Polymer Green Flame Retardants

Polymer Green Flame Retardants

Edited by

Constantine D. Papaspyrides

&

Pantelis Kiliaris

National Technical University of Athens
School of Chemical Engineering
Laboratory of Polymer Technology
Zographou Campus, Athens, Greece

ELSEVIER

AMSTERDAM • BOSTON • HEIDELBERG • LONDON
NEW YORK • OXFORD • PARIS • SAN DIEGO
SAN FRANCISCO • SINGAPORE • SYDNEY • TOKYO

Elsevier
Radarweg 29, PO Box 211, 1000 AE Amsterdam, Netherlands
The Boulevard, Langford Lane, Kidlington, Oxford OX5 1GB, UK
225 Wyman Street, Waltham, MA 02451, USA

British Library Cataloguing in Publication Data
A catalogue record for this book is available from the British Library

Library of Congress Cataloging-in-Publication Data
A catalog record for this book is available from the Library of Congress

ISBN: 978-0-444-53808-6

This book has been manufactured using Print On Demand technology. Each copy is produced to order and is limited to black ink. The online version of this book will show color figures where appropriate.

Printed in the United States of America

To Dida, my wife
-CDP

To my Alexandra, and to those who could not become giants
and thus tried to stand on the shoulders of giants
-PK

Contents

List of Contributors

V. Babrauskas
Fire Science and Technology Inc., Washington, USA

A.R. Bahramian
Tarbiat Modares University, Tehran, Islamic Republic of Iran

Åke Bergman
Stockholm University, Stockholm, Sweden

A. Blum
Green Science Policy Institute, Berkeley, California, USA

S. Bocchini
Istituto Italiano di Tecnologia, Torino, Italy

K. Bocz
Budapest University of Technology and Economics, Budapest, Hungary

S. Bourbigot
Ecole Nationale Supérieure de Chimie de Lille, Villeneuve d'Ascq, France

V. Cádiz
Universitat Rovira i Virgili, Tarragona, Spain

G. Camino
Politecnico di Torino, Dipartimento Scienza Applicata e Tecnologia, Alessandria, Italy

C. Campagne
Université Lille Nord de France, Ecole Nationale Supérieure des Arts et Industries Textiles, Roubaix, France

D. Deli
University of Bolton, Bolton, UK

E. Devaux
Université Lille Nord de France, Ecole Nationale Supérieure des Arts et Industries Textiles, Roubaix, France

Ph. Dubois
University of Mons & Materia Nova Research Center, Mons, Belgium

C.J. Ellison
McKetta Department of Chemical Engineering, The University of Texas at Austin, Austin, Texas, USA

Marta Escamilla
LEITAT Technological Center, Terrassa (Barcelona), Spain

M. Ferreira
Université Lille Nord de France, Ecole Nationale Supérieure des Arts et Industries Textiles, Roubaix, France

A. Fina
Politecnico di Torino, Dipartimento Scienza Applicata e Tecnologia, Alessandria, Italy

G. Fontaine
Ecole Nationale Supérieure de Chimie de Lille, Villeneuve d'Ascq, France

R. Fuoco
Green Science Policy Institute, Berkeley, California, USA

M. Galià
Universitat Rovira i Virgili, Tarragona, Spain

Z. Han
Harbin University of Science and Technology, Harbin, China

P. Hornsby
Queens University, Belfast, UK

S. Hörold
Clariant Produkte (Deutschland) GmbH, Gersthofen, Germany

T.R. Hull
Centre for Fire and Hazard Science, University of Central Lancashire, Preston, UK

B.K. Kandola
University of Bolton, Bolton, UK

P. Kiliaris
National Technical University of Athens, Athens, Greece

M. Kokabi
Tarbiat Modares University, Tehran, Islamic Republic of Iran

R.M. Kozlowski
Institute for Engineering of Polymers Materials and Dyes, Piastow and ESCORENA Focal Point at the Institute of Natural Fibres and Medicinal Plants, Poznan, Poland

T. Kuila
Surface Engineering and Tribology Division, CSIR-Central Mechanical Engineering Research Institute, Durgapur, India

F. Laoutid
University of Mons & Materia Nova Research Center, Mons, Belgium

R.J. Law
Centre for Environment, Fisheries and Aquaculture Science (CEFAS) Lowestoft, UK

G. Lligadas
Universitat Rovira i Virgili, Tarragona, Spain

S.M. Lomakin
Russian Academy of Sciences, Moscow, Russia

Gy. Marosi
Budapest University of Technology and Economics, Budapest, Hungary

M.S.S. Martins
University of Porto, Porto, Portugal

M. Murariu
University of Mons & Materia Nova Research Center, Mons, Belgium

M. Muzyczek
The Institute of Natural Fibres and Medicinal Plants, Poznan, Poland

S. Pack
Department of Materials Science and Engineering, Stony Brook University, Stony Brook, New York, USA

C.D. Papaspyrides
National Technical University of Athens, Athens, Greece

Roshan Paul
LEITAT Technological Center, Terrassa (Barcelona), Spain

C.M.C. Pereira
University of Porto, Porto, Portugal

R. Pfaendner
Fraunhofer Institute for Structural Durability and System Reliability
LBF Division Plastics, Darmstadt, Germany

D.A. Purser
Hartford Environmental Research, Hatfield, Hertfordshire, UK

M.H. Rafailovich
Department of Materials Science and Engineering, Stony Brook University, Stony Brook, New York, USA

J.C. Ronda
Universitat Rovira i Virgili, Tarragona, Spain

R. Rothon
Rothon Consultants and Manchester Metropolitan University, UK

A.M. Sakharov
Russian Academy of Sciences, Moscow, Russia

P.A. Sakharov
Russian Academy of Sciences, Moscow, Russia

M. Schlummer
Fraunhofer Institute for Process Engineering and Packaging, Freising, Germany

K. Shanmuganathan
McKetta Department of Chemical Engineering, The University of Texas at Austin, Austin, Texas, USA

Kelvin K. Shen
Rio Tinto Minerals (Consultant), Denver, USA

Marcia Simon
Professor and Director for Graduate Studies, Department of Oral Biology and Pathology School of Dental Medicine, Stony Brook University, Stony Brook, NY

S. Solarski
Université Lille Nord de France, Ecole Nationale Supérieure des Arts et Industries Textiles, Roubaix, France

S.K. Srivastava
Department of Chemistry, Indian Institute of Technology Kharagpur, Kharagpur, India

J.E.J. Staggs
University of Leeds, Leeds, United Kingdom

A.A. Stec
Centre for Fire and Hazard Science, University of Central Lancashire, Preston, UK

B. Szolnoki
Budapest University of Technology and Economics, Budapest, Hungary

A. Toldy
Budapest University of Technology and Economics, Budapest, Hungary

J. Walentowska
The Institute of Natural Fibres and Medicinal Plants, Poznan, Poland

Carl-Eric Wilén
Åbo Akademi University Åbo, Finland

G.E. Zaikov
Russian Academy of Sciences, Moscow, Russia

Preface

The ever-expanding use of polymers in a wide variety of applications results in a continuous demand for improved properties, so as to endure increasingly stringent conditions. The reduction in their propensity to ignite or burn efficiently is of major importance, since plastic materials comprise a large fraction of the fire load in houses, commercial environments, and transportation. Flame retardants are among the most common and varied plastic additives; there are literally hundreds of different substances on the market, destined for preventing or inhibiting fire spread in polymers. However, during the previous years, regulatory concerns about the human and environmental contamination caused by the use of traditional flame retardants have prompted a rapid progress in the research of effective "green" compounds. Fire safety legislation dictates, in particular, designing products of high-flame retardancy along with minimal environmental impact. Therefore, producers of flame retardants, compounders, and transformers are forced to abort traditional strategies of polymer flame retardation and develop "green" solutions. Investments on R&D activities are continuously increasing, because the rewards are believed to be enormous. Not only industries, but the vast majority of universities and research institutions around the world are engaged in the exploration of new flame retardant formulations. The flame retardancy of polymers is, hence, a hot topic, to which continuous and extensive research activity is dedicated. The intense research activity in this field is reflected by the significant amount of the corresponding literature that has been accumulated during the recent years. The growing interest and the continuously changing technology in the area of polymer flame retardancy are also revealed by the numerous national and international conferences dedicated on this subject.

In this context, the purpose of this book is to filter the accumulated literature and present a thorough treatise on polymer "green" flame retardancy, covering almost all of the key issues regarding the response of polymers during fire, the mechanisms of their flame retardation, the regulations imposed on their use, and the health hazards arising from their combustion. This book presents also the latest developments in the research conducted on this field, discussing the synthesis, properties, commercial potential, and environmental impact of the proposed fire-resistant plastic materials. Focus is placed on nanocomposites, as they are believed to be the most promising approach for producing materials of superior physical properties along with low flammability and ecological impact; the fire properties of nanocomposites of various matrixes and fillers are discussed, the toxicological characteristics of these materials are analyzed, addressing also their environmental sustainability. Moreover, for the first time in a book, important topics are explicitly analyzed such as the stabilization of fire retardant polymers and nanocomposites, the fire toxicity of these materials, their recycling potentials, and the LCA methodologies that can be applied to provide insights on their environmental impact. This book is designed to be both a handy textbook for people unfamiliar to these topics and a useful tool for scientists who expertise in this field. Furthermore, it is intended to comprise a handbook on "green" flame retardancy, including tables, in which promising environmentally benign fire retardant formulations can be found.

The book is divided into five parts. In the opening part, including four chapters, an introduction to the field of polymer fire retardancy is provided, analyzing the environmental reasons that render the use of greener flame retardants essential. More specifically, first the context is set by discussing the fundamentals of polymer combustion, the main fire tests, the modes of flame retardation, and the action of the commonly used flame retardants. Then, the fire-safety benefits against the health and environmental harm of the typically employed flame retardant chemicals are presented, highlighting the necessity for moving to alternative strategies that maintain fire protection without creating toxicity problems. In the next section (Chapters 5–13), the environmentally benign approaches that can be adopted are presented; the information provided is organized by flame retardant classes. Firstly, the mode of action of additives, belonging to the family of phosphorous and intumescent flame retardants, is discussed, presenting a case study dealing with the development of environmentally friendly intumescent/char forming coatings for wood and polymeric materials. Subsequently, a comprehensive insight into novel advances within the family of nitrogen-based chemicals and of inorganic flame retardants, precisely metal hydroxides, is given. Given that the latter comprises the largest segment of the flame retardant market, modelling their performance during fire is of major importance. Hence, the issues involved in modelling the endothermic decomposition of such fillers and generally of intrinsically hydrated solids are pointed out in a chapter. Apart from the above flame retardant categories, the recent advances and developments in the use of boron-based flame retardants and of silicon-based additives are reviewed. The second part of the book closes by analyzing the challenges of the stabilization of flame-retarded polymers, a topic not explicitly addressed in the past. Many of the commonly used flame retardants negatively affect the oxidative and photooxidative stability of polymers resulting in early loss of critical properties. Therefore, the selection of appropriate stabilizer systems (or even the development of novel ones) is of major significance when designing flame retardant formulation with a high potential of market success.

In the third section (Chapters 14–20), the focus is placed on nanocomposites, which, to the quest of environmentally friendly flame retardant formulations exhibiting enhanced physical performance, are believed to the holy grail. In particular, the performance during fire of thermoplastic polymer nanocomposites including the most thoroughly studied types of nanofillers is analyzed. Moreover, the behavior of nanocomposites in different fire scenarios, with correlations to the mechanisms of action of nanoparticles, is commented. Then, a mathematical model is presented able to give predictions about the thermal decomposition, thermophysical properties change, and ablation of polymer layered silicate nanocomposites; the validity of the model is tested on a thermoset nanocomposite. The flame retardancy of thermoset nanocomposites is addressed thoroughly in the next two chapters. To cover all types of polymer matrices, the behavior during fire of elastomer-based nanocomposites as well as of nanocomposites having a polymer blend as a matrix, is also discussed in this part of the book. Most studies performed so far are dedicated mostly to layered silicates and carbon nanotubes. Layered double hydroxides comprise an emerging class of nanofillers, whose mode of flame retardant action is analyzed in the last chapter of this section.

The scarcity of nonrenewable resources along with the negative environmental impact of petroleum-based materials renders the development of polymer products derived from nonpetrochemical feedstock indispensable. Thus, section four of the book (Chapters 21–23) deals with the advances and current research trends in the field of flame retardant bio-nanocomposites, the ones with environmentally biodegradable matrix derived from renewable resources. The analysis is extended to natural fibers making comparisons between their flammability properties and those of the artificial ones. The final section (Chapters 24–26) includes chapters dealing with topics, which to the best of our knowledge, are for the first time thoroughly analyzed in a book. More particularly, light is shed on the possible human health, safety risks, and environmental threats associated with nanomaterials. Recycling is another subject addressed in this final section; after illustrating the state-of-the-art recycling techniques, the processes believed to comprise economic and environmentally sound routes for the recycling of flame retarded polymers are presented. The book ends with a presentation of the LCA approaches that can be performed to gain quantitative information on the environmental profile of the commonly employed flame retardants, setting the context for similar studies on the utilization of nanocomposites in fireproof products.

Including contributions from an array of international industry and academia experts, this publication hopes to appeal to chemical, mechanical, environmental, material, and process engineers, to upper-level undergraduate and graduate students in these disciplines, and generally to researchers dedicated to developing commercially attractive and environmentally friendly fireproof products. Our objective is to provide a handy book, which will progress knowledge on the fire retardancy of polymers, triggering further research in this field. This goal would be unachievable without the thorough and intelligent contributions of the chapter authors, whom we cordially thank for the excellent and fruitful collaboration we had. Our appreciation is extended also to the staff of Elsevier for all their help and support.

Constantine D. Papaspyrides
Pantelis Kiliaris
National Technical University of Athens
School of Chemical Engineering
Laboratory of Polymer Technology
Zographou Campus, Athens, Greece

CHAPTER

Polymers on Fire

1

P. Kiliaris, C.D. Papaspyrides
National Technical University of Athens, Athens, Greece

CHAPTER OUTLINE

1. Introduction

Owing to their outstanding combination of properties—low weight and ease of processing—polymeric materials, being used in countless application areas and under the most demanding conditions, have added greatly to the quality of modern life and the progress of human civilization. However, one of the most important problems arising by their use is associated to the fact that the majority of the polymers, on which these materials are based, are organic and thus flammable. Unwanted fires account for considerable losses of life and property; therefore, fire hazards related to the use of polymeric materials are of particular concern among government regulatory bodies, consumers, and manufacturers, and there are great economic, sociological, and legislative pressures on plastics industries to produce materials with greatly reduced fire risk. The approach, typically followed for enhancing the performance of polymers against fire, involves the addition of the so-called *flame retardants*. The role of the flame retardant is to render the polymer formulation less flammable by interfering with the chemistry and/or the physics of the combustion process. In most cases, the

Polymer Green Flame Retardants. http://dx.doi.org/10.1016/B978-0-444-53808-6.00001-9

use of flame retardants to reduce polymer combustibility (along with smoke or toxic fume production) comprises a critical part of the plastics materials production. Among the major markets, in which flame retardants are required, are the industries dealing with construction, transportation, and electrical and electronics components [1–3].

The interest in reducing the flammability of polymeric materials goes back to the nineteenth century when highly flammable cellulose nitrate and celluloid were discovered. Until quite recently, however, the enhancement of polymer flame retardancy has been as much an art as it has been a science. The strategies commonly adopted involve incorporating nonflammable fillers, compounds that decompose endothermically, substances that provoke char formation, and/or materials that during fire liberate gases acting as radical traps in the gas phase. Typical of the last mentioned group of fire retardants are halogenated compounds (such as chlorinated and brominated aliphatics and aromatics), which evolve hydrogen halides when heated. Traditionally, the incorporation of halogen-based compounds comprised the most widely applied method for enhancing the flame retardancy of polymers, since, besides its low cost, it led to materials exhibiting mechanical performance similar to that of their Flame retardant-free (FR-free) analogs. However, despite these benefits, concerns were recently raised which were caused by findings of brominated flame retardants in the environment, biota, and humans, and evidence of formation of toxic dioxins and furans during the combustion of halogens. These concerns have led to restrictions to the marketing and use of certain flame retardants, e.g. the ban on penta- and octa-brominated diphenyl ethers. Moreover, some Scandinavian countries like Norway and Sweden have been considering national banning of some additional brominated compounds, even restricting some which had a positive European Union (EU) risk assessment. As a consequence, the market trend has been pushed to halogen-free flame retardants. Nevertheless, non-halogen compounds are also confronted with disadvantages associated with their inefficiency at low concentrations, which may lead to additional costs and inferior polymer mechanical properties. Therefore, the development of environmentally friendly and highly effective flame retardants is a hot topic in the plastics industry that has prompted extensive research into deeply understanding the action of existing flame retardants (so as to improve their performance and extend their application field), as well as into novel technologies [4–6].

This chapter provides a concise introduction to the flame retardancy of polymers; more specifically, the fundamentals of polymer combustion are discussed presenting the typical pattern of fire growth and evolution of fire hazards, the main tests applied to estimate the fire performance of polymeric materials are described, and the modes of flame retardation along with the mechanisms of (synergistic) action of the major classes of flame retardants are briefly reviewed. The main objective of this chapter is to make this book and, thus, the field of "(green) flame retardants for polymers" more accessible to the community of materials science and engineering. However, though designed with a "back to basics" orientation in order to comprise a useful tool for people new to these topics, scientists who expertise in this field may also find new interesting information.

2. Polymer combustion

Polymers are chemical compounds or mixtures of compounds consisting of a relatively large number of repeating structural units (of low relative molecular mass). The binding of these smaller molecules is achieved through the process of polymerization. The term derives from the ancient Greek words "πολύ" ("poly," meaning "many") and "μέρος" ("meros," meaning "part"), and refers to a process through which molecules of high molecular weight (exceeding 1500 g mol^{-1}) are formed exhibiting properties not shared by low-molecular-weight materials. Polymers are thus composed of "chains" of atoms linked to one another; the simplest (carbon—carbon) chain is that of polyethylene, in which the repeating unit is ethylene ($-CH_2-$). Polymers are neither purely crystalline nor purely amorphous, and thus they are not, strictly speaking, solid materials (even though they are normally solids in the sense that they are bounded by plane surfaces). When all the structural units correspond to the same monomer, the material is called a *homopolymer*; on the contrary, copolymers are compounds in the molecules of which different structural units exist [7,8].

Polymers can be natural or synthetic and they were initially classified, according to their response to temperature, into thermoplastics (whose deformation at elevated temperature is reversible) and thermosets (which when heated undergo irreversible changes). A categorization based on their physical/mechanical properties (in particular their elasticity and degree of elongation) is also used, classifying polymers into elastomers, plastics, and fibers. Elastomers (rubbers) are characterized by high extensibility and recovery, plastics exhibit intermediate properties, whereas fibers possess very high tensile strength but low extensibility. Polymers are also classified according to the type of the polymerization reaction through which they are formed: condensation and addition. However, the most useful classification method is the one based on their chemical composition since it gives indication of their reactivity, including thermal decomposition and response against fire. According to this method, polymers are basically categorized into carbon-containing (e.g. polyolefins), oxygen-containing (e.g. polyesters), nitrogen-containing (e.g. polyamides), chlorine-containing (e.g. poly(vinyl chloride)), and fluorine-containing polymers (e.g. polytetrafluoroethylene) [9,10].

Due to their chemical structure, polymers are generally highly combustible. When exposed to sufficient heat, they decompose (pyrolyze) generating combustible gases which mix with the oxygen of the ambient air to form an ignitable blend. Ignition occurs either impulsively if the temperature is adequate for autoignition (defined as the temperature at which the activation energy of the combustion reaction is achieved) or at a lower temperature (called the *flash point*) due to the presence of an external source (flame or spark). Upon combustion, heat is released, a part of which is transferred to the substrate promoting further decomposition. If the heat evolved is enough to keep the decomposition rate of the polymer above that required to maintain the concentration of the volatiles within the flammability limits, then a self-sustaining combustion cycle is established (Figure 1). It becomes, therefore,

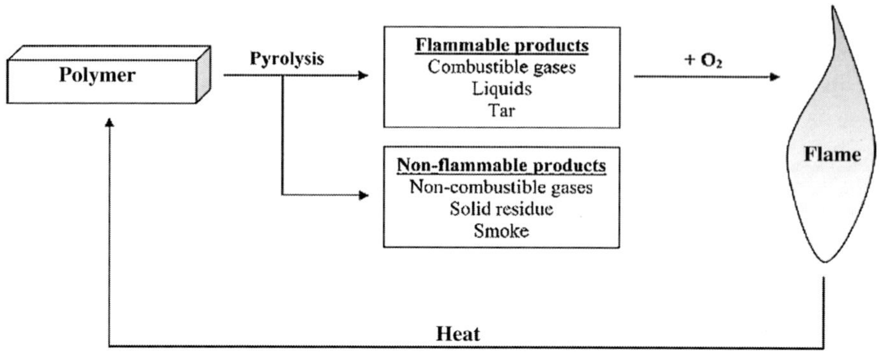

FIGURE 1

The combustion cycle of polymers [11].

apparent that for the development of a sustained fire, the so-called fire triangle should be established, which means that the contribution of three elements are essential: fuel (combustible volatiles evolved from a thermally degrading carbon rich substance), heat (supplied by an external source or by the exothermic oxidative decomposition of the fuel), and oxygen (provided usually by the air) [10,11].

The amount of heat needed to initiate combustion is mainly determined by the physical characteristics of the polymer. In particular, ignitability basically depends on the rate that the surface of the polymeric material reaches the temperature required for ignition. Accordingly, the energy stored by the polymer during its exposure to a source of heat correlates with its heat-storage capacity, enthalpy of fusion, and degree of crystallinity. For example, thermosetting polymeric materials do not melt and do not undergo simple phase changes upon heating; thus, heating leads directly to polymer decomposition. This is also the case for amorphous thermoplastics due to the absence of a melting point. Semicrystalline thermoplastics, on the other hand, on heating become soft, they melt and drip, withdrawing a part of heat from the surface. This behavior is beneficial for the fire performance of polymers that do not char upon combustion. Nevertheless, at a higher heat exposure, such polymers may ignite relatively easy since ignition may occur before the surface is heated to adequate depth for the melted material to drip [10,12].

Whereas ignition is principally related to the initial temperature of polymer decomposition, steady combustion is typically connected to the tendency of the polymer to yield char at the expense of combustible volatiles. Hence, the degradation pathway of a polymeric material is of major importance when designing the strategy of its flame retardancy enhancement. The most important mechanisms of polymer thermal decomposition are the following: (1) random chain scission, in which the polymer chains split at random locations into smaller fragments; (2) chain-end scission, where individual monomer units are successively removed from the chain ends; (3) chain stripping, during which elimination of atoms not part of the polymer chain or pendant groups are cleaved without breaking of the backbone; and (4) cross-linking, in which bonds are created between polymer chains. In most cases, the thermal decomposition of a polymer involves more than one of these reaction mechanisms [12–14].

Generally, the degradation pathway is the result of the combination of the effects of both heat and oxygen. Degradation can be therefore distinguished into nonoxidizing thermal degradation and oxidizing thermal degradation. In the first case, chain scission takes place due to the individual effect of temperature, leading to varying degrees of material depolymerization. The main factors governing initial scission include the presence of oxygen atoms in the polymer backbone, the existence of catalyst residues, and/or former residues of oxidation, and the presence of the chemical defects (e.g. weak bonds) in the polymer chains. In oxidizing conditions, the polymer reacts with the oxygen of the air, generating low molecular weight products (carboxylic acids, alcohols, aldehydes, etc.), and/or very reactive species, (H• and OH•, which confer a high velocity on the flame front resulting in rapid flame spread). Through recombination reactions of the macromolecular radicals, oxidative thermal degradation can lead also to cross-linking. In terms of flammability, polymers undergoing random scission and depolymerization are usually more flammable than polymers, which experience cross-linking or pendant group elimination upon heating. The latter two mechanisms (cross-linking and pendant group removal) result in the production of species, which acting as precursors of char, reduce flammability. The char development proceeds basically through four stages: (1) cross-linking, (2) aromatization, (3) fusion of aromatics, and (4) graphitization. The tendency of a polymer to form char during heating depends principally on its structure. For instance, polymers with short flexible bonds between aromatic rings are inclined to cross-linking and, thus, charring. Nonetheless, polymers, without inherent inclination to cross-link, can undergo charring through the addition of specific flame retardants (this topic is discussed later in this chapter) [15,16].

3. The stages of fire evolution and the corresponding hazards

Within a closed compartment, a fire can go through several phases of growth (Figure 2). In order, these are the following:

- *Ignition*: First, the fuel source is ignited and undergoes sustained combustion. As mentioned in the previous section, the ignition event occurs when flammable products (i.e. fuel) evolved from a decomposing (owing to its exposure to a heat source) material mix with oxygen from the surrounding air. When the lower flammability limit is reached (a flammable mixture has been formed), the mixture will either autoignite if the autoignition temperature has been reached, or "flash" ignite due to the presence of external flame or spark. As a consequence, a flame appears producing heat, some of which is transferred back to the surface of the fuel maintaining the generation of flammable volatiles. In fact, an

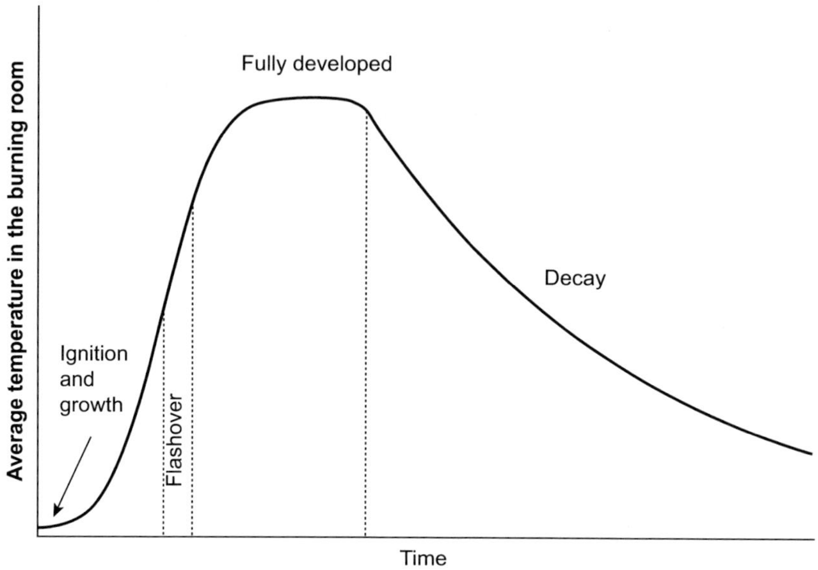

FIGURE 2

Schematic representation of a compartment fire versus time [18].

equilibrium is established between the rate of heat transfer back to the fuel, the heat losses from the fuel, and the evolution of degradation products.

- *Growth*: The fire grows if adequate fuel and oxygen are available, causing the continuous increase of room temperature. The initial growth of a fire mainly depends on the fuel itself; the combustible materials within the compartment have little or no influence. In this stage, materials exposed to the flame ignite when the temperature exceeds 350–500 °C.
- *Flashover*: Flashover takes place when all combustible items in the compartment are involved in fire. Hence, an unchecked fire eventually grows rapidly, resulting in an increase of temperature and heat, which motivates the ignition of nearby materials (by heat transfer). Flashover is not a precise term since a strict definition of this stage is not possible; typically it usually occurs when the average upper gas temperature in the room exceeds approximately 600 °C.
- *Fully developed fire*: Once the heat release rate (HRR) and temperature are at their prime, a fully developed fire is evident. The fire may spread to other parts of the building and (when severe enough) may induce structural damage to it. The peak temperature of a typical postflashover room fire is 900–1000 °C, even though temperature as high as 1200 °C can be reached. The fire in this stage is said to be ventilation controlled: if the compartment has openings, the unburnt fuel will escape (following the gas flow) and possibly burn outside of it.
- *Decay*: Decay eventually occurs due to the consumption of all combustible materials (or as a consequence of fire suppression systems, such as sprinklers,

operation). Thus, the HRR declines and the compartment temperature to falls. During this period, the fire may be changed from ventilation- to fuel-controlled [17–19].

In view of the above, it becomes apparent that during fire, the temperature inside a compartment reaches values so high that it would be impossible for the occupants remaining in the vicinity to survive. However, the vast majority of injuries and deaths, as well as the environmental hazards, arising from fires, result by far from the exposure to (toxic) products evolved by the decomposing burning materials. In the case of plastic materials, the nature of the fire products highly depends on polymer's chemical composition and the conditions under which the burning process takes place. Predominantly, during fire, smoke is produced, which is a combination of complete (CO_2, H_2O, and acid gases) and incomplete (CO, soot, and partially oxidized fuel gases) combustion species, and a solid residue, which is composed mostly of carbon and ash (oxidized metals). Some simplified generalizations about the structural factors determining smoke generation are the following: first, it has been found that aromatic and polyenic polymers exhibit greater tendency to produce smoke than the aliphatic or oxygenated polymers. Furthermore, polymers having aromatic units in their backbone present lower inclination to reduce smoke than polymers with pendant aromatic groups. On the other hand, low or intermediate halogenation leads to an increase of the amount of smoke, while highly halogenated polymers show reduced smoking tendencies. Generally, smoke quantity is related to thermal stability of the polymers and the type of fuel evolved upon polymer decomposition [20,21].

Depending on ventilation and the burning rate of the polymer, smoke can be the main fire hazard, containing asphyxiant gases which cause incapacitation, soot which absorbs light limiting visibility, and acid gases which are irritating to eyes and nasal passages causing respiratory tract pain and breathing difficulties. More specifically, the most hazardous toxic products generated during fire are organic irritants (such as formaldehyde, phenol, and styrene), inorganic irritants (such as hydrogen halides, nitrogen oxides, and phosphates), and asphyxiant gases (such as CO, HCN, and CO_2). Inadequate heat in the gas phase, quenching of the flame reactions (e.g. when halogens are present in the flame), the presence of stable molecules (such as aromatics) in well-ventilated conditions, and insufficient oxygen (e.g. in underventilated fires) are considered to be the main reasons of CO production during fire. Carbon monoxide is the principal toxic gas and, as concluded through postmortem studies, it comprises the major cause of death in fires. There is an exposure level to each toxic gas (especially CO and HCN), established by animal experimentation, which results in the loss of consciousness eliminating escape capability and, at a higher dose, in death. Under well-ventilated conditions and in the growth stage, limited visibility or radiant heat flux is often the first tenability criterion reached in the room of origin, followed by the toxic gas criteria. However, in several cases, even if the escape from the immediate area of flames and the high temperatures encountered there is achieved, the movement of smoke with growing levels of toxic and/or irritant compounds can cause injury or death far from the source of fire. Fire is constrained inside the room of origin until the flashover; after that point, fire fully develops and spreads to other parts of the building rendering death, as a result of burning or hyperthermia (due to the effect of heat), possible [18,20,22].

4. Fire testing

The performance against fire of polymers is mainly evaluated through ignitability, flame spread, and heat release. In order to assess (one or more of) these flammability characteristics, appropriate tests are conducted, depending on the intended application of the polymeric material in question. Numerous tests have been developed and performed either in laboratory scale on small representative samples or during industrial manufacturing on final products. Test methods can thus be small, intermediate, or full scale, varying in complexity from simple (e.g. ASTM D2863 and UL 94, for which 10-cm-long specimens are used) to very complex (e.g. ASTME 84, which require 750-cm-long specimens) [23]. However, the most widely used tests to characterize plastics flame retardancy are the UL 94 vertical burning test, the limiting oxygen index (LOI) test, and the cone calorimetry.

The UL 94 tests have been developed by Underwriters Laboratory Inc. for evaluating the flammability of plastic materials (intended for parts in devices and appliances). Several tests have been developed (small and large flame vertical tests, horizontal tests for bulk and foamed materials, and radiant panel flame-spread test), with the UL 94 vertical burning test being in practice the most common procedure to check the regulatory compliance of a product, with respect to flammability. In particular, the test measures the ignitability and flame spread of polymeric materials exposed to a small flame. Two sets of five bar-shaped specimens are employed, which depending on the intended use of the material may be 3.2, 1.6, or 0.8 mm thick. In this test, the specimen is mounted vertically and held from the top so that the lower end is located above a cotton layer to catch any flaming drip (Figure 3). The flame is applied at specimen's bottom for 10 s; if the specimen self-extinguishes, a subsequent application (10 s) is performed. Depending on its performance regarding the individual duration of burning for each specimen, the total burning time for all specimens, and the presence or absence of burning drips, the material is classified into three categories (V-0, V-1, and V-2); the specific burning characteristics required for each classification are summarized in Table 1 [24,25].

The other commonly used (mostly in laboratory scale) test is the LOI test. Standardized as ASTM D2863 or ISO 4589, this test is applied to quantify the resistance to ignition of a polymeric material. In particular, the minimum oxygen concentration in nitrogen, which supports the combustion of the material for at least 3 min or for the consumption of 5 cm of the sample, is evaluated. LOI is expressed as $LOI = 100\,([O_2]/([O_2] + [N_2]))$. Hence, the more the oxygen required (the higher the LOI value), the more fire resistant the material is considered to be. Given that air contains 21% oxygen, polymeric materials with an LOI value lower than 21

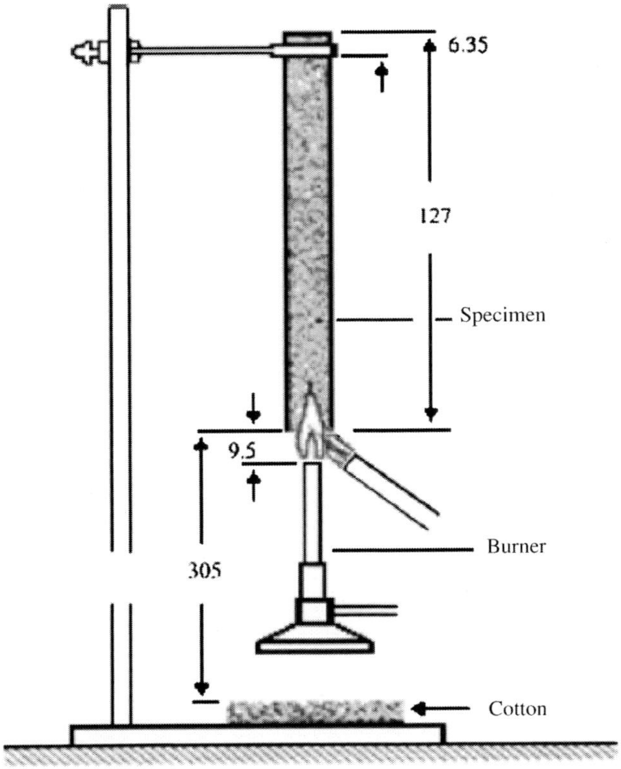

FIGURE 3

Schematic view of the UL94 vertical burning test [24].

Table 1 Criteria for UL 94 Classifications [24]

	V-0	V-1	V-2
Total flaming combustion for each specimen	≤10 s	≤30 s	≤30 s
Total flaming combustion for all five specimens of any set	≤50 s	≤250 s	≤250 s
Flaming and glowing combustion for each specimen after second burner flame application	≤30 s	≤60 s	≤60 s
Cotton ignited by flaming drips from any specimen	No	No	Yes
Glowing or flaming combustion of any specimen to holding clamp	No	No	No

are categorized as combustible, whereas those with LOI above 21 are classified as self-extinguishing since their combustion cannot be maintained at ambient temperature without the contribution of an external energy source. The apparatus utilized consists of a glass tube, in which the specimen is vertically mounted (Figure 4).

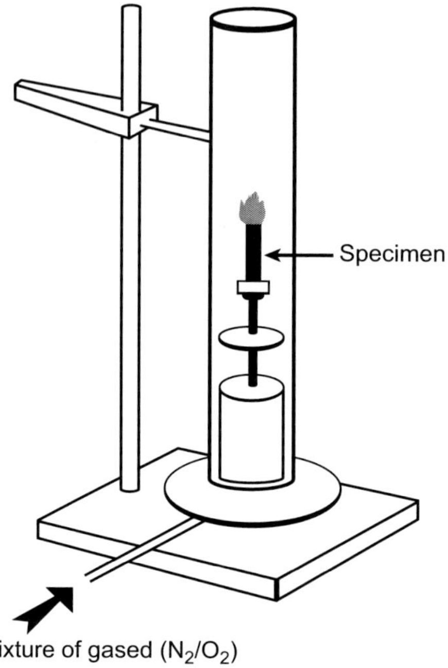

FIGURE 4

Schematic view of the limiting oxygen index test apparatus [24].

During the test, a slow stream of oxygen—nitrogen mix is supplied at the bottom of the tube and a small candle-like flame is applied to the top of the specimen in an attempt to ignite it. The specimen size and shape are not strictly specified in this test; usually bars of about 100 mm × 65 mm × 3 mm are generally employed when testing rigid plastics. Despite the fact that the LOI test is quite unsophisticated and does not represent a real fire scenario, it still remains one of the most important quality control tools used in the plastics industry because it gives a numerical value instead of a discrete classification [24,26].

Cone calorimetry (standardized as ASTME 1354 or ISO 5660), on the other hand, is a more elaborate (medium-sized bench-scale) method, which since its development by the National Institute of Standards and Technology (NIST) has gained popularity due to the numerous fire properties of a material that can be determined by applying this method. More specifically, in a cone calorimeter (Figure 5), the HRR, the peak of heat release rate (pHRR), the time to ignition (TTI), the total heat released (THR), the mass loss rate (MLR), the peak of mass loss rate (pMLR), and the specific extinction area reflecting the smoke production can be measured. Among the data gathered, HRR and pHRR comprise the most important parameters to evaluate fire safety: HRR is thought as the driving force of fire and pHRR represents the point in fire where heat is apt to propagate further or ignite adjacent objects.

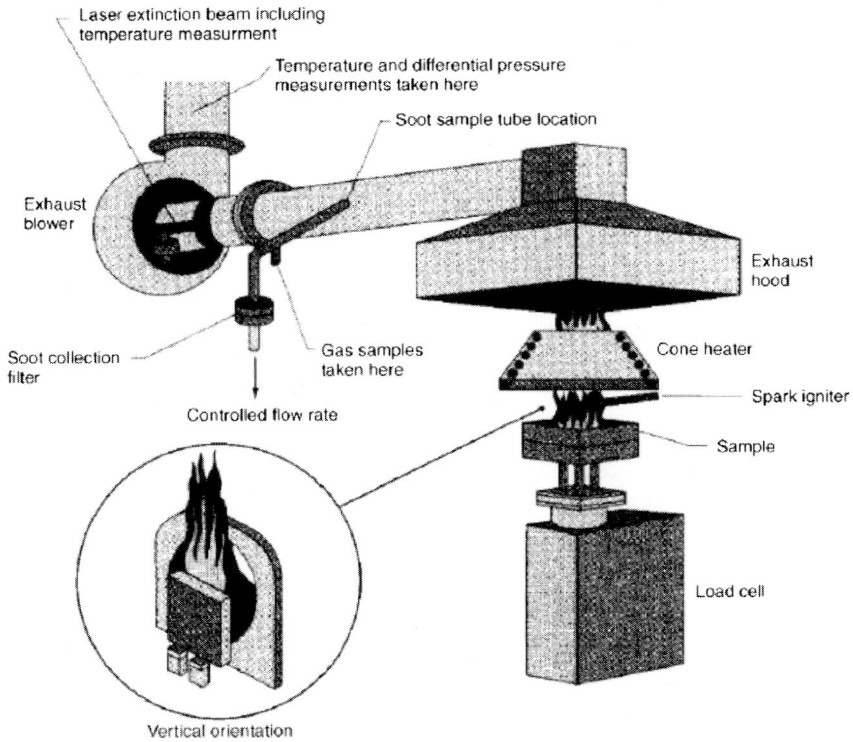

FIGURE 5

Schematic view of the cone calorimeter [28].

During this test, a sample (100 mm × 100 mm × up to 50 mm) is placed on a load cell allowing the evaluation of MLR. A conical radiant electrical heater uniformly irradiates (at a constant heat flux) the sample from above; the combustion is triggered by an electric spark, which ignites the gases evolved from the degrading (heated) specimen. The produced gases are collected in an exhaust duct system with centrifugal fan and a hood, and the HRR is calculated based on the fact that there is a constant relationship between the mass of oxygen consumed from the air and the amount of heat released throughout polymer combustion. This proportionality factor is constant from one material to another and is equal to 13.1 kJ g^{-1} consumed oxygen. The basic merit of cone calorimetry is that it can be applied to estimate the properties of a material that determine its behavior at a variety of fire scenarios (TTI: ignition, HRR: developing fire, THR: fully developed fire) [27–29].

Due to the fact that, in general, different fire tests try to simulate special fire scenarios, the correlation of the results obtained from these tests has proven to be problematic. Correlating the collected data has been found to be imperfect even in the case of different tests imitating the same scenario. For example, while both the cone calorimeter and the UL-94V test (or the LOI test) measure flammability, they do in a different way, giving in the quantitative correlation between them a high level of uncertainty. A great number of parameters such as dimension and geometry of the sample, source of ignition, ventilation, heat flux, and so on, are altered in these tests, preventing easy and straightforward correlations, mainly if different materials and fire-retardant mechanisms are examined. For instance, as a consequence of the fact that the samples inside the cone calorimeter are tested in a horizontal geometry under forced flaming conditions, cone calorimetry (contrary to the UL94 test) has a lack in characterizing the fire behavior of dripping materials [30]. However, in a comprehensive study of Morgan et al. [31], in which cone calorimeter analysis was conducted on 18 thermoplastics with different UL-94V ratings, it was shown that even though a universal model between UL-94 and cone calorimetry cannot be established, a more quantitative relationship can be developed when a specific polymer type and (may be even more narrowly) a flame retardant is examined.

5. Strategies of flame retardancy

Fire (or flame) retardation is a process through which the flammability of a polymeric material is reduced by some modification as measured by one of the accepted test methods. In the fire science community, the words "fire" and "flame" in the term "fire (flame) retardancy" are used interchangeably and it is mostly a matter of personal preference to use one or the other, rather than drawing a clear distinction between them. Sometimes, "flame" implies the result of smaller scale fire test (where as an ignition source, a small flame is used) and "fire" refers to a larger scale one (a fire test simulating a more realistic fire scenario). The absence of a clear distinction between "fire" and "flame" is at times further exacerbated by a lack of distinction among many in the field between the terms "fire (flame) retardant" and "fire (flame) resistant". The former term is applied to additives blended or otherwise incorporated into a polymer, endowing superior performance against fire. Often, the name "fire" retardant is given to a larger group of additives (including char formers, intumescent systems, etc.), while the term "flame" retardant refers to the compound acting mostly in the gas phase. However, throughout this book, these terms have the same meaning. On the other hand, as "fire (flame) resistant," rather than "fire (flame) retarded," is characterized a polymeric material possessing inherently better fire properties than ordinary polymers [32,33].

Successful strategies to improve the performance of a polymeric material against fire involve interrupting the combustion process at one or more of its complex stages, with the aim of inhibiting ignition, reducing the burning rate and/or changing the mechanism of combustion. In essence, the main objective is to exclude one of the three required components of fire: (1) the combustible substance or fuel, (2) the heat supplied either externally or from the combustion process itself, and (3) the oxidizing gas (primarily oxygen).

In practice, this is accomplished either (1) by the mechanical blending of a suitable flame-retardant compound with the polymeric material during the transformation process or (2) by the chemical incorporation of the flame retardant into the polymer during synthesis (by copolymerization) or through chemical modification of the preformed polymer (chemical grafting using a reactive component). The first category of flame retardants, the so-called additive flame retardants are designed not to react with the polymer at the stage of blending but only at higher temperatures, at the beginning of a fire. On the other hand, reactive flame retardants, which comprise the second category of flame retardants, are integrated in the polymer chains (covalently attached). This is why they offer several advantages over those that are merely additives. More specifically, owing to the fact that their incorporation takes place at the time of polymerization, they can be homogeneously dispersed preventing the formation of separate phase, which causes problems during the processing of the polymer into final plastic product. Moreover, since "mixing" occurs at the molecular level, reactive flame retardants are inherently immobile within the polymer, being thus less susceptible to loss during service through migration to the polymer surface (blooming) or due to solvent leaching. Additionally, lower concentrations (compared that of additives) are needed to achieve the desired level of flame retardancy; thus, the overall properties of the polymer are less likely to be adversely affected. Despite these merits of the reactive ones, the additive flame retardants are more generally used; the reasons behind their popularity are wider applicability and lower cost. More particularly, the reactive modification of chain-growth polymers is less readily accomplished than for step-growth polymers. Furthermore, extensive reactive modification of a partly crystalline polymer is likely to result in a considerable reduction of crystallinity (additive flame retardants end up in the amorphous phase). Besides, reactively modifying polymers for which commercially well-established methods for manufacturing already exist would lead to a significant increase of expenses [24,34,35].

Regardless of their type (additive or reactive), flame retardants, depending on their nature, interfere with the combustion cycle through physical or chemical modes of action (Figure 6). The main underlying physical mechanisms for hindering combustion are the following:

- Promotion of endothermic reactions (heat sink), which cool the substrate to a temperature below that required to maintain combustion.
- Generation of inert gases, which dilute the oxygen supply at the surface of the burning polymer, resulting in flame snuffing because of lack of oxidizing agent.
- Formation of a protective impermeable coating, which decreases the amount of heat transferred to the polymer, hampers the diffusion of oxygen to the area of decomposition, and impedes the escape of volatile flammable gases generated during polymer decomposition.

On the other hand, the following are the chemical modes of action by which flame retardants operate:

- Inhibition of the oxidation reactions, which occur in the gas phase, through trapping of free-radical species (especially, H• and OH•) evolved as a consequence of polymer degradation. Hydrogen radicals are responsible for the

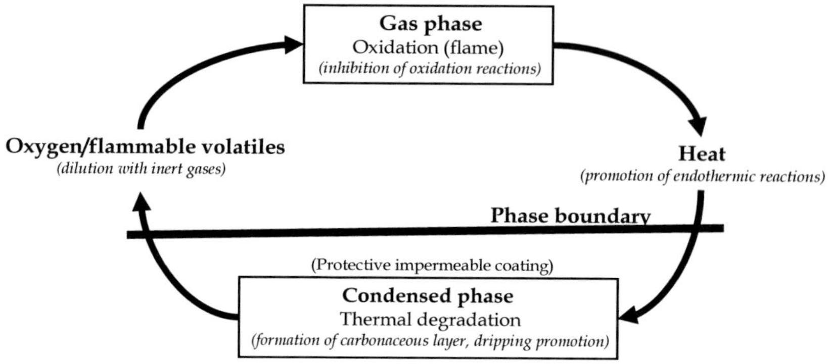

FIGURE 6

Combustion/flame retardation cycle.

chain-branching reaction which propagates the combustion of fuel ($H^{\bullet} + O_2 \rightarrow OH^{\bullet} + O^{\bullet}$) while hydroxyl radicals is involved with the most exothermic reaction ($OH^{\bullet} + CO \rightarrow H^{\bullet} + CO_2$) of polymer decomposition, which provides most of the energy needed to preserve the flame. These highly reactive species react with specific radicals released by the flame retardants to form less reactive or even inert molecules.

- Formation of a carbonaceous (or vitreous) layer on the polymer surface by promoting low-energy solid-state reactions, which lead to the carbonization of the polymer at the expense of volatiles production. This layer acts as a physical insulating barrier between the gas phase and the condensed phase.
- Acceleration of polymer degradation, causing pronounced dripping and, thus, the withdrawal of the fuel from the flame source. Additives operating through this mode of action are usually not preferable; they can be used however for the qualification of a polymeric material under the UL94 protocol [36–39].

Generally, the abovementioned mechanisms do not take place separately but in combination. Therefore, the way in which a flame retardant acts is usually a complex process consisting of individual stages, among which one is the dominating. The criteria for selecting a suitable additive for each case are based on a variety of factors, the most important of which are the following:

- The effectiveness of the flame retardant in the particular polymer matrix.
- The stability of the flame retardant in the processing conditions of the polymer.
- The compatibility and the ability of the additive to protect the physical properties of the polymer (mechanical properties, electrical performance, color, etc.).
- The cost efficiency.
- The toxicity of the flame retardant together with its tendency to migrate and/or to cause corrosion [29,34].

Speaking about toxicity, the public concern and the (increasingly) prohibitive legislation with regard to the persistent, bioaccumulative and toxic nature of several halogen-based compounds have divided flame retardants into two categories: halogenated and nonhalogenated. The class of halogen-free compounds is further disintegrated into smaller categories; the most frequently referred in literature are phosphorus-based flame retardants, nitrogen compounds, intumescent systems, mineral flame retardants (metal hydroxides and born compounds), silicon-based additives, and nanoparticles. Even though intumescent systems usually contain phosphorous and nitrogen substances, they perform differently than their individual counterparts; therefore, they comprise a separate class of flame retardants. The abovementioned classification is also applied in this chapter; the mode of action of each group is described in the following sections.

5.1 Halogen-based flame retardants

Even though a wide variety of flame retardants have been developed, halogenated compounds, due to their cost, availability, and extensive industry experience with this class of additives, have been the most frequently employed for enhancing the flame retardancy of polymers. Halogen-based flame retardants interfere with the combustion cycle primarily through free-radical scavenging mechanisms, inhibiting the flame propagation in the vapor phase. As already mentioned, (solid and liquid) materials burn only if they can be made to generate volatiles, which when mixed with oxygen, form a combustible mixture; hence, the phenomena that lead to ignition and the release of heat are those of the gas phase. There are literally thousands of chemical reactions occurring in the flame that convert oxygen and fuel to stable combustion products. However, kinetic modeling and experimental data have shown that the burning velocity is most sensitive to reactions (depicted in Figure 7) involving active radicals of fuel (R), hydroxyl (OH), hydrogen (H),

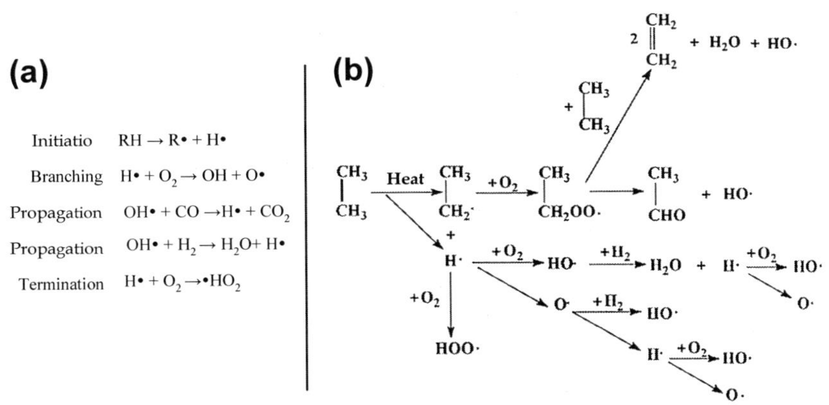

FIGURE 7

(a) Free-radical reaction mechanism during combustion, and (b) proliferation of OH• in the case of burning ethane.

and oxygen (O). Combustion, thus, proceeds via a free-radical mechanism, including a number of (chain-branching) steps. The most important one is propagated by the highly reactive OH• ($O\bullet + H_2 \rightarrow OH\bullet + H\bullet$) and H• ($H\bullet + O_2 \rightarrow OH\bullet + O\bullet$) radicals. As a simplified model of the flame chemistry to show the generation of the OH• radicals, the example of ethane is used in Figure 7; a similar contribution is also made by the H• radicals. Both OH• and H• radicals confer a high velocity on the flame front leading to rapid flame spread. Nevertheless, the exothermic reaction supplying most of the energy that maintains the flame is $OH\bullet + CO \rightarrow CO_2 + H\bullet$ [10,12,40].

Considering the abovementioned mechanism, it becomes apparent that in order to impede combustion, the chain-branching reactions, mainly $O\bullet + H_2 \rightarrow OH\bullet + H\bullet$ and $H\bullet + O_2 \rightarrow OH\bullet + O\bullet$, should be inhibited. Halogenated flame retardants are able to react with the highly radical species, OH• and H• (produced during polymer combustion), hampering the chain decomposition and, thus the burning of the polymer. This effect is a consequence of the generation of hydrogen halides ($MX \rightarrow HX + M\bullet$, where X is halogen and M is the residue of the flame-retardant molecule) or, if the flame-retardant molecule does not contain hydrogen, of halogen atoms ($MX \rightarrow X\bullet + M\bullet$) during the degradation (upon heating) of the halogenated compounds. Hydrogen halides (produced also through the reaction between halogen atoms and the polymer, $RH + X\bullet \rightarrow HX + R\bullet$), are considered to be the actual flame inhibitors as they react with the polymer degradation radicals ($HX + H\bullet \rightarrow X\bullet + H_2$ and $HX + OH\bullet \rightarrow X\bullet + H_2O$), leading to much less active species. Radical trapping is not the only activity of the halogen-based flame retardants. Physical factors, such as the heat capacity of halogens, have been shown to contribute to the flame-retarding effect; more specifically, halogens work as "heat sinks," reducing the heat evolved by the combustion of the polymer degradation gases, thus hampering further thermal decomposition. They also have a "blanketing" effect, excluding oxygen from the surface of the degrading polymer. In addition, halogens are believed to catalyze the oxidation of the solid phase, yielding products that tend to cyclize, forming eventually a solid protective layer on the surface of the burning polymer [41,42].

Typical representatives of the halogenated flame retardants are chlorine-, bromine-, fluorine-, and iodine-containing compounds, with the first two comprising in tonnage terms the most widely used members of this group. This is due to the fact that efficiency and stability, depending on the kind of the halogen, are the two primary factors that determine the type of the compound that can be used as flame retardant. Fluorine- and iodine-based compounds are not employed owing to their inability to interfere with the polymer combustion process. Fluorinated organics are more stable than most commercial polymers and do not release fluorine radicals or hydrogen fluoride at the temperature range, where the decomposition of the polymers takes place. On the contrary, iodinated compounds are less thermally stable than most polymers, thus releasing halogenated species during polymer processing. An additional reason that limits the development of flame retardants based on these two halogens is the higher cost of fluorine and iodine compared to that of chlorine or bromine. Bromine and chlorine, due to their low bonding energy with carbon atoms, are released at a somewhat lower temperature than the decomposition temperature of the polymer, inhibiting the combustion process in the gas phase, especially via the previously discussed free-radical mechanism. As a general rule, the thermal stability of halogenated flame retardants decreases as brominated aromatic > chlorinated aliphatic > brominated aliphatic (chlorinated aromatic products are relatively stable and therefore not very efficient to be used as flame retardants) [24,33,43].

Generally, halogen derivatives are used in combination with compounds, i.e. synergists, which improve their flame retarding performance. There exists a considerable number of such synergists, the most important one being antimony trioxide; formulations based on the halogen–antimony synergism are widely used for enhancing the fire retardancy of a variety of polymers, e.g. polyolefins, polystyrene, polyesters, polyamides, polyurethanes, etc. Antimony trioxide mainly operates in the gas phase. During pyrolysis, the hydrogen halides, which are released by the self-decomposition of the halogenated compound or by the interaction with antimony trioxide and/or with the polymer, react with the Sb_2O3 producing antimony halides. The latter react with atomic hydrogen producing SbX_2 ($SbX_3 + H\bullet \rightarrow SbX_2 + HX$), SbX ($SbX_2 + H\bullet \rightarrow SbX + HX$), and Sb ($SbX + H\bullet \rightarrow Sb + HX$). Antimony reacts with atomic oxygen, water, and hydroxyl radicals, producing SbOH ($Sb + OH\bullet \rightarrow SbOH$) and SbO ($Sb + O\bullet \rightarrow SbO$), which in turn scavenge OH• ($SbOH + OH\bullet \rightarrow SbO + H_2O$) and H• ($SbO + H\bullet \rightarrow SbOH$, $SbOH + H\bullet \rightarrow SbO + H_2$) radicals. In addition, antimony halides have been found to delay the escape of halogen from the flame, thus increasing its concentration and concurrently diluting flammable volatiles. Furthermore, they operate as a blanket hindering the oxygen penetration into the decomposing polymer. Apart from halogen–antimony mixtures, synergistic interactions, leading to the reduction of polymer flammability, have been also discovered by blending brominated and chlorinated flame retardants. Evidence of synergy has been given in several publications also in the case of bromine- and phosphorus-based derivatives [34,44–46].

Tetrabromobisphenol-A (TBBPA), polybromodiphenylether (PBDE) compounds (containing up to 10 bromine atoms attached to a diphenyl ether molecule), hexabromocyclododecane (HBCD, HBCDD), and chlorinated paraffins have been high production halogen-based flame retardants for many years. Some of them have been used either as additive (in one application) or as reactive compounds (in another), with TBBPA comprising the most notable example. Recently, concerns over the persistence, ability to bioaccumulate, and potential for toxicity of some of the most broadly employed halogenated flame retardants have led to increasing regulation and restrictions on their production and use. Producers in Europe and the USA have already ceased the production and use of polybrominated biphenyls (PBBs). Penta- and octabromodiphenylethers have been banned too since they have been also identified as probable dioxine precursors; only decabromodiphenylether is still used in a limited number of countries outside Europe. Along with regulations, there is increasing consumer pressure as downstream users and consumers are reluctant to use halogenated compounds in general, even those against which evidence of adverse environmental and health effects has not been yet discovered. However, despite restrictions, in several cases, the phased out halogenated flame retardants are replaced by chemicals of similar structure and unknown toxicity. Moreover, at the end of life, the hazardous flame retardant containing products are often exported to developing countries (mainly as electronic waste, second-hand electronics, and used cars), where they are recycled through primitive technologies. Therefore, further investigation on the wider group of halogenated compounds, which are currently used as alternatives to banned ones, is required in order to assess the long-term health and environmental effects of these products. Furthermore, additional regulations, clarifying the waste management policy that should be adopted, are needed [47—49].

5.2 Phosphorous-based flame retardants

The phosphorus-containing substances comprise, value-wise, one of the most commonly used groups and probably the fastest growing segment of flame retardants due to their versatility and the increasing environmental awareness of the problems arising from the use of halogen-based additives. Since the element exists in several oxidation states, the range of phosphorus-containing flame retardants is extremely broad, allowing a wide variety of applications. Phosphorus-based chemicals are employed for enhancing the fire performance of thermoplastics, thermosets, textiles, paper, coatings, and mastics. They can be added in the polymer by simple blending or they may be incorporated into the polymer chains through homopolymerization, or copolymerization. Their flame-retardant mechanism depends on the type of phosphorus compound, the chemical structure of the polymer, and the fire exposure conditions. Phosphorous compounds can function through various mechanisms, in the condensed polymer or in the gas phase, and possibly in both phases concurrently. Regarding the operation in the condensed phase, it is generally accepted that phosphorus flame retardants are notably more efficient in oxygen- or nitrogen-containing polymers, which could be either heterochain polymers or polymers with these elements in pendant groups. By generating anhydrides of phosphoric and related acids, which act as dehydrating agents, phosphorous compounds react with these polymers promoting char formation (dehydration reactions lead to the formation of double bonds, which at elevated temperatures result in cross-linked or carbonized structures). In general, the chemical reactions involved in the decomposition of the matrix are redirected in favor of those yielding carbon rather than CO or CO_2. The charring mode of action of phosphorus can be exploited even in the case of poor char forming polymers, such as polyolefins and styrenics, through the introduction of a char forming additive. The acids evolved during the degradation of phosphorous-based chemicals can also form a thin glassy or liquid protective coating on the condensed phase, lowering thus the oxygen diffusion, as well as the heat and mass transfer between the gas and the condensed phase.

This glassy residue may also coat the char making it more strong and cohesive, thus delaying its collapse. Even if the char does undergo oxidation (usually by smoldering), the presence of a phosphorus compound tends to inhibit complete oxidation of the carbon to carbon dioxide, thus decreasing the exothermic heat of combustion (heat sink). In some cases, phosphorus compounds have been reported to act under fire-exposure conditions by evolving acids, which catalyze the thermal degradation of the polymer matrix. Hence, the melt viscosity is reduced and the flow (dripping) of the molten polymer away from the fire zone is facilitated, withdrawing the fuel that feeds the flame [11,50–52].

Besides retarding the progress of fire in the condensed phase, phosphorus additives may well perform in the gas phase exhibiting free-radical trapping properties. Volatile phosphorus compounds are among the most efficient inhibitors of combustion. They liberate species that cause the hydrogen atom concentration in the flame to be reduced, thus quenching the flame. The step in the flame chemistry, which is inhibited, is the same step hampered by the scavenging effect of halogens: the rate-controlling branching step involving the reaction of a hydrogen atom with an oxygen molecule to yield a hydroxyl radical and an oxygen atom. In fact, phosphorus is, at the same molar concentration, much more effective radical scavenger than bromine and chlorine. Some key reactions, of the hundreds possible, that have been proposed as the ones governing the gas phase action of phosphorous-based additives are the following:

$$PO\bullet + H\bullet \rightarrow HPO$$

$$PO\bullet + OH\bullet \rightarrow HPO_2$$

$$HPO + H\bullet \rightarrow H_2 + PO\bullet$$

$$OH\bullet + H_2 + PO\bullet r \rightarrow H_2O + HPO$$

$$HPO_2\bullet + H\bullet \rightarrow H_2O + PO$$

$$HPO_2\bullet + H\bullet \rightarrow H_2 + PO_2$$

$$HPO_2\bullet + OH\bullet \rightarrow H_2O + PO_2$$

In spite of the fact that there is a scarcity of detailed investigations in the flame zone identifying the intermediate products and monitoring the concentrations of the different species generated, the main principle seems to have been understood: it is believed that the PO radical plays the major role. However, the vapor-phase action of phosphorous compounds does not necessarily have to involve flame chemistry. Flame inhibition via a physical means of action, based on heat capacity and heat of vaporization, and, probably, endothermic dissociation in the gas phase, may also be important during the flame retarding process. In particular, it has been shown that in certain cases phosphorus compounds, besides their modes of action mentioned above, result in flame retardation also by virtue of their heat of vaporization and heat capacity [42,53–55].

Phosphorous-based flame retardants can either be inorganic or organic. The most commonly used inorganic compounds are red phosphorous and ammonium polyphosphate. The latter is mainly employed in intumescent formulations; its mode of operation will be hence discussed in the corresponding section. Red phosphorus, typically manufactured as a red black powder, is the simplest and most concentrated source of phosphorus for flame retardancy. Added just in small quantities, it is very effective in a wide range of polymeric systems including polyolefins, polyamides, polyesters, and thermosetting resins. In Europe, it is used in electrical parts (e.g. electrical switches) molded from glass-filled nylon. Red phosphorous is believed to act principally in the solid phase through the formation of phosphoric acid, although some evidence also suggests free-radical scavenging. Except its red color that impairs the optical homogeneity of the polymer matrix, the main disadvantage of red phosphorus is its propensity (attributed to its poor thermal stability) to react with water (moisture) during melt compounding, releasing the highly toxic phosphine (PH_3). An efficient route of minimizing the trace amounts of phosphine involves the incorporation of metal oxides (such as copper oxide, cadmium oxide, or zinc oxide), which transform phosphine into phosphoric acid. Alternatively, prior to use, red phosphorus can be coated with a resin [50,56,57].

The organic phosphorus-containing compounds exhibiting flame retarding characteristics are numerous, yet only few of them have met commercial success. Three are the major groups of organic phosphorus fillers that are commonly used: (1) phosphinates, (2) phosphonates, and (3) phosphate esters (Figure 8). Triethyl phosphate is an organic phosphate used in practice for enhancing the flame retardancy of polyester laminates and in cellulosics. Cycloaliphatic phosphates, such as pentaerythritol phosphates, which take advantage of the char forming ability of the pentaerythritol structure, are also employed. Aryl phosphates comprise an additional group of organic phosphates. Triphenyl phosphate is a representative of this group; it is used as a flame-retardant additive for engineering thermoplastics such as polyphenylene oxide–high impact polystyrene and Acrylonitrile butadiene styrene(ABS)–polycarbonate blends. The improvement of the flame retardancy of ABS–polycarbonate blends (as well as of thermoplastic polyesters, polyamides,

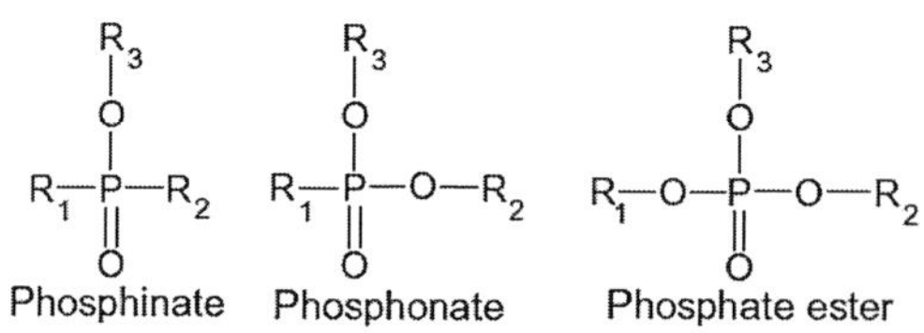

FIGURE 8

Chemical structures of commonly used organic phosphorus flame retardants [58].

and polycarbonates) can be achieved also with a member of the "aromatic diphosphates" family, precisely tetraphenyl resorcinol diphosphate. Turning to phosphonates, dimethyl methylphosphonate has been employed as a flame retardant in Aluminum trihydroxide (ATH)-filled polyester resins. On the other hand, aluminum diethylphosphinate, belonging to the class of phosphinates, has been used in combination with a melamine for improving the fire performance of polyamides, PBT, and epoxies. Apart from the abovementioned compounds, halogenated phosphorous flame retardants constitute a broadly used group, since they combine the properties of both the halogen and the phosphorus components. Two examples of compounds of this category are tris(chloropropyl)phosphate and tris(2-chloroethyl)phosphate. The phosphorous flame retardants, as mentioned, can be incorporated in the polymer either as additives or as reactive (co-)monomers. Simple reactive phosphate monomers, linear polyphosphazenes, and aromatic cyclic phosphazenes have attracted most attention as reactive flame retardants; however, none of them has been commercially exploited [4,51,58–60].

The growing need for the development of commercially viable and environmentally benign flame retardants, capable of competing the halogenated compounds, has rendered the approach of synthesizing formulations based on phosphorus-containing substances popular. In particular, it has been understood that currently the most feasible route to control or change the mechanisms of action and thus optimize the flame retardant efficiency of phosphorous compounds involves their combination with other (conventional) flame retardants. Numerous synergists or adjuvants have been examined for this purpose, proving that there is no universal synergist for phosphorus compounds as it is, for example, known for flame retardants containing halogen and antimony. In fact, various metal oxides, borates, inorganic fillers, and nitrogen-based compounds have been reported to exhibit synergistic actions with phosphorus flame retardants. In few cases, different phosphorus compounds were successfully used together as flame retardants, suggesting a "phosphorus–phosphorus" synergism [42,55,61].

5.3 Nitrogen-based flame retardants

Nitrogen-containing compounds comprise one of the most environmentally benign classes of flame retardants, producing low amounts of smoke, and no dioxin and/or halogen by-products during combustion. They also present several very important advantages over halogen-based compounds and red phosphorus owing to their chemistry, which is very similar to that of most polymers themselves. In addition, polymeric materials including nitrogen-based flame retardant have been shown to be suitable for recycling. Nitrogen compounds of principal interest as flame retardants are melamine, melamine derivatives, and related heterocyclic compounds. Nitrogen-containing flame retardants have been used for enhancing the fire performance primarily of polyamides and secondarily of polyolefins (and polyurethanes), but usually they are not very effective in other polymers alone [62,63].

Melamine (2,4,6-triamino-1,3,5-triazine, $C_3N_6H_6$) is a thermally stable crystalline product, produced by urea. Like cyanamide, it contains 67% nitrogen by mass and it is characterized by a melting point as high as 345 °C. The flame retardant actions of this inexpensive solid originates from the ability of melamine to sublimate (melt with vaporization) at about 350 °C, absorbing a significant amount of heat, thus decreasing the temperature of the degrading polymer's surface exposed to fire. At high temperature (or if evaporation is constrained), melamine decomposes evolving ammonia (which dilutes oxygen and combustible gases) and leads to the formation of the thermally stable condensates melem (2,5,8-triamino-1,3,4,6,7,9,9b-heptaazaphenalene), melam ((N-4,6-diamino-1,3,5-triazin-2-yl)-1,3,5-triazine-2,4,6-triamine), and melon (Poly(8-amino-1,3,4,6,7,9,9b-heptaazaphenalene-2,5-diyl)imino). The evolution of melam, melem, and melon is accompanied by the generation of residues in the condensed phase, resulting in the occurrence of endothermic processes, also effective for flame retardancy. Melamine is occasionally used by itself as a flame retardant (at levels of 3–20 phr, it has been found to endow nylon 6, 6.6, and 6/6.6 blends with V-0 rated performance), but more often it is employed as a blowing agent and flame-retardant adjuvant in intumescent coatings, elastomers, and plastic formulations as well as in flexible polyurethane foams [24,59,64].

Melamine is a weak base, which can form well-defined, thermally stable salts, with both organic and inorganic acids. The salts, having been found to possess flame retardant characteristics, and for that reason, commonly used in practice for improving polymer fire behavior, are melamine cyanurate, melamine phosphate, and melamine pyrophosphate. The latter two operate as intumescent compounds during combustion, and their mode of action is discussed in the following section. Melamine cyanurate (salt of melamine and cyanuric acid), the chemical structure of which is shown in Figure 9, on the other hand, has been reported to act in a completely different manner. In particular, it has been demonstrated that melamine cyanurate degrades polyamides at burning conditions promoting enhanced noncombustible flow dripping; hence, the polymer is withdrawn from the flame source not providing fuel to feed the fire. It also interferes with the combustion cycle in a physical manner, functioning as a heat sink and as an inert gas source. More specifically, its initial endothermic dissociation followed by the sublimation of the produced melamine and the subsequent degradation of melamine vapors absorb the generated

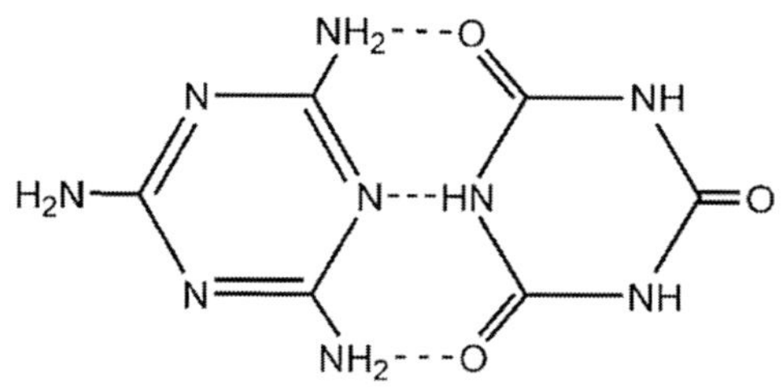

FIGURE 9

Chemical structure of melamine cyanurate.

heat, cooling the substrate. Furthermore, the inert gases produced contribute additionally to flame retardancy by diluting the oxygen and the flammable gases. Similar action to that of melamine cyanurate has been reported also in the case of melamine oxalate and melamine phthalate. However, melamine cyanurate is the leading flame retardant for unfilled PA6 and PA6,6, being quite more efficient (lower quantity of the flame retardant is needed) in the case of the PA6,6. This difference is associated to the degradation mechanisms of these two polymers: the degradation products of PA66 (cyclopentanone) may cross-link with the decomposition products (mainly NH_3) of melamine cyanurate, yielding in less flammable high molecular weight structures. On the contrary, PA6 decomposed to less reactive compounds that do not cross-link. Though efficient for nonreinforced polymers, melamine cyanurate is relatively ineffective for filled compositions like glass-fiber-reinforced polymeric materials; glass fibers act as a "wick" and retard melt flow thus holding the polymer in place to fuel the flame [65–69].

The majority of melamine salts operate in the condensed phase. A rather different mode of action was found by using *N*-alkoxy or cycloalkoxy derivatives as flame retardants forming radical species during decomposition. These compounds are effective for enhancing the performance against fire mainly of polypropylene and polyethylene films, and in several cases of polypropylene moldings, polyurethane adhesives and polystyrene. This novel family of nitrogen-based flame retardants was discovered by the inventors of Ciba Specialty Chemicals (now BASF), who were the first that realized that alkoxyamines can individually act as flame retardants without the assistance of other additives (synergists). Alkoxyamines are typically synthesized from the corresponding sterically hindered piperidine via the nitroxyl intermediate through reaction with a hydrocarbon; to facilitate this reaction, the presence of an organic peroxide (e.g. peracetic acid), a hydrogen peroxide, molybdenum oxide or iodide is needed. They were first introduced as UV stabilizers for providing protection against oxidation in automotive coatings and agricultural films. However, it was unexpectedly found that they do not only improve the UV stability of these materials but they also contribute to their fire performance, even when solely added. Their activity is based on the thermolysis of nitroxyl ethers, during which either alkoxy and aminyl radicals or alkyl and nitroxyl radicals are formed interfering with the free radical chemical reactions taking place during the combustion process. Alkoxyamines have also been reported to synergize with conventional flame retardants, including brominated, phosphorus, and inorganic compounds, enhancing further their flame retarding action. The first commercial product exploiting the fire retarding characteristics of alkoxyamines was put on the market in 2000 by Ciba under the trade name Flamestab NOR 116 (its structure includes an *N*-cyclohexyloxy-2,2,6,6-tetramethylpiperidinyl group attached to an oligomeric polymer chain). Since then, further research has been focused on a series of different *N*-alkoxy hindered amines as well as on nitrogen compounds able to generate free radicals during combustion like azoalkanes [59,70–72].

Another nitrogen-containing formulation, exhibiting flame retardant properties, is the oxazene resin; it can be used either as homopolymer or as a reactive additive in epoxies, presenting, compared to brominated epoxy resins, superior mechanical and electrical properties, good processing characteristics, and low toxicity and corrosiveness, in addition to elevated flame retardancy. Poly(isocyanurate)s (PICs) also present inherent flame retardancy, owing to their cyclic structure. Nevertheless, it is especially effective when combined with phosphorus containing epoxy resins. Epoxy resins generally exhibit an LOI of 20; quite recently, a route for producing phosphazene–triazine formulations achieving LOI between 32 and 38 was reported. These polymers, through a condense-phase mode of action, develop a charred coating on their surface during combustion preventing the substrate from further degradation. Formulations, produced via the polycondensation of macrocyclic chloric cyanure chloride with aromatic diamines or diols, have also been proposed to possess promising fire properties. It has been, accordingly, shown that the inclusion of –NH– linkages in the polymer chain comprises an effective means of enhancing polymer flame retardancy. Moreover, the co-presence of metals (Cu, Co, and Ni) results in further increase of thermal stability [4,73–76].

5.4 Intumescent systems

In French, the verb "tumere" means "to swell," while the word "intumescence" comes from the Latin word "intumescere," which means "to swell up". The verb "intumesce" was first used by John Webster (1580–1624) having two meanings: "to grow and to increase in volume against the heat" or "to show an expanding effect by bubbling". Nowadays, this term is used to describe a material (called intumescent), which when exposed to heat and beyond a critical temperature starts to swell and then to expand. The major historical use of this intumescent chemistry has been in water-based paints and coatings, since, as a consequence of the "intumescence" process, these materials can develop a superficial, foamed, cellular, charred layer during combustion protecting the substrate. They have been extensively employed in the construction industry for the protection of metallic materials, preventing the loss of their mechanical strength that leads eventually to the collapse of building structures. The early formulations were based on ammonium polyphosphate and nitrogen-rich compounds. The charred layer, whose density typically decreases as a function of temperature, acts thus as a physical barrier that impedes heat and mass transfer between the gas and condensed phases. In particular, by limiting heat transfer, results in the reduction of the substrate's temperature, causing a lag in the temperature rise of the surface beneath the char. The char hinders furthermore the diffusion of oxygen to the site of combustion, preventing also the flammable volatiles from reaching the flame. In the case of plastics, the dripping of the molten material is also reduced by the presence of the charred "shield," eliminating hence the possibility of the creation of another source that would result in further fire propagation. Furthermore, when the underlying polymer takes part in the formation of the charred layer, its charring occurs at the expense of the amount of the polymer that would be converted into flammable gases to feed the flames [77–79].

Considering the above, it becomes apparent that intumescent flame retardants operate thus in the condensed phase, interrupting the self-sustained combustion of the polymer at its earliest stage (i.e. the stage of thermal decomposition, during which the evolution of gaseous fuel takes place). Generally, intumescent systems consist of three components: (1) an acid source (e.g. ammonium polyphosphate, melamine polyphosphate), (2) a carbonization compound (e.g. pentaerythrithol, phenol—formaldehyde resins, starch), and (3) a blowing agent (e.g. melamine, guanidine, urea, chlorinated paraffin). The development of the swollen char, which occurs via a multistage physical/chemical process, is the outcome of the interaction of these components: the acid source decomposes to a mineral acid that dehydrates the polyol to produce the carbon char (mainly via a free radical process), which is in turn foamed by nonflammable gases emitted during the degradation of the blowing agent. The most critical step is the evolution of the gases; they should be released during the thermal decomposition of the carbonizing agent (after the melt forms, but before the char hardens) in order to trigger the expansion of the charred layer. Moreover, the generation of the acid source must occur at relatively low temperatures, specifically the temperature at which the decomposition of the polyhydric material begins. Initially, it was thought that the role of the carbonization (a.k.a. carbonific) compound and of the blowing (a.k.a. spumific) agent could be played only by additives. Nevertheless, studies showed that the polymer to be protected may also contribute. For example, polymers containing suitable functional groups, such as polyamide 6, when combined with melamine polyphosphate contribute to the formation of the charred layer as well as to its expansion. More specifically, in the case of PA6, polyphosphoric acid, generated by the elimination of melamine from the flame retardant, attacks the polymer forming phosphoric esters, which thermally degrade to carbon—carbon unsaturated species, yielding eventually charred structures. Simultaneously, ammonia, evolved by the decomposition PA6 (and of melamine), volatilizes to gases which inflate the char residue. The reactions between PA6 and melamine polyphosphate leading to char formation are summarized in Figure 10 [80—83].

In general, the intumescent char is a heterogeneous material consisting of gases "captured" in a phosphoro-carbonaceous cellular material; the latter comprises a mixture of a solid phase and liquid phase (acidic tars), capable of trapping the gaseous and liquid products of polymer degradation. This phosphoro-carbonaceous cellular material is composed of (1) polyaromatic species that are organized in stacks (pregraphitization stage) and are bridged by polymer links and phosphate groups, (2) crystalline additive particles, and (3) an amorphous phase that encapsulates the crystalline domains. The latter (made of small polyaromatic molecules, easily hydrolyzed phosphate species, alkyl chains generated due to the additives degradation, and polymer fragments) has to be voluminous enough to perfectly coat the crystalline domains. The final volume of the charred layer (as well as the shape and size of the cells) highly depends on (1) the amount of gases liberated during char formation and, mainly on, (2) the rheological characteristics (plasticity) of the charring mass. In fact, the rheological and mechanical properties of the intumescent "shield"

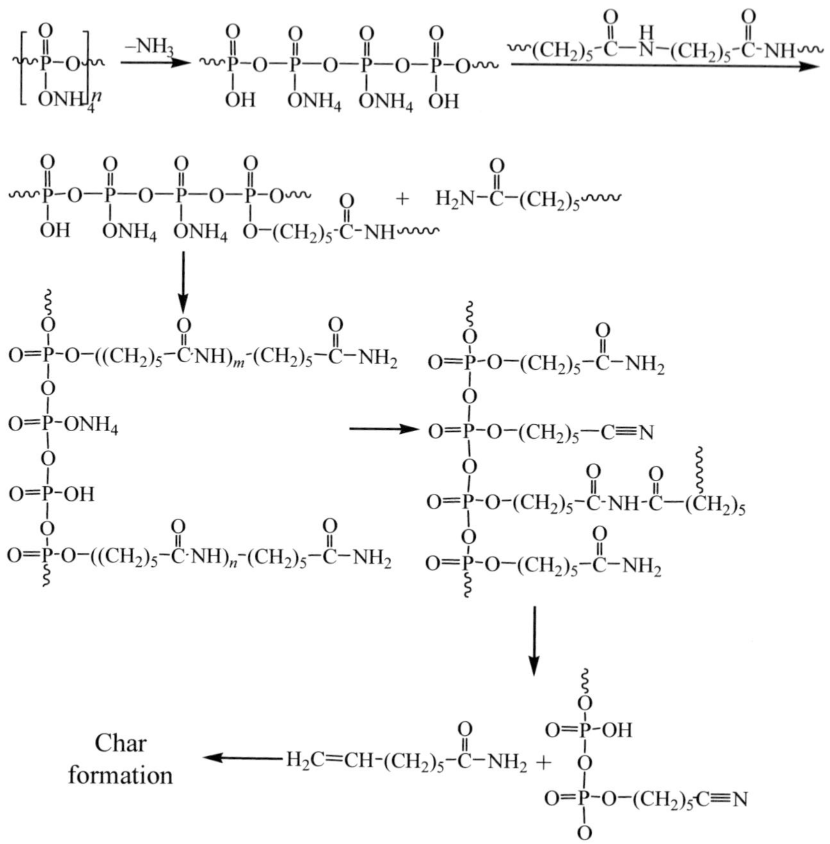

FIGURE 10

Reaction mechanism of melamine polyphosphate in PA6.

determine its fire retardant performance. The changes in these properties can be (roughly) explained by the chemistry of the intumescence process, which in the case of an intumescent polypropylene can be summed up in three steps: first (at 300–340 °C), due to the reduction with time of the apparent viscosity of the material under strain, the entanglements in the residual polymeric chains break up and (via a free radical process) the polymer chain length decreases. Then (at 340–390 °C), the apparent viscosity increases with time, and as a result of the reaction between the polymer chains and phosphate species, a stable layer (of high viscosity) is formed, able to trap the gaseous products emanating from polymer degradation and also to accommodate the stress induced by the solid particles and the high pressure applied by the trapped gases. Finally, at temperatures higher than 400 °C, the viscosity decreases dramatically leading to the formation of cracks in the material (breakage of the polyaliphatic links and evolution of the corresponding polyaliphatic chains) and, hence, the loss of its protective performance [36,79,81,84,85].

Even though intumescent systems have been evidently demonstrated to provide efficient fire protection to polymeric materials, a major problem that may be encountered when employing several intumescent flame retardants is their water solubility, owing to which they may leach out from the polymer matrix during normal use. Other disadvantages of these systems are related mainly to their thermal stability, high cost, and their inefficiency at low addition levels. To enhance the fire properties of intumescent formulations, several synergistic agents have been proposed. For example, it has been found that by adding, along with ammonium polyphosphate (APP)/pentaerythrithol (PER), small amounts of zeolite to polyolefins, their fire retardant properties can be drastically enhanced. Zinc borate, or at least a combination of zinc oxide and a borate, has been also claimed to improve the protection provided by typical intumescent systems. Moreover, intumescent coatings can be rendered more protective by incorporating Ti or Zr or other metal borides, nitrides, or carbides. Another route for improving the insulation effectiveness of an APP-containing intumescent (in PA6) involves incorporating additionally small amounts of talc and manganese dioxide. The boroxo-siloxane elastomer is also synergized effectively with the classic APP/PER intumescent system, improving the flame retardancy of polyolefins. However, quite recently it was found that large synergistic effects are evident when intumescent formulations are reinforced by nanofillers (such as layered silicates). The presence of nanofillers exerts a positive influence on the chemical (e.g. in the case of nanoclay, cohesive aluminophosphate ceramic structures are formed) and physical (superior char strength, better thermophysical characteristics) properties of the intumescent char enhancing its performance against fire [11,86–89].

5.5 Mineral flame retardants

Although the addition of any type of inorganic filler, even inert, influences the fire behavior of polymers (the content of combustible products is reduced, the thermal conductivity of the material is altered, and the melt viscosity is increased), due to stringent requirements (flame retarding effectiveness, ease of processing, no effect on other polymer properties, etc.) for the commercial use, very few have achieved market acceptance. The minerals most commonly used as flame retardants are metal hydroxides, borates, and hydroxycarbonates [24,90].

Metal hydroxides comprise the most important segment in the market of flame retardants; of the estimated 465,000 metric tonnes of flame retardants used in Europe in 2007, they accounted for 251,000 metric tonnes or 54% of the total (halogenated compounds: 17%, phosphorous-based: 20%, melamine-based: 3%, borates: 3%, antimony trioxide: 3% (source: CEFIC)). Their commercial prominence is mainly due to their relatively low cost, nontoxicity, noncorrosiveness and the fact that they can be handled easily. During combustion, metal hydroxides serve to remove heat by releasing copious quantities of water at the same temperature range or below the temperature at which polymer's decomposition takes place. Thus, by absorbing heat, they slow down the polymer pyrolysis process. Also, the generated water vapors dilute the combustible gases of polymer decomposition, thus preventing the exothermic radical reactions from taking place in the combustion zone. Furthermore, a nonflammable layer on the surface of the material is developed, protecting the substrate. Smoke suppression is another contribution of metal hydroxides to flame retardancy. Studies have shown that the amount of the carbon monoxide evolved during the combustion of polymers containing inorganic hydroxides is generally lower than that of the neat matrix. The exact mechanism has not been yet clearly elucidated; the most probable explanation seems to be that carbon, resulting from the degradation of the polymer, is deposited on the oxide that is produced by the decomposition of the hydroxide, and then it is volatilized as carbon dioxide, releasing no smoke. Furthermore, owing to the fact that water is the only product of their decomposition, inorganic hydroxides do not add significantly to the corrosive nature of the smoke liberated during the combustion of polymers containing them. The two most commonly employed mineral flame retardants are aluminum trihydroxide (ATH) and magnesium dihydroxide (MDH) [90,91].

ATH (a.k.a. alumina trihydrate or hydrated alumina) is the most popular inorganic flame retardant in current use. The dominating flame retarding effect of ATH is a consequence of its endothermic decomposition (occurring at 180–200 °C) to aluminum oxide and water, which absorbs around $1\ kJ\ g^{-1}$ of heat energy: ($2Al(OH)_3 \rightarrow Al_2O_3 + 3H_2O$). Thus, the heat feedback from the burning gases to the decomposing polymer is alleviated; the resulting aluminum oxide forms also a layer which further protects the unburned material beneath. Major applications of ATH are in vinyl and polyolefin cable jackets, in unsaturated polyester resins, and in elastomers used for the production of conveyer belts, and carpet backing. In general, the range of ATH applications is limited due to its relatively low decomposition temperature; thus it can be used only in polymers that are processed below its decomposition temperature (so that water evolution coincides with fuel release from the degrading polymer). By virtue of its higher decomposition temperature ($\sim$340 °C) compared to ATH, MDH is the second largest selling inorganic compound, extensively used as a stand-alone flame retardant in a wide range of polymers. Similar to ATH, the effectiveness of MDH (a.k.a magnesium hydroxide) derives from the endothermic ($1.3\ kJ\ g^{-1}$) water release, leaving MgO which serves as thermal barrier ($Mg(OH)_2 \rightarrow MgO + H_2O$). The largest volume uses are in wires and cables, and in thermoplastic polyolefin roofing membranes. The principal drawback of both ATH and MDH is that typically at least 60% by mass of them should be incorporated in the polymer in order to achieve good fire retardant properties. Such high loading levels, however, negatively affect the mechanical properties of the resulting material; to increase their effectiveness, metal hydroxides are often used in combination with other types of flame retardants. For example, ATH along with antimony oxide are employed to flame retard flexible PVC wire and cable formulations. PVC-based materials exhibit improved fire performance when mixed with inorganic hydroxides and molybdenum-containing compounds. ATH or MDH and red phosphorus have been reported to comprise another couple of effective synergists for enhancing polymer flame retardancy. Quite recently, it was shown that improved fire performance, at considerably lower additive loadings, can be also achieved by using MDH nanoparticles [59,92–94].

Borates, precisely zinc borates, comprise another group of inorganic additives that have been found to impart to polymeric materials enhanced performance during combustion. They are usually depicted as a combination of oxides and water, (*x*) $ZnO \cdot (y)B_2O_3 \cdot (z)H_2O$, and there are at least 25 distinct members of this family, but only two of them possess characteristics (flame retarding efficiency and good thermal stability) allowing their use as flame retardants: $4ZnO \cdot 6B_2O_3 \cdot 7H_2O$ and $2ZnO \cdot 2B_2O_3 \cdot 3H_2O$. Zinc borate belonging to the second category is the most used. It works predominantly in the condensed phase promoting the formation of char, but it also performs as smoke and afterglow suppressant. In addition, as a result of its endothermic decomposition (0.5 kJ g^{-1}) liberating water, boric acid, and boron oxide, it serves as a heat sink diluting also flammable volatiles. Moreover, boron oxide (B_2O_3) forms a superficial barrier further protecting the underlying material. The major use of zinc borate is in PVC and halogenated polyester, either as a complete or, more frequently, as a partial replacement for antimony oxide. Contrary to antimony oxide, which is a vapor phase flame retardant, zinc borate increases markedly the amount of char formed during polymer combustion. In particular, it reacts with hydrogen chloride, released during the degradation of PVC, forming zinc hydroxychloride and zinc chloride as well as boric oxide and boron trichloride. The latter functions in the gas phase, while boric oxide, comprising a low melting glass, covers the char, enhancing its stability. The formation of char is provoked the zinc species (remaining in the condensed phase), which alter the degradation pathway of the polymer by promoting dehydrohalogenation. Zinc borate is often used in combination with other flame retardants such as ATH, triggering the development of a porous ceramic layer during polymer combustion that thermally insulates the substrate. A ceramic structure is formed also by the reaction of zinc borate with MDH [29,95–97].

An alternative to the aforementioned conventional flame retardants has been reported to be hydroxycarbonates, specifically magnesium hydroxycarbonate, better known as hydromagnesite. Hydromagnesite ($3MgCO_3 \cdot Mg(OH)_2 \cdot 3H_2O$) is a mixture of magnesium carbonate and magnesium hydroxide. Hydromagnesite decomposes endothermically releasing water and carbon dioxide over a wider temperature range than ATH and MDH, implying superior flame retardant effectiveness. It can be synthetically prepared via a process involving the decomposition of magnesium bicarbonate. In nature, hydromagnesite is usually found in mixed deposits with huntite. This is the reason why such mixtures have attracted interest as fire retardants. In Figure 11, a natural mixture of hydromagnesite and huntite is shown: the larger hydromagnesite particles can be seen interspersed within the smaller platy huntite particles. Being a carbonate of magnesium and calcium, huntite ($Mg_3Ca(CO_3)_4$) is not really an inorganic hydroxide. Thus, it does not exhibit a flame-retardant performance similar to that hydroxides; it is used when superior thermal stability of the flame retardant is needed, e.g. in compounds where the addition of hydromagnesite alone becomes unsuitable due to higher processing

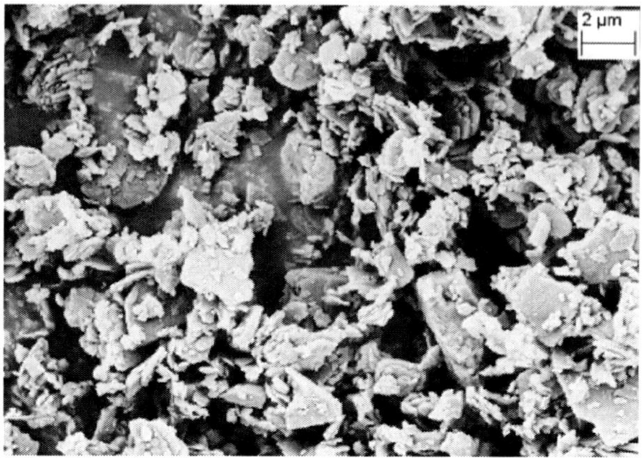

FIGURE 11

Natural mixture of hydromagnesite—huntite mixtures [100].

temperatures. Moreover, huntite has been found to possess good smoke-suppressant properties. In practice, hydromagnesite—huntite mixtures are employed in applications, especially in wire and cable manufacture, where ATH cannot survive processing and the use of MDH is too expensive. It has been also demonstrated that these mixtures endow with improved flame retardancy both halogen-containing (e.g. ethylene propylene copolymer containing antimony trioxide-decabromodiphenyl oxide) and halogen-free (ATH-doped ethylene propylene diene monomer rubber) polymeric materials [24,90,98—100].

5.6 Silicon-based flame retardants

Silicon-based compounds have long been reported to substantially reduce the flammability of polymers. For years, however, silicons were considered only as useful co-additives in flame retardant systems; quite recent research findings showed that in several cases, they can be also employed as stand-alone additives. In fact, they can be used as fillers in polymers, as components in copolymers, or as the main polymer matrix. Almost all forms of silicon have been explored as flame retardants, including silicones, silicas, silsesquioxanes, and silicates. The latter two have been mostly used also in the nanoscale form; their mechanism of action will be discussed in the following section [4,101,102].

Silicones have been commercialized since the beginning of the 1940s, growing over the years into a considerable segment of polymer market. They are used in many diverse applications in civil engineering, construction, electrical, transportation, aerospace, textiles, and cosmetics industries. During combustion, silicones burn at a slow rate, without generating dripping, and, when pure, without emission of toxic smoke (low yields of carbon monoxide). In general, a condensed-phase mechanism which involves the formation of a protective surface layer has been proposed to explain the flame retarding action of silicones. When exposed to heat, silicones, unlike organic polymers, decompose leaving an inorganic silica residue, which serves as a mass and heat transport barrier between the gas and the condensed phase. The most commonly used flame retardant based on silicone is a polyorganosiloxane, namely polydimethylsiloxane (PDMS). PDMS is among the polymers preferred for applications, at which high temperatures are encountered, such as in electrical wires and cables. The key structural elements of polysiloxanes that (directly or indirectly) influence their stability include the inherent strength of the siloxane (Si—O) bond, the flexibility of the —(Si—O)x— chain segments, and the stability of low-molecular-weight cyclic siloxanes compared to that of their high-molecular-weight linear counterparts. Extensive research, conducted on the fire retardancy of various types of PDMA block copolymers and blends (e.g. with polycarbonates, polyurethane, and polystyrene), has shown that in most cases, the flammability properties of polymer—silicone blends and of the copolymers are about the same but not sufficient to get a UL-94 V-0 rating. Therefore, other flame retardants should be additionally incorporated. Several types of fillers have been used to improve the flame retardancy of silicone, such as silica, calcium carbonate, wollastonite, mica, talc, kaolin, montmorillonite, carbon black, aluminum trihydrate, and magnesium dehydrate. Another promising approach for fire performance enhancement involves the use of polyborosiloxane (which is produced by PDMS through a polycondensation reaction). Contrary to PDMS which transforms into a fine silica powder during combustion, polyborosiloxane, due to the presence of the boron atom, forms a more cohesive ceramic layer on the surface of the burning polymer, thus offering better protection against fire [103—106].

Similarly to silicone-containing formulations, silica systems form a superficial layer upon exposure to heat, protecting the substrate from the action of fire. Hence, just like silicones, silicas have been found to operate in the condensed phase through a physical flame-retardant mechanism rather than a chemical one, meaning that they do not provoke the development of carbonaceous char, but they form an inorganic "shield" on polymer surface [24,103]. Various types of silica have been examined for reducing polymer flammability. For example, Kashiwagi et al. [107] investigated the effect of silica gel, fumed silica, and fused silica in a non-char-forming thermoplastic, precisely polypropylene and in a polar char-forming thermoplastic, namely polyethylene oxide. Accordingly, it was concluded that the effectiveness of the additive is determined by the balance between its density and surface area. Having large surface area and low density, fumed silica and silica gel increased polymer melt viscosity; so, they could accumulate near the regressing polymer surface without sinking, forming a barrier that resulted in superior fire performance compared to that of the meat matrix. On the contrary, fused silica, due to its high density and low surface area, tended to sink through the polymer melt during combustion (as the melt viscosity did not increase), did not accumulate near the surface, and hence, being unable to construct a superficial layer, was not effective for decreasing polymer flammability. An analogous conclusion, i.e. the larger the coverage over the polymer surface (associated with the level of silica accumulation) is, the better the fire behavior is, was derived by the authors in the case of PMMA. Two types of silica (fused silica and silica gel) were introduced in two different molecular weight PMMA samples; in so doing, it was found that both the level of accumulation and the surface coverage depended highly not only on the characteristics of silica but also on the melt viscosity of the matrix [108]. Another characteristic of silica influencing its flame retarding efficiency, apart from surface area and density, was highlighted in a previous study, by Gilman et al. [109], to be mainly its pore volume. By blending polypropylene with three silica gels of different pore volume, particle size, and surface silanol concentration, it was revealed that the higher pore volume silica led to greater reduction in the HRR. Larger pores could slow down volatilization and thus the evolution of degradation flammable gases, owing to their capability to accommodate polymer macromolecules impeding their mobility.

5.7 Nanoparticles

Polymer nanocomposites comprise one of the most promising developments in the area of flame retardancy, offering significant advantages over conventional formulations. They first attracted attention due their unique properties that are typically not shared by their conventional microscopic counterparts. Attributed to their nanometer size features and the extraordinarily high surface area of the dispersed nanoparticles, nanocomposites have been found to possess remarkably enhanced mechanical and thermal properties accompanied by an excellent barrier performance. What is more, it has been well established that this dramatic improvement in properties can be achieved with the addition (through processing techniques normally used for plastics like extrusion, injection molding, and casting) of only a small fraction of nanoparticles, thus without impairing the optical homogeneity of the matrix. The main factor determining the enhancement of properties is the degree of nanoparticles dispersion; better dispersion increases polymer—filler interactions, leading to the most significant enhancement in physical properties [110—112].

Among all potential nanocomposite precursors, those based on layered silicates have been most extensively investigated, probably due to the easy availability of the starting materials and the fact that their intercalation chemistry has been studied for a long time. However, over the years many different shapes and types of nanoparticles have been subject to examination. On the basis of dimensionality, the nanoparticles, which have been mostly investigated in terms of their ability to enhance polymer fire retardancy, can be sorted into the following three categories: (1) layered, (2) fibrous, and (3) particulate materials. The layered nanofillers (referred as 2D), which are characterized by one nanometric dimension and are present in the form of sheets of one to a few nanometers thick and hundreds to thousands nanometers in extent, include clays (e.g. montmorillonite), graphite, and layered double hydroxides (LDHs). The best-known representatives of the second category (particles, referred as 1D, whose two dimensions are on the nanometer scale and the third is larger, forming an elongated structure) are carbon nanotubes. Particulate materials (0D nanoparticles), the three dimensions of which are on the order of nanometers, that have been examined as potential flame retardants are polyhedral oligosilsesquioxane (POSS) and spherical silica [113–115].

Nanoparticles operate principally in the condensed phase promoting the formation of a continuous superficial layer comprising a network, which protects the underlying material. In the case of polymer-layered silicates (clay) nanocomposites, this structure is developed as a consequence of polymer ablation caused by pyrolysis, with the dewetted clay particles left behind. Rising bubbles (which are formed by the decomposing polymer, the clay surfactant, or even by clay itself inducing heterogeneous nucleation and subsequent bubbling) has been reported to be an additional reason for the movement and accumulation of silicates from the bulk to the surface. Clay particles are also claimed to play a catalytic role on the promotion of char forming reactions. According to another mechanism, paramagnetic iron, present in the clay structure, can induce radical trapping to prevent degradation. LDHs, on the other hand, provide the superficial layer mainly via the promotion of char and the formation of metal oxide residues. Additionally, the endothermic release of interlayer and structural water enables considerable heat absorption and dilution of combustible gases. Similarly to clays, graphite particles provoke char formation during combustion. In fact, the development of polymer–graphite nanocomposites was motivated by research findings that layered silicates form graphite-type char during the combustion of polymer nanocomposites. Two kinds of graphites have been employed for nanocomposites manufacture: (1) expanded graphite and (2) graphite oxide. The most widely studied carbon-based nanoparticles are, however, carbon nanotubes, existing in two different types: single-walled and multiwalled carbon nanotubes. The flame retardant action of nanotubes is credited to the formation of a structured char-nanotube network on the surface of the polymer, acting as a shield that reemits much of the incident radiation back into the gas phase. Moreover, carbon nanotubes increase the thermal conductivity of the polymer. Turning to particulate materials, POSS-containing nanocomposites form upon combustion oxidatively stable, nonpermeable surface chars reinforced by ceramic materials generated during POSS degradation. A physical, i.e. generation of inorganic residues, not chemical, i.e. promotion of charring reactions, effect is reported as well in the case of nanoscale silica. Apart from the abovementioned nanoparticles, the influence of metal oxides (2D, 1D, and 0D) on polymer fire performance has been also studied. Accordingly, it has been shown that metal oxides, such as titanium oxide (TiO_2) and ferric oxide (Fe_2O_3) particles, improve polymer flame retardancy by (1) acting as a heat shield that limits the thermal conduction into the sample and (2) hindering the release of (flammable) gases due to the increase in melt viscosity [11,59,116–121].

The most important contribution of nanocomposites to the flame retardancy of polymers is the decrease in the pHRR, reflecting reduction in flame spread. However, pHRR is not an intrinsic polymer property; it strongly depends not only on the nature of the material but also on the conditions at which it is measured (e.g. geometry and thickness of the samples, set-up of the test). For example, in thin samples, volatilization predominates, thus the char-promoting action of nanoparticles has not significant effect. On the other hand, the THR, which comprises an intrinsic property of polymers, is not altered much by the addition of nanofillers. THR gives an indication of material's performance in fully developed fires. The reduction of pHRR along with the nonalteration of THR implies that nanocomposites burn slowly for a longer time, but most of the material is converted to products that feed the flames. The time required for ignition (TTI) is another property that is not significantly influenced or, in most cases, worsened by the addition of nanoparticles. More specifically, the ignition delay times of nanocomposites (basically those based on carbons) measured in a cone calorimeter tend to be shorter than those of neat matrixes. It seems, therefore, that in the first stages of combustion, nanocomposites exhibit an earlier mass loss. Faster ignition is a consequence of the rapid temperature increase occurring on the surface of the nanocomposites; their higher melt viscosity compared with that of the unfilled matrix along with the labyrinths constructed by the nanoparticles hampers the diffusion of heat in the bulk of the material, accelerating the degradation (on the surface) of the polymer and thus the production of flammable volatiles. Other causes that may account for accelerated ignition are associated with by the exact type of the nanofillers used. For example, as shown in Figure 12 (where the flame retardant modes of actions of polymer—clay nanocomposites are schematically represented), layered silicates catalyze the decomposition of the polymer matrix (due to the hydrolysis by the water bound to clay, and due to the catalytic activity of silicates acidic sites), resulting in earlier weight loss. The

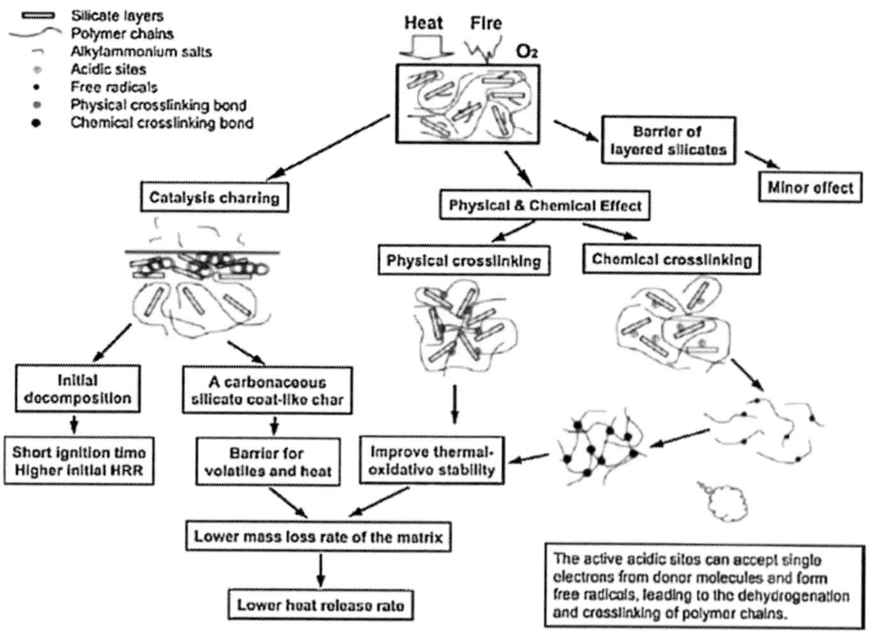

FIGURE 12

Flame retardant mechanism of polymer/montmorillonite(MMT) nanocomposites [122].

major drawback of nanofillers is, however, that they fail to perform better than the pristine polymer in industrially important tests, such as the LOI test, where the ignition resistance is under examination or the UL94 test, where a self-extinguishing performance is required for the material to qualify. In the fire scenarios replicated in these tests, barrier properties play a minor role, while changes in the dripping characteristics become crucial. Viscous melts, produced in the presence of nanoparticles, exhibit a limited ability to drip, thus holding the polymer in place to feed the flame [11,122–125].

Considering that a flame retardant additive is considered to be any compound or mixture of compounds that render polymers less likely to ignite or, once ignited, less prone to burn effectively, *whatever the extent of the effect*, nanofillers (at least the ones mentioned above), by definition, can be regarded as flame retardants (even though their main effect is the pHRR reduction). Nevertheless, fulfilling the "entry requirements" of a definition does not guarantee market acceptance. In fact, it has long been realized that nanofillers should be combined with conventional flame retardants, in order to develop synergistic systems with high efficiency, acceptable environmental impact, and hence a high potential of commercial exploitation. As a result, a great deal of effort has been spent and a large number of studies have been conducted for this purpose. Almost all possible ways of inserting flame retardants into nanocomposites have been followed involving direct blending, chemical attachment to the nanofillers, incorporation in the polymer backbone to form flame retardant copolymer nanocomposites, inclusion in the curing agent in the case of thermosetting nanocomposites, etc. Furthermore, the vast majority of conventional flame retardants operating through chemical and/or physical actions in either the gas or the condensed phase have been investigated for enhancing the fire retardant properties of nanocomposites. For example, synergistic phenomena were quested by exploiting the radical trapping ability of the halogenated compounds, the char promoting action of phosphorous-based flame retardants, the "intumescent" effect of melamine derivatives, etc. In certain cases, promising results were also obtained when combining different types of nanoparticles. Generally, research findings point out that flame retardants serve to optimize the fire properties of nanocomposites, suggesting that the proposed approaches offer novel means of developing flame retardant polymeric materials [11,24,124,125].

6. CONCLUSIONS

The growing use of polymers in many and varied applications with increasingly stringent requirements has resulted in a continuous demand for improved properties. Among these properties, low propensity of igniting or effective combustion is of major importance, given that plastic materials comprise a large fraction of the fire load in houses, commercial environments, and transportation. Most unwanted fires are fueled by synthetic polymers, which, as manufacturing technology has advanced, have substituted the vast majority of their, generally less flammable, natural counterparts (contained in wood, cotton, leather, and wool). The flammability of a polymer is not an intrinsic property, such as its density or heat capacity; its response against fire is dependent on the conditions. The behavior of two materials during fire may be reversed if tested under different conditions. Similarly, by altering the material composition, basically by adding a fire retardant, changes also the fire behavior of polymers. Traditionally, the incorporation of halogen-based fire retardants has been an economical method for improving the flame retardancy of polymers without sacrificing the quality of the final product. In recent years, however, concerns have been raised in regards to the environmental impact of (some of) the most commonly used halogenated compounds since they were found to be persistent in the environment, and bioaccumulative and toxic to humans and wildlife. This resulted in the banning of certain compounds, pushing the market trend to halogen-free flame retardants. Even though they are nontoxic and environmentally sound, halogen-free compounds also present disadvantages; inorganic substances, for example, which currently comprise the most used flame retardants, require high levels of loading, leading to additional costs, processing difficulties, and deterioration of polymer mechanical properties. On the other hand, intumescent systems are relatively expensive to be applied for the large-scale production of low-cost combustible materials. Generally, although potential alternatives do exist, they do not live up to the halogenated compounds on a performance or cost basis. Consequently, the objective of developing highly effective, "green" flame retardants has prompted a rapid progress during the last decade, extending research into novel technologies, with polymer nanocomposites being one of the most promising.

Research findings so far have shown that the reduction of polymer flammability remains a very complex problem, for which no universal solution can be given. Nevertheless, with the advent of new techniques, the improvement of the fire retardancy of plastic materials is becoming less empirical. Hence, it is believed that soon deeper knowledge sill be attained on the phenomena occurring during combustion and on mechanism of flame retardants action, promoting new technological breakthroughs. This book, presenting the state-of-the-art in polymer "green" flame retardation, is hoped to contribute to this direction. It is planned to comprise an inspiring and useful tool not only for those who are familiar with the topic of polymer flame retardancy but also to people new in this field. In this context, this introductory chapter provided a brief account of the phenomena of polymer combustion and flame retardancy, setting the background for understanding the topics which are considered in detail in the following chapters.

References

[1] Tesoro GC. Chemical modification of polymers with flame-retardant compounds. J Polym Sci Macromol Rev 1978;13:283–353.

[2] Hilado C. Flammability handbook for plastics. Pennsylvania, USA: Technomic Publishing Company; 1998. pp. 1–88.

[3] Landrock AH. Handbook of plastics flammability and combustion toxicology. New Jersey, USA: Noyes Publications; 1983.
[4] Lu SY, Hamerton I. Recent developments in the chemistry of halogen-free flame retardant polymers. Progr Polym Sci 2002;27:1661–712.
[5] Lomakin SM, Haslam E. Ecological aspects of polymer flame retardancy (new concepts in polymer science). Leiden, The Netherlands: Brill Academic Publishers; 1999.
[6] Crummett WB. Decades of dioxin. Princeton, NJ: Xlibris Corp; 2002.
[7] Elias HG. Plastics. In: Gerhartz W, editor. Ullman's encyclopedia of industrial chemistry, vol. 1. Weinheim, Germany: Wiley-VCH Verslag; 2002. pp. 1–123.
[8] Rosen SL. Polymers. In: Seidel A, editor. Kirk-Othmer encyclopedia of chemical technology, vol. 11. New Jersey, USA: John Wiley & Sons Inc.; 2004. pp. 1–20.
[9] Hirschler MM. Heat release from plastic materials. In: Brabauskas V, Grayson SJ, editors. Heat release in fires. London, UK: Elsevier Applied Science; 1992. pp. 375–8.
[10] Price D, Anthony G, Carty P. Introduction: polymer combustion, condensed phase pyrolysis and smoke formation. In: Horrocks AR, Price D, editors. Fire retardant materials. Cambridge, UK: Woodhead Publishing Ltd; 2001. pp. 1–30.
[11] Kiliaris P, Papaspyrides CD. Polymer/layered silicate (clay) nanocomposites: an overview of flame retardancy. Prog Polym Sci 2010;35:902–58.
[12] Hirschler MM. Chemical aspects of thermal decomposition of polymeric materials. In: Grand AF, Wilkie CA, editors. Fire retardancy of polymeric materials. New York, USA: Marcel Dekker Inc.; 2000. pp. 28–79.
[13] Bilingham NC. Degradation. In: Mark HF, editor. Encyclopedia of polymer science and technology, vol. 6. New Jersey, USA: John Wiley & Sons Inc.; 2005. pp. 1–49.
[14] Pielichowski K, Njuguna J. Thermal degradation of polymeric materials. Shropshire, UK: Rapra Technology Limited; 2005.
[15] Levchik SV, Wilkie CA. Char formation. In: Grand AF, Wilkie CA, editors. Fire retardancy of polymeric materials. New York, USA: Marcel Dekker Inc.; 2000. pp. 171–215.
[16] Wilkie CA, Levchik SV, Levchik GF. Is there a correlation between crosslinking and thermal stability? In: Al-Malaika S, Golovoy A, Wilkie CA, editors. Specialty polymer additives: principles and application. Oxford, UK: Blackwell Science; 2001. pp. 359–74.
[17] Drysdale D. An introduction to fire dynamics. Sussex, England: John Wiley & Sons; 1999.
[18] Irvine DJ, McCluskey JA, Robinson IM. Fire hazards and some common polymers. Polym Degrad Stab 2000;67:383–96.
[19] Mouritz AP, Gibson AG. Fire properties of polymer composite materials. Dordrecht, The Netherlands: Springer; 2006.
[20] Purser D. Toxicity of fire retardants in relation to life safety and environmental hazards. In: Horrocks AR, Price D, editors. Fire retardant materials. Cambridge, UK: Woodhead Publishing Ltd; 2001. pp. 69–127.
[21] Quintiere JG. Fundamentals of fire phenomena. Sussex, England: John Wiley & Sons; 2006.
[22] Schartel B, Hull TR. Development of fire-retarded materials-Interpretation of cone calorimeter data. Fire Mater 2007;31:327–54.
[23] Troitzsch J. International plastics flammability handbook. Munich, Germany: Hanser Publishers; 1990.

[24] Laoutid F, Bonnaud L, Alexandre M, Lopez-Cuesta JM, Dubois P. New prospects in flame retardant polymer materials: from fundamentals to nanocomposites. Mater Sci Eng R 2009;63:100−25.
[25] UL94-test for flammability of plastic materials for parts in devices and appliances. Northbrook, IL: Underwriters Laboratories Inc.; 1997.
[26] Babrauskas V, Peacock RD. Heat release rate: the single most important variable in fire hazard. Fire Saf J 1992;18:255−72.
[27] Kandola BK. Nanocomposites. In: Horrocks AR, Price D, editors. Fire retardant materials. Cambridge, UK: Woodhead Publishing Ltd; 2001. pp. 204−19.
[28] Beyer G. Nanocomposites: a new class of flame retardants for polymers. Plast Addit Compd; 2002:22−8.
[29] Levchik S. Introduction to flame retardancy and polymer flammability. In: Morgan AB, Wilkie CA, editors. Flame retardant polymer nanocomposites. Hoboken, USA: Wiley; 2007. pp. 1−29.
[30] Schartel B, Braun U. Comprehensive fire behaviour assessment of polymeric materials based on cone calorimeter investigations. e-Polymers. Article #13, http://www.e-polymers.org; 2003.
[31] Morgan AB, Bundy M. Cone calorimeter analysis of UL-94 V-rated plastics. Fire Mater 2007;31:257−83.
[32] Grand AF, Wilkie CA. Introduction: fire-retarded polymeric systems. In: Grand AF, Wilkie CA, editors. Fire retardancy of polymeric materials. New York, USA: Marcel Dekker Inc; 2000. pp. ix−xiii.
[33] Mack AG. Flame retardants, halogenated. In: Seidel A, editor. Kirk-Othmer encyclopedia of chemical technology, vol. 11. New Jersey, USA: John Wiley & Sons Inc.; 2004. pp. 454−83.
[34] Sutker BJ. Flame retardants. In: Gerhartz W, editor. Ullman's encyclopedia of industrial chemistry, vol. 14. Weinheim, Germany: Wiley−VCH Verslag; 2002. pp. 1−19.
[35] Joseph P, Tretsiakova-Mcnally S. Reactive modifications of some chain- and step-growth polymers with phosphorus containing compounds: effects on flame retardance-a review. Polym Adv Technol 2011;22:395−406.
[36] Camino G, Costa L, Luda di Cortemiglia MP. Overview of fire retardant mechanisms. Polym Degrad Stab 1991;33:131−54.
[37] Focke WW, Strydom CA, Bartie N. Thermal analysis of commercial inorganic flame retardants. SA J Chem Eng 1997;9:41−51.
[38] Skinner GA. Flame retardancy: the approaches available. In: Pritchard G, editor. Plastics additives, an A−Z reference. London, UK: Chapman and Hall; 1998. pp. 260−7.
[39] Bourbigot S, Le Bras M. Fundamentals: flame retardant plastics. In: Troitzsch J, editor. Plastics flammability handbook: principles, regulation, testing and approval. Munich, Germany: Hanser Publishers/Hanser Gardner Publications Inc.; 2004. pp. 134−48.
[40] Lyon RE, Janssens ML. Flammability. In: Mark HF, editor. Encyclopedia of polymer science and technology, vol. 6. New Jersey, USA: John Wiley & Sons Inc.; 2005. pp. 1−73.
[41] Georlette P, Simons J, Costa L. Halogen-containing fire-retardant compounds. In: Grand AF, Wilkie CA, editors. Fire retardancy of polymeric materials. New York, USA: Marcel Dekker Inc; 2000. pp. 246−84.
[42] Lewin M, Weil ED. Mechanisms and modes of action in flame retardancy of polymers. In: Horrocks AR, Price D, editors. Fire retardant materials. Cambridge, UK: Woodhead Publishing Ltd; 2001. pp. 31−68.

[43] Chanda M, Roy SK. Plastics fundamentals, properties, and testing. Florida, USA: CRC Press, Taylor & Francis Group; 2009.
[44] Ballistreri A, Montaudo G, Puglisi C, Scamporrino E, Vitallini D. Intumescent flame retardants for polymers. I. The poly(acrylonitrile)-ammonium polyphosphate-hexabromocyclododecane system. J Appl Polym Sci 1983;28:1743–50.
[45] O'Brien DD. Flame retarded polyethylene wire insulation. US Patent 5358991, 1994.
[46] Touval I. Flame retardants, antimony and other inorganic agents. In: Seidel A, editor. Kirk-Othmer encyclopedia of chemical technology, vol. 11. New Jersey, USA: John Wiley & Sons Inc.; 2004. pp. 1–19.
[47] Kemmlein S, Herzke D, Law RJ. Brominated flame retardants in the European chemicals policy of REACH-Regualtion and determination in materials. J Chromatogr A 2008;1216:320–33.
[48] Shaw SD, Blum A, Weber R, Kannan K, Rich D, Lucas D, et al. Halogenated flame retardants: do the fire safety benefits justify the risks? Rev Environ Health 2010;25: 261–305.
[49] Papachlimitzou A, Barber JL, Losada S, Bersuder P, Law RJ. A review of the analysis of novel brominated flame retardants. J Chrom A 2012;1219:15–28.
[50] Levchik SV, Weil ED. Combustion and fire retardancy of aliphatic nylons. Polym Int 2000;49:1033–73.
[51] Green J. Phosphorous containing flame retardants. In: Grand AF, Wilkie CA, editors. Fire retardancy of polymeric materials. New York, USA: Marcel Dekker Inc.; 2000. pp. 148–70.
[52] Schmitt E. Phosphorus-based flame retardants for thermoplastics. Plast Addit Compd 2007;9:26–30.
[53] Green J. Mechanisms for flame retardancy and smoke suppression – a review. J Fire Sci 1996;14:426–42.
[54] Babushok V, Tsang W. Inhibitor rankings for alkane combustion. Combust Flame 2000;124:488–506.
[55] Schartel B. Phosphorus-based flame retardancy mechanisms – old hat or a starting point for future development? Materials 2010;3:4710–45.
[56] Davis J. Flame retardants: halogen-free systems (including phosphorus additives). In: Pritchard G, editor. Plastics additives, an A–Z reference. London, UK: Chapman and Hall; 1998. pp. 276–86.
[57] Levchik SV, Weil ED. A review of recent progress in phosphorus-based flame retardants. J Fire Sci 2006;24:345–64.
[58] van der Veen I, de Boer J. Phosphorus flame retardants: properties, production, environmental occurrence, toxicity and analysis. Chemosphere 2012;88:1119–53.
[59] Zhang S, Horrocks AR. A review of flame retardant polypropylene fibres. Progr Polym Sci 2003;28:1517–38.
[60] Weil ED. Flame retardancy. In: Mark HF, editor. Encyclopedia of polymer science and technology, vol. 10. New Jersey, USA: John Wiley & Sons Inc.; 2005. pp. 21–54.
[61] Weil ED. Synergists, adjuvants, and antagonists in flame-retardant systems. In: Grand AF, Wilkie CA, editors. Fire retardancy of polymeric materials. New York, USA: Marcel Dekker Inc.; 2000. pp. 115–45.
[62] Lomakin SM, Zaikov GE. Ecological aspects of polymer flame retardancy. Utrecht, The Netherlands: VSP; 1999.
[63] Levchik SV, Weil ED. New developments in flame retardancy of styrene thermoplastics and foams. Polym Int 2008;57:431–48.

[64] Crews GM, Ripperger W, Kersebohm DB, Seeholzer J, Güthne T. Melamine and guanamines. In: Gerhartz W, editor. Ullman's encyclopedia of industrial chemistry, vol. 22. Weinheim, Germany: Wiley-VCH Verslag; 2002. pp. 1–18.

[65] Levchik SV, Levchik GF, Balabanovich AI, Camino G, Costa L. Mechanistic study of combustion performance and thermal decomposition behaviour of nylon 6 with added halogen-free fire retardants. Polym Degrad Stab 1996;54:217–22.

[66] Levchik SV, Balabanovich AI, Levchik GF, Costa L. Effect of melamine and its salts on combustion and thermal decomposition of polyamide 6. Fire Mater 1997;21: 75–83.

[67] Gijsman P, Steenbakkers R, Furst C, Kersjes J. Differences in the flame retardant mechanism of melamine cyanurate in polyamide 6 and polyamide 66. Polym Degrad Stab 2001;78:219–24.

[68] Weil ED, Levchik S. Current practice and recent commercial developments in flame retardancy of polyamides. J Fire Sci 2002;22:251–64.

[69] Kiliaris P, Papaspyrides CD, Pfaendner R. Polyamide 6 filled with melamine cyanurate and layered silicates: evaluation of flame retardancy and physical properties. Macromol Mater Eng 2008;293:740–51.

[70] Pfaendner R. Nitroxyl radicals and nitroxylethers beyond stabilization: radical generators for efficient polymer modification. CR Chim 2006;9:1338–44.

[71] Nicolas RC, Wilen CE, Roth M, Pfaendner R, King III RE. Azoalkanes: a novel class of flame retardants. Macromol Rapid Commun 2006;27:976–81.

[72] Pfaendner R. How will additives shape the future of plastics? Polym Degrad Stab 2006;91:2249–56.

[73] Schreiber H, Saur W. New fire retardant halogen free polymers. Makromol Chem Macromol Symp 1993;74:165–71.

[74] Aseeva RM, Zaikov GE. Flammability of macrocyclic polymers and their complexes with metals. Makromol Chem Macromol Symp 1993;74:335–8.

[75] Von Gentzkow W, Huber J, Kapitza H, Rogler W. Halogen-free flame-retardant plastics for electronic applications. J Vinyl Add Technol 1997;3:175–8.

[76] Mathew D, Nair CPR, Ninan KN. Phosphazene-triazine cyclomatrix network polymers: some aspects of synthesis thermal- and flame-retardant characteristics. Polym Int 2000;49:48–56.

[77] Horrocks AR. Developments in flame retardants for heat and fire resistant textiles – the role of char formation and intumescence. Polym Degrad Stab 1996;54:143–54.

[78] Antonov AV, Reshetnikov IS, Khalturinskij NA. Combustion of char forming polymeric systems. Russ Chem Rev 1999;68:605–14.

[79] Bourbigot S, Le Bras M, Duquesne S, Rochery M. Recent advances for intumescent polymers. Macromol Mater Eng 2004;289:499–551.

[80] Levchik SV, Camino G, Costa L, Levchik GF. Mechanism of action of phosphorus-based flame retardants in nylon 6. I. Ammonium polyphosphate. Fire Mater 1995; 19:1–10.

[81] Duquesne S, Jimenez M, Bourbigot S. Fire retardancy and fire protection of materials using intumescent coatings – a versatile solution? In: Hull TR, Kandola BK, editors. Fire retardancy of polymers – new strategies and mechanisms. Cambridge, UK: RSC Publishing; 2009. pp. 240–54.

[82] Kiliaris P, Papaspyrides CD, Pfaendner R. Polyamide 6 filled with melamine polyphosphate and layered silicates: evaluation of flame retardancy and physical properties. Macromol Mater Eng 2011;296:617–29.

[83] Kiliaris P, Papaspyrides CD, Xalter R, Pfaendner R. Study on the properties of polyamide 6 blended with melamine polyphosphate and layered silicates. Polym Degrad Stab 2012;97:1215–22.

[84] Camino G, Delobel R. Intumescence. In: Grand AF, Wilkie CA, editors. Fire retardancy of polymeric materials. New York, USA: Marcel Dekker Inc; 2000. pp. 218–43.

[85] Bourbigot S, Duquesne S. Intumescence and nanocomposites: a novel route for flame-retarding polymeric materials. In: Morgan AB, Wilkie CA, editors. Flame retardant polymer nanocomposites. Hoboken, USA: Wiley; 2007. pp. 131–62.

[86] Bourbigot S, Le Bras M, Tremillon JM, Breant P, Delobel R. Zeolites: new synergistic agents for intumescent thermoplastic formulations-criteria for the choice of the zeolite. Fire Mater 1996;20:145–58.

[87] Levchik SV, Levchik GF, Camino G, Costa L, Lesnikovich AI. Mechanism of action of phosphorus-based flame retardants in nylon 6, III: ammonium polyphosphate/manganese dioxide. Fire Mater 1996;20:183–90.

[88] Anna P, Gy Marosi, Bourbigot S, Le Bras M, Delobel R. Intumescent flame retardant systems of modified rheology. Polym Degrad Stab 2002;77:243–7.

[89] Weil ED. Fire-protective and flame-retardant coatings – a state-of-the-art review. J Fire Sci 2001;29:259–96.

[90] Horn WE. Inorganic hydroxides and hydroxycarbonates: their function and use as flame retardants. In: Grand AF, Wilkie CA, editors. Fire retardancy of polymeric materials. New York, USA: Marcel Dekker Inc.; 2000. pp. 285–352.

[91] Brown SC. Flame retardants: inorganic oxide and hydroxide systems. In: Pritchard G, editor. Plastics additives, an A–Z reference. London, UK: Chapman and Hall; 1998. pp. 288–96.

[92] Rothon RN. Effects of particulate fillers on flame retardant properties of composites. In: Rothon RN, editor. Particulate filled polymer composites. Shrewsbury, England: Rapra Technology Ltd; 2003. pp. 263–302.

[93] Papazoglou ES. Flame retardants for plastics. In: Harper CA, editor. Handbook of building materials for fire protection. Ohio, USA: McGraw-Hill; 2003. 4.1–4.88.

[94] Hornsby PR, Rothon RN. Fire retardant fillers for polymers. In: Le Bras M, Wilkie CA, Bourbigot S, Duquesne S, Jama C, editors. Fire retardancy of polymers: new applications of mineral fillers. London, UK: Royal Society of Chemistry; 2005. pp. 19–41.

[95] Yang Y, Shi X, Zhao R. Flame retardancy behavior of zinc borate. J Fire Sci 1999;17: 355–61.

[96] Shen KK, O'Connor R, Brown SC. Flame retardants: borates. In: Pritchard G, editor. Plastics additives, an A–Z reference. London, UK: Chapman and Hall; 1998. pp. 268–76.

[97] Shen KK, Kochesfahani S, Jouffret F. Zinc borates as multifunctional polymer additives. Polym Adv Technol 2008;19:469–74.

[98] Toure B, Cuesta J-ML, Benhassaine A, Crespy A. The combined action of huntite hydromagnesite for reducing flammability of an ethylene-propylene copolymer. Int J Polym Anal Char 1996;2:193–202.

[99] Toure B, Cuesta JL, Gaudon P, Benhassaine A, Crespy A. Fire resistance and mechanical properties of a huntite/hydromagnesite/antimony trioxide/decabromodiphenyl oxide filled PP-PE copolymer. Polym Degrad Stab 1996;53:371–9.

[100] Hollingbery LA, Hull TR. The fire retardant behaviour of huntite and hydromagnesite – a review. Polym Degrad Stab 2010;95:2213–25.

[101] Zaikov GE, Lomakin SM. Flame retardants: poly(vinyl alcohol) and silicon compounds. In: Pritchard G, editor. Plastics additives, an A–Z reference. London, UK: Chapman and Hall; 1998. pp. 315–26.
[102] Kashiwagi T, Gilman JW. Silicon-based flame retardants. In: Al-Malaika S, Golovoy A, Wilkie CA, editors. Specialty polymer additives: principles and applications. Oxford, England: Blackwell Science; 2001. pp. 3153–89.
[103] Kashiwagi T, Gilman JW. Silicon-based flame retardants. In: Grand AF, Wilkie CA, editors. Fire retardancy of polymeric materials. New York, USA: Marcel Dekker Inc.; 2000. pp. 353–89.
[104] Connell JE, Metcalfe E, Thomas MJK. Silicate–siloxane fire retardant composites. Polym Int 2000;49:1092–4.
[105] Marosfoi BB, Szabo A, Kiss K, Marosi G. Use of organosilicone composites as flame retardant additives and coatings for polypropylene. In: Hull TR, Kandola BK, editors. Fire retardancy of polymers – new strategies and mechanisms. Cambridge, UK: RSC Publishing; 2009. pp. 49–58.
[106] Hamdani S, Longuet C, Perrin D, JLopez-cuesta JM, Ganachaud F. Flame retardancy of silicone-based materials. Polym Degrad Stab 2009;94:465–95.
[107] Kashiwagi T, Gilman JW, Butler KM, Harris RH, Shields JR, Asano A. Flame retardant mechanism of silica gel/silica. Fire Mater 2000;24:27789.
[108] Kashiwagi T, Shields JR, Harris Jr RH, Davis RD. Flame-retardant mechanism of silica: effects of resin molecular weight. J Appl Polym Sci 2003;87:1541–53.
[109] Gilman JW, Kashiwagi T, Harris Jr RH, Lomakin S, Lichetenhan JD, Jones P, et al. Char enhancing approaches to flame retarding polymers. In: Al-Malaika S, Wilkie C, Golovoy CA, editors. Chemistry and technology of polymer additives. London, UK: Blackwell Science; 1999. pp. 135–50.
[110] Koo JH. Polymer nanocomposites: processing, characterization and applications. NewYork, USA: McGraw-Hill Companies Inc.; 2000. 125–234.
[111] Alexandre M, Dubois P. Polymer-layered silicate nanocomposites: preparation, properties and uses of a new class of materials. Mater Sci Eng 2000;28:1–63.
[112] Ray SS, Okamoto M. Polymer-layered silicate nanocomposite: a review from preparation to processing. Progr Polym Sci 2003;28:1539–641.
[113] Mai YW, Yu ZZ. Polymer nanocomposites. New York, USA, UK: WoodHead publishing; 2006.
[114] Pavlidou S, Papaspyrides CD. A review on polymer-layered silicate nanocomposites. Progr Polym Sci 2008;33:1119–98.
[115] Ma HY, Song PA, Fang ZP. Flame retarded polymer nanocomposites: development, trend and future perspective. Sci China Chem 2001;54:302–13.
[116] Uhl FM, Wilkie CA. Polystyrene/graphite nanocomposites: effect on thermal stability. Polym Degrad Stab 2002;76:111–22.
[117] Ke YC, Stroeve P. Polymer-layered silicate and silica nanocomposites. Amsterdam, The Netherlands: Elsevier BV; 2005.
[118] Bourbigot S, Le Bras M, Xavier F, Rochery M, Devaux E, Lichtenhan JD. Polyhedral oligomeric silsesquioxanes: application to flame retardant textiles. In: Le Bras M, Wilkie CA, Bourbigot S, Duquesne S, Jama C, editors. Fire retardancy of polymers: new applications of mineral fillers. London, UK: Royal Society of Chemistry; 2005. pp. 189–201.
[119] Laachachi A, Cochez M, Ferriol M, Leroy E, Lopez-Cuesta JM. Influence of TiO_2 and Fe_2O_3 fillers on the thermal properties of poly(methyl methacrylate) (PMMA). Mater Lett 2005;59:36–9.

[120] Du F, Winey KI. Nanotubes in multifunctional polymer nanocomposites. In: Gogotsi Y, editor. Nanotubes and nanofibers. Florida, USA: CRC Press, Taylor & Francis Group; 2006. pp. 179−98.
[121] Costa FR, Saphiannikova M, Wagenknecht U, Heinrich G. Layered double hydroxide based polymer nanocomposite. Adv Polym Sci 2008;210:101−68.
[122] Qin H, Zhang S, Zhao C, Hu G, Yang M. Flame retardant mechanism of polymer/clay nanocomposites based on polypropylene. Polymer 2005;46:8386−95.
[123] Schartel B, Bartholmai M, Knoll U. Some comments on the main fire retardancy mechanisms in polymer nanocomposites. Polym Adv Technol 2006;17:772−7.
[124] Kashiwagi T. Progress in flammability studies of nanocomposites with new types of nanoparticles. In: Morgan AB, Wilkie CA, editors. Flame retardant polymer nanocomposites. Hoboken, USA: Wiley; 2007. pp. 285−324.
[125] Zhu J, Wilkie CA. Flammability properties of polymer nanocomposites. In: Gupta RK, Kennel E, Kim KJ, editors. Polymer nanocomposites handbook. Florida, USA: CRC Press, Taylor & Francis Group; 2010. pp. 437−62.

CHAPTER

Fire Safety Performance of Flame Retardants Compared with Toxic and Environmental Hazards

2

D.A. Purser

Hartford Environmental Research, Hatfield, Hertfordshire, UK

CHAPTER OUTLINE

1. Introduction

As with many products and services, justification for the use of flame retardants (FRs) depends ultimately on the overall benefits and costs to society (Figure 1).

The principal benefits of FRs relate to the number of fires that are prevented or reduced in severity and the resulting savings in property loss, injuries, and deaths. Although fire injuries and deaths may be relatively small in number compared to other causes, the consequences to individuals, communities, and businesses can be severe when fires do occur, so that all societies place a high value on fire

Polymer Green Flame Retardants. http://dx.doi.org/10.1016/B978-0-444-53808-6.00002-0

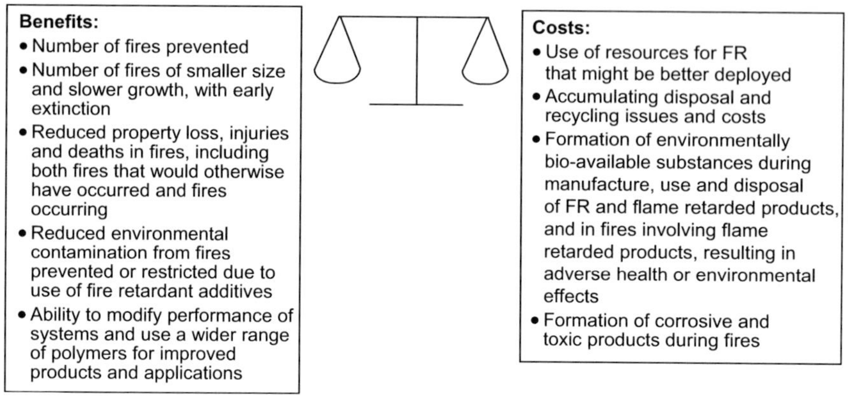

FIGURE 1

Benefits and costs of flame-retardants (FR).

prevention and mitigation. Another important benefit of the use of FRs lies in improving the fire performance of particular polymers to widen their range of safe applications, thereby reducing costs and facilitating the development of innovative products and applications. Arguably, since all fires result in contamination of the local environment (e.g. building interiors) and the outdoor environment (by airborne dispersal and contamination of water runoff), by preventing and limiting fires, FRs reduce environmental contamination from unwanted fires.

The principal disadvantages of FRs are that some FR additives have been found to leach or wear out of products during usage. They can then be released into the local indoor environment, coming into intimate contact with occupants, and in some cases they have been widely dispersed into the outdoor environment, even concentrating in remote regions such as the polar arctic [1]. While this release and dispersal may in itself be considered a cause of concern, far more serious aspects emerging for certain compounds are that they not only show a significant bioavailability but also have been shown to be biologically and toxicologically active in humans and wildlife, giving rise to significant health concerns for both. Another disadvantage of some FR systems is that while they may have some beneficial effects on some aspects of fire performance, they may also increase the release of toxic combustion products in fires, thereby increasing toxic hazards in some situations as well as increasing environmental contamination.

In weighing the benefits and costs of any specific FR usage, it is therefore necessary to compare the overall benefits of avoiding a relatively low incidence of very high consequence events (fires) with the potential costs of widespread exposure of almost the entire human population and planetary ecosystems to very low concentrations of products with risks that are difficult to quantify. This also needs to be considered in the context of the large array of non-FR chemicals also released into the environment that may show bioavailability and be toxicologically active.

Where such FR additives can be replaced by compounds or systems that are not released into the environment or are not bioaccumulative or toxicologically active, while showing equivalent costs and FR efficacy, then this is obviously advantageous.

2. FR applications

In the United Kingdom, fire safety in buildings is achieved by three main requirements of the Building Regulations [2].

- Means of warning and escape: this provision is designed to provide early warning and appropriate means of escape from a building to a place of safety capable of being safely and effectively used at all material times. In practice this means that the building and systems are constructed so that there will be sufficient time from when a fire is detected and a warning is then provided for occupants to escape in safety.
- Internal fire spread (linings): in order to inhibit the spread of fire within a building the internal linings are required adequately to resist the spread of flame over their surfaces and have if ignited a rate of heat release or a rate of fire growth which is reasonable in the circumstances.
- Internal and exterior fire spread (structure): there is a requirement to maintain the stability of a building, resist fire spread between buildings, and where necessary, inhibit the spread of fire within the building by subdivision with fire-resisting construction or suitable fire suppression systems. It is also required that unseen fire and smoke spread within concealed spaces in the structure and fabric be inhibited.
- In a separate legislation, the fire performance of contents are controlled in relation to fire safety of appliances, upholstered furniture and bedding [3], and some categories of clothing such as children's' nightwear.
- The elements of all these systems are subject to design requirements and/or fire tests with performance classifications and the performance of the overall systems may be subject to fire engineering assessments.

While it may be possible to achieve an adequate performance without FRs by using suitable materials and systems, the application of FR additives increases the range of materials and composites that can be used, particularly in relation to natural and synthesized polymer-based products. In certain applications, the use of FR polymers may be particularly advantageous. For example, fluoropolymers are used for aircraft electrical systems because they have good dielectric properties as well as good fire resistance, so that good fire performance can be achieved with thin insulation on cables, resulting in considerable weight and space saving. Light, high-performance FR polymer composites are also widely used in aircraft for seating and cabin linings.

While some materials with a very low organic content may be considered essentially noncombustible, most flame-retarded materials and products; in particular polymeric materials with high organic (carbon) content, are designed to withstand a particular level of thermal insult. If the thermal insult is below the critical design level, either they will not ignite or flaming will fail to be maintained or propagated, but if the thermal insult exceeds this value, a growing fire may result. An important aspect is then the subsequent fire development. For some systems a rapidly growing fire may result, with a growth rate and maximum fire size (heat release rate (HRR)) similar to that of non-FR versions of the polymer system. For other systems, the fire growth curve (t^2 curve) may be shallower, providing more time for escape or fire suppression activities, and the final fire size may be smaller, reducing the overall hazard. The main beneficial properties of the flame-retarded system are then:

- Lower probability of ignition from a range of ignition sources up to a particular degree of severity
- Slow subsequent fire growth and flame spread (depending on the system)
- Smaller ultimate fire (depending on the system and boundary conditions)

Once a fire occurs in a building, the outcome in terms of fire safety and hazard depends on two parallel time-based processes: the growing fire hazard and the escape performance of the occupants. In fire safety engineering terms it is necessary to ensure that the available safe escape time (ASET) is greater than the required safe escape time (RSET) by an acceptable margin of safety [4,5]. ASET is the time from ignition to that when conditions become untenable to occupants such that they are no longer able to escape without assistance. RSET is the time for ignition to that when affected occupants are able to reach a place of safety. In situations designed such that occupants are not expected to escape (defend-in-place strategy), the ASET must be essentially infinite.

2.1 Effect of application of FR systems to upholstered furniture, bedding, and televisions

The history of fire death and injury statistics in the United Kingdom provides an example of the application of FR systems to polymers used for furniture and bedding. Most fire injuries and deaths occur in domestic dwellings and result from fires primarily involving upholstered furniture or bedding. Figures 2 and 3 show the causes of deaths and injuries in fires per million population over the period from when the first national statistics are available (1955) to the latest set for 2010 [6]. Figure 2 shows that during the late 1950s the total annual number of deaths were approximately 9.7 per million, of which around 7.8 were caused by burns and 1.9 were due to exposure to toxic smoke (essentially carbon monoxide poisoning). Between the late 1950s and the early 1970s a profound change occurred. While the number of burn victims remained more or less constant, there was a fourfold increase in the number of persons dying as a result of exposure to toxic smoke, bringing the total number of deaths up to 14.3 per million per year (900 persons per year). This led to concern that something was causing fires in the 1970s to be more "toxic" than

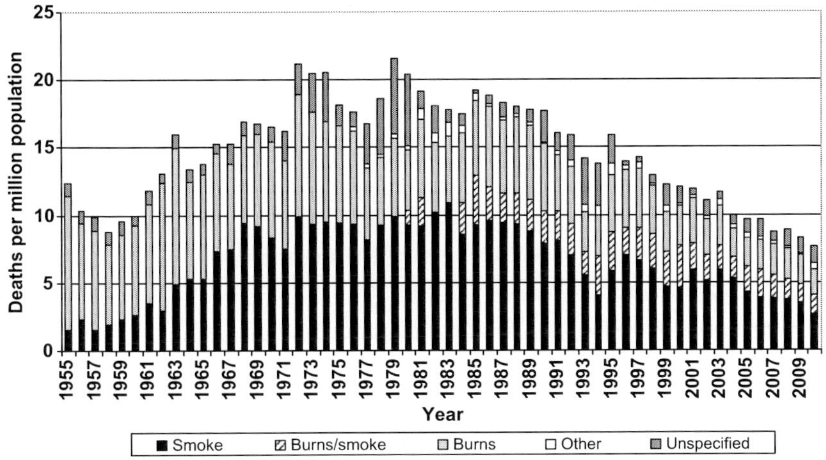

FIGURE 2

United Kingdom fire deaths per million population during 1955–2010.

had been the case in the 1950s, and it was considered that a major causative factor was changes in the materials used for coverings and fillings for upholstered furniture and bedding.

In the immediate post–World War II period most upholstered furniture was constructed from traditional, natural materials, with wooden frames; cotton, wool, or

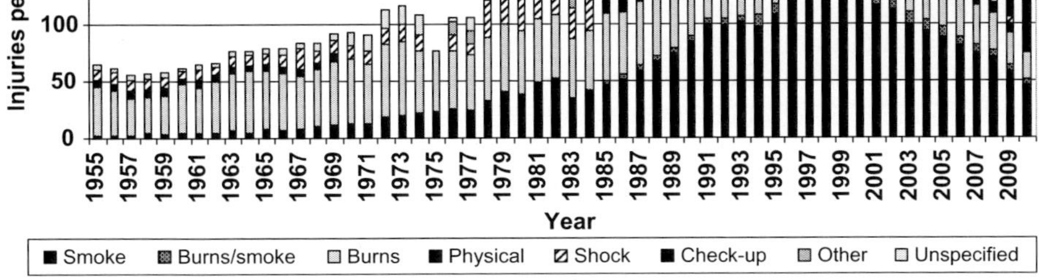

FIGURE 3

United Kingdom fire injuries during 1955–2010.

leather covers; and sprung interiors with fillings using materials such as coconut fiber (coir), kapok, or shoddy (mixed recycled textile fiber wadding). In those days fire fighters rarely used breathing apparatus. When they attended a fire, their first action would be to open all the doors and windows to vent the smoke, and then rescue the occupants, who would often be conscious. After sitting outside and coughing for a few minutes they would generally make a good recovery and were seldom taken to hospital or required treatment at the scene. By the early 1970s, the nature of furniture fires had changed, so that when fire fighters arrived and opened the exterior doors and windows, instead of the smoke clearing, the fire rapidly grew larger, producing much larger volumes of smoke. In addition, the smoke was found to be much more toxic, so that fire fighters now used breathing apparatus to enter a dwelling, and the rescued smoke victims were often found to be in a serious condition requiring hospitalization.

Although the traditional furniture covers and fillings were not generally treated with FR additives, up to a point at least some of them "naturally" showed some of the desirable properties of flame-retarded systems. Wool and to some extent leather upholstery covers are reasonably resistant to flaming ignition by smoldering (cigarette) or small flame (match) ignition sources, while cotton tends to smolder. Also, if the resistance to flaming ignition is overcome by the application of a larger flame source for a sufficient period, the typical combination of cover materials and fillings tended to result in slowly growing and spreading fires, with a shallow t^2 fire growth curve. The combustion conditions for these small fires were relatively efficient (well ventilated) so that the yields of smoke and toxic gases were low. Although domestic smoke detectors were nonexistent at this time, in most cases, occupants had sufficient time to become aware of the fire and were able to escape or were rescued before the exposure resulted in serious injury or death—so the ASET was sufficiently longer than the RSET for them to survive.

By the 1970s, the most important changes were that covering materials included polypropylene, viscose or acrylic fibers, or artificial leather (made from polyvinyl chloride (PVC) or polyurethane over a fiber such as polyester), as well as the more traditional cotton, wool and leather, and the fillings commonly consisted of a non-FR polyurethane foam or similar material. When furniture constructed using some of these materials is ignited, especially using a flaming ignition source such as a No. 7 crib (a test ignition source made from small sticks equivalent to a few sheets of crumpled newspaper [7]), a rapidly growing flaming fire can result, producing a relatively large fire within a few minutes with a high HRR. In domestic rooms, these fires rapidly become ventilation restricted, producing large quantities of thick black smoke and high yields of carbon monoxide and hydrogen cyanide. Increasing the ventilation to such fires by opening doors and windows can result in further rapid growth, sometimes resulting in flashover [8]. When fires grow rapidly postignition in this way, and produce large quantities of toxic smoke, the time available for escape before an occupant is incapacitated is very short, so that the ASET can often be shorter than the RSET [9]. Although the UK statistics can be interpreted in a number of ways, it was considered by many at the time that the large increase in smoke deaths was at least partly a result of this fire behavior.

In domestic buildings in the United Kingdom most fires tend to be confined to the item, or at least the room, of origin, although the fire effluents, including the smoke and toxic gases, usually spread throughout the building. Burn victims are almost always in the room of fire origin, often close to the fire, while toxic smoke victims are sometimes found in the room of fire origin, and also commonly in rooms beyond the fire enclosure, where they can rapidly be overcome by toxic smoke. It may be for this reason that while the incidence of burns was similar in the 1950s and the 1970s, the incidence of smoke deaths remote from the site of the fire was greatly increased.

Due to concerns about fire performance of upholstered furniture and the increasing incidence of fire deaths and injuries in the United Kingdom, a regulatory requirement was introduced in 1988 for domestic upholstered furniture covers, fillings, and composites to pass a cigarette and small flame ignition test [3,7]. Since in practice non-FR polyurethane foam and similar fillings failed this standard, flame-retarded foams were used subsequently, in combination with some FR treatment for common covering materials. The regulation was applied to sales of both new and second-hand furniture, and after a few years during which furniture already in use was gradually replaced, the fire death and injury statistics gradually improved, from the peak during the early 1990s of 900 deaths and 9000 smoke and burn injuries annually down to the current levels of around 350 deaths and 4000 injuries. The general adoption of domestic smoke alarms from around 1990 almost certainly contributed to the improvement, by providing earlier warning of fires so that occupants had more time available for escape during the early burning phase.

The use of brominated FRs in printed circuit boards and casings of television sets provides an example of the dilemma between fire safety and environmental hazard issues. The polybrominated ether FRs were first phased out in Europe due to the environmental concerns from release of the additives during use and of brominated dioxins and dibenzofurans when burned [10,11]. Following the reduction in the use of these and other FR treatments, there was evidence for an increase in the occurrence of fires in television sets in Europe [12] compared to that United States and Asian countries in which the use of the FR additives continued. A television fire can present a serious toxic hazard to building occupants, even if the fire does not spread to other room contents, as is often the case. If the fire spreads and destroys part of the building then a potential environmental hazard arises from the contamination of air and water by the total toxic effluent, including the very chlorinated and brominated dioxins and dibenzofurans that the cessation of use of the FRs was designed to prevent, so that over a product life cycle FR-treated televisions can have a lower environmental impact than non-FR-treated televisions [13]. Some halogenated FRs are now being replaced by "greener", more durable, halogen systems or phosphorus-based and other systems to obtain an acceptable performance [14,15].

2.2 Fire hazard development

To obtain the ASET time for any system it is necessary to determine:

- Time–concentration (or intensity) curves for the major toxic products, smoke, and heat in the fire at the breathing zone of the occupants, which in turn depend on:
  - Fire growth curve (mass loss rate of the burning fuel (kilograms per second) and its dispersal volume (kilograms per cubic meter) with time)
  - The yields of the major toxic products (kilograms per kilogram) and heat (kilojoules per kilogram) (for example, kilograms of CO per kilogram of material burned).
- Concentration/time/physiological effect relationships of these products in terms of the physiological/toxic potency of the products and heat (the exposure concentration (kilograms per cubic meter)), or exposure dose (kilograms·per cubic meter·minute or parts per million·minute) causing toxic effects (and the equivalent effects for heat and smoke obscuration) in terms of:
  - Concentrations or doses (or heat intensity) likely to impair escape efficiency due to behavioral and/or physiological effects
  - Exposure concentrations or doses likely to cause incapacitation or prevent escape due to behavioral and/or physiological effects
  - Lethal exposure concentrations or doses

The first term can be obtained from a full-scale reconstruction of the fire scenario of interest, including all the fuels in their end-use form and location, placed inside a full-scale fire compartment with appropriate dimensions and boundary conditions (especially with respect to ventilation). Fires can then be ignited using appropriate ignition sources and the subsequent fire development and conditions can be measured directly in terms of the time–concentration curves for the temperature, heat radiation, smoke density, and major irritant and asphyxiant toxic gases. The effects of the exposure on occupants can then be assessed using fractional effective dose (FED) methodology [9]. The FED of a toxic combustion product (of set of products) is calculated as the dose to which a person is exposed over a period of time during a fire, expressed as a fraction of the dose required to cause a given toxicity endpoint such as incapacitation. The FEDs for each successive short time period during a fire (for example, each minute) are summed until the fraction reaches unity, at which time it is predicted that the endpoint is reached. The FED concept is also applied to "exposure doses" of heat and fractional exposure concentrations (FIC) of smoke and irritants. The FED method is therefore used to calculate the effects and times of the different hazard endpoints [9,16]. The fire conditions (in terms of time–concentration curves for smoke, toxic products, and heat) can either be measured directly in a full-scale fire test or calculated using a zone or computational fluid dynamics fire model.

For assessing the effects of different FR systems and comparison with non-flame-retarded alternatives it is particularly important to consider what heat insult or ignition source should be applied to the main item of interest. The placing and separation of fuel items may also influence the fire spread and size. Items such as upholstered furniture or wall surface coverings are designed to a range of different fire performance standards depending on their intended application. Upholstered furniture for domestic use may have no ignition performance requirements, or may be required to resist ignition following application of a cigarette or a small flame, depending on the local performance requirements for different countries or states. Upholstered furniture intended for use in public buildings, for institutional use, or for transport applications such as aircraft seating may have much more stringent ignition requirements (be required to resist ignition from large ignition sources such as jet flame burners). The design system used to fulfill these ignitability requirements may also affect the rate of flame spread and subsequent fire growth involving the item, should the ignition resistance criteria be exceeded by application of a larger ignition source resulting in sustained flaming. Thus, in general a low resistance to ignition by small heat or flame sources is likely to be followed by subsequent rapid flame spread and involvement of the whole item, while a high ignition resistance may lead to slow subsequent fire growth. This is partly because fire spread of a flame across a surface covering depends to some degree on the ease with which heat radiation from a flaming area causes ignition of adjacent surface areas. This may not always be the case, so that for some systems (covering materials and interior foams) initial ignition, fire growth, and flame spread may be slow, but then increase very rapidly once a threshold HRR is achieved.

Similar considerations apply to FR surface lining materials. Some materials may have a good resistance to ignition, with slow fire spread following application of relatively small ignition sources and this may be maintained even when a fire has developed for some time. Other materials may also have a good resistance to ignition and local spread during the early stages of application of an ignition source, but once the fire has spread and the upper layer temperature in the fire compartment has increased to a critical level, a period of very rapid fire development may occur as the FR properties of the lining material are overwhelmed. One such example occurred in an experiment in which an International Organization for Standardization (ISO) room was lined with a material incorporating a brominated FR system. Brominated FRs are designed to inhibit gas phase flame reactions following the local release of bromine (or volatile bromine compounds). In bench-scale tests it was found that at a certain temperature the bromine compounds were released leaving a combustible residue. In the ISO room when the material was exposed to radiant heat from a burner in a corner test configuration, the initial performance was good, with slow local fire growth and flame spread, but after some minutes, as the upper layer temperature increased, the fire resistance of the system failed resulting in rapid involvement of the combustible lining and a sudden large flame issuing from the open room doorway to the consternation of the test observers [17].

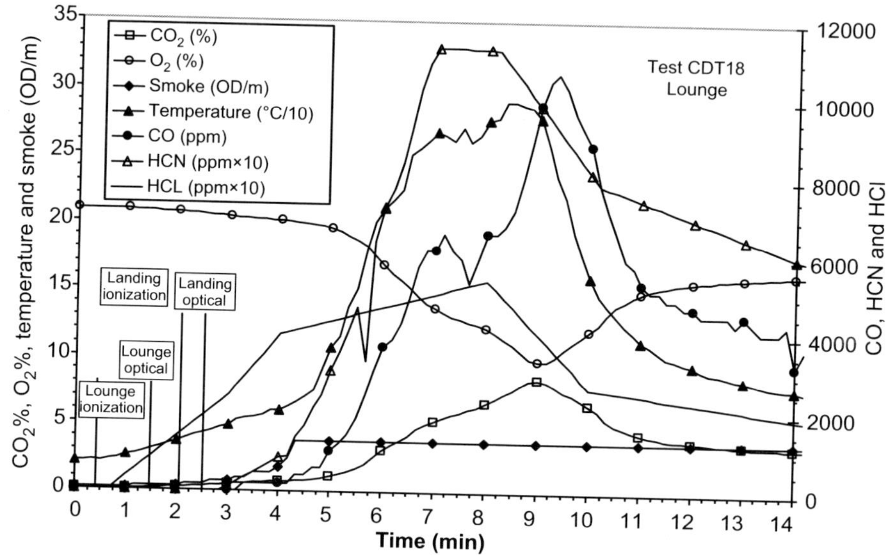

FIGURE 7

Conditions at head height in the lounge for a fire in an armchair with FR acrylic covers and combustion modified FR polyether foam [8].

carried out in an enclosed system, the fire self-extinguished after 5 min as the oxygen concentration in the lounge dropped to 4.4% O_2.

Figure 7 shows similar plots, for an identically constructed armchair, except that for this chair an FR-back-coated cotton layer containing nitrogen (9.8%), chlorine (3.64%), and bromine (2.38%) was applied to the acrylic fabric cover, over combustion modified melamine-containing polyurethane foam (10.75% nitrogen, 2.22% chlorine). This combination was compliant with the UK furniture regulations. This chair was burned in a similar lounge fire scenario with the lounge door open, this time in the two-story house. Although the overall pattern of fire development for this armchair is generally similar to the previous one, there are some important differences. The initial rate of fire growth is significantly slower than for the previous case, so that approximately 5 min elapsed before the period or rapid temperature increase occurred. The peak temperature was also considerably lower than for the previous case, with a maximum of 300 °C before the fire self-extinguished, and the smoke did not become dense until around 4 min after ignition. The peak CO was 11,000 ppm compared with 14,800 ppm, the peak HCN was 1297 ppm compared with 1606 ppm, and the oxygen minimum was 9.8%. The lower CO and HCN concentrations are particularly interesting because under specific combustion conditions the addition of halogenated FRs is known to increase the yields of these toxic gases. However, in this case the concentrations are lower because the total mass of material decomposed was less, due to the fire self-extinguishing with a lower mass of chair consumed (5.9 compared to 7.3 kg), thus involving less oxygen-depleted combustion conditions.

The effect on ASET shown in Figure 8 is that although the concentration of irritant smoke becomes problematic between approximately 1.5–2.5 min after ignition, collapse form asphyxia is predicted only after 5.5 min. Pain from heat is predicted after 5.5 min for a standing person, or 6 min for a person sitting. This means that an occupant of this room would experience vision difficulties from around 2 min but they would have a further 2.5 min in which to escape before asphyxia caused collapse and the heat became painful. Although the heat exposure would be painful and may affect escape capability, approximately a further 2- to 3-minute exposure is required for extensive, life-threatening (full-thickness or third-degree), burns (to both exposed and clothed skin). A person in the bedroom would also have approximately a further 2 min in which to descend the stairs and escape. In a further series of experiments in which different armchairs were burned in a totally enclosed room, the tenability times increased from 0.5 min for non-FR covers and foams to 10–12 min for CMHR and CM foams.

These examples illustrate that FR systems definitely can improve fire performance and tenability (ASET) times for upholstered furniture even when the design ignition resistance is overwhelmed by relatively large ignition sources. For some

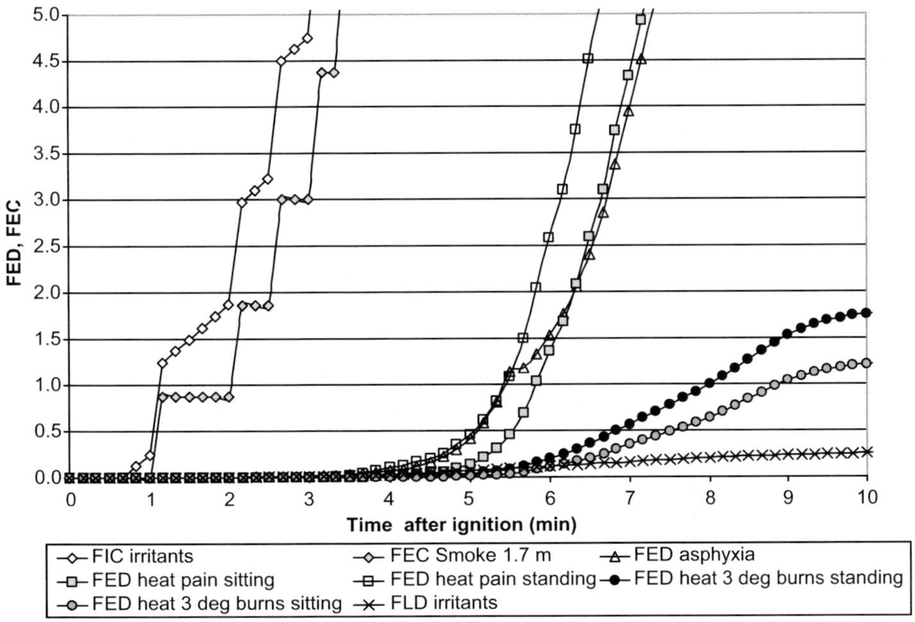

FIGURE 8

FED analysis for fire conditions in Figure 7, showing time to tenability limits for smoke, irritants, heat, and asphyxiant gases [8].

examples the ASET time is increased by only a few minutes, while for others a much greater time benefit of 10–20 min can be achieved. The question then arises as to how much improvement in postignition performance is required to significantly improve the probability of escape? The time required for escape (RSET) is the time from ignition to that when an occupant is able to reach a place of safety. For the example of a domestic dwelling this means that a lounge occupant would need to be able to leave the lounge and escape through the hallway exit. For a bedroom occupant they would need to be able to descend the stair (without significant hindrance from hot smoke) and exit via the hallway door before being overcome by asphyxiating gases. Or they would need to be alerted in sufficient time to close their bedroom door and either take refuge in the bedroom or escape via the bedroom window.

RSET consists of a set of additive terms [4,5,20]:

RSET (Δt_{esc}) $= \Delta t_{det} + \Delta t_a + \Delta t_{PTAT} + \Delta t_{trav}$

- Time from ignition to detection (t_{det})
- Time from detection to general alarm (Δt_a)
- Evacuation time, which has two phases
  - Pretravel activity time (or "premovement time")—time from alarm to when occupants begin to move toward the exits Δt_{PTAT}
  - Travel time (the time for occupants to travel to a place of safety) Δt_{trav}

The first term is the time from ignition to detection, and the second, the time from detection to alarm. For domestic smoke detectors, detection and alarm time are essentially the same, since the sounder is activated as soon as the detection threshold is achieved. For the experiment in Figure 7 the alarm times are shown. For the detectors placed in the fire room (lounge) the ionization detector was activated very quickly after 0.42 min and the optical detector after 1.5 min. For the detectors placed on upper floor landing ceiling (the normal location for a single detector in a house), the activation times were 2.1 and 2.5 min. For a flaming fire such as these the alarm time therefore varied between approximately 0.5 and 2.5 min depending on the detector type and location. The next term in the RSET equation, premovement recognition and response time, is much more variable and difficult to estimate. A sleeping occupant may require approximately up to a minute or so to awaken and recognize an alarm, and then a further minute or so to carry out a range of behaviors before attempting to escape or take refuge. The travel phase is likely to be very short—approximately 30 s from a bedroom to leaving a house via a door to the outside.

For the examples of the non-FR chair fires, the time available for escape was around 0.5–2.5 min. The activation time for a typical landing smoke alarm was 2.1–2.5 min. Adding a very short premovement time of 1 min plus 0.5 min travel time gives a very approximate minimum RSET of around 3.5–4 min for a sleeping occupant. For rapidly growing fires involving this kind of non-FR furniture the time available for escape is therefore likely to be shorter than, or very close to, the time required for escape, with little or no margin of safety. Although there are considerable uncertainties and variability associated with fire development, detection, and human response in such scenarios, these experiments demonstrate the considerable hazards associated with fires involving furniture of this type, which is commonly used worldwide. They also demonstrate that even a minute or so increase in time to rapid fire growth could significantly increase the probability of occupants escaping in safety, so that the UK-compliant FR-treated covers and foams, providing approximately 5—10 min escape time, present a definite safety advantage over the pre-1988 furniture. Of course, as shown in Figure 4, the use of FR additives is not essential to obtain a satisfactory fire performance and escape time, since a similar or superior performance can also be obtained by selection of appropriate alternative cover and filling materials and/or fire-resistant interliners.

3. Different FR systems and additives

The properties and performance of different FR systems and additives range from highly effective systems with environmentally friendly properties to less effective systems and additives, often with potential environmental hazards both in normal use and in fires as follows:

1. FR systems not releasing substances with significant bioavailability or toxicity during normal use:
   a. Product has minimal organic content, releases little or no fumes on heating, and is noncombustible.
   b. Product or system has minimal organic content, releases little or no fumes on heating, and encloses combustible components, preventing or reducing the probability of involvement of the combustible components and subsequent release of fumes or occurrence of fire.
   c. Product reduces the probability of fire up to defined ignition source, the rate of fire growth and the size of the subsequent fire, and the yields of toxic products.
   d. Product reduces the probability of fire up to defined ignition source, reduces the rate of fire growth and the size of the subsequent fire, but increases the yields of toxic products.
   e. Product reduces the probability of fire up to defined ignition source, but once ignited, fire growth is similar to that in a non-FR system and toxic product yields tend to be increased.
2. FR systems as above but may release substances during normal use, some of which may show significant bioavailability or toxicity

Although this classification covers the fire- and toxicity-related aspects of different systems and products, it does not address other important aspects such as effects on the performance of the product in use, costs, manufacturing, and recycling attributes. Table 1 lists a range of common FR systems and products and summarizes their properties.

Table 1 Summary of Different FR Systems, Additives, and Properties

System	Reduces Fire Probability	Reduces Fire Size	Bioavailability/ Environmental Toxic Hazard from FR System or Substance	Yields of Toxic Combustion Products	Yields of Environmental Toxic Products
Inert insulation panels, layers, fillers, interliners: glass fiber, mineral wool, ceramic fiber, aramid	Yes	Yes	Glass fiber and mineral wool have low bioavailability and toxicity, biodegradable if inhaled. Minor health hazards during installation or removal Ceramic fibers' inhalation hazard may be similar to asbestosis?	None or minimal toxic products	None or minimal environmental hazards
Intumescent coatings	Yes	Yes	No known issues	No	None
Aluminum trihydroxide Magnesium dihydroxide	Yes	Yes—by water release	No known issues	no	None
Boric acid	Yes	Yes, by glass formation	Easily released from substrate REACH classification *H360FD* (*May damage fertility. May damage the unborn child*) but exposure risk generally considered low. Regarded as a "green" FR	Reduced	Reduced
Phosphorus and phosphorus/nitrogen FRs	Yes	Yes, by char formation and in gas phase	Depends on compound and durability in application	Some reduced	Reduced

Phosphorus halogen	Yes	Yes, mainly gas phase	Depends on compound and durability in application	Increased	Increased
Ammonium polyphosphate	Yes	Yes, with char formation	No known issues	Reduced	Reduced
Organic halogens antimony salts	Yes	Yes in gas phases Sb aids char	Depends on the compound, some have been phased out	Increased	Increased
Inorganic halogens: PVC	Yes	To some extent	Not known to be an issue (other than VCM during manufacture)	Increased	Increased
Fluoropolymers	Yes	Yes	Not know to be an issue	Increased extreme toxicity nanoparticles under some conditions, HF, PFIB	Increased
Nanoparticle clays and fibers	Possibly	Possibly	Can nanoparticles be released during use? What toxicity occurs?	Some reduction of existing toxic products but potential toxic hazard from aerosolized nanoparticles	Potential issue

PFIB, perfluoroisobutylene; REACH, Registration, Evaluation and Authorisation of Chemicals; VMC, vinyl chloride monomer.

3.1 Noncombustible systems

The most effective systems and products from the perspective of fire performance, while minimizing release of bioavailable toxic products both during normal use and in fires, are the various noncombustible systems. These have a very limited organic content, so release little or no potentially toxic fumes when heated. They are relevant in the context of "Green FRs" where they exist as alternatives to commonly used combustible products. In the area of construction products, while many building materials are noncombustible, organic-based thermal insulation materials are widely used, such as fillings or panels in wall cavities or sandwich panels. Mineral fiber or glass fiber systems provide alternatives to polymer-based systems such as rigid polyisocyanurate foam or polystyrene, which may contain halogenated FRs. The main potential health hazards from mineral fiber systems are direct skin contact with fibers or inhalation of airborne fibers, especially during installation or removal. Fibers may also be released during large conflagrations as parts of building collapse. The potential dermal hazards from mineral and glass fibers can be avoided by the use of suitable protective clothing. The inhalation hazards can also be minimized by suitable respiratory protection, but under circumstances when fibers are inhaled the potential health hazards depend mainly on the size and aspect ratio of the fibers and their biodegradability.

Asbestos fibers are not biodegradable. When deposited in the lung they are engulfed by macrophages, but largely due to their shape, they penetrate the macrophage cell wall causing the cytotoxic contents to leak. The resulting inflammatory response leads to the fibrotic lung condition asbestosis. Fibers that enter the lymphatic system can end up as pleural plaques, and the chronic irritation and inflammation can eventually lead to mesothelioma. The initiation of mesothelioma is considered to be probabilistic, so that there is no "safe" dose and in theory a single inhaled fiber can result in cancer; however, the probability increases with the dose (total number of fibers) inhaled, the form of the fibers, and the number of years since exposure. Ceramic fibers, used in certain high-temperature insulation applications, are somewhat similar in form and persistence to asbestos fibers and so should be handled with care and protected from situations likely to release fibers. Mineral and glass fibers do not present a serious inhalation hazard due to the size and shape of the fibers (although basic precautions should be taken during installation and removal). Glass fibers in particular are slowly dissolved in the lung by macrophage activity, so are less persistent than asbestos fibers. These systems are not completely inert, however, since small amounts of organic binder oils are used to coat the fibers, which when heated have been found to produce small amounts of highly irritant fumes [21]. Carbon fibers are used in some FR systems and are considered to present a potential inhalation hazard if released, depending on their size and aspect ratio. Inhalation health problems (mostly sore throats) have been encountered by fire fighters following fires involving military aircraft constructed form carbon fiber composites.

Furniture can be constructed from noncombustible materials and noncombustible glass fiber curtains have been produced for domestic use, although dermal irritation sometimes resulted from careless handling due to the fibers penetrating the skin. Wire and cable is another area in which FRs are widely used to improve ignition and fire propagation performance. Cables have been included in the classification system under the European Construction Products Directive flame spread test including heat release and smoke production measurements according to prEN 50399. There are seven classes of which the best performance classification "ACA" is met only by cables with ceramic insulation [22].

3.2 Noncombustible protective layers and interliners

Another method for obtaining a good fire performance while limiting the release of toxic products, both in general use and in fires, involves applying a protective layer or interliner, to reduce the probability of involvement of a combustible inner layer. Two major applications are in sandwich panels and upholstered furniture. Sandwich panels typically consist of steel, aluminum, or mineral board outer layers with a polystyrene or rigid polyurethane (or polyisocyanurate) inner insulation layer. Such systems can be constructed to provide a good standard of fire performance, but this can sometimes be overcome in more severe fire scenarios, mainly due to delamination of the panels or failure at fixing points. When this occurs the combustible core materials can be exposed leading to large fires with release of considerable quantities of toxic products. A problem with untreated insulation filling materials is that these may also be vulnerable to ignition from small heat sources with subsequent fire spread throughout the cavity (a problem occurring in buildings and in transport applications such as aircraft fuselage insulation). FR treatment or other alternative materials may prevent this.

As described, interliners can be very effective in reducing postignition fire spread and development in upholstered furniture. A good performance can be obtained even when interliners are placed under relatively combustible cover materials such as non-FR cotton, due to prevention or delay of involvement of the combustible foam filling. Suitable combinations of cover, interliner, and filling are used to provide exceptionally high fire performance standard in aircraft seats [23]. Where interliners use noncombustible materials such as glass fiber, or high-performance aramid polymers, they do not require FR additives and so may be considered environmentally friendly.

Intumescent coating systems are often used to protect steel from overheating, but can also be applied to timber surfaces, thereby acting as an FR coating, inhibiting ignition, surface spread of flame, and heat propagation. These coatings consist of a carbon supplier (starch, polyol, or pentaerythritol), ammonium polyphosphate, and melamine as an expanding agent. These components are sealed into the paint, and so should have limited environmental release during use. When the surface is heated the coating forms a blown char, with approximately a 100-fold increase in thickness, thereby protecting the surface below for an extended period [24].

3.3 FR additives and toxic product yields during use and combustion

Where FR chemicals are applied as additives to products, their release to the environment depends on the exposure of the product to mechanical wear, weathering (especially washing), and the durability of the finish. FR treatments used for textiles provide an example of finish durability and its implications with respect to potential for environmental release [19]. The main approaches used, with increasing durability, are as follows:

- Application of an FR chemical finish
- Application of an FR chemical agent which either reacts with the fiber molecules or polymerizes within the fiber
- Inclusion of an FR additive during man-made fiber extrusion, which is thereby incorporated into the solid fiber matrix.
- Polymerization of special monomers or copolymerization of FR compounds with fiber-forming monomers to confer inherent flame redundancy to the resulting synthetic fiber
- Use of an inherently FR monomer such as vinyl chloride or aramid (or use of an inherently flame-retarded natural polymer such as wool or leather)

These approaches represent a generally increasing scale of durability and decreasing risk or potential level of release to the environment. While chemical finishes such as sodium borate are easily released by mechanical handling or washing, and other additives with significant vapor pressure may evaporate from the product over time, systems that are incorporated in, or form part, of the molecular structure of the solid product may exhibit minimal if any release to the environment.

The yields of toxic products from materials during fires depend on the following:

- The elemental composition
- The organic composition
- The combustion conditions

Of these, the combustion conditions are the most important determinants of toxic products. The main combustion conditions are:

- Nonflaming
- Well-ventilated flaming
- Underventilated flaming (pre- and postflashover)

Under nonflaming conditions, such as during smoldering or exposure to a radiant heat source, rates of decomposition and hazard development are slow, but the yields of toxic products including organic and inorganic irritants and carbon monoxide are high. The organic component consists of a wide range of aliphatic and aromatic compounds. Some of these are basically common to most decomposing polymers, with the highest yields often of low molecular weight aliphatic hydrocarbons such as methane and ethane and lower yields of higher molecular weight hydrocarbons such as hexane and octane [25,26]. Ring cyclization results in the formation of aromatic compounds including benzene, styrene, and methylstyrene. Partial oxidation results in the formation of aldehydes and ketones, which form the main organic irritant component, including the most potent lower molecular weight compounds (formaldehyde, acrolein, and crotonaldehyde) [9,21]. The yields of different components also depend to some extent on the organic composition of the polymer so that additional products are introduced, such as proteoglycans from cellulosic materials and isocyanate derivatives from polyurethanes. Cellulosic materials provide relatively high yields of formaldehyde and acrolein, while styrene produces high yields of styrene monomer. In addition to these immediately irritant compounds, systemically toxic compounds such as polycyclic aromatic hydrocarbons (PAHs) are also formed. For any material containing halogens the formation of halogenated benzenes and phenols acts as precursors of halogenated biphenyls, dibenzodioxins, and furans.

Once a transition to flaming combustion occurs, rates of growth and hazard development are often very rapid, but during the early stages of a flaming fire, before ventilation becomes limited, combustion of most untreated materials is very efficient. The main products are heat, water, and carbon dioxide, with very low yields of CO and organic irritants. When inorganic components such as halogens, sulfur, nitrogen, or phosphorus are present in the polymer, the acid gases tend to be released at high yields as hydrogen halides or sulfur oxides, with low yields of nitrogen oxides and phosphorus oxides. FR additives acting in the gas phase, mainly halogen systems, also inhibit efficient flame oxidation, increasing yields of CO from fuel carbon and HCN from fuel nitrogen [27]. Primary formation and yields of halogenated dibenzodioxins and furans are lowest under these conditions, but secondary formation can occur in hot plumes unless these are rapidly cooled (for example in incinerator stacks). Fires occurring inside enclosures such as buildings grow rapidly to a size at which they become underventilated, resulting in vitiated combustion conditions. Under these conditions, there is a considerable increase in the yields of toxic products as a function of the fuel/air equivalence ratio (ϕ). Figure 9 illustrates the increasing yields of toxic species (CO, HCN, total organics, and particulates) as a function of equivalence ratio for Polyamide 6 [27]. The yields are very low under well-ventilated flaming combustion conditions but increase dramatically once combustion becomes underventilated at ϕ values above 1.

Figure 10 shows the effects of different fire types on the overall toxicity of the combustion products from six materials. Toxicity is expressed as lethal toxic potency, which is $1/LC_{50}$ calculated for 30-min exposures according to the method of Purser [21,30]. The first four materials include pinewood and three cable materials (low-density polyethylene (LDPE), a low smoke and fume cable material (LSF) (containing 32% polyethylene, 8% vinyl acetate, and 60% aluminum hydroxide trihydrate), and a plasticized PVC (containing 28.5% calcium carbonate)) [28] decomposed using the ISO19700 (BS7990) tube furnace [31], while the cotton was decomposed in the DIN53436 furnace [29]. The densities of the four materials tested in the tube furnace were wood 0.52, LDPE 0.92, LSF 1.57, and PVC 1.55. The toxic product yields and toxic potency of a material or product

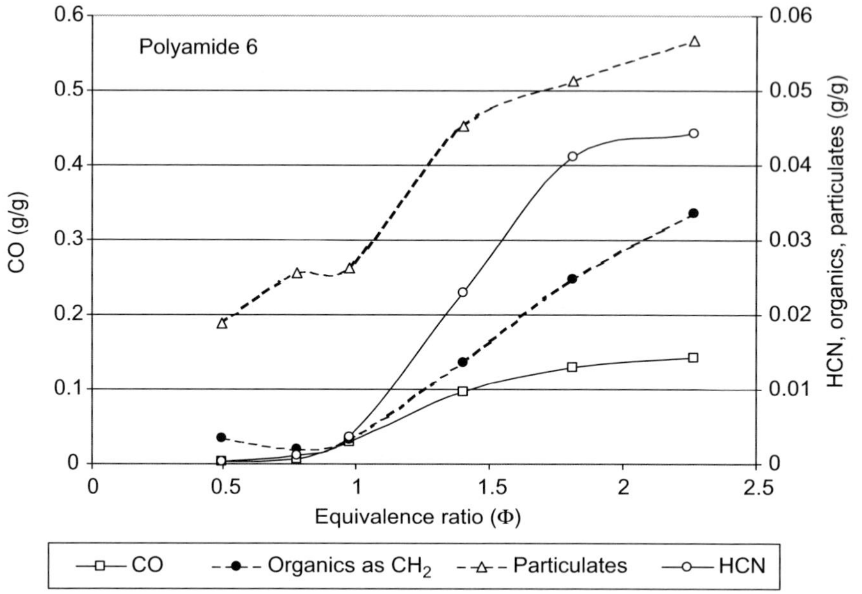

FIGURE 9

Increasing yields of toxic species as a function of equivalence ratio under flaming combustion (Polyamide 6, 650 °C in the ISO TS19700 tube furnace) [27].

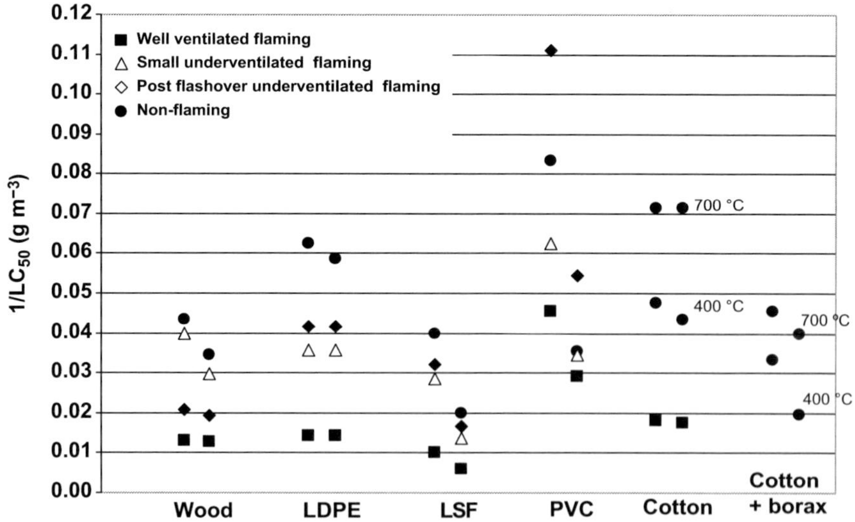

FIGURE 10

$1/LC_{50}$ concentrations for 30-minute exposures calculated from ISO 19700 tube furnace data for thermal decomposition and combustion products from three cable materials (LDPE, LSF and PVC), wood, and calculated from DIN 53436 data for cotton and FR-borax cotton (left of each pair expressed as mass loss, right as mass charge) [28,29].

can usefully be expressed in a number of different ways depending on the use to which the data are to be put. Expressing yield in terms of the mass of toxic product per unit mass of material decomposed (e.g. kilograms of CO/kilogram fuel mass loss) is useful in understanding the toxicity of the products released during a fire and for fire engineering hazard calculations. When considering the application of products in terms of fire load it is useful to express yield in terms of the mass of material used rather than the mass consumed (or in terms of material area or volume depending on the nature of the product). In Figure 10 the toxic potencies for each material are expressed both in terms of mass loss (left) and mass charge (right), with significant differences for materials not fully consumed during combustion.

The results show that well-ventilated flaming produces the lowest toxic potency and toxic product yields, while nonflaming decomposition and underventilated flaming combustion give considerably higher yields and toxic potencies. The rate of production of toxic products depends on the mass loss rate of the fuel during the fire multiplied by the toxic product yields, so although the yields may be similar under nonflaming and underventilated flaming conditions, the rates of production and hence the hazard are much greater in flaming fires.

3.3.1 Inert and active fillers

Inert fillers reduce the organic content of the fuel, which tends to reduce not only the combustibility but also the yields of toxic products. FR properties are enhanced when the filler is active, such as by releasing water or by char formation. Char formation has the added benefit of locking fuel carbon in the char, which reduces the yields of toxic products containing carbon. The borax additive to the cotton also reduces fuel carbon, but inhibits flaming by forming a glasslike layer.

In Figure 10, LDPE is a non-FR 100% organic polymer containing 86% carbon, which is completely consumed during flaming combustion, with a small residue under nonflaming decomposition. The mass loss and mass charge toxic potencies are therefore very similar. Figure 10 also shows three examples of the effects of fillers in reducing toxic product yields and toxicity (LSF, PVC, and cotton-borax). The LSF contains 60% aluminum hydroxide trihydrate as an active filler. The plasticized PVC cable material contained 28.5% by mass of $CaCO_3$, acting mainly as a passive filler.

For LDPE the yields of toxic products (Figure 11) are low under nonflaming conditions up to 350 °C but at 450 °C and above it ignites and burns. Well-ventilated combustion is efficient with small amounts of organics and particulates but CO and organic yields increase with underventilated combustion. LSF (Figure 12) was relatively stable up to 350 °C and nonflaming up to 450 °C, with some release of decomposition products. With well-ventilated flaming the main product was CO_2, while low yields to CO, organics, and particulates were formed with underventilated flaming. This mass loss toxic potency is considerably lower than that of LDPE under all fire conditions (Figure 10), and even lower in mass charge terms. This means that

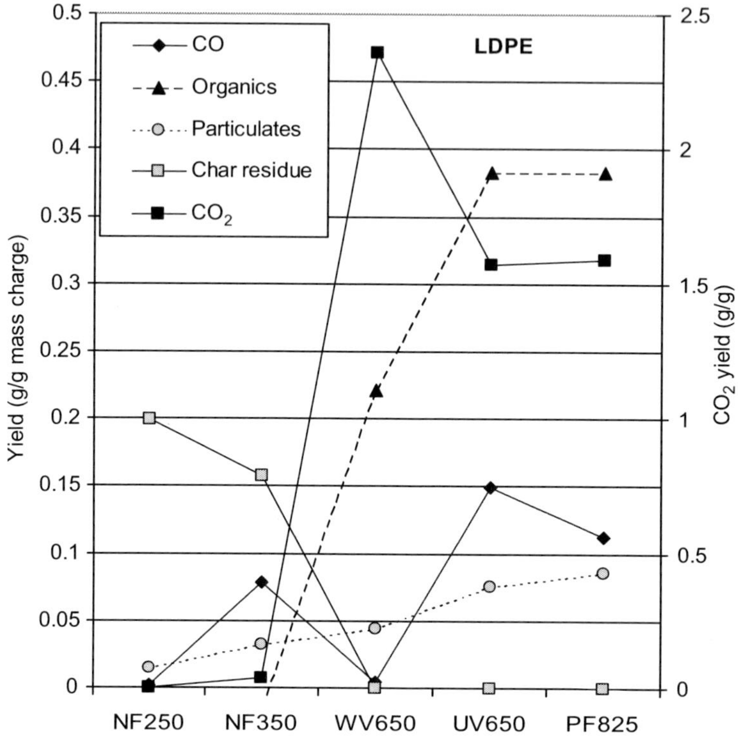

FIGURE 11

LDPE product yields under different fire conditions [28].

1 m of LSF cable with the same mass load as an LDPE cable would produce considerably less toxic products even if the FR treatment failed to limit fire propagation. For the plasticized PVC (Figure 13), the calcium carbonate filler, in addition to reducing the organic content of the material, also reduced the yield of HCl from the chlorine component. While the overall toxic potency under each fire condition is significantly greater than those of LDPE and LSF, the presence of inorganic residue means that the mass charge potency is significantly lower than the mass loss potency. The borax/boric acid treatment of cotton twill produces a glasslike layer on the solid inhibiting fuel loss when tested in the DIN 53436 tube furnace. The toxic potency (Figure 10) of untreated cotton was highest under nonflaming decomposition conditions, and greater at 700 °C than at 400 °C due to increased toxic product yields (Figure 14), but when flaming occurred at 700 °C the toxic potency was considerably reduced. The use of the borax approximately halved the toxic potency under nonflaming conditions and flaming did not occur. The untreated cotton produced significant yields of the irritant organics acrolein and formaldehyde, especially under nonflaming conditions, but neither was detected form the borax-treated cotton [29].

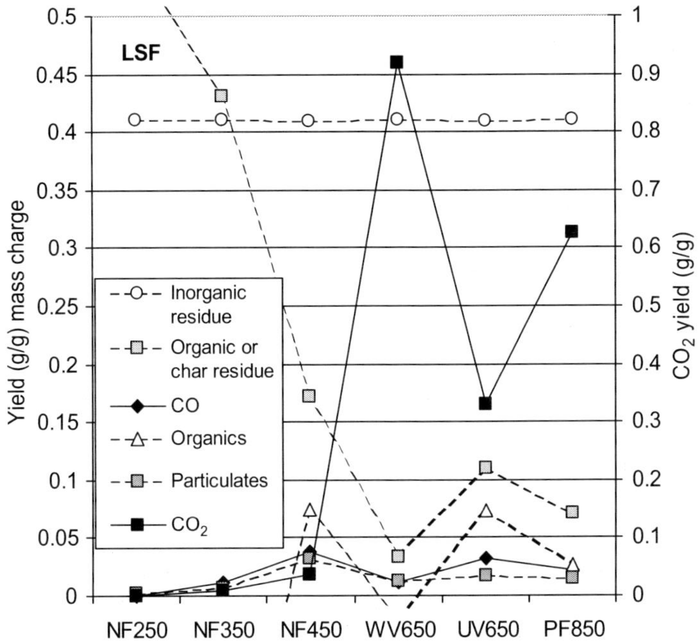

FIGURE 12

LSF product yields under different fire conditions [28].

3.3.2 Phosphorus-based and nitrogen, melamine, and melamine/chlorinated phosphate systems

3.3.2.1 Durability and potential environmental release issues

Many different substances containing phosphorus are used as FR additives including elemental red phosphorus as well as inorganic and organic phosphorus compounds. These may be used alone or in combination with bromine or chlorine and they may also be used with nitrogen-containing substances such as melamine.

With regard to environmental release issues these have different levels of durability. A common durable organophosphorus compound used with cotton is based on tetrakis(hydroxymethyl) phosphonium chloride (or sulfide). This is combined with ammonia to form a polymer containing both nitrogen and phosphorus within the cellulose fiber structure [19]. While direct dosing has produced some toxicity but not carcinogenicity [32], the bioavailability from garments is low. Pyrovatex, a formulation based on *N*-methylol dimethylphosphopropionamide is another durable N-P FR used with cotton, which is grafted onto the cellulose via a traditional crease-resistant finish [33]. For viscose, organophosphorus compounds can be incorporated within the spinning dope [19]. For polyester, phosphorus or halogen FR additives can be incorporated as comonomer, as additives in the melt spin, or as finishes or coatings, with varying degrees of durability.

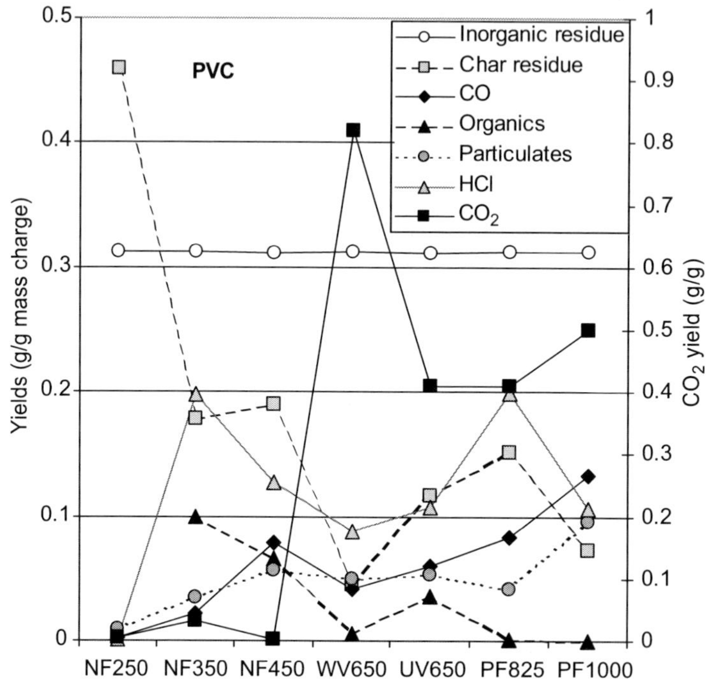

FIGURE 13

PVC product yields under different fire conditions [28].

The FR tris-(2,3, dibromopropyl) phosphate was found to be a very effective FR additive, such as for children's nightwear, but was subsequently withdrawn in 1977 due to concerns regarding the potential to cause genetic mutations and cancer, combined with bioavailability, since the chemical could be absorbed by children through the skin or by "mouthing" tris-treated children's clothing [34,35]. A durable alternative uses phosphinate groups as a comonomer with glycol and dimethyl terephthalate such that the phosphorus moieties are chemically linked onto the polymer main chain to impart flame resistance [36]. Polypropylene provides another example of the application of different FRs with different levels of durability and hence potential bioavailability. Most commercially available FRs are not compatible with polyolefins and they are slowly lost to the environment.

The general situation with respect to these different FR additive systems is therefore that irrespective of the potential toxicity of the parent compounds, the potential toxic and environmental hazard depends on the durability of the treatment–substrate system. Nondurable additives have the disadvantage of loss to the environment (either during normal wear or following washing), with accompanying loss of FR performance and potential bioavailability, to users in the domestic environment

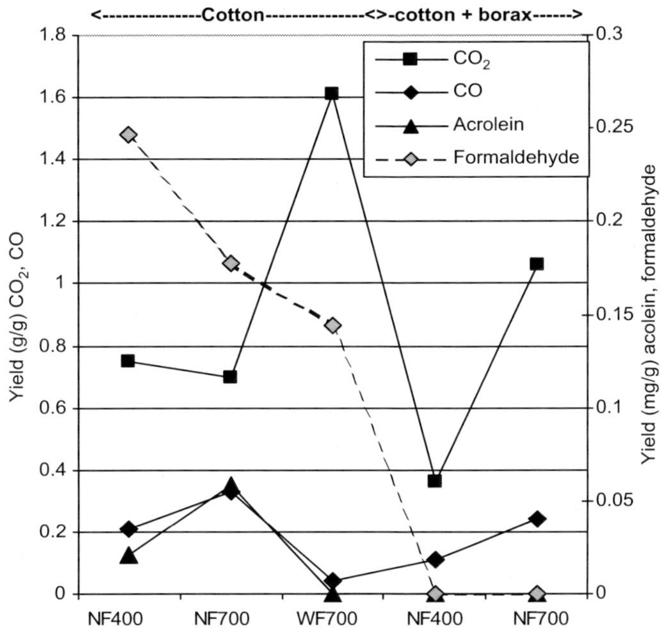

FIGURE 14

Cotton and cotton + borax product yields under different fire conditions [29].

and to the general environment through dispersal in waste. Alternatively, durable treatments can have very low environmental release, even after washing, thereby preserving their effectiveness and greatly reducing potential environmental contamination problems.

Phosphorus acts as a FR primarily in the solid phase, by promoting char formation, and also to some extent in the vapor phase. Vapor phase flame inhibition is enhanced by the presence of nitrogen and halogens. By promoting char formation, the release of potentially toxic carbon compounds is reduced, and some of the phosphorus is also retained. An important consideration is the form in which the vapor phase phosphorus is released during fires. This has been little studied, but in general phosphorus and organic phosphorous compounds are readily oxidized, so that the main product is P_2O_5, which then hydrolyzes to H_3PO_4. This is moderately toxic (1 h LC_{50} 1.2 g m^{-3}), and may make some small contribution to overall toxicity of fire effluents from treated materials. Phosphine (PH_3) is another toxic compound detected in some fire effluents (a highly toxic substance causing lung edema, with a 1 h rat LC_{50} of 44 ppm) [37,38]. Some traces of organophosphorus compounds have been detected in the combustion products from burning materials, but toxicity test protocols designed to detect neurotoxic compounds have not been used with combustion atmospheres [38]. One serious exception is the finding that any phosphorus source when combined with a trimethylol polyol can produce a potent neurotoxic class of bicyclophosphate esters in combustion products (Figure 15). In practice

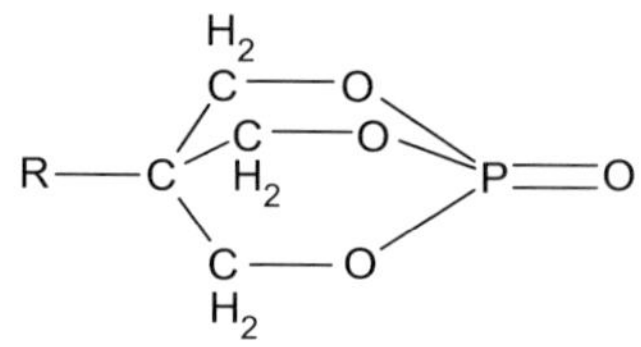

FIGURE 15

Bicyclophosphate ester.

this has been a potential problem only with certain turbine lubricants containing trimethylol (see reference [38]).

3.3.2.2 Toxicity of combustion products under different fire conditions

The toxicity of the combustion products from materials treated with phosphorus-containing additives depends on the additive combination used. The use of inorganic phosphorus or organic compounds containing phosphorus alone may reduce the yields of the normal range of toxic combustion products including CO and organic irritants, with the addition only of some slightly irritant phosphoric acid. However, phosphorus is most commonly used in combination with additives containing halogens and nitrogen. The effects of halogens are described in the next section, but the presence of nitrogen in the fuel, either in the base polymer or as an additive, is an important issue due to the release of highly toxic nitrogen-containing combustion products. The fate of fuel nitrogen during fires is illustrated in Figure 16, in terms of the yields of toxic nitrogen-containing products under flaming combustion conditions for Polyamide 6 [27]. Under the well-ventilated combustion conditions used in most fire tests, a small proportion of the fuel nitrogen is released as oxides of nitrogen, mostly in the form of NO, which slowly oxidizes further to produce very low yields of NO_2. NO is essentially nontoxic acutely, while NO_2 is a potent lung irritant. Since, as shown in Figure 16, the yields of NO and NO_2 are extremely low, this does not add significantly to the toxicity of the combustion products under these conditions. However, once a compartment fire develops larger and becomes underventilated, with the equivalence ratio exceeding unity, then hydrogen cyanide is produced at rapidly increasing yields. The yield depends on the equivalence ratio and the compartment upper layer temperature, so that higher yields can be obtained for underventilated conditions and even higher yields if flashover occurs. The effect of this is illustrated in Figures 6 and 8, where HCN is the main agent driving the predicted loss of consciousness from asphyxia after 3 and 5 min. If the HCN is removed from the calculation the time to asphyxia is increased by several minutes, by which point CO intoxication is predicted. In general it has been found that the efficiency of conversion of fuel nitrogen to HCN is similar to the efficiency of conversion of fuel carbon to CO, and as an approximate guide it is considered that HCN is likely to make a significant contribution to toxic hazard when the nitrogen content of the fuel exceeds 1% by mass.

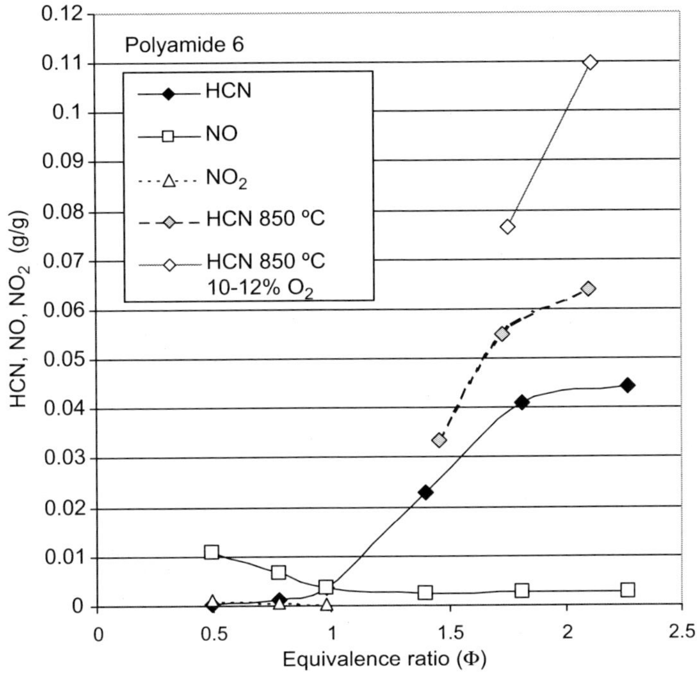

FIGURE 16

Yields of HCN and nitrogen oxides from Polyamide 6 under flaming conditions as a function of equivalence ratio and temperature [27].

3.3.3 Halogen acid vapor phase systems

3.3.3.1 Bioavailability and environmental toxicity

With FR systems containing halogens the same issues of durability, bioavailability, and toxicity apply as with phosphate-containing systems, and many additives contain both phosphorus and halogens. PVC is an example of an inherently flame-retarded material for which the active agent (Cl) is part of the monomer and hence durable in the polymeric structure. However, leaching of some vinyl chloride monomer, which is a known carcinogen, had been detected at low levels prior to 1977 from PVC water pipes, but is not a current concern [39,40]. Tetrabromobisphenol A (TBBPA), is a FR monomer which can be incorporated into the formation of epoxy polymers. It is widely used in electrical products and has also been used as a durable FR for polyesters in the form of 2,2-bis-(4,2′-hydroxy-3,5-dobromophenyl) propane (a TBBPA derivative) or as an additive. In the first two categories it is durable, with minimal releases, but some leaching can occur when it is used as an additive. Despite this, studies of leaching, bioavailability, toxicity, and levels in food and breast milk have found environmental hazards not to be of concern [41]. In contrast to these durable systems, polybrominated ether additives have been found to be released widely into the environment, and have been largely phased out (although decabrominated ether additives (deca) are still in use [42]).

3.3.3.2 Toxicity of combustion products from halogenated materials under different fire conditions

A major concern with all halogenated FRs is that the halogen component is designed to be released and act mainly in the vapor phase as a flame inhibitor. This not only results in the release of toxic halogenated combustion products but also, by reducing combustion efficiently, causes a significant increase in the yields of all "normal" toxic species, including CO, HCN (if nitrogen is present), organic irritants, particulates, and environmental contaminants (polyhalogenated biphenyls, dioxins, and furans). The presence of the halogens somewhat reduces the organic content of the material (especially for PVC and fluoropolymers) which to some extent offsets the increase in yields of carbonaceous products.

Under flaming conditions the presence of halogens leads to the production of relatively high yields of toxic products throughout the range from well ventilated to underventilated. The effect is illustrated in Figure 13 for plasticized PVC, and in Figure 17 for CO yields from two non-FR polymers PMMA and a bouclé fabric

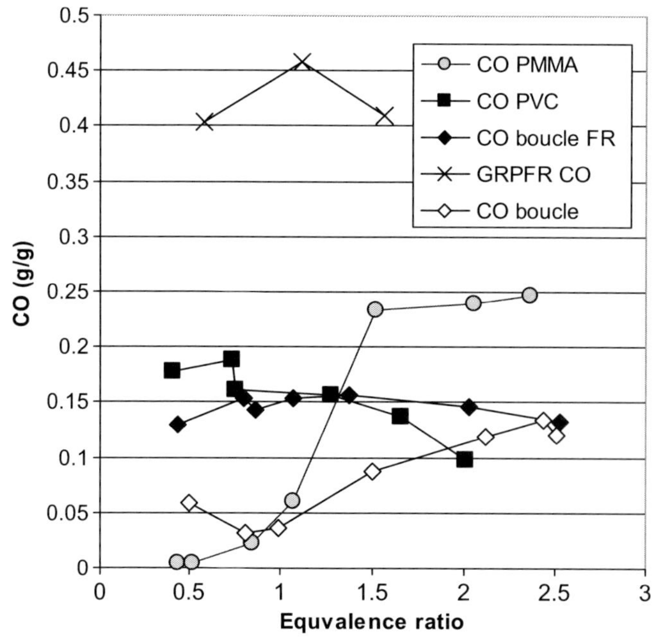

FIGURE 17

CO yields from nonhalogenated and halogenated polymers as a function of equivalence ratio [43].

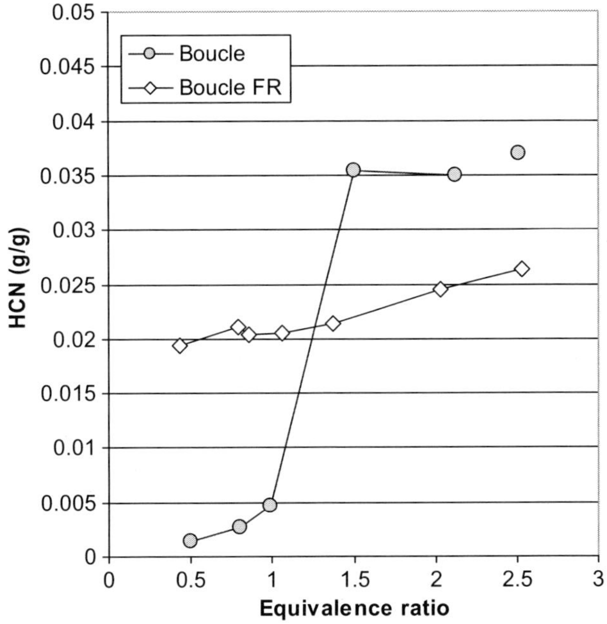

FIGURE 18

HCN yields from nonhalogenated and halogenated boucle as a function of equivalence ratio [27,43].

(38% acrylic, 38% wool, 24% polyester) and three polymers containing halogens (rigid PVC, a back-coated bouclé containing 6.09% Br and 0.95% Cl, and an FR-glass-reinforced plastic (GRP) containing 15.24%Br, 3.83% Cl and 1.3% P) [43]. The non-FR polymers show typical sigmoid yield curves with low CO and HCN yields under well-ventilated conditions ($\phi < 1$) increasing for underventilated conditions, while the FR polymers show high yields through the range. Figure 18 shows the same effect on HCN yields for the non-FR and FR bouclé fabrics. The conversion efficiency of polymer halogens to vapor phase acid gases is often high (often approaching 50–100%) under all combustion conditions (Figure 19), which adds to the toxicity of the combustion products.

These systems tend to provide a degree of ignition and flame-spread resistance up to a design level of performance. When subjected to local heating or ignition, flame spread may be slow or self-extinguishing, but if the general environment reaches high temperatures, or if the local heat source becomes large, so that fuel halogens are driven off, then the base polymer may burn more readily. This was observed with the bromine- and chlorine-containing GRPs shown in Figures 17 and 19. High yields of brominated fumes were released when the material was burned in the tube furnace, which then enabled the base polymer to flame. In a room-corner wall-lining configuration in the (open-roofed) Single Burning Item

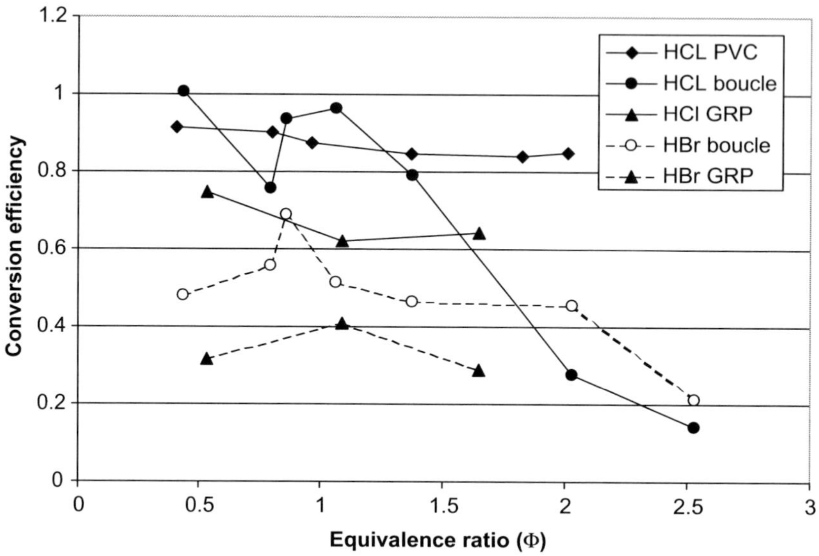

FIGURE 19

Conversion efficiency of fuel chlorine and bromine to gas phase halogens [43].

test the material performed quite well, with a slow flame-spread rate, but when the material was burned in a lined ISO 9705 room (which has walls and a ceiling), although it performed quite well in terms of flame spread for a period, once the fire size and upper layer temperature reach a critical stage, a sudden flashover occurred [17]. Findings such as these again emphasize the importance of ensuring that systems are tested in relevant full-scale scenarios. Under well-ventilated combustion conditions the yields and efficiencies with which carbon and nitrogen in the fuels are converted to CO, HCN, organics, and particulates are approximately proportional to the chlorine or bromine content, as illustrated for a variety of polymers in Figure 20.

3.3.3.3 Fluoropolymers and ultrafine particles (nanoparticles)

Fluoropolymers are inherently flame-retarded materials, but when thermally decomposed they produce toxic gases including carbonyl fluoride, hydrogen fluoride, and perfluoroisobutylene [44]. Inhalation of these at sufficiently high doses results in fatal lung edema and inflammation. The overall toxic potency of fluoropolymer thermal decomposition and combustion products is approximately 10 times that of most common polymers, but under a specific set of conditions and decomposition temperatures (450–650 °C with recirculation of the fume through the hot zone), perfluorinated polymers produce high concentrations of ultrafine (nano-) particles (10–150 nm diameter) which cause fatal lung inflammation with a very high toxic potency (LC_{50} 0.015 g m^{-3}). This is approximately 1000 times the toxic potency of combustion products from wood and similar materials [43].

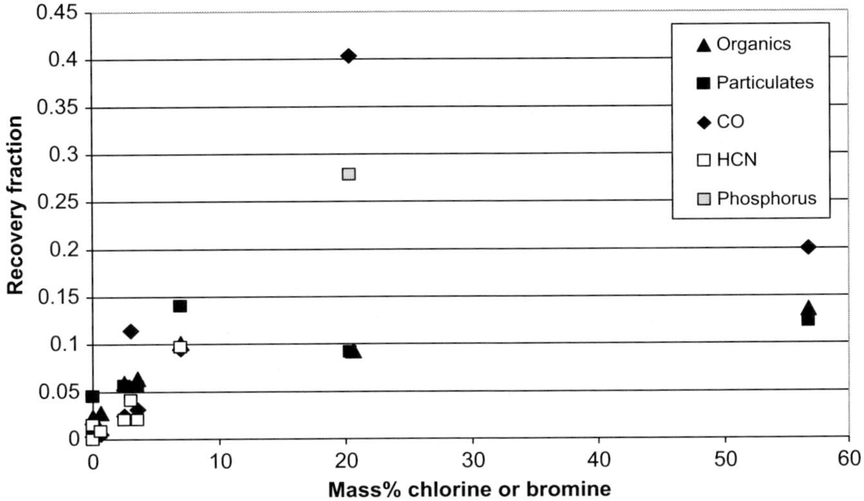

FIGURE 20

Conversion efficiency of fuel carbon, nitrogen, and phosphorus to toxic products as function of mass percentage chlorine or bromine in the original polymer [43].

Ultrafine particles are now known to form a significant fraction of the range of particles in most combustion product atmospheres, and are currently of considerable interest as a form of environmental air pollution considered responsible for a significant incidence of acute cardiovascular and respiratory deaths annually in the United Kingdom, the death rate increasing proportionally to the concentration of fine particles in the atmosphere [45]. These toxic effects occur directly as inflammation of the alveolar lung interstitium, and systemically following penetration into the blood circulation and when taken up by phagocytes. This may have implications for the development of nanoclay FR systems in that it may be important to ensure that combustion of these materials does not result in the release of ultrafine clay particles, or toxicity if released [46].

Some studies have been made of the effects of nanoclays additives on the yields of "normal" toxic combustion products (such as CO and particulates in the particulate matter 2.5–10 size range) [47,48] but there is a need for detailed studies of nanoparticle release and bioavailability from materials treated with such additives both during normal use and when burning.

3.3.3.4 Environmental contamination by dioxins and furans from halogenated materials

Combustion of any material containing chlorine or bromine produces a range of products which are not immediately hazardous to life safety in a fire but which may contaminate a building interior or the wider environment with toxic substances such as polychlorinated biphenyls, and halogenated dioxins and furans. These may be hazardous to health over prolonged periods.

With respect to unwanted fires there are two particular issues:

1. The extent to which the production, use, and disposal of organohalogen compounds leads to general levels of environmental contamination by dioxins, dibenzofurans, and phthalates, which constitutes a risk to public health
2. The extent to which exposure of people to combustion products from halogenated materials during and after fires in buildings constitutes a risk to their long-term health.

These issues are reviewed in Purser [37]. PVC disposal has not been a major factor in the general background level of soil and vegetation contamination by dioxins. Figure 21 shows dioxin and furan contamination of environment and food peaking in the 1960s, while annual PVC production has increased enormously since this time [37,49,50]. However, since other historical major sources of environmental dioxin contamination have now been eliminated, halogenated materials could become a problem for the future. Yields of dioxins and furans highly depend on the combustion conditions in a fire. Low-temperature (300–400 °C) vitiated fires produce high yields. An Environmental Protection Agency study showed very high yields of dioxins and furans from inefficiently burning fires involving the burning of household waste (4.5% PVC content) in a 55-gallon oil drum. In one experiment the total yield of polychlorinated dibenzodioxins (PCDs) and polychlorinated dibenzofurans (PCDFs) was 493 ng g^{-1} waste decomposed compared with 0.0035 ng g^{-1} waste decomposed in a modern incinerator, a factor of more

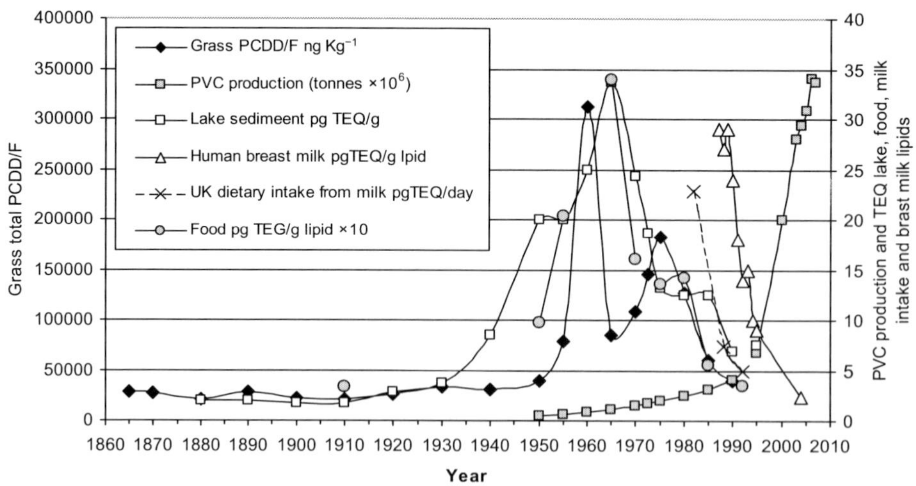

FIGURE 21

Trends in environmental dioxins compared with annual PVC production.

than 100,000 [51]. One household using this method could produce more dioxins than the entire output of a modern municipal waste incinerator. Using the tube furnace we have found high concentrations of dioxin precursors (halogenated benzenes and phenols) when materials are combusted under vitiated conditions and it is considered that this provides a useful method for measurement of the yields of dioxins and other environmental contaminants from materials decomposed over a range of combustion conditions.

Studies of fire scenes such as the Dusseldorf airport fire have revealed significant levels of dioxin contamination in soot residues. This raises issues for the health and safety of people working in such environments (such as fire investigators and fire brigade personnel), for methods of postfire decontamination and their effectiveness, and for health of building occupants after restoration. Table 2 summarizes findings from an examination of levels of contamination at several sites and calculations of possible uptake by exposed individuals at the scene through smoke inhalation, dust inhalation, or oral intake. The basic findings were that a person near or inside such a building for 1 h is unlikely to receive a significant dioxin and furan exposure dose (pg toxic equivalent TEQ) due to smoke inhalation, that there could be a minor hazard from dust inhalation, but that a potentially significant dose

Table 2 Possible Total Dioxin and Furan Intake (pgTEQ) for a Person Near or Inside a Building During or after a Fire

Dioxin Concentration of Smoke Particles and Soot (ngTEQ g^{-1})	Intake	Dose Received (pg[b] TEQ)	Fraction of Daily Intake Limit (700 pgTEQ day^{-1})
Old Swedish incinerator plume (assuming 1/100 plume dilution factor)	0.05–0.40 ngTEQ m^{-3} diluted smoke inhalation for 1 h (1 m^3)	50–400	0.07–0.57
EPA household waste study 700–7,000[a]	Inhalation 100 m visibility smoke dilution 2–20 ng TEQ m^{-3} for 1 h (1 m^3)	2000–20,000	2.9–29
EPA household waste study 700–7000[a]	Dust inhalation 1 mg m^3 for 5 h (5 m^3)	3300–33,000	4.7–47
German fire residue maximum 200	Dust inhalation 1 mg m^{-3} for 5 h (5 m^3)	1000	1.4
EPA household waste study 700–7000[a]	Oral intake 0.1 g soot	70,000–700,000	100–1000
German fire residue maximum 200	Oral intake 0.1 g soot	20,000	28.6

EPA, Environmental Protection Agency.
[a] Assuming TEQ = total dioxin and furan content × 0.1 (copyright HER).
[b] pg = picogram

(300–10,000 times a maximum acceptable daily intake of 700 pgTEQ day^{-1}) if 0.1 g of soot was ingested. The World Health Organization recommended maximum daily intake has since been decreased by a factor of 5 to 140 pqTEQ day^{-1}. These findings demonstrate that a potential hazard may exist at fire scenes and that personnel should take care to avoid dust inhalation or ingestion by the use of simple hygiene precautions (dust particulate masks and gloves—with careful hand washing after exposure).

Potential hazards from dioxins, furans, and PAHs in relation to health and safety of workers at the scenes of the Dusseldorf and Lengerich fires were examined in some details following these incidents [52–54]. The findings were that the cancer risk from PCDDs and PCDFs was substantially lower than that from PAHs, but as already stated, both types of pollutants are strongly bound to soot and have a low bioavailability. No chronic toxic effects have been reported from individuals accidentally involved in single fire incidents and a German study has shown professional fire fighters to have no higher dioxin blood levels than the dioxin background levels in the general population [55]. These findings support the conclusion that exposure at a single fire incident is unlikely to present a serious chronic health hazard from dioxins, furans, and PAHs, and that simple hygiene precautions can protect professionals from the potential effects of repeated exposures.

4. CONCLUSIONS

FR systems and additives reduce fire losses, fire injuries, and deaths by reducing the probability of serious fires developing from exposure to ignition sources of different intensities up to the design limits of the systems used. If the ignition resistance is overcome, they may also increase the time to rapid fire development and the size of the ultimate fire, thereby increasing the time available for escape and probability of survival for occupants. A wide variety of FR additive compounds are used with different products and systems. In many cases they enable materials with an otherwise poor fire performance to fulfill regulatory fire performance criteria, widening the choice and range of applications for different polymers and products. For some applications it may be difficult to develop effective alternatives to existing FR systems, but in most cases an adequate or improved fire performance can be obtained by either alternative materials or systems that do not require FR additives, or by the use of alternative FR systems with less environmental impact. With respect to overall advantages and disadvantages from the use of FRs it is essential that each individual product is examined in terms of the extent to which the FR is released to the environment, the extent of bioavailability if it is released, and any associated toxic effects. These aspects then need to be considered in terms of the benefits of use in terms of fire safety. Since individual FR additives and systems vary enormously in these terms it should be possible to choose candidates for all applications which are "environmentally friendly" while still providing a useful fire safety function.

Abbreviations and definitions

ASET = available safe escape time (interval between ignition of a fire and the time at which tenability criteria for escape are exceeded)
BRE = Building Research Establishment (UK)
CBUF = Combustion Behaviour of upholstered furniture project
CFD = computational fluid dynamics
CM and CMHR = combustion modified and combustion modified high resilience flexible polyurethane foams
CPD = Construction products directive (European Community)
EPA = Environmental Protection Agency (USA)
FEC = fractional effective concentration: the exposure concentration at any point during a fire expressed as a fraction of an exposure concentration required for a given toxicity endpoint (such as incapacitation) (see also FIC)
FED = fractional effective dose: the exposure dose at any point during a fire expressed as a fraction of the exposure dose required for a given toxicity endpoint (such as incapacitation), where exposure dose = concentration × time
FIC = fractional irritant concentration: the FEC of an irritant
FR = Flame-retardant
HRR = Heat release rate
ISO = International Organization for Standardization
LDPE = Low-density polyethylene
LSF = Low smoke and fume cable material
PAH = Polycyclic aromatic hydrocarbon
PCDD = Polychlorinated dibenzodioxin
PCDF = Polychlorinated dibenzofuran
PFIB = Perfluoroisobutylene
PVC = Polyvinyl chloride
REACH = Registration, Evaluation and Authorisation of CHemicals (in the European Community)
RSET = Required safe escape time (the interval between ignition and the time at which all occupants are able to reach a safe location)
TEQ = Toxic equivalent: mass of a set of polyhaloganeted dibenzodioxins ("doxins"), dibenzofurans ("furans), and diphenyls of differing toxic potencies normalized to an equivalent mass of the most toxic dioxin, 2,3,7,8-TCDD (2,3,7,8-Tetrachlorodibenzo-*p*-dioxin)
TBBPA = Tetrabromobisphenol A
VCM = Vinyl chloride monomer

References

[1] Ikonomou MG, Rayne S, Addison RF. Exponential increases of the brominated flame retardants, polybrominated diphenyl ethers, in the Canadian arctic from 1981 to 2000. Environ Sci Technol 2002;36:1886–92.

[2] The building regulations. Approved document B 2006 edition. Communities and local government; 2000.
[3] Furniture and furnishings (Fire) (Safety) regulations statutory instrument 1988/No. 1324, The Stationery Office Ltd.
[4] The application of fire safety engineering principles to fire safety design of buildings. PD7974–6: human factors: life safety strategies - occupant evacuation, behaviour and condition. British Standards Institution; 2004.
[5] ISO TR 16738 Fire safety engineering – technical information on methods for evaluating behaviour and movement of people. Geneva: International Organization for Standardization; 2009.
[6] United Kingdom Fire Statistics, Communities and local Government, London (published annually).
[7] BS 585 Methods of test for assessment of the ignitability of upholstered seating by smouldering and flaming ignition sources. British Standards Institution; 2006.
[8] Purser DA. Toxic product yield and hazard assessment for fully enclosed design fires involving fire retarded materials. Polym Int 2000;49:1232–55.
[9] Purser DA. Assessment of hazards to occupants from smoke, toxic gases and heat. In: DiNenno PJ, editor. The SFPE handbook of fire protection engineering. 4th ed. Massachusetts, USA: National Fire Protection Association; 2009. pp. 2/96–2/193.
[10] Buser H-R. Polybrominated dibenzofurans and dibenzo-p-dioxins: thermal reaction products of polybrominated diphenyl ether flame retardants. Environ Sci Technol 1986;20:404–8.
[11] Thoma H, Hauschulz G, Knorr E, Hutzinger O. Polybrominated dibenzofurans (PBDF) and dibenzodioxins (PBDD) from the pyrolysis of neat brominated diphenylethers biphenyls and plastics mixtures of these compounds. Chemosphere 1987;16: 277–85.
[12] Sperring K. Current state of regulations relating to brominated fire retardants. Society of Chemical Industry Fire Chemistry Discussion Group. University of Lancaster UK; 1998.
[13] Simonson M, Blomqvist P, Boldizar A, Möller K, Rosell L, Tullin C, et al. Fire life cycle analysis model: TV case study. SP report study; 2000. 13.
[14] Dietz M, Hörold S, Nass B, Schacker O, Schmitt E, Wanzke W. New environmentally friendly phosphorus based flame retardants for printed circuit boards as well as polyamides and polyesters in E&E applications. Fraunhofer IZM. Electronics goes green; 2004. pp. 771–776.
[15] Keeping fire in check, an introduction to flame retardants used in electrical and electronic devices. European Flame Retardants Association; 2010.
[16] Purser DA. Behavioural impairment in smoke environments. Toxicology 1996;115: 25–40.
[17] OPDM Building Regulations Division Project. The production of smoke and burning droplets from products used to form wall and ceiling linings. BRE report 213073, April 30, 2004.
[18] Fire safety of upholstered furniture – the final report on the CBUF research programme. In: Sundstrom B, editor. European commission measurement and testing report; June 1995. EUR 16477 EN.
[19] Heap JA. Studies of the toxic fume emissions from burning textiles [Ph.D. thesis]. University of Salford, 1990.
[20] Purser DA. ASET and RSET: addressing some issues in relation to occupant behaviour and tenability. In: 7th international symposium on fire safety science, Worcester Polytechnic Institute – Worcester Massachusetts, USA; June 16–21, 2002. Fire Safety

Science – Proceedings of the seventh international symposium. International Association for Fire Safety Science, 2003; 7:91–102. http://dx.doi.org/10.3801/IAFSS.FSS.7-91.
[21] Purser DA. Applications of human and animal exposure studies to human fire safety. In: Stec A, Hull R, editors. Fire toxicity. Cambridge, UK: Woodhead; 2010. pp. 282–345.
[22] prEN 50399-2-1 and prEN 50399-2-2 Common test methods for cables under fire conditions – heat release and smoke production measurement on cables during flame spread test; 2011.
[23] FAR 25.853 fire test to seat - fire test to aircraft material. Federal Acquisition Register; 2011.
[24] Camino G, Costa L, Martinasso G. Intumescent fire-retardant systems. Polym Degrad Stabil 1999;23:359–76.
[25] Woolley WD, Fardell PJ. Basic aspects of combustion toxicology. Fire Saf J 1982;5: 29–48.
[26] Purser DA, Stec AA, Hull TR. Effects of material and fire conditions on toxic product yields. In: Stec A, Hull R, editors. Fire toxicity. Cambridge, UK: Woodhead; 2010. pp. 515–40.
[27] Purser DA, Purser JA. HCN yields and fate of fuel nitrogen for materials under different combustion conditions in the ISO 19700 tube furnace and large-scale fires. In: Proceedings of 9th international symposium on fire safety science. Germany: University of Karlsruhe; September 21–26, 2008. pp. 1117–28. Fire Safety Science 2008; 9: 1117-28. http://dx.doi.org/10.3801/IAFSS.FSS.9-1117.
[28] Purser DA, Fardell PJ, Rowley J, Vollam S, Bridgeman B. An improved tube furnace method for the generation and measurement of toxic combustion products under a wide Range of fire conditions. In: Proceedings of the 6th international conference flame retardants '94. London, UK: Interscience Communications; 1994.
[29] Wright T. Environmentally friendlier flame retardant systems. PhD Dissertation. Leeds UK: Leeds University; 1997.
[30] Purser DA. Validation of additive models for lethal toxicity of fire effluent mixtures. Polym Degrad Stabil 2012;97:2552–61.
[31] ISO/TS 19700 Controlled equivalence ratio method for the determination of hazardous components of fire effluent. Geneva, Switzerland: International Organization for Standardization; 2007.
[32] Toxicology and carcinogenesis studies of tetrakis(hydroxymethyl)phosphonium sulfate (thps) (cas no. 55566-30-8) and tetrakis(hydroxymethyl)phosphoniumchloride (thpc)(cas no. 124-64-1) in f344/n rats and b6c3f1 mice. National toxicology program technical Report Series No. 296.
[33] Perfect JRW. Flame-resistant finishing with phosphorus-containing resins. J Soc Dye Colorists 1958;74:829–34.
[34] CPSC bans tris-treated children's garments. Consumer Product Safety Commission; April 7, 1977. Release # 77–030.
[35] Council Directive 79/663 EEC. Official J Eur Communities; 1979. L197/37.
[36] Chang S-J, Sheen Y-C, Chang R-S. Chang FC- the thermal degradation of phosphorus-containing copolyesters. Polym Degrad Stabil 1996;54:365–71.
[37] Purser DA. Toxicity of fire retardants in relation to life safety and environmental hazards. In: Horrocks AR, Price D, editors. Fire retardant materials. Cambridge, UK: Woodhead Publishing Ltd; 2001. pp. 69–127.
[38] Purser DA. Combustion toxicology of anticholinesterases. In: Ballantyne B, Marrs TC, editors. Clinical and experimental toxicology of organophosphates and carbamates. Oxford UK: Butterworth-Heinemann; 1992. pp. 386–95.

[39] Flournoy RE, Monroe D. History and risk assessment of vinyl chloride monomer leaching from PVC pipe: case history in Doniphan County, Kansas RWD # 5. Bridging the gap: meeting the World's water and environmental resources challenges. Proc World Water Environ Resour Congr; 2001.

[40] PVC-rRohre für trinkwasser-kontakt. Gutachten bezüglich möglicher Abgabe von monomeren Vinylchlorid an trinkwasser. A-1030 Wien: ÖKI (Osterreichisches Kunstoffinstitut); 1999.

[41] Scientific opinion on tetrabromobisphenol A (TBBPA) and its derivatives in food. EFSA panel on contaminants in the food chain (CONTAM). EFSA J 2011;9:2477−538.

[42] Fisk PR, Girling AE, Wildey RJ. Prioritisation of flame retardants for environmental risk assessment. UK Environment Agency; 2003.

[43] Purser DA, Purser JA. The potential for including fire chemistry and toxicity in fire safety engineering; March 28, 2003. BRE report no 202804.

[44] Purser DA. Recent developments in understanding the toxicity of PTFE thermal decomposition products. Fire Mater 1992;16:67−75.

[45] Handbook on air pollution and health. Department of health committee on the medical effects of air pollutants. London: The Stationery Office; 1997.

[46] Aitken RJ, Hankin SM, Ross B, Tran CL, Stone V, Fernandes TF, et al. EMERGNANO: a review of completed and near completed environment, health and safety research on nanomaterials and nanotechnology Defra Project CB0409. Institute of occupational Medicine; 2009.

[47] Stec AA, Hull TR. Toxic combustion products from fire retarded nanocomposite polymers. In: Proceedings of the 5th international seminar on fire and explosion hazards, Edinburgh, UK; 23−27 April 2007.

[48] Rhodes J, Smith C, Stec AA. Characterisation of soot particulates from fire retarded and nanocomposite materials, and their toxicological impact. Polym Degrad Stabil 2011;96: 277−84.

[49] Alcock RE, Jones KC. Dioxins in the environment: a review of trend data. Environ Sci Technol 1996;11:3133−43.

[50] Papke O. PCDD/PCDF: human background data for Germany, a 10-year experience. Environ Health Perspect 1998;106:723−31.

[51] Evaluation of emissions from the burning of household waste in barrels. Volume 1 Technical report. United States Environ Prot Agency Report EPA-600/R-97-134a, 1997, pp 1−69.

[52] Troitzsch JH. Fire gas toxicity and pollutants in fires: the role of flame retardants. In: FRPM07. 11th European meeting on fire retardant polymers. Bolton; July 3−6, 2007.

[53] Independent Experts Commission for the Minister President of the State of North Rhine Westphalia for Investigating the Consequences from the Dusseldorf Airport Fire. Report - Part I. Analysis April 11, 1996. Fire. Recomm Consequences Dusseldorf Airpt. April 14, 1997.

[54] Published by the Ministry for the Environment and the Ministry of the Interior of the State of North Rhine Westphalia and the town of Lengerich Major fire in a plastics warehouse in Lengerich, October 1992-Documentation; June 1994 [In German].

[55] Environmental medical studies on fire fighters. Ruhr-University Bochum and Heinrich-Heine University Dusseldorf on Behalf of the Ministry of Labour, Health and Social Affairs of the State of North Rhine Westphalia; October 1992 [In German].

CHAPTER

Flame Retardant Additives in Polymers: When do the Fire Safety Benefits Outweigh the Toxicity Risks?

3

V. Babrauskas
Fire Science and Technology Inc., Washington, USA

R. Fuoco, A. Blum
Green Science Policy Institute, Berkeley, California, USA

CHAPTER OUTLINE

Polymer Green Flame Retardants. http://dx.doi.org/10.1016/B978-0-444-53808-6.00003-2

1. Introduction

Since the 1970s, increasing flammability regulations continue to expand the global usage of organic brominated, chlorinated, and phosphate flame retardants (FRs). Manufacturers add these chemicals to a large range of products to increase their flame ignition resistance. Even though a wide variety of FR additives have been developed, for most man-made polymers, halogenated FR chemicals have been the most frequently used. This is due to their cost, availability, and extensive industry experience with this class of additives. Until recent years, regulatory bodies have considered only the potential benefits of their usage and not the potential drawbacks. For example, after decades of use and hundreds of studies detecting adverse health and environmental consequences, two polybrominated diphenyl ether (PBDE) commercial mixtures—pentaBDE and octaBDE—were banned in 2003 in California and in 2004 in the European Union and voluntarily withdrawn from production by the sole U.S. manufacturer. These two PBDE mixtures were subsequently listed as persistent organic pollutants (POPs) by the Stockholm Convention. The cause for concern with these and other PBDEs, their replacements, and other currently used halogenated and phosphate replacement chemicals is now well recognized [1,2].

The assumption held by much of the public, industry, and scientists is that any hazardous FR additives would be restricted from use in consumer products. However, in the United States, only chemicals in foods, drugs, and pesticides are regulated prior to reaching the marketplace. There is no requirement for data on potential human toxicity nor is there sufficient authority to regulate other chemicals. When a number of halogenated FRs received detailed study, they were found to be persistent when introduced into the environment and to have serious adverse health consequences [3]. In light of these findings, it seems necessary to consider the net outcome associated with the use of FR agents, instead of only evaluating their effects on improvement of fire safety. This could require a complex weighing of alternatives which lack a common basis for comparison, e.g. death or injury due to fire, versus damage to the environment or long-term health effects associated with direct ingestion, consumption of contaminated foodstuffs, and other modes of transfer. In cases where an evaluation determines a complete lack or near-lack of fire safety benefit, such a complex assessment may not be needed.

2. Innate fire performance characteristics of common polymers

Resins used for making commercial plastics are primarily composed of carbon (C) and hydrogen (H), with oxygen and nitrogen being relatively common and other heteroatoms only showing up in specialized application polymers. In many cases, resins consist solely of C and H. Polyethylene (PE) and polypropylene are two very common resins that are purely C and H with no heteroatoms. Consequently, they have exceptionally high heat of combustion values [4], with the net heat of combustion, $\Delta h_{c,net}$ being 43.1–44.6 MJ kg^{-1}. For comparison, gasoline is 43.7 MJ kg^{-1}. Other common polymers with no heteroatoms include poly-1,4-butadiene, poly-1-butene, and polystyrene, and these have heat of combustion values of 40–43 MJ kg^{-1}. Polymers which contain oxygen show a lower heat of combustion since the oxygen part of the molecule does not function as a fuel. Examples include liquid crystal polymer, polyester, polyethylene oxide, and polyvinyl acetate. These have heat of combustion values of 20–30 MJ kg^{-1}, and as low as 15.9 MJ kg^{-1} for polyoxymethylene, which has a notably high oxygen content. Similarly, polymers containing nitrogen in their structure tend to show lower heats of combustion although the effect is slight if the nitrogen content is small.

Heat of combustion, however, is only one of the many physicochemical properties of a polymer that influence its burning behavior. Whether a polymer is thermoplastic or thermoset, its ease of gasification, its thermal properties (especially thermal inertia, the product of conductivity, density, and heat capacity), and even properties such as diathermancy influence the fire behavior.

The actual fire behavior cannot be simply computed from knowing some basic chemical properties of the polymer. For fire safety purposes, the basic question is "How big is the fire?" In quantitative terms, this question leads to the concept of heat release rate (HRR), the most important variable in quantifying fire hazard [5]. HRR is measured in kilowatts (kW) and directly expresses the thermal hazard of combustion. The production of toxic gas species is highly correlated to HRR [6]. Thus, HRR quantifies human hazards of fire almost entirely. The two main fire science variables not encompassed by HRR are ignitability and flame spread. Since none of these hazard variables can be obtained from a chemical calculation, it is necessary to do testing instead. For bench-scale studies on materials, HRR and ignitability can be measured simultaneously in the same test apparatus, the Cone Calorimeter [7,8], ASTM E 1354 [9], or ISO 5660 [10]. In bench-scale testing, one would need to separately account for flame spread since HRR is measured on a per-unit-area basis (kW m^{-2}) and flame spread is not measured directly in the Cone Calorimeter or similar testing techniques. In many cases, the strong relationship between ignitability and flame spread can be utilized for this purpose. But the issue is obviated if large-scale HRR testing is done since, in that case, there is no need to separately characterize flame spread—the flame spread characteristics of the material intrinsically come into play in establishing its large-scale HRR behavior and do not need to be considered independently. To do such testing, a large-scale rig has to be built which is, to some extent, configured to deal specifically with the commodity in question. However, in general, such tests have often used the room configuration of ISO 9705 [11], along with the instrumentation details specified in ASTM E 2067 [12]. Results of diverse large-scale HRR tests have been compiled by Babrauskas [13].

Some polymers are innately FR and do not require additives to meet many test requirements. Poly(vinyl chloride) (PVC) is considered to be innately mildly FR, but the effect is small and FR grades of PVC are also used which do contain FR additives. Apart from PVC, other innately FR polymers tend to be expensive and many are only affordable for aerospace or military applications. Polytetrafluoroethylene (PTFE, Teflon) is probably the most commercially common of such intrinsically FR polymers. PTFE is fully fluorinated PE but several variants exist of partially fluorinated PE or copolymers, e.g. PE—tetrafluoroethylene (Tefzel), polytrifluoroethylene, polyvinylidenefluoride/hexafluoropropylene copolymer (Viton A), PTFE/PE copolymer (Tefzel), and also PE with both chlorine and fluorine substituents, e.g. polychlorotrifluoroethylene (CTFE, Kel-F). Other intrinsically FR polymers that show up in aerospace or military applications include polyamide-imide (Torlon), polybenzimidazole, polyetheretherketone (PEEK), polyetherimide (Ultem), polyimide (Kapton), poly(phenyl sulfone), polyphosphazenes (Eypel), and poly(vinylidene fluoride). Silicon-based intrinsically FR polymers include siloxanes and silphenylenes. For high-performance fabrics, poly(m-phenylene isophthalamide) (Nomex) or poly(p-phenylene isophthalamide) (Kevlar) is often used. The HRR properties determined in Cone Calorimeter tests were compiled by Hirschler for many common polymers [14]. For polymers with primarily aerospace applications, Lyon and coworkers at the FAA have provided collections of test results [15—18]. A compilation of Cone Calorimeter abstracts is also a useful resource [19].

3. Use of FR additives to improve the fire safety of polymers

FR chemicals are any form of additives which are used to impart an improved fire performance to combustible products. The only universally shared characteristics of such substances are that they are additives (i.e. not the innate substance itself) and that they are added to combustible substances to decrease, in some way, their combustibility. Products for enhancing the fire safety of noncombustible substances also exist, an example being protective coatings applied to steel structural members intended to make the member resist postflashover fire temperatures longer. Such substances are not called fire retardants and are not within the scope of this chapter. The main classes of combustible substances to which FR additives are added are plastics, wood and cellulosic products, and textiles. The chemical families used for these three main applications tend to be different although with some overlaps. For example, the FR additive for wool textiles is commonly potassium hexafluorozirconate (K_2ZrF_6), a substance not generally used for any other applications. However, only a tiny fraction of commercial textiles of any kind receive an FR treatment, with the majority being untreated. Treatments using ammonium phosphates are common for wood, but not for other substances. Similarly, borax and boric acid are common additives for cellulosic insulation and cotton batting, but do not find much usage with other categories of combustible substances. However, a distinction exists in that most cellulose insulation and cotton batting products are FR treated, whereas only a small fraction of whole wood products (dimensional lumber, plywood, etc.) receive an FR treatment.

FRs are most often added to products in response to a specific regulation; so they will refer to a specific test method to assess compliance. The test methods are rarely based on fundamental fire science variables (e.g. HRR). Instead, the test methods typically are ad hoc flame tests without a fundamental engineering basis; well-known examples include UL 94 [20] and MVSS 302 [21]. Babrauskas [22] has reviewed numerous US test methods while Troitzsch [23] published a compendium of test methods used in various countries around the world.

It is important to realize that the term "flame retardant" lacks any quantitative definition. ISO [24] defines *flame retardant* as "substance added, or a treatment applied, to a material in order to suppress or delay the appearance of a flame and/or reduce the flame-spread rate." *Fire retardant* is normally considered to be synonymous with "flame retardant," but ISO declares it to be either a synonym of "flame retardant," or a "substance added, or a treatment applied, to a material in order to delay ignition or to reduce the rate of combustion." ASTM [25] interestingly notes that *flame retardant* is "not a defined term," but defines *flame retardant chemical* as "a chemical, which when added to a combustible material, delays ignition and reduces flame spread of the resulting material when exposed to flame impingement." Thus, it is clear from these definitions that the qualities sought to be improved by adding FR chemicals include ignitability, flame spread, and rate of combustion. Scientifically, "rate of combustion" would be considered as referring to HRR, but the variables used in most regulatory mandated tests are inches or millimeters, with requirements such as "the burning rate shall not exceed X mm s^{-1}."

The largest fraction of FR chemicals produced is for treating of plastics, with the remaining markets being small in comparison. With three exceptions, however, FR chemicals are not designed for the treating of "plastics", but rather only specific families of polymers. The main exceptions are hydrated minerals, organohalogen compounds, and polymer–clay nanocomposites. Hydrated minerals are primarily alumina trihydrate and magnesium hydroxide and these achieve their FR action by endothermic decomposition, release of H_2O, and lowering of calorific value due to displacement of polymer by the mineral product. Organohalogen compounds are typically various ring compounds with Cl or Br substituents. Overall, organohalogen compounds have had a dominant role in the FR market. Antimony trioxide is often used as a synergist in formulations using an organohalogen FR additive. Polymer–clay nanocomposites are relatively new and, over the last two decades, there have been a huge number of academic studies [26] on the properties of these materials as FR additives for a very wide array of polymers. However, the FR effectiveness is often only modest, and it is not clear that this family will achieve widespread commercial use.

4. Toxicological behavior of common FR additives in laboratory studies

The four brominated FRs that have been used most extensively are tetrabromobisphenol A (TBBPA), hexabromocyclododecane (HBCD), decabromodiphenyl ether (decaBDE), and pentabromodiphenyl ether (pentaBDE) (Table 1).

Due to toxicological concerns, pentaBDE was withdrawn from U.S. commerce in 2004 and decaBDE will be phased out by the U.S. Environmental Protection Agency by the end of 2013. Despite these restrictions, discussion of the PBDEs is still relevant to this text as large amounts of upholstered furniture, juvenile products, carpet cushion, and plastics containing PBDEs are still in use and will eventually end up in landfills. Indeed, because of their high-volume use in consumer products over

Table 1 Heavily Used Brominated Flame Retardant Chemicals

Structure	Name(s) and Usage
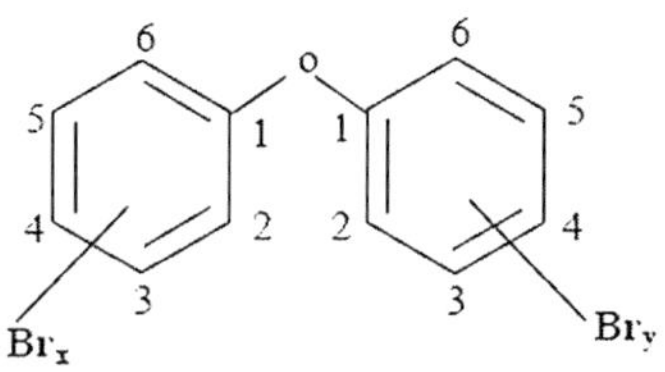 PBDEs	PentaBDE Polyurethane foam for use in upholstered furniture and baby products to comply with California Technical Bulletin 117 DecaBDE Polyvinyl chloride, polycarbonates, high-impact polystyrene, and back coating for textiles
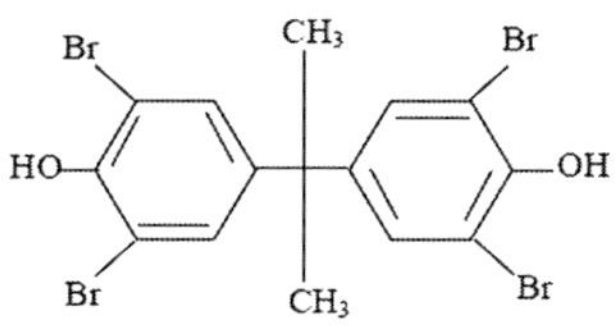 TBBPA	TBBPA Electronics including printed circuit boards
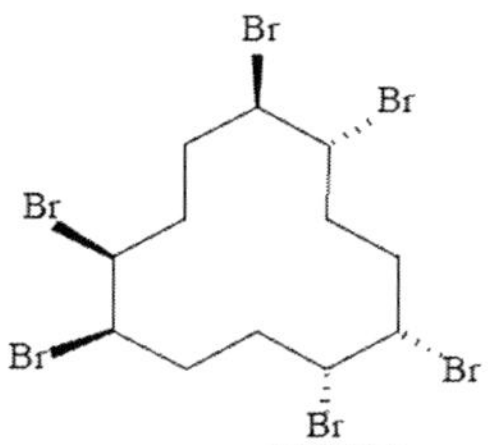 α-HBCD	HBCD Polystyrene foam insulation in buildings, upholstery, textile coatings, cable, latex binders, and electrical equipment

decades, the PBDEs have been considered unique among toxic chemicals in that they can be found at pound levels in homes, public places, and transportation. Moreover, large volumes of these materials are in the global recycling flow (e.g. plastic from waste electronics or polyurethane foam recycled to carpet padding) and will continue to be used in consumer products for a considerable time.

Major replacement chemicals for pentaBDE commercial mixtures include tris(1,3-dichloro-2-propyl) phosphate (TDCPP) and the mixture Firemaster 550®. Firemaster 550® contains triphenyl phosphate (TPP), mono-, di-, and tri-isopropylphenyl phosphates, 2-ethylhexyl-2,3,4,5-tetrabromobenzoate (TBB), and Bis (2-ethyl hexyl) tetrabromophthalate (TBPH). Replacements for decaBDE include 1,2-Bis(2,4,6-tribromophenoxy) ethane (BTBPE) and decabromodiphenylethane (DBDPE). As discussed below, these chemicals appear to have similar toxicological behavior as the FRs they have replaced.

4.1 PentaBDE and replacements

A large body of experimental data, both in vivo and in vitro, shows that pentaBDE can disrupt the endocrine system resulting in effects on thyroid, ovarian, and androgen function [27]. PentaBDE also disrupts thyroid hormone homeostasis which can cause neurologic impairments, including a reduction in the IQ of offspring [28]. Effects of prenatal or neonatal exposure include impacts on gene expression of liver enzymes [29] endocrine disruption [30], reproductive damage [31], immunotoxicity [32], and neurotoxic effects [33]. As an endocrine disruptor, pentaBDE exposure results in increased lipolysis and reduced insulin-stimulated metabolism in rat adipocytes [34], effects which have been associated with obesity, insulin resistance, and Type 2 diabetes.

The brominated Firemaster 550 components TBB and TBPH are genotoxic in fish, causing increased DNA strand breaks in orally exposed fish [35]. Triphenyl phosphate (TPP) is toxic to aquatic organisms including *Daphnia* [36], rainbow trout, and fathead minnows [37]. Mono-, di-, and tri-isopropylphenyl phosphates are reproductive and developmental toxins at mid- to high doses in rats [38]. Histopathologic changes are observed in female reproductive organs and adrenals at all doses. The offspring of rats exposed to low doses of Firemaster 550 showed weight gain and anxiety in a small pilot study [39].

TDCPP, or chlorinated Tris, was removed from use in children's pajamas in 1978 due to its mutagenicity [40] and has subsequently been found to be a probable human carcinogen in a study at the US Consumer Product Safety Commission (CPSC) [41]. The CPSC report estimates the lifetime cancer risk from tris-treated furniture foam is up to 300 cancer cases/million; their chronic hazard guidelines define a substance as hazardous if lifetime cancer risk exceeds one in a million. TDCPP is also absorbed by humans [42]. The U.S. EPA considers TDCPP a moderate hazard for cancer and reproductive/developmental effects [43]. TDCPP is listed as a carcinogen by the California Office of Environmental and Health Hazard Assessment [44].

4.2 DecaBDE and replacements

DecaBDE has been evaluated less often than the other common FRs, even though its component congener BDE 209 is the most widespread PBDE congener in use. Studies in mice and rats suggest evidence for thyroid hormone disruption and delayed developmental neurotoxicity [45], hyperthyroidism, and hepatotoxicity [46]. Chronic, oral exposure studies have demonstrated that very high doses of BDE 209 result in hepatic carcinogenicity in rodents [47].

DBDPE is acutely toxic to water fleas in vivo and reduces the hatching rates of exposed zebra-fish eggs and significantly raises the mortality of hatched larvae [48]. There has not yet been adequate toxicity testing for BDBPE.

4.3 TBBPA

TBBPA, a cytotoxicant, immunotoxicant, and thyroid hormone agonist, has the potential to disrupt estrogen signaling [49]. Like other brominated FRs, the central mechanism of TBBPA toxicity is thought to be through disruption of thyroid homeostasis. This is of primary importance during development when small thyroidal changes in the mother can result in cognitive defects in the children [50]. TBBPA has an even closer structural relation to thyroxin (T4) than PCBs, and TBBPA toxicity to the thyroid might be related to its effects on the transport of T4 in the blood. TBBPA is toxic to primary hepatocytes, destroys mitochondria, and results in membrane dysfunction in liver cells and inhibition of a key mixed function oxidase enzyme [51]. TBBPA is also neurotoxic in rat brain cells, where it causes oxidative stress, inhibits dopamine uptake, and generates free radicals [52]. The carcinogenic potential of TBBPA is currently being studied by the National Toxicology Program.

4.4 HBCD

Experimental evidence indicates that HBCD used in high volumes as an FR for polystyrene exerts a range of endocrine disrupting and reproductive developmental effects in animals [53]. Repeated-dose oral toxicity studies collectively point to the liver and thyroid as target organs for HBCD toxicity [54]. Neonatal HBCD exposure has been shown to have adverse effects on neurodevelopment [55]. HBCD has also been shown to block dopamine uptake in rat brain synaptosomes in vitro and inhibit neurotransmitter release [56]. Like other brominated FRs, HBCD may disrupt thyroid hormone homeostasis resulting in decreased levels of T4 and increased levels of thyroid stimulating hormone (TSH) [57] and affect thyroid hormone receptor-mediated gene expression [58]. HBCD is one of the first "Substances of Very High Concern" scheduled to be phased out by 2015 in the European Union [59]. Canada "is proposing to implement regulations to prohibit the manufacture, use, sale, offer for sale, import and export of HBCD and products containing HBCD [by the end of 2016]" [60]. HBCD is also under consideration for listing as a persistent organic pollutant by the 178 member countries belonging to the Stockholm

Convention on Persistent Organic Pollutants in 2013. In October 2012, an expert review committee recommended listing HBCD under Annex A (Elimination) with limited exemptions [61].

5. Routes of human exposure to FR additives

Most halogenated and phosphate FRs are semivolatile organic compounds that are mixed with the polymer resin as plastics and are not chemically bound. They therefore leach continuously out of products via volatilization and weathering of the plastic [62]. Over time, the chemicals accumulate in indoor air and dust [63]. The chemicals also enter the outdoor environment when products end up in landfills and when chemicals escape during the FR production of and manufacture of products containing them. Fugitive emissions and discharges from manufacturing or dismantling and disposal sites may contaminate the environment near the facilities, as well as farther away from them through long-range transport. Because of their persistence and bioaccumulative properties, some organohalogen FRs in the outdoor environment biomagnify up the food chain and end up in our food supply.

Most of what we know regarding exposure pathways of FRs comes from studies of PBDEs. However, studies show that TBBPA, HBCD, TDCPP, Firemaster 550, BTBPE, and DBDPE similarly migrate into indoor dust [64—68].

5.1 Ingestion of contaminated house dust

Associations between PBDEs found in office dust, hands, and serum suggest that potential exposure pathways may involve ingestion of PBDE-contaminated dust on hands [69]. PBDE and HBCD body burdens are associated with levels in house dust [70,71]. For North Americans, 82% of exposure to PBDEs is assumed to come mainly from this route of exposure [72]. Children's frequent hand-to-mouth activity and contact with indoor surfaces predispose them to particularly high exposures resulting from ingestion of contaminated dust [66]. Positive correlations between PBDE concentrations in house dust and umbilical cord plasma and between levels in house dust and levels in breast milk have been shown [73,74].

5.2 Ingestion of contaminated food

While ingestion of contaminated dust is considered the major route of human exposure to PBDEs in North America [75], about 20% of exposure to PBDEs in Americans (and more so in Europeans) is currently estimated to derive from diet, with the highest exposure from butter, seafood, and meat [76].

5.3 Inhalation of contaminated indoor air

Inhalation of airborne PBDEs has been estimated to be a small fraction of overall daily exposure except for in heavily contaminated office conditions where this route is as important as diet [77]. Airborne particulate matter collected in an office building's HVAC system contained high levels of PBDEs suggesting inhalation as a possible route of exposure [78].

5.4 Dermal absorption

Human dermal absorption of FRs has not been directly assessed. Laboratory studies, however, have shown that TDCPP is readily absorbed through skin in mice [79].

6. Human levels and health associations of FR additives from epidemiological studies

The pentaBDE concentrations in the North American general population are 10–40 times higher than the concentrations reported for populations in Europe and other parts of the world [80]. In California, populations have been shown to be disproportionately exposed to pentaBDE, likely due to the state's furniture flammability standard TB117 that has led to high usage of halogenated and phosphate FRs in furniture and baby products [81]. A recent study found that body burden levels in California children are 2–9 times higher than in similar aged children across the U.S., and 4–9 times higher than children in Mexico, and up to 100 times higher than those in children of similar ages in Europe and Mexico [82].

As a consequence of this widespread exposure, there is a growing body of human epidemiological literature finding associations between increased pentaBDE body burdens and health problems such as reduced IQ in children [83], endocrine and thyroid disruption [84], changes in male hormone levels and reduced fertility [85], increased time to pregnancy in women [86], adverse birth outcomes like low birth weight [87] and premature delivery [88,89], cryptorchidism [90], and impaired cognitive and motor development [91].

The current pentaBDE replacements appear to be similarly problematic. A recent study showed that men living in homes with high amounts of the organophosphate FRs TPP (a component of Firemaster 550) and TDCPP in household dust had reduced sperm counts and altered levels of hormones related to fertility and thyroid function [92]. High levels of TPP in dust were associated with a substantial reduction of sperm concentrations and an increase in prolactin levels. Increased prolactin is considered a marker of decreased neuroendocrine/dopamine activity and also may be associated with erectile dysfunction [93].

There is much less human biomonitoring and epidemiological research conducted on the other major FR chemicals.

One study of decaBDE levels in individuals living near electronic waste recycling facilities found a significantly inverse association between levels of BDE-209 in breast milk with infant neurodevelopment [94].

TBBPA can be rapidly metabolized by mammalian liver and eliminated in bile, urine, and feces and therefore has a lower potential for bioaccumulation [95].

However, in a study of 93 pregnant women in France, TBBPA was detected in 44% of breast milk samples and 30% of serum samples [96].

In recent years, the global demand for HBCDs has increased in Europe and Asia, and temporal studies show that the levels of the most persistent HBCD stereoisomer are increasing in humans [97].

7. Environmental contamination and wild animal exposure

Levels of FRs are rising in the environment as a result of their widespread use, their persistence, and their tendency to bioaccumulate in the food web. Since their introduction, halogenated FRs have become widespread global contaminants and have been detected throughout the world in air, water, soil, sediment, sludge, dust, bivalves, crustaceans, fish, amphibians, reptiles, birds, and mammals.

In North America, PBDE levels in wildlife are the highest in the world and are increasing [98]. In fish, PBDE exposure may affect thyroid hormone homeostasis, sperm production, disease resistance, and neurodevelopment [99,100]. PBDEs are detected at high concentrations in birds of prey, and studies have shown that PBDEs may exert DDT-like endocrine-disrupting and reproductive effects in several bird species [101,102]. Marine mammals from the California coast contain the highest reported PBDE levels in the world [103].

Two components of Firemaster 550, TBB and TBPH, have been detected in blubber of dolphins and porpoises near FR production facilities in south China [104] as well as in seven arctic species [105]. Both BTBPE and DBDPE have been reported in air, water, sewage sludge, sediment, mussels, fish, and birds [106]. Current concentrations of DBDPE (a replacement for decaBDE) in eggs of herring gulls from the Great Lakes are similar to or higher than those of decaBDE and appear to be increasing after a comparatively short period of usage [107] suggesting DBDPE may be more persistent and/or bioaccumulative than decaBDE. High concentrations of DBDPE have also been reported in captive pandas [108] and water birds [109] from China's Pearl River Delta, where many e-waste recycling facilities are located.

Despite its lower relative potential to bioaccumulate, TBBPA has been detected in various environmental media and biota including air, soils, water, sediment, and bird muscle from e-waste regions of China [110]; in water and sediment from English lakes [111]; and in bottlenose dolphins and bull sharks from the Florida Coast [112]. Also, there is concern about the increasing levels of HBCD in wildlife especially in Asia [113].

8. Overview of risk-benefit considerations

Prior to implementing new flammability standards, decision makers should evaluate the potential fire safety benefit as well as the health and environmental impacts of the chemicals, materials, or technologies likely to be used. Special scrutiny should be given to small open-flame standards that are likely to be met by adding organohalogen or phosphate FRs to foams or plastics in consumer products.

8.1 Fire safety evaluation

In general, use of FR additives in products would have a fire safety benefit if a positive answer could be obtained to at least one of the two questions:

1. Does use of materials or products with the FR additive significantly reduce the severity of fire, assuming the product gets ignited?
2. Does use of materials or products with the FR additive serve to prevent ignitions from small flame sources? (Products intended for consumers or the general commercial market normally would not be able to resist attack from large flame sources.)

8.2 Health and environmental evaluation

If indeed a fire safety evaluation of a chemical additive leads to the conclusion that its use in an application would provide a fire safety benefit, one should consider the following questions:

1. Does this chemical have the potential to persist in the environment and/or bioaccumulate in living organisms?
2. Is this chemical associated with health harm of any kind in laboratory or epidemiological studies of animals or humans?

The evaluation of risks and benefits can be complicated, or not, depending on the findings for fire safety and health aspects. Evaluation is simple if one or other of these factors is essentially nil. If the fire safety benefits are small, negligible, or dubious, then use should be discouraged if there is any harm to health or environment. Such examples are discussed below as Case 1 and Case 2. If, on the other hand, there is no significant health or environmental harm potential, and there are positive fire safety benefits, then the status is acceptable. An example is the use of boric acid for FR treatment of cotton batting mattresses. This has positive benefits in reducing the propensity toward smoldering mattress fires, but without a drawback of environmental health harm. The situation becomes complicated if there is fire safety benefit but health or environmental harm is also documented. Quantitative assessment of situations where fire safety benefits and environmental harm aspects are in direct conflict can be tackled if enough good quality data are available and certain social policy decisions are made, e.g. what is the dollar cost of human life. But even if a common dollar-cost metric can be applied, "trading" can be difficult when the question of "for whom" or "to whom" arises. In other words, if benefits accrue to one social group and harm to another, how acceptable is that? This chapter cannot answer such challenging social policy questions.

9. Case 1: FR additives in furniture foams

The California Furniture Flammability Standard Technical Bulletin 117 (TB117) was implemented in 1975 by the California Bureau of Electronic and Appliance Repair, Home Furnishings and Thermal Insulation (the Bureau). Its purpose is to reduce fire deaths and injuries associated with upholstered furniture. The standard is a small-flame ignition standard which requires polyurethane foam in juvenile products and upholstered furniture to withstand exposure to a small open flame for 12 s [114]. The standard also regulates smolder behavior of foams and provides fire tests for other, nonfoam components. But these aspects are not of relevance to the present chapter, since chemical fire retardants are not added to foams to pass the smolder test, while other components are either in very limited use or require no treatment, or both. Compliance with the standard is mandatory for all products sold in California, irrespective of where they are manufactured. Of the 50 US States, California has been the only state which issued such a standard, but increasingly many national furniture manufacturers have been using this standard for all their furniture sold across North America to avoid maintaining a double inventory and for defense against fire liability claims. Thus TB117 has become a de facto national standard, with the FRs typically used to meet it being found in most furniture and baby products containing polyurethane foam sold in the US and Canada [115,116].

From its implementation in the late 1970s until 2004, TB117 was primarily met with the addition of 3–5% pentabromodiphenyl ether (pentaBDE) to the foam in furniture and juvenile products (nursing pillows, strollers, baby carriers, etc.). PentaBDE is a commercial mixture of several congeners of polybrominated diphenyl ethers (PBDE). PentaBDE was banned in California in 2003 due to its persistence and toxicity; eight other states and the European Union (EU) followed suit. Subsequent to 2004, TB117 has generally been met by adding TDCPP, Firemaster 550™ or mixtures of phosphate FRs [114,115].

In order to assess the risk–benefit situation for TB117, the two main questions established in the previous section must be answered, as examined in a recent study [117].

9.1 Do TB117 compliant materials significantly reduce the severity of the fire?

Babrauskas [118] studied furniture where the peak HRR of identically constructed furniture passing TB117 was compared to furniture made with non-FR foam. The differences observed between the furniture with non-FR foam and furniture made with foam complying with TB117 were within the normal data scatter from this type of test. In addition, visually, the fire development over the furniture was seen to be identical. The Consumer Product Safety Commission conducted similar tests and reached the same conclusions [119]. The lack of effectiveness of the TB117 foams was due to amount of FR additive used, which has generally been 3–5%. When much higher FR concentrations are used, such foams can be almost nonburning [120]; however, this is not practical for consumer furniture. Thus, the answer to the first question is clearly No.

9.2 Does TB117 foam serve to prevent ignitions from small flame sources?

A severe fault of the TB117 test is that the foam alone is exposed to a small burner flame rather than the composite piece of furniture. Under such conditions, TB117 foam with 3–5% of FR additive can resist a small open flame. But actual upholstered furniture always consists of a composite of at least two layers, with a fabric cover on top of foam. Furthermore, fabrics are thin membranes of about 1 mm that do not serve as a barrier to the flow of heat from the outside to the foam and it is common for the upholstery fabric itself to ignite from small flame sources such as a candle or a lighter. Once the fabric is burning, the foam is presented with a flame challenge which is many times larger than the cigarette lighter flame which may have originally ignited the fabric. (Note that TB117 does not consider protection from ignition by large flame sources and that it is well established that targets which might resist a smaller flame attack may be unable to resist one from a larger flame [121,122]). It must additionally be noted that (a) the TB117 standard contains a separate test provision for fabrics. But this is a moot test since the procedure is the same as mandated in the Federal CS 191-53 test [123], which all fabrics sold in the U.S. are already required to meet by the Federal government. (b) While an ostensible, but moot fabric test exists within TB117, the standard never assesses the behavior of an actual composite, i.e. a fabric on top of a layer of foam. However, the latter question has been experimentally investigated by two groups. Talley [124] tested 15 different upholstery fabrics, each over non-FR and TB117 foams of matched density (24 kg m^{-3}). All specimens ignited except those using one of the 15 fabrics, and for that fabric neither of the foams used led to ignition. Talley also visually observed flame spread behavior of the specimens that did ignite, and his conclusion was that "The TB117 foam made no significant, consistent difference in either ignition or flame spread." Talley also ran additional tests which showed that TB117 foams did not offer any benefit in regards to resisting smoldering ignition from cigarettes. In addition, the U.S. Consumer Product Safety Commission (CPSC) conducted laboratory research on full-size chairs and also found the test to be ineffective [118]. Thus, the answer to the question is also No.

Consequently, with California furniture flammability standard TB117, the analysis is especially simple: there are no fire safety benefits, yet environmental harm has been documented. Thus, no further detailed benefit/harm ratio needs to be established or quantified. Because of the environmental harm issues, the TB117 standard is currently being revised by the State of California and a new version is expected in 2013.

10. Case 2: FR additives in plastic foam thermal insulation

Polystyrene, polyisocyanurate, and polyurethane foams are energy-efficient insulation materials whose use in buildings is increasing. To meet U.S. building codes, FR chemicals are added to these materials. FRs primarily used in building insulation are found at increasing levels in household dust, human, and animal body fluids and are becoming widespread in the environment. When tested, some of the building insulation FRs are found to be persistent, bioaccumulative, and may be toxic. (Being persistent means that they do not break down into safer chemicals in the environment. Being bioaccumulative means that they accumulate in plants and animals, becoming more concentrated as they move up the food chain.) Additionally, some of these FRs are carcinogens, mutagens, or reproductive, neurological, thyroid, or developmental toxicants. Finally, when these foams burn, often following a few seconds exposure to a heat source, the retardants produce highly toxic dioxins and furans.

Given the cost of adding these chemicals and their potential to adversely effect on health and the environment, an important question is whether the FRs in building insulation products provide a fire safety benefit. This benefit, if any, appears to be quite small. This is a consequence of the fact that building codes take a "belt and suspender" approach requiring two sorts of fire safety protection, where one could be equally effective. But, as we shall show, in this case, such a dual set of requirements does not enhance the likelihood of fire safety being achieved since one of the two requirements has no positive effect on enhancing fire safety [125].

Until 2000, the U.S. had three separate building code bodies and three separate "model" building codes. "Model" means that they are published by private organizations, and only acquire a regulatory status when States, Counties, or municipalities enact them into regulation. Eventually, the three organizations merged and in 2000 the successor organization, the International Code Council, issued the International Building Code (IBC). Since that time, most jurisdictions have been using one or another of the triennial IBC editions, occasionally supplemented by local provisions. In the current codes, the provisions governing foam plastic insulation in buildings are essentially identical to the provisions that were in place for some three decades within the prior model building codes.

In the early 1970s, there were instances where persons would install foam plastic insulations exposed to habitable spaces, e.g. in unfinished basement rooms, garages, etc. This resulted in some very serious fires and galvanized the code groups into action, so that the 1976 Uniform Building Code (UBC) acquired a new section (Sec. 1717) for controlling foam plastics [126]. This new section generally required foam plastics to be separated from the interior of a building by a "thermal barrier," most typically ½ in (12.7 mm)-thick gypsum wallboard. The UBC did allow foams to be used uncovered if they met certain large-scale (corner, or room fire test) test requirements, but foams sufficiently advanced to meet such requirements are expensive and typically will not show up in exposed uses in general purpose buildings. More problematically, the code maintained the preexisting requirement that "foam plastic insulation used in building construction shall have a flame spread index of not more than 75 and a smoke-developed rating of not more than 450 when tested in accordance with [the ASTM E 84 Steiner Tunnel [127]." The other model building codes generally set requirements which were very similar to the UBC, and these provisions then were continued when the first edition of the IBC [128] was published in 2000. Thus, for over three decades now, it was required in nearly all U.S. jurisdictions that foam plastic insulations both show a flame spread index (FSI) of $\leq$75 and be protected by a thermal barrier from sources of ignition, heat, or fire impinging on it from the room. There are some exceptions such as cold storage facilities and other limited, specialized situations, but these are a small fraction of the building insulation market.

The first requirement does not entail any usage of FR chemicals since the foam does not need to be modified, only a gypsum board, or equivalent, barrier provided. However, the second requirement, for limited FSI, is usually met by the addition of halogenated FR chemicals to the insulating material. So the question arises, is this requirement justified on the basis of fire safety need? The answer is clearly No, but to understand this requires that the research literature be considered.

A gypsum board barrier will keep fire, heat, and ignition sources originating in the occupied space of room from impinging onto foam insulation located in a wall or ceiling cavity for far longer than there is any opportunity for individuals to survive within that room. Thus, the only area where the insulation might be conceivably challenged is if fire gets into the void spaces. Now, all building codes have stringent firestopping provisions, requiring that openings to or from such void spaces be closed. Thus, even this premise is based on the assumption of a code violation. It is, of course, illogical to design code provisions whose only function is to act as an antidote to an assumed code violation occurring in regards to some other code provision. Nonetheless, researchers have conducted tests to determine if fire gets into the void space, will insulation materials there allow it sustain continued propagation. Choi and Taylor [129] ran large-scale tests at the National Research Council of Canada (NRCC) and concluded that, in the absence of proper firestopping, fire can spread vertically up in wall cavities. However, they found that this behavior was only dependent on the thickness of the gap between the insulation and the inside of the wall (assuming the insulation fails to fully fill the cavity). Gaps of more than 1 in (25 mm) showed propagation, while smaller ones did not. But, they found that this was only a geometry question and that "the flame spread rating of the materials used in the tests was not a significant factor." Thus, they demonstrated that having an improved fire spread test result for the insulation does not improve the fire safety of the cavities.

There might be perhaps an argument that individuals could erect buildings where, in flagrant violation of the codes, they would apply foam plastic insulations to exposed room surfaces and not install any form of barrier over them, and that use of FR-treated foams (that is, foams with FSI $\leq$ 75), would still make these constructions acceptably fire safe. This again seems to be a stretch of logic since if flagrant flouting of the codes is assumed, there would be nothing stopping such individuals from procuring insulation products meeting no standards whatsoever. But there is extensive research on this question too. Several studies have examined whether it is adequately safe to use foam insulation with FSI < 75 uncovered on room walls/ceiling surfaces. It turns out that the ASTM E 84 Steiner Tunnel test, while producing sensible results for some categories of building materials, e.g. wood products, is fundamentally unreliable with regards to assessing the hazard of plastic foams [130]. Williamson and Baron [131] demonstrated that rigid polyurethane insulation of FSI < 25 led to extremely rapid, severe fire development within a room, if applied naked on the walls and ceiling. Underwriters Laboratories (UL) ran tests similar to Williamson and Baron's, and got even more extreme results [132]. Using extruded PS foam having FSI = 3 (sic) caused an extremely serious room fire with the entire ceiling being burned out and much of the walls. There were some foams in their tests which performed much better, but there was no correlation of the room fire hazard to FSI. Another study at NRCC [133] demonstrated that foams of FSI in the range of 25–50 would lead to room flashover (the most extreme state of a room fire) in as little as 0.5 min. The National Bureau of Standards (NBS, now NIST) similarly showed [134] that uncovered polyisocyanurate and polystyrene foams having FSI < 25 showed very rapid flashover times when tested in full-scale room fire test, and a later NIST study [135] produced very similar findings.

Factory Mutual is one of the largest organizations in the U.S. doing fire safety research and testing. On the basis of in-house research, they already issued an advisory notice [136] in 1978 that "Flame spread ratings by ASTM-E84 tunnel test should be disregarded for foamed plastics."

So the answer to the question, "does this requirement leading to flame retardants in insulation improve fire safety?" is clearly "No".

In contrast to the lack of fire safety benefit from the addition of FRs to insulation used in wall cavities, the adverse impacts of exposure to FR chemicals upon workers, human and animal health, building occupants, wildlife, and the global environment have been documented in the peer-reviewed scientific literature.

For example, all polystyrene foam insulation used in building insulation (both XPS, such as Styrofoam™, and EPS) is currently (in 2013) treated with hexabromocyclododecane (HBCD), a persistent, bioaccumulative, and toxic fire retardant. This chemical is one of the first "Substances of Very High Concern" being phased out in the EU and under consideration for listing by the Stockholm convention as a persistent organic pollutant. HBCD is found in dust, sewage sludge, breast milk and body fluids, wildlife, and the environment. 90% of the use of HBCD is with polystyrene insulation, which is the probable primary source of the global contamination. The rest is also used with fabrics and plastic. The suggested replacement for HBCD is another persistent brominated FR lacking toxicological or health information.

Polyisocyanurate (polyurethane) board often contains tris (1-chloro-2-propyl) phosphate (TCPP). While its toxicity in mammals and the effects of long-term exposure are unknown, it is toxic in aquatic environments. Polyurethane spray foam formulations commonly contain TCPP or proprietary chemicals with unknown composition and health effects. For example, Dow THERMAX™ Insulation board contains a maximum of 10% of TCPP.

Given this fire safety, health, and environmental information, the IBC should be encouraged to reconsider their foam-plastic insulation provisions that mandate an FSI of 75 or less and a smoke-developed index of 450. Needed levels of fire safety are attained by existing fire barrier requirements and the use of added FR chemicals and with potential harm to health and environment does not enhance that safety.

11. Case 3: FR additives in electronics housings

Rigorous safety standards govern the internal functioning of electronic equipment and prevent fires, electric shock, and other harm. Over the last decade, a series of "candle flame" standards have been proposed requiring that the plastic enclosures around consumer electronics products resist external ignition from a small open flame such as a candle. Such requirements would result in the addition of up to 30% by weight of organohalogen, antimony, or other flame retarding chemicals to the housings [137].

In 2000, the U.S. National Association of State Fire Marshals (NASFM) first proposed [138] that the enclosures around home electronics such as computers and televisions should be required to be resistant to ignition by a candle flame. The International Electrotechnical Commission Technical Committee 108 (IEC TC 108) then voted to develop and incorporate such "candle resistance" requirements in upcoming revisions of a series of product safety standards for the enclosures of electronic housings.

11.1 What health and environmental risks would these standards carry?

Organohalogen and phosphate FRs are cost-effective and commonly used in plastic to meet small open flame standards. The annual increase with the IEC candle resistance requirement is estimated to range from a low of 250,000 metric tons (550 million pounds) of organohalogen compounds to a high of 790,000 metric tons (1.7 billion pounds) for the total increase [139]. The FR chemical can migrate out of products with potential adverse health impacts as described previously in this chapter.

The addition of a variety of flame retarding chemicals into product housings would also make responsible recycling of electronics more expensive and difficult. The rapid obsolescence of many consumer electronic products is leading to a huge expansion of the plastic waste stream. Increasing the FR chemical load in this waste stream and introducing a mixture of retardants may make certain types of recycling cost prohibitive or even impossible. The small open flame standards would be likely to cause more downcycling and energy recovery (burning) to occur. In addition, brominated and chlorinated fire retardants form highly toxic dioxins and furans during the controlled and uncontrolled combustion that is still the unfortunate end-of-life fate for much of the world's electronics enclosures [140].

11.2 Do these small open flame standards provide a fire safety benefit?

Four sources—The U.S. Consumer Product Safety Commission (CPSC), the US National Fire Protection Association (NFPA), the Consumer Electronics Association (CEA), and the Telecommunications Industry Association (TIA)—have concluded that the risk of candles igniting consumer electronics was low, the latter two stating it is, in fact, too low to justify this requirement [141].

A review of the fire statistics from the NFPA reveals that a very small number of open flame or candle fire injuries and deaths are associated with consumer electronics. In 2002, the CEA produced a report [142] based on data from the CPSC collected from 100 hospitals around the country. The CEA found that electronics fires accounted for only about 2% of incidents, but when the few specific incidents (five in all) were reviewed, improving the FR capability of plastic enclosures of electronics would not prevent the types of injuries seen. They concluded "... that there are a limited number of instances where candles ignite electronic products, but that these incidents are so few and of such a nature that they do not appear to warrant a change in the product safety standards for electronic equipment."

The situation is different for TVs versus other electronics, and also in the U.S. versus in Europe. In the U.S., TV set cabinets have been required to be V-0 rated since 1979. In Europe, meanwhile, HB ratings have been the norm. Apart from consumer TV sets, other audiovisual equipments, however, have typically only carried HB-rated housings, both in the U.S. and in Europe.

The primary justification for the proposed small flame requirement were analyses by Simonson and colleagues showing a large fire safety benefit from the use of FR chemicals in TV enclosures [143,144]. This work, based on eight TV fires in one suburb of Stockholm in 1994, was extrapolated to predict a large fire rate from TVs in all of Europe a decade or more later. However, Simonson et al. did not accurately represent the European or the U.S. fire data. As discussed in some detail below, these analyses, which were funded by the FR chemical manufacturers, are seriously flawed.

The Simonson life cycle model compares TV sets with and without FRs. It compares emissions from TV production, FR production, and TV incineration, but does not include health or environmental effects of the chemical retardants themselves. The model considers the chemicals emitted when a TV burns, but not the effects of exposures that are due to the migration of retardants from the plastic TV enclosures into household dust [145], and therefore into humans, animals, and the food chain. In view of the recent significant increase in knowledge about the health and ecologic effects of FRs, the Simonson model would benefit from considering these other inputs across the life cycle of the products. These effects would include health effects from exposures occurring during manufacturing and use of the product, as well as toxicity to humans and the environment during and after disposal.

As detailed in Table 2, the Simonson model also uses dissimilar data sources for comparing the annual fire rates in TV enclosures without FRs (Europe) and to those with FRs (U.S.). For Europe, the model input extrapolates from eight TV fire incidents during 1994 in one suburb of Stockholm, Sweden to an estimate of 165 TV fires per million TVs annually across all of Europe. The 1994 in-depth review used a broad definition of a TV fire, which included very small fires. Unjustifiably, her analysis ignored the U.K. Department of Trade and Industry (DTI) review of fire brigade data from across Europe, which found 12.2 European TV fires per million TVs by internal ignition plus 6 TV fires per million TVs by external ignition [146]. The Simonson estimate, based on data from one Stockholm suburb, was an order of magnitude higher than the European fire brigade data, which are clearly statistically more valid, and are comparable in magnitude to the U.S. experience (discussed below). A detailed critique of Simonson's extrapolating from such a small, unrepresentative sample to fire data for all of Europe has been previously published [147].

Annual U.S. TV fire data as reported by the NFPA was based on fire-department surveys and fire-incident data from the U.S. Fire Administration from across the country [148]. The data were normalized to 13 TV fires per million TV sets, from both internal and external ignition. Without justification, the Simonson model then assumed that only a fraction of the U.S. TV fires reported by the NFPA were due to internal ignition and restricted further analysis to only the internal ignition fires (reducing the number of annual U.S. fires from 13 TV fires per million TVs to 5 TV fires per million TVs). In contrast, the model does not restrict European TV fires to internal ignition fires. The dissimilar data sources are based on different definitions and sizes of TV fires, potentially leading to an

Table 2 Annual Estimated TV Fires per Million TVs in the U.S. and Europe that Breach the TV Enclosure

Simonson Estimate for Europe	Simonson Estimate for U.S.	DTI Estimate for Europe
165 per million (100 internal ignition, 65 external ignition)	13 per million (Internal and external ignition)	18.2 per million (12.2 internal ignition, 6 external ignition)
Based on 8 reported TV fires in a Stockholm suburb in 1994	NFPA National 1990–1994 data based on fire department surveys and fire incident data from the U.S. Fire Administration	U.K. Department of Trade and Industry (DTI) 1996 report based on fire brigade data and a review of fire statistics across Europe

unrealistically high estimate for European television fires and a correspondingly low one for the U.S.

The DTI review suggests a similar rate of TV fire incidents in Europe, where FR chemicals have not been used, as that in the U.S., where TV enclosures contain decaBDE and other FRs.

The end-of-life emissions calculations in the Simonson model also used different inputs and methodologies for the U.S. and Europe. The model assumes that all disassembled TV enclosures will be incinerated, which overestimates the extent of incineration of European TVs. The nontreated plastics in European TVs can be reused repeatedly without degradation and account for 87% of recycled electronics plastics [149]. Plastic enclosures manufactured with FRs (U.S. TVs) are not recommended for recycling, as combining flame-retardant-containing recycled plastics with new product in manufacturing can result in dioxin and furan concentrations in the plastic enclosures above legal limits [150]. The model underestimates incineration emissions by assuming that 100% of incinerated plastic enclosures are used for energy recovery. In Europe and the U.S., only a small percentage of plastic waste (31% in Europe and 14% in the U.S.) is incinerated with energy recovery [151]. Also not accounted for in the model are the emissions from uncontrolled incineration, such as landfill fires or open pit burning of plastic waste after disassembly. Such unregulated incineration of TVs treated with FRs emits toxic brominated dioxins and furans into the environment.

With technical and environmental arguments against the candle flame ignition requirement such as those discussed above as well as the lack of proven fire safety benefit, the majority of the IEC TC 108 committee members from 31 countries voted against incorporation of the requirement in each of three proposed standards votes (IEC 60065, IEC 60950, and IEC 62368.)

Other proposed candle flammability standards from Underwriters Laboratories and Canadian Standards Association were also voted down in 2008. Nonetheless, a candle flame ignition requirement for television housings in the EU (Amendment 11 to EN 60065) passed in 2008 and such a standard for TVs worldwide is under consideration.

The primary impetus for all seven small open flame standards for home electronics enclosures has come from the three major FR chemical producers Chemtura, Albermarle, and Israel Chemicals Limited and their affiliated organizations, such as U.S. National Association of State Fire Marshals (NASFM). NASFM is operated out of the same address and shares employees with Sparber & Associates, a firm which has provided lobbying services for the three chemical companies [152]. These candle standards, promoted by the manufacturers who would profit from their enactment, do not improve fire safety and have a large potential harm to health and environment worldwide.

The above example points to a serious weakness in the standard development procedures, as special interest groups can influence the process to achieve their marketing objectives.

12. Alternatives to the use of FR additives

12.1 Alternate design and materials

Design alternatives, such as making components of metal, glass, or ceramics instead of plastics, can reduce flammability without chemicals [153]. FR barriers, usually made of polymeric materials like those used in mattresses, could be investigated for use around foam in furniture and juvenile products. Avoiding low-density polyurethane foam, which is relatively flammable, will reduce the fire hazard from a small open flame. High-density foam and other filling materials such as polyester (which melts rather than flames) can sometimes meet standards such as TB117 without the addition of chemicals.

The evolution of drastic changes in product characteristics also must be carefully considered as concerns the hazards needing to be addressed. FRs were first added to TV enclosures in the mid-1970s to meet UL 1410, which protected against internally initiated fires [154]. The likelihood of such fires was far greater at that time with the very high voltages required by early cathode ray tube technology. Today's technology eliminates this hazard. High voltages are now contained in the ballast of fluorescent lamps that backlight the LCD display or within the individual cells of a plasma display. The voltages are so low that arcing hazard is small. The current technology also eliminates such hazards as internal heat sources like the tube filaments, rectification, and plate voltages that could reach as high as several hundred volts, and related high currents that were required in the printed circuit boards to deliver (particularly) filament power.

12.2 Safer chemicals

Green chemistry—the design and use of safer materials and processes with minimal adverse impact on human health and the environment—should be used to design a new generation of alternative FR chemicals that are not based on bromine or chlorine. Currently, several halogen-free compounds have been shown to be effective FRs, and almost all production polymers (such as nylon, polypropylene, polyester, or polyurethane) have halogen-free formulations that will meet current fire safety standards. The alternatives include mineral fillers, which absorb heat and give off water, such as aluminum trihydrate; low melting glasses and ceramics, which seal the surface from the flame; char promoters that protect the surfaces by the formation of a carbonaceous layer; and intumescent coatings, which swell to form a barrier some distance from the polymer surface.

It is argued that reactive FRs that are bonded to the substrate are safer substitutes than additive chemicals that are not bonded and can migrate out into the environment. TBBPA is found bonded to the substrate inside electronics to protect against internal fires. The use of reactive TBBPA inside electronics should therefore present a lower health and environmental hazard than TBBPA used as an additive FR in the plastic housing of electronics. However, brominated dioxins and furans are produced upon combustion of electronics containing TBPPA. A life-cycle approach should be used in evaluating replacement chemicals.

13. CONCLUSIONS

The rapidly increasing scientific research literature documenting long-term human, animal, and environmental harm from FR chemicals should be considered when flammability standards and regulations which lead to the use of these chemicals are developed. The regulatory process should include consideration of the improvement of fire safety as well as life-cycle impacts associated with their production, use, and disposal (including health, environmental, economic, and other types of direct and indirect consequences). Special scrutiny should be given to small open-flame standards that are likely to be met by adding organohalogen or phosphate FRs to foams or plastics in consumer products. Quantitative assessment of situations where fire safety benefits and health or environmental harm aspects are in direct conflict can be tackled if enough good quality data are available and social policy valuations are made. In cases where an evaluation determines a complete lack or near lack of fire safety benefit, further outcome assessment is not needed. This is the case with FRs added to furniture foam to meet California's flammability standard TB117, foam plastic building insulation to meet U.S. building codes, and plastic electronics housings to meet IEC requirements. Such regulations need to be modified to reduce harm and improve fire safety outcomes.

References

[1] DiGangi J, Blum A, Bergman A, de Wit CA, Lucas D, Mortimer D, et al. San Antonio statement on brominated and chlorinated flame retardants. Environ Health Persp 2010; 118:A516–8.

[2] Waaijers SL, Kong D, Hendriks HS, de Wit CA, Cousins IT, Westerink RHS, et al. Persistence, bioaccumulation, and toxicity of halogen-free flame retardants. Rev Environ Contam Toxicol 2012;222:1–71.

[3] Shaw SD, Blum A, Weber R, Kannan K, Rich D, Lucas D, et al. Halogenated flame retardants: do the fire safety benefits justify the risks? Rev Env Health 2010;25: 261–305.

[4] Babrauskas V. Tables and charts. In: Fire protection handbook. 20th ed. Massachusetts, USA: National Fire Protection Association; 2008. pp. 269–94.

[5] Babrauskas V, Peacock RD. Heat release rate: the single most important variable in fire hazard. Fire Saf J 1992;18:255–72.

[6] Babrauskas V. Quantifying the combustion product Hazard on the basis of test results. In: Babrauskas V, Gann RG, Grayson SJ, editors. Hazards of combustion products: toxicity, opacity, corrosivity and heat release. London: Interscience Communications Ltd; 2008. pp. 339–53.

[7] Babrauskas V. Development of the cone calorimeter-a bench scale heat release rate apparatus based on oxygen consumption. Fire Mater 1984;8:81–95.

[8] Babrauskas V. The cone calorimeter. In: The SFPE handbook of fire protection engineering. 4th ed. Quincy, MA: National Fire Protection Association; 2008. pp. 90–108.

[9] ASTM International. Standard test method for heat and visible smoke release rates for materials and products using an oxygen consumption calorimeter (E 1354).

[10] International Organization for Standardization. Reaction-to-fire tests — heat release, smoke production and mass loss rate-part 1: heat release (Cone calorimeter method), ISO 5660–1.

[11] International Organization for Standardization. Fire tests — full scale room test for surface products (ISO 9705).

[12] ASTM International. Standard practice for full-scale oxygen consumption calorimetry fire tests (ASTM e 2067).

[13] Babrauskas V. Heat release rates. In: The SFPE handbook of fire protection engineering. 4th ed. Massachusetts, USA: National Fire Protection Association; 2008. pp. 1–59.

[14] Hirschler MM. Heat release from plastic materials heat release from plastic materials. In: Babrauskas V, Grayson SJ, editors. Heat release in fires. London, UK: E&FN Spon; 1992. pp. 375–422.

[15] Lyon RE. Fire-resistant elastomers. New Jersey, USA: Federal Aviation Administration; 2002 (DOT/FAA/AR-TN01/104).

[16] Zhang H. Fire-safe polymers and polymer composites. New Jersey, USA: Federal Aviation Administration; 2004 (DOT/FAA/AR-04/11).

[17] Lyon RE, Janssens ML. Polymer flammability. New Jersey, USA: Federal Aviation Administration; 2005 (DOT/FAA/AR-05/14).

[18] Lyon RE. Nonhalogen fire-resistant plastics for aircraft interiors. New Jersey, USA: Federal Aviation Administration; 2008 (DOT/FAA/AR-TN08/5).

[19] Babrauskas V. Cone calorimeter annotated bibliography. Washington, USA: Fire Science Publishers; 2004.

[20] Underwriters Laboratories. Tests for flammability of plastic materials for parts in devices and appliances (UL 94).

[21] US Code of Federal Regulations. Standard No. 302; flammability of interior materials, 49 CFR 571.302.

[22] Babrauskas V. Ignition handbook. Washington, USA: Fire Science Publishers/Society of Fire Protection Engineers; 2003.

[23] Troitzsch J. Plastics flammability handbook: Principles, regulations, testing, and approval. 3rd ed. Cincinnati, OH: Hanser Gardener; 2004.

[24] International Organization for Standardization. Fire safety – vocabulary (ISO 13943).

[25] ASTM International. Standard terminology of fire standards (ASTM e 176).

[26] Morgan AB, Wilkie CA, editors. Flame retardant polymer nanocomposites. New Jersey, USA: Wiley; 2007.

[27] Legler J. New insights into the endocrine disrupting effects of brominated flame retardants. Chemosphere 2008;73:216–22.

[28] Costa LG, Giordano G. Developmental neurotoxicity of polybrominated diphenyl ether (PBDE) flame retardants. NeuroToxicol 2007;28:1047–67.

[29] Szabo DT, Richardson VM, Ross DG, Diliberto JJ, Kodavanti PRS, Birnbaum LS. Effects of perinatal PBDE exposure on hepatic phase I, phase II, phase III, and deiodinase 1 gene expression involved in thyroid hormone metabolism in male rat pups. Toxicol Sci 2009;107:27–39.

[30] Kuriyama SN, Wanner A, Fidalgo-Neto AA, Talsness CE, Koerner W, Chahoud I. Developmental exposure to low-dose PBDE-99: tissue distribution and thyroid hormone levels. Toxicology 2007;242:80–90.
[31] Talsness CE, Kuriyama SN, Sterner-Kock A, Schnitker P, Grande SW, Shakibael M, et al. In utero and lactational exposures to low doses of polybrominated diphenyl ether-47 alter the reproductive system and thyroid gland of female rat offspring. Environ Health Persp 2008;116:308–14.
[32] Arkoosh MR, Boylen D, Dietrich J, Anulacion BF, Ylitalo GM, Bravo CF, et al. Disease susceptibility of salmon exposed to polybrominated diphenyl ethers (PBDEs). Aquat Toxicol 2010;98:51–9.
[33] Eriksson P, Jakobsson E, Fredriksson A. Brominated flame retardants: a novel class of developmental neurotoxicants in our environment? Environ Health Persp 2001;109: 903–8.
[34] Hoppe AA, Carey GB. Polybrominated diphenyl ethers as endocrine disruptors of adipocyte metabolism. Obesity 2007;15:2942–50.
[35] Bearr JS, Stapleton HM, Mitchelmore CL. Accumulation and DNA damage in fat-head minnows (*Pimephales promelas*) exposed to 2 brominated flame-retardant mixtures, firemaster 550 and firemaster BZ-54. Environ Toxicol Chem 2010;29:722–9.
[36] Lin K. Joint acute toxicity of tributyl phosphate and triphenyl phosphate to Daphnia magna. Environ Chem Lett 2009;7:309–12.
[37] Chemtura. Material safety data sheet #694; 2006. Product: Durad 150.
[38] U.S. EPA. Furniture flame retardancy partnership: environmental profiles of chemical flame-retardant alternatives for low-density polyurethane foam (EPA 742-R-05-002A), pp. 4-2–4-5; 2005.
[39] Patisaul HB, Roberts SC, Mabrey N, McCaffrey KA, Gear RB, Braun J, et al. Accumulation and endocrine disrupting effects of the flame retardant mixture firemaster 550 in rats: an exploratory assessment. J Biochem Mol Toxicol; 2012. http://dx.doi.org/10.1002/jbt.21439.
[40] Gold MD, Blum A, Ames BN. Another flame retardant, tris-(1,3-dichloro-2-propyl)-phosphate, and its expected metabolites are mutagens. Science 1978;200(4343): 785–7.
[41] Babich MA. CPSC staff preliminary risk assessment of flame retardant (FR) chemicals in upholstered furniture foam; 2006.
[42] Hudec T, Thean J, Kuehl D, Dougherty RC. Tris(dichloropropyl)phosphate, a mutagenic flame retardant: frequent occurrence in human seminal plasma. Science 1981; 211:951–2.
[43] U.S. EPA. Furniture flame retardancy partnership: environmental profiles of chemical flame-retardant alternatives for low-density polyurethane foam; 2005.
[44] Faust JB, August LM. Evidence on the carcinogenicity of tris (1,3-dichloro-2-propyl) phosphate. OEHHA; 2011.
[45] U.S. EPA. IRIS health assessment of 2',3,3',4,4',5,5',6,6'-decabromodiphenyl ether; 2008 (BDE-209) CASRN 1163-19-5. EPA/635/R-07/008F.
[46] Lee E, Kim TH, Choi JS, Nabanata P, Kim NY, Ahn MY, et al. Evaluation of liver and thyroid toxicity in Sprague-Dawley rats after exposure to polybrominated diphenyl ether BDE-209. J Toxicol Sci 2010;3:535–45.
[47] U.S. National Toxicology Program. Toxicology and carcinogenesis studies of decabromodiphenyl oxide (CAS No 1163-19-5) in F344/N rats and B6C3F mice (feed studies); 1986. NTP TR 309.

[48] Nakari T, Huhtala S. In vivo and in vitro toxicity of decabromodiphenyl ethane, a flame retardant. Environ Toxicol 2010;25:333–8.
[49] Talsness CE, Andrade AJ, Kuriyama SN, Taylor JA, Vom Saal FS. Components of plastic: experimental studies in animals and relevance for human health. Philos T Roy Soc B 2009;364:2079–96.
[50] Haddow JE, Palomaki GE, Allan WC, Williams JR, Knight GJ, Gagnon J, et al. Maternal thyroid deficiency during pregnancy and subsequent neuropsychological development of the child. N Engl J Med 1999;341:549–55.
[51] Boecker RH, Schwind B, Kraus V, Pullen S, Tiegs G. Cellular disturbances by various brominated flame retardants (Abstract). In: Presented at the second international workshop on brominated flame retardants. Stockholm, Sweden; 14–16 May 2001.
[52] Reistad T, Mariussen E, Fonnum F. In vitro toxicity of tetrabromobisphenol-A on cerebellar granule cells: cell death, free radical formation, calcium influx and extracellular glutamate. Toxicol Sci 2007;96:268–78.
[53] Birnbaum LS, Staskal DF. Brominated flame retardants: cause for concern? Environ Health Persp 2004;112:9–17.
[54] Canton RF, Peijnenburg AA, Hoogenboom RL, Piersma AH, van der Ven LT, van den Berg M, et al. Subacute effects of hexabromocyclododecane (HBCD) on hepatic gene expression profiles in rats. Toxicol Appl Pharmacol 2008;231:267–72.
[55] Mariussen E, Fonnum F. The effect of pentabromodiphenyl ether, hexabromocyclododecane and tetrabromobisphenol- A on dopamine uptake into rat brain synaptosomes. Organohalogen Compd 2002;57:395–9.
[56] Dingemans MML, Heusinkveld HJ, de Groot A, Bergman A, van de Berg M, Westerink RHS. Hexabromocylcododecane inhibits depolarization-induced increase in intracellular calcium levels and neurotransmitter release in PC12 cells. Toxicol Sci 2009;107:490–7.
[57] Germer S, Piersma AH, van der Ven LT, Kamyschnikow A, Fery Y, Schmitz H-J, et al. Subacute effects of the brominated flame retardants hexabromocyclododecane and tetra-bromobisphenol A on hepatic cytochrome P450 levels in rats. Toxicology 2006;218:229–36.
[58] Yamada-Okabe T, Sakai H, Kashima Y, Yamada-Okabe H. Modulation at a cellular level of the thyroid hormone receptor-mediated gene expression by 1,2,5,6, 9,10-hexabromocyclo-dodecane (HBCD), 4,4′-diiodobiphenyl (DIB), and nitrofen (NIP). Toxicol Lett 2005;155:127–33.
[59] European Union. Chemicals/REACH: six dangerous substances to be phased out by the EU (IP/11/196); 2011.
[60] Environment Canada. Proposed risk management measure for hexabromocyclododecane (HBCD). Consultation Document; 2012.
[61] (UNEP/POPS/POPRC.8/16 2012). Stockholm Convention on Persistent Organic Pollutants Report of the persistent organic pollutants review committee on the work of its eighth meeting; 2012.
[62] Hutzinger O, Thoma H. Polybrominated dibezo-p-dioxins and dibenzofurans: the flame retardant issue. Chemosphere 1987;16:1877–80.
[63] Jones-Otazo H, Clarke JP, Diamond ML, Archbold AJ, Ferguson G, Harner T, et al. Is house dust the missing exposure pathway for PBDEs? an analysis of the urban fate and human exposure to PBDEs. Environ Sci Technol 2005;39: 5121–30.

[64] D'Hollander W, Roosens L, Covaci A, Cornelis C, Reynders H, Campenhout KV, et al. Brominated flame retardants and perfluorinated compounds in indoor dust from homes and offices in Flanders, Belgium. Chemosphere 2010;81:478–87.

[65] Roosens L, Abdallah M, Harrad S, Neels H, Covaci A. Exposure to hexabromocyclododecanes via dust ingestion, but not diet, correlates with concentrations in human serum. Environ Health Persp 2009;117:1707–12.

[66] Stapleton HM, Klosterhaus S, Eagle S, Fuh J, Meeker JD, Blum A, et al. Detection of organophosphate flame retardants in furniture foam and U.S. house dust. Environ Sci Technol 2009;43:7490–5.

[67] Stapleton HM, Allen JG, Kelly SM, Konstantinov A, Klosterhaus S, Watkins D, et al. Alternate and new brominated flame retardants detected in U.S. house dust. Environ Sci Technol 2008;42:6910–6.

[68] Ali N, Harrad S, Goosey E, Neels H, Covaci A. "Novel" brominated flame retardants in Belgian and UK indoor dust: implications for human exposure. Chemosphere 2011;83:1360–5.

[69] Watkins DJ, McClean MD, Fraser AJ, Weinberg J, Stapleton HM, Sjödin A, et al. Exposure to PBDEs in the office environment: evaluating the relationships between dust, handwipes, and serum. Environ Health Persp 2011;119:1247–52.

[70] Frederiksen M, Thomsen C, Froshaug M, Vorkamp K, Thomsen M, Becher G, et al. Polybrominated diphenyl ethers in paired samples of maternal and umbilical cord blood plasma and associations with house dust in a Danish cohort. Int J Hyg Environ Health 2010;213:233–42.

[71] Roosens L, Abdallah MA, Harrad S, Neels H, Covaci A. Exposure to hexabromocyclododecanes (HBCDs) via dust ingestion, but not diet, correlates with concentrations in human serum: preliminary results. Environ Health Persp 2009;117:1707–12.

[72] Lorber M. Exposure of Americans to polybrominated diphenyl ethers. J Expo Sci Env Epid 2008;18:2–19.

[73] Frederiksen M, Thomsen C, Frøshaug M, Vorkamp K, Thomsen M, Becher G, et al. Polybrominated diphenyl ethers in paired samples of maternal and umbilical cord blood plasma and associations with house dust in a Danish cohort. Int J Hyg Environ Health 2010;213:233–42.

[74] Wu N, Herrmann T, Paepke O, Tickner J, Hale R, Harvey E, et al. Human exposure to PBDEs: associations of PBDE body burdens with food consumption and house dust concentrations. Environ Sci Technol 2007;41:1584–9.

[75] Johnson-Restrepo B, Kannan K. An assessment of sources and pathways of human exposure to polybrominated diphenyl ethers in the United States. Chemosphere 2009;76:542–8.

[76] Schecter A, Harris TR, Shah N, Musumba A, Päpke O. Brominated flame retard-ants in US food. Mol Nutr Food Res 2008;52:266–72.

[77] Mandalakis M, Atsarou V, Stephanou EG. Airborne PBDEs in specialized occupational settings, houses and outdoor urban areas in Greece. Environ Pollut 2008;155:375–82.

[78] Batterman S, Godwin C, Chernyak S, Jia C, Charles S. Brominated flame retardants in offices in Michigan. U.S.A. Environ Int 2010;36:548–56.

[79] Hughes MF, Edwards BC, Mitchell CT, Bhooshan B. In vitro dermal absorption of flame retardant chemicals. Food Chem Toxicol 2001;39:1263–70.

[80] Sjödin A, Päpke O, McGahee EE, Focant J-F, Jones RS, Pless-Mulloli T, et al. Concentration of polybrominated diphenyl ethers (PBDEs) in household dust from various countries. Chemosphere 2008;73:131–6.

[81] Zota AR, Rudel RA, Morello-Frosch RA, Brody JG. Elevated house dust and serum concentrations of PBDEs in California: unintended consequences of furniture flammability standards? Environ Sci Technol 2008;42:8158–64.

[82] Rose M, Bennett DH, Bergman Å, Fangstrom BF, Pessah IN, Hertz-Picciotto I. PBDEs in 2-5 year-old children from California and associations with diet and indoor environment. Environ Sci Technol 2010;44:2648–53.

[83] Herbstman JB, Sjödin A, Kurzon M, Lederman SA, Jones RS, Rauh V, et al. Prenatal exposure to PBDEs and neurodevelopment. Environ Health Persp 2010; 118:712–9.

[84] Chevrier J, Harley KG, Bradman A, Gharbi M, Sjödin A, Eskenazi B. Polybrominated diphenylether (PBDE) flame retardants and thyroid hormone during pregnancy. Environ Health Persp 2010;118:1444–9.

[85] Meeker JD, Johnson PI, Camann D, Hauser R. Polybrominated diphenyl ether (PBDE) concentrations in house dust are related to hormone levels in men. Sci Total Environ 2009;407:3425–9.

[86] Harley KG, Marks AR, Chevrier J, Bradman A, Sjödin A, Eskenazi B. PBDE concentrations in women's serum and fecundability. Environ Health Persp 2010;118: 699–704.

[87] Harley KG, Chevrier J, Schall RA, Sjödin A, Bradman A, Eskenazi B. Association of prenatal exposure to polybrominated diphenyl ethers and infant birth weight. Am J Epidemiol 2011;174(8):885–92.

[88] Chao HR, Wang SL, Lee WJ, Wang YF, Päpke O. Levels of polybrominated diphenyl ethers (PBDEs) in breast milk from central Taiwan and their relation to infant birth outcome and maternal menstruation effects. Environ Int 2007;33:239–45.

[89] Wu K, Xu X, Liu J, Guo Y, Li Y, Huo X. Polybrominated diphenyl ethers in umbilical cord blood and relevant factors in neonates from Guiyu, China. Environ Sci Technol 2010;44:813–9.

[90] Main KM, Kiviranta H, Virtanen HE, Sundqvist E, Tuomisto JT, Tuomisto J, et al. Flame retardants in placenta and breast milk and cryptorchidism in newborn boys. Environ Health Persp 2007;115:1519–26.

[91] Roze E, Meijer L, Bakker A, Van Braeckel KNJA, Sauer PJJ, Bos AF. Prenatal exposure to organohalogens, including brominated flame retardants, influences motor, cognitive, and behavioral performance at school age. Environ Health Persp 2009; 117:1953–8.

[92] Meeker JD, Stapleton HM. House dust concentrations of organophosphate flame retard-ants in relation to hormone levels and semen quality parameters. Environ Health Persp 2010;118:318–23.

[93] Betts K. Endocrine Damper? Flame retardants linked to male hormone, sperm count changes. Environ Health Persp 2010;118:A130.

[94] Chao HR, Tsou TC, Huang HL, Chang-Chien GP. Levels of breast milk PBDEs from southern Taiwan and their potential impact on neurodevelopment. Pediatr Res 2011; 70:596–600.

[95] Schauer UMD, Volkel W, Dekant W. Toxicokinetics of tetrabromobisphenol A in humans and rats after oral administration. Toxicol Sci 2006;91:49–58.

[96] Cariou R, Antignac JP, Zalko D, Berrebi A, Cravedi JP, Maume D, et al. Exposure assessment of French women and their newborns to tetrabromobisphenol-A: occurrence measurements in maternal adipose tissue, serum, breast milk and cord serum. Chemosphere 2008;73:1036–41.

[97] Covaci A, Gerecke AC, Law RJ, Voorspoels S, Kohler M, Heeb NV, et al. Hexabromocyclododecanes (HBCDs) in the environment and humans: a review. Environ Sci Technol 2006;40:3679–88.

[98] Hites RA. Polybrominated diphenyl ethers in the environment and in people: a meta-analysis of concentrations. Environ Sci Technol 2004;38:945–56.

[99] Chou C-T, Hsiao Y-C, Ko F-C, Cheng J-O, Cheng Y-M, Chen T-H. Chronic exposure of 2,2',4,4'-tetrabromodiphenyl ether (PBDE-47) alters locomotion behavior in juvenile zebrafish (*Danio rerio*). Aquat Toxicol 2010;98:388–95.

[100] Muirhead EK, Skillman AD, Hook SE, Schultz IR. Oral exposure of PBDE-47 in fish: toxico-kinetics and reproductive effects in Japanese medaka (*Oryzias latipes*) and fathead minnows (*Pimephales promelas*). Environ Sci Technol 2006;40: 523–8.

[101] Fernie KJ, Shutt JL, Mayne GJ, Hoffman D, Letcher RJ, Drouillard KG, et al. Exposure to polybrominated diphenyl ethers (PBDEs): changes in thyroid, vitamin a, glutathione homeostasis, and oxidative stress in American kestrels (*Falco sparverius*). Toxicol Sci 2005;88:375–83.

[102] Fernie KJ, Shutt JL, Letcher RJ, Ritchie JI, Sullivan K, Bird DM. Changes in reproductive courtship behaviors of adult American kestrels (*Falco sparverius*) exposed to environmentally relevant levels of the polybrominated diphenyl ether mixture, DE-71. Toxicol Sci 2008;102:171–8.

[103] Shaw SD, Kannan K. Polybrominated diphenyl ethers in marine ecosystems of the American continents: foresight from current knowledge. Rev Environ Health 2009; 24:157–229.

[104] Lam JCW, Lau RKF, Murphy MB, Lam PKS. Temporal trends of hexabromocyclododecanes (HBCDs) and polybrominated diphenyl ethers (PBDEs) and detection of two novel flame retardants in marine mammals from Hong Kong, South China. Environ Sci Technol 2009;43:6944–9.

[105] Sagerup K, Herzke D, Harju M, Evenset A, Christensen GN, Routti H, et al. New brominated flame retardants in Arctic Biota. Sci Total Environ 2010;408:2885–918.

[106] Law RJ, Allchin CR, de Boer J, Covaci A, Herzke D, Lepom P, et al. Levels and trends of brominated flame retardants in the European environment. Chemosphere 2006;64: 187–208.

[107] Gauthier LT, Potter D, Hebert CE, Letcher RJ. Temporal trends and spatial distribution of non-polybrominated diphenyl ether flame retardants in the eggs of colonial populations of Great Lakes herring gulls. Environ Sci Technol 2009;43:312–7.

[108] Gao F, Luo X, Yang Z, Wang X, Mai B. Brominated flame retardants, polychlorinated biphenyls, and organochlorine pesticides in bird eggs from the Yellow River Delta, North China. Environ Sci Technol 2009;43:6956–62.

[109] Hu G, Luo X, Dai J, Zhang X, Wu H, Zhang CL, et al. Brominated flame retardants, polychlorinated biphenyls, and organochlorine pesticides in captive giant panda (*Ailuropoda melanoleuca*) and red panda (*Ailurus fulgens*) from China. Environ Sci Technol 2008;42:4704–9.

[110] Shi T, Chen S-J, Luo X-J, Zhang X-L, Tang C-M, Luo Y, et al. Occurrence of brominated flame retardants other than polybrominated diphenyl ethers in environmental and biota samples from southern China. Chemosphere 2009;74:910–6.

[111] Harrad S, Abdallah MA-E, Rose NL, Turner SD, Davidson TA. Current-use brominated flame retardants in water, sediment, and fish from English lakes. Environ Sci Technol 2009;43:9077–83.

[112] Johnson-Restrepo B, Adams DH, Kannan K. Tetrabromobisphenol A (TBBPA) and hexa-bromocyclododecanes (HBCDs) in tissues of humans, dolphins, and sharks from the United States. Chemosphere 2008;70:1935–44.
[113] Tanabe S, Ramu K, Isobe T, Takahashi S. Brominated flame retardants in the environment of Asia-Pacific: an overview of spatial and temporal trends. J Environ Monit 2008;10:188–97.
[114] Damant GH. California upholstered furniture combustibility standard: 1981 update. J Consum Prod Flamma 1981;8:209–20.
[115] Stapleton HM, Klosterhaus S, Keller A, Ferguson PL, van Bergen S, Cooper E, et al. Identification of flame retardants in polyurethane foam collected from baby products. Environ Sci Technol 2011;45:5323–31.
[116] Stapleton HM, Sharma S, Getzinger G, Ferguson PL, Gabriel M, Webster TF, et al. Novel and high volume use flame retardants in US couches reflective of the 2005 pentaBDE phase out. Environ Sci Technol 2012;46(24):13432–9.
[117] Babrauskas V, Blum A, Daley RE, Birnbaum LS. Flame retardants in furniture foam: benefits and risks. Fire Saf Sci 2011;10:265–78.
[118] Babrauskas V. Upholstered furniture heat release rates: measurements and estimation. J Fire Sci 1983;1:9–32.
[119] Consumer Product Safety Commission. Upholstered furniture validation memoranda; 2012.
[120] Babrauskas V, Harris RH, Gann RG, Levin BC, Lee BT, Peacock RD, et al. Fire hazard comparison of fire-retarded and non-fire-retarded products (Spec. Publ. SP 749). Gaithersburg, MD: U.S. Natl. Bur. Stand; 1988.
[121] Dembsey NA, Williamson RB. Effect of ignition source exposure and specimen configuration on the fire growth characteristics of a combustible interior finish material. Fire Saf J 1993;21:313–30.
[122] Babrauskas V. Ignition handbook. Issaquah, WA: Fire Science Publishers and Society of Fire Protection Engineers; 2003.
[123] Government Printing Office. Standard for the flammability of Clothing textiles (CS 191-53), code of federal regulations (16 CFR 1610).
[124] Talley TH. Phases 1&2, UFAC small open flame tests and cigarette ignition tests. In: Annual AFMA flammability Conf; 1995.
[125] Babrauskas V, Lucas D, Eisenberg D, Singla V, Dedeo M, Blum A. Flame retardants in building insulation: a case for re-evaluating building codes. Build Res Inf 2012;40: 738–55.
[126] Williamson RB, Mowrer FW. The role of interior finish in fire development. Fire Prot Eng; 2004:24–40.
[127] ASTM International. Standard test method for surface burning characteristics of building materials (ASTM E 84).
[128] International Code Council, Inc. International Building Code. 2000, 2003, 2006, and 2009 editions.
[129] Choi KK, Taylor W. Combustibility of insulation in cavity walls. J Fire Sci 1984;2: 179–88.
[130] Babrauskas V, White JA, Urbas J. Testing for surface spread of flame: new tests to come into use. Build Stand 1997;66:13–8.
[131] Williamson RB, Baron FM. A corner fire test to simulate residential fires. J Fire Flammabl 1973;4:99–105.

[132] Castino TG, Beyreis JR, Metes WS. Flammability studies of cellular plastics and other building materials used for interior finishes, subject 723. Northbrook, IL: Underwriters Laboratories; 1975.

[133] Rose A. Fire testing of rigid cellular plastics (IR-422). Ottawa: National Research Council of Canada; 1975.

[134] Lee BT. Standard room fire test development at the national Bureau of standards. In: Fire safety: science and engineering (ASTM STP 882). Philadelphia, PA: ASTM international; 1985. pp. 29−44.

[135] Dillon SE. Analysis of the ISO 9705 Room/Corner test: Simulations, correlations and heat flux measurements (NIST-GCR-98−756). Gaithersburg, MD: Nat Inst Stand and Technol; 1998.

[136] Foamed Polystyrene for Construction (Data Sheet 1-58). Norwood, MA: Factory Mutual, 1978.

[137] International Electrotechnical Commission. Study on the ignition of ICT and audio/video equipment by external ignition sources: Report of the International Electrotechnical Commission's Technical Committee 108 Special Fire Research Group. Document TC 108/AGS/SFRG (Report/chairman), 2003.

[138] NAFSM. Draft candle fire safety analysis and recommendations. Consumer product safety task force of the NASFM; 2000.

[139] Blum A. The case against candle resistant electronics. Berkeley, CA: Green Science Policy Institute; 2008. Available at: http://greensciencepolicy.org/sites/default/files/MASTERWhitepaper.pdf.

[140] Weber R, Kuch B. Relevance of BFRs and thermal conditions on the formation pathways of brominated and brominated-chlorinated dibenzo-dioxins and dibenzofurans. Environ Int 2003;29:699−710.

[141] TR41.7 letter to UL Objecting to the preliminary proposal for candle flame requirements UL 60950-1, 12/14/2007.

[142] Wilson D. Electronic equipment ignition by candle − recent incident reports. Consumer Electronics Association; 10/1/2002.

[143] Simonson M, Tullin C, Stripple H. Fire-LCA study of TV sets with V0 and HB enclosure material. Chemosphere 2002;46:737−44.

[144] Simonson M, Andersson P, van den Berg M. Cost benefit analysis model for fire safety: methodology and TV (DecaBDE) case study. SP Swed Natl Test Res Inst 2009;28: 1−65.

[145] Takigami H, Suzuki G, Hirai Y, Sakai S. Transfer of brominated flame retardants from components into dust inside television cabinets. Chemosphere 2008;73:161−9.

[146] UK Department of Trade and Industry. Consumer safety research - TV fires (Europe); 1996.

[147] Muir T. An attempt to replicate the risk management analysis of cost, benefits, and risks in the Deca-BDE and TV case study. Organohalogen Compd 2007;69:2611−4.

[148] Hall JR. The U.S. home product report, 1990-1994 (appliances and equipment). Quincy, MA: NFPA Fire Analysis and Research Division; 1997.

[149] Dawson RB, Landry SD. Recyclability of flame retardant HIPS, PC/ABS, and PPO/HIS used in electronic equipment. In: Proceedings of the international symposium on electronics and the environment (ISEE); 2005. pp. 77−82.

[150] Mark F. The characteristics of plastics-rich waste streams from end-of-life electrical and electronic equipment. Plast Eur; 2006.

[151] Fisher MM, Mark FE, Kingsbury T, Vehlow J, Yamawaki T. Energy recovery in the sustainable recycling of plastics from end-of-life electrical and electronic products. In: IEEE international symposium on electronic and the environment. Los Angeles, USA; 2005.
[152] State fire marshals endorse Gall for CPSC chairman. CPSC Monitor, January 1, 2001.
[153] United Nations Environment Programme. Stockholm convention on persistent organic pollutants; 2009.
[154] Underwriters Laboratories. UL 1410: television receivers and high-voltage video products; 1997.

CHAPTER

Environmental Drivers for Replacement of Halogenated Flame Retardants

4

T.R. Hull
Centre for Fire and Hazard Science, University of Central Lancashire, Preston, UK

R.J. Law
Centre for Environment, Fisheries and Aquaculture Science (CEFAS) Lowestoft, UK

Åke Bergman
Stockholm University, Stockholm, Sweden

CHAPTER OUTLINE

Polymer Green Flame Retardants. http://dx.doi.org/10.1016/B978-0-444-53808-6.00004-4

1. Introduction

The rapid development of synthetic polymers was accompanied by an increase in fires and fire fatalities. In the 1970s, the use of halogenated flame retardants (HFRs) grew rapidly. Halogen-based fire retardants are very effective in reducing fire risk, i.e. the probability of occurrence of a fire, but they show a high fire hazard, that is, the probability of producing toxic, corrosive, obscuring smokes while performing the fire-retardant action or when involved in a developed fire, when the fire can no longer be easily extinguished. HFRs act by releasing hydrogen bromide or hydrogen chloride, which interfere with the gas phase free radical reactions, typically producing more carbon monoxide, smoke, and other products of incomplete combustion. The dense smoke obscures escape routes and contaminates property, while the resultant hydrobromic and hydrochloric acids are highly corrosive—significantly increasing the costs of unwanted fires. The ensuing threat to people, structures, and goods involved in the fire may discourage the use of these fire retardants, despite their versatility and ease of incorporation.

In contrast, modern fire retardants prevent fuel from escaping to the gas phase, often by formation of a protective barrier layer, acting as a radiation shield, and inhibiting the flow of fuel and oxygen. Such barriers have been augmented in intumescent systems, where gas is released within the molten polymer, causing massive swelling, and so increasing the effectiveness of the thermal barrier. While the formulations required for char promotion and intumescence are generally specific to a particular polymer, HFRs tend to be nonspecific in their action, and one flame retardant (FR) can be incorporated into many polymers. This has increased their popularity, especially among smaller plastic compounders and product manufacturers, who do not have the expertise to devise such formulations. Unfortunately, this ease of use is matched by an ease of release, particularly when operating at elevated temperatures, such as in televisions (TVs) and other electronics; or during breakdown of the polymer, during use or at end of life; and leaching (especially in foam or textile products), allowing significant quantities to escape into the environment. In addition, losses of brominated flame retardants (BFRs) also occur during manufacture and from end-of-life processing.

Many HFRs are persistent and bioaccumulative, and are now ubiquitous throughout the built and natural environment [1,2]. In 2010 a group of over 100 eminent environmental scientists signed the San Antonio statement on HFRs [3] condemning their continued use and requesting urgent remedial action. These persistent organic pollutants (POPs) take several years to break down in the environment, are bioaccumulative (they accumulate in plants and animals, becoming more concentrated as they move up the food chain), and are toxic. All the 22 chemicals currently designated as POPs by the Stockholm Convention [4] belong to the family of halogenated chemicals, where carbon is bonded to a halogen such as fluorine, chlorine, bromine, or iodine. Studies have shown higher levels in house dust in California and the United Kingdom, where the most stringent furniture flammability regulations operate [5]; at lower levels they have been detected in flora and fauna, from the developed world to the Himalayan and Arctic regions. These are described in detail later in this chapter. Very high levels have been detected in the Pearl River delta in southern China, where many electronic recycling activities are based. The most detailed studies of health effects have concentrated on polybromodiphenyl ethers (PBDEs), proven endocrine disruptors in both experimental animals and humans. Elevated levels have been found in human blood serum in Californian children at 5 times the US average, and 10–100 times the European and Mexican averages. These elevated levels have been linked to infertility, hyperactivity and attention disorders in children, etc.

Several of the most widely used HFRs, including PBDEs and polybrominated biphenyls (PBBs), have already been banned in Europe and the United States. *Tris*-dibromopropyl phosphate was banned by the US Consumer Product Safety Commission in 1977 [6]. In some cases they have been replaced by similar organohalogen compounds with unknown eco- and human toxicity. The release of FR from a product, during or at the end of life; its stability and transport through the environment; its uptake and bioaccumulation by humans and other living organisms; and the resultant toxic effects are very difficult to predict; it took nearly 40 years of exposure to PBDEs to accumulate sufficient evidence for action to be taken, and it may take as long for environmental levels to fall below safe limits.

The market for fire retardants continues to grow, but there are very clear drivers in Europe, Japan, and to a lesser extent in the United States for development of sustainable products. In Europe, proposals to identify the presence of BFRs using "ecolabels" will provide consumers with additional information, and raise awareness of the potential hazards of certain materials. Inevitably, manufacturers will want to harmonize their global product range (so they are less likely to offer a choice of products with and without HFRs); to demonstrate their "green" credentials, they are therefore more likely to opt for halogen-free solutions. Already the major personal computer (PC) manufacturers are committed to producing products free of HFRs.

2. Fire statistics

Historically, the increased growth in the use of synthetic polymers was accompanied by an increase in the number of fire deaths and injuries, as shown in Figure 1 and Figure 2 [7]. After a peak around 1980, these have been in decline.

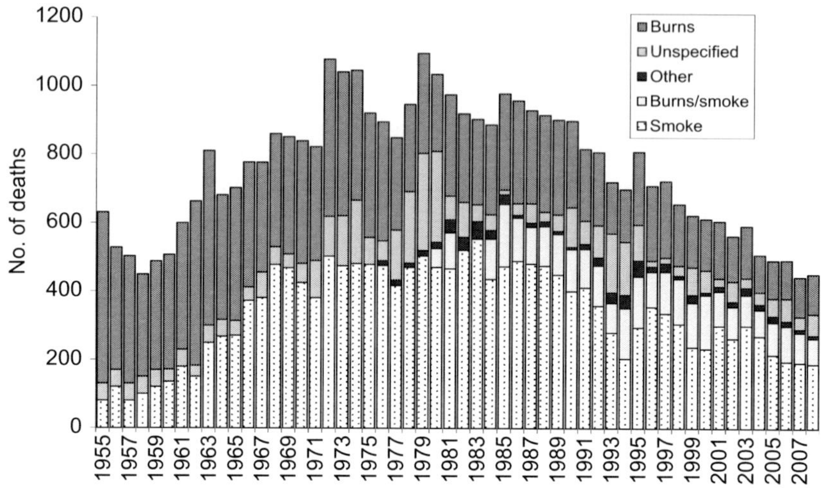

FIGURE 1

Causes of UK fire deaths during 1955–2008.

The shift in cause of death from burns to smoke inhalation is no less dramatic. The ratio of smoke inhalation to burns as cause of death increased by a factor of 20; in 1955 it was less than 0.16, in 1980 it was 2.1, and in 2004 it reached a peak of 2.5. Evidence for this change comes primarily from the easily detected

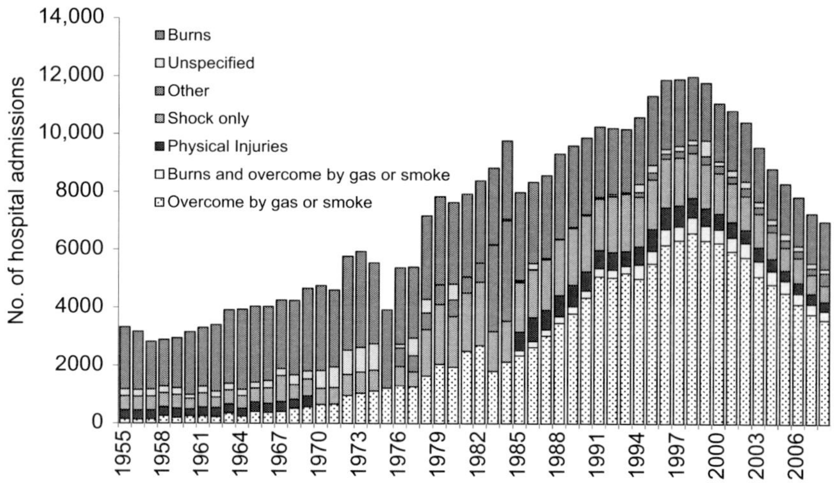

FIGURE 2

UK nonfatal fire injuries requiring hospital treatment, 1955–2008, excluding precautionary checkup.

carboxyhemoglobin concentrations in the blood, indicating deaths from smoke inhalation, since many victims continue to burn (but not breathe) after death. The higher toxicity is a combination of more lethal products, especially hydrogen cyanide in the fire effluents of certain (nitrogen containing) synthetic polymers, and the effect of gas phase FRs inhibiting the conversion of carbon monoxide (CO) to carbon dioxide (CO_2) and promoting hydrogen cyanide (HCN) formation over nitrogen oxides.

A similar but more pronounced trend is observed in the UK fire injury statistics. Injuries did not start to decline until around 2000, and the shift from burns to smoke and toxic gas inhalation is even more apparent. Again, this is explained at least in part by the higher fire toxicity of certain polymers, and particularly those containing gas phase FRs.

It was argued [8] in a report commissioned by the FR industry that "the introduction of fire-safe furniture [in the UK] from 1988 onwards is estimated to have resulted in at least 50% of the estimated 2002 savings in injuries and domestic fire deaths", the other 50% being attributed to low-cost smoke detectors. Factors such as changes in cigarette smoking habits, the change from exposed flame heating sources, and a general improvement in standard of living were disregarded. It seems unlikely that this often-quoted estimate would stand up to any serious scrutiny. More recently, a detailed study produced for the European Commission [9] on the risks and benefits of fire retardants analyzed the fire fatality data from individual European countries with different levels of flammability regulation. While they acknowledged the difficulty in comparing statistics from different countries, they concluded that "in some instances, drops in the number of fire deaths coincide with the introduction of non-flammability requirements for domestic consumer products. In other instances, however, there is no change in the on-going trend of fire deaths. This suggests that these numbers do not reflect the stringency of non-flammability requirements, respectively that non-flammability requirements do not visibly decrease the number of fire deaths."

On average the fire deaths in European states have fallen progressively over the period 1980–2008, although the decrease is less evident in some former Eastern European countries. Figure 3 compares the fire deaths for the United Kingdom and Ireland, which both introduced furniture flammability regulations (in 1988 and 1996, respectively) with those from Germany [9], which has no such regulation; Germany was chosen because of the completeness of the data, and the broad similarity to the United Kingdom and Ireland. Both UK and Irish statistics show clear evidence of a declining trend, both before and after the introduction of furniture flammability regulations, while German statistics indicate an increase between 1988 and 1990 (presumably associated with reunification with the poorer German Democratic Republic), followed by a very similar downward trend. It is difficult to draw conclusions from the higher numbers of fire deaths in different jurisdictions, because they depend on the statistical protocols for each country (e.g. deaths involving fire may instead be classified as crime—arson—or road traffic accidents, while fire survivors who subsequently die in hospital may not be recorded). There

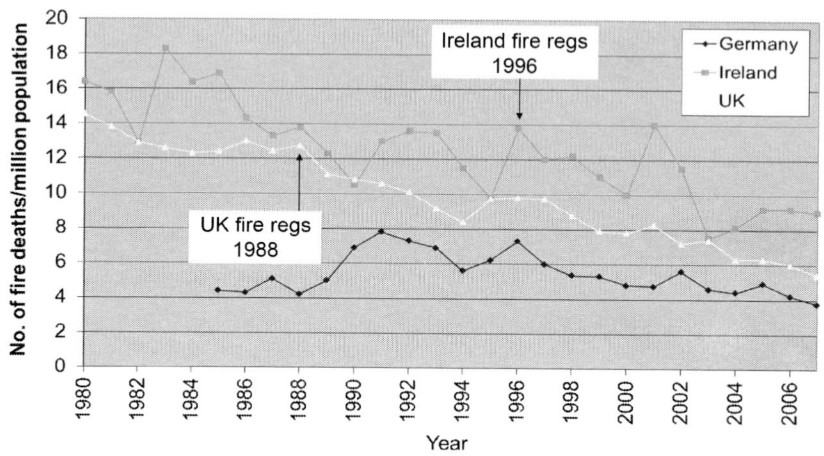

FIGURE 3

Fire deaths in the United Kingdom, Ireland, and Germany per million populations (1980–2007) where data is available.

does not appear to be any clear evidence that the furniture flammability regulations have had a quantifiable impact on the loss of life from fire.

Fire causes £1.7 billion of direct losses per year in the United Kingdom, or overall, 0.7% of gross domestic product [10], while fire safety measures represent 2.5% of the cost of a new building. Fires represent a complex sequence of chemical reactions, which are not well understood and, as synthetic polymers replace wood and wool, we are surrounded by more flammable materials. Recently, pressure has mounted to make buildings more energy efficient, and therefore traditional building materials such as brick, stone, and timber may be replaced by more flammable alternatives, such as polystyrene or polyurethane (PU) foams. Fortunately, the manufacturing process of synthetic polymers often allows the incorporation of fire-retardant additives, while the porosity of most naturally occurring polymeric materials permits access to fire-retarding substances. Globally, in 2010, the fire-retardant industry has been estimated to produce 1.5 million tons of product worth US $4.6 billion [11].

3. Fire retardants

3.1 Classification

The wealth of different fire-retardant strategies is a testament to the ingenuity of the chemists who developed them. Fire-retardant agents have been classified in many ways: place of action—gas or condensed phase; mode of action—physical or chemical; chemical nature of agent—halogen, phosphorus, metal hydroxide or carbonate, etc.; and means of incorporation of agent—additive or reactive (i.e. bound onto the

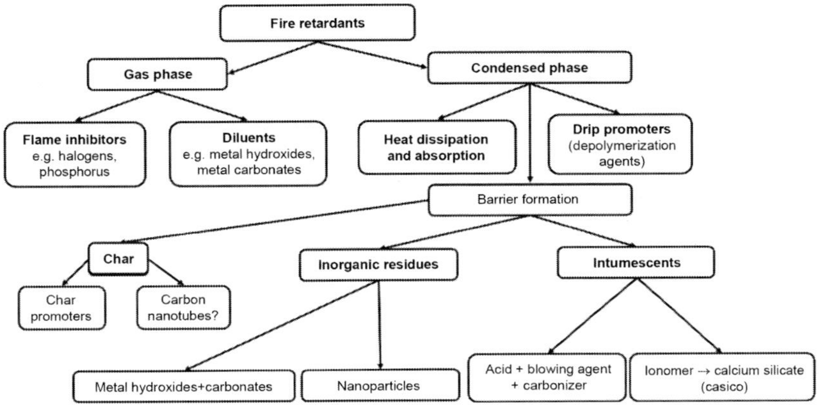

FIGURE 4

Classification of fire-retardant (FR) strategies.

polymer chain). Unfortunately, these classifications cannot be unambiguously applied to particular fire retardants—for example, the most widely used fire retardant, aluminum hydroxide (ATH) releases water, so acting in the gas phase, but in doing so absorbs heat and leaves a protective residue in the condensed phase [12]. It is useful to distinguish within the broadest category: fire retardants include any material which reduces the flammability of the polymer; FRs inhibit the gas phase free radical reactions responsible for flaming combustion, although less careful authors seem to use the terms interchangeably. Figure 4 illustrates how the major fire-retardant strategies could be subdivided. Fire retardants have evolved over the past four decades in order to meet the demands of industry and regulators, from halogen-based flame inhibitors to cleaner, char promoters, resulting in less smoke and toxic gas emissions. While HFRs continue to be used in a wide range of existing products, very little new work has been published on them. Instead, research has been heavily focused on finding suitable halogen-free replacement fire retardants, indicating that the industry has indeed recognized the need for change, in the face of increasing pressure, predominantly resulting from environmental concerns.

3.2 Drivers in fire-retardant development

The history of fire retardants goes back to Egyptian times when solutions of alum (hydrated potassium aluminum sulfate ($KAl(SO_4)_2 \cdot 12H_2O$)) were used to treat timber. Gay-Lussac protected theater fabrics from fire by treatment with mixtures of ammonium phosphate, ammonium chloride, and borax which formed a glassy layer on heating [13]. However, the main driver for development came with the growth of the plastics industries and the resultant widespread distribution of synthetic polymers across the anthrosphere. From the 1960 to the 1970s fires became more common and more severe. Anecdotally, fire fighters reported a change from fires with

Table 1 Events Driving the Deployment of Fire Retardants

Decade	Event	Demand
1960s	Widespread availability of cheap polymer products—more serious fires	Reduced ignitability
1970s	Smoke much worse (PU foam furniture and HFRs)	Reduced smoke
1980–1990s	Development of cone calorimeter (and emphasis on peak heat release rate, rather than ignitability); increase in deaths from smoke inhalation	Reduced peak heat release Reduced fire toxicity
2000s	Halogen FRs found across the ecosystem	Halogen-free FRs
2010s	Climate change and other environmental concerns become mainstream	Sustainable FRs

limited visibility, to those with almost no visibility due to dense smoke, primarily resulting from newly available low-cost PU foam furniture. In the 1960s–1970s this was accompanied by a shift from burns and, other injuries to a predominance of harm caused by toxic gas inhalation, as discussed later. The increasing severity of the fire problem led to the development from empirical tests for flammability to engineering models capable of providing data on burning behavior. This was accompanied by a shift in emphasis from ignitability to peak heat release rate. Unfortunately, the model chosen to quantify this behavior, the cone calorimeter, was only capable of replicating penetrative burning into a sample, not surface spread of flame. Predictive fire modeling is still at an early stage of development. A summary of the events that drove the deployment of fire retardants during the past 50 years is presented in Table 1.

4. Halogenated FRs

The first halogenated fire retardants were polychlorinated biphenyls (PCBs) (Figure 5), which were widely used as fire retardants before their very high toxicity was discovered in the 1960s [14]. Unfortunately, as fire retardants, PCBs were

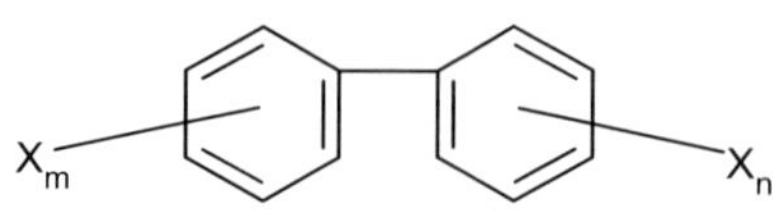

FIGURE 5

Structure of polyhalobiphenyls (X = Cl for PCB or Br for PBB).

replaced by chemically analogous polybrominated biphenyls (PBBs) (also Figure 5) without any prior toxicity evaluation. PBBs have a structure similar to PCBs; their use as FR additives for plastics began in the 1970s as mixtures of many different congeners. In 1973, 100 kg of 200–400 PBB-based fire retardants were accidentally mixed with livestock feed that was distributed to farms in West Central Michigan, United States [15,16]. Millions of domestic animals consumed this feed, and 85% of Michigan's population received some exposure to PBBs, inadvertently alerting the authorities to their toxicity. The immune systems of exposed farmers showed significant abnormalities, and the production of PBBs was rapidly discontinued. This is discussed in the final section *Harmful Effects of Halogenated Flame Retardants.*

The withdrawal of PCBs and PBBs led to a proliferation of different halogenated fire retardants. In 1972 there were significant sales of additives such as Dechlorane Plus, a chlorinated alicyclic (~1400 tons of polyolefins); *tris*-(dibromopropyl) phosphate (450 tons); brominated aromatics (700 tons); and a small amount of pentabromochlorocyclohexane and hexabromocyclododecane (HBCDD) for polystyrene foam [17]. There are more than 40 BFRs in current use and an additional number of chlorinated FRs [18]. These are used by simple melt blending with the polymer (additive FRs) or by incorporation into the polymer chain during polymerization (reactive FRs). Some additive FRs are small molecules, while others are oligomers or polymers. From a health and environmental perspective, small molecules will be easiest to release from the polymer matrix (by evaporation, leaching, end-of-life processing, etc.). Oligomeric or polymeric additives may be released during end-of-life processes, while reactive FRs, which have been successfully incorporated into the polymer, are most likely to be problematic during burning or incineration processes.

Of the small molecule BFRs, the PBDEs have been widely used in acrylonitrile butadiene styrene (ABS), high-impact polystyrene (HIPS), polyvinyl chloride (PVC), polybutylene terephthalate, polyethylene, polypropylene (PP), and flexible PU foam, and cited in more environmental studies, described later. Typical products are penta-PBDE, octa-PBDE, and deca-PBDE, although in practice each contains mixtures of the 209 possible congeners of polybrominated diphenyl ethers (Figure 6).

As the structure of PBDEs is similar to that of PCBs, the same system of abbreviation for individual congeners is used. As for PCBs, there are 209 brominated diphenyl ether (BDE) congeners and these are numbered from BDE1 to BDE209 (decabromodiphenyl ether). As an example, 2,2′,4,4′-tetrabromodiphenyl ether,

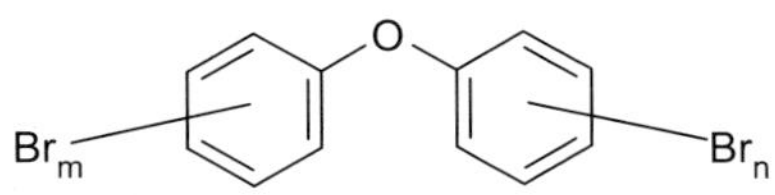

FIGURE 6

Structures of polybromodiphenyl ether (PBDE) (where, for example, $m+n=8$ for the major octa-PBDE congeners).

substituted at the 2 and 4 positions on both aromatic rings, is known as BDE47. When environmental data are summarized, all the congeners determined in a particular study are summed (see later in this chapter). ΣBDE_{10} then represents the sum of 10 congeners determined in environmental samples in one specific study.

Toxicity and environmental concerns led to submission of a proposal to the European Union to ban the use of PBDEs in 1989 (111-4301-89-EN Draft). The proposal was rejected on the basis of recommendations issued by a thorough debate between scientists, regulators, producers, and users of fire retardants, stating that banning would involve an unacceptable fire risk since alternatives were not available to replace HFRs with comparable effectiveness. Ever since, research on fire retardants has focused on the development of nonhalogenated replacements for the halogen fire retardants. Penta-PBDE and octa-PBDE have now been withdrawn from use in the United States and European Union, and alternatives are being introduced. In some instances, these are structurally similar to PBDEs. For example, one group of alternatives being recommended for facile substitution is the polybromodiphenylethanes (PBDPEs) (Figure 7).

Given the persistence of the PBDE molecule in the environment, and its toxicity, it is remarkable that the proposed replacement PBDPE has such a similar structure, and is without adequate environmental persistence or toxicological data.

HBCDD (Figure 8) is widely used, particularly in very large quantities in polystyrene foam for building insulation, PP, and textile applications. Bromination of cyclododeca-1,5,9-trienes theoretically results in 16 stereoisomers, six enantiomeric pairs, and four meso forms, since each bromine can occupy axial or equatorial positions, which will then affect the puckering of the ring. The first three pairs of enantiomers, the α-, β-, and γ-HBCDD stereoisomers, dominate in technical products [19]. The isomers have different environmental and toxicological properties, and

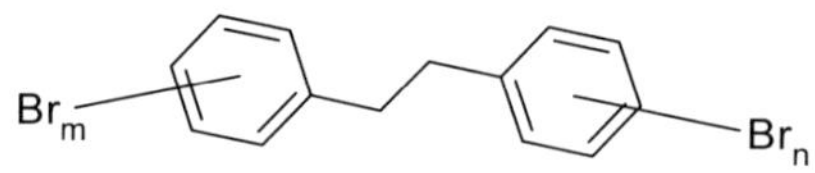

FIGURE 7

Structure of polybromodiphenylethanes (PBDPE).

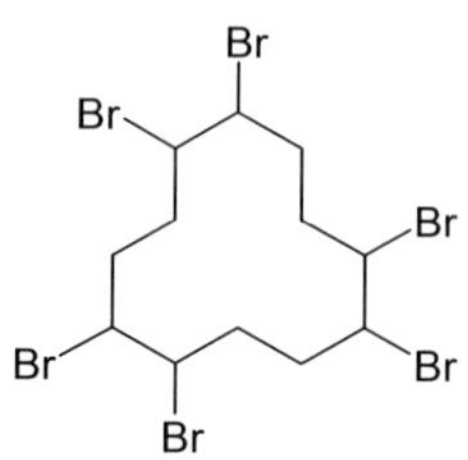

FIGURE 8

Structure of hexabromocyclododecane (HBCDD).

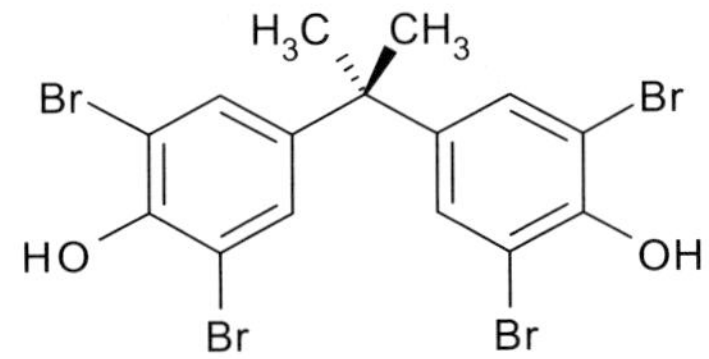

FIGURE 9

Structure of tetrabromobisphenol A (TBBP).

biological and thermal isomerization have been reported [20,21]; thus the ratio of molecular forms identified in the ecosystem may not correspond to those produced industrially.

Tetrabromobisphenol A (TBBPA) (Figure 9) is widely used in epoxy resins and polycarbonate where it is copolymerized into the polymer structure. Once bound into the polymer, it is much less likely to be released until the polymer itself is broken down. It is also used as an additive FR in ABS and HIPS [22], where losses to the environment may occur more readily. It is less stable than PBDEs or HBCDD and hence less persistent in the natural environment. Indeed, it might be expected to be below detectable limits, although in practice it is ubiquitous throughout the anthropogenic and natural environments [1], and has been identified in particularly high concentrations in Asia, near sites of manufacture and destruction of polymeric products [23].

4.1 Polymeric BFRs

A small number of polymeric BFRs [24] have recently been developed. They are considered better since they are too large to penetrate cell membranes. They also have very low solubility in water, so that once incorporated into the polymer matrix, leaching is unlikely. Examples include polybrominated acrylate, brominated epoxy polymers, and brominated polystyrene. They are used as FRs in a wide range of polymers, including polystyrene and copolymers, polycarbonates, polyesters, and polyamides.

4.2 Mode of action of HFRs

HFRs act by releasing hydrogen halides (HX) during thermal decomposition. If the hydrogen halide release coincides with fuel gasification, then HX can interfere with the gas phase combustion processes. Comparison of the energy (or temperature at which the polymer has acquired such energy) of decomposition of HX with, for example, decomposition of C—C bonds indicates which type of HFR is likely to be most suitable for the structural unit of a particular polymer.

Table 2 shows the average bond energies of carbon—hydrogen and carbon—carbon bonds, compared with the rather weak C—I bond, increasing progressively

Table 2 Bond Energies and Decomposition Temperatures for Polymer and Carbon–Halogen Bonds (C is Aliphatic Carbon, C_{ar} is Aromatic Carbon) [22]

Bond	Bond Energy (kJ mol^{-1})	Onset of Degradation (°C)
Polymer Bonds		
C–C	330–370	400
C–H	390–436	>500
C_{ar}–H	469	>500
C_{ar}–C_{ar}	518	>500
C=C	612	>500
Halogenated FR Bonds		
C–I	222–235	180
C–Br	285–293	290
C_{ar}–Br	335	360
C–Cl	339–352	370–380
C_{ar}–Cl	419	>500
C–F	443–450	>500

to the very high strength C—F bond. Since their mode of action depends on scission of the C—halogen bond, fluorine- and iodine-based FRs are not used in practice because neither type interferes with the combustion process at the right point. Fluorine cannot be effective as a radical interceptor in the gas phase because of its strong bond to carbon, whereas iodine is attached to carbon so weakly that it is liberated at polymer processing temperatures, and before polymer pyrolysis occurs.

Bromine is the most effective since its weaker bonding to carbon enables it to interfere at a more favorable point in the combustion process. Halogen atoms can be bound aliphatically or aromatically in FRs. Aliphatic bromine compounds are easier to breakdown than their aromatic counterparts, and are more suitable for polymers with low decomposition temperatures.

Flaming combustion involves a very small number of highly reactive free radicals to propagate the gas phase oxidation processes. For ignition to occur, the number of radicals must increase. This occurs in reaction 1 and 2 where each "•"represents an unpaired electron.

$$H\bullet + O_2 \rightarrow OH\bullet + O\bullet \tag{1}$$

$$\bullet O\bullet + H_2 \rightarrow OH\bullet + H\bullet \tag{2}$$

Halogen-containing FRs act by interfering with the radical chain mechanism taking place in the gas phase. The high-energy OH• and H• radicals formed by chain branching are removed by the halogen-containing FR.

At first the FR breaks down to

$$RX \rightarrow R\bullet + X\bullet \tag{3}$$

where X• is either Cl• or Br•.

The halogen radical reacts to form the hydrogen halide:

$$X\bullet + RH \rightarrow R\bullet + HX \tag{4}$$

which in turn interferes with the radical chain mechanism:

$$HX + H\bullet \rightarrow H_2 + X\bullet \tag{5}$$

The removal of H• is key to elimination of the main chain branching step (when one unpaired electron becomes three).

$$HX + OH\bullet \rightarrow H_2O + X\bullet \tag{6}$$

The removal of OH• blocks the main heat release step of hydrocarbon combustion, the conversion of CO to CO_2, by replacement with less reactive halogen free radicals in the gas phase [25]. The H• and OH• radicals are essential for many flame reactions and are involved in the main heat release in reaction Eqn 7.

$$CO + OH\bullet \rightarrow CO_2 + H\bullet \tag{7}$$

Loss of H• and OH• will increase the yield of toxic carbon monoxide and other products of incomplete combustion (hydrogen cyanide (HCN), organoirritants, and soot). The high-energy H• and OH• radicals are removed by reaction with HX and replaced with lower energy X• radicals. The actual FR effect is thus produced by HX.

Kinetic reaction schemes predict that HX must be recycled around seven times in order to account for the observed flame inhibition [26]. Thus the hydrogen halide is regenerated by reaction with hydrocarbon:

$$X\bullet + RH \rightarrow R\bullet + HX \tag{8}$$

In the condensed phase, the resulting unsaturated polyenes may act as char precursors, forming products with a tendency to cyclize and condense to yield carbonaceous products, which protect the condensed phase below the flame zone against attack by oxygen and radiant heat. In PVC, after loss of 60% mass as hydrogen chloride from the surface layers, char formation is a significant fire-retardant mechanism protecting the underlying polymer.

The halogen content in the polymer compound, and its chemical binding, will dictate the FR behavior. In the presence of antimony oxide (Sb_2O_3), the efficiency of HFRs is improved, although antimony has no FR effect on its own. This is believed to result from the formation of volatile SbX_3 and other species which are more effective halogen carriers than HX. Antimony oxide has similar toxicity to arsenic oxides, with a threshold limit value of 0.5 mg m^{-3}, in common with most antimony compounds [27]. Currently (in 2012), its high price (around €10 kg^{-1}) is also driving searches for its replacement.

4.3 Applications of BFRs

A frequent argument for continued use of HFRs is that they save lives [8]. Given that fire retardants add cost to a product, they are only deployed to meet regulatory requirements. Thus fire retardants actually allow more flammable materials to be used in high-risk situations (Table 3). Furthermore, suitable halogen-free alternatives have been demonstrated in almost all major application areas. In electrical and electronic (E & E) products, trade organizations [28,29] and manufacturers have identified suitable alternatives, and in the PC market, consumer pressure has persuaded all major manufacturers to commit themselves to make all their products halogen free (for example, Apple, Dell, HP, Toshiba, Sony, LG, and Acer). In upholstered furniture, a study of the response of 360 foam–fabric combinations to the United Kingdom's cigarette and match ignition test (BS 5852), in which 61% of combinations contained at least one fire-retarded component, 32% passed both tests; of these, 38% were without any FR, while 23%, 22%, and 17% had FRs in the cover fabric, the foam, and both, respectively [30]. Thus cigarette and match ignition tests on upholstered products do not require use of fire retardants at all. A recent demonstration of the efficacy of a low-cost halogen-free interliner (between fabric and foam) able to pass the severe No.5 burning wooden crib test of BS 5852 showed that the thousands of tons of HFRs in home furnishings in the United Kingdom and the United States were unnecessary [31]. In expanded polystyrene foam, used extensively with HBCDD for building insulation, suitable halogen-free alternatives are still under development, but the industry has yet to adopt them.

Although not strictly an additive fire retardant, it is appropriate to consider the use of PVC for electric cables, compared to halogen-free alternatives. PVC is a rigid polymer, containing 60% chlorine, which can only be made flexible by the incorporation of 30–50% flammable organic plasticizer. Typical PVC cable jacketing contains PVC polymer, phthalate ester plasticizer, and chalk in approximately equal proportions. In this case the PVC acts both as the polymer (giving both the strength and cohesion) and as the FR (releasing HCl on heating), while the plasticizer provides most of the fuel. Alternative cable systems are based on the rubberlike copolymers, such as ethylene vinyl acetate, capable of remaining flexible at loadings of up to 70% mineral filler fire retardant, such as ATH.

According to a survey carried out by SRI Consulting [32], the total market for FRs in the United States, Europe, and Asia in 2007 amounted to about 1.8 million metric tons (Figure 10). This market was expected to grow at an average annual rate of about 3.7% per year on a volume basis over the period 2007–2012. It is split roughly equally between Europe, America, and Asia, with HFRs and antimony oxide comprising around 37.5%.

In terms of tonnage, brominated FRs are second only to the inorganic compounds such as ATH, although this is primarily due to the large use of these compounds in Asia. In terms of value, the global market is worth between $4.2 and 4.25 billion, of which BFRs have the greatest value (Figure 11). About 44% by value of FRs are consumed in Asia, with 67% globally being halogenated or antimony oxide.

Table 3 Typical Uses of Halogenated FRs

Polymer	Additive		Reactive	Sb_2O_3 Compatibility
	Small Molecule	Oligomeric/Polymeric		
Polystyrene foam	HBCDD			✗
HIPS and ABS	Deca-PBDE, TBBPA, SCCP	Br-epoxy		✓
Epoxy resin	Deca-PBDE, BTBPE, bis	Br-PS,	TBBPA	✓
Polyamides	Deca-PBDE, DDC-CO	Br-PS, Br-epoxy, PPBBA		✓
Polyolefins	Deca-PBDE, TTBNPP, SCCP			
PUs	CAPs, Penta-PBDE (before prohibition)		TBNPA	✓
Aromatic polyesters		Br-PS,		✗?
Unsaturated polyesters	Deca-PBDE, SCCP	Br-epoxy	TBP-Anh	✗? (causes opacity)
Polycarbonate		Br-PS, Br-epoxy oligomers, PTC-TBBPA	TBBPA	✗
PVC	SCCP, Deca-PBDE (soft foams), TBPEs			✓
Rubber	SCCP			✓

BTBPE, bis(tribromophenoxy) ethane; Br-PS, brominated polystyrene; Br-epoxy, brominated epoxy (oligomeric or polymeric); CAP, chloroalkyl phosphates (e.g. tris(chloroethyl)phosphate (TCEP), tris(chloroisopropyl phosphate (TCPP), etc.); DDC-CO, Diels–Alder adducts of hexachlorocyclopentadiene and 1,5-cyclooctadiene (also referred to as Dechlorane Plus); PPBBA, poly(pentabromobenzyl acrylate); PTC-TBBPA, phenoxy-terminated carbonate oligomer of TBBPA; SCCP, short-chain chlorinated paraffins; TBP-Anh, tetrabromophthalic anhydride; TBPEs, tetrabromophthalate esters, e.g. bis(2-ethylhexyl) *tetrabromophthalate (BEH-TBP); TTBNPP, tris(tribromoneopentyl) phosphate.*

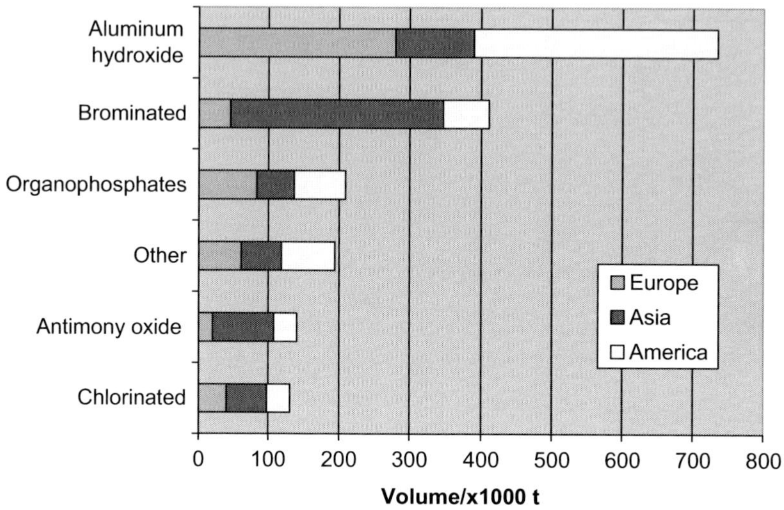

FIGURE 10

Volume consumption of flame retardants (FR) in the United States, Europe, and Asia [32].

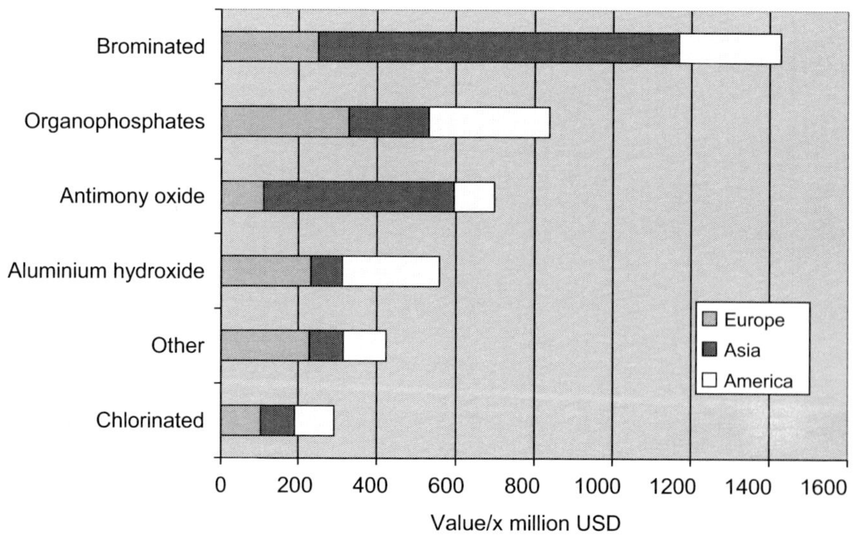

FIGURE 11

Value of flame retardants (FR) worldwide [32]. AATF, approved authorized treatment facility.

5. Impact of halogenated FRs on fire safety

5.1 Fire toxicity

Gas phase FRs interfere with the free radical reactions responsible for flaming combustion, described earlier. This results in incomplete oxidation of vapor phase fuel molecules, leading to higher yields of all products of incomplete combustion. These are all more toxic than the cleaner products of complete combustion (carbon dioxide and water), and include carbon monoxide, hydrogen cyanide, hydrocarbons, oxygenated organics (including organoirritants, such as acrolein and formaldehyde), and larger cyclic molecules such as polycyclic aromatic hydrocarbons and soot particulates. Fire toxicity increases as combustion becomes more incomplete, which can arise from chemical quenching (for example by gas phase FRs), insufficient heat (for example, during smoldering), or when the fire becomes ventilation controlled, and there is insufficient oxygen for complete combustion, as discussed in the chapter on the influence of fire retardants and nanofillers on fire toxicity.

Thus in a fire, HFRs will act to increase the gas phase products of incomplete combustion, and hence the yields of toxic gases and smoke, when compared to either the non-flame-retarded polymer or the same polymer incorporating a condensed phase fire retardant, which increases the char yield. As most fire deaths, and most fire injuries, result from toxic gas inhalation, the use of HFRs is a compromise, between suppressing ignition and increasing the fire toxicity (or decreasing the fire risk at the expense of increasing the fire hazard).

5.2 Fire retardancy and scaleup

The effects of fire retardants on large-scale fire behavior are less clear. Fire retardants can delay ignition, or reduce heat release rate in small-scale tests, and are most effective in the "first object ignited". However, they are only effective on a larger scale if they decrease the radiant component of heat transfer. Radiant heat allows the flames to spread horizontally or even downward. Soot particles absorb heat from the flame, and emit infrared radiation in all directions, which can heat adjacent objects causing flame spread, and is ultimately responsible for the devastating impact of unwanted fires. Gas phase free radical inhibitors, such as HFRs, which act by interfering with the combustion process, increase the soot yields and hence the severity of a large-scale fire, as soot particles convert more heat in flames into radiation, spreading the fire to other objects, alongside and below the flames [33].

6. Regulatory considerations

6.1 Flammability standards and regulations

Since fire retardants are generally more expensive than the base material they are used to treat, and their addition is at high-enough loadings to be detrimental to the physical properties of the material, few manufacturers would add fire retardants on a purely voluntary basis. Thus regulatory requirements generally drive the deployment of fire retardants and other strategies for fire safety. Standards typically define methods for quantifying performance, such as ignitability, heat release rate, flame spread, and smoke production. They may apply to materials or finished products and, from flammability perspective, should reflect likely fire scenarios in normal usage. There are a large number of industry standard tests for specific product types designed to achieve this. Regulations require standards in order to define minimum levels of performance, which can then be stipulated for particular applications or conditions (railway vehicles, high-rise buildings, electrical products, etc.).

European Parliament Directive 2001/95/EC on general product safety imposes a requirement on any consumer product to ensure that it is safe with regard to normal or reasonably foreseeable use, unless it is covered by specific sector legislation, but does not provide specific requirements for determining safety or "acceptable risk". The specific sectorial directives covering the safety aspects and nonflammability requirements of certain consumer products are:

- The Toys Safety Directive 2009/48/EC;
- Directive 2006/95/EC on electrical equipment designed for use within certain voltage limits;
- Directive 1999/5/EC on radio equipment and telecommunications terminal equipment;
- Directive 2006/42/EC on machinery; and
- The Construction Products Regulation (EU) No 305/2011, which comes into force in 2013.

At present, the fire safety of furniture and textiles is not regulated by specific European legislation.

6.2 Restrictions on the use of HFRs

6.2.1 European legislation

In 2003, the Restriction of Hazardous Substances Directive (RoHS) was adopted by the European Union. This restricts the use of certain hazardous substances in E & E equipment (Directive 2002/95/EC). It limits the use of six hazardous materials, including BFRs, in the manufacture of various types of electronic and electrical equipment, and is closely linked with the Waste Electrical and Electronic Equipment Directive (WEEE) 2002/96/EC, which sets collection, recycling, and recovery targets for electrical goods and is part of a legislative initiative to solve the problem of huge amounts of toxic electronic waste [34].The maximum concentrations of PBBs and PBDEs allowed are 0.1 wt% of homogeneous material, without separate collection and treatment.

The WEEE/RoHS Directives require the separation of plastics in E & E waste containing BFRs prior to recycling, followed by treatment at a specialist facility (Figure 12), rather than energy recovery or disposal. In practice this does not always occur [35]. Instead, WEEE containing BFRs has been shown to leave the approved

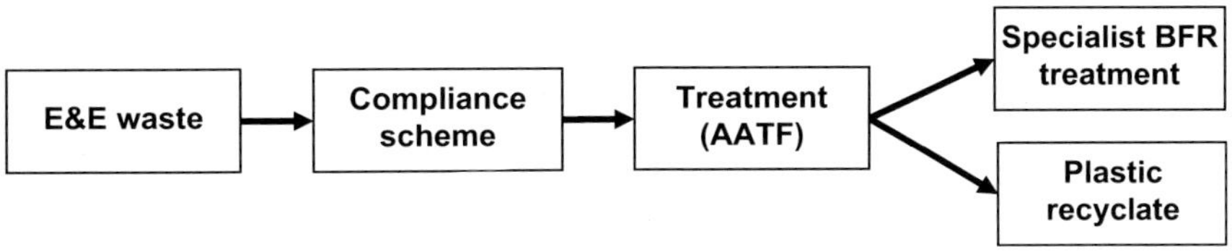

FIGURE 12

Schematic of WEEE process for dealing with E&E waste.

authorised treatment facility after separation of metals, and was then granulated and reclassified as halogen-free plastic recyclate.

In addition, it has been reported by the German environmental agency (UBA) that 155,000 tons of German WEEE were exported outside the European Union, including 50,000 tons TV and PC monitors which are highly likely to contain BFRs, possibly exploiting the loophole of waste being classified as being in working order, to permit their export [36].

The RoHS procedure has now combined with EU Registration, Evaluation, Authorisation and restriction of Chemicals (REACH), to become a new regulation EC/2006/1907. Four additional substances are listed that will be assessed as a priority; among these is HBCDD, an BFR widely used in expanded polystyrene for which few alternatives have so far been found. REACH addresses the production and use of chemical substances and their potential impacts on both human health and the environment, regulating manufacture and use of all man-made nonpolymeric chemical substances. The 10-year phased implementation of REACH started in 2007. BFRs were one of the first categories investigated by REACH and the investigation was almost complete within 2 years. The use of penta-PBDE and octa-PBDE was severely restricted, while deliberations on deca-PBDE still continue.

6.2.2 Voluntary regulation—ecolabels

Manufacturers can demonstrate compliance with environmental standards through the use of ecolabels. Although the framework for the operation of ecolabel schemes is governed by regulation, their actual use is voluntary. Individual schemes develop and publish environmental performance criteria for particular product groups (for example, TVs, furniture, etc.), which have to be met to qualify for an ecolabel.

The European Ecolabel is the most widespread scheme, established in 1992 to encourage businesses to market products and services that are more environmental friendly, and aims to substitute hazardous substances with safer ones. For example, the latest European Ecolabel proposals would prohibit the use of FRs which are banned in electronic appliances, including deca-PBDE, in Ecolabelled textiles, and bed mattresses [37].

The current proposals use "risk phrases" which ignore environmental and other nonoccupational effects, such as fire toxicity. They use a separate classification for additive and reactive "FRs" putting mineral fillers, such as ATH or magnesium hydroxide, or even naturally occurring mixtures of the minerals huntite and hydromagnesite, which show fire retardancy comparable to ATH [38], in same category as small molecule BFRs, while permitting the use of polymeric BFRs, such as brominated polystyrene, which may not leach into the environment, but will still increase fire toxicity, act as precursors for dioxins, and may increase the radiation and hence flame spread. Although Ecolabels are voluntary, they could be adopted into purchasing policy by local authorities and large institutions, with potentially greater impact on markets.

7. Environmental distribution of BFRs

After the first report of the environmental occurrence of PBDEs in Sweden in 1980 [39], interest in the actual and potential impacts of BFRs has grown steadily, with a surge in publications from the mid-1990s onward. Since then, knowledge of the environmental distribution of BFRs has been extensively reviewed [1,40–48]. The latest of these reviewed literature were published or available on journal websites up until the end of 2009 or early in 2010. These showed that BFRs are widely distributed in the environment across all matrices studied, and are present in samples from the Arctic [49] and Antarctic [50], and in other remote areas such as the Tibetan plateau [51], as a result of long-range atmospheric transport processes. Humans and wildlife species are exposed to BFRs via the atmosphere and diet (see [52], for a review of human dietary exposure to PBDEs) and, for humans, through their indoor environments (including houses and cars) and, in particular, indoor dust [53,54]. Indeed, Johnson et al. [55] have recently found a strong correlation ($r = 0.65–0.89$; $p < 0.05$) between PBDE concentrations in house dust and the blood serum levels of PBDEs in the residents. In addition these reviews have recently highlighted the issue of emissions from e-waste recycling, primarily in developing countries, and the importance of the exposure of the local populations to the emissions from these activities (see Ni et al. [56]). Transport of e-waste from developed countries continues, despite regulation [57]. It has also been noted that, in humans, levels of BFRs in human milk reflect varying exposures across the world according to local usage, with the highest BDE concentrations in North America [58]. In this section, we update this knowledge of the environmental distribution of BFRs, in regard to levels in humans and wildlife, using material published up to the end of December 2011.

As chemical names are often long and complicated, a series of abbreviations have been adopted in order to make referring to individual compounds easier. PCBs, a legacy contaminant, are composed of 209 individual congeners (compounds) which are numbered from CB1 to CB209 (the fully chlorinated decachlorobiphenyl), each number indicating an individual compound with a specific chloro-substitution. There is great structural similarity between PCBs and PBDEs, and so these latter compounds are numbered in the same manner (BDE1–BDE209). Concentrations can be expressed on the basis of either the wet weight of a tissue or its lipid (fat) content. The latter is often preferred for these compounds as they are lipophilic (fat soluble) and expressing concentrations on a lipid basis can normalize concentrations between tissues with differing lipid contents. Wet weight reflects the concentration expressed on a whole tissue basis. Different studies can also involve the analysis of different suites of BDE congeners. These are often quoted as, e.g. ΣBDE_3, which represents the sum of the concentrations of the three BDE congeners determined. This convention is used in the following section. Limits of detection and quantification can vary depending on the methods and instrumentation used, but are typically in the low to sub-microgram per kilogram range.

7.1 Brominated diphenyl ethers

7.1.1 Wildlife

7.1.1.1 Marine mammals and turtles

Vorkamp et al. [49] studied temporal trends of ΣBDE concentrations in ringed seals from East Greenland between 1986 and 2008. ΣBDE_3 concentrations in juvenile seals showed a slight upward trend, however, dependent on the low concentration seen in 1986. Removing this data point resulted in a downward trend, which was also found for adult seals for a time trend beginning in 1994.

In the blubber of 202 harbor seals (primarily young of the year ($n = 177$) from California, Greig et al. [59] determined ΣBDE_{10} concentrations to be 23–63,000 μg kg^{-1} lipid weight, up to about one-fifth of the total PCB concentration. Such broad concentration ranges occur as a result of differences in exposure via the diet, as some prey species are more highly contaminated than others, and maternal transfer to offspring.

Frouin et al. [60] studied ΣBDE_{15} concentrations in serum and blubber of three seal species from eastern Canada. In both sample types, concentrations from harbor seals from Bic Island in the St Lawrence estuary were higher than those in harbor seals from St Paul, grey seals and harp seals. ΣBDE_{15} concentrations in blubber and serum from harbor seals at Bic Island were 530 μg kg^{-1} lipid weight and 600 μg kg^{-1} lipid weight, respectively. Mean ΣBDE_{15} concentrations in the other samples ranged from 21 to 98 μg kg^{-1} lipid weight. ΣBDE_{14} was determined in the blubber of finless porpoises taken in 2000–2007 from Japan [61]. Concentrations ranged from 24 to 1000 μg kg^{-1} lipid weight. There were no obvious temporal trends during this period. In another study, ΣBDE_{23} concentrations in finless porpoise blubber from Korean coastal waters ($n = 52$) ranged from 180 to 1600 μg kg^{-1} lipid weight [62]. BDE49 and BDE66 comprised 4–16% of the total, which seemed to be associated with debrominated products from higher brominated congeners.

Nomiyama et al. [63] studied BDEs in archived whole blood samples from 11 species of cetaceans stranded on the Japanese coasts. ΣBDE_9 concentrations ranged from <1 to 54,000 μg kg^{-1} wet weight, with the highest concentrations being observed in a killer whale. High concentrations were also seen in Pacific white-sided dolphins and Stejneger's beaked whale, up to 10,000 μg kg^{-1} wet weight.

In blubber of harbor porpoises from the Black Sea, Weijs, et al. [64] reported ΣBDE_8 concentrations ranging from 43 to 85 μg kg^{-1} lipid weight in samples from 1998. These concentrations are low, indicating that PBDEs were not commonly used in the countries surrounding the Black Sea in the 1990s. In porpoises from the North Sea during 1990–2008, where PBDEs have been extensively used, ΣBDE_7 concentrations in blubber ranged from 42 to 4770 μg kg^{-1} lipid weight, the highest concentration occurring in a juvenile animal sampled during 1990–1998 [65].

Within a broader study of POPs in bottlenose dolphin tissues, Yordy et al. [66] determined ΣBDE_{10} concentrations in blubber. These ranged from 0.05 to 1.7 μg kg^{-1} wet weight in the four animals studied, from North Carolina, Virginia, and Florida. Whole body burden estimates ranged from 1.2 to 71 mg of ΣBDE_{10}. Quinete et al. [67] investigated BDEs in fish and tucuxi dolphins from the Paraíba do Sul River in Southeast (SE) Brazil. Mean ΣBDE_9 concentrations in liver were 10 μg kg^{-1} wet weight in scabbardfish and 3.5 μg kg^{-1} wet weight in whitemouth croaker, and 53 μg kg^{-1} wet weight in tucuxi dolphins. In muscle samples, mean ΣBDE_{15} concentrations were 3.9 and 2.1 μg kg^{-1} wet weight, in the two fish species, respectively.

Between 1996 and 2003, blubber samples from five dolphin species from the area around São Paulo, Brazil, were analyzed for PBDEs [68]. These were Guiana dolphin, franciscana, Atlantic spotted dolphin, rough-toothed dolphin, and bottlenose dolphin. Mean ΣBDE_{39} concentrations were highest in Atlantic spotted dolphin (770 μg kg^{-1} lipid weight), followed by rough-toothed dolphin (475 μg kg^{-1} lipid weight), Guiana dolphin (66 μg kg^{-1} lipid weight), bottlenose dolphin (64 μg kg^{-1} lipid weight), and franciscana (60 μg kg^{-1} lipid weight). Concentrations were highest in animals feeding over the continental shelf than in those feeding just offshore, possibly as they consume larger prey. The major source of BDEs was the penta-mix PBDE formulation, as seen also in porpoises from the United Kingdom [58].

Malarvannan et al. [230] determined ΣBDE_{14} in livers of three species of turtles (green, hawksbill, and loggerhead turtles) from Japan collected between 1998 and 2006. Concentrations ranged from 0.06 to 7.3, 1.4 to 140, and 0.7–4.5 μg kg^{-1} lipid weight, respectively.

7.1.1.2 Fish and shellfish

Kim & Stapleton [69] reported ΣBDE_{10} concentrations ranging from 21 to 292 μg kg^{-1} lipid weight in squid from Korean offshore waters.

In 12 species of deep-sea fish (20 samples) collected off the coast of Japan, Takahashi et al. [70] determined ΣBDE_{14} concentrations between 1.3 and 53 μg kg^{-1} lipid weight. Tri- to hexa-BDE congeners were predominant. In mussels and oysters from Japanese coastal waters, Ueno et al. [71] reported ΣBDE_{14} concentrations of 3.1–86 μg kg^{-1} lipid weight. The estimated dietary exposure via seafood was 0.05–6.8 ng kg body $weight^{-1}$ day^{-1}.

Nomiyama et al. [72] reported ΣBDE_7 concentrations in blood of Japanese amberjack of 15 ± 12 ng kg^{-1} wet weight in females and 12 ± 12 ng kg^{-1} wet weight in males. In scalloped hammerhead shark these were 8.3 ± 1.1 and 12 ± 8.8 ng kg^{-1} wet weight in females and males, respectively. So, similar levels were observed in both species of predatory fish.

In muscle tissue of wild Chinook salmon from Chile, Montory et al. [73] reported ΣBDE_{14} concentrations between 0.3 and 1.05 μg kg^{-1} wet weight, similar to levels reported for the Northern Hemisphere. In juvenile Chinook salmon from the Pacific Northwest United States, Sloan et al. [74] reported gutted body concentrations between not detected (nd) and 13,000 μg kg^{-1} lipid weight. Lipid contents averaged 1.7%, so the highest concentration would be (very roughly, as individual values could not be corrected) equivalent to 220 μg kg^{-1} wet weight. In brown trout from Andean Patagonia, Argentina, Ondarza et al. [75] reported mean ΣBDE_9 concentrations of 1.1 and 1.6 μg kg^{-1} wet weight in muscle and liver, respectively.

Skinner [76] reported concentrations of ΣBDE_{40} in smallmouth and striped bass from the Hudson River, NY, USA, taken in 2003. Mean concentrations ranged from 2.1 to 138 μg kg^{-1} wet weight in smallmouth bass and 35–169 μg kg^{-1} wet weight in striped bass, along a 385-km stretch of the river. Marine fish (five salmon species plus sablefish and walleye pollock) were sampled from several sites in the Northeast (NE) Pacific Ocean and the Bering Sea during 2002–2008 [77]. BDE47, BDE49, BDE99, and BDE100 dominated the congener distribution. Concentrations ranged from 0.03 μg kg^{-1} wet weight in walleye pollock from the Bering Sea in 2003 to 8.7 μg kg^{-1} wet weight in Chinook salmon from west coast Vancouver Island, also in 2003.

ΣBDE_{10} concentrations in muscle tissue of mullet and bass from Bizerte Lagoon, Tunisia, were reported by Ben Ameur et al. [78], along with concentrations of eight methoxy brominated diphenyl ethers (MeO-BDEs). Mean concentrations were 45 and 96 μg kg^{-1} lipid weight, respectively, similar to, or slightly lower than, those reported elsewhere and related to the trophic level at which the two fish species feed.

Munschy et al. [79] determined BDEs and four non-PBDE BFRs in muscle and liver of sole from the west coast of France. ΣBDE_{14} concentrations ranged from 0.01 to 0.16 μg kg^{-1} wet weight in muscle, and 0.07–2.8 μg kg^{-1} wet weight in liver. These concentrations were at the lower end of those reported in the literature at other European locations.

ΣBDE_{13} concentrations ranging from 81 to 1510 μg kg^{-1} wet weight were reported in livers of wild cyprinid fish (mostly carp) from the River Po in Italy [80]. BDE47 dominated the congener profile. The authors also interpreted the presence of BDE179, BDE188, and BDE202, and elevated levels of BDE154 versus BDE153, as evidence of metabolic debromination of BDE209 in these fish.

A nationwide study in Canada during 2007–2009 looked at concentrations of PBDEs in whole body fish homogenates (ΣBDE_{40}), with a focus on lake trout (*Salvelinus namaycush*) [81]. Concentrations are difficult to compare as they were presented on a homolog (degree of bromination) basis, but concentrations were higher in the Great Lakes and Lake Champlain than at other locations. These concentrations were comparable to those seen in other industrialized regions of the world, whereas those from the other sampling locations were similar to those of the Arctic and other remote locations, suggesting that long-range atmospheric transport may be the major input pathway at these sites.

Carlsson et al. [82] determined concentrations of ΣBDE_{19} in whole herring and sprat, eggs from common eider, and eggs and livers from herring gulls from the west coast of Sweden. Concentrations ranged from 0.7 to 2.0 $\mu g\ kg^{-1}$ wet weight in fish, from 0.3 to 0.6 $\mu g\ kg^{-1}$ wet weight in eider eggs, and from 2.2 to 21 and 1.3 to 30 $\mu g\ kg^{-1}$ wet weight in herring gull eggs and liver, respectively.

Li et al. [83] determined ΣBDE_8 concentrations in eight fish and one crab species from China. BDE209 was not detected in any of the fish species, but was found in clams and crabs at concentrations around 1 $\mu g\ kg^{-1}$ lipid weight.

Liu et al. [84] determined ΣBDE_8 concentrations in muscle tissue of marine fish from four areas of China. Mean concentrations ranged from 0.8 to 388 $\mu g\ kg^{-1}$ wet weight. Congener profiles indicated that BDE209 made up about 23–70% of ΣBDE (mean 54%), maybe reflecting the fact that the deca-mix PBDE product is the most used BFR in China [85]. Xia et al. [86] studied BDEs in two fish species (yellow croakers and silver pomfret) from nine coastal cities along the eastern China coastline in 2008. ΣBDE_9 concentrations averaged 3.0 $\mu g\ kg^{-1}$ lipid weight, a lower level than that seen in other regions of the world, especially North America. BDE47 and BDE154 were the dominant congeners in both species, accounting for >60% of ΣBDE_9. The authors suggested that the relatively high proportion of BDE154 may be due to the debromination of higher brominated congeners, such as BDE183 and BDE209, in these two fish species.

Since the 1980s, eel populations have been declining across Europe. Tapie et al. [87] have determined ΣBDE_4 concentrations in glass and silver eels from the Gironde estuary, France, during 2004–2005. Mean concentrations varied from 24 to 237 $\mu g\ kg^{-1}$ lipid weight, although PCB concentrations seemed to be of more importance in relation to their decline [88]. McHugh et al. [89] determined ΣBDE_{11} and ΣPBB_5 concentrations in eel muscle tissue from five catchments in Ireland. These ranged from 1.0 to 7.0 and 0.005 to 0.01 $\mu g\ kg^{-1}$ wet weight, respectively. In eels from 30 sites across Scotland, MacGregor et al. [90] determined ΣBDE_{13} concentrations in muscle tissue. These ranged from 1.3 to 132 $\mu g\ kg^{-1}$ wet weight. Roosens et al. [91] studied BDEs in eels from 50 sites in Flanders, Belgium, during 2000–2006. ΣBDE_{11} concentrations ranged from 10 to 5810 $\mu g\ kg^{-1}$ lipid weight. BDE47 dominated the profile in most samples, except for six samples in which BDE209 dominated. These samples may have had recent exposure to the deca-mix PBDE product.

7.1.1.3 Birds

Jaspers et al. [92] investigated the use of body, primary, and tail feathers in white-tailed sea eagles as a monitoring tool for organohalogen compounds, including ΣBDE_2 (BDE47 and BDE99). Concentrations ranged from nd to 22 $\mu g\ kg^{-1}$. Nordlöf et al. [93] determined concentrations of ΣBDE_6 in eggs of white-tailed sea eagles from four regions of Sweden in 1992–2005 (two freshwater: Lapland (LAP) and inland lakes (INL), and two brackish marine areas in the Baltic Sea: the south Bothnian Sea (SB) and the Baltic Proper (BP)). BDE209 was not detected in any of the samples. Geometric means were 720 (LAP), 1500 (INL), 4100 (SB), and 4300

(BP) $\mu g\,kg^{-1}$ lipid weight, suggesting higher concentrations in marine than in freshwater fish. Further study of sea eagles was undertaken by Eulaers et al. [94] in Norway in 2009, during which 20 nestlings, approximately 9 weeks old and so fledged, were captured on the nest prior to release in Ireland as part of a reintroduction program. Body feathers, blood, and preen oil samples were taken. In all, 20 feather, 19 preen oil, and 15 blood plasma samples were collected. In order that only individuals that could provide all three samples were considered, the final sample size was 14. Median ΣBDE_8 concentrations were 3.4 $\mu g\,kg^{-1}$ dry weight in feathers, 0.3 $\mu g\,kg^{-1}$ wet weight in plasma, and 31 $\mu g\,kg^{-1}$ wet weight in preen oil. Strong significant correlations between blood plasma and preen oil concentrations indicated that preen oil concentrations closely reflected the internal level of body contamination.

Van den Steen et al. [95] investigated the use of the eggs of blue tits (*Cyanistes caeruleus*) as a biomonitor for PCBs and PBDEs. Ten sampling locations in seven European countries were studied, including suburban, rural, and remote areas. ΣBDE_{22} concentrations ranged from 4.0 to 114 $\mu g\,kg^{-1}$ lipid weight. As expected, ΣBDE concentrations were highest in suburban locations, and profiles differed in the different locations, likely due to local sources. ΣBDE_{28} concentrations were determined in feathers of two colonial egret species from the Punjab, Pakistan, by Malik et al. [96]. Median concentrations were 2.4 and 1.9 $\mu g\,kg^{-1}$ wet weight in little and cattle egrets, respectively. BDE47, BDE100, BDE138, and BDE153 were abundant, and differences in profiles could be seen between little egrets (which eat fish) and cattle egrets (which feed on terrestrial species).

Van den Steen et al. [97] determined ΣBDE_{22} concentrations in eggs of two common bird species from the Falkland Islands in the south Atlantic Ocean, rockhopper penguins and imperial shags. Mean concentrations were 0.98 and 1.9 $\mu g\,kg^{-1}$ lipid weight, respectively. These values are significant, however, as this is both a remote and sparsely populated location, and in the Southern Hemisphere.

Lavoie et al. [98] determined ΣBDE_5 and HBCDD (discussed in detail in the following section) in the eggs of six seabird species (great black-backed gull, herring gull, black-legged kittiwake, common eider, razorbill, and black guillemot) from the Gulf of St Lawrence in Canada taken in 2006–2007. Mean ΣBDE_5 concentrations ranged from 134 to 3780 $\mu g\,kg^{-1}$ lipid weight and mean HBCDD concentrations from 6 to 116 $\mu g\,kg^{-1}$ lipid weight, with the highest concentrations (in both cases) observed in herring gulls.

Liver concentrations of BDEs in coastal herring gulls from the Hvaler Archipelago, Norway, in 1998 were determined [99]. ΣBDE_8 concentrations ranged from 135 to 985 $\mu g\,kg^{-1}$ lipid weight. $\delta^{13}C$ values suggest a higher abundance in terrestrial food sources. Inverse relationships of $\delta^{15}N$ with BDE99, BDE183, and BDE209 suggest greater bioaccumulation from lower trophic level prey species. This may be because those congeners are the products of debromination (of BDE209) in higher trophic level species, such as teleost fish. Levels of BDE209 (up to 95 $\mu g\,kg^{-1}$ lipid weight) were in the higher range reported for European birds, and higher than those previously reported for North Sea seabirds.

In eggs of great skua (*Stercorarius skua*) taken from the Shetland Islands (to the north of the Scottish mainland) in 1980 and 2008, ΣBDE_{14} concentrations ranged from 5.9 to 1550 μg kg^{-1} wet weight in 1980 and 55–734 μg kg^{-1} wet weight in 2008 [100]. Although the mean concentrations increased, the sample sizes were fairly small (c.25) and any apparent trend was unconvincing.

A temporal trend of BFRs in eggs of a terrestrial bird of prey, the peregrine falcon (*Falco peregrinus*), over the period 1974–2007 was presented by Johansson et al. [101]. ΣBDE_6 (BDE47, BDE99, BDE100, BDE153, BDE183, BDE209—so including markers for all three PBDE formulations) and HBCDD concentrations were determined. All BDE congeners were detected, with ΣBDE_6 concentrations ranging from 270 to 23,000 μg kg^{-1} lipid weight, BDE183 from 33 to 170 μg kg^{-1} lipid weight, and BDE209 from 10 to 870 μg kg^{-1} lipid weight. HBCDD concentrations ranged from 28 to 1200 μg kg^{-1} lipid weight. Fernie and Letcher [102] determined ΣBDE_{14} concentrations in plasma of nestling peregrine falcons from the Great Lakes area of Canada in 2004–2005. Concentrations ranged from 0.9 to 196 μg kg^{-1} wet weight. BDE congeners derived from the penta-mix and octa-mix PBDE formulations, and BDE47, BDE99, BDE100, BDE153, and BDE183 comprised 93% of the total. BDE209 concentrations were minimal ($<1\%$ of the total). A comparison between ΣBDE_{16} in the eggs of peregrine falcons from Canada and Spain was reported [103]. Overall, concentrations exceeded those of HBCDD and were an order of magnitude higher than those of seven novel BFRs (see below). ΣBDE_{16} concentrations in eggs from four sites in Canada and two sites in Spain ranged from 530 to 38,000 μg kg^{-1} lipid weight and 160 to 5300 μg kg^{-1} lipid weight, respectively. BDE183, BDE197, BDE207, and BDE209 were detected in all 25 eggs studied. BDE153 was the dominant congener in more than 60% of peregrine eggs—the congener pattern, regardless of the country of origin, was BDE153 > BDE99 > BDE47 > BDE100 > BDE183. These five congeners constituted $66 \pm 24\%$ and $85 \pm 7.8\%$ of the sum in Spanish and Canadian peregrines, respectively. The dominance of BDE153 may reflect its longer half-life compared to other BDE congeners as seen in kestrels [104], its accumulation as a possible debromination product [101], and the significant biomagnifications but low biotransformation of BDE153 in terrestrial birds of prey [45,105].

Chen et al. [105] compared PBDE contamination in eggs of osprey, double-crested cormorant, brown pelican, and peregrine falcon from the Chesapeake Bay area of the United States. Mean ΣBDE_{29} concentrations were 290 μg kg^{-1} wet weight in ospreys, 182 μg kg^{-1} wet weight in peregrine falcons, 28 μg kg^{-1} wet weight in brown pelicans, and 12 μg kg^{-1} wet weight in double-breasted cormorants. BDE47 dominated in piscivorous birds, while BDE153 was predominant in peregrines. BDE209 was only detected in peregrine falcons. Peregrine falcon eggs collected from urban and nonurban habitats in California were analyzed for BDEs by Newsome et al. [106]. Stable carbon ($\delta^{13}C$), hydrogen (δD), and nitrogen ($\delta^{5}N$) isotope compositions of the eggs were also determined. Differences in stable carbon values between animals indicate that they feed via different food webs and that in stable nitrogen values, the trophic level at which they feed. $\delta^{13}C$ values were significantly higher in urban than in nonurban eggs, suggesting that urban peregrines indirectly receive anthropogenic subsidies via their consumption of prey reliant on corn-based anthropogenic foods. $\delta^{15}N$ and δD values were significantly lower in eggs from urban than in nonurban eggs, reflecting differences in dietary diversity and food/water sources available to peregrines in each habitat. Study of bird feathers from nests from which eggs were collected represented 44 prey types in urban locations and 77 prey types in nonurban locations. Mean concentrations of ΣBDE_{15} were 11,600 μg kg^{-1} lipid weight in urban areas, and 4740 μg kg^{-1} lipid weight in nonurban areas. Concentrations of higher brominated congeners (BDE183–BDE209)—i.e. those derived from the octa-mix and deca-mix PBDE formulations and from the possible debromination of BDE209—were twofold higher than those of the lower brominated congeners in urban sites and 0.3 times as high in nonurban locations.

In a long-term study of BDE209 in wild birds, Leslie et al. [107] studied muscle tissue, liver, and eggs of 10 terrestrial and four aquatic bird species. BDE209 was detected in 47% of the terrestrial bird samples (nine species) but only in 9% of the aquatic bird samples (six species). For sparrowhawks, no significant changes in BDE209 concentrations between 1973 and 2001 were found. Peak BDE209 concentrations in peregrines were seen in 1995, followed by a decline until 2001. Concentrations ranged from nd to 412 μg kg^{-1} lipid weight in terrestrial and nd to 563 μg kg^{-1} lipid weight in aquatic species, the highest concentration being observed in the muscle of a grey heron from the United Kingdom.

Crosse et al. [108] used the eggs of Northern gannets to determine long-term trends (1977–2007) in BDE contamination in two colonies from the Eastern Atlantic Ocean (Ailsa Craig) and the North Sea (Bass Rock). ΣBDE_{25} concentrations in eggs from both colonies increased mainly from the late 1980s, peaked in 1994, and then declined rapidly so that concentrations in 2002 were similar to or lower than those observed in the 1970s or 1980s. The main source of contamination was the penta-mix PBDE formulation, as also seen for porpoises in the United Kingdom [58].

7.1.1.4 Other wildlife species

White-nose syndrome is a fungal growth seen on hibernating bats—the causes are not known. Kannan et al. [109] determined BDEs as part of a study of POPs in little brown bats in the NE United States. Concentrations of ΣBDE_{12} in fat tissues ranged from 520 to 10,900 μg kg^{-1} lipid weight in bats from New York, and 4300 to 13,000 μg kg^{-1} lipid weight in bats from Kentucky (a reference population unaffected by the disease). This similarity in maximum concentrations may indicate that BDEs are not a cause of this syndrome.

Vermeulen et al. [231] studied accumulation of BDEs through the soil–earthworm–hedgehog food chain. Biota-soil accumulation factors (2.0–5.7 for BDEs) indicated that animals feeding on earthworms are exposed to higher levels of accumulated BDEs, as suggested earlier [40], and that this provides a route into the higher levels of the food chain.

In whole wild rice frogs from close to an e-waste recycling site in China (*Rana limnocharis*), Liu et al. [110] reported ΣBDE_8 concentrations ranging from 5.8 to 54 μg kg^{-1} wet weight (mean 28 μg kg^{-1} wet weight) and also presented data for brain, liver, testes, and eggs. Congener profiles were dominated by BDE99 and BDE153, with BDE28, BDE47, and BDE209 as lesser congeners. Hen's egg and chicken tissues from an e-waste recycling area in SE China were studied by Qin et al. [111]. ΣBDE_7 concentrations in tissues ranged from 15 to 3140 μg kg^{-1} lipid weight and in egg was 564 μg kg^{-1} lipid weight. ΣBDE levels in tissue were up to two to three orders of magnitude higher than others reported in the literature. BDE209 was the predominant congener in all chicken tissues except brain. The dietary intake of BDEs from these sources, assuming that only local produce was eaten, ranged from 2.2 to 23 ng kg body weight^{-1} day $^{-1}$, with a mean value of 13.5 ng kg body weight^{-1} day^{-1}, one order of magnitude higher than the value reported in previous studies for consumption of all foodstuffs.

Hu et al. [112] determined ΣBDE9 concentrations in 16 aquatic species from Baiyangdian Lake, North China. ΣBDE9 concentrations ranged from 3.4 to 160 μg kg^{-1} lipid weight. BDE47 was the dominant congener in most samples, apart from river snails and swan mussels. BDE209 was detected in 50% of all samples. Correlation between lipid-normalized ΣBDE_9 concentrations and trophic levels as determined using nitrogen stable isotope measurements confirmed biomagnification in the freshwater food chain. Wild species studied were zooplankton, river snails, swan mussels, shrimps, crabs, carp (four species), northern snakehead, yellow catfish, rice field eel, and loach. Farmed species included oriental sheatfish, turtles, and ducks. Qiu et al. [113] determined ΣBDE_{15} concentrations in 19 species of benthic organisms taken from Deep Bay, South China, in 2004. Mean ΣBDE_{15} concentrations in fish (2.0 ± 1.1 μg kg^{-1} lipid weight) were generally higher than those in shrimps (0.9 μg kg^{-1} lipid weight) and crabs (1.4 μg kg^{-1} lipid weight). Positive correlations between ΣBDE_{15} concentrations (lipid basis) and $\delta^{15}N$ values were seen in planktivorous fish, whereas negative correlations were seen for small carnivorous fish. BDE47 was the dominant congener in these benthic organisms.

Zhang et al. [114] studied the trophodynamics of BDEs in a marine food web. ΣBDE_{14} concentrations were an order of magnitude higher in seabirds (184 ± 162 μg kg^{-1} lipid weight) than in invertebrates and fish (19 ± 14 μg kg^{-1} lipid weight); although the congener profiles in invertebrates, fish, and juvenile seabirds were dominated by BDE47, BDE99 and BDE153 were also significant in adult seabirds. Correlations between lipid-normalized BDE concentrations and trophic levels confirmed that seven BDE congeners were magnified in the invertebrate–fish–seabird food web and that BDE concentrations increased at a much greater rate across trophic levels in seabirds than in invertebrates and fish.

In order to determine the input sources of PBDEs in fish farming environments in South China, Zhang et al. [115] took samples of seven environmental matrices from cultured fish ponds during 2006–2007. Tri- to deca-BDEs were detected in all samples collected, with mean ΣBDE_{39} concentrations of 5.7 ± 3.6 ng l^{-1} in pond water, 15 ± 11 μg kg^{-1} dry weight in pond sediment, 12 ± 3.8 μg kg^{-1} dry weight in bank soil, 21 ± 20 μg kg^{-1} lipid weight in fish, and 93 ± 62 μg kg^{-1} lipid weight in fish feeds. BDE209 predominated in all samples except fish, while BDE47 predominated in fish. Relatively high abundances of BDE49 were detected in all samples compared to those in the penta-mix PBDE formulation. Fish feed may have been the major source of BDEs in fish, as seen in other studies.

In a laboratory study, Josefsson et al. [116] studied the influence of contaminant burial depth on the bioaccumulation of BDEs by two benthic invertebrate species. Three layers of sediment (2.0–2.5 cm, 5.0–5.5 cm, and 10–10.5 cm below the sediment surface) were spiked with different BDE congeners, uncommon in environmental samples. Uptake was higher from shallower depths, as bioturbation remobilized buried contaminants to the biologically active surface layer.

7.1.2 Humans

A study of human milk samples undertaken in New Hampshire in 2005–2006 found ΣBDE_8 concentrations ranging from 6.5 to 167 μg kg^{-1} lipid weight (median 30 μg kg^{-1} lipid weight) [117]. In this study, 40 women provided up to three samples at the end of their first, second, and third months of breastfeeding. BDE47 was generally the dominant congener, although BDE153 dominated in 20% of the samples. There was no significant variation in ΣBDE levels either day to day or month to month. BDE levels were influenced by both diet and the home environment.

Human milk samples collected during 2003–2005 from 82 first-time mothers in 24 communities located throughout California, where furniture flammability regulations drive higher levels of FR use, contained levels of ΣBDE_8 (tri- to hexa congeners) ranging from 9.6 to 1290 μg kg^{-1} lipid weight (median 53 μg kg^{-1} lipid weight), which are among the highest in the world [118]. Levels varied by a factor of 100 times. BDE47 was the dominant congener, and levels exceeded the US Environmental Protection Agency (EPA) reference dose for neurodevelopmental toxicity (100 ng kg^{-1} day^{-1}) in 60% of the samples. In some samples, BDE209 (2/82) or BDE153 (5/82) were the dominant congeners. Zota et al. [119] studied the relationship between BDE and hydroxyl-BDEs (OH-BDE) concentrations (OH-BDEs are products of metabolism) in pregnant women in California and thyroid function using serum samples collected from 25 second-trimester pregnant women during 2008–2009. Median concentrations of BDE47 and the sum of lower brominated congeners (BDE17–BDE154) were 43 and 86 μg kg^{-1} lipid weight, respectively. The sum of the higher brominated congeners (BDE183–BDE209), although defined, could not be located in the manuscript.

BDEs were determined in breast adipose tissues of women from California in the late 1990s. ΣBDE_5 concentrations ranged from 7.4 to 4500 μg kg^{-1} lipid weight [120]. Chevrier et al. [121] determined serum concentrations of ΣBDE_5 (BDE28, BDE47, BDE99, BDE100, and BDE153) and free thyroxine (T_4), total T_4, and thyroid-stimulating hormone (TSH) in 270 pregnant women around the 27th week of gestation. ΣBDE concentrations were inversely associated with TSH levels. Decreases in TSH ranged between 11% and 19% for every 10-fold increase in concentrations of individual congeners. Odds of subclinical hyperthyroidism (low TSH but normal T_4) were also significantly elevated in participants in the highest quartile of ΣBDE and congeners BDE100 and BDE153 relative to those in the first quartile. Statistically significant associations between ΣBDE and free and total T_4 were not observed. In the serum of 271 adolescent girls from the United States, during 2003–2004, the median ΣBDE_6 concentration was 45 μg kg^{-1} lipid weight [122]. The study related to the influence of PBDEs on the age at menarche, and will be discussed later in this chapter.

While young children are rarely included in biomonitoring studies, they are considered to be at greater risk of ingesting PBDEs than adults [123], from dust ingestion, for example. These authors reported blood ΣBDE_{20} concentrations for children in the United States (1.5–4 years old) typically 2.8 times higher than those in their mothers. Early life was indicated as an intense period of PBDE intake, and it was shown that young children have higher exposure to PBDEs (and likely other BFRs) than adults sharing the same living spaces.

Human milk samples were taken from 50 breastfeeding mothers in Shijiahuang, 60 in Tianjin, and 48 in Yantai, northern China (2006–2007), and 20 samples were randomly selected for BDE analysis [124]. The average concentrations of ΣBDE_{17} were 3.7, 3.4, and 4.2 μg kg^{-1} lipid weight in the three regions, respectively (lower than the range of concentrations reported above for the United States). BDE209, BDE207, BDE197, BDE153, BDE15, BDE28, and BDE47 were the dominant congeners, suggesting exposure to mainly the penta- and deca-mix PBDE formulations, and probably also some debromination of BDE209, although formulations manufactured in China will need to be examined for the presence of BDE207 and BDE197 in the deca-PBDE products on sale.

A total of 24 pooled human milk samples (from 1237 individual samples) collected in 12 Chinese provinces in 2007 were analyzed for BDEs [125]. ΣBDE_7 concentrations ranged from 0.9 to 3.0 μg kg^{-1} lipid weight and were dominated by BDE28, BDE47, and BDE153: BDE183 (considered as a marker for the octa-mix PBDE formulation) was also detected at concentrations ranging from 0.03 to 0.49 μg kg^{-1} lipid weight. BDE209 was not determined. Shen et al. [126] examined 22 pooled blood samples representing 66 children from e-waste recycling areas and control locations in Zhejiang province, China. Mean ΣBDE_{15} concentrations were 32 ± 18 μg kg^{-1} lipid weight, compared to 12 ± 7.8 and 8.4 ± 4.0 μg kg^{-1} lipid weight in control areas. Yu et al. [127] followed up on previous studies which had reported high serum concentrations of BDEs, especially BDE209, in the residents of an e-waste recycling site in Guiyu town, south China. They identified three OH-BDEs in human serum samples from this area, all substituted in the ortho-position: 6-OH-BDE196, 6-OH-BDE199, and 6′-OH-BDE206. This indicates that hydroxylated octa- and nona-BDEs can accumulate in human blood, and suggests that oxidative metabolism of BDE congeners occurs in humans.

Hair samples from people living in urban, e-waste recycling, and rural areas of South China were analyzed for BDEs [128]. ΣBDE_{21} concentrations in 173 human hair samples from sites in the Pearl River Delta ranged from 2.6 to 75 μg kg^{-1} dry weight in rural areas, 4,2–70 μg kg^{-1} dry weight in urban areas, 7.4–219 μg kg^{-1}

dry weight in residents from e-waste recycling areas, and 12–845 μg kg^{-1} dry weight in samples from e-waste recycling workers. Dust concentrations in the same four areas had ΣBDE_{21} concentrations ranging from 251 to 1890; 540 to 41,400; 859 to 5890; and 2700 to 28,000 μg kg^{-1} dry weight, respectively. BDE209 was the dominant congener in all four areas, c.60% of the total. Nona-BDEs were the second group of congeners (c.20%) followed by BDE47 (c.5%), BDE99, BDE183, and BDE153. This indicates the presence of all three PBDE products in the South China region, probably in the order deca-mix > penta-mix > octa-mix.

A study aimed at assessing dual body burdens of PCBs and PBDEs and potential health risk to local residents close to two e-waste recycling sites in SE China, one handling mainly each type of waste, was conducted [129]. PCB-containing waste was handled mainly at Luqiao, and PBDE-containing waste at Wenling. ΣBDE_8 concentrations in residents' blood were 40–396 μg kg^{-1} lipid weight in Luqiao and 132–714 μg kg^{-1} lipid weight in Wenling, higher than those observed in North America. BDE209 was the dominant congener in both locations.

Johnson et al. [55] reported concentrations of ΣBDE_{11} in 24 samples of blood serum from people in Massachusetts. Concentrations ranged from 13 to 225 μg kg^{-1} lipid weight. As noted above, the authors found a strong correlation ($r = 0.65–0.89$; $p < 0.05$) between PBDE concentrations in house dust and the blood serum levels of PBDEs in the residents.

Paired samples of human milk and maternal and cord blood were collected from 20 Korean women in 2007 [130]. In comparison to data from other countries and to previous data from Korea, relatively higher and gradually increasing concentrations for ΣBDE_{18} were found (nd to 590 μg kg^{-1} lipid weight; median 90 μg kg^{-1} lipid weight). PBDE concentrations in human milk were lower in the younger mothers and/or those with more than one child. Similarly, Frederiksen et al. [131] looked at paired samples of maternal and umbilical cord blood plasma. ΣBDE_{12} concentrations were 0.6–51.9 μg kg^{-1} lipid weight in maternal plasma and 0.2 and 54.3 μg kg^{-1} lipid weight in umbilical cord samples, and these were highly correlated.

During 2008–2009, 90 umbilical cord blood and 21 human milk samples were collected from a hospital in Korea [132]. ΣBDE_7 concentrations ranged from 2.8 to 95 and 1.8–8.7 μg kg^{-1} lipid weight, respectively. BDE47, BDE99, and BDE153 were the dominant congeners in cord blood and BDE47, BDE153, BDE28, and BDE100 in milk. BDE47 concentrations in blood were lower than those observed in US samples in earlier studies, but higher than those seen in Europe and other Asian regions. Similarly, BDE47 concentrations in milk were lower than those seen in North America and Australia, but similar to, or lower than, those seen in Europe and other Asian regions.

ΣBDE_{27} concentrations in fasting blood (taken after a period in which food was not eaten) were determined in a cross-sectional study of 72 participants aged 15–74 years in Japan [133]. The subjects lived in two urban areas and two fishing villages, but were not occupationally exposed to PBDEs. The median ΣBDE_{14}

concentration was 3.6 μg kg^{-1} lipid weight and more than half of the concentrations of 14 congeners were below the detection limit. The most abundant congener was BDE209 (median concentration 0.9 μg kg^{-1} lipid weight), followed by BDE153, BDE207, and BDE47, in that order. Most congeners with six bromine atoms or fewer had significant positive correlations with biomarkers of fish consumption, which were not evident for congeners with eight or more bromines. This suggests that, for the lower molecular weight BDEs, fish consumption is a major uptake route, while for higher congeners the main uptake route is different.

Roosens et al. [134] developed an exposure model to the Flemish population in Belgium to PBDEs. The model covered human exposure of infants, children, and adults through human milk, food, dust/soil ingestion, and air inhalation. Exposure levels through human milk were consistent with those of a background-exposed European population, whereas dust and food intake were at the low end of such estimates. Average ΣBDE_5 intakes by newborn infants were 12 μg kg^{-1} body weight per day, rising to 21 μg kg^{-1} body weight per day for those most exposed individuals and declining to 0.7 μg kg^{-1} body weight per day in adulthood. In a separate study, Roosens et al. [91], determined that intake of ΣBDE_{11} from eels for an average consumer ranged from 3 to 2300 ng day^{-1}.

In human adipose tissue samples collected in Siena, central Italy, during 2005–2006, Schiavone et al. [135] determined BDEs alongside other POPs (polychlorinated naphthalenes, PCBs, organochlorine pesticides, and polycyclic musks). ΣBDE_6 concentrations ranged from 3.2 to 36 μg kg^{-1} lipid weight.

Tue et al. [136] investigated the contamination status (using human milk samples) and possible exposure pathways in three Vietnamese e-waste recycling sites (Trang Minh, Dong Mai, and Bui Dau) and at one reference site (Hanoi). Levels were significantly higher in Trang Minh and Bui Dau than at the reference site. Recyclers from Bui Dau had the highest ΣBDE_{14} concentrations (20–250 μg kg^{-1} lipid weight), two orders of magnitude higher than those in the reference group. A specific accumulation of higher brominated BDEs, unrelated to diet, was observed in e-waste recyclers, suggesting extensive exposure to these compounds during recycling activities, possibly through inhalation and ingestion of dust. The estimated infant intake dose of PBDEs from the milk of some mothers involved in e-waste recycling were close to, or higher than, reference doses issued by the US EPA.

In a study conducted by Thomsen et al. [137], 393 human milk samples from mothers living in various regions of Norway were analyzed for BDEs. Ten congeners were analyzed in all samples, and BDE209 was analyzed in a subset of 46 samples. The median concentration of ΣBDE_7 (the most prominent congeners) was 2.1 μg kg^{-1} lipid weight, and of BDE209 was 0.3 μg kg^{-1} lipid weight. These concentrations were comparable to levels generally observed in European populations. The mothers' diet did not influence the concentrations observed. In another study, Thomsen et al. [138] studied elimination rates of BDEs in 10 mothers by collecting and analyzing milk samples monthly from about 2 weeks to 12 months after birth. A significant decrease of 2.5% per month was observed for ΣBDE_6. After a year of nursing this was c.30% in total.

In a study of BDEs in blood plasma of 111 Hong Kong residents undertaken in 2008 by Qin et al. [139], mean ΣBDE_{10} concentrations were 5.6 μg kg^{-1} lipid weight, approximately 10-fold less than those observed in North America. Concentrations of BDE209 were (very) approximately half those of BDE47.

Chovancová et al. [140] determined BDEs in 33 human milk samples collected from first-time mothers in 2006–2007 close to four industrial areas of Slovakia. ΣBDE_7 concentrations ranged from 0.2–1.6 μg kg^{-1} lipid weight, with median and mean values of 0.43 and 0.57 μg kg^{-1} lipid weight, respectively. No significant differences were observed between locations. In 204 samples of milk of randomly recruited first-time mothers from Sweden in 2000–2004, Glynn et al. [141] reported ΣBDE_4 concentrations of 0.2–1.4 μg kg^{-1} lipid weight in Uppsala, 0.2–1.3 μg kg^{-1} lipid weight in Gothenburg, 0.3–2.1 μg kg^{-1} lipid weight in Lund, and 0.4–1.8 μg kg^{-1} lipid weight in Lycksele.

Vorkamp et al. [142] studied possible associations between BDE levels in placenta samples and those in dust and indoor air from the mothers' homes. ΣBDE_{12} concentrations could not be easily calculated from the data presented; however, the authors indicated that BDE209 dominated in dust samples (with BDE206 the dominant nona-congener) but not in placenta.

Ten human adipose tissue samples from individuals aged 19–65 years in Örebro, Sweden, taken in 2007 were analyzed for BDEs [143]. ΣBDE_{20} concentrations ranged from 1.2 to 6.7 μg kg^{-1} lipid weight. Törnkvist et al. [144] estimated dietary intake of BDEs in four regions of Sweden in 2005. Food items from four groups (fish, meat, dairy products, as well as eggs and fats/oils) were analyzed for a range of POPs followed by per capita intake calculations. ΣBDE_9 intake was estimated to be 0.7 ng kg bodyweight^{-1} day^{-1}. Fish (38%) and dairy products (31%) were the major contributors.

Chen et al. [145] analyzed 29 plasma samples collected from children aged 0–11 years in Dalian, China, in 2008. ΣBDE_{16} concentrations ranged from nd to 188 μg kg^{-1} lipid weight, and no significant differences were observed between sexes. These levels were 9–30 times higher than those in nonoccupationally exposed people from Guangzhou.

Maximum concentrations of ΣBDE_5 in human milk samples from Norway taken from 2003 ranged from nd to 78 μg kg^{-1} lipid weight [146]. Concentrations of BDE209 were 0.005–5.8 μg kg^{-1} lipid weight.

Human exposure to PBDEs was evaluated in Ghana, Africa, using human milk samples collected in 2004 and 2009 [147]. Mean concentrations and ranges of ΣBDE_{13} were unexpectedly high (4.5; 0.9–18 μg kg^{-1} lipid weight), in spite of the fact that Ghana is relatively unindustrialized. A significant increase in concentrations was observed between the two study years.

The nonoccupational exposure to BFRs was studied using human milk samples collected in Limpopo province, South Africa [148]. ΣBDE_9 concentrations ranged from 0.7 to 6.3 μg kg^{-1} lipid weight.

Although exposure of children and adults to PBDEs is relatively well characterized, developmental exposure is less so. Foster et al. [149] studied BDEs in maternal serum and umbilical cord serum during 2004–2005 in Canada. ΣBDE_9 concentrations in maternal serum at 24–28 weeks' gestation ranged from 21–1290 μg kg^{-1} lipid weight and at delivery from 6.8 to 1100 μg kg^{-1} lipid weight. Concentrations in umbilical cord serum at delivery ranged from 51 to 1450 μg kg^{-1} lipid weight. The authors concluded that fetuses were exposed to PBDEs from at least the second trimester of pregnancy onward. Wu et al. [150] also studied umbilical cord blood from China in 2007. Umbilical cord blood was collected from 153 pregnant women in Guiyu (an e-waste recycling area) and Chaonan (a reference area) shortly after birth. ΣBDE_8 concentrations were higher in samples from Guiyu than from the control area (median concentrations 13.8 and 5.2 μg kg^{-1} lipid weight, respectively; ranges 1.1–505 and 0.3–364 μg kg^{-1} lipid weight, respectively).

Vizcaino et al. [151] analyzed maternal blood serum and cord blood serum from 174 mother/newborn pairs from Valencia, Spain, during 2003–2005. Median concentrations were the same in both, 9.6 μg kg^{-1} lipid weight, and the ranges were similar (nd to 120 μg kg^{-1} lipid weight and nd to 140 μg kg^{-1} lipid weight, respectively). Neonates (but not their mothers) from rural areas exhibited statistically significant lower concentrations than those from urban, semiurban, or metropolitan locations.

7.2 HBCDD and TBBPA

7.2.1 Wildlife

7.2.1.1 Marine mammals

Vorkamp et al. [49] studied temporal trends of HBCDD concentrations in ringed seals from East Greenland between 1986 and 2008. These showed a significant exponential increase from 2.0 to 8.7 μg kg^{-1} lipid weight (median concentrations) at a rate of +6.1% per year. HBCDD was determined in the blubber of finless porpoises taken in 2000–2007 from Japan [61]. Concentrations ranged from 7.2 to 2700 μg kg^{-1} lipid weight. There were no obvious temporal trends during this period.

7.2.1.2 Fish

In the study by Takahashi et al. [70] described earlier, HBCDD concentrations in deep-sea fish taken off the coast of Japan were <0.05–110 μg kg^{-1} lipid weight. The highest concentrations of most of the organohalogen compounds determined (PCBs, dichlorodiphenyltrichloroethane, chlordanes, BDEs, and HBCDD) were found in the same individual fish, a snubnosed eel. In mussels and oysters from Japanese coastal waters, Ueno et al. [71] reported HBCDD concentrations of 12 to 5200 μg kg^{-1} lipid weight, considerably higher than those of BDEs. The estimated dietary exposure via seafood was 0.45–34 ng kg body weight^{-1} day^{-1} the highest value about five times higher than that estimated for BDEs (see above).

HBCDD concentrations in marine fish (two species: large yellow croaker and silver pomfret) from nine coastal Chinese cities were reported by Xia et al. [152]. The average concentration was 3.7 μg kg^{-1} lipid weight (range 0.6–10 μg kg^{-1} lipid weight), which is lower than that seen in other parts of the world, especially Europe, where HBCDD is used to a greater degree than in China.

In eels from Flanders, Roosens et al. [91] found HBCD concentrations ranging from 16 to 4400 μg kg^{-1} lipid weight, quite similar to the range seen for ΣBDE_{11} in the same study (see above).

7.2.1.3 Birds

Esslinger et al. [153] studied HBCDD in the eggs of herring gulls from two colonies on the German North Sea coast and one on the Baltic Sea coast between 1998 and 2008. α-HBCDD was the dominant isomer and concentrations ranged from 3.7 to 107 μg kg^{-1} lipid weight. HBCDD concentrations in eggs from all three islands showed a peak around 2000, followed by a decline thereafter.

In the study of Sørmo et al. [99] reported above, liver concentrations of HBCDD in herring gulls from the Hvaler Archipelago in Norway taken in 1998 ranged from 10 to 698 μg kg^{-1} lipid weight. An inverse relationship between HBCDD concentrations and $\delta^{13}C$ values suggested a higher abundance of HBCDD in terrestrial diet sources.

In the study by Fernie and Letcher [102] described above, HBCDD concentrations were also determined in the plasma of nestling peregrine falcons in the Great Lakes area. These ranged from nd to 0.46 μg kg^{-1} wet weight. Guerra et al. [103] reported HBCDD concentrations in peregrine falcon eggs from Canada and Spain. ΣHBCDD concentrations ranged from 230 to 15,000 μg kg^{-1} lipid weight in eggs from Canada and 0.9 to 1600 μg kg^{-1} lipid weight in eggs from Spain, and α-HBCDD dominated in both cases. The highest level was in samples from Montreal. On the face of it, this is rather surprising, as HBCDD usage in Europe is generally held to be greater than that in North America, where the PBDE products have dominated the market.

In the study in white-tailed sea eagle eggs reported above, Nordlöf et al. [93] also determined concentrations of HBCDD. The geometric means were 60 (LAP), 90 (INL), 150 (SB), and 140 (BP) μg kg^{-1} lipid weight, again suggesting somewhat higher concentrations in marine rather than freshwater fish.

In the long-term study of BDE209 in birds referred to above, HBCDD was also determined in some samples [107]. ΣHBCDD concentrations in peregrine falcon eggs from the United Kingdom ranged from 8.9 to 1150 μg kg^{-1} lipid weight, and in sparrowhawk muscle tissue, also from the United Kingdom, from 84 to 2360 μg kg^{-1} lipid weight.

Numerous studies are now becoming available from e-waste recycling sites, mostly in Asia, as these have been recognized as significant local sources. He et al. [154] studied TBBPA and HBCDD in muscle tissue of six bird species from an e-waste recycling region in Qingyuan, South China. Stable isotope ratios ($\delta^{13}C$ and $\delta^{15}N$) were also determined in order to investigate the dietary sources and trophic levels of the birds. Median TBBPA and HBCDD concentrations in the birds ranged from 28 to 173 μg kg^{-1} lipid weight and from nd to 2000 μg kg^{-1} lipid weight, respectively. α-HBCDD predominated in birds feeding in an aquatic environment and γ-HBCDD in birds feeding in a terrestrial environment, whereas no clear dominance was evident in birds inhabiting freshwater wetland. A significant positive correlation between $\delta^{13}C$ and the percentage of α-HBCDD was seen, indicating the importance of dietary exposure pathways in the determination of HBCDD diastereoisomer patterns. The enantiomer fractions for α-HBCDD differed substantially between aquatic and terrestrial bird species with a significant enrichment of the (+) α-HBCDD enantiomer in aquatic birds and preferential enrichment of the (−) α-HBCDD in terrestrial birds. Trophic magnification (as defined by $\delta^{15}N$ levels) showed a strong positive relationship only for α-HBCDD ($p = 0.001$).

7.2.1.4 Food webs

Diastereoisomer-specific accumulation of HBCDD from a point source in five marine species representing a typical food web in a Norwegian coastal area (Åsefjorden) was studied by Haukås et al. [155]. Samples of mussels, polychaetes, crabs, and seabird eggs were analyzed for HBCDD, as well as stable isotopes of nitrogen to determine trophic level. HBCDD concentrations ranged from nd to 280 $\mu g\ kg^{-1}$ lipid weight. In contrast to β- and γ-HBCDD, α-HBCDD increased significantly with trophic level. Köppen et al. [156] used liquid chromatography-tandem mass spectrometry (LC-MS/MS) to quantify HBCDD enantiomers in six fish species from the Norwegian Etnefjorden. Concentrations of α-, β- and γ-HBCDD in fillets ranged from <5 to 11,100 $\mu g\ kg^{-1}$ lipid weight, with α-HBCDD enantiomers dominating throughout. In most cases, the preferential accumulation of (−) α-, (−) β-, and (+) γ-HBCDD was observed. In most cases the enantiomer fractions differed significantly from the racemic ratio.

Haukås et al. [157] studied the bioavailability of HBCDD in laboratory experiments through sediment and food to a polychaete worm in relation to assessment of an HBCDD-contaminated Norwegian Fjord (Åsefjorden). Although mean sediment concentrations were as high as 40,000 $\mu g\ kg^{-1}$ dry weight in the contaminated sediment, no HBCDD was detected in the exposed ragworm. In contrast, exposure to contaminated food (mussels) resulted in elevated worm tissue concentrations of 9.1 $\mu g\ kg^{-1}$ wet weight. α-HBCDD showed a relative increase during transfer from mussel to ragworm. Hence, in the contaminated Fjord, trophic transfer seemed to be a more important mechanism for the entry of HBCDD into the local benthic food web than sediment to biota accumulation.

7.2.2 Humans

In the study by Roosens et al. [134] described above, average intake of ΣHBCDD was 3.1 $\mu g\ kg^{-1}$ body weight per day for newborn infants, rising to 15 μg kg body $weight^{-1}\ day^{-1}$ for the most exposed individuals. For median exposed individuals, HBCDD intake peaked at 6.6 $\mu g\ kg^{-1}$ body weight per day at age 3−6 years, and declined to ~1 $\mu g\ kg^{-1}$ body weight per day at later age. In a separate study by Roosens et al. [91], intake of HBCDD from eels for an average consumer ranged from 3 to 1110 ng day^{-1}.

In the study by Tue et al. [136] reported above, recyclers from Bui Dau were the only group with significant exposure to HBCDD (human milk ΣHBCDD concentrations from 1.4–7.6 μg kg^{-1} lipid weight).

The median ΣHBCDD concentration in Norwegian human milk samples in the study described earlier [137] was 0.86 μg kg^{-1} lipid weight. The mothers' diet did not influence these concentrations. The concentrations of HBCDD in human milk samples from Norway taken from 2003 ranged from 0.1 to 31 μg kg^{-1} lipid weight [146].

In Sweden in 2000–2004, human milk concentrations were <0.2–10 μg kg^{-1} lipid weight [141]. The study is described above. Törnkvist et al. [144] determined HBCDD in samples from a market basket sample taken in Sweden in 2005. The intake of HBCDD in the human diet was estimated to be 0.14 ng kg body weight^{-1} day^{-1}, with fish (65%), dairy products (24%), and meat (10%) contributing almost the entire dietary intake. The food groups studied are described above.

As above for PBDEs, human exposure to HBCDD was evaluated in human milk samples from Ghana collected in 2004 and 2009 [147]. Concentrations of ΣHBCDD ranged from <0.01–18 μg kg^{-1} lipid weight and showed no increase between the two study years. In the study undertaken using human milk samples collected in Limpopo province, South Africa, reported above [148], HBCDD was also determined. HBCDD concentrations ranged from <0.2 to 1.4 μg kg^{-1} lipid weight.

7.3 Novel BFRs

7.3.1 Wildlife

Munschy et al. [79] determined four non-PBDE BFRs in muscle and liver of sole from the west coast of France. Concentrations of 1,2-*bis*(2,4,6-tribromophenoxy) ethane (BTBPE), decabromodiphenylethane (DBDPE), hexabromobenzene (HBB), and 2,2′,4,4′,5,5′-hexabromobiphenyl (BB153) ranged from nd to 0.001, nd to 0.004, nd to 0.005, and 0.1 to 0.05 μg kg^{-1} wet weight, respectively, in muscle; and nd to 0.07, nd to 0.02, nd to 0.2, and nd to 0.1 μg kg^{-1} wet weight, respectively, in liver. The slightly higher levels of BB153 may be related to the fact that technical decabromobiphenyl was produced by one company in France until 2000 [158].

Guerra et al. [103] determined concentrations of seven novel BFRs in peregrine falcon eggs from Canada and Spain. The concentration ranges were: 5,6-dibromo-1,10,11,12,13,13-hexachloro-11-tricyclo[8.2.1.02,9]tridecene), 17–22 μg kg^{-1} lipid weight in Canada and not detected in Spain; *bis*(2-ethylhexyl)tetrabromophthalate, 1.1–4.5 μg kg^{-1} lipid weight and not detected in Spain; octabromotrimethylphenylindane, 1.8–29 μg kg^{-1} lipid weight in Canada and 0.8–1.0 μg kg^{-1} lipid weight in Spain; DBDPE, nd to 8.2 μg kg^{-1} lipid weight in Canada and not detected in Spain; BTBPE, 13 μg kg^{-1} lipid weight in Canada and 3.3–7.4 μg kg^{-1} lipid weight in Spain; BB153 (a polybrominated biphenyl congener), 7.2 to 1300 μg kg^{-1} lipid weight in Canada and 3.6–67 μg kg^{-1} lipid weight; and HBB, 4.1–7.2 μg kg^{-1} lipid weight in Canada and 0.6–3.1 μg kg^{-1} lipid weight in Spain, presumably reflecting the different usage patterns of these compounds in and around the two countries.

Relevant information on the known uses and time spans of these compounds is given in Guerra et al. [103].

In the study in white-tailed sea eagle eggs reported above, Nordlöf et al. [93] also determined concentrations of BB153. The geometric means were 20 (LAP), 30 (INL), 100 (SB), and 120 (BP) $\mu g\ kg^{-1}$ lipid weight, once more suggesting somewhat higher concentrations in marine rather than freshwater fish.

7.3.2 Humans

Zheng et al. [128] also determined HBB, PBBs, BTBPE, and DBDPE in human hair samples from South China. Levels declined in the order: occupationally exposed workers > residents in e-waste recycling areas > residents in urban and rural areas. Maximum concentrations were 19, 14, 29, and 365 $\mu g\ kg^{-1}$ dry weight, respectively.

8. FRs in food

Schecter et al. [159] provided an update to earlier US market basket surveys of BDE levels and dietary intake estimates, and also quantified HBCDD for the first time. A total of 31 food types and 310 samples were studied. ΣBDE_{24} concentrations in food varied by food type, ranging from 12 $ng\ kg^{-1}$ wet weight in whole milk to 1550 $ng\ kg^{-1}$ wet weight in tinned sardines and 6210 $ng\ kg^{-1}$ wet weight in butter. ΣHBCDD concentrations ranged from 23 $ng\ kg^{-1}$ wet weight in tinned beef chilli to 593 $ng\ kg^{-1}$ wet weight in tinned sardines. HBCDD was not detected in any dairy sample. Estimated daily dietary intakes were 50 ng for ΣBDE (not only from dairy consumption but also from meat and fish) and 16 ng for HBCDD (primarily from meat).

In the James Bay region of sub-Arctic Canada, Liberda et al. [160] studied BDEs in traditional diets of wild game and fish. A total of 147 samples were analyzed and dietary exposure assessed and compared to the US EPA reference doses. ΣBDE_4 concentrations were up to 230 $\mu g\ kg^{-1}$ lipid weight in birds (mallards) and 595 $\mu g\ kg^{-1}$ lipid weight (whitefish). The reference doses were not exceeded, but modeling of body burdens showed that ΣBDE body burdens increased as access to manufactured goods increased, suggesting household dust as a major exposure route.

Contamination of 10 butter samples from the United States and their wrapping paper was studied by Schecter et al. [161]. They found that one of the samples and its wrapping paper contained very high levels of ΣBDE_{24}, with total concentrations ranging from <0.001 to 42.2 $\mu g\ kg^{-1}$ wet weight. In the wrapper of the most contaminated sample, the ΣBDE_{24} concentration was c.750 $\mu g\ kg^{-1}$ wet weight, with BDE209 predominating, followed by BDE206, BDE196, and BDE207, in that order.

In muscle fillets (edible tissues) of turbot, from sites of Spanish aquaculture, Blanco et al. [162] reported ΣBDE_{27} concentrations 0.5–2.1 $\mu g\ kg^{-1}$ lipid weight, reflecting concentrations in their feed (2.4–4.8 $\mu g\ kg^{-1}$ lipid weight). Predominant congeners were BDE47, BDE99, and BDE100. The proportion of BDE49 increased from 2% in feed to 15% of ΣBDE in fillets, probably as a result of metabolism of higher brominated congeners in these fish. The need to source fish feed with lower levels of contamination in order to reduce human dietary intake was stressed.

Edible marine products ($n = 24$) collected from fish markets in Adelaide, Australia, were analyzed for ΣBDE_{11}. Concentrations ranged from 1.0 to 45 μg kg^{-1} wet weight. BDE209 was the dominant congener, followed by BDE47, BDE99, and BDE100, and the highest levels were seen in imported silverfish and prawns from Vietnam and Thailand [163]. ΣBDE_{11} concentrations in 11 species of fish, shellfish, and shrimps taken from markets close to Laizhou Bay, Shandong province, China, in a PBDE production area ranged from 2.7 to 42 μg kg^{-1} wet weight [164]. The mean dietary intake of PBDEs in this area was 218 ng day^{-1}, while the daily intake via inhalation was 612 ng for men and 455 ng for women. BDE209 contributed, on average, 80% of this intake. Dietary intake and inhalation contributed 29 and 71%, respectively, to the total intake, and human exposure to PBDEs by both routes was the highest reported in the world to date.

Concentrations of ΣBDE_{13} were determined in 299 vegetable and animal-based food samples of 31 species from Shanghai, China, collected in 2008–2009 [165]. ΣBDE_{13} concentrations ranged from nd to 1.25 μg kg^{-1} wet weight, with higher concentrations in animal-based foods than in vegetables, particularly in some fish and shellfish.

Guo et al. [166] analyzed 602 seafood samples from Guandong Province, South China, collected on a market basis in 2004–2005. ΣBDE_{10} concentrations ranged from nd in all sample types to maxima of 5.9 (fish), 1.1 (shrimp), 1.6 (crab), and 2.2 μg kg^{-1} wet weight (mollusk). Dietary intakes were dominated by fish and molluscs, each contributing 45% to the total.

In a study conducted in order to assess exposure of the Belgian population to HBCDD, Goscinny et al. [167] determined HBCDD diastereoisomers in 45 composite food samples from five major food groups. The estimated average daily intake was 0.99 ng kg^{-1} body weight per day (67% γ-HBCDD, 25% α-HBCDD and 8% β-HBCDD). Meat products and "other" products were the major contributors. These concentrations were low in comparison to other EU countries and the estimated daily intake was substantially below proposed threshold values.

Nakagawa et al. [168] determined HBCDD in seafood samples collected in coastal areas of Japan. HBCDD concentrations in Anguilliformes, Perciformes, Clupeiformes, and farmed Salmoniformes ranged from 0.05 to 37, nd to 26, 0.09 to 77, and 1.1–1.3 μg kg^{-1} wet weight, respectively. However, HBCDD was not detected in samples classified as Crustacea, Mollusca, Pleuronectiformes, or Scorpaeniformes, or, if detected, levels were very low. For fish samples with concentrations >20 μg kg^{-1} wet weight, γ-HBCDD was dominant, suggesting exposure to local industrial discharge, while in other wild fish, α-HBCDD was dominant. Based on the median concentrations for Anguilliformes, farmed Salmoniformes, and Perciformes, the daily intake of HBCDD from fish by an average Japanese adult was tentatively calculated to be 3.7, 2.3, and 1.3 ng kg body weight^{-1} day^{-1}, respectively.

9. Dioxin formation

Another major concern over the use of HFRs is the production of polyhalogenated dibenzodioxins and dibenzofurans (Figure 13). Characterization of the toxicity of dioxins followed the "Agent Orange" scandal in Vietnam, when a broad-leaf defoliant contaminated with 2,3,7,8-tetrachlorodibenzo-*p*-dioxin (TCDD) was sprayed by US aircraft to prevent the Vietcong using foliage for cover, resulting in an estimated 1,000,000 deaths and injuries and 50,000 children born with birth defects [169]. The US the National Academy of Sciences has set out a list of diseases that are found to have "sufficient" evidence of an association to Agent Orange, including soft tissue sarcoma, non–Hodgkin lymphoma, Hodgkin disease, and chloracne, and "suggestive" evidence of respiratory cancers (lung, bronchus, larynx, trachea), prostate cancer, multiple myeloma, acute and subacute transient peripheral neuropathy, porphyria cutanea tarda, type 2 diabetes, spina bifida, and acute myelogenous leukemia [170]. This was followed by the industrial accident known as the "Seveso disaster" which focused public attention on the very high toxicity of "dioxins". In Seveso, an aerosol cloud containing between a few hundred grams and a few kilograms of TCDD was released over an 18 km^2 area, resulting in short- and long-term ill-health effects in the exposed population [232].

In the late 1980s, pyrolysis experiments were performed on BFRs and FR systems and the breakdown products analyzed. PBBs, PBDEs, 2,4,6-tribromophenol, pentabromophenol, TBBPA, and tetrabromophthalic anhydride all produced quantifiable yields of polybrominated dibenzodioxins (PBDD) and polybrominated furans (PBDF). Pyrolysis of the FRs alone [171], as well as in combination with polymers [172], was investigated. Although the results are generally not comparable to actual fire situations, and dioxin formation is known to vary widely depending on conditions of heating, the wide variety of conditions found in unwanted fires suggests that the dioxin problem will always be associated with all HFRs, while controlled incineration of waste of polymer materials containing HFRs will always require special treatment (See section on Regulatory Considerations).

More recently it has been shown that BFRs, especially PBDEs, and PBBs, are a major source of toxic tetra- to octa-brominated dioxin and furan contamination [173,174]. Brominated dioxins and furans have been shown to have toxicities similar to their chlorinated counterparts in human cell lines and mammalian species

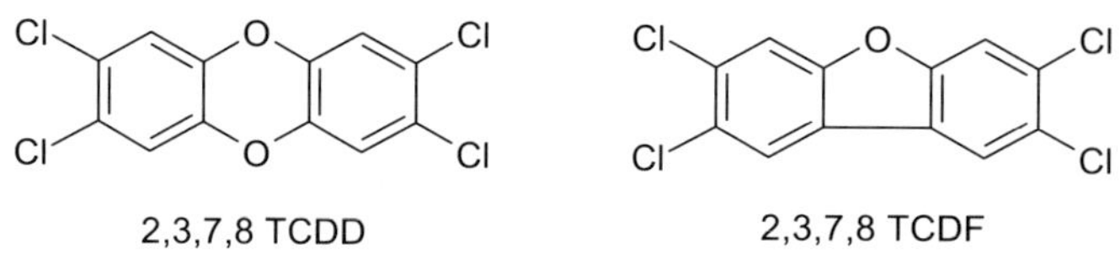

FIGURE 13

Structures of 2,3,7,8-tetrachlorodibenzo-*p*-dioxin (TCDD) and 2,3,7,8-tetrachlorodibenzofuran (TCDF).

[175–177]. Brominated dioxins and furans have been found as contaminants in commercial BFRs, such as PBDEs, TBBPA, and bromophenols [178,179], and following thermal processing of plastics containing BFRs, such as molding and extrusion [173,180,181]. PBDEs in plastics [182] or textiles [183] are converted into PBDD/PBDF by sunlight. Primitive recycling of electronic waste containing PBDEs has been estimated to release up to 0.1% by mass of PBDD/PBDF [174] resulting in excessively high levels of these compounds in soil near such sites.

Personnel involved in the production and recycling of BFRs [184], people living near primitive electronic waste recycling facilities [185–188]; firefighters, and cleanup workers [175,189,190] are all exposed to elevated levels of brominated and chlorinated dioxins and furans. Unwanted fires involving BFRs also produce high levels of PBDD/PBDFs [173,180]. Fire fighters have significantly elevated rates of four types of cancer that are potentially related to exposure to dioxins and furans: multiple myeloma, non–Hodgkin lymphoma, prostate cancer, and testicular cancer [189,191].

10. Harmful effects of halogenated FRs

About 7 years after the discovery of PCBs as an environmental contaminant [192], and 5 years after the accidental rice oil PCB toxification in Japan in 1968 [193], in 1973 PBBs were accidentally distributed as a feed supplement to farm animals in Michigan [16,194]. The PBB brand Firemaster BP-6 was accidentally packed into sacks identified as a farm animal feed supplement and distributed to farms in Michigan. The mistake took nearly a year to discover. Farm animals became increasingly sick, with symptoms which included lethargy, malformation of newborn lambs, reduced milk production, and many more [16,194]. This incident resulted in the emergency slaughter of 1.5 million chickens, more than 30,000 cattle, almost 6000 pigs, and about 1500 sheep. The acute toxic effects of PBB were observed directly in the farm animals, while the effects on the human population of Michigan took longer to appear, and have been the subject of debate ever since. Recent data link PBBs to cancer, neurobehavior, as well as thyroidogenic, embryotoxic, and teratogenic effects [195]. As a result of the toxic effects PBBs had on the farm animals, it is beyond doubt that similar effects would occur in humans. Furthermore, the endocrine disruption effects, such as immunological dysfunction and effects on thyroid functions, related to PBBs, may occur at very low doses. Unfortunately, PBB exposure is not limited to the population of Michigan, but to humans all over the world, despite the restrictions and phasing out of PBBs in commercial products. Very recently, Norrgran and coworkers reported high levels of decabromobiphenyl in cat blood [196], a potential reason for feline hyperthyroidism [197]. This observation has no obvious explanation since it is more than a decade ago that the production of the compound ceased, calling for further work to identify the source of this BFR.

Following the Michigan PBB incident, the human exposure to *tris*(2,3-dibromopropyl) phosphate, often referred to as "tris-BP", caused major turbulence in health science in the later half of the 1970s [198]. Tris-BP is known both as a mutagen and as a carcinogen, but was still used as an additive FR in infant's pajamas. The scientific data on this compound led to changes in the application areas of this compound. However, based on information given by the Chemical Abstracts database, Scifinder, it is still a compound in use as a BFR. Interestingly, the corresponding chlorinated compound, *tris*(2,3-dichloropropyl) phosphate, is also used as a FR. Both the compounds are highly questionable for this use due to their toxic properties [198,199].

The two examples above illustrate the serious adverse health effects of some HFRs on farm animals and humans, respectively. At the same time, other HFRs, such as pentabromotoluene and polybrominated diphenyl ethers gave cause for concern, but the absence of toxicological data meant that the severity of the problem remained unknown [200,201]. This is generally true for most BFRs; although these compounds have been identified in biota, including humans, the gaps in the data make any assessment of their potential harm inconclusive.

Of all the types of HFRs, the most detailed investigations have focused on PBDEs. The effects of both the three commercially produced groups—penta-mix, octa-mix and deca-mix PBDE—and of individual PBDE congeners (specific isomers) have been studied. The most recent update on PBDEs and their toxicological profile is given in a European Food Safety Authority (EFSA) opinion from 2011 [202]. Numerous other reviews have addressed the toxicity issues of PBDEs [203–206]. The neurodevelopmental effects of PBDEs have been reported by several researchers, such as causing learning disabilities [202]. PBDEs are also linked to thyroid and liver cell effects [202,207], and to reprotoxic effects, including effects on sperm. BDE209 was shown to induce cancer in male rats as determined in a major US National Toxicity Program study [208].

Approximately 15 years after the first report on PBDEs in biota (fresh water fish from the Sweden) [39] a report came out showing almost exponential increases in the levels of PBDEs in mothers' milk [209]. Soon after that, far higher concentrations were determined in humans living in the United States [210–212]. In fact, the levels of individual PBDE congeners exceeded that of the common environmental pollutant, PCB. Human exposure, in particular of young children, together with HFR toxicity, led to the banning of penta-mix and octa-mix PBDE within the European Union [213] and voluntary actions being taken in the United States [214]. In addition, the permitted uses of deca-BDE were restricted [215]. Penta- and octa-PBDEs have been listed for elimination by the Stockholm Convention on Persistent Organic Pollutants confirming their persistence, bioaccumulativity, toxicity (PBT), and long-range transport [216]. Additional matters of concern are the transfer of PBDEs through the placenta to the growing fetus [217] and the formation of strongly thyroidogenic metabolites of PBDEs that are retained in human blood [218]. Of all HFRs, the PBDEs are the most well researched.

HBCDDs are presently under discussion for inclusion in the list of POPs scheduled for elimination by the Stockholm Convention. The background for this is the PBT characteristics of the HBCDD isomers. Again, a very recent EFSA opinion is reviewing all scientific data regarding their properties, environmental fate, occurrence in food, and toxicity [2] of commercial mixtures of HBCDD and individual HBCDD isomers. In summary, exposure to technical-grade HBCDD during development in rodents affects the nervous system, with subsequent behavioral changes [219]. The resultant changes in thyroid hormone regulation may play a critical role in the onset of the observed effects [220], while effects on the immune system have also been reported [221]. Further, a two-generation reproduction study indicated increased pup mortality in rats, during lactation [222]. However, HBCDD has a very low acute toxicity. It is important to recognize that the toxicology of HBCDD isomers is rudimentary. In particular, technical grades of HBCDD, in which the γ-isomer predominates, differ from the environmental composition of HBCDD, where the α-isomer predominates. Accordingly, isomer-specific toxicological data are required, but do not currently exist.

The environmental concentration of HBCDD varies, but is, in general, still lower than for several PBDE congeners. In humans, HBCDDs are present in the low nanogram per gram fat levels, but HBCDDs have been shown to increase over time in mothers' milk among in nursing mothers in Stockholm [223], with an indication that they may level off during the 2000s.

The most abundant BFR is TBBPA. This compound is bioavailable and readily taken up in the gastrointestinal tract. It has a short half-life in humans (a couple of days) and presumably also in other mammals. Accordingly, the compound may be unlikely to undergo environmental transformations, but is readily metabolized and excreted. The widespread use of TBBPA may, however, lead to pseudopersistency of the compound, i.e. continuous exposure rather than the persistency of TBBPA leads to the establishment of steady-state levels in human blood. So far TBBPA has been of less concern than many of the other HFRs, although it has been linked to several toxicological effects, as presented by EFSA 2011 [224]. It has very low acute toxicity, and does not seem to produce the neurodevelopmental effects caused by PBDEs and HBCDDDs [225]. The major toxicologically relevant effect relates to thyroid hormone disruption, which is due to the chemical structure similarity of TBBPA and thyroxin. TBBPA may be immunotoxic, while no reprotoxic effects have hitherto been reported for the compound [225]. No studies of the potential carcinogenicity of the compound have been performed, but it seems not to be genotoxic, based on study results so far.

TBBPA is used for preparation of a number of derivatives, e.g. TBBPA *bis*(2-hydroxyethyl) ether, TBBPA bis-allyl ether, TBBPA *bis*(2,3-dibromopropyl ether) (TBBPA-BDBPE), and TBBPA *bis*(glycidyl ether). For other TBBPA derivatives there is some uncertainty about their commercial use as FRs: TBBPA bisacrylate, 3,3′,5,5′-TBBPA bis-acetate, and TBBPA bis(2-hydroxyethyl)ether bisacrylate. Almost no toxicological information has been obtained for any of these new BFRs, making hazard identification impossible. However, based on structure–activity comparisons, it is reasonable to believe that TBBPA-BDBPE, used as an additive in polymers, has toxicological properties of some concern. This is due to the 2,3-dibromopropyl group in which the bromine substituents will readily undergo substitution reactions, facilitating reaction with both deoxyribonucleic acid and proteins (CAS no 1163-19-5).

In total, more than 50 BFRs are listed today as being in common use [226] but among these there are still several for which the actual applications as FRs are unknown. However, many have been identified in the environment, either in air, soil, sediments, or in water or in biota (wildlife or humans). These chemicals are regarded as emerging BFRs, in contrast to the established BFRs for which the database of environmental occurrence and health and environmental effects is extensive (cf. above). Others have been indicated as novel BFRs and these have only been indicated through chemical analysis of materials and products [227]. Typically, the majority of BFRs have very limited, if any, retrievable toxicological and/or ecotoxicological information in the scientific literature. This means that it is not possible to assess either their environmental or their health risks. This is a typical situation for a very large number of chemicals in use. In general terms, a man-made chemical is regarded as safe until data identify any adverse effects (so the absence of toxicity studies is taken to indicate the safety of a compound). The new European REACH regulations will eventually require each man-made chemical to undergo toxicological assessment once established on the market. However, it is questionable if REACH will be able to change this situation, at least not in the near future.

Based on the present knowledge we can see that several BFRs are both persistent and bioaccumulative. In the future, these chemicals will cause problems if they are used as BFRs, since environmental levels and concentrations in biota will potentially increase to a point where adverse effects will be observed. For others, toxicological data is required to assess their environmental effects or health hazard. The risk for low-dose effects on hormones and hormone-mediated actions call for such testing. Endocrine effects include the effects on reproduction (sperm counts, sex organ malformation), neurodevelopment, thyroid, puberty, cancer, and numerous others [228]. Hence it is not a matter of acute toxicity but of chronic intoxication. This is particularly serious since most of these chemicals are transferred via the placenta to the fetus, with the risks associated with exposure during early development. A summary of the serious concern regarding both brominated and chlorinated FRs and their environmental and health effects were presented and signed by more than 100 scientists around the world, as the "San Antonio Statement 2010 on Brominated and Chlorinated Flame Retardants" [3]. This document states that many brominated and chlorinated FRs appear to be persistent and bioaccumulative, resulting in food web contamination including human milk; lack adequate toxicity information, but the available data raise concerns; have been incorporated into products, even though comprehensive toxicological information is lacking; are released to the indoor and outdoor environments; and currently cannot be handled safely at end of life. It goes on to recommend improved availability of and access to information on HFRs, the need to make consumers aware of their presence, the need to identify safer alternatives, and particularly to ensure that current and alternative chemical FRs do not have hazardous properties such as mutagenicity, carcinogenicity, or adverse effects on the reproductive, developmental, endocrine, immune, or nervous systems.

11. CONCLUSIONS

The widespread use of synthetic polymers has increased the flammability of our built environment. The demand to reduce energy consumption by replacing traditional building materials with flammable polymeric foam insulation materials looks set to exacerbate the problem. To ensure fire safety, regulations place limitations on the flammability of materials which may be present in prescribed high-risk/high-hazard situations, such as mass transport, high-rise buildings, furnishings, and E & E equipment. The incorporation of fire retardants into synthetic polymers reduces their flammability; in effect, this means that more inherently flammable polymers are able to meet the regulatory requirements for use in such high-risk applications.

Some fire retardants, such as hydrated mineral fillers, nanoclays, or alums are naturally occurring, and have low toxicity. Others, such as ammonium polyphosphate, are also of low toxicity and are rapidly broken down in the environment. The problem group, including many of those described here, are those which are environmentally persistent, bioaccumulative, and toxic (i.e. have PBT properties).

HFRs release hydrogen chloride or hydrogen bromide, which inhibits the gas phase flame reactions delaying ignition, and/or reducing the heat release, while increasing the yields of products of incomplete combustion, such as carbon monoxide, hydrogen cyanide, organoirritants, polycyclic aromatic hydrocarbons, and soot. If the item containing the FR was always the only item involved in the fire, and its FR treatment led to extinguishment, this would not be a problem. However, most fire deaths result from inhalation of toxic fire effluents, and HFRs increase the toxicity of these effluents.

At loadings of 5–20% halogen, sufficient flame inhibition is observed to alter the performance in standard tests from a fail to a pass. Thus, finding an HFR that can be incorporated into the flammable polymer at a suitable loading without detriment to the physical properties provides a quick and easy solution. Unfortunately, this ease of incorporation is matched by an ease of release of the FR into the surroundings, increasing human exposure through direct contact, inhalation, and especially through household dust (ingestion and inhalation).

In contrast, the halogen-free fire retardants are often polymer specific—melamine is an excellent fire retardant for PU foam, but useless in PP—in general, specialist knowledge is required to develop and optimize halogen-free fire- retardant formulations.

In addition to fire-retardant action often being specific to particular classes of polymers, it can also be specific to particular aspects of flammability, such as ignitability, flame spread, or heat release rate. For example, the new generation of nanoparticulate fire retardants often reduce the time to ignition, but, in doing so, they also reduce the heat release rate, slowing down burning. Regulatory tests often emphasize just one of these aspects.

In the future, as petrochemical resources become more scarce, an additional attribute of any "green fire retardant" must be that the polymer and FR additive should be able to be recycled, preferably retaining its fire retardancy over several cycles of reuse. At present, many fire retardants are incompatible with the recycling processes, and most of the world's stock of polymer recyclate is contaminated with BFRs and their breakdown products [229].

A huge volume of painstaking analytical chemistry has been undertaken to map the presence of BFRs across human and natural environments. This indicates the extraordinary stability (or persistence) of these molecules. It is unclear whether the leveling off observed in some studies is a consequence of better emission controls and slow dispersion processes, or gradual breakdown. The distribution of BDE congeners certainly seems to be indicative of some debromination of the higher brominated congeners, while the patterns of global distribution indicate the effectiveness of the Earth's transport processes, particularly via the atmosphere. The routes from manufacture to the environment remain the subject of conjecture. Careless manufacture and processing leading to environmental discharges during production are believed to be in decline, particularly in Europe and America. The prevalence of BFRs in house dust suggests breakdown of domestic articles (FR furniture and fabrics, evaporation/pyrolysis from heated electrical electronic devices, etc.), while the presence of BFRs in the food chain and in humans confirms the severity of the problem.

Epidemiology of nonfatal illness in diverse human populations is extremely challenging. Diet, lifestyle, exposure, and metabolism all differ so greatly from individual to individual, that establishing clear trends is problematic. Such problems allowed the tobacco industry to get away with defending the indefensible for decades. Human exposure to BFRs has been correlated to neurodevelopmental effects, leading to learning disabilities and effects on reproduction.

The large-scale exposure of domestic animals and humans to PBBs in Michigan categorically demonstrated their potential to cause sickness and birth defects, while they were also suspected of causing cancer as well as neurobehavioral, thyroidogenic, embryotoxic, and teratogenic effects. Tris-BP was also identified as a mutagen and carcinogen, although since it was voluntarily withdrawn, not banned outright, it seems that both *tris* and its homologs continue to be in use as FRs today.

Recently, HBCDD has been considered for inclusion by the Stockholm Convention on Persistent Organic Pollutants. Studies have shown that it causes behavioral changes in mice and disrupts the thyroid hormone regulation and the immune system in rats.

The most abundant BFR is TBBPA. However, despite its ready bioavailability, its short half-life in humans and low acute toxicity suggest that it does not pose a major problem. Further, with two reactive functional groups, it is readily incorporated into the polymer backbone, where it is believed to be retained for the lifetime of the polymer. It may contribute to dioxin release during fires.

Millions of tons of HFRs are now distributed throughout the environment, originating in the anthrosphere, but accumulating in the biosphere at the top of the food chain, in fatty tissue. It seems that the most effective means mammals have for eliminating BFRs is to express them through breast milk due to its high lipid content, unwittingly passing them on to their offspring. Their deployment has allowed cheaper, more flammable polymers to be used in higher risk situations. They undoubtedly present hazards to human health, and may continue to do so until their eventual decomposition. It is evident that some of the highest discharges to the environment result from recycling processes in the Far East and Africa. Since the largest quantities of HFRs are probably incorporated into products currently in use, unless urgent action is taken to properly enforce regulations such as the European WEEE directive, levels are set to rise before they start to fall, and the legacy of BFRs may be with us longer than the 40 years they have currently been on the planet.

References

[1] De Wit CA. An overview of brominated flame retardants in the environment. Chemosphere 2002;46:583.

[2] Shaw SD, Blum A, Weber R, Kannan K, Rich D, Lucas D, et al. Halogenated flame retardants: do the fire safety benefits justify the risks? Rev Environ Health 2010;25:261–305.

[3] DiGangi J, Blum A, Bergman Å, de Wit CA, Lucas D, Mortimer D, et al. San Antonio statement on brominated and chlorinated flame retardants. Environ Health Perspect 2010;118:A 516–36.

[4] Stockholm convention on persistent organic pollutants, Stockholm, 22 May 2001 (depositary notification). United Nations; 2009. Adoption of amendments to annexes a, b and c.

[5] Sjödin A, Päpke O, McGahee E, Focant J-, Jones RS, Pless-Mulloli T, et al. Concentration of polybrominated diphenyl ethers (PBDEs) in household dust from various countries. Chemosphere 2008;73:S131–6.

[6] U.S. Consumer Product Safety Commission. NEWS from CPSC - CPSC bans TRIS-treated children's garments. U.S.Consumer Product Safety Commission; 1977, http://www.cpsc.gov/cpscpub/prerel/prhtml77/77030.html [accessed 12.10.10].

[7] Stec A, Hull R, editors. Fire toxicity. Cambridge: Woodhead Publishing; 2010.

[8] Emsley A, Lim L, Stevens G, Williams P. International fire statistics and the potential benefits of fire counter-measures. Report commissioned by the European Flame Retardants Association. UK: University of Surrey; 2005.

[9] Arcadis EBRC. Report for European Commission (DG health and consumers) - evaluation of data on flame retardants in consumer products – final report 17.020200/09/549040. Brussels, http://ec.europa.eu/consumers/safety/news/flame_retardant_substances_study_en.pdf; 2011.

[10] World fire statistics No. 27. Geneva: Geneva Association; 2011.

[11] (Market Research Report) Flame retardant chemicals. Stamford CT: BCC Research; 2006.

[12] Hull TR, Witkowski A, Hollingbery L. Fire retardant action of mineral fillers. Polym Degrad Stab 2011;96:1462–9.

[13] Gay-Lussac JL. Ann Chim Phys 1821;18:211.

[14] Jensen S. Report of a new chemical hazard. New Sci 1966;32:612.

[15] Carter LJ. Michigan's PBB incident: chemical mix up leads to disaster. Science 1976;192:240–3.

[16] Fries GF. The PBB episode in Michigan: an overall appraisal. Crit Rev Toxicol 1985; 16(2):105–56.
[17] Green J. 25 years of flame retarding plastics. J Fire Sci 1997;15:52.
[18] EFSA Panel on Contaminants in the Food Chain (CONTAM). Scientific opinion on brominated flame retardants (BFRs) in food: emerging and novel BFRs in food. EFSA J 10, 2908 (125 pp.).
[19] Law RJ, Kohler M, Heeb NV, Gerecke AC, Schmid P, Voorspoels S, et al. Hexabromocyclododecane challenges scientists and regulators. Environ Sci Technol 2005;39: 281A–7A.
[20] Morris S, Allchin CR, Zegers BN, Haftka JJH, Boon JP, Belpaire C, et al. Distribution and fate of HBCD and TBBPA flame retardants in North sea estuaries and aquatic food webs. Environ Sci Technol 2004;38:5497–504.
[21] Law K, Palace VP, Halldorson T, Danell R, Wautier K, Evans B, et al. Dietary accumulation of hexabromocyclododecanediastereoisomers in juvenile rainbow trout (*Oncorhynchus mykiss*) I: bioaccumulation parameters and evidence of bioisomerization. Environ Toxicol Chem 2006;25:1757–61.
[22] Bocchini S, Camino Halogenated G. Flame retardants in fire retardancy of polymeric materials. In: Wilkie CA, Morgan AB, editors. 2nd ed. Boco Raton: CRC Press; 2010.
[23] Zhang Xiu-Lan, Xiao-Jun Luo, She-Jun Chen, Jiang-Ping Wu, Bi-Xian Mai. Spatial distribution and vertical profile of polybrominated diphenyl ethers, tetrabromobisphenol A, and decabromodiphenylethane in river sediment from an industrialized region of South China. Environ Pollut 2009;157:1917–23.
[24] Tange L, Borms R, Goebelbecker S, Georlette P, Reznick G, Yaakov YB. Sustainable use of brominated flame retardants in plastics. In: Proceedings of the 5th European conference on additives and colors; 2007.
[25] Schnipper A, Smith-Hansen L, Thomsen SE. Reduced combustion efficiency of chlorinated compounds, resulting in higher yields of carbon monoxide. Fire Mater 1995; 19:61–4.
[26] Babushok V, Tsang W, Linteris GT, Reinelt D. Chemical limits to flame inhibition. Combust Flame 1998;115:551–60.
[27] Newton PE, Schroeder RE, Zwick L, Serex T. Inhalation developmental toxicity studies in rats with antimony trioxide (Sb2O3). Toxicologist 2004;78(1-S):38.
[28] Non-halogen fire safety solutions - phosphorus, inorganic and nitrogen flame retardants association, Cefic, Brussels.http://www.pinfa.org/.
[29] High Density Electronic Packaging Users Group (HDPUG) (trade association for electronic component packaging) providing guidance on halogen-free alternatives. http://www.hdpug.org/content/completed-projects#HalogenFree.
[30] Chivas C, Guillaume E, Sainrat A, Barbosa V. Fire Saf J 2009;44:801.
[31] Martini P, Spearpoint MJ, Ingham PE. Low-cost wool-based fire blocking inter-liners for upholstered furniture. Fire Saf J 2010;45:238.
[32] SRI Consulting, Report on flame retardants, Published December 2008.
[33] Tewarson A. Combustion efficiency and its radiative component. Fire Saf J 2004;39: 131.
[34] Castrovinci A, Lavaselli M, Camino G. Recycling and disposal of flame retardant materials. In: Horrocks AR, Price D, editors. Advances in fire retardant materials. Cambridge: Woodhead Publishing; 2008. pp. 213–54.

[35] Reade A. The treatment of WEEE plastics containing brominated flame retardants by United Kingdom approved authorised treatment facilities. MSc dissertation. University of Central Lancashire; 2010.
[36] Transboundary shipment of waste electrical and electronic equipment/electronic scrap – optimization of material flows and control, Federal Environmental Agency (Germany) (UBA), http://www.umweltbundesamt.de/uba-info-medien/mysql_medien.php?anfrage=Kennummer&Suchwort=3769.
[37] European Environment Bureau (EEB) and the European Consumers' Organisation (BEUC) joint letter to Environment Ministers on European Ecolabel voting decisions. March 2009. Ref: BMO/MGO/2009077/go.
[38] Hollingbery LA, Hull TR. The fire retardant effects of huntite in natural mixtures with hydromagnesite. Polym Degrad Stab 2012;97:504–12.
[39] Andersson Ö, Blomkvist G. Polybrominated aromatic pollutants found in fish in Sweden. Chemosphere 1981;10:1051–60.
[40] Law R, Alaee M, Allchin C, Boon J, Lebeuf M, Lepom P, et al. Levels and trends of polybrominated diphenylethers and other brominated flame retardants in wildlife. Environ Int 2003;29:757–70.
[41] Law RJ, Allchin CR, de Boer J, Covaci A, Herzke D, Lepom P, et al. Levels and trends of brominated flame retardants in the European environment. Chemosphere 2006;64: 187–208.
[42] Law RJ, Herzke D, Harrad S, Morris S, Bersuder P, Allchin CR. Levels and trends of HBCD and BDEs in the European and Asian environments, with some information for other BFRs. Chemosphere 2008;73:223–41.
[43] Tanabe S, Ramu K, Isobe T, Takahashi S. Brominated flame retardants in the environment of Asia-Pacific: an overview of spatial and temporal trends. J Environ Monit 2008;10:188–97.
[44] Law RJ. Chapter 2: brominated flame retardants. In: Harrad S, editor. Persistent organic pollutants. John Wiley and Sons; 2010, ISBN 1-4051-6930-3. p. 288. pp. 5–24.
[45] Chen D, Hale RC. A global review of polybrominated diphenyl ether flame retardant contamination in birds. Environ Int 2010;36:800–11.
[46] Law RJ, Herzke D. Current levels and trends of brominated flame retardants in the environment. In: Eljarrat E, Barceló D, editors. Brominated flame retardants. HdbEnvChem, vol. 16. Heidelberg, Germany: Springer-Verlag Berlin; 2011. pp. 123–40. http://dx.doi.org/10.1007/698_2010_82.
[47] Domínguez AA, Law RJ, Herzke D, de Boer J. Bioaccumulation of brominated flame retardants. In: Eljarrat E, Barceló D, editors. Brominated flame retardants. HdbEnvChem, vol. 16. Heidelberg, Germany: Springer-Verlag Berlin; 2011. pp. 141–85. http://dx.doi.org/10.1007/698_2010_95.
[48] Covaci A, Harrad S, Abdallah MA-, Ali N, Law RJ, Herzke D, et al. Novel brominated flame retardants: a review of their analysis, environmental fate and behaviour. Environ Int 2011;37:532–56.
[49] Vorkamp K, Rigét FF, Bossi R, Dietz R. Temporal trends of hexabromocyclododecane, polybrominated diphenyl ethers and polychlorinated biphenyls in ringed seals from East Greenland. Environ Sci Technol 2011;45:1243–9.
[50] Yogui GT, Sericano JL, Montone RC. Accumulation of semivolatile organic compounds in Antarctic vegetation: a case study of polybrominated diphenyl ethers. Sci Total Environ 2011;409:3902–8.

[51] Wang Y, Yang R, Wang T, Zhang Q, Li Y, Jiang G. Assessment of polychlorinated biphenyls and polybrominated diphenyl ethers in Tibetan butter. Chemosphere 2010;78: 772–7.

[52] Domingo JL. Polybrominated diphenyl ethers in food and human dietary exposure: a review of the recent scientific literature. Food Chem Toxicol 2012;50:238–49.

[53] Abdallah MA-E, HarradS. Personal exposure to HBCDs and its degradation products via ingestion of indoor dust. Environ Int 2009;35:870–6.

[54] Harrad S, de Wit CA, Abdallah MA-E, Bergh C, Björklund JA, Covaci A, et al. Indoor contamination with hexabromocyclododecanes, polybrominated diphenyl ethers, and perfluoroalkyl compounds: an important exposure pathway for people. Environ Sci Technol 2010;44:3221–31.

[55] Johnson PI, Stapleton HM, Sjödin A, Meeker JD. Relationships between polybrominated diphenyl ether concentrations in house dust and serum. Environ Sci Technol 2010;44:5627–32.

[56] Ni H-G, Zeng H, Tao S, Zeng EY. Environmental and human exposure to persistent halogenated compounds derived from e-waste in China. Environ Toxicol Chem 2010;29:1237–47.

[57] BBC. Europe still exporting electronic waste, www.bbc.co.uk/news/world-europe-10846395; 2010 [accessed 04.08.10].

[58] Law RJ, Barry J, Bersuder P, Barber JL, Deaville R, Reid RJ, et al. Levels and trends of brominated diphenyl ethers in blubber of harbor porpoises (*Phocoena phocoena*) from the UK, 1992-2008. Environ Sci Technol 2010;44:4447–51, http://dx.doi.org/10.1021/es100140q.

[59] Greig DJ, Ylitalo GM, Wheeler EA, Boyd D, Gulland FMD, Yanagida GK, et al. Geography and stage of development affect persistent organic pollutants in stranded and wild-caught harbour seal pups from central California. Sci Total Environ 2011;409: 3537–47.

[60] Frouin H, Lebeuf M, Hammill M, Sjare B, Fournier M. PBDEs in serum and blubber of harbour, grey and harp seal pups from Eastern Canada. Chemosphere 2011;82:663–9.

[61] Isobe T, Oshihoi T, Hamada H, Nakayama K, Yamada TK, Tajima Y, et al. Contamination status of POPs and BFRs and relationship with parasitic infection in finless porpoises (*Neophocoena phocaenoides*) from Seto Island Inland Sea and Omura Bay, Japan. Mar Pollut Bull 2011;63:564–71.

[62] Park B-K, Park G-J, An Y-R, Choi H-G, Kim GB, Moon H-B. Organohalogen contaminants in finless porpoises (*Neophocaena phocaenoides*) from Korean coastal water: contamination status, maternal transfer and ecotoxicological implications. Mar Pollut Bull 2010;60:768–74.

[63] Nomiyama K, Eguchi A, Mizukawa H, Ochiai M, Murata S, Sorreya M, et al. Anthropogenic and naturally occurring polybrominated phenolic compounds in the blood of cetaceans stranded along Japanese coastal waters. Environ Pollut 2011;159:3364–73.

[64] Weijs L, Das K, Neels H, Blust R, Covaci A. Occurrence of anthropogenic and naturally-produced organohalogenated compounds in tissues of Black sea harbour porpoises. Mar Pollut Bull 2010;60:725–31.

[65] Weijs L, van Elk C, Das K, Blust R, Covaci A. Persistent organic pollutants and methoxylated PBDEs in harbour porpoises from the North Sea from 1990 to 2008: young wildlife at risk ? Sci Total Environ 2010b;409:228–37.

[66] Yordy JE, Pabst DA, McLellan WA, Wells RS, Rowles TK, Kucklick JR. Tissue-specific distribution and whole-body burden estimates of persistent organic pollutants in the bottlenose dolphin (*Tursiops truncatus*). Environ Toxicol Chem 2010;29:1263–73.

[67] Quinete N, Lavandier R, Dias P, Taniguchi S, Montone R, Moreira I. Specific profiles of polybrominated diphenylethers (PBDEs) and polychlorinated biphenyls (PCBs) in fish and tucuxi dolphins from the estuary of Paraíba do Sul River, Southeastern Brazil. Mar Pollut Bull 2011;62:440–6.

[68] Yogui GT, Santos MCO, Bertozzi CP, Sericano JL, Montone RC. PBDEs in the blubber of marine mammals from coastal areas of São Paulo, Brazil, southwestern Atlantic. Mar Pollut Bull 2011;62:2666–70.

[69] Kim GB, Stapleton HM. PBDEs, methoxylated PBDEs and HBCDs in Japanese common squid (*Todarodes pacificus*) from Korean offshore waters. Mar Pollut Bull 2010; 60:935–40.

[70] Takahashi S, Oshihoi T, Ramu K, Isobe T, Ohmori K, Kubodera T, et al. Organohalogen compounds in deep-sea fishes from the western North Pacific, off-Tohoku, Japan: contamination status and bioaccumulation profiles. Mar Pollut Bull 2010;60:187–96.

[71] Ueno D, Isobe T, Ramu K, Tanabe S, Alaee M, Marvin C, et al. Spatial distribution of hexabromocyclododecanes (HBCDs), polybrominated diphenyl ethers (PBDEs) and organochlorines in bivalves from Japanese coastal waters. Chemosphere 2010;78: 1213–9.

[72] Nomiyama K, Uchiyama Y, Horiuchi S, Eguchi A, Mizukawa H, Hirata SH, et al. Organohalogen compounds and their metabolites in the blood of Japanese amberjack (*Seriola quinqueradiata*) and scalloped hammerhead shark (*Sphyrna lewini*) from Japanese coastal waters. Chemosphere 2011;85:315–21.

[73] Montory M, Habit E, Fernandez P, Grimalt JO, Barra R. PCBs and PBDEs in wild Chinook salmon (*Oncorhynchus tshawytscha*) in the Northern Patagonia, Chile. Chemosphere 2010;78:1193–9.

[74] Sloan CA, Anulacion BF, Bolton JL, Boyd D, Olson OP, Sol SY, et al. Polybrominated diphenyl ethers in outmigrant juvenile Chinook salmon from the lower Columbia River and estuary and Puget Sound, Washington. Archives Environ Contam Toxicol 2010; 58:403–14.

[75] Ondarza PM, Gonzalez M, Fillmann G, Miglioranza KSB. Polybrominated diphenyl ethers and organochlorine compound levels in brown trout (*Salmo trutta*) from Andean Patagonia, Argentina. Chemosphere 2011;83:1597–602.

[76] Skinner LC. Distributions of polyhalogenated compounds in Hudson River (New York, USA) fish in relation to human uses along the river. Environ Pollut 2011;159:2565–74.

[77] Ikonomou MG, Teas HJ, Gerlach R, Higgs D, Addison RF. Residues of PBDEs in northeastern Pacific marine fish: evidence for spatial and temporal trends. Environ Toxicol Chem 2011;30:1261–71.

[78] Ben Ameur W, Hassine SB, Eljarrat E, El Megdiche Y, Trabelsi S, Hammami B, et al. Polybrominated diphenyl ethers and their methoxylated analogs in mullet (*Mugil cephalus*) and sea bass (*Dicentrarchus labrax*) from Bizerte Lagoon, Tunisia. Mar Environ Res 2011;72:258–64.

[79] Munschy C, Héas-Moisan K, Tixier C, Boulesteix L, Morin J. Classic and novel brominated flame retardants (BFRs) in common sole (*Solea solea*) from main nursery zones along the French coasts. Sci Total Environ 2011;409:4618–27.

[80] Viganò L, Rocioli C, Guzzella L. Decabromodiphenyl ether (BDE-209) enters the food web of the River Po and is metabolically debrominated in resident cyprinid fishes. Sci Total Environ 2011;409:4966–72.
[81] Gewurtz SB, McGoldrick DJ, Clark MG, Keir MJ, Malecki MM, Gledhill M, et al. Spatial trends of polybrominated diphenyl ethers in Canadian fish and implications for long-term monitoring. Environ Toxicol Chem 2011;30:1564–75.
[82] Carlsson P, Herzke D, Wedborg M, Gabrielsen GW. Environmental pollutants in the Swedish marine ecosystem, with special emphasis on polybrominated diphenyl ethers. Chemosphere 2011;82:1286–92.
[83] Li Q, Yan C, Luo Z, Zhang X. Occurrence and levels of polybrominated diphenyl ethers (PBDEs) in recent sediments and marine organisms from Xiamen offshore areas, China. Mar Pollut Bull 2010;60:464–9.
[84] Liu Y-P, Li J-G, Zhao Y-F, Wen S, Huang F-F, Wu Y-N. Polybrominated diphenyl ethers (PBDEs) and indicator polychlorinated biphenyls (PCBs) in marine fish from four areas of China. Chemosphere 2011;83:168–74.
[85] Zhou ZM. Implement of administrative measure on the control of pollution caused by electronic information products and the exemption of deca-BDE mixture. Flame Retard Mater Technol 2006;4:15–6 [in Chinese].
[86] Xia C, Lam JCW, Wu X, Sun L, Xie Z, Lam PKS. Levels and distribution of polybrominated diphenyl ethers (PBDEs) in marine fishes from Chinese coastal waters. Chemosphere 2011;82:18–24.
[87] Tapie N, Le Menach K, Pasquaud S, Elie P, Devier MH, Budzinski H. PBDE and PCB contamination of eels from the Gironde estuary: from glass eels to silver eels. Chemosphere 2011;83:175–85.
[88] de Boer J, Dao QT, van Leeuwen SPJ, Kotterman MJJ, Schobben JHS. Thirty year monitoring of PCBs, organochlorine pesticides and tetrabromodiphenylether in eels from the Netherlands. Environ Pollut 2010;158:1228–36.
[89] McHugh B, Poole R, Corcoran J, Anninou P, Boyle B, Joyce E, et al. The occurrence of persistent chlorinated and brominated organic contaminants in the European eel (*Anguilla anguilla*) in Irish waters. Chemosphere 2010;79:305–13.
[90] MacGregor K, Oliver IW, Harris L, Ridgway IM. Persistent organic pollutants (PCB, DDT, HCH, HCB & BDE) in eels (*Anguilla anguilla*) in Scotland: current levels and temporal trends. Environ Pollut 2010;158:2402–11.
[91] Roosens L, Geeraerts C, Belpaire C, Van Pelt I, Neels H, Covaci A. Spatial variations in the levels and isomeric patterns of PBDEs and HBCD in the European eel in Flanders. Environ Int 2010;36:415–23.
[92] Jaspers VLB, Rodriguez FS, Boertmann D, Sonne C, Dietz R, Rasmussen LM, et al. Body feathers as a potential new biomonitoring tool in raptors: a study on organohalogenated contaminants in different feather types and preen oil of West Greenland white-tailed eagles (*Haliaeetus albicilla*). Environ Int 2011;37:1349–56.
[93] Nordlöf U, Helander B, Bignert A, Asplund L. Levels of brominated flame retardants and methoxylated diphenyl ethers in eggs of white-tailed sea eagles breeding in different regions of Sweden. Sci Total Environ 2010;409:238–46.
[94] Eulaers I, Covaci A, Hofman J, Nygård T, Halley DJ, Pinxten R, et al. A comparison of non-destructive sampling strategies to assess the exposure of white-tailed sea eagle nestlings (*Haliaeetus albicilla*) to persistent organic pollutants. Sci Total Environ 2011;410:258–65.

[95] Van den Steen E, Pinxten R, Covaci A, Carere C, Eeva T, Heeb P, et al. The use of blue tit eggs as a biomonitoring tool for organohalogenated pollutants in the European environment. Sci Total Environ 2010;408:1451–7.

[96] Malik RN, Moeckel C, Jones KC, Hughes D. Polybrominated diphenyl ethers (PBDEs) in feathers of colonial water-bird species from Pakistan. Environ Pollut 2011;159: 3044–50.

[97] Van den Steen E, Poisbleau M, Demongin L, Covaci A, Dirtu AC, Pinxten R, et al. Organohalogenated contaminants in eggs of rockhopper penguins (*Eudyptes chrysocome*) and imperial shags (*Phalacrocorax atriceps*) from the Falkland Islands. Sci Total Environ 2011;409:2838–44.

[98] Lavoie RA, Champoux L, Rail J-F, Lean DRS. Organochlorines, brominated flame retardants and mercury levels in six seabird species from the Gulf of St. Lawrence (Canada) relationships with feeding ecology, migration and molt. Environ Pollut 2010;158: 2189–99.

[99] Sørmo EG, Lie E, Ruus A, Gaustad H, Skaare JU, Jenssen BM. Trophic level determines levels of brominated flame-retardants in coastal herring gulls. Ecotoxicol Environ Saf 2011;74:2091–8.

[100] Leat EHK, Bourgeon S, Borgå K, Strøm H, Hanssen SA, Gabrielsen GW, et al. Effects of environmental exposure and diet on levels of persistent organic pollutants (POPs) in eggs of a top predator in the North Atlantic in 1980 and 2008. Environ Pollut 2011; 159:1222–8.

[101] Johansson A-K, Sellström U, Lindberg P, Bignert A, de Wit CA. Temporal trends of polybrominated diphenyl ethers and hexabromocyclododecane in Swedish peregrine falcon (*Falco peregrinus peregrinus*) eggs. Environ Int 2011;37:678–86.

[102] Fernie KJ, Letcher RJ. Historical contaminants, flame retardants, and halogenated phenolic compounds in peregrine falcon (*Falco peregrinus*) nestlings in the Canadian Great Lakes basin. Environ Sci Technol 2010;44:3520–6.

[103] Guerra P, Alaee M, Jiménez B, Pacepavicius G, Marvin C, MacInnis G, et al. Emerging and historical brominated flame retardants in peregrine falcon (*Falco peregrinus*) eggs from Canada and Spain. Environ Int 2012;40:179–86.

[104] Drouillard K, Fernie K, Letcher R, Shutt L, Ryan M, Gebink W, et al. Bioaccumulation and biotransformation of sixty-one polychlorinated biphenyl (PCB) and four polybrominated diphenyl ether (PBDE) congeners in juvenile American kestrels (*Falco sparverius*). Environ Toxicol Chem 2007;26:313–24.

[105] Chen D, Hale RC, Watts BD, La Guardia MJ, Harvey E, Mojica EK. Species-specific accumulation of polybrominated diphenyl ether flame retardants in birds of prey from the Chesapeake Bay region, USA. Environ Pollut 2010;158:1883–9. 3.

[106] Newsome SD, Park J-S, Henry BW, Holden A, Fogel ML, Linthicum J, et al. Polybrominated diphenyl ether (PBDE) levels in peregrine falcon (*Falco peregrinus*) eggs from California correlate with diet and human population density. Environ Sci Technol 2010;44:5248–55.

[107] Leslie HA, Leonards PEG, Shore RF, Walker LA, Bersuder PRC, Morris S, et al. Decadibromodiphenylether and hexabromocyclododecane in wild birds from the United Kingdom, Sweden and the Netherlands: screening and time trends. Chemosphere 2011;82:88–95.

[108] Crosse JD, Shore RF, Jones KC, Pereira MG. Long term trends in PBDE concentrations in gannet (*Morus bassanus*) eggs from two UK colonies. Environ Pollut 2012; 161:93–100.

[109] Kannan K, Yun SH, Rudd RJ, Behr M. High concentrations of persistent organic pollutants including PCBs, DDT, PBDEs and PFOS in little brown bats with white-nose syndrome in New York, USA. Chemosphere 2010;80:613–8.

[110] Liu P-Y, Du G-D, Zhao Y-X, Mu Y-S, Zhang A-Q, Qin Z-F, et al. Bioaccumulation, maternal transfer and elimination of polybrominated diphenyl ethers in wild frogs. Chemosphere 2011;84:972–8.

[111] Qin X, Qin Z, Li Y, Zhao Y, Xia X, Yan S, et al. Polybrominated diphenyl ethers in chicken tissues and eggs from an electronic waste recycling area in Southeast China. J Environ Sci 2011;23:133–8.

[112] Hu G-c, Dai J-y, Xu Z-c, Luo X-j, Cao H, Wang J-s. Bioaccumulation behaviour of polybrominated diphenyl ethers (PBDEs) in the freshwater food chain of Baiyangdian Lake, North China. Environ Int 2010;36:309–15.

[113] Qiu Y-W, Zhang G, Guo L-L, Zheng GJ, Cai S-Q. Bioaccumulation and historical deposition of polybrominated diphenyl ethers (PBDEs) in Deep Bay, South China. Mar Environ Res 2010;70:219–26.

[114] Zhang K, Wan Y, An L, Hu J. Trophodynamics of polybrominated diphenyl ethers and methoxylated polybrominated diphenyl ethers in a marine food web. Environ Toxicol Chem 2010;29:2792–9.

[115] Zhang B-Z, Ni H-G, Guan Y-F, Zeng EY. Occurrence, bioaccumulation and potential sources of polybrominated diphenyl ethers in typical freshwater cultured fish ponds of South China. Environ Pollut 2010a;158:1876–82.

[116] Josefsson S, Leonardsson K, Gunnarsson JS, Wiberg K. Influence of contaminant burial depth on the bioaccumulation depth of PCBs and PBDEs by two benthic invertebrates (*Monoporeia affinis* and *Marenzelleria* spp.). Chemosphere 2011;85: 1444–51.

[117] Dunn RL, Huwe JK, Carey GB. Biomonitoring polybrominated diphenyl ethers in human milk as a function of environment, dietary intake, and demographics in New Hampshire. Chemosphere 2010;80:1175–82.

[118] Park J, She J, Holden A, Sharp M, Gephart R, Souders-Mason G, et al. High postnatal exposures to polybrominated diphenyl ethers (PBDEs) and polychlorinated biphenyls (PCBs) via breast milk in California: does BDE-209 transfer to breast milk ? Environ Sci Technol 2011;45:4579–85.

[119] Zota AR, Park J, Wang Y, Petreas M, Zoeller RT, Woodruff TJ. Polybrominated diphenyl ethers, hydroxylated polybrominated diphenyl ethers, and measures of thyroid function in second trimester pregnant women in California. Environ Sci Technol 2011;45:7896–905.

[120] Petreas M, Nelson D, Brown FR, Goldberg D, Hurley S, Reynolds P. High concentrations of polybrominated diphenyl ethers (PBDEs) in breast adipose tissue of California women. Environ Int 2011;37:190–7.

[121] Chevrier J, Harley KG, Bradman A, Gharbi M, Sjödin A, Eskenazi B. Polybrominated diphenyl ether (PBDE) flame retardants and thyroid hormone during pregnancy. Environ Health Perspect 2010;118:1444–9.

[122] Chen A, Chung E, DeFranco EA, Pinney SM, Dietrich KN. Serum PBDEs and age at menarche in adolescent girls: analysis of the National Health and Nutrition Examination Survey 2003-2004. Environ Res 2011;111:831–7.

[123] Lunder S, Hovander L, Athanassiadis I, Bergman Å. Significantly higher polybrominated diphenyl ether concentrations in young U.S. children than in their mothers. Environ Sci Technol 2010;44:5256–62.

[124] Sun S, Zhao J, Leng J, Wang P, Wang Y, Fukatsu H, et al. Levels of dioxins and polybrominated diphenyl ethers in human milk from three regions of northern China and potential dietary risk factors. Chemosphere 2010;80:1151–9.

[125] Zhang L, Li J, Zhao Y, Li X, Yang X, Wen S, et al. A national survey of polybrominated diphenyl ethers (PBDEs) and indicator polychlorinated biphenyls (PCBs) in Chinese mothers' milk. Chemosphere 2011;84:625–33.

[126] Shen H, Ding G, Han G, Wang X, Xu X, Han J, et al. Distribution of PCDD/Fs, PCBs, PBDEs and organochlorine residues in children's blood from Zhejiang, China. Chemosphere 2010;80:170–5.

[127] Yu Z, Zheng K, Ren G, Zheng Y, Ma S, Peng P, et al. Identification of hydroxylated octa- and nona-bromodiphenyl ethers in human serum from electronic waste recycling workers. Environ Sci Technol 2010;44:3979–85.

[128] Zheng J, Luo X-J, Yuan J,-G, Weng J, Wang Y-T, Chen S-J, et al. Levels and sources of brominated flame retardants in human hair from urban, e-waste, and rural areas in South China. Environ Pollut 2011;159:3706–13.

[129] Zhao X-R, Qin Z-F, Yang Z-Z, Zhao Q, Zhao Y-X, Qin X-F, et al. Dual body burdens of polychlorinated biphenyls and polybrominated diphenyl ethers among local residents in an e-waste recycling region in Southeast China. Chemosphere 2010;78: 659–66.

[130] Kang CS, Lee J-H, Kim S-K, Lee K-T, Lee JS, Park PS, et al. Polybrominated diphenyl ethers and synthetic musks in umbilical cord serum, maternal serum, and breast milk from Seoul, South Korea. Chemosphere 2010;80:116–22.

[131] Frederiksen M, Thomsen C, Frøshaug M, Vorkamp K, Thomsen M, Becher G, et al. Polybrominated diphenyl ethers in paired samples of maternal and umbilical cord blood plasma and associations with house dust in a Danish cohort. Int J Hyg Environ Health 2010;213:233–42.

[132] Kim U-J, Lee I-S, Kim HS, Oh J-E. Monitoring of PBDEs concentrations in umbilical cord blood and breast milk from Korean population and estimating the effects of various parameters on accumulation in humans. Chemosphere 2011;85:487–93.

[133] Uemura H, Arisawa K, Hiyoshi M, Dakeshita S, Kitayama A, Takami H, et al. Congener-specific body burden levels and possible determinants of polybrominated diphenyl ethers in the general Japanese population. Chemosphere 2010;79:706–12.

[134] Roosens L, Cornelis C, D'Hollander W, Bervoets L, Reynders H, Van Campenhout K, et al. Exposure of the Flemish population to brominated flame retardants: model and risk assessment. Environ Int 2010;36:368–76.

[135] Schiavone A, Kannan K, Horii Y, Focardi S, Corsolini S. Polybrominated diphenyl ethers, polychlorinated naphthalenes and polycyclic musks in human fat from Italy: comparison to polychlorinated biphenyls and organochlorine pesticides. Environ Pollut 2010;158:599–606.

[136] Tue NM, Sudaryanto A, Minh TB, Isobe T, Takahashi S, Viet PH, et al. Accumulation of polychlorinated biphenyls and brominated flame retardants in breast milk from women living in Vietnamese e-waste recycling sites. Sci Total Environ 2010;408: 2155–62.

[137] Thomsen C, Stigum H, Frøshaug M, Broadwell SL, Becher G, Eggesbø M. Determinants of brominated flame retardants in breast milk from a large scale Norwegian study. Environ Int 2010;36:68–74.

[138] Thomsen C, Haug LS, Stigum H, Frøshaug M, Broadwell SL, Becher G. Changes in concentrations of perfluorinated compounds, polybrominated diphenyl ethers, and

polychlorinated biphenyls in Norwegian breast-milk during twelve months of lactation. Environ Sci Technol 2010;44:9550–6.
[139] Qin YY, Leung CKM, Lin CK, Leung AOW, Wang HS, Giesy JP, et al. Halogenated POPs and PAHs in blood plasma of Hong Kong residents. Environ Sci Technol 2011; 45:1630–7.
[140] Chovancová J, Čonka K, Kočan A, Sejáková ZS. PCDD, PCDF, PCB and PBDE concentrations in breast milk of mothers residing in selected areas of Slovakia. Chemosphere 2011;83:1383–90. 3.
[141] Glynn A, Lignell S, Darnerud PO, Aune M, Ankarberg EH, Bergdahl IA, et al. Regional differences in levels of chlorinated and brominated pollutants in mother's milk from primiparous women in Sweden. Environ Int 2011;37:71–9.
[142] Vorkamp K, Thomsen M, Frederiksen M, Pedersen M, Knudsen LE. Polybrominated diphenyl ethers (PBDEs) in the indoor environment and associations with prenatal exposure. Environ Int 2011b;37:1–10.
[143] Jogsten IE, Hagberg J, Lindström G, van Bavel B. Analysis of POPs in human samples reveal a contribution of brominated dioxin of up to 15% of the total dioxin TEQ. Chemosphere 2010;78:113–20.
[144] Törnkvist A, Glynn A, Aune M, Darnerud PO, Ankarberg EH. PCDD/F, PCB, PBDE, HBCD and chlorinated pesticides in a Swedish market basket from 2005 – levels and dietary intake estimations. Chemosphere 2011;83:193–9.
[145] Chen C, Chen J, Zhao H, Xie Q, Yin Z, Ge L. Levels and patterns of polybrominated diphenyl ethers in children's plasma from Dalian, China. Environ Int 2010;36:163–7.
[146] Eggesbø M, Thomsen C, Jørgensen JV, Becher G, Odland JØ, Longnecker MP. Associations between brominated flame retardants in human milk and thyroid-stimulating hormone (TSH) in neonates. Environ Res 2011;111:737–43.
[147] Asante KA, Adu-Kumi S, Nakahiro K, Takahashi S, Isobe T, Sudaryanto A, et al. Human exposure to PCBs, PBDEs and HBCDs in Ghana: temporal variation, sources of exposure and estimation of daily intakes by infants. Environ Int 2011;37:921–8.
[148] Darnerud PO, Aune M, Larsson L, Lignell S, Mutshatshi T, Okonkwo J, et al. Levels of brominated flame retardants and other persistent organic pollutants in breast milk samples from Limpopo province, South Africa. Sci Total Environ 2011;409:4048–53.
[149] Foster WG, Gregorovich S, Morrison KM, Atkinson SA, Kubwabo C, Stewart B, et al. Human maternal and umbilical cord blood concentrations of polybrominated diphenyl ethers. Chemosphere 2011;84:1301–9.
[150] Wu K, Xu X, Liu J, Guo Y, Li Y, Huo X. Polybrominated diphenyl ethers in umbilical cord blood and relevant factors in neonates from Guiyu, China. Environ Sci Technol 2010;44:813–9.
[151] Vizcaino E, Grimalt JO, Lopez-Espinosa M-J, Llop S, Rebagliato M, Ballester F. Polybromodiphenyl ethers in mothers and their newborns from a non-occupationally exposed population (Valencia, Spain). Environ Int 2011;37:152–7.
[152] Xia C, Lam JCW, Wu X, Sun L, Xie Z, Lam PKS. Hexabromocyclododecanes (HBCDs) in marine fishes along the Chinese coastline. Chemosphere 2011;82: 1662–8.
[153] Esslinger S, Becker R, Jung C, Schröter-Kermani C, Bremser W, Nehls I. Temporal trend (1988-2008) of hexabromocyclododecane enantiomers in herring gull eggs from the German coastal region. Chemosphere 2011;83:161–7.
[154] He M-J, Luo X-J, Yu L-H, Liu J, Zhang X-L, Chen S-J, et al. Tetrabromobisphenol-A and hexabromocyclododecane in birds from an e-waste region in South

China: influence of diet on diastereoisomer- and enantiomer-specific distribution and trophodynamics. Environ Sci Technol 2010;44:5748−54.

[155] Haukås M, Hylland K, Nygård T, Berge JA, Mariussen E. Diastereomer-specific bioaccumulation of hexabromocyclododecane (HBCD) in a coastal food web, Western Norway. Sci Total Environ 2010;408:5910−6.

[156] Köppen R, Becker R, Esslinger S, Nehls I. Enantiomer-specific analysis of hexabromocyclododecane in fish from Etnefjorden (Norway). Chemosphere 2010;80:1241−5.

[157] Haukås M, Ruus A, Hylland K, Berge JA, Mariussen E. Bioavailability of hexabromocyclododecane to the polychaete (*Hediste diversicolor*): exposure through sediment and food from a contaminated Fjord. Environ Toxicol Chem 2010;29:1709−15.

[158] Hardy ML. A comparison of the properties of the major commercial PBDPO/PBDE product to those of major PBB and PCB products. Chemosphere 2002;46:717−28.

[159] Schecter A, Haffner D, Colacino J, Patel K, Päpke O, Opel M, et al. Polybrominated diphenyl ethers (PBDEs) and hexabromocyclododecane (HBCD) in composite U.S. food samples. Environ Health Perspect 2010;118:357−62.

[160] Liberda EN, Wainman BC, LeBlanc A, Dumas P, Martin I, Tsuji LJS. Dietary exposure of PBDEs resulting from a subsistence diet in three first nation communities in the James Bay region of Canada. Environ Int 2011;37:631−6.

[161] Schecter A, Smith S, Colacino J, Malik N, Paepke O, Birnbaum L. Contamination of U.S. butter with polybrominated diphenyl ethers from wrapping paper. Environ Health Perspect 2011;119:151−4.

[162] Blanco SL, Martínez A, Porro C, Vieites JM. Dietary uptake of polybrominated diphenyl ethers (PBDEs), occurrence and profiles, in aquacultured turbots (*Psetta maxima*) from Galicia, Spain. Chemosphere 2011;85:441−7.

[163] Shanmuganathan D, Megharaj M, Chen Z, Naidu R. Polybrominated diphenyl ethers (PBDEs) in marine foodstuffs in Australia: residue levels and contamination status of PBDEs. Mar Pollut Bull 2011;63:154−9.

[164] Jin J, Wang Y, Yang C, Hu J, Liu W, Cui J. Human exposure to polybrominated diphenyl ethers at production area, China. Environ Toxicol Chem 2010;29:1031−5.

[165] Yu Y-X, Huang N-B, Zhang X-Y, Li J-L, Yu Z-Q, Han S-Y, et al. Polybrominated diphenyl ethers in food and associated human daily intake assessment considering bioaccessibility measured by simulated gastrointestinal ingestion. Chemosphere 2011;83: 152−60.

[166] Guo J, Wu F, Shen R, Zeng EY. Dietary intake and potential health risk of DDTs and PBDEs via seafood consumption in South China. Ecotoxicol Environ Saf 2010;73: 1812−9.

[167] Goscinny S, Vandevijvere S, Maleki M, Van Overmeire I, Windal I, Hanot V, et al. Dietary intake of hexabromocyclododecane diastereoisomers (α-, β- and γ-HBCD) in the Belgian adult population. Chemosphere 2011;84:279−88.

[168] Nakagawa R, Murata S, Ashizuka Y, Shintani Y, Hori T, Tsutsumi T. Hexabromocyclododecane determination in seafood samples collected from Japanese coastal areas. Chemosphere 2010;81:445−52.

[169] Palmer MG. The legacy of agent orange: empirical evidence from central Vietnam. Soc Sci Med 2005;60:1061−70.

[170] Institute of Medicine, US. Updated summary of findings in occupational, environmental, and veterans studies regarding the association between specific health outcomes and exposure to herbicides, http://books.nap.edu/html/veterans2000/reportbrief.pdf; 2000.

[171] Addink R, Olie K. Mechanisms of formation and destruction of polychlorinated dibenzo-p-dioxins and dibenzofurans in heterogeneous systems. Environ Sci Technol 1995;29:1425–35.
[172] Blumenstock M, Zimmermann R, Schraumm L-W, Kaune A, Henkelmann B, Kettrup A. Presence of polychlorinated dibenzo-p-dioxins (PCDD), dibenzofurans (PCDFs), biphenyls (PCB), chlorinated benzenes (PCBz) and polycyclic aromatic hydrocarbons (PAH) under various combustion conditions in a post combustion chamber. Organohalogen Compd 1998;36:59–63.
[173] Weber R, Kuch B. Relevance of BFRs and thermal conditions on the formation pathways of brominated and brominated–chlorinated dibenzodioxins and dibenzofurans. Environ Int 2003;29:699–710.
[174] Zennegg M, Yu X, Wong MH, Weber RR. Fingerprints of chlorinated, brominated and mixed halogenated dioxins at two e-waste recycling sites in Guiyu, China. Organohalogen Compd 2009;71:2263–7.
[175] Birnbaum LS, Staskal DF, Diliberto JJ. Health effects of polybrominated dibenzo-p-dioxins (PBDDs) and dibenzofurans (PBDFs). Environ Int 2003;29:855–60.
[176] Olsman H, Engwall M, Kammann U, Klempt M, Otte J, van Bavel B, et al. Relative differences in aryl hydrocarbon receptor-mediated response for 18 polybrominated and mixed halogenated dibenzo-p-dioxins and -furans in cell lines from four different species. Environ Toxicol Chem 2007;26:2448–54.
[177] Samara F, Gullett BK, Harrison RO, Chu A, Clark GC. Determination of relative assay response factors for toxic chlorinated and brominated dioxins/furans using an enzyme immunoassay (EIA) and a chemically-activated luciferase gene expression cell bioassay (CALUX). Environ Int 2009;35:588–93.
[178] Polybrominated dibenzo-p-dioxins and dibenzofurans. Environmental health criteria 205. Geneva: World Health Organization; 1998, http://www.inchem.org/documents/ehc/ehc/ehc205.htm.
[179] Hanari N, Kannan K, Miyake Y, Okazawa T, Kodavanti PRS, Aldous KM, et al. Occurrence of polybrominated biphenyls, polybrominated dibenzo-p-dioxins, and polybrominated dibenzofurans as impurities in commercial polybrominated diphenyl ether mixtures. Environ Sci Technol 2006;40:4400–5.
[180] Ebert J, Bahadir M. Formation of PBDD/F from flame-retarded plastic materials under thermal stress. Environ Int 2003;29:711–6.
[181] McAllister DL, Mazac CJ, Gorsich R, Freiberg M, Tondeur Y. Analysis of polymers containing brominated diphenyl ethers as flame retardants after molding under various conditions. Chemosphere 1990;20:1537–41.
[182] Kajiwara N, Noma, Takigami H. Photolysis studies of technical decabromodiphenyl ether (DecaBDE) and ethane (DeBDethane) in plastics under natural sunlight. Environ Sci Technol 2008;42:4404–9.
[183] Kajiwara N, Takigami H. Behavior of additive brominated flame retardants in textile products. Proceedings 5th international symposium on brominated flame retardants. 2010, Kyoto, Japan, http://www.bfr2010.com/abstract-download/2010/90074.pdf.
[184] Ota S, Aitawa H, Kondo Y, Takigami H, Hirai Y, Sakai S. Current status o fpolybrominated dibenzo-p-dioxins and furans (PBDD/DFs) emission in Japan. Organohalogen Compd 2009;71:1340–5.
[185] Wong MH, Wu SC, Deng WJ, Yu XZ, Luo Q, Leung AOW, et al. Export of toxic chemicals - a review of the case of uncontrolled electronic-waste recycling. Environ Pollut 2007;149:131–40.

[186] Yu X, Zennegg M, Engwall M, Rotander A, Larsson M, Wong MH, et al. E-waste recycling heavily contaminates a Chinese city with chlorinated, brominated and mixed halogenated dioxins. Organohalogen Compd 2008;70:813–6.
[187] Bi X, Thomas G, Jones KC, Qu W, Sheng G, Martin FL, et al. Exposure of electronics dismantling workers to polybrominated diphenyl ethers, polychlorinated biphenyls, and organochlorine pesticides in South China. Environ Sci Technol 2007; 41:5647–53.
[188] Jin J, Wang Y, Yang C, Hu J, Liu W, Cui J, et al. Polybrominated diphenyl ethers in the serum and breast milk of the resident population from production area, China. Environ Int 2009;35:1048–52.
[189] LeMasters GK, Genaidy AM, Succop P, Deddens JA, Sobeih T, Berriera-Viruet H, et al. Cancer risk among firefighters: a review and meta-analysis of 32 studies. J Occup Environ Med 2006;48:1189–202.
[190] Brandt-Rauf PW, Fallon LFJ, Tarantini T, Idema C, Andrews L. Health hazards of fire fighters: exposure assessment. Brit J Ind Med 1988;45:606–12.
[191] Kang D, Davis LK, Hunt P, Kriebel D. Cancer incidence among male Massachusetts firefighters, 1987–2003. Am J Industrial Med 2008;51:329–35.
[192] Anonymous. Report of a new chemical hazard. New Sci 1966;32:612.
[193] Kuratsune M, Yoshimura H, Hori Y, Okumura M, Mauda Y. YUSHO: a human disaster caused by PCBs and related compounds. Fukuoka, Japan: Kyushu University Press; 1996.
[194] Eggington J. The poisoning of Michigan. Michigan State University Press; 2009, ISBN 978-0-87013-867-6.
[195] EFSA Panel on Contaminants in the Food Chain (CONTAM). Scientific opinion on polybrominated biphenyls (PBBs) in food. EFSA J 2010;8(10):1789 [151 pp.].
[196] Norrgran J, Jones B, Lindqvist N, Bergman Å. Decabromobiphenyl, polybrominated diphenyl ethers and brominated phenolic compounds in serum of cats diagnosed with the endocrine disease Feline Hyperthyroidism. Arch Environ Contam Toxicol, in press. doi: 10.1007/s00244-012-9750-y.
[197] Peterson M, Johnson G, Andrews L. Proc Am Coll Intern Med Forum 1979;108.
[198] Blum A, Gold MD, Ames BN, Jones FR, Hett EA, Dougherty RC. Children adsorb tris-BP flame retardant from sleepwear: urine contains the mutagenic metabolite, 2,3-dibromopropanol. Science 1978;201(4360):1020–3.
[199] Gold MD, Blum A, Ames BN. Another flame retardant, tris-(1,3-Dichloro-2-Propyl)-phosphate, and its expected metabolites are mutagens. Science 1978;200:785–7.
[200] Norström Å, Andersson K, Rappe C. Major components of some brominated aromatics used as flame retardants. Chemosphere 1976;5:255–61.
[201] Sundström G, Hutzinger O. Environmental chemistry of flame retardants V. The composition of bromkal 70-5 DE - a pentabromodiphenyl ether preparation. Chemosphere 1976;5:187–90.
[202] EFSA Panel on Contaminants in the Food Chain (CONTAM). Scientific opinion on polybrominated diphenyl ethers (PBDEs) in food. EFSA J 2011;9(5):2156 [274 pp.].
[203] ECB (European Chemicals Bureau). European union risk assessment report. Diphenyl ether, pentabromo derivative. Office for Official Publications of the European Communities; 2001. CAS No: 32534-81-9. EINECS No: 251-084-2.
[204] ECB (European Chemicals Bureau). European union risk assessment report. Bis(pentabromophenyl) ether; 2002. CAS No: 1163-19-5. EINECS No: 214-604-9. Final draft. June 2002.

[205] ECB (European Chemicals Bureau). European union risk assessment report. Diphenyl ether, octabromo derivative. Office for Official Publications of the European Communities; 2003. CAS No: 32536-52-0. EINECS No: 251-087-9.
[206] ECB (European Chemicals Bureau). European union risk assessment report. Update of the risk assessment of bis(pentabromophenyl) ether (decabromodiphenyl ether); 2004. CAS No: 1163-19-5. EINECS No: 214-604-9. Final Environmental Draft of May 2004.
[207] Pearce EN, Braverman LE. Environmental pollutants and the thyroid. Best practice & research. Clin Endocrinol Metabol 2009;23:801–13.
[208] NTP. Toxicology and carcinogenesis studies of decabromodiphenyl oxide (CAS no 1163-19-5) in F344/N and B6C3F1 mice (feed studies). NTP TR 1986;309.
[209] Meironyté D, Norén K, Bergman Å. Analysis of polybrominated diphenyl ethers in Swedish human milk. A time-related trend study, 1972-1997. J Toxicol Environ Health, Part a 1999;58:329–41.
[210] Mazdai A, Dodder NG, Abernathy MP, Hits R, Bigsby RM. Environ Health Perspect 2003;111:1249–52.
[211] Hites R. Polybrominated diphenyl ethers in the environment and in people: a meta-analysis of concentrations. Environ Sci Technol 2004;38:945–56.
[212] Fischer D, Hooper K, Athanasiadou M, Athanassiadis I, Bergman Å. Children show highest levels of polybrominated diphenyl ethers in a California family of four: a case study. Environ Health Perspect 2006;114:1581–4.
[213] Cox P, Efthymiou P. Directive 2003/11/EC of the European parliament and of the council of February 6, 2003 amending for the 24th time council directive 76/669/EEC relating to restrictions on the marketing and use of certain dangerous substances and preparations (pentabromodiphenyl ether, octabromodiphenyl ether). Official J Eur Union, OJ L 2003;42:45–6.
[214] Great lakes flame retardants: regulatory updates - United States, http://www.e1.greatlakes.com/fr/common/jsp/index.jsp; 2005.
[215] EU T.C.o.J.o.t.E.C.: directive 2002/95/EC-Electrical and electronic equipment - restriction of use of certain hazardous substances - decabromodiphenyl ether ('Deca-BDE') - Commission Decision 2005/717/EC - exemption of DecaBDE from the prohibition on use - actions for annulment - Commission's implementing powers - infringement of the enabling provision; 2008, http://curia.europa.eu.
[216] http://chm.pops.int/Convention/ThePOPs/tabid/673/Default.aspx.
[217] Meijer L, Weiss J, van Velzen M, Brouwer A, Bergman Å, Sauer PJJ. Serum concentrations of neutral and phenolic organohalogens in pregnant women and some of their infants in the Netherlands. Environ Sci Technol 2008;42:3428–33.
[218] Athanasiadou M, Cuadra SN, Marsh G, Bergman A, Jakobsson K. Polybrominated diphenyl ethers (PBDEs) and bioaccumulative hydroxylated PBDE metabolites in young humans from Managua, Nicaragua. Environ Health Perspect 2008;116:400–8.
[219] Eriksson P, Fischer C, Wallin M, Jakobsson E, Fredriksson A. Impaired behaviour, learning and memory, in adult mice neonatally exposed to hexabromocyclododecane (HBCDD). Environ Toxicol Pharmacol 2006;21:317–22.
[220] van der Ven L, Verhoef A, van de Kuil T, Slob W, Leonards PEG, Visser TJ, et al. A 28-day oral dose toxicity study enhanced to detect endocrine effects of hexabromocyclododecane in Wistar rats. Toxicol Sci 2006;94:281–92.
[221] van der Ven L, van de Kuil T, Leonards PEG, Slob W, Lilienthal H, Litens H, et al. Endocrine effects of hexabromocyclododecane (HBCD) in a one-generation reproduction study in Wistar rats. Toxicol Lett 2009;185:51–62.

[222] Ema M, Fujii S, Hirata-Koizumi, Matsumoto M. Two-generation reproductive toxicity study of the flame retardant hexabromocyclododecane in rats. Reprod Toxicol 2008; 25:335–51.

[223] Fängström B, Athanassiadis I, Odsjö T, Noren K, Bergman Å. Temporal trends of polybrominated diphenyl ethers and hexabromocyclododecane in milk from Stockholm mothers, 1980-2004. Mol Nutrition Food Res 2008;52:187–93.

[224] EFSA Panel on Contaminants in the Food Chain (CONTAM). Scientific opinion on tetrabromobisphenol A (TBBPA) and its derivatives in food. EFSA J 2011;9(12): 2477 [61 pp.].

[225] Viberg H, Eriksson P. Differences in neonatal neurotoxicity of brominated flame retardants, PBDE 99 and TBBPA, in mice. Toxicology 2011;289:59–65.

[226] Bergman A, Ryden A, Law RJ, de Boer J, Covaci A, Alaee M, et al. A novel abbreviation standard for organobromine, organochlorine and organophosphorus flame retardants and some characteristics of the chemicals. Environ Int 2012;49:57–82.

[227] EFSA Panel on Contaminants in the Food Chain (CONTAM). Scientific opinion on emerging and novel brominated flame retardants (BFRs) in food. EFSA J 2012; 10(10):2908. 125 pp.

[228] UNEP/WHO. State-of-the-science of endocrine disrupting chemicals - 2012. Geneva, Switzerland: United Nations Environment Programme; 2012.

[229] Bantelmann E, Ammann A, Naef U, Tremp J. Brominated flame retardants in products: results of the Swiss market survey. In: Abstracts from the 5th international symposium on brominated flame retardants (BFR2010), Kyoto, Japan; 2008.

[230] Malarvannan G, Takahashi S, Isobe T, Kunisue T, Sudaryanto A, Miyagi T, et al. Levels and distribution of polybrominated diphenyl ethers and organochlorine compounds in sea turtles from Japan. Mar Pollut Bull 2011;63:172–8.

[231] Vermeulen F, Covaci A, D'Have H, Van den Brink NW, Blust R, De Coen W, et al. Accumulation of background levels of persistent organochlorine and organobromine pollutants through the soil-earthworm-hedgehog food chain. Environ Int 2010;36: 721–7.

[232] Bertazzi PA. Long-term effects of chemical disasters. Lessons and results from Seveso. Sci Total Environ 1991;106:5–20.

CHAPTER

5

Reactive and Additive Phosphorus-based Flame Retardants of Reduced Environmental Impact

Gy. Marosi, B. Szolnoki, K. Bocz, A. Toldy

Budapest University of Technology and Economics, Budapest, Hungary

CHAPTER OUTLINE

1. Introduction

In spite of recent development in the performance of fire-fighting equipment no significant decrease in the danger of fire can be seen because the use of polymer-based products grows even more rapidly. These objects are responsible for 25% of fire-caused injuries and for 40% of deaths. The time available for building evacuation was around 17 min in the seventies, while it is around 3 min nowadays. Electrical/electronic, building/construction, and transportation application fields require more and more strict benchmarks for flame retardancy (FR) of polymer systems. In order to meet the complex requirements as well as the social expectations, the FR of polymers should be improved in a way that

Polymer Green Flame Retardants. http://dx.doi.org/10.1016/B978-0-444-53808-6.00005-6

allows the maintenance of other important (mechanical, insulation, conductivity, etc.) characteristics and of the environment. On the other hand flame retardants allow the use of lightweight structures, resulting, for example, in new passenger cars with reduction of their energy consumption and CO_2 emission (the current aim is to reduce this emission to 120 g km^{-1} in the European Union). The global consumption of fire retardants is expected to grow with an annual growth rate of 4.9% and reach 1.95 million tons, while their sales are supposed to reach $6.1 billion in 2014 worldwide.[1] It is essential that the huge amount of such fire retardants should involve low impact on health and environment during their entire life cycle including recycling and disposal.

The RoHS directive (Restriction of the use of certain Hazardous Substances in electrical and electronic equipment, 2002/95/EC) defines two categories of halogenated flame retardant as toxic and bioaccumulative: the polybromobiphenyles and the polybromodiphenylethers; therefore, decabromodiphenylether is rapidly being phased out. At the moment the polymeric brominated flame retardants are considered to be acceptable environmentally [1]. However, the increasing focus on the health and environmental compatibility of flame retardants has drawn the attention to the halogen-free additives. At the same time, the rising cost pressure in industry requires cost-effective solutions. The water releasing inorganic additives, such as metal hydroxides, meet the health, environmental, and cost requirements but the very high concentration, needed for appropriate flame retardant effect, restricts their use to polymers of high rubbery amorphous content (e.g. ethylene vinylacetate copolymers). Even if combined with synergistic (catalytic) nanofillers or other coadditives (montmorillonite, attapulgit, zinc hydroxystannates, metal molybdates, etc.) their application field can be hardly extended [2–4]. Nonetheless, halloysite is an exception; halloysite is a mineral nanotube of kaolin origin, releases structurally bound water similar to aluminum trihydrate or MDH but, as the release temperature is above 400 °C, it can be used in a wider range of polymers (EVA, PO, PA, PET, epoxy, etc.) [5]. Due to its high surface area and high aspect ratio, it is both char former and reinforcing agent. Its fire retardant efficiency alone needs to be improved but it is advantageously applicable as a char-forming synergist [6].

As the application field of the mentioned (most environmental) group of flame retardants is limited, intensive research is focused on the development of phosphorus-containing flame retardants of reduced ecological impact. The range of phosphorus containing flame retardants is extremely wide and versatile since the P element exists in several oxidation states [7–9]. Additives being degradable during processing circumstances or low molecular weight substances are, however, too mobile and thus leach out from the polymer. A well-designed system does not migrate to the surface of the matrix during its normal use, minimizing thus the emission of toxic gases, in case of fire. This can be achieved either with

[1]The Global Market for Flame-retardant Chemicals (CHM014K), BCC Research, 2009.

reactive or macromolecular flame retardants or with the incorporation of phosphorus additives in the interfaces of a polymer system. Another effective route involves altering the mechanism of decomposition in favor of carbonaceous char; in so doing, the emission of CO and CO_2 is reduced and less flame retardant is needed to achieve the same level of safety if the efficiency is increased. In view of the above, it becomes apparent that understanding the mechanisms in gas/solid phase and increasing the efficiency leads indirectly to protection of the environment.

Phosphorus-containing flame retarded systems, which are produced according to green chemistry, are favored environmentally because they are safe during their production and use; furthermore, they are not harmful at the end of their life. Examples for these advantages are discussed in this study showing the intensive current research work worldwide aiming to answer the challenges. At first, reactive P-containing epoxy resins are discussed as examples of thermosetting materials. It is followed by two examples of thermoplastics demonstrating the different applicable fire retardancy strategies of such systems highlighting the role of interphases (IPs) [10]. Striving for the development of even more environmentally friendly flame retardant systems, the enormous amount of end-of-life (EOL) plastics cannot be left out of consideration. The European legislation (2008/98/EC Directive) aims at supporting the environmentally friendly recycling technologies and promoting the use of polymers from renewable resources, thus urgent need exists for development of viable recycled and green products. Accordingly in the last part of this study flame retarded composites made of recycled and biodegradable components are discussed.

2. Reactive flame retardancy of epoxy resins

Epoxy resins that found wide use in various industrial application fields (for surface coatings, castings, laminates, adhesives, composites, etc.) can be considered as typical representatives of thermosetting polymers. They possess excellent characteristics, such as toughness; chemical, mechanical, and electrical resistance; low shrinkage on cure; and high adhesion to many substrates; they also exhibit great disadvantages related mainly to flammability and EOL issues. Furthermore, the cost of their use is higher compared to other thermosetting polymers. Hence, their thermal and flame resistance needs to be improved for many application areas, most preferably in environmental and economic way. The thermal stability of epoxy resins, as well as their flammability, depends on the structure of the epoxy monomer, of that of the curing agent, and on the cross-link density.

Epoxy resins that are fire retarded with conventional additives are of poorer physical properties than the unmodified ones; therefore, in many cases, the use of reactive comonomers is preferred. Despite their disadvantages, the additive fire retardants dominate the market because most of the available reactive solutions are too complicated and expensive. The most versatile method involves incorporating phosphorus-containing compounds that react easily with the OH groups of the resin, resulting during fire in high char yield. The composition of the epoxy system (e.g. the type of hardener and the presence or absence of fibers or fillers) and its application determine the amount of phosphorus needed to meet the flammability requirements (e.g. V-0 rating UL-94 standard). If anhydride hardeners are used, up to 5% P is required while usually 3% is enough with amines. For laminates with 60% fiber content even 2% P can be sufficient. Hence an iterative optimization must be carried out for every system. Extensive reviews on phosphorus-containing flame retardants for epoxy resins have been previously published by Jain et al. [11] in 2002, by Levchik et al. [12] in 2004, and by Döring et al. [13] in 2010.

The phosphorus-containing chemical unit can be incorporated into the epoxy component, into the cross-linking agent, or both. The understanding of their mode of action in the gas and solid phase needs in most cases further investigations, which is one of the reasons why the fulfillment of the strict requirements of electronic industry is still quite rare.

2.1 Phosphorus-containing epoxy monomers

The synthesis methods of phosphorus-containing epoxy components can be categorized into the following four main groups:

- reaction of phosphorus-containing comonomers with epoxy components

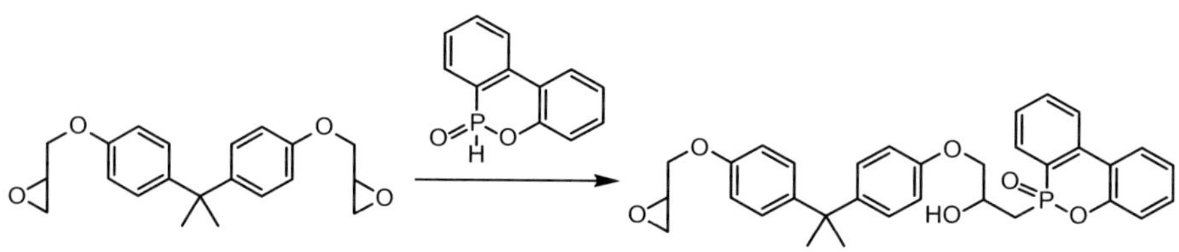

The most common comonomer, used to react with the oxirane ring of the epoxy component, is 9,10-dihydro-9-oxa-10-phosphaphenantrene-10-oxide (DOPO). Epoxy resin of 1−3% phosphorus content has been synthesized by reacting DOPO with diglycidyl ether of bisphenol-A (DGEBA) [14]. At 3% P content (using 4,4′-diaminodiphenylmethane (DDM) hardener) the limiting oxygen index (LOI) increased to 30 and V-0 rating was reached [15]. With the application of 2,8-dimethyl-phenoxaphosphin-10-oxide (DPPO) in a novolac resin cured with DDM, 0.75% P content was necessary for the V-0 rating [16]. Another reactive comonomer, 4-[(5,5-dimethyl-2-oxide-1,3,2-dioxaphosphorinan-4-yl)oxy]-phenol (DODPP), could provide at 2.5% P-content V-0 rating, LOI value of 30, and 80% cut of heat release rate (HRR) in DGEBA cured with low molecular weight polyamide [17].

- modification of phosphorus-containing comonomers with phenols followed by reaction with epichlorohydrin

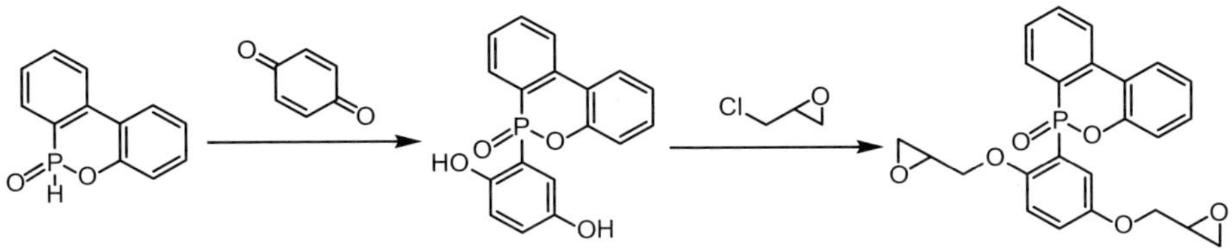

The comonomers decreasing cross-link density reduce the T_g of the resin, which is the main drawback of their application. Therefore, new DOPO derivatives were prepared by the reaction with benzoquinone (DOPO-BQ) [18] or naphthoquinone (DOPO-NQ) [19]. Both systems led to V-0 rating with 2% P.

- Reaction of phosphorus-containing alcohols/phenols with epichlorohydrin

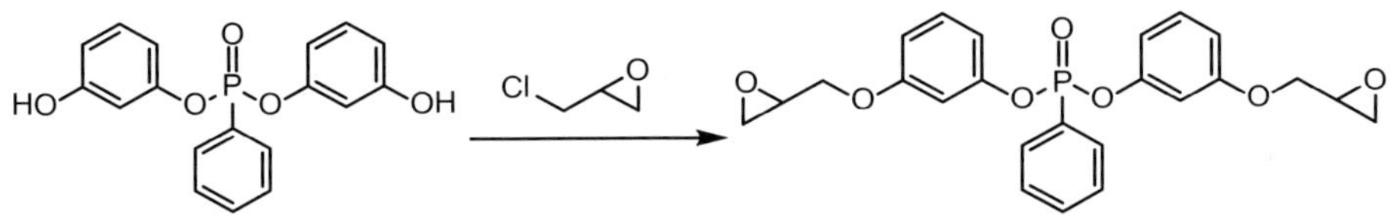

Epichlorohydrin, widely used as industrial reactant, easily reacts with hydroxyl groups, which allows the synthesis of diglycidyl ether of (2,5-dihydroxyphenyl) diphenyl phosphine oxide (Gly-HPO). In a study of Cadiz et al. [20], it was cured with the phosphorus-containing bis(3-aminophenyl)methylphosphinoxide resulting in 8.5% of P but still reached only an LOI value of 32. The same result could be achieved by curing it with benzoxazine with only 3.5% P content [21]. The product of the reaction between resorcinol and phenyl phosphonic dichloride was reacted with epichlorohydrin to form the diglycidyl ether of bis-phenoxy (3-hydroxy)phenyl phosphine oxide (BPHPPO) [22]. Cured with diamino-4,4′-diphenylsulphone (DDS), this system reached only LOI of 34, although the phosphorus content was 7.8%.

- reaction of phosphorus (oxy)chlorides with glycidyl alcohol

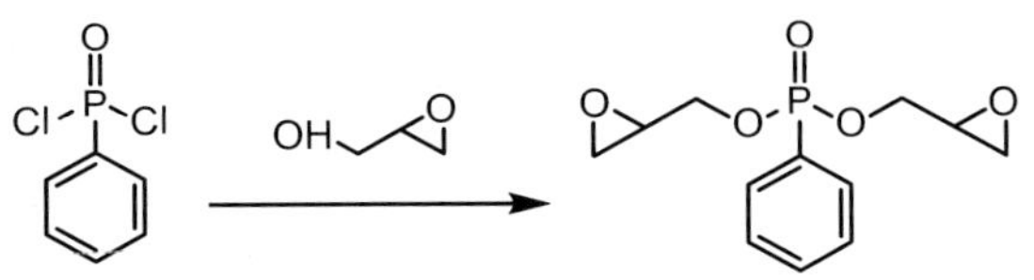

An easy method to synthesize phosphorus-containing epoxy components is through the reaction of phosphorus oxychlorides with glycidyl alcohol. A series of simple epoxy components prepared via this method was incorporated by Hergenrother et al. [23] into a *N,N,N′,N′*-tetraglycidyl-4,4′-methylenedianiline (TGMDA)–DDS system in different concentrations. Low phosphorus content (3%) was enough to fulfill the strict requirements of the aircraft industry. Recently, cyclophosphazene-based epoxy components were also synthesized [24,25]. By replacing 20% of DGEBA with the new component, a V-0 composition was achieved. In the case of diglycidylphenylphosphate (DGPP) [26] cured with 2,5-bis(p-aminophenyl)-1,3,4-oxadiazole the phosphorus content of the system was 7.4% and it resulted in an LOI of 47.

2.2 Phosphorus-containing curing agents

Incorporation of phosphorus-containing groups into aliphatic or aromatic diamine curing agents is also a widely used method. Aliphatic diamines cure epoxies at room temperature, while for aromatic diamines, elevated temperatures are required. However, the presence of aromatic groups is promising with respect to FR. The methods described for the synthesis of phosphorus-containing amines can be categorized into the following main groups:

- Reaction of phosphorus oxychlorides with aminophenols or aminoalcohols, or with nitrophenols followed by reduction to obtain the amino group

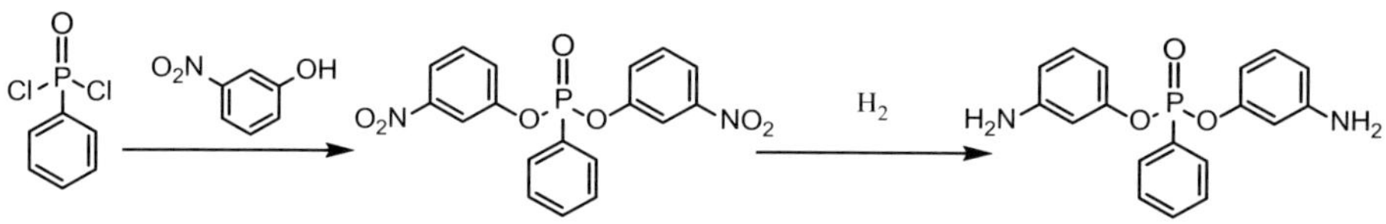

The reaction of *p*-nitrophenol with methylphosphonic dichloride followed by reduction results in the formation of bis(4-aminophenyl)methylphosphonate [22]. This compound has been used as curing agent of TGMDA and resulted in 3.9% P content, yielding an immediately extinguishing resin. Bis(4-aminophenyl)phenylphosphonate (BAPP) has been prepared in the same way [27] and applied in DGEBA and TGDDM epoxy resins. In the case of DGEBA (2.82% P content) the maximum of heat release was found to be reduced by 30% compared to the reference, while in case of TGDDM the reduction was more than 50% [28]. In another study BAPP was used to cure DGEBA mixed in different ratios of siliconized DGEBA [29]. When the mixing ratio of the two epoxies was 100:15, the LOI increased from 32 to 42 v/v%.

- Reaction of phosphorus oxychlorides with amines

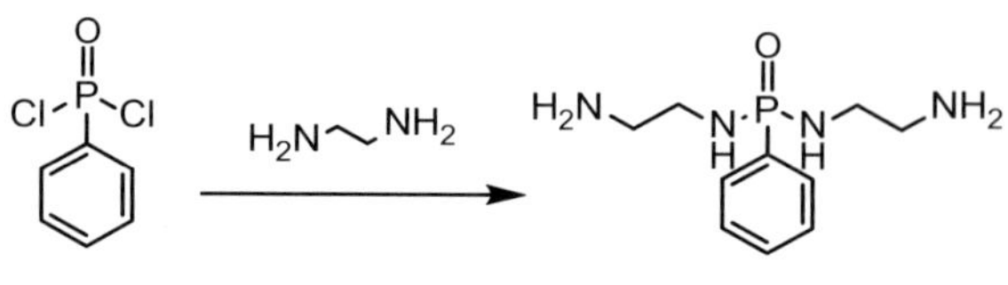

Series of phosphorus-containing poly(alkylene) amines with or without aromatic groups have been synthesized by reacting phosphoryl chloride derivatives with commercially available polyetheramines, ethylenediamine, and *N*-phenyl-1,4- phenylenediamine [30]. The prepared amines reacted with DGEBA and their flame retardant properties were determined. The best results were achieved when the highest phosphorus-containing amine, the reaction product of ethylenediamine and phenylphosphonicdichloride (4.07% P) was used. This system reached an LOI of 31, and the char yield was 12.2% in air at 850 °C. Combined with phosphorus-containing epoxy component, the increased P content did not increase the former values significantly [31].

- Transesterification of phosphate esters with aminophenols or aminoalcohols

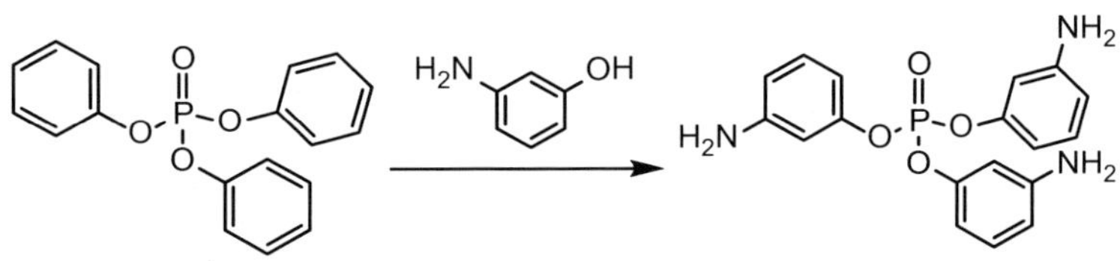

Tris-(3-aminophenyl)-phosphate (TAPP) is prepared by the transesterification of triphenyl phosphate with 3-aminophenol [32]. Incompletely replaced compounds and oligomers can be found among the by-products, which are as effective as TAPP but they decrease the cross-link density. In novolac-type epoxy resin laminates, it has been shown that a V-1 rating can be achieved (according to the vertical UL-94) test with the application of TAPP.

- Nitration of aromatic phosphine oxides followed by reduction to obtain the amino group

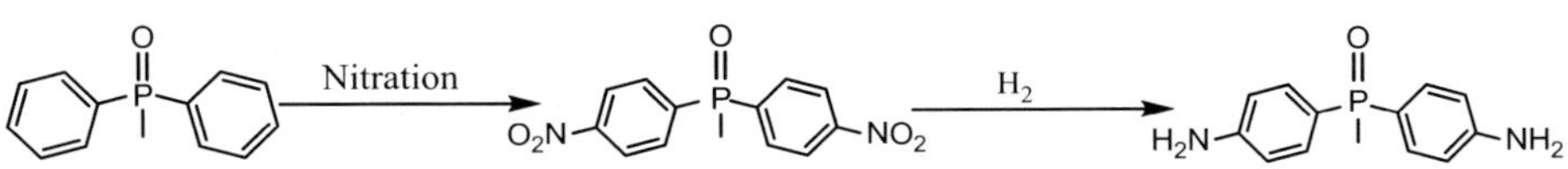

According to the extensive comparative research of Braun et al. [33] about the effect of different oxidation states of phosphorus on FR, the best results were achieved when aromatic phosphinate was applied. Bis(4-aminophenyl)methylphosphine oxide was synthesized by the nitration of diphenylmethylphosphine oxide followed by reduction of the nitro groups [23]. This phosphorus-containing compound was used as curing agent of TGMDA and resulted in 4% P content, which led to an immediately extinguishing resin, with 23% char yield at 800 °C in N_2.

- Reaction of DOPO with reagents containing amine groups

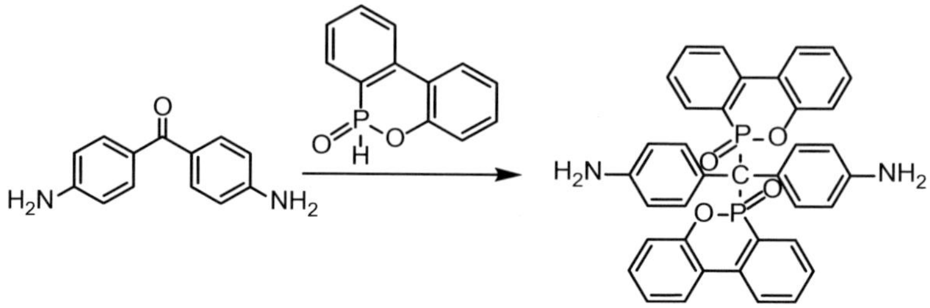

DOPO-based diamines can be prepared by the additional reaction of DOPO with different amine-containing substituents. The addition can occur on oxo [34] or imine groups [35,36]. The reaction product of DOPO and 4,4′-diaminobenzophenone (2DOPO-A) in siliconized DGEBA has been reported to result in LOI of 35% with 2.35% P content and 4.57% Si content [37]. By the reaction of DOPO with an aromatic diimine, a symmetric diamine can be gained. By its application as co-curing agent beside DDM in DGEBA at 1.5% P content, the LOI value was found to be 37 while a UL-94 V-0 rating was reached. The rings of DOPO can also be nitrated and then reduced to form phosphorus-containing hardener [38]. The synthesized diamine increased the LOI from 22 to 35% at 4.5% P content. Wang et al. [39] synthesized an oligomeric form of a DOPO derivative. The new organophosphorus oligomer, poly(DOPO-substituted hydroxyphenyl methanol pentaerythritol diphosphonate), was incorporated into epoxy resin, cured by DDM. The reactive modification resulted in higher glass transition temperature and significantly enhanced the char yield compared to the reference resin.

2.3 Novel P-N bonded structures

There has always been a strong interest in P-N bonded structures. Most of these P-N bonded structures, like phosphoramidates, can exhibit condensed-phase or gas-phase action depending on their structures [40,41].

An environmentally friendly, cost-effective method for the synthesis of both aliphatic and aromatic phosphorylated amines with high phosphorus content has been recently published by the authors. The materials applied in the synthesis were ethylenediamine (MM 60.1; ρ 0.899 g ml^{-1}; bp. 118 °C), and triethylphosphate (MM 182.2; ρ 1.071 g ml^{-1}; bp. 215 °C) supplied by Aldrich Chemical Co. The aliphatic epoxy resin matrix consisted of the epoxy component type ipox MR 3016 (tetraglycidyl ether of pentaerythritol, properties: nonmodified, resinlike reactive dilutant, epoxy equivalent: 156–170; viscosity at 25 °C: 850–1200 mPa·s; density at 25 °C: 1.24 g cm^{-3}) applied with ipox MH 3122 curing agent (3,3′-dimethyl-4,4′-diamino-dicyclohexyl-methane, properties: amine number:

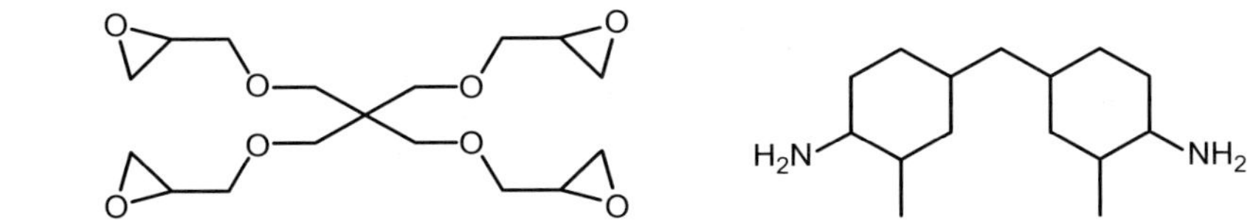

SCHEME 1

Main components of ipox MR 3016 and ipox MH 3122.

464–490 mg KOH g^{-1}; viscosity at 25 °C: 80–120 mPa·s; density at 20 °C: 0.944 g cm^{-3}) supplied by ipox chemicals Ltd., Hungary (Scheme 1).

The aromatic epoxy resin matrix consisted of epoxy component type D.E.R. 330 (diglycidyl ether of bisphenol A, properties: epoxy equivalent: 176–185; viscosity at 25 °C: 7000–10,000 mPa·s; density at 25 °C: 1.16 g cm^{-3}) applied with D.E.H. 24 curing agent (triethylene-tetramine, properties: amine number: 1443 mg KOH g^{-1}; viscosity at 25 °C: 19.5–22.5 mPa·s; density at 25 °C: 0.981 g cm^{-3}) supplied by DOW Chemical Company (Scheme 2).

Using the mentioned components a novel, easy-to-scale-up method was developed and patented [42] in order to provide an environmentally friendly alternative route to the reaction of phosphorous oxychlorides with aminophenols or aminoalcohols (Figure 1) [43]. The phosphorus-containing hardener is the product of the reaction between an alkyl phosphate and a diamine, in this particular case triethyl phosphate and ethylenediamine. These compounds are commercially available and produced in large quantities for other purposes, they are not harmful to the environment, and during the reaction no harmful by-products are formed.

In order to investigate the flame retardant effect of TEDAP (tris(ethylenediamino)phosphate, systematic name: *N,N′,N″*-tris(2-aminoethyl)-phosphoric acid triamide), MH 3122 curing agent was replaced by TEDAP in 20, 40, 60, 80, and finally in 100 mass% in the pentaerythritol-based aliphatic epoxy resin. The appropriate amount of TEDAP to replace MH 3122 was calculated according to its amine number in order to reach the same level of curing. The fire resistance was characterized by LOI measurement (according to ASTM D-2863), UL-94 test (according to ASTM D3801, ASTM D635), and mass loss type cone calorimeter (according to ISO 13927, Fire Testing Technology, heat flux of 50 kW m^{-2}).

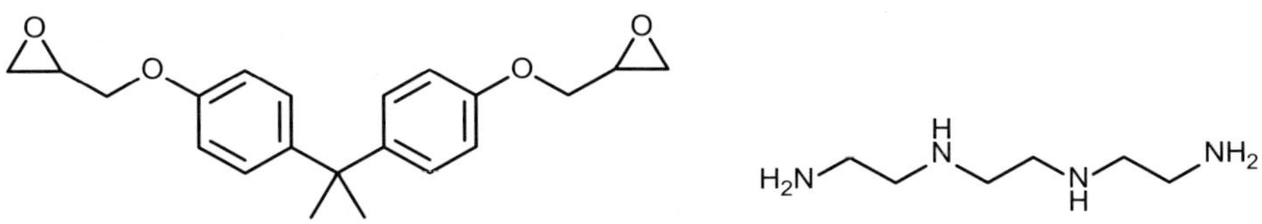

SCHEME 2

Main components of D.E.R. 330 and D.E.H. 24.

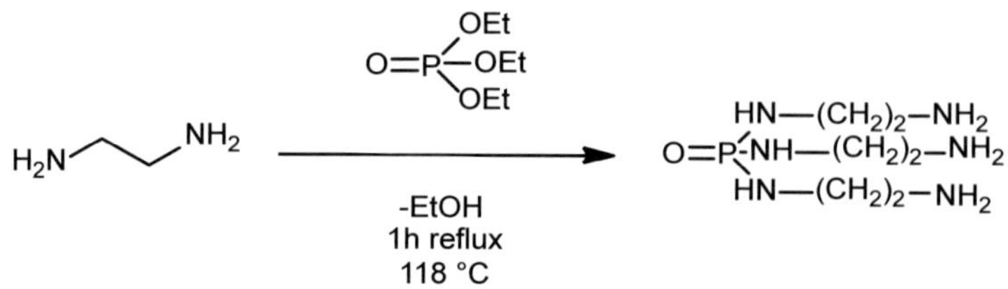

FIGURE 1

Synthesis of TEDAP [42].

With the proportion of TEDAP and so the P concentration in the system increasing, a clear improvement of FR is observable (Table 1). To achieve the V-0 UL-94 classification, 60 mass% of the original cross-linking agent was replaced by TEDAP. Best results (LOI of 33 and V-0 UL-94 classification) were reached when the original amine was completely replaced by TEDAP, which means 3.5 mass% of P in the epoxy resin [43].

According to the HRR results the greater the portion of TEDAP (and so the phosphorus content), the smaller peak value of HRR, which shows a good correlation with LOI and UL-94 results. The largest decrease in peak HRR was observed when the original amine was completely replaced by TEDAP; the peak HRR was reduced approximately to the tenth of its original value. A significant time delay was observed too, which increases the time to escape in case of fire.

As epoxy resins are mostly used in the aircraft industry, carbon fiber-reinforced composites (CCs) were made as well (Table 2). Beside the aliphatic one, the experiments were extended to the commonly used DGEBA, referred to as aromatic epoxy resin in the following. With the application of the flame retardant curing agent in the aromatic matrix, an LOI of 31 and V-0 could be reached, while the peak of heat

Table 1 FR Values as a Function of TEDAP Proportion (MR 3016/MH 3122 System)

Ratio of Hardener Replaced by TEDAP (%)	P Content (mass%)	LOI	UL-94	TTI (sec)	pk HRR (kW m^{-2})	Total HR (MJ m^{-2})
Reference	-	21	HB	45	962	87.6
20% TEDAP	0.7	26	HB	44	562	55.7
40% TEDAP	1.4	29	HB	51	373	54.3
60% TEDAP	2.1	29	V-0	83	383	41.6
80% TEDAP	2.8	29	V-0	90	272	24.8
100% TEDAP	3.5	33	V-0	97	111	12.0

Table 2 Flame Retardancy Results of the Matrices and Composites

Sample	LOI	UL-94	pk HRR ($kW\,m^{-2}$)	Total HR ($MJ\,m^{-2}$)	Residue (%)
Matrices					
Aliphatic reference	21	HB	962	87.6	0.4
Aromatic reference	26	HB	1022	103.6	1.0
Aliphatic FR	33	V-0	111	12.0	36.6
Aromatic FR	31	V-0	166	48.1	25.1
Composites					
Aliphatic CC	29	HB	445	46.5	52.6
Aromatic CC	33	HB	363	49.5	50.3
Aliphatic FR CC	33	V-0	195	16.0	59.3
Aromatic FR CC	31	V-0	307	39.1	55.4

release decreased by more than 80%. Moreover, it could be concluded that in case of the reference systems (both aliphatic and aromatic), the inclusion of carbon fibers itself resulted in significant increase in flame retardant performance; however, in the case of the aliphatic system the effect was somewhat poor. In case of the flame retarded systems, the outstanding LOI and UL-94 results could be maintained.

Concerning the heat release results, the simultaneous application of flame retardant and carbon fibers did not lead to further improvement (expected, assuming additivity); this is probably explained by the hindered intumescence caused by the plies of the reinforcing carbon fibers.

The mechanical characteristics (presented in Table 3) of the composite specimens were determined with an instrumented tensile tester type Zwick Z050 and Z020. Tensile strength measurements were carried out according to ISO 527-4 in

Table 3 Tensile Properties of the Matrices and Composites

MATRICES	Elasticity modulus (GPa)	Tensile strength (MPa)	Composites	Elasticity modulus (GPa)	Tensile strength (MPa)
Aliphatic reference	2.1	30	Aliphatic CC	21	523
Aromatic reference	2.6	53	Aromatic CC	18	424
Aliphatic FR	1.7	23	Aliphatic FR CC	17	394
Aromatic FR	2.4	46	Aromatic FR CC	16	375

case of composites and ISO 527-2 in case of the reference resins. Regarding the mechanical properties, it could be concluded that the inclusion of the carbon fibers resulted in a significant amelioration, as expected. However, the application of the polar flame retardant in composites deteriorated the mechanical properties to some extent in most cases [44].

Two kinds of coated systems were developed in order to combine the enhanced mechanical properties of fiber-reinforced composites and the outstanding flame retardant properties of the intumescent systems (eliminating the negative effect of the fibers on the intumescence and that of the flame retardant on the interlaminar adhesion). Both coated systems consisted of 2-mm-thick D.E.R. 330–D.E.H. 24 aromatic composite core with five plies of woven carbon fibers; one of them had a 2-mm-thick MR 3016–TEDAP aliphatic intumescent FR coating on the top while the other had two 1-mm-thick MR 3016–TEDAP aliphatic intumescent coatings—one above and one below the core composite.

This combination seemed to be synergistic in FR in both cases; an increase in LOI from 33 to 37 was measured in both coated composites compared to their constituent layers. According to the HRR results, a difference between the two types of coated systems was found. In case of the composite with 2 × 1 mm layers, the peak of HRR remained between the values of the constituent layers (318 versus 363 kW m^{-2} of the aromatic core and 111 kW m^{-2} of the aliphatic coating) but it caught fire 84 s later. In case of the composite with 2 mm layer, the improvement was even more significant; the HRR value fluctuated around zero as no ignition occurred at all at 50 kW m^{-2} heat flux.

Summarizing the above, the environmental and efficiency features of the fire retarded epoxy resins can be improved by applying reactive phosphorus compounds. Amines are especially preferred because of the P-N synergism, although their synthesis applies often hazardous and expensive reagents. The industrial production of such compounds is doubtful. However, new effective flame retardants can be easily synthesized from harmless commercial components in an economic and environmentally friendly way, even in the case of CCs, whose mechanical properties deteriorate in the presence of the polar flame retardant segments. A simultaneous increase both in FR and mechanical performance could be achieved by the formation of a multilayer composite structure consisting of a reference composite core and an intumescent epoxy resin coating layer. This synergistic combination led to a composite that does not ignite when subjected to 50 kW m^{-2} heat flux, owing to combined flame retardant effect in the gas and solid phases and to the heat distribution effect of the carbon fiber reinforcement.

3. Additive and reactive flame retardant systems in thermoplastic polyurethanes

Thermoplastic polyurethanes (TPUs) exhibit many favorable properties (such as high tensile and flexural fatigue strength, excellent low-temperature flexibility, etc.);

however, being an organic polymer, their FR needs to be improved in order to fulfill the safety standards. Environmental halogen-free methods of fire retardancy—reported mostly in the patent literature—include a wide range of phosphorus-, nitrogen-, silicon-, and boron-containing additives. These are used usually as part of intumescent systems. Siddhamalli and Brown [45] used a combination of 15–25 mass % phosphinate, 0–5 mass% phosphate, 2.5–10 mass% pentaerythritol or dipentaerythritol, 0–5 mass% inorganic filler (talc, ammonium phosphate, zinc borate, calcium carbonate, montmorillonite clay, etc.), and 0.05–2 mass% antioxidant to reach V-0 UL-94 rate and LOI of 35% in jackets for production of a wire and cable. Peerlings and Winkler [46] used phosphine oxide(s) in combination with melamine derivate(s), while Lee et al. [47] applied 0.5–25 mass% red phosphorus, 0.01–20 mass% trialkyl phosphate and triaryl phosphate, and 5–50 mass% melamine cyanurate, melamine phosphate, melamine polyphosphate, and/or melamine borate to avoid dripping through formation of char. Kim et al. [48] used, besides 1–20 mass% organic phosphate-based flame retardant and 5–50 mass% melamine or melamine derivative, 0.1–1 mass% of active isocyanate groups to avoid dripping. Henze et al. [49] also used isocyanate, as a cross-linking agent in addition to conventional flame retardants. The combination of phosphorus-containing flame retardants with metal hydroxides was also reported: Kuwasaki and Nishiguchi [50] combined red phosphorus with metal hydrates, Yoon et al. [51] applied a mixture of red phosphorus and melamine cyanurate, and magnesium hydroxide or aluminum hydroxide as flame retardant. Application of metal hydroxides alone is not preferred given that only V-2 UL-94 could be achieved even with as high rates as 80 phr (which causes difficulties in processing) [52].

Silicon-containing additives, mainly different types of clays, are also applied in combination with other flame retardants (see the above examples). Hydroxyl silicone oil was used by Chen et al. [53] for microencapsulation of ammonium polyphosphate (APP) in order to give better water resistance and FR in TPU; at 20 mass% additive level, the previously V-2 material achieved UL-94 V-0 classification. High phosphorus content is, however, not preferred in case of TPU used for cable insulation because phosphoric acid evolving during an incidental fire may cause corrosion problems in copper-containing cables. Significant reduction in HRR can be achieved even without phosphorus additive when polyhedral oligomeric silsesquioxane is applied individually in TPU, as reported by Bourbigot et al. [54]; however, other flame retardant classifications as UL-94 and LOI cannot be considerably ameliorated.

A systematic research work was published comparing the ideas of these promising additive combinations [55]. Based on this comparison the two main routes are compared here: application of phosphorus-containing intumescent additives and use of flame retardant system containing reactive components, with as low content of phosphorus as possible. Polyether-type TPU, Elastollan 1185A10 from BASF, was used as base polymer matrix (density: 1.12 g cm^{-3}, Shore A type hardness: 86, extrusion temperature: 180–205 °C). A reference flame retarded benchmark material, Elastollan 1185A10FHF from BASF

Table 4 Types of the Applied Flame-Retardant (FR) Component

Commercial Name	Abbreviation	Producer	Type
Reogard 2000	Reogard	Chemtura	Pentaerythritol phosphate alcohol- and melamine phosphate-containing intumescent FR
–	PER	Montedison	Pentaerythritol
Exolit OP 550	OP550	Clariant	Phosphorus-containing polyol type FR
Elastan 6568/103	MDI	BASF	4,4′-diphenylmethane-diisocyanate
Firebrake ZB	ZnB	Chemarco	Zinc borate
Melapur MC	MC	Ciba	Melamine-cyanurate
Hostavin N30	HALS	Clariant	Polymer of hindered amine and epichlorohydrin
HDK H15	SiO_2	Wacker	Silica

(density: 1.23 g cm^{-3}, Shore A type hardness: 88, LOI: 25%, UL-94 classification: V-0 in case of 3-mm-thick sample, V-2 in case of 1 mm thick sample), was also used in the comparison. The types and properties of the applied flame retardants are summarized in Table 4.

In order to reach V-0 classification on 1-mm-thick samples, necessary for cable applications of TPU, extended screening of various additive and reactive combinations were performed [55]. Based on the screening of potentially synergistic flame retardant systems, an additive system of 4.5 mass% P content providing V-0 rating and of highest LOI (containing 30% Reogard and 5% PER, referred to as TPU additive FR in the following), and the best reactive system of 0.4 mass% P content (containing 1.92% MDI, 2.08% OP550, 5% ZnB, 19% MC, 1% HALS, and 1% SiO_2, referred to as TPU reactive FR) were subjected to detailed analysis and compared to the reference TPU and flame retarded benchmark TPU system (Table 5).

Thermogravimetric analysis results (given in Table 2) revealed that the introduction of chemically bound FR caused a significant shift in temperature, which means that it successfully delayed the main degradation of TPU. Comparing the amounts of residue at 600 °C, it was noticeable, that the residue of the FR benchmark material has appropriate heat stability at the beginning of degradation but at higher temperatures it is less stable than the additive or reactive FR systems and even than the reference TPU.

Comparing the HRR curves (Figure 2), it can be concluded that the additive system was the most successful in reducing the HRR (to 20% of the original value) and total heat released (tHR) of TPU. However, the thermal degradation, expressed by time to ignition, begins earlier (as suggested by TG results as

Table 5 LOI And UL-94 and Thermogravimetric Results of the Chosen Flame Retarded Systems Compared to Reference TPU and Flame Retarded Benchmark TPU System

	P-content Mass (%)	LOI (%)	UL-94	$T_{5\%}$ (°C)	$T_{50\%}$ (°C)	v_{max} (% min^{-1})	T_{max}	Residue (%)
TPU reference	0	23	HB	331	429	−6.9	344	26.4
TPU FR benchmark	n.a.	25	V-2	321	382	−12.8	348	10.8
TPU additive FR	4.5	32	V-0	290	449	−5.2	320	37.8
TPU reactive FR	0.4	23	V-0	306	421	−6.7	415	26.7

$T_{5\%}$: temperature at 5% mass loss; $T_{50\%}$: temperature at 50% mass loss; v_{max}: maximum of the degradation speed at 10 °C min^{-1} heating rate; T_{max}: temperature at the maximal degradation speed, residue: amount of residue at 600 °C.

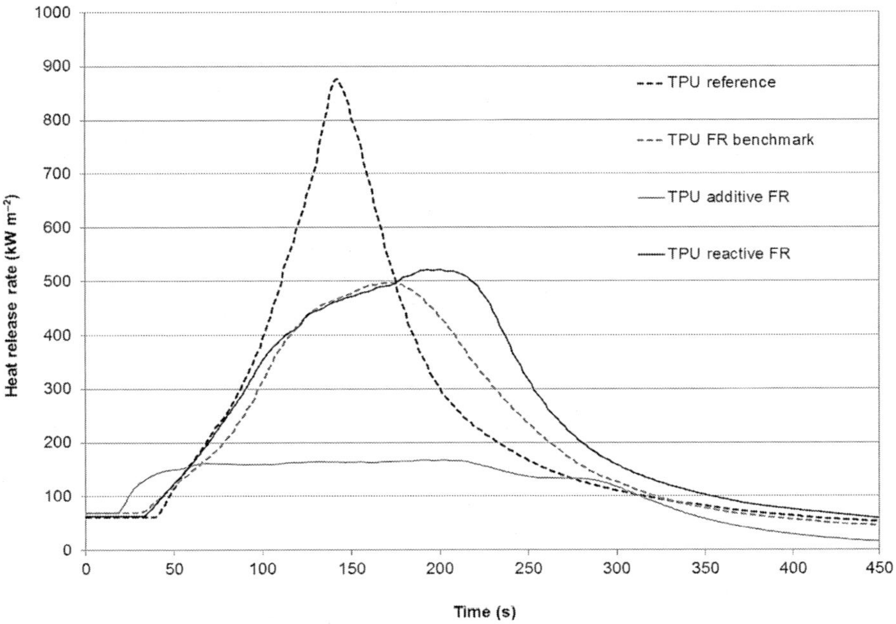

FIGURE 2

Heat release rate results of the chosen flame retarded systems compared to reference TPU and flame retarded benchmark TPU system.

well). The HRR curve of the reactive FR system follows the pattern of the TPU reference until 90 °C (no earlier degradation occurs as in case of the additive FR). The peak of HRR is reduced in this case to 60% of the original value and was shifted in time by 60 s. This result is comparable to the reduction reached by the TPU FR benchmark system.

The mechanical characteristics were compared before and after aging in order the estimate the applicability of the developed flame retarded systems as cable material. As expected the mechanical properties were deteriorated by the inclusion of flame retardants but these decreased values are still within the range acceptable for the cable-coating application. Furthermore, the FR system containing reactive components resulted in better tensile properties than the additive system. All values measured after aging remained at least 70% of the original value, which is a requirement for cable-coating materials. (Further details and results are published elsewhere [55].)

Summarizing, the application of both additive and reactive phosphorus-based FR system led to V-0 UL-94 rate of 1-mm-thick samples in TPU but with different flame retardant concentrations. The studied reactive system met this requirement at a phosphorus content as low as 0.4 mass%. The additive system reduced the peak of HRR to 20% of the original value; however, the time to ignition became shorter. The reactive FR system did not ignite earlier than the reference and the peak of HRR was reduced to 60% of the original value (shifted in time by 60 s) with a phosphorus content of 0.4 mass%. Environmental advantage of the reactive method is the in situ formation of the active component, which provides flexibility in the composition design according to the requirements of the given application area and allows lower P content than the additive method probably because of the better distribution of the P atom in the polymer.

4. Interphase modifications in fire retarded polypropylene

Concerning thermoplastic polymers, apart from some examples of reactive modification, the application of additive fire retardants is by far the most dominant approach. The "reactive" method is hardly applicable, especially in the case of polyolefins where well-designed interfacial structures and macromolecular components can serve for controlling the migration of additives. The phosphorus-based fire retardant additives act as fillers or plasticizers in the polymer matrix. These flame retardants allow the use of low-cost, recyclable thermoplastic polymer systems, which would otherwise be too flammable, to meet fire safety standards for many applications. Polyolefins, especially the most widely studied polypropylenes, are a good example to demonstrate the environmental gain of interface/IP modification in flame retarded thermoplastics. The environmental and health issues that are to overcome by modification are generally accompanied with undesirable migration effects. The small molecule FR additives can migrate out of the polymer matrix during their application and thereby lose their efficacy and expose humans to these (often) toxic chemicals. The intumescent fire retardant (IFR) systems containing APP and polyol mostly start to degrade during the polymer processing, causing bleed-out of the degradation products. The aim of modification in these cases is to preserve or enhance the efficiency of additives in case of fire and to ensure environmental production, use, and recycling of FR thermoplastics.

Modification of interfacial interactions can affect the degree of crystallinity and hinder or enhance the segmental mobility in the amorphous phase resulting in increased or reduced diffusion coefficient. Weak interaction at the interfaces enhances water diffusion, initiating hydrolytic degradation of flame retardants.

To overcome the chemical and thermal sensitivity of fire retardants, *insulating interlayers* can be employed [56]. APP, for instance, has to be protected against hydrolytic degradation at higher temperatures; for this purpose several insulating polymer layers have been developed [57,58]. In the case of a shear stress-resistant silicone elastomer the protecting effect (preserved under processing conditions) has been confirmed by the decrease in the conductivity of water used for extracting the FR product (from 200 μS of the reference sample to 50 μS of protected APP) [56].

For improving the compatibility of the phases, *connecting interlayers* are needed in all multiphase polymer systems including fire retarded polypropylene (PP). However, the use of conventional interface modifiers, such as surfactants (e.g. quaternary ammonium salts), functionalized (e.g. maleated) polyolefins, and coupling agents (e.g. vinyl silanes) decrease the fire retardancy, respectively, due to their high flammability, less ordered structure, and the restriction that they impose to the mobility of FR species [59,60]. In the case of nanofiller additives of very high surface area, the amount of interface modifier is comparable with that of the filler. Montmorillonite, the most extensively studied nanofiller in FR systems, is generally coated by quaternary ammonium salt of fatty acids. Replacement of this interfacial agent with phosphonium salt-type ionic liquid increases the fire retardant effect to some extent but substantial improvement was not detected because the P content of the thin interfacial layer was still low.

In order to increase the P content of fire retarded PP, many attempts have been published focusing mostly on the combination of nanofillers with intumescent flame retardants [61]. One of the most recent examples is the synergistic combination of halloysite and IFR, in which maleic anhydride grafted PP (PP-g-MA) was employed as interface modifier. Uniform dispersion was achieved only when compatibilizer was combined with water-assisted extrusion process. The formation of a compact, mechanically reinforced, and stable inorganic barrier at the surface of the char is considered to be the reason of the synergistic effect [62].

Another synergistic but even more complex combination is the application of a novel organosilicone derivative of cyclotriphosphazene with APP in PP. Thermogravimetric analysis revealed the increase in residues' weight. The cone calorimeter results showed considerable decrease in HRR compared to the reference composites. The interfacial interaction of the two components causes a better dispersion and more homogeneous distribution of APP particles in the PP composites [63].

Hierarchic multilayer interfacial structures, including polymeric interlayer, comprises another promising flame retarding approach, proposed originally for polymer composites [64,65] in which particles coated with surfactant were surrounded also with polymer interlayer. This concept can be extended to fire retarded systems. Thick IP, with polymer layer of relatively high phosphorus content, can potentially fulfill the functions of both fire retardant and mechanical compatibilizer additives. Being oligomers or polymers, such additives would hardly deteriorate the physical properties of thermoplastics. There are several phosphorus flame-retardant additives that can be suitable for this purpose. These include, for example, the bisphenol A/bis(diphenyl phosphate) oligomer, which is commercially available; nevertheless, one of its degradation products, the bisphenol A, is potent endocrine disruptor. Another one, the resorcinol bis(diphenyl phosphate) (RDP), is considered as one of the most promising alternative to brominated flame retardants. These, mostly liquid, oligomers require the addition of PTFE to prevent dripping. Nanostructures, owing to their high surface area, can adsorb these oligomers forming a complex, stable structure. Such a nanostructure, called nanosponge, can be formed by building cyclodextrin into a polymer network. Active substances can be included into the internal and external cavities of cyclodextrine-polymer network. The method was elaborated originally for drug delivery purposes and adapted later to fire retardancy [66]. Intumescent system can be prepared in polyolefins (ethylene vinyl acetate copolymer, PP, etc.) by their melt blending with the cyclodextrin-phosphorus (e.g. RDP) complex. Applying the adduct in EVA alone, only V-2 classification could be achieved with UL-94 test for every formulation regardless of the type and the amount of complex used [67]. However, this phosphorus-containing nanosponge can be applied advantageously in combination with conventional IFR in polyolefins.

While cyclodextrin is a relatively expensive cyclic oligosaccharide, the polysaccharide chitosan (derivative of chitin, which is the second most abundant naturally occurring biopolymer) is an ideal base for forming green phosphorus flame retardants. Treatment of chitosan has been performed, for example, with DOPO in carbon tetrachloride forming a phosphorus-containing polymer, which functions as a flame-retarding additive for a number of polymers [68].

Thermoplastic polymers can be prepared through much more environmental-friendly, solvent-free, melt polymerization process through the reaction of a high-purity phosphonate monomer with aromatic dihydroxy compounds [69]. The achievable phosphorus content in the formed polyphosphonate is larger than 10%. The by-product of the polymerization reaction is separated and reused to produce one or more of the starting materials, so little or no waste is generated. The LOI value of polyphosphonates is 65%, which is the highest number measured for a thermoplastic material. Showing good miscibility with a number of polymers (such as polycarbonate, polyester, polyurethane, epoxy, and polyurea), this polymer can be applied as nonmigrating flame retardant additive. Compared to halogenated flame retardants, this additive would allow better recyclability of polyolefins after their use if the compatibility between the two thermoplastic polymers were better.

An alternative solvent-free method for the synthesis of phosphorus-containing polymers is the polymer analogous modification [70]. Incorporation of phosphorous- and nitrogen-containing functionalities into polymers can be done through classical polymer analogous modification or by click chemistry.

Nanocompatibilization is another way to enhance the miscibility of such incompatible polymers. One of the approaches of nanocompatibilization is the introduction of organomodified nanofillers into an immiscible polymer system leading to the formation of a thermodynamically stable organic–inorganic multilayer IP Organoclays and other nanoparticles arranged at the phase border between the two polymers proved to improve the performance of incompatible polymer blends. Polymer chains of both polymer phases can be absorbed to the clay surface to form a more compatible nanocomposite [71]. Organoclays of high aspect ratio (500–1000:1) and large surface area (700–800 $m^2\,g^{-1}$) create sites for interaction with polymer chains. Apart from the thermodynamic factors, the increased viscosity and local shear forces, evidenced in presence of nanofillers, contribute to the improvement of the miscibility of incompatible polymer phases. The large surface to volume ratio is efficient in the reduction of the interfacial tension between the phases [71]. Application of this compatibilization concept to fire retardancy proved to be advantageous in PP containing phosphorylated epoxy resin (PEP). Montmorillonite nanoclay was introduced to improve dispersion by nanocompatibilization. The result is shown in Figure 3.

Without nanoclay the distribution of epoxy in PP matrix was poor, while homogeneous dispersion could be achieved after attachment of PEP to clay nanoparticles as interlayer. The improved FR performance is ascribed to homogeneous char-forming action of PEP interlayer [72]. Nonleaching intumescent flame retardant system can be produced this way using additives in REACH-compliant

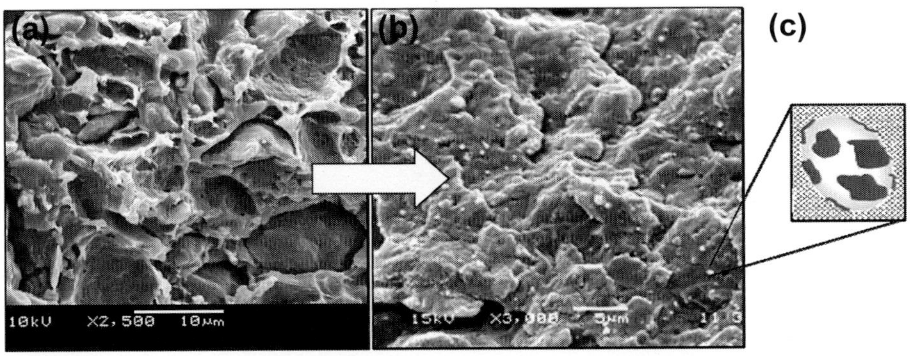

FIGURE 3

Compatibilization of immiscible PP and PEP polymer phases with nanoclay: (a) scanning electron microscope (SEM) image of the blend without montmorillonite, (b) SEM image of the blend with montmorillonite, (c) scheme of the interfacial region with montmorillonite nanolayers.

oligomeric, polymeric, and/or polymerizable form, combining the coupling and/or cross-linking functions with FR action.

Responsive interlayers represent, however, the most challenging design of a phase border. A signal from its environment initiates irreversible or reversible structural change in such an interlayer. The inspiration for designing responsive structures came from the biological IPs where this behavior is common. Regarding fire retardancy, the application of this idea involves the development of nonreversibly adaptive IP, the transformation of which is induced by a trigger signal (e.g. high temperature) resulting in increased protecting capability [73]. The transporter IP, for example, acts as a carrier layer, which is activated only by a signal of environmental change. Fire retardancy of PP was improved via this way [10]. The role of this IP at normal environment, similarly to other multiphase systems, is to enhance the compatibility, facilitate the processability, and enhance the mechanical properties of the formulation [74]. However, at elevated temperatures caused by fire attack, the IPs around the particles promotes the development of a heat-insulating protective char-foam, delivering the particles to the surface where a fire-protective barrier IP is formed. Thermally induced detachment of a compatibilizing unit from the surface of fire retardant makes the underlying incompatible polymer uncovered, which, acting as a carrier layer, delivers the flame retardants to the surface [60]. Above the critical temperature the whole interlayer expands forming a fire-protective foam. On the other hand, the converting IP plays an adaptive role by promoting chemical conversion, induced by thermal signal. Temperature-induced ceramization (Figure 4), for example, can transform a boron-containing polysiloxane surface layer that becomes this way intumescent white layer fairly resistant against the heat and oxidative effect of fire [75–78].

FIGURE 4

Melamine borate and boron-containing polysiloxane white ceramizing intumescent layer protecting the underlying polypropylene (photo taken after cone calorimeter treatment).

Silicone-modified clay nanoparticles, utilizing similar mechanism, contribute to the fire retardancy in intumescent PP nanocomposites [56]. Expandable interlayers play their role by causing volume change in the material. Irreversible expansion is initiated by chemical reactions, i.e. curing or decomposition. This effect can be utilized for the enhancement of the efficiency of fire retarded systems. The particles expanded owing to heat and catalytic effects may reach each other, leading to percolation of the layer-coated particles. Compared to the conventional intumescent fire retarded systems, in which continuous char layer is formed only after the decomposition of a considerable amount of matrix polymer, the expanding particles of a "percolating intumescent system" reach each other in earlier stage of the process providing rapid protection to the underlying polymer phase [79]. Parameters influencing the percolation are the interparticle distance, the volume increase, and the rate of action [80]. Expandable interlayers also provide a way for rapid delivery of active substances to the surface [10,81]. A material utilizing this mechanism contains interlayer of relatively low decomposition temperature intercalated between nanolayers. At the early stage of fire, the interlayer starts to decompose and the gaseous degradation products rapidly drive the separated nanolayers to the surface. This mechanism is known for expandable graphite and for montmorillonite intercalated by FR-active agent.

Skin technology comprises a simple alternative to the responsive interlayers. When the processing line and the composition allow it, the fire protective surface layer can be formed not only in case of fire, as described in the previous paragraph, but also in advance by forming a coating layer already during processing. Laminating, coinjection molding, and coextrusion processing methods are generally applied for this purpose [44,78,82]. The conversion of the outmost layer of materials to a more fire-resistant one can also be achieved by plasma treatment of their surface coating (e.g. acrylate, polysiloxane) layer [75,83,84]. The most recently proposed method for performing skin technology, the application of electrospun nanofiber mats coating, originates from the fields of tissue engineering and drug delivery (Figure 5) [85].

Electrospinning of 4,4′-bishydroxydeoxybenzoin-polyphosphonate was performed from 65 mass% dimethylformamide (DMF) solution. The heat release capacity and char yield of the formed nanofibers were $70 \pm 3\,J\,g^{-1}\,K^{-1}$, and 53%, respectively, in close agreement with the values obtained from the bulk materials [86]. Polyimide (PI) nanofiber mats were deposited on PA66 through electrospinning. Because of their low thermal conductivity, electrospun nanofiber mats were found to act as a protective surface that delays ignition [87].

As a conclusion of the described examples of the IP modification in fire retarded polyolefins, the environmental gain of this approach can be summarized. As reactive fire retardancy is hardly applicable in the case of some thermoplastic polymers the modification of IPs is the right route for controlling the migration of additives in such systems. Several phosphorous additives are available for the replacement of the halogeneous additives, the most favored of which are the oligomeric/polymeric ones prepared by methods of green chemistry. After successful

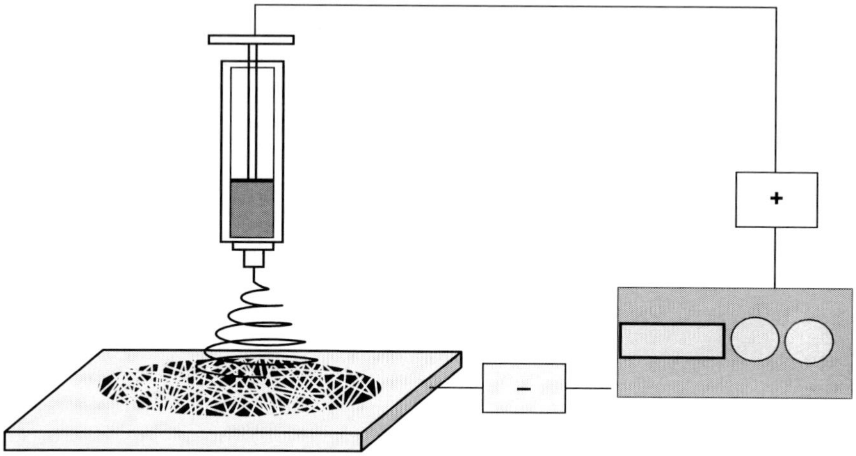

FIGURE 5

Method for the formation of electrospun nanofiber mat coating.

nanocompatibilization, increased efficiency, thus reduced concentration, can be achieved. The recyclability of these fire retarded polymers is an obvious environmental advantage compared to the thermosetting systems.

5. Going more green: recycled and biodegradable matrices

Steps toward more green FR can be made by extending the view to matrices of secondary and renewable resources. Combination of recycling and fire retardancy concepts allows upgrading recycling (upcycling) of polymer wastes [88]. However, the systems using recycled polymers (mostly polyolefins) as matrix are under economic pressure; therefore, these must include the most common phosphorus compounds at the lowest efficient concentration. This concept can be accomplished in an attractively and economically feasible way only by reducing the cost of FR as much as possible and increasing the technical value of the polymer system simultaneously. To achieve economically effective recycling rates, the application of effective and controlled separation techniques is required. Nevertheless, the compensation of the negative effects of the remaining impurity and inhomogeneity is indispensable. For this purpose the application of self-reinforced (SR) composites was considered in the present work. Biodegradable matrices from renewable resources are of increasing importance but up to now mainly in packaging industry. Extension of these matrices to engineering application fields needs various modifications including FR. In this chapter polylactic acid (PLA) and starch matrices as well as their combinations and composites are discussed.

5.1 Flame retarded recyclable self-reinforced composite from secondary plastics

The polyolefin fraction of a mixed plastic waste is the most valuable for thermomechanical recycling. Recyclable polyolefin fractions can mostly be obtained from automotive, electrical and electronic equipments, and packaging waste. The results discussed in this chapter were obtained using recycled density-separated polyolefin fraction ($\rho < 0.92\ \mathrm{g\ cm^{-3}}$) of automotive shredder waste (REC) from Alcufer Ltd. (Hungary). According to thermogravimetry (TG), differential scanning calorimetry (DSC), infrared (IR), and Raman spectroscopic measurements the main component (approximately 92 mass%) of this polyolefin fraction was PP (the minor polymeric components are PE and PS), while its inorganic filler (mainly silicates and calcium carbonate) content was lower than 0.75 mass%. (Similar experiments using real polymer wastes, instead of virgin "waste-modeling" polymer blends, are quite rarely discussed in the literature.)

In order to enhance the mechanical properties of polyolefins conventional glass or natural fiber reinforcement is generally employed. However, the glass or natural fiber-reinforced composites are multicomponent systems, whose reprocessing through the known waste treatment technologies is cost-defective. In contrast their novel alternatives, the SR polymer composites [89], are easy to recycle [90]. In the literature, there are several papers reporting comparison of commercial glass fiber reinforced and SR composites (Kmetty et al. [89], Alcock et al. [91,92] and Cabrera et al. [93]), and reveal that the most important advantages of SR composites are the potential of weight reduction, high tensile strength and impact resistance, excellent viscoelasticity, and recyclability. Thus the SR composites show continually increasing application potential. This potential initiated the development of environmentally friendly composites from secondary raw materials with improved mechanical properties and reduced flammability at the same time (Bocz et al. [94]). To reach this goal multilayered composites were prepared from the polyolefin fraction of automotive shredder waste and reinforcing PP fabrics (Stradom S.A., Czestochowa, Poland; identification code: T-PPT-181, areal weight of $171.5\ \mathrm{g\ m^{-2}}$). The flammability was reduced by modifying the recycled matrix layers with an APP (Exolit AP766, Clariant GmbH)-based intumescent flame retardant additive.

Two types of flame retarded recycled samples, simple compounds and multilayered SR composites, have been prepared with the same IFR content of 18 mass%. In the first case the specimens (REC-IFR18) were prepared by simple mixing of 18 mass% IFR with the recycled matrix (REC) in internal-mixer (Brabender Plasti Corder PL 2000) at 190 °C, with a rotor speed of 30 rpm, for 10 min and then hot pressed to 4-mm-thick specimens in laboratory press (Collin P200E). In case of the multilayered composites, a higher amount (34%) of IFR was introduced into the recycled matrix and 180-μm-thick layers were compression molded at 190 °C in laboratory press (Schwabenthan Polystat 300 S) from the obtained compounds. The multilayered fabric-reinforced composites were manufactured by lamination of 11 recycled matrix layers and 10 woven PP fabrics according to

Table 6 The LOI Values (ASTM D 2863) and UL-94 Ratings (ASTM 1356-90) of the Flame Retarded Recycled Samples

Sample	LOI (v/v%)	UL-94	Burning Rate (mm min^{-1})
REC	20	HB	31.5
REC-IFR18	26	V-1	–
SR-REC-IFR18	30	V-0	–

the film stacking method [95]. The film-stacked packages were compressed to composite in the same press at 169 °C with a pressure of 6 MPa, for 2 min. The flame retarded content of the matrix layers (34%) with the additive-free reinforcing PP fabrics adds up to 18 mass% considering the total mass of the multi-layered SR composite (SR-REC-IFR18). The nominal thickness of the prepared SR composites was 3.78 mm.

In the case of APP-based intumescent flame retardants (IFR) at least 20–22% IFR is required in order to achieve the best (V-0) classification, according to the UL-94 standard, of polyolefins [96]. It is also the same in the case of polyolefin-type recycled matrices when preparing simple compounds. In Table 6, however, it can be seen that with the preparation of multilayered SR composites (SR-REC-IFR18) the concentration of IFR needed for V-0 level can be reduced to 18 mass %. It is also presented, that despite the same IFR content, 4 v/v% higher LOI can be reached by preparing SR composites. This beneficial effect is assumed to be attributed to the physical interaction between the intumescent expanding structure and the highly oriented shrinking tapes. It is presumed that this system exposed to heat forms a special compact charred layer on the surface of the sample owing to the shrinkage of the PP fabrics. In this case the vaporization of phosphorus compounds and the gas release of the nitrogen-containing foaming agent occurs in a much smaller bulk, being as a consequence more effective in the fire extinction. This mechanism has been published by Bocz et al. [97].

Regarding the heat release and mass loss results, the different fire behaviors of the two types of recycled samples is also pronounced. (The mass loss type cone calorimeter tests were carried out by an instrument made by FTT Inc., according to ASTM E 906 standard, 100 × 100 mm specimens were exposed to constant heat flux of 50 kW m^{-2} and ignited.) The calculated experimental results are summarized in Table 7. Although the type of their HRR curves is similar (Figure 6(a)), 30% less total heat released (tHR) was recorded in the case of the SR-REC-IFR18 sample compared to that of the simple compound (REC-FR18). The peak of HRR of the SR composite is also somewhat lower, but it appears almost 2 min earlier. The prolonged burning of REC-FR18 is expressed by the smaller slope of its mass loss rate curve in Figure 6(b); moreover, the residual mass of the SR composite was 5 mass% more than that of the sample prepared by simple compounding.

Table 7 Results of Mass Loss Calorimeter Tests

Sample	Mass (g)	TTI (s)	pk HRR ($kW\,m^{-2}$)	pk HRR_{time} (s)	tHR ($MJ\,m^{-2}$)	Residue (%)
REC	34.1	23	875.0	182	109.5	0.3
REC-IFR18	39.8	24	253.1	728	108.8	3.7
SR-REC-IFR18	36.9	25	213.1	613	78.0	8.1

TTI: time to ignition, pk HRR: peak of heat release rate, pk HRR_{time}*: time of peak heat release rate, tHR: total heat released.*

This difference is more obvious when comparing the photographs taken of the residues (Figure 7).

The most significant benefit of the SR materials, besides the observed enhanced FR effectiveness, is their mechanical performance, such as tensile and flexural strength and impact resistance. (The tensile tests were performed on specimens 20 × 150 mm using a Zwick Z020 machine with a crosshead speed of 5 mm min, while the instrumented falling weight impact tests were performed on Ceast-Fractovis 6785 machine using the following settings: maximal energy: 228.64 J; diameter of the dart: 20 mm; diameter of the support rig: 40 mm; weight of the dart: 23.62 kg; and drop height: 1 m. Specimens 70 × 70 mm were subjected to the impact tests, from which the specific perforation energy, i.e. the thickness-related total energy (joules per millimeter) was determined. These properties could be significantly improved compared to the secondary polymer matrix. For example, the tensile strength of the SR flame retarded recycled composite is

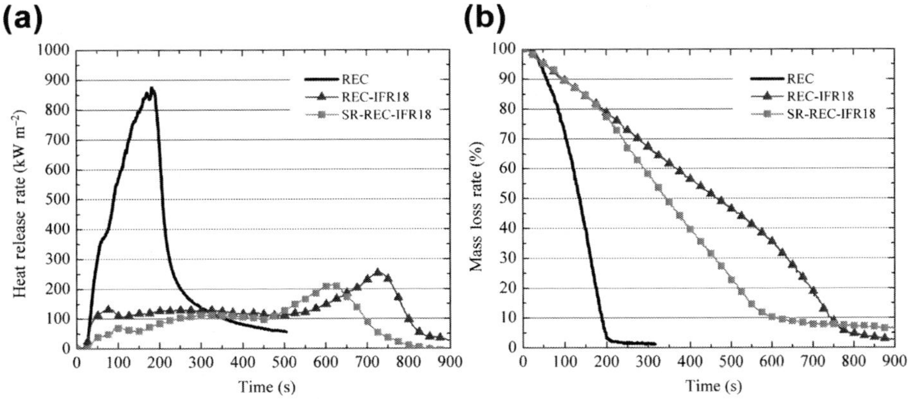

FIGURE 6

Heat release rate (a) and mass loss rate (b) measured during the burning of flame retarded recycled samples.

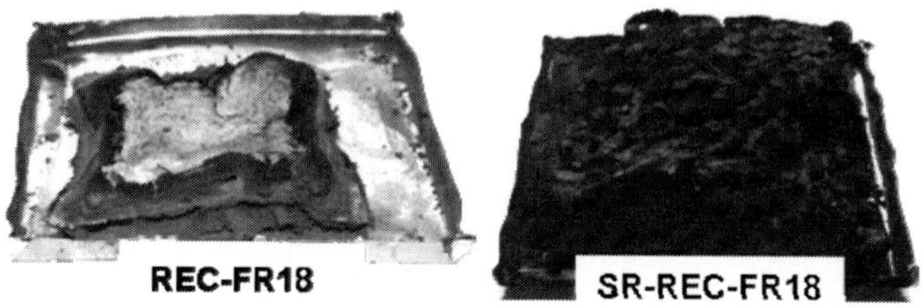

FIGURE 7

Photographs taken of the charred residues obtained after mass loss calorimetric tests.

fivefold higher than that of the reference secondary matrix, as also presented in Figure 8. The obtained prominent perforation energy was more than four times higher than that of the recycled reference matrix) of the prepared recycled SR composite (Figure 9), suggesting that these materials can be advantageous in many fields of application where shock-resistant protection (covers, equipment housing, etc.) is required.

Therefore, it can be concluded that a special synergistic effect was observed between an APP-based intumescent flame retardant additive and highly oriented PP fabrics in recycled matrix. This interaction results in improved FR such as 4 v/v% higher LOI value, and better UL-94 classification (V-0) than expected. Furthermore, in the case of the recycled multilayered composites with reduced flammability, excellent mechanical properties can be guaranteed, thus the preparation of such composites can be considered as an excellent example of upgrading recycling of plastic waste [94]. Consequently, required levels of FR and increased strength can be achieved with lower amount of flame retardants resulting in cost-effective, more environmental, and higher market value secondary products.

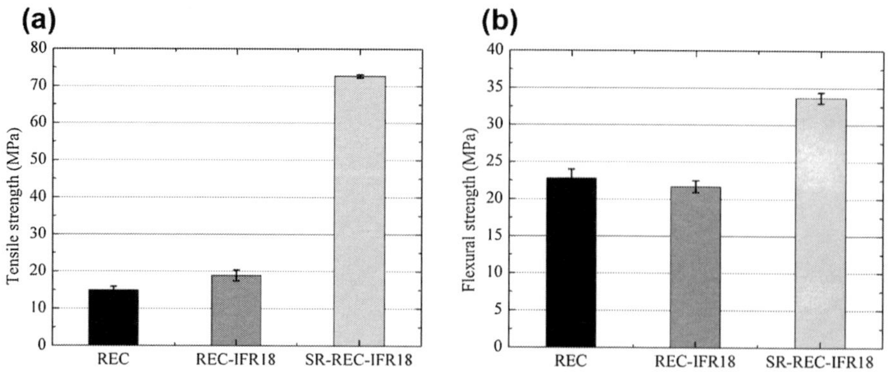

FIGURE 8

Tensile (a) and flexural strength (b) of the recycled composites.

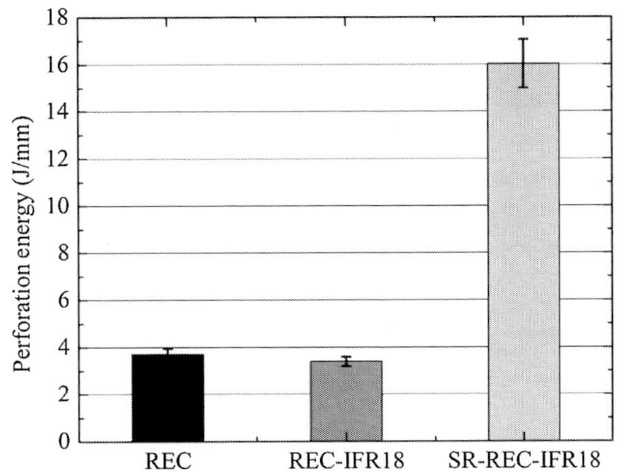

FIGURE 9

Specific perforation energy (Ep) of the recycled composites.

5.2 Flame retarded polymer systems from renewable resources

New flame retarded polymers should be as "green" as possible, thus the systems based on biodegradable renewable resources releasing no toxic products in the environment have to be discussed in this context as well. All the strategies mentioned up to now can be applied on the biomatrices.

Reactive modification applied to soybean oil copolymers, PLA, and lignin resulted in phosphorus-containing polymers showing significantly reduced flammability (Sacristan et al. [98] and Yuan et al. [99]). Biodegradable epoxy resins from renewable sources such as plant oils (e.g. soybean, linseed, and cottonseed oils (these are triglycerides of unsaturated fatty acids and can be easily epoxidized on the C—C double bonds)) are the most commonly used biobased epoxy resins [100]. Due to the high amount of glycerol ester bonds, which are easily hydrolyzable in soil, these thermosetting polymers are potentially (bio)degradable [101], especially when cured with anhydrides. According to Lligadas et al. [102,103] these epoxidized plant oils can react with the commonly used flame retardant DOPO, to form fire retarded, biodegradable epoxy resins from renewable resources.

Several examples for FR of biobased polymers, such as thermoplastic starch (TPS) (Matkó et al. [104]), PLA (Bourbigot et al. [105]), etc., incorporating conventional flame retardants and nanofillers have been published. In particular, it has been found by Wu et al. [106] that in PVA/glycerol-plasticized TPS and its intumescent flame retardant composites, microencapsulated form of APP is necessary in order to restrain the reaction between APP and starch during processing; these composites with only 2 mass% APP can pass V-0 in UL 94 test. Most of the starch-based thermoplastic polymer blends are, however, mechanically not acceptable when sufficient amounts of flame retardants are introduced. Furthermore, melt-blending of starch with other biodegradable polymers is also difficult since very few polymers are compatible with starches. Nonetheless, incorporating nanoparticles, coated with RDPs, can modify the interfacial interactions promoting good dispersion and synergism. Various combinations of RDP-modified starch, RDP-modified clays, and RDP-halloysite tubes have been reported by Yang et al. [107] to form flame retardant biodegradable blends with PLA passing the UL-94 V-0 test. The RDP-coated clays are localized at the polymer interfaces, reducing the interfacial tension and contributing to compatibilization. As the total amount of added particles is less than 10%, the embrittlement can be minimized.

Achieving a good balance of the fire safety and mechanical properties is a challenge in the case of biopolymers as well. A possible solution is the combination of fire retarded biopolymer with reinforcing fabrics. Figure 10 shows the heat release curves of composites consisting of flax fiber and modified starch-containing phosphorous plasticizer. For achieving V-0 rating in this system, the presence of 2% P atom is needed.

Hence, the TPS applied in reinforced and flame retarded forms seems to be promising to certain purposes as cost-effective fully biodegradable alternative of synthetic polymers.

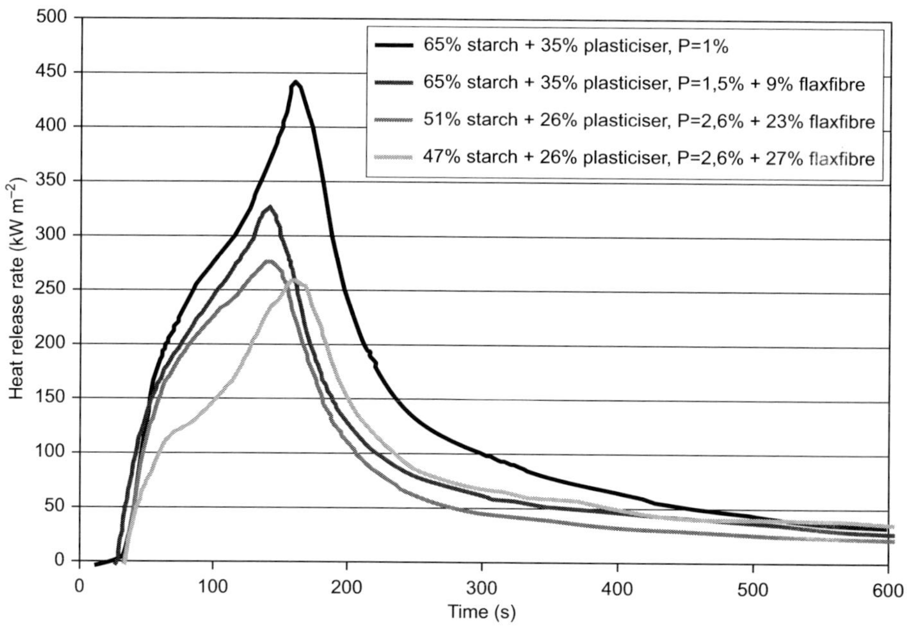

FIGURE 10

Heat release rate of flame retarded thermoplastic starch reinforced with flax fiber (heat flux: 50 kW m^{-2} in mass loss type cone calorimeter).

6. CONCLUSIONS

Green flame retardants are mostly expected to be free of halogen. The applicability of the metal hydroxides is restricted by the high concentration required for their efficiency, while the phosphorous flame retardants can be used in all polymer types. As P element exists in several oxidation states, a wide variety of FR molecules can be synthesized. Their green application requires enhanced efficiency and nonmigrating character during use. In the case of epoxy resins the best solutions are provided by reactive fire retardants, the most recent examples of which have been discussed in this chapter (the most important ones are summarized in the appendix). New details are presented about the effect of a flame retardant cross-linking agent made from alkyl phosphate and a diamine and applied here in multilayer composite form.

Comparison of reactive and additive fire retardant types in TPU has also been presented with the conclusion that the reactive one meets even the complex requirements (including good processability) of TPU cable coating. The types of relevant interface modifications promoting the fire retardancy of thermoplastics are discussed in the case of PP, highlighting the increasing importance of polymeric additives and nanocompatibilization. Nanodispersion of PEP in PP could be achieved via this way.

Two even "greener" approaches, the use of recycled and renewable resources in fire retarded systems, have been proposed in the last part of this chapter. The self-reinforced fire retarded polypropylene (SRFRPP), by significantly improving both the mechanical and fire retardancy performance of the material, provides an economically attractive route for recycling. The examples shown for successful formation of flame retardant systems from fully renewable resources seem promising for the future. These phosphorus-containing biodegradable polymers can act in the soil as fertilizers.

7. Appendix: resume of the results of studies on phosphorus-based flame retardants

Matrix	FR Composition	FR Result	Other Remarks	References
Modified DGEBA epoxy	DOPO 0–3.5% P	V-0 at 3% P cured with DDM and at 2.6% P cured with DETDA	Decreasing cross-link density	[15]
Modified novolac resin	DPPO 0.5–2.0% P	V-0 at 0.6%P cured with DDM	LOI: 26.6	[16]
Modified DGEBA epoxy	DODPP 0–3.0% P	V-0 at 2.5% P cured with low molecular weight poliamide	Char yield at 700 °C in N_2 21.4%	[17]

Continued

—Cont'd

Matrix	FR Composition	FR Result	Other Remarks	References
Modified DGEBA epoxy	DOPO-NQ	V-0 at 2% P cured with dicyanate ester	Char yields at 700 °C in N_2 ~ 20%	[19]
Gly-HPO P-containing epoxy	BAMPO P-containing curing agent 8.5% P	LOI: 32	Char yield at 800% in N_2 25%; in air 29%	[20]
BPHPPO P-containing epoxy	P content 7.8% cured with DDS	LOI: 34	Char yield at 800% in N_2 51.8%; in air 52.1%	[22]
TGDDM epoxy, partially replaced	P-containing epoxies: Glycidyl phosphonates and phosphates	Immediate extinguishing, cured with DDS	Char yields: 28–44% in N_2 at 800 °C	[23]
TGDDM epoxy	Cured with BAMPO	Immediate extinguishing at 4%P	Char yield at 800 °C in N_2 23%	[23]
DGEBA epoxy, partially replaced	HGCP cyclotriphosphazene-based epoxy component	V-0 by replacing 20% of DGEBA, cured with DDM	Char yield at 800 °C in air 20%	[24]
DGPP P-containing epoxy	P content 7.4% cured with BPOD	LOI: 47	Char yield in N_2 48%	[26]
TGDDM epoxy	Cured with BAPP	Peak of heat release reduced by 50% (reference hardener DETDA)	Char yield at 800 °C in N_2 40.5%, in air 8.5%	[28]
DGEBA epoxy	Cured with P-containing poly(ethylene diamine) 4.07%P	LOI:31	Char yield at 850 °C in air 12.2%	[30]
Novolac resin	Cured with TAPP	V-1 at 3.2% P		[32]
Siliconized DGEBA epoxy	Cured with 2DOPO-A	LOI: 35 at 2.35% P and 4.57% Si	Char yield at 700 °C in N_2 24.9%, in air 14.6%	[37]
DGEBA epoxy	Cured with amino-functionalized DOPO 4.5%P	LOI: 35	Char yield at 750 °C in N_2 35.2%	[38]
PER-based epoxy	Cured with TEDAP	V-0 at 2.5%P	Peak of heat release reduced by 90% at 3.5% P	[43]
TPU	15–25% Phosphinate, 0–5% phosphate, 2.5–10% pentaerythritol or dipentaerythritol, 0–5% inorganic filler, 0.05–2% antioxidant	UL-94: V-0 LOI: 35%	Good melt processing characteristics and mechanical properties	[45]

—Cont'd

Matrix	FR Composition	FR Result	Other Remarks	References
TPU	2–20% Phosphine oxide(s), 5–50% melamine derivate(s),	UL-1581 Self-extinguishing	Extrusion quality	[46]
TPU	0.5–25% red phosphorus, 0.01–20% trialkyl phosphate and triaryl phosphate 5–50% melamine cyanurate, melamine phosphate, melamine polyphosphate and/or melamine borate	No dripping	Without lowering mechanical properties	[47]
TPU	1–20% Organic phosphate-based FR, 5–50% melamine or melamine derivative, 0.1–1% active isocyanate groups	No dripping		[48]
TPU	Conventional FR, 0.1–30% isocyanate dissolved in polyurethane	UL-94: V-2, V-1 or V-0	Increased mechanical stability	[49]
Thermoplastic elastomer and/or ethylene-based copolymer	0–13% red phosphorus, 33–67% metal hydrates	Not specified	Moisture resistant	[50]
TPU	Red phosphorus, melamine cyanurate, magnesium or aluminum hydroxide	Self-extinguishing		[51]
TPU	20% ammonium polyphosphate microencapsulated with hydroxyl silicone oil	UL-94: V-0		[53]
TPU	30% pentaerythritol phosphate alcohol- and melamine phosphate-containing intumescent FR 5% pentaerythritol	UL-94: V-0	4.5 mass% P	[55]
TPU	1.92% isocyanate 2.08% phosphorus-containing polyol-type FR 5% zinc borate 19% melamine-cyanurate 1% hindered amine stabilizer 1% silica	UL-94: V-0	0.4 mass% P	[55]
PP	21% APP 7% PER 1.5% SIL 0.5% Clay	UL-94: V-0		[56]
PS	5% Anionic surfactant modified MMT	64% Reduction of pk HRR		[59]

—Cont'd

Matrix	FR Composition	FR Result	Other Remarks	References
EVA	Complex of cyclodextrin nanosponge-phosphorus compounds (TPP, TEP, APP, APb, PhEt)	UL-94:V-0, pk HRR decreased up to 84%		[67]
PMMA	20% DB 5% AO 5% Cloisite 20A	UL-94:v-0, LOI: 25.6, pkHRR reduced by 75%		[71]
PP	31% IFR 2% BSil 1% MMT	UL-94: V-0, LOI: 37		[76]
PA-6	Coatings Obtained from polymerization of 1.1.3.3-tetramethyldisiloxane (TMDS) monomer mixed with oxygen	LOI: 25, pkHRR decreased by 28%		[83]
PA-6	Coated by AC8	50% Decrease in pk HRR		[84]
PA66	PI nanofiber mat	Increase in TTI up to 60 s		[87]
Self-reinforced recycled automotives shredder polyolefin waste	18% intumescent FR	UL94:V-0, LOI: 30, pkHRR reduced by 75%	FR self-reinforced composites outperform the nonreinforced recycled matrices of identical FR contents	[94]
Self-reinforced PP	9% intumescent FR	UL94:V-0, LOI: 26, pkHRR reduced by 50%	FR self-reinforced composites outperform the nonreinforced PP mixtures of identical FR contents	[97]
Soybean oil copolymers	P-containing styrene derivatives (diphenyl styryl phosphine oxide and dimethyl-p-vinylbenzylphosphonate) as comonomers	LOI: 24	1% P	[98]
PLA	HQ-DOPO chemically incorporated into the PLA backbone	UL94: V-0, LOI: 33	2% P	[99]

—Cont'd

Matrix	FR Composition	FR Result	Other Remarks	References
Fatty acid diepoxide	Cured with DOPO/BAMPO and DOPO/DDM	LOI: 32 for DOPO/BAMPO and LOI: 31 for DOPO/DDM	Char yield at 800 °C in N_2 16%; in air 16% for DOPO/BAMPO and in 18% N_2 and in air 10% for DOPO/DDM	[102,103]
TPS	10% APP	UL94: V-0, LOI: 33		[104]
TPS	Microencapsulated APP	UL94: V-0, LOI: 21	Char yield at 700 °C in N_2 21.8%	[106]

Abbreviations

BAMPO bis(3-aminophenyl)methylphosphinoxide
BAPP bis(4-aminophenyl)phenylphosphate
BPOD 2,5-bis(p-aminophenyl)-1,3,4-oxadiazole
DDM 4,4′-diaminodiphenylmethane
DDS diamino-4,4′-diphenylsulphone
DETDA diethyldiaminotoluene
DGEBA diglycidyl ether of bisphenol-A
DODPP 4-[(5,5-dimethyl-2-oxide-1,3,2-dioxaphosphorinane-4-yl)oxy]-phenol
DOPO 9,10-dihydro-9-oxa-10-phosphaphenantrene-10-oxide
DPPO 2,8-dimethyl-phenoxaphosphin-10-oxide
EVA ethylene vinyl acetate
HGCP hexaglycidyl cyclotriphosphazene
MDH magnesium dihydroxide
PA polyamide
PE polyethylene
PET polyethylene terephtalate
PMMA poly (methyl methacrylate)
PO polyolefine
PS polystyrene
PTFE poly(tetrafluoroethylene)
PVA poly(vinylalcohol)
REACH Registration, Evaluation, Authorisation and Restriction of Chemicals
TAPP tris(3-aminophenyl)phosphate
TEDAP N,N′,N″-tris(2-aminoethyl)-phosphoric acid triamide
TGDDM tetraglycidyl-4,4′-diaminodiphenylmethane
TGMDA N,N,N′,N′-tetraglycidyl-4,4′-methylenedianiline

Acknowledgments

This work is connected to the scientific program of the "Development of quality-oriented and harmonized R + D + I strategy and functional model at BME" project supported by the New Széchenyi Plan (Project ID: TÁMOP-4.2.1/B-09/1/KMR-2010-0002). The project was supported by an EU7 Framework Program No. 212782 (W2PLASTICS), by RECYTECH, TECH_08-A4/2-2008-0142 project, and by the Hungarian Scientific Research Fund (OTKA K73346 and OTKA NN82462). A. Toldy is thankful to the János Bolyai Scholarship of the Hungarian Academy of Science.

References

[1] Narayan S, Moore M. A new, versatile flame retardant for olefinic and styrenic polymers. Spec Chem 2011;11:24–5.

[2] Marosfői B, Garas S, Bodzay B, Zubonyai F, Marosi G. Flame retardancy study on magnesium hydroxide associated with clays of different morphology in polypropylene matrix. Polym Adv Technol 2008;19:693–700.

[3] Hornsby PR, Cusack PA, Cross M, Toth A, Zelei B, Marosi G. Zinc hydroxystannate-coated metal hydroxide fire retardants: fire performance and substrate-coating interactions. J Mater Sci 2003;38:2893–9.

[4] Isarov A, Temples D, Chen T, Walker JK. Molybdate/borate complexes for enhanced cable compound fire performance. In: Proceedings of the 60th international wire & cable symposium (IWCS). North Carolina, USA: Charlotte; November 6–9, 2011. pp. 250–6.

[5] Khunová V, Kristóf J, Kozánková J. Polypropylene nanocomposites based on halloysite and montmorillonite: the effect of dimaleinimide reactive modifiers on structure and properties. Chem Listy 2011;105:277–9.

[6] Du M, Guo B, Jia D. Thermal stability and flame retardant effects of halloysite nanotubes on poly(propylene). Eur Polym J 2006;42:1362–9.

[7] Lu SY, Hamerton I. Recent developments in the chemistry of halogen-free flame retardant polymers. Prog Polym Sci 2002;27:1661–712.

[8] Weil ED. Phosphorus-based flame retardants. In: Engel RE, editor. Handbook of organophosphorus chemistry. New York, USA: Marcel Dekker; 1992.

[9] Aaronson AM. In: Walsh EN, Griffith EJ, Parry RW, Quin LD, editors. Phosphorus chemistry: developments in American science, ACS symposium series 468. American Chemical Society; 1992. pp. 218–28.

[10] Gy Marosi, Marosfői BB, Bodzay B, Igricz T, Gy Bertalan. Design of interlayers for fire-retarded polymeric systems. In: Wilkie CA, Morgan AB, editors. Fire retardancy of polymeric materials. 2nd ed. Florida, USA: CRC Press; 2009. pp. 329–47.

[11] Jain P, Choudhary V, Varma IK. Flame retarding epoxies with phosphorus. J Macromol Sci-Pol R 2002;C42:139–83.

[12] Levchik SV, Weil ED. Thermal decomposition, combustion and flame-retardancy of epoxy resins – a review of the recent literature. Polym Int 2004;53:1901–29.

[13] Rakotomala M, Wagner S, Döring M. Recent developments in halogen free flame retardants for epoxy resins for electrical and electronic applications. Materials 2010; 3:4300–27.

[14] Wang CS, Lin CH. Synthesis and properties of phosphorus-containing epoxy resin by novel method. J Polym Sci Pol Chem 1999;37:3903–9.

[15] Schäfer A, Seibold S, Walter O, Döring M. Novel high T_g flame retardancy approach for epoxy resins. Polym Degrad Stabil 2008;93:557–60.

[16] Schäfer A, Seibold S, Lohstroh W, Walter O, Döring M. Synthesis and properties of flame-retardant epoxy resin based on DOPO and one of its analog DPPO. J Appl Polym Sci 2007;105:685–96.

[17] Gao LP, Wang DY, Wang YZ, Wang JS, Yang B. A flame-retardant epoxy resin based on a reactive phosphorus-containing monomer of DODPP and its thermal and flame-retardant properties. Polym Degrad Stabil 2008;93:1308–15.

[18] Ho TH, Hwang HJ, Shieh JY, Chung MC. Thermal, physical and flame-retardant properties of phosphorus-containing epoxy cured with cyanate ester. React Funct Polym 2009;69:176–82.

[19] Ho TH, Hwang HJ, Shieh JY, Chung MC. Thermal and physical properties of flame-retardant epoxy resins containing 2-(6-oxido-6H-dibenz <c,e> <1,2> oxaphosphorin-6-yl)-1,4-naphthalenediol and cured with dicyanate ester. Polym Degrad Stabil 2008;93:2077–83.

[20] Spontón M, Mercado LA, Ronda JC, Galià M, Cádiz V. Preparation, thermal properties and flame retardancy of phosphorus- and silicon-containing epoxy resins. Polym Degrad Stabil 2008;93:2025–31.

[21] Spontón M, Ronda JC, Galià M, Cádiz V. Cone calorimetry studies of benzoxazine-epoxy systems flame retarded by chemically bonded phosphorus or silicon. Polym Degrad Stabil 2009;94:102–6.

[22] Ren H, Sun J, Wu B, Zhou Q. Synthesis and properties of a phosphorus-containing flame retardant epoxy resin based on bis-phenoxy (3-hydroxy) phenyl phosphine oxide. Polym Degrad Stabil 2007;92:956–61.

[23] Hergenrother PM, Thompson CM, Smith Jr JG, Connell JW, Hinkley JA, Lyon RE, et al. Flame retardant aircraft epoxy resins containing phosphorus. Polymer 2005; 46:5012–24.

[24] El Gouri M, El Bachiri A, Hegazi SE, Rafik M, El Harfi A. Thermal degradation of a reactive flame retardant based on cyclotriphosphazene and its blend with DGEBA epoxy resin. Polym Degrad Stabil 2009;94:2101–6.

[25] Liu R, Wang X. Synthesis, characterization, thermal properties and flame retardancy of a novel non-flammable phosphazene-based epoxy resin. Polym Degrad Stabil 2009;94: 617–24.

[26] Sudhakara P, Kannan P. Diglycidylphenylphosphate based fire retardant liquid crystalline thermosets. Polym Degrad Stabil 2009;94:610–6.

[27] Liu W, Varley RJ, Simon GP. A phosphorus-containing diamine for flame-retardant, high-functionality epoxy resins. I. Synthesis, reactivity, and thermal degradation properties. J Appl Polym Sci 2004;92:2093–100.

[28] Liu W, Varley RJ, Simon GP. Understanding the decomposition and fire performance processes in phosphorus and nanomodified high performance epoxy resins and composites. Polymer 2007;48:2345–54.

[29] Ananda kumar S, Denchev Z, Alagar M. Synthesis and thermal characterization of phosphorus containing siliconized epoxy resins. Eur Polym J 2006;42:2419–29.

[30] Jeng RJ, Wang JR, Lin JJ, Liu YL, Chiu YS, Su WC. Flame retardant epoxy polymers using phosphorus-containing polyalkylene amines as curing agents. J Appl Polym Sci 2001;82:3526–38.
[31] Jeng RJ, Shau SM, Lin JJ, Su WC, Chiu YS. Flame retardant epoxy polymers based on all phosphorus-containing components. Eur Polym J 2002;38:683–93.
[32] Mauerer O. New reactive, halogen-free flame retardant system for epoxy resins. Polym Degrad Stabil 2005;88:70–3.
[33] Braun U, Balabanovich AI, Schartel B, Knoll U, Artner J, Ciesielski M, et al. Influence of the oxidation state of phosphorus on the decomposition and fire behaviour of flame-retarded epoxy resin composites. Polymer 2006;47: 8495–508.
[34] Wu CS, Liu YL, Chiu YS. Synthesis and characterization of new organosoluble polyaspartimides containing phosphorus. Polymer 2002;43:1773–9.
[35] Artner J. Darstellung und erprobung aminofunktionalisierter, phosphororganischer flammschutzmittel [Ph.D. thesis]. Heidelberg: Ruprecht-Karls-Universität; 2008.
[36] Sun D, Yao Y. Synthesis of three novel phosphorus-containing flame retardants and their application in epoxy resins. Polym Degrad Stabil 2011;96:1720–4.
[37] Wu CS, Liu YL, Chiu YS. Epoxy resins possessing flame retardant elements from silicon incorporated epoxy compounds cured with phosphorus or nitrogen containing curing agents. Polymer 2002;43:4277–84.
[38] Schartel B, Braun U, Balabanovich AI, Artner J, Ciesielski M, Döring M, et al. Pyrolysis and fire behaviour of epoxy systems containing a novel 9,10-dihydro-9-oxa-10-phosphaphenanthrene-10-oxide-(DOPO)-based diamino hardener. Eur Polym J 2008;44:704–15.
[39] Wang X, Hu Y, Song L, Xing W, Lu H. Preparation, mechanical properties, and thermal degradation of flame retarded epoxy resins with an organophosphorus oligomer. Polym Bull 2011;67:859–73.
[40] Rupper P, Gaan S, Salimova V, Heuberger M. Characterization of chars obtained from cellulose treated with phosphoramidate flame retardants. J Anal Appl Pyrol 2010;87: 93–8.
[41] Gaan S, Mauclaire L, Rupper P, Salimova V, Tran TT, Heuberger M. Thermal degradation of cellulose acetate in presence of bis-phosphoramidates. J Anal Appl Pyrol 2011;90:33–41.
[42] Toldy A, Anna P, Marosi GJ. Process for the preparation of amine functional phosphoric amides and their use as flame retardant and curing agent for epoxy resins; 2009. WO 2009/07779.
[43] Toldy A, Anna P, Csontos I, Szabó A, Marosi G. Intrinsically flame retardant epoxy resin - fire performance and background – part 1. Polym Degrad Stabil 2007;92: 2223–30.
[44] Toldy A, Szolnoki B, Marosi G. Flame retardancy of fibre-reinforced epoxy resin composites for aerospace applications. Polym Degrad Stabil 2011;96:371–6.
[45] Siddhamalli SK, Brown CA. Non halogen flame retardant thermoplastic polyurethane; 2007. WO/2006/121549.
[46] Peerlings H, Winkler J. Self-extinguishing thermoplastic polyurethanes, a process for their production and their use. 2007. US Patent 20080132615.
[47] Lee HY, Lee TW, Kim DS, Kim SK. Halogen-free flame retardant thermoplastic polyurethane composite resin composition prepared from thermoplastic polyurethane, red phosphorus and melamine derivative; 2008. KR2007055886-A.

[48] Kim DS, Kim NJ, Kim YW. Halogen-free flame retardant thermoplastic polyurethane composition having excellent mechanical properties without containing a halogen-based flame retardant; 2008. KR1020080057636.
[49] Henze OS, Hansen M, Hackmann K, Meier D, Beckmann C, Mühren O. Halogen-free flame-retardant TPU; 2009. WO/2009/103765.
[50] Kuwasaki Y, Nishiguchi M. Flame retardant optical-fiber strand for wire-cable, has secondary coating layer consisting of inner layer and outer layer comprising resin composition containing preset amount of metal hydrate, triazine derivative compound and clay; 2007. JP2007219325-A.
[51] Yoon SH, Eom YB, Sung IK. Thermoplastic polyurethane having good self-extinguishing property or excellent flame retardancy, and a preparation method thereof. 2008. Korean Patent 1020080095097.
[52] Pinto UA, Yuan Visconte LL, Gallo J, Reis Nunes RC. Flame retardancy in thermoplastic polyurethane elastomers (TPU) with mica and aluminum trihydrate (ATH). Polym Degrad Stabil 2000;69:257–60.
[53] Chen X, Jiao C, Zhang J. Microencapsulation of ammonium polyphosphate with hydroxyl silicone oil and its flame retardance in thermoplastic polyurethane. J Therm Anal Calorim 2011;104:1037–43.
[54] Bourbigot S, Turf T, Bellayer S, Duquesne S. Polyhedral oligomeric silsesquioxane as flame retardant for thermoplastic polyurethane. Polym Degrad Stabil 2009;94:1230–7.
[55] Toldy A, Harakály Gy, Szolnoki B, Zimonyi E, Marosi Gy. Flame retardancy of thermoplastics polyurethanes. Polym Degrad Stabil 2012;97:2524–30.
[56] Marosi Gy, Anna P, Márton A, Bertalan Gy, Bóta A, Tóth A, et al. Flame retarded polyolefin systems of controlled interphase. Polym Advan Technol 2002;13:1103–11.
[57] Jimenez M, Duquesne S, Bourbigot S. Intumescent fire protective coating : toward a better understanding of their mechanism of action. Thermochim Acta 2006;449:16–26.
[58] Anna P, Marosi Gy, Csontos I, Bourbigot S, Le Bras M, Delobel R. Influence of modified rheology on the efficiency of intumescent flame retardant systems. Polym Degrad Stabil 2001;74:423–6.
[59] Wang Z, Du X, Yu H, Jiang Z, Liu J, Tang T. Mechanism on flame retardancy of polystyrene/clay composites - the effect of surfactants and aggregate state of organoclay. Polymer 2009;50:5794–802.
[60] Marosi G, Ravadits I, Bertalan Gy, Anna P, Maatoug MA, Tóth A, et al. Role of migration process in the efficiency of intumescent flame retardant additives in polypropylene. In: Camino G, Le Bras M, Bourbigot S, Delobel R, editors. Fire retardancy of polymers: the use of intumescence. London: Royal Society of Chemistry; 1998. pp. 325–40.
[61] Bourbigot S, Duquesne S. Intumescence and nanocomposites: a novel route for flame-retarding polymeric materials. In: Morgan AB, Wilkie CA, editors. Flame retardant polymer nanocomposites. Hoboken: John Wiley & Sons; 2007. p. 131.
[62] Lecouvet B, Sclavons M, Bourbigot S, Bailly C. Fire retardancy of polypropylene nanocomposites based on halloysite nanotubes: synergistic effects with intumescent flame retardant. In: Wilkie CA, editor. Recent advances in flame retardancy of polymeric materials XXIII. Stamford: BCC. Inc; 2012. p. 225.
[63] He L, Li D, Yang R. Flammability and thermal stability studies of cyclotriphosphazene derivative containing silicon in halogen-free flame retardant polypropylene. In: Wilkie CA, editor. Recent advances in flame retardancy of polymeric materials XXII. Stamford: BCC. Inc; 2012. p. 305.

[64] Kardos JL. In: Ishida H, Kumar G, editors. Molecular characterisation of composites interfaces. New York: Plenum Publishing Corporation; 1985.

[65] Marosi G, Bertalan G. Role of interfaces in multicomponent polymer systems in modification and blending of synthetic and natural macromolecules for preparing multiphase structure and functional materials. In: Ciardelli F, Penczek S, editors. NATO advanced study institute series. Pisa, vol. 175; 2004. pp. 135–54.

[66] Cavalli R, Trotta F, Tumiatti W. Cyclodextrin-based nanosponges for drug delivery. J Incl Phenom Macro 2006;56:209–13.

[67] Alongi J, Pošković M, Frache A, Trotta F. Novel flame retardants containing cyclodextrin nanosponges and phosphorus compounds to enhance EVA combustion properties. Polym Degrad Stabil 2010;95:2093–100.

[68] Howell BA, Dumitrascu A, Al-Omari M. Chitosan as a base for the development of green phosphorus flame retardants. In: Wilkie CA, editor. Recent advances in flame retardancy of polymeric materials XXIII. Stamford: BCC. Inc; 2012. p. 142.

[69] Chen L, Wang YZ. Aryl polyphosphonates: useful halogen-free flame retardants for polymers. Materials 2010;3:4746–60.

[70] Paul DR, Robeson LM. Polymer nanotechnology: nanocomposites. Polymer 2008;49: 3187–204.

[71] Zaitsev MSV, Goldman M, Frenkel A, Peiffer DG, Weil E, Sokolov JC, et al. Self-extinguishing polymer/organoclay nanocomposites. Polym Degrad Stabil 2007;92: 86–93.

[72] Szabó A, Marosfői B, Anna P, Marosi G. Complex micro-analysis assisted design of fire retardant nanocomposites - contribution to the nano-mechanism. In: Kandola K, Hull R, editors. Fire retardancy of polymers, new strategies and mechanisms, vol. 4. Cambridge: The Royal Society of Chemistry; 2008. pp. 74–91.

[73] Keszei S, Matkó Sz, Bertalan Gy, Anna P, Marosi Gy, Tóth A. Progress in interface modifications: from compatibilization to adaptive and smart interphases. Eur Polym J 2005;41:697–705.

[74] Marosi Gy, Anna P, Bertalan Gy, Szabó Sz, Ravadits I, Papp J. Role of interface modification in flame retarded multiphase polyolefin systems. In: Nelson G, Wilkie C, editors. Fire and polymers, vol. 797. Washington: ACS Series; 2001. pp. 161–71.

[75] Ravadits I, Tóth A, Marosi G, Márton A, Száp A. Organosilicon surface layer on polyolefins to achieve improved flame retardancy through oxygen barrier effect. Polym Degrad Stabil 2001;74:414–22.

[76] Marosi Gy, Márton A, Szép A, Csontos I, Keszei S, Zimonyi E, et al. Fire retardancy effect of migration in polypropylene nanocomposites induced by modified interlayer. Polym Degrad Stabil 2003;82:379–85.

[77] Marosi Gy, Márton A, Anna P, Bertalan Gy, Marosfői B, Szép A. Ceramic precursor in flame retardant systems. Polym Degrad Stabil 2002;77:259–65.

[78] Marosfői B, Szabó A, Kiss K, Marosi Gy. Use of organosilicone composites as flame retardant additives and coating for polypropylene. In: Kandola K, Hull R, editors. Fire retardancy of polymers, new strategies and mechanisms, vol 4. Cambridge: The Royal Society of Chemistry; 2008. pp. 49–58.

[79] Guo Z, Fang Z, Tong L. Application of percolation model on the brittle to ductile transition for polystyrene and polyolefin elastomer blends. Express Polym Lett 2007;1: 37–43.

[80] Zhang F, Zhang J, Wang Y. Modeling study on the combustion of intumescent fire-retardant polypropylene. Express Polym Lett 2007;1:157–65.

[81] Marosi G, Anna P, Márton A, Matkó S, Szép A, Keszei S, et al. Mechanism of interactions in flame retarded polymer nanocomposites. In: Wilkie C, Al-Malaika S, editors. International conference on additives, vol. 12. San Francisco: ECM Ltd; 2003. p. 203.
[82] Márton A, Anna P, Marosi G, Szép A, Matkó S, Rácz I, et al. Use of layered structures in recycling of polymers. Prog Rubber Plast Re 2004;20:97–104.
[83] Quédé A, Jama C, Supiot P, Le Bras M, Delobel R, Dessaux O, et al. Elaboration of fire retardant coatings on polyamide-6 using a cold plasma polymerization process. Surf Coat Tech 2002;151-152:424–8.
[84] Errifai I, Jama C, Le Bras M, Delobel R, Gengembre L, Mazzah A, et al. Elaboration of a fire retardant coating for polyamide-6 using cold plasma polymerization of a fluorinated acrylate. Surf Coat Tech 2004;180-181:297–301.
[85] Nagy ZsK, Nyúl K, Wagner I, Molnár K, Marosi Gy. Electrospun water soluble polymer mat for ultrafast release of Donepezil HCl. Express Polym Lett 2010;4:763–72.
[86] Moon S, Ku BC, Emrick T, Coughlin BE, Farris RJ. Flame resistant electrospun polymer nanofibers from deoxybenzoin-based polymers. J Appl Polym Sci 2009;111: 301–7.
[87] Gallo E, Fan Z, Schartel B, Greiner A. Electrospun nanofiber mats coating—new route to flame retardancy. Polym Advan Technol 2011;22:1205–10.
[88] Marosi Gy. Interfaces in multiphase polymer systems. In: Boudenne A, Ibos LY, Candau Y, Thomas S, editors. Handbook of multiphase polymer systems. John Wiley and Sons; 2011. pp. 81–121.
[89] Kmetty Á, Bárány T, Karger-Kocsis J. Self-reinforced polymeric materials: a review. Progr Polym Sci 2010;35:1288–310.
[90] Alcock B, Cabrera NO, Barkoula N-M, Peijs T. Low velocity impact performance of recyclable all-polypropylene composites. Comp Sci Technol 2006;66:1724–37.
[91] Alcock B, Cabrera NO, Barkoula N-M, Loos J, Peijs T. The mechanical properties of unidirectional all-polypropylene composites. Compos Part A-Appl S 2006;37: 716–26.
[92] Alcock B, Cabrera NO, Barkoula N-M, Spoelstra AB, Loos J, Peijs T. The mechanical properties of woven tape all-polypropylene composites. Compos Part A-Appl S 2007; 38:147–61.
[93] Cabrera NO, Alcock B, Peijs T. Design and manufacture of all-PP sandwich panels based on co-extruded polypropylene tapes. Compos Part B-Eng 2008;39:1183–95.
[94] Bocz K, Toldy A, Kmetty Á, Bárány T, Igricz T, Marosi Gy. Development of flame retarded self-reinforced composites from automotive shredder plastic waste. Polym Degrad Stabil 2012;97:221–7.
[95] Izer A, Bárány T, Varga J. Development of woven fabric reinforced all-polypropylene composites with beta nucleated homo- and copolymer matrices. Comp Sci Technol 2009;69:2185–92.
[96] Zhang S, Horrocks AR. A review of flame retardant polypropylene fibres. Progr Polym Sci 2003;28:1517–38.
[97] Bocz K, Bárány T, Toldy A, Bodzay B, Csontos I, Madi K, et al. Self-extinguishing polypropylene with a mass fraction of 9% intumescent additive - a new physical way for enhancing the fire retardant efficiency. Polym Degrad Stabil 2013;98:79–86.
[98] Sacristan M, Ronda RC, Galià M, Cádiz V. Phosphorus-containing soybean-oil copolymers: cross-metathesis of fatty acid derivatives as an alternative to phosphorus-containing reactive flame retardants. J Appl Polym Sci 2011;122:1649–58.

[99] Yuan XY, Wang DY, Chen L, Wang XL, Wang YZ. Inherent flame retardation of bio-based poly(lactic acid) by incorporating phosphorus linked pendent group into the backbone. Polym Degrad Stabil 2011;96:1669−75.
[100] Raqueza JM, Deléglise M, Lacrampe MF, Krawczak P. Thermosetting (bio)materials derived from renewable resources: a critical review. Progr Polym Sci 2010;35: 487−509.
[101] Uyama H, Kuwabara M, Tsujimoto T, Kobayashi S. Enzymatic synthesis and curing of biodegradable epoxide-containing polyesters from renewable resources. Biomacromolecules 2003;4:211−5.
[102] Lligadas G, Ronda JC, Galià M, Cádiz V. Synthesis and properties of thermosetting polymers from a phosphorous-containing fatty acid derivative. J Polym Sci Pol Chem 2006;44:5630−44.
[103] Lligadas G, Ronda JC, Galià M, Cádiz V. Development of novel phosphorus-containing epoxy resins from renewable resources. J Polym Sci Pol Chem 2006;44: 6717−27.
[104] Matkó Sz, Toldy A, Keszei S, Anna P, Bertalan Gy, Marosi Gy. Flame retardancy of biodegradable polymers and biocomposites. Polym Degrad Stabil 2005;88:138−45.
[105] Bourbigot S, Fontaine G. Flame retardancy of polylactide: an overview. Polym Chem 2010;1:1413−22.
[106] Wu K, Hu Y, Song L, Lu H, Wang Z. Flame retardancy and thermal degradation of intumescent flame retardant starch-based biodegradable composites. Ind Eng Chem Res 2009;48:3150−7.
[107] Yang K, He S, Pack S, Muir N, Davis R, Juang C, et al. Engineering flame retardant biodegradable nanocomposites. In: Wilkie CA, editor. Recent advances in flame retardancy of polymeric materials XXIII. Stamford: BCC. Inc; 2012. p. 224.

CHAPTER

Phosphorus-based and Intumescent Flame Retardants

6

S. Hörold

Clariant Produkte (Deutschland) GmbH, Gersthofen, Germany

CHAPTER OUTLINE

1. Introduction

Phosphorus-based flame retardants (FRs) have been exploited commercially for a long time. Intumescent FRs consist of a phosphorus FR combined with a nitrogen-containing additive and a further synergist which forms char when ignited. The first use of phosphorus FRs was ammonium phosphate on canvas already in the eighteenth century. The market for phosphorus FRs is strongly increasing, and a lot of new products and new formulations are under development. This review covers the recent developments concerning the use of phosphorus FRs in their different oxidation states.

Phosphorus is a multivalent nonmetal element of the nitrogen group. As a mineral, it is present as inorganic phosphate rocks. Elemental phosphorus exists in two major forms—white phosphorus and red phosphorus, but due to its high reactivity, phosphorus is never found as a free element on the Earth. Phosphates come from ore beds rich in fossils of ancient marine life. Phosphate rock, mainly made of apatite

Polymer Green Flame Retardants. http://dx.doi.org/10.1016/B978-0-444-53808-6.00006-8

(an impure tricalcium phosphate mineral), is an important commercial source of this element. Large deposits of apatite are located in China, Russia, Morocco, Florida, Idaho, Tennessee, Utah, and elsewhere [1]. It is expected that a "Peak Phosphorus" will occur in 30 years and that "at current rates, reserves will be depleted in the next 50–100 years" [2].

Phosphorus is an essential element for all living cells, and organisms tend to accumulate and concentrate it. For example, elemental phosphorus was historically first isolated from the sediment in human urine, and bone ash was an important early phosphate source. Low phosphate levels are an important limit to growth in some aquatic systems. Today, the most important commercial use of phosphorus-based chemicals is the production of fertilizers, to replace the phosphorus that plants remove from the soil. Phosphates are also essential ingredients in consumer goods such as detergents and cleaners, metal treatment, water treatment, toothpaste, food, and beverages such as cola drinks [3].

Phosphorus-based compounds also comprise a class of widely used FRs, next being halogen-containing organics, inorganic minerals, nitrogen compounds, and others. Phosphorus-based FRs include organic and inorganic phosphates, phosphonates, and phosphinates as well as red phosphorus, thus covering a wide range of phosphorus compounds in all possible oxidation states between 0 and +5. Additionally, the phosphorus content in these molecules varies largely from 100% (red P) to about 10% (phosphates) and does not correlate to the FR efficiency of the specific substance. This indicates that the FR effect of the P products cannot be described by a single mechanism as it is done with halogen-based FRs. Phosphorus-based FRs can act in both the condensed and the gaseous phases. In the gaseous phase, phosphorus results in flame inhibition through radical trapping; in the condensed phase, it initiates the formation of carbon char or inorganic residue inhibiting the pyrolysis process necessary to feed the flames. The activity of phosphorus depends on the FR used as well as on the chemical structure of the polymer matrix and on its interaction with other additives.

Phosphorus FRs can either be mixed into plastics or be reactive, that is, chemically bound into the plastic molecules at polymerization, depending on the properties required of the plastic in terms of finished product performance. Reactive phosphorus FRs are mainly used in polyethylene terephthalate (PET) fibers, polyurethane (PU) foams, and epoxy resins. Additive phosphorus FRs are common in engineering plastics and polyolefins. However, it is a particular property of phosphorus-based FRs that they have to be formulated individually for each polymer system. Therefore, the supplier must have extensive expertise in phosphorus chemistry and the mechanisms of flame retardancy. The decomposition of the polymer in a fire and the reaction of the phosphorus FR have to be synchronous in order to be effective [4]. Typically, a phosphorus-based FR is designed to develop its activity in combination with the starting decomposition of the specific polymer it is used for. It can offer a partial gas phase contribution to the flame-extinguishing effect (comparable to bromine- or chlorine-containing FRs), but more often char-forming activity (intumescence) and formation of protective top layers on the plastic surface are the main features (Figure 1).

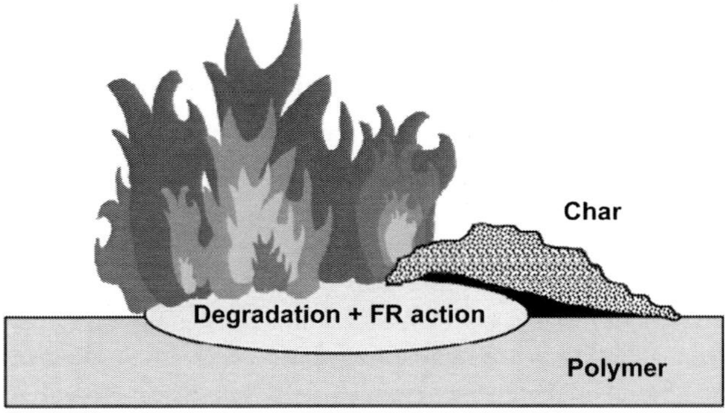

FIGURE 1

Solid phase action of P-based FRs.

The obvious advantage of such a "solid phase mechanism" is that it causes less release of smoke and off-gases in a rising fire situation, thus keeping secondary fire damages as low as possible.

Phosphate esters constitute the largest volume phosphorus FRs, and can be divided into halogenated and nonhalogenated types. Halogenated phosphorus compounds, usually chloroalkyl phosphate esters, are used predominantly in PU. Brominated organophosphate esters are used in PP fibers and provide the vapor-phase flame retardancy characteristics of brominated compounds with condensed phase control by the phosphorus. Aryl and alkylaryl phosphate esters represent the largest class of nonhalogenated phosphorus FRs, and are commonly used in PVC and engineering plastics. Organophosphorus compounds with P–C bonds are more stable under certain circumstances than phosphates (e.g. acid hydrolysis), but are more expensive and are used only when the additional cost can be justified [4].

In this chapter, the recent work and progress will be described for four groups of phosphorus compounds used as FRs (Figure 2):

- Phosphates—$OP(OR)_3$
- Phosphonates—$OP(OR)_2R$
- Phosphinates—$OP(OR)R_2$, and
- Red phosphorus as the pure element itself

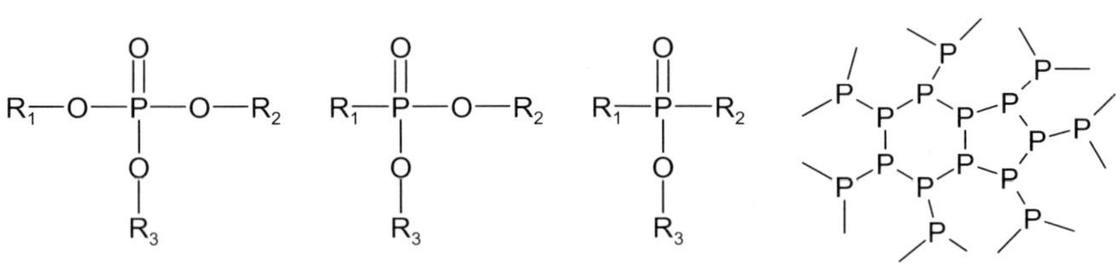

FIGURE 2

General structures of phosphorus flame-retardants (FR).

2. Phosphate flame retardants

2.1 Phosphoric acid aryl esters for the application enclosures

Electronic enclosures are a market segment which includes primarily the enclosures of consumer and information technology equipment such as televisions (TVs), desktop or notebook computers, monitors, printers, copiers, household appliances, etc. These housings are made of different types of polymer resins such as high-impact polystyrene (HIPS), acrylonitrile butadiene styrene (ABS) copolymers, polycarbonate (PC)/ABS blends (PC/ABS), and polyphenylene ether (PPE)/HIPS blends (PPE/HIPS). Halogen-free phosphate-based FRs are applied in PC, PC/ABS, and PPO/HIPS blends, which are extensively used for housings of business computers (mostly laptops) and other business devices. Recently these resins, in spite of being more expensive than HIPS or ABS, are also finding their way to consumer electronics as they are mostly used by original equipment manufacturers (OEMs) who are marketing halogen-free products [5,6]. Phosphates are rarely used in plain PC because of partial loss of clarity, tendency to stress cracking and reduced hydrolytic stability [7].

Aryl phosphates (Figure 3) are suitable halogen-free FRs for PC/ABS blends. Resorcinol bis(diphenyl phosphate) (RDP) and bisphenol A bis(diphenyl phosphate) (BDP) are mixtures of oligomers with two to five phosphorus atoms, but with their molecular weight distribution shifted mainly toward the diphosphates. RDP has higher phosphorus content than BDP and is therefore more efficient as a FR. On the other hand, BDP is more thermally and hydrolytically stable than RDP [6]. Resorcinol bis(di-2,6-xylenyl phosphate) (RXP) is exclusively made in Japan and has a higher hydrolytic stability than RDP and BDP due to the steric hindrance by the 2,6-xylyl groups. Triphenyl phosphate (TPP) is the most volatile arylphosphate.

The required loadings depend on the ratio of PC and ABS in the blend. In commercial PC/ABS blends where the ABS content normally does not exceed 25% it is possible to achieve a UL 94 V-0 rating with these products at 8–15 wt% loading in combination with a coadditive. The coadditive usually retards dripping. A common antidripping agent is polytetrafluoroethylene (PTFE) with loadings up to 0.5 wt%. These phosphate esters also function as plasticizers; their use is limited by the ability of the polymer to tolerate the plasticization and other changes impacting mechanical properties [5].

Levchik et al. [8,9] studied the fire behavior and the physical properties of PC/ABS blends fire retarded with TPP, RDP, and BDP, and proposed that the main fire retardant action of BDP takes place in the condensed phase. According to Lyons [10] this condensed phase mechanism results from char formation through esterification and dehydration, wherein phosphorus compounds act like acid precursors. It has also been reported that phosphates form volatile PO, P, and P2 species that act in the gas phase through flame poisoning, which could be identified spectroscopically [11–13]. Pawlowski and Schartel studied the detailed thermal

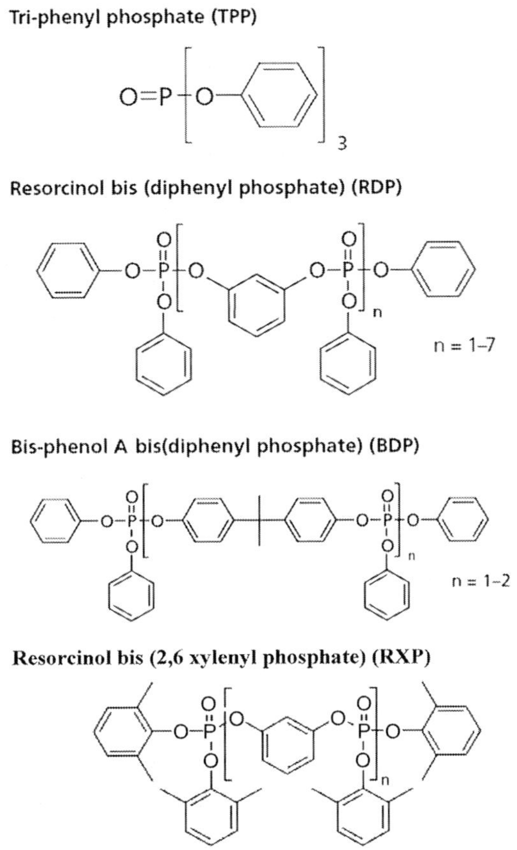

FIGURE 3

Structures of aryl phosphate flame-retardants (FR) [5].

decomposition, fire behavior, and flame retardancy mechanisms of BDP, RDP, and TPP in PC/ABS + PTFE [14]. It was revealed that TPP shows only a gas phase action (flame inhibition), whereas RDP shows some and BDP a crucial condensed-phase action (enhancing char formation) in addition to the gas-phase action. RDP and BDP induced cross-linking not only by acting like acid precursors but also by taking part in the cross-linking network. Table 1 shows dosages of TPP, RDP, and BDP in PC/ABS and the resulting compound properties. RDP gives the best impact and heat distortion temperature (HDT) performance.

PC is sensitive to acidic conditions. Different phosphate esters have different hydrolytic stabilities. All phosphate esters decompose to acid species especially after exposure to high heat and humidity. Bisphenol-A-bridged phosphate FRs are less susceptible to hydrolysis but will show PC polymer degradation eventually. RDP analogs displayed instability earlier in the testing exposure regime but with

Table 1 Properties of PC/ABS UL 94 V-0 with Different Phosphates [5]

Flame Retardant		TPP	RDP	BDP
Dosage (V-0)	(%)	14	9	12.3
Tensile strength	(N mm^{-2})	46.0	53.3	55.6
Elastic modulus	(MPa)	2620	1586	1655
Flexural strength	(N mm^{-2})	82.7	81.4	84.8
Notched impact strength	(J m^{-2})	123	651	496
HDT 1.80 MPa	(°C)	67.8	83.3	81.4

the addition of stabilizers, substantial improvements are seen [5]. PPO/HIPS blends can be also flame retarded with the aromatic phosphates TPP, RDP, and BDP. Typical levels of PPO used to achieve a UL 94 V-0 rating are 30–70%; these blends also contain 10–20% of TPP, RDP, or BDP [5].

Both RDP and BDP are very viscous liquids (BDP is much more viscous than RDP) and therefore need special dosing systems like heat-traced lines to liquid metering units and extruders or are even storage at elevated temperature. A proprietary new aromatic bisphosphate, a solid powder melting at 105–108 °C, is currently introduced under the trade name Fyrolflex Sol-DP [6]. It can be handled by standard plastics processing equipment, equipped with solids feeding conventional systems. The product is thermally stable up to 300 °C and its phosphorus content is 10.8 wt%. According to the current patent literature [7], it is a hydroquinone-based oligomeric aryl phosphate. The oligomers of the invention are suitable to flame retard PC/ABS composites, with better ultraviolet (UV) stability, hydrolytic stability, and higher HDT as compared with resorcinol-based phosphates [7].

Another commercially available bisphosphate is RXP. It is a solid melting at 95 °C, which makes it sometimes easier to add it to the polymer, and has a phosphorus content of 9%. It is mainly used in Asia and is said to be more expensive than the liquid bisphosphates, but has a better hydrolytic stability due to the steric hindrance of the 2,6-xylyl groups. A brominated phosphate, precisely tris(tribromoneopentyl)phosphate, is commercially available as Reoflam FR 370 and has 3% phosphorus and 70% bromine. It is melt blendable with polypropylene and does not require antimony trioxide as synergist, so that it can be used for fiber spinning. It also has an excellent UV stability and can be used together with light stabilizers in stadium seats and polypropylene carpets.

Arylphosphates are also used as plasticizing FRs in PVC (flooring, wire, and cable) and certain elastomers, such as neoprene rubber used for conveyor belts in the mining industry. Furthermore, these phosphates improve the compatibility of chlorinated paraffins with PVC [4]. The total consumption of aryl phosphate esters was estimated at 111.5000 metric tons, valued at $488 million (SRI 2008). While the consumption of monoarylphosphates over the next 5 years will track the global gross domestic product, consumption of bisarylphosphates will grow at 5% per year as a result of the replacement of halogen FRs in engineering resins (PPO/HIPS, PC/ABS blend) used for the production of housings for computer monitors, TV sets, and mobile phones [4].

2.2 Intumescent FRs

2.2.1 Ammonium polyphosphate

Ammonium polyphosphate (APP) (Figure 4) is an inorganic salt of polyphosphoric acid and ammonia with the chemical formula $[NH_4 \cdot PO_3]_n$. The properties of APP depend on the number of monomers in each molecule and the structure of the polymeric chains. Shorter chains ($n < 100$) are more water sensitive, more soluble, and less thermally stable than longer chains ($n > 1000$) [15]. Commercial products differ in molecular weight, particle size, solubility, and surface coating. Processing of APP in plastics is limited by its thermal stability of about 250 °C. Encapsulated grades, normally with thermosetting aminoplast coatings, prevent hydrolysis during formulation, processing, or use, and simultaneously provide better dispersibility.

APP can be prepared by reacting concentrated phosphoric acid with ammonia. As a fire retardant, APPs decompose to form ammonia and phosphoric acid. The phosphoric acid acts as an acid catalyst in the dehydration of carbon-based polyalcohols, such as cellulose in wood. The phosphoric acid reacts with alcohol groups to form heat-unstable phosphate esters. The esters decompose to release carbon dioxide and regenerate the phosphoric acid catalyst. The release of nonflammable carbon dioxide helps to dilute the air of oxygen and flammable decomposition products of the material that is burning. The resultant carbonaceous char is less flammable than before [16].

APP is currently used primarily as a component in intumescent paints and coatings. Intumescence refers to the formation on heating of a stable foamed char that insulates a material from further heat and prevents access of oxygen to the burning material. Intumescent formulations consist of an acid source, such as APP; a polyhydric source of carbon to form the char, such as pentaerythrithol; a blowing agent, such as melamine or similar compounds; and resin binders, such as vinyl copolymers, epoxies, polyhydroxy compounds, and melamine formaldehydes. It not only acts as a FR but also retains the dimensional stability of the coated substrates in a fire [4]. Intumescent coatings are applied to protect combustible materials such as wood, rubber, and plastics, as well as steel structures in buildings and ships. Steel loses its strength when exposed to high temperatures in a fire and intumescent

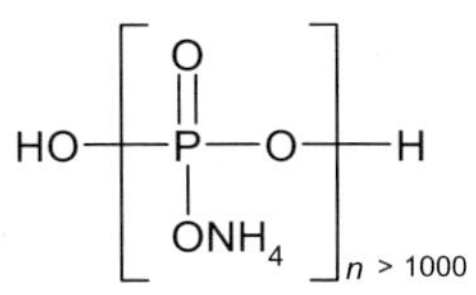

FIGURE 4

Chemical structure of ammonium polyphosphate (APP).

coatings can contribute to preventing the collapse of a structure exposed to severe heat. These functional coatings are applied at thicknesses in excess of 0.5 mm and sometimes up to 20 mm (epoxies), whereas conventional paints are usually applied at less than 1 mm thicknesses. Aside from applications in buildings and ships, these coatings are also used in the oil refining and petrochemical industries, for offshore structures for oil and gas exploration and production, and for protective coatings of liquid petroleum gas containers. The intumescent paints swell in heat to form on the substrate a char layer about 10—100 times the thickness of the original coating. Trends in formulating intumescent paints include the replacement of polyols with polymeric hydroxylic compounds, as well as their combination with nanoclays in order to increase the melt viscosity of the paint in case of fire and to reinforce the barrier effect of the coating against oil and fuel.

Beside its use in intumescent coatings, APP is used in FR systems for installation equipment (cables, cable ducts, sealants), timber (boxes, doors), and polymer applications (E & E, polyolefins, thermoset resins) [4]. Long-chain APP of crystallinic form II is used in intumescent systems for polyolefins and starts to decompose slowly in the temperature range between 240 and 280 °C. Organic nitrogen compounds, e.g. melamine or melamine phosphate, and polyol components, e.g. dipentaerythritol or sorbitol, are employed as synergists [16,17]. However, these additives often create problems, coming from their volatility and water solubility. Furthermore, they are not compatible with the polymeric matrix, and this weakens the mechanical properties of the materials. Therefore, a lot of work has recently been done on applications of different kinds of carbonization agents. The currently commercially available APP-based intumescent FRs are used in polypropylene in approximately 25% dosage to pass UL 94 V-0 at 1.6 mm thickness (e.g. Exolit AP 765 from Clariant or Budit 3167 from Chemische Fabrik Budenheim). APP-based FRs can normally not be used in applications with permanent water contact or prolonged contact with warm water due to their relative hydrophilic nature. In very thin film applications below 0.5 mm thickness intumescent FRs are not effective because the char and barrier formation cannot work sufficient.

Wanzke et al. [15] showed that a beginning water release at processing temperature reduces the stability of APP and synergist blends (Figure 5). Mixtures of APP

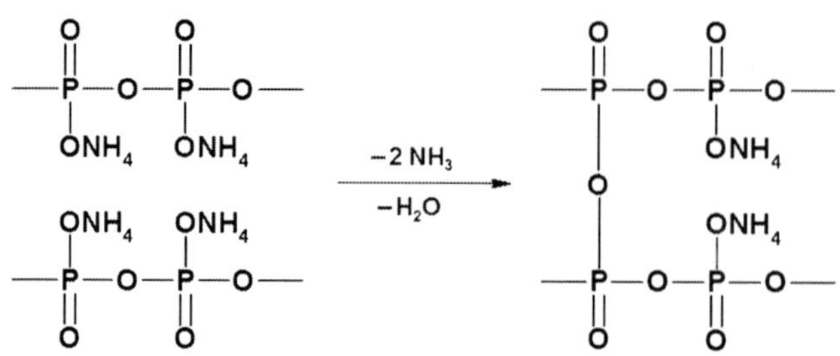

FIGURE 5

Release of water and ammonia at the beginning decomposition of APP [15,20].

and polyol components release water which partially hydrolyzes the APP chains leading to more acidic polyphosphate and shorter chain products with phosphoric acid terminal groups [18—20]. This kind of interaction was found with several synergists in APP-based systems, but particularly with polyol components. It is basically the effect of that earlier starting decomposition in connection with water release that makes the intumescent blends more sensitive in processing than the APP alone. If the intumescent FR is dispersed in a polyolefin compound, the effect is much less drastic than with the pure powder blend but the more decomposition takes place during processing, the more negative effects will be observed. A "soapy" surface on the polymer strands after running through a cooling bath and a high water content of the granules after compounding are very good indicators for partial decomposition during processing.

As alternatives to polyols, triazines and their derivatives have been proved to be good charring agents. These agents have excellent charring effect because they contain abundant nitrogen and possess a structure of tertiary nitrogen. Hu et al. [21] synthesized a derivative of triazines and used it as a charring agent with APP in polyethylene. When the charring agent (8 wt%) and APP (22 wt%) were incorporated into polyethylene, the limiting oxygen index (LOI) reached the maximum of 31.2, and class V0 in the UL 94 test was obtained. Li and Xu [22] also combined a novel triazine charring—foaming agent based on cyanuric chloride, ethanolamine, ethylene diamine, and piperazine or 1,2-diamino-4-methylcyclopentane with APP and used them in polypropylene. The results indicated that polypropylene with an FR concentration of only 18 wt% could obtain a V0 rating with a high LOI value of 30.2. In addition, polyamides (Pas) such as PA 6, PA 66, and PA 11 can be used as charring agents. PA 6 is a well-known char-forming polymer, and the interaction between the amide groups of PA 6 and an acid catalyst at a high temperature can form some cross-linking chemical structures, which strengthen the char layer. Recently, Jun, Yi, and Cai described a novel PA-based charring agent, poly (*p*-propane terephthalamide) (PPTA), which was synthesized and used together with APP to prepare a novel intumescent FR for ABS. With 22.5% APP and 7.5% PPTA V-0 was observed [23].

Owing to the mode of action of APP (formation of a protective layer), only very little smoke is formed in the early stages of a fire, a fact that can give those involved extra time to escape or be rescued. Figure 6 shows the increase in smoke density in the first few minutes of a fire involving PP materials containing various FRs (measured in the NBS Smoke Chamber). Whereas dense black smoke is formed very rapidly with brominated or chlorinated FRs, the smoke density value for plastic containing an APP-based intumescent system (®Exolit AP 750) remains low for a long time. Valuable time for escape or rescue is thus gained in the event of a fire [15].

In view of their high asset values, electronic installations in particular are at risk from even small volumes of smoke. Even if the fire is small, corrosive substances that are released by, for example, halogenated FRs or PVC, can destroy entire installations by attacking the sensitive electronic components. The PP containing the

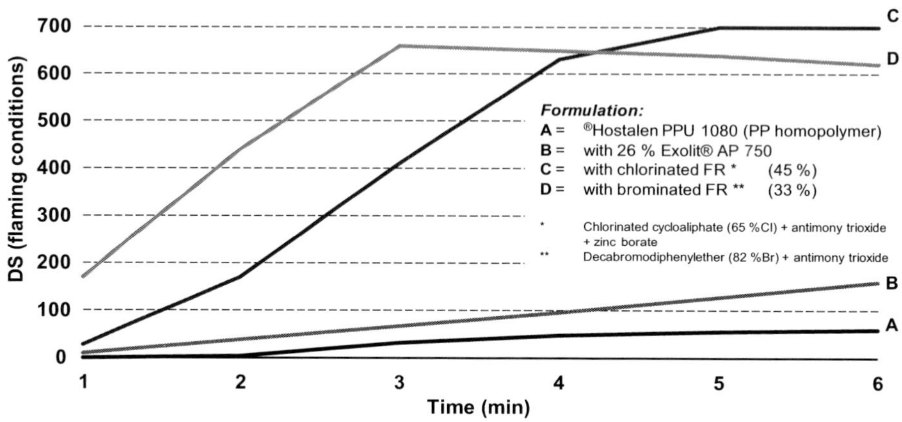

FIGURE 6

Optical smoke density of flame-retardant (FR)-modified PP compounds [15].

intumescent system does not release any corrosive halogen hydrides and generates only insignificant amounts of toxic fumes such as HCN or NO_x.

The release of heat from plastics components in a fire is also an important criterion, since the rate of flame spread depends quite fundamentally on this factor. Figure 7 shows data from measurements in a cone calorimeter, which again make clear the significant difference between a halogenated FR and an intumescent system. The halogenated FR cannot prevent a relatively large amount of heat being released from the PE/EVA specimen within a short time, which promotes rapid flame spread in the immediate vicinity of the ignition source. The intumescent system, by contrast, ensures relatively slow, constant heat dissipation at a significantly lower level that delays flame spread [15].

Consumption of APP is estimated at 29,000 metric tons, valued at $130 million (SRI 2008 [4],). The largest market segment for its FR use, with 80–85%, is represented by intumescent coatings, which are used almost exclusively in building and construction to protect steel structures. APP is reported to be biodegradable and could be used or regarded as a fertilizer, which makes it an attractive FR for bio-based polymers like PLA and biodegradable polymers. The general use of intumescent FRs in polyolefins and other polymers is discussed in a book by Le Bras et al. [24] and an additional review [25] by the same author.

2.2.2 Ethylene diamine phosphate

Ethylene diamine phosphate (Figure 8), a white powder with a melting point of 250 °C, is, unlike APP, self-intumescent and does not need any char-forming synergists [26]. It was introduced as FR for polyolefins, EVA, and PVC. Products are commercially available, e.g. under the trade name Jjazz from JJI Technologies, USA. Limits of its use are the low thermal stability with decomposition beginning

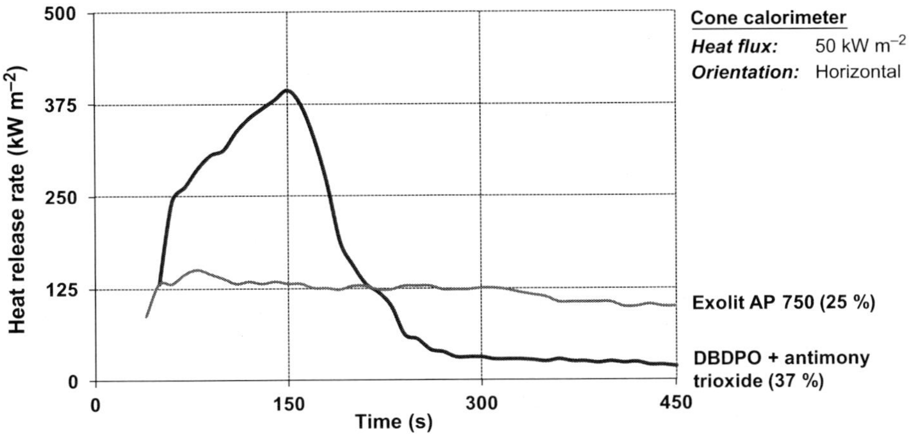

FIGURE 7

Heat dissipation from flame-retardant (FR) PE/EVA materials (UL 94 V-0, 1.6 mm) [15].

at 240 °C and its water solubility. Synergistic effects with melamine and melamine phosphates are reported [27].

2.2.3 Melamine polyphosphate

Melamine derivatives are salts with organic or inorganic acids such as boric acid, cyanuric acid, and phosphoric or polyphosphoric acid. These derivatives have higher decomposition temperatures (melamine cyanurate (MC), 310 °C; melamine polyphosphate (MPP), 360 °C) in comparison to melamine (250 °C) alone, are less soluble in water, and can add other features to performance such as nondripping (cyanuric acid) or char formation (boric acid, phosphoric acid). Melamine homologs such as melam, melem, and melon have even higher decomposition temperatures with 400 °C and above, but they have found only experimental use so far.

MC is used in the production of halogen-free flame-retarded unfilled and mineral-filled PAs for electrical and electronic applications, in thermoformable

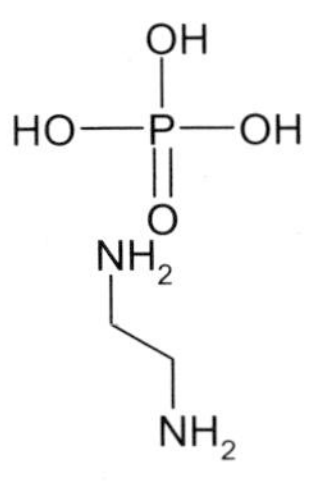

FIGURE 8

Structure of ethylene diamine phosphate (EDAP).

polyurethane foams, and in polypropylene intumescent formulations in conjunction with APP. Melamine phosphate and melamine pyrophosphate are mainly used in intumescent coatings and in glass-filled polyolefins and PAs. Melamine phosphate and melamine pyrophosphate can also be used in paper, textiles, and wood laminates to effect fire-retardant barriers. Melamine borate is mainly used in intumescent coatings, in various polyolefins, in PVC as an antimony-free synergist and smoke suppressant, and in textile backcoatings [4].

Like APPs, melamine phosphates are also substances combining the synergistic effect of nitrogen- with phosphorus-containing components in one salt. Based on increasing thermal stability the melamine phosphates can be ranked as follows: melamine phosphate < melamine pyrophosphate < melamine polyphosphate [28]. Melamine monophosphate is a salt of melamine and phosphoric acid. Above ~200 °C melamine phosphate will react with melamine pyrophosphate with release of reaction water, which will result in a heat sink. Above ~260 °C melamine pyrophosphate will react with release of reaction water with melamine polyphosphates (Figure 9) which again results in a heat sink effect.

Above 350 °C, melamine polyphosphate undergoes endothermic decomposition thus acting as a heat sink and cooling the combustion source. The released phosphoric acid acts to coat and therefore shield the condensed combustible polymer. The phosphoric acid along with the polymer also works to form a char around the fuel source (polymer), thus reducing the amount of oxygen present at the combustion source. The melamine released is also a blowing source, which blows up the char, resulting in an intumescent behavior.

Compared to APP, MPP is more thermally stable and less sensitive to hydrolysis, but its FR effect is lower than that of APP, so it could not replace APP in intumescent coatings and polyolefin FRs. It was originally developed for glass fiber-reinforced (GF) PA 66 and gives a V-0 at 25% dosage, but it fails to improve the flammability characteristics of PA 6 at reasonable levels. Jahromi demonstrated that MPP depolymerizes at above 350 °C, inducing significant cross-linking (as evidenced by increase in viscosity) in PA 66 and leading to a drastic depolymerization of PA 6 (as evidenced by a decrease in viscosity) [29].

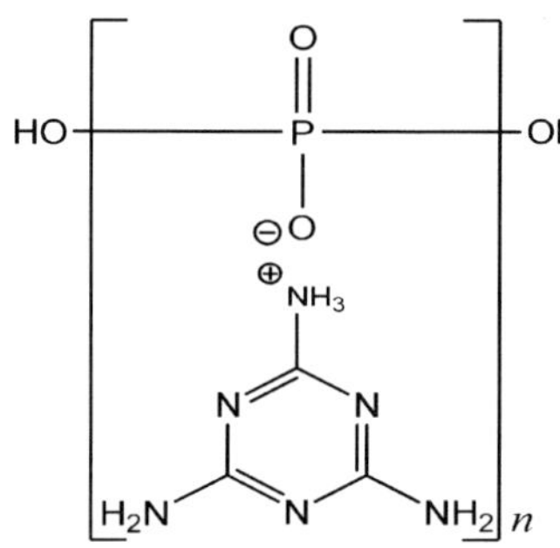

FIGURE 9

Structure of melamine polyphosphate (MPP).

MPP is today mostly used in combination with other FRs, such as metal phosphinates (see phosphinate chapter), metal hydroxides, and phosphates [28]. It is characterized by its good thermal stability and a low impact on the glass-transition temperature (Tg). As mentioned, under thermal stress, melamine derivatives decompose endothermically (heat sink) and release inert nitrogen gases (e.g. ammonia) that dilute oxygen and the flammable gases in the flame. Often phosphoric acid is also formed as a decomposition product and promotes the formation of insulating char on the surface of the polymer [30,31].

2.2.4 Piperazine pyrophosphate

In addition to ammonium and melamine phosphate, piperazine-based phosphates are also commercially available. In polyolefins, piperazine pyrophosphates are effective FRs. Piperazine pyrophosphate (Figure 10) can be obtained through condensation of two piperazine phosphate units. Impurities with piperazine phosphate lead to a lowering of the decomposition temperature of the pyrophosphate. The purest piperazine pyrophosphates have a decomposition temperature of 324 °C [32], making it more stable than APP and less stable than MPP. Piperazine pyrophosphate is used together with melamine salts like melamine pyrophosphate and antidripping agents like PTFE. With 22–24% a V-0 classification in polypropylene can be achieved [33].

3. Red phosphorus as a flame retardant

Red phosphorus is, unlike white phosphorus, nontoxic, thermally stable up to 450 °C and is not spontaneously flammable [34–36]. Red phosphorus is made by heating yellow phosphorus to 270–300 °C in the absence of air. When incorporated in a polymer, red phosphorus breaks down during the process of fire to form P2 molecules that are active species in the gas phase. Red phosphorus is a very effective FR for glass fiber-reinforced PAs and is used in thermosets and polyolefins for wire and cable applications as well. In reinforced PA 6.6 only 5.5–7% by weight of red phosphorus is needed to achieve a V-0 classification. Scherzer et al. [36] claimed flame-proofed PA molding compounds reinforced with natural fibers and red phosphorus as FR. UL 94 V-0 is achieved with just 5% red phosphorus in PA

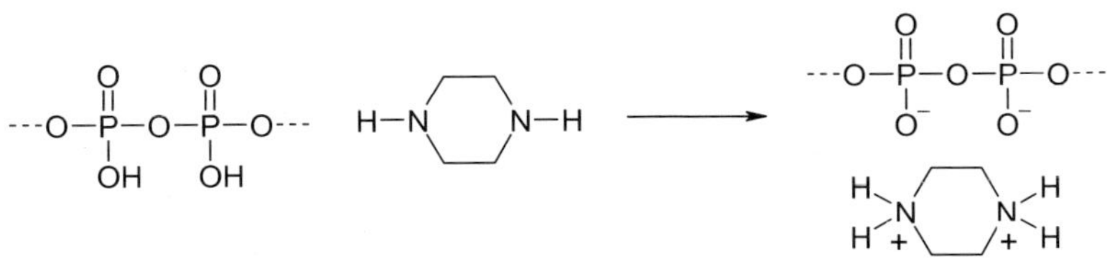

FIGURE 10

Structure of piperazine pyrophosphate.

6, 66, and 6.10. Obviously the natural fibers (Kenaf and Cordenka) improve the FR effect of red phosphorus [37]. Red phosphorus can be also used as a FR in polybutylene terephthalate (PBT); UL 94 V-0 can be attained at a level of 5–6% stabilized and coated red phosphorus. These compounds are said to be used in casings for electronic appliances.

Toshiba described an adhesive formulation made from a blend of bisphenol A and cresol novolac epoxies using red phosphorus as a FR. They achieved UL 94 V-0 rating with 4 wt% red phosphorus (encapsulated in a phenolic resin and coated with ATH) and 25 wt% ATH as a synergist [36]. Due to its effectiveness and the good mechanical and electrical properties of the resulting compounds, red P is widely used in industrial E & E applications in Europe and to some extent in Asia but only to a very limited extent in the United States [38]. Due to its high phosphorus content, red phosphorus may well offer some advantages over organophosphorus compounds, but its use is currently subject to several limitations. Red phosphorus is not very reactive but the powder can burn. The main safety issue during compounding is to prevent the ignition of red phosphorus at hot machine parts of the compounding line. Direct compounding of red phosphorous powder into the resin is the most demanding process and requires an investment because an inert gas blanket is needed for the whole extruder and the feeding units. In order to guarantee safe handling by the customer, suppliers deliver red phosphorus in dispersions with carriers such as water, epoxy resin, castor oil, TCPP, or diphenyl octyl phosphate, and as concentrates or masterbatches in granule or pastille form based on different polymeric or wax-type carrier materials. In this case an inert gas blanket is not required and the control of phosphine gas emissions at the workplace remains as the main safety issue. The emission of toxic phosphine gas can be created by the reaction of phosphorus with traces of water and oxygen. Therefore, suppliers provide red phosphorus in encapsulated or chemically stabilized compounds and recommend that dry base resin and FR be dried prior to processing and that processing temperatures be limited to 300 °C. Furthermore, the long-term evolution of phosphine and its oxidation to corrosive and conductive phosphorus acids at metal contact on the surface of the plastics can cause problems in electrical and electronic applications. Additionally, red phosphorus imparts color, and can at present be used only in dark-colored products.

The current consumption of red phosphorus as a FR in Western Europe is about 1200 metric tons per year valued at $13.2 million; consumption is stagnating to slowly declining (0% to −2%). In contrast to Europe, the other regions use this inorganic FR only in very small quantities [4].

4. Phosphonate FRs

Phosphonates or phosphonic acids are organic compounds containing $C{-}PO(OH)_2$ or $C{-}PO(OR)_2$ groups (where R = alkyl, aryl). Most phosphonates are highly water soluble, while the phosphonic acids are only sparingly soluble. Phosphonates are not

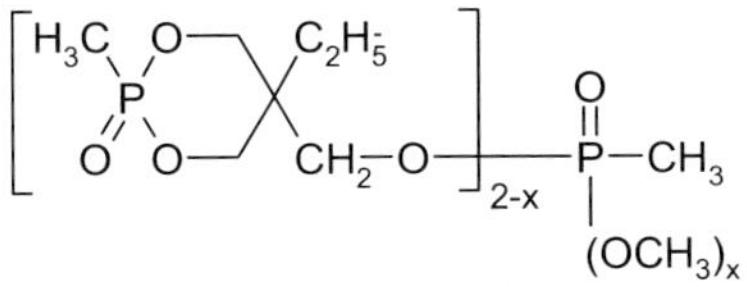

FIGURE 11

Cyclic phosphonates for the thermosol process.

volatile and are poorly soluble in organic solvents. Normally they are more efficient FRs than phosphates, but raw materials for phosphonates are more expensive than for phosphates. Phosphonates can be synthesized using the Michaelis–Arbuzov reaction from phosphites $P(OR)_3$. Some phosphonates are produced on industrial scale, and a few are also applied as FRs for polymers. A cyclic phosphinate, Antiblaze 19 or Antiblaze 1045 from Rhodia, with high phosphorus content is used in PET fibers. The product is applied as an aqueous solution to the surface of the fiber and then heated to a temperature at which the fiber becomes soft and dissolves the FR (thermosol process, Figure 11 [39]). Applications in injection molding PET and PBT are also reported in the literature. With 1% phosphorus in PET resin, UL 94 V-2 is achieved, and with 3% phosphorus even V-0. Here the handling of the high-viscosity liquid is still a challenge.

A polymeric phosphonate, Fyrol PMP (Figure 12) from ICL-IP, is designed for epoxy resins in printed circuit boards. The product is a solid, which melts at about 50 °C and has curing agent properties for epoxy resins by the phenolic groups. It is thermally stable up to 300 °C. The reactivity of Fyrol PMP's aromatic hydroxyl groups is said to be similar to other common curing agents, so there is no need to change the curing cycles.

Recently, the US Company FRX (Figure 13) introduced a phosphorus-containing PC designed for the use in PC and PC/ABS, epoxy resins, and polyesters [40,41].

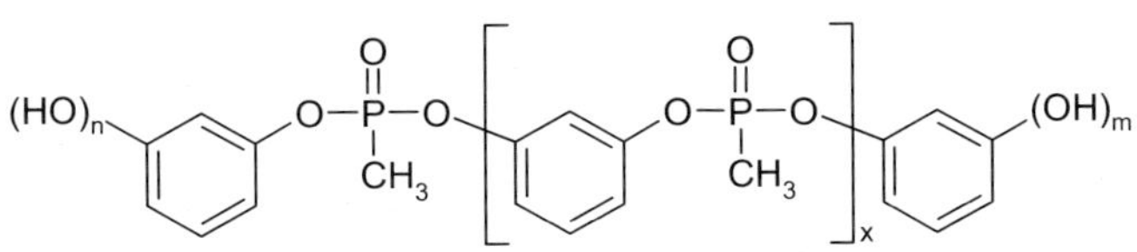

FIGURE 12

Structure of fyrol PMP.

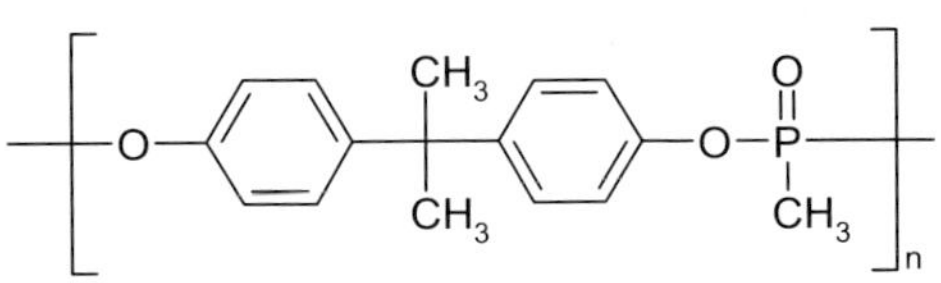

FIGURE 13

Structure of FRX polymer [40,41].

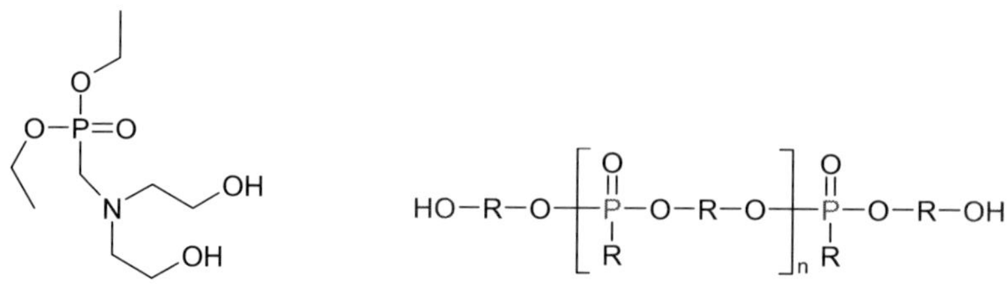

FIGURE 14

Structure of diethyl (*N,N*-bis(2-hydroxyethyl)amino) methane phosphonate (Fyrol 6, Levagard 4090) (left) and of phosphorus diols, e.g. Exolit OP 560 (right).

The product is synthesized by transesterification of phosphonic acid diarylester with bisphenol A, giving a polymeric material with a phosphorus content of 10.8% and being inherently FR. This polymer can be also melt blended with other polymer resins like ABS or PBT. Transparent flame-retarded PC could be made with FRX polymer as well as PC/ABS with high HDTs. In PBT, UL 94 V-0 can be passed with >25% addition. Oligomers with reactive phenolic end groups can be made as well and can be incorporated in epoxy and other thermoset resins. Wanzke has reported synergistic effects of FRX polymer with phosphonates in polyester-based elastomers [42].

A reaction product of diethyl phosphite and formaldehyde and diethanolamine is commercially available under the trade name Fyrol 6 (ICL-IP) and Levagard 4090 (Lanxess) (Figure 14). It is used as an FR for rigid PU foams and can react in the PU network with its hydroxyl groups. The mode of action of this molecule was found to be the promotion of increased char yield [43,44]. It is used in spray foams and pour-in-place foams as well as in some elastomers. An advantage of a reactive FR as such is its good retention, especially in applications with long-term heating requirements, e.g. in refrigerators [43].

Fyrol 6 and Levagard 4090 can not be used in flexible PU foams (except in small amounts), since their hydroxyl number is too high. Phosphorus-containing diols suitable for flexible foams, e.g. for automotive seats, have been a subject of research for many years. A phosphonate polyol with 12% phosphorus and a hydroxyl number of 400 was introduced by Clariant as Exolit OP 560 (Figure 14). FMVSS 302 class SE is attained with 2.5–12.5 php Exolit OP 560, depending on the density of the foam (php = parts per hundred polyole). It meets low fogging requirements and is mainly used in automotive foams [45,46]. Phosphonates are also known for their good efficiency and good scorch performance in PU foams.

More recently Thor has introduced three phosphonate-based FRs. The pentaerythritol ester of methane phosphonic acid is available under the trade name Aflammit TL 1260 (Figure 15) and is claimed to be an efficient FR for PA 6. At a dosage of 4–8% UL 94 V-0 was passed. Due to its melting behavior, the product can be used for fiber applications as well [47].

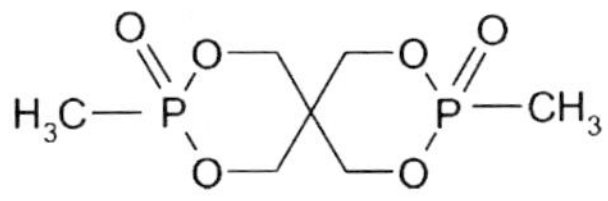

FIGURE 15

Structure of Aflammit TL 1260 [47].

It was also found that some phosphonates can have synergistic effects with special hindered amine light stabilizers, so-called NOR-HALS molecules, for flame-retarding polyethylene films [48]. NOR-HALS structures are FRs for polyolefin films and fibers and are effective even at about 1% dosage, but the FR effect is limited to very thin films (normally up to 200 μm). Combining NOR-HALS with special phosphonates, fire tests can be passed in all thicknesses [48]. The dimethyl spirophosphonate TL 1260 is mentioned here as well as an ethylene diamine methyl phosphonate and a melamine methyl phosphonate. Both products are also currently introduced to the market.

5. Phosphinate FRs

Phosphinates are organophosphorus compounds with the formula $OP(OR)R_2$.The use of metal phosphinates for PAs and polyesters was first described by Sandler in 1979 [49]. Zinc phosphinates were found to be polymeric and show an increase in LOI in PA 66 in 10–20% dosage. Kleiner in 1994 found aluminum and calcium salts of phosphinic acid as being very effective FRs for polyesters and PAs [50,51]. Calcium and aluminum phosphinates with R standing for short-chain alkyl groups are hydrophobic fine powders with a phosphorus content of 23–24% with a very low residual solubility in water and they do not melt before decomposing at >350 °C. Hypophosphites like aluminum or calcium hypophosphite with R standing for hydrogen are (also) white powders with phosphorus contents of about 40% and beginning decomposition above 300 °C. Hypophosphite salts (Figure 16) are claimed to be FRs in PAs and polyesters as well [52]. The thermal stability of hypophosphites is in most cases lower than the stability of alkyl phosphinates (Table 2).

The thermal decomposition of aluminum hypophosphite is characterized by two steps with maximal mass loss rates at 332 °C and 450 °C and can be represented by the two equations below [53]. During polymer processing the release of phosphine gas (PH_3, CAS-No. 7803-51-2), which is very toxic and highly flammable with garliclike smell, can occur at elevated temperatures. The FR effect of hypophosphites is also in most cases lower than that of short-chain alkyl phosphinates.

With 20% loading of calcium methyl ethyl phosphinate in a reinforced PBT, UL 94 was found at acceptable mechanical properties. In PAs, the addition of about 30% of aluminum or calcium phosphinates was necessary to pass UL 94 V-0 [51].

Jenewein and Kleiner have found that some nitrogen compounds are synergistic with phosphinates in polyesters so that the compounding properties could be

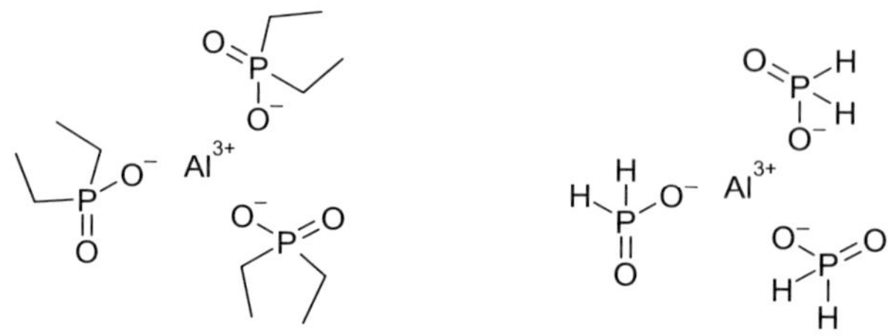

FIGURE 16

Structure of phosphinic acid salts, Al-diethylphosphinate (left), and Al-hypophosphite (right).

improved and the costs of the flame-retarded polyesters could be reduced [54]. The combination of phosphinates with MPP could reduce the dosage for an UL 94 V-0 classification in PAs from 30% to less than 20%; this finding opened reinforced PA 6 and 66 compounds as applications for phosphinates [55]. Flame-retarded PAs are used mainly for electrical and electronic applications. Phosphinates are efficient in unreinforced PAs as well, but here MC is more economic.

The combination of phosphinates with MPP in reinforced PA has been also investigated by cone calorimeter. The combination of the FRs in PA 6-GF reduced the peak HRR by about 250 kW m^{-2}, the THR by about 20 MJ m^{-2}, and THR/TML by about 0.4 MJ m^{-2}g compared to PA 6-GF. The residue increased by about 6 wt%. Thus the results clearly show that the combination of both additives results in greater flame retardancy than expected from a simple superposition of both additive effects. Taking a UL 94 V-0 classification at 1.6 mm or 0.8 mm as a basis for the comparison, a compounder needs about 20% of brominated polystyrene and 6% of antimony trioxide in a PA 6-GF or PA 66-GF to fulfill the requirement compared to 20% of a mixture of aluminum phosphinate and MPP, preferred in a ratio of 2:1. The formulations and compound properties are given in Table 3.

Table 4 shows that with the new phosphinate flame-retarded PA 66-GF compounds with a balanced range of properties can be obtained. A V-0 rating in combination with passing the glow wire flammability index at 960 °C demonstrates good FR efficiency at an attractive dosage level. Especially the high comparative tracking index (CTI) value and the low compound density are very favorable. In contrast to the formulation based on red phosphorus, the compound based on the phosphinate allows bright colors. Physical properties are as well on a satisfying level.

The fire retardancy mechanism of aluminum diethylphosphinates as single FR and in combination with MPP in glass–fiber-reinforced PA 6 (PA 6-GF)

Table 2 Thermal Stability of Phosphinate and Hypophosphite Salt

	Maximum Decomposition in N_2	Maximum Decomposition in Air
Al-diethylphosphinate	479 °C	403 °C
Al-hypophosphite	310 °C	300 °C

Table 3 Synergistic Effect of Phosphinate and Melamine Polyphosphate in Polyamide (PA) [55]

PA 6	55	55	55
Glass fiber	25	25	25
Aluminum phosphinate	20		13.3
MPP		20	6.7
UL 94 0.8 mm	V-1	No rating	V-0
UL 94 1.6 mm	V-0	No rating	V-0

Table 4 Formulations and Properties of PA 66-GF Compounds

	PA 66-GF	PA 66-GF with BrPS	PA 66-GF with Red Phosphorus	PA 66-GF with Phosphinate
PA 66	70 wt%	43.6 wt%	56 wt%	52 wt%
Glass fibers	30 wt%	30 wt%	30 wt%	30 wt%
Brominated polystyrene	–	20 wt%	–	–
Antimony trioxide (80% in PA 6)	–	6.0 wt%	–	–
PTFE	–	0.4 wt%	–	–
Red phosphorus (50% in PA 6)	–	–	14 wt%	–
Aluminum phosphinate + MPP	–	–	–	18 wt%
UL 94 class (0.8 mm) (UL 94 vertical test)	No rating	V-0	V-0	V-0
CTI [V] (DIN IEC 112)	600	400	600	600
Tensile strength (N mm^{-2}) (ISO 527-2)	177	164	174	132
Elongation at break (%) (ISO 527-2)	3.7	2.7	3.2	2.8
Impact strength (kJ m^{-2}) (ISO 179/1eU)	79	64	78	64
Notched impact strength (kJ m^{-2}) (ISO 179/1eA)	15.3	14.3	11.6	12.5

has been investigated by analyzing the pyrolysis, flammability, and fire behavior [56]. The phosphinate alone partly vaporizes and partly decomposes to volatile diethylphosphinic acid and aluminum phosphate residue. In fire, a predominant gas phase action was observed, but the material did not reach a V-0 classification for the moderate additive content used. For the combination of both phosphinate and MPP in PA 6-GF a synergistic effect occurred, because of the reaction of MPP with phosphinate. Aluminum phosphate is formed in the residue and melamine and phosphinic acid are released in the gas phase. The aluminum phosphate acts as a barrier for fuel and heat transport, whereas the melamine results in fuel dilution; the phosphinic acid results in flame inhibition. The higher amount of aluminum phosphate in the phosphinate + MPP system stabilized the residue in flammability tests in comparison to the phosphinate alone, so that this material achieved a V-0 classification in the UL 94 test. MPP and phosphinate in PA 6-GF also showed synergy with respect to HRR and THR. The observed aluminum phosphate formation in pyrolysis experiments has been proposed to improve the fire behavior through an optimized barrier effect, whereas the melamine decomposition still results in fuel dilution and phosphinic acid decomposition in flame inhibition (Figure 17).

The four PA formulations shown in Table 4 were also studied with cone calorimeter and NBS smoke chamber. Similar heat release curves for all the three flame-retarded PAs were observed, whereas the compound flame retarded with phosphinate and MPP show the lowest smoke density (Figure 18).

The European Union's RoHS directive forced E & E manufacturers to switch to lead-free solder systems running at about 30 °C higher temperatures than traditional systems, prompting the rapid growth of polymers like high-temperature polyamides (HTPAs). Recently, commercialized HTPA includes PA 9T (306 °C), PA 4T, and PA 10T. Traditional PAs such as PA 66 are processed at around 290–300 °C, whereas HTPAs are normally processed at 320 °C or above. Due to their higher thermal stability, HTPAs can be used in the so-called surface mount technology of semiconductor components. Facing the requirement of lead-free reflow soldering, these materials have to withstand peak temperatures of 260 °C and more during the soldering process. HT-PAs can only compete with more expensive materials like liquid

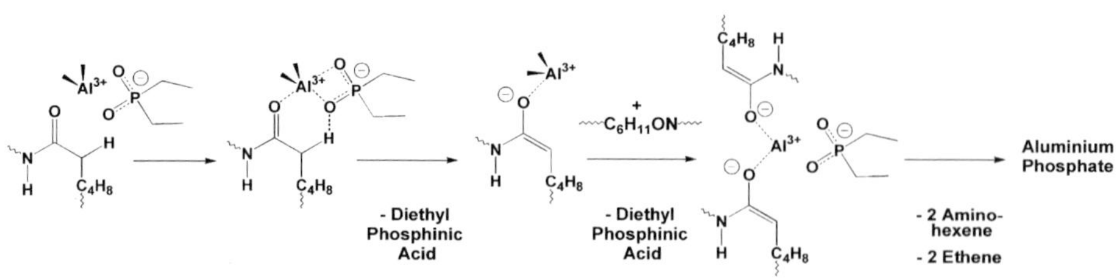

FIGURE 17

Interaction model between Al-diethylphosphinate and PA 6 (dark wedges at Al ion indicate diethylphosphinate units) [56].

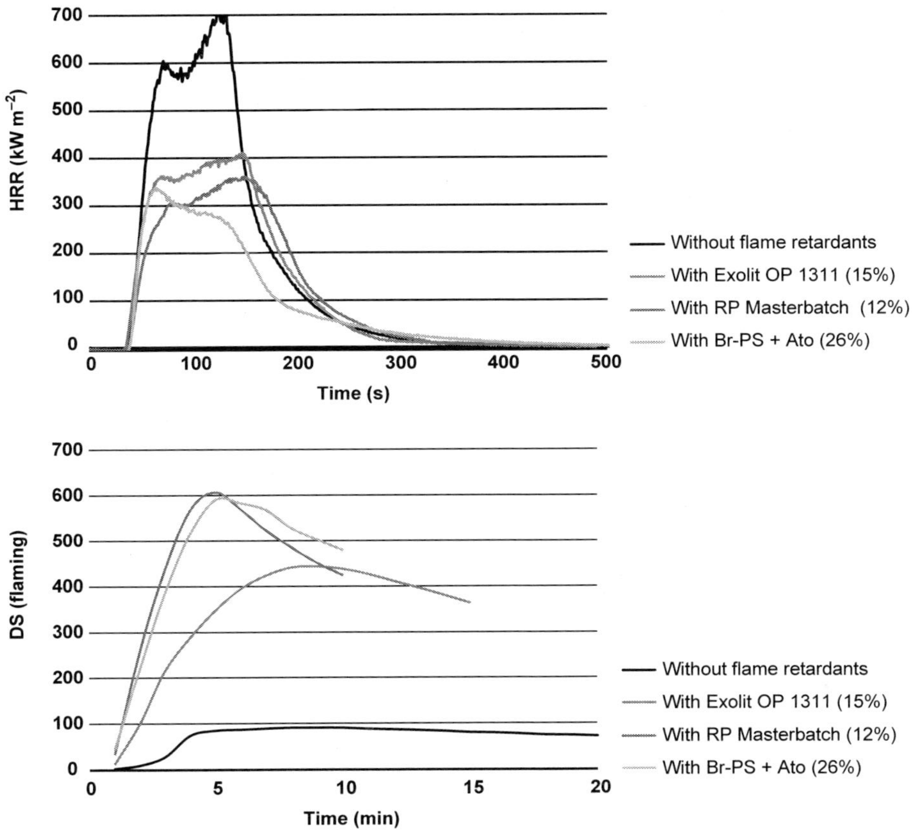

FIGURE 18

Cone calorimeter and smoke chamber results of flame-retarded PAs.

crystal polymers, if the FR system keeps their properties on the high level required by the application [57]. Aluminum phosphinate has proven to be stable enough, providing very efficient flame retardancy for HT-PAs. At a dosage of 12–15%, UL 94 V-0 is achieved up to 0.4 mm thickness. In contrast to other FR materials, a phosphinate-based compound shows no blistering effects in the so-called solder dip test at 260 °C. Moreover, an excellent moisture sensitivity level Class 2 is reached according to the standard IPC/Jedec J-STD-020-C.

Quite recently, Yao and Levchik et al. [58] found that the aluminum salt of diisobutylphosphinic acid (Figure 19) is a more efficient FR, especially in polyamides, than the salt of diethylphosphinic acid. The diisobutylphosphinate has a lower phosphorus content, but a better volatility compared to the diethylphosphinate. This result also indicates a strong gas phase mode of action of phosphinates in PAs and polyesters. Commercial applications of diisobutylphosphinates are not known.

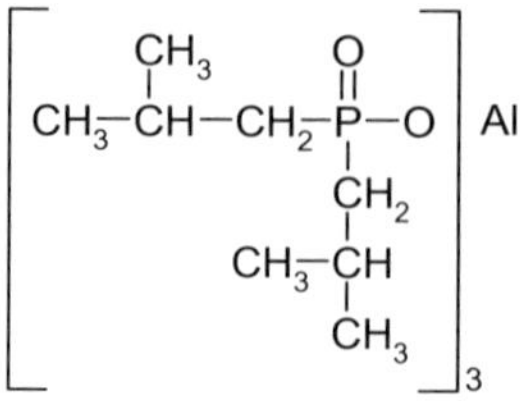

FIGURE 19

Structure of the aluminum salt of diisobutylphosphinic acid.

Higher volatility might cause problems in compound and injection molding with outgassing and mold deposits.

Except PAs, another important group of engineering plastics that can be flame retarded by phosphinates are thermoplastic polyesters obtained from aliphatic glycols and aromatic dicarboxylic acids. The largest volume is represented by PET and PBT. PBT has a lower melting point and HDT than PET, but is a better electrical insulator and can be molded at lower temperatures. Thermoplastic polyesters are often used for insulating parts in the electrical and electronic industry and are competing with PA 6 and 66. PBT has good dimensional stability and thermal and chemical resistance. The strength and stiffness of reinforced PBT allow its use in thinner wall sections than is possible with unreinforced materials. The high heat deflection temperatures of reinforced grades make PBT able to perform in high-temperature environments. The FR systems which have been most commonly used in PBT resins consist of polymeric brominated compounds and an antimony source.

In glass fiber-reinforced PBT for E & E applications a total FR dosage of approximately 20 % is needed to safely achieve the most required classification, a V-0 according to the UL 94 standard. When using aluminum diethylphosphinate, there are a few options for the compounder to achieve the V-0 class. One can simply add the neat phosphinate or combine it with a nitrogen-containing co-FR like MC or MPP. The class UL 94 V-0 is passed even at 0.4 mm thickness. Using aluminum hypophosphite, V-0 can be passed down to a thickness of 1.6 mm, but not below (Table 5).

Compared to a flame-retarded PBT with a brominated FR, e.g. the widely used pentabromobenzylacrylate plus antimony trioxide, a halogen-free PBT with aluminum diethylphosphinate and MPP shows higher CTI, but reduced tensile and impact properties due to the fillerlike character of the halogen-free FRs. Therefore, it is preferable to reduce the glass fiber content from 30% to 25% in these materials. Tensile and impact properties of the flame-retarded compound with 25% glass are more balanced and less different from the values of a non-FR material with 30% glass (Table 6). Adding MPP to aluminum diethylphosphinate offers different interesting benefits. It causes less PBT degradation at processing temperatures and allows better physical properties of the final compound but might need fine-tuned formulation to counteract a potential corrosion issue of tool steels during injection molding. Recently, Narayanan et al. [59] found that some organometallic

Table 5 V-0 Formulations for PBT GR

PBT	55	55
Glass fibers	25	25
Aluminum diethylphosphinate	13.3	
Aluminum hypophosphite		10
MC	6.7	10
UL 94 0.4 mm	V-0	n.c.
UL 94 0.8 mm	V-0	n.c.
UL 94 1.6 mm	V-0	V-0

n.c.: not classified

phosphorus compounds, preferably titanates, can prevent the polymer degradation of phosphinates and thus provide improved mechanical properties of the polyester compounds.

Turning to the effect of MC, cone calorimeter testing of PBT with phosphinate and MC compared to PBT with PBBA + Ato (antimony trioxide) have shown

Table 6 Halogen-Free PBT versus V-0 with Br/Ato

	PBT with Phosphinate + MPP	PBT with PBBA + Ato
PBT	50	55
Glass fibers	30 wt%	30 wt%
Aluminum phosphinate	13.3	
MPP	6.7	
Pentabromobenzylacrylate		9
Ato (80% in PBT)		6
UL 94 class (0.8 mm) (UL 94 vertical test)	V-0	V-0
CTI (V) (DIN IEC 112)	550	225
Tensile strength (N mm^{-2}) (ISO 527-2)	105	128
Elongation at break (%) (ISO 527-2)	2.0	2.4
Impact strength (kJ m^{-2}) (ISO 179/1eU)	40	56
Notched impact strength (kJ m^{-2}) (ISO 179/1eA)	6.7	8.1

Ato, Antimony trioxide; PBBA.

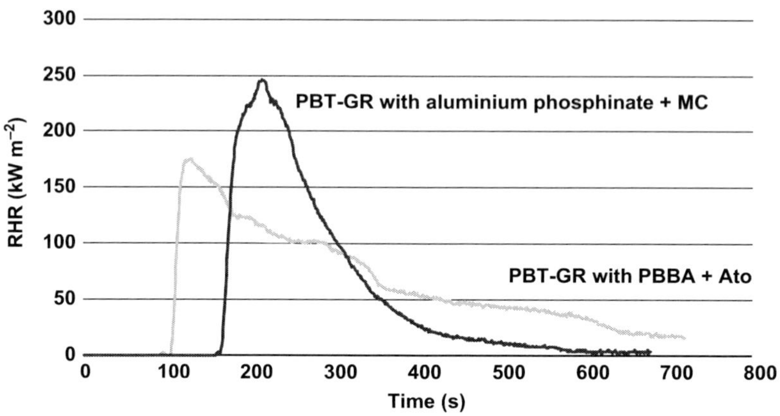

FIGURE 20

Rate of heat release of FR PBT [59].

some differences in the burning behavior: the PBT with phosphinate gets ignited significantly later than the brominated one, but the peak heat release is higher (Figure 20). Braun and Schartel showed that MC decomposes mainly at the beginning of polymer decomposition. So the shift in ignition time is mainly attributed to the fuel dilution effect of MC [60].

A detailed analysis of the fire retardancy mechanisms of aluminum phosphinate alone and in combination with MC in PBT-GF was published by Braun and Schartel [61] recently. A predominantly gas phase action of phosphorus occurred due to the release of phosphinic acid into the gas phase, with a probable additional effect due to the formation of a small amount of carbonaceous char. So the fire retardancy mechanism of aluminum phosphinate in PBT is different from the mode of action in PA 66, where the aluminum phosphinate is combined with MPP. In these polymers, the formation of an aluminum phosphate barrier dominates the fire retardancy mechanism [60]. Aluminum phosphinate in PBT-GF decomposed through the diethylphosphinic acid release (Figure 21). Through interaction of phosphinate anions with terephthalic acid, aluminum phosphinate terephthalate was formed. These derivates decomposed during the second decomposition step, evolving benzene, carbon dioxide, and ethene. Aluminum orthophosphate remained in residue [60]. When aluminum orthophosphate was formed at the end of decomposition, two-thirds of the phosphinate groups were released to the gas phase by diethylphosphinic acid formation. The presence of MC did not influence the pyrolysis behavior of phosphinates significantly. However, it influenced the decomposition characteristics of the polymer and increased the LOI through fuel dilution.

Apart from the FR criteria, the main advantages of phosphinate-based PBT compounds are their low density and the good electrical properties. In contrast to halogen-based FR materials, a much higher level of CTI can be achieved.

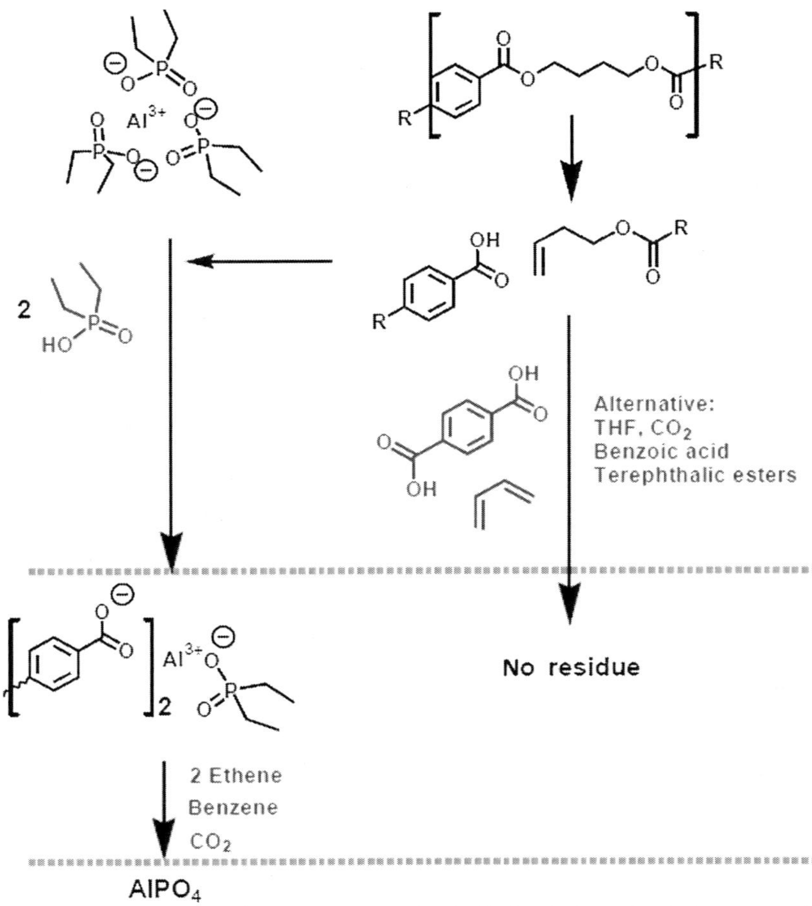

FIGURE 21

Decomposition pathway of PBT with phosphinates.

Reinforced PBT molding compounds based on phosphinates are typically in the 550 or 600 V range, whereas halogen-containing FRs in combination with antimony trioxide usually show results at the 200–300 V level. The density difference of phosphinate-based PBT vs halogen-based PBT is in the range of 10% (depending on specific recipe). It allows molders to produce more parts per weight of compound and it contributes to a typical goal of a part designer in the E&E industry which is lighter and smaller equipment.

Due to the fillerlike behavior of aluminum phosphinate and MPP or MC, PBT with high glass fiber contents and sufficient mechanical properties is still a challenge. One approach is to combine fillerlike phosphorus FRs with polymer or

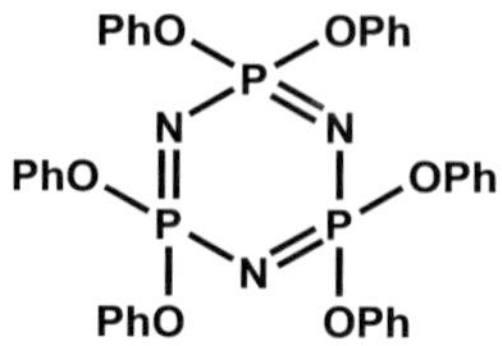

FIGURE 22

Structure of a cyclic phenoxy phosphazene.

liquid ones, e.g. arylphosphates or phosphazenes. Phosphazenes are a class of chemical compounds in which a phosphorus atom is covalently linked to a nitrogen atom by a double bond and to three other atoms or radicals by single bonds. Polyphosphazenes are a class of inorganic polymers with the repeat unit (−RR'P=N−). The substituents R,R′ are usually alkoxy or phenoxy groups (Figure 22). For applications as polymer additive, phenoxy-substituted phosphazenes are preferred due to their better hydrolytic stability. Phenoxy phosphazenes are commercially available [62] and are thermally stable up to 320 °C making them suitable for PAs and polyesters. They are melting under processing conditions.

Combinations of phosphazenes and phosphinates are described as FRs in the recent patent literature, especially for partial aromatic PAs with melting points between 280 and 320 °C [63]. The use of phosphazenes in polyesters is advantageous when used together with liquid crystalline polymers as synergists [64]. Combinations of aluminum phosphinate, phosphazene, and MPP used in PBT GF, give a compound with V-0 classification and around 2.5% elongation at break (Table 7). Such materials are useful for electrical and electronic equipment and have a very balanced property profile.

The aluminum diethylphosphinate selected for the use in plastics have been extensively studied with regard to their toxicology, safety at the workplace, and their possible release into the environment during the life cycle of the corresponding E&E products [65]. Phosphinates used as FRs in PAs will usually end up as phosphates, especially aluminum phosphate, in the ashes of incineration or metal refining processes. The possibility of mechanical recycling has been tested as well with good results for the properties of the reworked materials. Phosphinate-based additives are also widely used as reactive FRs in PET textile applications, giving some kind of inherently flame-retarded fibers and fabrics. These materials are used for upholstery and draperies in hotels, hospitals, institutions, and public transports like ships and airplanes, and to some extent in apparel for high-risk situations. Two main products are widely used. One is based on the reaction product of dihydrooxaphosphaphenanthrene oxide (DOPO) and itaconic diester to make a phosphinate diester which can be reacted in the PET chain (Figure 23) [66]. This fiber shows good hydrolytic stability, since the PET backbone is not affected by the phosphorus molecule. However, addition of the bulky phosphorus group decreases the crystallinity and melting point of the PET [67].

Table 7 Combination of Phosphinate and Phosphazene in PBT GR

	PBT with Phosphinate + MPP	PBT with Phosphinate + Phosphazene	PBT with Phosphinate + Phosphazene + MPP
PBT	50	50	50
Glass fibers	30 wt%	30 wt%	30 wt%
Aluminum phosphinate	13.3	13.3	13.3
Phenoxy phosphazene[a]		6.7	3.3
MPP	6.7		3.4
UL 94 class (0.8 mm) (UL 94 vertical test)	V-0	V-1	V-0
Tensile modulus [MPa] (ISO 527-2)	11,710	9840	10,533
Tensile strength ($N\,mm^{-2}$) (ISO 527-2)	105	104	108
Elongation at break (%) (ISO 527-2)	2.0	2.9	2.5
Impact strength ($kJ\,m^{-2}$) (ISO 179/1eU)	40	56	54
Notched impact strength ($kJ\,m^{-2}$) (ISO 179/1eA)	6.7	7.6	8.1

[a] *Grade SPB 100 from Otsuka Chemicals or FP 110 from Fushimi Pharmaceuticals.*

The other type of product contains the phosphorus in the PET backbone. A monophosphinic acid is reacted with acrylic acid to produce a carboxyethyl phosphinic acid. Products based on methylphosphinic acids and phenylphosphinic acids are commercially available. The diacids are first reacted to ethylene glycol diesters and then reacted in the PET esterifications [68,69]. These flame-retarded PET fibers show lower hydrolytic and thermal stability compared to the DOPO-based products, but are, nevertheless, widely used.

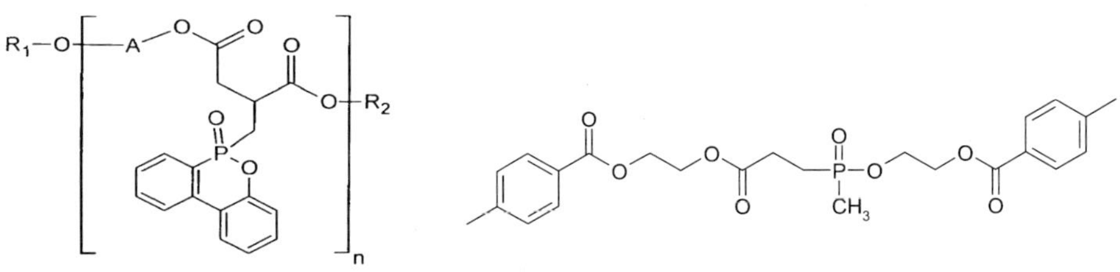

FIGURE 23

Polyesters with phosphorus reacted in DOPO (left) and carboxyethyl phosphinic acid (right).

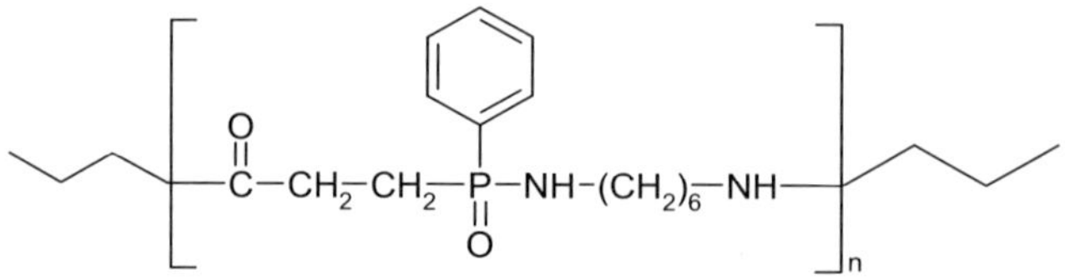

FIGURE 24

A Carboxyethyl-phenyl phosphinic acid is reacted in a Polyamide (PA) 66 chain [70].

The mode of action of these phosphinates in PET fibers seems to be mainly flow enhancement [67]. The fabric melts away from the flames and gets self-extinguishing. Therefore, blends of FR PET with cotton cannot be made, since this mode of action does not work anymore.

The incorporation of carboxyethyl phosphinic acids in a PA 66 backbone is also described in the literature [70]. A so-called AH salt analog is made from the diacid and a diamine, which is then condensed with the normal AH salts to a PA 66 chain (Figure 24). Commercial applications of these kinds of fiber are not known. Presumably, the phosphorus—nitrogen bonds formed are very sensitive to hydrolysis.

DOPO is also incorporated in epoxy resins, mainly as a reaction product of DOPO with benzoquinone. This difunctional phenolic product reacts similar to the commonly used tetrabromobisphenol A and has been commercially used already for some years, but its higher price has limited the market share so far. DOPO can be reacted by the P—H bond to epoxy groups as well, but due to its monofunctionality it can be used only with special curing agents at a sufficient Tg. An excellent review on reactive modifications of polymers with phosphorus compounds in general is recently published by Joseph and Tretsiakova-Mcnally and also covers also polyolefins, acrylics, PCs, epoxies, and PUs [71].

6. CONCLUSIONS

Phosphorus FRs are currently in a very active state of development driven by their good performance and also by environmental and regulatory problems with the brominated FRs [72]. New industrial production capacities for phosphorus-based FRs are under construction and new molecules are subjects of research. While the WEEE legislation is being reinforced in the European Union, requiring separation of bromine FR-containing components before recycling, many OEMs in the E&E industry such as Intel, HP, Dell, and Apple are executing road maps to phase out PVC and halogenated additives first from enclosures, and then from PCBs, cables, and connectors. Phosphorus FRs can act as char formers in the condensed phase and to some extent also as radical scavengers in the gas phase. The existence of these two different modes of actions suggests many synergistic combinations, e.g. phosphinates and polyphosphates in PAs. Flame-retarded PAs and polyesters with phosphinates are very attractive materials, especially for electrical and electronic applications. The property profile of the flame-retarded compounds is very balanced with good electrical and mechanical properties. For halogen-free polyesters, phosphinates are the FRs with the best performance on the market. For PAs, they combine the advantages of brominated FRs with the ones of red phosphorus and are now an established system in many applications. The use of DOPO-based FRs in epoxy resins is also strongly increasing.

In this chapter, the recent work and progress regarding the use of phosphorous-based FRs was analyzed. Table 8 gives a summary of the cited literature, the kind of phosphorus FR and the polymer, and the main topic of the chapter.

Table 8 Resume of Literature

Flame Retardant		Polymer	Literature	Content
Phosphate	Arylphosphate	PC/ABS	[6,7]	UL 94 V-0, comparison of products
			[8–14]	Mode of action, cone calorimeter
	Intumescent: APP	PP	[15]	UL 94 V-0, cone, compound properties
			[16,17]	Synergists: Polyols
			[18–20]	Stability: Interactions APP + synergists
			[21–23]	Triazine and PA synergists
			[24,25]	Review on intumescent FRs
	Intumescent: EDAP	PP	[26]	Recipes UL 94 V-0
			[27]	Synergists: melamine and phosphates
	Intumescent: MPP	PA 66	[28]	Thermal stability, comparison of products
			[29]	MPP in PA 66 vs PA 6
			[30,31]	Mode of action, cone calorimeter
	Intumescent: PIP	PP	[32,33]	UL 94 V-0, thermal stability
Red phosphorus		PA 6	[35]	Mode of action, cone calorimeter
		Epoxies	[36]	UL 94 V-0 with ATH
		PA 6, 66, 610	[37]	UL 94 V-0, with natural fiber reinforcement

Continued

[14] Pawlowski KH, Schartel B. Flame retardancy mechanisms of triphenyl phosphate, resorcinol bis(diphenyl phosphate) and bisphenol A bis(diphenylphosphate) in polycarbonate/acrylonitrile butadiene styrene blends. Polym Int 2007;56:1404–14.
[15] Wanzke W. Phosphorus containing flame retardants – compounding and material properties. London, UK: Interscience Communications Ltd.; 2002. 63–74.
[16] Young DC. Ammonium polyphosphate production. US Patent 4041133 (Union Oil) 1977.
[17] Maurer A, Staendeke H. Activated ammonium polyphosphate, a process for making it, and its use. US Pat 4515632 (Hoechst AG) 1985.
[18] Camino G, Costa L, Trossarelli L. Study of the mechanism of intumecence in fire retardant polymers: part V – mechanism of the formation of gaseous product in the thermal degradation of ammonium polyphosphate. Polym Degrad Stabil 1985;12: 203–12.
[19] Camino G, Grassie N, McNeill IC. Influence of the fire retardant, ammonium polyphosphate, on the thermal degradation of poly(methyl methacrylate). J Polym Sci Polym Chem Ed 1978;16:95–106.
[20] Camino G, Costa L, Trossarelli L. Study of the mechanism of intumescence in fire retardant polymers: part I – thermal degradation of ammonium polyphosphate – pentaerythritol mixtures. Polym Degrad Stabil 1984;6:243–52.
[21] Hu X, Li YL, Wang YZ. Synergistic effect of the charring agent on the thermal and flame retardant properties of polyethylene. Macromol Mater Eng 2004;289:208–12.
[22] Li B, Xu M. Effect of a novel charring – foaming agent on the flame retardancy and thermal degradation of intumescent flame retardant polypropylene. Polym Degrad Stabil 2006;91:1380–6.
[23] Jun W, Yi J, Cai X-F. Synergistic effect of a novel charring agent and ammonium polyphosphate on the flame retardancy of acrylonitrile butadiene styrene. J Appl Polym Sci 2011;120:968–73.
[24] Le Bras M, Camino G, Burbigot S, Delobel R, editors. Fire retardancy of polymers: the use of intumescence. Cambridge, UK: Royal Society of Chemistry Special Publication, Springer Verlag; 1998.
[25] Bourbigot S, Le Bras M, Duquesne S, Rochery M. Recent advances for intumescent polymers. Macromol Mater Eng 2004;289:490–511.
[26] Goin CL, Huggard MT. Paper presented at 2nd annual BCC conference on recent advantages in flame retardancy of polymeric materials. Connecticut, USA; May 1991.
[27] Rhodes MS, Izrailev L. Activated flame retardants and their applications. US Patent 6733697 (Broadview Technologies) 2004.
[28] Salemis P. EFRA, flame retardants fact sheet—Melamine phosphates http://www.cefic-efra.eu; 2007.
[29] Jahromi S, Gabrielse W, Braam A. Effect of melamine polyphosphate on thermal degradation of polyamides: a combined X-ray diffraction and solid-state NMR study. Polymer 2003;44:25–37.
[30] Bras ML, Bourbigot S, Delporte C, Siat C, Tallec YL. New intumescent formulations of fire-retardant polypropylene – discussion of the free radical mechanism of the formation of carbonaceous protective material during the thermo-oxidative treatment of additives. Fire Mater 1996;20:191–203.
[31] Bras ML, Bourbigot S, Tallec YL, Laureyns J. Synergy in intumescence- application to beta-cyclodextrin carbonization agent in intumescent additives for fire retardant polyethylene formulations. J Polym Degrad Stabil 1997;56:11–21.

[32] Ke CH, Li J, Fang KY, Zhun QL, Yan Q, Wang YZ. Synergistic effect between a novel hyperbranched charring agent and ammonium polyphosphate on the flame retardant and anti-dripping properties of polylactide. Polym Degrad Stabil 2010;95:763−70.

[33] Dai JF, Li B. Synthesis, thermal degradation, and flame retardance of novel triazine ring-containing macromolecules for intumescent flame retardant polypropylene. J Appl Polym Sci 2010;116:2157−65.

[34] Weil ED. Mechanisms and modes of action in flame retardancy of polymers. In: Proceedings of the conference on recent advances in flame retardancy of polymeric materials, Stamford. Connecticut, USA; May 22−24, 2000.

[35] Levchik SV, Levchik GF, Balanovich AI, Camino G, Coast L. Mechanistic study of combustion performance and thermal decomposition behaviour of nylon 6 with added halogen-free fire retardants. Polym Degrad Stabil 1996;54:217−22.

[36] Honda N, Sugiama T. Halogen-free flame-retardant epoxy resin composition (Toshiba). U.S. Patent 5994429; 1999.

[37] Scherzer D, Scheibitz M. Flameproofed polyamide molding compounds. PCT Int Appl; 2011. WO2011009877 (BASF SE).

[38] Weil ED. Recent advances in flame retardancy of polymeric material. In: Proceedings of the 11th annual BCC conference on flame retardancy; May 2000.

[39] Weil ED. Flame retardants based on phosphorus. In: Proceedings of the conference on recent advances in flame retardancy. Connecticut, USA: Business Communications Corp; 1991.

[40] Freitag D. Preparation of polyphosphonates via transesterification without a catalyst. PCT Int Appl; 2005. WO2005121220 (Triton Systems Inc).

[41] Freitag D. Method for the production of block copolycarbonate/phosphonates and compositions therefrom. PCT Int Appl; 2007. WO2007065094 (Triton Systems Inc).

[42] Wanzke W. Flame retardant synergies with organic phosphinates. In: 21st annual recent advances in flame retardancy of polymer materials. BCC Research; 2010.

[43] Weil ED, Levchik SV. Flame retardants for plastics and textiles. Munich, Germany: Hanser Publishers; 2009. pp. 153−78.

[44] Wang XL,Yang KK, Wang YZ. Physical and chemical effects of diethyl N,N-diethanolaminemethylphosphonate on flame retardancy of rigid polyurethane foam. J Appl Polym Sci 2001;82:276−82.

[45] Eilbracht, C, Sicken, M. Process for producing flame-retardant flexible polyurethane foams. US Patent 6380273 (Clariant) 1999.

[46] Witte, A, Krieger, W. Halogen-free, water-blown, flame-retardant rigid polyurethane foam and a process for its production. US Patent 6518325 (Clariant) 2000.

[47] Butz V. Flame-protected polyamide moulding materials. PCT Int Appl; 2009. WO 2010 146033 (Thor).

[48] Butz V. Flame-retardant composition comprising a phosphonic acid derivative. PCT Int Appl; 2010. WO 2010 026230 (Thor).

[49] Sandler SR. Polyamide resins flame retarded by poly(metal phosphinate)s. US Patent 4208321 (Pennwalt Corp.) 1978.

[50] Kleiner H-J, Budzinsky W. Flame retarded polyester molding compositions. Eur. Patent 699708 (Hoechst AG) 1996.

[51] Kleiner H-J, Budzinsky W. Flame retarded polyamide molding compositions. Eur Pat 792912 (Hoechst AG) 1997.

[52] Costanzi S, Leonardi M. Polyester compositions flame retarded with halogen-free additives. PCT Int Appl; 2005. WO 2005121232 (Italmatch).

[53] Yang W, Song L, Hu Y, Lu H, Yuen R. Enhancement of fire performance of glass-fibre reinforced poly(ethylene terephthalate) composites with the incorporation of aluminium hypophosphite and melamine cyanurate. Compos Part B 2011;42:1057–65.

[54] Jenewein E, Kleiner H-J. Synergistic flame protection agent combination for thermoplastic polymers. US Patent 6,365,071 (Clariant) 2002.

[55] Schlosser E, Nass B. Flame retardant combination. US Patent 6,255,371 (Clariant) 2001.

[56] Braun U, Schartel B, Fichera MA, Jäger C. Flame retardancy mechanisms of aluminium, phosphinate in combination with melamine polyphosphate and zinc borate in glass-fibre reinforced polyamide 6,6. Polym Degrad Stabil 2007;92:1528–45.

[57] Steffner K-J. Lead-free soldering. Kunstst Plast Europe 2005;9:195–8.

[58] Yao Q, Levchik SV, Alessio R. Phosphorus-containing flame retardant for thermoplastic polymers; 2008. US 2008 01326619 (Supresta).

[59] Narayanan R. Flame retardant polymer compositions. PCT Int Appl; 2010. WO 2010009026 (Ticona).

[60] Braun U, Bahr H, Sturm H, Schartel B. Flame retardancy mechanisms of metal phosphinates and metal phosphinates in combination with melamine cyanurate in glass-fiber reinforced poly(1,4-butylene terephthalate): the influence of metal cation. Polym Adv Technol 2008;19:680–92.

[61] Braun U, Schartel B. Flame retardancy mechanisms of aluminum phosphinate in combination with melamine cyanurate in glass-fiber reinforced poly(1,4-butylene terephthalate). Macromol Mater Eng 2008;293:206–17.

[62] Nakacho Y, Nishioka Y, Tada Y, Yabuhara T. Crosslinked phenoxyphosphazene compounds, flame retardant, flame-retardant resin compositions, and moldings of flame-retardant resins. Eur Patent 0945478 (Otsuka Chem Corp.) 1999.

[63] Seki M. Flame-retardant polyamide composition. US Patent 2010 0261818 (Mitsui Chemicals) 2010.

[64] Yabuhara T, Tada Y. Flame-retardant resin composition. DE Patent 69907251 (Otsuka Chemical Corp.) 2004.

[65] Marzi T, Beard A. The ecological footprint of flame retardants over their life cycle – a case study. Flame Retard 2006;12:21–32.

[66] Seiji E, Takao K, Akitada O, Tatsuhiko S, Tadashi I. Phosphorus-containing compounds. US Patent 4127590 (Toyo Boseki) 1978.

[67] Weil ED, Levchik SV. Flame retardants for plastics and textiles. Munich, Germany: Carl Hanser Verlag GmbH & Co. KG; 2009. 105–108.

[68] Kleiner H-J, Finke M, Bollert U, Herwig W. Flame retarding linear polyesters and shaped articles thereof. US Patent 3941752 (Hoechst Aktiengesellschaft) 1973.

[69] Asrar J. Flame retardant polyester copolymers. US Patent 5399428 (Monsanto Company) 1994.

[70] Asrar J. Polymer-bound non-halogen fire resistant compositions. Eur Patent EP 0983311 (Solutia Inc.) 2000.

[71] Joseph J, Tretsiakova-Mcnally S. Reactive modifications of some chain- and step-growth polymers with phosphorus containing compounds. Polym Adv Technol 2011;22:395–406.

[72] Levchik SV, Weil ED. A review of recent progress in phosphorus based flame retardants. J Fire Sci 2006;24:345–63.

[73] Levchik SV, Weil ED. A review of recent progress in phosphorus based flame retardants. J Fire Sci 2006;24:345–63.

CHAPTER

Novel Class of Eco-Flame Retardants Based on the Renewable Raw Materials

7

A.M. Sakharov, P.A. Sakharov, S.M. Lomakin, G.E. Zaikov
Russian Academy of Sciences, Moscow, Russia

CHAPTER OUTLINE

1. Introduction

Prevention and extinguishing of fires has always been one of the global problems facing humanity. Annually thousands of new reagents appear dedicated to the solution of this problem. However, problems such as the extinguishment of forest fires or of low polymeric materials are still far from the solution.

The combustion of natural and polymeric materials, like the combustion process of any other fuel material, is a combination of complex physical and chemical processes, which include the transformation of initial products. This whole conversion process may be divided to several stages, with specific physical and chemical processes occurring in each of these stages. In contrast to the combustion of gases, the combustion process of condensed substances has a multiphase character. Each stage of the initial transformation of a substance correlates with a corresponding value (combustion wave) with specific physical and chemical properties (e.g. state of aggregation, temperature range, concentration of the reacting substances, and kinetic parameters of the reaction) [1,2].

One category of flame retardants widely used for protecting polymers against fire is that of char-forming materials. These materials swell (intumesce) during degradation (combustion), creating a shield that protects the substrate. Thus, research and development of new intumescent flame retardants has been focused on finding ways to increase this char-forming tendency, since there is a strong correlation between char yield and fire resistance. The tendency of a polymer to char can be increased with chemical additives and/or by altering its molecular structure [3–5]. These polymeric additives usually produce a highly conjugated system of

Polymer Green Flame Retardants. http://dx.doi.org/10.1016/B978-0-444-53808-6.00007-X

aromatic structures, which char during thermal degradation and/or transform into cross-linking agents at high temperatures.

Char, formed during thermal degradation, can play several roles in fire retardancy. The formation of char has a significant effect on the degradation because it occurs at the expense of other reactions that may form volatiles; thus, char formation may limit the amount of fuel available. For example, cellulose degrades either by a series of dehydration reactions that yield water, carbon dioxide, and char, or by a process in which levoglucosan is produced, leading eventually to the formation of volatiles.

It is believed that the temperature at the surface of a burning polymer is close to the temperature at which extensive thermal degradation occurs (usually 300−600 °C). The bottom layer of char, near the polymer surface, is at the same temperature, whereas the upper surface, exposed to the flame, can be as hot as 1500 °C [6]. Therefore, the chemistry of fire retardancy concerns chars, which are produced at temperatures between 300 and 1500 °C [7]. Furthermore, the objective of a flame retardant approach is to promote the formation of such intumescent chars. Intumescent systems, on heating, give a swollen multicellular char capable of protecting the underlying material from the action of the flame. Fire-protective coatings with intumescent properties have been in use for over 50 years, whereas incorporation of intumescent additives in polymeric materials is a relatively recent approach [8−10]. On burning, these additives develop the foamed char on the surface of degraded material, which acts as a physical barrier against heat transmission and diffusion of oxygen toward the polymer and of combustible degradation products of the polymer toward the flame. Thus, the rate of pyrolysis of the natural and polymeric materials eventually decreases below flame-feeding requirements, leading to flame extinguishment.

The intumescent approach, i.e. charring and foaming of the surface of the burning polymers is being widely adopted for application where fire retardancy is required, because it is characterized by a low environmental impact. Moreover, intumescent coatings comprise one of the most effective methods of protecting wood and natural materials from fire. When heated, they form a thick, porous carbonaceous layer (upon heating, intumescent compounds expand up to 200 times their volume). This provides an ideal insulation to the substrate against an excessive increase in temperature and oxygen availability; thus thermal decomposition, which plays a decisive role in retarding the combustion process, is prevented [11]. Commercially available intumescent systems include paints, lacquers, mastics, and linings; these products contain ingredients that, when heated at elevated temperatures, form large amounts of nonflammable residues. These residues, in turn, under the influence of emitted gases produce foam with good insulating properties [11−13].

Heating of cellulosic materials and wood at 105−110 °C results in the removal of moisture. Reactions occurring at this stage proceed slowly and are mostly endothermic. After exceeding the above temperature, a slow thermal decomposition of the components of natural polymers begins and at 150−200 °C gas products of the decomposition start to be released. At 160 °C begins the decomposition of lignin. Lignin under thermal degradation yields phenols from cleavage of ether and carbon–carbon linkages, resulting in more char than that produced during the decomposition of cellulose. Most of the fixed carbon in charcoal originates from lignin. At 180 °C, the hydrolysis of low molecular weight polysaccharides (hemicelluloses) takes place [13]. Thermal stability of hemicelluloses is lower than that of cellulose; more incombustible gases and fewer tarry substances are released [12]. The most abundant gaseous products contain about 70% of incombustible CO_2 and about 30% of combustible CO. Depending on the availability of oxygen, subsequent reactions can be exothermic or endothermic. In the temperature between 220 and 260 °C, exothermic reactions begin. They are characterized by evolution of gaseous products, release of tarry substances, and appearance of ignition areas of hydrocarbons with low boiling points. Cellulose decomposes between 260 and 350 °C, and it is primarily responsible for the formation of flammable volatile compounds [12].

At temperatures of 275–280 °C, accelerated release of considerable quantities of heat begins and increased amounts of liquid and gaseous products (280–300 °C), including methanol and acetic acid, are formed. The amount of evolving carbon monoxide and dioxide decreases, while mechanical slackening of wood structure proceeds, and ignition occurs; the mass loss of wood reaches about 39% [12]. Tar begins to appear at 290 °C. The release of gases still increases and rapid formation of charcoal takes place. This reaction is highly exothermic and proceeds at 280–320 °C. Secondary reactions of pyrolysis become predominant, resulting in an increased amount of gaseous products. Combustion proceeds in the gas phase at a small distance from the surface rather than on the wood surface itself. The ignition of wood occurs at 300–400 °C, depending on its origin and lignin content [13]. At the final stage of wood's combustion (above 500 °C), the formation of combustible compounds is small, while the formation of charcoal increases. Charcoal makes an insulating layer that hinders heat transfer, thus preventing the temperature from increasing to when the pyrolysis of wood occurs.

Mechanisms of flame retardancy depend on the action of compounds present in fire retardants. In their comprehensive review on wood flame retardants, R. Kozlowski and M. Wladyka-Przybylak proposed that there are two general groups of fire retardants for wood: the additive and the reactive ones [12]. Additive compounds are those whose interaction with a substrate is only physical in its nature, whereas reactive compounds interact chemically with wood (frankly with cellulose, hemicellulose, or lignin) [12,14]. The additive compounds include mono- and diammonium phosphate, halogenated phosphate esters, phosphonates, inorganic compounds (such as antimony oxide and halogens), ammonium salts (ammonium bromide, ammonium fluoroborate, ammonium polyphosphate, and ammonium chloride), amino resins (compounds used for their manufacture are dicyandiamide, phosphoric acid, formaldehyde, melamine, and urea), hydrated alumina, stannic oxide hydrate, zinc chloride, and boron compounds (boric acid, borax, zinc borate, triammonium borate, and ethyl and methyl borates) [12].

Reactive-type flame retardants (aliphatic diglycidyl phosphonates, glycidyloxy diphenylphosphine oxide, phosphorus-containing epoxy resins, etc.) appear to be most versatile in thermosetting plastic materials. They are chemically bound to polymer products either by incorporating them into the polymer backbone during the polymerization reaction or by grafting them onto it.

As mentioned, one of the most effective methods of protecting wood from fire is the use of fire retardant intumescent coatings. However, the existing fire retardants present some disadvantages related to efficiency and environmental impact. Therefore, research is focused on the chemical modification of natural polymers and wood, and on the development of more efficient intumescent systems and fire protectors. Here, the results of research conducted regarding the use of nonpolluting reagents from renewed vegetative raw materials is presented; the materials can perform as highly effective flame retardants intumescent/char-forming systems for cellulosic materials and wood, presenting a high potential for use in a wide range of applications, due to their cheapness and simple technology of production.

Starch, cellulose, and lignin have been found to be easy oxidized by oxygen in the presence of copper salts and alkali. The basic products of oxidation of polysaccharides and lignin are salts of polyoxyacids [15]. Such salts can be used as components of washing powders, components of boring solutions, gums for use in building, etc. However, unexpectedly it was found that the salts of polyoxyacids can also prevent several materials from burning [16]. In the present study, the data obtained from the evaluation of the flame retardant properties of oxidized polysaccharides and a lignin, are presented.

2. Oxidized polysaccharides and lignin

The oxidative modification of polysaccharides, leading to polyacids as main products, is widely applied in practice due to the availability of raw materials and high consumer-attractive properties of oxidized polysaccharides. Salts of polyacids are used as water-soluble glues in the production of paper and cardboard, as components of drilling agents, etc. The oxidizing agent, used until recently for the oxidation of polysaccharides, was hydrogen peroxide since gaseous oxygen presents low activity in such procedures [17,18]. However, when copper complexes and bases are present, not only the structurally simple alcohols and ketones, but also polysaccharides (e.g. starches, dextranes, and cellulose) can be oxidized by oxygen, achieving high rates. These high rates of oxidation by oxygen, exceeding the ones achieved by oxidation with hypochlorites, are reached at temperatures 40–90 °C [19].

The oxidation of polysaccharides proceeds with the greatest efficiency in the presence of copper and alkali salts. In Figure 1, the kinetic curve of oxygen absorption (curve 1) and of NaOH consumption (curve 2) are presented along with the change of viscosity during oxidation of potato starch (curve 3). It can be observed that at the first 10–20 min of starch oxidation, the viscosity of the (initial) gel decreases by hundred times, reaching finally (after 3–4 h) a value similar to that

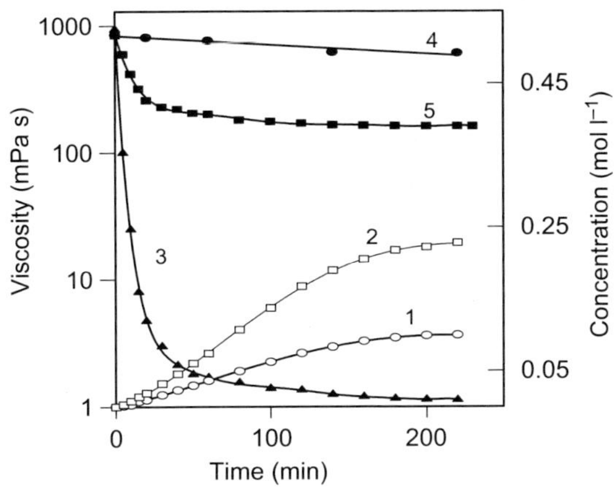

FIGURE 1

(1): O_2 consumption, (2): NaOH consumption, (3): viscosity change of initial gel during starch oxidation in the present of copper ions, (4): viscosity change of gel in the absence of catalyst, (5): viscosity change of initial gel of starch (5 g starch and 0.5 g NaOH in 50 ml H_2O) under inert atmosphere after addition of 1 g dry powder of deep oxidized starch. ($[CuCl_2]_o = 5.10^{-3}$ M, 5 g potato starch, 0.5 g NaOH, 50 ml H_2O, 338 K).

of water. On the contrary, the molecular weight of an initial polysaccharide exhibits a slight decrease during the first minutes [20].

Without the copper salt, the oxidation of starch does not proceed with measurable rate and viscosity of gel practically does not change (curve 4). But the addition of oxidized starch to the native gel of starch (in the absence of oxygen) drastically (by a factor of ten) diminishes viscosity (curve 5).

The most probable formula of the oxidized polysaccharide (starch) is presented below (Figure 2). In this structure, unmodified α-D-glucopyranosyl cycles and the oxidized links of polysaccharides containing carboxyl groups are included [20].

Copper salts are the most effective catalysts of oxidation of polysaccharides in the presence of alkali. In the absence of the catalyst, starch and cellulose practically cannot be oxidized at temperatures below 373 K. However, lignin, which contains many high-reactive phenolic groups (under high rates), can be oxidized by oxygen at measurable rates, even without the catalyst.

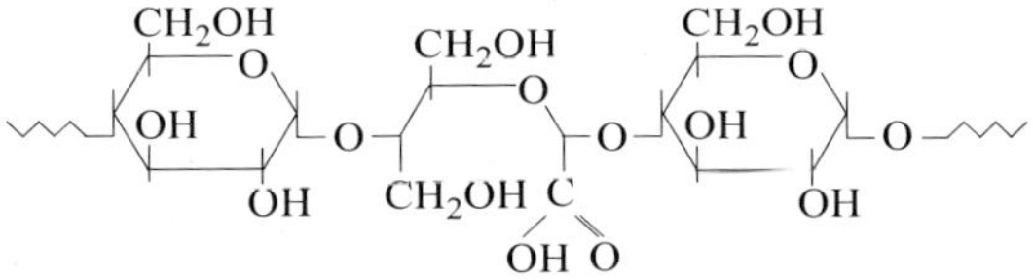

FIGURE 2

Chemical formula of oxidized polysaccharide.

Similar to what occurs during the oxidation of polyols with low molecular weights [21], polysaccharides react with Cu^{2+} ions with $[Cu^{2+}...A^{-}]$ adducts formation (A^{-}-deprotonated polysaccharide). The role of the catalyst (Cu^{2+} ions) is to activate the deprotonated forms of substrate to react with oxygen. In other words, the role of bivalent copper ions is supposed to be the facilitation of the oxidation of anion forms of substrate (interaction of intermediate radicals or anion radicals with O_2). However, high-molecular polysaccharides with a low content of end aldehyde groups are redox inactive in anaerobic medium and the rate of reduction of Cu^{2+} to Cu^{+} ions (due to electron transfer from substrate anion form to Cu^{2+}) is 100 times lower than the rate of oxygen adsorption during the process of polysaccharides oxidation in alkaline mediums. Therefore, it seems that it is just the direct interaction of oxygen with anions A^{-} in coordination sphere of copper ion that leads to the formation of hydroxycarboxylates (as the main primary products of oxidation). Absorption of oxygen, on the other hand, is completely stagnated after neutralization of introduced alkali by the polyoxyacids, formed during the oxidation process. Thus, by varying of the amount of introduced alkali the degree of polysaccharides oxidation into polyacids can be altered, giving the possibility to tailor the properties of the final product (such as viscosity, bonding ability, and solubility in water).

As initial raw materials for receiving salts of polyoxyacids, not only starch can be used, but also any of other starch-containing raw materials (like corns of maize, oats, rice, etc.), including ill-conditioned raw materials (such as grain crops affected by various fungus diseases, waste of rice slashing, and waste of mill houses). In so doing, the prime cost of the final product can be decreased significantly. Moreover, behavior similar to the polysaccharides, under oxidation in alkali media with/without the catalyst, has been found to be exhibited by lignin. Lignin is among the most abundant biopolymers on earth. It is estimated that the planet currently contains 3×10^{11} metric tons of lignin with an annual biosynthetic rate of approximately 2×10^{10} tons [12]. Lignin constitutes approximately 30% (of the dry weight) of softwoods and about 20% of hardwoods [22]. It is absent from primitive plants such as algae and fungi, which lack a vascular system and mechanical reinforcement. The presence of lignin within the cellulosic fiber wall, mixed with hemicelluloses, creates a naturally occurring composite material that imparts strength and rigidity to trees and plants. Lignin, moreover, is one of the most important bioresources for obtaining raw materials needed for the synthesis of environmentally compatible polymers, since it is nontoxic and extremely versatile in performance. Lignins are derived from renewable resources such as trees, grasses, and agricultural crops. About 30% of wood constituents are lignin. Furthermore, the production of lignin as a by-product of pulping process is worldwide over 30 million tons per year [22].

Lignin is a random copolymer consisting of phenylpropane units having characteristic side chains. Lignin slightly cross-links and takes an amorphous structure in the solid state. The molecular motion is observed as glass transition by thermal, viscoelastic, and spectroscopic measurements. The hydroxyl group of lignin plays

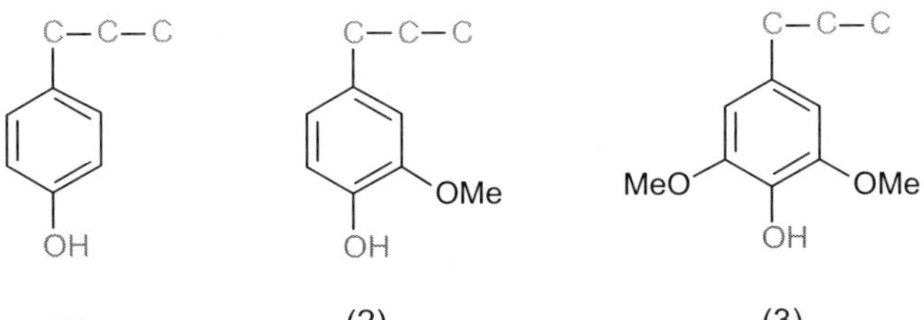

FIGURE 3

Chemical formula of (1) 4-hydroxyphenyl, (2) guaiacyl, and (3) syringyl molecules.

a crucial role during its interaction with water. Lignin is usually considered as a polyphenolic material having an amorphous structure, which arises from an enzyme-initiated dehydrogenative polymerization of *p*-coumaryl, coniferyl, and sinapyl alcohols. The basic lignin structure consists of two components; one is the aromatic part and the other is the C3 chain. The only usable reaction site in lignin is the OH group, which is the case for both phenolic and alcoholic hydroxyl groups. In the process of radical-based biosynthesis, the types of reactants employed to produce the lignin polymer are three: (1) 4-hydroxyphenyl, (2) guaiacyl, and (3) syringyl structures (Figure 3).

The estimation of fire-protecting action of coverings based on oxidized starch-containing raw materials with different amounts of applied reagent was carried out by performing experiments in a vertical tube furnace (ASTM E136-09) (Figure 4). Accordingly, it was found that all pine samples treated with "one-layer covering" of the oxidized starch reagent (OSR), 100 g m^{-2}, provided good fire-protection activity, reaching the rank of hardly inflammable wood (II group). In the case of "multilayer" covering, three to four covering, (OSR consumption 300—400 g m^{-2}), the mass loss during fire testing was significantly decreased,

FIGURE 4

Formation of coke after the heat treatment of a metal plate with an oxidized starch covering.

offering to a fire protection of the highest ranking (group I—hardly inflammable wood).

3. Mode of flame retarding action of oxidized polysaccharides

Studies of thermal decomposition of oxidized starch (at dynamic heating between 25 and 950 °C) have shown that the formation of foam coke occurs at the very early stages of decomposition (at 150–280 °C) as a result of synchronous processes taking place: the reduction of polymer viscosity (transition from the glassy state to that of a viscous fluid) and the chemical reactions of decarboxylation and dehydration. Reactions of intermolecular dehydration/decarboxylation promote the formation of the sewed spatially mesh structure, and the stabilization and hardening of the foam coke. In Figure 4, the formation of intumescent coke from oxidized starch under the heating at 400 °C in air, is shown.

The efficiency of this intumescent additive was also confirmed when studying the combustion behavior of pinewood samples treated with an aqueous solution of oxidized starch (Figures 5 and 6).

The fireproof action of the high-molecular oxidized starch is connected with formation on a surface of wood of a foamed coke layer. This foam coke shows excellent heat-shielding barrier properties, impeding heat access to the wood surface and hindering the evolution of combustible gaseous products of decomposition of wood. Figure 7 shows the actual mass loss of pinewood samples treated with aqueous solution of different oxidized polysaccharides after fire tests (ASTM E136-09).

The combustibility of pristine pinewood sample and that treated by oxidized lignin and starch was also evaluated by the mass loss calorimeter (ISO 13927) at an incident heat flux of 35 kW m^{-2} condition [23]. Mass loss rates (MLRs) of pinewood samples acquired at 35 kW m^{-2} are presented in Figure 8.

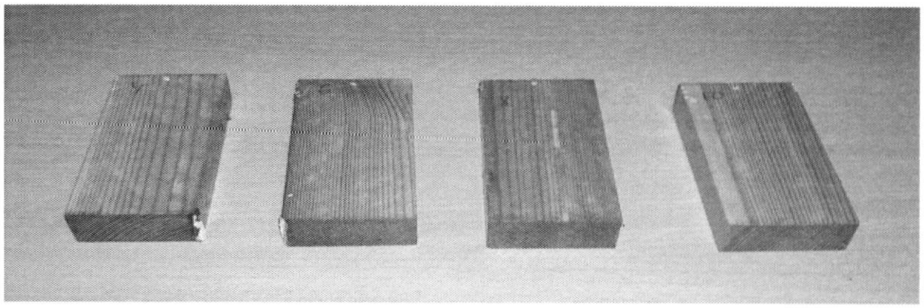

FIGURE 5

Pinewood samples, treated with an aqueous solution of oxidized starch, before fire tests (ASTM E136-09).

FIGURE 6

Pinewood samples, treated with an aqueous solution of oxidized starch, after fire tests (ASTM E136-09).

It is clearly shown that under the conditions of initial surface combustion, the mass loss rate of the pinewood samples treated with oxidized lignin (15% by wt) and that of the pinewood sample treated with oxidized starch are similar and noticeably lower than for the rate of the neat pinewood samples. The peak of MLR of the pinewood sample treated with oxidized lignin is 41% lower than that of pure pinewood sample, while that of the pinewood sample treated with oxidized starch is 33% lower. Due to the predominance of the solid-phase mode of action of the oxidized

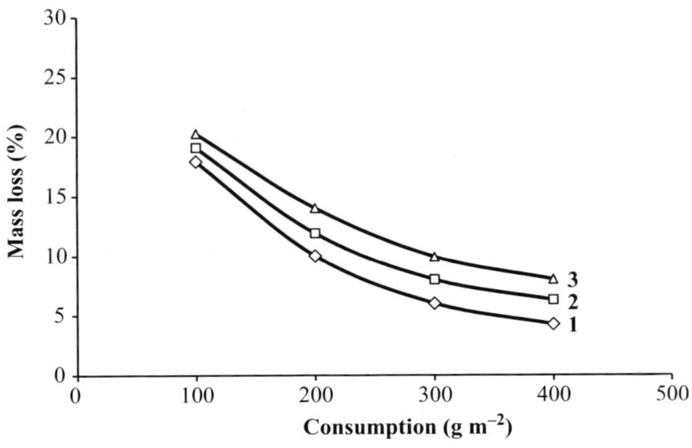

FIGURE 7

Dependence of mass loss of wood samples on the amount of applied fire-protecting coverings based on modified polysaccharides. (1) Oxidized starch (high degree of oxidation), (2) oxidized rice (high degree of oxidation), (3) oxidized starch (average degree of oxidation).

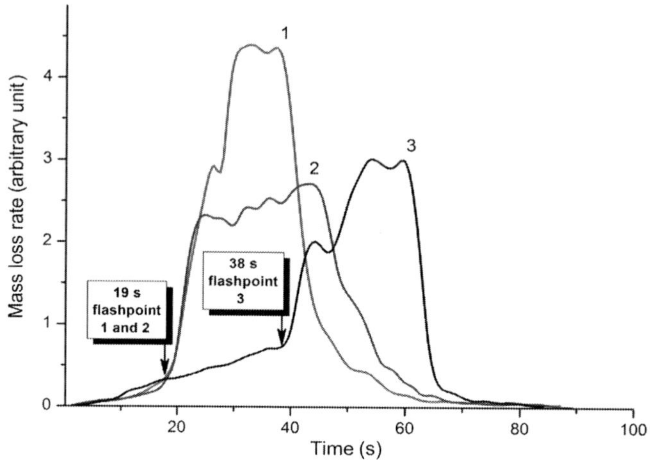

FIGURE 8

Mass loss rate versus time of plain pinewood samples. (1) Plain Pinewood sample, (2) Pinewood sample treated with oxidized lignin (15% by wt), (3) pinewood sample treated with oxidized starch (15% by wt).

lignin and starch, the MLR curves are supposed to be similar to the rate of heat release (RHR) curves; hence, the reduction of the MLR is evidently the primary factor responsible for the lower RHR of the pinewood samples. The improvement in flame resistance of the pinewood samples treated with oxidized lignin is a result of the char formation, provided by the transient protective barrier of oxidized lignin (as polyphenolic carbonizing structure), whereas the pinewood sample treated with oxidized starch exhibits a strong intumescent action. An additional observation obtained from the MLR plots is the substantial increase in induction period time of self-ignition (flash point) of the pinewood sample treated with oxidized starch as compared with that of the plain pinewood sample and that of the pinewood sample treated with oxidized lignin due to intumescent behavior of oxidized starch.

4. CONCLUSIONS

Industrial fire retardants present some disadvantages related to efficiency and environmental impact. Therefore, our research was focused on the chemical modification of natural polymers and wood, and on the development of more efficient intumescent systems and fire protectors.

These materials can perform as highly effective flame retardants intumescent/char-forming systems for cellulosic materials and wood, presenting a high potential for use in a wide range of applications, due to their cheapness and simple technology of production.

References

[1] Lomakin SM, Brown JE, Breese RS, Nyden MR. Investigation of the thermal stability and char-forming tendency of cross-linked poly(methyl methacrylate). Polym Degrad Stab 1993;41:229–43.

[2] Factor A. Char formation in aromatic engineering polymers. In: Nelson GL, editor. Fire and polymers. ACS symposium series, vol. 425; 1990. p. 274. Washington, USA.

[3] Lomakin SM, Zaikov GE, Artsis MI. Advances in nylon 6,6 flame retardancy. Int J Pol Mater 1996;32:173–202.

[4] Lomakin SM, Zaikov GE, Artsis MI, Ruban LV, Aseeva RM. New pathway of polymers flammability depression with poly(vinyl alcohol) oxidative system. Oxid Comm 1995; 18:105–12.

[5] Lomakin SM, Zaikov GE, Artsis MI. Thermal and oxidative stability of PVA and nylon-6,6. Int J Polym Mater 1994;26:187–94.

[6] Aseeva RM, Zaikov GE. Combustion of polymeric materials. Moscow, Russia: Nauka; 1981. pp. 84–135 (in Russian).

[7] Levchik S, Wilkie CA. Char formation. In: Grand AF, Wilkie CA, editors. Fire retardancy of polymeric materials. New York, USA: Marcel Dekker Inc.; 2000. pp. 171–216.

[8] Camino G. Flame retardants: intumescent systems. In: Pritchard G, editor. Plastics additives. London, UK: Chapman Hall; 1998. pp. 297–306.

[9] Camino G, Delobel R. Intumescence. In: Grand AF, Wilkie CA, editors. Fire retardancy of polymeric materials. New York, USA: Marcel Dekker Inc.; 2000. pp. 217–44.

[10] Camino G, Lomakin S. Intumescent materials. In: Horrocks AR, Price D, editors. Fire retardant materials. Washington, USA: CRC Press; 2001. pp. 318–35.

[11] Vandersall HL. Intumescent coating systems, their development and chemistry. J Fire Flammability 1971;2:97–140.

[12] Kozlowski R, Wladyka-Przybylak M. Natural polymers, wood and lignocellulosic materials. In: Horrocks AR, Price D, editors. Fire retardant materials. Washington, USA: CRC Press; 2001. pp. 293–317.

[13] Hurst NW, Jones TA. A review of products evolved from heated coal, wood and PVC. Fire Mater 1985;9:1–9.

[14] Lewin M, Atlas SM, Pearce EM. Flame-retardant polymeric materials. New York, USA and London, UK: Plenum Press; 1975.

[15] Sakharov AM. One stage method of catalytic oxidation of vegetal raw materials by oxygen: novel ecologically pure products and perspectives of their practical use. In: Zaikov GE, editor. Chemical and biochemical physics. New York, USA: Nova Science Publishers Inc.; 2006. pp. 113–25.

[16] Sivenkov AB, Serkov BB, Aseeva RM, Sakharov AM, Sakharov PA, Skibida IP. Fire proof fire-retardant coatings based on the modified polysaccharides. Fire Expl Saf 2002;1:39 [in Russian].

[17] Floor M, Schenk KM, Kieboom APG, Van Bekkum H. Oxidation of maltodextrins and starch by the system tungstate-hydrogen peroxide. Starch 1989;41:303–9.

[18] Santacesria E, Trully F, Brussani GF, Gelosa D, Serio MD. Oxidized glucosidic oligomers: a new class of sequestering agents-preparation and properties. Carbohydr Polym 1994;23:35–46.

[19] Skibida IP, Sakharov AM, Sakharov AM. Pat EP 91122164.6; 1996.

[20] Sakharov AM, Skibida IP. Mechanism of catalytic oxidation of polysaccharides in the alkaline gels. Chem Phys 2001;20:101 [in Russian].
[21] Run-Cang S. In: Cereal straw as a resource for sustainable biomaterials and biofuels: chemistry, extractives, lignins, hemicelluloses and cellulose. Amsterdam, The Netherlands: Elsevier; 1996.
[22] Hatakeyama H, Hatakeyama T. Lignin structure, properties, and applications. In: Abe A, Dusek K, Kobayashi S, editors. Biopolymers – lignin, proteins, bioactive nanocomposites. New York, USA: Springer; 2010. pp. 2–64.
[23] ASTM E2102: standard test method for screening test for mass loss and ignitability of materials. Philadelphia, USA: American Society for Testing and Materials. Standard published 04/01/2013 by ASTM International

CHAPTER

8

Design and Utilization of Nitrogen Containing Flame Retardants Based on *N*-Alkoxyamines, Azoalkanes and Related Compounds

Carl-Eric Wilén

Åbo Akademi University Åbo, Finland

R. Pfaendner

Fraunhofer Institute for Structural Durability and System Reliability LBF Division Plastics, Darmstadt, Germany

CHAPTER OUTLINE

1. Introduction

Flame retardants containing nitrogen are a class of various materials among which are well-known halogen-free commercial products, namely ammonium polyphosphate (APP), melamine cyanurate (MC), melamine borate, and melamine (poly)phosphate.

APP, which is available in different crystal modifications and with coatings (e.g. melamine and silicone) or in an encapsulated form, decomposes by exposition to fire or heat to ammonia, polyphosphoric acid, and phosphorus oxides. In the presence of a char-forming agent, e.g. a polyol, an intumescent barrier via phosphoric ester

Polymer Green Flame Retardants. http://dx.doi.org/10.1016/B978-0-444-53808-6.00008-1

intermediates is formed [1]. The main applications of APP are coatings for steel and wood and within the plastic field unsaturated polyesters, epoxides, polyurethanes, and polyolefins.

MC is a 1:1 adduct of melamine and cyanuric acid. MC decomposes endothermically into its components, and melamine decomposes further to nitrogen-containing gases such as ammonia. MC is mainly used as a flame retardant component in polyurethanes, polyesters, epoxides, and polyamides, specifically in unfilled polyamide-6.

Melamine borate decomposes endothermically into its components and promotes glassy char formation mainly in combination with phosphates.

Melamine phosphates cover melamine orthophosphate (1:1 salt of melamine and phosphoric acid), dimelamine orthophosphate (2:1), dimelamine pyrophosphate, melamine pyrophosphate, and melamine polyphosphate (MPP). Because of thermal stability reasons, the most interesting material in this class of plastic applications is MPP. The endothermic decomposition of MPP produces ammonia, other nitrogen species, and poly(phosphoric acid); the latter dehydrates the polymer under char formation. Application fields are polyamides (PA6, PA6.6 glass fiber reinforced) and polyesters. For example, 25% MPP provides UL 94 V-0 classification at 1.6 mm in PA6.6/20% glass fibers [2]. In combination with other phosphorus flame retardants, a synergistic effect of MPP is found, which allows one to reduce the overall flame retardant loading, e.g. combinations of MPP and a phosphinate to achieve a UL-94 V-0 classification in glass fiber-reinforced PA6.6 with an overall loading of 15–20% [3–5].

Furthermore, melamine salts other than phosphates, higher condensation products of melamine (melam, melon, melem), and closely related substances (ammeline, ammelide, urea, guanamine, guanidine, and derivatives thereof) as well as all kind of salts thereof have been proposed as flame retardants for plastic applications. For example, mixed salts of melamine–melam–melem and polyphosphate or pyrosulfate have been synthesized at condensation temperatures >300 °C [6]. Phosphorus derivatives of nitrogen-containing species other than phosphates have been described, e.g. in the form of nitrilotrismethylenephosphonic acid [7,8] or as guanidine phenylphosphinate [9].

Most of the above-mentioned salt structures of melamine and related substances act in the condensed phase through endothermic decomposition of the flame retardant, dilution of the burning gases through the formation of inert gaseous molecules, and through char formation. A somewhat different activity mechanism was found by using alkoxyamines (NOR) as flame retardants forming radical species during decomposition. The use of radical generators as flame retardant components was already proposed in the 1960s [10,11], where it was shown that peroxides, azo, sulfur, bibenzyl compounds, and hydrazones act synergistically with brominated compounds. Consequently, combinations of hexabromocyclododecane and peroxides have been used as standard flame retardant (FR) systems in polystyrene foam for many years [12]. This system, however, has to be replaced due to environmental reasons in the near future. Furthermore, it was as also stated that radical generators of peroxide, hydroperoxide, and polysulfide types alone do not act as flame retardants [13].

The first who realized that alkoxyamines act as potential flame retardants without any other flame retardant were inventors from Ciba Specialty Chemicals [14]. Alkoxyamines (or hydroxylamine ethers, NOR) were first introduced as noninteracting ultraviolet (UV) stabilizers with a low basicity. Such stabilizers provide protection for automotive coatings and for agricultural films even under severe conditions, e.g. in the presence of pesticides. Flame retardants such as halogens or phosphorus compounds generate thermally or photochemically acidic species, which result in inferior UV stability by deactivating hindered amine light stabilizers. This problem was solved by using alkoxyamines instead of amines in combination with brominated flame retardants. Surprisingly, it was found that the alkoxyamines do not only improve the UV stability but they also contribute to flame retardancy. Even without any additional flame retardant, it was demonstrated that alkoxyamines alone can achieve flame retardancy, e.g. in polypropylene fibers and films [15,16]. Meanwhile, various structures and process patents have been published that claim alkoxyamines to be efficient flame retardants [17,18]. The synthesis of alkoxyamines is performed usually from the corresponding sterically hindered piperidine via the nitroxyl intermediate through reaction with a hydrocarbon in the presence of an organic peroxide such as peracetic acid, hydrogen peroxide under catalysis of copper, molybdenum oxide, or iodide. A more recent process uses the copper-catalyzed reaction of aldehydes with the nitroxyl intermediate [17].

In addition, several other chemical structures related to alkoxyamines are claimed to provide flame retardancy, namely hydroxylamine esters [19], hydroxylamines, and nitroxyl radicals [20].

The first commercial product on the basis of alkoxyamines was introduced by Ciba (now BASF) in 2000 under the trade name Flamestab NOR 116 (Figure 1).

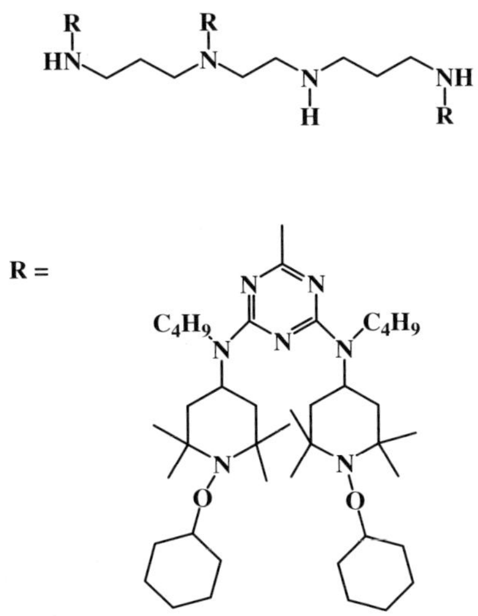

FIGURE 1

Structure of Flamestab NOR 116.

It was demonstrated that Flamestab NOR 116 provides flame retardancy at a concentration of only 0.5% passing the NFPA 701 fiber test [21] on polypropylene knitted socks [15].

Moreover, the addition of alkoxyamines as an FR synergist can improve the efficiency of conventional flame retardants, whereas all kinds of flame retardants and synergistic mixtures, e.g. brominated compounds [22–25], phosphorus compounds [26,27], phosphacenes, melamines [28,29], and inorganics, fluorolefins [30] are claimed. For example, the combination of NOR and decabromodiphenyloxide surpasses the flame retardants alone [31]. Therefore, it is possible with alkoxyamines to design flame retardant polyolefin molding compositions with lower levels of halogenated flame retardants, and in addition to eliminate antimony trioxide, which is often used as an FR synergist together with brominated compounds. Low FR concentrations and no antimony trioxide result in polymers with improved processability, better mechanical properties, and reduced smoke density. For example, UL 94 V-2 rating is achieved in PP injection-molded plaques at concentrations as low as 3.5% of formulations containing Flamestab NOR 116 [32].

The activity of the alkoxyamine as a flame retardant is based on the thermolysis of nitroxyl ethers, which leads to the formation of either alkoxy and aminyl radicals or alkyl and nitroxyl radicals (Figure 2). Aminyl and alkoxy radicals are very reactive and cause, on the one hand, degradation of polypropylene (and crosslinking of polyethylene). On the other hand, they are involved in the free radical chemical reactions during the combustion process [31].

In synergistic combinations, alkoxyamines can interact with brominated flame retardants and facilitate the release of bromine, consequently increasing the overall FR performance. It has been experimentally shown that alkoxyamine acts in the condensed and gas phase, modifies the decomposition pathway of phosphoric acid tris-(3-bromo-2,2-bis-bromomethylpropyl)ester and polypropylene while lowering the temperature to release HBr as flame quenching species and also the temperature

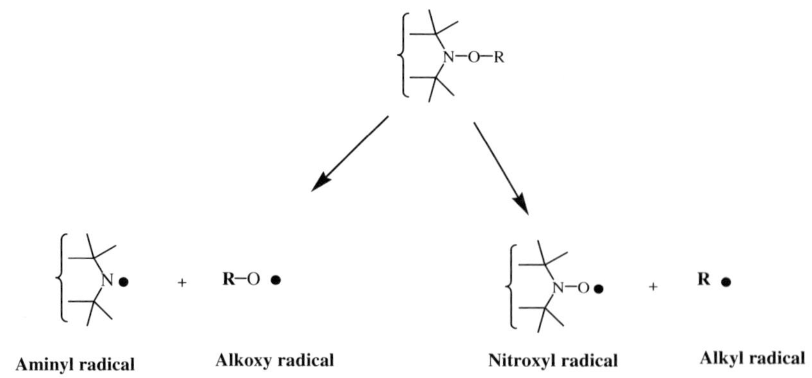

FIGURE 2

Thermolysis of alkoxyamines.

of dripping of the polymer. Decomposition experiments [33,34] of tris(3-bromo-2,2-bis(bromomethyl)propyl)phosphate (TBBPP) and analysis via nuclear magnetic resonance, electron spin resonance, and gas chromatography/mass spectroscopy combination showed that there is an interaction of NOR and TBBPP below the decomposition temperature of polypropylene. In the first step, the nitroxyl radical is formed from NOR, which abstracts a bromomethyl group from TBBPP. As further intermediates, a cyclic phosphate ester and 1,3-dibromo-2,3-bis(bromomethyl)propane (mainly) and 2,3-dibromo-2-butene are generated via a radical process. These newly formed bromo compounds are thermally less stable than TBBPP and immediately release HBr as a gas phase active flame retardant, as shown in Scheme 1.

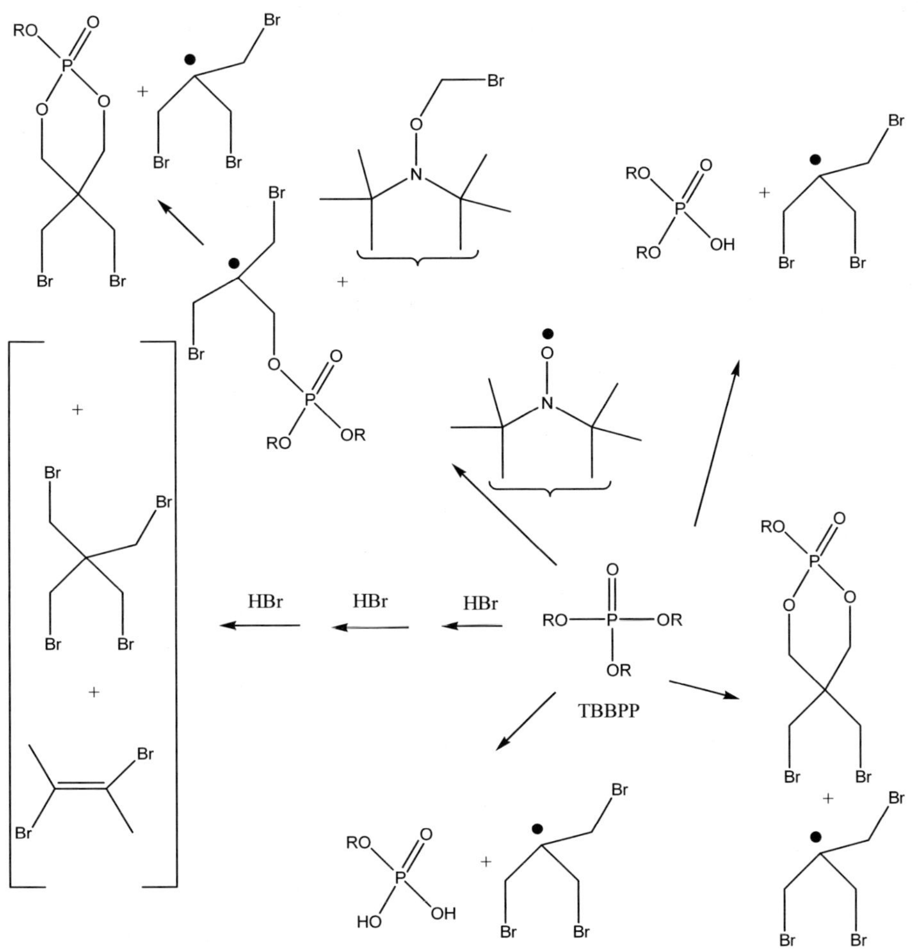

SCHEME 1

Various nitrogen-based classes of flame retardants

The decomposition of TBBPP to the brominated species takes place in any case; however, it is shifted to lower temperatures in the presence of NOR. Obviously, this effect depends on the structure of the brominated flame retardant; however, so far, there are no other published results in this respect.

Applications of alkoxyamines such as Flamestab NOR 116 are according to the manufacturer in polyolefin fibers, nonwovens, and films [35]. In addition, the hindered amine light stabilizer (HALS) molecule provides light and long-term thermal stability.

Overall, nitrogen-based flame retardants still represent a small but fast growing segment of flame retardants that meet the requirements of sustainable production and products. As a consequence of this, researchers have recently started to examine the mode and action of these compounds. This review focuses on some new developments within the family of nitrogen-based flame retardants and their structure–property relationship. The general structures of these nitrogen-based flame retardant families are depicted in Figure 3.

2. Mechanistic considerations and studies on structure–activity relationship of *N*-alkoxyamines as flame retardants

Inspired by the flame retardant efficacy of Flamestab NOR 116, a series of NOR analogs have been prepared, and their property–structure relationship has been examined. As earlier mentioned, synthetic routes for a large variety of *N*-alkoxyamine structures have been developed in recent years. In particularly, a great number of various NOR derivatives for nitroxide-mediated "living-free" radical polymerizations have been synthesized. The mechanisms of nitroxide-mediated "living" free radical polymerizations and their structure–property relationship have been well documented [36–38]. The literature available on the role of various NOR structures on fire retardant efficacy is still very limited and mostly in the form of patents [14,15]. However, a clear understanding of the mechanism of action of NOR fire retardants would be essential for their further optimization. It has been well established that the active part in NOR flame retardants is centered around the nitrogen–oxygen–carbon (N–O–C) segment of the molecule, as shown in Figure 2 [39,40]. As a consequence of this, it was investigated by GC/MS analysis as to what type of radicals are prevailing after thermally induced breakdown of NOR compounds and what is the relative extent of the two homolysis pathways [41].

Thus, a synthesized model NOR molecule (1-cyclohexyloxy-2,2,6,6-tetramethylpiperidine) was highly diluted in a hydrogen donor solvent (9,10-dihydroanthracene) in a closed vessel under vacuum. The solution was kept for 1 h at 200 °C. The radicals generated from the thermal decomposition of the NOR compound were hydrogenated by the solvent. The solution was then analyzed by GC and GC/MS, and the results are shown in Figure 4. In the case of model NOR compounds, the preliminary results indicate that both types of thermal homolysis reactions take place to a similar extent.

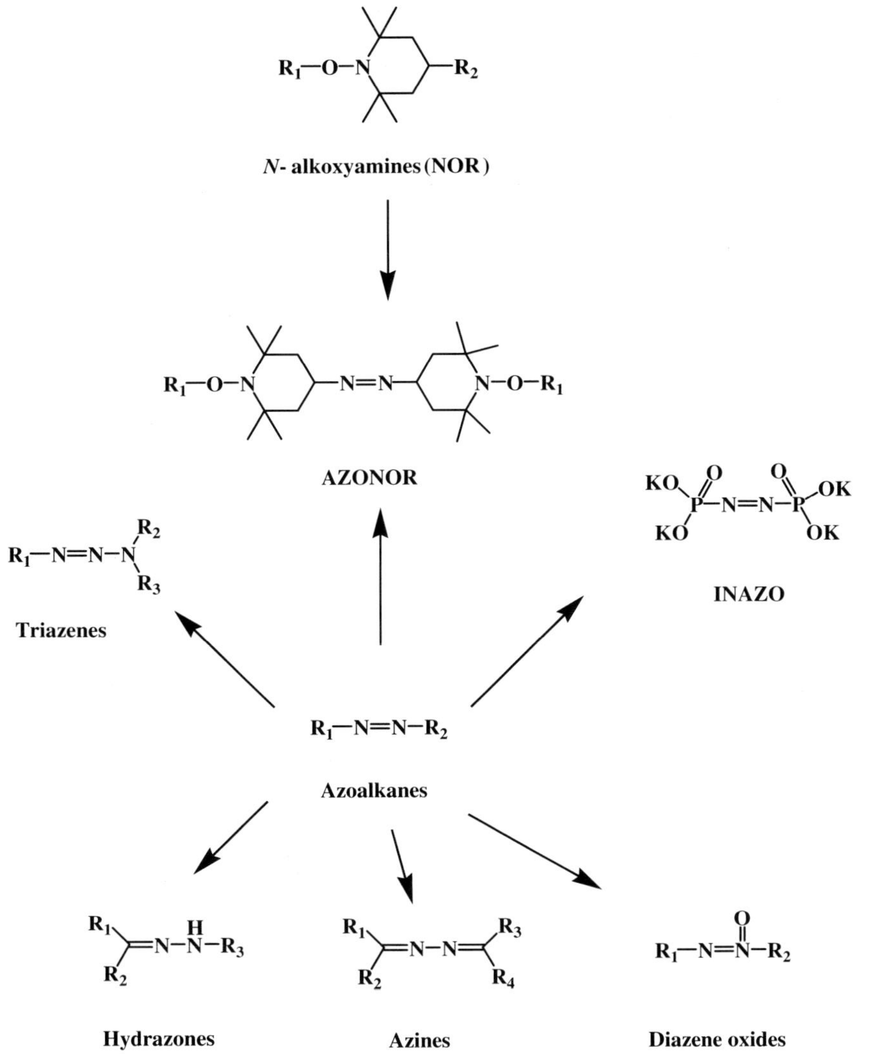

FIGURE 3

Nitrogen-based classes of flame retardants.

In order to define the key parameters in the NOR structure for an optimal flame retardancy performance, a diverse library of *N*-alkoxyamine additives was prepared as shown in Figure 5. Most of these chemicals are 1-alkoxy-2,2,6,6-tetramethylpiperidine-based compounds that differ from each other either in terms of backbone structure, ring size, degree of ring substitution, or 1-alkoxy group. NOR derivatives in which any of the four methyl groups had been replaced by larger alkyl groups such as ethyl or cyclohexyl moieties were excluded from this study due to the

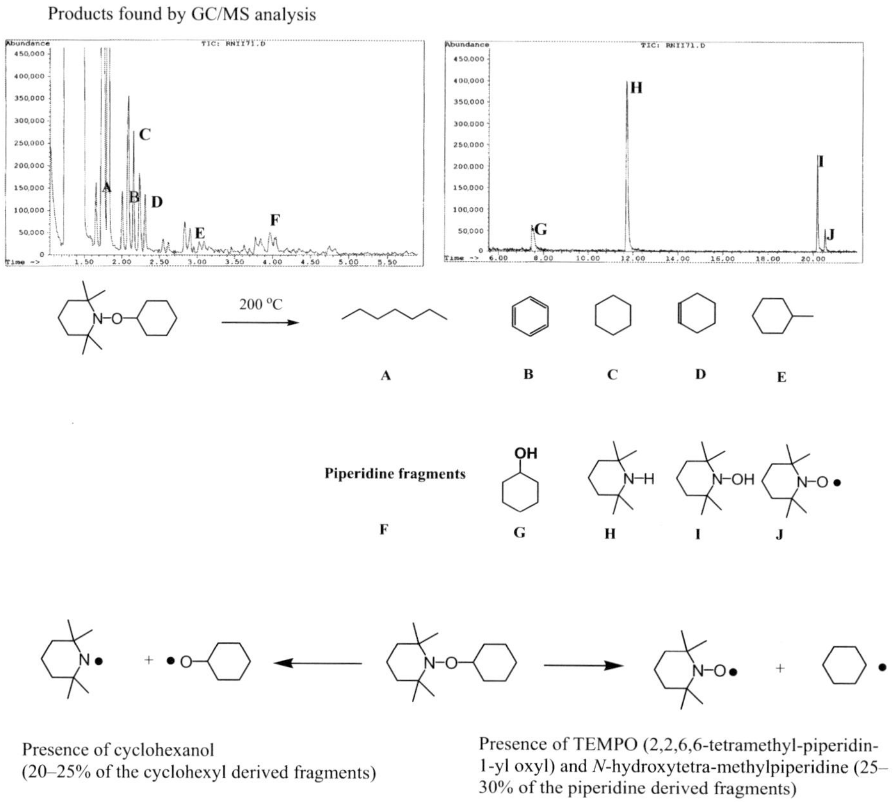

FIGURE 4

Homolysis products detected by GC—GC/MS from thermal decomposition of 1-cyclohexyloxy-2,2,6,6-tetramethylpiperidine.

fact that sterically highly hindered nitroxide derivatives are generally known to exhibit rather low thermal stabilities [42,43] and are thus more suitable for living free radical polymerizations than for flame retardancy per se.

The flame retardant properties were evaluated according to the UL94 protocol using rectangular 108 mm × 10 mm × 4-mm polypropylene bars and thermal stabilities were studied by differential scanning calorimetry analysis (Table 1). From early on, it was clear that a UL rating could not be achieved for any of the synthesized NOR compounds alone using thick section polypropylene bars. Therefore, the relative flame retardant efficacy of the various NOR additives have been based on the burning time of polypropylene plaques using 1 wt% of Flamestab NOR 116 as standard formulation. Virgin polypropylene samples without any flame retardant burned down in 132 s under these experimental conditions, whereas polypropylene bars containing 1, 2, and 10 wt% of Flamestab NOR 116 burned down in 238, 237, and 205 s, respectively. (The flame retardant experiments verify the inverse

Effect of nitroxyl radicals

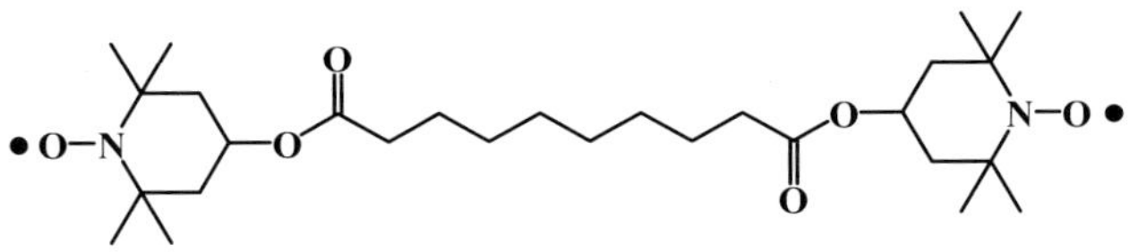

TEMPO1

Effect of 4-substituent

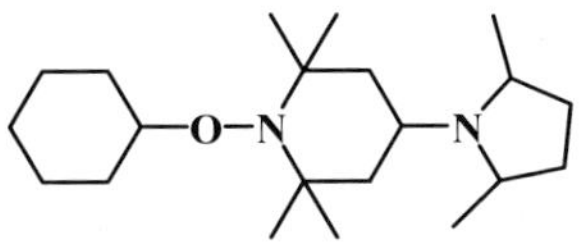

4PyNO6R

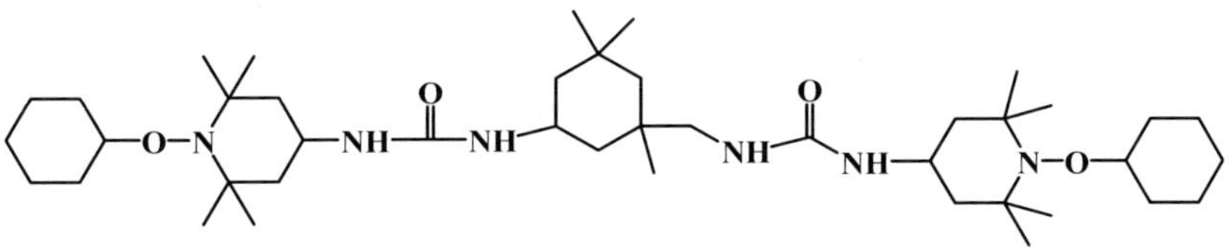

4-UREANO6R

Effect of alkoxy ring size

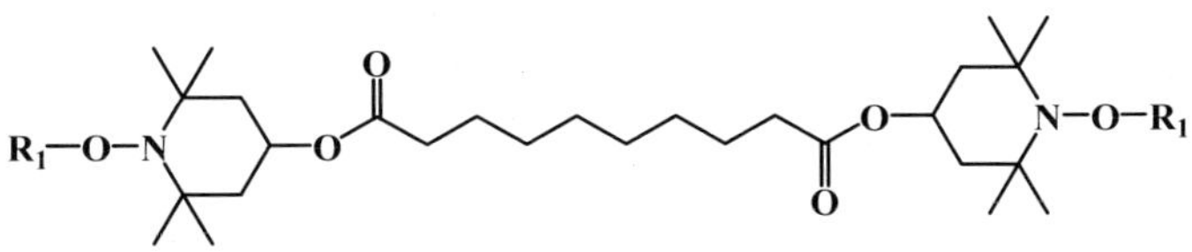

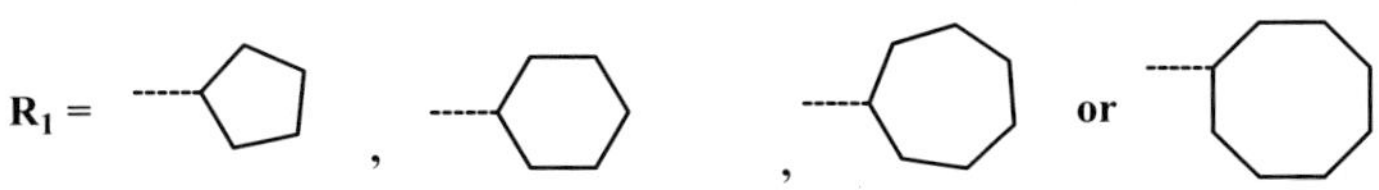

NO5R **NO6R** **NO7R** **NO8R**

FIGURE 5

Various NOR additives tested as flame retardants for polypropylene plaques.

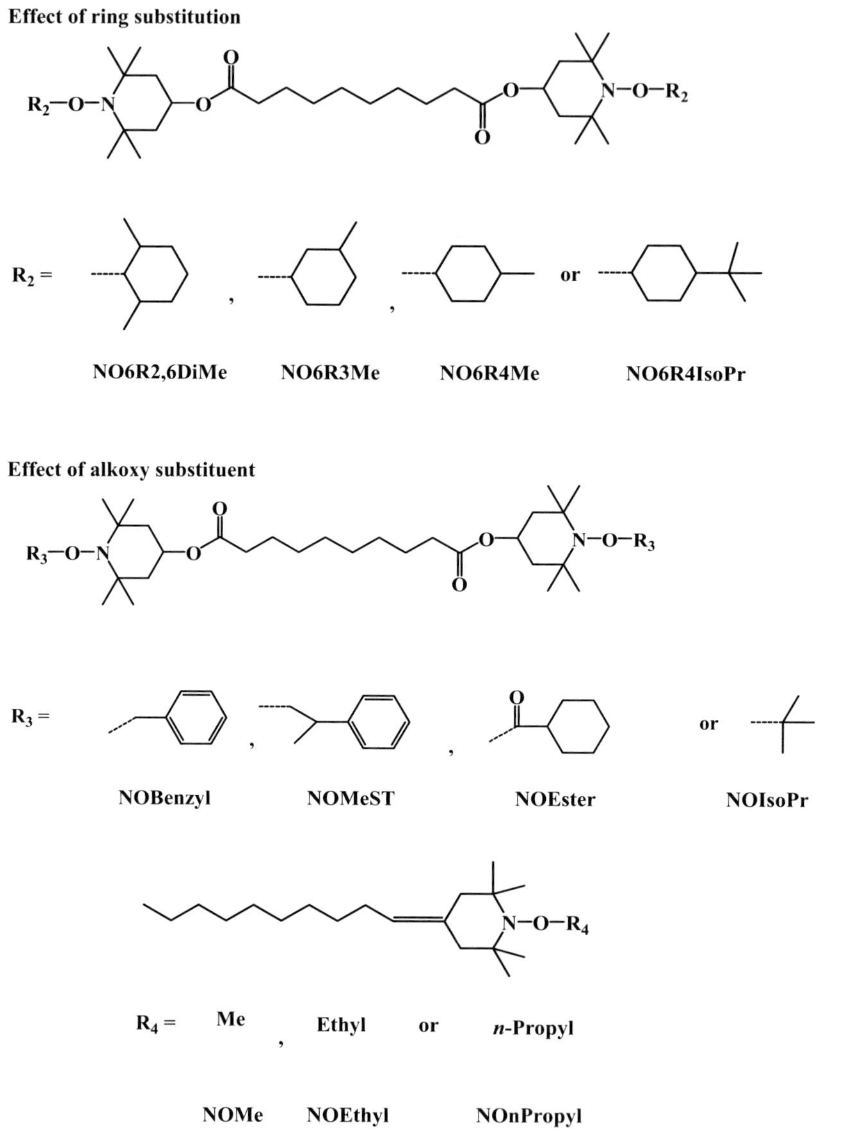

FIGURE 5 cont'd.

concentration effect of Flamestab NOR 116 [15].) All other burning times are given relative to 1 wt% of Flamestab NOR 116, i.e. a value of >1 and <1 means a better or worse performance than using 1 wt% of Flamestab NOR 116, respectively.

The polypropylene formulations containing 1, 2, and 3 wt% of TBBPA as a brominated reference flame retardant exhibited relative burning times of 0.95,

Table 1 Flame Retardant Efficacies for a Series of Different *N*-Alkoxyamines

Additive Code	Additive %wt	DSC—Decomposition Temperature (°C)	UL 94 t_1 (s)/t_1 (s, NOR 116)
No FR	–	–	0.55
TBBPA	1	–	0.95
TBBPA	2	–	1.07
TBBPA	5	–	1.32
NOR 116	1	284	1
NOR 116	2	284	1
NOR 116	3	284	0.92
Nitroxyl Radical			
TEMPO1	1	–	0.85
4-Substituted Cyclohexyloxytetramethylpiperidines			
4PyNO6R	1	276	1.25
4UREANO6R	1	–	1.13
Cycloalkyloxytetramethylpiperidine Diesters Having Various Ring Sizes			
NO5R	1	278	0.85
NO6R	1	295	1.13
NO7R	1	261	0.88
NOR8R	1	236	0.86
NORs with Different Alkyl Substituents on the Cyclohexyl Ring			
NO6R2,6DiMe	1	248	0.71
NO6R3Me	1	288	0.81
NO6R4Me	1	288	0.93
NO6R4IsoPr	1	277	1.15
Various Alkoxy Substituted Tetramethylpiperidines			
NOBenzyl	1	262	0.86
NOMeST	1	234	0.81
NOEster	1	197	0.8
NOIsoPr	1	234	0.76
NOMe	1	300	(t_1 = 1 s) 1.13[a]
NOEthyl	1	304	1.02
NOnPropyl	1	295	0.90

TEMPO, 2,2,6,6-tetramethylpiperidine-1-oxyl; TBBPA, tetrabromobisphenol A.
[a] *Based on the burning time after second ignition.*

1.07, and 1.32, respectively. Even if one needs to admit that this method is not the ideal way to compare flame retardant efficacies, it is still a feasible method for the preliminary evaluation of the potential of various NOR structures as flame retardants for polypropylene. Overall, the flame retardant properties of different alkoxyamines mainly depend on three parameters: (1) thermal stability (homolysis temperature), (2) products of decomposition and mechanisms thereof, and (3) reactivity of the radicals generated during decomposition.

The results indeed reveal that nitroxyl radicals (TEMPO1) as such also exhibit some flame retardant properties although significantly more weaker than Flamestab NOR 116. One can readily see that two 4-substituted NOR compounds (4PyNO6R and 4UREANO6R) containing the 1-cyclohexyloxy-2,2,6,6-tetramethylpiperidine moiety showed similar thermal stabilities. Surprisingly, 4PyNO6R and 4UREA-NO6R exhibited even higher flame retardant properties than Flamestab NOR 116 under these experimental conditions. In general, one would expect that the substituent in the 4-position would not directly affect flame retardant properties. However, the differences in recorded flame retardant properties may arise from differences in compatibility, ease of dispersion, and volatility of these NOR compounds due to their variations of backbone structures. In order to avoid the problem of dispersion in the subsequent NOR series, compounds having the same NOR backbone were prepared (either a diester or a long alkenyl chain in the 4-position).

Next, the effect of ring size on flame retardancy was investigated by comparing the flame test results obtained using 5-, 6-, 7-, and 8-membered ring 1-cycloalkyloxy-2,2,6,6-tetramethyl piperidine-4-diesters. Although, the cycloalkyloxytetramethylpiperidine diesters NO5R, NO6R, NO7R, and NO8R had relatively comparable electronic and steric properties, their flame retardancy potentials differed quite significantly. The best result was recorded for the six-membered ring *N*-alkoxyamine (NO6R), whereas larger or smaller ring sizes showed significantly lower flame retardant efficacies. As a consequence of this, 1-cyclohexyloxy-2,2,6,6-tetra-methylpiperidine appeared to be a promising template for further optimization of the NOR structure. Therefore, as a next step, NOR additives with alkyl substitutions on the cyclohexyl ring were synthesized, and their flame retardant properties were studied.

Compound NO6R2,6DiMe having methyl groups in 2 and 6 positions of the cyclohexyl group as well as NO6R3Me having a methyl group in 3 position exhibited inferior flame retardant properties compared to Flamestab NOR 116. This can be attributed to the lower thermal stability of NO6R2,6DiMe and the fact that the substituted cyclohexyl radical liberated during thermal decomposition is also less reactive than the unsubstituted cyclohexyl radical. Thus, it is understandable that NO6R2,6DiMe presented worse flame retardant properties than Flamestab NOR 116 or NO6R. Compounds NO6R4Me and NO6RIsoPr are similar additives as NO6R with alkyl substitutions of the cyclohexyl ring in the 4-position. The flame retardant performances of these two molecules are close to NO6R. Thus, the presence of alkyl substituents in the 4-position of the cyclohexyl ring does not significantly affect the flame retardant properties. However, the adjustment of the cyclohexyl radical volatility and diffusion properties may be fine tuned by the optimization of alkyl substitution of the cyclohexyl ring.

In this study, alkoxyamines with other alkoxy substituents than cyclohexyl or related alkyl substituted cyclohexyl derivatives were also investigated. Compounds bearing isopropyloxy (NOIsoPr), ester (NOEster), or α-methyl styrene (NOMeST) exhibited fairly low potential as flame retardants since they lack sufficient thermal stabilities for processing. The processing of polypropylene was conducted at 230 °C and, therefore, this type of "low thermal stability" NOR compounds may have undergone partial decomposition already during the polymer-processing step itself. This fact, explains the poor efficacies recorded for these NOR series.

In contrast, in the series of 1-*n*-alkoxy-4-(dodec-2-enyl)-2,2,6,6-tetramethylpiperidines, all the compounds showed a comparable or even higher efficiency than Flamestab NOR 116. The results indicate that the flame retardant properties within the family of NOR compounds increased as a function of the thermal stability of the *N*-alkoxyamine functionality.

One of the most thermally stable additives prepared was 1-methoxy-4-(dodec-2-enyl)-2,2,6,6,tetra-methylpiperidine (NOMe), and it was shown to be the most efficient FR additive within the synthesized library of NOR compounds. This study clearly indicated that the formed nitroxyl and aminyl radicals of NOR thermolysis are perhaps not the only radical species involved in the retardation of the combustion process. Thus, the simultaneously formed alkyl radicals might also play a vital role as flame retardant species (Section 3).

In addition, NOMe showed an extraordinarily high synergistic effect in conjunction with brominated flame retardants. Thus, the UL94 tests showed that V-0 rating in combination with an already 11 wt% of brominated flame retardant was reached when employing NOMe, whereas 14 wt% of the brominated flame retardant was needed when using Flamestab NOR 116 in order to achieve the V-O rating. The result suggests that methyl-TEMPO additives also have a higher tendency to interact/react with brominated flame retardants than those based on the more sterically hindered cyclohexyl-TEMPO analogs.

3. Azoalkanes (AZO) as flame retardants for polypropylene films and plaques

The conclusion derived from studies of the structure–activity relationship of NOR flame retardants and their breakdown was that two distinctly different thermolysis reaction pathways clearly take place (Figure 4). Thus, the fragmentation of the NOR molecule may occur at both the N–O and O–R bonds whereby in theory either alkyl and nitroxyl radicals or alkoxy and aminyl radicals are formed. Given the fact that these four types of radicals clearly differ from each other in terms of chemical reactivity and stability, the significance of their presence in the flame retardant process was not expected to be equivalent. In the original Flamestab NOR 116 flame retardant publications, it was thought that aminyl and alkoxy radicals would be the most important species for interrupting the free radical reactions during the polymer combustion process [39]. Whereas the very high synergistic effect of NOR additives with conventional brominated flame retardants was ascribed to the generation of aminyl, alkoxy, and nitroxyl radicals that in turn interact with brominated compounds and thereby facilitate the release of bromine, thus improving flame retardant performance [40]. The potential role of generated alkyl radicals was not discussed.

The apparently complex decomposition of *N*-alkoxyamines made it worthwhile to investigate the flame retardancy properties of each of the aforementioned radical families separately. The investigation was focused on the least studied radicals, i.e. the role of alkyl radicals on flame retardancy of polypropylene. Azoalkanes were selected as the cleanest and most convenient source of alkyl free radicals [44] (Figure 6). As in the case of *N*-alkoxy hindered amines, the thermal stability of azoalkanes is very closely related to their chemical structure. Their thermal stability is mainly governed by the ground-state energy, stability of the incipient radicals, and orbital symmetry factors. Aromatic azo compounds and azotriphenylmethane represent extreme cases: the former decomposes at 500 °C, and the latter has a transient existence only at −40 °C. The Polanyi plot of activation energy for thermolysis of both symmetrical azoalkanes and *N*-alkoxy hindered amines versus the C−H bond dissociation energy of the corresponding hydrocarbons are in both cases essentially linear [38,45]. As a consequence of this, the thermal stability of the azoalkane can be optimized for various polymers in terms of their processing temperatures and decomposition temperatures.

In analogy to *N*-cyclohexyloxy-2,2,6,6-tetramethylpiperidinyl-functionalized flame retardants, three azocyclohexyl compounds were synthesized, i.e. 4,4′-bis(cyclohexylazo-cyclohexyl)methane (ABO AA1), cyclohexylazooctadecane (ABO AA 2), and azocyclohexane (ABO AA3) as sources of cyclohexyl radicals [46]. For the evaluation of the flame retardancy of azocyclohexyl, ignitability tests were performed according to the DIN 4102-1/B2 standards [47]. First, polypropylene was blended with 0.1% by weight of calcium stearate, 0.2% of Ciba® Irganox® B 501 (stabilizer), and with 0, 0.25, or 0.5% of the radical precursor. The blends were melt processed into fibers, spun into socks, and finally compression molded into 200 mm films. Surface exposure of the samples to a gas flame was separately carried out, and the results are shown in Table 2 [48]. Slight differences in burn times and damage length were observed, and overall, the formulation containing the ABO AA1 additive gave a marginally better performance than the others in thin PP sections. When tested, all the samples containing azoalkanes gave a clean burn residue after the flame application. In contrast, the Flamestab NOR 116 samples exhibited

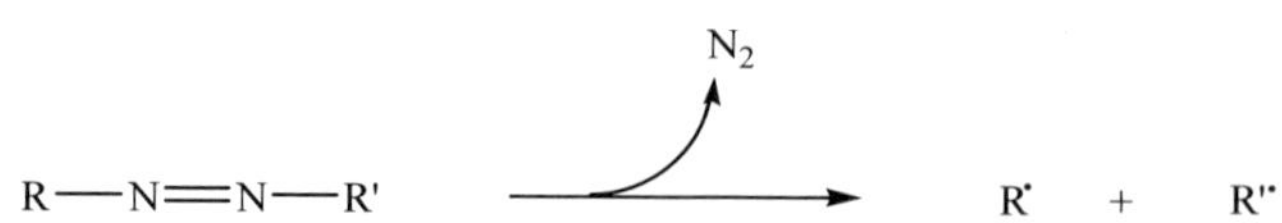

FIGURE 6

Azoalkanes: "The cleanest and most convenient source of alkyl radicals".

Table 2 Flammability Results According to DIN 4102 Part-1 Classification B2 Test Method: Face Ignition Test of 230 mm × 90-mm samples[a]

Formulation	Average Damage Area (cm^2)	Total Burning Time (s)	Pass/Fail
Blank	100	100	Fail
0.25 wt% of Flamestab NOR 116	n.d.[b]	20	Pass
0.50 wt% of Flamestab NOR 116	n.d.	22	Pass
0.25 wt% of ABO AA1	28	13	Pass
0.5 wt% of ABO AA1	21	8	Pass
0.25 wt% of ABO AA2	39	18	Pass
0.5 wt% of ABO AA2	31	13	Pass
0.25 wt% of ABO AA3	39	16	Pass
0.5 wt% of ABO AA3	22	10	Pass

[a] *Substrate: polypropylene is blended with 0.1 wt% of calcium stearate; 0.2 wt% of Ciba Irganox B501; and 0 wt%, 0.25 wt%, or 0.5 wt% of the test FRs. The blends are melt processed into fibers, spun into socks, and subsequently compression molded into thin films.*
[b] *n.d. = not determined.*

yellow chars around the burn damage area. The most striking difference in flame retardant efficacy was observed in thick polypropylene plaques of 1 mm, i.e. azo compounds (e.g. ABO AA3) had a much better FR performance than did Flamestab NOR 116 in thick PP sections (Table 3.) In addition, azo compounds resulted in less discoloration and ABO AA1 exhibited an even nonburning dripping behavior.

3.1 The potential of symmetric versus unsymmetric azoalkanes

Most interestingly, many critical properties of azoalkanes of the general formula R′—N=N—R, such as physical state, toxicity, volatility, initiator efficiency, and in particular thermal stability (in contrast to peroxides (R′—O—O—R)) can be adjusted over a very wide range by selection of groups R′ and R. According to Engel, as a rule of thumb, the thermal stability of unsymmetrical azoalkanes (R = R′) is intermediate between the stabilities of the symmetrical azoalkanes R—N=N—R and R′—N=

Table 3 FR Performance in PP Moldings (1 mm)[a]

FR Compound 0.5 wt%	Weight Loss %	Burn Length	Pass/Fail
Control no FR	100	100	Fail
NOR 116	49.4	80	Fail
ABO AA3	12.9	47	Pass

Compounding: W&P ZSK25 WLE, 230 °C, 100 rpm, 4 kg h $^{-1}$.Molding: Injection molding Arburg 320S, 1-mm plaque.
[a] *Substrate: Moplen HF500N, PP homo, 0.3% Ix B225 + 0.05% CaSt.*

N—R′ [49]. In addition, a consensus exists that most symmetrical azoalkanes lose nitrogen by the simultaneous breakage of both C—N bonds, whereas unsymmetrical azoalkanes may undergo one-bond cleavage [49]. As a natural extension of the work with ABO AA azoalkane structures, systematic structural changes of the aforementioned basic azoalkane compound were screened in terms of variations in ring size, electronic (resonance and inductive) and steric hindrance with regard to flame retardant efficiency. Therefore, a number of symmetrical and unsymmetrical azoalkanes of the general formula R′—N=N—R were prepared, as shown in Figure 7 [46]. The experimental results show that in the series of different-sized azocycloalkanes, the flame retardant efficacy decreased in the following order: R = cyclohexyl > cyclopentyl > cyclobutyl > cyclooctanyl ≫ cyclododecanyl. However, in the series of aliphatic azoalkanes compounds, the efficacy decreased in the following order: R = *n*-alkyl > tert-butyl > tert-octyl. As a spin-off, it was observed that unsymmetrical azoalkanes, in particularly *N*-cyclohexyl-*N*′-*t*-octyl-diazene, can successfully be used for manufacturing of controlled-rheology polypropylene and of crosslinked high-density polyethylene (PEX) [50,51].

3.2 Azoxy, azine, hydrazone, and triazene compounds

Furthermore, structures related to azo compounds were synthesized in the form of azoxy, hydrazone, triazene as well as azine derivatives in order to assess their potential as novel flame retardants for polypropylene (Figure 7). The prepared azoxy (R—N=NO—R), azine (R=N—N=R), and hydrazone (R—N=NH—R) derivatives provided flame retardancy to polypropylene films at already very low concentrations (0.25—1 wt%). More recently, various triazene ($R—N^1=N^2—N^3R'R'$)-based flame retardants were prepared. For example, polypropylene samples containing a very low concentration of only 0.5 wt% of bis-4,4′-(3,3′-dimethyltriazene)diphenyl ether passed the test with B2 classification. It is noteworthy that no burning dripping could be detected and the average burning times were very short with exceptionally low weight losses [52]. Therefore, triazene compounds constitute a new and interesting family of radical generators for flame retarding of polymeric materials. The high flame retardant potential of triazenes can be attributed to their ability to generate various types of radicals during their thermal decomposition. According to thermogravimetric analysis/Fourier transform infrared spectroscopy/MS analysis, triazene units are homolytically cleaved into various aminyl, resonance-stabilized aryl radicals, and different CH fragments with simultaneous evolution of elemental nitrogen, as shown in Figure 8. These radicals can effectively provide flame retardant properties.

3.3 Tetrapotassium azo diphosphonate (INAZO) as flame retardants

With an effort to further extend the scope of azoalkane-based flame retardants, inorganic azo phosphonate compounds were synthesized, i.e. $(KO)_2(O)P—N=N—P(O)(OK)_2 \cdot 4H_2O$, (INAZO), and test its potential as a novel type of flame retardant additive for two-component polyurethane adhesives. The action mechanism of

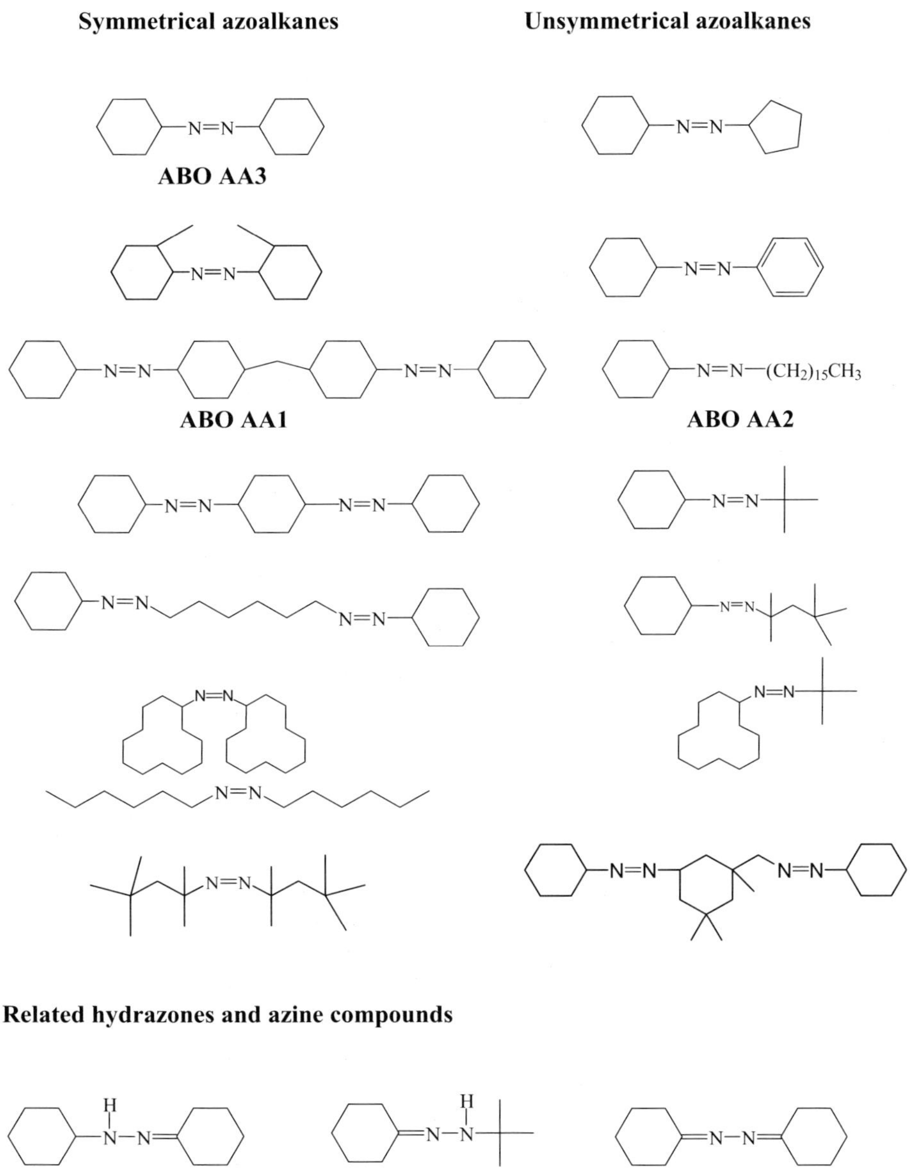

FIGURE 7

Chemical structures of symmetrical and unsymmetrical azoalkanes and related compounds.

the developed INAZO flame retardant was suggested to be mainly in the condensed phase. UL 94 V-0 rating was achieved in the vertical burning test when 10 wt% loading of INAZO was used, whereas the reference flame retardant APP required a loading of 20 wt% to reach the V-0 classification [53].

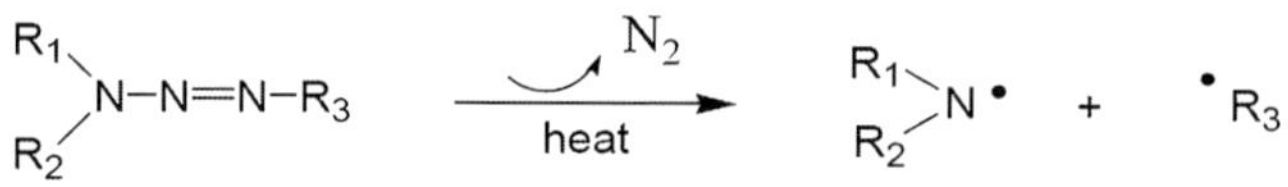

TRIAZENE — Aminyl radical — Alkyl radical

FIGURE 8

Thermolysis of triazene compounds.

3.4 Multifunctional bis(1-propyloxy-2,2,6,6-tetramethylpiperidyl)-4-diazene

A disadvantage of the azoalkane-based compounds in comparison to NOR 116 is that they do not contribute to light stability. Although selective azo compounds have both the advantage of nonburning dripping properties during fire tests and excellent optical properties, they also importantly function well alone in thick polypropylene sections such as 1-mm plaques. As a consequence of this, the hypothesis of creating an optimized multifunctional flame retardant family without any of the aforementioned (as 1) lack of light stability, (2) low synergistic effect with brominated halogens, and (3) lack of flame retardant performance in thick

Table 4 Flame Test on 1-mm Injection Molded PP Plaques According to the DIN 4102/B2 Test

FR Compound 0.5 wt%	Weight Loss %	Burn Length	Pass/Fail
Control	100	100	Fail
NOR 116	49.4	80	Fail
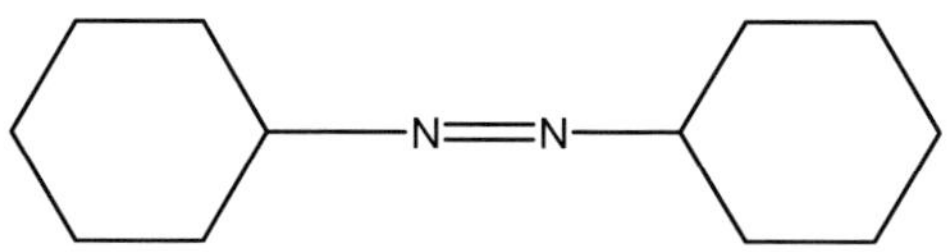	12.9	47	Pass
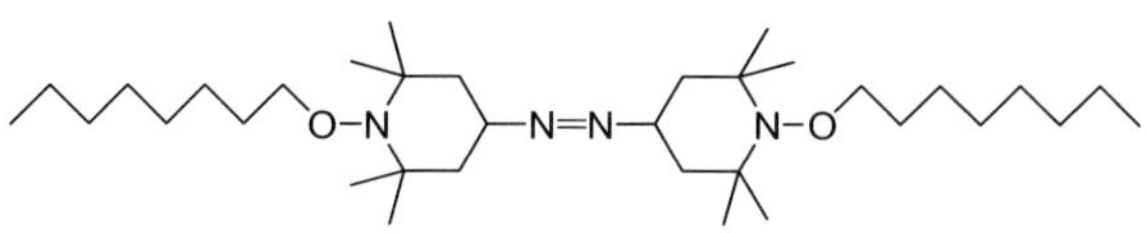	14	43	Pass
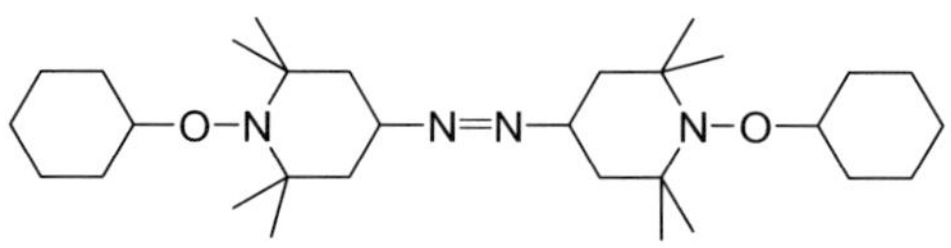	8.9	37	Pass
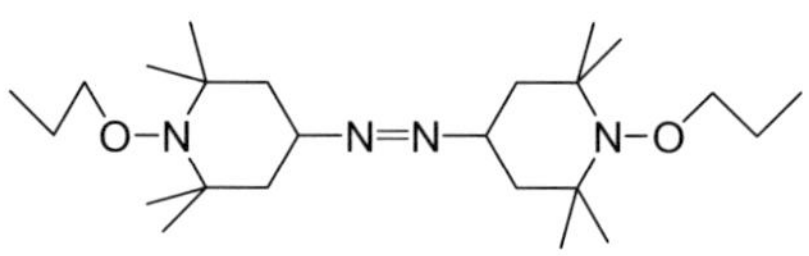	5.2	27	Pass

polymer plaques was explored. This was done by combining the diazene and *N*-alkoxyamine chemistries into a new molecule, i.e. bis(1-propyloxy-2,2,6,6-tetramethylpiperidyl)-4-diazene (AZONOR) containing both of the favorable constituents [54]. Polypropylene samples containing very low concentrations of 0.25−1 wt% of this additive successfully passed not only the fire standard tests of DIN 4102 B2 and NF P92-505 but also the more challenging UL94 VTM-2 standard. By further refining the AZONOR structure, excellent flame retardant properties could be recorded in polypropylene moldings of 1 mm that clearly outperform NOR 116, as shown in Table 4. Besides relatively low levels of addition and having no detrimental effect on polypropylene appearance or its mechanical and processing properties, another great advantage offered by this flame retardant is its multifunctionality, i.e. high flame retardant durability in combination with light stability. Thus, even after 2000 h of artificial weathering, no significant decrease in flame retardant efficacy could be observed [55]. In addition, we could observe a strong synergistic effect between AZONOR and e.g. brominated flame retardants and aluminum trihydrate (ATH) for polypropylene plaques.

4. CONCLUSIONS

Different nitrogen-based compounds alone or in combination with conventional flame retardant systems have demonstrated their versatility and usefulness as flame retardants for both thermoplastics and thermosets. Their main common advantages are they are effective at low concentrations, they cannot form dioxans and furans during combustion, they are noncorrosive, they exhibit good UV stability, they are recyclable due to their high thermal stabilities, some of them exhibit excellent synergistic effects with conventional flame retardants, and many of the nitrogen compounds exhibit multiple modes of action. One of the major drawbacks of today's nitrogen flame retardants is that they require a good fit between the polymer and type of nitrogen-based flame retardant, i.e. nitrogen-based flame retardants are polymer specific. The key drivers for future research of flame retardants will be increased fire safety in combination with flame retardant formulations that fulfill the criteria of sustainable production and products. In recent years, important questions about the environmental safety of antimony, phosphate, and in particular, brominated flame retardants have been raised. All this information suggests a reappraisal of nitrogen-based flame retardants, and many improvements are expected when chemistries have been further refined and fire characterization of various polymers have been elucidated in detail. It would now appear that there are still plenty of nitrogen-based flame retardants to be discovered. Thus, continued research efforts are definitely warranted.

Acknowledgments

The authors are indebted to coworkers at BASF (former Ciba) and Åbo Akademi University for flame retardant synthesis and for all the fruitful discussions related to flame retardancy of polymers. They would especially like to thank R. Nicholas, M.Sc., Dr M. Roth, Dr H. Hoppe, Dr S. Kniesel, Dr R. Xalter, Dr R. King III, Dr R. Drewes, M. Aubert, M.Sc., T. Tirri, M.Sc., and W. Pawelec, M.Sc.

References

[1] Camino G, Costa L, Martinasso G. Intumescent fire-retardant. Polym Degrad Stab 1989;23:359–76.

[2] Kersjes JG, Kierkels RHM. Polyphosphate derivative of a 1,3,5-triazine compound, method for producing the same and its use. PCT Int Appl; 2000. WO 00/02869, (DSM N.V.).

[3] Steenbakkers-Menting HNA, Tijssen J, Tummers DJ. Halogen-free flame retarder composition and flame retardant polyamide composition. PCT Int Appl; 2002. WO 02/28953 (DSM N.V).

[4] Schlosser E, Nass B, Wanzke W. Combined flame retardant. Clariant GmbH; 2004. EP 1386942.

[5] Schlosser E, Nass B, Wanzke W. Flame retardant combination. Clariant Produkte GmbH; 2001. EP 1070754.

[6] Suzuki K, Shindo M, Iljima M. 1,3,5-Triazine derivative salts of polyacids comprising phosphorus sulfur, and oxygen and process for production the same. EP 0974588. Nissan Chemical Industries; 2000.

[7] Wehner W, Dave T. Triazene compounds containing phosphorous as flame retardant. PCT Int Appl; 2009. WO 2009/015772 (Catena Additives GmbH).

[8] Hara Y, Tamura K, Nishimura T, Matsumoto N. Flame retardant composition and flame-retardant resin composition. EP 1116773. Nippon Chemical Industrial Company Ltd; 2001.

[9] Xalter R, Weiss T. Guanidine phenylphosphinate flame retardant composition. PCT Int Appl; 2010. WO 2010/133570 (BASF Schweiz AG).

[10] Eichhorn J. Synergism of free radical initiators with self-extinguishing additives in vinyl aromatic polymers. J Appl Pol Sci 1964;8:2497–524.

[11] Eichhorn J, Bates SI. Foamed self-extinguishing alkenyl aromatic resin composition containing and organic bromide and peroxide, and method of preparation. US Pat 3058928 (Dow Chemical Corp.); 1962.

[12] Modesti M, Lorenzetti A. FR design for foam materials. In: Wilkie CA, Morgan AB, editors. Fire retardancy of polymeric materials. 2nd ed. 2010. pp. 763–81. Boca Raton.

[13] Chiotis A, Clouet G, Brossas J. Effect of free radicals generating compounds towards the flammability of phosphonated polystyrenes-V. Polym Bull 1984;11:275–80.

[14] Horsey DW, Andrews SM, Davies LH, Dyas DD, Gray RL, Gupta A, et al. Flame retardant compositions. PCT Int Appl; 1999. WO 99/00450 (Ciba Specialty Chemicals Holding Inc.).

[15] Srinivasan R, Rotzinger B. Presentation, polyolefins 2000. In: International conference on polyolefins, Houston, USA; 2000. pp. 571–81.

[16] Pfaendner R. Nitroxyl radicals and nitroxylethers beyond stabilization: radical generators for efficient polymer modification. Chimie 2006;9:1338–44.

[17] Basbas AI, Alvisi D, Cordova R, DiFazio MP, Fischer W, Kotrola JA, et al. Process for the preparation of sterically hindered nitroxyl amines. PCT Int Appl; 2008. WO 2008003605 (Ciba Specialty Chemicals Holding Inc.).

[18] Pfaendner R, Roth M, Schoening KU, Weiss T, Hindalekar CB. Phospho-substituted alkoxyamine compounds. PCT Int Appl; 2011. WO 2011/086114 (BASF SE).
[19] Roth M, Simon D, Leslie G, Nesvadba P, King RE. Flame retardant polymer composition containing hydroxylamine. PCT Int Appl; 2003. WO 2003087211 (Ciba Specialty Chemical Holdings Inc.).
[20] Troutman MV, Ravichandran R, Srinivasan R, King RE, Horsey DV. Flame retardant composition. PCT Int Appl; 2002. WO 02/074847 (Ciba Specialty Chemicals Holdings Inc.).
[21] Standard methods of fire tests for flame-resistant textiles and films. Quincy, MA, USA: National Fire Protection Association; 1996.
[22] Wilen C-E, Pfaendner R. Improving weathering resistance of flame-retarded polymers. J Appl Polym Sci 2013;129:925—44.
[23] Roth M. Flame retardant composition comprising sterically hindered amines. PCT Int Appl; 2009. WO 2009080554 (Ciba Specialty Chemicals Holding Inc.).
[24] Kaprinidis N, Lelli N. Flame retardant polymer electrical parts. PCT Int Appl; 2004. WO 2004035673 (Ciba Specialty Chemicals Holding Inc.).
[25] Kaprinidis N, Lelli N. Flame retardant composition. PCT Int Appl; 2004. WO 2004035671 (Ciba Specialty Chemicals Holding Inc.).
[26] Butz V. Flame retardant composition comprising a phosphonic acid derivative. PCT Int Appl; 2010. WO 2010/026230 (Thor GmbH).
[27] Xalter R, Weiss T, Roth M, Hoppe H. Phenylphosphonate flame retardant composition. PCT Int Appl; 2011. WO 2011003773 (BASF SE).
[28] Kaprinidis N, Lelli N, Kierkels RHM. Flame retardant composition. PCT Int Appl; 2004. WO 2004041919 (Ciba Specialty Chemicals Holding Inc.).
[29] Kaprinidis N, Chandrika G, Zingg J. Flame retardant composition. PCT Int Appl; 2004. WO 2004031286 (Ciba Specialty Chemicals Holding Inc.).
[30] Kaprinidis N, Horsey DW. Flame retardant composition. PCT Int Appl; 2003. WO 03/016388 (Ciba Specialty Chemicals Holding Inc.).
[31] Kaprinidis N, Zingg J. Overview of flame retardant compositions UV stable flame retardant products for polyolefins; halogen free melamine based flame retardants for polyamides. Presentation, Additives 2003, Executive Conference Management, San Francisco, USA; 2003.
[32] Troutman MV, Ravichandran R, Srinivasan R, King RE, Horsey DV. US Pat 2006084731 (Ciba Specialty Chemicals Holdings Inc.), 2002.
[33] Marney DCO, Russell LJ, Stark TM. The influence of N-alkoxy HALS on the decomposition of a brominated fire retardant. Polym Degr Stab 2008;93:714—22.
[34] Marney DCO, Russell LJ, Soegang TM, Dowling VP. Mechanistic analysis of the fire performance of a fire retardant system. J Fire Sci 2007;25:471—97.
[35] http://www.specialchem4polymers.com/tds/flamestab-nor-116/ciba-now-part-of-basf/2672/index.aspx.
[36] Salomon DH, Rizzardo E, Cacioli P. Polymerization process and polymers produced thereby. US Pat 4581 429 (Commonwealth Scientific and Industrial Research Organization); 1985.
[37] Hawker CJ, Bosman AW, Harth E. New polymer synthesis by nitroxide mediated living radical polymerization. Chem Rev 2001;101:3661—88 [and references therein].
[38] Ciriano MV, Korth HG, van Scheppingen WP, Mulder P. Thermal stability of 2,2,6,6-tetramethylpiperidine-1-oxyl (TEMPO) and related *N*-alkoxyamines. J Am Chem Soc 1999;121:6375—81.

[39] Kaprinidis N, Zingg J. International fire safety. Conference; papers presented at Hotel Monetleon; 2003.

[40] Kaprinidis N, Shields P, Leslie G. Flame retardants; 2002. Queen Elizabeth II Conference, London, England.

[41] Nicolas R, Wilen C-E. NOR structure—flame retardant property relationship. Internal Abo Akademi University Report; 2003.

[42] Chauvin F, Dufils PE, Gigmes D, Guillaneuf Y, Marque SRA, Tordo P, et al. Molecular disorder and mesoscopic order in polydisperse acrylic block copolymers prepared by controlled radical polymerization. Macromolecules 2006;39:5238—50.

[43] Miura Y, Nakamura N, Taniguchi I. Low-temperature living radical polymerization of styrene in the presence of nitroxides with spiro structures. Macromolecules 2001;34: 447—55.

[44] Nicolas R, Wilen C-E, Roth M, Pfaendner R, King III RE. Azoalkanes: a novel class of flame retardants. Macromol Rapid Commun 2006;27:976—81.

[45] Danneberg JJ, Rocklin DJ. A theoretical study of the mechanism of the thermal decomposition of azoalkanes and 1,1-diazenes. J Org Chem 1982;47:4529—34.

[46] Aubert M, Nicolas R, Pawelec W, Wilen C-E, Roth M, Pfaendner R. Azoalkanes—novel flame retardants and their structure property relationship. Polym Adv Technol 2011;22:1529—38.

[47] German standard DIN 4102-1: fire behaviour of building materials and building components—part 1: building materials, concepts, requirements and tests; 1998.

[48] Nicolas R, Wilen C-E. PCT Int Appl; 2005. WO 2005/030852A2 (Ciba Specialty Chemicals Holding Inc.).

[49] Engel PS. Mechanism of the thermal and photochemical decomposition of azoalkanes. Chem Rev 1980;2:99—150.

[50] Roth M, Pfaendner R, Nicolas R, Wilen C-E. Azo compounds for polymer degradation. PCT Int Appl; 2006. WO 2006/106059 (Ciba Specialty Chemicals Holding Inc.).

[51] Aubert M, Roth M, Pfaendner R, Wilen C-E. Azoalkanes: a novel class of additives for cross-linking and controlled degradation of polyolefins. Macromol Mater Eng 2007; 292:707—14.

[52] Pawelec W, Aubert M, Pfaendner R, Hoppe H, Wilén C-E. Triazene compounds as and effective class of flame retardants for polypropylene. Polym Deg Stab 2012;97:948—54.

[53] Tirri T, Aubert M, Wilen C-E, Pfaendner R, Hoppe H. Novel tetrapotassium azo diphosphonate (INAZO) as flame retardant for polyurethane adhesives. Polym Deg Stab 2012;97:375—82.

[54] Aubert M, Wilen C-E, Pfaendner R, Kniesel S, Hoppe H, Roth M. Bis(1-propyloxy-2,2,6,6-tetramethylpiperidine-4-yl)-diazene — an innovative multifunctional radical generator providing flame retardancy to polypropylene even after extended artificial weathering. Polym Deg Stab 2011;96:328—33.

[55] Roth M, Pfaendner R, Aubert M, Wilen C-E. Symmetric azo compounds in flame retardant compositions. PCT Int Appl; 2008. PCT Int Appl WO 2008/101845 (Ciba Specialty Chemicals Holding Inc.).

CHAPTER

Fire Retardant Fillers for Polymers

9

R. Rothon

Rothon Consultants and Manchester Metropolitan University, UK

P. Hornsby

Queens University, Belfast, UK

CHAPTER OUTLINE

Polymer Green Flame Retardants. http://dx.doi.org/10.1016/B978-0-444-53808-6.00009-3

1. Introduction

Although there are a number of potential compounds that could be considered as fire retardant fillers, there are stringent requirements for their successful commercial use. To inhibit combustion, they must possess a significant endothermic decomposition at a temperature exceeding the polymer melt processing temperature, typically between 180 and 300 °C, together with the release of at least 25% by weight of water and/or carbon dioxide. Additionally, they should have a small particle size and surface area with optimum particle morphology to enable their use at high filler levels without unduly affecting the polymer rheology during processing or ultimate mechanical properties of the composite. Also, they should be colorless, free from impurities, which might compromise electrical insulation properties or initiate premature polymer degradation, be nontoxic and available in large quantities, ideally at a low cost.

Although there are several fillers that meet these criteria (Table 1), very few have achieved commercial prominence. They mostly require extensive chemical modification or synthesis, but some occur naturally and can be used with minimal preprocessing. A key requirement for any fire retardant filler is the need to be thermally stable during polymer melt processing, which is a distinguishing feature between the filler types.

Table 1 Decomposition Temperatures, Enthalpies, and Volatile Emission Levels for Some Candidate Metal Hydroxide Flame Retardants

Metal Hydroxide	Approximate Onset of Decomposition (°C)	Approximate Enthalpy of Decomposition ($kJ\ kg^{-1}$)	Volatile Content (%)
Aluminum hydroxide $Al(OH)_3$	180	1300	34.5
Boehmite $AlO(OH)$	340	560	15
Magnesium hydroxide $Mg(OH)_2$	300	1450	31
Basic magnesium carbonate $4MgCO_3 \cdot Mg(OH)_2 \cdot 4H_2O$	220	1300	57
Calcium hydroxide $Ca(OH)_2$	440	1150	24

Metal hydroxides, in particular those of aluminum and magnesium, are very effective flame retardant fillers for polymers with strong "green credentials" and have a significant market presence. Originally introduced commercially in the 1960s, the first metal hydroxide of this type to achieve any commercial significance was aluminum hydroxide, which is still the main product in use today. Its main competitor is magnesium hydroxide, which has an advantage in some polymers of a higher decomposition temperature. Despite its potentially lower cost, calcium hydroxide is relatively ineffective, and has found little use. Some related materials such as basic magnesium carbonates (including layer double hydroxides or hydrotalcites) are also used commercially and have been included in this chapter.

Aluminum hydroxide is used in a wide range of elastomers, thermoplastics, and thermosetting resins processed at temperatures $<200\,^{\circ}C$. However, magnesium hydroxide has a much greater thermal stability ($>300\,^{\circ}C$), and can be used in polymers processed at higher temperatures. Both these hydroxides are available in a number of forms that can critically determine their suitability as fire retardants and general acceptability in polymer parts.

Fire retardant fillers differ from conventional inert fillers added to plastics and rubbers in that they possess additional functionality, such that when they are exposed to heat they decompose evolving water or inert gasses. As mentioned earlier, in the most commercially significant examples, this compositional change results from an endothermic reaction, which together with the by-products evolved, imparts fire retardancy to the host polymer. However, a number of other factors may also contribute to the overall mechanism of action, including dilution of the matrix phase and the thermally insulating oxide barrier formed. Unlike many other commonly used fire retardant types, notably halogen-based systems, these decomposition products can be considered benign, without contributing to the hazards of the fire through emission of toxic gases and additional smoke. In fact, fire retardant fillers generally possess smoke suppressing properties, which has the important benefit of reducing smoke obscuration from the burning polymer.

These attributes have resulted in the widespread industrial use of hydrated fillers, especially aluminum and magnesium hydroxides, particularly in applications such as wire and cable covering and electronic goods, where fire safety is at a premium or legislation demands low smoke emission from the polymer. Therefore, they are regarded as among the most environmentally friendly types of flame retardants. Importantly, from a commercial viewpoint, they are also relatively inexpensive.

Despite these advantageous qualities, fire retardant fillers have several drawbacks, notably their relative inefficiency, requiring high addition levels to the polymer in order to meet acceptable performance levels required in fire testing specifications. This in turn can adversely affect the physical properties of the polymer, including its rheological, mechanical, and, sometimes, electrical properties. Various approaches have been adopted to overcome these issues, including the use of surface modifiers to the filler to influence filler—matrix interactions yielding enhanced toughness and strength, and combinations of hydrated fillers and other additives, which together may act synergistically. In this regard, the most important recent development has been the use of nanoclays to boost the performance of the metal hydroxides. There is also interest in the use of nanofiller metal hydroxide variants, which can reduce the level of additive required and in some cases can also act as reinforcements.

This chapter will provide an overview of fire retardant filler types and their application in polymers, the general mechanistic principles that govern their fire retardant and smoke suppressing action, together with the means for potentially enhancing their effectiveness through the use of synergists and potentially nanoscale filler variants. It will also consider the effects of fire retardant fillers on other properties of the host polymer and means for modifying the surface chemistry of the filler.

2. Market background

Metal hydroxides have about 40% by weight of the commercial fire retardant market today, which amounts to about 700,000 tons worldwide [1]. The main commercial products range in price from €200–€1800 per ton, and the market value is estimated as being about €400 million.

For reasons that will be discussed later, aluminum hydroxide dominates the market, accounting for >600,000 of the 700,000 tons. The next largest volume is for magnesium hydroxide at about 50,000 tons. (Note that there is considerable uncertainty in the magnesium hydroxide figure due to the possible internal use of natural brucite in China.)

Because of their green credentials, the use of metal hydroxides has been growing faster than the overall fire retardant market, and this is expected to continue with aluminum hydroxide expected to continue to dominate this sector of the market, at least in the short to medium term.

Although its use is limited to polymers processed below 200 °C, aluminum hydroxide finds application in a wide range of elastomers, thermosets, and some thermoplastics. The greater thermal stability of magnesium hydroxide enables its use in many engineering polymers processed at higher temperatures however, although its use is more selective and consequently has a much lower market share. Other hydrated fillers, such as hydromagnesite/huntite mixtures also find some use due to their lower cost. Halloysite has also shown potential in a variety of polymers, including those processed at high temperatures (e.g. PA12, polyethylene terephthalate, polypropylene sulfide, and polyether ether ketone), due to its greater thermal stability. The smoke-suppressing characteristics of hydrated fillers have important implications in their selection in especially demanding applications, such as in cables and in transportation, and where low smoke emissions are generally paramount. Table 2 gives a summary of the application areas for the principal fire retardant types.

Table 2 Application Areas for Fire Retardant Fillers

Hydrated Filler	Polymers	Applications
Aluminum hydroxide (ATH)	*Rubbers* (e.g. SBR)	Carpet backing, cables, hoses, Conveyor belting
	Thermosets Unsaturated polyester resins	Molded sheet and dough molding compounds (e.g. laminated countertops and wall coverings, electronic circuit boards, pultruded profiles for construction
	Epoxy and phenolic resins	Adhesives, laminates, circuit boards, potting compounds
	Crosslinked acrylic resins	Sinks, countertops, bathroom panels
	Thermoplastics Flexible and rigid PVC, ethylene–propylene rubber and EPDM, EVA copolymer, polyethylene (LDPE, HDPE, LLDPE)	Wire and cable insulation, conduit, piping, appliance housings, adhesives, insulating foams
Magnesium hydroxide (MH)	EVA copolymer, PP, polyamides, ABS, modified PPO, elastomers	Wire and cable, appliance housings, construction laminates, electrical components
Hydromagnesite (usually in combination with huntite)	Ethylene–propylene copolymers, PVC	Wire and cable

PVC, poly(vinyl chloride); SBR, styrene–butadiene rubber; EPDM, ethylene propylene diene monomer; EVA, ethylene vinyl acetate; LPDE, low-density polyethylene; HDPE, high-density polyethylene; LLDPE, linear low-density polyethylene; ABS, acrylonitrile–butadiene–styrene; PPO, poly(p-phenylene oxide); ATH, alumina trihydrate.

3. Environmental benefits as fire retardants and smoke suppressants

A selection of data from the literature follows, showing the general features of metal hydroxide flame retardants. It must be stressed that most of the published results refer to small-scale laboratory tests, of which there are several. These small-scale tests do not necessarily agree with each other, or with all real fire situations. Results are thus quoted from most of the main tests in use and where evidence exists, correlation between the tests is discussed.

The general effectiveness of metal hydroxide flame retardants in typical small-scale tests is illustrated by the comparison data for aluminum hydroxide and calcium carbonate given in Table 3.

Table 3 Typical Results in Different Small-Scale Fire Tests for Aluminum Hydroxide and Calcium Carbonate at Equivalent Weight Loadings (60%) in an Ethylene Vinyl Acetate Copolymer

Fire Test	Typical Aluminum Hydroxide	Typical Calcium Carbonate
Oxygen index (%)	30	20
UL 94 vertical burn	V0	Unclassified
Cone calorimeter ASTM E1354-90 (35 kw m^{-2})	Ignition time 150 s Peak heat release 125 kW m^{-2}	Ignition time 125 s Peak heat release 310 kW m^{-2}

High addition levels are necessary before commercially useful levels of flame retardancy can be achieved, and this is well exemplified in the work of Hughes and coworkers using a variety of flame retardant tests on aluminum hydroxide-filled poly(methyl methacrylate) (PMMA) [2]. This work is summarized in Table 4.

The need for high addition levels is one of the key features of the use of this type of additive and influences much of the formulation work needed to arrive at commercially viable final products. It should be noted that some applications where the very highest flame retardancy is required can use loadings as high as 80%w/w (~60%v/v).

Although the green credentials of the metal hydroxides are widely accepted, it is worth examining them in more detail here. There are several aspects to consider, namely production, handling, use, and disposal.

The impact of production on the environment relative to other types of flame retardant is obviously important, but has received little attention. The vast majority of the metal hydroxide market is taken by synthetic products, produced chemically from mined natural minerals. The manufacturing processes are often relatively energy intensive and can produce significant amounts of waste. This is especially true of aluminum hydroxide where large quantities of residue (usually referred to as red mud) are generated and have to be safely stored or disposed of. The question then is

Table 4 Effect of Flame Retardant Aluminum Hydroxide Loading on the Performance of a PMMA Compound in a Variety of Small-Scale Fire Tests

Addition Level (%w/w)	Oxygen Index (%)	UL 94 Vertical Burn Rating	Ignition Time ISO 5657 (s)
0	17.5	Unclassified	99
33	20.1	Unclassified	108
50	23.5	Unclassified	166
60	28.2	V0	213

PMMA, poly(methyl methacrylate).

how this compares with the manufacture of other flame retardants, not forgetting to allow for the higher addition levels used. To the authors' knowledge, these comparisons have not been reported.

Being nontoxic and noncorrosive, metal hydroxides are certainly benign to handle. The main focus of the green debate for flame retardants is usually on their performance when there is a fire situation. The prime demand is, of course, that they make ignition harder and flame propagation slower. By definition, all flame retardants do this. The green issues are around the levels of smoke and of corrosive and possible toxic gasses given off. Here, the metal hydroxides score highly, especially by the standard tests employed. Some examples taken from the literature follow.

The data for smoke production strongly support the environmental friendliness of metal hydroxides. Herbert [3] shows how the addition of alumina trihydrate (ATH) reduces the smoke generation of crosslinked EVA (40% VA) from a Dmax of 250 to <100 (BS6401 flaming mode). Rothon and Hornsby [4] have also reported that the use of 37%v/v (60%w/w) of an magnesium hydroxide reduced the smoke level for a polypropylene compound and showed that this was more than was observed for an inert filler (Table 5).

When looking at the smoke levels, it must also be remembered that many other flame retardant formulations would increase the smoke above that of the polymer itself, so the advantage of the hydroxides is greater than may appear simply by comparison with the base polymer.

Corrosion is most relevant to electrical applications and, as they only generate water, the metal hydroxides would not be expected to contribute to corrosive gas emissions, and there is evidence to support this (Table 6) [6].

Likewise, toxic gas emissions would not be expected to be enhanced by metal hydroxides. Most published information on this topic refers to carbon monoxide generation in cone calorimetry and supports this assumption. Thus, Landemaine found that the addition of magnesium hydroxide to polypropylene reduced carbon monoxide formation to about half that of the unfilled polymer [5].

Of course, many factors have to be taken into consideration in comparing the environmental impact of different fire retardant approaches. Perhaps the most comprehensive relevant study has been that of O'Neill [6] who argues against too simplistic an interpretation of small-scale test data on model compounds. He used

Table 5 The Effect of Magnesium Hydroxide on the Smoke Release from Burning Polypropylene (Cone Calorimeter-35 kW m^{-2})

Compound	Average Specific Extinction Area $m^2\,kg^{-2}$ After 6 min
Unfilled polypropylene	610
Polypropylene with 60%w/w of magnesium hydroxide	250
Polypropylene with 60%w/w of glass beads	600

Table 6 A Comparison of the Acidity, Electrical Conductivity, and Corrosivity of the Combustion Gasses from Typical Commercial Flame Retardant Cable Insulations

Material	Acidity (pH)	Conductivity ($\mu S\ cm^{-1}$)	Corrosion CNET Test (%)
PVC	2.08	3350	14.2
Fire retarded PVC	2.51	1500	8.8
Chlorosulfonated polyethylene	3.58	50	5.6
ATH/EVA (thermoplastic)	3.84	32	0.6
ATH/EVA (crosslinked)	4.44	20	0.9

PVC, poly(vinyl chloride); ATH, alumina trihydrate; EVA, ethylene vinyl acetate; CNET, Centre National d'Etudes des Telecommunications.

a number of different fire retardant polymer formulations, halogenated and nonhalogenated, typical of those found in commercial electrical cable applications. This included standard and fire retarded PVC and aluminum hydroxide in both thermoplastic and silane crosslinked EVA. Table 6 summarizes his results regarding smoke, toxicity, and corrosion. They show the clear advantage of the ATH-based formulations over PVC ones and to a lesser extent over chlorosulfonated polyethylene. The smoke results are presented in Table 7 and again confirm the smoke advantage of the ATH-based formulation.

Finally, one has to consider disposal. Here, the metal hydroxides again seem to score well. Certainly, they do not interfere with incineration, unlike many other types of flame retardants. Also, they should not lead to any leaching of hazardous chemicals if the composite finds its way into landfill. The authors would like to see efforts made to recover and reuse the fillers, but there is little evidence of this so far. Overall then, with the possible exception of the manufacturing process, the green badge worn by these products seems well justified.

Table 7 A Comparison of Smoke Generation by Typical Commercial Flame Retardant Cable Insulations

Material	Specific Extinction Area ($m^2 kg^{-1}$) Cone Calorimeter-50 $kW\ m^{-2}$
PVC	1660
Fire retarded PVC	1240
Chlorosulfonated polyethylene	405
ATH/EVA (thermoplastic)	334
ATH/EVA (crosslinked)	326

PVC, poly(vinyl chloride); ATH, alumina trihydrate; EVA, ethylene vinyl acetate.

4. Fire retardant mechanisms

4.1 General considerations

Fire retardant fillers are relatively inefficient compared to alternative phosphorous-based intumescent and halogen-containing additives, necessitating the use of high filler levels, to achieve commercially acceptable levels of combustion resistance. The latter point has important consequences due to the high filler levels required to achieve adequate flame retardancy. For example, with magnesium hydroxide, this may require 60 wt% of filler to be added to the polymer (equivalent to around 35 vol%), which significantly reduces the amount of combustible matrix present. However, particular loadings depend on the host polymer used, its degradation mechanism, its char forming tendencies, the specific application requirements determining the level of fire retardancy needed and the extent to which detrimental changes to physical properties, resulting from the presence of the filler, can be tolerated.

The general mechanism governing the flame retardant action of hydrated fillers is governed by a number of factors. Of key importance is the endothermic decomposition of the filler together with its heat capacity, causing a cooling of the underlying matrix, associated release of volatiles (specifically water) diluting and cooling the vapor phase and providing a barrier to heat flow in the condensed phase, through the formation of residual oxide ash.

As shown in Figure 1 and in Table 1, the temperature at which this commences and the magnitude of the effect varies between the different filler types, although

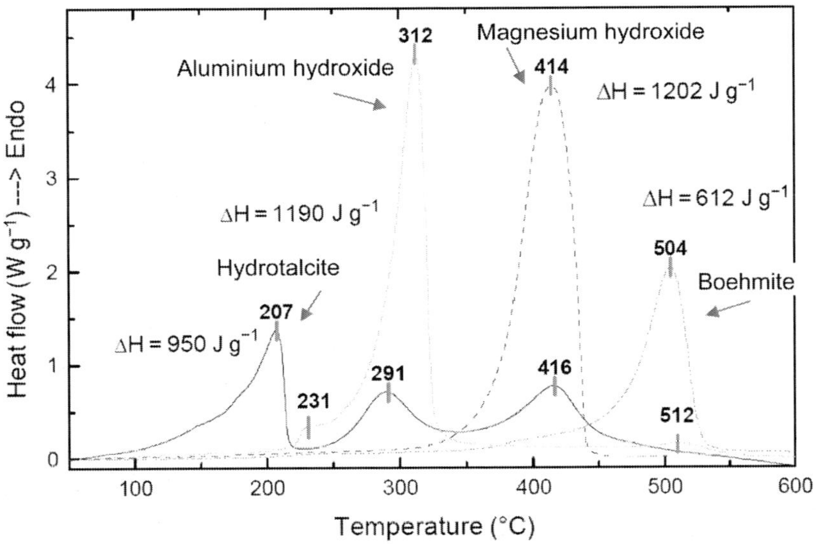

FIGURE 1

Endothermic decomposition of selected hydrated fillers.

From Ref. [7].

these values may vary depending on the purity and other factors associated with the particles, surface area and morphology of the filler and the analysis conditions used. It should be noted that thermal stability, as indicated by the onset of decomposition, is not a fundamental material property and varies with external factors such as the heating rate and ability of gasses to escape from the system. Surface treatments can also affect the stability [8]. This variability is particularly important with aluminum hydroxide, as its stability is borderline for some applications, and small improvements in stability can then be very useful.

With aluminum and magnesium hydroxides, there is normally one decomposition event with the release of water; however, other fillers, for example hydromagnesite, may exhibit multiple decomposition stages evolving both CO_2 and water. Generally, the determination of thermal decomposition behavior is undertaken using differential scanning calorimetry (DSC); however, experimental factors, such as the sample size, heating rate, gas flow rate, and pressurization within the pan depending on the extent of closure, can influence the resulting observations [9]. The particle size of the filler may also be a factor influencing stability and decomposition kinetics. In this regard, ATH is reported to be less thermally stable with increasing particle size [10], with finer precipitated grades being significantly more thermally stable than ground material [11].

Boehmite, AlO(OH), is a monohydroxide found at an intermediate stage during the decomposition of aluminum hydroxide, with a thermal stability of around 450 °C, potentially making it suitable for use at high melt temperatures during processing. However, being a partial decomposition product, its effectiveness as a fire retardant is significantly limited.

More recently, halloysite is being marketed as a promising "green" fire retardant filler in engineering polymers [12]. It is an aluminosilicate material with the general formula $Al_2Si_2O_5(OH)_4 \cdot nH_2O$, the same as kaolin, but with the sheets rolled up into tubes. Like other hydrated fillers, it evolves water of hydration through endothermic decomposition, but at much higher temperatures (>400 °C), and being a high surface area fibrous filler, it also promotes char formation. In this regard, there is reported to be synergy between glass fiber reinforcements and halloysite enhancing the contiguous nature of the char produced [12]. Attractive features of this material are its relatively low cost being mined from natural deposits, combined with effective mechanical reinforcement, due to its fibrous form (Figure 2). Interestingly, being tubular in nature, it has been suggested that the lumens can host synergists (discussed later), without compromising mechanical properties of the polymer, which can occur when introduced separately.

Whereas effects from the filler decomposition endotherm occur in the condensed phase, release of volatiles, in particular water, can serve to dilute flammable vapor formed from polymer decomposition products and also exert a cooling effect. Furthermore, the filler residue remaining in the polymer following decomposition has a significant insulating role, protecting and reducing the rate of heat transmission to the underlying polymer [13]. The so-called forced combustion studies have been used to demonstrate the significance of this effect for a variety of filler–polymer

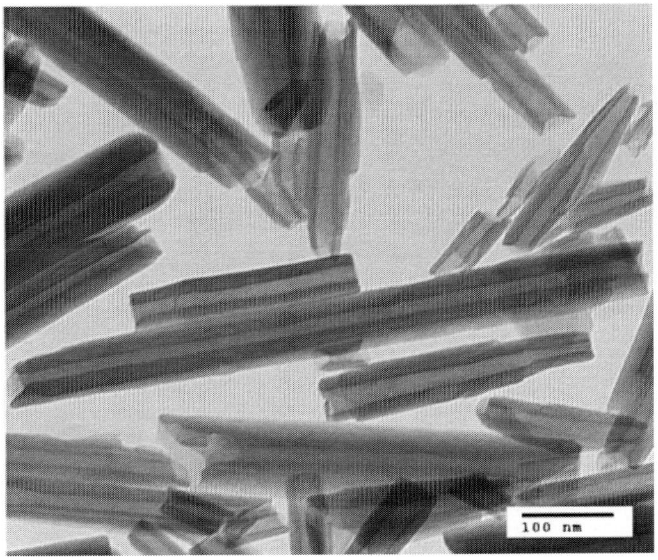

FIGURE 2

Transmission electron microscopy image of Dragonite™ halloysite tubes.

Courtesy Applied Minerals Inc., New York, USA.

combinations [14]. With aluminum and magnesium hydroxides, this results in a metal oxide residue creating a thermal barrier combined with carbonaceous char from the polymer. It has been found that the integrity and stability of this char/filler residue can lead to differences in overall flame retardancy from nominally the same filler type. With magnesium hydroxide, this appears to be related to the particle morphology and whether or not it is platy [15].

The relative importance of these factors is difficult to ascertain and undoubtedly varies with the polymer and especially with the flame retardant test. Case and Jackson developed a model for the behavior of endothermic fillers in the oxygen index test, which is based on the heat feedback from the flame [9]. This model fits experimental data well except where very small size fillers are used. Here, their model breaks down with very small filler sizes giving anomalously high values, the reason for which has not been fully established. It has been found that, unlike the coarser sizes that fit the model, these fillers lead to the formation of a firmly attached ash. Continual removal of this ash restores the oxygen index to the predicted value, so the discrepancy would appear to originate from this source. It seems reasonable to speculate that the ash is insulating the underlying polymer and reducing heat feedback, an effect not included in the model. This heat balance approach has been developed and refined by Hull and coworkers [16], whose results for the relative contribution that various potential fire retardant fillers could make are reproduced in Table 8. Their analysis gives very similar results to those of Case and Jackson for the two fillers that they studied, aluminum hydroxide and nesquehonite.

Table 8 Calculated Total Energy Absorption and Contributions from Different Mechanisms for Various Candidate Flame Retardant Metal Hydroxides

Metal Hydroxide	Total Energy Absorbed by All Mechanisms ($J\ g^{-1}$)	% From Mineral Heat Capacity	% From Endothermic Decomposition	% From Residue Heat Capacity	% From Evolved Gasses
Aluminum hydroxide	2300	9	55	13	23
Magnesium hydroxide	2500	19	56	9	15
Calcium hydroxide	2000	29	55	5	11
Hydromagnesite $4MgCO_3 \cdot Mg(OH)_2 \cdot 4H_2O$	2750	10	56	14	21
Boehmite AlO(OH)	1250	18	46	20	15

Hull and coworkers also comment on the mechanisms involved in Bunsen burner tests such as the Underwriters UL 94 tests and in cone calorimetry.

It is not only the magnitude of the endotherm that is important. It must also occur at an appropriate temperature, close to that of the pyrolyzing polymer. If it is too low, then there may be processing difficulties, and although it may delay combustion (increase ignition time), the polymer will burn freely once the additive has been used up. If it is too high, then the polymer will combust before the endotherm comes into effect. This latter point is illustrated by calcium carbonate. This has a large endotherm but only at temperatures of >600 °C, so it has little fire retardant effect. For most polymers, a decomposition starting in the range 150–350 °C is acceptable, but there is some evidence that magnesium carbonate with a decomposition starting at about 450 °C is effective in some polymers.

The earlier consideration of thermal decomposition of hydrated fillers assumes that there is no interaction between the filler and the polymer, which might influence their relative degradation behavior. This is important since hydrated fillers, in particular magnesium hydroxide, may promote polymer degradation, through hydrolysis from water release. For example, this has been observed with polyamide 6 and polyamide 6,6 containing magnesium hydroxide using thermal analysis (thermogravimetric analysis and DSC) combined with evolved gas analysis (mass spectroscopy and Fourier transform infrared spectroscopy (FTIR)) [17]. In these systems, it was found that the decomposition profile of MH/PA-6 overlapped more closely than with MH/PA-6,6 where the polymer decomposed before the filler (Figure 3). This resulted in superior fire retardancy in PA-6 than in PA-6,6 at an equivalent filler loading.

Similar effects have been seen in EVA copolymer compositions (with 30% vinyl acetate content) containing ATH and MH at the same filler contents [18]. A much higher oxygen index value was seen using MH (46%) relative to ATH (37%), which was attributed to acetic acid evolved from the degrading polymer causing earlier decomposition and water release from the MH filler.

However, care must be taken when assessing the relative performance of filler type flame retardants when using small-scale laboratory tests. For example, additives that show an endothermic decomposition can sometimes give high values in the oxygen index test in particular, without it being reproduced in other small-scale tests or, more importantly, in large-scale ones.

As an example of this, Ashley and Rothon [19] reported that 125 phr of a precipitated calcium carbonate raised the oxygen index of a crosslinked EVA compound from 18.5 to 29.0 similar to ATH (30.5) and much higher than glass Ballotini (21.5). This indicates that the precipitated calcium carbonate is acting as an effective flame retardant. However other tests, including UL 94, showed this not to be the case. They went on to analyze the ash residue from the oxygen index test and found it to contain significant amounts of calcium oxide, showing that decomposition had occurred under the conditions of the test. Unlike ground calcium carbonate, the precipitated form gives a coherent ash in the test, which remains attached to the sample. The flame can be seen to travel down through this ash, raising it to red heat and causing the filler to decompose. This effect will only be observed in tests where the

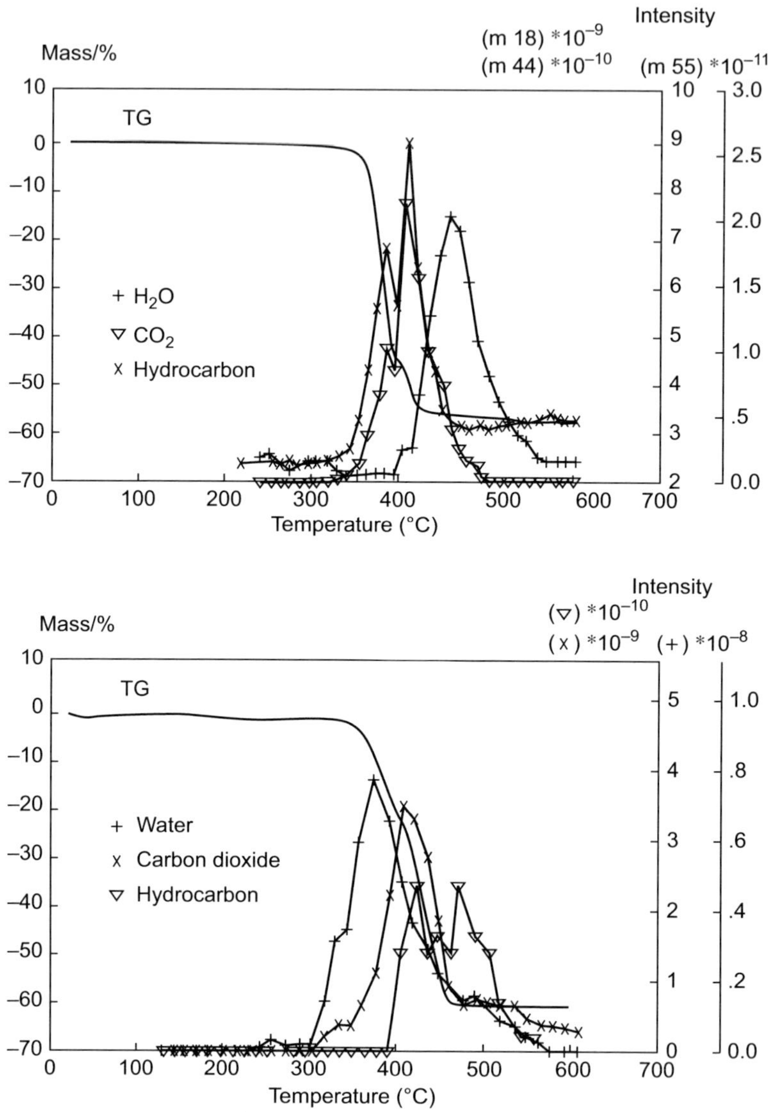

FIGURE 3

FTIR/EGA (evolved gas analysis) profiles for water, carbon dioxide, and hydrocarbon in (top) magnesium hydroxide/PA-6,6 and (bottom) magnesium hydroxide/PA-6 compounds.

From Ref. [17].

ash is able to contribute in this way, but in most other tests (e.g. UL 94), it falls off the specimen and is ineffective.

The same authors reported a similar effect for another endothermic filler, magnesium carbonate subhydrate. This appeared to be an effective flame retardant in polyamide 66 with an oxygen index of 31% (at 50%w/w). In support of this, analysis of the ash showed that decomposition to magnesium oxide had occurred. Despite this, it gave an unclassified rating in the UL 94 vertical burn test. In crosslinked EVA, the same filler at 125 phr gave an oxygen index of 33%, and the ash again contained magnesium oxide. Again, the UL 94 vertical burn test gave an unrated classification. In order to demonstrate the effect of the ash, they removed this during the oxygen index test, and then, the rating fell to 22%.

As discussed later, metal hydroxides are often surface treated to enhance processing and to improve mechanical and other properties. The common coatings are fatty acids and organosilanes. As Rothon has shown, such treatments can affect the thermal stability of metal hydroxides and their effect on some flame retardancy tests [20]. An example is given in Table 9. The fatty acid treatment is seen to have no effect on oxygen index, but to markedly reduce ignition time, while the reverse is true for the silane coupling agent.

4.2 Particle size and shape

This is a much neglected, but very significant, area. There is abundant evidence that these properties can have a significant effect that probably varies with the nature of the polymer matrix and most importantly with the tests used, but the nature of all the factors involved has not been fully elucidated. Because of this variation with particle properties, it is not possible to answer fully the question as to how one material, say ATH, does compare with another as a flame retardant. Any comparison should be at equal particle size and shape distribution, but the manufacturing processes make this almost impossible to achieve.

The magnitude of the differences that can be seen is illustrated in the work of Hughes and coworkers referred to earlier [2]. Results from this work for two aluminum hydroxides of differing particle sizes are presented in Table 10. These results show that there is a significant particle size effect on the UL rating and on ignition time, but not on oxygen index, in keeping with Case and Jackson's results

Table 9 The Effect of Some Common Surface Treatments on Performance in Some Small-Scale Tests

Surface Treatment of Magnesium Hydroxide	Oxygen Index (%)	Ignition Time (s)
None	42.5	20.5
Fatty acid	43.0	13.5
Organosilane	28.5	19.0

Table 10 Effect of Aluminum Hydroxide Particle Size on Fire Retardant Performance in a Number of Small-Scale Tests (60% w/w in PMMA)

Aluminum Hydroxide Average Particle Size	Oxygen Index (%)	UL 94 Vertical Burn Rating	Ignition Time ISO 5657 (s)
69 μm	27.5	Unclassified	135
8.4 μm	28.8	V0	203

PMMA, poly(methyl methacrylate).

referred to earlier. It is only with much smaller sizes than those used in this work that deviation from their model is seen.

There is little published information on the effect of particle size in the cone calorimeter test, especially for the larger particle sizes. The most comprehensive data are reported by Herbert for aluminum hydroxide in EVA copolymer [21] and indicates that the particle size effect may be quite small. However, the particle size range used did not include the larger sizes found to give reduced performance in the work referred to above.

Particle size distribution is as important as particle size in many instances. This is especially true when very high filler loadings are used to maximize flame retardancy. Such high loadings can give rise to unacceptable viscosities unless special particle size controlled grades are used, often in conjunction with high efficiency dispersants.

Particle shape almost certainly also plays a role, but there is little hard evidence on this, as it is difficult to vary independently of other particle properties. One of the few attempts to study this was by Hughes and coworkers with aluminum hydroxides in PMMA [2]. The results for two samples with a similar particle size, but different shape are summarized in Table 11. While the indications are that platy particles may be better than spherical ones, care must be taken as the specific surface areas were inevitably different. It must also be remembered that platy particles cause more processing difficulties.

Extensive work has been undertaken using nanofillers for polymers, including silicate layer structures, carbon nanotubes, cellulose nanofibers, etc., describing benefits to mechanical and other physical properties obtainable at very low addition levels. Hence, bearing in mind the size effect highlighted above, it is logical to consider the use of nanosize fire retardant fillers in polymers in order to obtain

Table 11 Effect of Particle Shape on Fire Retardant Performance in a Number of Small-Scale Tests

Aluminum Hydroxide	Oxygen Index at 60%w/w	Amount Needed for V0 (%w/w)	Ignition Time ISO 5657 (s) 50%w/w
None	17.5	Not relevant	99
12.3-μm blocky	28	>50	166
11.1-μm platy	32	<50	187

potential improvements in fire retarding performance, but at lower additive levels than are currently used.

A limited amount of work has been published adopting this approach, and the discussion below considers examples of reported observations using magnesium hydroxide, magnesium aluminum, layered double hydroxides, and aluminum hydroxide. In general, however, the findings are not completely convincing, considering the greater complexity and cost in preparing these systems and in some cases, the questionable benefits achievable compared to micron-scale filler variants.

Commercially available grades of nano-magnesium hydroxide (hexagonal sheets with dimensions of a 100-nm width by <50-nm thickness) and three forms of micro-magnesium hydroxide (particle size of around 2.5 μm) were combined with EPDM rubber, although at relatively high filler levels (60–100 phr), and the mechanical properties and fire resistance of these composites were compared [22]. The nanocomposite material gave a significantly lower heat release rate, a greater time to ignition and less tendency to drip during the limited oxygen index (LOI) test (although LOI values were similar), perhaps reflecting greater stability of the char formed. Additions of only 12 wt% magnesium hydroxide with a particle size of 25 nm (synthesized from $MgCl_2$) to polypropylene gave a 35% reduction in burning rate compared to unfilled material combined with a very large 433% increase in Young's modulus [23].

Similarly, magnesium hydroxide nanocrystals with different geometries (needle, lamellar, and rod-like) were synthesized from $MgCl_2$ and added to EVA in order to compare their flammability and mechanical properties with micron-sized filler composites [24]. It was shown that tensile strength for the larger particles decreased with increasing filler level, whereas the values for the nano-MH/EVA composite increased. A more compact char structure giving an enhanced gas barrier obtained with the nanoscale fillers was considered to explain the superior LOI results reported for the nano-MH. However, relatively high filler levels were still required to achieve acceptable fire resistance.

Magnesium hydroxide with a mean particle size of 10 nm has been made with the particles surrounded by alkyl chains in a core–shell morphology [25]. It was shown that the peak heat release rate (HRR) of a resulting EVA/nano-magnesium hydroxide composite was reduced by 60% compared to pure EVA at only 5 wt% filler addition, together with a significant reduction in smoke emission, although LOI was not significantly influenced. The nano-magnesium hydroxide gave improved fire performance to conventional 1-μm magnesium hydroxide filler at equivalent loadings between 5 and 60 wt%. A similar approach was adopted for the synthesis of nano-aluminum hydroxide with an alkyl hydrocarbon chain surrounding shell structure [26]. At a 10 phr filler loading, this gave a much lower heat release rate than did conventional 1-μm ATH.

Hydrotalcites are a form of hydrated magnesium–aluminum (Mg–Al) hydroxycarbonate with a double hydroxide lamellar structure and general formula $Mg_6Al_2(OH)_{16}CO_3 \cdot 4H_2O$. They are of interest as nanofire retardant fillers on account of their potential to be exfoliated into nanoplates after chemical treatment and intercalation by polymer chains. Furthermore, hydrotalcite shows three thermal decomposition peaks at 207, 291, and 416 °C with corresponding endothermic decomposition and weight loss behavior similar to that of magnesium and aluminum hydroxides (Figure 1) [7].

Various studies using micron-sized hydrotalcite have been reported. Using relatively course hydrotalcite (200-μm) particles, additions of only 30% by weight gave an LOI value of 29 and UL94 V0 rating in unplasticized PVC [27]. A comparison in EVA between hydrotalcite with a median particle size of 2.26 μm and a surface area of 17 $m^2\,g^{-1}$, and MH and ATH with a similar particle size, showed that EVA filled with 50 wt% hydrotalcite gave the slowest heat release rate, longest ignition time, and lowest evolved gas temperature [7]. This was attributed to the wide temperature range (200–500 °C) for water loss and associated heat absorption, a delay in deacetylation of the EVA due to interlayer water loss at about 207 °C and possible intercalation of acetate anions in the layer structure. In epoxy resin containing microhydrotalcite and nanohydrotalcite (the latter made by intercalation with citric acid), UL94 horizontal burn rates were compared. The unmodified resin burned at a rate of 22 mm min^{-1}, whereas with 5 wt% addition of filler, this was reduced to 18 mm min^{-1} and <5 mm min^{-1}, for the microhydrotalcite and nanohydrotalcite, respectively [28].

4.3 Smoke reduction

Effect on smoke is a very important property of any flame retardant, and as mentioned previously, metal hydroxides not only do not contribute to this but they also actually reduce it compared to the unfilled polymer. This effect has been demonstrated in many polymers including styrene–butadiene copolymer [29], ethylene–propylene–diene elastomers [30], polypropylene [14], polystyrene [31], modified polyphenylene oxide, polybutylene terephthalate, and ABS [13].

As well as diluting the amount of flammable material present, the fire retardant filler also reduces the rate of combustion and hence the amount of smoke produced. Additionally, on thermal decomposition, MH and ATH, in particular, generate very high surface area oxide particles, which can have catalytic properties [32], promoting decomposition of carbon particles from the smoke with their resulting oxidation [33]. A consequence of this action is the so-called afterglow or incandescence phenomenon, which is commonly associated with the use of hydrated fillers [34,35].

Afterglow, smoldering, and reignition are important properties in some applications, but seem to be largely overlooked in the recent scientific literature relating to metal hydroxide flame retardants. This is despite evidence that it can be a real issue with some of these materials. Delfosse and coworkers [35] and Hornsby and Watson [14] have conducted some useful studies in these areas, and Rothon has again summarized some of the work [36]. Overall, it seems that afterglow, smoldering, and reignition certainly can be a problem with metal hydroxide flame retardants and are related to the smoke reducing mechanism described above. The oxides formed, when these additives decompose, trap soot particles on their surfaces during the main burning process, and it is these residues that cause the later problems as they oxidize and burn away. Indeed, in some tests, the ash is seen to be black or gray initially and to change to white later as the carbonaceous matter is oxidized. It would seem prudent to use afterglow suppressors if this issue is likely to be of importance. Afterglow is one of the important parameters included in the UL 94 test.

4.4 Synergistic additives

There is clear evidence from the literature that a number of additives can boost the fire retardant performance of metal hydroxides and therefore reduce the overall additive levels required. Most of this is in the commercial and patent literature, and some of the additives are proprietary, so the chemical nature is not revealed. In some instances, hydrated fillers have been added to alternative fire retardants in order to reduce overall smoke emission.

This includes combinations of hydrated fillers such as MH and ATH [37,38], and incorporation of varying amounts of red phosphorus [39], melamine and novolacs [40], transition metal oxides [41], metal nitrates [42], zinc stannate and zinc hydroxystannate [43–45], carbon fibers [46], and polyacrylonitrile fibers [47].

In particular, organoclays or nanoclays have received a great deal of attention as polymer additives over the last decade, including as environmentally friendly flame retardants [48]. They are particularly effective in cone calorimeter tests but have some limitations in other tests. It has been found that they can give very good results when mixed with other flame retardant types, especially metal hydroxides. Work by Beyer [49] demonstrates this well and products where such clays are combined with aluminum or magnesium hydroxides have become commercially available. The main benefit of the combined products is the ability to significantly reduce loading levels, without comprising flame retardancy. This, in turn, allows other properties to be better optimized. As an example, Herbiet [50] has reported that 50%w/w of an aluminum hydroxide/nanoparticle combination gives a similar flame retardancy as that of 61.5% of aluminum hydroxide alone. Similar results are reported for an magnesium hydroxide/nanoparticle combination. Some results from cone calorimetry (EVA/LLDPE mixture 35 kW m^{-2}) are summarized in Table 12.

A very interesting development in recent years has been the use of metals such as zinc, chromium, and nickel coprecipitated with a metal hydroxide. If correctly

Table 12 Effect of Nanoparticle/ATH Combination on Fire Performance in Cone Calorimetry

Additive System	Level (phr)	Time to Ignition (s)	Peak Heat Release Rate (kW m^{-2})	Smoke Maximum (m^2 m^{-2} s^{-1})
Aluminum hydroxide	50	95	215	2.0
Aluminum hydroxide	61.5	121	100	0.75
Aluminum hydroxide nanoparticle combination	50	121	105	1.0

carried out, the metal is in solid solution in the hydroxide, and by doing so, it appears to be able to vary the decomposition temperature and also to increase flame retardant effectiveness by increasing char formation. Magnesium hydroxide doped with nickel has been introduced commercially in Japan, and this is discussed further in the section dealing with that hydroxide.

5. Effects on other important composite properties

While primarily described as flame retardants, the high addition levels necessary for commercially useful effects are typical of mineral fillers, and the metal hydroxides need to be treated as such in any discussion of their overall effects on polymer composites. These affects have to be allowed for when choosing products and are briefly discussed below.

5.1 Processing

The high loadings involved mean that viscosity effects can become very significant. As a result, particle size, shape, and especially particle size distribution, become important factors. Unfortunately, while small particle size generally gives the best flame retardancy, it also gives the highest viscosities, and this can lead to compromises having to be made. Additives (sometimes present as coatings) are often necessary to reduce viscosities, and their effect on fire retardant performance also has to be taken into account.

Recent work demonstrates how small changes in particle size distribution can significantly improve melt flow without compromising other properties [51].

5.2 Mechanical properties

Although used as flame retardants, the high addition levels means that the hydroxides and related materials will inevitably have a very significant effect on the mechanical properties of the composite. These effects are not always helpful to the final product, and the main ones are briefly discussed below.

Strength is usually significantly reduced by uncoated metal hydroxide. Coatings can either restore some of the loss, or even make it worse, depending on their nature. Those that give a weak interface, such as fatty acids, usually reduce strength while those such as coupling agents that give a strong one usually minimize any losses. An example is given in Table 13 [52].

Modulus (stiffness) is always increased, sometimes by very significant amounts. As an example, the elastic modulus of a polypropylene was increased from 1000 to 3200 MPa by the addition of 65%w/w of a stearate-coated magnesium hydroxide. This property is little affected by coating in most polymers uncoated products usually significantly reduce elongation to break. Some coatings minimize these effects while other may increase them. An example is given in Table 13.

Table 13 An Example of the Effect of a Flame Retardant Filler and of Surface Modification on Strength and Elongation (60%w/w Filler in EVA with 18% VA)

Filler	Tensile Strength (MPa)	Elongation at Break (%)
None	20.1	>250
Uncoated	14.7	48
Fatty acid coated	4.0	234
Organosilane coated	15.3	123

Toughness (including impact strength) is frequently significantly reduced by the uncoated products. Surface treatments or toughened polymer compositions are then needed to offset this. This is a key area for many thermoplastic applications and is one where surface treatments are often critical to success. This is a very complex area with the optimum surface treatment often varying with the exact nature of the polymer (type of polypropylene copolymer for instance). Unfortunately, from a scientific point of view, many of the best performing treatments are proprietary, and most of the useful information is to be found in the trade literature. As an example of the effects, it has been reported (Magnifin® brochure) that while addition of 65%w/w of an uncoated magnesium hydroxide to a random copolymer polypropylene reduced the impact strength (DIN 53453) from no break to only 7 kJ m^{-2}, the "no break" performance could be restored by a number of different proprietary coatings.

5.3 Polymer stability

Metal hydroxides can significantly reduce the thermal stability of polymer systems, and this has to be guarded against. The first way this can happen is by adsorption of stabilizers present in the formulation. The high loadings and relatively small particle size means that there is a lot of surface present for this adsorption to occur on. This problem can be countered by increasing stabilizer levels, or by the use of surface treatments. The second way is through the presence of active forms of some trace impurity metals such as copper. A few parts per million of such metals can be very detrimental. Again extra stabilizer can be used to overcome the problem, but the best way is to eliminate or minimize such impurities. In an interesting twist to the stability issue; dopants such as nickel and chromium have been promoted as char-enhancing agents.

5.4 Electrical properties

Because of the lack of corrosive gas emissions, many of the applications for metal hydroxides are in electrical applications, especially cables. Electrical stability, especially in wet conditions, is particularly important here. This is significantly reduced by the presence of certain trace impurities, notably sodium. As a result, conductivity of the extract is often specified, and special electrical grades are produced, often by washing to extract as much sodium as possible.

Table 14 Summary of the Main Effects on Composite Properties Associated with Commonly Used Surface Modifiers

Property	Effect Relative to Untreated	Comments
Filler equilibrium water content	Reduced	Improves processing but increases dust problems
Filler reactivity	Reduced	Mainly applies to magnesium hydroxide where carbonation can be a problem
Water absorption of composite	Reduced	Less swelling and more stable electricals
Leaching of additive	Reduced	Mainly an issue with some magnesium hydroxides in especially adverse conditions; can lead to surface chalking and loss of fire retardant properties
Composite viscosity	Reduced	Improves processing, especially with very high additive levels
Composite strength	Can be reduced or improved depending on the additive	–
Composite elongation and toughness	Can be reduced or improved depending on the additive	–
Flame retardancy	Can be reduced, stays the same or be improved depending on the additive	–

5.5 Use of surface modifiers to improve composite properties

Many of the surface treatments typical of the mineral filler industry are utilized to control the detrimental effects on processing and composite properties.

The most common reasons for using surface treatments are listed in Table 14. Some examples of the utility of these additives have been given above, together with a mention of their potential effect of flame retardancy itself. Many treatments show a combination of the various effects.

6. Principal commercial products

6.1 Aluminum hydroxide (also widely referred to as ATH)

Aluminum hydroxide has dominated the market since its first commercial introduction and continues to do so today. This is due to it exhibiting similar flame retardancy and smoke suppression to the other candidates (principally magnesium hydroxide) while being significantly less expensive to manufacture in useful form.

The flame retardant ability of aluminum hydroxide was recognized almost a century ago and a patent on its use as a flame retardant for elastomers appeared in 1921 [53]. Despite this, nothing much happened until the mid-1960s, although there was a patent on its use to improve arc resistance of various polymers in 1960 [54], an effect also related to its endothermic decomposition. The first significant commercial use as a flame retardant started in unsaturated polyesters in the mid-1960s, following the work by Connolly and Thornton [55]. This was followed by an interest in elastomer systems in the early 1970s when workers such as Lawson et al. [29] and Nelson [56] recognized the low smoke benefits associated with the use of ATH as a flame retardant. They also recognized the effectiveness of other fillers, especially magnesium hydroxide.

Today, ATH is widely used as a flame retardant in many types of polymer and a host of grades varying in particle size distribution and particle shape are available to suit particular applications.

6.1.1 Production methods

Aluminum hydroxide occurs widely in nature, mainly as gibbsite, a mineral that is the principal component of the ore bauxite. Unfortunately, this is not pure enough for direct use as a flame retardant, even with physical beneficiation. Chemical processing is therefore used to increase the purity and control particle shape and size.

The main chemical process is carried out by extracting the aluminum in natural ores such as bauxite with sodium hydroxide to form a solution of sodium aluminate. After purification, aluminum hydroxide is precipitated from this solution by dilution and pH adjustment. Seeds of aluminum hydroxide crystals are also often used to control particle size. This extraction and precipitation is also the basis for manufacture of aluminum oxide and aluminum by the Bayer process, which is carried out on a vast scale. The Bayer process is operated to give reasonably pure, blocky, particles of an about 70-μm diameter. These are known as Bayer hydrates and are the lowest cost form of aluminum hydroxide available for fire retardant use. Where low cost smaller size particles are required, these can be made by grinding or milling of Bayer hydrates. The grinding process converts the particles into a platy shape however.

Higher quality grades are made by using better purification procedures and different precipitation conditions (especially with regard to seed addition). Today, they often start by reextracting Bayer hydrate. Because they can be directly precipitated at the desired size, no milling is required, and they are usually blocky in shape. These grades are often referred to as precipitated or fine precipitated aluminum hydroxide.

6.1.2 Endothermic decomposition

The decomposition of aluminum hydroxide starts to occur at about 180 °C. While it looks simple on paper, the decomposition has some unexpected subtleties. Under certain conditions, it can take place in two stages, with the intermediate formation of the monohydroxide (boehmite) phase.

$$2Al(OH)_3 \rightarrow 2AlO(OH) + 2H_2O$$

$$2AlO(OH) \rightarrow Al_2O_3 + H_2O$$

Once formed, this intermediate phase is stable to much higher temperatures and only starts to decompose at about 340 °C. Formation of boehmite is favored under conditions where escape of the water is hindered and such conditions have been reported to exist with larger particle sizes [57] and may also exist for the finer sizes when they are encapsulated in relatively impermeable polymer.

The important question is whether the formation of the intermediate phase affects the flame retardancy and in what way. There is no definitive answer to this, but it is probable that it does; with the direction and magnitude of any affect depending on the polymer and the test. If the intermediate phase does not decompose under the test conditions, then one would expect a negative effect, but where it does, then there may be no effect or even an improvement due to a spreading out of the decomposition.

Synthetic boehmite itself is marketed as a flame retardant additive, and this is briefly discussed in Section 7.2.

6.1.3 Grades of aluminum hydroxide

The main difference between aluminum hydroxides is that between milled or ground Bayer hydrates and the specially purified and precipitated forms. The principal differences between these two types are in purity and particle shape (platy versus spherical). There is also some evidence that the ground grades have a slightly lower onset of decomposition than the precipitated ones [58]. There is also a report by Ashton that the ground types have superior flame retardancy than the precipitated types, at least in some circumstances [59]. At first sight, this seems unexpected, but may be due to their platy nature, compared to the more spherical precipitated ones. This is in line with the results of Hughes et al. referred to earlier.

Particle sizes of commercial grades vary from an average of 70 μm down to one of about 1 μm with specific surface areas (Brunauer–Emmett–Teller method) ranging from under 1 to about 15 $m^2\ g^{-1}$. Particle size and shape distributions significantly affect the packing ability of the particles and hence the maximum loading that can be achieved. This is usually indicated by oil absorption values that range from 53 down to as low as 15 (cubic centimeters of oil per 100 g of filler). Packing ability is maximum at the lowest oil absorptions, and special grades are made to maximize packing (minimize oil absorption). Many applications are in electrical equipment where the stability of electrical properties, especially under humid conditions, is important. Traces of residual soluble species, especially sodium, can have a marked deleterious effect on this, and hence a property called the conductivity of aqueous extract is often quoted, with special electrical grades being produced, which have been carefully extracted to minimize this property. Conductivity of the extract varies from 800 down to under 50 μS cm^{-1}.

6.1.4 Surface treatments

Although surface modification is frequently required in final applications, most grades are sold uncoated with any surface treatments being applied separately at the compounding stage. Common surface treatments are organosilane coupling agents and various dispersants.

6.2 Magnesium hydroxide

Magnesium hydroxide is very similar to the aluminum compound in fire retardant performance and is the second most widely used product for this type of application. Even so, its market share is only about 10% of that of aluminum hydroxide. This is mainly due to the fact that equivalent quality grades are more expensive to produce. The principal advantage of magnesium hydroxide is its greater thermal stability, with decomposition only starting at about 300 °C or at approximately 100 °C higher than that of the aluminum compound. This makes it more suitable where high processing temperatures are required, principally in thermoplastics such as polypropylene and polyamides. It is also much more chemically reactive than ATH, and this limits its suitability in one of the main markets, unsaturated polyesters. Indeed, magnesium hydroxide is so reactive in this type of polymer that small amounts are used as thickeners.

6.2.1 Production methods

Unlike aluminum hydroxide, there are natural sources of magnesium hydroxide that are pure enough to be used with only grinding and physical beneficiation; these are usually referred to by the mineral name brucite. There are also deposits where the brucite is mixed with significant amounts of calcium carbonate (brucitic marbles), and these are also being exploited. The main processing given to the natural products is drying and grinding, although it is reported that natural brucite is not easy to grind.

Despite its smaller market share, magnesium hydroxide is produced by more varied synthetic processes than aluminum hydroxide. These largely depend on the raw material source used.

The simplest processes are probably precipitation of magnesium hydroxide by adding base to natural solutions of magnesium chloride such as in seawater or some brines. In some regions, natural deposits of magnesium chloride are solution mined to provide a suitable brine. Such processes have been carried out on a significant scale in the past as an intermediate step in the production of magnesium oxide for use in refractory applications, and there is still some manufacture using this route today. For this purpose, the oxide is produced by calcination of the magnesium hydroxide precipitate. For cost purposes, the alkali used is lime or dolomitic lime.

There is a direct analogy here with the production of aluminum hydroxide in the Bayer alumina process. Unfortunately, unlike the Bayer hydrate, the magnesium hydroxide precipitate is not in a very useful form for flame retardant applications, so the same economy of scale is not forthcoming. There have been many efforts to upgrade this hydroxide and even to change the alkali to the more expensive sodium or ammonium hydroxides, but none of these has proved to be completely successful.

Another route that has been extensively explored is to use hydration of magnesium oxide. The oxide can be obtained by calcination of magnesium carbonate (the mineral magnesite is abundant and can be obtained relatively pure) or by calcination of the seawater or brine precipitates. Such processes can give satisfactory products for some, but not all applications, and the calcination significantly increases costs.

The most successful commercial process is probably the route known as hydropyrolysis, in which an aqueous solution of magnesium chloride is sprayed into a hot reactor, producing steam, hydrogen chloride, and magnesium oxide. The magnesium oxide is collected in water where it hydrates. If the starting solution is pure enough and the hydropyrolysis and hydration conditions are carefully controlled, then very good particle forms for flame retardancy can be produced. These products are satisfactory for almost all applications. As can be imagined, the process is quite expensive to operate and requires a low cost source of magnesium chloride solution and a commercial use for the hydrochloric acid by-product. In some processes, this hydrochloric acid is recycled to leach a magnesium mineral and obtain the magnesium chloride.

The other significant commercial process is referred to as the hydrothermal process and is an adaptation of the precipitation of magnesium hydroxide by adding a base to a magnesium chloride solution. In this process, ammonia or pure sodium hydroxide is the alkali source, and the poor morphology of the magnesium hydroxide precipitate is corrected by hydrothermal growth using an autoclave. Although there are important trace impurity effects, this type of process can give excellent products and was used for the first significant commercial production of fire retardant magnesium hydroxide. It tends to be even more expensive than the hydropyrolysis process however and, although still in commercial use, has largely been superseded by hydropyrolysis.

One variant of the hydrothermal process incorporates nickel into the precipitate in order to increase char formation. While very interesting results were published, and a commercial plant was opened in Japan, this does not appear to have been a commercial success.

6.2.2 Endothermic decomposition

This starts at about 300 °C and, unlike ATH, does not involve any intermediate phase. As mentioned earlier, surface treatments have been reported as affecting the onset of decomposition, with fatty acids reducing it and silica increasing it.

6.2.3 Grades of magnesium hydroxide

Most magnesium hydroxide flame retardants available today are either ground natural brucite, or products made by the hydropyrolysis process.

The natural brucite derived products typically contain from 85 to 95% magnesium hydroxide, most of the impurities are usually magnesium and calcium carbonates. Average particle size ranges from 1 to >15 μm and specific surface areas from 2 to 10 $m^2 g^{-1}$.

The synthetic grades made by hydropyrolysis are much purer, with >99% magnesium hydroxide (excluding any coating that may be present). The average particle size is usually from 0.5 to 2.5 μm, and the specific surface area is from 2 to 2 $m^2 g^{-1}$. As mentioned earlier, there are several reports of nanosize grades of magnesium hydroxide, but to date, these have not achieved commercial significance.

6.2.4 Surface treatments

Pretreatment with surface coatings is much more widely used with magnesium hydroxides than with aluminum hydroxides; with probably about half of the commercial products being pretreated in some way. This is partly due to the higher surface reactivity and partly reflects the different polymer applications involved.

The principal conventional surface treatments used are fatty acids such as stearates and oleates, and organosilanes. Proprietary treatments are also commonly used.

6.2.5 Metal doping

It is relatively easy to introduce other metals into solid solution in magnesium hydroxide, and this has been widely studied as a way to improve the fire retardant product. Unfortunately, the information on this is in the patent rather than scientific literature [60]. According to the patent claims, doping with metals such as nickel has a number of useful effects. These are listed below.

Alteration of decomposition temperature. This can be lowered significantly depending on the metal dopant and the amount used. It is claimed, with some level of justification, that the fire retardant performance of magnesium hydroxide can be improved (in some tests at least) by lowering the onset of decomposition.

Improved acid resistance. Magnesium hydroxide is quite reactive chemically and readily solubilized by even mild acids such as carbonic. While this does not seem to have been of great concern in Europe and the United States, this issue has attracted some criticism in Japan. The incorporation of metal dopants appears to help significantly with this problem.

Increased char formation and hence improved fire retardant performance. This is the principal reason for adding the dopant allowing lower filler levels to be used.

7. Other hydroxides and related materials

7.1 Calcium hydroxide

Calcium hydroxide is the most available and lowest cost of the metal hydroxides and has a very significant endothermic decomposition at a temperature which, although a little on the high side, ought to be suitable for some polymers. As such, it has attracted some interest as a possible flame retardant additive.

Despite these apparently favorable properties, conventional calcium hydroxide does not appear to have a sufficient fire retardant effect. Ashley and Rothon [19] have reported testing at 125 phr in crosslinked EVA where it only achieved an oxygen index of 24.5% compared to that of aluminum hydroxide at 29.0%. Even less effect was found in polypropylene and polyamide. The polyamide result was especially disappointing as its decomposition temperature is similar to that of calcium hydroxide, so a good effect was expected. A strong ash was observed in the EVA case, but analysis showed this to be calcium carbonate not oxide. This is somewhat surprising in view of the same workers reporting that precipitated calcium carbonate formed an oxide ash in the same test and polymer. Nishimoto and coworkers also reported the lack of any significant effect in polypropylene [61].

There have been claims that calcium hydroxide can be made into an effective flame retardant by making it into a composite hydroxide with other metals in solid solution [62]. This does not seem to have been converted into any significant commercial production though.

7.2 Boehmite

Boehmite has the composition AlO(OH) and has already been mentioned as an intermediate in the decomposition of aluminum hydroxide and is being promoted commercially as a flame retardant. As demonstrated in Table 6 and Figure 1, the onset of decomposition is significantly higher while the enthalpy of decomposition and volatile release are significantly less than for aluminum and magnesium hydroxides. The fire retardant effect is thus not expected to be as marked as these other hydroxides in most low-temperature decomposing polymers and boehmite finds most application where the high thermal stability is important. A good example is in thermoset printed circuit boards where soldering can create significant problems for other environmentally friendly flame retardants.

7.3 Basic magnesium carbonates

Several hydrated double salts of magnesium hydroxide and magnesium carbonate are known and have some place in the history of flame retardants. The only one of any significance today is hydromagnesite ($3MgCO_3 \cdot Mg(OH)_2 \cdot 3H_2O$), which is exploited in both natural (mineral) form and as a synthetic product. The first mention of the flame retardant use of hydromagnesite was at least as early as 1967 in a patent to Bakelite Xylonite Limited [63].

Initially, the synthetic form was of most interest and was produced commercially in the United States. This form has quite a high specific surface area and oil absorption and cannot be added at very high loadings. Its main use is thus as a smoke suppressor in PVC formulations.

More recently, exploitable natural deposits have been discovered, mainly in Greece and Turkey. In these deposits, the hydromagnesite is mixed with another mineral (huntite $3MgCO_3 \cdot CaCO_3$), and it is these mixtures that are marketed for flame retardant applications. The particle shapes are more suitable for polymer use than is synthetic hydromagnesite and high enough addition levels to achieve good fire retardant performance are possible.

Hydromagnesite decomposition is quite complex involving at least two steps and the evolution of water and carbon dioxide, with a possible intermediate conversion to magnesium carbonate depending on the conditions [64,65]. There is some uncertainty over the role, if any, that the huntite plays in the natural product. Some attempt to unravel this was reported by Doyle and coworkers [66] and more recently by Hollingbery [65].

7.4 Layer double hydroxides

Magnesium hydroxide is a layer structured mineral, and there are a family of double hydroxides, where the magnesium hydroxide layers alternate with trivalent metal hydroxide layers. Water, plus anions such as carbonate, is found between the layers. When the trivalent metal is aluminum, the double hydroxides are known as hydrotalcites and have the general formula:

$$Mg_{1-x} \cdot Al_x(OH)_2, \text{etc.}$$

Hydrotalcites have large decomposition endotherms occurring over a range of temperatures that makes them potential fire retardant additives for polymers. Despite some patent literature, there does not seem to have been much commercial activity, possibly because they are more expensive to produce than aluminum hydroxide and do not offer a great deal more in terms of flame retardant performance. They have, however, found significant use as stabilizers for PVC. As indicated earlier, there is also growing interest in them for nanoparticle applications, and this may revive interest in their use as flame retardants. More information on their potential use as flame retardants can be found in Ref. [67].

8. CONCLUSIONS

Certain metal hydroxides, notably those of aluminum and magnesium and related compounds, are effective flame retardants for polymers and also possess good green credentials relative to many other types of flame retardant. In particular, they have a low toxicity and give low smoke and low corrosion formulations. As a result, they have a very significant share of the current market and are expected to continue to be one of the fastest growing sectors.

Since they generally require use at high filler loadings to confer adequate flame retardancy, this can result in a deleterious effect on mechanical properties and processability. However, these drawbacks can be overcome by judicious application of surface treatments, control of particle size, particle size distribution and shape, and by combination with other flame retardants in order to lower overall filler content. In this regard, nanoclays also possess good green credentials and are useful synergists with hydrated fillers and many other flame retardants. Halloysite is a recent newcomer to the nanoclay family and is showing some promise.

References

[1] Rothon R. Minerals flame again. Ind Min; August 2011:53–7.

[2] Hughes P, Jackson GV, Rothon RN. Particle morphology effects on the performance of PMMA filled with aluminium hydroxide in a variety of fire tests. Makromol Chem Macromol Symp 1993;74:179–83.

[3] Herbert MJ. Paper presented at cables and fire protection conference. ERA Technology Limited; January 1992.

[4] Rothon RN, Hornsby PR. Flame retardant effects of magnesium hydroxide. Polym Degrad Stab 1996;54:383–5.

[5] Landemaine B. [Ph.D. thesis]. Universite du Maine; Submitted December 2001; 119.

[6] O'Neill T. The heat & flame response of flame retardant halogenated and non-halogenated polymers and compounds. In: British Plastics Federation, editor. Proceedings of flame retardants '90. Elsevier Applied Science; 1990, ISBN 1-85166-461-0. pp. 43–52.

[7] Camino G, Maffezzoli A, Braglia M, De Lazzaro M, Zammarano M. Effect of hydroxides and hydroxycarbonate structure on fire retardant effectiveness and mechanical properties in ethylene–vinyl acetate copolymer. Polym Degrad Stab 2001;74: 457–64.

[8] Ziegan G, Hondesberg H. Magnesium hydroxide: effective flame retardant for rubber and thermoplastics. In: The Plastics and Rubber Institute, editor. Proceedings of flame retardants '92. Elsevier Applied Science; 1992, ISBN 1-85166-758-X. pp. 120–3.

[9] Rothon RN. Chapter 6. In: Rothon RN, editor. Particulate filled polymer composites. 2nd ed. Rapra Technology Limited; 2003, ISBN 1-85957-382-7. pp. 277–90.

[10] Hancock M, Rothon RN. Chapter 2 'Principal types of particulate fillers'. In: Rothon RN, editor. Particulate filled polymer composites. 2nd ed. Rapra Technology Limited; 2003. pp. 85–7.

[11] Dando NR, Clever TR, Pearson A, Stinson JM, Kolok PL, Martin ES. Aluminium trihydroxide (ATH) as a filler for polymer composites: improvements in thermal stability by controlled precipitation. In: Proceedings from 50th annual technical conference. Composite Institute, Society of Plastics Industry Inc, Session 1-D; 1995. pp. 1–4.

[12] Halloysite handbook. New York, USA: Applied Chemicals Inc.; 2012.

[13] Hornsby PR, Watson CL. A study of the mechanism of flame retardancy and smoke suppression in polymers filled with magnesium hydroxide. Polym Degrad Stab 1990;30: 73–87.

[14] Hornsby PR, Watson CL. Mechanistic aspects of smoke suppression and fire retardancy in polymers containing magnesium hydroxide filler. Plast Rubber Process Appl 1989; 11:45–51.

[15] Hornsby PR, Mthupha A. Analysis of fire retardancy in magnesium hydroxide-filled polypropylene composites. Plast Rubber Compos Process Appl 1996;25:347–55.

[16] Hull TR, Witkowski A, Hollingbery L. Fire retardant action of mineral fillers. Polym Degrad Stab 2011;96:1462–9.

[17] Hornsby PR, Wang J, Rothon R, Jackson G, Wilkinson G, Cosstick K. Thermal decomposition behaviour of polyamide fire retardant compositions containing magnesium hydroxide filler. Polym Degrad Stab 1996;51:235–49.
[18] Rychlý J, Vesely K, Gal E, Kummer M, Jancar J, Rychla L. Use of thermal methods in the characterization of the high-temperature decomposition and ignition of polyolefins and EVA copolymers filled with $Mg(OH)_2$, $Al(OH)_3$ and $CaCO_3$. Polym Degrad Stab 1990;30:57.
[19] Ashley RJ, Rothon RN. Use of inorganic fillers to reduce the flammability of polymers. Plast Rubber Compos 1991;15:19–21.
[20] Rothon RN. Chapter 2. In: Rothon RN, editor. Particulate filled polymer composites. 2nd ed. Rapra Technology Limited; 2003, ISBN 1-85957-382-7. p. 90.
[21] Herbert MJ. Cone calorimetry evaluation of ATH filled EVA. In: Proceedings of flame retardants '94. Interscience Communications; 1994, ISBN 0-9516320-5-1. pp. 59–72.
[22] Zhang Q, Tian M, Wu Y, Lin G, Zhang L. Effect of particle size on the properties of $Mg(OH)_2$-filled rubber composites. J Appl Polym Sci 2004;94:2341–6.
[23] Mishra S, Sonawane SH, Singh R, Bendale A, Patil K. Effect of nano-$Mg(OH)_2$ on the mechanical and flame-retarding properties of polypropylene composites. J Appl Polym Sci 2004;94:116–22.
[24] Qiu L, Xie R, Ding P, Qu B. Preparation and characterization of $Mg(OH)_2$ nanoparticles and flame retardant property of its nanocomposites with EVA. Compos Struct 2003;62: 391–5.
[25] Okobo N, Okumura H, Okoshi M, Hamada H. Flame retardancy of EVA/nano magnesium hydroxide hybrid. In: Proceedings from additives 2004, Clearwater Beach, Florida; 22–24h March, 2004.
[26] Okoshi M, Nishizawa H. Flame retardancy of nanocomposites. Fire Mater 2004;28: 423–9.
[27] Wang X, Zhang Q. Effect of hydrotalcite on the thermal stability, mechanical properties, rheology and flame retardance of poly(vinyl chloride). Polym Int 2004;53:698–707.
[28] Torno O. Paper 19, Nanocomposites based on boehmite alumina. In: Proceedings from conference on functional fillers for plastics, 15–17 September. Hamburg: Interscience; 2004.
[29] Lawson DF, Kay EL, Roberts DT. Mechanism of smoke inhibition by hydrated fillers. Rubber Chem Technol 1975;48:124.
[30] Moseman M, Ingham JD. Smoke properties of highly filled ethylene-propylene–diene terpolymer rubbers. Rubber Chem Technol 1978;51:970.
[31] Hirschler MM, Thevaranjan TR. Effects of magnesium oxide/hydroxide on flammability and smoke production tendency of polystyrene. Eur Polym J 1985;21:371.
[32] Krylov OV. Catalysis by non-metals. London: Academic Press; 1970.
[33] McKee DW. The catalyzed gasification reactions of carbon. In: Walker PL, Thrower PA, editors. Chemistry and physics of carbon, vol. 16. New York: Marcel Dekker; 1981.
[34] Delfosse L, Baillet C, Brault A, Brault D. Combustion of ethylene-vinyl acetate copolymer filled with aluminium and magnesium hydroxides. Polym Degrad Stab 1989;23:337.
[35] Baillet C, Delfosse L. The combustion of polyolefins filled with metallic hydroxides and antimony trioxide. Polym Degrad Stab 1990;30:89–99.
[36] Rothon RN. Chapter 6. In: Rothon RN, editor. Particulate filled polymer composites. 2nd ed. Rapra Technology Limited; 2003, ISBN 1-85957-382-7. pp. 290–4.

[37] Kirschbaum GS. Recent developments in ATH and magnesium hydroxides—a challenge to traditional materials. In: Proceedings of the 1995 fall conference of the fire retardant chemical association. Rancho Mirage, CA: Technomic Publishing Co.; October 29–November 1, 1995. p. 145.
[38] Kirschbaum GS. Halogenfreier flammschutz. Kunststoffe 1989;79:62–4.
[39] Staendeke H. In: Proceedings of the spring conference of the fire retardant chemical association, vol. 32. Grenelefe, FL: Technomic Publishing Co.; 1988. March 20–23.
[40] Weil EW, Lewin M, Lin HS. Enhanced flame retardancy of polypropylene with magnesium hydroxide, melamine and novolac. J Fire Sci 1998;16:383–404.
[41] Imahasi T, Okada A, Abe T. Flame retardant aid, flame retardant and flame-retardant composition U.S. Patent 5,583,172; December 10, 1996.
[42] Zhu W, Weil ED. A paradoxical flame-retardant effect of nitrates in ATH-filled ethylene-vinyl acetate copolymer. J Appl Polym Sci 1995;56:925–33.
[43] Baggaley RG, Hornsby PR, Yahya R, Cusack PA, Monk AW. The influence of novel zinc hydroxystannate-coated fillers on the fire properties of flexible PVC. Fire Mater 1997;21:179.
[44] Cusack PA, Hornsby PR. Zinc hydroxystannate/zinc stannate-coated fillers: novel flame retardants and smoke suppressants for polymeric materials. J Vinyl Addit Technol 1999; 5:21–30.
[45] Mohai M, Tóth A, Hornsby PR, Cusack PA, Cross M, Marosi G. XPS analysis of zinc hydroxystannate coated fillers. Surf Interface Anal 2002;34:735–9.
[46] Namike Y, Kato Y, Kitano Y, Kuriso H, Yokata Y. Flame retardant polyolefin compound having low smoking and toxicity. US Patent No. 5,654,356; August 5, 1997.
[47] Miyata S, Imahasi T. Fire-retardant resin composition, US Patent 5,094,781; March 10, 1992.
[48] Gilman JW, Jackson CL, Morgan AB, Harris R, Maniac E, Giannelis EP, et al. Flammability properties of polymer layered-silicate nanocomposites. In: Proceedings of flame retardants 2000. Interscience Communications; 2000, ISBN 0-9532312-4-0. pp. 49–68.
[49] Beyer G. Nano-composites—a new flame retardant system for polymers. In: Proceedings of flame retardants 2002. Interscience Communications; 2002, ISBN 0-9541216-0-0. pp. 209–16.
[50] Herbiet R. Mineral flame retardants; market outlook and latest developments. In: Paper 4, proceedings of high performance fillers 2005. Cologne, Germany: Rapra Technology; March 2005, ISBN 1-859-57-500-5.
[51] Kotian V. Optimised aluminium hydroxide grades with enhanced properties. Minerals in compounding conference, Atlanta, GA, USA; 2010.
[52] Rothon RN, Liauw CM, Lees GC, Schofield WCE. Magnesium hydroxide filled EVA: the effects of filler surface modification. J Adhesion 2002;78:603–28.
[53] Frood H, Alger HP. GB Patent 183,922; 1921.
[54] General Electric Company. GB Patent 831,490; 1960.
[55] Connolly WJ, Thornton AM. Proceedings of the SPI 20th reinforced plastics technology and management conference, Chicago, IL, USA; 1965.
[56] Nelson GL. J Fire Flammabl 1974;5:125.
[57] Sobolev I, Woychesin EA. In: Katz HS, Milewski JV, editors. Handbook of fillers for plastics. New York, USA: Van Nostrand Reinhold; 1987, ISBN 0-442-26024-5. pp. 294–6.

[58] Dando NR, et al. Aluminium trihydroxide (ATH) as a filler for polymer composites. In: Proceedings of SPI Composites Institute 51st annual conference, paper 19-A. Cincinnati, OH, USA; 1996.
[59] Ashton HC. In: Xanthos M, editor. Functional fillers for plastics. Wiley −VCH; 2010, ISBN 978-3-527-32361-6. pp. 319−22.
[60] Kurisu H, et al. Tateho Chemical Industries Co. Ltd. US Patent 5,766,568; 1998.
[61] Nishimoto K, Muraida T, Murata T. Natl Tech Report Jpn 1975;21(3):367−80.
[62] Miyata S. US Patent 5,326,548; 1992.
[63] Millane JJ. Bakelite Xylonite Limited, British Patent BP 1,213,625; 1967.
[64] Rothon RN. Chapter 2. In: Rothon RN, editor. Particulate filled polymer composites. 2nd ed. Rapra Technology Limited; 2003, ISBN 1-85957-382-7. pp. 91−5.
[65] Hollingbery LA, Hull TR. The fire retardant behaviour of huntite and hydromagnesite—a review. Polm Degrad Stab 2010;95(12):2213−25.
[66] Doyle M. Non-halogentated flame retardant for polypropylene. In: Proceedings of flame retardants '94. London: Interscience Communications; 1994, ISBN 0-9516320-2-7. pp. 193−202.
[67] Lefebre J, Le Bras M, Bourbigot S. In: Le Bras M, et al., editors. Chapter 3, Fire retardancy of polymers. R.S.C; 2005, ISBN 0-85404-582-1.

CHAPTER

10

Modeling the Endothermic Decomposition of Hydrated Solids

J.E.J. Staggs

University of Leeds, Leeds, United Kingdom

CHAPTER OUTLINE

1. Introduction

The recent environmental concerns over halogenated flame retardants (FRs) and toxicity (and smoke production in general) have provided considerable momentum to the search for new, greener FRs. Interest has naturally refocused on hydrated mineral fillers, particularly in conjunction with other additives. Metal hydroxides such as alumina trihydroxide (ATH) and magnesium hydroxide (MH) have long been known to impart FR properties to host polymers, incurring no toxic or smoke production overhead (smoke, in fact, is reduced). These materials act as FRs mainly through their endothermic liberation of water on heating. They also have an added benefit of leaving behind a protective microporous layer of the metal oxide after dehydration. Hornsby [1] neatly summarizes the FR properties of hydrated mineral fillers and also briefly explores their compatibility in polymeric systems.

Polymer Green Flame Retardants. http://dx.doi.org/10.1016/B978-0-444-53808-6.00010-X

More recently, there have been a number of authors who have explored using ATH or MH in conjunction with other fillers in specific polymer systems in order to obtain a beneficial FR effect. For example, Zhang-Yu Wang et al. [2] looked at a combination of ATH with melamine and Novolac resin (originally called Novolak, a phenol–formaldehyde resin) in polyoxymethylene using a variety of quantitative techniques. Their cone calorimeter results showed a significant reduction in heat release rate and observations of the sample showed the development of a compact char. Gul et al. [3] studied a combination of MH with sepiolite (a clay mineral that is a complex of magnesium silicate) in a low-density polyethylene (PE) host, although unfortunately their flammability measurements were limited to LOI.

In a slightly different vein, Bellayer et al. [4] reported results for low-density PE filled with 50% by weight of calcium carbonate, magnesium carbonate, and talc, either treated with stearic acid or untreated. Their cone calorimeter results showed significant decrease in heat release rate for the untreated fillers and even better performance when the fillers were coated with stearic acid. In fact, they report intumescent behavior for the $CaCO_3$–stearic acid combination. The FR mechanism of these fillers remains poorly understood, but the effect of filler particle shape and size are thought to be important.

Given the re-emergence of mineral fillers as FRs, it is important to have a strong theoretical model against which comparisons may be made. Of especial importance is the ability to then use the model to test claimed observed phenomena such as synergistic interactions. For example, is there some simpler thermal or thermochemical explanation for the observation? In this chapter we consider various modeling approaches to problems involving the endothermic decomposition of solids. Starting with a discussion of decomposition kinetics, the centrally important characteristic kinetic temperature (CKT) and characteristic kinetic temperature range (CKTR) are defined. We shall see that reformulating kinetic rate equations in terms of the CKT and CKTR yields a more intuitive parameter set rather than the traditional Arrhenius parameters, which consequently makes parameter estimation easier. We then progress to develop statements for conservation of energy in simple thermally thin situations and also when temperature gradients are not negligible. These, combined with statements for conservation of mass, form the basis of a mathematical model that is discussed at some length in the penultimate section. The final model is valid for endothermic decomposition of a filled polymer or a hydrated solid. It may also be extended to hydrated mineral fillers in a polymer host.

2. Decomposition kinetics

2.1 Introduction

Thermal decomposition of a solid is usually quantified in terms of one or more *rate equations*, which are ordinary differential equations expressing the rate of change of a mass fraction (or concentration) in terms of temperature. A rate equation may be assigned to a particular conversion step in a decomposition process, where some chemical reaction occurs, or may also be used to model a gasification step where volatile species are generated. Consider the simplest possible case, where a solid decomposes under the action of heat to produce gas in a single step. Let $\mu = m/m_0$ be the ratio of sample mass m to initial mass m_0. A rate equation describing the gasification process is anything of the form $d\mu/dt = f(\mu,T)$, where t is the time and T is the absolute temperature. Traditionally, two critical assumptions are usually employed:

- The function f may be expressed as $f(\mu,T) = -k(T)F(\mu)$, where $k(T)$ is a function of temperature only (with units s^{-1}) and $F(\mu)$ is a dimensionless function of mass fraction only.
- The function k is usually assumed to be of the same form as the Arrhenius rate function $k(T) = A\exp(-T_A/T)$.

The parameters A and T_A in the definition of k are referred to as the *pre-exponential factor* (with units s^{-1}) and the *activation temperature* (with units K), respectively. Note that T_A is sometimes replaced by E_A/R, where E_A is the *activation energy* and R is the universal gas constant. It should be appreciated that there is little or no physical justification for these assumptions regarding the functional form of f. The parameters on which f depends are invariably found from a curve-fitting process where the optimal solution of the rate equation is estimated from comparison with appropriate experimental data.

Taking the separable form of f admits a simple implicit solution of the rate equation: $\int_{\mu}^{1} d\mu/F(\mu) = \int_{0}^{t} k(T)dt$. The form $F(\mu) = \mu^n$ is often assumed for thermal decomposition of solids and if this is adopted, then it follows from the separable solution that

$$\mu = \begin{cases} \exp(-K), & n = 1 \\ [1-(1-n)K]^{\frac{1}{1-n}}, & n \neq 1 \end{cases} \quad \text{where } K = \int_0^t k(T)dt \tag{1}$$

For isothermal applications, $K = k(T)t$, but if a linear heating regime is imposed, then $dt = dT/H$, where H is the constant heating rate, and so for the Arrhenius rate function

$$K = \frac{A}{H}\int_{T_0}^{T} \exp\left(-\frac{T_A}{T}\right)dT \tag{2}$$

Here T_0 is the initial temperature. For most applications, the activation temperature T_A is large enough such that $T_A/T_0 \gg 1$, implying that $\exp(-T_A/T_0)$ is negligible. Consequently, we may safely assume that $\int_0^{T_0} \exp(-T_A/T)dT$ is also negligible and so the lower bound in the integral in Eqn (2) may be replaced by 0 without significant loss of accuracy. If z is defined by $z = T_A/T$, then integrating by parts gives $K = AT_A e^{-z}[1 - ze^z E_1(z)]/(Hz)$, where $E_1(z)$ is the *exponential integral* [5] defined by $E_1(z) = \int_z^{\infty} \exp(-z)\mathrm{d}z/z$. Furthermore, if T_A is large enough such that $T_A/T \gg 1$ for all temperatures of relevance, then the asymptotic relation $z\exp(z)E_1(z) = 1 - 1/z + O(1/z^2)$ may be employed [5] and so

$$K \approx \frac{AT^2}{T_A H} \exp\left(-\frac{T_A}{T}\right) \tag{3}$$

The kinetic parameters n, A, and T_A are normally estimated from experimental thermogravimetric (TG) data. There are a number of methods available for this, ranging from Freeman & Carroll's method [6] to techniques based on nonlinear optimization where the kinetic parameters are adjusted to optimize the fit of a numerical solution of the rate equation to observation (a selection of popular methods is summarized in Ref. [7]). Here, for illustration an optimization technique for $F(\mu) = \mu^n$ is presented based on initially finding a reaction order n that gives a best correlation coefficient in a constant heating rate TG experiment (the *maximum* r^2 *method*). Let $y = \ln(-\mu^{-n}\mathrm{d}\mu/\mathrm{d}T)$. Then since $y = \ln(A/H) - T_A/T$, it follows that a plot of y against $x = -1/T$ should be linear with gradient T_A and y-intercept $\ln(A/H)$. Let the correlation coefficient for the (x,y) plot be r. Then the closer r^2 is to 1, the more linear the plot. It therefore follows that if we choose the reaction order n to minimize the quantity $S = 1 - r^2$, then the fit will be as linear as possible, giving the best possible fit of the kinetic model to the experimental data. Therefore, in order to find the optimal kinetic parameters, the following scheme is suggested.

1. Fix some initial value of n (say $n = 1.0$).
2. From the experimental data, compute $x = -1/T$ and $y = \ln(-\mu^{-n}\mathrm{d}\mu/\mathrm{d}T)$.
3. Fit a straight line to the (x,y) data and compute the y-intercept, gradient, and r^2 value for the fit. The y-intercept will be an estimate of $\ln(A/H)$ and the gradient will be an estimate of T_A.
4. Compute $S = 1 - r^2$.
5. Repeat steps 2–4 for a range of n-values and identify the optimal value that gives the smallest S. The estimates of $\ln(A/H)$ and T_A at the optimal n then give the best possible kinetic parameters.

It is relatively straightforward to use a spreadsheet to automate this process and also compute the optimal n-value using a built-in minimization function.

The graph in Figure 1 shows the plots of S (multiplied by 10^4), $\ln(A/H)$, and T_A derived as above for a constant heating rate TG experiment (at $H = 10\ \mathrm{K\ min^{-1}}$) on a sample of water over the temperature range 20–100 °C. The S-curve clearly shows a minimum close to $n = 0.8$. Further calculation shows that the optimal reaction order is $n = 0.83$ with $T_A = 4165$ K and $\ln(A/H) = 6.677$. Figure 2 shows the

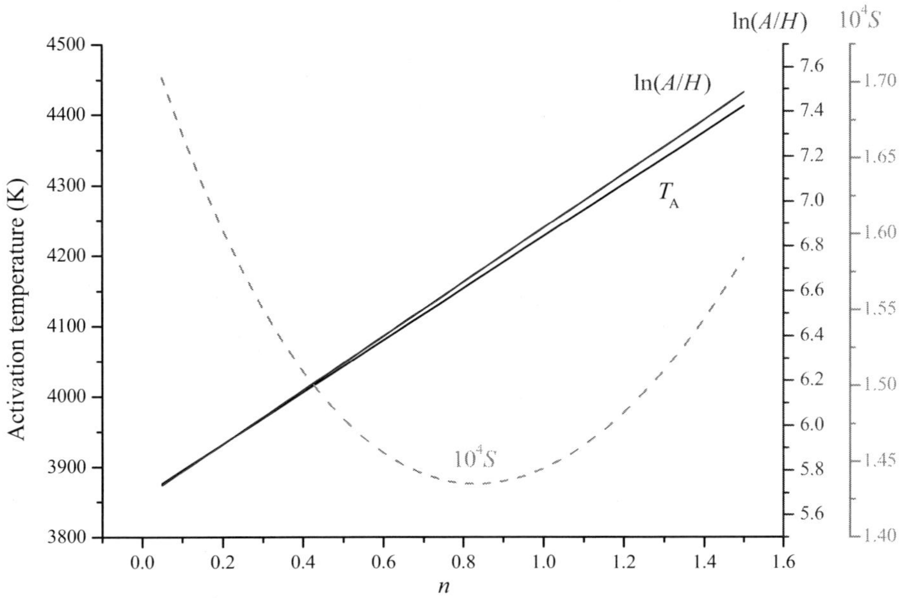

FIGURE 1

Derived kinetic parameters for water.

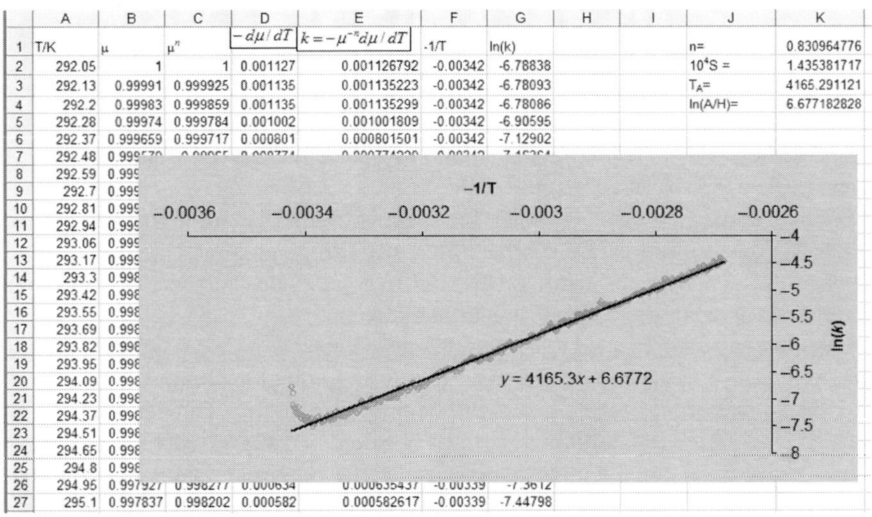

FIGURE 2

Spreadsheet used to determine optimal kinetic parameters.

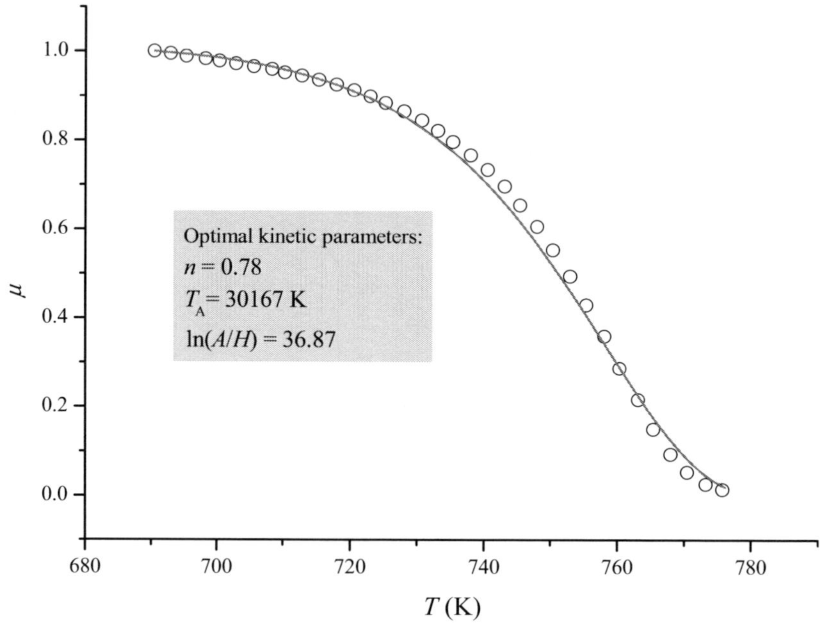

FIGURE 3

Maximum r^2 method applied to thermogravimetric data for polyethylene.

MS-Excel spreadsheet used for the calculations. The optimal n-value was obtained by using the Solver built-in function to minimize S (in cell K2) by adjusting n (in cell K1). The correlation coefficient (used to find S) is computed using the built-in function COREL. The activation temperature in cell K3 is found from the ratio of the covariance of the (x,y) data (columns F and G) divided by the variance of x using the built-in functions COVAR and VARP. The value for $\ln(A/H)$ is found from $\bar{y} - T_A\bar{x}$, where $\bar{x}$ is the average of x (found using the AVERAGE function) and $\bar{y}$ is the average of y.

Figure 3 shows the results of applying the maximum r^2 method to constant heating rate TG data for a standard sample of PE in a nitrogen atmosphere. The symbols represent the experimental data and the curve corresponds to the calculated mass fraction values using the optimal kinetic parameters.

If isothermal data are available, then there is no need to impose a specific form on $F(\mu)$, as it can be constructed directly. If the rate equation $d\mu/dt = -k(T)F(\mu)$ is valid, then a plot of $-d\mu/dt$ against μ will be proportional to F. One has to be careful about precisely how the derivative $-d\mu/dt$ is calculated from experimental data as there is likely to be considerable noise present. If isothermal data are available for a number of different temperatures, then this may be used to check the validity of the assumption that $f(\mu,T) = -k(T)F(\mu)$. Plots of $-d\mu/dt$ against μ for different temperatures will all have the same shape if this assumption is true. However, if they do

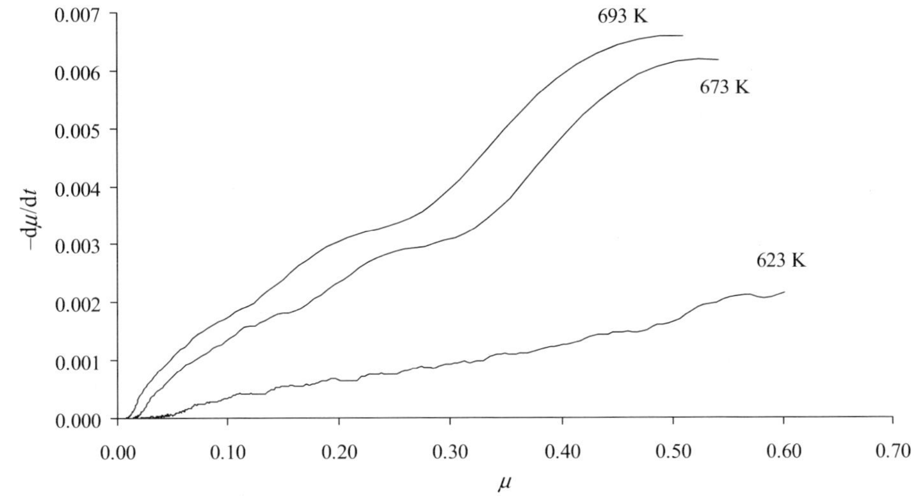

FIGURE 4

Isothermal thermogravimetric data for polyethylene.

not have the same shape then the assumed form of f is definitely incorrect. Using the Savitsky–Golay technique detailed in Ref. [8] to determine the mass fraction derivative, Figure 4 shows the plot of the mass fraction derivative against mass fraction for isothermal TG experiments on a standard sample of PE conducted in a nitrogen atmosphere. The curves at 693 and 673 K appear to be convincingly similar in shape; however, the 623 K curve lacks the characteristic shape of the other two, shedding doubt on the assumed functional dependence of the rate function.

2.2 The characteristic kinetic temperature

The Arrhenius rate equation for a single-step degradation process in a linear heating regime,

$$d\mu/dT = -A\mu^n \exp(-T_A/T)/H, \tag{4}$$

implies that the curve of μ versus T will have a characteristic ogive shape. Given this fact, it is possible to identify a temperature and temperature range that characterizes the curve, as Figure 5 shows. The *characteristic kinetic temperature* T_k is defined as the temperature at which the mass fraction μ is 0.5. Now consider a tangent drawn on the TG curve at T_k produced so that it meets the temperature axis at one end (at $T = T_1$ say) and the line $\mu = 1$ at the other (at $T = T_2$ say). The *characteristic kinetic temperature range* ΔT_k is then defined as the temperature difference $T_1 - T_2$.

If the reaction order n is known, then it can be shown [9] that there is a unique correspondence between the characteristic kinetic parameters and the Arrhenius parameters determined from the solutions of the equations

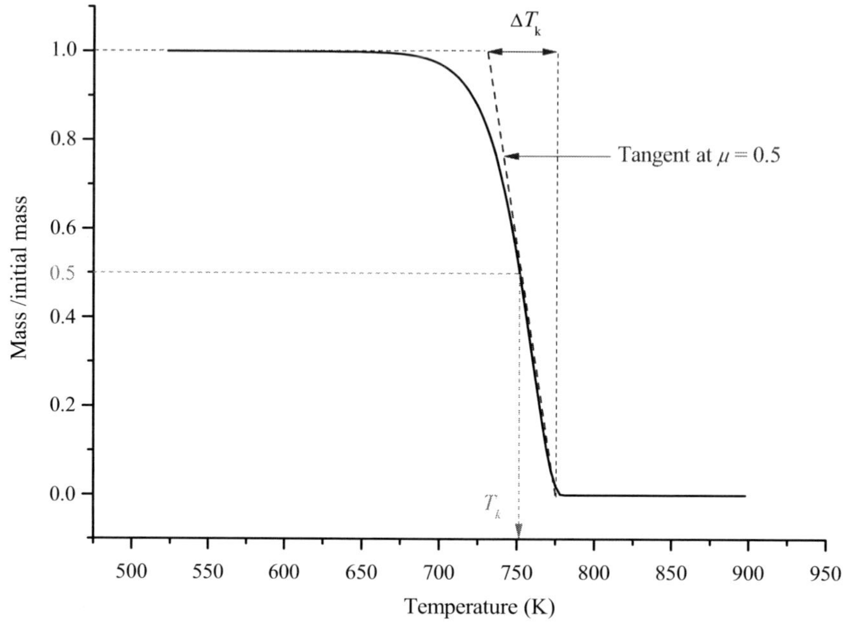

FIGURE 5

Characteristic kinetic temperature and temperature range.

$$1 - z_k e^{z_k} E_1(z_k) = \varphi_n \Delta T_k / T_k, \quad A/H = 2^n e^{z_k} / \Delta T_k, \tag{5}$$

where $z_k = T_A/T_k$, E_1 is the exponential integral defined in the preceding section and

$$\varphi_n = \begin{cases} \dfrac{2^{1-n} - 1}{2(1-n)} & n \neq 1, \\ \dfrac{\ln 2}{2}, & n = 1. \end{cases} \tag{6}$$

If $\Delta T_k / T_k \ll 1$, then it follows that z_k will be large and so the asymptotic relation for E_1 given above may be used to obtain the approximations

$$\frac{T_A}{T_k} \approx \frac{T_k}{\varphi_n \Delta T_k}, \quad \frac{A}{H} \approx \frac{2^n}{\Delta T_k} \exp\left(\frac{T_k}{\varphi_n \Delta T_k}\right) \tag{7}$$

The expressions above may be used with $n = 1$ to obtain convenient estimates for the kinetic parameters from TG curves. For example, from the data for PE in Figure 3, we have $T_k \approx 751.3$ K and $\Delta T_k \approx 52.5$ K. Substitution of these values into Eqn (7) gives $T_A \approx 31{,}009$ K and $\ln(A/H) \approx 38.0$, which compare well with the optimal Arrhenius parameters found above.

It is now apparent that a sufficient condition for the approximate solution to the rate equation for constant heating rates (Eqn (1) above with K given by Eqn (3)) to be valid is that T_A/T_k is large, which is equivalent to $\Delta T_k/T_k$ being small. Therefore, a good approximation for first-order decomposition with constant heating rate H is

$$\mu = \exp\left(-\frac{AT^2}{T_A H} e^{-T_A/T}\right), \tag{8}$$

provided that the ratio of characteristic kinetic temperature range (CKTR) to characteristic kinetic temperature (CKT) is small. In practical terms, this means that the constant heating rate TG curve should show a relatively sharp transition in the mass-loss region.

Another way of writing the expression for approximate mass fraction using the relations for the CKT and CKTR (Eqn (7)) is

$$\mu = \exp\left(-\frac{T^2 \ln 2}{T_k^2} e^{\beta_k(1-T_k/T)}\right), \quad \beta_k = \frac{2T_k}{\Delta T_k \ln 2} \tag{9}$$

The expression above naturally lends itself to a convenient method to estimate the Arrhenius parameters for a single-step decomposition process since the CKT and temperature range are easily estimable from a constant heating rate TG curve. One may simply compare the predicted mass fraction from Eqn (9) with the experimental curve and adjust T_k and ΔT_k until the curves are acceptably close. The Arrhenius parameters may then be estimated from Eqn (7) with $n = 1$. A more precise method would be to evaluate a residual error, defined in terms of the estimated mass fraction and the observed mass fraction, and to minimize this by adjusting T_k and ΔT_k.

Finally in this section, it is worth noting that the CKT and CKTR are a more natural parameter choice than the traditional Arrhenius parameters when quantifying thermal decomposition. Suppose that TG data is used at a predefined standard heating rate H^θ (such as 10 K min^{-1}) to determine the characteristic kinetic parameters, and let these be denoted by T_k^θ and ΔT_k^θ. Then the standard rate equation may be rewritten in terms of the new parameters as

$$\frac{d\mu}{dt} = -\frac{\mu^n}{t_k^\theta}\exp\left\{\beta_k^\theta\left(1-\frac{T_k^\theta}{T}\right)\right\}, \quad t_k^\theta = \frac{\Delta T_k^\theta}{2^n H^\theta}, \quad \beta_k^\theta = \frac{T_k^\theta}{\varphi_n \Delta T_k^\theta}. \tag{10}$$

Note that t_k^θ may be interpreted as a natural timescale for the decomposition process.

2.3 Parallel reactions: hydrated fillers

Hydrated mineral oxides such as alumina trihydrate or MH comprise a nonvolatile metal oxide together with chemically bound water. On heating, water is liberated leaving a porous metal oxide residue. If the residue yield is μ_s, a straightforward extension of the standard nth order Arrhenius rate equation may be used to model the decomposition:

$$H\mathrm{d}\mu/\mathrm{d}T = -A(\mu - \mu_s)^n \exp(-T_A/T). \tag{11}$$

Now setting $M = (\mu - \mu_s)/(1 - \mu_s)$, so that M varies between 1 and 0 as μ varies between 1 and μ_s, it follows that the rate equation for M is identical to that for a simple nonfilled solid with the modification that A is replaced by $A(1 - \mu_s)^{n-1}$. Hence, if the CKT and CKTR are evaluated at the mass-fraction midpoint $\mu = (1 + \mu_s)/2$ and using $\mu = \mu_s$ as the baseline rather than $\mu = 0$ (Figure 6), then provided that $\Delta T_k/T_k$ is small, the following expressions will be true approximately:

$$\frac{T_A}{T_k} \approx \frac{T_k}{\varphi_n \Delta T_k}, \quad \frac{A}{H} \approx \frac{2^n}{(1 - \mu_s)^{n-1} \Delta T_k} \exp\left(\frac{T_k}{\varphi_n \Delta T_k}\right). \tag{12}$$

When hydrated materials are used as flame retardant fillers in conjunction with a polymer, the simplest way to model the thermal decomposition of the composite is via a pair of parallel reactions. Implicit in this approach is the assumption that the filler and polymer do not interact during the decomposition. For simplicity, we shall assume that the reaction orders for both components are unity. Let the mass of filler

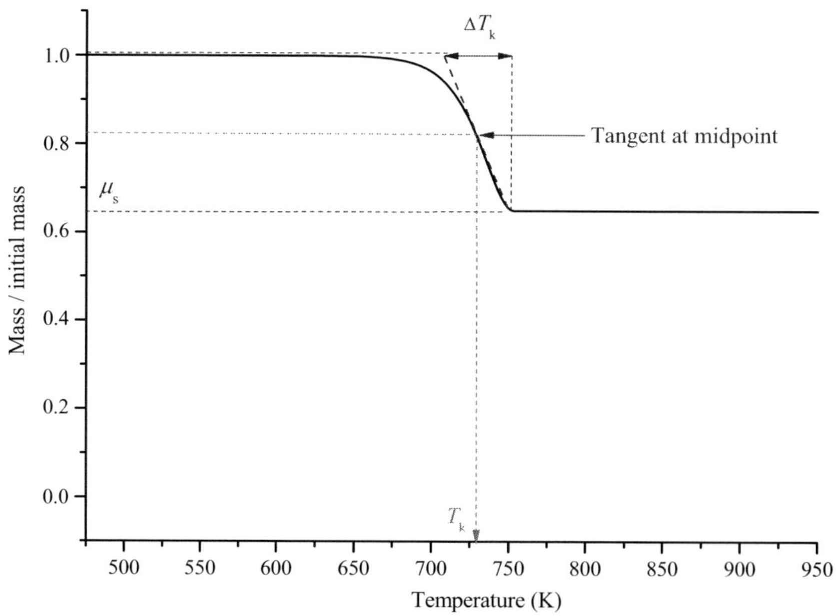

FIGURE 6

Characteristic kinetic temperature and temperature range for decomposition with finite residue.

be m_f, with solid residue mass m_s, and the mass of polymer be m_p. Then if the initial mass of the composite is m_0 and the component mass fractions are $\mu_j = m_j/m_0$ ($j = \text{f, p, s}$), the corresponding rate equations are

$$d\mu_f/dt = -k_f(\mu_f - \mu_s), \quad d\mu_p/dt = -k_p\mu_p, \tag{13}$$

where $k_j = A_j \exp(-T_{A,j}/T)$, $j = \text{f, p}$. If ϕ is the initial mass fraction of filler in the composite, then the initial conditions are $\mu_f(0) = \phi$ and $\mu_p(0) = 1 - \phi$. The total mass fraction of the composite, i.e. the ratio of composite mass to m_0, is $\mu = \mu_f + \mu_p$.

Under isothermal conditions, the solution of the rate equations is straightforward and we have that the composite mass fraction is

$$\mu = \mu_s + (\phi - \mu_s)e^{-k_f t} + (1 - \phi)e^{-k_p t}. \tag{14}$$

For constant heating rate conditions, the theory of preceding sections may be applied. Each rate equation is independent and so a CKT and CKTR may be associated with each decomposition step. Using the approximate solution (Eqn (8)), we have

$$\mu \approx \mu_s + (\phi - \mu_s)\exp\left(-\frac{A_f T^2}{T_{A,f}H}e^{-T_{A,f}/T}\right) + (1 - \phi)\exp\left(-\frac{A_p T^2}{T_{A,p}H}e^{-T_{A,p}/T}\right). \tag{15}$$

Naturally this approach can be extended easily to the case of a polymer that consists of N components, each of which decomposes independently via parallel reactions. In terms of the CKT parameters, let $T_{k,f}$ and $\Delta T_{k,f}$ denote the CKT and CKTR of the filler, respectively, and let $T_{k,p}^{(j)}$ and $\Delta T_{k,p}^{(j)}$ denote the CKT and CKTR of component j of the polymer, respectively. Then if ν_j is the mass fraction of component j of the polymer (so that $\sum_{j=1}^{N}\nu_j = 1$), the expression for total mass fraction becomes

$$\mu \approx \mu_s + (\phi - \mu_s)\exp\left(-\frac{T^2 \ln 2}{T_{k,f}^2}e^{\beta_{k,f}(1 - T_{k,f}/T)}\right)$$
$$+ (1 - \phi)\sum_{j=1}^{N}\nu_j \exp\left(-\frac{T^2 \ln 2}{\left(T_{k,p}^{(j)}\right)^2}e^{\beta_{k,p}^{(j)}\left(1 - T_{k,p}^{(j)}/T\right)}\right),$$

$$\beta_{k,f} = \frac{2T_{k,f}}{\Delta T_{k,f} \ln 2}, \quad \beta_{k,p}^{(j)} = \frac{2T_{k,p}^{(j)}}{\Delta T_{k,p}^{(j)} \ln 2}. \tag{16}$$

Figure 7 shows the constant heating rate TG data for ethylene vinyl acetate (EVA) filled with ATH (symbols). The mass derivative $-d\mu/dT$ shows that there are three main mass loss processes—the filler at ~525–600 K and the two steps

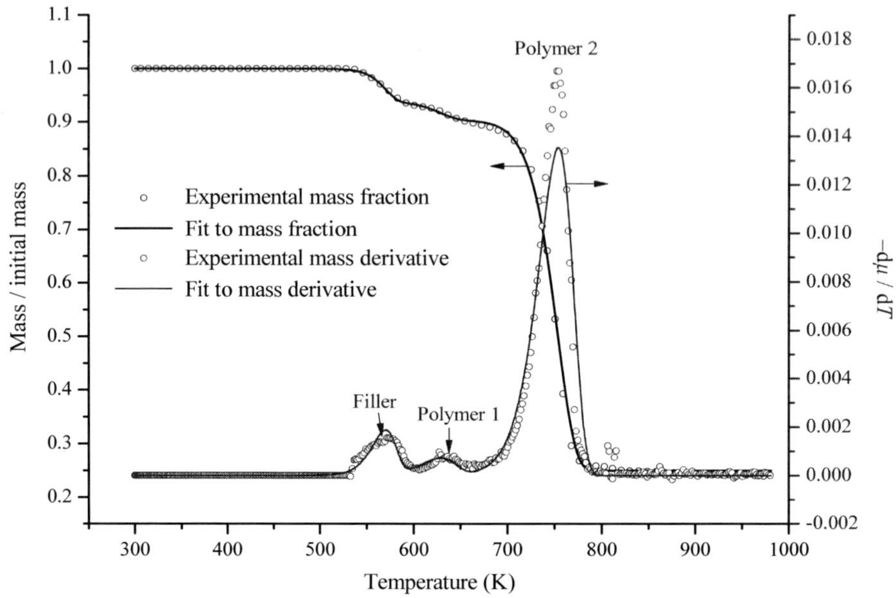

FIGURE 7

Application to constant heating rate thermogravimetric for ethylene vinyl acetate–alumina trihydroxide sample.

of the EVA decomposition at ~600–660 K and ~700–800 K. The solid curves correspond to the model fit given by Eqn (16) with $N = 2$ where the CKT parameters were obtained by minimizing the total residual between the fit and experimental data. Initial guesses for the CKT parameters were easily obtained from the mass fraction derivative curve. The optimized parameters are shown in Table 1.

Table 1 Optimal Characteristic Kinetic Temperature Parameters for Ethylene Vinyl Acetate–Alumina Trihydroxide Sample

	Characteristic Kinetic Temperature (K)	Characteristic Kinetic Temperature Range (K)	T_A (K)	ln(*A*/*H*)
Filler	565.4	36.6	25,176	41.6
Polymer step 1	622.8	53.9	20,761	30.0
Polymer step 2	747.6	53.0	29,766	36.6

3. Conservation of energy with no heat diffusion

3.1 Simple endothermic solid

In this section, the endothermic decomposition of a simple solid under the action of an external heat source is considered. General features of the decomposition process are discussed and an approximate solution is developed.

Consider the case of a thin sample of solid heated by an external heat flux. By "thin" here, we mean thin enough such that temperature gradients may be neglected. For simplicity, we consider a simple solid that undergoes an endothermic decomposition to liberate gas according to $\mathrm{d}\mu/\mathrm{d}t = -k(\mu - \mu_\mathrm{s})$, where $\mu = m/m_0$ is the ratio of mass to initial mass as before, μ_s is the yield of nonvolatile residue, and k is the Arrhenius rate function. Let external heat be supplied to the sample at a rate $\dot{q}$. Then in a small time step δt, the change in internal energy δU of the sample will be given by

$$\delta U = \dot{q}\delta t + \delta m c_\mathrm{g}(T - T_\mathrm{a}) + \delta m \Delta H \qquad (17)$$

Here δm is the change of mass of the solid in the time step, $\dot{q}\delta t$ is the external heat supplied to the solid, T_a is some reference temperature (taken as ambient temperature for convenience), $\delta m c_\mathrm{g}(T - T_\mathrm{a})$ is the heat lost convectively by the flow of gas (with specific heat capacity c_g) out of the solid and ΔH is the heat of the decomposition reaction in J kg^{-1}. Now, since $U = mc(T - T_\mathrm{a})$, where c is the specific heat capacity of the solid, it follows that the expression for conservation of energy can be written as the ordinary differential equation:

$$\mu c \mathrm{d}T/\mathrm{d}t = \dot{q}/m_0 + \left[(c_\mathrm{g} - c)(T - T_\mathrm{a}) + \Delta H\right]\mathrm{d}\mu/\mathrm{d}t. \qquad (18)$$

Equation (18) admits a quasi-steady solution (when the temperature is approximately constant) provided that

$$\frac{\mathrm{d}\mu}{\mathrm{d}t} = -\frac{\dot{q}/m_0}{(c_\mathrm{g} - c)(T - T_\mathrm{a}) + \Delta H}, \qquad (19)$$

i.e. when

$$k(T) \approx \frac{\dot{q}/(m_0 - m_\mathrm{s})}{(c_\mathrm{g} - c)(T - T_\mathrm{a}) + \Delta H}, \qquad (20)$$

where m_s is the mass of solid residue. Let the constant temperature during the approximately steady period be T_∞. The solution of Eqn (20) gives a good estimate of T_∞. For strongly endothermic reactions such that $\Delta H >> |c_g - c|(T_\infty - T_a)$ and constant $\dot{q}$, then the steady temperature is given in terms of the Arrhenius parameters by $\exp(J - T_\mathrm{A}/T_\infty) \approx 1$, where $J = \ln(A(m_0 - m_s)\Delta H/\dot{q})$. This implies that

$$T_\infty / T_\mathrm{A} \approx 1/J. \qquad (21)$$

The variation of quasi-steady temperature is shown in Figure 8.

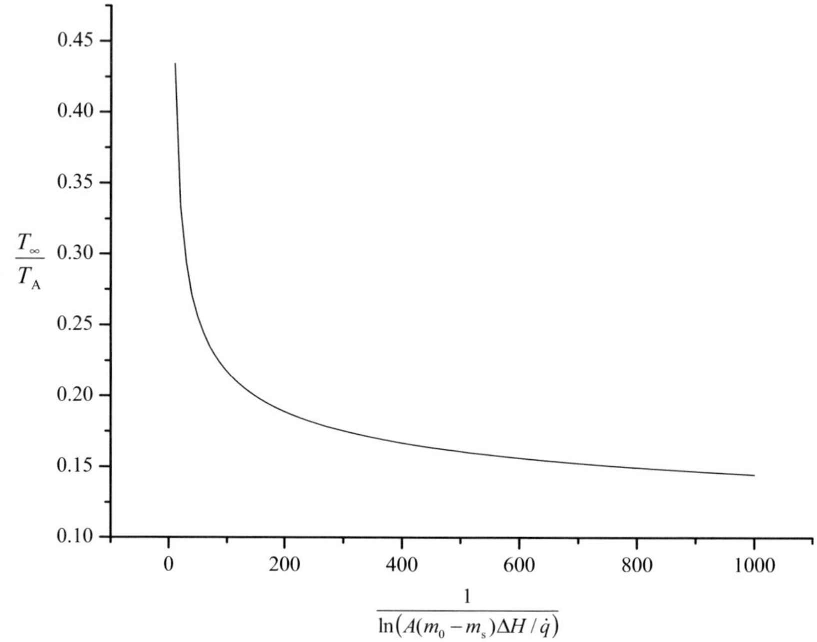

FIGURE 8

Approximate steady temperature for endothermic solid.

It is likely that $A >> \dot{q}/(m_0 - m_s)\Delta H$ and consequently $\ln(A(m_0 - m_s)\Delta H/\dot{q}) = \ln(A) - \ln(\dot{q}/(m_0 - m_s)\Delta H)$ will be dominated by the value of ln$A$. Now in terms of the CKT parameters, using Eqn (5) gives

$$\ln(A) = \ln\left(2H^\theta/\Delta T_k^\theta\right) + T_A/T_k^\theta \approx T_A/T_k^\theta, \tag{22}$$

when $T_A >> T_k^\theta$. Consequently, it follows from Eqn (21) that to a leading order of approximation, $T_\infty/T_A \approx T_k^\theta/T_A$, i.e. the steady temperature will be approximately the same as the CKT.

This last result follows directly from the rate equation in terms of the characteristic kinetic parameters (Eqn (10)). Note that for large β_k^θ, when $T < T_k^\theta$, $\beta_k^\theta(1 - T_k^\theta/T)$ is large in absolute terms and negative, implying that $d\mu/dt$ is small. As time progresses, T will increase until it gets close to T_k^θ, so that $\beta_k^\theta|1 - T_k^\theta/T|$ is no longer large and so $-d\mu/dt$ increases. As this happens, the term $-\Delta H d\mu/dt$ will come to balance the term $\dot{q}/m_0$ in the energy Eqn (18), resulting in quasi-steady temperature.

Hence it follows that, to a reasonable order, the approximate behavior for the thermal decomposition of a strongly endothermic solid under the action of external heating is given by the solution of $m_0 c dT/dt = \dot{q}$, for $T < T_\infty$ and subsequently

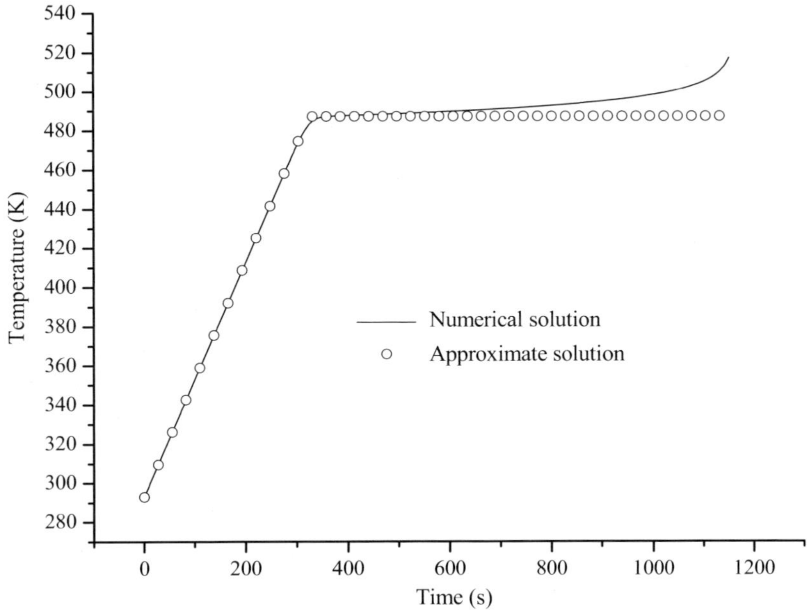

FIGURE 9

Comparison of approximate behavior (symbols) with numerical solution (curves) for simple endothermic solid.

$T \approx T_\infty$ until the decomposition ceases. If $\dot{q}$ is constant, then the time for T to reach T_∞ will be approximately

$$t_\infty = m_0 c(T_\infty - T_a)/\dot{q}. \tag{23}$$

The latent heat required to convert mass $m_0 - m_s$ of solid to vapor is $\Delta H(m_0 - m_s)$ and so the time required for the decomposition t_g will be

$$t_g \approx \Delta H(m_0 - m_s)/\dot{q}. \tag{24}$$

Thus for constant $\dot{q}$, the approximate thermal behavior will be given by

$$T = \begin{cases} T_a + \dfrac{\dot{q}t}{m_0 c}, & 0 \le t \le t_\infty, \\ T_\infty, & t_\infty < t \le t_\infty + t_g. \end{cases} \tag{25}$$

Figure 9 compares the approximate solution with a numerical solution of the full problem with the parameters shown in Table 2. Note that as material is lost during decomposition, the agreement becomes less good, which is to be expected.

When heat losses are considered, $\dot{q}$ itself becomes a function of T. For convection and radiation losses, the net heat flux $\dot{q}'' = \dot{q}/S$ is $\dot{q}''(T) = \varepsilon\dot{q}''_{ext} + h(T_a - T) + \varepsilon\sigma(T_a^4 - T^4)$, where S is the area of the exposed

Table 2 Parameter Values for Numerical Comparison with Approximate Solution

m_0 (kg)	0.2
ΔH (J kg^{-1})	2.28E + 06
T_a (K)	293
T_A (K)	36,067
A (s^{-1})	1.773×10^{29}
$\dot{q}$ (W)	500
m_s/m_0	0.1

surface of the solid, ε is the emissivity of the exposed surface, $\dot{q}''_{\text{ext}}$ is the external heat flux, h is the convection heat transfer coefficient, and σ is the Stefan–Boltzmann constant. When heat losses are included, a maximum temperature that the solid can attain exists, termed the *adiabatic temperature* T_{ad}, given by the solution of $\dot{q}''(T) = 0$. The steady temperature for the decomposition process T_∞ still exists, but is now given approximately by the solution of

$$\theta(T) = \exp(J - T_A/T), \quad \text{where} \ \ \theta(T) = \frac{\dot{q}''(T)}{\varepsilon \dot{q}''_{\text{ext}}}, \quad J = \ln\left(\frac{A(m_0 - m_s)\Delta H}{\varepsilon \dot{q}''_{\text{ext}} S}\right). \tag{26}$$

Given the discussion above, we expect that the steady temperature T_∞ will still be close to the CKT provided that $T_k^\theta < T_{\text{ad}}$ and if $T_k^\theta > T_{\text{ad}}$, then $T_\infty \approx T_{\text{ad}}$. Therefore, the expected qualitative thermal behavior will be as follows. Given that $T_k^\theta < T_{\text{ad}}$, the temperature of the solid will increase broadly according to the solution of

$$m_0 c \mathrm{d}T/\mathrm{d}t = S\dot{q}''(T), \quad T(0) = T_a, \tag{27}$$

until the temperature reaches T_k^θ at time t_∞. The temperature will then remain quasi-steady while the solid decomposes. When the volatile component of the solid is exhausted, after an approximate time $t_g \approx \Delta H(m_0 - m_s)/S\dot{q}''(T_k^\theta)$, the temperature will then increase to the adiabatic temperature T_{ad} according to the solution of

$$m_s c \mathrm{d}T/\mathrm{d}t = S\dot{q}''(T), \quad t > t_\infty + t_g, \quad T(t_\infty + t_g) = T_k^\theta. \tag{28}$$

This behavior is summarized in Figure 10, where the solid curve represents a numerical solution of the full energy equation.

A slightly more detailed view of the decomposition of a hydrated solid is to consider the solid as a homogeneous mixture of a volatile component m_v that degrades according to $\mathrm{d}m_v/\mathrm{d}t = -km_v$ and a nonvolatile component m_s. If $m = m_v + m_s$ is the total mass, then the rate equation in terms of m is $\mathrm{d}m/\mathrm{d}t = -km_v$ since component m_s is nonvolatile and since $m_v = m - m_s$, this may be

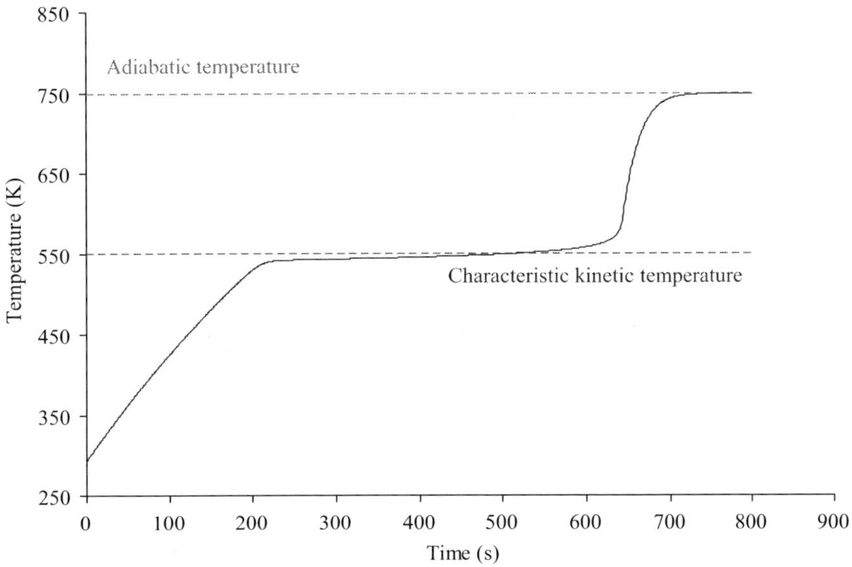

FIGURE 10

Illustration of thermal behavior for simple endothermic solid with heat losses.

written as $dm/dt = -k(m - m_s)$, which has the same form as considered above. However, applying conservation of energy gives the equation

$$(c_v\mu_v + c_s\mu_s)dT/dt = \dot{q}/m_0 - \left[(c_g - c_v)(T - T_a) + \Delta H_v\right]k\mu_v, \qquad (29)$$

which is similar to the energy Eqn (18) above, except for the more general form of composite specific heat capacity on the left-hand side. Here c_j is the specific heat capacity of species j (j = v, s, g) $\mu_j = m_j/m_0$, and ΔH_v is the heat of the reaction $m_v \rightarrow$ gas. The qualitative behavior of the new energy equation will be the same as discussed above.

3.2 Application to fire-resistant glazing

It transpires that the simple model presented in the previous section can be applied reasonably successfully to provide an approximate description of the performance of materials such as hydrated gels in glazing systems. This form of fire-resistant glazing consists of two planes of glass separated by a layer of hydrated gel. The thermal properties of the hydrated gel may be taken to all intents and purposes to be the same as liquid water. The thermal performance of the glazing system may be quantified in a standard test furnace environment, where thermocouples are attached to the unexposed pane of glass and the composite is exposed to a prescribed standard heating regime.

Two important heating regimes used in practice are the cellulosic and hydrocarbon thermal histories [10,11], where the temperature of the test furnace T_f is adjusted as a function of time to follow specific curves given by

$$T_f(t) = 293 + 345\,\log_{10}(1 + 2t/15), \tag{30}$$

$$T_f(t) = 293 + 1080\left(1 - 0.325e^{-0.167t/60} - 0.675e^{-2.5t/60}\right), \tag{31}$$

for the cellulosic and hydrocarbon curves, respectively. Note that time t is taken in seconds here.

Following the approach in Refs [12,13], the net heat flux to the sample is now

$$\dot{q}'' = h_a(T_a - T) + h_f(T_f - T) + \varepsilon_f\sigma\left(T_f^4 - T^4\right) + \varepsilon_{ext}\sigma\left(T_a^4 - T^4\right), \tag{32}$$

where h_a is the convection heat transfer coefficient for heat loss through the unexposed face of the test sample, h_f is the convection heat transfer coefficient for the furnace interior, ε_f is the effective emissivity of the furnace interior [13,14], and ε_{ext} is the effective emissivity for radiation losses from the unexposed surface.

The graph in Figure 11 shows the results of a suitably modified version of the model (Eqn (29)) with heat transfer rate per unit area (Eqn (32)) compared with

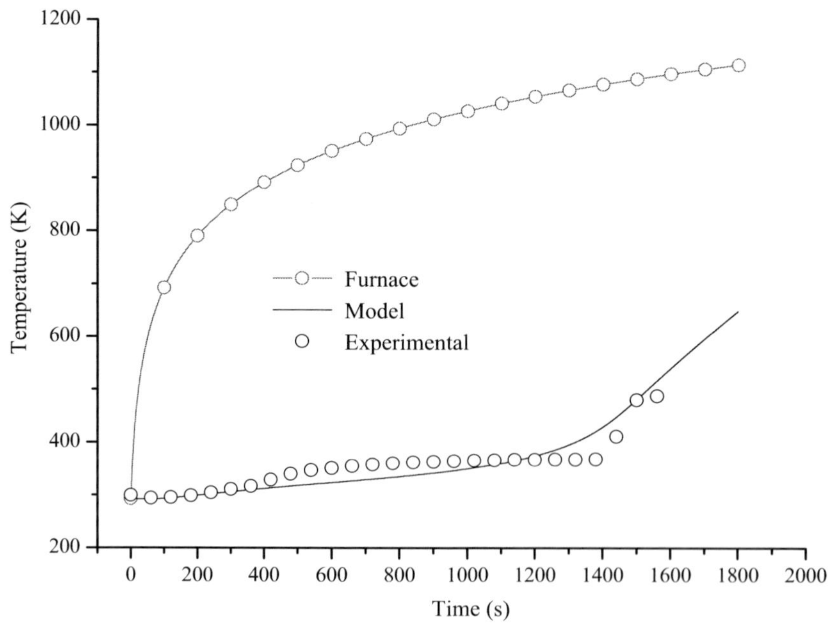

FIGURE 11

Comparison of experimental temperature data with model predictions for a hydrated gel sample.

observed temperatures (on the unexposed face) for a sample of experimental hydrated gel in a standard cellulosic furnace test.

4. Conservation of energy with heat diffusion

4.1 Heat conduction

An effective method of reducing flammability is to reduce the amount of heat absorbed by combustible material. One way that this may be achieved is by building up a thermally insulating layer between the external heat source and the flammable component. Perhaps the most important parameter of the insulating layer is the *conduction heat transfer coefficient* $\lambda' = \lambda/l$, where λ is the thermal conductivity and l is the thickness of the layer. The importance of λ' may be demonstrated by considering the temperature distribution through the thickness of a solid exposed to a constant heat flux $\dot{q}''$, insulated on its unexposed face. In this case, the difference between the temperature at the exposed surface and unexposed surface ΔT is given by

$$\Delta T = \frac{\dot{q}''}{\lambda'}\left\{\frac{1}{2} - \frac{2}{\pi^2}\sum_{n=1}^{\infty} e^{-n^2\pi^2\tau}\left(\frac{1-(-1)^n}{n^2}\right)\right\} \tag{33}$$

Here τ is the ratio of time to the diffusion timescale l^2/α, where α is the thermal diffusivity. The graph in Figure 12 shows the dimensionless temperature difference $\Delta T\lambda'/\dot{q}''$ as a function of τ for illustration.

For large time, if y is the distance through the solid from the unexposed surface, the temperature distribution across the solid is such that

$$\frac{T\lambda'}{\dot{q}''} \sim \tau - \frac{1}{2}\left(1 - \frac{y^2}{l^2}\right), \tag{34}$$

which implies that $\Delta T\lambda'/\dot{q}'' \to 1/2$ as $\tau \to \infty$. It therefore follows that for a fixed heat flux, the most effective thermal protection is provided by minimizing λ', i.e. minimizing thermal conductivity λ and maximizing layer thickness l.

When convection heat losses are included, it is again possible to construct an exact solution for the temperature distribution. If h is the convection heat transfer coefficient and $\dot{q}''$ is again the external heat flux, then it is possible to

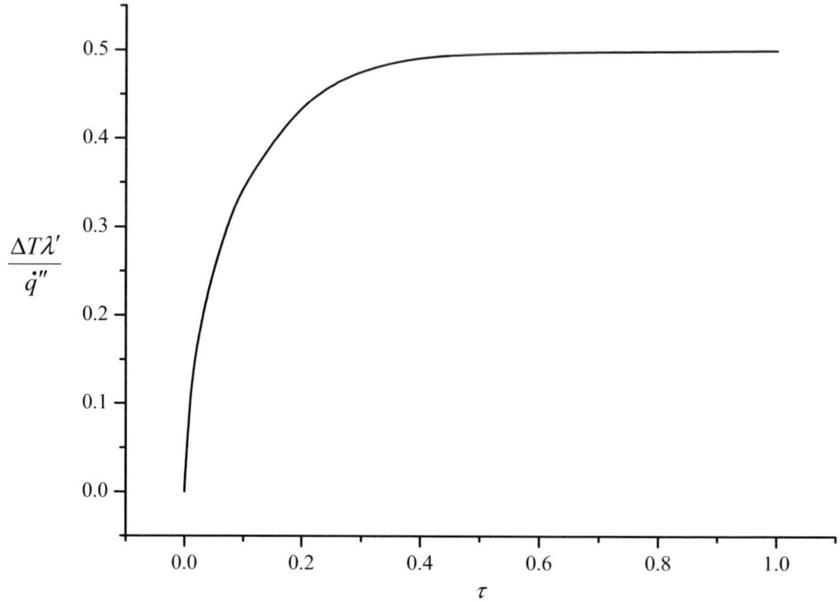

FIGURE 12

Variation of dimensionless temperature difference with τ.

show that the temperatures at the exposed surface T_1 and the unexposed surface T_0 are given by

$$\frac{h(T_1 - T_a)}{\dot{q}''} = 1 - 2\sum_{n=1}^{\infty} \frac{e^{-\xi_n^2\tau} \sin^2\xi_n}{\beta + \sin^2\xi_n}, \quad \frac{h(T_0 - T_a)}{\dot{q}''} = 1 - 2\beta\sum_{n=1}^{\infty} \frac{e^{-\xi_n^2\tau} \sin\xi_n}{\xi_n(\beta + \sin^2\xi_n)}, \tag{35}$$

where T_a is the ambient temperature, $\beta = h/\lambda'$ is the Biot number, τ is again the ratio of time to diffusion timescale l^2/α, and the eigenvalues ξ_n are given by the solutions of $\xi_n \sin\xi_n = \beta\cos\xi_n$.

Figure 13 shows the variation of exposed surface temperature, unexposed surface temperature, and temperature difference $\Delta T = T_1 - T_0$ with dimensionless time τ for $\beta = 10$.

Suppose that the purpose of the insulating layer is to maximize the time taken for the unexposed surface to reach a given temperature T_{crit}, which is less than the adiabatic temperature $T_a + \dot{q}''/h$. The dimensionless time τ_{crit} taken for the unexposed surface to reach T_{crit} is given by the solution of Eqn (35) with $T = T_{crit}$ and $\tau = \tau_{crit}$. Thus it is apparent that τ_{crit} will be a function of the two dimensionless parameters β and $\theta_{crit} = h(T_{crit} - T_a)/\dot{q}''$. Consequently, it follows that t_{crit} must be of the form $t_{crit} = l^2F(\theta_{crit},\beta)/\alpha$.

The graph in Figure 14 shows the contours of $\tau_{crit} = F(\theta_{crit},\beta)$. The hatched parts of this plot represent areas where $\tau_{crit} > 3$.

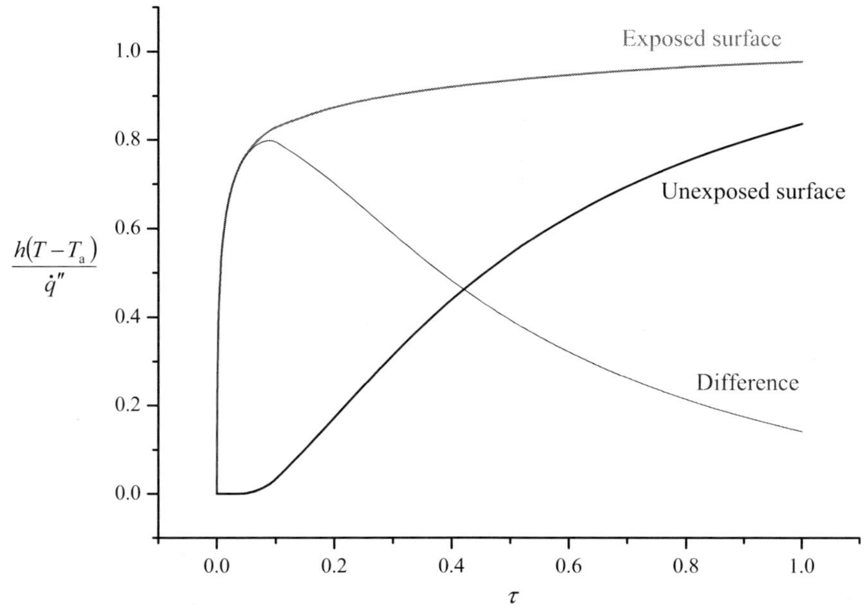

FIGURE 13

Illustration of temperature variation for external heating with convection.

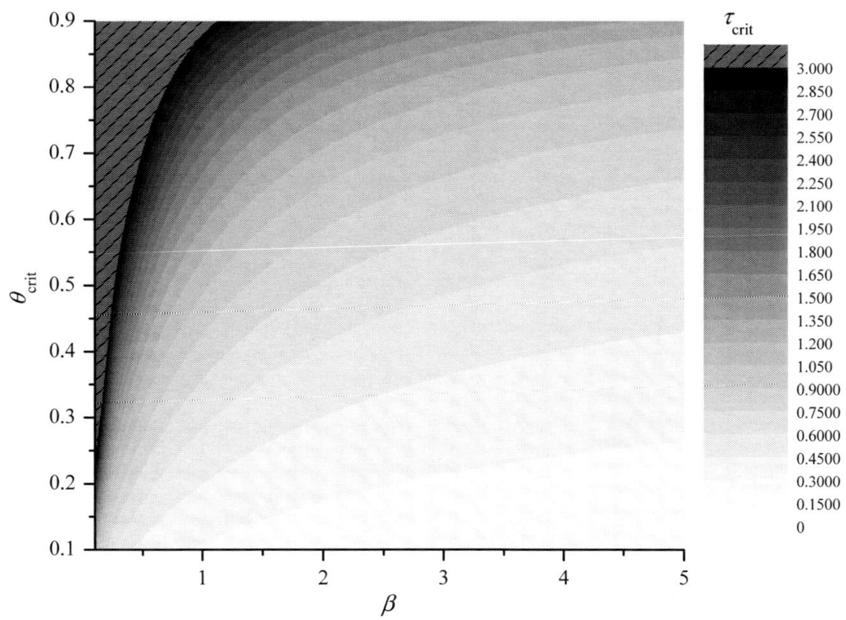

FIGURE 14

Contours of $\tau_{crit} = F(\theta_{crit}, \beta)$.

4.2 Composite thermal conductivity

The most effective insulators are porous solids, where gas-filled pores are embedded in a solid matrix and so it is important to consider the thermal properties (particularly the thermal conductivity) of these materials. The *porosity* is defined as the ratio of pore volume to the total volume of the solid. The composite density ρ and composite specific heat capacity c of a porous solid are easy to define. If the subscript s refers to a solid property and * to a pore property, then

$$\rho = (1-\varphi)\rho_s + \varphi\rho_* \approx (1-\varphi)\rho_s, \quad c = (1-\varphi\rho_*/\rho)c_s + \varphi\rho_* c_*/\rho \approx c_s, \quad (36)$$

to good approximations.

The composite thermal conductivity λ, however, is more problematic. It transpires that the composite thermal conductivity lies between two bounds, given by

$$\frac{\Lambda}{\Lambda + \varphi(1-\Lambda)} \leq \frac{\lambda}{\lambda_s} \leq 1 - \varphi(1-\Lambda), \quad (37)$$

where $\Lambda = \lambda_*/\lambda_s$. In the rather likely event that the porous solid does not have all its pores arranged together in series with the direction of heat transfer (resulting in the upper bound) or in parallel (resulting in the lower bound), then the actual thermal conductivity will depend on pore shape, distribution, and to a lesser extent on pore size as well as porosity. The maximum range of thermal conductivity is $\lambda_s(1-\Lambda^{1/2})^2$, which occurs at porosity $\Lambda^{1/2}(1-\Lambda^{1/2})/(1-\Lambda)$. See Figure15 for clarification.

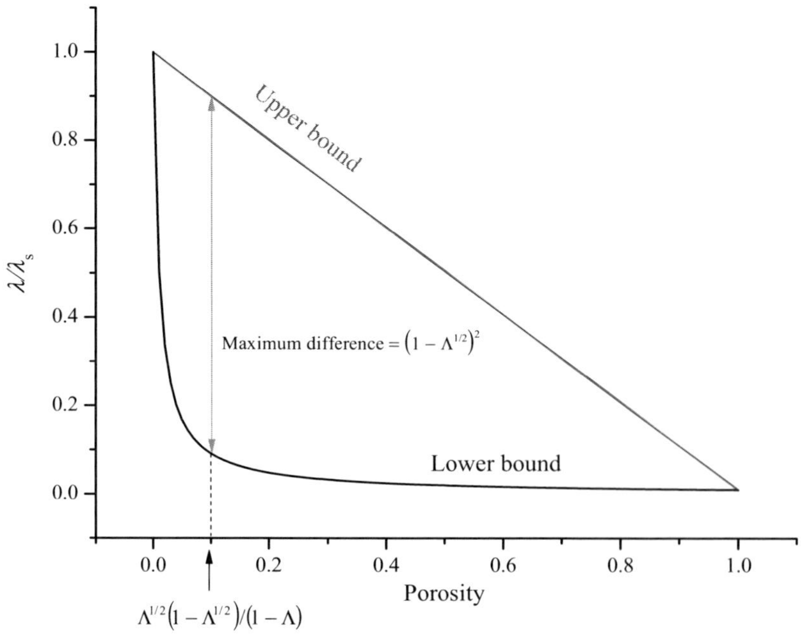

FIGURE 15

Bounds for composite thermal conductivity.

Correlations exist for a number of idealized situations featuring pores of specific shapes. An obvious starting point for a correlation would be to use the conductivity bounds themselves to obtain

$$\frac{\lambda}{\lambda_S} = \xi\{1 - \varphi(1 - \Lambda)\} + (1 - \xi)\left\{\frac{\Lambda}{\Lambda + \varphi(1 - \Lambda)}\right\}. \tag{38}$$

Here ξ is a parameter between 0 and 1 such that $\xi = 0$ gives the lower conductivity bound and $\xi = 1$ gives the upper conductivity bound [15]. Of the standard correlations, perhaps the most well known is Maxwell's expression for spherical pores of modest porosity,

$$\frac{\lambda}{\lambda_S} \approx \frac{2\varphi(\Lambda - 1) + \Lambda + 2}{2 + \Lambda - \varphi(\Lambda - 1)}, \tag{39}$$

which is $\sim 1 - 3\varphi/2$ for small φ and Λ. Arguably more realistic, but less convenient, are expressions for randomly distributed circular pores in 2D and spherical pores in 3D [16]:

$$1 - \varphi = \begin{cases} \dfrac{(\lambda/\lambda_S) - \Lambda}{(\lambda/\lambda_S)^{1/2}(1 - \Lambda)} & \text{for circular pores,} \\ \dfrac{(\lambda/\lambda_S) - \Lambda}{(\lambda/\lambda_S)^{1/3}(1 - \Lambda)} & \text{for spherical pores.} \end{cases} \tag{40}$$

Note that these are *implicit* relations and for a given φ, must be solved for λ. For the spherical pores case, it is possible to show that for small Λ,

$$\lambda/\lambda_S = (1 - \varphi)^{3/2} + 3(1 - \varphi)\Lambda\left\{1 - (1 - \varphi)^{1/2}\right\}/2 + O(\Lambda^2) \tag{41}$$

Here $O(\Lambda^2)$ means terms involving powers of Λ of order 2 and greater.

Direct numerical calculation is an alternative to using a correlation. It has the distinct advantage that pore sizes and shapes are modeled explicitly. However, implementation is difficult and computational overheads can be high. Two main numerical methods are available, namely, finite differences and finite elements. In both cases, the pores are digitally represented and the steady temperature distribution is computed on a finite grid of nodes with fixed temperature boundary conditions. The composite thermal conductivity is found from the resulting steady heat flux [17,18]. Figure 16 shows the results of finite element calculations for randomly distributed circular pores compared with Bruggeman's 2D result for $\Lambda = 0.01$.

If pore shape has such a large effect on the composite thermal conductivity, then this begs the question: is there some criterion for composite thermal conductivity that gives the best thermal resistance? It is well known that if pores are flattened relative to the direction of heat flow, creating a more disc-like shape with the axis of the disk aligned with the heat flow, then better thermal resistance results [18]. Figure 17 shows the results of direct numerical calculations of composite thermal conductivity for two composite solids with the same porosity. In both cases, the direction of heat

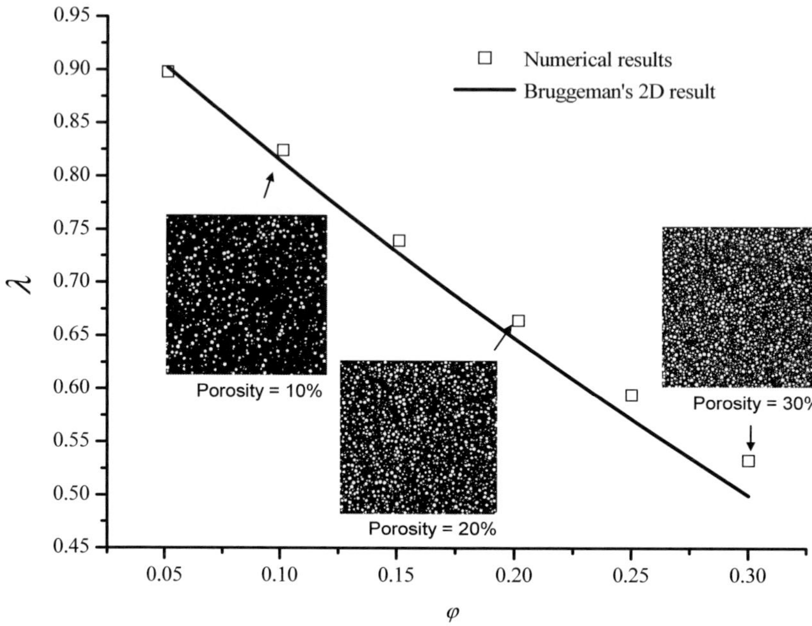

FIGURE 16

Comparison between finite-element results for randomly distributed circular pores and Bruggeman's expression.

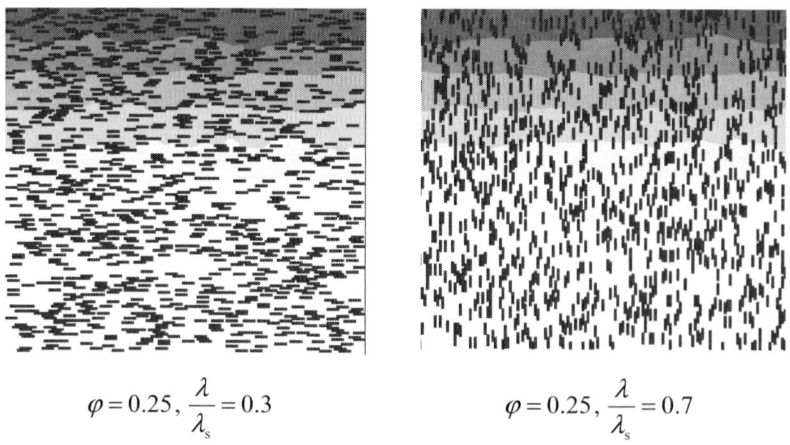

FIGURE 17

Effect of pore shape on composite thermal conductivity.

flow is from top to bottom and the shading corresponds to local temperature. The solid on the left has plate-like pores of identical size and the solid on the right has rod-like pores of the same size. The calculation demonstrates that the composite thermal conductivity of the solid with plate-like pores is approximately 43% lower than the solid with rod-like pores.

For external heating with convection, it was demonstrated above that the time taken for the temperature on an unexposed face to reach a particular value $T_{crit} < T_{ad}$ may be written as

$$\alpha_s t_{crit}/l^2 = (\rho c/\rho_s c_s)(\lambda_s/\lambda)F(\theta_{crit}, \beta), \tag{42}$$

where $\theta_{crit} = h(T_{crit} - T_a)/\dot{q}''$, $\alpha_s = \lambda_s/\rho_s c_s$, and $\beta = hl/\lambda$. Now if we use the expressions for composite density and specific heat capacity (Eqn (36)) above, we have that

$$\alpha_s t_{crit}/l^2 \approx (1-\varphi)(\lambda_s/\lambda)F(\theta_{crit}, \beta), \quad \varphi < 1. \tag{43}$$

Now, ideal thermal insulation performance will be given by a porosity dependence such that t_{crit} is always an increasing function of φ. Now, since θ_{crit} does not depend on porosity, this implies that

$$F\frac{\mathrm{d}}{\mathrm{d}\varphi}\left(\frac{1-\varphi}{\lambda}\right) + \frac{1-\varphi}{\lambda}\frac{\mathrm{d}\beta}{\mathrm{d}\varphi}\frac{\partial F}{\partial \beta} > 0. \tag{44}$$

Using the fact that $\mathrm{d}\beta/\mathrm{d}\varphi = -(\beta/\lambda)\mathrm{d}\lambda/\mathrm{d}\varphi$, the last inequality may be written as

$$-\frac{1-\varphi}{\lambda}\frac{\mathrm{d}\lambda}{\mathrm{d}\varphi} > \frac{F}{F + \beta\,\partial F/\partial\beta} = \frac{1}{1 + \partial(\ln F)/\partial(\ln\beta)}. \tag{45}$$

Looking at Figure 18, it is apparent that F is a decreasing function of β and so $\partial F/\partial\beta < 0$. Consequently, it follows that the right-hand side of the inequality directly above is always >1. Hence any composite thermal conductivity such that

$$-\frac{1-\varphi}{\lambda}\frac{\mathrm{d}\lambda}{\mathrm{d}\varphi} < 1 \tag{46}$$

could not provide ideal insulating performance.

If we look at the thermal conductivity upper bound $\lambda/\lambda_s = 1 - \varphi(1-\Lambda)$, we find that

$$-\frac{1-\varphi}{\lambda}\frac{\mathrm{d}\lambda}{\mathrm{d}\varphi} = \frac{(1-\Lambda)(1-\varphi)}{1-\varphi(1-\Lambda)}, \tag{47}$$

which is always <1. Therefore any materials with composite thermal conductivity close to the upper bound will not make effective insulators. In fact, increasing the porosity for such a layer will actually make thermal resistance worse. Alternatively if we look at the lower thermal conductivity bound, we find that

$$-\frac{1-\varphi}{\lambda}\frac{\mathrm{d}\lambda}{\mathrm{d}\phi} = \frac{(1-\Lambda)(1-\varphi)}{\Lambda + \varphi(1-\Lambda)} \tag{48}$$

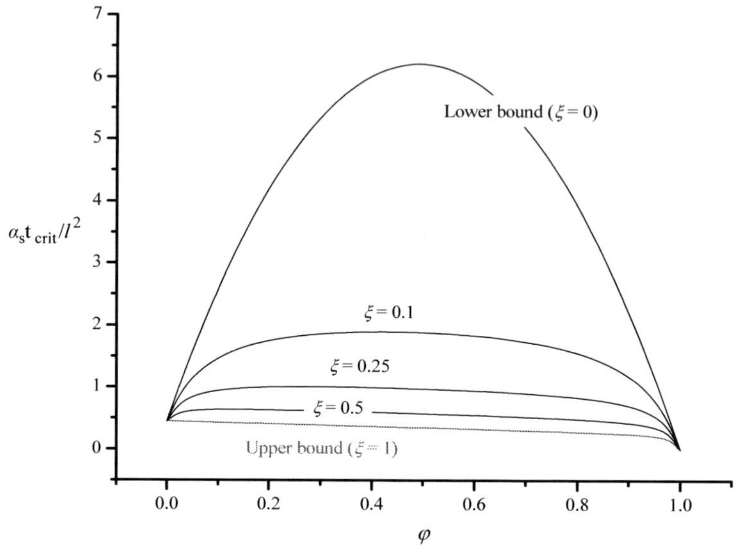

FIGURE 18

Illustration of insulation effectiveness as a function of porosity.

which is <1 when

$$\varphi > \frac{1 - 2\Lambda}{2(1 - \Lambda)} \tag{49}$$

When Λ is small, this inequality approximately reduces to $\varphi > 1/2$.

By way of illustration, the graph in Figure 18 shows the variation of $\alpha_s t_{crit}/l^2$ with porosity, computed from expression (35), for the composite thermal conductivity model (Eqn (38)). This figure was computed for $\theta_{crit} = 0.3$, $hl/\lambda_s = 2$, and $\Lambda = 0.01$. Note that as shown above, there is no improvement in thermal resistance with increasing φ for the upper thermal conductivity bound. For the lower thermal conductivity bound, thermal resistance increases as φ increases for φ approximately <0.5, and then decreases with increasing porosity.

4.3 Effect of radiation

As temperature increases, the composite thermal conductivity is augmented by radiation heat transfer across the pores. If the pore size δ is small, there will be no internal convection and the additional heat transfer rate due to radiation is approximately $-4\varepsilon\sigma\delta^3 T^3 \partial T/\partial y$, where $\partial T/\partial y$ is the temperature gradient across the pore, ε is the emissivity of the interior surface, and σ is the Stefan–Boltzmann constant. Thus, if a layer of porous solid of thickness δ and cross-sectional area S

(normal to the overall direction of heat transfer) is considered, the total additional heat transferred by radiation will be approximately $-4\varepsilon\sigma\varphi\delta\, ST^3\partial T/\partial y$. Hence if λ is the observed composite thermal conductivity at ambient temperature T_a, the net heat flux due to both radiation and conduction across the layer is $\dot{q}'' \approx -\left\{\lambda + 4\varepsilon\sigma\varphi\delta\left(T^3 - T_a^3\right)\right\}\partial T/\partial y$. Note that the term $4\varepsilon\sigma\varphi\delta T_a^3$ will probably be small in practice. Thus, to leading order, the effect of radiation across small pores may be accounted for by defining an effective thermal conductivity

$$\lambda_{\mathrm{eff}} = \lambda + \varphi\lambda_{\mathrm{R}}\left(\frac{T^3}{T_a^3} - 1\right), \tag{50}$$

where $\lambda_{\mathrm{R}} \sim 4\varepsilon\sigma\delta\, T_a^3$. If one is able to measure effective thermal conductivity as a function of temperature, then $\varphi\lambda_{\mathrm{R}}$ may be easily determined from the gradient of a plot of λ_{eff} against $x = (T/T_a)^3$.

The effect of radiation enhancement of thermal conductivity clearly depends on porosity, but it also depends on pore size. The analysis above suggests that radiation enhancement is an increasing function of pore size. Hence to minimize the effect of radiation, it is better for a porous layer of given porosity to have many small pores rather than few large pores.

4.4 A basic mathematical model for decomposition of a filled solid

As above, consider a solid with a reactive volatile component and a solid, inert, component. For simplicity, let us suppose that heat transfer is in one direction through the solid, normal to an exposed surface. Let the overall thickness of the solid be l and let y denotes the distance through the solid. The surface $y = 0$ is assumed to be well insulated for convenience and the surface $y = l$ is taken as the exposed surface where the solid is subject to heating from an external source.

Consider a small control volume (CV) of solid of thickness Δy and volume ΔV. In general, this volume will consist of reactive component, inert component, and gas. Let the concentrations in kg m^{-3} of the volatile and solid components in the CV be $\overline{\rho}_v$ and $\overline{\rho}_s$, respectively, and let the (constant) skeletal densities be ρ_v and ρ_s, respectively. Note that since the solid component is inert, $\overline{\rho}_s$ will be constant. In order to formulate a model, it is necessary to impose some assumptions and chief among these are the following:

- As the reactive component degrades, gas is generated and pores develop.
- There is no overall change in volume during the decomposition and the increase in pore volume is equal to the decrease in reactive component volume.
- As gas is generated, compressibility effects are negligible and it flows through the porous residue toward the exposed surface.
- The components of the CV are in thermal equilibrium.

If ρ is the overall composite density, then $\rho = \varphi\rho_g + \overline{\rho}_v + \overline{\rho}_s$, where φ is the porosity and ρ_g is the gas density. Let the initial mass fraction of inert solid in the composite be $\mu_{s,0}$, then the initial composite density ρ_0 is given by

$$1/\rho_0 = \mu_{s,0}/\rho_s + (1 - \mu_{s,0})/\rho_v. \tag{51}$$

Consequently, the initial concentration of reactive component $\rho_{v,0}$ and the constant concentration of inert solid are given by

$$\frac{\overline{\rho}_{v,0}}{\rho_v} = \frac{(1 - \mu_{s,0})}{(1 - \mu_{s,0}) + \mu_{s,0}P}, \quad \frac{\overline{\rho}_s}{\rho_s} = \frac{\mu_{s,0}P}{(1 - \mu_{s,0}) + \mu_{s,0}P}. \tag{52}$$

Here $P = \rho_v/\rho_s$. Note that it then follows that $\overline{\rho}_{v,0}/\rho_v + \overline{\rho}_s/\rho_s = 1$. In reality, it is probably likely that $P \approx 1$, in which case $\overline{\rho}_{v,0}/\rho_v \approx 1 - \mu_{s,0}$ and $\overline{\rho}_s/\rho_s \approx \mu_{s,0}$. The volume of reactive component in the CV is $\overline{\rho}_v \Delta V/\rho_v$ and so the pore volume is $(\overline{\rho}_{v,0}/\rho_v - \overline{\rho}_v/\rho_v)\Delta V$. Consequently, the porosity will be given by

$$\varphi = (\overline{\rho}_{v,0} - \overline{\rho}_v)/\rho_v. \tag{53}$$

Since the control volume is fixed, the mass of reactive component in ΔV is $\overline{\rho}_v \Delta V$ and the rate equation for thermal decomposition may be written as

$$\partial\overline{\rho}_v/\partial t = -k\overline{\rho}_v, \quad \overline{\rho}_v(y, 0) = \overline{\rho}_{v,0}. \tag{54}$$

Note that this implies that the mass generation rate of gas per unit volume is $k\overline{\rho}_v$. Let the mass flow rate of gas per unit cross-sectional area of the CV be $\dot{m}''_g$. Then, by conservation of mass, the difference between the mass flow rate out of the control volume $\dot{m}''_g(y + \Delta y, t)$ and into the control volume $\dot{m}''_g(y, t)$ will be equal to the mass generation rate of gas within the CV per unit area minus the accumulation rate of gas due to the changing pore volume. Hence if ρ_g is the gas density, conservation of mass implies that

$$\dot{m}''_g(y + \Delta y, t) = \dot{m}''_g(y, t) + k\overline{\rho}_v\Delta y - k\overline{\rho}_v\frac{\rho_g}{\rho_v}\Delta y \Rightarrow \frac{\partial \dot{m}''_g}{\partial y} = k\overline{\rho}_v\left(1 - \frac{\rho_g}{\rho_v}\right). \tag{55}$$

This equation may be integrated with respect to y to obtain (noting that there is no flow of gas at the insulated boundary),

$$\dot{m}''_g(y, t) = \int_0^y k\overline{\rho}_v\left(1 - \frac{\rho_g}{\rho_v}\right)dy. \tag{56}$$

The total mass lost per unit cross-sectional area from the solid due to decomposition of the reactive component is given by the gas mass flux at the exposed surface, i.e. $\dot{m}''_g(l, t)$.

Applying conservation of energy to the control volume, we have that the net change of internal energy per unit cross-sectional area $\delta U''$ is given by

$$\delta U'' = -\left[\left(c_g(T - T_a)\dot{m}''_g - \lambda_{\text{eff}}\frac{\partial T}{\partial y}\right)\right]_y^{y+\Delta y}\delta t + \Delta H\delta\overline{\rho}_v\Delta y. \tag{57}$$

Here c_g is the specific heat capacity of the gas generated by decomposition of the reactive component, λ_{eff} is the effective composite thermal conductivity defined by Eqn (50), and ΔH is the heat of the decomposition reaction. Transport of heat per unit area into, and out of, the CV by the flow of gas generated from decomposition of the reactive component is accounted for by the term $-\left[c_g(T-T_a)\dot{m}''_g\right]_y^{y+\Delta y}$. Since we are assuming that the components of the composite are in thermal equilibrium, $U=(\rho_g\varphi c_g+\overline{\rho}_v c_v+\overline{\rho}_s c_s)(T-T_a)\Delta V$, where c_v and c_s are the specific heat capacities of the reactive component and the inert solid component, respectively. Substitution into the statement of energy conservation yields

$$\frac{\partial}{\partial t}\left\{(\rho_g\varphi c_g+\overline{\rho}_v c_v+\overline{\rho}_s c_s)(T-T_a)\right\}=\frac{\partial}{\partial y}\left(\lambda_{eff}\frac{\partial T}{\partial y}-c_g(T-T_a)\dot{m}''_g\right)-\Delta Hk\overline{\rho}_v \tag{58}$$

The boundary condition on $y=0$ is $\partial T/\partial y=0$ since we assume that this surface is well insulated. The boundary condition on the exposed surface $y=l$ depends on the physical circumstances. Perhaps the two most common situations are where the surface is exposed to a prescribed heating regime, such as a standard furnace test, or where the surface is exposed to a prescribed heat flux with heat losses. Since the second situation is more often encountered when testing the effectiveness of flame retardant, we shall assume that

$$\lambda_{eff}\partial T/\partial y=\varepsilon\dot{q}''_{ext}+h(T_a-T)+\varepsilon\sigma\left(T_a^4-T^4\right) \tag{59}$$

on $y=l$. Here $\dot{q}''_{ext}$ is the external heat flux, h is the convection heat transfer coefficient, T_a is ambient temperature, ε is the emissivity of the exposed surface, and σ is the Stefan–Boltzmann constant. As noted above, this boundary condition implies that there is a maximum temperature that the solid can achieve, the adiabatic temperature T_{ad}, given by the solution of $\varepsilon\dot{q}''_{ext}+h(T_a-T_{ad})+\varepsilon\sigma(T_a^4-T_{ad}^4)=0$. Unfortunately it is not possible to write down the explicit form for T_{ad} when radiation losses are included. In general, $\theta_{ad}=T_{ad}/T_a$ is a function of the two dimensionless parameters $Q=\dot{q}''_{ext}/(\sigma T_a^4)$ and $H=h/(\varepsilon\sigma T_a^3)$. If we fix θ_{ad}, then Q and H will lie on the straight line:

$$H=\frac{Q}{\theta_{ad}-1}-\frac{\theta_{ad}^4-1}{\theta_{ad}-1}. \tag{60}$$

In other words, the contours of T_{ad}/T_a as functions of Q and H will be straight lines with gradients $1/(\theta_{ad}-1)$. Figure 19 confirms this observation and shows a contour plot of T_{ad}/T_a as a function of Q on the independent axis and H on the dependent axis.

From the definition of T_{ad}, it is possible to rewrite the boundary condition as

$$\lambda_{eff}\partial T/\partial y=h(T_{ad}-T)+\varepsilon\sigma\left(T_{ad}^4-T^4\right), \tag{61}$$

which will prove more convenient below.

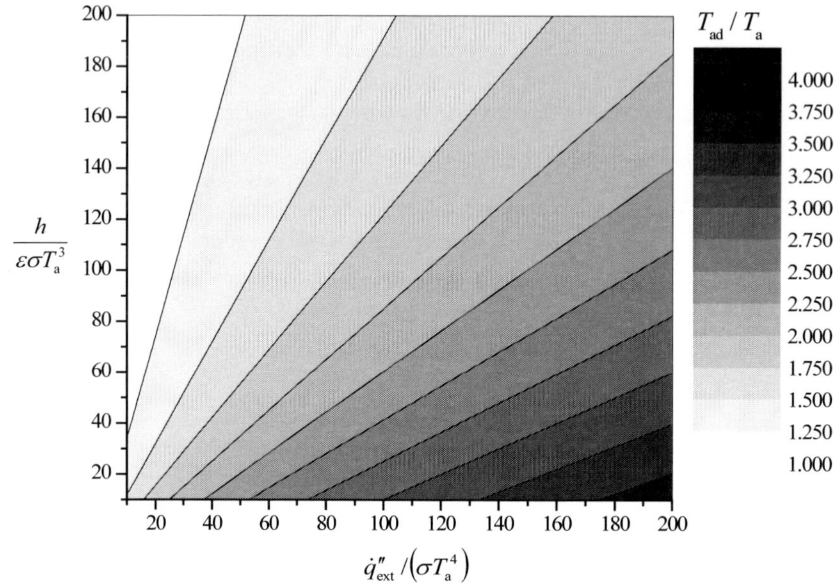

FIGURE 19

Contour plot of T_{ad}/T_a.

Using the equations above for conservation of mass and porosity, Eqn (58) can be written in the more accommodating form:

$$\left(\rho_g \varphi c_g + \overline{\rho}_v c_v + \overline{\rho}_s c_s\right)\frac{\partial T}{\partial t} + c_g \frac{\partial T}{\partial y}\int_0^y k\overline{\rho}_v\left(1 - \frac{\rho_g}{\rho_v}\right)\mathrm{d}y$$

$$= \frac{\partial}{\partial y}\left(\lambda_{\mathrm{eff}}\frac{\partial T}{\partial y}\right) - k\overline{\rho}_v\left[\left(c_g - c_v\right)\left(T - T_a\right) + \Delta H\right] \tag{62}$$

The effective composite thermal conductivity λ_{eff} includes a correction for radiation enhancement across pores as they develop. However, an important detail concerns the form of the purely conductive term λ. In the present case, the composite consists of reactive, inert, and pore components each with volume fraction $\eta_v = \overline{\rho}_v/\rho_v$, $\eta_s = \overline{\rho}_s/\rho_s$, and $\varphi = (\overline{\rho}_{v,0} - \overline{\rho}_v)/\rho_v$, respectively. There are a number of ways that we might proceed, but here we shall divide the porous solid into just two parts: a solid part consisting of a mixture of reactive and inert components and a porous part. The volume fraction of the solid mixture is $\eta_{v+s} = \eta_v + \eta_s$. Let λ_{v+s} be the thermal conductivity of the solid mixture and λ_*, the thermal conductivity of the pores. Then, again for simplicity, we shall assume that the composite thermal conductivity of the solid mixture has the same functional dependence as the thermal conductivity of the porous solid. In other words, for a two-part composite with species 1 having volume fraction η and thermal conductivity λ_1 and species 2 having volume fraction $1-\eta$ and thermal conductivity λ_2, we assume that the composite thermal conductivity λ_{1+2} has the form $\lambda_{1+2}=\lambda_2\kappa(\eta,\Lambda)$, where κ has some specific functional form such as discussed in Section 4.2 above with $\Lambda=\lambda_1/\lambda_2$ and $\kappa(0,\Lambda)=1$, $\kappa(1,\Lambda)=\Lambda$. Hence, the solid mixture will have thermal conductivity $\lambda_{v+s}=\lambda_s\kappa(\eta_v/\eta_{v+s},\lambda_v/\lambda_s)$ and the overall composite will have thermal conductivity

$$\lambda=\lambda_{v+s}\kappa(\varphi,\lambda_*/\lambda_{v+s}). \tag{63}$$

If we take $\lambda_s/\lambda_v=1$ for simplicity, then it follows that $\lambda_{v+s}/\lambda_v=1$ and consequently λ/λ_v will be given simply by $\lambda/\lambda_v=\kappa(\varphi,\Lambda_*)$, where $\Lambda_*=\lambda_*/\lambda_v$.

The term $r=\rho_g\varphi c_g+\overline{\rho}_v c_v+\overline{\rho}_s c_s$ corresponds to the composite volumetric heat capacity in J K^{-1} m^{-3}. In practice, it is likely that we can assume $\rho_g/\rho_v\ll 1$ and if we also assume that $\rho_v/\rho_s\approx 1$, then provided that c_g/c_v is not large,

$$r\approx\rho_v c_v\left(\eta_v+\mu_{s,0}\frac{c_s}{c_v}\right) \tag{64}$$

Hence when $\rho_g/\rho_v\ll 1$, and setting $x=y/l$, $\alpha_v=\lambda_v/\rho_v c_v$, $t_v=l^2/\alpha_v$, $\tau=t/t_v$, $\widehat{k}=t_v k$, $\widehat{\lambda}=\lambda_{eff}/\lambda_v$, $\theta=T/T_{ad}$ and taking all of the specific heat capacities to be equal for simplicity, the energy and mass equations with the boundary and initial conditions become

$$(\eta_v+\mu_{s,0})\frac{\partial\theta}{\partial\tau}+\frac{\partial\theta}{\partial x}\int_0^x\widehat{k}\eta_v\mathrm{d}x=\frac{\partial}{\partial x}\left(\widehat{\lambda}\frac{\partial\theta}{\partial x}\right)-H_v\widehat{k}\eta_v, \tag{65}$$

$$\frac{\partial\theta}{\partial x}=0\quad\text{on } x=0,\quad \widehat{\lambda}\frac{\partial\theta}{\partial x}=B_v(1-\theta)+E_v(1-\theta^4)\quad\text{on } x=1,\quad \theta=\theta_a$$
$$=T_a/T_{ad}\quad\text{at } \tau=0, \tag{66}$$

$$\widehat{\lambda}=\kappa(\varphi,\Lambda_*)+\varphi\widehat{\lambda}_R\left(\frac{\theta^3}{\theta_a^3}-1\right),\quad \Lambda_*=\frac{\lambda_*}{\lambda_v},\quad \Lambda_R=\frac{\lambda_R}{\lambda_v}, \tag{67}$$

$$\partial\eta_v/\partial\tau=-\widehat{k}\eta_v,\quad \eta_v(x,0)=1-\mu_{s,0}, \tag{68}$$

$$\varphi=1-\mu_{s,0}-\eta_v, \tag{69}$$

where $H_v=\Delta H/(c_v T_{ad})$, $B_v=hl/\lambda_v$, and $E_v=\varepsilon\sigma l T_{ad}^3/\lambda_v$.

In terms of the CKT, the Arrhenius rate function $\widehat{k}$ may be written as $\widehat{k}=\exp\{J_v+\beta_k^\theta(1-\theta_k^\theta/\theta)\}$, where

$$J_v=\ln\left(\frac{2l^2H^\theta}{\alpha_v T_{ad}\Delta\theta_k^\theta}\right),\quad \beta_k^\theta=\frac{2\theta_k^\theta}{\Delta\theta_k^\theta\ \ln 2},\quad \theta_k^\theta=\frac{T_k^\theta}{T_{ad}},\quad \Delta\theta_k^\theta=\frac{\Delta T_k^\theta}{T_{ad}} \tag{70}$$

Hence, it transpires that the solution depends on nine dimensionless parameters:

- The ratio of ambient temperature to adiabatic temperature θ_a.
- Two heating parameters B_v and E_v, corresponding to a Biot number and a ratio of radiation heat transfer rate to conduction heat transfer rate.
- A so-called Stefan parameter H_v corresponding to a ratio of latent heat of reaction to sensible heat.
- The log of a Damköhler number J_v (representing the ratio of diffusion timescale to chemically reactive timescale).
- Two characteristic kinetic parameters θ_k^θ and $\Delta\theta_k^\theta$.
- Two thermal conductivity parameters Λ_* and Λ_R.

The dimensional parameters in Table 3 result in $\alpha_v = 1.44 \times 10^{-7}\,\mathrm{m^2\,s^{-1}}$ and $T_{ad} = 880.6$ K. The resulting dimensionless parameters are shown in Table 4.

Figure 20 shows the solution distributions for dimensionless temperature θ and volume fraction η_v at regular time intervals ($\Delta\tau = 0.25$) between $\tau = 0.25$ and 2.25 for the variables in Table 4. As time progresses, the reaction front progresses through the material from the exposed face at $x = 1$ to the unexposed face at $x = 0$. Heat diffusion causes the reaction zone to initially broaden as it advances and the width of the reaction zone is shown as a function of τ in Figure 21. Note that the local temperature in the vicinity of the reaction zone is close to the CKT.

The graph in Figure 22 shows the effect of the amount of reactive solid on the temperature at the unexposed face and total porosity and Figure 23 shows the total evolved dimensionless mass flux. As the amount of inert solid reduces, the effectiveness of the reactive component becomes more apparent. The general effect of the reactive component is to maintain the local temperature in the vicinity of the CKT until the component is exhausted. Once exhausted, the local temperature

Table 3 Example Dimensional Parameters

ΔH (MJ kg^{-1})	2.28
ρ_v (kg m^{-3})	1000
c_v (J K^{-1} kg^{-1})	4180
λ_v (W m^{-1} K^{-1})	0.6
λ_R (W m^{-1} K^{-1})	0.009
Λ_* (W m^{-1} K^{-1})	0.0273
ε	0.9
$\dot{q}''_{ext}$ (kW m^{-2})	50
h (W m^{-2} K^{-1})	25
T_a (K)	293
l (mm)	20
T_k^θ (K)	600[a]
ΔT_k^θ (K)	50[a]

[a] *Taken at heating rate $H^\theta = 10\,\mathrm{K\,min^{-1}}$.*

Table 4 Corresponding Dimensionless Parameters

θ_a	0.3327
H_v	0.6194
B_v	0.8333
E_v	1.1616
Λ_R	0.0150
Λ_*	0.0455
θ_k^θ	0.6814
$\Delta\theta_k^\theta$	0.0568
J_v	2.9220

increases to the adiabatic value. The total evolved mass flux has a characteristic double-peaked shape. The first peak corresponds to rapid gasification near the exposed surface and the second peak occurs after a diffusive lag and corresponds to gasification from subsurface regions.

The graph in Figure 24 shows the solutions of the model for the parameter values shown in Table 4 as Λ_R is varied for $\mu_{0,s} = 0.3$ (solid gray curves) and $\mu_{0,s} = 0.7$ (solid black curves). Also shown on the figure are total porosity curves for $\Lambda_R = 0$ for each

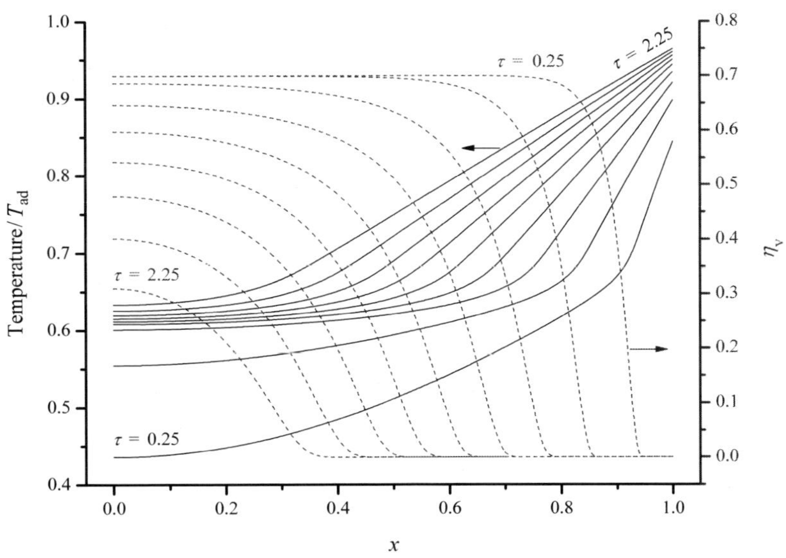

FIGURE 20

Solution distributions (dashed curves: volume fraction; solid curves: temperature) at regular intervals $\Delta\tau = 0.25$.

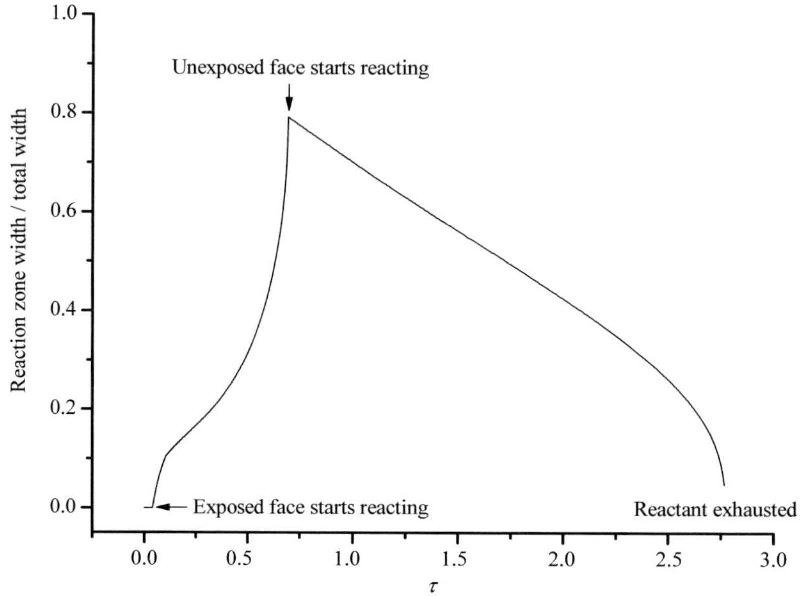

FIGURE 21

Width of reaction zone as a function of dimensionless time.

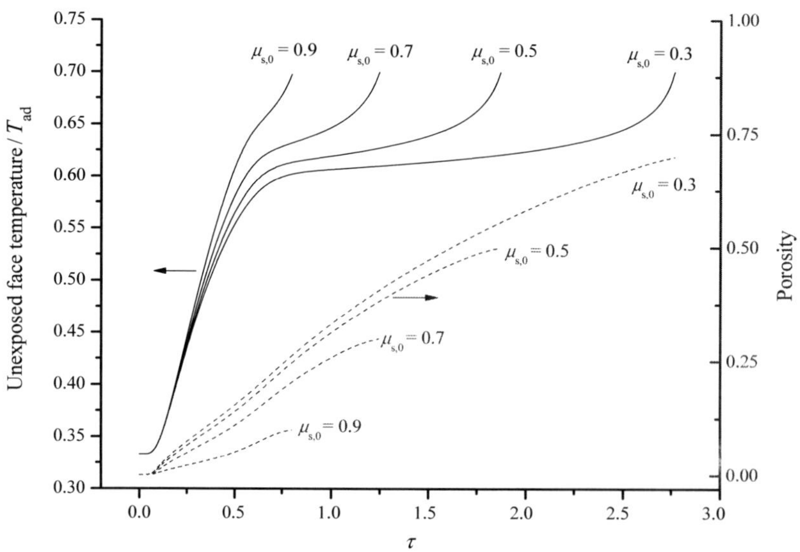

FIGURE 22

Effect of reactive component on temperature at unexposed face (solid curves) and total porosity (dashed curves).

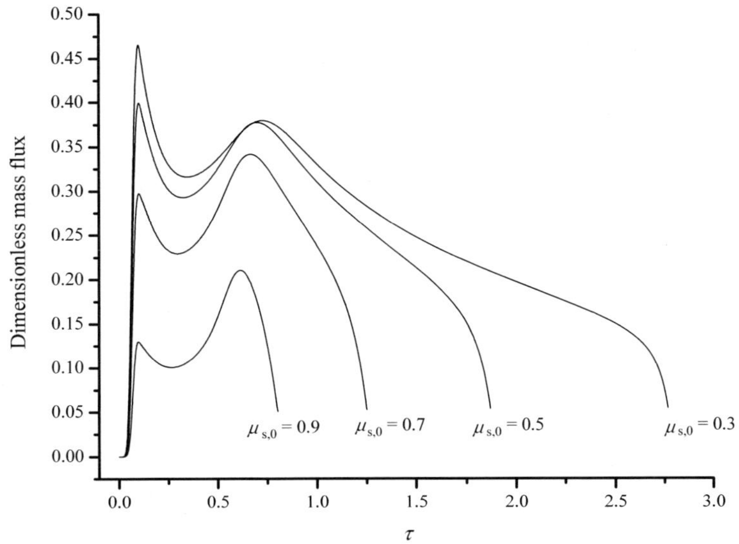

FIGURE 23

Effect of reactive component on total dimensionless mass flux $\widehat{\dot{m}}'' = \int_0^1 \widehat{k}\eta_V dx$.

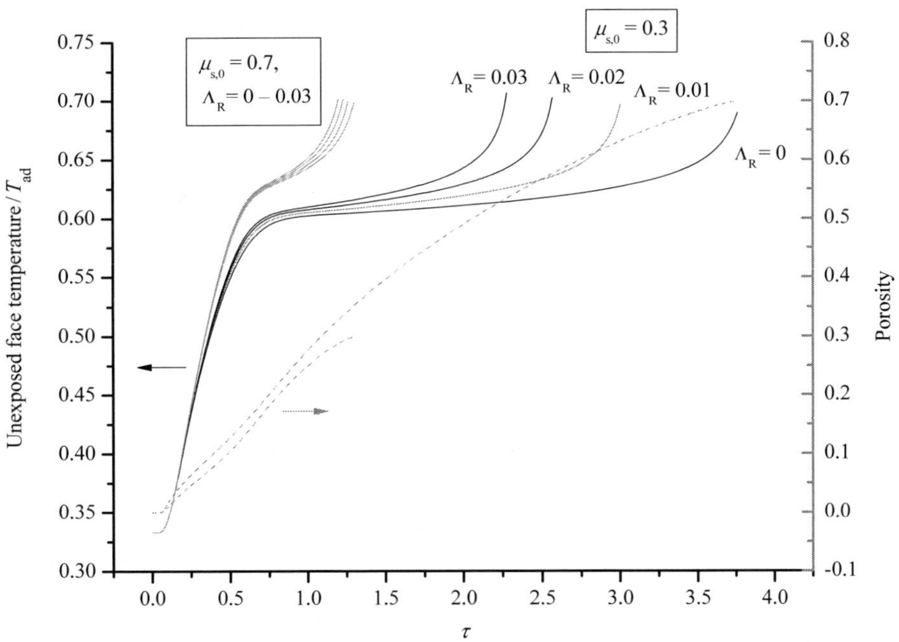

FIGURE 24

Effect of pore radiation parameter on temperature at the unexposed face.

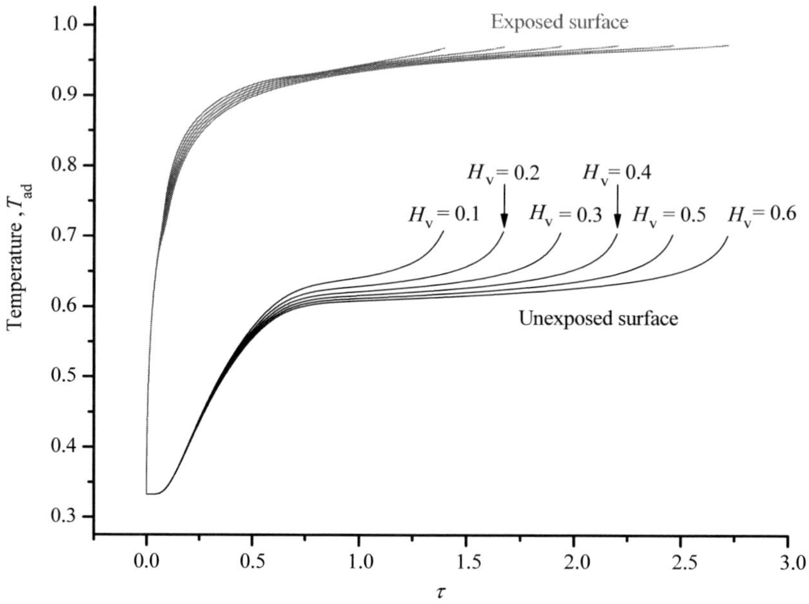

FIGURE 25

Effect of endothermicity on temperature.

$\mu_{0,s}$ value (dashed curves). Note that as the solid mass fraction reduces, radiation augmentation becomes increasingly important due to increasing porosity.

The graph in Figure 25 shows the effect of changing the heat of reaction on the exposed surface temperature ($x = 1$) and the unexposed surface temperature ($x = 0$). Note that this term does not greatly affect the exposed surface temperature but does have a large effect on the time that the unexposed face remains below the CKT. Since the exposed surface temperature does not very greatly during decomposition, the net heat flux at the exposed surface will remain approximately constant. Thus we should expect that the time taken to exhaust the reactant should be approximately a linear function of ΔH_V. This is confirmed in Figure 26, where the dimensionless time τ for the reactant to be exhausted is plotted as a function of H_V.

The graph in Figure 27 shows the effect of reducing the CKTR on the solution distributions at $\tau = 1.25$. As the CKTR reduces, the reaction zone reduces in size and the local temperature becomes increasingly controlled by the reaction kinetics during gasification. Figure 28 confirms the effect of reducing CKTR on local temperature, which shows the temperatures on the unexposed face as functions of τ. It becomes increasingly difficult for a numerical code implementing standard algorithms to calculate the solution for very small CKTR and the code implemented herein has effectively reached its computational limit for a CKTR ~ 2 K.

Theoretically, as the CKTR approaches zero, the solution tends to the *ablation limit*, given by the solution of a two-phase Stefan problem analogous to the melting

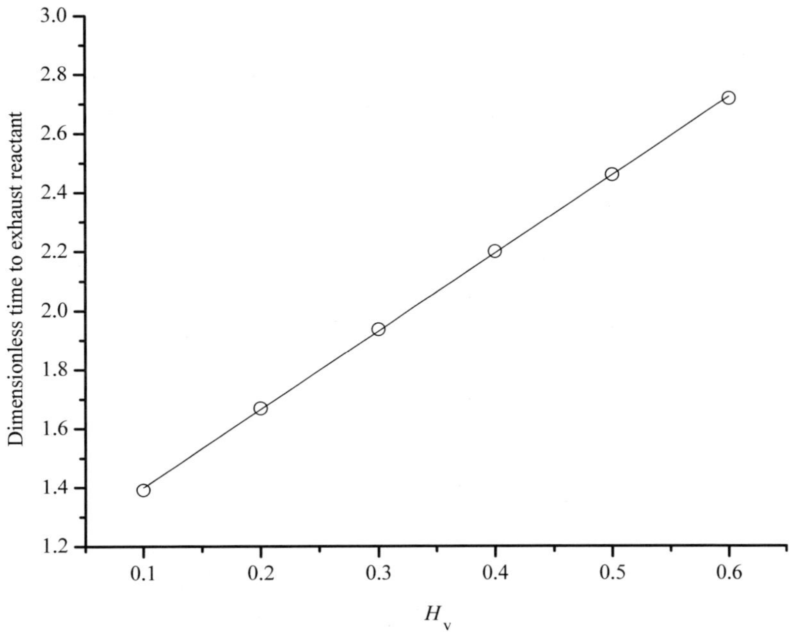

FIGURE 26

Time taken to exhaust reactant as a function of endothermicity.

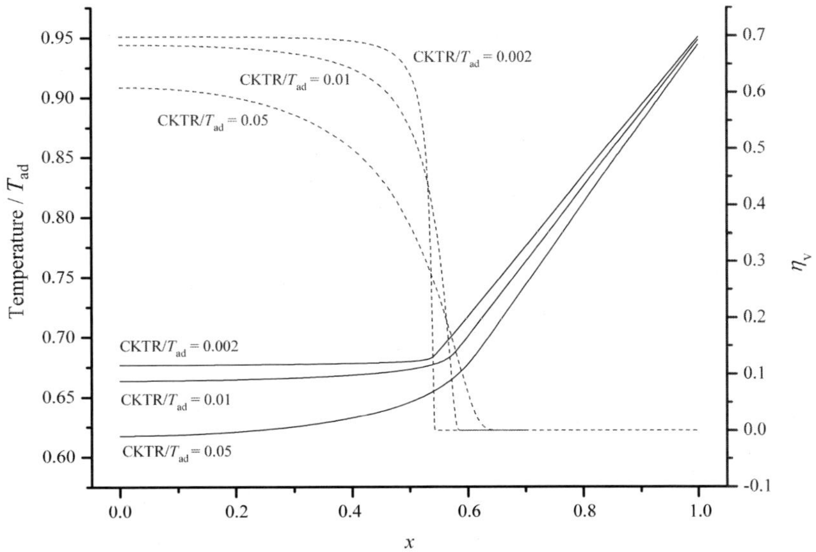

FIGURE 27

Effect of characteristic temperature range on solution distributions.

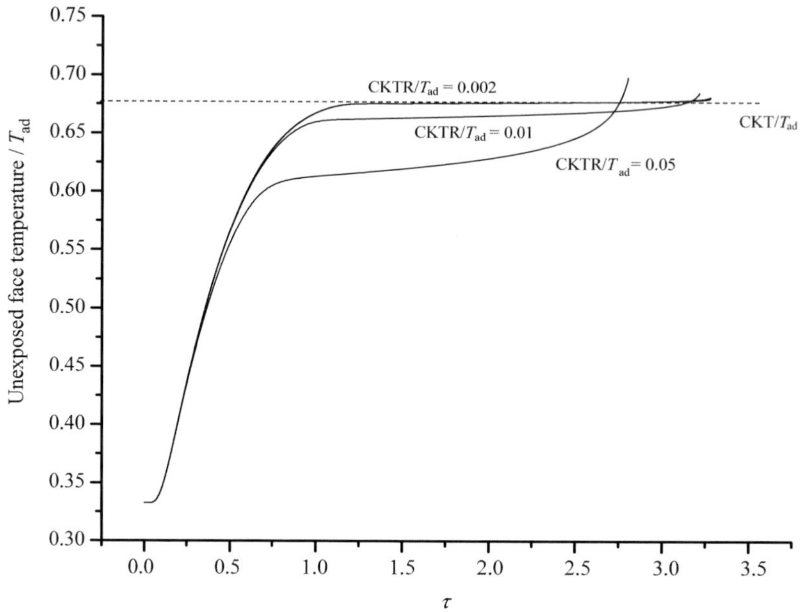

FIGURE 28

Effect of characteristic temperature range on temperature of unexposed face.

problem with different thermal properties either side of the melting transition [19–21].

In a thick solid, the through-flow of gas resulting from decomposition of material has a local cooling effect. The reason for this is that gas flow will be in the direction of increasing temperature gradient (since the decomposition reaction will be more advanced in hotter regions resulting in greater porosity). The magnitude of this effect is determined by the term $c_g(T - T_a)\dot{m}''_g$, where $\dot{m}''_g$ is the local mass flux of gas, given by Eqn (56). This contributes to the net local heat flux $\dot{q}''$, so that

$$\dot{q}'' = -\lambda_{eff}\frac{\partial T}{\partial y} + c_g(T - T_a)\int_0^y k\bar{\rho}_v\left(1 - \frac{\rho_g}{\rho_v}\right)dy \tag{71}$$

The magnitude of the effect of outgassing on local temperature is directly proportional to the specific heat capacity of the gas, but assuming that this is not too different from that of the surrounding solid, Figure 29 shows the effect that this may have on local temperature. The dashed curves correspond to no thermal effect of outgassing ($c_g = 0$) and the solid curves correspond to $c_g = c_v$, with all other parameters as in Table 4.

Finally, the effect of the porosity dependence of composite thermal conductivity λ on local temperature is shown in Figure 30. The solid curves correspond to the

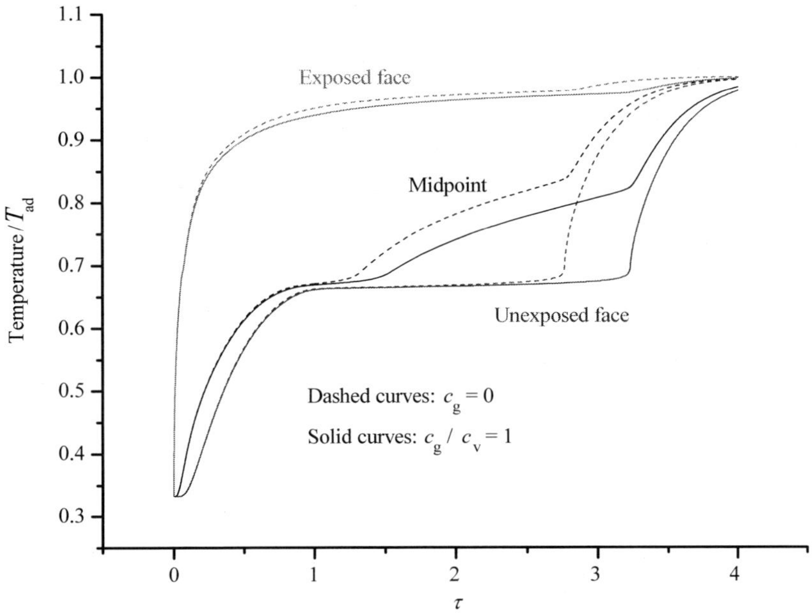

FIGURE 29

Effect of outgassing on local temperature.

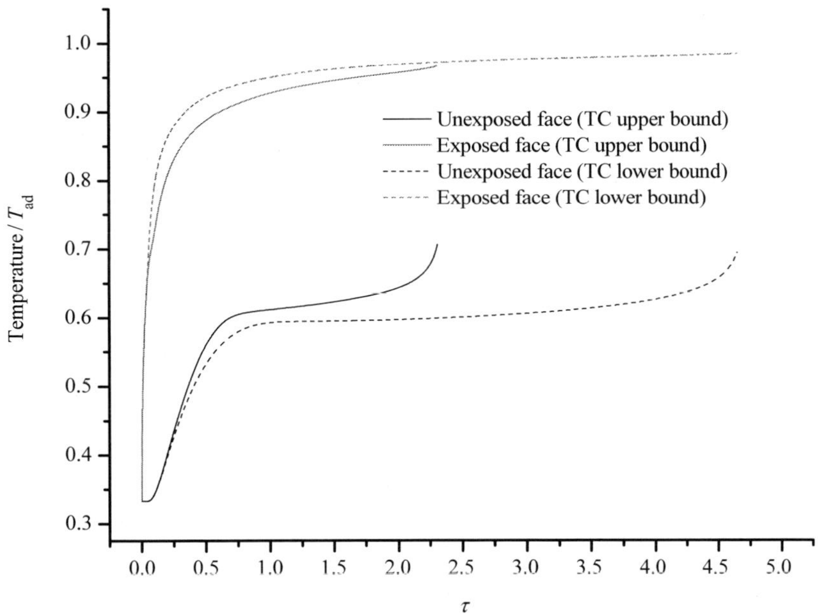

FIGURE 30

Effect of composite thermal conductivity on temperature.

thermal conductivity upper bound and the dashed curves to the thermal conductivity lower bound (defined in Eqn (37) above). Thus we see that for the range of possible composite thermal conductivities, there is a large effect on the time taken for the reactant to be exhausted.

5. CONCLUSION

The main thermal effect of endothermic gasification is to control the local temperature in a decomposing solid for a certain period of time and this chapter has demonstrated the central role of the CKT in this regard. During decomposition, the local temperature is determined from an interaction between the local heat transfer rate and the decomposition kinetics of the solid. However, the effect of local heat transfer rate (assuming that it is sufficient to drive the decomposition) is of a lower order than the kinetics and to a leading order of approximation, the local temperature will be close to the CKT. Broadly speaking, for a thin piece of material with no significant temperature gradients, the period for which local temperature is below some target temperature (assumed to be greater than the CKT) is illustrated in Figure 31 and given by the sum of the three terms $t^* = t_{i1} + t_k + t_{i2}$. Here, t_{i1} is an induction period

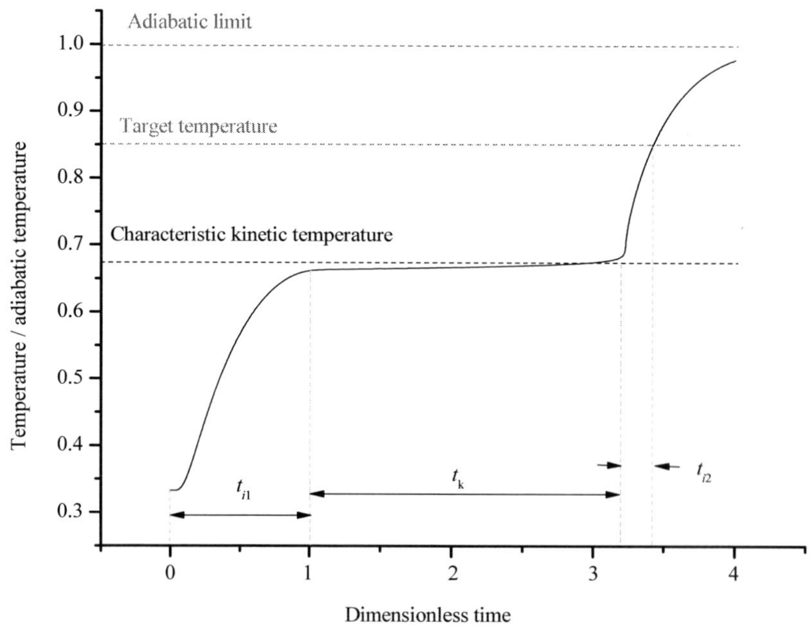

FIGURE 31

Total characteristic thermal time (no diffusion).

where the local temperature increases to the CKT. It is determined purely by the net heat flux through the exposed surface and the volumetric heat capacity. Once attained, the temperature remains in a neighborhood of the CKT for a period t_k, which is defined mainly by the local heat transfer rate, the amount of reactive component, and the endothermicity of the reactant. Once the reactant has been exhausted, the temperature increases again until the target is reached. This final period t_{i2} is again specified by the net heat transfer rate and the volumetric heat capacity. In general, the volumetric heat capacity during the initial induction period will be different to that in the final period t_{i2}.

When the solid is thick enough for thermal diffusion to play a part, the situation is more complicated. Figure 32 shows the numerical solutions of the model presented above and illustrates the variation of temperature and reactant volume fractions with time at the exposed surface, midway through the solid and at the unexposed surface of a thick solid. The gasification at the exposed surface occurs as soon as the local temperature gets sufficiently close to the CKT. At the midpoint, there is an induction period (labeled I on the figure) caused by diffusive thermal lag and then the temperature remains close to the CKT until all reactant has been consumed (region II). After this, there is an interplay between heat diffusing from above and also being diffused away to lower regions where further gasification is still occurring (region III). When all reactant throughout the entire solid has been consumed, the temperature increases purely by diffusion to the adiabatic limit

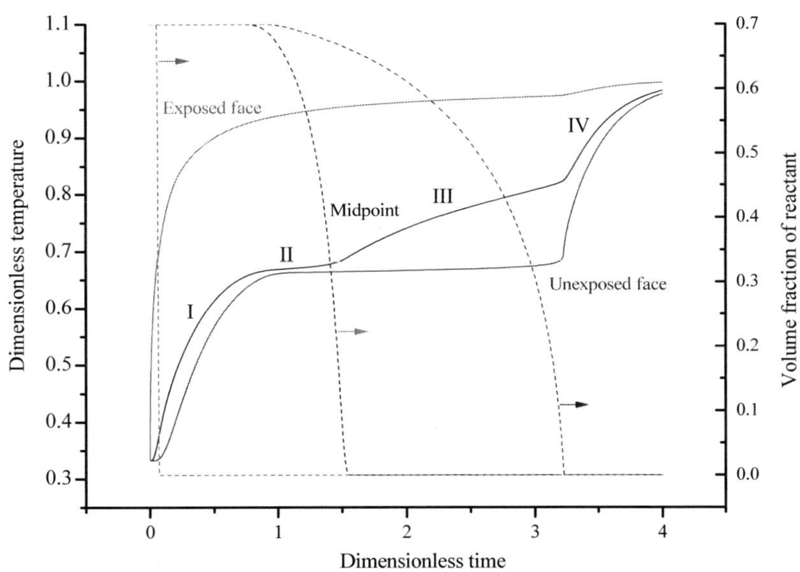

FIGURE 32

Total characteristic thermal time (with diffusion).

(region IV). The reactant adjacent to the unexposed face is the last to be consumed and since this face is assumed to be perfectly insulated, there can be no further gasification from regions below it and so region III of the thermal history of the midpoint is not present.

In a thick solid, there is an additional beneficial effect due to venting gas that is not commonly associated with hydrated fillers. Through-flow of gas during decomposition has a local cooling effect, which may contribute to the flame retardant effectiveness (assuming that the gas is inert). The net result of this effect is to modify the local heat flux by a term that is proportional to the local temperature, the specific heat capacity of the gas, and also the net production rate of gas. Numerical results show that this effect is by no means negligible and can significantly reduce the local temperature, which will result in lower decomposition rate of combustible material and hence lower heat release rate.

If pores form as gasification occurs, then two effects are likely. The basic composite thermal conductivity λ will reduce and radiation enhancement will cause the overall effective thermal conductivity λ_{eff} to increase with temperature approximately like T^3. The discussion above also highlights the important effect of pore shape and distribution on λ. It is apparent that as the dependency of λ on porosity φ should be such that as φ increases, situations such that $-\text{d}\lambda/\text{d}\varphi < \lambda/(1-\varphi)$ should be avoided. Radiation enhancement increases with both pore size and porosity. It transpires from the discussion above that in order to provide the best thermal protection, for a given porosity, it is better to form a large number of small pores rather than a small number of large pores.

References

[1] Hornsby PR. The application of hydrated mineral fillers as fire retardant and smoke suppressing additives for polymers. Macromol Symp 1996;108:203.

[2] Wang Z-Y, Liu Y, Wang Q. Flame retardant polyoxymethylene with aluminium hydroxide/melamine/novolac resin synergistic system. Poly Degrad Stab 2010;95:945.

[3] Gul R, Islam A, Yasin T, Mir S. Flame-retardant synergism of sepiolite and magnesium hydroxide in a linear low-density polyethylene composite. J Appl Polym Sci 2011;121:2772.

[4] Bellayer SP, Tavard E, Duquesne S, Piechaczyk A, Bourbigot S. Natural mineral fire retardant fillers for polyethylene. Fire Mater 2011;35:183.

[5] Abramowitz M, Stegun IA, editors. Handbook of mathematical functions. 9th ed. New York, USA: Dover Publications Inc; 1972. § 5.1.

[6] Freeman ES, Carroll B. The application of thermoanalytical techniques to reaction kinetics: the thermogravimetric evaluation of the kinetics of the decomposition of calcium oxalate monohydrate. J Phys Chem 1958;62:394.

[7] Hatakeyama T, Quinn FX. Thermal analysis: fundamentals and applications to polymer science. Chichester, UK: Wiley; 1994. § 5.3.

[8] Staggs JEJ. Savitsky–Golay smoothing and numerical differentiation of cone calorimeter MLR data. Fire Saf J 2005;40:493.

[9] Staggs JEJ. Thermal degradation of solids using single-step first-order kinetics. Fire Saf J 1999;32:17.

[10] BS 476—20. Fire tests on building materials and structures. 1987.

[11] BS EN 1991-1-2. Eurocode 1: actions on structures. 2002.

[12] Staggs JEJ, Phylaktou HN. The effects of emissivity on the performance of steel in furnace tests. Fire Saf J 2008;43:1.

[13] Staggs JEJ. Numerical characterisation of the thermal performance of static porous insulation layers on steel substrates in furnace tests. J Fire Sci 2011;29:177.

[14] Staggs JEJ, Crewe RJ, Butler R. A theoretical and experimental investigation of intumescent behaviour in protective coatings for structural steel. Chem Eng Sci 2012;71:239.

[15] Maxwell JC. A treatise on electricity and magnetism. 3rd ed. New York, USA: Dover Publications Inc; 1954.

[16] Bruggeman DAG. Berechnung verschiedener physikalischer konstanten von heterogenen substanzen. Ann Phys 1935;24:636.

[17] Staggs JEJ. Modelling the thermal conductivity of porous materials using thermal resistor networks. Fire Saf J 2002;37:107.

[18] Staggs JEJ. Thermal conductivity estimates of intumescent chars by direct numerical simulation. Fire Saf J 2010;45:238.

[19] Landau HG. Heat conduction in a melting solid. Q J Appl Math 1950;8:81.

[20] Leung CH, Staggs JEJ, McIntosh A, Brindley J. Modelling of a one-dimensional ablation process with char formation. Comb Sci Tech 1996;119:301.

[21] Staggs JEJ. A discussion of modelling idealised ablative materials with particular reference to fire testing. Fire Saf J 1997;28:47.

CHAPTER

11

Review of Recent Advances on the Use of Boron-based Flame Retardants

Kelvin K. Shen
Rio Tinto Minerals (Consultant), Denver, USA

CHAPTER OUTLINE

1. Introduction

Borates function as flame retardants, smoke suppressants, afterglow suppressants, and antitracking agents in both halogen-containing and halogen-free polymers. Boron compounds such as boric acid (BA) and sodium borates are well-known flame retardants for cellulosic products. However, the use of boron compounds such as zinc borates, ammonium pentaborate, melamine borate, boron phosphate, metal borophosphate, and borosiloxane as flame retardant in plastics and rubbers has become prominent only in recent years (Table 1).

The use of boron-based compounds in both halogen-containing and halogen-free polymers (up till 2008) was previously reviewed [1–3]. This chapter focuses only on recent advances and those developments that were not covered in the previous reviews, particularly on halogen-free polymers. Chemical–physical properties,

Polymer Green Flame Retardants. http://dx.doi.org/10.1016/B978-0-444-53808-6.00011-1

Table 1 Major Commercial Boron-based Flame Retardants

Chemical Name	Formula (Typical B_2O_3 wt%)	Starting Dehydration Temperature (°C)	Water Solubility (wt%, ~25 °C)	Applications
Borax pentahydrate	$Na_2O \cdot 2B_2O_3 \cdot 5H_2O$ (49.0%)	65	4.4	Wood/cellulose/cotton, urethane, coating
Borax decahydrate	$Na_2O \cdot 2B_2O_3 \cdot 10H_2O$ (37.5%)	~45	5.8	Wood/cellulose
Boric acid	$B_2O_3 \cdot 3H_2O$ (56.6%)	70	5.5	Wood/cellulose/cotton, urethane, coating
Boric oxide	B_2O_3 (98.5%)	–	3.1	Engineering plastics
Anhydrous borax	$Na_2O \cdot 2B_2O_3$ (68.8%)	–	3.1	Urethane, wire and cable
Disodium octaborate tetrahydrate	$Na_2O \cdot 4B_2O_3 \cdot 4H_2O$ (67.3%)	40	9.7 (20 °C)	Wood products, cotton
Calcium borate (colemanite)	$2CaO \cdot 3B_2O_3 \cdot 5H_2O$ (44–48%)	290	0.2	Rubber-modified roofing membrane
Calcium metaborate	$CaO \cdot B_2O_3 \cdot 4H_2O$	~115	Soluble	Coating
Barium metaborate	$BaO \cdot B_2O_3 \cdot H_2O$ (26% with ~90% purity)	200	0.3	PVC, coating
Zinc borates (Table 2)	$xZnO \cdot yB_2O_3 \cdot zH_2O$	–	–	Polymers, elastomers, coatings, sealants/caulks
Ammonium pentaborate	$(NH_4)_2O \cdot 5B_2O_3 \cdot 8H_2O$ (64.6%)	120	10.9	Epoxy, urethane, coating
Melamine diborate	$(C_3H_8N_6)O \cdot B_2O_3 \cdot 2H_2O$ (22.0%)	130	0.7	Epoxy intumescent coating, cotton textile
Boron phosphate	BPO_4 (18.7% as B)	NA	Low	PPE/polyamide, PPE/HIPS, PO
Boron phosphate (single molecule)	BPO_4 (18.7% as B)	Divetrify at around 1000 °C	Low	Coating
Magnesium borate (whisker)	$2MgO \cdot B_2O_3$ (46.4%)	None	Insoluble	Engineering polymers

Table 2 Major Commercial Zinc Borates		
Formula	**Approx. Starting Dehydration Temperature (°C)**	**Trade Name**
$2ZnO \cdot 3B_2O_3 \cdot 7H_2O$	170	ZB-237
$2ZnO \cdot 2B_2O_3 \cdot 3H_2O$	200	ZB-223
$2ZnO \cdot 3B_2O_3 \cdot 3.5H_2O$	290	*Firebrake*®ZB, ZB-467
$2ZnO \cdot 3B_2O_3$	NA	*Firebrake*®500
$4ZnO \cdot B_2O_3 \cdot H_2O$	>415	*Firebrake*®415

end-use applications, fire test performance, as well as mode of actions of boron-based flame retardants are reviewed. In particular, synergistic interaction between "boron and metal hydroxide," "boron and nitrogen," "boron and phosphorus," "boron and carbon," and "boron and silicon" in fire test performances will be illustrated.

2. Products and applications

2.1 Zinc borate

2.1.1 Zinc borates

Among all the commercial zinc borates, the multifunctional *Firebrake*®ZB ($2ZnO \cdot 3B_2O_3 \cdot 3.5H_2O$) is of the most commercial importance (Figures 1–3 and

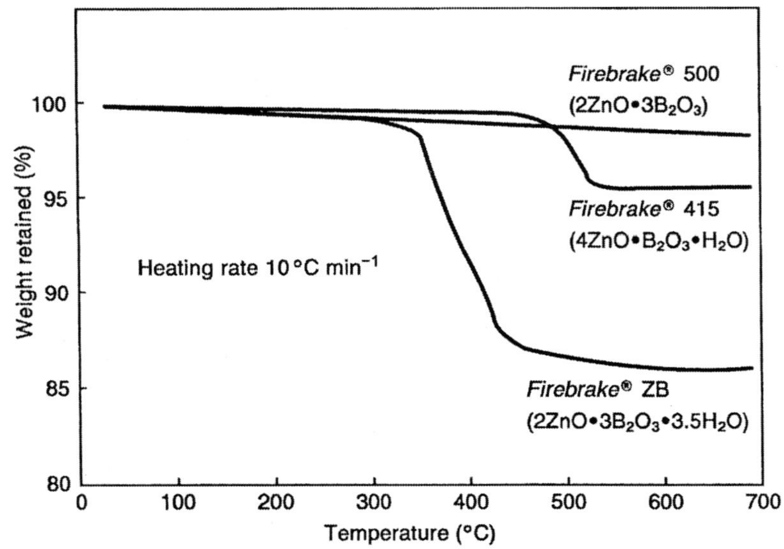

FIGURE 1

Thermogravimetric analyses (TGA) of *Firebrake*® zinc borates.

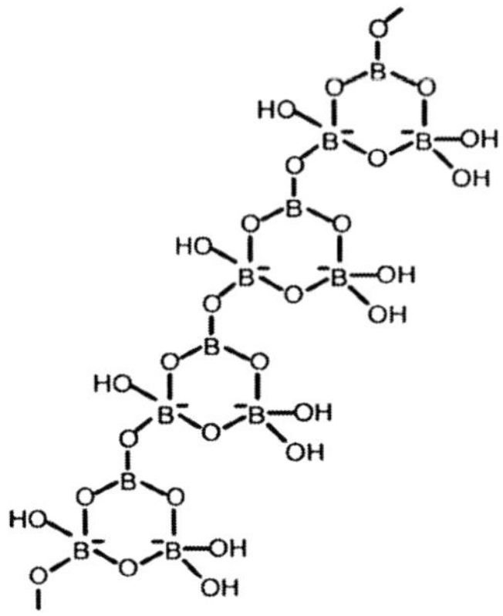

FIGURE 2

Molecular structure of *Firebrake*®ZB (zinc atoms complexing with oxygen atoms are not displayed).

Table 2) [1–3]. Figure 1 displays the thermogravimetric analysis of different zinc borates. A recent single-crystal X-ray crystallography study showed that *Firebrake* ZB has a structure of $Zn[B_3O_4(OH)_3]$ (Figure 2) [4]. Figure 3 (differential thermal analysis) illustrates the dehydration/crystallization/melting temperatures of *Firebrake* ZB. These zinc borates have been used as flame retardants in both halogen-containing and halogen-free polymers such as PVC [5], polyamide [6], polyolefin [7,8], epoxy [9], phenolics, and various elastomers [10]. *Firebrake* 500

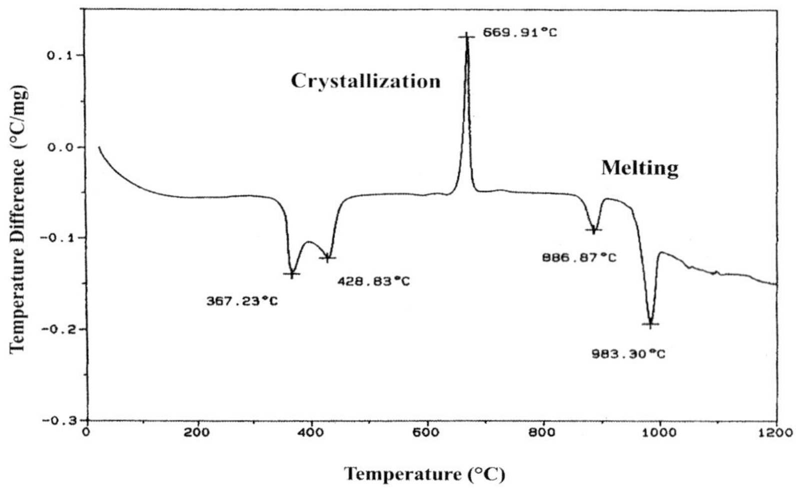

FIGURE 3

Differential thermal analysis (10 °C/min) of *Firebrake*®ZB.

($2ZnO \cdot 3B_2O_3$) and *Firebrake* 415 ($2ZnO \cdot 3B_2O_3 \cdot 3.5H_2O$) are used mostly in engineering polymers such as polyamides and polyetherketones that are processed at 300 °C or above.

In halogen-containing systems, zinc borates can provide the following benefits:

- Synergist of most halogen sources—Zinc borates are synergists of both chlorine- and bromine-containing flame retardants. They are particularly effective toward aliphatic halogens. Zinc borates can react with hydrogen halide to form zinc chloride, zinc hydroxychloride, boric oxide, and water. The zinc species remaining in the condensed phase can alter the pyrolysis chemistry by catalyzing the dehydrohalogenation and promoting cross-linking between polymer chains, resulting in increased char formation and a decrease in both smoke production and faming combustion. The released boric oxide, a low melting glass, can stabilize the char and inhibit afterglow through glass formation on the char surface.
- Synergist of antimony oxide—In certain polymer system, *Firebrake* ZB can display synergistic effects with antimony oxide in fire retardancy.
- Can either partially or completely replace antimony oxide; her efficacy depend on both the polymer used and halogen source used (i.e. aliphatic halogen versus aromatic halogen).
- Function as smoke suppressants—In contrast to antimony oxide, *Firebrake* ZB is predominately a condensed phase flame retardant and can function as a smoke suppressant.
- Function an afterglow suppressants—The B_2O_3 moiety in *Firebrake* ZB is responsible for its afterglow suppression effect.
- Promote char formation and prevent dripping—The zinc oxide moiety can promote char formation with halogen sources; the B_2O_3 moiety can stabilize the char.
- Improve thermal stability of some bromine/antimony trioxide fire retardant package.
- Improve the comparative tracking index (CTI) that is used to measure the electrical breakdown (tracking) properties of surface of polymer—The B_2O_3 moiety can improve antitracking and antiarcing property of electrical/electronic parts.
- Can form a porous, ceramic-like residue such as in flexible PVC-containing zinc borate, antimony trioxide, and aluminum trihydrate (ATH)—see also in halogen-free polymers below.

In halogen-free systems, zinc borates can provide the following benefits:

- *Firebrake* zinc borates reduce the heat release rate and smoke evolution in most metal hydroxide-containing polymers. The main benefit of *Firebrake* zinc borates is promoting the formation of a strong char/ceramic residue that prevents burning drips, delays oxidative pyrolysis, and protects underlying polymers.
- The use of co-additives such as silicone/silica, melamine polyphosphate, phosphate ester, red phosphorous, ammonium polyphosphate (APP), melamine

cyanurate, and nanoclay can augment the fire test performance of zinc borate/metal hydroxide system.

- When metal hydroxides and *Firebrake* zinc borates are used together, generally it is recommended that the ratios of zinc borate to ATH or magnesium hydroxide (MDH) are about 1:10 to about 2:10.
- In some polymers and in the absence of metal hydroxides, zinc borates can display synergy with phosphorus compounds, nitrogen-containing compounds such as melamine cyanurate, and silicone/silica.
- In some polymer system such as polyethersulfone, zinc borates alone can function as effective flame retardants.

Recently several research groups reported their efforts in developing zinc borate with nanoparticle size, different morphologies, different surface treatment, and even different coatings. The most interesting examples are as follows.

Nanosized zinc borate: Wu et al. [11] reported the preparation of nanosized zinc borate ($2ZnO \cdot 3B_2O_3 \cdot 3.5H_2O$) by reacting sodium borate and zinc chloride in the presence of a surfactant. They evaluated the use of this zinc borate in polyolefin/APP/pentaerythritol (PER). It can improve flame retardancy as evidenced by the UL-94 performance and Cone Calorimeter (heat release rate (HRR)). It is claimed that the zinc borate can cause the LDPE/APP/PER to form stable graphite-like char with aromatic hydrocarbon. In addition, boron phosphate was also detected in char. Both graphite and boron phosphate are believed to promote the formation of a stable char.

Nanowhisker zinc borate: Zheng et al. [12] reported the preparation of nanowhisker of zinc borate ($4ZnO \cdot B_2O_3 \cdot H_2O$) with the use of borax, zinc nitrate, and a phosphate ester template. They found the nanowhisker zinc borate can increase thermal stability of polyethylene significantly.

Nanoflake zinc borate: Li et al. [13] reported the preparation of oleic acid treated zinc borate nanoflakes ($2ZnO \cdot 3B_2O_3 \cdot 3.5H_2O$). The addition of this hydrophobic zinc borate nanoflake can increase the thermal stability of polyethylene at only 2% loading.

Nanodisc zinc borate: Zinc borate ($2ZnO \cdot 3B_2O_3 \cdot 3H_2O$) nanodiscs were prepared by using sodium borate/zinc sulfate and oleic acid [14]. The nanodiscs have average diameter from 100 to 500 nm and thickness about 30 nm. No evaluation of this material was carried out for fire retardancy performance.

Zinc molybdate-coated zinc borate: Recently Isarov et al. reported the preparation of zinc molybdate (high surface area shell) precipitated onto zinc borate core [15]. It was claimed that the resulting product enhances the borate/molybdate synergy and provides smoke suppression in flexible PVC (ASTM E662 smoke test) and flame retardancy in ethylene-vinyl acetate (EVA) (UL-94 test).

Zinc borate-coated silica: Spherical SiO_2 particles were coated with zinc borate ($5ZnO \cdot 2B_2O_3$) through a self-assembly process [16]. The obtained SiO_2—zinc borate core—shell nanospheres are claimed to have perfect spherical shape (core with a mean diameter of 40 nm and shell thickness 5 nm). No fire retardancy evaluation of this material was reported.

2.1.2 Zinc borate and metal hydroxides

Polyolefin: The use of *Firebrake* ZB ($2ZnO \cdot 3B_2O_3 \cdot 3.5H_2O$), *Firebrake* 415, and *Firebrake* 500 with metal hydroxides in polyolefin was extensively reported by Shen, Bourbigot, and Duquesne [17] and others. When *Firebrake* ZB is used in conjunction with metal hydroxide, it can function as a sintering aid to promote the formation of a strong char/ceramic residue. As a result, partial replacement of metal hydroxide with *Firebrake* zinc borates in polyolefin can reduce rate of heat release, smoke evolution, and carbon monoxide generation. In addition, *Firebrake* zinc borates can improve mechanical properties (except tensile strength), electrical properties, and processability of EVA/metal hydroxide systems.

Figure 4 illustrates the effect of *Firebrake* zinc borates on HRR of MDH/silica system. In addition to the significant reduction in the HRR, *Firebrake* zinc borate promotes the formation of an expanded char and vitreous layer. It is interesting to note that, compared to *Firebrake* ZB and ZB-Fine, *Firebrake* 500 gives the lowest HRR in the EVA/MDH/silica system. It is believed that *Firebrake* 500 (an anhydrous zinc borate) can promote sintering (ceramic formation) much more readily than regular *Firebrake* ZB (i.e. without the need of undergoing endothermic dehydration prior to the promotion of sintering) [8].

Basfar et al. [18] recently claimed the use of MDH and magnesium carbonate (huntite/hydromagnesite) in a radiation cross-linked EVA [VA (vinyl acetate) 25%]/LLDPE for ECO wire and cable applications. It was claimed that *Firebrake* ZB provides synergistic effect on flame retardancy when used in combination with metal

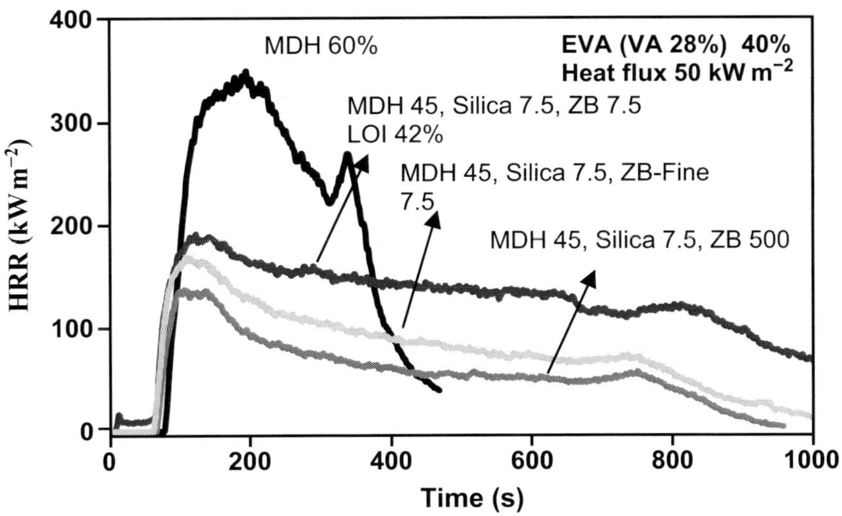

FIGURE 4

HRR curves of MDH/silica/*Firebrake* zinc borates in EVA (28%VA) [8]. HRR, heat release rate.

hydroxide. The typical formulation consists of EVA (90 parts), LLDPE, and PEgMA compatibilizer (10), MDH (70), Mg carbonate (65), *Firebrake* ZB (5), ammonium octamolybdate (2), talc (2), PTFE (0.5), coupling agent (2), and stabilizers. Interestingly enough, the zinc borate was claimed to improve physical properties of cross-linked product (i.e. with much better Hot Set Test performance (IEC60811)).

Epoxy: The use of a combination of zinc borate and ATH was reported being used in ablative coatings for rocket launcher [9]. Formicola et al. [19] recently reported the use of a combination of ATH and zinc borate in epoxy for aerospace application. Their cone calorimeter results demonstrated that zinc borate promotes the formation of a strong char via sintering with alumina and results in substantial decrease of pHRR, aveHRR of epoxy resin. Synergy was observed when the total loading reaches 20% or higher at 1:1 ratio of zinc borate:ATH.

Polyurethane: Henkel (Thiele et al. [20]) claims the use of ATH (72%) in a reactive PU adhesive that has a specific caloric values of not more that 12 $MJ\,kg^{-1}$ (EN ISO 1716). The adhesive can thus be applied in a maximum density of 333 $g\,m^{-2}$ so that the sandwich bonds have greater strength. In addition, sandwich bonds of mineral wool panels and metal outer layers can meet the requirements of A2 classification of EN 13501-1. Upon partial replacement of ATH with zinc borate (68% ATH and 4% zinc borate), the combination gives the same caloric value but it also meets the more stringent requirements of tube test (DIN 4102-1).

2.1.3 Zinc borate and nitrogen compounds

A combination of boron-based and nitrogen-based compounds can provide synergy in fire retardancy [1]. For example, a combination of BA and nitrogen compounds, such as urea or guanidine, is commonly used as flame retardant in wood product. Ammonium pentaborate [$(NH_4)_2O \cdot 5B_2O_3 \cdot 8H_2O$] (APB) is an effective flame retardant in intumescent coating, polyurethane, and paper products. Upon thermal decomposition, APB first gives off a large amount of water starting at about 120 °C. At about 200 °C, it starts to give off ammonia and, at about 450 °C, it is all converted to boric oxide which is a glass former. APB functions as both an inorganic blowing agent and a glass forming fire retardant. PPG industries patented the use of zinc borate and APB in epoxy intumescent coatings.

Melamine diborate ($C_3H_6N_6 \cdot 2H_3BO_3$) is commercially available from Budenheim for fire retarding textile and coatings. The mode of action between B and N is not well understood presumably due to the formation of boron nitride at high temperature.

Thermoplastic polyester: Until recently, the interaction between zinc borate and nitrogen compound is not widely reported. Sony reported the use of a combination of melamine cyanurate (MC) and zinc borate in a halogen-free PET flat cable adhesive (Table 3) [21]. This combination provides not only good fire retardancy but also good electrical resistance and adhesion; whereas the APP/MC combination performed poorly in electrical resistance and adhesion. In addition, zinc borate also works well with melamine or benzoquanamine in this system. The mode of action is probably due to the release of NH_3/melamine and possibly some boron nitride formation.

Table 3 Halogen-free Flat Cable (PET) [21]

	Examples (pbw)					
	1	2	3	4	5	6
Components						
PET	100	100	100	100	100	100
Decabromodiphenyl oxide	60	–	–	–	–	–
Melamine cyanurate	–	–	100	100	–	–
Ammonium polyphosphate	–	–	–	–	100	–
Antimony trioxide	20	–	–	–	–	8
Zinc borate	–	–	10	50	–	–
ATH	–	–	–	–	–	–
Properties						
UL-VW1 (SUB 758)	–	X	O	☺	☺	X
Combustion gas	X	O	O	O	O	O
Electrical resistance (Ω)	$>10^{11}$	$>10^{11}$	$>10^{11}$	$>10^{11}$	$<10^{10}$	$>10^{11}$
Adhesion (kg cm^{-1})	>0.5	>0.5	>0.5	>0.5	<0.3	>0.5

☺ *implies the best performance*

Epoxy: Henkel reported the benefit of using a combination of *Firebrake* ZB and melamine cyanurate in halogen-free epoxy encapsulant for semiconductors (Table 4) [22]. The usage level of MC is limited by its negative impact on the flowability of the epoxy compound.

Polyurethane (nitrogen-containing polymer): Yildiz et al. [23] reported that zinc borate ($2ZnO \cdot 3B_2O_3 \cdot 7H_2O$) at 0.5—7.0% loading can increase the flame retardancy of polyurethane as evidenced by the horizontal burn test. Interestingly enough, the zinc borate can also improve oxidative stability of polyurethane significantly as evidenced by the oxygen induction time test and weathering chamber test. The zinc borate actually outperforms commercial light stabilizer (Tinuvin B75).

2.1.4 Zinc borate and phosphorus compounds

Polypropylene: The use of zinc borate and APP (Figure 5) in epoxy is well documented [9]. Interestingly enough, Bourbigot et al. [24] reported the preparation of a neutralized intumescent flame retardant (NIFR) by reacting petaerythritol, phosphoryl trichloride, and melamine (Figure 6). The addition of a small amount of *Firebrake* 500 (2 wt% at the expense of NIFR) in PP leads to a dramatic improvement in the fire test performances [Limiting Oxygen Index (LOI) and V-0 rating]. The zinc borate promotes the formation of a cohesive intumescent barrier.

Polyphenylene oxide: The synergy between zinc borate and phosphate ester in flexible PVC has been well documented [5]. Recently, the use of phosphate ester [resorcinol bis[bis(2,6-biphenylphspate)]] (Figure 7) and *Firebrake* ZB-Fine (fine grade of $2ZnO \cdot 3B_2O_3 \cdot 3.5H_2O$) in cross-linked polyphenyleneoxide (PPO) was

Table 4 Epoxy Encapsulant for Semiconductors [22]

	Examples (wt%)	
	1	2
Composition		
Bisphenol-A epoxy (Epon 1001F)	40.04	40.04
Bisphenol-A 2004	7.46	7.46
Amine binder	8.69	8.69
Melamine cyanurate	–	1.0
Firebrake ZB (zinc borate)	-	6.0
Magnesium hydroxide	-	4.0
Fused silica	37.06	26.06
Triphenyl phosphine	0.2	0.2
Other additives	6.55	6.55
Properties		
UL-94 (1/4 inch) Total burn time (s)	26	Burn to clamp
Rating	Fail	V-0

reported by Kyocera [25]. This combination exhibits good halogen-free flame retardancy and excellent characteristics such as dielectric property, heat resistance, moisture resistance, and corrosion resistance for printed wire board application (Table 5). In a subsequent patent, Kyocera claimed the use of 9,10-dihydro-9-oxa-10-phosphaphenanthrene-10-oxide (DOPO) derivative (Figure 8) and *Firebrake* ZB-XF (extra fine grade $2ZnO \cdot 3B_2O_3 \cdot 3.5H_2O$) is PPO for PWB application [26].

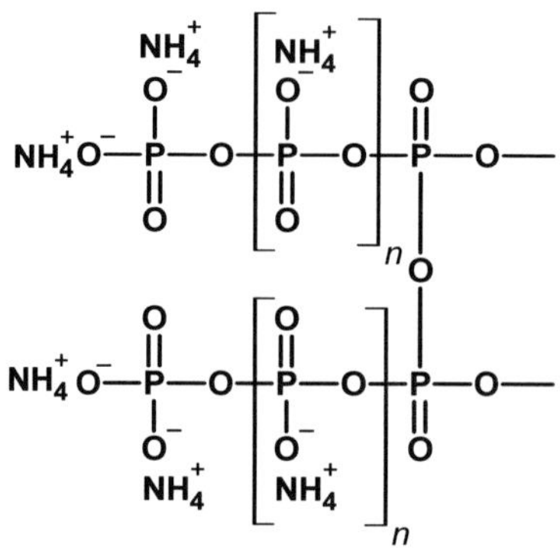

FIGURE 5

Ammonium polyphosphate.

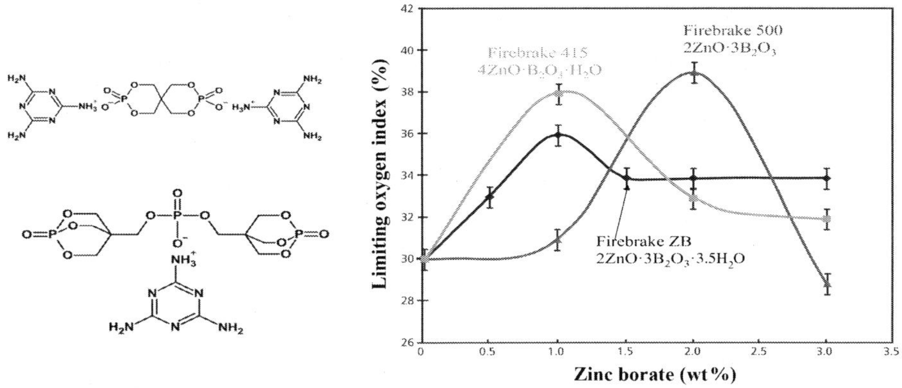

FIGURE 6

Firebrake® zinc borates and N/P intumescent retardants in PP (30%) [24].

Polyamides (PA): Zinc borates are used in both halogen-containing and halogen-free PA and high-temperature PA (HTPA) [2,27]. Melamine polyphosphate (MPP) and melamine pyrophosphate (MPyP) (Figure 9) have been used to fire retard PAs. MPP works well in PA 66 but fails to improve flammability characteristics of PA 6 at reasonable levels. Martens et al. [28] reported the use of unspecified zinc borate in conjunction with MPyP in achieving V-0 at 1.6 mm. Partial replacement of MPyP with zinc borate can increase the CTI from 275 to 600 V.

Clariant developed a unique product-based aluminum diethylphosphinate (AlDPi) (Figure 10) mostly for PAs and PBT. It is the dominant fire retardant on the market for halogen-free PAs. AlDPi is a hydrophilic fine powder with a very low residual solubility in water. It does not melt before it starts to decompose at temperatures >350 °C [28]. Clariant offers three major products for the PA industry—Exolit OP 1230 (AlDPi), Exolit OP 1311 (blend of AlDPi and MPP), and Exolit OP 1312 (a blend of AlDPi, MPP, and zinc borate). In 25% glass-reinforced PA 66, OP 1311 and OP 1312 can achieve V-0 at 1.6 mm with 18%

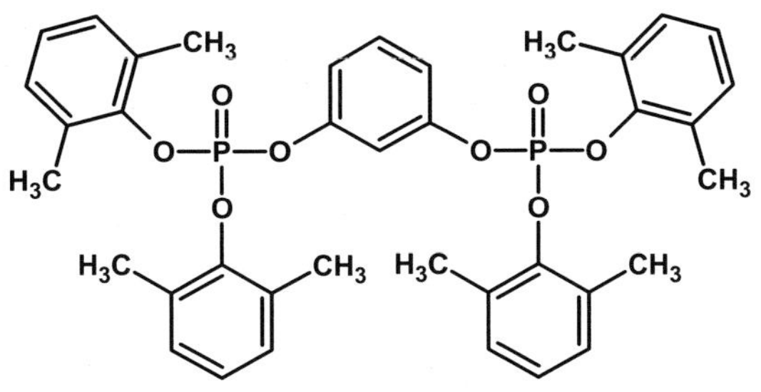

FIGURE 7

Resorcinol-bis[bis(2,6-biphenylphosphate)].

Table 5 Printed Wire Board Based on PPO/PS [25]

	Examples (parts by wt)		
	1	2	3
Components			
PPO (MA-modified)	50	50	50
Triallyl isocyanurate	46	46	46
Polystyrene (GP)	4	4	4
Bis(1,2-pentabromophenyl)ethane	20	–	–
Sb_2O_3	4	–	–
Phosphate esters (PX-200)	–	15	15
Silica (7 μm)	15	15	-
Firebrake ZB-Fine	–	–	15
Peroxide	6	6	6
Properties			
UL-94 (1.6 mm)	V-0	V-0	V-0
Peel strength ($KN\,M^{-1}$)	1.2	1.1	1.0
Insulation resistance (Ω)	1×10^{14}	4×10^{14}	4×10^{14}
Glass transition (°C)	170	160	165
Dielectric constant (MHz)	3.6	3.6	3.5
Soldering heat resistance (260 °C)	Good	Good	Good

and 16% loading, respectively. Both formulations can achieve even UL 94 5VA at 20% and 18% loadings, respectively.

Kaynak et al. [29] reported a comparison of fire test performance of OP 1312 versus brominated epoxy oligomer/antimony trioxide in PA 66/GF. They showed that the OP 1312 package outperforms the brominated system in terms of peak heat release, time to ignition, and fire growth index. The zinc borate in OP 1312 is a thermal stabilizer. OP 1312 in PA is highlighted by a low loading level (compared to about 27% with brominated polystyrene/Sb_2O_3 system), low material

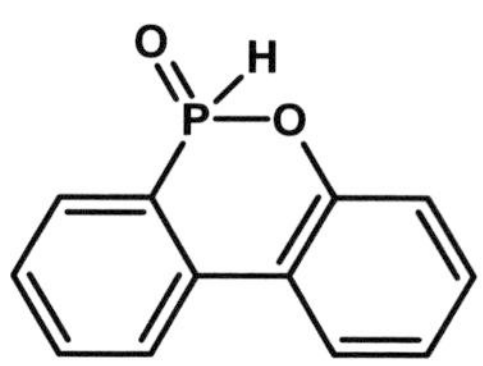

FIGURE 8

DOPO (9,10-dihydro-9-oxa-10-phosphaphenanthrene-10-oxide).

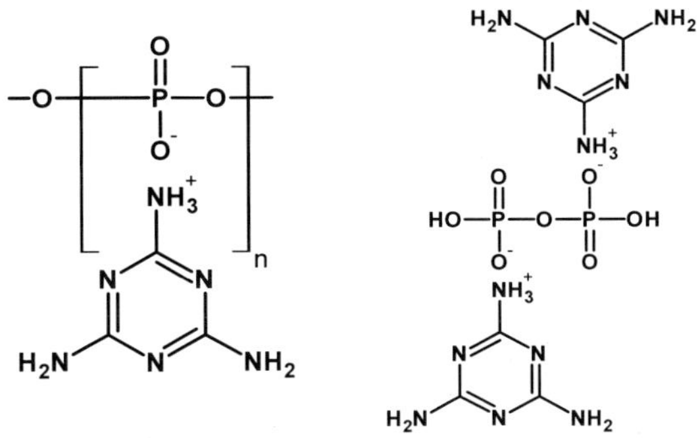

FIGURE 9

Melamine polyphosphate (left) and melamine pyrophosphate (right).

density, and a high CTI (>600 V). OP 1312 is not thermally stable enough for use in HTPA.

Martens et al. [31] reported the use of a combination of OP 1230 (AlDPi only) and *Firebrake* 500 (2%) in a HTPA—PA6T/66 (alloy of polyhexamethylene terephthalate and polyhexamethylene adipamide) with 30% glass fiber. It was claimed that this formulation possesses excellent flame retardancy, good mechanical properties, high CTI, and low density. In addition, the composition can be molded without generating significant mold deposit. The thermal stabilizing effect of *Firebrake* zinc borate in OP 1312 is also illustrated by the color stability in Table 6. Formulation 4 in the table is a halogen-containing MXD6 (poly metaxylene adipamide) from Solvay (IXEF 1521 0008) [30]. Clariant recently reported that PA 66 containing AlDPi (11.4%), MPP (5.7), talc (10), and *Firebrake* 500, and glass fiber (20%) can achieve UL 94 V-0 at 0.6 mm and GWIT of 775 °C at 1 mm and 800 °C at 2 mm [32].

More recently, Yin of DuPont claimed the use of *Firebrake* 500 in a HTPA 6T/66 containing AlDPi (12%) can reduce corrosion of processing equipment [33].

Braun and Schartel et al. [34] reported the mode of actions of AlDPi, MPP, and zinc borate in PA 66/GF. When used alone, most AlDPi vaporizes (as phosphinic acid) and acts effectively in the gas phase. When MPP is added, the mode of action shifts to condensed phase by forming aluminum phosphate. The

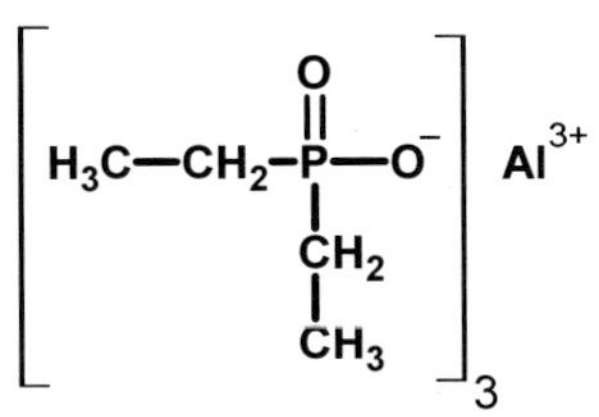

FIGURE 10

Aluminum diethylphosphinate.

Table 6 Flame Retardant Halogen-free Polyamide MXD6 [30]

	Examples (wt%)			
	1	2	3	4[a]
Component				
PA MXD6 (IXEF 6002)	42	42	42	42
Fiberglass	50	50	50	50
Bromine/antimony	–	–	–	?
Exolit OP 1230	–	–	8.0	–
Exolit OP 1311	–	8.0	–	–
Exolit OP 1312	8.0	–	–	–
Properties				
UL 94 (0.8 mm)	V-0	V-0	V-1	V-0
Degradation onset temperature (°C)	387	387	386	372
Discoloration during molding	No	Yes	No	Yes
Acid out gas during molding	No	No	No	Yes

[a] *Formulation 4 is a halogenated fire retardant MXD6.*

insulating char barrier becomes more pronounced with the addition of zinc borate (via the formation of boron–aluminum phosphate layer) (Figure 11). The formation of boron–aluminum phosphate layer is believed to be indispensable for the sample to meet the UL-94 5V test.

2.1.5 Zinc borate and carbon

Afterglow/smoldering combustion is a nonflaming combustion of carbon (char). Some flame retardant polymers systems are prone to afterglow/smoldering combustion; borates are known to be afterglow/smoldering combustion inhibitors. It is generally believed that borates can form a glassy layer on the surface of the char and inhibit the nonflaming oxidation. It was also proposed that borates can remove free electrons from the active sites with the result that the sites are no longer capable

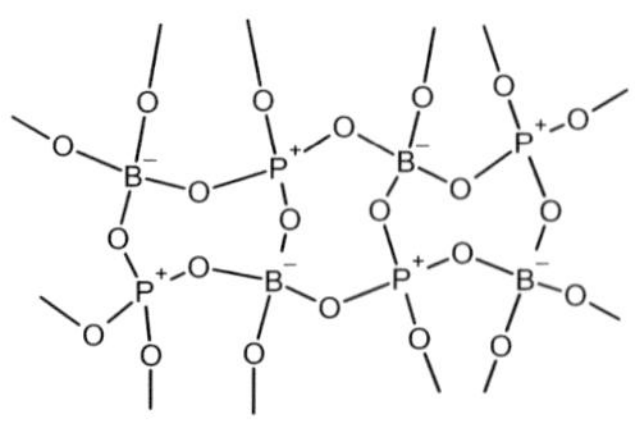

FIGURE 11

Boron phosphate.

Table 7 Flame EVA with Expandable Graphite [35]

	Examples (parts by wt)			
	1	2	3	4
Components				
EVA (15% VA, MFR 1.5 9/10 min)	100	100	100	100
Expandable graphite	0	10	0	10
Zinc borate	0	0	20	20
Properties				
UL-94 (3.2 mm)	NR	NR	NR	V-0
Oxygen index (%)	21.0	25.0	20.0	27.5

of reacting. In addition, the inhibiting borate molecule may be covering the active sites so that, geometrically, oxygen is not able to attack the surface [35].

Polyolefin: Nakasaki [36] first reported the benefit of using a combination of *Firebrake* ZB and expandable graphite in EVA (Table 7).

In the USA, approximately 225 million new wooden pallets enter the supply chain each year. Recently, a significant amount of wooden pallets are replaced by plastic pallets. High-density polyethylene (HDPE) is the preferred material due to ease of processing. Muirhead [36] claimed the use of HDPE layer containing mixed metal hydroxides, expandable graphite, zinc borate, and antimony trioxide for the body of radiofrequency transparent wooden pallet. The fire retardant layer is claimed to have a peak HRR between 300 and 500 $kW\,m^{-2}$ at heat flux of 35 $kW\,m^{-2}$ (ASTM E1354).

Syndiotactic polystyrene (S-PS): Table 8 illustrates the effect of zinc borate on afterglow combustion of carbon black in S-PS [37]. In the presence of carbon black, S-PS plus bromine/Sb can result in high afterglow time. Partial replacement of antimony trioxide with zinc borate results in considerable reduction of afterglow combustion and V-0 performance.

2.1.6 Zinc borate and silicon-containing compound

In recent years, silicon-based flame retardants have attracted considerable attention by many research groups and industrial fire retardant suppliers. Recently Shen [38] reviewed the use of zinc borate in conjunction with different types of silicon-based flame retardants such as silica (fumed silica, fused silica, and silica gel), silicone (polydimethylsiloxane and its derivatives), metallosilicate/silicate minerals, silicate glass/frits, and so on.

Silicone: Sharp (Raychem) claimed the use of zinc borate (and/or BA) and ATH in a silicone cable sheathing composition [39]. The use of borate and ATH provides much better thermal insulation during fire testing. It is believed that boric oxide from zinc borate may react with silica (derived from silicone combustion) to form borosilicate. Recently, Czubarow (Saint Gobain) reported the use of a combination of

Table 8 Ignition Resistance of Syndiotactic Polystyrene (S-PS) [37]

	Examples (wt%)			
	1	2	3	4
Components				
S-PS, brominated FR				
Plus other additives	96.7	96.0	97.0	97.0
Sb_2O_3	3.3	4.0	1.5	2.0
Firebrake®ZB	–	–	1.5	1.0
Properties				
Uncolored sample				
UL-94 (1.6 mm)	V-0	V-0	V-0	–
Avg. afterglow time (s)	8.6	9.0	9.4	–
Sample with carbon black				
UL-94 (1.6 mm)	V-1	V-1	V-0	V-0
Average afterglow time (s)	32.8	34.8	13.8	18

alumina (80%)/Fe_2O_3(2)/*Firebrake* ZB(2)/ATH(2) in a Pt-catalyzed RTV (room temperature vulcanization) silicone to achieve V-0 [40]. The electrical insulating sheeting is thermally conductive and electrically insulative.

Alexander [41] reported that the use of a combination of mica (25%)/glass frit (1.5)/MDH (10) in peroxide cured silicone to produce a self-supporting ceramic material at 1050 °C. Addition of zinc borate (1.5%) resulted in not only better mechanical properties but also much better fire resistance properties as evidenced by the mass loss measurement. The composition is suitable for partitions, wall linings, fire door inserts, door seals, or cables.

Polyolefin: Shen et al. [38] reported the benefit on HRR when SFR 100 (PDMS-a linear silicone and silicone resin blend) is used in conjunction with MDH and *Firebrake* ZB-Fine in EVA. The peak HRR can be reduced drastically from 257 to 131 kW m^{-2}. SEM of the pyrolysis residues of ATH/*Firebrake* ZB/SFR100 in EVA displayed a network of particles (presumably silica) on the surface of the residue. IEC 332-1 single wire burn test also confirmed a stronger char formation.

Recently Paul (PP Polymers) [42] claimed the use of a combination of silicone, *Firebrake* ZB, ATH, and a zinc or sodium ionomer in PP to achieve low HRR in PP. Cogen et al. [43] reported the use of ethylene polymers, metal hydroxides, ultra-high-molecular-weight polysiloxane, and zinc borate for wire and cable applications with excellent processability.

Silicone–polyetherimide: Borates are known to be effective flame retardants in both silicone and polyetherimide. Recently Bhandari and Gallucci [44] reported the use of *Firebrake* 500 or *Firebrake* 415 (about 5%) and PTFE in silicone–polyetherimide to meet the Flame Spread Index requirement (<35, ASTM E162) for transportation application.

2.2 Boric acid ($B(OH)_3$)/trialkyl borate ($B(OR)_3$)/sodium borate

BA/trialkyl borate/sodium borate are widely used in many different fire retardant applications. Some recent examples are presented as follows.

Cellulose: A flame retardant composition consisting of an oligomeric phosphate ester and a silica sol–gel in an alcohol solvent is claimed for fire retarding lignocellulosic-based material such as honeycomb paper structures that are frequently used in doors, walls, furniture, and so on [45]. More importantly, addition of BA derivatives (trialkyl borates) to this fire retardant composition provides positive effect on the durability of the cellulosic products, protection against biological degradation, and improvement in fire resistance.

Polypropylene: Wang et al. [46] reported the use of BA as a catalyst in PP containing a novel intumescent flame retardant (IFR) based on bis (2, 6, 7-trioxa-1-phosphabicyclo [2,2,2] octane-1-oxo-4-hydroxymethyl) phenylphosphonate. They demonstrated that BA has a synergistic effect with the IFR. For example, addition of 1 wt% of BA to 29 wt% of IFR in PP gives an LOI of 34.2% and UL-94 V-0 rating, whereas 30% loading of IFR give an LOI of only 30.3. In addition, the incorporation of BA can improve the thermal stability and promote the formation of carbonaceous charred layer. This is probably due to the fact that BA can cause cross-linking of polyphosphoric acid.

Epoxy: BA, zinc borate, or combination of these two boron compounds have been used in conjunction with APP and other nitrogen compounds in epoxy intumescent coatings [1]. Duquesne and Bourbigot et al. [47,48] reported that when APP is used in conjunction with BA in epoxy, the char developed remains well stuck on the steel plate. The backside of the steel plate reaches 400 °C in 29.5 min versus 4 min for epoxy resin alone. They also observed that in the presence of APP, BA is a key ingredient to reduce the drop of viscosity around 350 °C during epoxy degradation.

Polyurethane foam: Nishyama et al. reported the preparation of flame retardant. PU insulation panel. Open-cell rigid PU foam was vacuum impregnated with a mixture of BA and borax (44:56 ratio in water—23 wt%) at 90 °C [49]. The treated foam is sandwiched between polyethylene-coated aluminum sheets. The product has a Gross Calorific Value (ISO5660 Cone Calorimeter test) of 0.1 MJ m^{-2} after 10 cycles of impact test.

Elastomers: Saito et al. [50] reported the use of a combination of expandable graphite, ATH, and BA in butyl rubber/styrene rubber as a sealing material. BA is claimed as a shape retention agent after thermal expansion. This patent again illustrates the benefit of expandable graphite/borate interaction (see also "zinc borate-carbon" section).

2.3 Metal borophosphate

In the presence of phosphorus compounds such as APP and BA or zinc borate, one can generate boron phosphate or metal borophosphate that can help to maintain the integrity of the char [47,48]. Schartel et al. [51] reported that addition of MPP/zinc borate to aluminum diethylphosphinate in polyamide causes the change of mode of

Table 9 Flame Retardant PA 6,6 with AlDpi and Metal Borophoshate [52]

	Examples (wt%)						
	1	2	3	4	5	6	7
Components							
PA 6,6	60	60	60	60	60	60	60
Glass fiber	20	20	20	20	20	20	20
Exolit 1312 (al diethylphosphinate: MPP 2:1 plus zinc borate)	20	17.5	15	–	17.5	15	–
Exolit 1240 (al diethylphosphinate)	–	–	–	13.4	–	–	13.4
Mg borophophate	–	2.5	5	6.7	–	–	–
Zn borophosphate	–	–	–	–	2.5	5	6.7
Properties							
UL-94 (0.8 mm)	V-0	V-0	V-0	V-1	V-0	V-0	V-1

action from gas phase to a mixed gas/condensed phase. They detected the formation of boron phosphate and aluminum phosphate in the condensed phase.

Recently Ewald et al. [52] of BASF reported the preparation of different metal borophosphates. These compounds were found to exhibit high thermal stability and are useful as flame retardants for plastics. Table 9 illustrates the use of magnesium borophosphate and zinc borophosphate in conjunction with aluminum diethylphosphinate in polyamides. Partial replacement of AlDPI with the supposedly less expensive metal borophosphate can still maintain the UL-94 performance. The advantage of this additive is that it contains both boron and phosphorus. For easy processing, it could replace both melamine polyphosphate and zinc borate.

2.4 Benzeneboronic acid ($(C_6H_5)B(OH)_2$)

It was proposed that boronic acid group in a burning polymer is active in the condensed phase by assisting intermolecular cross-linking, forming the six-membered boroxine ring and promoting the formation of a protective char layer [1]. Morgan et al. reported the syntheses of 1,4-benzenediboronic acid and 1,3,5-benzenetriboronic acid and evaluated them as fire retardants in ABS and polycarbonate [53]. For example, 5% by weight of 1,4-benzenediboronic acid in PC can provide a V-0 performance. In ABS (a much more flammable polymer), this diboronic acid failed to give a quantifiable UL-94 result, but the observation of char that could not be ignited is noteworthy.

Recently Morgan et al. [54] proposed the possibility of using benzeneboronic acid in PU foam. They showed that addition of boronic acid moieties to terephthalic acid can reduce the total HR by 50–96% and increase char yield from 0% to 66 wt% of the potentially promising additives.

Mosurkal et al. [55] reported a novel class of polyborosiloxane terpolymers prepared from phenyl boronic acid, aminopropyl-terminated polydimethylsiloxane, and phthalic anhydride. It was concluded that increase in boron concentration in these polymers lowers their heat release due to the formation of a glassy intumescent that hinders heat transfer to the underlying polymer. For example, the polyborosiloxane/1,1,4,4-tetracarboxylicphenyldianhydride (DAH) terpolymer with boron content of 0.5% versus polysiloxane/DAH with no boron showed heat release capacity of 171 and 266 $J\ g^{-1}\ K$, respectively, and total heat release of 12.1 and 15.6, respectively.

3. CONCLUSIONS

New market demand of improved thermal stability, better processability (better flow and recyclability), and miniaturized parts (better mechanical properties) will continue to push the envelope of fire retardant technology. While product safety of halogenated flame retardants will continue to be scrutinized, brominated flame retardants will still be the major player in the foreseeable future. New halogen-free technologies such as aluminum diethylphosphinate and DOPO are steadily gaining more market share. In addition to developing new halogen-free compounds, future endeavors will be mostly focused on developing a fire retardant package of more than one or two additives.

Boron-based flame retardants have very broad applications. They are multifunctional fire retardants and can function as flame retardants, smoke suppressants, afterglow suppressants, antitracking/antiarcing agents. In plastics and elastomers, zinc borates have been very commercially successful in both halogen-containing and halogen-free polymers. The fire test performance of zinc borates can be enhanced with the use of co-additives such as metal hydroxides, N, P, and Si-containing compounds.

In halogen-containing polymer, the zinc moiety in zinc borate can cause cross-linking and promote char formation and the boron moiety can stabilize the char and improve char hardness. In halogen-free polymers, zinc borate can function as a flux and form eutectic mixture (lower melting point) at the particle—particle interface between metal oxide at medium high temperatures. At high temperatures, borates such as zinc borate and BA can undergo fusion with metal oxide such as silica or silicate minerals. It is believed that borosilicate glass or ceramics is formed. With N and P compounds, borate can form boron phosphate (or metal borophosphate) and boron nitride, respectively, at high temperatures.

References

[1] Shen KK, Kochefahani S, Jouffret F. Boron-based flame retardants and flame retardancy. In: Wilkie CA, Morgan AB, editors. Fire retardancy of polymeric materials. 2nd ed. Florida, USA: Taylor and Francis; 2010. p. 207.

[2] Shen KK, Kochefahani S, Jouffret F. Zinc borates as multifunctional additives. Polym Adv Technol 2008;19:469–74.
[3] Shen KK, Olson E. Borate as fire retardants in polymers. In: Wilkie CA, Nelson GL, editors. Fire and polymers IV-ACS symposium series 922. Washington, USA: American Chemical Society; 2005. p. 224.
[4] Schubert DM, Alam F, Visi MZ, Knobler C. Structural characterization and chemistry of the industrially important zinc borate, $Zn[B_3O_4(OH)_3]$. Chem Mater 2003;15: 866–71.
[5] Shen KK. Overview of flame retardancy and smoke suppressant in flexible PVC. In: Society of plastics engineering vinyltech conference. Atlanta, USA; October 2006.
[6] Shen KK. Fire retardant polyamides – present technology and challenges. Zinc borates as multifunctional fire retardants in polyamides. In: Twenty first annual BCC conference on flame retardancy. Connecticut, USA; May 2010.
[7] Shen KK. Zinc borate as a flame retardant in halogen-free wire and cable systems. Plastics compounding. Cleveland, USA: Edgell Communication; November/December 1998.
[8] Shen KK, Olson E, Amigouet P, Tong C. Recent advances on the use of metal hydroxides and borates as fire retardants in halogen-free polyolefins. In: Seventeenth annual BCC conference on flame retardancy. Connecticut, USA; May 2006.
[9] [a] Shen KK. Zinc borates as multifunctional fire retardants in epoxies. In: Thirteenth annual bcc conference on flame retardancy. Connecticut, USA; June 2002;
[b] Shen KK, Sprague RW. Zinc borate as a flame retardant and smoke suppressant in epoxy systems. J Fire Retard Chem 1982;9:161.
[10] Shen KK, Schultz D. Flame retardants. In: Rubber technology-compounding and testing for performance. Munich, Germany: Hanser Publishers; 2001. p. 248.
[11] Wu Z, Hu Y, Shu W. Influence of ultrafine zinc borate on the thermal degradation behavior of a low-density polyethylene/intumescent flame retardant system. J Vinyl Add Technol; 2009:260–5.
[12] Zheng Y, Tian Y, Ma H, Qu Y, Wang Z, An D. Synthesis and performance study of zinc borate nano-whiskers. Colloid Surface A: Physicochem Eng Aspect 2009;339:178–84.
[13] Li S, Long B, Wang Z, Tian Y, Zheng Y, Zhang Q. Synthesis of hydrophobic zinc borate nano-flakes and its effect on flame retardant properties of polyethylene. J Solid State Chem 2010;183:957–62.
[14] Tian Y. Synthesis of hydrophilic nanodiscs zinc borate for lubrication. Mater Lett 2006; 60:2511–5.
[15] Isarov A, Temples D, Chen T, Walker J. Molybdate/borate complexes for enhanced cable compound fire performance. In: Proceedings of the 60th IWCS conference. North Carolina, USA; 2011.
[16] Tian Y, Zhou W, Yu L, Meng F, Yu K, Ding X, et al. Self assembly of monodisperse SiO_2-zinc borate core-shell nanosphere for lubrication. Mater Lett 2007;61:506–10.
[17] Bourbigot S, Duqusene S. Fire retardant polymers: recent development and opportunities. J Mater Chem 2007;17:2283–300.
[18] Basfar AA, Nolreaux P, Coudreuse A, Noblat R. ECO friendly cross-linked flame retardant composition for wire and cable. US Pat 20110278039; 2011.
[19] Formicola C, Fenzo A, Zarrelli M, Frache A, Giordano M, Camino G. Synergistic effects of zinc borate and aluminum trihydroxide on flammability behavior of aerospace epoxy system. eXPRESS Polym Lett 2009;3:376–84.

[20] Thiele L, Garmann H, Gorzny M, Sanch F. Highly resistant polyurethane adhesives. US Pat 20050080218; 2005.
[21] Ueno Y, Kumakura M. Fire retardant adhesive, fire-retardant adhesive film using the same, and flat cable. US Pat 6,764,765; 2004.
[22] Heschke JE, Ahsan T, Volante CN, Bischof CS. Flame Retardant composition for coating powder. US Pat 20060036006; 2006.
[23] Yildiz B, Seydibeyoglu MO, Guner F. Polyurethane-zinc borate composites with high oxidative stability and flame retardancy. Polym Degrad Stab 2009;94:1072–5.
[24] Fontaine G, Bourbigot S, Duquesne S. Neutralized flame retardant phosphorus agent: facile synthesis, reaction to fire in PP and synergy with zinc borate. Polym Degrad Stab 2008;93:68–76.
[25] Satoshi M, Kzuhiko O. Polyphenylene ether resin composition, prepreg containing the same, laminated plate, copper-clad laminated plate and printed circuit board. Jap Pat 2005008829; 2005.
[26] Kazumichi U. Halogen-free flame retardant resin composition, prepreg using the same and application of prepreg. Jap Pat 2006063258; 2006.
[27] Martens MM, Kasowsk R, Cosstick KB, Penn RE. Jap Pat 2005272851; 2005.
[28] Hoerold S, Dietz M, Schmitt S, DeBoysere J. Seventeenth BCC conference on flame retardancy of polymeric materials. Connecticut, USA; May 2006.
[29] Isitma NA, Gunduz HO, Kaynak C. Halogen-free flame retardants that outperform halogenated counterparts in glass fibber reinforced polyamides. J Fire Sci 2010;28:87–98.
[30] Desio GP, Norfolk LM. Flame retarded polymer composition with improved thermal stability. PCT Int Appl WO2005/121234; 2005.
[31] Martens MM, Saga Y. Flame resistant aromatic polyamide resin and articles there from. PCT Int Appl WO 2005/033192; 2005.
[32] Nass B, Hoerold S, Schaker O. Polymer molding compositions based on thermoplastic polyamides. US Pat Appl 2007/0072967; 2007.
[33] Yin Y. Flame resistant semi-aromatic polyamide resin composition and article. US Pat Appl 2009/0030124; 2009.
[34] Rakszawski JF, Parker WE. The effect of group IIIA–VIA elements and their oxides on graphite oxidation. Carbon 1964;2:53–63.
[35] Nakasaki F. Proceedings of fire retardant chemical association. California, USA: Spring Conference; 1997.
[36] Muirhead SAW. Flame retardant-containing plastic shipping container. US Pat Appl 2011/0187022; 2011.
[37] Anonymous. Ignition resistance SPS formulation containing zinc borate. Res Discl 1998;737.
[38] Shen KK. Review of recent advances on the use of silicon-based flame retardants – silicon/boron synergy. In: 22nd annual BCC conference on flame retardancy. Connecticut, USA; May 2011.
[39] Sharp DW. Article for protecting substrates. GB Pat 2,140,325A; 1984.
[40] Czubarow P. Flame resistant thermal interface material. US Pat Appl 2005/0197436; 2005.
[41] Alexander G, Cheng YB, Burford RP, Shanks R, Mansouri J, Hodzic A, et al. Fire-resistant silicone polymer compositions. US Pat Appl 2006/0155039; 2006.
[42] Paul S. Flame retardant additive for polymers, free of halogen, antimony oxide, and phosphorus containing substance. US Pat Appl 20100004370; 2010.

[43] Cogen JM, Jow J, Whaley PD. Flame retardant composition with excellent processability. US Pat Appl 2007/0010615; 2007.

[44] Bhandari YJ, Gallucci RR. Silicone polyimide compositions with improved flame retardance. US Pat Appl 2012/0021153; 2012.

[45] Sailer MF. Compositions useful as fire retardant. PCT Int Appl WO 2008085055; 2008.

[46] Peng HQ, Wang YZ. Effects of boric acid on flame retardancy of intumescent flame retardant polypropylene system containing a caged bicyclic phosphate. In: Wilkie CA, Morgan A, Nelson GL, editors. Fire and polymers V, chapter 14-ACS symposium, vol. 1013. Washington, USA: American Chemical Society; 2009. p. 225.

[47] Jimenez M, Duquesne S, Bourbigot S. Characterization of the performance of an intumescent fire protective coating. Sur Coat Tech 2006;201:979–87.

[48] Jimenez M, Duquesne S, Bourbigot S. Kinetic analysis of the thermal degradation of an epoxy-based intumescent coating. Polym Degrad Stab 2009;94:404–9.

[49] Nishiyama I, Ono T, Nishino T. Flame retardant polyurethane foam. Jap Pat 2011063655; 2011.

[50] Saito K, Wada S. Intumescent fireproofing rubber-based sealing materials. Jap Pat 4,669,573; 2010.

[51] Braun U, Schartel B, Ficherra MA, Jaeger C. Flame retardancy mechanisms of aluminium phosphinate in combination with melamine polyphosphate and zinc borate in glass-fibre reinforced polyamide 6,6. Polym Degrad Stab 2007;92:1528–45.

[52] Ewald B, Hibst H. Borophosphate, borate phosphate, and metal borophosphate as flame proofing additives for plastics. PCT Int Appl WO 2011/000913; 2011.

[53] Morgan AB, Jurs JL, Tour JM. Synthesis, flame retardancy testing, and preliminary mechanism studies of non-halogenated aromatic boronic acids. J Appl Polym Sci 2000;76:1257–68.

[54] Durganala S, Morgan AB, Benin V. Reactive flame retardants for polyurethane current technology and new directions. In: Proceedings of fire and materials conference. California, USA; 2011.

[55] Mosurkal R, Kirby R, Muller WS, Soares JW, Kumar J. Simple green synthesis of phenylborosiloxane as environmentally-safe, non-halogenated flame retardant polymers. Green Chem 2011;13:659–65.

CHAPTER

12

Organosilicon Compounds as Polymer Fire Retardants

Z. Han
Harbin University of Science and Technology, Harbin, China

A. Fina, G. Camino
Politecnico di Torino, Dipartimento Scienza Applicata e Tecnologia, Alessandria, Italy

CHAPTER OUTLINE

1. Introduction

The urgent need for novel methods and/or formulations to protect polymers from fires continuously increased since the first evidences arose that universally used halogen-based fire retardants, although they are characterized by high effectiveness and versatility in reducing fire risk, may enhance fire hazard. Indeed, smoke and CO production increase when halogenated fire retardants suppress the flame which is the result of gas phase radicals trapping by halogenidric acids produced by the fire retardants. Moreover, corrosion may occur due to halogenidric volatile moieties and environmental impact may be enhanced by occasional production of halogenated supertoxicants

Polymer Green Flame Retardants. http://dx.doi.org/10.1016/B978-0-444-53808-6.00012-3

such as halogenated dibenzodioxins and dibenzofurans during use and/or postconsumer recycling [1]. The increasing safety awareness and environmental issues are therefore driving the development of environmentally friendly flame retardants [2]; the early stages of these attempts have been reviewed by Lomakin and Zaikov [3].

Inorganic additives such as aluminum and magnesium hydroxides were first proposed as alternative low fire hazard and environmentally friendly fire retardants because on heating, they supply to the flame inert diluting gases, such as water from hydroxides decomposition and have been shown to reduce smoke production in polymer combustion. However, their application is limited since very high loading (40–70%) of inorganics are needed for improving physical–mechanical properties of polymers, such as rubber compounds for electrical cables sheeting or epoxy composites in which the inert fillers may be partially or totally replaced by inorganic fire retardants.

The general attention was thus turned to phosphorous organic compounds and ammonium polyphosphate-based intumescent mixtures. Results are however not completely satisfactory because these fire retardants have a variable effectiveness, depending on the polymer matrix. Moreover, some concerns are being expressed about toxicity of volatile phosphorous moieties that may be produced during polymer combustion, although evidence is not conclusive so far [4].

Polymer nanocomposites raised a great interest for their ability to decrease the rate of polymer combustion through surface protection by inorganic coalescing nanoparticles and polymer char. However, in nanocomposites developed so far, this surface shielding cannot extinguish the flame in the short time required by fire safety standards.

In this fire scenario, silicon-based compounds have recently attracted research attention [5]; it has been shown that the addition of relatively small amounts of silicon-containing compounds to polymers can significantly improve their fire retardancy without adding to the toxicity and environmental hazards of polymer fires [6,7]. Silicon-containing compounds can be used as fire-retardant additives, smoke suppressing agents, and synergistic compounds, being helpful in depressing the release of combustible gas, smoke, or poison compounds during burning, reducing thus the fire harmful impact on environment [8,9]. Several forms of silicon compounds have been explored as fire retardants. This chapter will present recent developments in the understanding of the combustion behavior and fire-retardant mechanism of some organosilicon compounds, with particular attention to silanes, siloxanes, and silselquioxanes.

2. Definitions and classification

Organosilicon compounds are organic compounds containing carbon–silicon bonds; a considerable number of organosilicon compounds is currently available and growing every year. Similarly to sp^3 carbon, silicon shows a tetrahedral bonding structure in organosilicon compounds. There are several proposed criteria for the classification and nomenclature of these compounds. For example, they can be classified according to spatial structure as linear, branched, ladder, and cross-linked (i.e. two- or three-dimensional). Also, they can be classified according to their molecular structures.

Those used as fire retardants, can be basically classified into two groups: (1) low molecular weight and (2) macromolecular compounds. Low molecular weight organosilicon compounds can be viewed as the result of hydrogen atoms substitution in silane (SiH_4) with organic radicals. Most of the widely applied macromolecular organosilicon compounds are based on Si—O bond, termed siloxane which can be polymerized to form polysiloxanes with different geometrical structures. Hereinafter, organosilicon compounds definition and classification for polymers fire retardancy will be described.

2.1 Silane

Silane is the silicon analog of methane in which a hydrogen is substituted by an organic functional group such as vinyl, amino, chloro, epoxy, mercapto, alkoxy, etc. Usually, silanes possess a hydrolytically sensitive center (e.g. —OR) that can react with hydroxyls of inorganic substrates to form stable covalent bonds, thus providing the inorganics with organic tails (e.g. R) that act as coupling agents between inorganic particles and polymer matrices. Silanes are used in fire-retardant materials taking advantage of the fairly good thermal stability and the coupling action to link inorganic flame retardants and polymer matrixes. While obtaining a good balance of strength and toughness [10], improved fire-retardant properties is obtained thanks to the improved dispersion of fire-retardant particles in the polymer [11] and due to the enhanced interaction between fire-retardant components [12], possibly leading to enhanced stability of combustion residues acting as a thermal shield toward the flame [13]. An active fire-retardant role is played by silanes which bear reactive functional groups that enable them to be incorporated in the polymer structure. Examples of silanes that have been reported in literature as fire retardants in different polymers are mentioned in Table 1.

2.2 Siloxanes

Siloxane is a kind of organosilicon compounds which is composed of —Si—O—Si—O— backbones with side chains R attached to the silicon atoms (R_2SiO), where R is a hydrogen atom or an organic radical possibly bearing functional groups. Polymers of siloxanes are referred as silicones, such as polydimethylsiloxane (PDMS), which is the dominant polymer in the silicone industry. Different kinds of siloxanes and polysiloxanes have been synthesized, aiming at fire-retardant applications. Silicones can be used as fire-retardant agents through direct blending within the polymer matrix, incorporation into porous fillers, or by synthesizing block/graft copolymers including silicone segments. Examples of siloxanes and polysiloxanes that have been used as fire retardants and have been reported in literature are presented in Table 2.

2.3 Silsesquioxanes

Silsesquioxanes are a family of organic—inorganic hybrid compounds with general formula ($RSiO_{1.5}$) where R is an organic group. Polysilsesquioxanes $(RSiO_{1.5})_n$ are obtained by hydrolytic condensation of trifunctional monomers $RSiX_3$, where X is a highly reactive substituent, such as Cl or alkoxy [28,29]. The commercial availability of several different $RSiX_3$ precursors and the possibility to tailor the reaction

Table 1 Silanes in Fire-Retardant Materials

Silane	Structure	Matrix	Role
Vinyl trimethoxy silane [14]	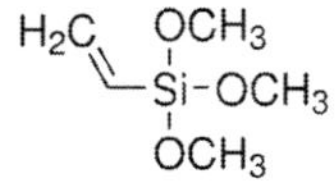	Ethylene–vinyl acetate copolymer	Coupling agent
3-Aminopropyl ethoxy silane [15]	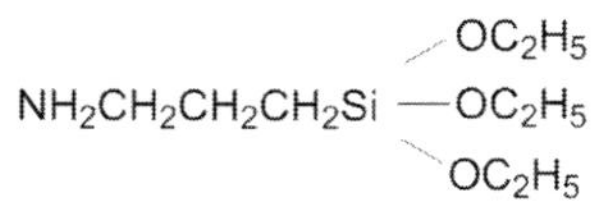	Cotton fabric	Fire-retardant component
Tri(acryloyloxyethyloxy) phenyl silane [16]	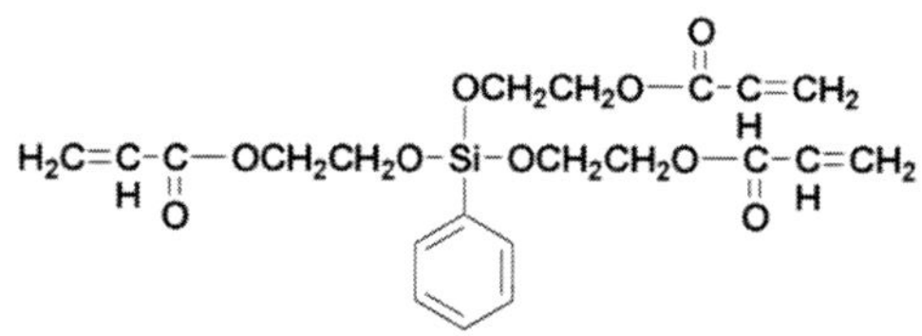	Epoxy acrylate	Monomer
Triglycidyloxy phenyl silane [17]	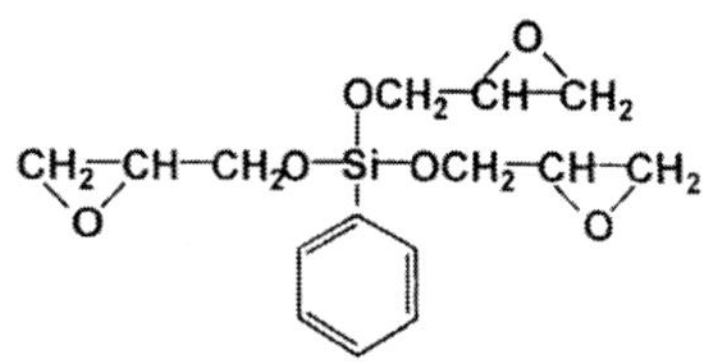	Epoxy	Monomer
N-maleimido phenoxy silane [18]	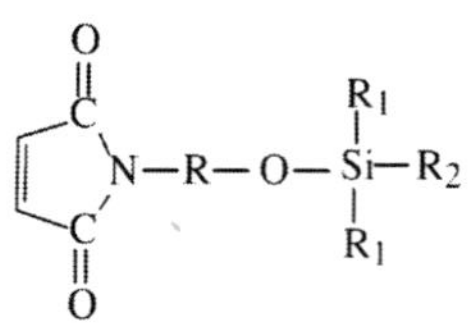	Maleimide polymer	Monomer

3-Isocyantaopropyl triethoxy silane [19]	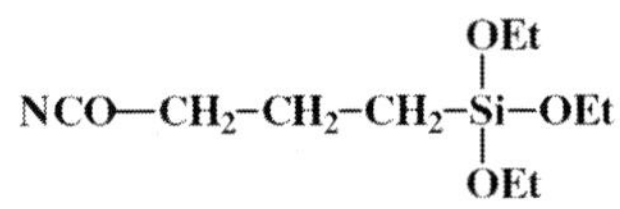	Epoxy	Reactive component
Silicon and phosphate modified acrylate [20]	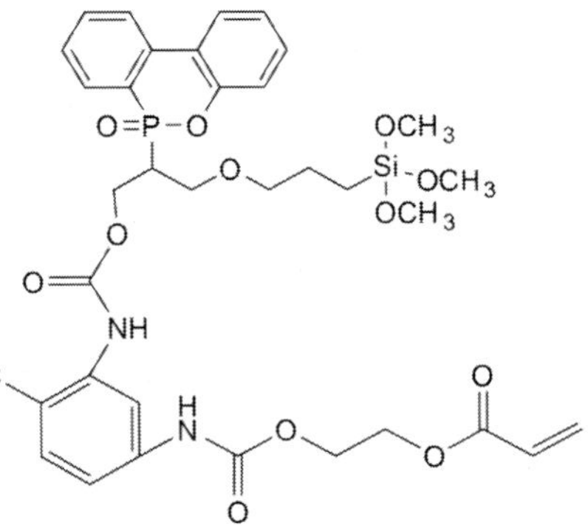	UV-curable coating	Monomer
Polysilastyrene [21,22]	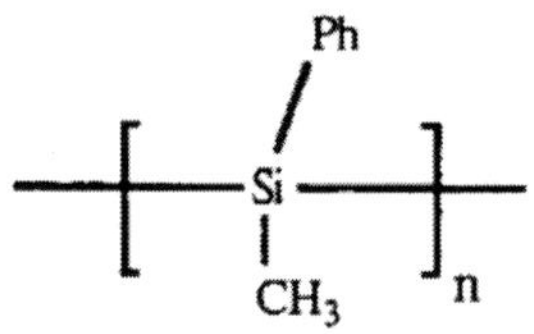	Polystyrene-polybutadyene-polystyrene copolymer Polytetramethileneylether-glicol-b-polyamide 12 copolymer	Additive

Table 2 Siloxanes in Fire-Retardant Materials

Siloxane	Structure	Matrix	Role
1,3-Bis(glycidyloxypropyl)-1,1,3,3-tetramethyldisiloxane [23]	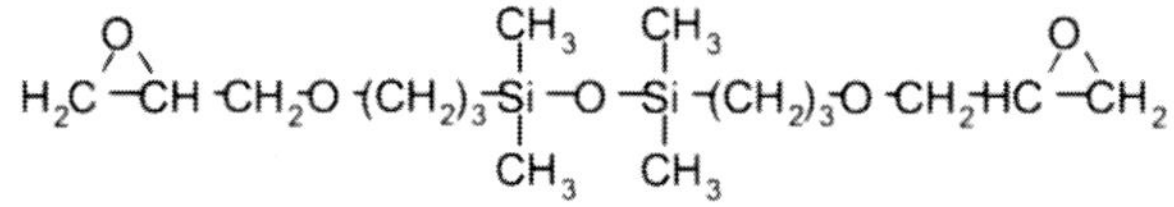	Epoxy	Monomer
Polymethyl(3-glycidyloxypropyl) siloxane [24]	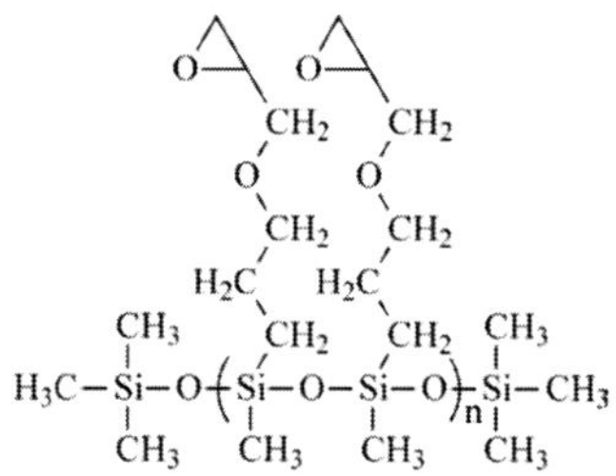	Epoxy	Monomer
Boroxo siloxanes [25]	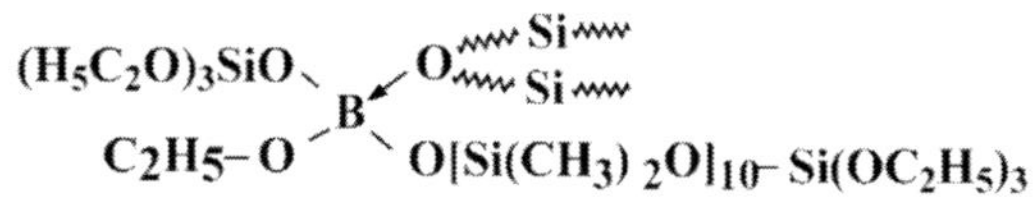	Polypropylene	Additive
Poly (9,10-dihydro-9-oxa-10-phosphaphenanthrene-10-oxide siloxane) [26]	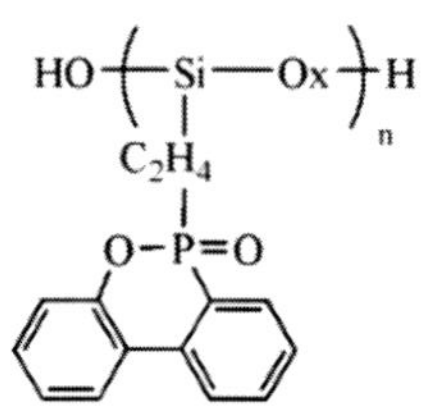	Polycarbonate	Additive
Oligomeric siloxane-containing triphenylphosphonium phosphate [27]	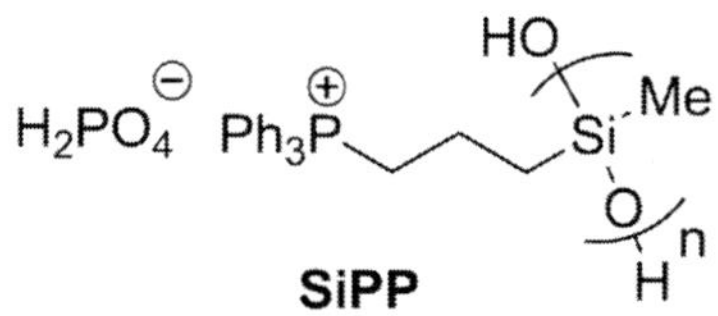 **SiPP**	Polycarbonate	Additive

conditions allow the preparation of very different chemical and geometrical structures, including random, ladder, and cage structures [30], as reported in Figure 1.

Polyhedral structures are the most widely studied silsesquioxanes, being also known as POSS, which is the acronym of polyhedral oligomeric silsesquioxanes.[1]

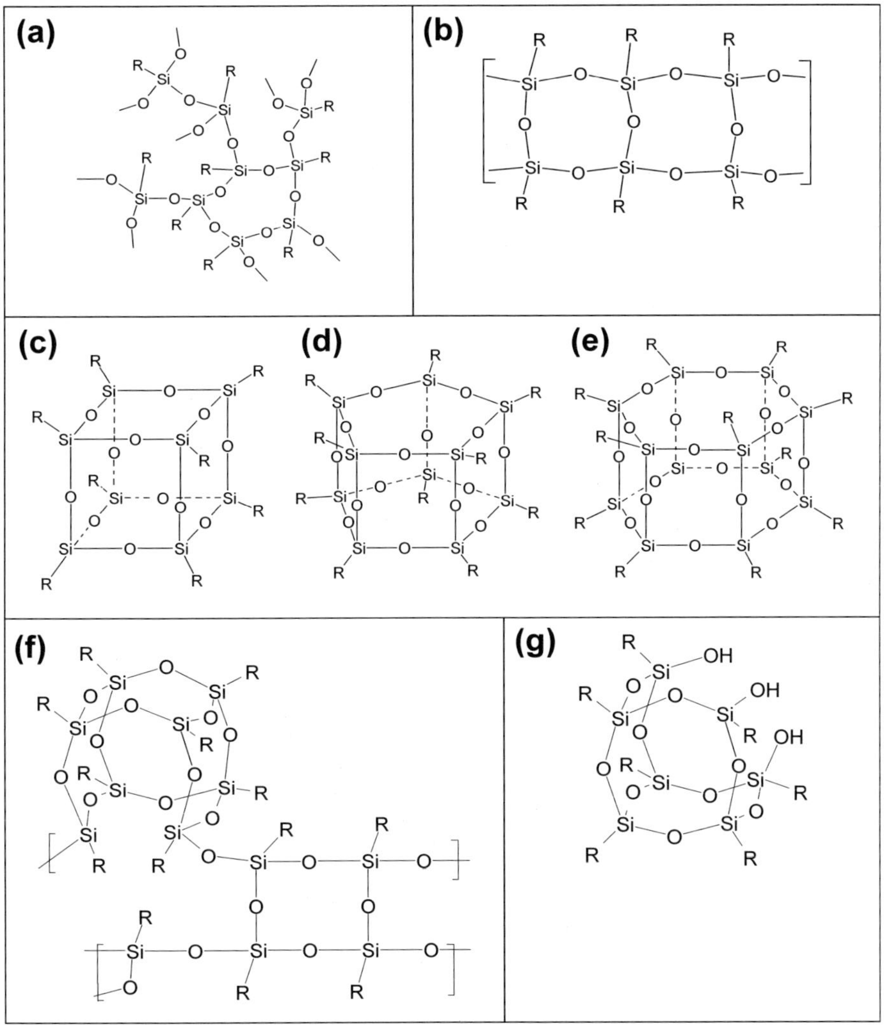

FIGURE 1

Silsesquioxanes geometrical structures: (a) random, (b) ladder, (c) T_8 POSS, (d) T_{10} POSS, (e) T_{12} POSS, (f) mixed cage-network structure, and (g) incompletely condensed POSS. POSS, polyhedral oligomeric silsesquioxanes.

[1]POSS is a trademark from Hybrid Plastics.

Since each silicon atom is bound to one-and-a-half oxygen (sesqui-) and to a hydrocarbon (-ane), POSS are "in the half way" between silica (SiO_2) and silicones $(R_2SiO)_n$. POSS are indeed particularly interesting when aiming to tailor the properties of a polymer-based material by the modification of the macromolecular structure, using POSS as nanoscale building blocks for new materials.

POSS are based on a Si—O—Si thermally and chemically robust inorganic core, which is the smallest possible silica particle, being also known as molecular silica. At the same time, the possibility to choose the group on POSS surface permits the solubility within the organic matrixes, while the possible presence of reactive groups allows chemical bonding onto the organic macromolecules. Polysilsesquioxanes and POSS have been used in fire retardancy of commodity and engineering polymers [5], generally leading to a decrease in combustion rate. Examples of polysilsesquioxanes reported in literature as fire retardants for different polymers are mentioned in Table 3.

Modification of POSS structures to include metal moieties have also been widely studied [28,37] for improving the fire retardancy of polymers; accordingly, clear positive effects have been reported on the thermoxidative decomposition and combustion of polypropylene [38,39].

2.4 Others

Polycarbosilane and polysilazane polymers consist of alternating Si—C and Si—N bonds, respectively, with pendent carbon-containing groups (Figure 2). These polymers are widely used as precursors of silicon carbide and silicon carbonitride ceramics.

Both polycarbosilane and polysilazane have been reported in a few papers as preceramic [40] compounds: examples are reported in Table 4.

3. Mechanistic aspects in fire-retardancy action

Fire retardants' role is to increase the time to "flashover", when fire can no longer be extinguished. This can be accomplished by increasing the time to ignition and/or by decreasing the rate of combustion i.e. rate of fire growth. The process involved, in both cases, depends on chemical reactions and physical phenomena occurring in condensed and gas phase. Recent literature is reviewed here concerning the fire-retardant mechanism of organosilicon compounds, which play an important role in the condensed phase of burning polymers.

3.1 Surface accumulation and ceramization

In order to build a ceramic protective layer from the thermal evolution of organosilicon compounds, a sufficient concentration of such compounds has to be reached on the top layer of a decomposing polymer. This can be reached by progressive ablation of the polymer or by migration of the preceramic compound. Recent literature suggests that migration exerts an important effect on surface accumulation. Owing to the low surface energy of silicones and similar products, these can indeed move

Table 3 POSS in Fire-Retardant Materials

POSS	Structure	R
Octamethyl POSS [31]	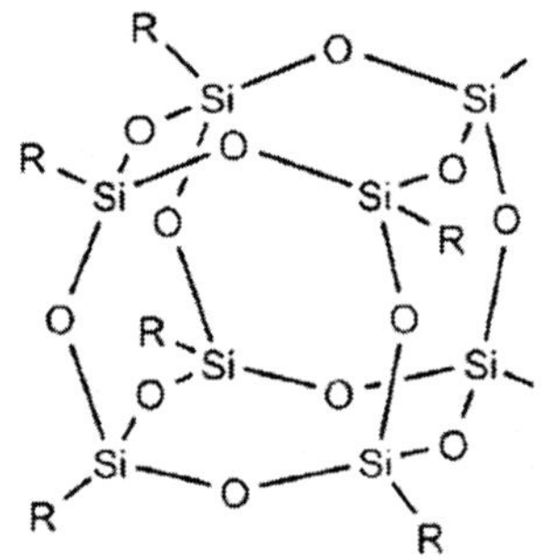	Methyl
Trisilanolphenyl-POSS [32]	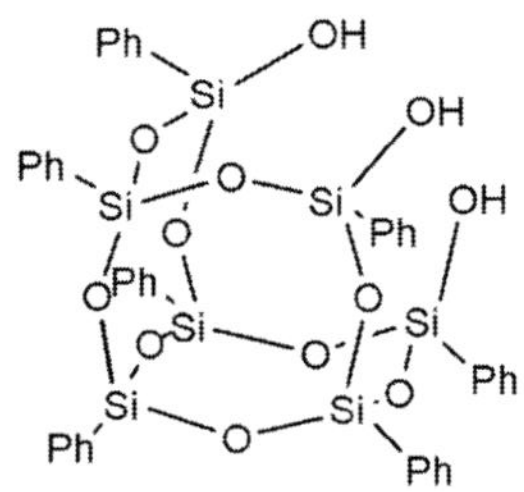	Phenyl
Octa(tetramethylammonium) POSS [33]	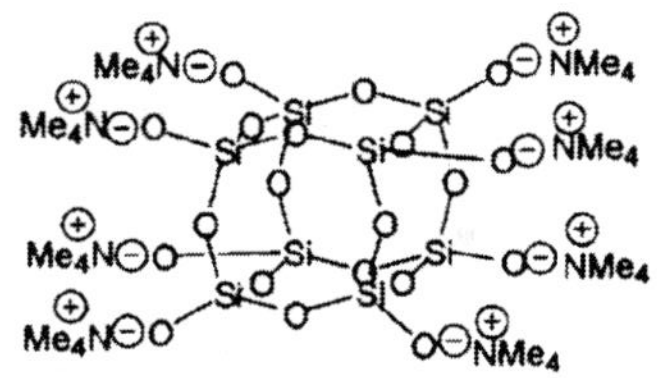	Methyl ammonium

Continued

Table 3 POSS in Fire-Retardant Materials—cont'd

POSS	Structure	R
POSS containing 9,10-dihydro-9-oxa-10-phosphaphenanthrene-10-oxide [34]	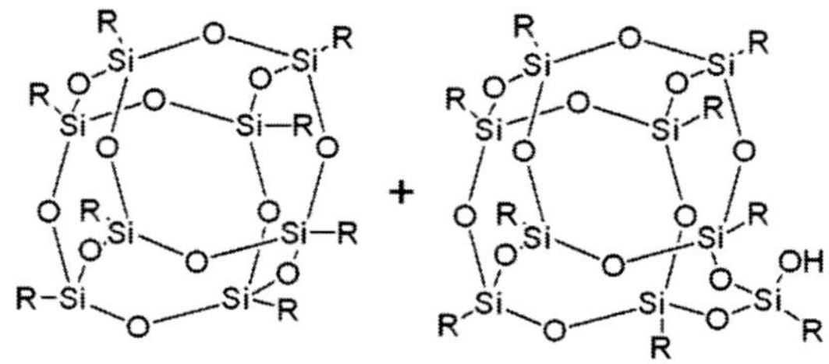	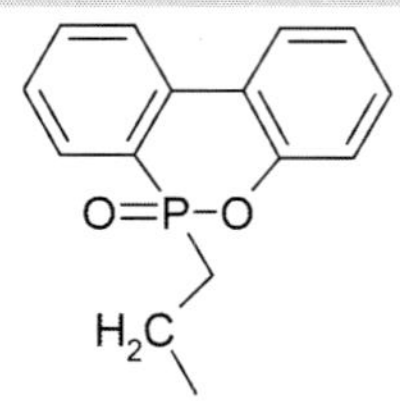
1,3,5,7,9,11,13,15-Octakis(di-phenylphosphine oxide- 2,1-ethanediyl)pentacyclo [9.5.1.13,9.15,15.17,13] octasiloxane [35]	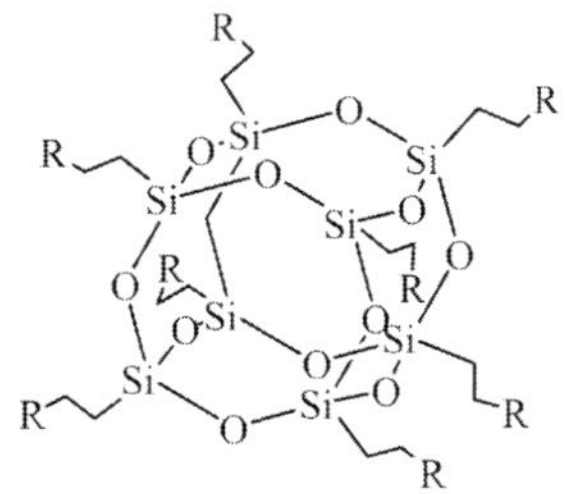	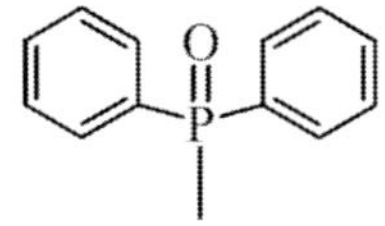
Poly(vinylsilsesquioxane) [36]	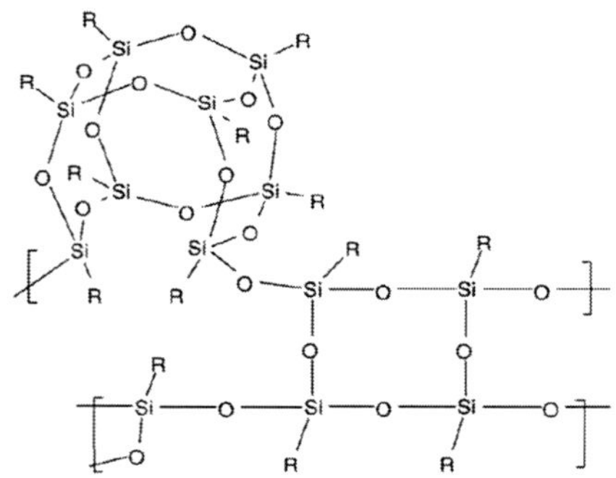	$-CH{=}CH_2$

POSS, polyhedral oligomeric silsesquioxanes.

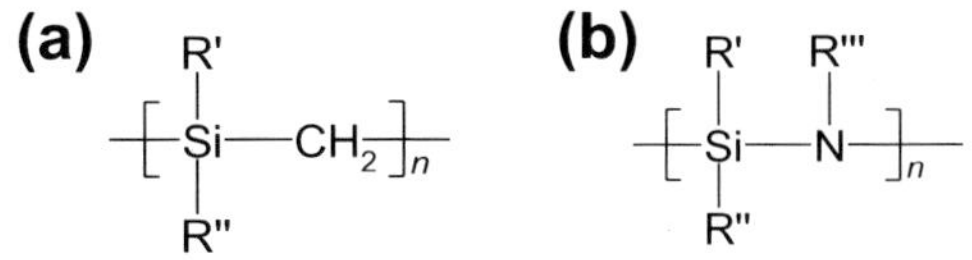

FIGURE 2

General structures for (a) polycarbosilanes and (b) polisilazanes.

toward the surface where they are available to convert to ceramic species, acting as a protective layer. Tang and Lewin investigated the migration behavior of POSS by annealing the composites at temperatures above the melting point of PP matrix or by heating it below the melting point in a microwave oven [41]. The distribution of POSS in the polymeric matrix after the migration process was reported. It was found that in all cases migration occurs simultaneously to the top and bottom surfaces. In samples originally containing 5 wt% of POSS, the concentration of POSS above 50 wt% on the surface was obtained.

Silicones and other silicon derivatives that do not volatilize on heating produce silica residues during burning which accumulate on the burning surface (while the polymer decomposes to combustible volatiles) and can serve as a surface barrier, hindering the mass and heat transportation between condensed and gas phase. The silica surface barrier can indeed delay the volatilization of decomposition products, thus reducing the volatiles concentration in the gas phase; it also insulates the polymer from external heat. These barrier effects provide the fundamentals for the development of silicone-based fire retardants. In particular, it was reported by Hshieh [42] that the silica ash layer integrity governs the efficiency of the diffusion barrier that restricts the diffusion of fuels into the combustion zone and the access of oxygen to the unburned fuels.

Silicone derivatives with a branched chain structure containing methyl and phenyl groups as chain silicon substituents and methyl groups as chain terminal silicon substituents were found to be effective in flame-retarding aromatic thermoplastics, particularly PC and PC/ABS alloy [6,43].

Polycarbonate (PC) containing methylphenyl-silicone with a branched structure (SFR-PC) was investigated by Zhou and Yang [43]. The addition of methylphenyl-silicone enhanced the thermal stability of PC and promoted the formation of char. The silicone was found effective in retarding the combustion of the PC. The limited oxygen index of the PC containing 5 wt% of methylphenyl-silicone was increased from 26% to 34%. The proposed fire-retardant mechanism was based on the formation of a cross-linked structured char consisting of condensed aromatic compounds and siloxane units on the surface of the burning material (Figure 3).

Polysilsesquioxane or polycarbosilane resins were also shown to be effective fire retardants in thermoplastic polymers such as polypropylene (PP), styrene–butadiene–styrene block copolymers (SBS), and polyether–polyamide copolymers (PTME-PA), reducing the heat release rate (HRR) during cone calorimeter tests [21]. In PP, a 40% reduction on the peak heat release rate (pHRR) was obtained

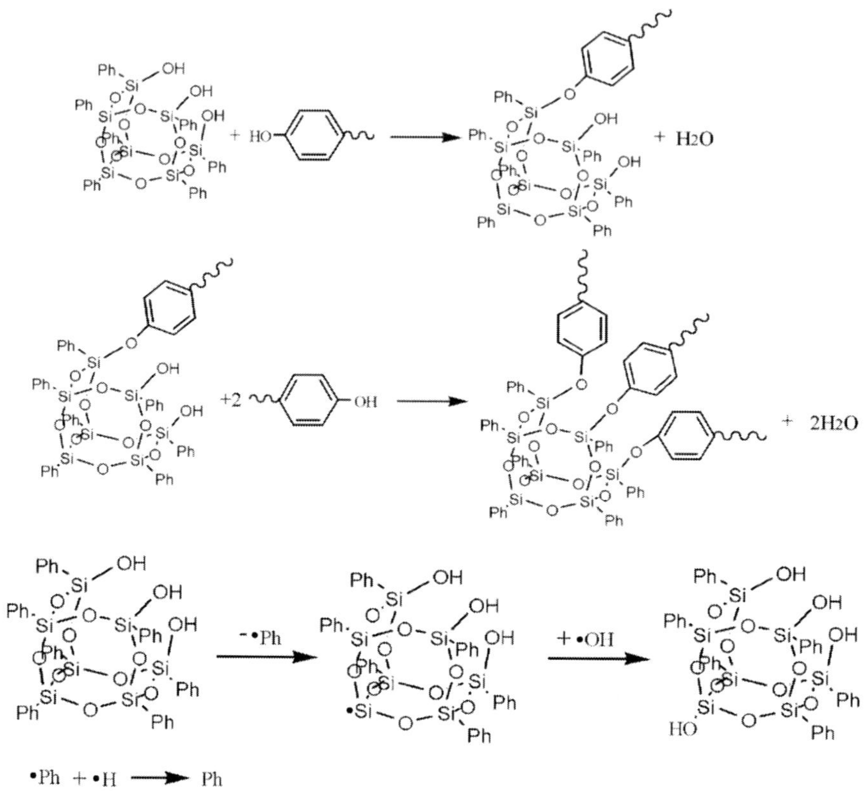

FIGURE 4

The reaction of trisilanolphenyl-polyhedral oligomeric silsesquioxanes with the degradation products of polycarbonate [32].

mechanism of bisphenol A polycarbonate (PC) and trisilanolphenyl-POSS (TPOSS). Song el al. [32]. proposed that the reactions of TPOSS with alkyl and aromatic substituted phenols of the PC matrix (shown in Figure 4) interfere with the decomposing and cross-linking processes of PC. With increase in temperature, TPOSS retards the oxidation of char residue and enhances the amount and thermal oxidative stability of the protective char layer, which builds up on the surface of the burning polymer.

A second example was reported for polyurethane (TPU) containing polyhedral oligomeric silsesquioxanes; a large reduction of PHRR compared to virgin TPU was obtained, ascribed to the protection from an intumescent layer composed of ceramified char made of silicon network in a polyaromatic structure [36]. The foaming process and char residues left are reported in Figure 5. The intumescent ceramified char layer acts as a thermal barrier at the surface of the substrate limiting heat and mass transfer thus leading to a limited HRR.

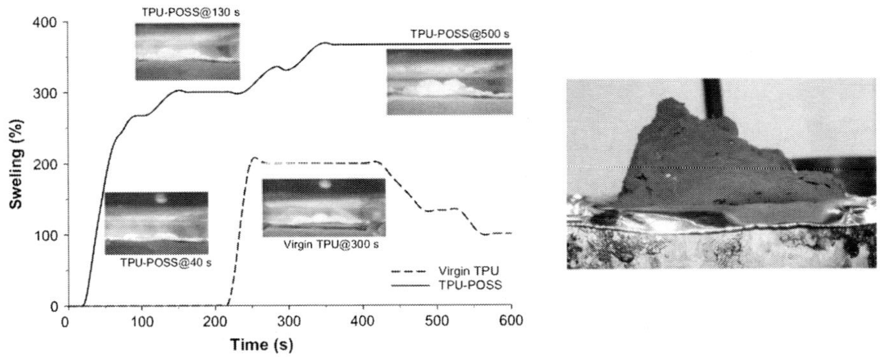

FIGURE 5

Foaming as a function of time of virgin TPU compared to TPU—POSS composite during a mass loss calorimetry experiment with infrared pictures showing the swelling at different characteristic times and residue of TPU—POSS composite after a mass loss calorimetry experiment (external heat flux 35 kW m^{-2}) [36]. TPU, polyurethane; POSS, polyhedral oligomeric silsesquioxanes.

3.2 Combination with other FRs

In order to obtain superior performances in fire retardancy, total heat released should also be reduced in addition to the reduction in combustion rate. Extensive charring of the polymer upon heating is aimed to reduce the amount of volatile fuel produced by the polymer decomposition. Therefore, the combination of organosilicon species with fire-retardant moieties able to promote charring of polymers is of very high interest. For example, the incorporation of silicones often provokes slight intumescent char formation in polypropylene, while the polymer itself burns without a char [49]. This solid phase reaction appears to be the main mode for the silicone fire-retardant action, making it possible that a combination of a linear polydimethylsililoxane and other optional components can effectively fire-retard polypropylene. The use of organosilicon derivatives is particularly suitable for the combination with intumescent FRs. Indeed, the fire-retardant mechanism of a halogen-free material based on calcium carbonate, silicone elastomer, and an ethylene acrylate copolymer was described by Hermansson et al. [50]. The fire-retardant mechanism was shown to take place in the condensed phase: polymer decomposition products form intumescent gases while calcium ions interact with acrylate pyrolysis—derived acidic groups on the polymer and result in (ionomer) cross-linking leading to an increased melt viscosity and reduced gases transportation rate in and out of the material. The type of copolymer used has been shown to be important for fire-retardant effectiveness [51].

Combination with phosphorous compounds was studied by several authors. Silane-derivatives used as a coupling agent was also found to be helpful in charring, based on the possible interaction between ammonium polyphosphate (APP), which is typically used as polyphosphoric acid source in intumescent fire-retardant formulations, and silane (3-aminopropyltriethoxy silane) [52], as shown in Figure 6.

The effects of hydroxy silicone oil (HSO) as a synergistic agent on the fire retardancy of intumescent fire-retardant polypropylene composites (IFR-PP) have also been studied [53]. The heat release and mass loss rate, total heat release, and carbon monoxide and carbon dioxide formation of PP/APP/PER/HSO composites were found to decrease in comparison with the PP/APP/PER ternary composite because HSO could promote the formation of a homogenous and compact intumescent char layer. On the other hand, the synergistic effect of organic silicon on the thermal stability and the fire-retardant properties of PC and potassium-4-(phenylsulfonyl)

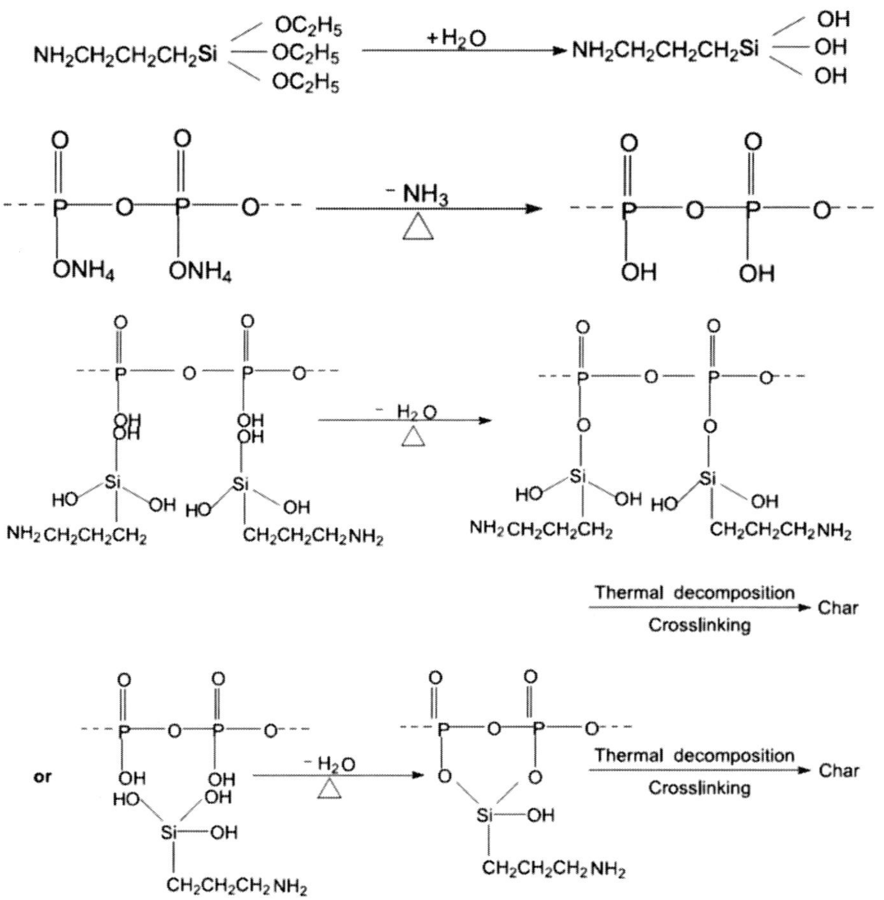

FIGURE 6

Proposed reaction mechanism of char formation in silane/ammonium polyphosphate systems [52].

benzenesulfonate (KSS) was reported by Yuan et al. [54]. The synergistic interactions between organic silicon and KSS enhanced the char formation ability with 39.4 wt% char residues of this system at 700 °C in comparison with 33.4 wt% according to calculation.

In another study, Kim and his coworkers reported the fire-retardant properties of the ABS/epoxy resin/triphenyl phosphate intercalated nanosilicate (Nano TPP)/ silane mixture [55]. Increase in the LOI value of up to 44 was observed with increasing the silane contents, which was related to the char morphologies after burning. As shown in Figure 7, crevices were hardly found and only small holes gushing flammable gas were observed in the system of ABS/epoxy resin/Nano TPP/silane.

Synergistic effect between octamethyl polyhedral oligomeric silsesquioxane (OMPOSS) and a phosphinate-based compound (Exolit OP950) was found in fire-retardant polyethylene terephthalate (PET) [31]. The addition of OMPOSS leads to a considerable increase in the fire-retarding performances of the polymer in terms of cone calorimetry as shown in Figure 8, even if a very poor dispersion of OMPOSS and OP950 into the matrix has been observed.

FIGURE 7

Optical micrographs of the chars formed after Limiting Oxygen Index (LOI) test: (a) ABS/Clay 99/1; (b) ABS/Nano TPP 85/15; (c) ABS/(Nano TPP/silane) 85/15; and (d) ABS/Epoxy/(Nano TPP/silane) 85/9/6 [55].

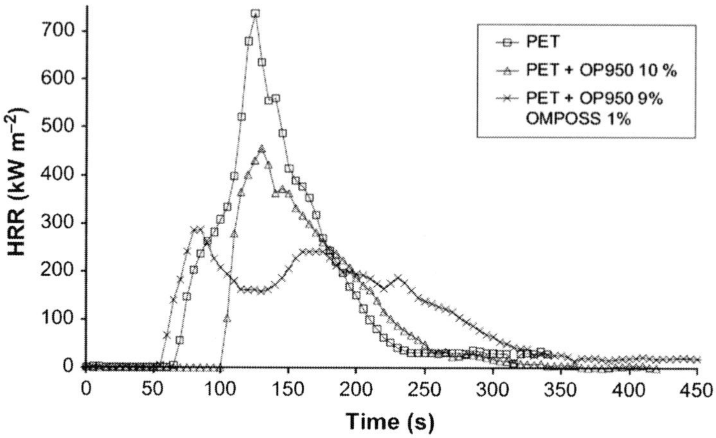

FIGURE 8

HRR curve of PET/OP950/OMPOSS (89/10/1 wt%) formulations [31]. HRR, heat release rate.

Polyamide 6 with 10 phr silicone elastomeric nanoparticles shows a 60% lower PHRR compared to pure polyamide 6, while with additional 0.5% clay, polyamide 6 PHRR decreases by 68% [56]. The silicone elastomeric nanoparticle and exfoliated clay platelets exhibit a synergistic fire-retardant effect, which is a result of the formation of two barriers, coat-like layer and island-like floccules, on the nanocomposite residue surface at the end of combustion as shown in Figure 9.

Similarly, MMT was observed to have a catalytic effect on promoting the barrier formation in PC containing DOPO-functionalized polysiloxane [26], as shown in Figure 10.

Silicon-based structures and phosphorous-based FR functions can also be conveniently combined at molecular level in the same chemical structure. As an example, oligomeric siloxane containing triphenylphosphonium phosphate (SiPP) was found to promote the formation of char layers at low temperature in PC, resulting in much lower HRR and total heat release (THR) compared to pristine PC as shown in Figure 11 [27]. The char layers in the condensed phase contributed to better fire retardation; the LOI value of PC increased from 22.8 to 30.9 with 20 phr SiPP content and a V-0 rating was reached with only 5 phr SiPP.

A silicon- and phosphorus-containing fire retardant, poly (9,10-dihydro-9-oxa-10-phosphaphenanthrene-10-oxide siloxane) was reported to have good thermal stability and to be able to be used as an efficient fire retardant for PC [26]. The LOI value of the flame-retardant PC systems could reach a maximal value of 32.8 when the content of the fire retardant was 5 wt%. When 2 wt% MMT was added, the UL94 rating reached V-0.

Combination with boron moieties have been also addressed: Marosi et al. synthesized boroxo siloxanes (BSil) which were used as a ceramic precursor and combined them with an intumescent fire retardant in polypropylene [25,57,58].

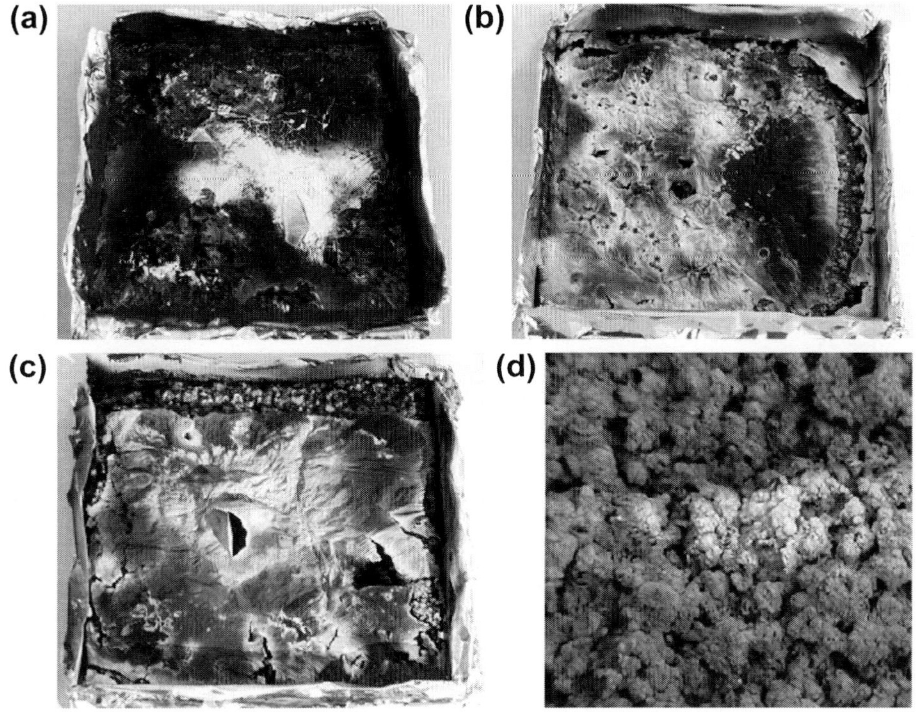

FIGURE 9

Surface of the residues at the end of the combustion of pure polyamide 6 (a), polyamide 6 with 10 phr silicon nanoparticle (b), with 10 phr mixtures of silicon nanoparticle and clay 4:1 wt/wt (c, d): surface layer (c) underneath island-like floccules (d) [56].

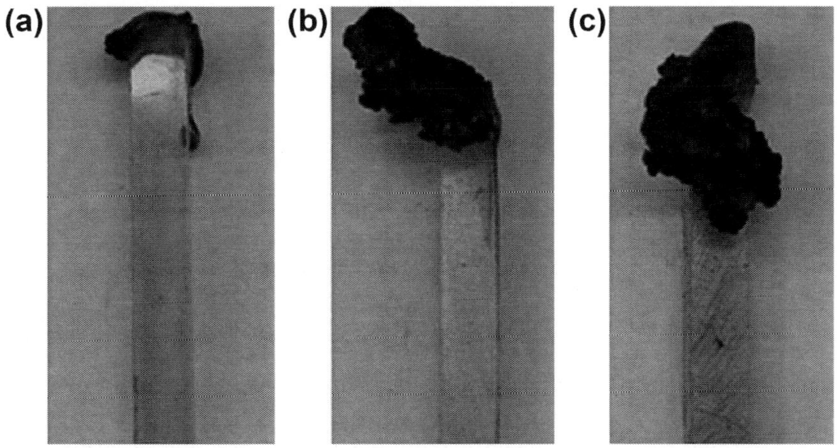

FIGURE 10

Char layer of the samples after LOI test (a) virgin PC, (b) PC/5%P(DOPO-VTES), and (c) PC/5%P(DOPO-VTES)/2%MMT [26].

Reprinted from [26], with permission from Elsevier.

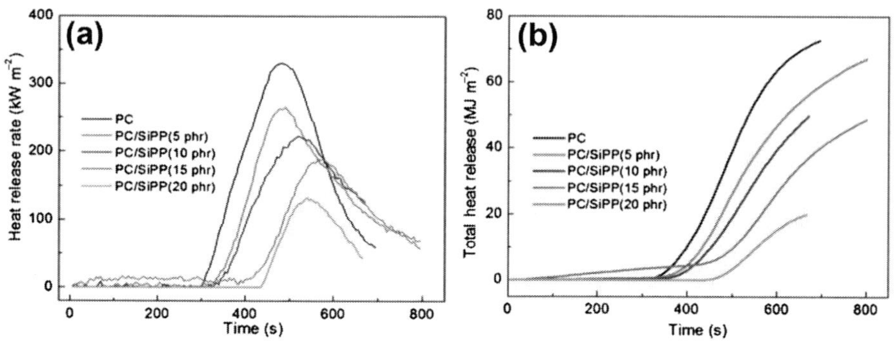

FIGURE 11

Heat release rates (a) and total heat releases (b) versus burning time for PC/SiPP composites [27].

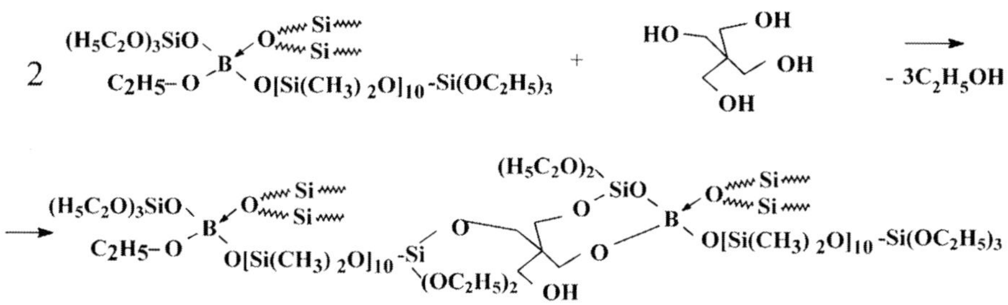

FIGURE 12

The reaction between boroxo siloxane and pentaerythritol [25].

BSil changed the melt viscosity by reacting with pentaerythritol and ammonium polyphosphate [57]. The proposed reaction between BSil and pentaerythritol is shown Figure 12 [25]. Transformation of the applied precursor to inorganic layer is not complete at the temperature of fire and the organic residue improves the plasticity of the char. As shown in Figure 13, the charred residue of FR-PP without BSil has a flat surface full of cracks, while the sample with BSil has a dome-like blown formation, manifesting the satisfactory plastic, barrier character of the charred layer [58].

Boroxo siloxanes (BSil) were also used as a coating for nanoparticles by Marosi et al. [25], to be used in combination with intumescent FR. It was found that BSil-coated nanoparticles accumulate on the surface in the case of fire and act as a skeleton in the transformation of the precursor to the barrier layer. BSil, in turn, may act as a glue keeping the particles together. Nanoparticles coated with BSil and the residue of APP may form a consistent ceramic layer even on ignition in the vertical position (Figure 14).

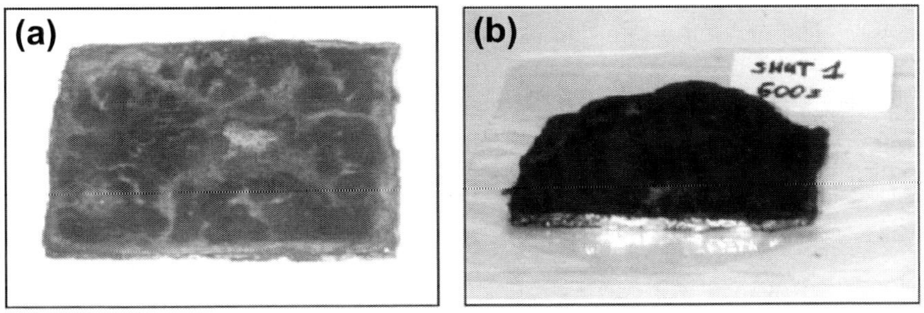

FIGURE 13

Charred layer of intumescent fire-retardant polypropylene (a) without and (b) with boroxo siloxane obtained in cone calorimetric test [58].

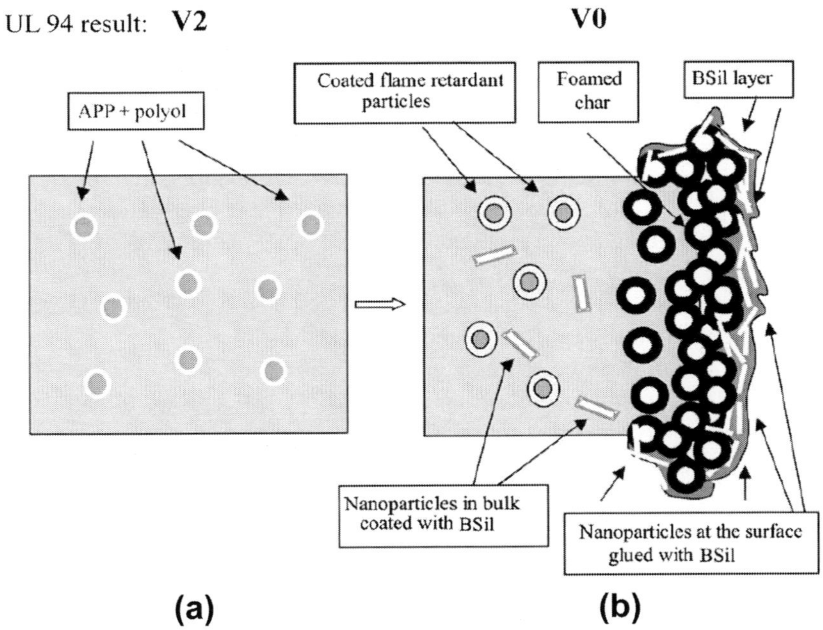

FIGURE 14

UL94 rating and scheme of action of intumescent fire-retarded polypropylene system without (a) and with (b) Boroxo siloxanes (BSil)-coated nanoparticles (additives: APP: 21%, PER: 7%, BSil: 1.5%, Cloisite 20A: 0.5%) [25]. APP, ammonium polyphosphate.

4. Applications

Different strategies can be applied for the inclusion of organosilicon compounds into polymers. The traditional technique in preparing fire-retardant polymers is to blend fire-retardant additives with polymers. However, since no chemical bonds form between the additive and the polymer, the technique suffers from low fire-retardant dispersion, leaking of fire-retardant additives, impaired polymer properties, and processing problems. Alternatively, fire retardants may be chemically bound to the polymer backbone, by copolymerization or grafting processes. The two different approaches are described in this section, presenting examples taken from the recent literature published in this field. Furthermore, applications of silicon-based compounds in surface protective coatings are briefly reviewed.

4.1 Reactive component

Reactive type fire retardants, which can be permanently bonded to the polymer, currently attract a great deal of attention. Bonding of fire retardant to polymer is widely used in thermoset polymers, such as epoxy resin, because the fire-retardant components can be incorporated into the backbone through the epoxy or the curing agent [59]. Preparation routes of silicon-containing reactive type compounds with reactive functional groups are widely reported.

Mercado et al. [60] reported fire-retardant epoxy resins with silicon-containing epoxides or silicon-containing prepolymers as shown in Figure 15. The presence of silicon increased the LOI values and the char yield at 700 °C for the silicon-containing resins as compared to the unmodified resin, indicating that the formation of a silicon-rich protective layer plays an important role in fire retardation.

A silicon-containing epoxy monomer, triglycidyloxy phenyl silane (TGPS), was used to prepare a fire-retardant epoxy resin which showed substantially higher solid char yield (40%) at 800 °C than that of the conventional epoxy (12%), indicating that the silicone-containing group favors the carbonization mechanism [17]. The LOI value of 35 for TGPS can be advantageous in its use as a reactive fire retardant.

Diglycidyloxy methylphenyl silane (DGMPS) as epoxy monomers and bis(4-aminophenoxy)dimethyl silane (APDS) as curing agents both showing

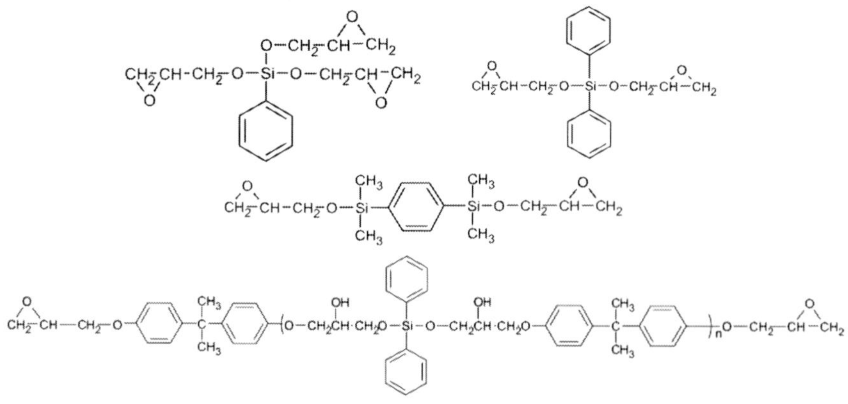

FIGURE 15

Silicon-containing epoxides and silicon-containing prepolymers [60].

suitable reactivity were used to form phosphorus- and silicon-containing epoxy resins with LOI values up to 37, with 10% Si concentration [61].

Siloxanes were also applied as reactive components in epoxy resins. Kang et al. [24] reported improved fire retardancy of epoxy resins by polysiloxanes. When the polysiloxanes content was 20 wt%, the LOI value reached 29.5 in contrast to 23.2 of the neat system. However, the replacement of a part of the epoxy resin or the hardener by their siloxane counterparts just slightly improved the fire resistance of resin according to the results from Murias et al. [23]. Hydroxyl-terminated polydimethylsiloxane (silicone) was used to prepare a siliconized epoxy matrix resin by reacting with diglycidyl ethers of bisphenol A (DGEBA) type epoxy resin [62]. When cured with bis (4-aminophenyl) phenylphosphate (BAPP), the thermal stability and flame-retardant properties were improved by the incorporation of both silicone and phosphorus moieties and a synergistic effect of silicone and phosphorus was observed.

Silicon-containing acrylate monomers were blended with epoxy acrylate oligomer to form UV-curable flame-retardant films with improved fire retardancy, increased char yield, and enhanced toughness [16]. Reactive polysilsesquioxanes and particularly POSS have also been widely used for inclusion in epoxy resins [63]. Fire retardancy of POSS/epoxy hybrids showed both reduction in HRR in cone calorimetry and inhibition of dripping in vertical UL94 tests [64].

4.2 Fire-retardant additive

Several papers report the use of organosilicon compounds as additives in fire-retarded formulations, either as such or in combination with other fire retardants. A few examples were given above in Sections 3.1 and 3.2.

Silicones show relatively low HRR, mostly within the range of 60–150 kW m^{-2} [65], minimal sensitivity to external heat flux and low yields of carbon monoxide release. In flammability tests, silicones generally show a slow burning rate without flaming dripping. Based on these fire properties, silicones offer advantages to be exploited in fire retardancy of other polymers. Furthermore, incorporation of silicones may also improve mechanical properties of the fire-retarded polymers [66].

Pape and Romenesko, on the basis of their results from cone calorimeter studies, reported that powdered silicone additives are helpful in improving fire-retardant properties of different polymers including polyolefins, polycarbonate, polystyrene, polyamide, etc. by reducing PHRR, carbon monoxide generation, and smoke evolution [67].

Karlsson et al. [68] investigated the behavior of an ethylene–methacrylic acid copolymer (EMAA) compounded with calcium carbonate and silicone elastomer (CaSiMAA) compared to a butyl acrylate copolymer (EBA) in terms of fire retardant, structural, and rheological properties. It was shown that the earlier formation of interactions in CaSiEMAA results in higher melt viscosity and lower foaming i.e. the overall fire-retardant performance is improved. In the EBA case, butene has to be released by ester pyrolysis before acid groups are available for cross-linking, which postpones the fire-retardant effect compared to EMAA.

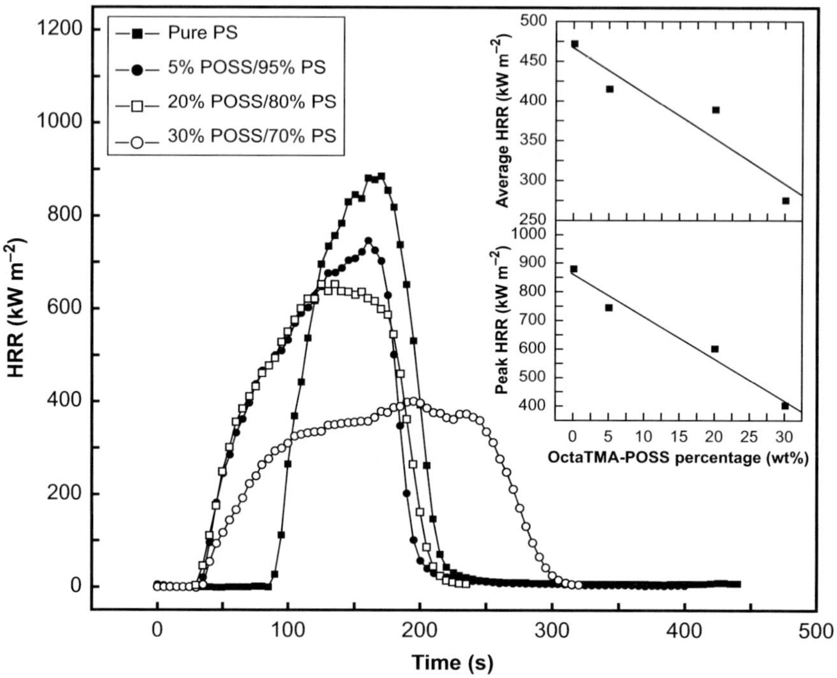

FIGURE 16

The heat release rate (HRR) results of OctaTMA-POSS/PS composites with high POSS weight ratios (from 5% to 30%). The radiant power is 35 kW m^{-2}. POSS, polyhedral oligomeric silsesquioxanes; PS, polystyrene.

Reprinted from Ref. [33], with kind permission from Springer Science and Business Media.

In another study, octa(tetramethylammonium) polyhedral oligomeric silsesquioxane (OctaTMA-POSS) was melt-blended as a fire-retardant additive, with polystyrene (PS) to prepare organic–inorganic composites [33]. Lower HRR and larger char yield were reported (Figure 16). Compared to pure PS, pHRR of the 40/60 wt% OctaTMA-POSS/PS composite decreased by 54.5% and the average HRR of the composites was reduced by 41.5%.

4.3 Protective coatings

Surface modification appears to be an attractive fire-retardant technology as it allows concentration of fireproof properties at the polymer surface and preserves the bulk properties of the material. Quédé et al. [69] observed an efficient fire-retardant effect of organosilicon thin films polymerized by 1,1,3,3-tetramethyldisiloxane when coated on polyamide-6 (PA-6) and polyamide-6 clay nanocomposite (PA-6 nano) substrates. The clay addition combined with the coating leads to an increase of LOI by 130% and a decrease of HRR and the total heat evolved by 41% and

33%, respectively, thanks to reinforcement in the formation of a silicon and carbonaceous protective layer during combustion.

Ravadits et al. [70] designed a model system consisting of a polyethylene substrate surface treated by vinyltriethoxysilane and by organoboroxo-siloxane (OBSi), and investigated an OBSi-containing intumescent flame-retarded compound (IFR-OBSi) based on polypropylene, ammonium polyphosphate, and pentaerythritol. The oxygen permeability decreased by about one order of magnitude after a radio-frequency plasma treatment. A small portion of OBSi remained unreacted in the surface layer of the flame-treated IFR-OBSi, which may ensure the plasticity necessary to prevent it from cracking and to ensure improved fire retardancy.

Gardelle et al. [40] investigated the thermal degradation and fire performance of polysilazane-based coatings (Figure 17). The degradation of polysilazane contributed to the formation of a ceramic char at high temperature. Moreover, it was shown that the performance of the polysilazane coating can be improved incorporating some fire retardant such as aluminum trihydroxide (ATH) as shown in Figure 17 for backside temperature of a polisilazane-coated steel plate exposed to a flame.

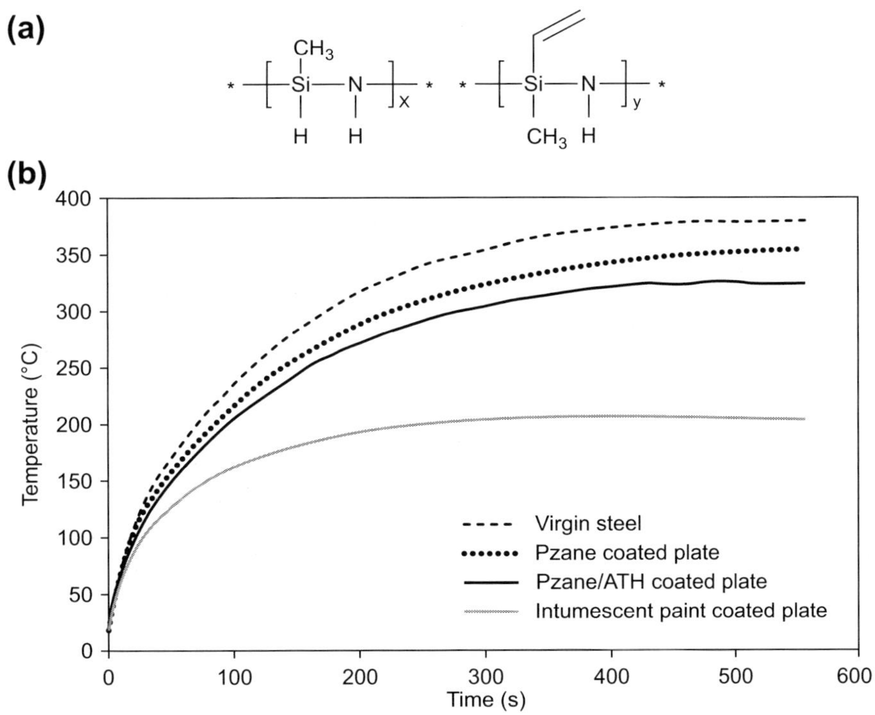

FIGURE 17

(a) Structure of vinyl polysilzane and (b) backside temperature profile for the virgin steel plate and for the polysilazane coatings plate [40].

5. CONCLUSION AND PERSPECTIVES

In the research effort to develop environmentally friendly and effective fire retardants, silicon-based compounds may play a decisive role because their general destiny when exposed to heat is transformation into inert unvolatile silica or silicon oxycarbide/silicon carbide which create a protective shielding on the polymer burning surface. Thus evolution of silicon-containing volatiles, which might be toxic or interact negatively with the environment, should be prevented.

Furthermore, silicon fire retardants, in combination with either inorganic or heteroatom-containing organic fire retardants, can produce on heating, ceramic residues on the surface of the burning polymer (e.g. Si–N, Si–B, Si–P, etc.). Silica and ceramics act as a thermally stable barrier to diffusion of combustible volatiles produced by the underneath polymer pyrolysis, toward the flame. Moreover, the silicon-derived inorganic/ceramic surface is characterized by heat reradiation properties which provide heat insulation to the condensed phase from the heat of the flame. Indeed, when the surface fire retarded polymer is completely consumed by thermal degradation in the range 250–500 °C in the burning process, the resulting inorganic/ceramic surface temperature can greatly increase by radiant heat from the flame, which can reach significantly higher temperatures. In these conditions, reradiation from the surface becomes an efficient thermal protection because it goes with the 4th temperature power. Thus the pyrolysis rate of the underlying polymer, which supplies the fuel to the flame by diffusion through the inorganic/ceramic surface, could be decreased below the flame self-feeding level.

The literature data so far available, reviewed here, although still relatively limited, allow an optimistic forecast on a very positive contribution of silicon-based fire retardants to limiting satisfactorily the fire risk and hazard in polymer materials. There is also evidence that this should be possibly done with an environmentally friendly approach, complying with the most safe fire standards and with relatively low concentration, to preserve or improve material physical properties.

References

[1] Covaci A, Harrad S, Abdallah MAE, Ali N, Law RJ, Herzke D, et al. Novel brominated flame retardants: a review of their analysis, environmental fate and behaviour. Environ Int 2011;37:532–56.

[2] Sjödin A, Patterson DG, Bergman A. A review on human exposure to brominated flame retardants-particularly polybrominated diphenyl ethers. Environ Int 2003;29:829–39.

[3] Lomakin SM, Zaikov GE. Ecological aspects of polymer flame retardancy. Zeist, NL: VSP; 1999.

[4] Van der Veen I, De Boer J. Phosphorus flame retardants: properties, production, environmental occurrence, toxicity and analysis. Chemosphere 2012;88:1119–53.

[5] Lu SY, Hamerton I. Recent developments in the chemistry of halogen-free flame retardant polymers. Prog Polym Sci 2002;27:1661–712.

[6] Iji M, Serizawa S. Silicone derivatives as new flame retardants for aromatic thermoplastics used in electronic devices. Polym Adv Technol 1998;9:593–600.
[7] Zhong H, Wei P, Jiang P, Wu D, Wang G. Synthesis and characteristics of a novel silicon-containing flame retardant and its application in poly[2,2-propane-(bisphenol) carbonate]/acrylonitrile butadiene styrene. J Polym Sci Part B Polym Phys 2007;45: 1542–51.
[8] Liu S, Ye H, Zhou Y, He J, Jiang Z, Zhao J, et al. Study on flame-retardant mechanism of polycarbonate containing sulfonate-silsesquioxane-fluoro retardants by TGA and FTIR. Polym Degrad Stab 2006;91:1808–14.
[9] Zhong H, Wei P, Jiang P, Wang G. Thermal degradation behaviors and flame retardancy of PC/ABS with novel silicon-containing flame retardant. Fire Mater 2007;31:411–23.
[10] Liauw CM, Lees GC, Hurst SJ, Rothon RN, Ali S. Effect of silane-based filler surface treatment formulation on the interfacial properties of impact modified polypropylene/magnesium hydroxide composites. Compos Appl Sci Manuf 1998;29:1313–8.
[11] Zaharri ND, Othman N, Mohd Ishak ZA. Thermal and mechanical properties of zeolite filled ethylene vinyl acetate composites. Procedia Chem 2012;4:95–100.
[12] Demir H, Balköse D, Ülkü S. Influence of surface modification of fillers and polymer on flammability and tensile behaviour of polypropylene-composites. Polym Degrad Stab 2006;91:1079–85.
[13] Cárdenas MA, García-López D, Gobernado-Mitre I, Merino JC, Pastor JM, de Martínez JD, et al. Mechanical and fire retardant properties of EVA/clay/ATH nanocomposites – effect of particle size and surface treatment of ATH filler. Polym Degrad Stab 2008;93:2032–7.
[14] Wang B, Wang X, Tang G, Shi Y, Hu W, Lua H, et al. Preparation of silane precursor microencapsulated intumescent flame retardant and its enhancement on the properties of ethylene–vinyl acetate copolymer cable. Compos Sci Tech 2012;72:1042–8.
[15] Lecoeur E, Vroman I, Bourbigot S, Delobel R. Optimization of monoguanidine dihydrogen phosphate and amino propylethoxysilane based flame retardant formulations for cotton. Polym Degrad Stab 2006;91:1909–14.
[16] Cheng X, Shi W. UV-curing behavior and properties of tri/di(acryloyloxyethyloxy) phenyl silane used for flame-retardant coatings. Progr Org Coat 2010;69:252–9.
[17] Wang WJ, Perng LH, Hsiue GH, Chang FC. Characterization and properties of new silicone-containing epoxy resin. Polymer 2000;41:6113–22.
[18] Shu WJ, Ho JC, Perng LH. Studies of silicon-containing maleimide polymers: 1. Synthesis and characteristics of model compounds. Eur Polym J 2005;41:149–56.
[19] Chiang CL, Chang RC. Synthesis, characterization and properties of novel self-extinguishing organic–inorganic nanocomposites containing nitrogen, silicon and phosphorus via sol–gel method. Compos Sci Tech 2008;68:2849–57.
[20] Chen X, Hu Y, Jiao C, Song L. Preparation and thermal properties of a novel flame-retardant coating. Polym Degrad Stab 2007;92:1141–50.
[21] Gilman JW, Kashiwagi T, Harris RH, Lomakin S, Lichtenhan JD, Bolf A, et al. Char enhancing approaches to flame retarding polymers. In: Al-Malaika S, Golovoy A, Wilkie CA, editors. Chemistry and technology of polymer additives. Malden: Blackwell Science Inc.; 1999. pp. 135–50.
[22] Lichtenhan JD, Gilman JW. Preceramic additives as fire retardants for plastics. US Patent 6,362,279; 2002.
[23] Murias P, Maciejewski H, Galina H. Epoxy resins modified with reactive low molecular weight siloxanes. Eur Polym J 2012;48:769–73.

[24] Kang N, Du Z, Li H, Zhang C. Synthesis of polysiloxane-type multifunctional flame retardant and its application in epoxy systems. J Appl Polym Sci 2012;124:4915–9.
[25] Marosi G, Márton A, Anna P, Bertalan G, Marosföi B, Szép A. Ceramic precursor in flame retardant systems. Polym Degrad Stab 2002;77:259–65.
[26] Hu Z, Chen L, Zhao B, Luo Y, Wang D-Y, Wang Y-Z. A novel efficient halogen-free flame retardant system for polycarbonate. Polym Degrad Stab 2011;96:320–7.
[27] Tang Z, Li Y, Zhang YJ, Jiang P. Oligomeric siloxane containing triphenylphosphonium phosphate as a novel flame retardant for polycarbonate. Polym Degrad Stab 2012;97: 638–44.
[28] Harrison PG. Silicate cages: precursors to new materials. J Organomet Chem 1997;542: 141–83.
[29] Voronkov MG, Lavrent'yev VI. Polyhedral oligosilsesquioxanes and their homo derivatives. Top Curr Chem 1982;102:199–236.
[30] Li G, Wang L, Ni H, Pittman Jr CU. Polyhedral oligomeric silsesquioxane (POSS) polymers and copolymers: a review. J Inorg Organomet Polym 2001;11:123–54.
[31] Vannier A, Duquesne S, Bourbigot S, Castrovinci A, Camino G, Delobel R. The use of POSS as synergist in intumescent recycled poly(ethylene terephthalate). Polym Degrad Stab 2008;93:818–26.
[32] Song L, He Q, Hu Y, Chen H, Liu L. Study on thermal degradation and combustion behaviors of PC/POSS hybrids. Polym Degrad Stab 2008;93:627–39.
[33] Liu L, Hu Y, Song L, Nazare S, He S, Hull R. Combustion and thermal properties of OctaTMA-POSS/PS composites. J Mater Sci 2007;42:4325–33.
[34] Zhang W, Li X, Yang R. Novel flame retardancy effects of DOPO-POSS on epoxy resins. Polym Degrad Stab 2011;96:2167–73.
[35] Song T, Li ZS, Liu JG, Yang SY. Novel phosphorus–silicon synergistic flame retardants: synthesis and characterization. Chin Chem Lett 2012;23:793–6.
[36] Bourbigot S, Turf T, Bellayer S, Duquesne S. Polyhedral oligomeric silsesquioxane as flame retardant for thermoplastic polyurethane. Polym Degrad Stab 2009;94: 1230–7.
[37] Feher FJ, Budzichowski TA. Silasesquioxanes as ligands in inorganic and organometallic chemistry. Polyhedron 1995;14:3239–53.
[38] Fina A, Abbenhuis HCL, Tabuani D, Camino G. Metal functionalized POSS as fire retardants in polypropylene. Polym Degrad Stab 2006;91:2275–81.
[39] Fina A, Bocchini S, Camino G. Catalytic fire retardant nanocomposites. Polym Degrad Stab 2008;93:1647–55.
[40] Gardelle B, Duquesne S, Vu C, Bourbigot S. Thermal degradation and fire performance of polysilazane-based coatings. Thermochim Acta 2011;519:28–37.
[41] Tang Y, Lewin M. Migration and surface modification in polypropylene (PP)/polyhedral oligomeric silsequioxane (POSS) nanocompositesy. Polym Adv Technol 2009;20: 1–15.
[42] Hshieh F-Y. Shielding effects of silica-ash layer on the combustion of silicones and their possible applications on the fire retardancy of organic polymers. Fire Mater 1998;22: 69–76.
[43] Zhou W, Yang H. Flame retarding mechanism of polycarbonate containing methylphenyl-silicone. Thermochim Acta 2007;452:43–8.
[44] Fina A, Tabuani D, Carniato F, Frache A, Boccaleri E, Camino G. Polyhedral oligomeric silsesquioxanes (POSS) thermal degradation. Thermochim Acta 2006;440: 36–42.

[45] Bolln C, Tsuchida A, Frey H, Muelhaupt R. Thermal properties of the homologous series of 8-fold alkyl-substituted octasilsesquioxanes. Chem Mater 1997;9:1475–9.

[46] Mantz RA, Jones PF, Chaffee KP, Lichtenhan JD, Gilman JW. Thermolysis of polyhedral oligomeric silsesquioxane (POSS) macromers and POSS-siloxane copolymers. Chem Mater 1996;8:1250–9.

[47] Fina A, Tabuani D, Camino G. Polypropylene–polysilsesquioxane blends. Eur Polym J 2010;46:14–23.

[48] Laoutid F, Bonnaud L, Alexandre M, Lopez-Cuesta J-M, Dubois P. New prospects in flame retardant polymer materials: from fundamentals to nanocomposites. Mater Sci Eng Rep 2009;63:100–25.

[49] Zhang S, Horrocks AR. A review of flame retardant polypropylene fibres. Prog Polym Sci 2003;28:1517–38.

[50] Hermansson A, Hjertberg T, Sultan B-A. The flame retardant mechanism of polyolefins modified with chalk and silicone elastomer. Fire Mater 2003;27:51–70.

[51] Krämer RH, Blomqvist P, Van Hees P, Gedde UW. On the intumescence of ethylene-acrylate copolymers blended with chalk and silicone. Polym Degrad Stab 2007;92:1899–910.

[52] Lin H, Yan H, Liu B, Wei L, Xu B. The influence of KH-550 on properties of ammonium polyphosphate and polypropylene flame retardant composites. Polym Degrad Stab 2011;96:1382–8.

[53] Chen X, Jiao C. Synergistic effects of hydroxy silicone oil on intumescent flame retardant polypropylene system. Fire Saf J 2009;44:1010–4.

[54] Yuan D-D, Deng L-T, Yin H-Q, Cai X-F. Synergistic effect of organic silicon on the flame retardancy and thermal properties of polycarbonate/potassium-4-(phenylsulfonyl) benzenesulfonate systems. J Appl Polym Sci 2013;127:2095–101.

[55] Kim J, Lee K, Lee K, Bae J, Yang J, Hong S. Studies on the thermal stabilization enhancement of ABS; synergistic effect of triphenyl phosphate nanocomposite, epoxy resin, and silane coupling agent mixtures. Polym Degrad Stab 2003;79:201–7.

[56] Dong W, Zhang X, Liu Y, Wang Q, Gui H, Gao J, et al. Flame retardant nanocomposites of polyamide 6/clay/silicone rubber with high toughness and good flowability. Polymer 2006;47:6874–9.

[57] Anna P, Marosi G, Bourbigot S, Le Bras M, Delobel R. Intumescent flame retardant systems of modified rheology. Polym Degrad Stab 2002;77:243–7.

[58] Anna P, Marosi G, Csontos I, Bourbigot S, Le Bras M, Delobel R. Influence of modified rheology on the efficiency of intumescent flame retardant systems. Polym Degrad Stab 2001;74:423–6.

[59] Weil ED, Levchik S. A review of current flame retardant systems for epoxy resins. J Fire Sci 2004;22:25–40.

[60] Mercado LA, Galià M, Reina JA. Silicon-containing flame retardant epoxy resins: synthesis, characterization and properties. Polym Degrad Stab 2006;91:2588–94.

[61] Spontón M, Mercado LA, Ronda JC, Galià M, Cádiz V. Preparation, thermal properties and flame retardancy of phosphorus- and silicon-containing epoxy resins. Polym Degrad Stab 2008;93:2025–31.

[62] Ananda kumar S, Denchev Z, Alagar M. Synthesis and thermal characterization of phosphorus containing siliconized epoxy resins. Eur Polym J 2006;42:2419–29.

[63] Kuo S-W, Chang F-C. POSS related polymer nanocomposites. Progr Polym Sci 2011;36:1649–96.

[64] Franchini E, Galy J, Gerard J-F, Tabuani D, Medici A. Influence of POSS structure on the fire retardant properties of epoxy hybrid networks. Polym Degrad Stab 2009;94: 1728−36.

[65] Buch RR. Rates of heat release and related fire parameters for silicones. Fire Saf J 1991; 17:1−12.

[66] Perret B, Schartel B. The effect of different impact modifiers in halogen-free flame retarded polycarbonate blends − II. Fire behaviour. Polym Degrad Stab 2009;94: 2204−12.

[67] Pape PG, Romenesko DJ. The role of silicone powders in reducing the heat release rate and evolution of smoke in flame retardant thermoplastics. J Vinyl Addit Tech 1997;3: 225−32.

[68] Karlsson L, Lundgren A, Jungqvist J, Hjertberg T. Influence of melt behaviour on the flame retardant properties of ethylene copolymers modified with calcium carbonate and silicone elastomer. Polym Degrad Stab 2009;94:527−32.

[69] Quédé A, Mutel B, Supiot P, Jama C, Dessaux O, Delobel R. Characterization of organosilicon films synthesized by N 2-PACVD. Application to fire retardant properties of coated polymers. Surf Coat Tech 2004;180−181:265−70.

[70] Ravadits I, Tóth A, Marosi G, Márton A, Szép A. Organosilicon surface layer on polyolefins to achieve improved flame retardancy through an oxygen barrier effect. Polym Degrad Stab 2001;74:419−22.

CHAPTER

13

(Photo)oxidative Stabilization of Flame-Retarded Polymers

R. Pfaendner

Fraunhofer Institute for Structural Durability and System Reliability

LBF Division Plastics, Darmstadt, Germany

CHAPTER OUTLINE

1. Introduction

Polymers like all kinds of organic materials are sensitive to oxidation. Oxidation takes place at any time of the lifetime of the polymer (and of the flame-retarded polymer) during thermal transformation, storage, or application. Furthermore, the combined action of light and oxygen results in photooxidation. Degradation of the polymer and irreversible changes of the polymer structure are the consequence of photooxidation resulting finally in insufficient mechanical properties, cracking, failure of the part, change in visual appearance (e.g. discoloration), and loss in value. Fortunately suitable stabilizers such as antioxidants and UV/light stabilizers inhibit the oxidation process, protect polymers against both oxidative and photooxidative damages, and maintain available properties during processing as well as during the product life cycle in the intended application. Flame-retarded polymers are mainly used in long-term applications; therefore, the role of stabilizers is decisive as many flame retardants influence the oxidative and photooxidative stability of polymers often in a negative way.

Polymer Green Flame Retardants. http://dx.doi.org/10.1016/B978-0-444-53808-6.00013-5

Furthermore, many flame retardants act as a typical polymer filler whereas it is known that filler interactions influence the stability of the polymer, which is even more critical if nanosized materials are used. Depending on the type of flame retardant used, on the polymer substrate, and on the intended application, adjusted stabilizer systems have to be selected or developed.

Challenges of the stabilization of flame-retarded polymers are analyzed below with regard to the components and mutual interactions with focus on "green" flame retardants. Processing and long-term thermal stabilization of flame retarded polymers are discussed and strategies of improving the light stability of flame-retarded polymers are provided. Additionally, the specific requirements of the stabilization of nanocomposites as potential flame-retardant components are covered.

2. (Photo)oxidative degradation and stabilization of polymers

Degradation of the polymer and irreversible changes of the polymer structure are the consequence of photooxidation resulting finally in insufficient mechanical properties, cracking, failure of the part, change in visual appearance (e.g. discoloration), and loss in value. The sensitivity to oxidation depends on the chemical structure of the polymer and of all other ingredients which may accelerate or delay any (photo)oxidation. For example, polystyrene, polymethylmethacrylate, or polyester are quite stable; polypropylene and other polyolefins such as polybutadiene are very sensitive to oxidation.

The so-called autoxidation of polymers starts with an initiation step forming radicals e.g. from defect structures in the polymer chain (Figure 1). Reaction with

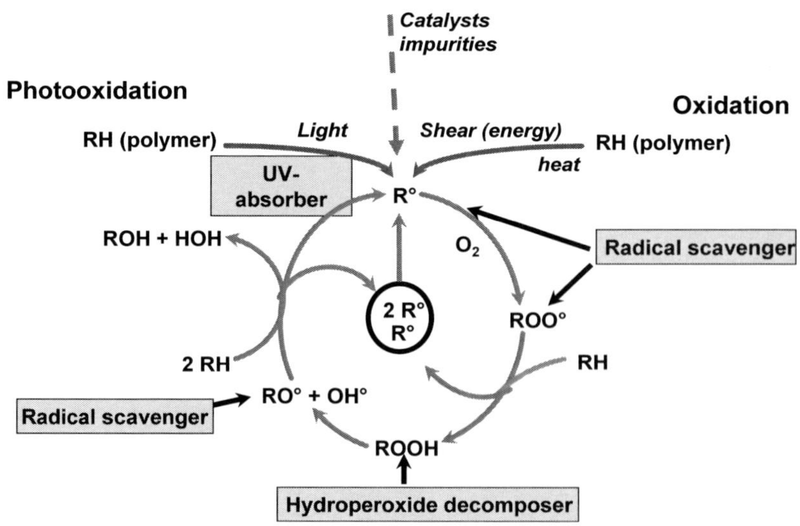

FIGURE 1

Autoxidation of polymers.

molecular oxygen results in peroxy radicals followed by hydroperoxide formation through hydrogen abstraction. Hydroperoxides decompose into alkoxy and hydroxyl radicals. Furthermore, the polymer chain is modified through forming oxygen-containing structures (carbonyl, alcohol, carboxyl) and unsaturated groups (vinyl, vinylidene, allylic). Disproportionation and depolymerization result in low molecular weight products. Cross-linked polymer chains originate from the radical recombination of lower molecular weight fragments. Through the action of light, free radicals are formed depending on the polymer structure and the energy of the light. These radicals can react with oxygen via peroxy radicals to hydroperoxides and initiate the autoxidation cycle. Fillers, metal ions, recycled polymers, some pigments, and flame retardants may accelerate the degradation; stabilizers such as antioxidants or light stabilizers protect the polymer and extend the lifetime [1].

Antioxidants inhibit or reduce the thermo-oxidative degradation in polymers and maintain inherent properties during processing as well as during the product life cycle in the intended application. Antioxidants interrupt the autoxidation cycle by chemical reactions with the formed intermediates. Typical antioxidants are sterically hindered phenols (e.g. AO-1, AO-2, AO-3), amines, phosphites (e.g. P-1), and sulfur compounds. Phosphites (or alternatively phosphonites, e.g. P-2) are mainly used as processing stabilizers; hindered phenols and amines provide processing and long-term thermal stability. Sulfur compounds (e.g. S-1) act in combination with phenolic antioxidants as long-term thermal stabilizers. To protect polymers from the negative influence of light, UV absorbers e.g. benzophenones (e.g. UV-1), benzotriazoles (e.g. UV-2, UV-3), cinnamates, hydroxyphenyltriazines (e.g. UV-4), and free radical scavengers in the form of hindered amine light stabilizers (HALS, e.g. LS-1, LS-2, LS-3) are used. In combination with phenolic antioxidants HALS provide in addition efficient long-term thermal stability to polymers. The chemical structures of the aforementioned stabilizers can be found in the appendix.

Standard test methods to analyze stability of polymers cover multiple extrusion, accelerated heat aging, and UV light exposure. Multiple extrusion evaluates the melt processing stability of a polymer or polymer formulation as high temperature, shear forces, and oxygen in the polymer transformation process cause degradation. The melt properties are usually characterized by melt flow rate (MFR) or melt volume rate (MVR) e.g. according to ISO 1133. Accelerated heat aging (e.g. ISO 188, ISO 2578) is simulating long-term thermal stability. Samples are oven-aged at defined temperatures and conditions, and the change of the properties, e.g. visual appearance, color, and mechanical values, is recorded over aging time. A simple version of long-term stability mainly for quality control purposes is to determine the oxygen induction time (OIT, e.g. ISO 11357), whereas with differential scanning calorimetry (DSC) or differential thermoanalysis (DTA) the induction time for the onset of the oxidation process is measured. Photooxidation is tested by UV exposure and artificial or natural weathering (e.g. ISO 4892). Change of the properties e.g. visual appearance such as chalking, crazes, gloss, and/or mechanical properties such as tensile strength, elongation, and impact strength are measured. More detailed information on testing methods, mechanisms of degradation, activities of stabilizers, and technical aspects of stabilization for different polymers can be found elsewhere [1–3].

3. Challenges of the stabilization of flame-retarded polymers

Flame retardants reduce the flammability of polymers. The concentrations needed are often quite high; therefore, these molecules "disturb" the polymer compound and influence e.g. the rheological properties during processing and the mechanical properties and visual appearance of the final product. Obviously, the stability of the polymer can be influenced as well. With regard to (photo)oxidation, the polymer and the flame retardant may degrade independently. There might be a direct interaction between the polymer and the flame retardant causing acceleration or retardation, and there might be an indirect interaction through the influence on the stabilizing additives. Moreover, secondary chemical processes during use like hydrolysis may result in degradation products, influencing the overall stability of the polymer formulation.

3.1 Processing and long-term thermal and light stabilization of halogenated flame retardants

Classical bromine-containing flame retardants such as tetrabromobisphenol A or brominated diphenyl ethers are less thermally stable than most polymers. Therefore, the dehydrobromination reaction may already occur during processing. However, it can be suppressed by the addition of heat stabilizers from the PVC processing range, e.g. organotin compounds such as dialkyltinmaleates [4–7], epoxides [8] or hydrotalcites [9], phosphites, polyols, and mixed metal soaps [10].

The photooxidation of polypropylene is accelerated in the presence of decabromodiphenylether and antimony oxide although a UV filter effect is present through the flame retardants in comparison to the neat sample [11,12]. It has been found that the fire retardant decomposes quickly under irradiation into free radicals, which attack the polymer. Although the main decomposition products of the polymer are unchanged, there is evidence of increased hydroperoxide decomposition [13]. Experiments with polystyrene [12] and polyethylene (HDPE, LDPE) showed similar acceleration of photooxidation like in the case of polypropylene [14].

Moreover, hindered amine stabilizers are deactivated through the formation of an amine salt (Figure 2), preventing the oxidation of HALS to the nitroxyl radical

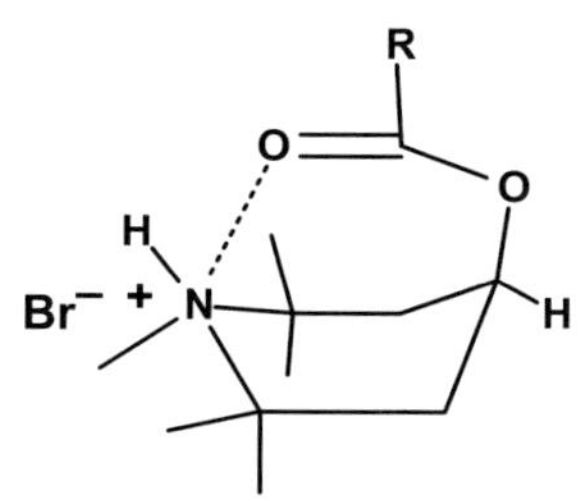

FIGURE 2

Potential structure of deactivated hindered amine light stabilizers.

which is mandatory in the stabilization process. Furthermore, the resulting aminium salt is less thermally stable than the corresponding amine and may be further degraded during processing [15]. Aliphatic bromine compounds are in this respect less detrimental than aromatic ones [16].

To eliminate the antagonistic effect of brominated flame retardants on HALS, HBr scavengers like tin maleates or antiacids such as ZnO, calcium stearate, and $Mg(OH)_2$ have been tested; yet, they were not successful [17]. However, increase in the HALS stabilizer concentration, additional pigment (TiO_2), and use of an alkoxyamine HALS or *N*-acylated HALS [18] with reduced basicity are potential solutions to increase the light stability [16,17]. Meanwhile, a number of different structures with low basicity HALS have been published in patent applications [19–24]. Microencapsulation of the halogenated flame retardant was proposed as an alternative to avoid impact of brominated flame retardants on light stability [25].

3.2 Processing and long-term thermal and light stabilization of phosphorus and nitrogen-based halogen free flame retardants

The published data on the photooxidative stability of phosphorus-based (ammoniumpolyphosphate, red phosphorus, and melamine(poly, pyro)phosphates) and of nitrogen-based (melamine cyanurate, alkoxyamines) flame retardants are not very numerous. Intumescent flame retardants on the basis of ammonium polyphosphate and pentaerythritol do not modify the degradation of unstabilized PP [26]; the photodegradation of the polymer and of the flame retardant are independent processes [27]. However, the intumescent flame retardant may influence the light stabilized material. There, the activity of HALS is somewhat reduced likely through protonation by the polyphosphoric acid [28]; however, it is less severe than with brominated compounds. On the other side, the efficiency of an UV-absorber (UV-1) is increased through the polyol present in the flame-retardant combination.

In a patent application comparison, it was demonstrated that halogen-containing flame retardants like octabromodiphenylether (OBDE) show significant surface degradation already after 400 h, even in the presence of HALS, while ammonium polyphosphate flame-retarded systems are stable up to 1800 h of artificial light exposure in the presence of 0.5% HALS (LS-4), which is nevertheless lower than that achieved by the stabilized polymer without APP (Table 1 [29]). However, in case of artificial weathering, hydrolysis of ammonium polyphosphate may occur and result in reduced fire retardancy as the formation of the intumescent network is disturbed already after exposure with an incident energy of 210 kJ cm^{-2}, a value which corresponds to less than 1 year of outside use in mid Europe [30]. A similar result is found in EVA/PA-6/PP-blends [31].

Discoloration of ammonium polyphosphate flame-retarded polypropylene during heat aging was reduced by adding heterocyclic basic additives such as piperazines, benzimidazole, or thiouracyl [32]. On the other hand, bisphenol-A-diphosphate PC/ABS blend showed, without UV stabilizer, discoloration and surface cracks; however, no influence on the flame retardancy after 210 kJ cm^{-2} [30].

Table 1 Light Stability of Flame-Retarded Polypropylene [29]

Composition	Flame Retardant (%)	HALS (%)	UL-94 Test	Hours until Severe Surface Degradation
PP	0	–	Not classified	400
PP	0	0.5	Not classified	>2500
PP OBDE/Sb_2O_3	4.27/1.22	–	V-2	400
PP OBDE/Sb_2O_3	4.27/1.22	0.5	V-2	400
PP APP	39.0	–	V-0	600
PP APP	39.0	0.5	V-0	1800

HALS, hindered amine light stabilizers.

Red phosphorus is mainly used in engineering plastics (polycarbonate, polyamide, polyester), therefore its influence on light and long-term thermal stability is less critical. From the stabilization point of view, it is more important to stabilize red phosphorus against phosphine formation through moisture, which is achieved by microencapsulation [33] or by adding salts such as copper acetate [34].

The influence of melamine phosphate on stabilized polypropylene in multiple extrusion experiments has been found to be negligible; moreover, there was no antagonistic effect with HALS in artificial weathering experiments [35]; even, a slight improvement could be detected due to a pigment effect of the flame retardant.

Turning to nitrogen-based compounds, stabilized polyamide melamine cyanurate formulations have been found to not loose their fire retardancy after artificial weathering (210 kJ cm^{-2}); however, some blooming of melamine cyanurate was detected [30]. Furthermore, the processing stability of melamine cyanurate-containing polyamide is claimed to be improved by addition of sulfates, e.g. magnesium sulfate [36] or metal acetates e.g. magnesium acetate [36].

Recycling of melamine cyanurate polyamide compounds, APP polypropylene, and melamine phosphate—polypropylene seems to be feasible with minor properties loss [31,35,37]. On the other side, recycling of PC/ABS flame-retarded with organic phosphate ester resulted in the loss of flame retardancy already after the first recycling step, although molecular weight and mechanical properties were still close to the starting level in most cases [38]. Contrary to this result, ABS with brominated flame retardants passed the UL 94 5VB test without failing, even after five extrusion steps.

More recent flame retardant systems mainly for polypropylene/polyolefins thin section applications, such as films or fibers, consist of alkoxyamine structures, which are related to hindered amine light stabilizers. These alkoxyamine flame retardants can be used alone [39] or in combinations with other flame retardants [40—43]. It is even possible to incorporate more flame retardant active structures chemically in one molecule e.g. in phosphorus-containing alkoxyamines [44]. In any case, alkoxyamines of selected structures can combine flame retardancy, inherent light stability, and long-term thermal stability in one molecule.

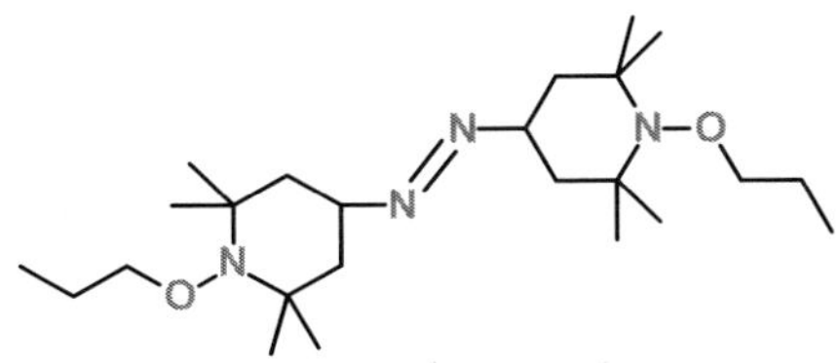

FIGURE 3

Example of azo alkoxyamine flame retardants (bis(1-propoxy-2,2,6,6-tetramethylpiperidin-4-yl)diazene, CA: 1050647-62-5).

An even more efficient molecule with combined flame retardancy and light stabilizer activity was identified in Azo-NOR compounds (e.g. Figure 3); very low concentrations (0.5%) provide flame retardancy to polyolefin films even after extended weathering time [45,46]. In Table 2, it is shown that the Azo-NOR compound is the only one which provides flame retardancy in polypropylene films for more than 2000 h of artificial weathering.

3.3 Processing and long-term thermal and light stabilization of inorganic flame retardants and of nanocomposites

A substantial flame retardant class is inorganic fillers such as $Al(OH)_3$, $AlO(OH)$, $Mg(OH)_2$, layered silicates, and nanocomposites which act often like filler-type materials as these are insoluble in the polymer and are distributed in particles throughout the matrix (melamine cyanurate, melamine polyphosphate, etc.). It is well known that inorganic fillers show, independent of the structures, experimentally a negative effect on the (photo)oxidative stability of the polymer; however, to a various extent. Interactions between the stabilizer and the filler and adsorption/desorption mechanisms are mainly responsible for this influence. The surface area of the filler and pore volumes, surface functionality, hydrophilicity, thermal and photosensitization properties of the filler, and transition metal ion content (manganese, iron, titanium) have been found as potential elements of the interaction [47]. Mechano- degradation through the filler forms additional radicals during processing resulting in additional consumption of stabilizers. Furthermore, efficient stabilization of polymers becomes even more complicated when additionally to natural inorganic fillers other materials, such as carbon black, or pigments, such as TiO_2, are incorporated in the formulation. The impact of inorganic fillers on the stability of polymers is illustrated in the following examples:

Calcium carbonate and talc decreased dramatically the photostabilizing efficiencies of all light stabilizers (hindered amines, benzophenones) and antioxidants. Polypropylene, in presence of 0.05% of a typical phenolic antioxidant (AO-1) withstood at 120 °C, a time, until 50% retention of tensile strength, of 3508 h versus only 470 h in the presence of 10% $CaCO_3$ (talc), resulted in 2180 h [48].

Polypropylene films, stabilized with 2,6-ditert-butyl-4-methylphenol and a low molecular weight hindered amine light stabilizer (LS-1) showed at 130 °C heat

Table 2 Properties of Flame-Retarded Polypropylene Films after Artificial Weathering [46]

Flame Retardant (wt%)	Before Artificial Weathering			After 400 h of Artificial Weathering			After 2000 h of Artificial Weathering		
	Weight Loss (%)	Burn Length (mm)	DIN 4102 B2 (Pass/Fail)	Weight Loss (%)	Burn Length (mm)	DIN 4102 B2 (Pass/Fail)	Weight Loss (%)	Burn Length (mm)	DIN 4102 B2 (Pass/Fail)
No FR	100	100	Fail	100	100	Fail	Plaques destroyed after 400 h		
0.5% Azo	12.9	47	Pass	39	93	Pass	Plaques destroyed after 800 h		
0.5% Azo + 0.25% NOR	14.5	44	Pass	21.1	75	Pass	Plaques destroyed after 1200 h		
0.5% NOR	49.4	80	Fail	42.9	100	Fail	Test failed before 400 h		
0.5% Azo-NOR	5.2	27	Pass	5.4	27	Pass	7	30	Pass

Azo = Dicyclohexylazo, NOR = Flamestab NOR 116, AZONOR = Figure 3.

aging a stability of 181 days until embrittlement i.e. until complete loss of mechanical properties. The same formulation containing 40% of talc (metal impurities: 60 ppm Mn, 5 ppm Cu) degraded already after 4 days only [49]. Photooxidation was even more severe: A typical stabilized polypropylene film (0.2% LS-1) achieved 2660 h to 50% retention of tensile strength compared to only 364 h in the presence of 10% $CaCO_3$ [50].

The influence of silica (SiO_2) on the oxidative and photooxidative stability of polypropylene is correlated to the metal ion (Fe, Al, Ti) content of impurities present in natural silicas. Most of the silicas act as photosensitizers and accelerate the degradation of the polymer even at low concentrations. Thermal oxidation at 130 °C (shown as time to embrittlement/0.06 carbonyl units of PP films) mainly resulted in reduced lifetime if SiO_2 is present (0.1% AO-1 achieved 456 h without and 336 h with SiO_2). Some HALS e.g. LS-2 or LS-1 seem to be an exception as these can increase the thermal stability of SiO_2 containing PP (e.g. with 0.1% LS-2, 312 h in the presence of SiO_2 and only 91 h without SiO_2 were measured). With the exception of LS-1 used as light stabilizer, photooxidation resulted in a considerable reduction of lifetime of all tested individual stabilizers and combinations, if SiO_2 was present, e.g. 0.1% of LS-2 achieved 1512 h without and 540 h in the presence of SiO_2. The results with different HALS stabilizers indicated that not only adsorption but as well desorption mechanisms and release of stabilizers are important [47].

Therefore, in order to achieve the requested processing and long-term properties, the adsorption effect of the filler and the effect of filler impurities have to be compensated. One simple solution is to increase the stabilizer level to the necessary extent. However, this might be an insufficient or a costly approach and it is limited because of compatibility reasons between the additives and the polymer. As mentioned in the examples above, especially light stabilizers of the HALS type show considerable performance differences in the presence of fillers, depending on the molecular weight and structure. Therefore, the selection of an adjusted stabilizer or stabilizer combination for the selected filler polymer pair is decisive. Furthermore, a number of chemical compounds, the so called filler deactivators or coupling agents, have been suggested to modify the filler surface and to improve the stability of the polymer [51]. Suitable additives range from typical filler coatings (stearic acid, stearates), oligomeric epoxides, silanes, titanates to functional polymers (e.g. polypropylene-graft-maleic anhydride or polypropylene-graft-acrylic acid). Excellent results with regard to the oxidative stability of filled polymers can be achieved by using amphiphilic modifiers, such as acrylates with long carbon chain side groups [52,53].

Layered silicates, another family of inorganic materials, have been extensively evaluated as flame retardant components [54,55], but the stabilization of the resulting "nano"fillers implies additional challenges. Layered silicates from natural sources (e.g. montmorillonite) contain metal ions as contaminants, which act in the same way as in other fillers but might be more homogeneously distributed and, therefore, their presence is more crucial to the polymer stability than metal ions in micro sized fillers.

Layered silicates as such are only compatible to hydrophilic polymers such as EVA, PA or TPU. In order to improve the compatibility of layered silicates to most polymers to allow intercalation or exfoliation, a modification step is necessary to make the filler organophilic. Modification of the clay is usually achieved by cation exchange with a long chain amine or a quaternary ammonium salt. The amount of organic material within the clay can be up to 40% and, therefore, the stability of this organic part cannot be neglected. At processing temperatures during extrusion or for injection molding of most polymers the thermal stability of the ammonium salts is very limited. Most of the ammonium structures tend to undergo Hofmann elimination resulting in volatile olefins and amines. The thermal degradation of ammonium salts starts already at 180 °C and is furthermore reduced by catalytically active sites on the alumosilicate layer [56].

Some nanocomposites manufactured by hydrophobic polymers such as polyolefins often require on top of the organic modification of the filler an additional compatibilizer e.g. polypropylene-g-maleic anhydride (PP-g-MAH), which is used in concentrations of 5–25%. These lower molecular weight modified polypropylene materials possess usually an inferior stability compared to the parent polymer and, therefore, comprise a further reason for a potential insufficient long-term performance of polymer nanocomposites.

As expected, nanocomposites based on montmorillonite show compared to unfilled materials a dramatic loss in processing and long-term stability [57–59], e.g. the long-term thermal stability of a stabilized polypropylene at 135 °C has been found to decrease from more than 40 days to only 15 days in the presence of 5% organically modified clay. An even more serious problem is the photooxidative stability of nanocomposites. Polypropylene montmorillonite based nanocomposites (stabilized with antioxidant) degrade much faster under photooxidative conditions than pure polypropylene [60,61]. This behavior is attributed to the active species of the clay generated by photolysis or photooxidation e.g. from alkylammonium modifiers and by interactions between antioxidant, montmorillonite and maleic anhydride modified PP. Furthermore, the photooxidation of ethylene propylene diene monomer (EPDM) montmorillonite nanocomposites drastically reduced the induction time of photooxidation in the presence of nanoclay, in particular if an exfoliated structure is present [62]. The photooxidation of syndiotactic PP/synthetic clay (fluorohectorite modified by octadecylammonium) nanocomposites showed that the nanoparticles catalyzed the decomposition process [63]. The presence of PP-g-MAH as compatibilizer accelerated the degradation further. In natural clays iron impurities also play an active role in the dramatic modification of the oxidation kinetics [64].

Photooxidation of LDPE and of LLDPE nanocomposites (modified organoclay) also resulted in a much shorter lifetime than the virgin polymer [65–67]. For example complete loss of mechanical properties was found already after 40 h of irradiation at 8% nanoclay loading compared to 100 h for the pure polymer in the Q-UV test [68]. The influence of the nanocomposite substrate is furthermore of importance, e.g. EVA or polystyrene show less acceleration of degradation than LDPE [69,70]. In some exceptional cases such as polyamide-6 [71] and PE/PA-6 blends [72] even an improvement in photooxidative stability in the presence of nanoclays could be detected.

Apart from layered silicates similar photooxidation acceleration was found e.g. for nanodispersed hydrotalcite in polypropylene [73], for $CaCO_3$ and SiO_2 nanoparticles [74], for EPDM nanocomposites synthesized from layered double hydroxides (LDH) [75] and for boehmite, modified by long-chain alkyl benzene sulfonic acid in isotactic PP and in syndiotactic PP [76].

An indication of the long-term thermal stability of stabilized nanocomposites based on layered silicates may be gained from OIT (oxygen induction time) measurements. It is quite obvious that conventionally stabilized nano materials are rather limited in thermal stability, e.g. PP stabilized with 0.05% phenolic antioxidant (AO-1) and 0.05% phosphite (PS-1) achieved only a lifetime of 1.8 min (fluorohectorite) or 2.3 min (montmorillonite) at 190 °C. With a proper stabilization including filler deactivators, the OIT value can be prolonged to more than 90 min. Key factor for this improvement is the combination of phenolic antioxidants, phosphite processing stabilizers and reactive molecules of selected oxirane or dianhydride structures [77]. With similar systems the long-term thermal stability can be raised again to values of unfilled materials, however a higher stabilizer loading is mandatory for this purpose [58].

In a more recent patent application it was demonstrated that by choosing a polymeric glycidyl group containing copolymer as filler deactivator not only the thermal stability (OIT) but even more the mechanical properties of nanocomposite formulations can be improved [78] (Table 3). At somewhat increased antioxidant level, the oxidative stability can be multiplied by adding the selected copolymer.

Because of the inferior light stability of polymer nanocomposites, it is obvious to evaluate the effect of light stabilizers on the stability, a topic addressed in several patents and papers. As expected the influence of antioxidants on the photostability is not very pronounced. Nevertheless the data show a certain delay of photooxidation [79]. In LLDPE nanocomposites it could be clearly demonstrated that the addition of UV absorbers of benzotriazole, benzophenone or hydroxyphenyltriazine structures extended decisively the lifetime of the nanocomposite film. However, a metal

Table 3 Stabilization of Nanocomposites (Polypropylene, 5% Organically Modified Clay, 15% PP-g-MAHah) [78]

Stabilization System	Styrene-Acrylate-Glycidylmethacrylate Copolymer (wt%)	OIT (ASTM D 3895-80) at 190 °C (min)	Tensile Impact Strength (ISO 8256) (kJ m^{-2})
0.1% AO-1/0.1% P-1	0	13.3	128
0.25% AO-1/0.25% P-1	0	30.3	
0.1% AO-1/0.1% P-1	0.3	19.9	
0.25% AO-1/0.25% P-1	0.3	61.5	128
0.25% AO-1/0.25% P-1	1.0	82.4	132
0.25% AO-1/0.25% P-1	2.0	96.6	178

deactivator (MD-1) alone outperformed the UV absorber in these experiments, indicating that the influence of metal impurities is very crucial [68,80]. A combination of UV-absorber with metal deactivator showed only a minor additional improvement. When MD-1 was added to the LLDPE nanocomposite, the elongations at break and tensile strength of the tested film samples were still at a level of 70% of the starting values after 300 h of Q-UV; complete loss of mechanical properties of the unstabilized sample was recorded after 70 h. An improvement of the photostability of syndiotactic PP/synthetic organo-clay nanocomposites could be shown by using a combination of phenolic antioxidant (AO-1), phosphite (PS-1) processing stabilizer, and oligomeric HALS (LS-2). However, the stabilization effect was reduced with increasing filler concentration [63]. At 5% organoclay level, the stability of the nanocomposite was lowered to one fourth in comparison to unfilled sPP at an LS-2 loading of 0.3% and the compatibilizer (PP-g-MAH) supported the photooxidation further. The stabilization of boehmite based PP nanocomposites with the same system (AO-1, PS-1, LS-2) did not show any improvement of photostability at higher filler concentrations; however a low molecular weight HALS was more efficient [76]. It was proven by adsorption experiments in solution that the oligomeric HALS interacts with the boehmite and, therefore, is not available any longer for protecting the polymer.

The latter results are in line with the beneficial use of combinations of low and high molecular weight light stabilizers to balance availability and migration to the surface of the plastic part. In addition to polyolefins, it has been experimentally demonstrated that polyamide nanocomposites can be synergistically stabilized by low and high molecular weight light stabilizers (LS-1, LS-2) in combination with processing and long-term heat stabilizers. In this case the criterion was yellowing after UV exposure [81].

Another solution to achieve sufficient light stability of nanocomposites is to include in the formulation suitable filler deactivators (Table 4). Artificial weathering of these stabilizer compositions revealed that the filler deactivator contributes by itself to the extended light stability and retention of mechanical properties [51,58]. Further improvement of light stability is feasible by selection of the optimized light stabilizer combination.

Table 4 Improved Light Stability of Nanocomposites through Filler Deactivators and Light Stabilizers [58]

Stabilization System for Processing/Aging	Light Stabilization	Time until 50% Retained Tensile Impact Strength (h)
0.25% AO-1 + 0.25% P-1	0.20% LS-1 + 0.20% LS-2	1100
0.50% Stabilizer system containing filler deactivator	0.20% LS-1 + 0.20% LS-2	2100
0.50% Stabilizer system containing filler deactivators	0.40% UV-4	4600

4. CONCLUSIONS

Most flame retardants influence the photooxidative stability of the polymer substrate directly through acceleration of the degradation process or indirectly by interacting with the antioxidants and light stabilizers. Whereas the behavior of brominated flame retardants is well understood, the impact of "green" halogen free flame retardants on the photooxidation of polymers is investigated in very few examples only. Phosphorus and nitrogen based flame retardants seem to influence the (photo)oxidative stability of polymers less than brominated compounds and they show no or less antagonism with hindered amine light stabilizers. Some recently developed flame retardants combine flame retardancy and light stabilizer functionality. Adjustment of stabilization of plastics containing fillers either used for mechanical reasons or as flame retardant is necessary to achieve sufficient long-term properties. Filler deactivators can play a decisive role in this respect.

Overall, the (photo)oxidative stabilization of "green" flame-retarded polymers stays a complex field as the polymer, the flame retardant system, and the additives (processing stabilizers, antioxidants, light stabilizers a.o.) have to be considered together. As not much is known about the interactions of the different components, there is still a vast field for further research and for cost performance optimization of flame-retarded formulations.

5. Appendix

Structures of stabilizers.

AO-1 Benzenepropanoic acid, 3,5-bis(1,1-dimethylethyl)-4-hydroxy-, 1,1′-[2,2-bis [[3-[3,5-bis(1,1-dimethylethyl)-4-hydroxyphenyl]-1-oxopropoxy]methyl]-1,3-propanediyl] ester (CA: 6683-19-8)

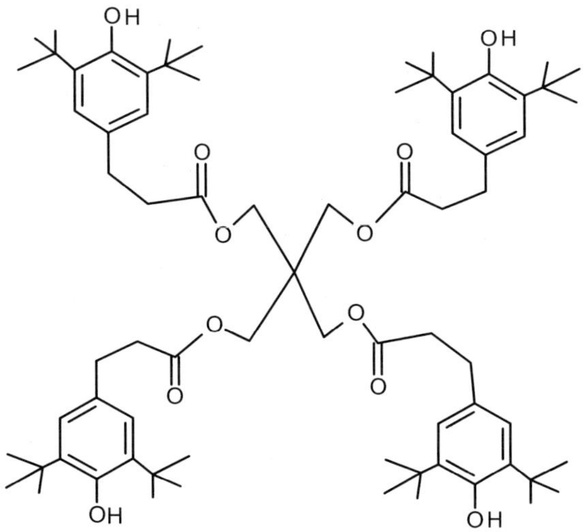

AO-2 Benzenepropanoic acid, 3,5-bis(1,1-dimethylethyl)-4-hydroxy-, octadecyl ester (CA: 2082-79-3)

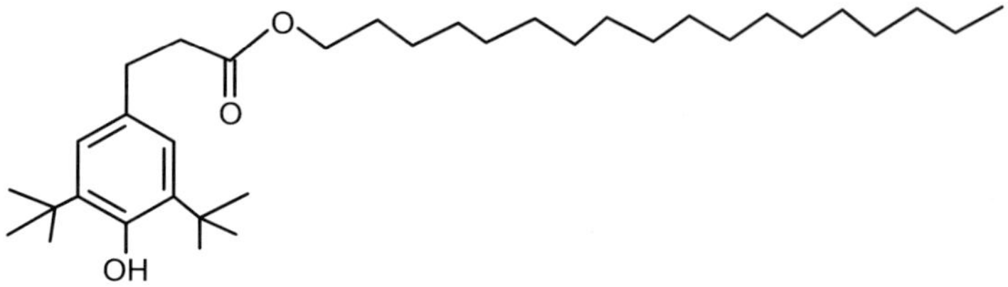

AO-3 Phenol, 4,4′,4″-[(2,4,6-trimethyl-1,3,5-benzenetriyl)tris(methylene)]tris[2,6-bis(1,1-dimethylethyl)- (CA: 1709-70-2)

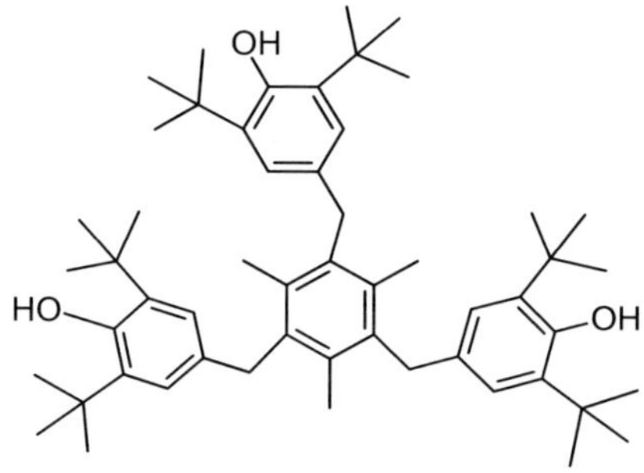

LS-1 Decanedioic acid, 1,10-bis(2,2,6,6-tetramethyl-4-piperidinyl) ester (CA: 52829-07-9)

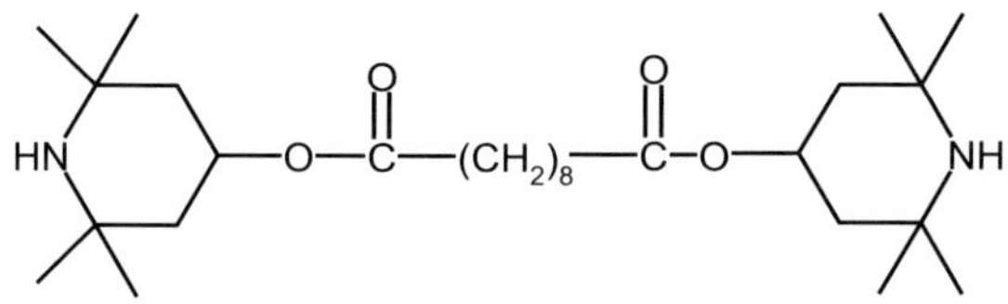

LS-2 Poly[[6-[(1,1,3,3-tetramethylbutyl)amino]-1,3,5-triazine-2,4-diyl][(2,2,6,6-tetramethyl-4-piperidinyl)imino]-1,6-hexanediyl[(2,2,6,6-tetramethyl-4-piperidinyl)imino]] (CA: 71878-19-8)

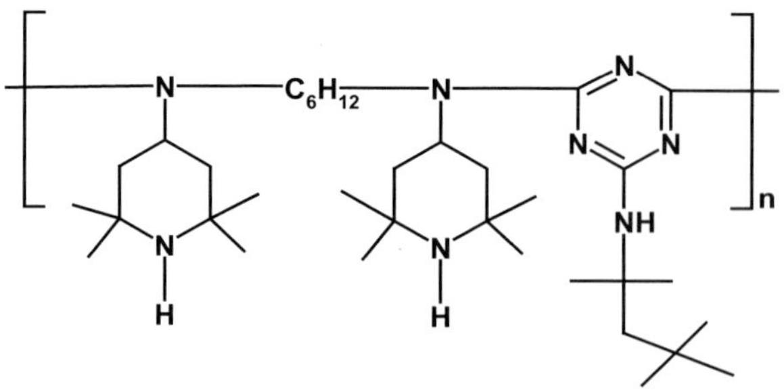

LS-3 Butanedioic acid, polymer with 4-hydroxy-2,2,6,6-tetramethyl-1-piperidineethanol (CA: 52722-86-8)

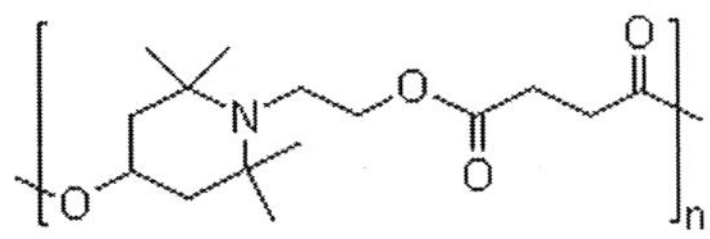

LS-4 7-Oxa-3,20-diazadispiro[5.1.11.2]heneicosan-21-one, 2,2,4,4-tetramethyl-20-(2-oxiranylmethyl)-, polymer with 2-(chloromethyl)oxirane (CA: 237081-56-0)

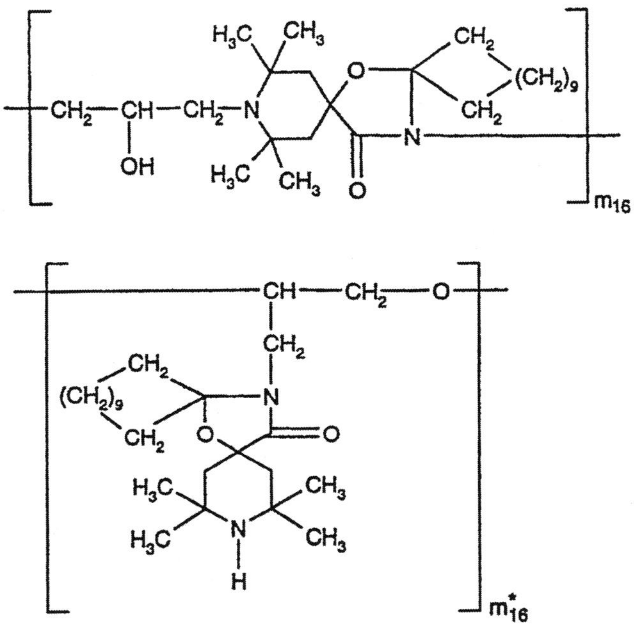

MD-1 Benzenepropanoic acid, 3,5-bis(1,1-dimethylethyl)-4-hydroxy-, 2-[3-[3,5-bis(1,1-dimethylethyl)-4-hydroxyphenyl]-1-oxopropyl]hydrazide (CA: 32687-78-8)

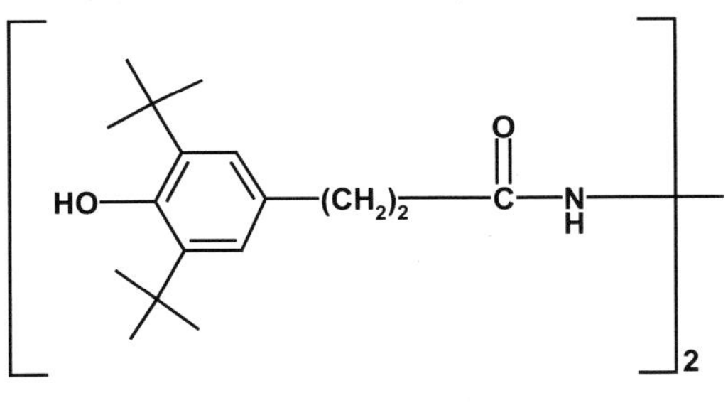

PS-1 Phenol, 2,4-bis(1,1-dimethylethyl)-, 1,1′,1″-phosphite (CA: 31570-04-4)

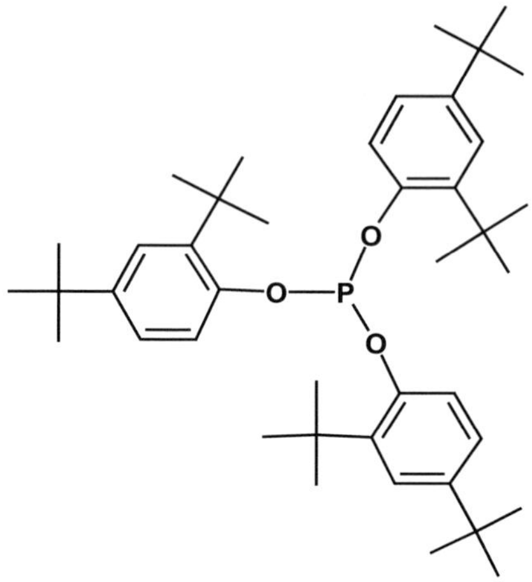

PS-2 Phosphonous acid, P,P′-[[1,1′-biphenyl]-4,4′-diyl]bis-, P,P,P′,P′-tetrakis [2,4-bis(1,1-dimethylethyl)phenyl] ester (CA: 38613-77-3)

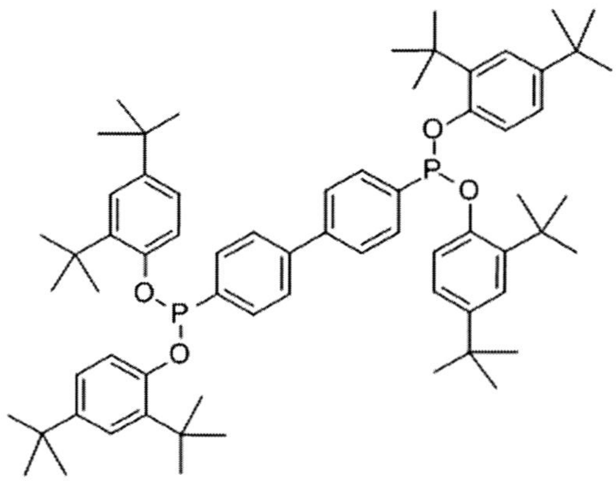

S-1 Disulfide, dioctadecyl (CA: 2500-88-1)

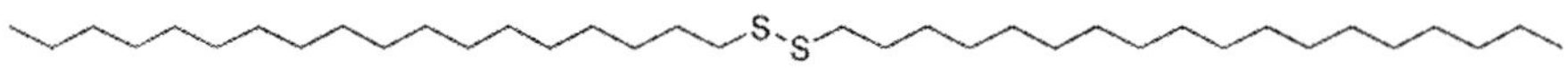

UV-1 Methanone, [2-hydroxy-4-(octyloxy)phenyl]phenyl- (CA: 1843-05-6)

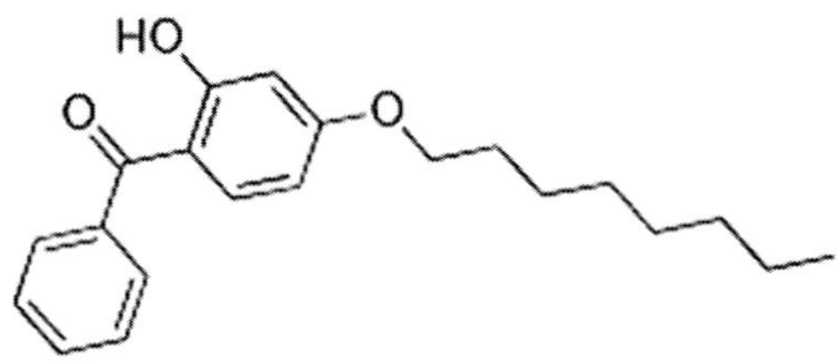

UV-2 Phenol, 2-(2H-benzotriazol-2-yl)-4-methyl- (CA: 2440-22-4)

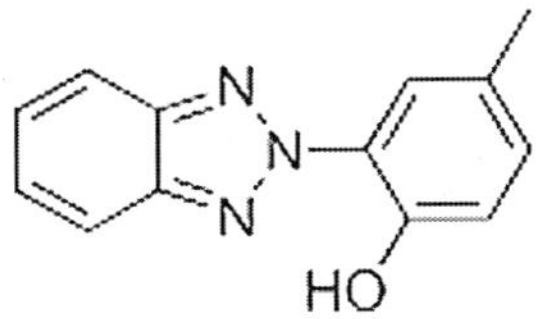

UV-3 Phenol, 2-(2H-benzotriazol-2-yl)-4,6-bis(1,1-dimethylpropyl)- (CA: 25973-55-1)

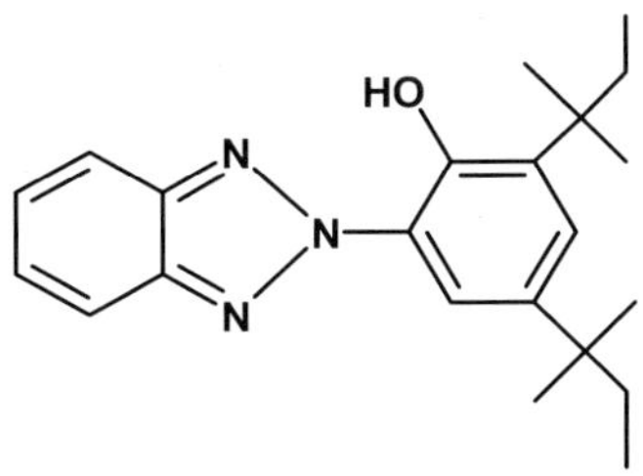

UV-4 Phenol, 2-(4,6-diphenyl-1,3,5-triazin-2-yl)-5-(hexyloxy)- (CA: 147315-50-2)

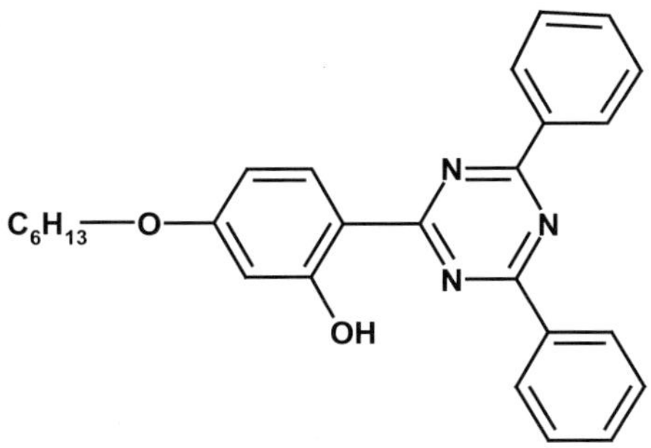

References

[1] Zweifel H, Maier RD, Schiller M. Plastic Additives Handbook. 6th ed. Munich, Germany: Hanser Publishers; 2009.

[2] Murphy J. Additives for Plastics Handbook. 2nd ed. New York, USA: Elsevier; 2001.

[3] Ash M, Ash I. Handbook of Plastics and Rubber Additives. 2nd ed. Shropshire, UK: Rapra Technology; 2005.

[4] Hindersinn RR, Marciniak HW. Fire retardant polymer composition containing halogenated cyclopentadiene adducts. US 3418263. Hooker Chemical Corp; 1968.

[5] Pillar WO. Molding expandable styrene polymer particles. US 4113672. Arco Polymers; 1978.

[6] Levek RP, Williams DO. Stabilized flame-retardant styrenic polymer compositions and the stabilizing flame-retardant mixtures used therein. US 4089912. Great Lakes Chemical Corporation; 1978.

[7] Dix JS, Mathis R. Organotin compounds as color and corrosion inhibitors in thermoplastic polymer compositions containing flame. US 3764579. Phillips Petroleum C; 1973.

[8] Mahling D, Meier L, Stange K, Steden F. Schwer entflammbare stabilisierte Formmassen auf Basis von Monoolefinpolymerisaten. AT 253210. BASF; 1967.

[9] Goossens DF. Flame retardant composition exhibiting superior thermal stability and flame retarding properties and use thereof. WO 2006132900. Albemarle Corp; 2006.

[10] Mathis RD. Flame retardant polyolefin composition of improved color and melt stability. US 3821141. Phillips Petroleum Comp; 1974.

[11] Sinturel C, Philippart JL, Lemaire L, Gardette JL. Photooxidation of fire retarded polypropylene, I Photoageing in accelerated conditions. Eur Polym J 1999;35:1773−81.

[12] Torikai A, Kato H, Fueki K, Suzuki Y, Okisaki F, Nagata M. Photodegradation of polymer materials containing flame-cut agents. J Appl Polym Sci 1993;50:2185−90.

[13] Sinturel C, Lemaire J, Gardette JL. Photooxidation of fire retarded polypropylene, II photooxidation mechanism. Eur Polym J 1999;35:1783−90.

[14] Torikai A, Chigita KI, Okisaki F, Nagata M. Photooxidative degradation of polyethylene containing flame retardants by monochromatic light. J Appl Polym Sci 1995;58:685−90.

[15] Antos K, Sedlar J. Influence of brominated flame retardant thermal decomposition products on HALS. Polym Degrad Stab 2005;90:188−94.

[16] Antos K, Tochacek J. Polypropylene with decreased flammability and increased resistance to UV radiation. Plasty a Kauc 2007;44:136−41.

[17] Sinturel C, Lemaire J, Gardette JL. Photooxidation of fire retarded polypropylene, III Mechanism of HAS inactivation. Eur Polym J 2000;36:1431−43.

[18] Carlsson DJ, Can Z, Wiles DM. Polypropylene photostabilization by hindered amines in the presence of acidic species. J Appl Polym Sci 1987;33:875−84.

[19] Hauck S, Fischer W, Schoening KU. Bi- or tricyclic sterically hindered alkoxyamines and process for their preparation. WO 2011018384. BASF SE; 2011.

[20] Schoening KU, Fischer W, Basbas AI, Dichtl A. Process for the preparation of sterically hindered nitroxyl ethers. WO 2008003602. Ciba Specialty Chemicals Holding; 2009.

[21] Basbas AI, Alvisi D, Cordova R, Difazio MP, Fischer W, Kotrola JA, et al. Process for the preparation of sterically hindered nitroxyl ethers. WO 2008003605. Ciba Specialty Chemicals Holding Inc; 2008.

[22] Zedda A, Ferri G, Sala M. Preparation of sterically hindered amine ethers. US 6117995. Ciba Specialty Chemicals Corp; 2000.

[23] Galbo JP, Grace HC, Horsey D, Solera P, Srinivasan R. N,N′,N″-[tris(2,4-bis(hydrocarbyloxy-2,2,6,6,-tetramethylpiperidin-4yl)alkylamino]-s-triazin-6-yl)-3,3′-ethylenediiminodipropylamines, their isomers and bridged derivatives and polymer compositions stabilized therewith. US 5844026. Ciba Specialty Chemicals Corp; 1998.

[24] Negishi Y, Ayabe T, Tobita E. Weakly basic hindered amines having carbonate skeletons, synthetic resin compostions and coating compositions. US 5844026. Asahi Denka C; 2006.

[25] Pietersen AH, Wolf R, Bill R. Flammschutzmittel-Kombination. DE3543414. Sandoz Patent GmbH; 1986.

[26] Camino G, Costa L, Martinasso G. Intumescent fire-retardant systems. Polym Degrad Stab 1989;23:359−76.

[27] Chantegraille D, Morlat-Therias S, Gardette JL. Photochemical behaviour of fire-retarded polymers. Polym Degrad Stab 2010;95:274−7.

[28] Camino G, Arnaud R, Costa L, Lemaire L. Photooxidative Behaviour of polypropylene fire retarded with intumescent additive. Angew Makromol Chem 1988;160:203−9.

[29] Nass B, Wanzke W. Flame-resistant moulding composition having improved light stability. EP 0718357. Hoechst AG; 1996.
[30] Braun U, Wachtendorf V, Geburtig A, Bahr H, Schartel B. Weathering resistance of halogen-free flame retardance in thermoplastics. Polym Degrad Stab 2010;95: 2421−9.
[31] Almeras X, LeBras M, Hornsby P, Bourbigot S, Marosi G, Anna P, et al. Artificial weathering and recycling effect on intumescent polypropylene-based blends. J Fire Sci 2004;22:143−61.
[32] Bertelli G, Locatelli R. Self-extinguishing polyolefinic compositions having improved flame resistant characteristics and a high thermal stability. EP 0125652. Montedison S.p.A; 1984.
[33] Liu Y, Wang Q. Preparation of microencapsulated red phosphorus through melamine cyanurate self-assembly and its performance in flame retardant polyamide 6. Polym Eng Sci 2006;46:1548−53.
[34] Largman T. Phosphine suppressants for polymeric compositions including red phosphorus as a flame retardant. US 4356282. Allied Corp; 1982.
[35] Horacek H, Grabner R. Advantages of flame retardants based on nitrogen compounds. Polym Degrad Stab 1996;54:205−15.
[36] Tomomura H, Harita K, Oonishi K. Flame-retardant polyamide resin composition. JP 06145506, JP 06145507. Kanebo Ltd; 1994.
[37] Grabner R, Dekker E. Melamine cyanurate: a simple but complicated flame retardant. Spec Chem Mag 2004;24:18−9.
[38] Takaretu I, Hamm S, Rothenbacher KP. Comparison of the recyclability of flame-retarded plastics. Environ Sci Technol 2003;37:652−6.
[39] Horsey DW, Andrews SM, Davis LH, Dyas DD, Gray RL, Gupta A, et al. Flame retardant compositions. WO 99/00450. Ciba Specialty Chem; 1999.
[40] Troutman MV, Ravichandran R, Srinivasan R, King RE, Horsey DW. Flame retardant compositions. WO02/074847. Ciba Specialty Chem; 2002.
[41] Kaprinidis N, Horsey DW. Flame retardant compositions. WO 03/016388. Ciba Specialty Chem; 2003.
[42] Butz V. Flame retardant composition comprising a phosphonic acid derivative. WO 2010026230. Thor GmbH; 2010.
[43] Sala M, Cocco G, Bassi A, Roth M, Schoening KU. Stabilizers. WO 2010089230. BASF SE; 2010.
[44] Pfaendner R, Roth M, Schoening KU, Weiss T, Hindalekar SB. Phospho-substituted alkoxyamine compounds. WO 2011086114. BASF SE; 2011.
[45] Roth M, Pfaendner R, Wilen CE, Aubert M. Symmetric azo compounds in flame retardant compositions. WO 2008101845. Ciba Holding; 2008.
[46] Aubert M, Wilen CE, Pfaendner R, Kniesel S, Hoppe H, Roth M. Bis(1-propyloxy-2,2,6,6-tetramethylpiperidin-4-yl)-diazene − an innovative multifunctional radical generator providing flame retardancy to polypropylene even after extended artificial weathering. Polym Degrad Stab 2011;96:328−33.
[47] Allen NS, Edge M, Corrales T, Childs A, Liauw CM, Catalina F, et al. Ageing and stabilisation of filled polymers: an overview. Polym Degrad Stab 1998;61:183−99.
[48] Hu X, Xu H, Zhang Z. Influence of fillers on the effectiveness of stabilizers. Polym Degrad Stab 1994;43:225−8.
[49] Rysavy D, Tkadleckova H. Thermooxidative Stabilität von gefülltem Polypropylen. Plaste Kautsch 1991;38:273−5.

[50] Pan J, Xu H, Qi J, Cen J, Ma Z. Study of CaCO3 filled polypropylene composite with long use life time. Polym Degrad Stab 1991;33:67–75.
[51] Pfaendner R. Stabilization of nanocomposites. In: Beyer G, editor. Industry guide to nanocomposites. Bristol, UK: Applied Market Information Ltd; 2009. pp. 117–35.
[52] Pfaendner R, Hoffmann K, Meyer F, Rotzinger B. Synthetic polymers comprising additive blends with enhanced effect. US 6610765. Ciba Specialty Chemicals Corp; 2003.
[53] Rotzinger B. Talc-filled PP: a new concept to maintain long term heat stability. Polym Degrad Stab 2006;91:2884–7.
[54] Beyer G. Industry guide to nanocomposites. Bristol, UK: Plastics Information Direct; 2009.
[55] Pavlidou S, Papaspyrides CD. A review on polymer-layered silicate nanocomposites. Prog Polym Sci 2008;33:1119–98.
[56] Xie W, Gao Z, Pan WP, Hunter D, Singh A, Vaia R. Thermal degradation chemistry of alkyl quaternary ammonium montmorillonite. Chem Mater 2001;13:2979–90.
[57] Kumar AP, Depan D, Tomer NS, Singh RP. Nanoscale particles for polymer degradation and stabilization – trends and future perspectives. Prog Polym Sci 2009;34: 479–515.
[58] Pfaendner R. Nanocomposites: industrial opportunity or challenge? Polym Degrad Stab 2010;95:369–73.
[59] Dominkovics Z, Hari J, Fekete E, Pukanszky B. Thermo-oxidative stability of polypropylene/layered silicate nanocomposites. Polym Degrad Stab 2001;96:581–7.
[60] Morlat-Therias S, Mailhot B, Gonzalez D, Gardette JL. Photooxidation of polypropylene/montmorillonite nanocomposites. 2. Interactions with antioxidants. Chem Mater 2005;17:1072–8.
[61] Tidjani A, Wilkie CA. Photo-oxidation of polymeric-inorganic nanocomposites: chemical, thermal stability and fire retardancy investigations. Polym Degrad Stab 2001;74: 33–7.
[62] Morlat-Therias S, Mailhot B, Gardette JL, Da Silva C, Haidar B, Vidal A. Photooxidation of ethylene-propylene-diene/montmorillonite nanocomposites. Polym Degrad Stab 2005;90:78–85.
[63] Chmela S, Kleinova A, Fiedlerova A, Borsig E, Kaempfer D, Thomann R, et al. Photooxidation of sPP/organoclay nanocomposites. J Macromol Sci Part a: Pure Appl Chem 2005;42:821–9.
[64] Morlat-Therias S, Mailhot B, Gonzalez D, Gardette JL. Photo-oxidation of polypropylene/mointmorillonite nanocomposites. 1. Influence of nanoclay and compatibilizing agent. Chem Mater 2004;16:377–83.
[65] Qin H, Zhao C, Zhang C, Chen G, Yang M. Photo-oxidative degradation of polyethylene/ montmorillonite nanocomposite. Polym Degrad Stab 2003;81:497–500.
[66] Dintcheva NT, Al-Malaika S, La Mantia FP. Effect of extrusion and photo-oxidation on polyethylene/clay nanocomposites. Polym Degrad Stab 2009;94:1571–88.
[67] Kumanayaka TO, Parthasarathy R, Jollands M. Accelerating effect of montmorillonite on oxidative degradation of polyethylene nanocomposites. Polym Degrad Stab 2010;95: 672–6.
[68] La Mantia FP, Tzankova Dintcheva N, Malatesta V, Pagani F. Improvement of photostability of LLDPE-based nanocomposites. Polym Degrad Stab 2006;91:3208–13.
[69] Botta L, Dintcheva NT, La Mantia FP. The role of organoclay and matrix type in photooxidation of polyolefin/clay nanocomposite films. Polym Degrad Stab 2009;942:712–8.

[70] Bottino FA, Pasquale GD, Fabbri E, Orestano A, Pollicino A. Influence of montmorillonite nano-dispersion on polystyrene photo-oxidation. Polym Degrad Stab 2009;94: 369–74.

[71] Kiliaris P, Papaspyrides CD, Pfaendner R. Influence of accelerated aging on clay-reinforced polyamide 6. Polym Degrad Stab 2009;94:389–96.

[72] Dintcheva NT, Filippone G, La Mantia FP, Acierno D. Photo-oxidation behaviour of polyethylene/polyamide 6 blends filled with organomodified clay: Improvement of the photo-resistance through morphology modification. Polym Degrad Stab 2010;95: 527–35.

[73] Bocchini S, Morlat-Therias S, Gardette JL, Camino G. Influence of nanodispersed hydrotalcite on polypropylene photooxidation. Eur Pol J 2008;44:3473–81.

[74] Li J, Yang R, Yu J, Liu Y. Natural photo-aging degradation of polypropylene nanocomposites. Polym Degrad Stab 2008;93:84–9.

[75] Kumar B, Rana S, Singh RP. Photo-oxidation of EPDM/layered double hydroxides composites: Influence of layered hydroxides and stabilizers. eXPRESS Polym Let 2001;1:748–54.

[76] Chmela S, Fiedlerova A, Borsig E, Erler J, Mülhaupt R. Photo-oxidation and stabilization of sPP and iPP/boehmite-disperal nanocomposites. J Macromol Sci Part a: Pure Appl Chem 2007;44:1027–34.

[77] Wermter H, Pfaendner R. Stabilization of thermoplastic nanocomposites. EP 1592741. Ciba Specialty Chem; 2004.

[78] Pfaendner R. Stabilizer composition for filled polymers and nanocomposites. WO 2011067222. BASF SE; 2011.

[79] Mailhot B, Morlat S, Gardette JL, Boucard S, Duchet J, Gerard JF. Photooxidation of polypropylene nanocomposites. Polym Degrad Stab 2003;82:163–7.

[80] Morlat-Therias S, Fanton E, Gardette L, Dintcheva NT, La Mantia FP, Malatesta V. Photochemical stabilization of linear low-density polyethylene/clay nanocomposites: towards durable nanocomposites. Polym Degrad Stab 2008;93:1776–80.

[81] Chin H, Solera PS, Horsey DW, Kaprinidis N, Sitzmann EV. Synergistic combinations of nano-scaled fillers and hindered amine light stabilizers. US 7084197. Ciba Corp; 2006.

CHAPTER

Comprehensive Approach to Flame-Retardancy Evaluation of Layered Silicate Nanocomposites

14

A. Fina, G. Camino

Politecnico di Torino, Dipartimento Scienza Applicata e Tecnologia, Alessandria, Italy

S. Bocchini

Istituto Italiano di Tecnologia, Torino, Italy

CHAPTER OUTLINE

1. Introduction

In the field of polymer nanocomposites, several classes of inorganic fillers have been explored in the last decades, including natural and synthetic nanoclays [1,2], carbon nanotubes [3,4], and graphenes [5,6]. However, the most widely studied nanofillers are certainly the natural nanoclays, particularly the layered silicates such as montmorillonite (MMT). For this reason, in this chapter we will focus on layered silicates, namely silicate mineral in which flat sheets are formed by the sharing of three of the four oxygen atoms in each SiO_4^{4-} tetrahedron with neighboring tetrahedrons (Figure 1).

Different cations (e.g. Fe, Mg, or Al) can bind to the silicate sheets, sharing the apical oxygen in their octahedral coordination. Two of the simplest structures are

Polymer Green Flame Retardants. http://dx.doi.org/10.1016/B978-0-444-53808-6.00014-7

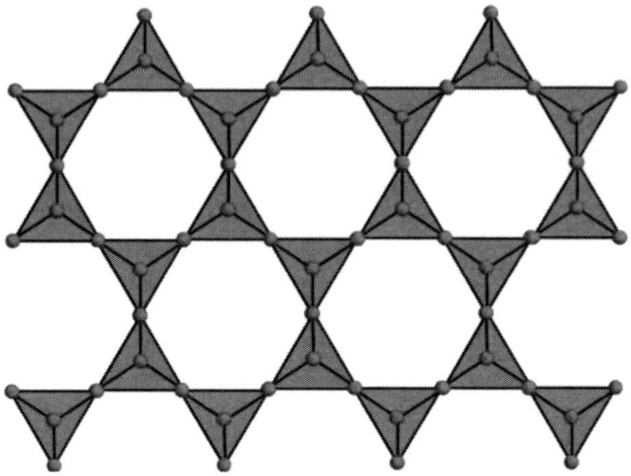

FIGURE 1

Silicate sheet: it is formed by tetrahedral SiO_4^{4-} linked together by three corners to form continuous sheets.

kaolinite and lizardite in which a silica tetrahedral sheet is fused respectively to an Al^{3+} or Mg^{2+} octahedral sheet in the 1:1 ratio [7].

The layered silicates commonly employed for the preparation of polymer nanocomposites, such as MMT and hectorite (HT), belong to the family of 2:1 phyllosilicates, more specifically smectites. Their crystal structure consists of stacked layers made of two silica tetrahedrons fused to an edge-shared octahedral sheet of alumina (Figure 2), sizing in the order of about 1 nm in thickness, whereas the lateral dimension may vary in a wide range from about 100 nm to a few microns [8]. Both in tetrahedrons and octahedrons, substitutions of Si^{4+} and Al^{3+} are possible with cations of lower charges (e.g. Al^{3+} or Mg^{2+}, respectively), inducing a negative charge on the silicate layer. This requires counterions to balance the electrical charges, which is typically obtained by sodium or calcium ions, generally carrying their coordination water molecules, in the interlayer space.

The incorporation of layered silicates organically modified into polymer matrices was first reported in 1950 by Carter et al. [9]. However, the interest increased in the 1990s thanks to the pioneering works of the Toyota research group [10–13] when nanocomposites exhibiting markedly improved thermal and mechanical properties were produced incorporating layered silicates into polyamide by in-situ polymerization. Nevertheless, layered silicate nanocomposites would not be so popular if it was not for the discover of Vaia that these can be obtained for direct melt intercalation, an approach that can be combined with conventional processing techniques such as extrusion offering a versatile and environmental method to produce nanocomposites [14–17].

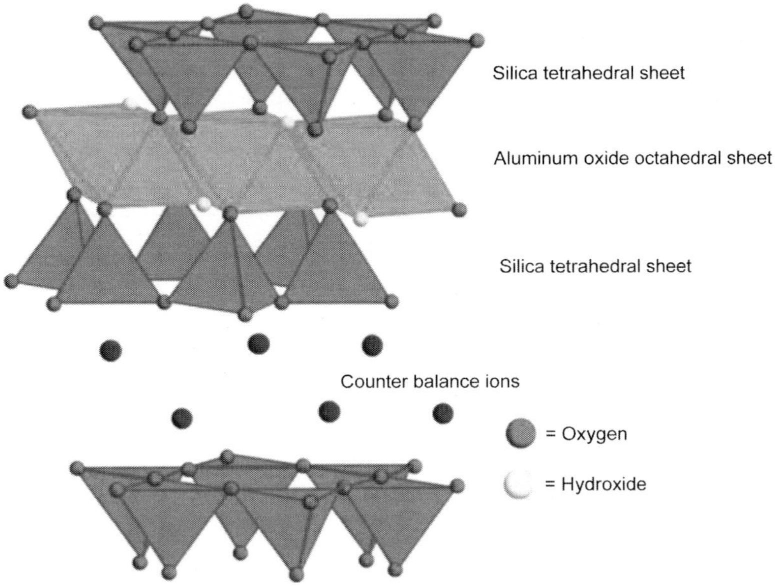

FIGURE 2

Basic structure of smectite.

Spontaneous intercalation between the layers of phyllosilicates is limited to small molecules [18,19] and hydrophilic polymers such as polyethyleneoxide. In order to render these hydrophilic phyllosilicates more organophilic, the hydrated cations in the interlayer can be exchanged with cationic surfactants such as alkylammonium or alkylphosphonium (onium) to lower the surface energy, thus improving the wetting by the polymer matrix. The phyllosilicates modified with organic surfactant is usually referred to as modified clay or organoclay.

There are mainly three different degrees of dispersion of layered silicates inside polymers, depending on the chemical nature (polymer matrix, clay filler and organic surfactant, relative concentrations) and processing conditions. When a polymer is unable to intercalate into the interlayer space polymer and clay forms two different phases and the properties are similar to traditional composites containing inorganic fillers. Higher compatibility between layered silicates and polymer matrix allows the intercalation of the polymer macromolecules between the clay layers, increasing the interlayer distance but keeping a certain regularity in stacking of clay platelets. Finally, when the silicate layers are separated and uniformly dispersed inside polymer matrix an exfoliated nanocomposite is formed. In most real cases, none of these ideal types of dispersion is observed, but a variable dispersion is often found, with coexistence of exfoliated and intercalated nanoparticles.

The first detailed study on flame-retardant properties of layered silicate nanocomposites was reported by Gilman et al. in 1997 [20]. On the basis of these findings, further studies were performed on a wide variety of different polymers: a comprehensive review on the argument was recently published by Kiliaris and Papaspyrides [21].

2. Mechanisms of nanoclays action during polymer decomposition and combustion

The currently available knowledge on the mechanisms for the formation of the physical shield during clay nanocomposite burning is the result of many years of work by different research groups and went through the accumulation of evidences on reorganization of the clay nanoparticles inside the polymer during thermal degradation/combustion. Gilman et al. [20,22–24] showed that exfoliated polyamide 6 (PA6)/clay nanocomposites significantly reduced the heat release rate in combustion, compared to neat PA6 and explained this result with the formation of protective floccules on the sample surface. The analysis of the protective floccules evidenced that these are mainly formed by clay particles, in which lamellae are stacked with a d-spacing larger than the original clay without organic modifier. This higher interlayer distance was explained by the presence of thermally stable organic components possibly with graphitic-like structure, produced by the partial charring of the polymer matrix or clay surfactant.

It is now generally agreed that the combustion behavior of polymer nanocomposites strictly depends on the interface between polymer condensed phase and the gas phase. Since the first studies on the flame-retardancy properties of nanocomposites [20,22], it was pointed out that the behavior under forced combustion for different nanocomposites were quite similar. In particular, a reduction of the heat release rate is generally observed as a consequence to a lower fuel feed rate, often without substantial modifications of the total heat evolved. Such a behavior is related to the formation of a physical shield built up by the inorganic nanoparticles left behind by polymer ablation or nanoparticles migration, which acts as a mass barrier, slowing down the release of generated gas fuel and/or as a heat shield, by re-radiating the external imposed heat flux.

Attempts of correlating flame-retardancy performance with the degree of dispersion of nanoclays did not bring to a general rule: despite homogeneous distribution of nanoparticles is needed to allow formation of a well structured surface residue during the early stage of combustion, exfoliated or intercalated structures may both result in significant decrease of combustion rate [25]. This fact relates to the thermal evolution of polymer nanocomposite during heating, thermal decomposition and the early combustion stages, changing both chemical and physical structure of the nanocomposite [26] and leading to a different morphology compared to the initial intercalated or exfoliated structure.

Two mechanisms were proposed for accumulation of the initially well-dispersed clay particles. The first is the classical accumulation of clay particles on the surface of the burning material by polymer ablation. De-wetted clay lamellae, left behind by polymer pyrolysis, tend to aggregate and stack against each other. The second mechanism involves the thermal degradation of the clay organic treatment in the molten polymer and subsequent aggregation in particles stacks in the polymer bulk. These aggregates of nanoparticles may be moved to the surface by the rising bubbles of volatile products obtained by organic modifier/polymer degradation and/or convection flow in the high temperature melt. Lewin [27] pointed out that the clay platelets inside the nanocomposites can also migrate at lower temperatures because of the thermal decomposition of the organic modifier and subsequent migration to the surface, driven by surface energy minimization. Pastore et al. [28] demonstrated a high-temperature structural rearrangement of intercalated clay nanocomposites based on poly[ethylene-co-(vinyl acetate)] (EVA). De-intercalation and decrease of the silicate interlayer space were found under nitrogen as a consequence of thermal degradation of the organic modifier; whereas, increase of the silicate interlayer space and exfoliation were the main processes found under air, by the partial polymer oxidation which favor the affinity with hydrophilic inorganic nanoparticles.

The surface accumulation is not the only way in which layered silicates can influence the polymer degradation: indeed, nanoparticles also provide a confined environment (cage effect) in which the trapped decomposing polymer chains could undergo additional degradation pathways [29]. Intermolecular reactions are favored, leading to the formation of complex compounds. Thus, the volatilization rate is lowered for the profit of char formation.

In the cases described so far, a purely physical effect of the nanofiller has been reported on the structure of the burning composite, related to the formation of an inorganic layer by the accumulation and coalescence of nanoparticles on the surface, with no significant effect on the polymer degradation chemistry in the condensed phase. However, nanoparticles may also play a chemical role in specific cases. Studies on PA6/layered silicates nanocomposites conducted by Vaia et al. [30] have revealed increased carbonaceous char yields in the presence of silicates. A detailed study of the flame-retardant mechanism of polymer-clay nanocomposites [31] showed that a clay reinforced carbonaceous char is formed during combustion also for nanocomposites which base polymer matrix normally produce little or no char when burned alone, such as polyethylene (PE) [32], polypropylene (PP) [33], poly[ethylene-co-(vinyl acetate)] (EVA) [34], polystyrene (PS) [35], and acrylonitrile-butadiene-styrene (ABS) copolymer [36].

Zanetti et al. [33] proposed a chemical oxidative dehydrogenation catalytic action of strong acid sites created by thermal decomposition of the organic modifier, in PP/clay nanocomposites. The resulting conjugated polyenes are typical precursor of aromatic chars as it is well known for example from studies of thermal degradation of polyvinylchloride and polyvinylacetate [37]. In EVA/MMT nanocomposites, the acceleration of deacetylation by the catalytic effect of acidic clay sites was shown, with formation unconjugated polyenes [34], which may evolve, through further dehydrogenation, to conjugated polyenes and then to polyaromatic compounds and carbonaceous char.

During combustion, the formation of strongly acidic proton sites created during the thermal decomposition of organic surfactant plays an important role in the formation of a stable char residue by their catalysis, which may result in delaying heat release [14,38]. One of the most important chemical characteristic which can strongly affect the combustion behavior is the clay acidity. In fact, weakly acidic SiOH and strongly acidic hydroxyl groups bridging two silicon atoms, present at the layer edges act as Brønsted acid sites. Lewis acid centers, instead, arise from the isomorphic substitutions with ions such as Fe^{2+} and Fe^{3+} as well as from crystallographic defect sites within the layers. They can act as catalytic sites accepting electrons from donor molecules with low ionization potential, coordinating organic radicals, abstracting electrons from vinyl monomers or, in the case of transition metals, working as peroxide decomposers [19,39–41].

One type of organic surfactant most often used is alkylammonium salts, which thermal degradation takes place in one step in a relatively narrow temperature range (250–350 °C) [42], whereas the alkylammonium ions intercalated between the clay layers decompose to volatile compounds in more than one step and in a larger temperature range (e.g. 200–500 °C) [42–45]. The degradation onset temperatures of alkylammoniums intercalated in the clay interlayer space are usually lower than those of the pure alkylammonium salt. The first event of decomposition is usually due to the unconfined fraction of alkylammonium, that consists of alkylammonium surfactant which has not undergone the ion exchange reaction and is either complexed on the outer surface of the clay, within void spaces between primary particles or inside the interlayer space, in a peripheral position. This behavior shows that the Lewis and/or Brønsted acid sites behave as catalytic sites affecting the initial stage of decomposition of the alkylammonium molecules accelerating the carbon–carbon bond scission at high temperature. The following degradation steps involve the alkylammonium molecules that have undergone ion exchange, whose weight loss is delayed by the barrier effect of the clay layers to the diffusion of the volatile compounds formed by thermal degradation of the organic moiety.

The importance of clay type, i.e. its chemical structure, in the degradation mechanism was proved by the comparison of MMT and fluorohectorite ion exchanged with the same octadecylammonium ion [42]. In thermal degradation under oxidative conditions the clays promote the charring of the organic surfactant with the formation of a char residue which oxidizes at higher temperatures with the production of CO_2. This effect could be partially due to the physical containment of the clay layers that favors recombination instead of volatilization of fragments. However, a major role is played by the catalytic action of the clays, depending on the kind of clay. The thermal degradation in air of the two clays show different amount of organic surfactant charred residue: the char yields of octadecylammonium decomposition are respectively 60 and 32.5 wt% for MMT or fluorohectorite. Acidity of MMT, due to the presence of −OH groups at the edges of clay lamellae, is enhanced by the cations in the clay surface [46], promoting charring catalysis more effectively than fluorohectorite, in which a large portion of edge −OH groups are substituted by fluorination. The clays catalytic action, shown for the organic modifier, is obviously occurring also on heating polymer nanocomposites and affects their behavior in combustion.

3. Nanocomposites flame-retardancy performance

The assessment of flame-retardancy performance of polymer nanocomposites is typically carried out by one or more of the three fire tests: flammability test UL 94, limiting oxygen index (LOI), and Cone Calorimeter. However, significance and relevancy to real fire scenarios of these different tests are often neglected in material science literature. In fact, flammability tests such as UL 94 and LOI reflect the exposure to a small flame, in terms of capability to ignite and to self-sustain a flame, thus representing a scenario in which the material is at the origin of a fire. On the other hand, cone calorimeter test is representative of a forced combustion, in which the material is forced to burn under controlled heat flux, corresponding to the development stage of a real fire.

A comprehensive approach to the flame retardancy of polymer nanocomposites should take into account all the different aspects of flaming behavior, including ignition, flammability tests, forced combustion, and flame spread. Smoke release and toxic products evolved during fire should also be taken into account, as discussed elsewhere in this book.

3.1 Ignition

Ignition of polymers under external irradiation has been extensively studied since the 1960s [47], but fundamental aspects concerning their chemical and physical evolution, prior and during the combustion process are not yet understood in detail. Indeed, especially in the case of complex polymer-based formulations containing fire retardants and/or nanoparticles, the thermophysical evolution of the condensed phase before ignition in a fire test has been rarely studied, probably owing to experimental difficulties in such properties measurement. On the other hand, polymers containing additives or nanoparticles often exhibit unpredictable time to ignition if blind reference is made only to the material thermal behavior, such as that shown in thermogravimetry. As a result, variable trends for time to ignition have been reported for polymer/layered silicate nanocomposites: a reduction of Time To Ignition (TTI) compared to neat polymers is often observed [48−50], but the opposite effect is also reported in other cases [34,51−54].

Despite the fact that different proposals have been made to interpret modification of ignition time in nanocomposites (including thermal instability of nanofiller alkylammonium organic modifiers [55–58], triggering polymer degradation by catalytic effects [51,59–63] and molten polymer viscosity [53,54]), little work has been devoted to the experimental investigation of materials evolution during heat exposure in preignition step. The use of temperature measurements in the condensed phase has been recently shown [64,65] to help in understanding phenomena controlling ignition in polymer nanocomposites, providing insights in the melting, which controls the time to reach the decomposition temperature in semicrystalline polymers, and decomposition, which controls the production of volatile fuel.

In a study of ours, intercalated polyethylene terephthalate (PET)/5 wt% organo-modified MMT (Cloisite 30B) nanocomposites was compared with correspondent microcomposite prepared with unmodified sodium MMT (Cloisite Na^+) and pristine PET [64]. PET shows an average TTI at 730; whereas, both composites exhibit lower ignition time, namely 630 s for PET microcomposite and 380 s for PET nanocomposite (Figure 3).

Comparing the surface temperature for microcomposite with PET, similar trend is observed: after complete melting of PET, surface temperature rapidly increases to the decomposition temperature and remains constant until ignition occurs. The comparison of average surface temperature at ignition shows that no difference is induced by the presence of microaggregates of MMT on the ignition temperature. This is reasonably explained by the fact that surface temperature is still controlled by the latent heat of polymer decomposition, taking into account that microdispersed MMT has negligible effects on the decomposition of PET. However, the shorter ignition time has to be correlated with some change in material physical properties induced by the microparticles, leading to the earlier achievement of critical fuel

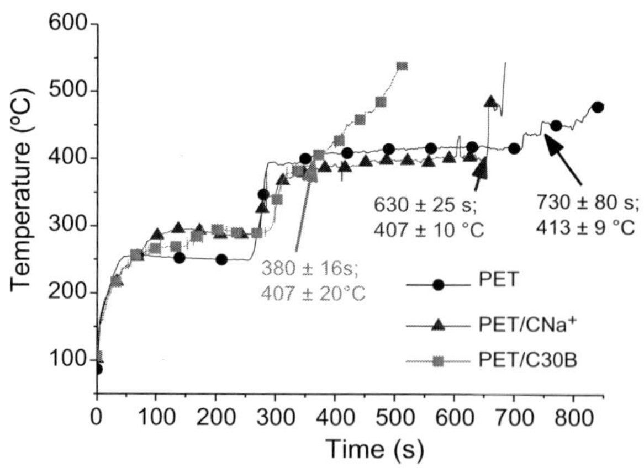

FIGURE 3

Surface temperature plots for PET and its nanoclay composites. Arrows indicate occurrence of ignition [64]. PET, polyethylene terephthalate.

production rate to reach the lower flammability limit. In the PET nanocomposite, ignition occurs as soon as polymer volatilization temperature is reached in the surface layer (as if gases evolved from the top layer of the specimen were sufficient for ignition). This was confirmed by the examination of the specimens obtained by the interruption of the test just after TTI, evidencing that the bottom part of the PET nanocomposite specimen was not yet decomposing at the ignition time.

Similar results were obtained on PA6 clay nanocomposites, in which ignition occurs earlier in time compared to pristine PA6, associated with a lower thickness of decomposing material at the time of ignition [64].

Despite the fact that ignition behavior is not understood in detail yet, the results available support for an interplay of several physical and chemical effects related to the presence of nanoparticles during polymer decomposition, including polymer viscosity control on heat transfer by convection, filler effect on heat reradiation, and clay barrier oxygen diffusion controlling polymer thermal oxidation as well as chain fragments volatilization.

3.2 Flammability

Flammability tests such as UL 94 and LOI on polymer nanocomposites usually evidence limited or no dripping from the ignited specimen, significantly lower rate of combustion and higher residue at the end of test, compared to the reference polymer.

As an example, Bartholmai and Schartel reported no reduction of flammability for maleic anhydride-grafted-polypropylene (PPgMA)/organoclay nanocomposites, the pristine polymer being classed V2 and the nanocomposites being nonclassified in the vertical UL 94 test [51]. This was explained by the different dripping behavior, as the unfilled PPgMA extinguishes thanks to dripping of flaming parts of the specimen, thus lowering heat flux to the polymer; whereas, the nanocomposites showed decreased dripping, so that the specimens did not extinguish. However, the fire behavior observed was different in terms of burning velocity measured in horizontal UL 94 tests, with a reduction of about 50% for the nanocomposites compared to reference PPgMA.

A visual example of hindered dripping for polybutylene terephthalate (PBT) in the presence of nanoclays is given in Figure 4. Ten seconds after ignition by the small flame, PBT already drips extensively, whereas the corresponding nanocomposite does not drip for the whole test. Furthermore, while both specimens are burned up to the clamp (nonclassifiable accordingly to the ASTM D 3801) a residue is left at the end of the test for PBT/nanoclay. Similar results were also obtained with PA6/organoclay nanocomposites (5% Cloisite 30B, mainly exfoliated), exhibiting reduced dripping compared to PA6 and preventing self-extinguishment by the loss of flaming parts of the specimen.

In order to properly take into account the material performance behind the bare UL 94 ranking, complementary description of polymer materials burning process should be considered, such as times of combustion, amount of material burned, and rate of combustion. As an example, a much smaller flame and reduced rate of

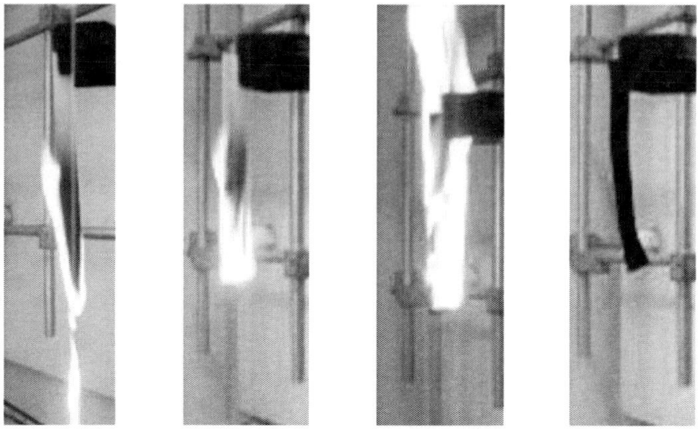

FIGURE 4

Snapshots taken from UL 94 test videos for PBT and PBT/fluorohectorite clay. PBT, polybutylene terephthalate.

combustion are often observed for nanocomposites compared to pristine polymers, which is not taken into account when using the UL 94 classification, as long as the persistency of a small flame prevents self-extinguishment of the specimen.

LOI test represents the opposite configuration compared to vertical UL 94, the specimen being ignited from the top, as to simulate a burning candle. Relatively few papers reported LOI test for nanocomposite materials [51,66,67], generally showing either decrease or no significant increase of LOI value compared to the pristine polymers. Indeed, LOI setup condition appears unfavorable for the production of an effective ceramic protective layer on the material surface, as the specimen below the flame front is not heated by the flame to allow nanoparticles accumulation on the surface, before being reached by the flame front. However, the oxygen index test (LOI) has the advantage of ranking materials by the LOI number, which is the minimum concentration of oxygen in a flowing mixture of oxygen and nitrogen that supports candle-like combustion of a specimen.

3.3 Forced combustion

Forced combustion tests are typically carried out by cone calorimetry, which has been by far the most used test for the study of nanocomposite's fire behavior, allowing quantitative testing of materials in controlled and repeatable conditions.

Among the many parameters supplied by cone calorimetry, attention has been focused primarily on heat release rate (HRR) and its peak value (pkHRR) during combustion because of its relevance to fire risk related to time to flashover. Typical layered silicate nanocomposites show a decrease in the pkHRR of about 50–70%

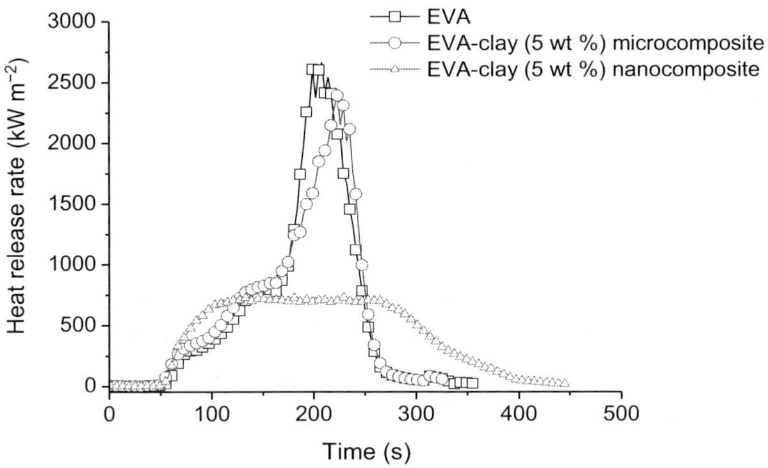

FIGURE 5

Forced combustion: heat release rate (HRR) for pure EVA; EVA with 5 wt% of fluorohectorite exchanged with aminododecanoic acid, (microcomposite); and EVA with 5 wt% of montmorillonite exchanged with methyl, tallow, bis-2-dihydroxyethylammonium, (nanocomposites), at a heat flux of 50 kW m^{-2} [69].

compared to reference unfilled polymer [20,68–70]. This behavior is usually associated only with clay well-dispersed inside the polymer as clearly shown in Figure 5, where pristine EVA is compared with layered silicate microcomposite and nanocomposite.

The primary parameter responsible for the decreased HRR of the nanocomposites is the development of a multilayered carbonaceous-silicate structure on the surface of nanocomposites during combustion switching the typical noncharring behavior of most thermoplastic polymers to that of charring materials. The structure is formed on the basis of the mechanism explained above, with migration of the clay toward the surface in the molten polymer at high temperature and accumulation caused by polymer ablation. These clay particles reassemble to form stacks. The carbonaceous char produced superficially by the clay catalysis during combustion is thus reinforced by silicates, creating an excellent physical barrier which protects the substrate from heat and oxygen, and slows down the escape of flammable volatiles generated during polymer degradation. This char generally grows by the progressive ablation of polymer from the composite leading to the formation of a structure with variable degree of compactness. The compactness of the residue during burning is generally related to the efficiency in HRR reduction; the higher the compactness, the lower HRR (Figure 6).

However, the simple comparison of peak of HRR values of polymer materials characterized by very different combustion behavior could be misleading. In particular, charring strongly affects the shape of the HRR curve: char-forming materials show the heat release rate peak at the early stage of the combustion

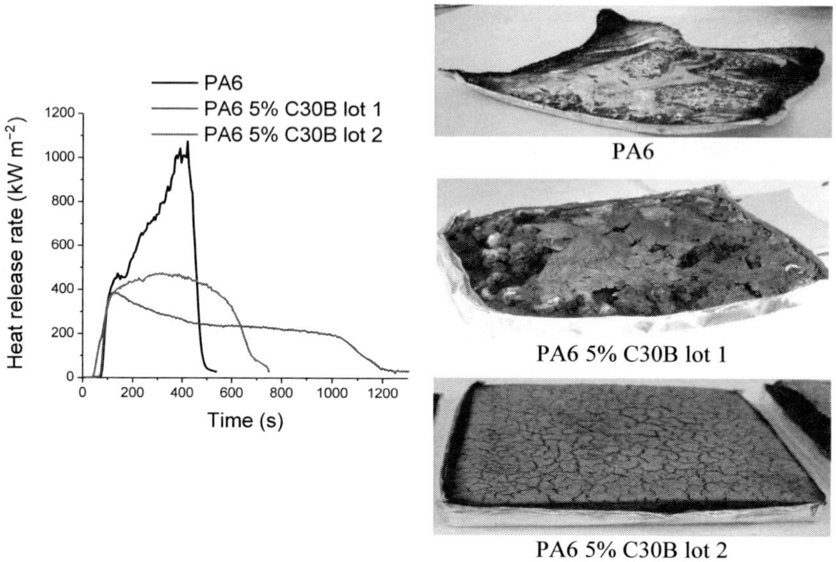

FIGURE 6

Comparison between pristine PA6 and two different PA6/organomontmorillonite lots (Cloisite 30B, mainly exfoliated). Left: HRR plots and right: residues at the end of the test for the different plots, showing correlation between compactness of the char and HRR reduction. PA6, polyamide 6; HRR, heat release rate.

because of the protective action of the char building up on the surface of the burning sample, whereas noncharring materials (including most of the thermoplastic polymers) show the peak of heat release rate at the end of the burning process [71–73].

This particular behavior is determined by the standard insulated sample holder setup in the cone calorimeter: indeed, the large pkHHR at the end of burning of noncharring materials has been demonstrated to disappear using a modified sample holder that conducts away the heat from the back of the sample by a "heat sink" effect [73–75]. On the other hand, HRR curve for charring materials is not significantly affected by thermal feedback, so that experimental setup plays a minor role on pkHRR for these materials.

Based on these observations, the evaluation of fire performance in terms of pkHRR for charring materials in comparison with noncharring polymer may lead to an overestimation of HRR reduction. To avoid this, comparison of HRR values should be done in terms of a "steady value of heat release rate" which can be observed for thermally thick materials[1] [76]. With this method, the effectiveness

[1]In the case of noncharring pristine polymers, thermally thick condition is generally obtained for thicknesses above 10 mm.

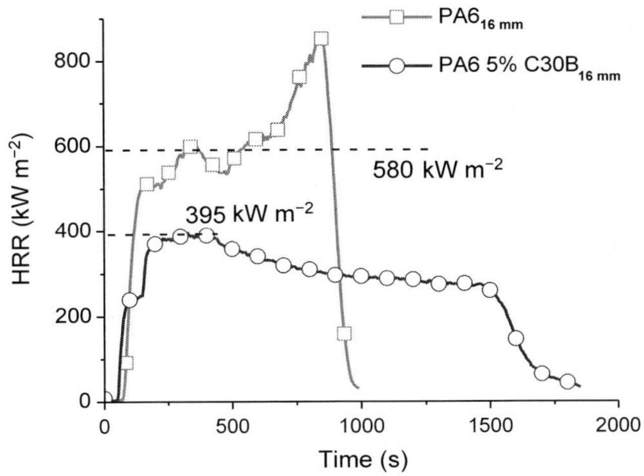

FIGURE 7

Cone calorimeter HRR curves for 16 mm thickness samples of PA6 and PA6/organomontmorillonite (Cloisite 30B, mainly exfoliated). Horizontal lines: steady value of heat release rate [76]. HRR, heat release rate; PA6, polyamide 6.

Reproduced by permission of The Royal Society of Chemistry.

of 5% nanoclay in the reduction of the HRR of PA6 was assessed and quantified to be about 33%, whereas about 50% reductions are observed considering pkHRR values (Figure 7).

Even though HRR reduction has been often used to claim superior flame retardancy of polymer nanocomposites, it has to be noted that the condition used in cone calorimetry in the standard horizontal configuration represents a single and specific testing condition. In particular, the horizontal and confined specimen setup implies that no material macroscopic flow occurs; therefore, progressive accumulation of char on the surface allows the ceramic-charred residue to efficiently protect the underlying polymer. These conditions are clearly specific of the horizontal configuration, whereas no direct correlation can be drawn for vertical configuration. Unfortunately, no results have been reported for cone calorimetry of polymer nanocomposites in vertical configuration, likely due to the experimental difficulties related to dripping of molten material.

3.4 Flame spread

Flame spread tests aim at measuring the tendency of a flame to propagate over a substrate and directly correlate to surface flame propagation in a real fire scenario. However, material scientists rarely used this kind of tests for the development of flame-retarded formulation, likely owing to the relatively large specimens needed

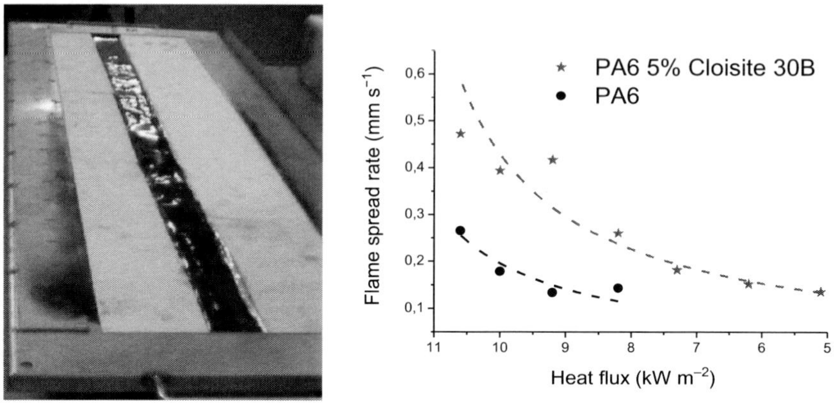

FIGURE 8

Modified specimen setup for the use of 50 mm wide strips (left) and flame spread rate vs heat flux (right) for PA6/Cloisite 30B compared with pristine PA6. PA6, polyamide 6.

for standard testing, such as flooring radiant panel (ISO 9239) or vertical flame spread (ISO 5658).

A modified approach using reduced width extruded strips in the flooring radiant panel was proposed by Gilman et al. [77]. The same approach was applied in this paper for the comparison between PA6 and PA6/organomontmorillonite (Cloisite 30B, mainly exfoliated). Both specimens showed complete combustion; however, PA6/clay nanocomposite showed significantly higher flame spread rate compared to pristine PA6 (Figure 8). This negative result might be correlated with the reduction in ignition time observed in the cone calorimeter for the same formulation, assuming flame spread over a surface as a succession of ignition steps. Indeed, in horizontal flame spread, each single portion of the specimens becomes covered by the flame as soon as the fuel concentration reaches the lower flammability limit, as a consequence of polymer decomposition by the external heat flux, similarly to the ignition scenario in cone calorimeter. This implies that flame spread is a completely different and complementary scenario as compared to both forced combustion and flammability test, which complete the assessment of flame retardancy of new materials.

4. CONCLUSIONS

Despite the fact that material scientists have been evaluating flame-retardancy properties mostly based on cone calorimetry and UL 94 tests, the comprehensive assessment of FR behavior of nanocomposites should use different tests to correlate with the different fire scenarios possible. Indeed, every fire test represents a specific condition which should not be generalized for other fire conditions.

Nanoparticles are well known to be very effective in reducing HRR in forced combustion, owing to the accumulation of a heat resistant layer on the material surface, by polymer ablation and/or as nanoparticle migration. This layer acts as a physical barrier for fuel diffusion toward the flame and/or thermal shielding. On the other hand, minor effects are generally observed in terms of reduction of total heat release and effective heat of combustion. Ignition time under external irradiation has also been addressed in recent literature, often showing reduced time to ignition for polymer nanocomposites, compared to pristine polymers; however, cases of delayed ignition have been also reported. The phenomena explaining these modifications in ignition time are not yet clarified and require further research efforts.

A second noticeable effect generally induced by the presence of dispersed nanoparticles is the reduction of polymer dripping during vertical flammability test. Though dripping could be beneficial or not in the real fire scenario, from the strict point of view of UL 94 material classification reduction of dripping often correspond to a worsened material classification. To describe the real material behavior beyond the bare standard ranking, other parameters such as the burning time, burning rate, and the specimen residue left at the end of test should also be considered.

On the other hand, studies on flame spread tests are insufficient; in fact, flame spread tests for nanocomposites are reported in this paper for the first time: the results show complementary information can be obtained to complete the assessment of polymer nanocomposites fire behavior in fires.

Based on these conclusions, nanoclays clearly induce a strong modification in the flame performance of polymers and can be used as flame retardants to slow down combustion and to reduce material dripping. However, the use of nanoclays alone is generally insufficient to reach the performance required in real applications. For this reasons, nanoclays are increasingly used in combination with other fire-retardant additives, either to reduce the total amount of fire retardants or to improve the flame retardancy of traditional materials.

The final target in flame retardancy is however not just to develop polymer materials complying with prescriptive codes regulating their specific applications. A more comprehensive and safer approach involves fire performance, assessed by quantitative tests such as the cone calorimeter which would supply material parameters for feeding real scenarios fire and hazard modeling.

Further development of flame-retarded polymer nanocomposites is expected in the next years, through the understanding of fundamental mechanisms controlling ignition and fuel/heat barrier. Knowledge-based development will hopefully allow to downsize heat/fuel flux below self-sustaining supply, to reach the final goal of self-extinguishing polymer nanocomposite materials.

Acknowledgments

The authors wish to thank Dr Alessandro Gallo and the company MARIS (www.mariscorp.com) for extrusion of strip specimens as well as Mr Giovanni Ortale and the fire laboratory IMAST (www.imast.biz) for flooring radiant panel testing.

References

[1] Ray SS, Okamoto M. Polymer/layered silicate nanocomposites: a review from preparation to processing. Prog Polym Sci 2003;28:1539–641.
[2] Pavlidou S, Papaspyrides CD. A review on polymer–layered silicate nanocomposites. Prog Polym Sci 2008;33:1119–98.
[3] Spitalsky Z, Tasis D, Papagelis K, Galiotis C. Carbon nanotube–polymer composites: chemistry, processing, mechanical and electrical properties. Prog Polym Sci 2010;35: 357–401.
[4] Moniruzzaman M, Winey KI. Polymer nanocomposites containing carbon nanotubes. Macromolecules 2006;39:5194–205.
[5] Kim H, Abdal AA, Macosko CW. Graphene/polymer nanocomposites. Macromolecules 2010;43:6515–30.
[6] Potts JR, Dreyer DR, Bielawski CW, Ruoff RS. Graphene-based polymer nanocomposites. Polymer 2011;52:5–25.
[7] Bailey SW. Polytypism of trioctahedral 1:1 layer silicates. Clays Clay Miner 1969;17: 355–71.
[8] Alexandre M, Dubois P. Polymer-layered silicate nanocomposites: preparation, properties and uses of a new class of materials. Mater Sci Eng R 2000;28:1–63.
[9] Carter LW, Hendricks JG, Bolley DS. United States Patent No. 2531396; 1950.
[10] Fukushima Y, Inagaki S. Synthesis of an intercalated compound of montmorillonite and 6-polyamide. J Inclusion Phenom Mol Recognit Chem 1987;5:473–82.
[11] Okada A, Fukushima Y, Kawasumi M, Inagaki S, Usuki A, Sugiyama S, et al. (assigned to Toyota Motor Co., Japan). United States Patent No. 4739007; 1988.
[12] Kawasumi M, Kohzaki M, Kojima Y, Okada A, Kamigaito O. United States Patent No. 4810734; 1989.
[13] Usuki A, Kojima Y, Kawasumi M, Okada A, Fukushima Y, Kurauch T, et al. Synthesis of nylon 6-clay hybrid. J Mater Res 1993;8:1179–84.
[14] Vaia RA, Ishii H, Giannelis EP. Synthesis and properties of two-dimensional nanostructures by direct intercalation of polymer melts in layered silicates. Chem Mater 1993;5: 1694–6.
[15] Vaia RA, Jandt KD, Kramer EJ, Giannelis EP. Kinetics of polymer melt intercalation. Macromolecules 1995;28:8080–5.
[16] Vaia RA, Giannelis EP. Lattice model of polymer melt intercalation in organically-modified layered silicates. Macromolecules 1997;30:7990–9.
[17] Vaia RA, Giannelis EP. Polymer melt intercalation in organically-modified layered silicates: model predictions and experiment. Macromolecules 1997;30:8000–9.
[18] Theng BKG. The chemistry of clay-organic reactions. New York: Wiley; 1974.
[19] Sherman JD. Synthetic zeolites and other microporous oxide molecular sieves. Proc Natl Acad Sci USA 1996;96:3471–8.
[20] Gilman JW, Kashiwagi T, Lichtenhan JD. Nanocomposites: a revolutionary new flame retardant approach. SAMPE J 1997;33:40–6.
[21] Kiliaris P, Papaspyrides CD. Polymer/layered silicate (clay) nanocomposites: an overview of flame retardancy. Prog Polym Sci 2010;35:902–58.
[22] Kashiwagi T, Gilman JW, Nyden MR, Lomakin SM. Polymer combustion and new flame retardants. In: Le Bras M, Camino G, Bourbigot S, Delobel R, editors. Fire retardancy of polymers: the use of intumescence. Cambridge, UK: The Royal Society of Chemistry; 1998. pp. 175–202.

[23] Gilman JW, Kashiwagi T, Giannelis EP, Manias E, Lomakin S, Lichtenhan JD, et al. Nanocomposites: radiative gasification and vinyl polymer flammability. In: Le Bras M, Camino G, Bourbigot S, Delobel R, editors. Fire retardancy of polymers: the use of intumescence. Cambridge, UK: The Royal Society of Chemistry; 1998. pp. 203–21.

[24] Kashiwagi T, Harris RH, Zhang X, Briber RM, Cipriano BH, Raghavan SR, et al. Flame retardant mechanism of polyamide 6–clay nanocomposites. Polymer 2004; 45:881–91.

[25] Samyn F, Bourbigot S, Jama C, Bellayer S. Fire retardancy of polymer clay nanocomposites: is there an influence of the nanomorphology? Polym Degrad Stab 2008;93: 2019–24.

[26] Fina A, Bocchini S, Camino G. Thermal behavior of nanocomposites and fire testing performance. In: Wilkie CA, Morgan AB, Nelson GL, editors. Fire and polymers V, materials and concepts for fire retardancy. Washington (USA): American Chemical Society; 2009. pp. 10–24.

[27] Lewin M. Some comments on the modes of action of nanocomposites in the flame retardancy of polymers. Fire Mater 2003;27:1–7.

[28] Pastore HO, Frache A, Boccaleri E, Marchese L, Camino G. Heat induced structure modifications in polymer-layered silicate nanocomposites. Macromol Mater Eng 2004;289:783–6.

[29] Jang BN, Costache M, Wilkie CA. The relationship between thermal degradation behavior of polymer and the fire retardancy of polymer/clay nanocomposites. Polymer 2005;46:10678–87.

[30] Vaia RA, Price G, Ruth PN, Nguyen HT, Lichtenhan J. Polymer-layered silicate nanocomposites as high performance ablative materials. Appl Clay Sci 1999;15:67–92.

[31] Gilman JW. Flame retardant mechanism of polymer–clay nanocomposites. In: Morgan AB, Wilkie CA, editors. Flame retardant polymer nanocomposites. Hoboken, NJ: Wiley; 2007. pp. 67–87.

[32] Zanetti M, Bracco P, Costa L. Thermal degradation behavior of PE/clay nanocomposites. Polym Degrad Stab 2004;85:657–65.

[33] Zanetti M, Camino G, Reichert P, Mulhaupt R. Thermal behaviour of poly(propylene) layered silicate nanocomposites. Macromol Rapid Comm 2001;22:176–80.

[34] Zanetti M, Camino G, Thomann R, Mulhaupt R. Synthesis and thermal behaviour of layered silicate–EVA nanocomposites. Polymer 2001;42:4501–7.

[35] Bourbigot S, Gilman JW, Wilkie CA. Kinetic analysis of the thermal degradation of polystyrene–montmorillonite nanocomposite. Polym Degrad Stab 2004;84: 483–92.

[36] Wang S, Hu Y, Lin Z, Gui Z, Wang Z, Chen Z, et al. Flammability and thermal stability studies of ABS/montmorillonite nanocomposite. Polym Int 2003;52:1045–9.

[37] McNeill IC. In: Allen G, Bevington J, Eastmond GC, Ledwith A, Russo S, Sigwalt P, editors. Comprehensive polymer science, vol. 6. Oxford, UK: Pergamon Press; 1989. pp. 452–97.

[38] Zanetti M, Camino G, Mülhaupt R. Combustion behaviour of EVA/fluorohectorite nanocomposites. Polym Degrad Stab 2001;74:413–7.

[39] Solomon DH, Hawthorne DG. Chemistry of pigments and fillers. Malabar, FL: Krieger; 1991.

[40] Sposito G, Skipper NT, Sutton R, Park SH, Soper AK, Greathouse JA. Surface geo-chemistry of the clay minerals. Proc Natl Acad Sci USA 1999;96:3358–64.

[41] Bocchini S, Fukushima K, Di Blasio A, Fina A, Frache A, Geobaldo F. Polylactic acid and polylactic acid-based nanocomposites photooxidation. Biomacromolecules 2010; 11:2919–26.
[42] Bellucci F, Camino G, Frache A, Sarra A. Catalytic charring-volatilization competition in organoclay nanocomposites. Polym Degrad Stab 2007;92:425–36.
[43] Xie W, Gao Z, Pan W, Hunter D, Singh A, Vaia R. Thermal degradation chemistry of alkyl quaternary ammonium montmorillonite. Chem Mater 2001;2001(13):2979–90.
[44] Davis RD, Gilman JW, Sutto TE, Callahan JH, Trulove PC, De Long HC. Improved thermal stability of organically modified layered silicates clays. Clays Clay Miner 2004;52:171–9.
[45] Bellucci F, Camino G, Frache A, Ristori V, Sorrentino L, Iannace S, et al. Effect of organoclay impurities on mechanical properties of EVA-layered silicate nanocomposites. e-Polym; 2006. n014.
[46] Park SJ, Seo DI, Lee JR. Surface modification of montmorillonite on surface acid-base characteristics of clay and thermal stability of epoxy/clay nanocomposites. J Colloid Interface Sci 2002;251:160–5.
[47] Babrauskas V. Ignition of common solids. In: Ignition handbook. Issaquah, USA: Fire Science Publishers; 2003. pp. 234–351.
[48] Zhu J, Start P, Mauritz KA, Wilkie CA. Thermal stability and flame retardancy of poly(methyl methacrylate)–clay nanocomposites. Polym Degrad Stab 2002;77:253–8.
[49] Liang ZM, Wan CY, Zhang Y, Wei P, Yin J. PVC/montmorillonite nanocomposites based on a thermally stable, rigid-rod aromatic amine modifier. J Appl Polym Sci 2004;92:567–75.
[50] Wu Y, Huang H, Zhao W, Zhang H, Wang Y, Zhang L. Flame retardance of montmorillonite/rubber composites. J Appl Polym Sci 2008;107:3318–24.
[51] Bartholmai M, Schartel B. Layered silicate polymer nanocomposites: new approach or illusion for fire retardancy? Investigations of the potentials and the tasks using a model system. Polym Adv Technol 2004;15:355–64.
[52] Bourbigot S, Samyn F, Turf T, Duquesne S. Nanomorphology and reaction to fire of polyurethane and polyamide nanocomposites containing flame retardants. Polym Degrad Stab 2010;95:320–6.
[53] Hull TR, Stec AA, Nazare S. Fire retardant effects of polymer nanocomposites. J Nanosci Nanotechnol 2009;9:4478–86.
[54] Nazare S, Hull TR, Biswas B, Samyn F, Bourbigot S, Jama C, et al. Study of the relationship between flammability and melt rheological properties of flame-retarded poly(butylene terephthalate) containing nanoclays. In: Hull TR, Kandola BK, editors. Fire retardancy of polymers: new strategies and mechanisms. Cambridge: Royal Society of Chemistry; 2009. pp. 168–83.
[55] Tang Y, Hu Y, Wang S, Gui Z, Chen Z, Fan W. Preparation and flammability of ethylene-vinyl acetate copolymer/montmorillonite nanocomposites. Polym Degrad Stab 2002;78:555–9.
[56] Zong R, Hu Y, Liu N, Li S, Liao G. Investigation of thermal degradation and flammability of polyamide-6 and polyamide-6 nanocomposites. J Appl Polym Sci 2007;104:2297–303.
[57] Zanetti M, Luigi Costa L. Preparation and combustion behaviour of polymer/layered silicate nanocomposites based upon PE and EVA. Polymer 2004;45:4367–73.
[58] Solarski S, Mahjoubi F, Ferreira M, Devaux E, Bachelet P, Bourbigot S, et al. (Plasticized) polylactide/clay nanocomposite textile: thermal, mechanical, shrinkage and fire properties. J Mater Sci 2007;42:5105–17.

[59] Zhao C, Qin H, Gong F, Feng M, Zhang S, Yang M. Mechanical, thermal and flammability properties of polyethylene/clay nanocomposites. Polym Degrad Stab 2005;87:183–9.

[60] Qin H, Zhang S, Zhao C, Hu G, Yang M. Flame retardant mechanism of polymer/clay nanocomposites based on polypropylene. Polymer 2005;46:8386–95.

[61] Qin H, Su Q, Zhang S, Zhao B, Yang M. Thermal stability and flammability of polyamide 66/montmorillonite nanocomposites. Polymer 2003;44:7533–8.

[62] Sahoo PK, Samal R. Fire retardancy and biodegradability of poly(methyl methacrylate)/montmorillonite nanocomposite. Polym Degrad Stab 2007;92:1700–7.

[63] Camino G, Tartaglione G, Frache A, Manferti C, Costa G. Thermal and combustion behaviour of layered silicate–epoxy nanocomposites. Polym Degrad Stab 2005;90:354–62.

[64] Fina A, Camino G. Ignition mechanisms in polymers and polymer nanocomposites. Polym Adv Technol 2011;22:1147–55.

[65] Fina A, Cuttica F, Camino G. Ignition of polypropylene/montmorillonite nanocomposites. Polym Degrad Stab 2012;97:2619–26.

[66] Kandola BK. Nanocomposites. In: Horrocks AR, Price D, editors. Fire retardant materials. Cambridge, UK: Woodhead Publishing Limited; 2001. pp. 204–19.

[67] Hartwig A, Putz D, Schartel B, Bartholmai M, Wendschuh-Josties M. Combustion behaviour of epoxide based nanocomposites with ammonium and phosphonium bentonites. Macromol Chem Phys 2003;204:2247–57.

[68] Bourbigot S, Duquesne S. Fire retardant polymers: recent developments and opportunities. J Mater Chem 2007;17:2283–300.

[69] Zanetti M, Kashiwagi T, Falqui L, Camino G. Cone calorimeter combustion and gasification studies of polymer layered silicate nanocomposites. Chem Mater 2002;14:881–7.

[70] Zheng X, Jiang DD, Wang D, Wilkie CA. Flammability of styrenic polymer clay nanocomposites based on a methyl methacrylate oligomerically-modified clay. Polym Degrad Stab 2006;91:289–97.

[71] Lyon RE, Janssens ML. Polymer flammability: U.S. Department of Transportation Federal Aviation Administration technical report. 1982 Technical Report DOT/FAA/AR05/14, FAA. Available online at: http://www.fire.tc.faa.gov; May 2005.

[72] Babrauskas V. Development of the cone calorimeter a bench scale heat release rate apparatus based on oxygen consumption. Fire Mater 1984;8:81–94.

[73] Schartel B, Hull TR. Development of fire-retarded materials-interpretation of cone calorimeter data. Fire Mater 2007;31:327–54.

[74] Schartel B, Bartholmai M, Knoll U. Some comments on the use of cone calorimeter data. Polym Degrad Stab 2005;88:540–7.

[75] Carvel R, Steinhaus T, Rein G, Torero JL. Determination of the flammability properties of polymeric materials: a novel method. Polym Degrad Stab 2011;96:314–9.

[76] Fina A, Canta F, Castrovinci A, Camino G. Significant assessment of nanocomposites combustion behaviour by proper use of cone calorimeter. In: Hull TR, Kandola BK, editors. Fire retardancy of polymers: new strategies and mechanisms. Royal Society of Chemistry; 2009. pp. 147–59.

[77] Gilman JW, Bourbigot S, Shields JR, Nyden M, Kashiwagi T, Davis RD, et al. High throughput methods for polymer nanocomposites research: extrusion, NMR characterization and flammability property screening. J Mater Sci 2003;38:4451–60.

CHAPTER

Polymer Nanocomposites as Ablative Materials 15

A.R. Bahramian, M. Kokabi

Tarbiat Modares University, Tehran, Islamic Republic of Iran

CHAPTER OUTLINE

1. Introduction

Advances in ultra–high-temperature conditions such as hypersonic atmospheric flight, rocket propulsion, and other areas of high-temperature technology have resulted in environments of extremely high heat flux and temperature gradients [1–3]. Combustion flames of chemical propulsion systems are of the order of 3000 °C and higher [1,4]. Similarly, boundary layer temperatures in excess of 5000 °C are characteristic of vehicles entering hypersonically into a planetary atmosphere [1,5]. In these and other very high-temperature environments, the use of conventional engineering materials is greatly restricted because of possible thermal degradation and catastrophic failure [3,6–8]. However, the materials may serve their intended purpose if they are protected from the intense heat and temperature of the environment [9,10]. One convenient means for accomplishing this form of environmental protection is by the proper use of a new class of engineering materials known as thermal protection systems (TPS) [11–14].

Polymer Green Flame Retardants. http://dx.doi.org/10.1016/B978-0-444-53808-6.00015-9

Novel materials based on nanotechnology creating nontraditional ablators are rapidly changing the technology applied for TPS. The introduction of inorganic nanomaterials as additives into polymer systems has resulted in polymer nanostructured materials exhibiting multifunctional, high-performance polymer characteristics beyond what traditional polymer composites possess. Multifunctional features attributable to polymer nanocomposites consist of improved thermal and flame resistance. Through control or alternation of the additives at the nanoscale level, one is able to maximize property enhancement of selected polymer systems to meet or exceed the requirements of current defense, aerospace, and commercial applications.

In this chapter, resole-type phenolic resin/asbestos cloth ablative composite and its layered silicate nanocomposite counterparts as new TPS are studied, and the influence of the experimental conditions in estimating the thermophysical properties of ablative composite and nanocomposites materials is examined. A mathematical model of transient processes of heat and mass transfer in composite and nanocomposites is finally proposed and confirmed by experimental data.

2. Thermal protection systems

TPS protect a body or substrate from the severe heating encountered during high-temperature conditions such as in a hypersonic flight through a planetary atmosphere or in a propulsion chamber of solid rocket motor engine [15]. Since TPS is a single point of failure subsystem, it is critical and its performance needs to be validated through modeling and simulation, ground test, and analysis. In general, there are two classes of TPS [16–19]:

(i) *Reusable TPS* [16,19,20], where after exposure to the external heat flux, there are no changes in the mass, geometry, or properties of the TPS materials. Typically, reusable TPS applications are limited to relatively mild entry environments (e.g. Shuttle) [16–18] or other long-time applications. The characteristics of a reusable TPS are shown in Figure 1. It is observed that radiative and convective heating results in a significant amount of energy being reradiated from the heated surface with the remainder conducted into the TPS material [16]. It is advantageous if the surface coating has high emissivity, to maximize the amount of energy reradiated, and low surface catalycity, to minimize convective heating by suppressing recombination of dissociated boundary layer species at the heated surface. It is also advantageous if the primary insulation has low thermal conductivity since that will minimize the mass of material required to insulate the primary structure (backup material) [16,17].

Therefore, reusable TPS systems are designed to reject as much heat as possible and conduct as little heat as possible in order to meet the bond-line temperature requirements with minimal heat-shield mass. Low catalytic efficiency, high emissivity, and low thermal diffusivity are desired in designing a TPS system.

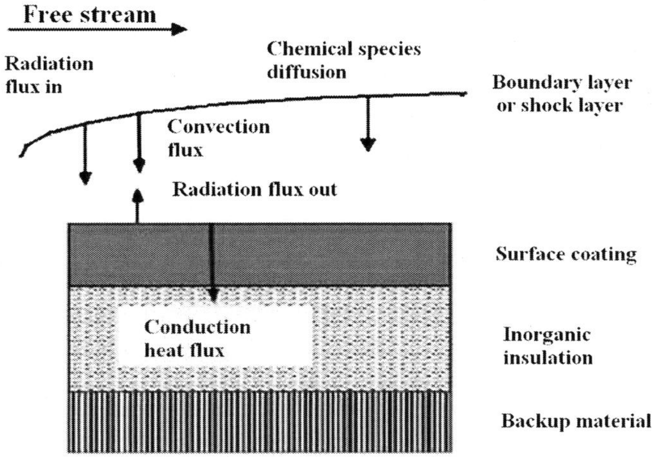

FIGURE 1

Energy transfer in reusable TPS materials [16].

(ii) *Ablative TPS* [16,19,20], in contrast, accommodate high heating rates and heat loads through phase change and mass loss. Ablative materials have been the classical approach to TPS used for over 40 years in a broad range of applications [16,18].

The word ablation is derived from the suppletive past particle of the Latin *auferre*, which means to remove. It was originally used in the geologic sense to describe the combined, predominantly thermal, processes by which a glacier wastes. The present use of the word maintains the thermal aspect and describes the absorption, dissipation, and blockage of heat associated with high-speed entry into the atmosphere. Thus ablative thermal protection materials are used to protect vehicles from damage during atmospheric reentry. Ablative materials are also used to protect rocket nozzles and ship hulls from propellant gas erosion, as protection from laser beams, and to protect land-based structures from high-heat environments.

To clarify further the objectives of this work, the principle of thermal ablation process will be explained in detail.

2.1 Ablative materials

2.1.1 Ablative composite

Ablation is an effective and reliable method largely used in aerospace structures to protect the pay load from the damaging effects of external high temperatures. In the ablation process, the high heat fluxes are dissipated by the material through a series of endothermic processes that finally lead to the loss and the consumption of the material itself [4–8]. The ablative material keeps the surface temperature within a certain range, and as a consequence, an increase of the heat flux will not cause a consistent temperature rise, but will cause an increase of the surface recession rate. Therefore, important parameters in the choice of a suitable ablator are the ablation temperature (the temperature at which the material degradation begins) and the density [9—11,14]. Charring ablators produce char as an effect of the degradation reaction, subsequently providing an insulation layer. During the ablation reaction, these materials do not reduce their volume significantly. Moreover, charring ablators are the most widely used thermal protection shields [10]. As the charring ablator is heated, the temperature increases until the surface reaches the degradation temperature and starts to release gaseous products, leaving a porous, carbonaceous residue [9,11]. Thermal protection via ablation is achieved through a self-regulating heat and mass transfer process (involving an insulator with low thermal conductivity) and concomitant formation of a tough refractory char on the insulator surface. Above the decomposition temperature, the insulator produces pyrolysis gases in the reaction zone and degrades to a char layer at higher temperatures. The majority of mass loss occurs in the reaction zone. The presence of the char layer regulates penetration of heat from the surface and produces a steep temperature gradient [12—14].

While the process of ablation is rather complex, many of its important physicochemical aspects have been identified. At the onset of heating, energy is absorbed by the TPS and then conducted internally [9]. The rate of heat penetration is dependent on the surface temperature, but in all cases, thermal diffusion is slow because of the low conductivity of the polymer. Maintaining the heat in the surface region causes its temperature to rise rapidly until material vaporization takes place. The initial volatiles are generally water, residual diluents, or polymers of low molecular weight. At still higher temperatures, thermal agitation becomes severe enough to split pendant and side groups from the polymer backbone. Eventually, the chemical bonds in the backbone are ruptured [3,4]. The polymer thus undergoes competitive reactions involving reversible depolymerization and/or irreversible pyrolysis. The rate of decomposition follows an *n*th-order Arrhenius-type reaction rate law and varies with the temperature and amount of undecomposed material present. If elimination of groups along the polymer chain predominates over chain cleavage, much of the original chain structure will remain as carbon [20]. Gaseous products formed during pyrolysis will vary greatly in composition and molecular weight. For example, they may be composed of water-gas constituents like hydrogen, carbon monoxide, carbon dioxide, and water vapor, saturated hydrocarbons like methane and ethane, and unsaturated hydrocarbons such as ethylene and acetylene [19,20]. Those pyrolytic species are injected into the boundary layer, lower the enthalpy of this layer, and undergo further thermochemical reactions [21,22].

If the polymer melts or the polymeric composite contains a meltable component, additional heat is absorbed at the melting point (latent heat of fusion). Subsequent fusion occurs with the formation of a liquid layer [13—19]. When the molten material is of low viscosity, a very thin layer of the liquid may be removed immediately by gas-dynamic shear forces acting at the gas—liquid interface [20]. On the other extreme, a highly viscous melt tends to remain on the surface until sufficient heat is absorbed to vaporize it. In this case, an additional amount of heat is absorbed (the latent heat of vaporization), and the surface temperature corresponds to the liquids vaporization temperature [16,18]. Most cases of melting ablation, however, involve partial flow and vaporization of the liquid layer. The extent of each competing reaction depends upon the environmental parameters (shear, heat flux, etc.) and the temperature-dependent viscosity of the melted material [13,23].

The presence of fibers or fillers in charring polymeric composites alters the ablation process; voids are thus created in the region originally occupied by the fibers. A highly porous char is formed, which is relatively susceptible to mechanical erosion [22]. Pyrolyzed species are reduced to lower molecular weight products in passing through the reactive char layer. Like the vapors arising from the organic resin, they also deposit carbon in the char zone. If the ablating surface is rich in carbon, its temperature will be dictated essentially by the incident heating rate and may reach values in excess of 3200 °C. With respect to melting oxide fibers like silica, the mechanism of ablation is quite different. The organic resin or its residual char is removed by the environment at a faster rate than the fibers, thus leaving them exposed and unsupported [17,20]. The vitreous fiber then melts, and the molten material covers the surface as liquid droplets, irregular globules, or a thin film. On the other hand, carbon fibers provide exceptional dimensional stability to charring composites. They remain in the char layer during pyrolysis, provide high strength to the char zone, and firmly attach it to the substrate. Unlike the nylon and silica fibers, the carbon fibers are little affected by high temperatures. Fiber removal occurs primarily, by thermochemical reactions such as oxidation [16,20].

From a thermophysical point of view, ablation is an eight-stage heat and mass transfer process in which a large amount of thermal energy is expended by sacrificial loss of surface region material [16,20–22,24,25]. The heat input from the environment is absorbed, blocked, dissipated, and generated by numerous mechanisms. They involve (1) heat conduction into the material substrate and storage by its effective heat capacity; (2) materials phase changes such as melting, vaporization, and sublimation; (3) heat absorption by gases in the substrate as they percolate to the surface; (4) convection of heat in a liquid layer, if one exists; (5) transpiration of gases from the ablating surface into the boundary layer with attendant heat absorption; (6) surface and bulk radiation; (7) endothermic and exothermic chemical reactions; and (8) possible yet unrecognized reactions. Such energy absorptive processes take place automatically, simultaneously control the surface temperature, and greatly restrict the inward flow of heat. The characteristics of ablative TPS materials are illustrated in Figure 2.

2.1.2 Ablative nanocomposite

Several limitations associated with composite ablatives have motivated current efforts to identify next generation materials [26,27]. Many times, the resulting chars are structurally weak and are susceptible to mechanical erosion, severely reducing the lifetime of the insulator or necessitating additional insulation thickness. Additionally, provided that the char remains intact, resistance to thermal and chemical

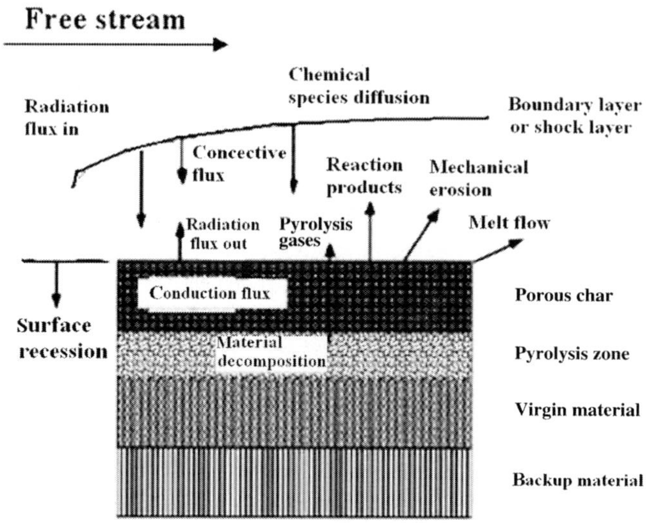

FIGURE 2

Energy transfer mechanisms of ablative TPS materials [16].

ablation ultimately limits lifetime and performance [26,27]. Recent advances showing that polymer-layered silicate nanocomposites present especially improved flammability resistance encourage the examination of this unique class of evolving materials as potential ablatives [28—33]. Polymer-layered silicate nanocomposites show excellent potential as ablative materials because upon pyrolysis, the organic—inorganic nanostructure reinforcing the polymer can be converted into a uniform ceramic layer, which may lead to significantly increased resistance to oxidation and mechanical erosion compared to composite ablative materials [34—37]. In contrast to conventional composites containing micron-scale or larger reinforcing constituents, nanocomposites contain ultrafine phase dimensions, typically on the order of a few nanometers, and thus exhibit unique combination of properties typically not found in traditional composites [38,39].

Polymer nanocomposites are of great industrial as well as scientific interest. From the scientific side, a new scale is available intermediate between molecular and microscales; the potential for more sophisticated tailoring of properties is apparent [16—18]. From the industrial point of view, the major interest is in the dramatically improved properties. In the present work, the focus is placed on polymer-layered silicate nanocomposites which have been developed over the past 10 years. A wide range of layered silicates, compatiblizing agents, and polymers have been used successfully [40—46]. A very low amount of layered silicate improves mechanical and physical properties of nanocomposites. Improved mechanical and thermal properties are of interest for under-the-hood applications in the aerospace industry [40—42]. Recently, environmental regulation has restricted the use of some halogenated flame-retardant (FR) additives, initiating a search for alternative FR additives. Nanoparticle fillers are highly attractive for this purpose since they can simultaneously improve both the physical and flammability properties of the polymer nanocomposite [47].

Since the most general thermal response model was first introduced, we began studying the effects of high heat flux on ablative composite and polymer-layered silicate nanocomposite performance. In this work, the principle of design and evaluation of the performance of composite and polymer-layered silicate nanocomposite TPSs are explained as follows:

1. The physical and thermodynamical aspects of ablative polymer-layered silicate nanocomposite systems are described.
2. The thermal degradation process of composite and nanocomposites are modeled using thermogravimetry analysis (TGA) and thermal degradation kinetic parameters are evaluated using analysis of thermogravimetry data.
3. Thermophysical properties and coefficient of ablation equation (mass and energy balance) are calculated using thermal degradation kinetic parameters.
4. Ablation equation is solved numerically using finite difference method, and inverse analyses of thermal diffusivity estimation are done. Finally, all the theoretical results are confirmed with experimental data.
5. Burning phenomena of polymer-layered silicate nanocomposite surface due to external heat flux is modeled using energy and mass conservation equations.

2.2 Mathematical model

The focus of this section is placed on the improvement and compliment of previous modeling work and investigation and modeling of thermal degradation and high-temperature ablation and burning of nanocomposites as a high-performance heat shield. The results of this work will clarify the mechanism of the ablation and thermal diffusivity of the layered silicate nanocomposite heat shields due to the high-temperature degradation in comparison with its composite counterpart.

2.2.1 Kinetic models of thermal degradation

The kinetic parameters of the thermal degradation of composite and polymer-layered silicate nanocomposite are determined using TGA. The knowledge of these parameters is required to calculate the variation of thermophysical properties of the thermoset composite and nanocomposite, and usage of equations which will be mentioned in the following sections.

2.2.1.1 Chemical reactions

Thermal degradation consists of three processes: heating of the polymer, thermal decomposition, and ignition of the gaseous decomposed products in air [23,47,48].

Pyrolysis gases mix and react with air in the combustion zone above the surface releasing heat and producing carbon dioxide, water, and incomplete combustion products such as carbon monoxide. The basic thermal degradation mechanism leading to volatile fuel generation in char-forming polymers has been described as a generalized chemical bond scission process consisting of primary and secondary decomposition events. The primary decomposition step is side-chain scission of the polymer to form intermediates. The primary char decomposes by dehydrogenation to form the secondary gas and thermally stable secondary char [48]. In essence, the thermal decomposition of polymeric materials is assumed to comprise parallel series reaction processes which include irreversible thermal degradation reactions [23,47,48].

A simple solution for the mass loss history of a polymeric material allows estimation of the generation rate under isothermal and nonisothermal heating, and can be verified using standard laboratory thermogravimetric techniques. By considering some assumptions, the process of thermal degradation will be reduced and simplified. The following assumptions are considered [23,48]:

A: The breaking of primary chemical bonds in the polymer is the rate-limiting step.
B: The reactive intermediate is in dynamic equilibrium with the parent polymer.
C: Thermal degradation of primary char to secondary char and gas is slow compared to the formation of the primary char.
D: The oxidative environment in the pyrolysis zone of a thermal degradation solid polymer is anaerobic.
E: All the chemical reactions are of first order.

The assumptions A–E lead to a simplified mass loss model as follows [23,47,48]:

$$P \underset{k_{-\mathrm{p}}}{\overset{k_\mathrm{p}}{\rightleftarrows}} I^* \tag{1}$$

$$I^* \xrightarrow{k_\mathrm{g}} \text{Volatiles} \tag{2}$$

$$I^* \xrightarrow{k_\mathrm{c}} \text{Gases} \tag{3}$$

By assuming the first-order reaction for all the chemical reactions, the systems of rate equations for the species are as follows:

$$\frac{\mathrm{d}P}{\mathrm{d}t} = -k_\mathrm{p}P + k_{-\mathrm{p}}I^* \tag{4}$$

$$\frac{\mathrm{d}I^*}{\mathrm{d}t} = k_\mathrm{p}P - \left(k_{-\mathrm{p}} + k_\mathrm{g} + k_\mathrm{c}\right)I^* \tag{5}$$

$$\frac{\mathrm{d}G}{\mathrm{d}t} = k_\mathrm{g}I^* \tag{6}$$

$$\frac{\mathrm{d}C}{\mathrm{d}t} = k_\mathrm{c}I^* \tag{7}$$

Using these equations systems and those assumptions, the principle equations of mass change due to thermal degradation are computed as shown below [23,47]:

$$\frac{m(t)}{m_0} = 1 - \left[\frac{k_g}{k_g + k_c}\right]\{1 - \exp(-k_p t)\} \tag{8}$$

Equation (8) shows that as $t \rightarrow \infty$, the residual mass approaches an equilibrium value at constant temperature given by [23,49,50]

$$\frac{m(\infty)}{m_0} = \left[\frac{k_c}{k_g + k_c}\right] = y_c \tag{9}$$

Assuming Arrhenius forms for k_g ($k_g = A_g \exp(-E_g/RT)$) and k_c ($k_c = A_c \exp(-E_c/RT)$) in Eqn (9),

$$y_c = \left[1 + \frac{A_g}{A_c}\exp\left[\frac{-(E_g - E_c)}{RT}\right]\right]^{-1} \tag{10.a}$$

$$\ln\left[\frac{1 - y_c}{y_c}\right] = \ln\left[\frac{A_g}{A_c}\right] - \left[\frac{E_g - E_c}{R}\right]\frac{1}{T} \tag{10.b}$$

Substituting Eqn (9) into Eqn (8) yields the final results for the isothermal mass loss history in terms of the rate constant for thermolysis of primary chemical bonds, k_p, as [47,48,51]

$$\frac{m(t)}{m_0} = y_c + (1 - y_c)\exp(-k_p t) \tag{11}$$

For a constant heating rate, $\beta = \frac{dT}{dt}$, the independent variable, i.e. $k_p t$, in Eqn (11) can be transformed from time to a dimensionless variable. The exact solution of Eqn (11) for the fractional mass as a function of temperature at constant heating rate is [23,47,48]

$$\frac{m(T)}{m_0} = y_c + (1 - y_c)\exp\left[-\frac{A}{\beta}(T - T_0)e^{-E/RT}\right] \tag{12}$$

The fractional mass loss rate during a linear temperature program is obtained by differentiating Eqn (12) with respect to time. No simple solution is possible when $y_c = y_c(T)$, so we make the approximation that the char yield is constant and not a function of temperature, thus with $y_c = \mu$, we obtain [23,52]

$$\frac{-1}{m_0}\frac{dm}{dt} = (1 - \mu)k_p(T)\left(1 + \frac{(T - T_0)E}{RT^2}\right)\exp\left[\frac{-(T - T_0)k_p(T)}{\beta}\right] \tag{13}$$

By this approximation, the rate of gas and char formation are equal; $k_g = k_c$, and from Eqn (13), char yield is 0.5 [23,48].

2.2.1.2 Mass loss kinetic

For TGA, the degree of weight loss may be defined as follows [23,47,49,50]:

$$\alpha = \frac{(m_0 - m)}{(m_0 - m_f)} \tag{14}$$

The rate of weight loss is dependent on temperature, and can be expressed as

$$\frac{d\alpha}{dt} = k \cdot f(\alpha) \tag{15}$$

The rate constant could be described in an Arrhenius form as

$$k = A \cdot \exp\left(\frac{-E}{RT}\right) \tag{16}$$

As mentioned earlier, according to the assumptions made, all the chemical reactions are of first order, but the mass loss of composite and nanocomposite system are n-th order. Using the phase boundary reaction method for textile polymeric system as one of the most general model ($f(\alpha) = (1 - \alpha)^n$), the final form of Eqn (15) could be expressed as [23,51]

$$\frac{d\alpha/dt}{(1-\alpha)^n} = A \cdot \exp\left(-\frac{E}{RT}\right)$$

For a case study of thermal degradation analysis and kinetics parameters evaluation, in this work, STA 625 from the Polymer Laboratories was employed to evaluate the thermal performance of the cured samples. Moreover, differential scanning calorimeter (DSC) has been used to calculate the heat of thermal degradation and specific heat.

In this work, a resole-type phenolic resin was used as the polymeric matrix. Asbestos cloth (Grade AAA) was added as textile reinforcing to the polymeric matrix. The kaolinite powder employed in this work was separated from Semirom Natural Clay Co. Iran and its surface was modified with dimethylsulfoxide in our research group [48]. The montmorillonite sample employed in this work was cloisite 15A from Southern Clay Products, Inc., U.S.A. The combination of solution and in situ intercalation was employed for nanocomposite sample preparations, as follows.

Ethyl alcohol was used to disperse the layered silicates (kaolinite & cloisite 15A) and at the same time to dissolve the phenolic resin as the polymeric matrix. To stack the layers, the crystallite was delaminated in ethyl alcohol. This occurs due to the weak van der Waals force. Phenolic resin then can be adsorbed onto the delaminated individual layer. However, upon ethyl alcohol removal, the layers can reassemble to reform the stack with phenolic chains sandwich in between, forming a well-ordered intercalated nanocomposites. Then, asbestos cloth was added as reinforcement to the phenolic resin matrix (asbestos cloth is very dangerous if it becomes dry and broken; therefore, before cutting, the cut line should be wet by using ethyl alcohol). The cloth was impregnated by phenolic resin/layered silicate intercalated mixture. The sample was dried by ethyl alcohol removal, then precured at 120 °C for 10 min,

Table 1 The Formulations of Samples

Sample	Component	W_{resin} (%)	W_{clay} (%)	$W_{asbestos}$ (%)
Composite	Asbestos cloth/phenolic resin composite	0.50	0.00	0.5
NKA*	Asbestos cloth/phenolic resin/kaolinite nanocomposite	0.42	0.06	0.52
NMA**	Asbestos cloth/phenolic resin/montmorillonite nanocomposite	0.42	0.06	0.52

** Nanocomposite, Kaolinite, Asbestos*
*** Nanocomposite, Montmorillonite, Asbestos*

and finally cured at 160 °C and 3 bar for 1 h in autoclave. After curing, the sample was postcured for 0.5 h at 150 °C. So, the nanocomposite samples are formed. The formulation of samples is given in Table 1.

The $\ln\left(\frac{d\alpha/dt}{(1-\alpha)^n}\right)$ versus the reciprocal of temperature can be given as a linear dependency at correct n. From this correlation, the activation energies, E, frequency factor, A, and degree of thermal degradation, n, are evaluated by linear regression analysis of the data points. Figure 3 shows this curve for asbestos cloth/phenolic resin composite, NKA3 and NMA3 nanocomposites in air, too. These curves can be plotted for these systems and net phenolic resin in nitrogen, alike.

Therefore, the activation energies of composite, NKA3, and NMA3 nanocomposites are 9.57×10^7, 12×10^7 and 10.5×10^7 J mol^{-1} in air, respectively, which is

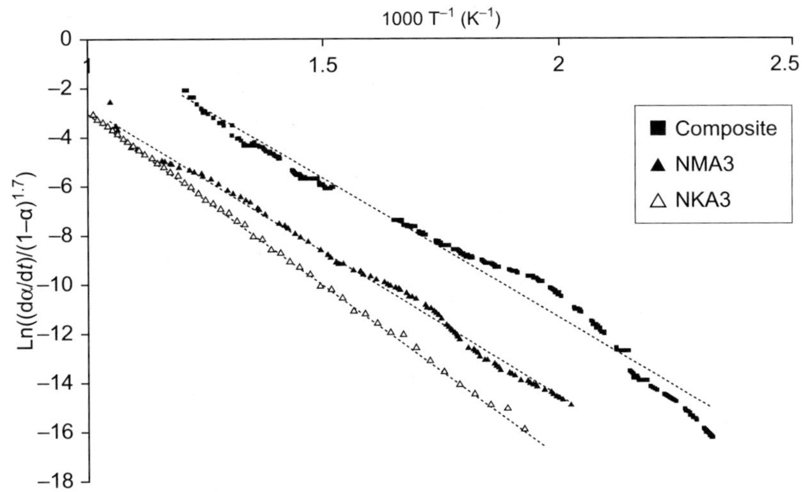

FIGURE 3

Plot of the natural logarithm of $(d\alpha/dt)/(1-\alpha)^n$ versus reciprocal temperature of asbestos cloth/phenolic matrix composite, NKA3, and NMA3 nanocomposites in air [52,53].

nearly three orders of magnitude larger than the activation energy of net phenolic resin calculated as $1.43 \times 10^5 \, J \, mol^{-1}$ in nitrogen. This is due to asbestos cloth chemical reactions and its interaction with resin. Similar data have been reported by the other researchers (asbestos cloth phenolic resin composite [23] and net phenolic resin [49,52]).

The relative rate constants for volatile and char formation are obtained by plotting Eqn (10.b). Thus, a plot of $\ln[(1 - y_c)/y_c]$ versus $1/T$ has a slope proportional to the difference in activation energies for volatile and char formation ($E_g - E_c$) and an intercept which is the natural logarithm of the ratio of the frequencies factors (A_g/A_c). Figure 4 shows the plot of Eqn (10.b) using the isothermal char yields for asbestos cloth/phenolic resin composite, NKA3, and NMA3 nanocomposites in air.

All the kinetic parameters of thermal degradation of phenolic resin, asbestos cloth/phenolic resin composite, NKA3, and NMA3 nanocomposites are given in Table 2.

2.2.2 Thermophysical properties

As said before, under the action of high temperatures and external heat flux, the heat shield's thermophysical properties, such as density, thermal conductivity, and specific heat, all change [23,54–57]. Therefore, to exactly solve the ablation equation (conservation equations of mass and energy), it is necessary to consider the behavior of its properties during thermal degradation and ablation.

2.2.2.1 Density

On heating a polymeric material to the pyrolysis temperature of its matrix, density decreases due to intensive gas generation in the matrix. It is assumed that the

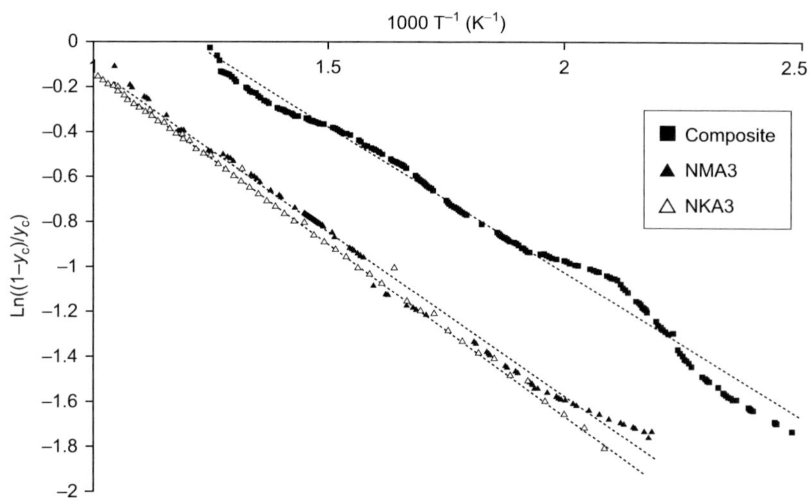

FIGURE 4

Plot of the natural logarithm of $(1 - y_c)/y_c$ versus reciprocal temperature of asbestos cloth/phenolic matrix composite, NKA3, and NMA3 nanocomposites in air [52,53].

Table 2 The Kinetic Parameters of the Samples [23,48,52,53]

Kinetic Parameter	Unit	Phenolic Resin (in Nitrogen)	Composite (in Nitrogen)	Composite (in Air)	NKA3 (in Air)	NMA3 (in Air)
Activation energy, *E*	$J\,mol^{-1}$	1.43×10^5	1.75×10^8	9.57×10^7	12×10^7	10.5×10^7
The difference in activation energy for volatile and char formation, $E_g - E_c$	$J\,mol^{-1}$	4.1×10^4	3.92×10^7	1.01×10^7	1.3×10^7	1.3×10^7
Frequency factor, *A*	s^{-1}	3×10^9	1.26×10^{10}	8.82×10^4	8.9×10^4	8.9×10^4
The fraction of frequency factor for volatile and char formation, A_g/A_c	–	13	46.5	4	4.1	4.3
Degree of thermal degradation reaction, *n*	–	1	6.5	1.7	1.7	1.7

density under thermal degradation conditions depends on the mass fraction of polymer remaining in the solid, as char, and the porosity created during decomposition. The change in density upon heating is described by the below equations [23,57]:

$$\rho = \sum_{i=1}^{3} \rho_i \nu_i \tag{17}$$

$$\sum_{i=1}^{3} \nu_i = 1 - \nu_4 \tag{18}$$

The porosity (ν_4) forms during the thermal degradation. At the initial state before heating, they are

$$\nu_1 = \nu_1^0; \quad \nu_2 = \nu_2^0; \quad \nu_3 = 0; \quad \nu_4 = \nu_4^0 \tag{19}$$

and

$$\rho = \rho_0 = \rho_1 \nu_1 + \rho_2 \nu_2^0 \tag{20}$$

The volume fraction of fibers and inorganic fillers, ν_1, does not change upon heating and the changes in the volumetric concentration of polymer and coke phases are described by the following equations [23,56,57]:

$$\rho_2 \frac{\partial \nu_2}{\partial t} = -J \tag{21}$$

$$\nu_3 = \left(\nu_2^0 - \nu_2\right)(1 - \Gamma)\frac{\rho_2}{\rho_3} \tag{22}$$

The rate of pyrolysis, J, depends on the temperature and gas pressure in the pores and is described by the modified Arrhenius equation [23,57]:

$$J = J_0\left[1 - \frac{p}{p_0}\exp\left(-\frac{2E}{RT}\right)\right]^{0.5}\exp\left(-\frac{E}{RT}\right) \tag{23}$$

Gaseous products of pyrolysis in pores are assumed ideal and perfect gas, therefore, [23,55]

$$p = R\rho_g T \tag{24}$$

The density and porosity of the composite and nanocomposites laminated at ambient temperature were determined with submerging in water according to standard test methods for density and specific gravity of plastics by displacement (ASTM D792).

2.2.2.2 Thermal conductivity

The change in thermal conductivity coefficient under high temperature is described by the following formula [23,57]:

$$K = \frac{K_0}{\nu_2^0}\left(\frac{T}{T_0}\right)^{0.5}(\nu_2 + n_K \nu_3) \tag{25}$$

K is directly determined experimentally by considering a specific value for T. Then n_K can be calculated from Eqn (23).

In Eqn (23), the change of the thermal conductivity coefficient is connected with the growing thermal conductivity of the polymeric matrix at the initial stage of heating occurring before the thermal degradation of polymeric matrix ($\nu_2 = \nu_2^0$; $\nu_3 = 0$), and with the decrease of the thermal conductivity with the formation of porosity (decrease of ν_2 and increase of ν_3). After completion of the pyrolysis of polymeric matrix ($\nu_2 = 0$), thermal conductivity grows again due to increasing thermal conductivity coefficient of the coke [23].

The thermal conductivity of the laminate through its thickness was measured by employing a comparative steady-state method against a reference sample according to ASTM E1225-87. At a steady-state condition, the thermal conductivity was derived by comparing the temperature gradient between a reference sample and a target one. The thermal conductivity of a target sample, K_t, was calculated using the following equation derived from the Fourier's law of the heat conduction:

$$K_t = K_r \frac{\Delta T_r}{\Delta T_t} \cdot \frac{\delta_t}{\delta_r} \cdot \frac{d_r^2}{d_t^2} \tag{26}$$

2.2.2.3 Specific heat

The specific heat during heating is described as the average value of the specific heats of its components [57]. Obviously, the specific heat of the sample and the volume fractions of its components are functions of temperature [23,57].

$$C = \frac{\sum_{i=1}^{3} \rho_i C_i \nu_i}{\sum_{i=1}^{3} \rho_i \nu_i} \tag{27}$$

In Eqn (27), the change of the specific heat is connected with the decreasing specific heat during thermal degradation of polymeric matrix (decrease of ν_2 and increase of ν_3). After completion of the pyrolysis of polymeric matrix ($\nu_2 = 0$), specific heat fixes constant due to the specific heat of the coke.

The specific heat capacity measurements were performed with a differential scanning calorimeter (DSC PL, Polymer Laboratory) according to ASTM D-1269. A heating rate of 10 °C/min was used. The physical and thermophysical properties of composite and nanocomposites samples at standard condition are given in Table 3.

2.2.3 Mathematical model of ablation process

Under the external high heat flux (*oxyacetylene flame test*) conditions, complex processes of heat and mass transfer take place in the ablative sample. When the sample is exposed to high-temperature and high-velocity fluid stream, decomposition of resin begins at a pyrolysis temperature and char layer ablation creates at a higher temperature. The free surface of the sample under the influence of high-temperature stream is continuously spilled and depends on the oxygen content in

Table 3 Physical and Thermophysical Properties of Asbestos Cloth/Phenolic Resin Composite and Nanocomposites at Standard Condition [23,52,53]

Properties	Dimension	Composite	NMA3	NKA3
Density	kg m^{-3}	1450	1520	1510
Density of pyrolysis gas	kg m^{-3}	0.3	0.3	0.3
Density of char	kg m^{-3}	1089	1174	1123
Specific heat of char	J kg^{-1} K^{-1}	773	791	786
Specific heat	J kg^{-1} K^{-1}	1270	3562	2947
Thermal conductivity	J m^{-1} s^{-1} K^{-1}	0.5	0.38	0.43
Thermal conductivity constant	–	1.5	1.5	1.5
Heat of ablation	kJ kg^{-1}	952	41.7	52.3

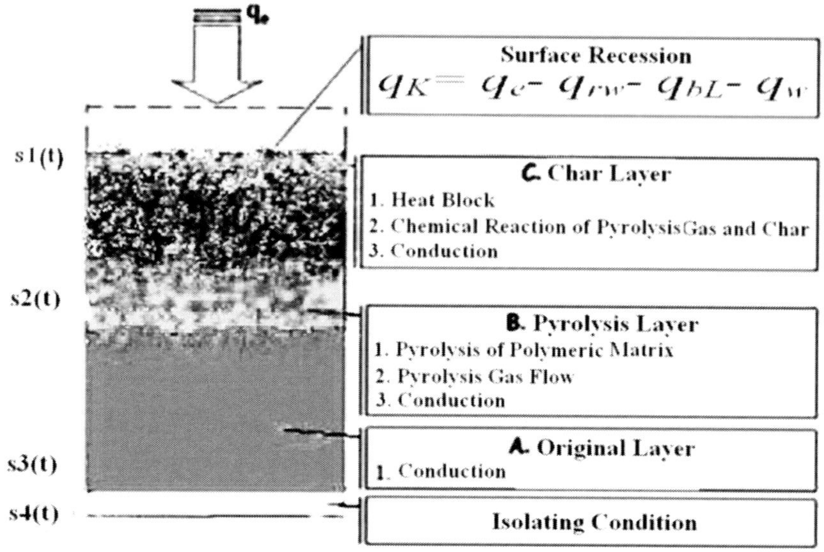

FIGURE 5

The zones within the ablating sample [23,53].

the gases, oxidized. Therefore, three zones are formed (Figure 5): the virgin material, the pyrolysis zone, and the porous char layer [8,9]. The char layer in nanocomposite samples has a dense ceramic layer [20,48,52,53].

Boundary Conditions

Ablation is a heat and mass transfer process in which a large amount of thermal energy is expended by sacrificial loss of surface material, thus restricting high environmental temperature to the surface region. The heat input from the environment is absorbed, dissipated, and blocked by numerous mechanisms. These are as follows [23,54–57]:

a. Heat conduction into the material substrate and storage by its effective heat capacity.
b. Material phase change.
c. Heat absorption by gases in the material substrate as they percolate to the surface.
d. Transpiration of gases from the ablating surface into the boundary layer.
e. Surface and bulk radiation.
f. Endothermic chemical reactions.

In its simplest form, the energy balance at the ablating surface (boundary condition) is [23,54–56]

$$q_{\mathrm{K}} = q_{\mathrm{e}} - q_{\mathrm{rw}} - q_{\mathrm{bL}} - q_{\mathrm{w}} \quad \text{at} \quad y = s_1(t) \tag{28}$$

where

$$q_K = -K\frac{\partial T}{\partial y}; \quad q_e = h(T_e - T_0); \quad q_{rw} = \varepsilon\sigma T_w^4;$$

$$q_{bL} = \gamma_{bL} G I^{-1}(q_e - q_w); \quad q_w = C_p I(T_w - T_0)$$

In this condition, q_{bl} is a transpiration cooling that is created as follows [24,53,54]. Gaseous products formed by material ablation are injected into the hot boundary layer. In diffusing through this high-temperature environment, the gases absorb heat by sensible temperature rise. The boundary layer is thus increased in thickness, and its original enthalpy or temperature is lowered. Consequently, less heat is transferred from the environment to the ablating surface.

From the theoretical and experimental results on air-to-air injection, a simple correlation formula has been derived for the transpiration number. This formula has been modified for the injection of gaseous products other than air and is given as [21,54] $\gamma_{bl} = N(M_a/M_v)^p$. In oxyacetylene flame test, N, M_v, and p are equal to 0.68–0.72, 16, and 0.55, respectively.

$$T(s_2(t), t) = T_p \tag{29}$$

The applied initial conditions for the ablative sample take the general form [23,54]:

$$T(y,t)|_{t=0} = T(y,0) = T_0 \tag{30}$$

2.2.3.1 Direct solution

In this section, the mathematical model illustrating the thermal decomposition and ablation of the polymeric heat shield material is described. The main assumptions on which the model rests are listed below [14,23,53,58–60]:

1. Heat flow only in the direction normal to the surface (y-axis).
2. The gases coming out of the bulk are in thermal equilibrium with the char.
3. Chemical reactions between the char and the gases coming out are neglected.
4. The cross-sectional area must be considered constant with the normal distance y.
5. No energy is transferred by mass diffusion.
6. There is thermal equilibrium between the decomposition gases and the solid material but there is no accumulation of these volatile gases in the solid material.
7. The feedback of the heat released by the flaming combustible volatiles in a small-scale fire resistance test (or indeed in a real postflashover fire) may be neglected owing to its relatively small contribution when compared to the heat flux created by the furnace or real fire.
8. Movement of liquid is assumed negligible compared to pyrolysis gases.
9. Pyrolysis gases may be considered "ideal gases" but their properties will be kept constant.

As mentioned earlier, it is assumed that the rate of decomposition (weight loss) is expressed by the first-order reaction and that the mechanism of polymer degradation is a single-step one [14,23]:

$$-\frac{\partial\left[\rho\left(\nu_2 X_g - V_{gas}\right)\right]}{\partial t} = k\rho\left(\nu_2 X_g - V_{gas}\right) \tag{39}$$

where $(\nu_2 X_g - V_{gas})$ represents the mass of decomposed resin with respect to mass of virgin sample. Using the Arrhenius type reaction rate constant, equations for calculating V_{gas} and $\dot{m}_g$ are obtained as follows [14,23]:

$$\frac{\partial V_{gas}}{\partial t} = k\left(\nu_2 X_g - V_{gas}\right) = A\left(\nu_2 X_g - V_{gas}\right)e^{-\frac{E}{RT}} \tag{40}$$

$$\dot{m}_g(t) = \int_0^{\delta_B} \frac{\partial\left(\rho V_{gas}\right)}{\partial t} = \int_0^{\delta_B} \rho A\left(\nu_2 X_g - V_{gas}(y,t)\right)e^{-\frac{E}{RT(y,t)}}dy \tag{41}$$

In Eqn (39), the following relations were experimentally determined by Kanevce [14,23] for X_g of resole-type phenolic resin matrix [14]:

$$\begin{aligned} X_g(T) &= -0.21322 \times 10^{-6} \times T^2 + 0.80560 \times 10^{-3} \\ &\quad \times T - 0.12527 \qquad \text{for} \quad T < 1255\ \text{K} \\ X_g(T) &= 0.55 \qquad \text{for} \quad T \geq 1255\ \text{K} \end{aligned} \tag{42}$$

For the case of a horizontal specimen exposed to a uniform heat flux from above, production of bubbles is restricted to a thin surface layer of thickness δ_B. The approximate size of δ_B may be estimated by matching the net heat flux at the top surface to the temperature gradient. The result is that [23,56]

$$\delta_B \approx \frac{K\Delta T_p}{\varepsilon q_0 + h\left(T_0 - T_p\right) + \varepsilon\sigma\left(T_0^4 - T_p^4\right)} \tag{43}$$

The last section of this work is dedicated to the numerical testing of the simulation model. All the experimental data and model equations obtained are now used in an ablation computer program in order to predict the temperature distribution profile through the thickness of the ablative material as a function of a given heat flux imposed on its external surface. A finite difference method is used to solve the conservation equations of mass and energy. While the parameters used in the solution were calculated using the method proposed.

To evaluate the thermal behavior and ablation performance of the ablative insulators, the oxyacetylene flame test is carried out according to ASTM-E-285-80. This test can be creating hot gas with 3400 K and 10×10^6 W m^{-2} external heat flux. Hot combustion gases are directed along the normal to the specimen. This test method covers the screening of the ablative materials to determine the relative thermal insulation effectiveness when tested as a flat panel, also the surface erosion when tested as a cylindrical element in an environment of a steady flow of hot gas provided by an oxyacetylene burner.

To analyze the temperature at the back of the sample (back surface temperature), a 6 × 100 × 100 mm flat sandwich panel structure made of 4 mm in thickness of sample and an aluminum substrate (2 mm in thickness) is prepared. For the oxyacetylene flame test and recording the back temperature, the nickel–chrome thermocouple was placed on the back of the sample.

The sample for surface erosion measurement has a cylindrical shape with 10 and 25 mm in diameter and height, respectively. This sample before and after oxyacetylene flame test is shown in Figure 7. $s_1(t)$ is equal to the height of cylindrical sample after flame test. Therefore, it can be measured directly. $s_2(t)$ is also an equivalent height of the sample after flame test which is experimentally determined as follows [23]:

$$s_2(t) = \frac{(m_0 - m(t))}{\rho \cdot S} \tag{44}$$

Figure 8 shows the zero and net heat flux on composite and nanocomposite sample in oxyacetylene flame test. In this test, temperature of flame and convection coefficient is 3000 K and 2960 W m^{-2} K, respectively. Therefore, the initial or zero heat flux is equal to 8.99 × 10^6 W m^{-2}. During the test time, net heat flux decreases because of surface boundary condition (Eqn (48)). Therefore, net heat flux for composite and nanocomposite are equal to 4.31 × 10^6 and 3.92 × 10^6 W m^{-2} at steady condition, respectively. This figure shows that the nanocomposite net heat flux is relatively 23.7% lower than composite net heat flux.

Figure 9 shows the char and pyrolysis surface erosion of asbestos/phenolic composite and NMA3 nanocomposite under oxyacetylene flame test condition. After 20 s, the theoretical values of the char and pyrolysis surface erosion are 1.35 and 1.77 mm for composite and 0.62 and 0.98 mm for NMA3 nanocomposite, respectively. The experimental values of these parameters were 1.9 and 2.6 mm for composite and 1.3 and 2.1 mm for NMA3 nanocomposite, respectively. The most

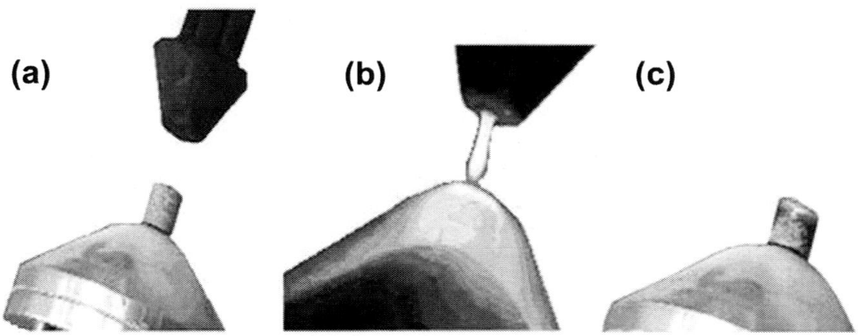

FIGURE 7

Measuring the surface erosion by oxyacetylene flame test (a) before, (b) due to, and (c) after surface erosion test [23].

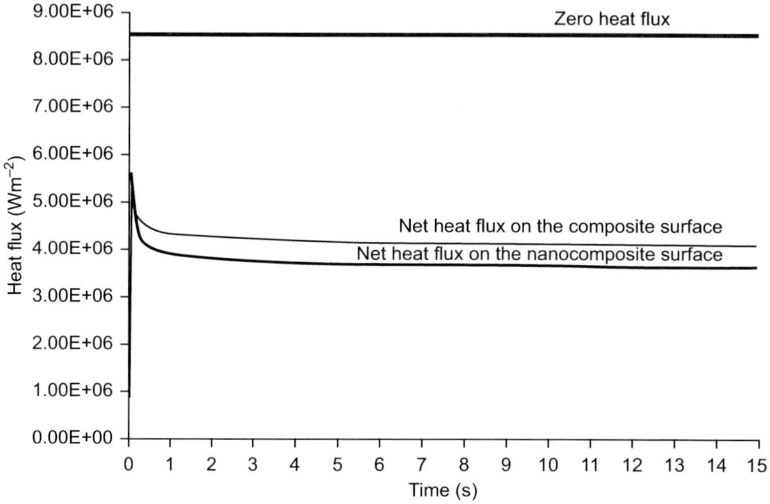

FIGURE 8

Zero and net heat fluxes which were applied on the surface of asbestos/phenolic composite and its nanocomposite counterpart.

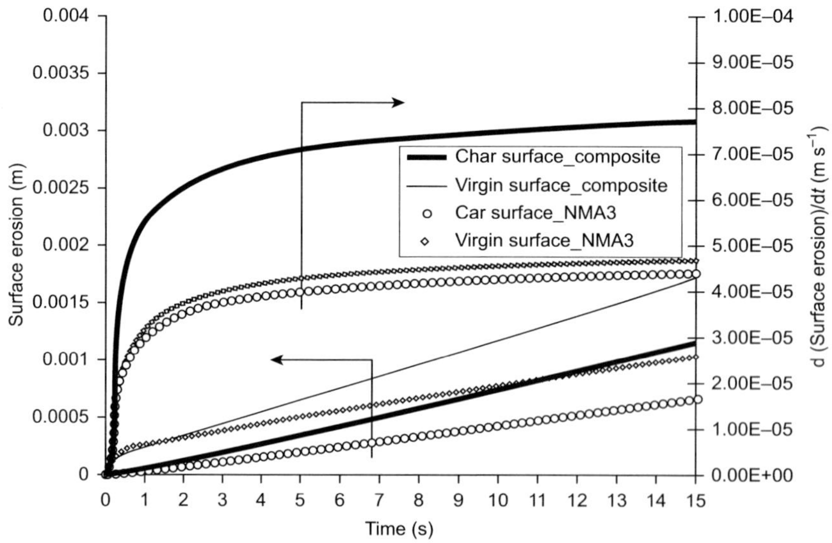

FIGURE 9

Char and pyrolysis surface erosion of composite and nanocomposite under oxyacetylene flame test condition.

important reason for this difference is the omission of the effect of fluid stream force of oxyacetylene flame in the calculation of the surface erosion.

Figure 10 shows the temperature distribution through the thickness of asbestos/phenolic composite based on ablation model prediction. Dot points are back surface temperature that is measured experimentally in oxyacetylene flame test. Test time duration is 20 s, front surface temperature rises to 1887 K, and external convectional heat flux is 8×10^6 W m^{-2}. After 20 s, the model predicts the back surface temperature of a 4 mm thickness asbestos/phenolic composite of 468 K, while oxyacetylene flame tests show these temperatures of 431, 438, and 423 K. This figure shows a difference between the model and the experimental results. The main reason of this difference is also the omission of the effect of fluid stream force of oxyacetylene flame that causes the mechanical erosion of the surface. Creation of the melting phase from the asbestos cloth and the assumptions in the evaluation of the thermophysical properties with increase of temperature in ablation model could also contribute to this difference.

2.2.3.2 Inverse analysis technique

To model the nanocomposite ablation process and estimate the effective thermal diffusivity, we used the inverse solution analysis technique. For the solution of the inverse problem of parameter estimation used for the identification of the effective thermal diffusivity of a resole-type phenolic resin—asbestos cloth composite and its layered silicate nanocomposites, we identified that [53]

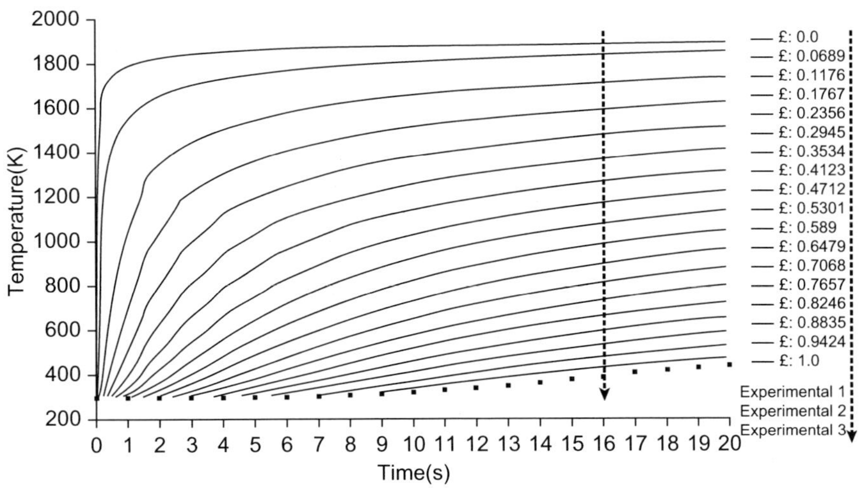

FIGURE 10

Temperature distributions through the thickness of asbestos/phenolic composite; dot points are back temperature that is measured in oxyacetylene flame test [23].

$$\frac{\mathrm{d}}{\mathrm{d}t}=\frac{\partial}{\partial t}+v_s\frac{\partial}{\partial y}$$

$$K = \text{constant (in each section of figure 5).} \tag{45}$$

$$\frac{\mathrm{d}\rho}{\mathrm{d}t}=\frac{\mathrm{d}\rho}{\mathrm{d}T}\cdot\frac{\mathrm{d}T}{\mathrm{d}t}$$

By the combination of Eqns (29) and (43), we have

$$\left(\rho C+\Delta H_p\frac{\mathrm{d}\rho}{\mathrm{d}T}\right)\frac{\partial T}{\partial t}+\left[C_g\dot{m}_g+v_s\left(\rho C+\Delta H_p\frac{\mathrm{d}\rho}{\mathrm{d}T}\right)\right]\frac{\partial T}{\partial y}=K\frac{\partial^2 T}{\partial y^2} \tag{46}$$

It can be defined that

$$\Psi(T)=\rho C+\Delta H_p\frac{\mathrm{d}\rho}{\mathrm{d}T}$$

$$H(T,v_s)=C_g\dot{m}_g+v_s\Psi(T)$$

$$\frac{K}{\Psi(T)}=\alpha_{\mathrm{eff}}(T) \tag{47}$$

$$\mathrm{Bi}=\frac{H(T,v_s)\cdot s(t)}{K}$$

Hence, we can rewrite Eqn (43) as follows [53]:

$$\frac{1}{\alpha_{\mathrm{eff}}}\frac{\partial T}{\partial t}+\frac{\mathrm{Bi}}{s(t)}\frac{\partial T}{\partial y}=\frac{\partial^2 T}{\partial y^2} \tag{48}$$

The sample for measuring the temperature distribution through the thickness and data fitting to Eqn (46) has a sandwich structure with a thickness of 8 mm for either composite or nanocomposite. Therefore, the sample is a flat panel with the dimensions $8 \times 100 \times 100$ mm. For oxyacetylene flame test, the nickel–chrome thermocouple was placed in along the thickness of sample. The setup of oxyacetylene flame test and its sample are shown in Figures 11 and 12.

Figure 13 shows the temperature distribution of top surface and through the thickness of asbestos cloth/phenolic resin composite and NMA3 nanocomposites in oxyacetylene flame test. Test time duration is 20 s. As observed in Figure 13, we named the regions between the temperature profile from top surface to bottom surface as sections one, two, and three, respectively. By comparing the temperature distribution in these sections, we can see the following:

1. The steady-state surface temperature of asbestos cloth/phenolic resin composite and NMA3 nanocomposite are around 1500 and 2100 °C at the same condition, respectively. This difference clearly shows that the nature of the top surface of the composite and that of the nanocomposite are different. It is known that the greater value of top surface temperature of the NMA3 nanocomposite is one of

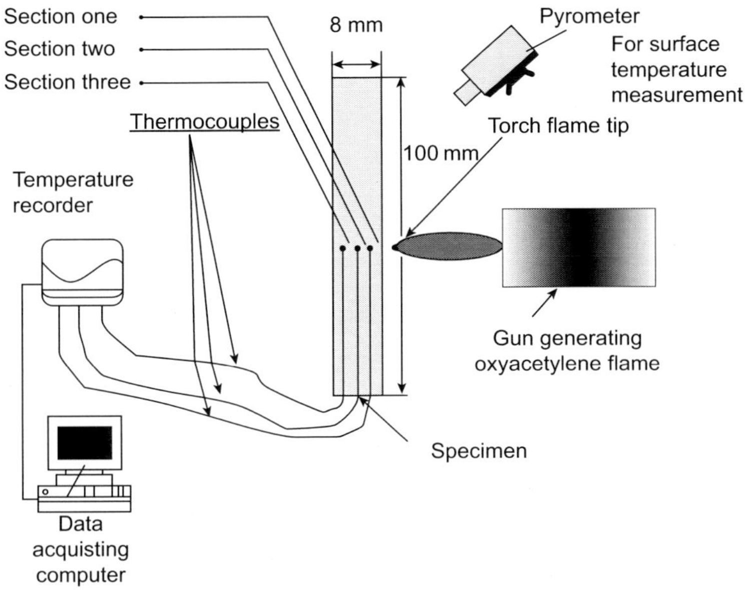

FIGURE 11

The oxyacetylene flame test setup [53].

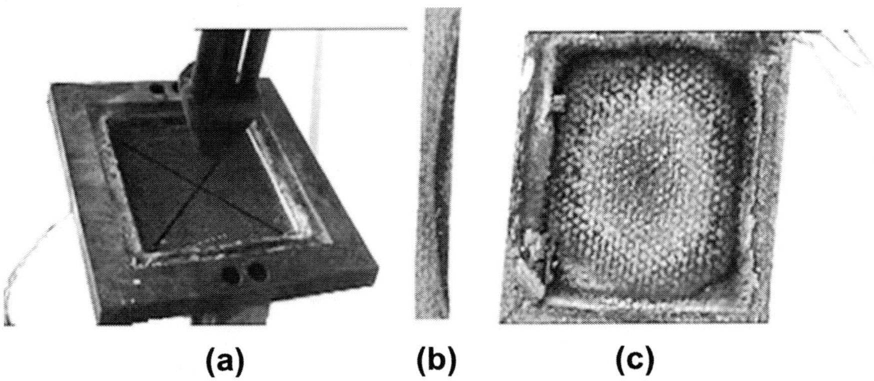

FIGURE 12

The sample for oxyacetylene flame test (a) before, (b) during, and (c) after test [53].

the key factors causing the higher insulation performance compared to asbestos cloth/phenolic resin composite. The radiation heat flux from surface increases by increasing the top surface temperature.

2. The temperature profile difference between section one of the composite and nanocomposite is more than sections two and three in each sample. On the other hand, as shown in Figure 13, section one of the composite and nanocomposite

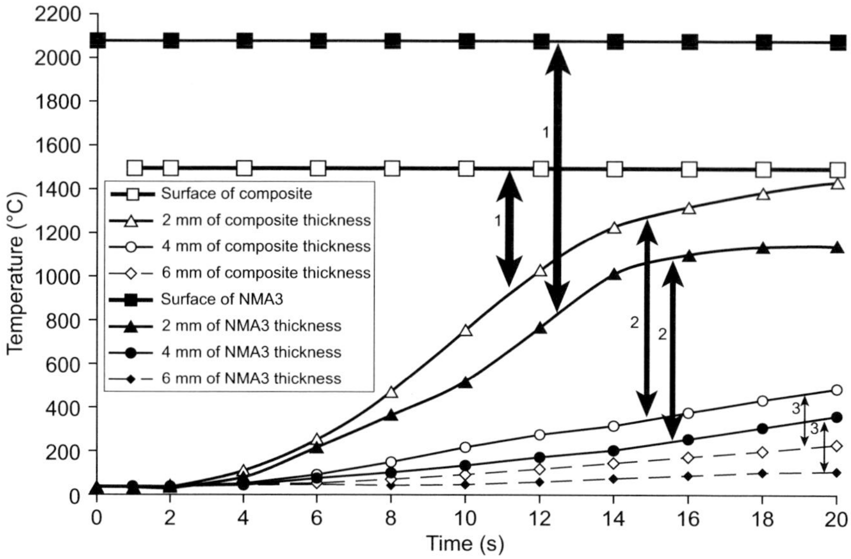

FIGURE 13

Experimental temperature distributions of top surface and along a thickness of composite and nanocomposite sample (NMA3) in oxyacetylene flame test [53].

sample suffers most thermal degradation after oxyacetylene flame test. It means that the major difference between ablation performance of the composite and of the nanocomposite is a consequence of thermal degradation, char forming, and nature of nanocomposite char layer.

Figure 14 shows the change of effective thermal diffusivity in the three sections of composite and NMA3 nanocomposite samples. These curves were obtained from numerical solution of Eqn (46) by considering the curves of Figure 13 as boundary conditions. From this figure, the following are observed:

1. In section one in which char layer is created due to high-temperature ablation, the effective thermal diffusivity of composite is twice than the NMA3 nanocomposite. This result is attributed to difference in char layer nature and ablation mechanism of composite and nanocomposite at high temperature.
2. The effective thermal diffusivity difference for the sections one and two of the composite and of the nanocomposite is significant while it is not so much for sections three of aforementioned samples. As shown in Figure 12, under oxyacetylene flame test, thermal degradation mostly occurs in sections one and two of samples. It means that the major difference between ablation performance of the composite and the nanocomposite is a consequence of the nature of nanocomposite char and pyrolysis layers.

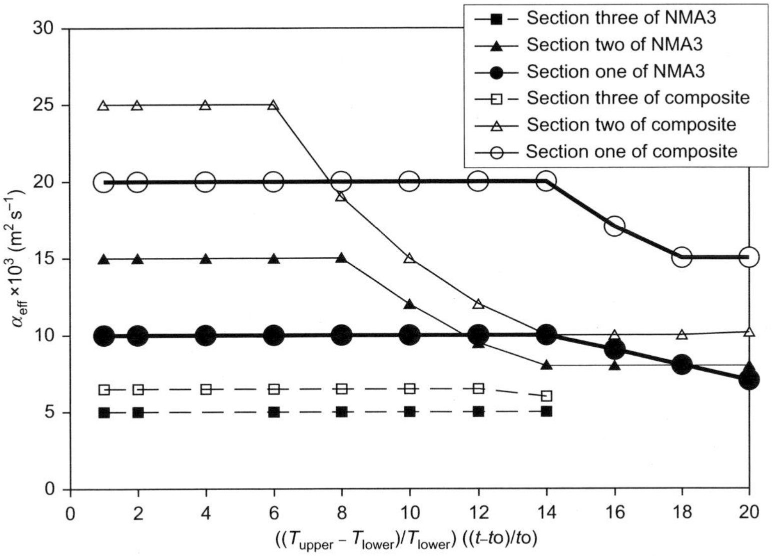

FIGURE 14

The change of effective thermal diffusivity in three sections of composite and nanocomposite (NMA3) samples [53].

3. As observed in Figure 14, in section three of samples which remained intact during oxyacetylene flame test, the effective thermal diffusivity is constant and lower than that of the others. The effective thermal diffusivities of composite and NMA3 nanocomposite in this section are close enough and equal to 6.5×10^{-8} ($m^2 s^{-1}$) and 5×10^{-8} ($m^2 s^{-1}$), respectively.

Figures 15 and 16 show the temperature distribution (as a function of the thickness) of asbestos cloth/phenolic resin composite, and NMA3 nanocomposite samples in oxyacetylene flame test, respectively. In these figures, dot points and lines are the experimental and theoretical results, respectively. Because of a good agreement between the theoretical and experimental results, the values of effective thermal diffusivity are reliable.

Generally, by considering the results of the oxyacetylene flame test, it can be concluded that the nanocomposite has a higher ablative performance than the composite sample. But this test cannot give proper evidence for this feature. The cone calorimetry and scanning electron microscopy are complementary techniques for this purpose. Experiments were conducted using the cone calorimeter which is an internationally recognized testing apparatus. Cone calorimetery was performed on an Atlas CONE2 according to ASTM E 1354 at a heat flux of 8×10^4 W m^{-2}, which is a normal irradiance level for the evaluation of the flammability, using a cone-shaped heater.

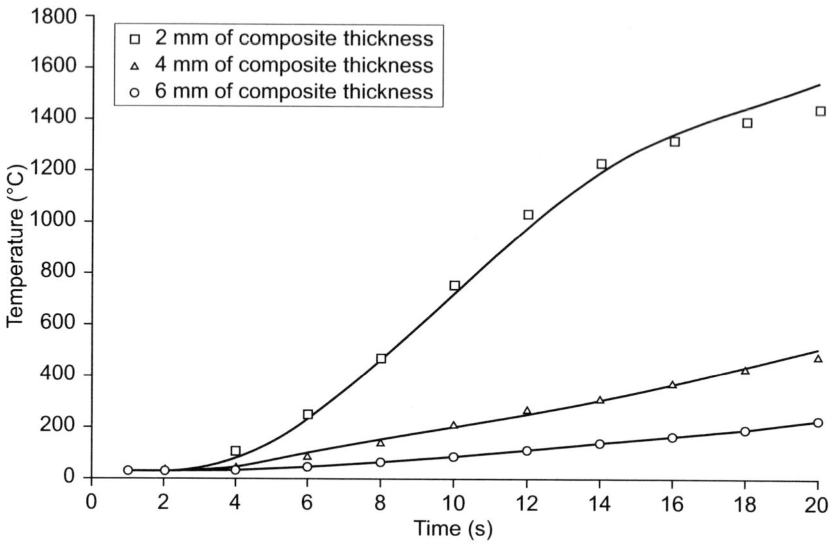

FIGURE 15

Temperature distribution through the thickness of composite sample in oxyacetylene flame test: dot points and lines are experimental and theoretical results, respectively [53].

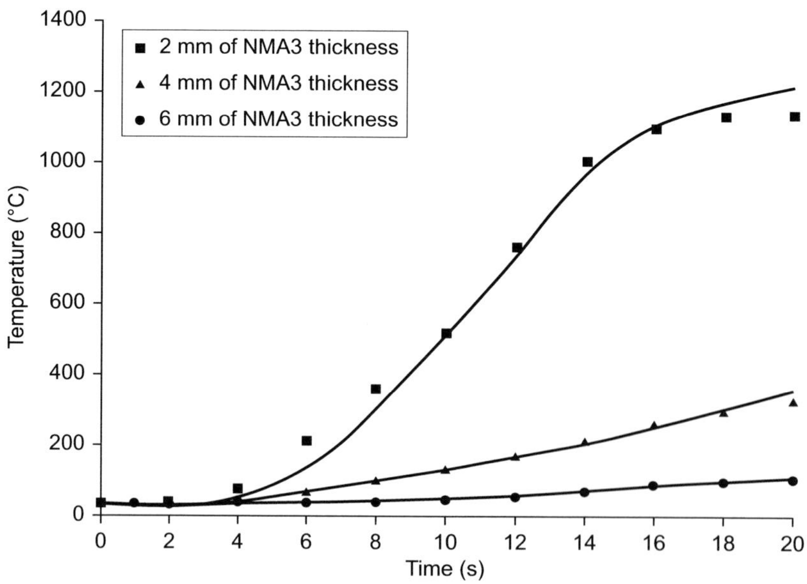

FIGURE 16

Temperature distribution through the thickness of nanocomposite sample (NMA3) in oxyacetylene flame test: dot points and lines are experimental and theoretical results, respectively [53].

2.2.4 Mathematical model of burning process

The cone calorimeter is considered to be the most significant bench scale instrument in fire testing. This apparatus has been adopted by the International Organization for Standardization for measuring the heat release rate (HRR) of a sample [61–63]. HRR is based on the fact that the consumed oxygen during combustion is proportional to the released heat. This device analyses the combustion gases and measures the produced smoke from a test specimen being exposed to a certain heat flux; the oxygen concentration is analyzed to calculate the released heat. The data collected from this bench scale real fire test can be used for fire modeling and the prediction of real-scale fire behavior [64].

To clarify further the objective of this section, it is required to completely explain the principle of burning and flammability process. Flaming combustion can be divided into physical and chemical processes taking place in each of three separate phases: gas, mesophase, and condensed phase [61,62]. Figure 17 shows the schematic diagram of a horizontal polymer slab that is burning with a flame. The physical processes are shown on the left-hand side, which include (1) energy transport by radiation and convection between the gas phase and the mesophase and (2) energy loss from the mesophase by mass transfer and conduction into the solid.

The important chemical processes are indicated on the right-hand side, which include (1) thermal degradation of the polymer in a thin surface layer as a consequence of the physical processes, (2) mixing of the volatile pyrolysis products with air by diffusion, and (3) combustion of the fuel–air mixture in a combustion zone that produces radiant energy over a spectrum of wavelengths including visible. The combustion zone is bounded by a fuel-rich region on the inside and a fuel-lean region on the outside [61].

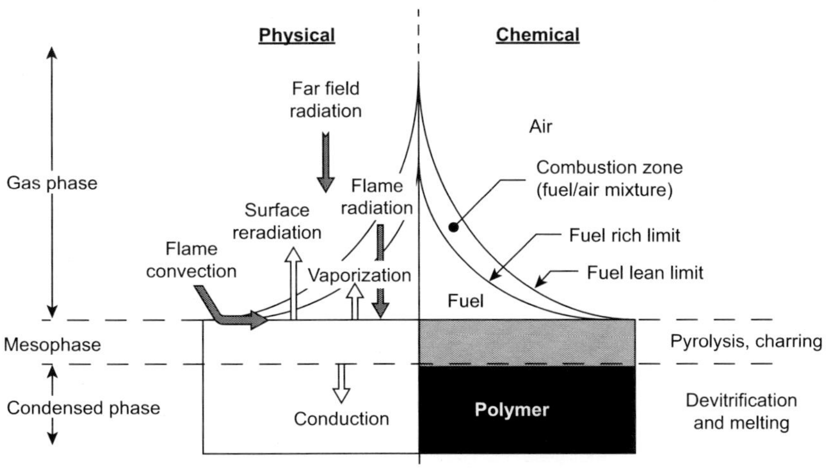

FIGURE 17

Physical and chemical processes in the flaming combustion of polymers [61].

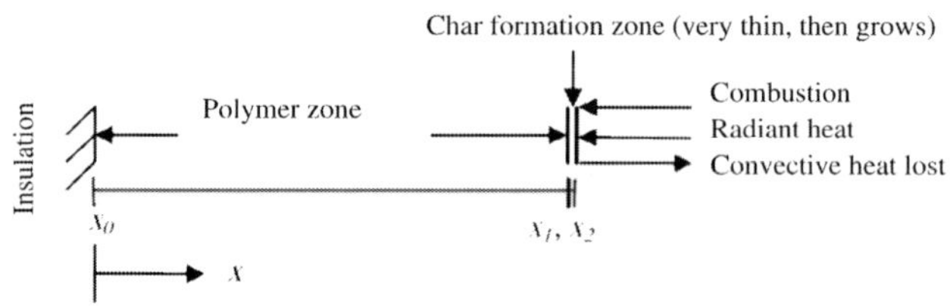

FIGURE 18

Geometry of physical model [62].

Due to burning, the polymer thermally degraded into gas and char. To model the problem, two distinct regions have been selected to account for the physics involved. The polymer zone accounts for the pyrolysis reaction; polymer is consumed and gas is produced [62].

Figure 18 shows the physical situation involved, where $x_0 - x_1$ is the polymer zone, and $x_1 - x_2$ is the char formation zone.

During this reaction, the mass of polymer is consumed and produces a fraction, α, of gas and, the remaining char. The first-order reaction rate for polymer mass consumption is [61–64]

$$\frac{\partial m}{\partial t} = -k \cdot m \tag{49}$$

The rate constant for the pyrolysis reaction, k, is a function of temperature described by the Arrhenius relationship:

$$k = A \cdot \exp\left(\frac{-E}{RT}\right) \tag{50}$$

In this problem, in reality, the boundary, x_1, would move to the left as polymer is consumed. It is assumed that polymer zone, $x_0 - x_1$, is constant, thus the volume is constant, while the concentration changes with time. Hence, the Eqn (47) [65] can be rewritten:

$$r = \frac{\partial c}{\partial t} = -k \cdot c \tag{51}$$

The production of gas in the polymer zone ($x_0 - x_1$) must also diffuse through this zone to reach the char formation zone ($x_1 - x_2$). Thus, the mass balance on the gas species that is produced [21,22] is

$$r_{\mathrm{Gp}} = \frac{\partial c_{\mathrm{G}}}{\partial t} - D_{\mathrm{polymer}} \cdot \frac{\partial^2 c_{\mathrm{G}}}{\partial x^2} = \alpha \cdot k \cdot c \tag{52}$$

When the pyrolysis gas reaches the char formation zone [$x_1 - x_2$], it must diffuse to the outside boundary. Writing the mass balance in the char formation zone [65],

$$\frac{\partial c_G}{\partial t} - D_{\text{char}} \cdot \frac{\partial^2 c_G}{\partial x^2} = 0 \tag{53}$$

In the polymer zone, heat is transferred by conduction. Heat of volatilization from pyrolysis and heat from other chemical processes are also shown [65–67]:

$$\rho_{\text{polymer}} \cdot Cp_{\text{polymer}} \cdot \frac{\partial T}{\partial t} - K \cdot \frac{\partial^2 T}{\partial x^2} = -\Delta H \cdot k \cdot c + \Delta H_e \tag{54}$$

In the char zone ($x_1 - x_2$), heat is transferred by conduction. The energy balance leads to

$$\rho_{\text{char}} \cdot Cp_{\text{char}} \cdot \frac{\partial T}{\partial t} - K_{\text{char}} \cdot \frac{\partial^2 T}{\partial x^2} = 0 \tag{55}$$

At the char surface, heat is introduced by radiation. Some of this heat is reflected back. Combustion is taking place on the surface and a fraction of this heat of reaction is fed back to the boundary. Heat also leaves the surface to the outside by convection. At this boundary, the heat flux entering the surface is described as

$$Q_{\text{surface}} = \varphi \cdot \Delta H_1 \cdot \left(-\frac{\partial c_G}{\partial x} \right)\bigg|_{\text{surface}} \cdot D_{\text{char}} - h \cdot (T - T_{\text{atm}}) + \varepsilon \cdot \sigma \cdot \left(T_{\text{cone}}^4 - T^4 \right) \tag{56}$$

The heat of combustion term, ΔH_1, can be found from the cone calorimeter experimental data by integrating the HRR versus time.

The HRR versus time is measured by heat of combustion multiplied by the flux of gas diffusing out of the surface [66,67]:

$$\text{HRR} = \Delta H_1 \cdot \left(-\frac{\partial \dot{c}_G}{\partial x} \right)\bigg|_{\text{surface}} \cdot D_{\text{char}} \tag{57}$$

The rate of mass growth of char is equal to the same polymer pyrolysis kinetics which can be equated as [65–67]

$$\frac{\partial m_c}{\partial t} = -(1 - \alpha) \cdot \frac{\partial m}{\partial t} = (1 - \alpha) \cdot k \cdot m \tag{58}$$

Dividing by ρ_{char}, the left-hand side becomes

$$\frac{1}{\rho_{\text{char}}} \frac{\partial m_c}{\partial t} = \frac{\partial V}{\partial t} = \text{Area} \cdot \frac{\partial x}{\partial t} \tag{59}$$

where V is the volume, Area is the area, and x is the displacement of the char. By dividing both sides by *Area* and multiplying the right-hand side top and bottom by a constant thickness, χ, which is the distance between x_0 and x_1:

$$\frac{\partial x}{\partial t} = -\frac{(1 - \alpha) \cdot \chi}{\rho_{\text{char}} \cdot \text{Area} \cdot \chi} \frac{\partial m}{\partial t} = \frac{(1 - \alpha) \cdot k \cdot \chi}{\rho_{\text{char}}} \frac{m}{\text{Area} \cdot \chi} \tag{60}$$

By considering that (Area χ) is a constant volume, $m/(\text{Area } \chi) = c$ and $m_G/(\text{Area } \chi) = c_G$. Therefore, we can consider Eqn (60) as a pyrolysis velocity.

Initial conditions for each zone of the problem are as follows:

$$\begin{aligned} c_G(x,0) &= 0 \\ T(x,0) &= T_{\text{initial}} \end{aligned}$$

Boundary conditions were set at the x_0 and x_2 boundaries. At x_1, all boundary conditions are assumed to be constant interfacial boundary conditions where temperature and concentration are held constant at the interface (pyrolysis temperature). Boundary condition at x_2 is Eqn (56).

Experimental and modeling results of the mass loss and HRR for asbestos cloth/phenolic composite and NKA3 and NMA3 nanocomposites at 8×10^4 W m^{-2} external heat flux are shown in Figures 19 and 20, respectively. These figures show that composite and nanocomposites exhibit the same thermal behavior before hundredth second of flame test. After that, by increasing the surface temperature around 1000 °C, rate of mass loss and HRR peak of nanocomposite were decreased. Generally, the NKA3 and NMA3 nanocomposites have a 33% and 51% lower HRR and 19% and 40% lower mass loss than the asbestos cloth/phenolic composite, respectively. Good agreements between experimental and theoretical results are observed in these figures. Data in Tables 2–4 were used to theoretical modeling of mass loss and HRR in cone colorimeter test conditions.

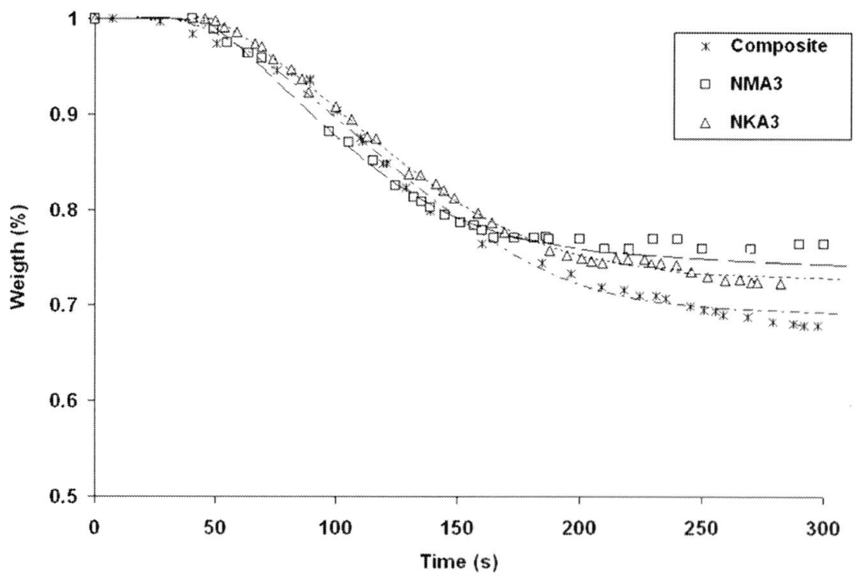

FIGURE 19

Experiment and modeling results of the mass loss for asbestos cloth/phenolic composite and NKA3 and NMA3 nanocomposites [67].

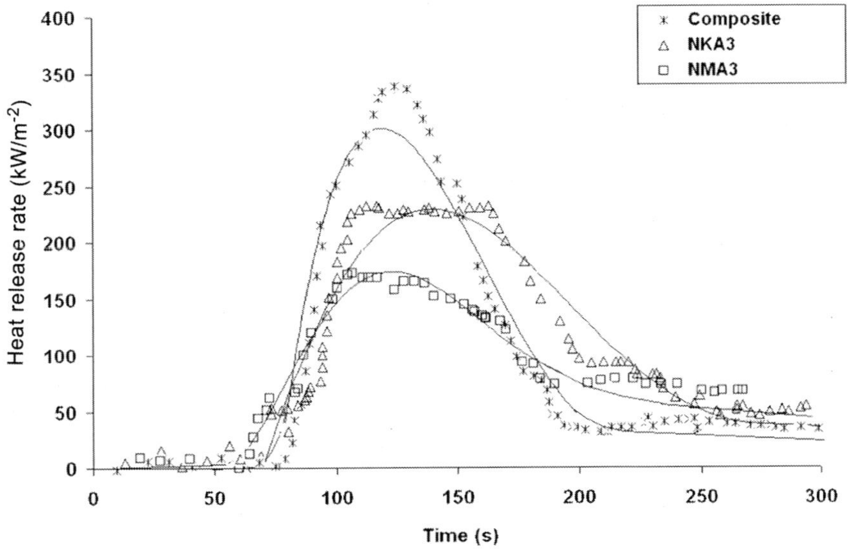

FIGURE 20

Experiment and modeling results of the heat release rate HRR for asbestos cloth/phenolic composite and NKA3 and NMA3 nanocomposites [67].

Figures 21–23 show the scanning electron micrographs of the asbestos–phenolic composite, and NKA3 and NMA3 nanocomposite samples after the cone calorimeter test at 8×10^4 W m^{-2} external heat flux, respectively. These figures are illustrating top surface (surface of char region), lateral surfaces, and heat flux direction. The characteristic ablation regions—virgin material, porous reaction zone, and dense char layer—are apparent in all of the samples. Therefore, the microstructure of nanocomposite chars implies that increasing clay content results in a tougher char. In these

Table 4 Thermophysical Properties of Asbestos Cloth/Phenolic Resin Composite and Nanocomposites at Cone Colorimeter Test Conditions [67]

Properties	Dimension	Composite	NMA3	NKA3
$D_{polymer}$	m^2 s^{-1}	8	8	8
D_{char}	m^2 s^{-1}	1.5×10^{-6}	1.5×10^{-6}	1.5×10^{-6}
ΔH	J kg^{-1}	2.4×10^4	1.3×10^4	1.8×10^4
Cone temperature	K	1000	1000	1000
Initial temperature	K	300	300	300
Time to ignition	s	67	72	69
Peak of HRR values	kW m^{-2}	342	164	228
Time of HRR peak	s	123	108	112

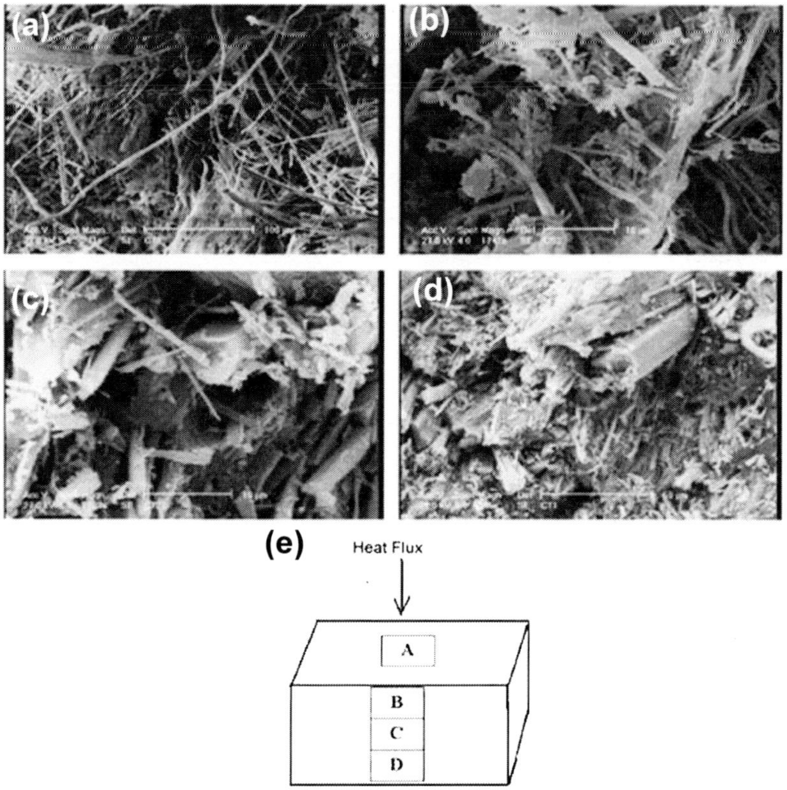

FIGURE 21

Scanning electron micrographs of the asbestos–phenolic composite sample after cone calorimetry test: (a) top surface; (b), (c), and (d) lateral surfaces; (e) illustration of heat flux direction [67].

chars, a nanoscopic packed plate-like morphology are observed (Figures 22-a and 23-a). In general, the spatially uniform arrangement of the silicate layers on an ultrafine nanometer level facilitates the formation of a spatially uniform inorganic char. This dense clay char caused the superior performance of nanocomposites at high temperature.

The XRD analysis shows that the mineral compositions of composite due to flame test change from clinochrysotile ($Mg_3Si_2O_5(OH)_4$) to forsterite (Mg_2SiO_4). The mineral compositions of NKA3 nanocomposite after cone calorimetry flame test are containing forsterite (Mg_2SiO_4) and aluminum silicate ($Al_2O_3 \cdot SiO_2$). The compositions of NMA3 nanocomposite after flame test are containing forsterite (Mg_2SiO_4) and sodium aluminum silicate ($NaAlSiO_4$).Therefore, it can be concluded that the char layer of ablative nanocomposite contains the ceramic based on aluminum silicate.

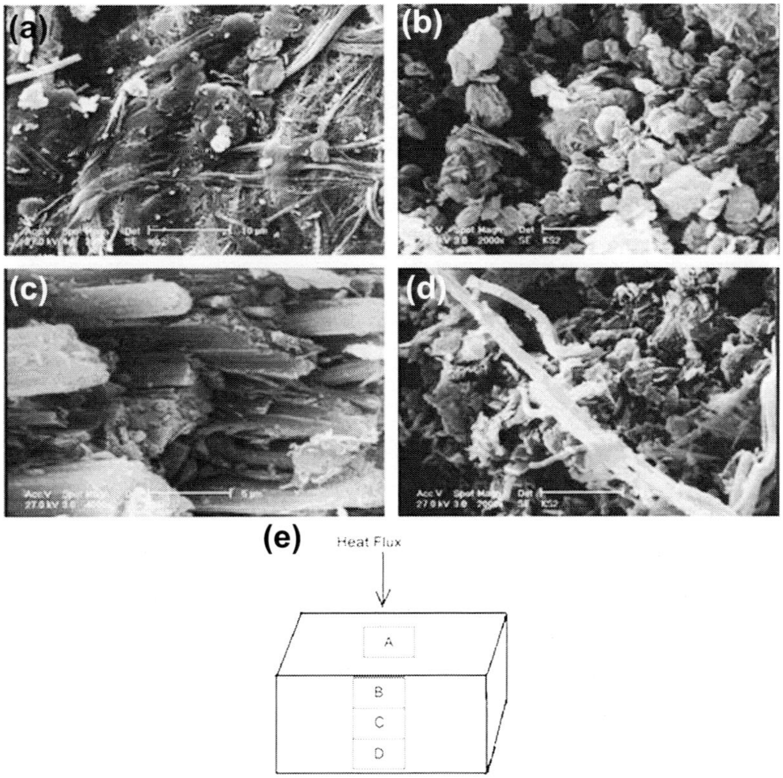

FIGURE 22

Scanning electron micrographs of NKA3 nanocomposite sample after cone calorimetry test: (a) top surface; (b), (c), and (d) lateral surfaces; (e) illustration of heat flux direction [67].

3. CONCLUSIONS

Polymer clay nanocomposites represent a new class of ablative materials that possesses unique thermal properties arising from the nanoscale dispersion of nanoparticles in the polymer matrices. A relatively tough, inorganic ceramic layer forms during the ablation of the nanocomposites, which improves ablative performance of nanocomposites relative to the net polymer or traditional filled systems. This enhanced ablative performance is not related to an alteration of decomposition kinetics associated with the presence of the silicate layers. Thus, the role of the dispersed silicate depends critically on the specific mechanisms associated with the ablation process due to nanocomposites surface uniform inorganic ceramic layer.

In this chapter, a mathematical model describing the peculiarities of the high-temperature properties of ablative heat shield was proposed. The model demonstrates the variation of thermophysical properties such as effective thermal

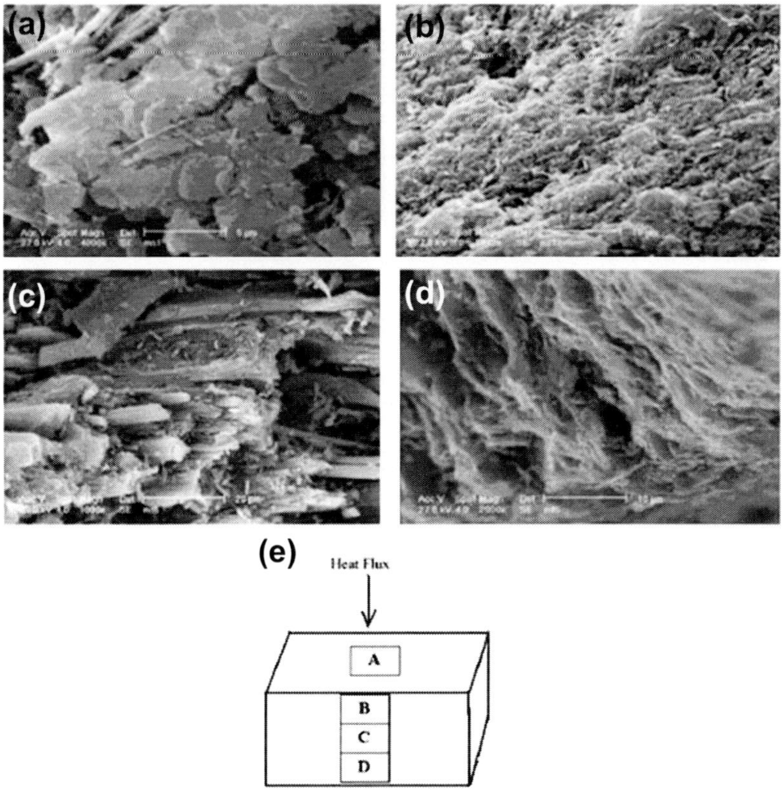

FIGURE 23

Scanning electron micrographs of NMA3 nanocomposite sample after cone calorimetry test: (a) top surface; (b), (c), and (d) lateral surfaces; (e) illustration of heat flux direction.

diffusivity, thermal conductivity, density, and specific heat at different temperatures. This model propounds a simple tool which allows ablation process simulation in designing the thickness of a heat shield.

Nomenclature

A	Frequency factor (s^{-1})
A_c	Frequency factor for char formation (s^{-1})
A_g	Frequency factor for gas formation (s^{-1})
Bi	Biot number
C	Heat capacity ($J\,kg^{-1}\,K^{-1}$)
c	Concentration of polymer ($kg\,m^{-3}$)
c_G	Concentration of pyrolysis gases ($kg\,m^{-3}$)
C_1	Heat capacity of the fiber ($J\,kg^{-1}\,K^{-1}$)

C_2	Heat capacity of the polymer (J kg^{-1} K^{-1})
C_3	Heat capacity of the coke (J kg^{-1} K^{-1})
C_g	Heat capacity of the gas (J kg^{-1} K^{-1})
C_p	Heat capacity of the air (J kg^{-1} K^{-1})
Cp	Heat capacity of unreacted material (J kg^{-1} K^{-1})
C1	Primary char
C2	Secondary char
d_r	Diameter of a reference sample (m)
d_t	Diameter of a target sample (m)
D_{char}	Diffusion coefficient of pyrolysis gases through char (m^2 s^{-1})
$D_{polymer}$	Diffusion coefficient of pyrolysis gases through polymer (m^2 s^{-1})
E	Activation energy of thermal degradation (J mol^{-1})
E_c	Activation energy for char formation (J mol^{-1})
E_g	Activation energy for gas formation (J mol^{-1})
$f(\alpha)$	Conversion functional relationship
G	Mass rate of gaseous products of thermodecomposition (kg s^{-1} m^{-2})
G1	Primary gas
G2	Secondary gas
h	Heat convection coefficient (J m^{-2} K s^{-1})
h_0	Enthalpy of virgin material (J kg^{-1})
h_1	Enthalpy of fiber (J kg^{-1})
h_3	Enthalpy of char (J kg^{-1})
h_g	Enthalpy of gas (J kg^{-1})
h_p	Enthalpy of polymer (J kg^{-1})
HRR	Heat Release Rate (W m^{-2})
I	Enthalpy of surrounding hot gas (J kg^{-1})
I^*	Reactive intermediate
J	Mass rate of pyrolysis of the matrix (kg m^{-3} s^{-1})
J_0	Intensity of volumetric ablation (kg m^{-3} s^{-1})
K	Thermal conductivity (J m^{-1} s^{-1} K^{-1})
K_0	Zero thermal conductivity (J m^{-1} s^{-1} K^{-1})
k	Rate constant (s^{-1})
k_C	Rate constant for char formation (s^{-1})
k_g	Rate constant for gas formation (s^{-1})
k_p	Rate constant for polymer formation (s^{-1})
L_0	The initial thickness (m)
m	Weight of sample (kg)
m_c	Mass of char (kg)
m_G	Mass of pyrolysis gas (kg)
m_0	Initial weight of the sample (kg)
m_f	Final weight of the sample (kg)
$\dot{m}_g$	Specific pyrolysis mass flow rate (kg m^{-2} s^{-1})
M_{wg}	Average molecular weight of hot gas
n	Degree of thermal degradation reaction
n_K	Experimental constant of thermal conductivity
p	Gas pressure (Pa)
p_0	Standard pressure (equal to 0.1 MPa)
Q	Heat of combustion per unit volume per unit time (W m^{-3})

q_0	Zero heat flux ($J\,s^{-1}\,m^{-2}$)
q_{bl}	The convective heat flux carried off from the heated surface at the expense of injection into the boundary layer of gaseous products of the thermodecomposition ($J\,s^{-1}\,m^{-2}$)
q_e	The convective heat flux carried off from the surrounding into the surface ($J\,s^{-1}\,m^{-2}$)
q_K	The heat flux consumed for warming the sample due to heat conduction ($J\,s^{-1}\,m^{-2}$)
Q_p	Internal heat source
q_{rw}	The radiation heat flux carried off from the surface at the expense of own radiation of the surface ($J\,s^{-1}\,m^{-2}$)
$\dot{Q}_{surface}$	Heat flux at surface ($J\,m^{-2}\,s^{-1}$)
q_w	The convective heat flux carried off from the heated surface heated up to temperature T_w into the surrounding ($J\,s^{-1}\,m^{-2}$)
R	Universal gas constant ($J\,kg^{-1}\,mol^{-1}\,K^{-1}$)
r	Rate of polymer consumption during pyrolysis ($kg\,m^{-3}\,s^{-1}$)
r_{GP}	Rate of gas evolution during pyrolysis ($kg\,m^{-3}\,s^{-1}$)
s	Char moving boundary (m)
s_2	Pyrolysis moving boundary (m)
s_3	Back of insulator fixed boundary (m)
s_4	Back of substrate fixed boundary of (m)
S	Surface area of cylindrical sample (m^2)
t	Time (s)
T	Absolute temperature (K)
T_{atm}	Temperature of atmosphere (K)
T_{Cone}	Temperature of cone heater (K)
T_g	Glass transition temperature (K)
T_e	Hot gas temperature (K)
T_w	Surface temperature (K)
V	Volatile
V_{gas}	The mass fraction of the resin transformed to the gas in respect to the mass of sample
X_g	The mass fraction of the resin which may be decomposed in respect to the total mass of resin
y	Distance (m)
y_c	Yield of charring
α	Degree of weight loss
α_{eff}	The effective thermal diffusivity ($m\,s^{-1}$)
β	Heating rate ($K\,s^{-1}$)
λ	Local mass fraction of polymer
ΔH_p	Heat of ablation ($J\,kg^{-1}$)
ΔH_e	Heat released or absorbed in other chemical processes in sample ($J\,kg^{-1}$)
ΔH_0	Heat of volatilization of polymer ($J\,kg^{-1}$)
ΔH_1	Heat of combustion ($J\,kg^{-1}$)
ΔT_p	Temperature difference between initial and final thermal degradation in thermogravimeter analysis (K)
ΔT_r	Temperature difference between two junction of thermocouples used in reference sample (K)

ΔT_t	Temperature difference between two junction of thermocouples used in target sample (K)
Δy	The distance differentiate (m)
δ_B	A thin surface layer of thickness that produces the bubbles (m)
δ_r	Distance between two junction of thermocouples used in reference sample (m)
δ_t	Distance between two junction of thermocouples used in target sample (m)
Φ	Percent heat transferred by heat of combustion
ε	Emission factor
ϕ	Expansion coefficient
ζ	Landau transformation
ν_1	Volumetric concentration of fibers in sample
ν_2	Volumetric concentration of polymer in sample
ν_3	Volumetric concentration of coke that forms in the matrix during pyrolysis
ν_4	Volumetric concentration of pores in the sample
ν_4^0	Volumetric concentration of initial porosity at the initial temperature (293 K)
ν_p	Volumetric concentration of unreacted material
Γ	Gasification coefficient of the matrix
γ_{bl}	Pyrolysis gas blowing coefficient
ρ	Density of sample (kg m^{-3})
ρ_g	Density of the pyrolysis gas (kg m^{-3})
ρ_1	Density of the fiber (kg m^{-3})
ρ_2	Density of the polymer (kg m^{-3})
ρ_3	Density of the coke (kg m^{-3})
ρ_p	Density of unreacted material (kg m^{-3})
σ	Stefan Boltzmann constant (J m^{-2} K^{-4})

Acknowledgments

The authors would like to appreciate Tarbiat Modares University for the support of the present work.

References

[1] Rasky DJ, Tran HK. Low-cost entry system for future planetary exploration missions. Acta Astronaut 1999;45:347–55.

[2] Kline GM. Behaviour of reinforced plastics at very high temperatures-Part 2. Mod Plast 1959;21:137–9.

[3] Resch CL. Ablation models of thermal protection materials. Johns Hopkins APL Tech D 1992;13:426–31.

[4] Cho D, Yoon BI. Microstructural interpretation of the effect of various matrices on the ablation properties of carbon-fibre-reinforced composites. Compos Sci Technol 2001; 61:271–80.

[5] Shen HS, Zheng JJ, Huang XL. Dynamic response of shear deformable laminated plates under thermomechanical loading and resting on elastic foundations. Compos Struct 2003;60:57–66.

[6] Beusman CC, Wesman J. Comparison of transpiration and ablation cooling. ARS J 1960;14:573–5.
[7] Atsios MD, Paros B. Flammability properties for charring materials. Fire Saf J 2003;38: 219–28.
[8] Reejith PS, Kirshnamurthy R, Narayanasamy K. Studies on the machining of carbon phenolic ablative composite. J Mater Process Technol 1999;88:43–50.
[9] Torre L, Kenny JM, Maffezzoli AM. Degradation behaviour of a composite material for thermal protection systems, part II process simulation. J Mater Sci 1998;33: 3145–9.
[10] Torre L, Kenny JM, Maffezzoli AM. Degradation behaviour of a composite material for thermal protection systems, part I experimental characterization. J Mater Sci 1998;33: 3137–43.
[11] Torre L, Kenny JM. Degradation behaviour of a composite material for thermal protection systems, part III char characterization. J Mater Sci 2000;35:4563–5.
[12] Gibson AC, Hume J. Fire performance of composite panels for large marine structures. Plast Rubber Compos Process 1995;23:175–83.
[13] Hiltz AA, Florence DE, Lowe DL. Thermal protection systems for a mars-entery vehicle. J Spacecr 1968;5:1278–84.
[14] Kanevce LD, Kanevece GH. Comparison of two kinds of experiments for estimation of thermal properties of ablative composite. In: 3rd international conference on Inverse Problems in Engineering. Port Ludlow. Wash USA; June 1999. pp. 13–8.
[15] Chen YK, Milos FS. Ablation and thermal response program for spacecraft heat shield analysis. J Spacecr Rockets 1999;36:475–83.
[16] Bradford JE, Olds JR. Thermal protection system sizing and selection for RLVs using the sentry code. In: 24nd AIAA/ASME/ASE/ASEE Propulsion Conference & Exhibit. California: Sacramento; 2006. July 9–12.
[17] Ko WL, Gong L, Quinn RD. Hypothetical reentry thermostructural performance of space shuttle orbiter with missing or eroded thermal protection tiles. California: NASA Dryden Flight Research Center; 2004. NASA/TM-212850.
[18] NASA space vehicle design criteria structures, entry thermal protection. NASA SP-8014; August 1968.
[19] EADS, Space transportation, Competence Center. Thermal & mechanical engineering unit re-entry system & technology department; 2005. June 10.
[20] Laub B, Venkatapathy E. Thermal protection system technology and facility needs for demanding future planetary missions. In: Presented at the international workshop on Planetary Probe Atmospheric Entry and Descent Trajectory Analysis and Science, Lisbon, Portugal; 2003. October 6–9.
[21] Alelio GFD, Parker JA, Marcel Dekker. Ablative plastics. New York: Dekker, Inc; 1971.
[22] Pering GA, Farrell PV, Springer GS. Degradation of tensile and shear properties of composites exposed to fire or high temperature. J Compos Mater January 1980;14: 54–68.
[23] Bahramian AR, Kokabi M, Famili MHN, Beheshty MH. Ablation and thermal degradation behaviour of a composite based on resole type phenolic resin; process modeling and experimental. Polymer 2006;47:3661–73.
[24] Griffis CA, Masumura RA, Chang CI. Thermal response of graphite epoxy composite subjected to rapid heating. J Compos Mater 1981;15:427–42.
[25] Henderson JB, Wiecek TE. A Mathematical model to predict the thermal response of decomposing, expanding polymer composites. J Compos Mater 1987;21:373–93.

[26] Vaia RA, Price G, Ruth PN, Nguyen HT, Lichtenhan J. Polymer/layered silicate nanocomposites as high performance ablative materials. Appl Clay Sci 1999;15:67–92.
[27] Zanetti M, Camino G, Thomann R, Mulhaupt R. Synthesis and thermal behaviour of layered silicate-EVA nanocomposites. Polymer 2001;42:4501–7.
[28] Gilman JW. Flammability and thermal stability studies of polymer layered-silicate (clay) nanocomposites. Appl Clay Sci 1999;15:31–49.
[29] Morgan AB, Gilman JW, Harris RH, Catheryn J. Flammability of polystyrene-clay nanocomposites. Repr Polym Mater Sci Eng (ACS) 2000;83:53–4.
[30] Morgan AB, Gilman JW. Characterisation of polymer-layered silicate (clay) nanocomposite by transmission electron microscopy and X-ray diffraction: a comparative Study. J Appl Polym Sci 2003;87:1329–38.
[31] Alexandre M, Dubois P. Polymer-layered silicate nanocomposites: preparation, properties and uses of a new class of materials. Mater Sci Eng 2000;28:1–63.
[32] Zanetti M, Camino G. Combustion behaviour of EVA/fluorohactorite nanocomposites. Polym Degrad Stab 2001;74:413–7.
[33] Uhl FM, Wilkie CA. Polystyrene/graphite nanocomposites: effect on thermal stability. Polym Degrad Stabil 2002;76:111–22.
[34] Lee D. Thermal degradation behaviour of polyaniline in polyaniline/Na^{+}montmorillonite nanocomposites. Polym Degrad Stabil 2002;75:555–60.
[35] Zhua J, Startb P, Mauritzb KA, Wilkiea CA. Thermal stability and flame retardancy of poly(methyl methacrylate)-clay nanocomposites. Polym Degrad Stabil 2002;77:253–8.
[36] Tanga Y, Hua Y, Wanga SF, Guia Z, Chenb Z, Fana WC. Preparation and flammability of ethylene-vinyl acetate copolymer/montmorillonite nanocomposites. Polym Degrad Stabil 2002;78:555–9.
[37] Bourbigot S, Devaux E, Flambard X. Flammability of polyamide-6/clay hybrid nanocomposite textiles. Polym Degrad Stabil 2002;75:397–402.
[38] Carrado KA. Synthetic organo-clay polymer: preparation, characterization, and materials applications. Appl Clay Sci 2000;17:1–23.
[39] Pinnavaia TJ, Beall GW. Polymer clay nanocomposites. New York: John Wiley & Sons, Ltd; 2001.
[40] Vaia RA. AMPTIAC Newsl 2006;6:17–24.
[41] Beyer G. Nanocomposites: a new class of flame retardants for polymers. Plast Addit Compos. Compd 2002;4:22–8.
[42] Capitaneo JL, Silva FT, Caffarena VR. Aluminosilicate nanocomposite-material. Appl Mineral 2004;1:45–8.
[43] Sheng N, Boyce MC, Parks DM. Multiscale micromechanical modeling of polymer/clay nanocomposites and the effective clay particle. Polymer 2004;45:487–506.
[44] Gultek A, Seckin T. Poly(methacrylic)acid and methacryloxy propyltrimethoxy silane/clay nanocomposites prepared by in-situ polymerisation. Turk J Chem 2002;26:925–37.
[45] Deana D, Walkerb R, Theodoreb M, Hamptonb E, Nyairo E. Chemorheology and properties of epoxy/layered silicate nanocomposites. Polymer 2005;46:3014–21.
[46] Tolle TB, Anderson DP. Morphology development in layered silicate thermoset nanocomposites. Compos Sci Technol 2002;62:1033–41.
[47] Lyon RE. Pyrolysis kinetics of char forming polymers. Polym Degrad Stabil 1998;61:201–10.
[48] Bahramian AR, Kokabi M, Famili MHN, Beheshty MH. Thermal degradation process of resole type phenolic resin kaolinite layered silicate nanocomposite. Iran Polym J 2007;16:375–87.

[49] Pektas I. High-temperature degradation of reinforced phenolic insulation. J Appl Polym Sci 1998;68:1337–42.
[50] Nam JD, Seferis JC. Generalized composite degradation kinetics for polymeric systems under isothermal and nonisothermal condition. J Polym Sci Part B Polym Phys 1992;30:455–63.
[51] Naval Surface Weapon Center (N43) Dahlgren. Determination of kinetic parameters for thermal decomposition of phenolic ablative materials by multiple heating rate method; 1990. VA 22448.
[52] Bahramian AR, Kokabi M, Famili MHN, Beheshty MH. High temperature ablation of kaolinite layered silicate nanocomposite. J Hazard Mater 2008;150:136–45.
[53] Bahramian AR, Kokabi M. Ablation mechanism of polymer layered silicate nanocomposite heat shield. J Hazard Mater 2009;166:445–54.
[54] Dimitrienko YI, Dimitrienkov ID. Effect of thermo-mechanical erosion on heterogeneous combustion of composite materials in high-speed flows. Combust Flame 2000;122:211–26.
[55] Dimitrienkov YI. Effect of finite deformations on internal heat-mass transfer in elastomer ablating materials. Int J Heat Mass Transfer 1997;40:699–704.
[56] Dimitrienko YI. Thermal stress in ablative composite thin walled structures under intensive heat flows. Int J Eng Sci 1997;1:15–31.
[57] Dimitrienkov YI. Thermomechanical behaviour of composite materials and structures under high temperatures. Mater Compos Part a 1997;28A. 453–61.
[58] Staggs J. Simple mathematical models of char forming polymers. Polym Int 2000;49:1147–52.
[59] Staggs JEJ. A simplified mathematical model for the pyrolysis of polymers with inert additives. Fire Saf J 1999;32:221–40.
[60] Staggs JEJ. A simple model of polymer pyrolysis including transport of volatiles. Fire Saf J 2000;34:69–80.
[61] Office of Aviation Research. Polymer flammability. Washington, D.C. 20591: DOT/FAA/AR-05/14; 2005.
[62] DOT/FAA/AR-07/2 Thermal analysis of polymer flammability. Washington, DC 20591: Air Traffic Organization Operations Planning Office of Aviation Research and Development; 2007.
[63] Zanetti M, Kashiwagi T, Falqui L, Camino G. Cone calorimeter combustion and gasification studies of polymer layered silicate nanocomposites. Chem Mater 2002;14:881–7.
[64] Konecki M, Galaj J. Correlation model for heat release rate of lignocellulosic materials in a full scale from small scale experimental data. J Appl Sci Res 2009;5:1558–67.
[65] Zhang J, Michael A, Delichatsios SB. Experimental and numerical study of the effects of nanoparticles on pyrolysis of a polyamide 6 (PA6) nanocomposite in the cone calorimeter. Combust Flame 2009;156:2056–62.
[66] Lautenberger C, Pello CF. Generalized pyrolysis model for combustible solids. Fire Saf J 2009;44:819–39.
[67] Bahramian AR, Kokabi M. Numerical and experimental evaluation of the flammability and pyrolysis of a resole based nanocomposite by cone calorimeter. Iran Polym J 2011;20:399–411.

CHAPTER

Flame-Retardant Thermoset Nanocomposites for Engineering Applications

16

B.K. Kandola, D. Deli

University of Bolton, Bolton, UK

CHAPTER OUTLINE

1. Introduction

The technological progress in aerospace, automotive, and marine industries of the last few decades has required materials with superior properties and functionalities

Polymer Green Flame Retardants. http://dx.doi.org/10.1016/B978-0-444-53808-6.00016-0

for the new applications and the higher demands. The tendency of nowadays advanced materials for engineering applications, in particular structural materials, is undoubtedly to posses high mechanical strength and light weight as this results in energy saving. For this reason in the new generation of aero-vehicles, automotives, boats, ships, etc. engineers have been trying to replace, when possible, conventional single phase materials such as metals with thermosetting based materials (glass/carbon fiber-reinforced composites) as they not only have good mechanical properties at reduced density but also lower cost and ease of processability.

Thermosets are essentially polymers able to inter and intracrosslink their chains to form a hard three-dimensional polymer network. Such a polymer network is covalently cross-linked and so once shaped into a permanent form cannot be remelted or reshaped as the basic polymeric component has undergone an irreversible chemical change. The procedure by which the raw material is converted to a hard, insoluble, and infusible product is referred to as cure (or curing) and corresponds to the final step of a polymerization reaction. As such materials are irreversibly cross-linked, their strength is higher than thermoplastics; and unlike the latter they also have better physical resistance to heat. For these reasons, they are always preferred for structural applications to thermoplastics. Although such materials are strong and their mechanical properties can resist to a certain extent to heat, they still own a glass transition temperature above which the Young's modulus of the material decreases of magnitude as the energy of the system increases the mobility of the polymeric chains. Hence the material softens and becomes more rubbery-like. There are several types of thermosets, each of them with different properties that depend upon their structures and components.

Epoxy is a class of resins that contain at least two reactive epoxide groups. The higher the number of these epoxide groups, the more functional is the epoxy resin. Therefore, various types of epoxies can be derived with respect to the number of their epoxide groups, the most common examples are: diglycidyl ether of bisphenol-A (DGEBA) and bisphenol-F (DGEBF), triglycidyl resins of *p*-aminophenol (TGAP), and tetraglycidyl diamino diphenyl methane (TGDDM). However, the most widely used is the DGEBA (see Scheme 1(a)) based. The main cure process takes place by the reaction between a di-epoxide and a primary diamine or an anhydride often referred as hardeners. Further cross-linking can then take place by the reaction of the resulting hydroxyl groups that generate by the opening of the epoxy ring. Reactants (epoxy prepolymers or hardeners) with higher reactive functionalities result in a highly cross-linked, stiff and tough epoxy network suitable for aerospace applications [1] Epoxy resins are also light, have increased adhesion, reduced degradation from water ingress, and increased resistance to osmosis making them ideal for marine type applications. However, despite their enhanced functionalities and properties, their use is limited due to high cost when compared to unsaturated polyester resins as it can be several times higher than the latter. Moreover, processing of these resins is less straightforward because of their high viscosity and as the curing reactions are slower and conditions are much more severe (often 150–180 °C).

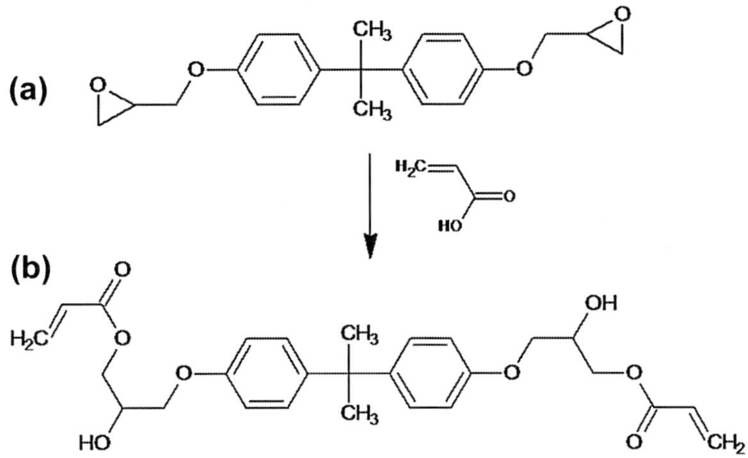

SCHEME 1

Chemical structure of (a) diglycidyl ether of bisphenol-A (DGEBA) and (b) vinyl ester.

Unsaturated polyester resins are mixtures containing polyester prepolymers bearing unsaturated groups and styrene with the latter serving as reactive solvent that solubilize the prepolymers and ensure radical cross-linking during the polymerization process [2,3]. A typical structure is presented in Scheme 2.

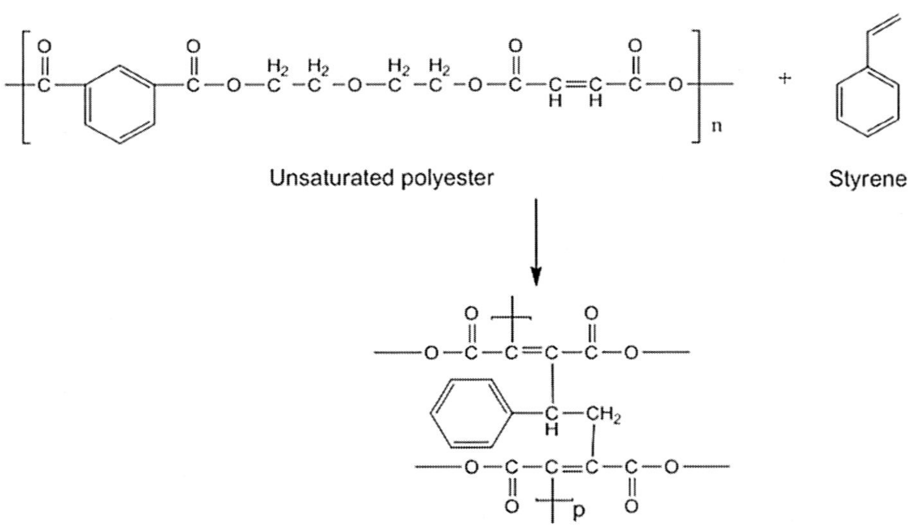

SCHEME 2

Chemical structure of an unsaturated polyester and cross-linking reaction with styrene.

The unsaturated polyester chain is generally composed of a glycol (commonly di-ethylene or propylene glycol), maleic anhydride which adds the unsaturation moieties onto the backbone and react with the glycol to give esters and a anhydride (commonly *ortho*-phthalic or iso-phthalic) which, similar to the maleic anhydride, reacts with the glycol to give esters. Easy processability, good mechanical, water and corrosion resistance properties, and relatively low cost make them ideal for the marine, automotive, and construction applications and hence one of the most commonly used resin types.

Vinyl esters, like polyesters, are solvent monomer systems with typically 40–50% of styrene content. They are prepared by a reaction between an epoxy resin prepolymer with acrylic acid and have similar structure as unsaturated polyesters differing only for the location of the unsaturation moieties. In fact, such unsaturations in a vinyl ester are located at the ends of the monomeric chains as shown in Scheme 1(b). Vinyl esters are extensively used in the marine industry [4] as a replacement for unsaturated polyesters as they offer better mechanical properties and are less prone to hydrolysis damage since they contain less ester linkages. Other applications include sewage pipelines, water and chemical storage tanks.

Phenolics are obtained by the reaction of phenols and formalaldehyde, and are used for the production of molded products, coatings, and adhesives. Their chemistry is well-established and in general there are two different types of materials: novolacs which are thermoplastic phenol based prepolymers that can be cross-linked using hexamethylene tetra amine as cross-linking agent and resoles, which are prepolymer phenol based mixtures bearing methylol pendants groups. Such groups can promote self-crosslinking under sufficient heat. Both types of phenolics react by condensation reaction to give cross-links of the type shown in Scheme 3. Generally speaking, phenolics are brittle and do not have high mechanical properties. In addition, they require severe conditions in order to cross-link and this drastically narrows down their possible applications.

Besides these, other resin systems such as cyanate esters, polyimides, and bismaleimides are also used in similar applications [5]. However, their use is limited compared to that of epoxy, unsaturated polyesters, vinyl esters, and phenolic resins. The ranking of the major properties of epoxy, polyester, vinyl ester, and phenolics are summarized in Table 1 [6,7], which undoubtedly place polyesters and vinyl esters as the most cost-effective structural materials for engineering applications.

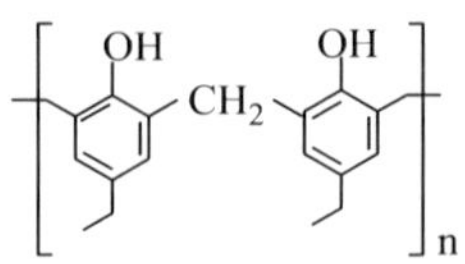

SCHEME 3

Chemical structure of a phenolic resin.

Table 1 Ranking of the Properties of Epoxy, Polyester, Vinyl Ester, and Phenolics

Property	Ranking
Mechanical property	Epoxy > polyester and vinyl ester > phenolics
Processability	Polyester & vinyl ester > epoxy > phenolics
Cost-effectiveness	Polyester & vinyl ester > phenolics > epoxy
Flammability	Phenolics < epoxy < polyester & vinyl ester

2. Thermosets and key flammability issues

2.1 Flammability of different thermosets

As discussed earlier, thermosetting resins are a very important group of materials as they have a wide range of applications; therefore, it is practically inevitable that they will experience fire situations. As shown in Table 1, the flame retardancy of thermoset resins in general can be ranked as [6—11]:

$$\text{Phenolic} > \text{Epoxy} > \text{Polyester \& vinyl ester}$$

In general, polymers that contain aromatic and heterocyclic groups in their chains are less combustible than polymers with aliphatic backbones [12]. On the other hand, polymers with relatively long flexible (aliphatic) linkages are still relatively combustible despite the aromatic groups.

After the ignition of the polymer, the combustion depends on the mechanism of thermal degradation of the polymer as it is related with the amount of char that is promoted by the particular mechanism. Char normally acts not only as a physical barrier for the heat flux coming from the flame to the surface of the thermoset but also as a diffusion barrier of the fuel flux from the polymer to the flame and therefore char formation is a required property of a thermoset. Generally, charring of polymers proceeds through different stages: cross-linking, aromatization, fusion of aromatics, and graphitization [13]. The chemical structure of the polymer determines whether the thermoset is able or not to perform one or several of these stages and produce char. Although many polymers tend to cross-link at the beginning of thermal decomposition, this does not necessarily result in char formation. Char is formed only if the cross-linked polymer contains aromatic fragments linked by short and flexible linkages that are prone to aromatization during thermal decomposition [14]. Examples of this are phenol formaldehyde resins that normally display good thermal stability, relatively good flame retardancy, and self-extinguishing property ($\text{LOI} = \geq 25$) [6,7,11] due to the high aromatic content and their tendency to promote fused aromatics on the polymer surface. Such condensed rings on the surface tend to assemble into small stacks, which are precursor of graphite. These pregraphitic domains can assemble together at about 600/900 °C (temperature typically found on the surface of burning polymers) forming an amorphous char. Chars containing pregraphitic domains are stable to thermal oxidation and therefore less likely burn and expose the polymer surface to the heat of the flame. Highly graphitized chars are rigid and have often cracks, which do not retard diffusion of combustible materials to the flame [13]. Despite having good fire-retardant properties, the phenolics are not decisive thermosets in applications where weight, mechanical and fire properties are needed because of their brittleness. Moreover, the severe conditions needed to fully cure the resin can be an issue in large-scale applications.

The thermal and flammability behavior of an epoxy is determined by the final aromatics that have been used. Moreover, the nitrogen content present in the hardener plays a fundamental role. Epoxy resins cured with amines tend to produce more char than anhydride-cured resin [15]. Decomposition of epoxies takes place via random chain-scissions reactions around temperatures of about 380 and 450 °C (~130 °C after polyesters) when the C—N bonds, having the lower bond dissociation energy, start to split. Such reaction decomposes 80—90% of the original polymer generating various volatile and flammable compounds that provide fuel source for the final decomposition of the epoxy [16]. As a consequence, their LOI values may be no lower than about 23 [6,7,11]. However, the improved fire performance property epoxies are limited in special applications because of their high cost and difficult processability.

Unsaturated polyesters display poor fire performances being composed of aromatics attached by relatively long aliphatic chains. The emission of styrene during the processing and the thermal combustion is a serious environmental issue due to the styrene toxicity. Moreover, the high flammability of styrene is the cause for the ignition of the polyester as the ignition event begins when a source of heat comes into contact with the styrene (fuel) generated by the thermal degradation of the polyester under the heat. This flow of flammable styrene can then react with the oxygen in the air producing flame and more heat. Some of this heat is then transferred to the surface of the polymer, maintaining the flow of flammable styrene that feeds the flame [17]. For this reason, currently there is considerable research that aims to reduce the styrene levels without compromising the mechanical properties of the polymer. In thermal performance terms, most polyesters start to decompose above 250 °C, with the main step of weight loss occurring between 300 and 400 °C and most importantly no char left at 600 °C [18].

Vinyl esters, as polyesters, display a thermal degradation mainly governed by the decomposition of the styrene content. As the amount of styrene in vinyl esters is generally higher than in polyesters, thermal degradation of vinyl esters tend to begin at a lower temperature and slightly higher smoke generation, heat release rate and time-to-ignition are often observed.

2.2 Flammability of glass fiber-reinforced thermoset composites

When a thermoset resin is incorporated in a composite structure with fibers, the overall burning behavior of a composite will be the sum of its component fibers and resin plus any positive (synergistic) or negative (antagonistic) interactive effects [6,7,11]. Except for some particular fibers (such as UHMW polyethylene or *para*-aramid), the most commonly used ones (such as glass and carbon) add little to the fuel content of a composite. The glass/carbon fiber-reinforced composites hence are less flammable than cast neat resins of similar thicknesses due to lower resin contents in the former and also the fibers present acting as fillers.

Apart from fiber types, other factors that affect the flammability of a composite include thickness of the composite, intensity of the fire, composite physical structure and presence, and type of core materials, discussed in detail in our previous reviews [6,7,11]. In context with polymer nanocomposites, the thickness of the cast resin or a composites laminate and the intensity of fire are important parameters. Thermally and physically thin samples ignite early, show large peak heat release values, burn for shorter time, and produce less total heat release compared to thermally and physically thick samples, which burn slowly but for a longer time [19,20]. "Thermally thick" means the heat wave penetration depth is less than the physical depth, whereas in "thermally thin", the composite has the same temperature through this limiting thickness. The transition from thermally thin to thermally thick is not a constant since it depends on material thermal properties including fiber and resin thermal conductivities.

Similarly with increasing fire intensity or incident external heat flux, the samples ignite early, show large peak heat release values, burn for shorter time, and produce less total heat release compared to those at lower heat fluxes. Polymer nanocomposites are shown to have more effect on reducing the flammability of the polymer in thick samples and when tested at lower heat fluxes, which is due to the fact that the nanoclays/nanoparticles will have more chance to form thermal barriers on the surface under these conditions. Specific examples in this context are reported in later sections.

2.3 Fire testing and performance requirements

Thermal stability of polymer nanocomposites has widely been studied by thermal analytical techniques such as differential thermal analysis (DTA), differential scanning calorimetry (DSC), and thermogravimetric analysis (TGA). These give information about the effect of nanoparticle on polymer's physical and chemical changes as a function of temperature, the important parameters being the onset of decomposition temperature, onset and maximum temperatures of different physical and chemical transitions, mass loss rate, and mass of charred residue left at the end of the test. Dynamic mechanical thermal analyzer (DMTA) is also very useful for thermosets, in particular to determine their glass transition temperatures. Similar to other polymers, limiting oxygen index [21], UL-94 [22], and cone calorimetry [23] are the common laboratory methods of quantifying the flammability of thermosets.

In order to use polymer-based materials as structural materials for commercial applications, such as in construction or transport sector, they have to conform to certain specified regulations for a particular application. There are no international fire safety standards for construction sector and most of the countries have their own requirements for fire performance. On the other hand, the European Union established its own fire safety regulations for composite materials based on thermosetting polymers used in constructions such as hospitals, schools, shops, clubs, leisure centers, stadiums, factories, etc. [24,25]. The European System evaluates the fire hazard of any material, including thermosets, by the reaction to fire and the fire-resistant properties. Heat release rate, time-to-ignition, flame spread, and oxygen index are recognized as the most important reaction to fire properties of any material. Such properties are used to describe the toxicity of a combustible material and how its flammability and combustion influence the early stage of a fire, generally from the ignition to the flashover.

On the other hand, the transport sector has formal fire regulations at both national and international levels. Since this subject is very complex, the reader is directed elsewhere for more detailed regulation and test information [26] and only key international fire performance codes and standards will be highlighted here. In brief, for the aerospace industry all products are required to comply with the US Federal Aviation Regulations (FAR), which govern the requirements for materials used for such applications [27,28]. The most important test is the Heat Release Rate Test [29], according to which when the sample is tested at 35 kW m^{-2} external heat in an Ohio State University calorimeter, both the peak heat release rate and total heat release rates measured over 2 min should not exceed 65 kW m^{-2}.

For automotive sector, in the United States the Federal Motor Vehicle Safety Standard 302 (FMVSS 302) [30] regulates the flammability of materials used in the interiors of cars and trucks. The FMVSS 302 test is virtually an international standard, as it has been harmonized with many equivalent designations.

For rail industry, the rate of fire spread and toxicity of the fumes produced during their combustion are of vital importance. In the United States the Federal Railroad Administration (FRA) has a standard, Federal Register 49:192 (1984), which gives requirements for all materials used in rail carriages. In Europe there is not a common standard across the different countries. Each European country has its own regulations, e.g. DIN 5510 (Germany), BS 6853 (Great Britain), and NF F16-101 (France). Most of these require low rate of fire spread and toxicity of the fumes produced during combustion as in railways and metro systems in particular, a large number of the fires occur in underground tunnels where longitudinal flame spread is rapid (coinciding with escape routes) and ventilation is poor [31]. For this reason, smoke generation is also important as the smoke reduces visibility thereby impeding evacuation.

In the marine industry, the regulations fall into two groups: commercial passenger/cargo vessels and naval surface vessels/submarines. All vessels that belong to the first group in European countries have to comply with the fire performance requirements contained in the International Convention for the Safety of Life at Sea (SOLAS) and Codes of Safety for High Speed craft of the International Maritime Organisation (IMO/HSC). The fire tests to be carried out and the acceptance criteria are defined in the International Code for Application of Fire Test Procedures (IMO/FTP), which have been mandatory since 1998 [32]. In this same geographic region, because there are no specific naval regulations concerning use of composite materials for manufacturing structural parts for vessels belonging to the second group, they also comply to IMO/HSC codes. In other countries outside of the immediate European region and especially in the United States, naval vessels and submarines are dealt under national regulations and codes [33,34].

3. Flame-retardant strategies for thermosets

3.1 Current flame-retardant strategies used for thermosets

As discussed above, to identify a thermosetting resin with good mechanical properties and fire retardancy, which is easy to process and produced at low price is a real challenge. For applications where the fire performance of the products is an important criterion, inherently flame-retardant resins such as phenolics are used. Alternatively, flame-retardant solutions are applied to resins or composites [6,7,11]. In this review, only possible solutions for resins are discussed.

Polymeric resins can be flame retarded either by adding flame-retardant chemicals into it or by chemical modification of the resin. For thermosets, chemical modification of the resin is very easy due to a vast choice of resins, curing agents, and hardeners. One method is chemical modification of the polymer backbone by introducing flame-retardant elements into it. For example, in unsaturated polyester the use of halogenated resin or replacement of curing agent from styrene to bromostyrene; introduction of the halogen elements in the epoxy backbone such as chlorine in diglycidyl ether of bisphenol C (DGEBC), fluorine in diglycidyl ether of bisphenol-F (DGEBF), bromine in tetrabromobisphenol A (TBBA). For detailed discussion on flame retardancy of thermosets, the reader is referred to our previous reviews [6,7,11]. The other way is the proper choice of curing agents and hardeners. As discussed before, epoxy resins cured with di-amines are less flammable than anhydride-cured resin [15].

Most commonly used flame-retardant additives for unsaturated polyester, vinyl ester, and epoxy resins include aluminum trihydrate, magnesium hydroxide, expandable graphite, brominated compounds, ammonium polyphosphate, and intumescent systems. Metal hydroxides such as alumina trihydrate and magnesium hydroxide act as flame retardants by undergoing endothermic reactions at elevated temperatures reducing polymer temperature and releasing water vapors, which effectively dilute the volatile species evolved during polymer degradation. However, metal hydroxides are effective only at very high percent loadings (typically >50 wt%) which might have an adverse effect on the mechanical properties of the polymer matrix [35]. These are quite often used in conjunction with halogenated thermosets and such mixtures are now considered to be ecologically undesirable and, in a fire, increase the amount of smoke and toxic fumes are given off by the burning material [6]. Nitrogen- and phosphorus-containing additives such as

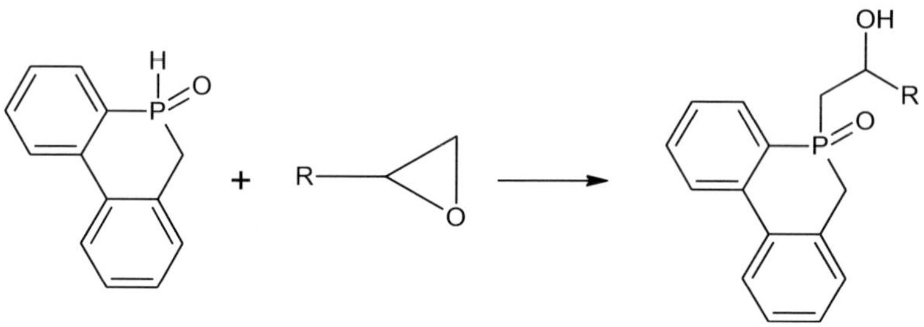

SCHEME 4

Reaction of DOPO with epoxy functionality.

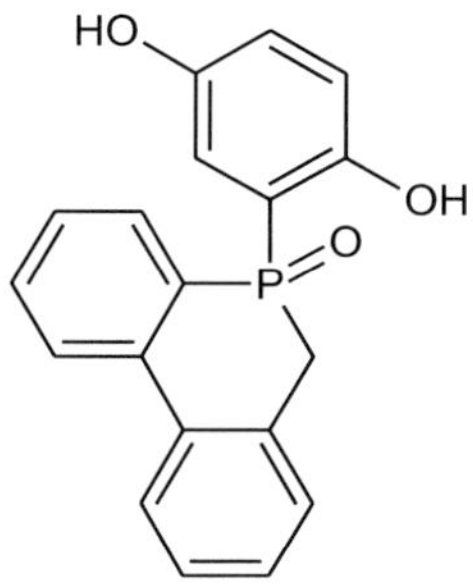

SCHEME 5

Chemical structure of DOPO-HQ.

ammonium polyphosphate, intumescent chemicals etc., on the other hand chemically interfere with polymer decomposition and promote char formation. Chemically reactive type flame retardants are required at comparatively lower levels (20–30 wt%). The red phosphorus alone has also shown an improvement in flame retardancy of the epoxy resin [36].

A recent development in chemically reactive type flame retardants for epoxy resin is an organophosphorus flame retardant, 9,10-dihydro-9-oxa-10-phosphaphenanthrene10-oxide (DOPO) [37]. DOPO is a reactive monofunctional flame retardant that can react with epoxy groups following Scheme 4. However, the problem on doing so is that the monofunctionality of DOPO reduces the cross-link density of the cured epoxy. For this reason, some researchers have been using modified DOPO with quinone groups (DOPO HQ in Scheme 5) [38].

3.2 Role and significance of nanocomposites

Nanocomposites are a relatively new development in the area of flame retardancy and offer significant advantages over the conventional flame-retardant formulations described above. They are materials composed of two or more physically and chemically distinct phases, where one of the components has a dimension on the submicron scale. This can give rise to intrinsically new properties that are not displayed by the pure components. Such new properties typically originate by the change of the polymer nature in the surrounding of the filler: the polymer can be absorbed by the filler surface or trapped between fillers. As a result, such new properties strongly depend upon the surface area of the filler and so the dispersion. Good dispersion allows smaller loading of filler as it increases the surface area of the filler.

This area has received significant attention since Toyota researchers [39,40] found that the addition of a low load of clay ($\sim$5%) to poly-amide 6 (PA-6) led to superior mechanical properties and much higher heat distortion temperature (87 °C higher than pure PA-6). Other general advantages of such type of materials are lightweight (compared to filled polymers), high modulus and strength, and decreased gas permeability. Nanocomposites exhibit better physical, mechanical, and other performance properties due to unique interfacial effects resulting from the dispersion of nanoparticles of high specific surface area and high aspect ratio [41–43]. Strong nanoparticle–polymer interactions increase the viscosity of the polymer melt, which increases with increasing nanoparticle concentration. This though poses processing problems (it is difficult to process above 10 wt% nanoparticle concentration) is helpful from flame retardancy point of view as melt dripping [44,45] and flame spread are significantly reduced. Their main mechanism of action is believed to be by physical means where during combustions nanoparticles, in particular nanosilica and polymer-layered silicates aggregate on the polymer surface due to low surface potential energy of silicon [46–48], forming a thermally insulative surface layer, which slows down the diffusion of pyrolysis gases and hence slowing down flame propagation. The accumulation of silicate layers on the surface is due to gradual degradation and gasification of the polymer. Or according to other theory the migration of silicates to the surface is due to the lower surface free energy of the clays and by convection forces, arising from the temperature gradients, perhaps aided by movement of gas bubbles present during melting of the thermoplastic polymers [49]. The larger aspect ratio particles lead to a greater reduction in mass loss rate, i.e. they are more effective at reducing flammability [50].

Nanoparticles when used in combination with other char forming conventional flame retardants, show synergistic action by catalyzing charring reactions [6,45,51–53]. They also have shown positive effect with vapor phase active flame retardants [45,54].

4. Synthesis of thermoset nanocomposites

4.1 Type of nanoparticles

Since the discovery of polymer nanocomposites by Toyota research group, major interest has laid in polymer-layered silicates, typically the montmorillonites, hectorites, and saponites, the montmorillonites being studied most. However, there are many different shapes and types of nanoparticles. Nanoparticle is defined as having at least one dimension on the nanoscale.

One-dimensional (1D) particles are sheets or layers, such as layered silicates or graphites, which are one to a few nanometers thick and hundreds to thousands nanometers in the other dimension. Due to very high aspect ratio, their reinforcement effect is better than any other nanoparticles. Nanoclays can be cationic (montmorillonite hectorites and saponites) or anionic (layered double hydroxides (LDHs)). Both types of clays have been used for epoxy, UP, and VE resins.

Nanoparticles with two dimensions on nanoscale (2D) and the third being larger have high aspect ratio, so also act as reinforcing element. Examples include carbon nanotubes, carbon nanofibers, and sepiolite. Carbon nanotubes have been widely studied in thermosets. When all three dimensions (3D) are of nanometer order, these are true nanoparticles with aspect ratio of 1. Examples include polyhedral oligomeric silsesquioxane (POSS) and nanometallic oxides (silica, titania, alumina) and carbides. Derivatives of POSS are hybrid inorganic/organic chemical composites that possess an inner inorganic silicon and oxygen core $(SiO_{1.5})_n$ and external organic substituents that can feature a range of polar or non-polar functional groups. The organic substituents on their outer surfaces make them compatible or miscible with most polymers. POSS nanostructures having diameters ranging from 1 to 3 nm can be considered as the smallest possible particles of silica, i.e. molecular silica. Examples of their usage in thermoset resins are available in literature as discussed in a latter section. Graphene is another nano-additive with a unique two-dimensional (2D) atomic carbon sheet structure [55] and is been increasingly used for producing polymer nanocomposites [56,57], including epoxy nanocomposites [58,59].

4.2 Methods of synthesis and characterization

Thermoset polymer nanocomposites are prepared by in situ polymerization process, where nanoparticles are first dispersed in the monomer using mechanical stirring, heating, or sonification followed by polymerization using the method suitable for that particular resin system. Alternatively, nanoparticles are dispersed in a solvent (compatible with the resin) to obtain a stable suspension before starting the polymerization process. Some interfacial agents can be added to promote a stable suspension [39].

It is well known that improvements in mechanical and thermal properties of polymer nanocomposites are largely associated with the extent of dispersion of the nanoparticles in the polymeric matrix [60–64]. In the case of polymer-layered silicate nanocomposites, three different types of nanocomposite morphologies can be formed depending on the degree and homogeneity of particle dispersion. In other words, the groups are determined according to the expansion of the interlayer space when the polymer diffuses into the gallery between the clay layers [62,65]. Figure 1 [66,67] illustrates the two structures of polymer-layered silicate nanocomposites. When the interlayer spacing is partially enlarged, the resulting nanocomposite is considered to be intercalated. Alternatively, the full separation of the clay layers results to an exfoliated structure [68]. If however the interlayer distance in the

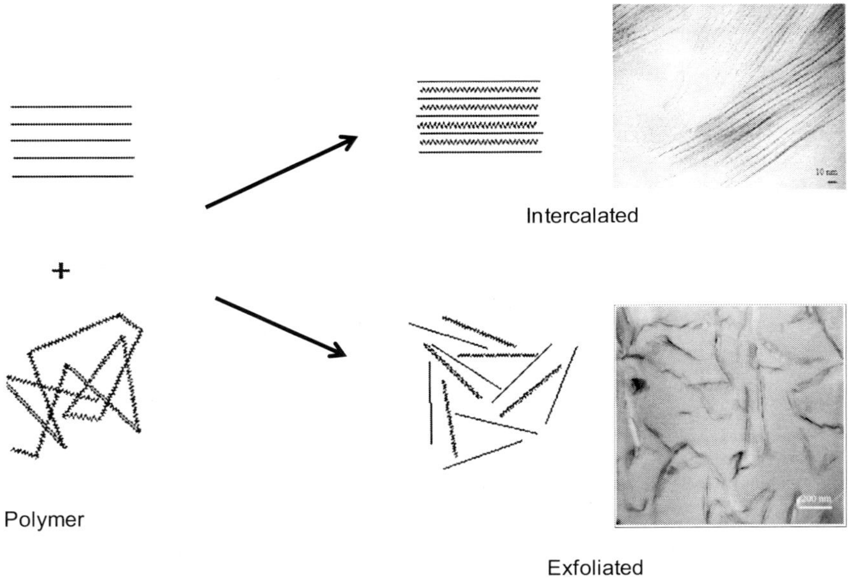

FIGURE 1

Structures of epoxy-layered silicate nanocomposites.

TEM images reproduced from Refs [66,67] by permission of the American Chemical Society.

clay remains at its minimum, a microcomposite is said to have been obtained. The properties of such material are similar to those of filled polymers. On the other hand, exfoliated systems have improved properties compared to intercalated nanocomposites and conventional composites due to homogeneous dispersion of the clays in the polymer matrix [43,69]. However, the strong tendency of nanoparticles to form agglomerates, due to their strong interfacial bonding can result in poor dispersion in the polymer, which consequently has a negative effect on the properties of the resulting nanocomposite [42]. Also, failure to achieve uniform dispersion usually leads to a phase separation [42,70,71].

The most commonly used techniques for characterization of nanocomposites are X-ray diffraction (XRD) and transmission electron microscopy (TEM). X-ray diffraction allows the determination of the spaces between structural layers of silicate utilizing Bragg's law: $\sin\theta = n\lambda/2d$. A reduction in the diffraction angle corresponds to an increase in the silicate layer distance in the low angle region ($2\theta = 3-9°$), indicating formation of intercalated structure, whereas if the nanocomposites are disordered (exfoliated), no peaks are observed due to loss of structural registry of the layers. With TEM, when nanocomposites have formed, the intersections of the silicate sheets are seen as dark lines which are the cross-sections of the silicate layers, measuring 1 nm thick. Sometimes, other analytical techniques like X-ray diffraction, DSC, TGA, and FTIR are also used to characterize polymer–nanocomposite structures by comparing the results for polymer alone and polymer—nanocomposite structures. In case of epoxy or other thermoset polymer—clay nanocomposites the organic modifier of the clay in addition to promoting intercalation/exfoliation can also react and have an effect on the polymerization process. Reactive organic modifiers promote intra gallery polymerization. This has been reviewed in detail by Zammarano [46] for epoxy composites. Both intercalated and exfoliated structures formed are reported in literature. With in situ polymerization mostly intercalated structures are formed [72] whereas exfoliated nanostructures have been reported to be obtained by exfoliation—adsorption process (organo-modified clay dispersed in a solvent prior to polymerization) [67,73]. In this method, the organo-modified clay when dispersed in a solvent can be easily delaminated. The polymer then adsorbs onto the delaminated sheets and when the solvent is evaporated a highly disordered structure is obtained before starting the standard in situ polymerization [46]. Chen et al. [67] and Ma et al. [73] have used this method to produce disordered and fully exfoliated nanocomposite structures.

Physical properties including the polarity, viscosity and functionality of the base resin, the nature of the nanoclay, cure conditions, and the type of curing agents are postulated to have an effect on nanocomposite structures. Resins and hardeners exhibiting low viscosities at the curing temperatures are expected to diffuse into the galleries at a faster rate leading to a high reactivity. We have recently investigated [74] the effect of nanoclays and double-walled carbon nanotubes (DWCNTs) on selected physical properties of two types of epoxy resins; a tetrafunctional epoxy resin, tetraglycidyl-4,4′ diaminodiphenylmethane (MY721) with 4,4′-diaminodiphenyl sulphone (Aradur 976-1) used as a curing agent and a bifunctional epoxy resin, 1,4-butanediol diglycidyl ether (LY5052) with a modified cycloaliphatic amine-based hardener (HY5052). Montmorillonite clays modified with different organic modifiers, double-walled nanotubes and nanosilica have been used, the details of which are provided in Table 2. High temperature curing and tetrafunctional epoxy resin (MY721) nanocomposites showed to have well intercalated polymer clay microstructures by XRD (Figure 2) and TEM (Figure 3) analysis.

On the other hand, a low temperature bifunctional epoxy resin (LY5020) nanocomposites exhibited moderately intercalated microstructures. At elevated cure temperatures the MY721 resin/hardener mixture has a lower viscosity or enhanced molecular mobility leading to improved mass transfer into the clay galleries. Intergallery reaction rates may be increased by rapid diffusion of the resin mixture into galleries which may promote exfoliation. LY5052, a room temperature curing resin will reach gelation faster than a high temperature curing resin (MY721) system. The rigid cross-linked network that forms much earlier for LY5052 resin systems within the layers of the clay at gelation points may prevent any further expansion leading to the formation of moderately intercalated microstructures. In contrast, gelation is delayed at elevated temperatures for MY721 allowing significant expansion of the clay layers before the polymer structure is fully cross-linked. The dispersion of carbon nanotubes was found to be poor although significantly better in the high temperature curing resin systems compared to the low temperature curing epoxy resin

Table 2 Composition, Cone Colorimetry Data, LOI and UL-94 Results for Tetrafunctional (MY721) and Bifunctional Epoxy Resins and Their Composites

	Nanoparticles		Cone Calorimetry				UL-94	
Sample	Type	Mass (%)	TTI (s)	PHRR (kW m^{-2})	THR (MJ kg^{-1})	LOI	TBT (s)[a]	Performance
EPR 486	Silica	30	34	822	43	29.4	20[b]	V-1
MY721	–	–	24	823	56	27.8	88	HB
MY-30B-5	30B	5	25	867	52	30.8	82	HB
MY-VTP-5	VTP	5	28	616	49	31.7	34[b]	V-1
MY-1.30E-5	1.30E	5	24	712	56	30.0	110	HB
MY-DWNT-0.5	DWNTs	0.5	21	727	53	29.6	21[b]	V-1
LY5052	–	–	34	1861	77	21.5	68	HB
LY-30B-5	30B	5	37	1780	77	21.9	63	HB
LY-VTP-5	VTP	5	37	1737	70	23.1	62	HB
LY-1.30E-5	1.30E	5	35	1636	79	20.5	65	HB
LY-DWNT-0.5	DWNTs	0.5	30	1805	69	20.1	65	HB

MY721, a high temperature curing tetrafunctional epoxy resin; LY5052, a low temperature curing bifunctional epoxy resin; EPR486, MY721 modified with 30% of nanosilica particles; 1.30E, Octadecyl ammonium ion-modified montmorillonite clay; 30B, Alkyl quaternary ammonium modified montmorillonite clay; VTP, Vinyl triphenyl phosphonium bromide modified montmorillonite clay; DWCNTs, double-walled carbon nanotubes.

[a] *TBT is the total burn time.*

[b] *Samples self-extinguished after this time.*

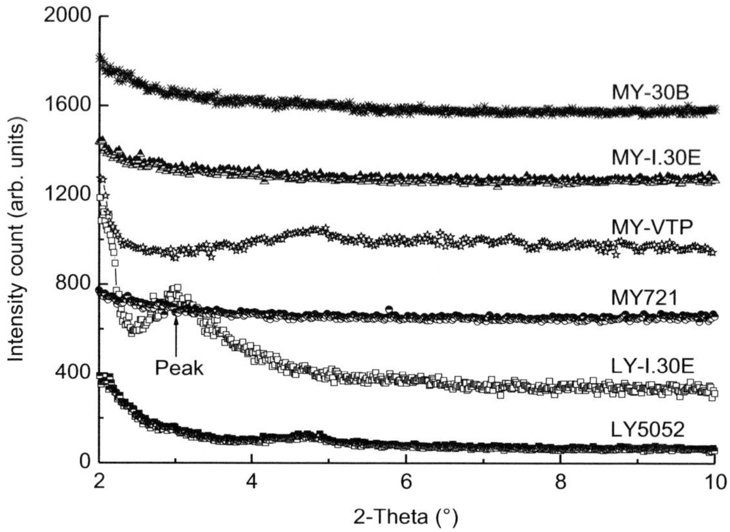

FIGURE 2

XRD patterns of epoxy resins (MY721 and LY5052) and their nanocomposites.

Reproduced from Ref. [74] by permission of Elsevier.

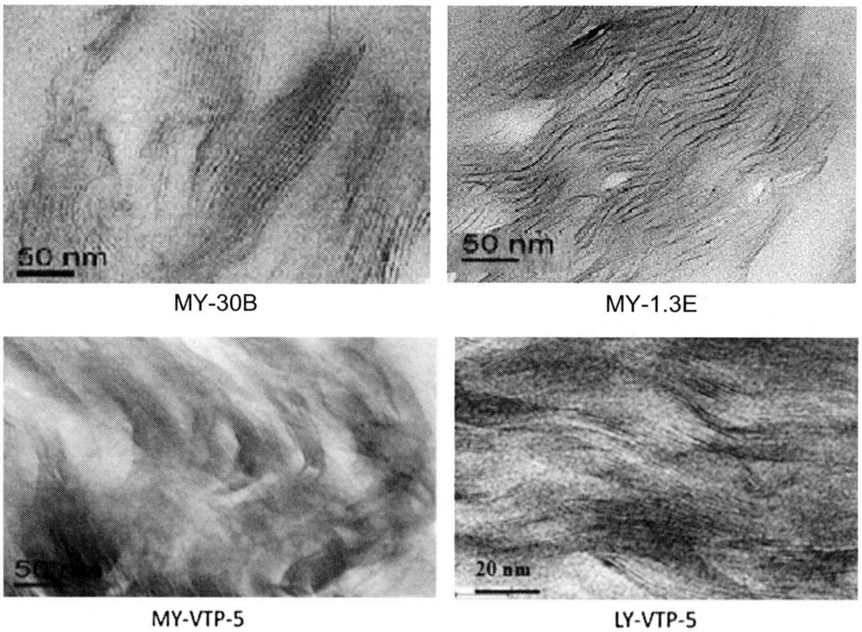

FIGURE 3

TEM images of MT-30B, MY-1.3E, MY-VTP-5, and LY-VTP-5.

Reproduced from Ref. [74] by permission of Elsevier.

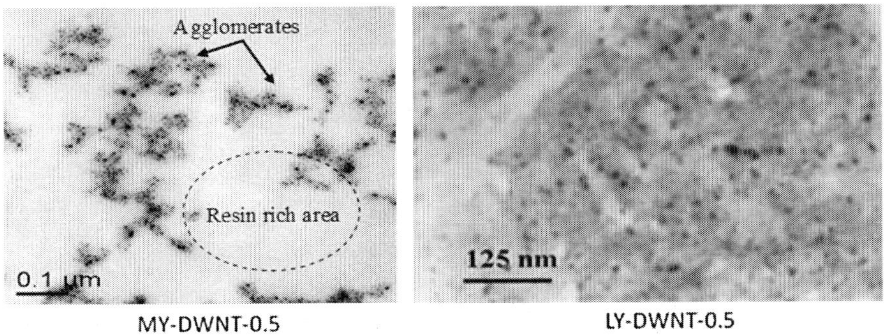

FIGURE 4

TEM images of MY-DWNT-0.5 and LY-DWNT-0.5 [74].

(Figure 4). Different rates of reactivity and variations in polarities, chemical interactions, and viscosities are some of the factors that could lead to differences in the morphology of the resultant nanocomposites when the same type of additives are used with different types of thermosetting resins. Layered double hydroxides (LDHs) and anionic clays are also reported to have been used in epoxy resins, mainly by in situ polymerization. They mostly form intercalated structures [46,75].

Unsaturated polyester and vinyl ester nanocomposites are prepared only by in situ polymerization. In these resins, intercalated nanocomposite structures are easily obtained by short mixing times of only 5 min [76,77]. Longer mixing time have no additional effect on *d*-spacings of the nanocomposites. This is contrary to that observed for epoxies where longer mixing times are required [46]. Suh et al. [78] have explained this due to high diffusion coefficient of styrene. In our laboratory, we have studied dispersion of various unsaturated polyester samples containing different organically modified clays using XRD [53,79,80]; selected results are shown in Figure 5. The clays were of montmorillonites types (Southern Clay Products, USA), modified with different organic modifiers, Cloisite 10A (modified by dimethyl benzyl hydrogenated tallow quaternary ammonium chloride), 15A (modified by dimethyl dehydrogenated tallow quaternary ammonium chloride), 25A (modified by dimethyl hydrogenated tallow 2-ethylhexyl quaternary ammonium methyl sulfate), and 30B (modified with methyl tallow bis-2-hydroxyethyl quaternary ammonium chloride). These results show that dispersion of functionalized clays in the polymer matrix depends on the type of functional group of the organic modifier used and so intercalation of polyester chains and formation of a possible nanocomposite structure also depend on this. Exfoliated nanocomposite structures may be obtained by appropriate selection of functionalized nanoclay. It is proposed that cyclohexyl groups in Cloisite 25A interact with phenyl groups of the unsaturated polyester resin, resulting in exfoliated nanocomposite structures whereas linear aliphatic chains in Cloisite 15A and Cloisite 30B give possible intercalated nanocomposite structures. It is also suggested that the benzyl group in Cloisite 10A causes steric hindrance with polyester resin to give only a microcomposite structure.

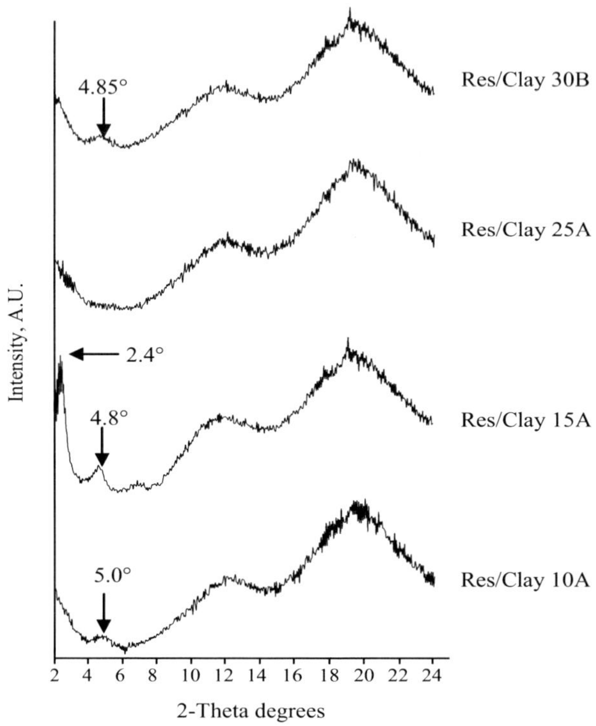

FIGURE 5

XRD curves for various resin/clay hybrids.

Reproduced from Ref. [53] with permission from John Wiley and Sons.

Regarding phenolic-clay nanocomposites, the published literature is quite limited. On the other hand, in the case of epoxy, apart from nanoclays and nanotubes, other nanoparticles such as POSS [81], graphene [82,83], nanosilica, etc. have been added to the resin during polymerization process. Their characterization, usually with high resolution SEM or TEM, has shown a uniform dispersion. As an example, TEM image of a commercially available tetraglycidyl-4,4′ diaminodiphenylmethane epoxy resin containing 30 wt% of nanosilica particles (Bakelite EPR 486) (Bakelite AG) is shown in Figure 6 [74]. Despite a higher loading fraction (30 wt%), nanosilica particles are uniformly dispersed throughout the epoxy matrix.

5. Thermal, fire, and mechanical performance of thermoset nanocomposites

The thermal decomposition of polymers and the effect of additives on enhancing their thermal stability are usually studied by thermogravimetric analysis (TGA), which is a good indicator of their flammability behavior. Generally, the addition

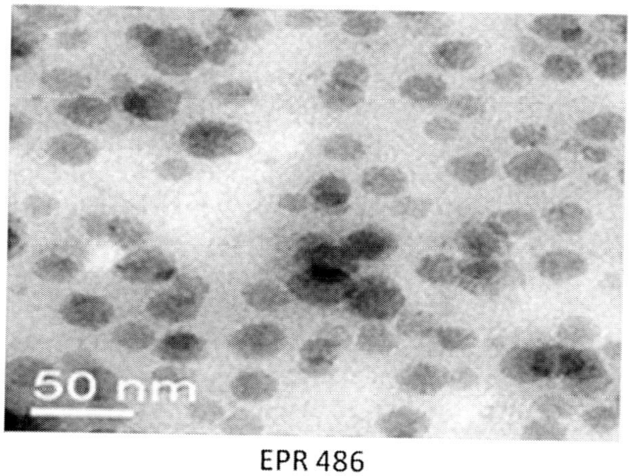

FIGURE 6

TEM image of epoxy-nanosilica (EPR 486).

Reproduced from Ref. [74] with permission from Elsevier.

of nanoclays, nanotubes, etc. to the polymer reduces the rate of mass loss of the polymer and hence, enhances the thermal stability by acting as a superior insulator and mass transport barrier to volatile by-products generated during decomposition, thereby suppressing combustion. In terms of fire performance based on cone calorimetric test, generally the presence of nanoclays or nanotubes alone reduces peak heat release rates, they most often reduce time-to-ignition and extend total burning periods while affecting the overall heat release of the polymeric substrate a little [84,85]. A typical effect of layered silicate nanocomposites as imparting flame retardancy to thermoset resins is shown by taking an example from Gilman's work [76] in Figure 7, where HRR versus time curves are plotted from cone experiment. As can be seen TTI is reduced, PHRR is reduced, however the nanocomposite sample burns for a longer time. While slowing down the burning process but encouraging more rapid ignition, they also encourage increased char formation. For fire-retardant properties, it is believed that the presence of clay in a polymer promotes carbonaceous-silicate build up on the surface during burning, which insulates the underlying material [47–49] as discussed earlier.

In general, the nano-additives have shown more promising thermal and fire-reaction properties in thermoplastics than in thermosetting resin systems. The nanoparticles can disrupt the cross-linking density of the thermosetting resin and this effect is more pronounced as the dispersion of the particles improves [86]. The reduced cross-link density may promote thermal decomposition and reduce the char yield [87]. Nanoparticles can also affect the cure chemistry of the resin. As demonstrated by Lan et al. [88], the organic modifier catalyzes the curing and the epoxy monomer (diglycidyl ether of bisphenol A-DGEBA) can self polymerize leading to linear rather than three-dimensional cross-linked

30 and 250 °C, 42% at 250 and 475 °C, followed by 56% at 475 and 700 °C. Respective values for LY 5021 are 8% between 30 and 350 °C, 55% at 350 and 475 °C, and 55% at 475 and 650 °C [74].

Most of the work reported in literature on epoxy nanocomposites is on cast resins [44,46,74,86,92], the effect shown by nanoclays varies depending on the cross-linking, affected by curing conditions. The presence of nanoclays usually reduces the onset of decomposition temperature of the resin [74,93,94], which is partly due to decomposition the organic modifier of the nanoparticles and partly due to catalytic effect of the modifier in sensitizing the thermal degradation of the resin. In nitrogen atmosphere, the presence of nanoclays and nanotubes does not show much effect on the char yield, once the presence of the silicate in the residue is accounted for [95]. However in air, the char yield increases [74,93], which is indicative of their effect as oxidation resistance of the char by acting as insulating sheets. The effect of five different nanoparticles on thermal stability of the two types of resins discussed above, on TGA behavior of the epoxy resins is shown in Figure 8, where the percent mass difference between MY721/LY5052 and their nanocomposites as a function of the degradation temperatures are also plotted (Figure 8(b) and (d)).

The addition of nanoclays and carbon nanotubes to both types of resins leads to thermal destabilization of the resin in the lower temperature region (280–400 °C). The effect is more pronounced in LY5052 than in MY721, which leads to significant reduction in the onset of thermal degradation of the former. Of the five different types of nanoparticles studied for the high temperature curing resin, only nanosilica particles at a loading fraction of 30 wt% showed significant improvements in the thermal stability of the matrix at temperatures below 400 °C. Cloisite® 30B showed some improvement on the thermal stability of MY721 at temperatures above 530 °C eventually giving a residual char of about 2% at 800 °C. The DWNTs showed an adverse effect on the thermal degradation of MY721 over a wider temperature range (280–800 °C) while I.30E and VTP resulted in slight improvements in char formation at temperatures above 600 °C. The thermal stability of LY5052 was significantly enhanced at temperatures above 400 °C when VTP and I.30E are used. The same effect was observed for LY-30B-5 albeit at temperatures above 570 °C. The addition of nanoclays enhanced char formation (up to 5%) at temperatures above 570 °C (final oxidation stage) for all epoxy-clay composites. The flammability results for these samples are shown in Table 2. In terms of limiting oxygen index, a typical increase between 1% and 3% in the LOI values is seen with 5 wt% organoclay [74]. Similar results are reported by Hartwig et al. [44]. As can be seen from Table 2, highly cross-linked resin MY721 has high LOI value compared to less cross-linked resin, LY5052. In both the cases, the presence of nanoclays increased the LOI compared to the neat resin, though the effect is more pronounced in MY 721. In both the cases, VTP clays have shown the best result, which is due to the presence of phosphorus in the organic modifier. The UL-94 test results also show that while none of the nanoparticles could improve the UL-94 test requirements, nanosilica, VTP containing nanoclay and nanotubes could help improving the performance from HB to V-1 rating. The cone results for all of these samples presented in Table 2 indicate that all of these nanoclays could reduce the peak heat release by 5–12%, except for MY-VTP sample where 25% reduction in PHRR was observed. These nanoparticles do not show much effect on the total heat release rate. These results show that these nanoparticles on their own are not very effective in reducing the flammability of these resins. The best result in literature is reported by Gilman et al. [76], where DGEBA resin containing 6 wt% nanoclay and cured with methylenedianiline (MDA) or benzyldimethylamine (BDMA) showed 40% reduction in PHRR and mass loss rate, however TTI was also reduced.

Another important group of nanoparticles is layered double hydroxide or anionic clays, which function in a different manner than the cationic clays such as montmorillonites discussed above. When subjected to heat, the LDHs first release adsorbed and intercalated water, followed by endothermic decomposition of hydroxyl layers and decomposition of organic anions [46]. LDHs contribute to the flame retardancy of polymers by producing a refractory oxide residue on the surface of the material and releasing aqueous vapor and carbon dioxide during the decomposition. The endothermic nature of these processes and the dilution of combustible gases of pyrolysis increase the ignition time and reduce the heat release during the combustion [46]. Hsueh and Chen [90] and Zammarano et al. [46,96] have reported the use of layered double hydroxide (LDH) nanoclays in epoxy and reported that in TGA experiments they show similar effect as montmorillonites, i.e. lowering of onset of decomposition temperature and increase in char formation in air.

In particular, Zammarano [46] has reported the work done at NIST on the effect of LDH nanoclays on burning behavior of epoxy resin and compared with that of montmorillonite type clays by UL-94 and cone calorimetry. The sample details and cone results are reproduced here in Table 3. The UL-94 horizontal burning test showed that epoxy-LDH samples had reduced burning rate and self extinguished

Table 3 Cone Calorimeter Results of Epoxy Nanocomposite Samples at 35 kW m^{-2} [46]

Sample	Nanoclay	TTI (s)	PHRR (kW m^{-2}) (Δ%)	AvHRR (kW m^{-2}) (Δ%)	Residue Yield (%)
Epoxy	–	109	1181	533	3.3
Epoxy-MMT	C14–C18 Primary alkylamine montmorillonite (Laviosa)	110	862 (27)	477 (11)	8.6
Epoxy-LDH1	3-Aminobenzenesulfonate modified LDH	98	715 (40)	382 (28)	8.4
Epoxy-LDH2	4-Toluenesulfonate modified LDH	112	584 (51)	347 (35)	9.5

TTI, time-to-ignition; PHRR, peak heat release rate; AvHRR, average HRR. The data are reproducible to within ±10%. Sample size = square plates 100 mm large and 8 mm thick.

whereas epoxy-MMT sample burned completely. Cone results shown in Table 3 also confirm the higher performance of LDH nanocomposites; PHRR for epoxy-LDH1 (aminobenzenesulfonate modified LDH) and epoxy-LDH2 (4-toluenesulfonate modified LDH) reduced by 51% and 40%, respectively as compared to the neat epoxy resin, whereas for epoxy-MMT only 27% reduction was observed. Zammarano et al. [96] also observed that LDH clays are more effective in thin samples as compared to thick samples, which is contrary to the action of MMT based cationic clays. This can be explained based on their different mechanism of action. LDHs function by releasing water vapors and CO_2, which is more effective when the polymer mass is less in the matrix, whereas MMTs function by physical means by aggregating on the surface and get more chance of aggregating on the surface when samples are thick. They also observed that when LDH clays are used in epoxy based coatings, the coating intumesced, which further supports this hypothesis. This work has shown that there is a potential of using these LDH nanoclays in flame-retardant coatings.

As discussed before and seen from the above discussion, nanoparticles though reduce the flammability of polymers or thermosets (in context of this chapter), are not effective on their own to help the product to pass commercial flammability tests such as UL-94, flame spread, etc. Hence, their use in combination with conventional flame retardants is widespread. Thermoplastics nanocomposites quite often show synergistic effect with conventional flame retardants [45,46,51,52]. Some of the results for epoxy resin are discussed here to explore any such action.

Hussain et al. [94] have studied the effect of nanoclay on DGEBA and tetra-functional, tetraglycidyl diamino diphenyl methane (TGDDM) epoxy resins, containing chemically reactive flame retardant DOPO (9,10-dihydro-9-oxa-10-phosphaphenanthrene-10-oxide) at 3% P level. The nanoclay presence showed anatogonistic effect in terms of reduction in PHRR values. The reduction of PHRR, as compared to the neat DGEBA, was 40%, 50%, and 38% for the clay–DGEBA nanocomposite, the 3 wt% P-modified epoxy and the 3 wt% P-modified epoxy nanocomposite, respectively. The cone results concluded that incorporating both clay and DOPO in epoxy resins provided no synergistic improvement in flame retardancy, even more an antagonistic effect was observed in most properties. Similar observations were made by Liu et al. [97] that by adding nanoclays in DGEBA and TGDDM resins, flame retarded by curing with diethyl toluene diamine (DETDA) and bis(4-aminophenoxy)phenyl phosphonate (BAPP), there was no enhancement in flame-retardant properties. Toldy et al. [98] also used phosphorus-based curing agents to flame-retard epoxy resins. However, they found that by adding 1 wt% nanoclay, the LOI of the resin increased from 33% to 36%.

Schartel et al. [45,99] have used a phosphorus-containing organic modifier (tetraphenyl-phosphonium) for the montmorillonite based clay. This clay in DGEBA resin showed improved flame-retardant properties compared to commercial Nanomer 1.28E (Nanocor, USA) clay. However, with additional presence of organophosphorus flame retardants, the LOI values observed were lower than those for samples containing flame retardants only. These antagonistic effects could be due to the effect of nanoclays or nanotubes on the cross-linking density of the resin [74,86,87]. Torre et al. [100] however, have reported that the addition of nanoclays and conventional phosphorus-based flame retardants do not affect the cure chemistry or cross-linking, on the contrary had catalytic effect in curing reactions.

In 2011, in our laboratory, we studied the effect of nanoclays and nanotubes without or in the presence of conventional flame-retardant additives on the flammability of epoxy resin and glass fiber-reinforced composites produced from these modified formulations [101]. From a series of nanoparticles discussed in Table 2, VTP clay and DWCNTs, and from two resins, LY5052 was chosen for this work. Flame retardants chosen included resorcinol bis-(diphenyl phosphate) (RDP), ammonium polyphosphate (APP), and tris-(tribromoneopentyl) phosphate (TBP). The effect of flame retardants and nanoparticles on the thermal stability of the resin, studied by TGA, is shown in Figure 9. When FR additives (15 wt%) or VTP containing nanoclay (5 wt%) were used independently, the overall thermal stability enhancement of the epoxy resin follows the order APP > nanoclay VTP > RDP > TBP. The

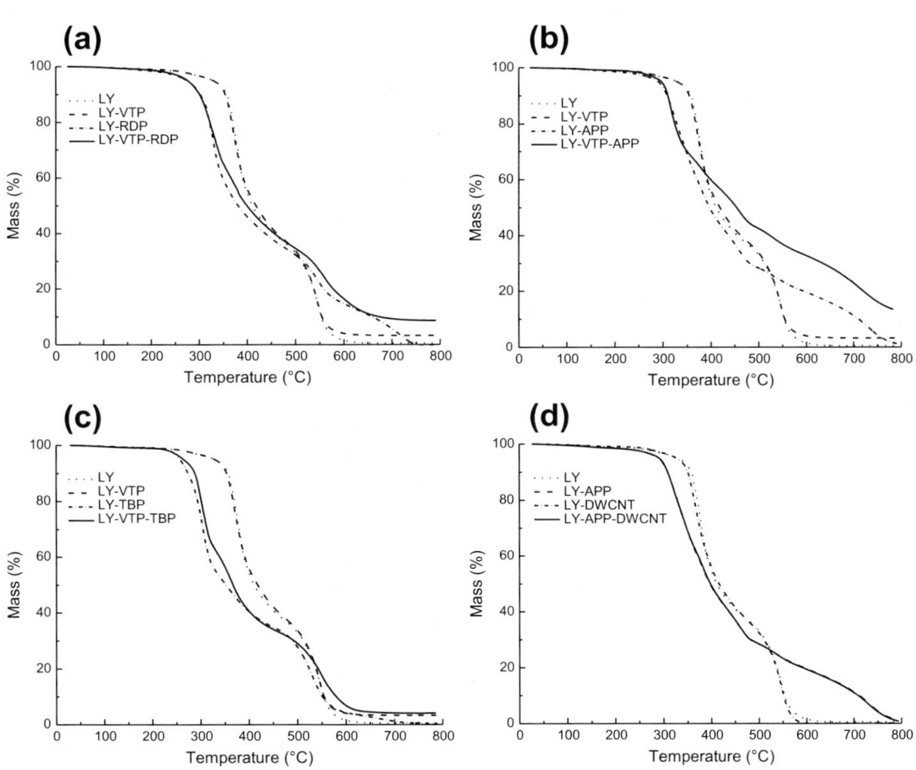

FIGURE 9

TGA mass loss—temperature profiles for resin, resin-FR, resin-VTP/DWCNT, and resin-FR-VTP/DWCNT formulations a) VTP and RDP, b) VTP and APP, c) VTP and TBP and d) DWCNT and APP combinations.

Reproduced from Ref. [101] with the permission from Sage Publications.

addition of DWCNTs alone did not significantly alter the thermal stability of the epoxy resin matrix. The concomitant addition of VTP together with RDP, APP, and TBP at a cumulative weight fraction of 20% led to synergistic improvements in the thermal stability of the resin. The reaction to fire properties of the glass fiber-reinforced laminates prepared from modified resin formulations were measured under cone calorimetry conditions at an incident heat flux of 50 kW m^{-2} and the results are shown in Table 4. The addition of conventional flame retardants significantly reduced the PHRR and THR relative to the control sample. The clay alone also reduced both the PHRR and THR compared to the control resin. However, when VTP clay and flame retardants were present, there was a little further improvement except for with RDP. With APP and TBP there was antagonistic effect, as also observed by other researchers [44,94,97,99]. As an example, HRR and mass loss versus time curves for formulations containing APP are shown in Figure 10. DWCNTs alone did not show a significant effect on the reaction to fire properties of the epoxy resin composites, whereas in the presence of APP, there was a significant reduction in both PHRR and THR.

Table 4 Composition of Resin Formulation (LY-Series) and Cone Colorimetric Data for LY5052 and Nanocomposites-Based Fiber-Reinforced Composites (GF Series) [101]

	Composition of the Resin (50 wt% of the Composite[a])			Cone Calorimetry			
Sample	**Epoxy**	**FR**	**Nanoclay/ DWCNT**	**TTI (s)**	**PHRR (kW m^{-2})**	**THR (MJ kg^{-1})**	**Char Yield (%)**
LY	100	0	0/0	47	713	40	54
LY-VTP	95	0	5/0	52	571	34	56
LY-RDP	85	15	0/0	36	474	34	55
LY-APP	85	15	0/0	34	190	32	61
LY-TBP	85	15	0/0	29	493	27	53
LY-VTP-RDP	80	15	5/0	30	440	34	56
LY-VTP-APP	80	15	5/0	35	262	32	63
LY-VTP-TBP	80	15	5/0	32	586	30	54
LY-DWCNT	99.5	0	0/0.5	44	682	40	56
LY-DWCNT-APP	84.5	15	0/0.5	26	230	30	61

LY5052, a low temperature curing bifunctional epoxy resin; VTP, vinyl triphenyl phosphonium bromide nanoclay; RDP, resorcinol bis-(diphenyl phosphate); APP, ammonium polyphosphate; TBP, tris-(tribromoneopentyl) phosphate; DWCNTs, double-walled carbon nanotubes.
[a] *Composite composition; glass:resin = 1:1.*

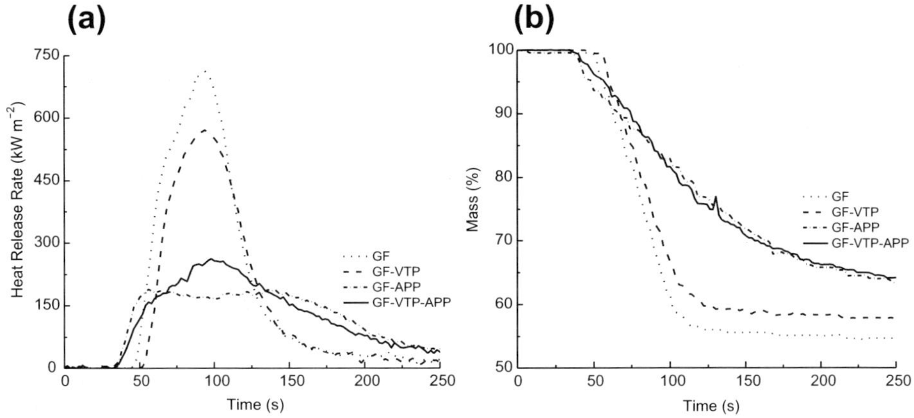

FIGURE 10

The (a) HRR and (b) mass profiles for GF, GF-VTP, GF-APP, and GF-VTP-APP at 50 kW m^{-2}, respectively.

Reproduced from Ref. [101] with the permission from Elsevier.

5.1.2 Carbon nanotubes: novel methods of usage

Wu et al. [102] have used a novel approach of using carbon nanotubes in the surface layer of the fiber-reinforced composites. Single-walled carbon nanotube (SWCNT) and multi-walled carbon nanotube (MWCNT) membranes (bucky paper) and carbon nanofiber (CNF) paper were incorporated onto the surface of epoxy carbon fiber composites, as proposed fire shields. Bucky papers (carbon nanotube membranes) are freestanding mats of tangled carbon nanotube ropes, which can be fabricated by the filtration of carbon nanotube suspensions. Dense nanotube networks and small pore size within the bucky paper provide low gas and mass permeability, which means bucky paper may act as an inherent flame-retardant shield when applied to the polymeric material surface. Their flammability behavior was investigated by a cone calorimeter. SWCNT bucky paper and CNF paper did not show notable improvement on fire retardancy. However, MWCNT bucky paper acted as an effective flame-retardant shield, reducing the peak heat release rate by more than 60% and reducing smoke generation by 50% during combustion. The results are shown in Table 5. The thermo-oxidation stability and low gas permeability of bucky papers or CNF nanofiber are key roles in improving flame-retardant properties of composites. In the cone calorimeter test condition, the MWCNT-based bucky paper, due to its high temperature thermo-oxidation and dense network, acted as an effective fire shield to reduce heat, smoke, and toxic gases generated during fire combustion. SWCNT-based bucky paper was burnt out after combustion in the Epoxy-SWCNT-BP composite and did not affect the flammability of the composite. In the case of Epoxy-CNF paper composite, the big pore size of the CNF paper network resulted in high gas permeability, which was the reason for its low flame-retardant efficiency.

Table 5 Cone Calorimeter Results of Carbon Fiber-Reinforced Epoxy Composites Samples Containing Carbon Nanotube and Nanofiber Membranes as Surface Layers at 50 kW m^{-2} [102]

	Bucky Paper/CNF Paper					
Sample	**Thickness (μm)**	**Surface Density (mg cm^{-2})**	**TTI (s)**	**PHRR (kW m^{-2})**	**THR (MJ m^{-2})**	**TSR (m^2 m^{-2})**
Epoxy	–	–	46	568	23.2	1124
Epoxy-SWCNT-BP	15–20	1.29	50	526	24.5	1180
Epoxy-MWCNT-BP	20–25	1.54	64	258	13.2	526
Epoxy-CF paper	55–75	3.08	59	508	24.8	1165

Control composite sample contains six carbon fiber fabric layers; the others contain one extra bucky paper/CNF paper. TTI, time-to-ignition; PHRR, peak heat release rate; THR, total HRR; TSR, total smoke release. The data are reproducible to within ±10%. Sample size = 100 × 100 × ~2.5 mm.

Yu et al. [103] grafted molybdenum—phenolic resin (Mo—PR) onto the surface of multi-walled carbon nanotubes (MWCNTs), and used the modified MWCNTs (CNT-PR) to prepare epoxy nanocomposites by *in situ* polymerization process. Mo—PR or CNT-PR were used with 8 wt% melamine. Selected results for LOI, UL-94, and cone calorimetric tests are shown in Table 6. The LOI value increased with increasing concentration of Mo—PR or CNT-PR and in UL-94 test V-0 could be obtained with 5 wt% CNT-PR or 2%Mo—PR. For these two samples, PHRR also decreased from 900 kW m^{-2} in the control to 579 and 468 kW m^{-2}, respectively. Similarly, THR decreased and char yield increased. The better performance of samples containing CNT-PR and melamine was due to more dense and integrated residual char. The microstructure observation showed that the decomposition of melamine could prevent the aggregation of CNTs during combustion. As a result, the strength of char layer was strengthened, and the structure of the char layer was closely knitted.

5.1.3 POSS

Although there is extensive work done in the area of using POSS with epoxy, phenolic, polyester, and VE resins to improve the physical and mechanical properties as compiled in Kuo and Chang's [81] comprehensive review, there are very limited papers related to flame retardancy of thermoset resins. Notable work is by Zhang et al. [104], who prepared epoxy_POSS composites and studied the effect of varying concentrations of PSS (0, 5, 10, and 15 wt%) on thermal stability by TGA. With increasing POSS concentration, the onset of decomposition temperatures increased by 24—30 °C, the decomposition temperature by 28—36 °C, and the

Table 6 Flammability Results of Epoxy Resin Containing Molybdenum–Phenolic (Mo–Pr) Resin and Mo–Pr Grafted Carbon Nanotubes at 35 kW m^{-2} [103]

Sample	Composition			LOI (%)	UL-94	Cone Calorimetry Results at 35 kW m^{-2}		
	Mo–PR (wt%)	CNT-PR (wt%)	Melamine (wt%)			PHRR (kW m^{-2})	TSP (m^2 kg^{-1})	Residual char (%)
Epoxy	–	–	–	21.5	No rating	900	26.7	2.9
1Mo–PR-8Ma	1	–	8	28	V-2	-	-	-
2Mo–PR-8Ma	2	–	8	29.5	V-0	579	21.8	11.4
1CNT-PR-8Ma	–	1	8	27.7	V-2	527	25.2	11.6
3CNT-PR-8Ma	–	3	8	28.6	V-2	535	22.5	12.0
5CNT-PR-8Ma	–	5	8	29.5	V-0	468	20.8	15.0

Mo–PR, molybdenum phenolic resin; CNT-PR, molybdenum–phenolic resin grafted carbon nanotubes; Ma, melamine.

Table 7 TGA and Mechanical Property Retention after Thermo-Oxidation of Epoxy-POSS Composites [81]

	TGA Results			Mechanical Property Retention after Thermo-Oxidation[a] (%)	
Conc of POSS in Epoxy (wt%)	T_{Onset} (°C)	T_{Dec} (°C)	Char Yield at 800 °C (%)	Flexural Strength	Impact Strength
0	309	354	13.0	57	57
5	334	384	14.7	78	78
10	334	398	22.0	81	83
15	341	401	24.8	87	90

T_{Onset}, *temperature of onset of decomposition;* T_{Dec}, *temperature of decomposition.*
[a] *Thermo-oxidation performed by heating in a muffle furnace at 500 °C for 5 min.*

char yield by 1.7—11.8% (see Table 7). Mechanical properties were used as the index to show the effect of POSS on the thermo-oxygen degradation resistance of the resin. For this all the samples were placed in a muffle furnace in air at 500 °C for 5 min. After removal of the samples from the furnace, they were quickly tested for their flexural strength. The results shown in Table 7 indicate that the incorporation of POSS into epoxy networks enhanced the thermo-oxidation resistance of epoxy. POSS led to the formation of inert layer on the surface of materials which could protect the internal structure from decomposition. As a result, the retention of mechanical properties of EP/POSS hybrids increased with increasing POSS content. However, more than 10 wt% POSS showed no beneficial effect.

Wu et al. [105] synthesized a functional POSS (NPOSS) and studied its effect on flame retardancy of epoxy resin. The functional POSS, NPOSS was synthesized by the reaction between trisilanolisobutyl-POSS (TPOSS), which contains a partial T8 cage with one corner Si atom missing and triglycidyl isocyanurate (TGIC). The results of microscale combustion calorimeter indicated that the presence of NPOSS (10% weight ratio) in epoxy resin (EP) can decrease its peak heat release rate by about 30%. The TGA results however, indicated that the NPOSS reduced the onset of decomposition temperature of the epoxy resin, but increased its char yield from 0.2% to 1.8%. In a subsequent work [106], POSS and TGIC were added to an epoxy resin individually or in combinations at low concentrations. The epoxy resin samples containing additives were cast into resin laminates and also used to produce glass fiber-reinforced composite laminates. The flammability of the two types of laminates was studied by limiting oxygen index, UL-94, and cone calorimetry and the results are presented in Table 8. The incorporation of POSS can increase the LOI value of neat cast resin slightly from 19.8% to 20.7%. With further addition of TGIC (sample EP/TGIC/POSS), there is a little effect on LOI. TGIC alone is also not effective in changing the LOI of the resin (LOI for EP/TGIC = 19.9%). When tested with UL-94, all samples burned completely and could not achieve any rating. However, the

Table 8 Composition and Flammability Results of Epoxy-POSS/TGIC Cast Resins and Glass Fiber-Reinforced Composites [106]

Sample	Sample Composition (mass%)				LOI (%)	UL-94[a] Burning[b] Rate (mm s^{-1})	Cone Calorimetry Results at 50 kW m^{-2}		
	Fiber	Epoxy	POSS	TGIC			TTI (s)	PHRR (kW m^{-2})	THR (MJ m^{-2})
EP	0	100	0	0	19.8	25.0	108	1634	78
EP/POSS	0	90	10	0	20.7	22.7	99	774	56
EP/TGIC	0	90	0	10	19.9	25.3	86	1190	67
EP/POSS/TGIC	0	90	5	5	20.9	21.7	88	944	58
FEP	50	50	0	0	–	–	125	857	50
FEP/POSS	50	45	5	0	–	–	121	420	32
FEP/TGIC	50	45	0	5	–	–	108	620	47
FEP/POSS/TGIC	50	45	2.5	2.5	–	–	114	385	32

'–' means test not performed.
[a] *In UL-94 tests all samples failed the test and had no rating.*
[b] *Burning rate = time taken to burn 50 mm length in horizontal mode.*

rate of burning (total burn length, 50 mm/time taken to burn) given in Table 8 shows that TGIC had no effect on rate of burning of the epoxy resin. POSS reduced the rate of burning slightly from 25.0 to 22.7 mm min^{-1}, which was further reduced to 21.7 mm min^{-1} by additional presence of TGIC. The cone results indicated that the use of POSS (10 wt%) in epoxy resin (EP) can significantly decrease resin/composite's peak heat release rate, total heat release, and CO production. The mechanical performance in flexural modes of the fiber-reinforced composite laminates was not adversely affected by the addition of POSS and TGIC in the resin [106].

5.1.4 Graphene

Graphene, graphene oxide (GO), and organic phosphate functionalized graphite oxides (FGO) have been reported to be used as flame retardants in epoxy (Ep) by Guo et al. [107]. The thermogravimetric results showed that Ep-graphene had the highest onset of decomposition temperature and maximum weight loss temperature, compared with those containing GO and FGO. The flame-retardant properties were investigated by micro-combustion calorimeter, where both EP/graphene and EP/FGO composites seemed to perform better than EP/GO composites. Ávila et al. [82,108] have studied the effect of nanoclay Cloisite 30B and surface modified graphene nanosheets on post-fire performance under bending load of carbon fiber-reinforced composites. The samples were exposed to 800 kW m^{-2} heat flux, for a period up to 120 s. The addition of nanoparticles (nanoclay and graphene nanosheets) increased the unburned thickness from 0.16 mm (original) to 2.63 mm and 2.74 mm, respectively. The impact resistance in post-fire low velocity impact tests, however, degraded as a function of heat exposure. Only nanoclays could increase the impact peak force slightly, i.e. by 12%.

In our own work, the effect of flame retardants and graphene on the flammability of glass fiber-reinforced epoxy composites was studied [109]. Table 9 shows the formulation of the samples prepared. As expected, incorporation of fire retardants (DOPO and MP) into epoxy resins decreases the PHRR and PHRR as shown in the cone results reported in Table 9. However, time-to-ignition (TTI) was also reduced; such phenomenon was probably due to an earlier decomposition of the phosphorus-containing compounds (DOPO and MP) that released some flammable gases accelerating the combustion of the material. Incorporation of graphene promoted further reduction of the PHRR and THR. The effect however, was less than that observed in a similar study on PBS (polyhedral oligomeric silsesquioxanes) [110], where graphene and melamine phosphate were added by melt blending. With 18 wt% MP and 2 wt% graphene, PBS could achieve the LOI value of 34.0% and UL-94 V0 grade, whereas with 20 wt% MP, the LOI was 31.5% and V1 UL-94 rating. The PHRR in the cone calorimetric test was also much lower (425 kW m^{-2}) in the former compared to MP only containing sample (691 kW m^{-2}).

5.1.5 Other nanoparticles

He et al. [111] have used nano-$CaCO_3$ particles to improve thermal and mechanical properties of DGEBA epoxy cast resin. $CaCO_3$ nanoparticles were surface treated

Table 9 Composition and Cone Colorimetric Data for Fiber-Reinforced Epoxy-Graphene-Flame-Retardant Composites [109]

	Composition of the Resin (50 wt% of the Composite)					Cone Calorimetry			
Sample	**Epoxy**	**FR**	**Graphene**	**MP**	**DOPO**	**TTI (s)**	**PHRR (kW m^{-2})**	**THR (MJ kg^{-1})**	**EHC (MJ kg^{-1})**
EP	100	0	0	0	0	44	853	51.9	25.7
EP-MP	90	0	0	10	0	38	528	48.8	22.2
EP-MP-G	90	15	1	9	1	36	483	47.9	21.7
EP-DOPO	90	15	0	0	0	34	624	41.3	20.5
EP-DOPO-G	90	15	1	0	1	32	538	36.5	20.0

EP, epoxy resin; MP, melamine phosphate; G, graphene; DOPO, 9,10-dihydro-9-oxa-10-phosphaphenanthrene-10-oxide.

with a silane coupling agent to improve interfacial properties. The thermal stability was studied by TGA. The results indicated that with 4 wt% nano-$CaCO_3$, $T_{50\%}$ (temp where 50% mass loss occurs) increased from 419 in the control resin to 425 °C, whereas the T_{max} (where max. mass loss occurs) increased from 410 to 422 °C. No flammability properties have been studied for these samples.

5.2 Unsaturated polyester

In our own laboratories [79,80], investigation has been made on the thermal degradation effects of introducing functionalized nanoclays along with phosphorus-containing flame retardants in unsaturated polyester resin. Subsequent work has reported the effect of nanoclays and flame retardants on cone calorimetric properties [53,112]. In this work, a typical polyester resin—(Crystic 471 PALV (Scott Bader)) was investigated by DTA-TGA in the presence of a range of clays (Cloisite Na^+, 10A, 15A, 25A, and 30B (Southern Clay Products, USA) same as in Table 2) and phosphorus-containing compounds [ammonium polyphosphate (Antiblaze MCM, Rhodia Specialities (now Solvay)), melamine phosphate (Antiblaze NH, Rhodia Specialities (now Solvay)), dipentaerythritol/melamine phosphate intumescent mixture (Antiblaze NW, Rhodia Specialities (now Solvay))], and alumina trihydrate. Smoke suppressants were included as well, particularly zinc borate (ZB, Firebrake ZB, Borax), zinc stannate (ZS, Flamtard S, William Blythe Ltd.), and zinc hydroxy stannate (ZHS, Flamtard H, William Blythe Ltd.). The details of the formulations are given in Table 10.

Polyester resin starts to decompose above 200 °C, the main decomposition occurs between 300 and 400 °C. Above 400 °C, solid phase oxidation of the char occurs, leaving very little char residue at higher temperatures (1% at 800 °C, see Table 10). For polyester nanocomposites, thermal stability of the resin is reduced below 400 °C; above 600 °C, thermal stability is increased slightly as can be seen from Figure 11(a), where the difference between TGA experimental and calculated (from weighted average component responses) masses versus temperature are plotted. Above 600 °C, char formations are similar to the expected from respectively calculated values and the type of clay has no effect on residual char formation at high temperatures. This gives indication that nanoclays on their own are not effective in increasing char formation and hence, reducing flammability of the resins. TGA results with flame retardants indicate that all flame retardants enhance char residues at higher temperatures as seen from Table 10. Additional presence of clays increases their thermal stability further. As an example in Figure 11(b), the actual and calculated TGA responses for Res/Clay 25A/APP formulations are plotted, where it can be seen that the thermal stability of this formulation is less than expected below 500 °C, but after that it is more stable, suggesting synergistic effect of clay and FR combination. Smoke suppressants also reduced thermal stability of Res/Clay/FR samples at lower temperatures, but at 800 °C, there was not much difference in char yields (see Table 10).

Table 10 TGA and Cone Calorimetric Data for Polyester–Clay Nanocomposites with and without FRs and Smoke Suppressants

Sample	TGA Results % Mass Residue		Cone Calorimetry Data at 50 kW m^{-2}				
	600 °C	800 °C	TTI (s)	FO (s)	PHRR (kW m^{-2})	THR (MJ m^{-2})	Smoke (l)
Resin	1.1	1.0	34	136	1153	79.0	761
Resin/Nanoclays							
Res/Clay10A	4	4	41	165	843	74.1	790
Res/Clay15A	4	4	36	160	886	75.5	823
Res/Clay25A	6	5	36	147	902	70.7	763
Res/Clay30B	6	5	37	158	853	72.7	781
Resin/FR/Smoke Suppressants							
Res/APP	30	3	31	190	478	52.5	754
Res/NH	16	7	33	205	526	62.9	684
Res/NW	16	7	30	159	740	58.0	1183
Res/ATH	14	10	38	198	636	66.3	761
Res/APP/ZB	15	14.4	37	234	404	54.9	730
Res/APP/ZS	14	13.5	28	176	578	60.6	610
Res/APP/ZHS	14	12.8	35	209	615	67.0	749
Resin/Nanoclay/FR							
Res/Clay10A/APP	23.9	16.4	35	251	357	59.7	717
Res/Clay10A/NH	17.8	8.9	36	280	493	70.0	1172
Res/Clay10A/NW	15.6	7.8	33	313	528	71.8	1443
Res/Clay10A/ATH	15.3	14.8	42	471	460	68.8	1212
Res/Clay10A/APP	23	15	34	224	454	59.2	764
Resin/Nanoclay/FR/Smoke Suppressants							
Res/Clay25A/APP/ZB	17	16.6	35	235	531	65.7	720
Res/Clay25A/APP/ZS	16.5	15.8	35	211	586	67.8	778
Res/Clay25A/APP/ZHS	16.7	15.9	38	228	521	65.9	717

Clay10A, Closite10A (modified by benzyl hydrogenated tallow quaternary ammonium chloride); Clay15A, Cloisite 15A (modified by dimethyl dehydrogenated tallow quaternary ammonium chloride); Clay25A, Cloisite 25A (modified by dimethyl hydrogenated tallow 2-ethylhexyl quaternary ammonium methyl sulfate); Clay 30B, Cloisite 30B (clay modified with methyl tallow bis-2-hydroxyethyl quaternary ammonium chloride); APP, ammonium polyphosphate; NH, melamine phosphate; NW, dipentaerythritol/melamine phosphate; ATH, alumina trihydrate; ZB, zinc borate; ZS, zinc stannate; ZHS, zinc hydroxy stannate.

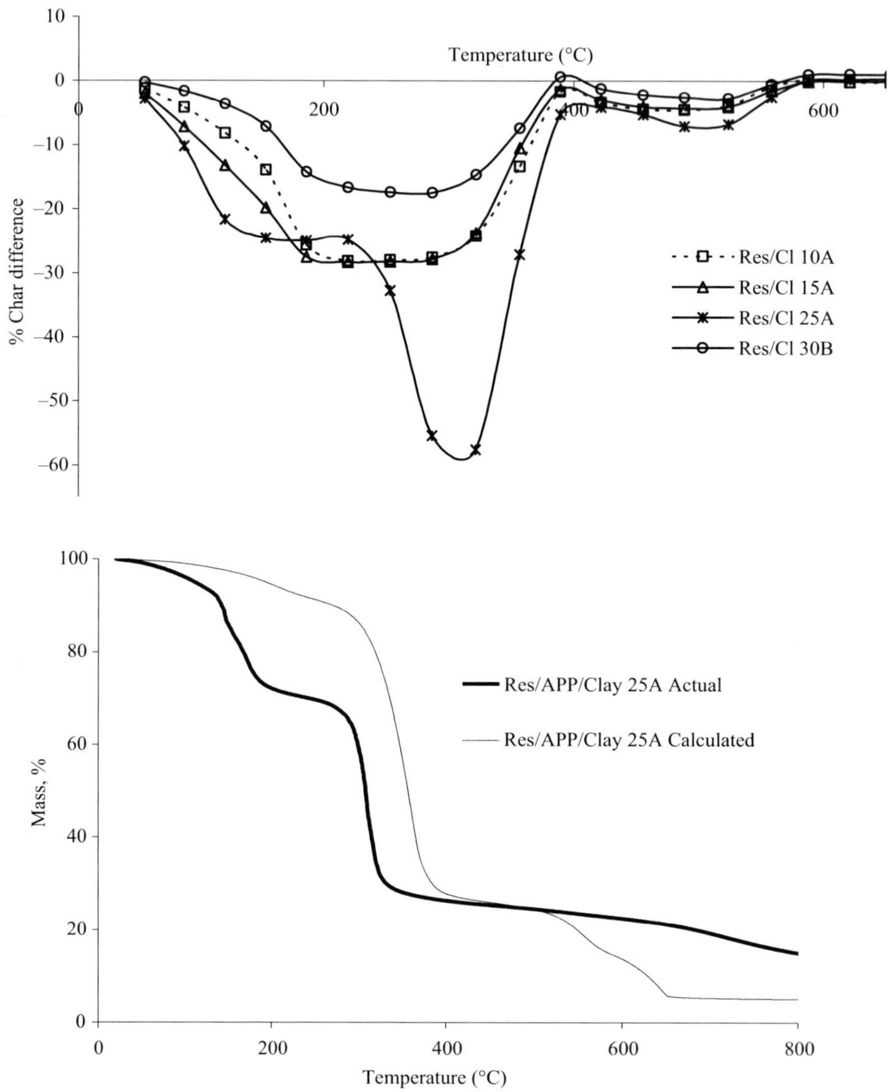

FIGURE 11

(a) Mass difference between a sample and the control resin-temperature for polyester nanocomposites and (b) TGA mass–temperature curve for Res/APP/25A sample.

Reproduced from Ref. [79] with the permission of RSC Publications.

The fire performance was derived by cone calorimetry at 50 kW m^{-2} incident flux and expressed in terms of peak heat release (PHRR), total heat release (THR), and smoke evolution as shown in Table 10. The presence of nanoclays slightly increased the ignition time of the resin but reduced the PHRR values

(22–27%). Longer burning times, however, resulted in higher total heat release values of the samples containing nanoclays. The amount of smoke also increased in resin/clay samples. All the flame retardants on their own lowered the ignition times and PHRR values of the resin. APP shows highest reduction (60%) in PHRR compared to that of pure resin. Total heat release (THR) and smoke production were also reduced in the presence of the FRs except for sample containing NW, where the total amount of smoke released is increased compared to resin sample. However, when both clay and FRs were present, the TTI values increased and the PHRR values further decreased as compared to respective Res/FR samples. Presence of clay in flame-retarded polyester resin, in general, increased the burning time of the samples and hence increased THR values of Res/Clay/FR samples compared to Res/FR samples. Incorporation of clays in the flame-retarded resin significantly increases the amount of smoke production during burning of specimens. This may be probably due to incomplete combustion of the specimen in presence of insufficient oxygen. Further cooling of non-flammable gases, flammable gases, and micro-solid particles may also lead to generation of heavy smoke. Presence of smoke suppressant zinc borate (ZB) reduced PHRR and slightly reduced smoke values in Res/APP/ZB compared to the Res/APP sample. Zinc stannate (sample Res/APP/ZS) reduced smoke generation but increased PHRR values. However, zinc hydroxy stannate increased both PHRR and smoke generation compared to the Res/APP sample. In conclusion, while their appears to be no general improvement in fire performance when nanoclays are added to conventionally flame-retarded resins, there is evidence that in certain formulations such as those containing APP and ATH, some benefits are observed and this opens opportunities for favorable exploitation of introducing nanoclays and other nanoparticles into flame retardant resin formulations for use in reinforced composites having improved fire properties.

In a subsequent work 10A and VTP clay were selected to prepare glass fiber-reinforced polyester composite laminates with all types of flame retardants. The sample compositions and their mechanical and cone results are shown in Table 11. Flexural modulus results generally show that the inclusion of the functionalized nanoclays enhances the flexural performance of the resin due to nano-level reinforcement whereas the inclusion of flame retardants adversely affect the flexural moduli of the composite laminates. The presence of clay in Res/Clay/FR samples show small but not significant improvement compared to Res/FR samples. Cone results for glass-reinforced composites in general show slight increase in TTI and flameout (FO) values of samples with and without flame retardants. PHRR and smoke production are reduced with regard to the control sample. Despite the higher TTI values and longer FO times, all the samples with Res/Clay/FR formulation show reduced THR values suggesting reduced flammability. Composite laminate Res/Clay (i, VTP)/APP shows the highest reduction (43%) in PHRR value as compared to control sample and this is probably due to presence of phosphonium group in the organic modifier of functionalized clay. HRR versus time curves of resin, Res/Clay (i) with and without APP are graphically shown in Figure 12 [6].

Most of the work on unsaturated polyester is on nanoclays, however, some work has been done on other nanoparticles. Zhuge et al. [113] have used exfoliated

Table 11 Physical, Flexural, and Flammability Properties of Glass Reinforced Polyester-Nanocomposites Samples

Samples	Composition (mass %)				Thickness (mm)	Flexural Modulus E (GPa)	Cone Calorimetry Data at 50 kW m^{-2}				
	Glass	Resin	Clay	FR			TTI (s)	FO (s)	PHRR (kW m^{-2})	THR (MJ m^{-2})	Smoke (l)
Res	67	33	–	–	2.42	14.91	36	251	401	31.0	359
Res/Clay10A	62	36.1	1.9	–	2.43	16.35	41	270	370	35.3	357
Res/Clay(i)	62	36.1	1.9	–	2.44	15.59	39	237	358	33.8	320
Res/APP	67	27.4	–	5.6	2.73	8.82	37	260	245	28.4	281
Res/ATH	63	30.7	–	6.3	2.79	8.32	40	231	260	33.8	423
Res/NH	64	29.9	–	6.1	2.67	9.94	42	238	340	30.1	422
Res/NW	64	29.9	–	6.1	–	–	39	296	239	22.5	389
Res/Clay10A/APP	67	26.1	1.3	5.6	2.72	9.50	39	238	287	26.5	335
Res/Clay(i)[a]/APP	67	26.1	1.3	5.6	2.74	8.63	38	295	229	31.6	413
Res/Clay10A/ATH	67	26.1	1.3	5.6	2.59	12.30	40	265	288	30.2	408
Res/Clay10A/NH	67	26.1	1.3	5.6	2.72	73.99	40	240	291	30.8	409
Res/Clay10A/NW	67	26.1	1.3	5.6	–	–	42	272	224	26.2	378

[a] *Clay (i): VTP clay*

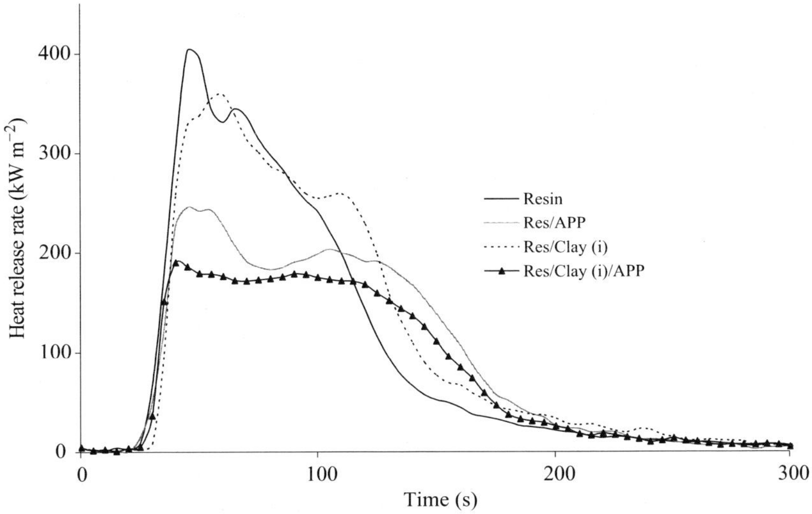

FIGURE 12

HRR vs time curves for glass fiber-reinforced polyester composites containing functionalized nanoclays with and without APP.

Reproduced from Ref. [6] with the permission from Woodhead Publishing Ltd.

graphite nanoplatelets (xGnP) in the presence of ammonium polyphosphate (APP) to improve the flame resistance of glass fiber-reinforced polyester composites. The cone calorimeter test results indicated that there was a synergistic effect between xGnP and APP at the optimal loading levels of 3 and 17 wt%, respectively. The flame resistant performance of the nanocomposites was further improved by applying xGnP-dominant carbon nanofiber (CNF)/xGnP hybrid nanopaper onto the surface of the samples.

Tibiletti et al. [114] incorporated nano-alumina (median diameter of 13 nm) and submicronic alumina trihydrate (median diameter of 300 nm) particles, individually or mixed together in an unsaturated polyester resin at various loadings. Synergistic effects on thermal stability by TGA and heat release rate by cone calorimetry were observed for combinations between both kinds of nanoparticles. The best result for fire behavior was obtained for a global loading of 10 wt% with 50:50 wt% ratio. Mass loss curves also exhibited increased char yield for mixed compositions. Moreover, the replacement of nano-alumina by a pyrogenated hydrophilic silica with the same particle size also lead to synergistic effects on fire behavior observed by cone calorimetry. These synergistic effects were ascribed to physical effects resulting from the arrangement of both kinds of mineral particles of very different median size at the surface of the composite during polymer combustion and ablation. The formation of this mineral barrier can also promote catalytic effects ascribed to the huge specific surface area of oxide nanoparticles.

5.3 Vinyl ester

Gilman et al. [76] have studied the effect of nanoclays on a nitrile rubber modified bisphenol-A epoxy based vinyl ester (Mod-bis-A), or a combination of bisphenol-A and novolac epoxy based vinyl ester (Bis-A/novolac). With 6 wt% Cloisite 15A, 25% and 39% reduction in PHRR for Mod-bis-A and Bis-A/novolac, respectively were observed by cone calorimetry. The clay also helped in improving char formation.

Another notable work in this area is by Chigwada et al. [77], who have investigated the effect of different types of nanoclays (i.e. modified montmorillonite, (Cloisite 15A), and magadiite) and POSS on thermal stability and flammability of VE resin. TGA results indicated that all of these nanoparticles did not affect the onset of decomposition temperature, however the char formation was increased. Both nanoclay Cloisite 15A at 6 wt% and POSS at 5 wt% level could reduce PHRR by about 30% compared to neat resin. They also studied the effect of these nanoparticles in combination with commercial flame retardants, tricesylphosphate (TCP), and resorcinol di-phosphate (RDP) [77]. With RDP at 30 wt%, there was no effect on PHRR with further nanoclay presence. With TCP, however, much lower PHRR was observed. Similar trend was seen for glass fiber-reinforced composites.

6. CONCLUSIONS AND FUTURE TRENDS

Thermosetting nanocomposites, despite showing increased thermal stability by TGA and reduction in peak heat release rate (PHRR) by cone calorimetry, are not as good as their thermoplastic counterparts. On their own, nanocomposites are incapable of passing commercial fire tests such as UL-94 and passing fire test standards. For this reason, traditional flame retardants are currently used in combination with nanoparticles, which enables a reduction of concentration of the latter to be made in the resin formulations. The epoxy nanocomposites have shown to have antagonistic effect with conventional flame retardants. This behavior is due to the effect of nanoparticles on the cross-linking of the resin. The results shown by unsaturated polyester and vinyl ester nanocomposites are better than for the epoxies. Future trends involve development of new kinds of nanoparticles and particularly those having flame-retardant chemicals incorporated in their structure, for example nano sized magnesium and aluminum hydroxides. Other potential nanoparticles include fluorinated synthetic mica, magadiite, and layered double hydroxides (LDHs). There is always a possibility of the use of multiple types of nanofillers in the same polymer to yield a multi-component nanocomposite. Another possible way is the use of nanoparticles as a surface layer, as shown by some researchers using thin films made of carbon nanotubes. Innovative FR systems can be designed by using optimized mixtures of nano- and microparticles or using microparticles in the resin and nanoparticles on the surface to maximize the barrier effect.

References

[1] Marsh G. GKN aerospace extends composites boundaries. Reinf Plast 2006;50:24–6.

[2] Baudry A, Dufay J, Regneir N, Mortaigne B. Thermal degradation and fire behaviour of unsaturated polyester with chain ends modified by dicyclopentadiene. Polym Degdn Stab 1998;61:441–52.

[3] Anon. Fire safety aspects of polymeric materials. In: Materials state of art, chapter 6, a report by national materials advisory board, national academy of sciences, vol. 1. Washington: Technomic Publ; 1977.

[4] Marsh G, Jacob A. Trends in marine composites. Reinf Plast 2007;51:22–7.

[5] Marsh G. Composites fight for share of military applications. Reinf Plast 2005;49: 18–22.

[6] Kandola BK, Kandare E. Composites having improved fire resistance. In: Horrocks AR, Price D, editors. Advances in fire retardant materials, chapter 5. Cambridge, UK: Woodhead Publishing Ltd; 2008. pp. 398–442.

[7] Horrocks AR, Kandola BK. Flammability and fire resistance of composites. In: Long AC, editor. Design and manufacture of textile composites, chapter 9. Cambridge, UK: Woodhead Publishing Ltd; 2005.

[8] Scudamore MJ. Fire performance studies on glass-reinforced plastic laminates. Fire Mater 1994;18:313–25.

[9] Brown JE, Loftus JJ, Dipert RA. Fire characteristics of composite materials - a review of the literature. Report; 1986. NBSIR: 85–3226.

[10] Kourtides DA, Gelwee VJ, Parker JA. Thermochemical characterisation of some thermally stable thermoplastic and thermoset polymers. Polym Eng Sci 1979;19(1): 24–9.

[11] Kandola BK, Horrocks AR. Composites. In: Horrocks AR, Price D, editors. Fire retardant materials, chapter 5. Cambridge, UK: Woodhead Publishing Ltd; 2001.

[12] Aseeva RM, Zaikov GE. Combustion of polymer materials. Munich, Germany: Carl Hanser Verlag; 1986. p. 149.

[13] Levchik S, Wilkie CA. Char formation. In: Grand AF, Wilkie CA, editors. Fire retardancy of polymeric materials. New York: Marcel Dekker; 2000. pp. 171–215.

[14] Wilkie CA, Levchik SV, Levchik GF. Is there a correlation between crosslinking and thermal stability? In: Al Malaika S, Golovoy A, Wilkie CA, editors. Specialty polymer additives: principles and application. Oxford, England: Blackwell Science; 2001. pp. 359–74.

[15] Bishop DP, Smith DA. Combined pyrolysis and radiochemical gas chromatography for studying the thermal degradation of epoxy resins and polyimides, the degradation of epoxy resins in nitrogen between 400 and 700° C. J Appl Polym Sci 1970;14:205.

[16] Vogt J. Thermal analysis of epoxy-resins: Identification of decomposition products. Thermochem Acta 1985;85:407–10.

[17] Irvine DJ, McCluskey JA, Robinson IM. Fire hazard and some common polymers. Polym Degdn Stab 2000;67:383–96.

[18] Levchik SV. Thermal degradation of thermosetting polymers. In: Troitzsch J, editor. Plastics flammability handbook. 3rd ed. Munich: Hanser; 2004. pp. 83–98.

[19] Kandola BK, Horrocks AR, Myler P, Blair D. The effect of intumescents on the burning behaviour of polyester-resin-containing composite. Compos Part A 2002;33: 805–17.

[20] Mouritz AP, Gibson AG. Fire properties of polymer composite materials. Dordrecht Netherlands: Springer; 2006.
[21] BS EN ISO 4589 – 2. Plastics – determination of burning behaviour by oxygen index – part 2: ambient temperature test; 1999.
[22] BS 2782-1: Method 140A. ISO 1210:1992, Plastics — Part 1: Thermal properties — Method 140A: determination of the burning behaviour of horizontal and vertical specimens in contact with a small-flame ignition source; 1992.
[23] ISO 5660–1. Fire tests on building materials and structures — part 15: method for measuring the rate of heat release of products; 1993.
[24] Murrel J, Briggs P. Developments in European and international fire test methods for composites used in building and transport applications. In: Proceedings of the 2nd conference of composites in fire, Newcastle-upon-Tyne; September 2001.
[25] Briggs P. Fire performance of composites in European construction applications. In: Proceedings of the 3rd conference on composites in fire, Newcastle-upon-Tyne; September 2003.
[26] Troitzsch J. Plastics flammability handbook. 3rd ed. Munich: Hanser; 2004.
[27] FAA fire testing; http://www.faafiretesting.com.
[28] Lyon RE. Materials with reduced flammability in aerospace and aviation. In: Horrocks AR, Price D, editors. Advances in fire retardant materials. Cambridge, UK: Woodhead Publishing Ltd; 2008. pp. 573–98.
[29] Federal Aviation Regulation (FAR), Airworthiness standards, department of transportation, federal aviation administration, FAA specification, JAR 25.853 Part IV Appendix F.
[30] FMVSS 302. Motor Vehicle safety standard No. 302, flammability of materials – passenger cars, multipurpose passenger vehicles, trucks and buses, National Highway Traffic Safety Administration, Washington, DC.
[31] Ingason H. Model scale railcar fire tests. Fire Saf J 2007;42:271–82.
[32] SOLAS. consolidated text of the international convention for the safety of life at sea, 1974, and its protocol of 1978: articles, annexes and certificates. Consolidated Edition 2004.
[33] Sorathia U. Flame retardant materials for maritime and naval applications. In: Horrocks AR, Price D, editors. Advances in fire retardant materials. Cambridge, UK: Woodhead Publishing Ltd; 2008. pp. 527–72.
[34] Title 49: Transportation US Code of Federal Regulations. Chapter II federal railroad administration, Part 238 passenger equipment safety standards, vol. 4; October 1, 2006. Standard 103, Fire Safety (FRA 238.103).
[35] Hörold S. Phosphorus flame retardants in thermoset resins. Polym Degdn Stab 1999; 64:427–31.
[36] Kim J, Yooa S, Baea JY, Yunb HC, Hwangb H, Kongb BS. Thermal stabilities and techanical properties of epoxy molding compounds (EMC) containing encapsulated red phosphorous. Polym Degdn Stab 2003;81:207–13.
[37] Hussain M, Varley RJ, Mathus M, Burchill P, Simon GP. Development and characterization of a fire retardant epoxy resin using an organo-phosphorus compound. J Mater Sci Lett 2003;22:455.
[38] Wang X, Zhang Q. Synthesis, characterisation and cure properties of phosphorus-containing epoxy resins for flame retardance. Eur Polym J 2004;40:385–95.
[39] Kojima Y, Usuki A, Kawasumi M, Okada A, Fukushima Y, Karauchi T, et al. Synthesis of nylon 6-clay hybrid. J Mater Res 1993;8:1179–84.

[40] Kojima Y, Usuki A, Kawasumi M, Okada A, Fukushima Y, Karauchi T, et al. Mechanical properties of nylon 6-clay hydrid. J Mater Res 1993;8:1185−9.
[41] Pinnavaia TJ. Intercalated clay catalysts. Science 1983;220:365−71.
[42] Giannelis EP. Polymer layered silicate nanocomposites. Adv Mater 1996;8:29−35.
[43] Vaia RA, Giannelis EP. Polymer nanocomposites: status and opportunities. MRS Bull 2001;26:394−401.
[44] Hartwig A, Putz D, Chartel B, Bartholmai M, Wendschuh-Josties M. Combustion behaviour of epoxide based nanocomposites with ammonium and phosphonium bentonites. Macro Chem Phys 2003;204:2247−57.
[45] Kandola BK, Yenilmez A, Horrocks AR, Smart G. Effect of flame retardants on the thermal, burning and char formation behaviour of polypropylene − nanoclay compounded polymers. In: Morgan A, Nelson GL, Wilkie CA, editors. Fire and polymers. American chemical society symposium series 1013, chapter 5; 2009. pp. 47−69.
[46] Zammarano M. Thermoset fire retardant nanocomposites. In: Morgan AB, Wilkie CA, editors. Fire retardant poymer nanocomposites, chapter 9. Wiley - Interscience; 2007. p. 235.
[47] Bourbigot S, Gilman JW, Wilkie CA. Kinetic analysis of the thermal degradation of polystyrene−montmorillonite nanocomposite. Polym Degdn Stab 2004;84(3):483−92.
[48] Gong F, Feng M, Zhao C, Zhang S, Yang M. Thermal properties of poly(vinyl chloride)/montmorillonite nanocomposites. Polym Degdn Stab 2004;84(2):289−94.
[49] Lewin M. Some comments on the modes of action of nanocomposites in the flame retardancy of polymers. Fire Mater 2003;27:1−7.
[50] Cipiriano BH, Kashiwagi T, Raghavan SR, Yang Y, Grulke EA, Yamamoto K, et al. Effects of aspect ratio of MWNT on the flammability properties of polymer nanocomposites. Polymer 2007;48:6086−96.
[51] Zanetti M, Kashiwagi T, Falqui L, Camino G. Cone calorimeter combustion and gasification studies of polymer layered silicate nanocomposites. Chem Mater 2002;14:881−7.
[52] Tang T, Chen X, Chen H, Meng X, Jiang Z, Bi W. Catalyzing carbonization of polypropylene itself by supported nickel catalyst during combustion of polypropylene/clay nanocomposite for improving fire retardancy. Chem Mater 2005;17:2802.
[53] Nazare S, Kandola BK, Horrocks AR. Cone calorimetric study of flame-retardant unsaturated polyester resin incorporating functionalised nanoclays. Polym Adv Technol 2006;17:294−303.
[54] Zhu J, Uhl FM, Morgan AB, Wilkie CA. Studies on the mechanism by which the formation of nanocomposites enhances thermal stability. Chem Mater 2001;13:4649−54.
[55] Fan XB, Peng WC, Li Y, Li XY, Wang SL, Zhang GL, et al. Deoxygenation of exfoliated graphite oxide under alkaline conditions: a green route to graphene preparation. Adv Mater 2008;20:4490−3.
[56] Higginbotham AL, Lomeda JR, Morgan AB. Graphite oxide flame retardant polymer nanocomposites. ACS Appl Mater Interfaces 2009;1:2256−61.
[57] Wang X, Song L, Yang H, Lu H, Hu Y. Synergistic effect of grapheme on antidripping and fire resistance of intumescent flame retardant poly(butylenes succinate) composites. Ind Eng Chem Res 2011;50:5376−83.
[58] Zaman I, Phan TT, Kuan HC, Meng Q, La TB, Luong L, et al. Epoxy/graphene platelets nanocomposites with two levels of interface strength. Polymer 2011;52:1603−11.
[59] Martin-Gallego M, Verdejo R, Lopez-Manchado MA, Sangermano M. Epoxy-graphene UV-cured nanocomposites. Polymer 2011;52:4664−9.

[60] Luo JJ, Daniel IM. Characteriszation and modelling of mechanical behaviour of polymer/clay nanocomposites. Comp Sci Technol 2003;63:1607–16.

[61] Vu YT, Mark JE, Pharm LH, Engelhardt M. Clay nanolayer reinforcement of cis-1,4-polyisoprene and epoxidized natural rubber. J Appl Polym Sci 2001;82:1391–403.

[62] Zhang YQ, Lee LH, Jang HJ, Nah CW. Preparing PP/clay nanocomposites using a swelling agent. Compos Part B 2004;35:133–8.

[63] Park CII, Park OO, Lim JG, Kim HJ. The fabrication of syndiotactic polystyrene/organophilic clay nanocomposites and their properties. Polymer 2001;42:7465–75.

[64] Thostenson ET, Ren Z, Chou TW. Advances in the science and technology of carbon nanotubes and their composites: a review. Comp Sci Technol 2001;61:1899–912.

[65] Morgan AB, Gilman JW. Characterization of polymer-layered silicate (clay) nanocomposites by transmission electron microscopy and X-ray diffraction: a comparative study. J Appl Polym Sci 2003;87:1329–38.

[66] Messersmith PB, Giannelis EP. Synthesis and characterization of layered silicate-epoxy nanocomposites. Chem Mater 1994;6:1719–25.

[67] Chen B, Liu J, Chen H, Wu J. Synthesis of disordered and highly exfoliated epoxy/clay nanocomposites using organoclay with catalytic function via acetone-clay slurry method. Chem Mater 2004;16:4864–6.

[68] Ray SS, Okamoto M. Polymer/layered silicate nanocomposites: a review from preparation to processing. Prog Polym Sci 2003;28:1539–641.

[69] Zeng QH, Yu AB, Lu GQ, Paul DR. Clay-based polymer nanocomposites: research and commercial development. J Nanosci Nanotechnol 2005;5:1574–92.

[70] Alexandre M, Dubois P. Polymer-layered silicate nanocomposites: preparation, properties and uses of a new class of materials. Mater Sci Eng Reports 2000;28:1–63.

[71] Wypych F, Satyanarayana KG. Functionalization of single layers and nanofibres: a new strategy to produce polymer nancomposites with optimized properties. J Colloid Interface Sci 2005;285:532–43.

[72] Lan T, Kadiratna KD, Pinnavaia TJ. Mechanism of clay tactoid exfoliation in epoxy-clay nanocomposites. Chem Mater 1995;7:2144–50.

[73] Ma J, Yu ZZ, Zhang QZ, Xie XL, Mai YW, Luck I. A novel method for preparation of disorderly exfoliated epoxy/clay nanocomposite. Chem Mater 2004;16:757–9.

[74] Katsoulis C, Kandare E, Kandola BK. The effect of nanoparticles on structural morphology, thermal and flammability properties of two epoxy resins with different functionalities. Polym Degdn Stab 2011;96:529–40.

[75] Leroux F, Adachi-Pagano M, Intissar M, Chauviere S, Forano C, Besse JP. Delamination and restacking of layered double hydroxides. J Mater Chem 2001;11:105–12.

[76] Gilman J, Kashiwagi T, Nyden M, Brown JET, Jackson CL, Lomakin S, et al. Flammability studies of polymer layered silicate nanocomposites: polyolefin, epoxy, and vinyl ester resins. In: Ak-Malaika S, Colovoy A, Wilkie CA, editors. Chemistry and technology of polymer additivies, chapter 14. Malden: Blackwell Science Inc.; 1990. pp. 249–65.

[77] Chigwada G, Jash P, Jiang DD, Wilkie CA. Fire retardancy of vinyl ester nanocomposites: synergy with phosphorous-based fire retardants. Polym Degdn Stab 2005;89: 85–100.

[78] Suh DJ, Lim YT, Park OO. The property and formation mechanism of unsaturated polyester-layered silicate nanocomposites depending on the fabrication methods. Polymer 2001;41:8557–63.

[79] Kandola BK, Nazare S, Horrocks AR. Thermal degradation behaviour of flame - retardant unsaturated polyester resins incorporating functionalised nanoclays. In: Le Bras M, Bourbigot S, Duquesne S, Jama C, Wilkie C, editors. Fire retardancy of polymers: the use of mineral fillers in micro- and nano-composites. Cambridge, UK: Royal Chemical Society; 2005.

[80] Kandola BK, Nazare S, Horrocks AR, Myler P. Effect of layered silicate nanocomposites on burning behaviour of conventionally flame-retarded unsaturated polyesters. In: Nelson GL, Wilkie CA, editors. Fire and polymers. American chemical society symposium series 922, chapter 13; 2006. pp. 155−71.

[81] Kuo SW, Chang FC. POSS related polymer nanocomposites. Prog Polym Sci 2011;36: 1649−96.

[82] Ávila AF, Dias EC, Lopes da Cruz DT, Yoshida MI, Bracarense AQ, Carvalho MGR, et al. An investigation on graphene and nanoclay effects on hybrid nanocomposites post fire dynamic behaviour. Mater Res 2010;13(2):143−50.

[83] Wang Z, Tang XZ, Yu ZZ, Guo P, Song H, Du X. Dispersion of graphene oxide and its flame retardancy effect on epoxy nanocomposites. J Polym Sci 2011;3:368−76.

[84] Gilman JW, Kashiwagi T. Polymer-layered silicate nanocomposites with conventional flame retardants. In: Pinnavia TJ, Beall GW, editors. Polymer-clay nanocomposites in wiley series in polymer science. Chichester and New York: Wiley; 2001. pp. 193−206.

[85] Morgan AB, Wilkie CA, editors. Flame retardant polymer nanocomposites. Hokoben, NJ, USA: Wiley Interscience; 2007.

[86] Becker OC, Cheng YB, Varley RJ, Simon GP. Layered silicate nanocomposites based on various high-functionality epoxy resins: the influence of cure temperature on morphology, mechanical properties and free volume. Macromol 2003;36(5):1616−25.

[87] Bharadwaj RK, Mehrabi AR, Hamiltion C, Trujillo C, Murga M, Fan R, et al. Structure-property relationships in cross-linked polyester-clay nanocomposites. Polymer 2002;43(13):3699−705.

[88] Lan T, Kaviratna PD, Pinnavaia TJ. Epoxy self-polymerization in smectite clays. J Phys Chem Solids 1996;57:1005−10.

[89] Lewis AF, Doyle MJ, Gillham JK. Effect of cure history on dynamic mechanical-properties of an epoxy-resin. Polym Eng Sci 1979;19:683−6.

[90] Hsueh HB, Chen CY. Preparation and properties of LDHs/epoxy nanocomposites. Polymer 2003;44:5275−83.

[91] Kandola BK, Horrocks AR, Myler P, Blair D. Thermal characterization of thermoset matrix resin. In: Nelson GL, Wilkie CA, editors. Fire and polymers. American chemical society symposium series, vol. 797; 2001. pp. 344−60.

[92] Zanetti M, Lomakis LS, Camino G. Polymer layered silicate nanocomposites. Macromol Mater Eng 2000;279:1−9.

[93] Camino G, Tartaglione G, Frache A, Manferti C, Costa G. Thermal and combustion behaviour of layered silicate−epoxy nanocomposites. Polym Degdn Stab 2005;90: 354−62.

[94] Hussain M, Varley RJ, Mathys Z, Cheng YB, Simon GP. Effect of organo-phosphorus and nano-clay materials on the thermal and fire performance of epoxy resins. J Appl Polym Sci 2004;91:1233−53.

[95] Wang Z, Pinnavaia TJ. Hybrid organic-inorganic nanocomposites: exfoliation of magadiite nanolayers in an elastomeric epoxy polymer. Chem Mater 1998;10: 1820−6.

[96] Zammarano M, Franceschi M, Bellayer S, Gilman JW, Meriani S. Preparation and flame resistance properties of revolutionary self-extinguishing epoxy nanocomposites based on layered double hydroxides. Polymer 2005;46:9314–28.
[97] Liu W, Varley RJ, Simon GP. Understanding the decomposition and fire performance processes in phosphorus and nanomodified high performance epoxy resins and composites. Polymer 2007;48:2345–54.
[98] Toldy A, Toth N, Anna P, Keglevich G, Kiss K, Marosi G. Flame retardancy of epoxy resin with phosphorus-containing reactive amine and clay minerals. Polym Adv Technol 2006;17:778–81.
[99] Schartel B, Knoll U, Hartwig A, Putz D. Phosphonium-modified layered silicate epoxy resins nanocomposites and their combinations with ATH and organo-phosphorus fire retardants. Polym Adv Technol 2006;17:281–93.
[100] Torre L, Lelli G, Kenny JM. Cure kinetics of epoxy/anhydride nanocomposite systems with added reactive flame retardants. J Appl Polym Sci 2004;94:1676–89.
[101] Katsoulis C, Kandare E, Kandola BK. The combined effect of epoxy nanocomposites and phosphorus flame retardant additives on thermal and fire reaction properties of fibre-reinforced composites. J Fire Sci 2011;29(4):361–83.
[102] Wu Q, Zhu W, Zhang C, Liang Z, Wang B. Study of fire retardant behavior of carbon nanotube membranes and carbon nanofiber paper in carbon fiber reinforced epoxy composites. Carbon 2010;48:1799–806.
[103] Yu H, Liu J, Wen X, Jiang Z, Wang Y, Wang L, et al. Charing polymer wrapped carbon nanotubes for simultaneously improving the flame retardancy and mechanical properties of epoxy resin. Polymer 2011;52:4891–8.
[104] Zhang Z, Gu A, Liang G, Ren P, Xie J, Wang X. Thermo-oxygen degradation mechanisms of POSS/epoxy nanocomposites. Polym Degdn Stab 2007;92:1986–93.
[105] Wu K, Song L, Hu Y, Lu H, Kandola BK, Kandare E. Synthesis and characterization of a functional polyhedral oligomeric silsesquioxane and its flame retardancy in epoxy resin. Prog Org Coatings 2009;65(4):490–7.
[106] Wu K, Kandola BK, Kandare E, Hu Y. Flame retardant effect of polyhedral oligomeric silsesquioxane and triglycidyl isocyanurate on glass fibre-reinforced epoxy composites. Polym Compos 2011;32(3):378–89.
[107] Guo Y, Bao C, Song L, Yuan B, Hu Y. In situ polymerization of graphene, graphite oxide, and functionalized graphite oxide into epoxy resin and comparison study of on-the-flame behavior. Ind Eng Chem Res 2011;50:7772–83.
[108] Ávila AF, Yoshida MI, Gabriela M, Carvalho R, Dias EC, de Á Junior J. An investigation on post-fire behaviour of hybrid nanocomposites under bending loads. Compos Part B 2010;41:380–7.
[109] Wang X, Song L, Pornwanchai W, Hu Y, Kandola B. The effect of graphene presence in flame retarded epoxy resin matrix on the mechanical and flammability properties of glass fiber-reinforced composites. Composites Part A 2013;53:88–96.
[110] Wang X, Hu Y, Song L, Yang H, Yu B, Kandola B, Deli D. Comparative study on the synergistic effect of POSS and graphene with melamine phosphate on the flame retardance of poly(butylene succinate). Thermochim Acta 2012;543(10):156–64.
[111] He H, Li K, Wang J, Sun G, Li Y, Wang J. Study on thermal and mechanical properties of nano-calcium carbonate/epoxy composites. Mater Des 2011;31:4521–7.
[112] Nazaré S, Kandola BK, Horrocks AR. Smoke, CO and CO_2 measurements and evaluation using different fire testing techniques for flame retardant unsaturated polyester resin formulations. J Fire Sci 2008;26:215–42.

[113] Zhuge J, Gou J, Ibeh C. Flame resistant performance of nanocomposites coated with exfoliated graphite nanoplatelets/carbon nanofiber hybrid nanopapers. Fire Mater 2012;36(4):241–53. http://dx.doi.org/10.1002/fam.1104. Article first published online.

[114] Tibiletti L, Longuet C, Ferry L, Coutelen P, Mas A, Robin JJ, et al. Thermal degradation and fire behaviour of unsaturated polyesters filled with metallic oxides. Polym Degdn Stab 2011;96:67–75.

CHAPTER

Flame Retardancy of Fiber-Reinforced Polymer Composites Based on Nanoclays and Carbon Nanotubes

17

C.M.C. Pereira, M.S.S. Martins
University of Porto, Porto, Portugal

CHAPTER OUTLINE

Polymer Green Flame Retardants. http://dx.doi.org/10.1016/B978-0-444-53808-6.00017-2

1. Introduction

Fiber-reinforced polymer (FRP) composites have been in a sustained growth use since the 1960s. Their advantages over other materials for high-performance and lightweight applications have increased their exploration and usage in many industries such as aerospace, automobile, civil infrastructures, sports, and marine. It is expected that the use of composite materials continues to grow in coming years with emerging applications in large bridge structures, wind blades, offshore platforms, engine machinery, computer hardware, and medical devices. Key advantages of FRP composites over many metal alloys include low density, high specific stiffness and specific strength, good fatigue endurance, excellent corrosion resistance, outstanding thermal insulation, and low thermal expansion.

However, composites have several disadvantages such as anisotropic properties, low through-thickness mechanical properties, poor impact damage tolerance, and poor performance in fire, mainly due to their polymeric or organic constitution. Polymers are in general highly flammable and produce large quantities of smoke and toxic gases. With the increasing usage of composites, these polymers may be brought into fire situations at any time, adding urgency for the development of effective and environmental friendly flame-retardant systems to reduce fire hazards.

Commercially available flame retardants for polymers matrices can be highly effective, but have some important disadvantages, such as a negative influence on the resin workability (e.g. viscosity, reactivity), the quality (physical and mechanical properties) of the final product, in addition to being harmful (e.g. halogenated-based flame retardants). The concept of nanoscale reinforcement and flame retardancy provides opportunity for synthesis of new polymer materials with unique properties. Several nanoparticles, like nanoclays and carbon nanotubes, can also represent a rational alternative to the conventionally filled polymers because they have distinct advantages over traditional flame retardants on its related problem areas.

An overview of the effects of different nano- and microscale materials (e.g. glass and carbon fibers) on the flammability of advanced fiber-reinforced polymer composites will be detailed and discussed in this chapter. Its goal is to bring together the recent bibliographic reviews and experimental results obtained for different materials. The synergetic effect of inorganic-based traditional flame retardants will also be mentioned.

2. Fiber-reinforced polymer composites

2.1 Introduction

A composite material can be defined as a heterogeneous mixture of two or more materials. The term composite was originally used in the engineering field when two or more distinct materials were combined into one with the intention of overcoming the individual limitations and favor individual advantages [1,2]. The final properties of a composite material depend on both the nature of the constituent materials, and on their physical or chemical affinity. In fiber reinforced polymer (FRP) composites, the major constituents are the polymer matrix and the reinforcing fibers. The role of the polymer matrix, which is also called the continuous phase of the composite, is to primarily bind the fibers together by keeping them in the desired location and orientation, to act as a load transfer medium between fibers and to protect the fibers from environmental damages [3]. The reinforcing fibers carry out the structural load, reduce thermal stresses, and provide macroscopic stiffness and strength.

Composite materials can be classified into three types based on reinforcing filler organization: particulate, fibrous (short or long fiber), and laminate. Particulate composites consist of a matrix reinforced by a dispersed phase in the form of particles, which can have a random or preferred orientation. Composite laminates consist of several layers of fibrous materials dispersed in a polymeric, ceramic, or metallic matrix, which can be combined in different fiber orientations (multilayer composite). A sandwich structure is a special type of composite material, composed of two or more thin layered composite materials attached to a relatively thick core material.

The type of polymer matrix, thermoset, or thermoplastic, plays an important role in FRP composites processing manufacturing. Thermoset materials have in general low viscosity, 50–500 times more than water, and can easily impregnate the empty spaces between the fibers. These types of resins require an additional processing step known as curing, which involves the exothermic chemical reaction of cross-linking between the polymer chains. The high density of cross-linking within the thermoset structure is responsible for the superior thermal stability and mechanical properties. Epoxies, polyesters, vinyl esters, and phenolic resins are the thermoset polymers commonly used as matrix in continuous or long fiber-reinforced composites. Unsaturated polyester resins are mostly used as matrix for glass fiber-reinforced composites, due to their lower costs and mechanical properties. Epoxies, on the other hand, are more expensive than polyester resins but have superior mechanical properties and good resistance to alkaline conditions. Epoxies are by far the most widely used polymer matrix for carbon fiber composites. The vinyl ester resins have properties that are intermediate between polyesters and epoxies, are easier to process than epoxides, coupled with a better chemical resistance than polyester resins. Phenolic resins are relatively cheap, have good high temperature performance, excellent fire and smoke resistance, and good resistance to acids [4]. Phenolic resins are cured by a condensation reaction, creating voids in their structure, which makes their mechanical properties far below other thermoset resins.

On the other hand, thermoplastic polymer chains do not cross-link and are highly viscous. Their viscosity can be high as 10^6 times higher than that of water. They flow after being heated-up and with effort (e.g. under pressure) fill the empty spaces between the reinforcing fibers. After the cooling-down step, the thermoplastic polymer chains keep their initial size, however the repetitive molecules (monomers) can be reorganized in different ways. Thermoplastic polymers are more commonly used in short fiber-reinforced composites that are injection-moulded; however, the interest in continuous or long fiber-reinforced thermoplastic (LFT) matrix is growing. Polyamides and polycarbonates are examples of thermoplastic polymers used as matrix in injection-moulded composites.

Various types of fibers are used as reinforcements for polymer composites. The fiber type, length, orientation, and volume fraction are extremely important, since they influence the density, the mechanical and fire reaction properties, the electrical and thermal conductivities, and the cost of the material. Several types of fibers are commercially available; however, the most commonly used are glass and carbon fibers. Glass fibers are the most widely used reinforcement in polymeric matrix composite. The principal advantages of glass fibers are their low cost, high tensile strength, high chemical resistance, and excellent insulating properties [2]. The main disadvantages are their relatively low tensile modulus, fatigue resistance, and high density. Glass fibers are chemically inert in fire and retain their chemical and physical stability even at high temperatures and heat fluxes [5]. Glass fibers remain unaffected by the fire until heated up to 830 and 1050 °C in the case of E-glass and S-glass fiber types, respectively. Carbon fibers are available in various types, depending on their production method. The two most common types are polyacrylonitrile-based (PAN) and pitch-derived carbon fibers. The main advantages of carbon fibers are their high specific tensile strength and modulus, very low coefficient of linear thermal expansion, high fatigue strength, and high thermal conductivity. The disadvantages are their high cost, low strain-to-failure, low impact resistance, and high electrical conductivity, which may cause undesirable electrical discharges in unprotected equipment. When exposed directly to fire, carbon fiber's surface can be oxidized and the onset degradation temperatures are about 350 and 450 °C, respectively, for PAN and pitch-based fibers.

A key ingredient in the successful production of a material or a component is a cost-effective and reliable manufacturing method [3]. Hand layup, compression moulding, pultrusion, resin transfer moulding (RTM), and filament winding are examples of manufacturing processes for thermoset FRP composites. For thermoplastic-based composites, the two most commonly used manufacturing techniques are extrusion and injection moulding.

2.2 Applications

Commercial and industrial applications of fiber-reinforced polymer composites are diverse in both structural and nonstructural application areas due to their many outstanding physical, thermal, chemical, and mechanical properties. The major structural application areas include aircraft, space, automotive, sporting goods, marine, and infrastructures. In nonstructural applications, fiber-reinforced polymer composites are also used in electronics (e.g. printed circuit boards), building construction (e.g. floor beams), furniture (e.g. chair springs), power industry (e.g. transformer housing), oil industry (e.g. offshore oil platforms and oil sucker rods used in lifting underground oil), medical industry (e.g. bone plates for fracture fixation, implants, and prosthetics), and in many industrial products, such as step ladders, oxygen tanks, and power transmission shafts [3]. Their use requires careful design practice and appropriate process development based on the understanding of their unique characteristics.

Commercial aircrafts are one of the major applications for FRP composites. Airbus was the first commercial manufacturer to make an extensive use of composites in their A310 aircraft, introduced in 1987. Figure 1 shows Airbus A380, introduced in 2006, where about 25% of its weight is made of composites, including 23% of carbon fiber-reinforced polymer. Some of the components in composites are the central torsion box, the tail, the flaps, spoilers, and ailerons.

According to Airbus, the use of composite materials has increased the service intervals for the aircraft from 6 to 12 years, significantly reducing maintenance costs. The high percentage of composites also reduces the need for fatigue-related inspections, usually required on more traditional aluminum jetliners, and lessens the requirement for corrosion-related maintenance checks. Boeing has started the use of composites in 1995, with the incorporation of about 10% of carbon fiber-reinforced epoxy. The next line of airplanes, called Boeing 787 Dreamliner is pushing the envelope with a total composite of 50% by weight, including the integration of an all-composite fuselage, wings, and tail [6]. In aerospace industry, materials used in external structures are usually carbon fiber epoxy

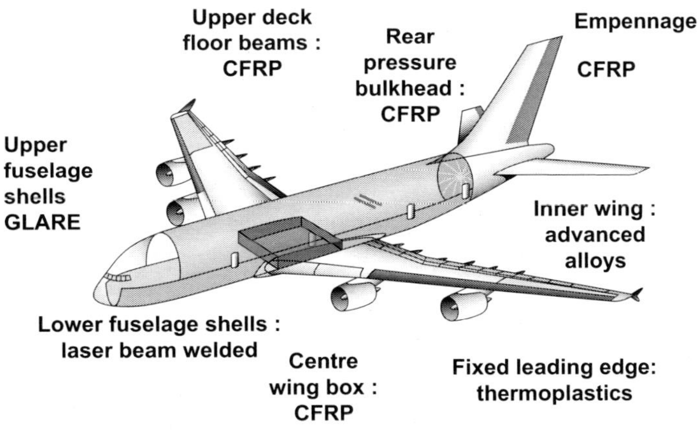

FIGURE 1

Use of fiber-reinforced polymer composites in airbus A380.

Adapted from Ref. [6].

composites, while glass/phenolic composites are mostly used in internal structures such as walls and ceilings. In all cases, mostly thin sections of composite laminates are used. The main reasons for using these materials in aerospace are: weight savings, with consequent fuel saving and payload increasing; reduction in the number of components and fasteners; higher fatigue and corrosion resistance; furthermore, FRP fabrication allows the possibility of aeroelasticity tailoring the stiffness of the airframe structure. The key limiting factors in using CFRP in aircraft structures are the high cost, relatively low impact damage tolerance, susceptibility to lightning damage and low fire resistance.

Applications in automotive industry can be classified into three groups of components: body, chassis, and engine. Glass fiber-reinforced polymer (GFRP) are the primary composite materials used in automobiles, however, it is recognized that the vehicle weight reduction for improved fuel efficiency can only be achieved with carbon fiber composites because of their higher specific strength and stiffness. Cars can start a fire in several ways, namely through mechanical failures, breakages, leaks, and electrical faults. Fire tests have shown that a fire starting in the engine compartment of a modern car results in flames reaching the passenger cabin in 10–20 min. Plastics in the engine compartment and body panels represent a real fire hazard and are not (yet) subject to any particular performance test [7]. Fire safety requirements for cars are lower than for public transportations; however, legislation in this field of application is becoming more restricted.

Composites also find several applications in the railway industry. They are commonly used in passenger coaches due to their excellent structural properties and improved aesthetics. In this particular case, glass fiber reinforced with polyesters, epoxies, and phenolic resins are the most commonly used materials. Weight savings in the order of 50% for structural and 75% for nonstructural applications benefits a high-speed transportation, reduces power consumption, and increases the ability to carry greater pay-loads. Composites also provide greater versatility in train design and optimization of train performance, such as lowering the center of gravity for enhanced stability. In transportation industries, glass fibers with unsaturated polyester or vinyl ester resins are the most commonly used FRP composites. These materials are also used in marine industries in the form of thick sections of composite laminates.

Although advanced polymer composite materials have been used mostly in transportation industries, in the last decades there has been a growing awareness amongst civil engineers of the importance of the unique mechanical and in-service properties of these materials, along with their customized fabrication technologies. FRP composites have been therefore employed not only in new construction of structures, but also as reinforcement materials in concrete, bridge decks, modular structures, formwork, and external reinforcement for strengthening and seismic upgrade [8,9].

2.3 Main concerns

Composites have several disadvantages and problems, such as low through-thickness mechanical properties, poor impact damage tolerance, anisotropic properties and significantly poor performance in fire, that have hindered their growth on some markets such as aircrafts and naval applications. As example, aircraft fires are extremely hazardous due the difficulty to escape from the fire scenario. In addition, aircraft composites must have excellent fire resistant properties to prevent structural failure during and after a fire [5].

Many polymer composites ignite when exposed to high heat fluxes, releasing heat that can, in some circumstances, contribute to the growth of the fire. Significant quantities of smoke and toxic fumes may also be released, limiting visibility and posing a serious health hazard. For these reasons, stringent fire regulations govern the use of composites in aircraft, ships, buildings, land transport, oil and gas facilities, and other applications. These regulations require that the fire reaction properties of these materials meet specified levels. The reaction properties that are often used to define the fire hazard include heat release rate, time-to-ignition, and flame spread rate.

In spite of such unfavorable properties, FRP materials present other attributes that are advantageous in a fire situation. In contrast with other traditional materials, such as steel, composites are very good heat insulators, and this feature is important for slowing the spread of fire from room to room [10]. Also when compared to steel, composites present better burn-through resistance, providing an effective barrier against flame, heat, smoke, and toxic fumes. Furthermore, several procedures can be applied to improve the fire performance of FRP materials [11].

3. Fire behavior, test methods, and fire reaction properties

3.1 Fire behavior

Fire is a complex event that can be developed in stages of increasing temperature and size before decaying. This event becomes even more complex when polymer composite materials are involved because they contribute (as a rich source of fuel) to temperature, size, and spread of the flame. The initiation and growth of a fire is determined by several factors: type, load and area of the fuel, oxygen content, wind speed, and closed or open space [5]. The fire hazard increases in enclosed spaces, such as in aircraft cabins and railway compartments, due to the heat, smoke, and toxic gases entrapment. Within a closed compartment and according to the respective conditions, a fire can be distributed into four main stages:

- *Ignition*. Materials are exposed to an ignition source, good exposure to air, low temperatures;
- *Developing fire*. The fire will grow and the environment temperature will continue to rise, resulting in external heat fluxes in the range of 20–60 kW m^{-2} and ventilated temperatures to 300–800 °C;
- *Fully developed fire*. This stage occurs when the heat release rate and temperature of a fire are at their maximum, resulting in high external heat fluxes, >40 kW m^{-2}, low ventilation, and temperatures >800 °C;
- Decay occurs after the consumption of all combustible materials.

Different fire properties are discernible in each of these stages. Materials ignitability and flammability is determined in an ignition scenario. Heat release and flame spread control the fire behavior in a developing fire. Fire load and fire penetration are the critical aspects of fire behavior in fully developed fires [12].

3.2 Test methods

The fire behavior of polymeric materials is primarily characterized by their ignitability, flame spread rate, and heat release. The appropriate tests needed for measuring the fire behavior depend on the application of the target materials. Each experimental setup or test defines a specific fire scenario and differs from others in their heat and mass transport characteristics such as applied heat flux, temperature, horizontal/vertical position, length scale, and ventilation [12–14]. Therefore, each fire test characterizes the performance resulting from an interaction of material properties, specimen, and defined fire scenario. The obtained results may have little relevance for other fire scenarios or fire tests.

There are numerous flammability tests used in industrial or academic laboratories, either for screening materials during product development or for testing manufactured products. The three more commonly used laboratory test methods for studies of polymer flammability are: UL94 vertical burning test, limiting oxygen index (LOI), and cone calorimeter.

Underwriters' laboratory UL94 test is designed to assess the ignitability and flame spread of polymeric materials exposed to a small flame. It is accepted for standardization in many countries and provides three major classifications of materials flammability. The three categories, V-0, V-1, and V-2, depend on the performance of the material, regarding the burning duration of individual and all specimens and also on the presence or absence of burning drips.

The LOI test is a screening tool and does not represent a real fire scenario. It is applied to quantify the resistance to ignition of material. The purpose of this test is to assess the minimum oxygen concentration in an oxygen/nitrogen mixture that is able to support the combustion of the material for at least 3 min or for the consumption of 5 cm of the same sample. The more the oxygen required, the more flame retarded the material is [15].

Cone calorimeter has become one of the most important and widely used instruments for the research and development of fire retarded polymeric materials [13]. The popularity of this method is due to the fact that it allows the determination of several parameters in one single test, using samples with reduced dimensions, plates with 100×100 mm^2 in size [5]. The sample size is of the smallest order of magnitude discussed in fire engineering and of the largest used in polymer analysis. In addition, the cone setup was developed thoroughly to target the properties of materials rather than to correspond to a special full-scale scenario of a real fire [13]. Cone calorimeter investigations can be used as a universal approach to ranking and comparing the fire performance of different materials, pass or fail a material according to a certain fire criteria, assess the likely response of a material when exposed to a large fire, and generate fire reaction data to validate fire models [5,13,16].

Fire tests using the cone calorimeter involve exposing a flat specimen to an incident heat flux generated by the cone shaped heater. It consists of an electric heater rod that is tightly wound into the shape of the truncated cone. Heating is controlled using a thermostatically controlled radiant heater that is capable of submitting a specimen to any heat flux usually up to 100 kW m^{-2}. Fire tests can be performed with the specimen in a horizontal or vertical direction. Tests are usually performed in the horizontal orientation because in this mode the convective component of heat transfer is almost negligible, although vertical tests may be preferred if the actual fire scenario involves a vertical face [5]. The combustion gases produced pass through the heating cone and are captured by means of an exhaust duct system with a centrifugal fan and a hood. The gas flow, oxygen, CO and CO_2 concentrations, and smoke density are measured in the exhaust duct. The gas flow and oxygen concentration measurements are used to calculate the quantity of heat released per unit of time and surface area: the HRR (heat release rate), expressed in kW m^{-2}. In addition, the cone calorimeter test also enables characterization of time-to-ignition, mass loss, and total smoke released [17].

3.3 Fire reaction properties

Of the many fire reaction properties, it is generally recognized that *heat release rate* (HRR) is the single most important parameter in controlling fire hazard. The amount of heat released from a polymer composite is controlled by the combustion of flammable gas products resulting from the decomposition of the organic components. The heat release rate is a complex property determined by the balance of the heat released by the flammable volatiles and the heat absorbed by the endothermic decomposition reactions occurring in the condensed phase or solid composite. Mouritz et al. [18] correlated smoke density, carbon monoxide, and carbon dioxide yield, mass loss rate, total mass loss, and flame spread with the heat release rate.

The *peak of heat release rate* (PHRR) is not only the cone calorimeter result probably most communicated to the polymer science community, but also a characteristic that displays a very strong and quite complex dependence on the fire scenario defined by the cone calorimeter setup. The PHRR represents the maximum amount of heat released by the material during the combustion process and frequently occurs shortly after the ignition. This is also considered a critical parameter as it controls the maximum temperature achieved during combustion and the rate of flame spread [5]. The peak heat release rate is strongly dependent not only on the properties of the material, but also on the geometry and the conditions of the fire test [13], as will be discussed in Chapter 4.

From the curve of HRR versus time, the *total heat released* (THR) during a cone calorimeter test can also be determined. It is calculated as the integral of the HHR curve with respect to time, representing the total heat output up to that time. The

THR depends strongly on the total mass loss, the effective heat of combustion of the volatiles and the combustion efficiency in the flame zone [14].

Time-to-ignition is defined as the time a combustible material can support, when exposed to a constant radiant heat flux, before initiating ignition and be subjected to combustion with continuous flame. It can be used as a simple measure of the material's resistance to fire [18]. The ignition time depends on several factors, namely the oxygen availability, the temperature, and the chemical and thermo-physical properties of the polymeric matrix and the reinforcement.

Mass loss is another important parameter when characterizing the fire reaction properties of a material, since it represents a quantitative measurement of the materials that decompose in a fire. The amount and rate of decomposition of the organic substances of a composite material can be determined by measuring the weight variation of the sample during cone calorimeter testing. The total mass loss is largely conditioned by the char yield of the organic constituents. Composites that yield a small amount of char experience a greater mass loss than materials with a high char yield. The mass loss rate is not constant, because it is controlled by diverse thermal processes that change with time, including ignition, combustion, char formation, and the transition of the material's behavior from thermally thick to thermally thin [5]. A sample is considered to be thermally thin when its thickness is below 1–2 mm.

Smoke is a combination of complete and incomplete combustions species, which can be composed of water, carbon dioxide and monoxide, acid gases, and soot. Additionally, it can also contain small fiber fragments [19]. Depending on the ventilation and burning rate, smoke can be the main fire hazard; it can be extremely dense and consequently reduce visibility and cause disorientation during a fire. Furthermore, carbon monoxide is toxic and acid gases are irritating to eyes. In fact, it is known that most deaths in a fire are caused by inhalation of smoke and toxic combustion gases, carbon monoxide being the most common cause. Smoke formation during burning of a material is not an inherent property. The level of smoke actually measured in a test depends on the burning conditions (heat flux, oxidant supply, specimen geometry, presence or absence of flame) as well as the test conditions (ambient temperature, test chamber volume, ventilation) [2]. The *specific extinction area* (SEA) is the parameter measured during cone calorimeter tests, which characterizes the density of the released smokes. It describes how effectively a given mass of volatiles generated by the combustible material is converted into smoke [5]. SEA is expressed in $m^2\ kg^{-1}$, representing the amount of smoke released per mass unit of sample.

4. FRP composites in fire

4.1 Introduction

Fire performance is one of the most significant factors affecting the wider use of composites. When heated to moderate temperatures (100–200 °C), FRP materials soften and creep, causing a considerable reduction of both strength and stiffness. Furthermore, when FRP materials are exposed to high temperatures (300–500 °C), their organic matrix decomposes, releasing heat smoke, soot, and toxic volatiles [5]. The behavior of FRP composite materials in fire is governed largely by the chemical processes involved in the thermal decomposition of the polymer matrix. When present, organic fibers such as aramid, polyethylene, flax, and hemp are also important contributors in the material fire response. A description of the main processes involved in the thermal decomposition is presented in Section 4.2.

The effects of the fire intensity, composite structure (e.g. resin composition and content, fiber type and organization, fiber sizing agents), and specimen thickness on the most critical fire reaction properties of FRP composites are described in Section 4.3. Despite the large number of polymers and fibers combinations used in composites, only the properties of the most commonly used thermoset matrices—epoxy and unsaturated polyester resins—and reinforcing fibers—glass and carbon—will be addressed here. The thermal stability and fire response of combustible organic fibers are out of the scope of this revision paper, even if some of them are extensively used. The flame retardancy of these composites will be addressed in a separate work.

4.2 Thermal decomposition

The events involved in the decomposition of an FRP composite exposed to fire conditions are represented in Figure 2. When the composite material is exposed to a sufficiently large source of heat radiated from a fire, the organic components will start to thermally decompose, releasing volatile gases and airborne soot particles (smoke) and yielding liquid products, tar, and solid carbonaceous char [5]. The volatiles consist of a variety of vapors and gases, both flammable (e.g. carbon monoxide, methane, low molecular weight molecules, or polymer fragments) and nonflammable (carbon dioxide, water). The flammable volatiles diffuse from the decomposing composite into the air creating a combustible gaseous mixture. These volatiles will react with the oxygen from air, leading to ignition, at low temperature when submitted to an external source of intense energy (spark, flame, etc.) or when the auto-ignition temperature is reached. After ignition, the combustion products are carbon dioxide and monoxide, soot, and heat. When the amount of heat liberated reaches a certain level, new decomposition reactions are induced in the solid phase. The combustion cycle is thus maintained [17]. As it is represented in Figure 2, this global process is complex and involves several reactions and transport phenomena in the solid, gaseous, and interfacial phases.

Due to their chemical structure, mainly made up of carbon and hydrogen, polymers are the main source of flammable volatiles in an FRP composite. Thus, the amount of heat released is controlled by the combustion of flammable gas products resulting from the decomposition of the matrix when inorganic fibers are present.

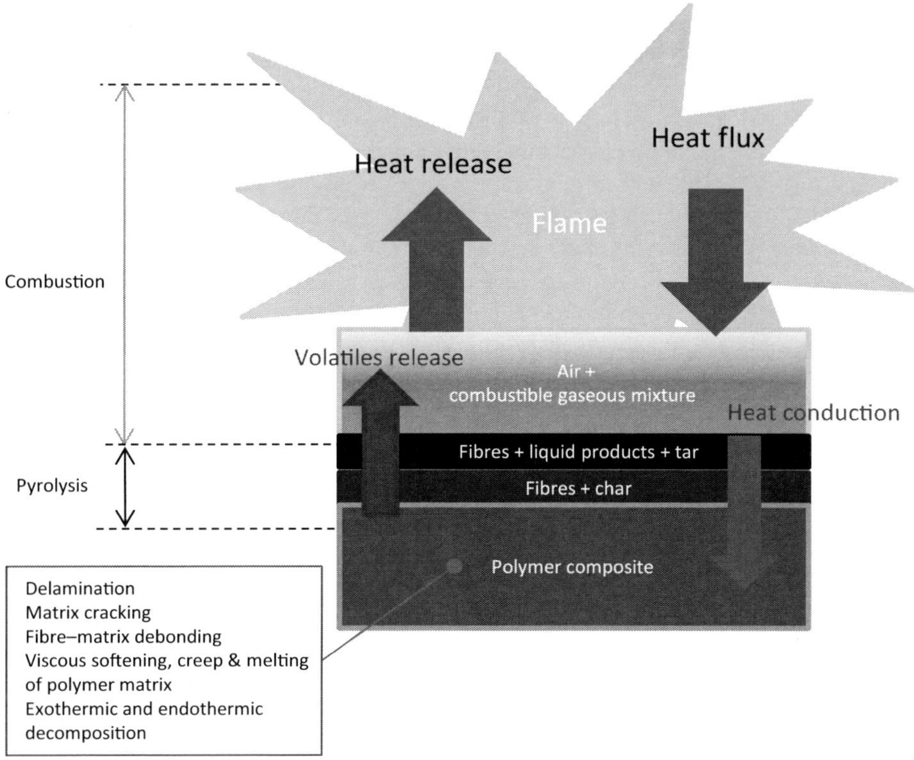

FIGURE 2

Schematic representation of the different phases, mechanisms, and reaction processes involved during the exposure to fire of a polymer composite.

The thermal decomposition of a polymer (i.e. covalent bond dissociation) is endothermic and therefore requires an input of energy. The amount of energy required to initiate combustion depends on the physical characteristics of the material. In the particular case of thermoset polymers, due to the absence of a melting point, the polymer immediately starts decomposing which occurs due to the dissociation of covalent bonds. Generally, thermal decomposition is the result of a combination of the effects of heat and oxygen where nonoxidizing and/or oxidizing thermal degradation take place. Polymers decompose through a series of chemical reaction mechanisms (mainly chain scissions) when heated to a sufficiently high temperature. The bonds requiring less energy are the first to break, with the liberation of volatile low molecular weight molecules (polymer fragments) that diffuse into the air. The thermal decomposition reactions of polymers may proceed by oxidative processes or simply by the action of heat (pyrolysis) [5]. The decomposition process is often accelerated by oxygen, but in thick composite sections it is generally only the surface region that decomposes in the presence of oxygen.

The outgassing of volatiles from the decomposition zone obstructs the ability of oxygen to diffuse much beyond the surface layers of the composite. In oxidizing thermal conditions, the polymer reacts with oxygen in the air and generates a variety of low molecular weight products: carboxylic acids, alcohols, ketones, aldehydes, etc. Oxidation can lead to cross-linking through recombination reactions of the macromolecular radicals. However, bond scission usually remains the dominant reaction. In literature it is referred that at combustion temperatures below 300 °C, polymer degradation takes place via nonoxidizing thermal decomposition [17]. Under these conditions, the rate of pyrolysis is much faster than the diffusion of oxygen in the solid phase. Oxidation therefore only occurs in the gas phase due to the presence of low molecular weight compounds produced by thermal decomposition.

The processes occurring within a polymer laminate exposed to fire conditions, and illustrated in Figure 2, are described by Mouritz et al. [16] and can be resumed as follows:

- *Thermal processes*. Heat conduction from the fire into the composite; heat generated or absorbed from decomposition of the polymer matrix; convective heat transported by hot reaction gases and water vapors coming from the composite; heat generated through char or fiber oxidation; and heat generated by the ignition of flammable reaction gases at the composite/fire boundary.
- *Chemical processes*. This involves decomposition and volatilization of the polymer matrix; formation, growth, and oxidation of char; oxidation of carbon fibers; and char/fiber reactions.
- *Physical processes*. This involves thermal expansion and contraction, viscous softening and melting of the matrix; thermally induced strains; internal pressure build up due to the formation of volatile gases and vaporization of water; formation of gas-filled pores; matrix cracking; fiber—matrix interfacial debonding; delamination damage; surface ablation; and softening, melting and fusion of fibers.
- *Failure processes*. The failure mode depends on the temperature, heat flux, and duration of the fire; magnitude and type of load (e.g. tension, compression, bending, torsion); and geometry of the structure.

4.3 Fire reaction properties versus composite structure and fire intensity

Fire reaction properties of FRP composites used in aerospace, marine, and infrastructure applications have received an increasing interest. As a result, the fire reaction behavior of these materials have been extensively characterized, and a large data of information is available regarding their time-to-ignition, heat release rate, limiting oxygen index, flame spread, smoke density, and smoke toxicity properties [5,18]. However, fire reaction behavior of composite materials is greatly dependent on different parameters. In this section, the effects of fire intensity, resin composition, resin content, fiber type and organization, fiber sizing agents, and specimen thickness on the most critical fire reaction properties of FRP composites are described.

4.3.1 Fire intensity

External heat flux or fire intensity is one of the fire conditions that greatly affect the fire reaction properties of a composite. Figure 3 shows the effect of heat flux on the ignition times of two different composite laminates. One can observe from the figure that the fire intensity shows a more pronounced effect on the flammability of glass/polyester composite, rather than for carbon/epoxy composite. The composites do not ignite below a threshold heat flux, even after exposure to fire for a very long time [5]. The onset heat flux is $\sim$13 kW m^{-2} for polyester- and epoxy-based composites, and $\sim$25 kW m^{-2} for phenolic-based composites. An increase in heat flux (or environment temperature) increases rapidly the matrix pyrolysis with consequent production of volatiles that flows out of the composite, thereby promoting the ignition.

The heat flux has also a significant influence on the heat release rate of a composite material. Figure 4 shows the effect of increasing heat flux on the PHRR values for a variety of glass/thermoset laminates. Glass/epoxy composite is the more flammable material and thus the heat released rises more rapidly with heat flux. The peak heat release rate of high-performance thermoset laminates is less sensitive to increases in the external heat flux.

Fire intensity also affects other fire reaction properties, namely the amount of smoke produced [5]. Smoke density may also be considered as an indirect

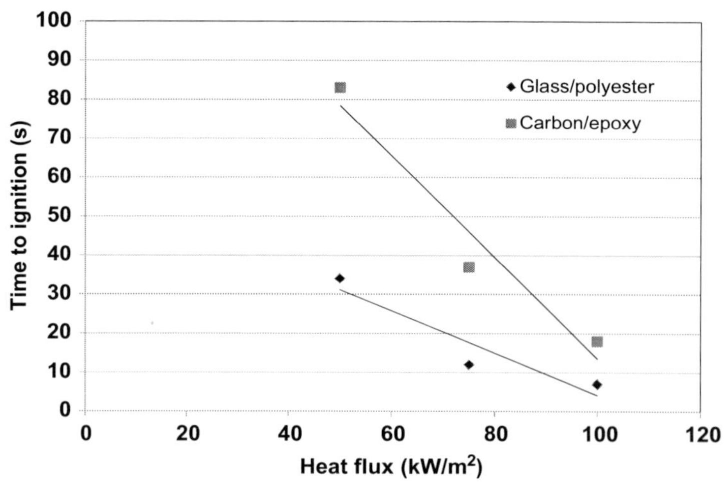

FIGURE 3

The effect of external heat fluxes in the ignition times for two different composites.

Graph constructed with data published on Ref. [5].

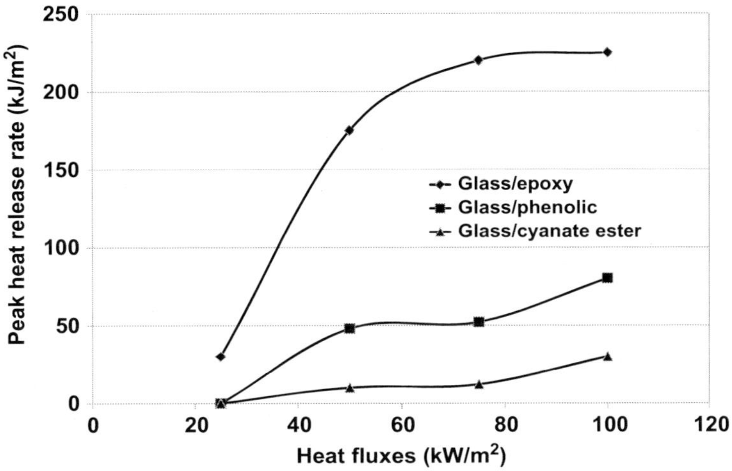

FIGURE 4

Effect of heat flux on the peak HRR for various glass/epoxy laminate.

Graph constructed with data published on Ref. [5].

measure of the toxic potency of the fumes. While there is not necessarily a direct correlation between the density and the toxicity of smoke, the smoke density is still an indirect measure of the concentration of potentially hazardous gases and airborne particles [19].

4.3.2 Resin type

As previously mentioned, the behavior of FRP composites is mainly governed by the thermal decomposition of the polymeric matrix and, as can be confirmed by the results reported in literature [5,20,21], the resin type has an important effect on the composite flammability. Cone calorimeter data reported by Brown and Mathys [20], particularly time-to-ignition, rate of heat release, and smoke obscuration, clearly emphasize the superior combustion properties of phenolic composites when compared to epoxy or polyester resins.

Several authors [22—24] have reported the thermal decomposition of polyesters or the fire behavior of polyester composites. The thermal decomposition process of all unsaturated polyesters is governed in the initial stages by scission of highly strained portions of the polystyrene cross-links, with the formation of free radicals that then go on to promote further decomposition, including some accompanying scission of the polyester backbone. This results in a variety of low molecular weight volatiles (CO, CO_2, methane, ethylene, propylene, butadiene, naphthalene, benzene, and toluene). These styrene-based polymers have another important disadvantage: the tendency for smoke production due to the presence of styrene monomer. Additionally these resins tend, during decomposition, to form an important quantity of liquid products and tar, which can result in the loss of flaming droplets. These tendencies can be lightened to a certain extent through the use of composites with high inorganic content, by employing, where possible, a high glass or filler content [5].

Decomposition of most epoxies occurs via random chain scission reactions over the temperature range of about 380–450 °C [25]. In amine- or amide-cured epoxies, the nitrogen linkages have lower bond dissociation energies than the ether or ester linkages, and therefore chain scission occurs at the C–N bonds. The hydroxyl groups are also vulnerable to degradation at high temperature. The scission reactions decompose 80–90% of the original polymer weight into almost 100 different volatile compounds, which are mainly various types of substituted alkylated phenols, aromatic ether derivations, and other flammable organic species [26]. These compounds provide a fuel source for the decomposition reaction to continue until the epoxy is completely degraded. Between 10% and 20% of the original polymer weight is transformed into a highly porous char, and in the presence of air this will start to oxidize above 550 °C. As with polyester composites, the high yield of flammable volatiles produced in the decomposition reaction is the main reason for the relatively poor fire performance of epoxy matrix composites.

4.3.3 Fiber type, content, and organization

The fiber reinforcement influences the composite fire reaction properties. Noncombustible fibers such as glass and carbon fibers show a diluting effect and have influence on the velocity of combustion through the formation of a protective layer [27]. Additionally, fiber reinforcement influences the combustion mechanism such as the radiation heat that penetrates into the material, the migration of the matrix degradation products, as well as the formation of char in the burning surface. Reinforcement acts as a barrier for these processes.

Glass fibers are inert to fire when the heat flux is below 100–125 kW m^{-2} but affect the ignition process and the heat released by composite materials. On the other hand, carbon fiber's surface can be oxidized at 350 and 450 °C, respectively, for PAN and pitch-based, when exposed directly to fire. However, in most type of fires the extent of oxidation is small because only the fibers in the surface exposed to fire are also exposed to oxygen-rich environment. The majority of carbon fibers within a composite are protected by char [5].

One of the first works reporting that the amount of fiber changes the time-to-ignition and the amount of heat released by a burning composite was presented by Le Bras et al. [28]. They found that time-to-ignition increases with glass fiber content and heat released rate declines with glass fiber content as it is shown in Figure 5. This behavior occurs because less polymer matrix material is available to generate heat during thermal decomposition.

In 2008, Avila et al. [21] reported a study about the effect of resin type and glass content on the overall fire performance of composites. The resin type was found to affect the resultant fire performance; however, the effect of glass content was a little more subtle. PHRR and burning time decrease with glass fiber content, time-to-ignition increases with glass fiber content, but no significant difference in the average heat release was observed. These authors observed that the minimum heat

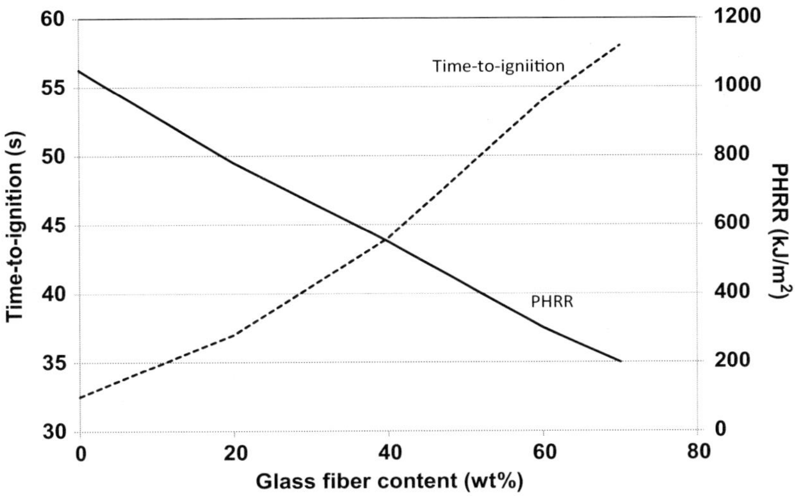

FIGURE 5

Schematic representation of the different phases, mechanisms, and reaction processes involved during the exposure to fire of a polymer composite. PHRR, peak of heat release rate.

Graph constructed with data published on Ref. [5].

flux for proper ignition increased with glass content, with exception for the polyester-based composite.

The presence of binder and the different physical constructions of chopped strand and woven roving reinforcements have significant effects on the combustion properties of their glass-reinforced polymer composites. Chopped strand-based composites have shorter ignition times, lower maximum heat release rates, lower and broader rate of heat release vs. time profiles and higher total heat release values than the corresponding woven roving based composites [20]. Eibl [27] observed from cone calorimetry and temperature distributions measurements through a laminate that the velocity and degree of combustion are dominated by fiber orientation for a given resin. They concluded that in general a quasi-isotropic fiber orientation leads to faster ignition, because of preferred delaminations, but delays combustion processes more effectively than a unidirectional layup.

4.3.4 Specimen thickness

Besides the chemical composition of a material, the dimensions of a component also influence its combustion properties. It is known that increasing specimen thickness retards ignition because of heat conduction into the bulk material. The total heat emitted during combustion rises with specimen thickness but can diminish when expressed as specific heat released per volume unit for thicker composite specimens [27]. The ignition time generally increases with thickness, as the thermal response changes from thermally thin to thermally thick behavior. The total HRR decreases with laminate thickness as reported in literature by Mouritz and Gibson [5,14].

Despite their combustibility, organic matrix composites have considerable potential for use in structures that may be subject to fire. When used in sufficiently thick sections, typical polyester/glass systems are capable of withstanding severe temperatures and ablative effects by virtue of their low intrinsic thermal conductivity and the endothermic nature of the resin decomposition process [10].

Delamination is one of the main concerns of thick structural composite laminates induced by fire conditions. Delamination is defined as the mechanical separation of interlaminar layers or plies. Sources of delamination can be attributed to internal pressure build up resulting from expansion of hot gases. These gases consist of combustion gases, i.e. combustion byproducts, and evolution of absorbed water or other liquids (e.g. residual solvents or low molecular weight components in the matrix). Because of the rapid heating conditions of fires the hot gases are incapable of quickly diffusing through the matrix material, thus producing a pressure build up. Due to this fact, initial moisture content is a major factor in the development of delaminations in thick composites exposed to fire.

5. Flame retardancy

Flame-retardant systems are intended to inhibit or stop polymer pyrolysis and combustion processes by interfering with the various mechanisms involved—heating, ignition, and propagation of thermal degradation [17]. In this chapter, the most prevalent principles, mechanisms, and modes of action in the field of flame retardancy of polymers are resumed.

Flame retardants can be divided into two main types: gas-phase-active and condensed-phase-active, and in function of their nature, they can either act physically or chemically. Gas-phase-active flame retardants disrupt the combustion in the flame and act either by:

- *Physical mechanisms*. Some flame retardants generate large amounts of noncombustible gases (H_2O, CO_2, NH_3, etc.), which dilute flammable gases; additionally, sometimes they dissociate endothermically causing a decrease in temperature by absorbing heat.
- *Chemical mechanisms*. The free-radical mechanism of the combustion process can be stopped by the incorporation of flame-retardant additives that preferentially release specific radicals (e.g. Cl*, Br*, P*) in the gas phase [17]. These radicals can react with highly reactive species (such as H* and OH*) to form less reactive or even inert molecules.

Condensed-phase-active mechanisms of action (physical and chemical) are numerous, as listed below:

- Dilution of the amount of combustible organic material by addition of inert particles.
- Reduction of the temperature of the composite by the addition of filler that acts as heat sink by absorbing heat due to their high specific heat capacity.

- Reduction of the temperature by the addition of fillers that decompose endothermically (heat consumption) to yield water or other noncombustible products.
- Reduction of the heat release rate by using polymers that decompose via endothermic reactions.
- Increase the aromaticity of the polymer matrix in order that it decomposes into an insulating surface layer of carbonaceous char (charring) that slows conduction into composite and reduces flammable gas emissions. Charring is the most common mode of action in the condensed-phase [29].

In addition, flame retardants can be classified into two categories: additive flame retardants (generally incorporated during the transformation process and do not react with the polymer at this stage) and reactive flame retardants (introduced into the polymer during synthesis or in a postreaction process). Reactive flame retardants are mainly based on halogen (bromine and chlorine), phosphorus, inorganic, and melamine compounds. These compounds act by liberating bromine and chlorine atoms into the flame where they disrupt the combustion reactions. However, these halogenated materials have severe health and environment concerns, mainly due to their high toxicity, and for that a new European regulation on chemical products (REACH) hinders their use.

Traditional flame-retardant additives for polymer composites include aluminum trihydroxide (ATH), magnesium hydroxide, and phosphorous compounds such as ammonium polyphosphate (APP) or red phosphorous. Hydrated metal salts like ATH act mainly by the absorption of heat from the polymer when it decomposes endothermically and by the dilution of the concentration of flammable gases due to the release of water vapor during decomposition. ATH suffers decomposition between 210 and 400 °C. The endothermic decomposition behavior of ATH is shown in Figure 6.

ATH is widely used in polymer and polymer composites mainly due to its notably low cost, good flame-retardant properties, and nontoxic smoke [5]. The efficacy of ATH as flame retardant in FRP composites is presented by several authors [5,30]. However, it is often necessary to add 50 wt% or more of ATH to many resin systems to achieve good flammability resistance and these high loadings affect negatively the polymer processability—which is particularly negative in the case of liquid infusion techniques for thermoset resins—and degrade the mechanical and durability properties of most types of polymer composites. ATH particles are usually 1 μm or bigger, and must be uniformly dispersed in the matrix in order to ensure even flame retardancy of the composite. This uniform distribution within the composite could be difficult to obtain mainly due to potential particles filtration in the fibers. As far as we know, this phenomenon is not reported in literature for flame retardants, but is referred several times in the case of processing FRP composites by using nano-doped resins and liquid infusion techniques. These phenomena must be also discarded and investigated in the case of micro-particulate flame retardants.

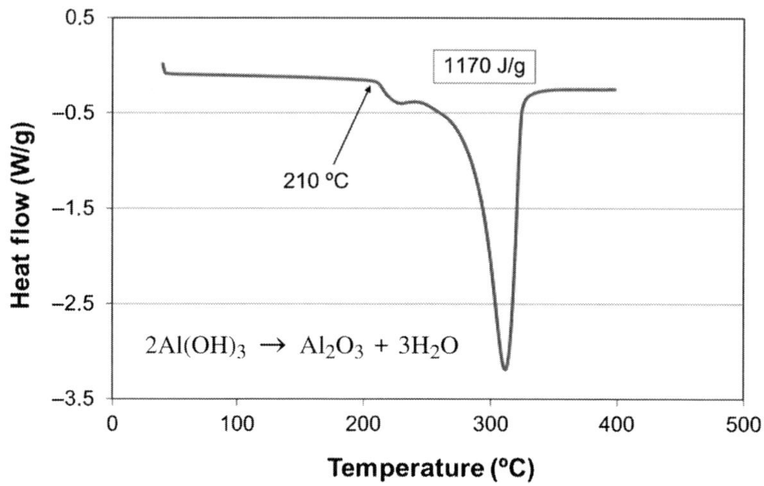

FIGURE 6

Differential scanning calorimeter test performed at INEGI showing the endothermic decomposition behavior of aluminum trihydroxide.

Phosphorous-based flame retardants act in the gas phase by a radical mechanism that interrupts the exothermic processes and suppresses combustion. Additionally, act in the condensed phase, promoting char formation, therefore creating a barrier to inhibit gaseous products from diffusing to the flame and shielding the polymer from heat and air. The efficacy of phosphorous-based flame retardants depends on the chemical structure of the polymer; they are usually mostly effective in oxygen containing polymers such as epoxy resin [31]. Traditional flame-retardant ammonium polyphosphate (APP), $(NH_4PO_3)_n$ is a nonvolatile, relatively water insoluble powder, with a phosphorous content of about 30%. When exposed to heat, ammonium polyphosphate is converted into phosphoric acid through thermal decomposition, as presented in the first step of Eqn (1). The degradation continues with the formation of polyphosphoric acid, which dehydrates the polymer in the pyrolysis stage, therefore creating unsaturated compounds that form a carbonaceous structure known as intumescence. This protecting layer is resistant to high temperatures and protects the polymer from the attack of heat and flame.

$$(NH_4PO_3)_n \xrightarrow{>250\ ^\circ C} \underset{\text{polyphosphoric acid}}{(HPO_3)_n} + nNH_3$$

$$(HPO_3)_n + \text{Polymer} \xrightarrow{-H_2O} H_3PO_4 + \text{Char}$$

Mouritz and Gibson [5] refer the reductions in both the peak and average heat release rate of an epoxy/carbon fiber composite with increasing phosphorous contents. A study about the influence of the oxidation state of different phosphorous-based hardeners on the fire behavior of epoxy/carbon fiber composites was reported by Braun et al. [32]. Phosphine-based composites showed significant reductions in PHRR, V-1 classification in UL94 tests and an LOI value of 39. However, these authors do not report about the possible delamination in the composite structure during the fire exposure. As explained in Section 4, delamination can occur when the internal pressure builds up as a result of the formation of hot gases, such as NH_3 gas molecules when APP is present in the matrix.

6. Polymer/clay nano- and microcomposites

Over the past decade, clay-based polymer nanocomposites have attracted considerable attention from scientific community. One of the first reviews describing the properties of polymer/clay nanocomposites was made by Alexandre and Dubois [33]. They presented a complete study on the methods of preparation of polymer/clay nanocomposites using different matrices, as well as their mechanical, thermal, and flame-retardant properties. This work, however, had a general focus, and flame retardancy was not exhaustively discussed. Furthermore, it has been more than 10 years since this review was published, and much work has been done since. In 2008, Paul and Robeson [34] presented a review study on clay-based polymeric nanocomposites. Attention was given to the reinforcement potential of clays, while other properties such as thermal, barrier, and flammability resistance were briefly mentioned. One of the first comprehensive overview of polymer nanocomposites for flame-retardancy applications was presented by Morgan and Wilkie [29]. The book covered themes as nanocomposite fundamentals, design, synthesis, characterization, and flammability fundaments, emphasizing the effects of nanocomposites on polymer flammability. They have focused their review mainly on thermoplastic-based nanocomposites and slight references were made to epoxy flame retarded nanocomposites.

In the following years, a joint revision work from the teams of Lopez-Cuesta and Dubois [17] put-out the fundaments of the polymer combustion theory, the main flame-retardant properties and tests used to describe fire behavior, together with the nature and modes of action of the most representative flame retardants (including nanometric materials such as nanoclays and other nanoparticles) and the synergistic effects that can be achieved by combining them. Again, they have focused their review on thermoplastic-based nanocomposites and have shown that nanoparticles alone or in combination with other traditional flame retardants are very promising flame-retardant solutions. However, it was demonstrated that the flammability of plastic materials remains a very complex scientific problem for which no single solution can be found, mainly due to the extensive diversity of polymers.

The team of Papaspyrides [15,35] reviewed extensively the recent advances in the field of polymer-layered silicate nanocomposites. One of these papers [15] was focused on the flame-retardancy enhancement. The combination of clays with halogenated and nonhalogenated flame retardants was analyzed, either for thermoplastic or thermoset polymers. They concluded that nanoclays are not effective on ignition and in fully developed fires stages, however, helpful in retarding flame spread in developing fires. The combination with traditional flame retardants is pointed out as solution for obtaining fully retarded materials. Interesting is the fact that authors figure out several questions such as, what is the effect of reactive sites of silicates during fire; what is the correlation (theoretical fundaments) between the dispersion degree and the reduction in PHRR; and how important is the lack of thermal stability of the clay's organic modifier (alkyl-ammoniums decompose above 200 °C) on the fire reaction properties (mainly earlier time-to-ignition) and (catalytic) degradation of the polymer matrix.

Although some promising works have recently emerged presenting nanoclays as promising flame retardants for FRP composites, no reference was done in all these reviews. The present work intends to overcome this gap and will present an overview of the current state of the art on these materials. Focus will be given to thermoset matrices (epoxy and UP) and their glass and carbon fiber-reinforced composites.

6.1 Nanoclays

Nanoclays are the most commonly used nanoparticles for flame retardancy; they can be natural or synthetic, cationic or anionic materials. While metal and metal oxide nanoparticles in use are typically spherical, nanoclays have flaky shapes with two of their dimensions within the range of 100−1000 nm. One of the main advantages of nanoclays is that the lamellas on their structure can be exfoliated, increasing the interfacial contact with the polymer matrix, and this is one of their major advantages. In this regard, montmorillonite (MMT) clays are being used in order to obtain polymeric nanocomposites with enhanced properties [33−35]. MMT clays are the most extensively studied [15]. MMT is a crystalline 2:1 layered cationic clay mineral with a hydrophilic surface in nature (see Figure 7), which can be organically modified to increase the compatibility with the organic polymer matrix. Alkylammonium cations type surfactants are the most widely used in the ion-exchange reaction to render the hydrophilic clay surface organophilic characteristics. The silicate layers are highly anisotropic, having a platelet thickness of 1 nm and a lateral size of about 30 nm. MMT is the basic mineral constituent of Bentonite clays.

Layered Double Hydroxides, LDH, also known as hydrotalcite-like compounds or anionic clays, are brucite-like layered materials with anionic counterions in the gallery space [36], as schematically represented in Figure 7. The positive layer charge is originated by the partial isomorphic substitution of the divalent cation for a trivalent one. In order to attain electroneutrality, an appropriate number of anions must be incorporated into the interlamellar domain. Their general chemical formula is $[M^{2+}{}_{1-x}M^{3+}{}_x(OH)_2^{x+}(A^{m-})]_{x/m}\cdot nH_2O$, where M^{2+} is a divalent cation, M^{3+} is a trivalent cation, and A is an interlamellar anion with formal charge m^-. The identities of M^{2+}, M^{3+}, and the interlayer anion, along with the value of the stoichiometric coefficient (x) may vary over a wide range, giving rise to a large class of isostructural materials. This flexibility in composition allows LDHs with a large

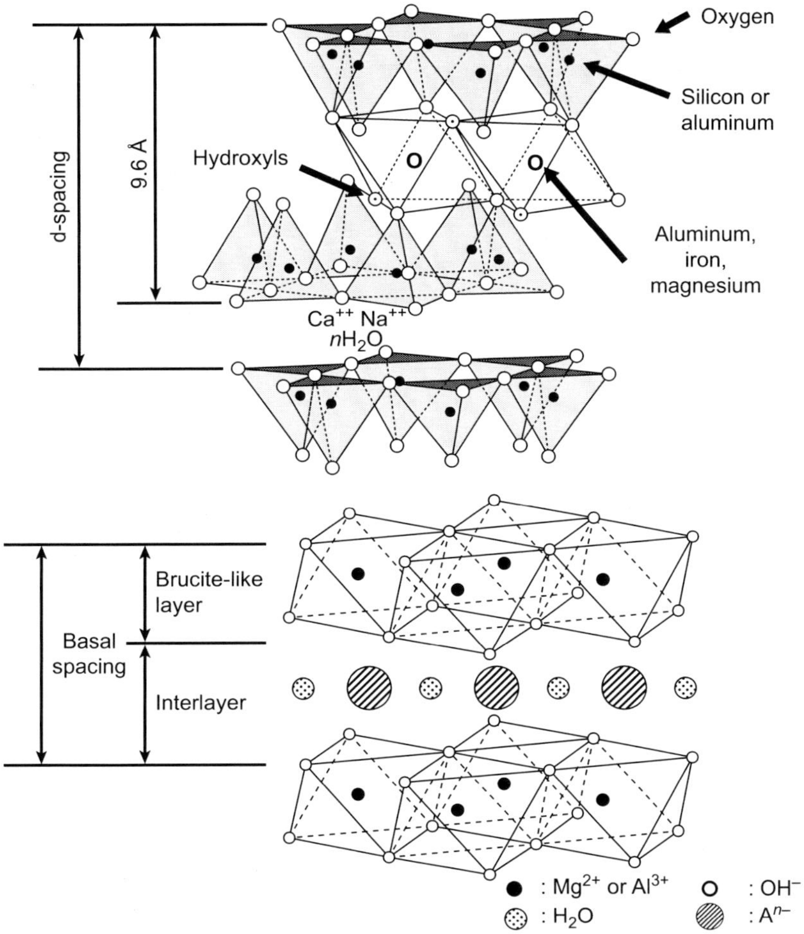

FIGURE 7

Schematic illustrations of montmorillonite (above) [37] and LDH (below) clays [38].

variety of properties to be prepared and this is one of their most attractive features [39]. When LDHs are used as additives, the exchange for organic molecules leads to materials with improved interactions with the polymeric chain and increased interlayer spacing [40].

6.2 Flame-retardant polymer/clay nano- and microcomposites

By far, the most common applications of nanoclays as flame retardants report their use in thermoplastic matrices. Enhanced fire reaction behavior properties have been observed in different types of thermoplastic polymers, namely on poly(methyl methacrylate)

(PMMA) [41], Ethylene Vinyl Acetate (EVA) [42], Polypropylene (PP) [43], Polyamide 6 (PA6) [44] and Polyethylene (PE) [45]. Nevertheless, it is intended with this review to focus on thermoset polymers. The number of papers reporting the fire reaction improvements of thermoset polymers with the incorporation of nanoclays is already extensive, as we can see in Table 1. However, the number of those reporting the use of these clay/polymer systems as matrices of fiber-reinforced composites is for now very limited.

From the analysis of table, one may conclude that cationic clays are the most widely studied, while the use of LDHs clays has been slightly approached on both epoxy and unsaturated polyester resins.

6.2.1 Epoxy resin

Concerning epoxy resins with LDHs, one of the first works was published in 2005 by Zammarano et al. [46]. They have reported the use of three different organo-modified LDHs and compared with alkylamine modified MMT and conventional flame retardants such as ATH and APP. All tested formulations had 5% additive. Cone calorimeter revealed intumescent and self-extinguishing behavior for LDH nanocomposites and a higher reduction in the peak heat release rate compared to MMT nanocomposites. They observed that the level of dispersion of LDH clays have a positive effect on the fire reaction behavior. In 2011, Becker et al. [47] have also reported a self-extinguishing behavior for LHD/epoxy nanocomposites prepared by solvent dispersion methods.

Several works [48–56] report the flame-retardant behavior of cationic clays (bentonite and montmorillonite) in epoxy resins. Hartwig et al. [48] observed that the incorporation of 5 wt% organo-montmorillonite—ammonium or phosphonium modified—enhanced the LOI of epoxy resin from 18.3% to 19.7%. Nanocomposites showed higher ignition times and a considerable decrease in PHRR. The presence of layered silicates on the sample surface, causing a barrier effect, was indicated by the authors as the main mechanism of flame retardancy for these nanocomposites, therefore resulting in a reduction of the fire growth rate.

In 2005, Camino et al. [50] claimed, for the first time, that besides the physical effects (formation of a protective layer), there is also a chemical effect of MMT clays in the flame retardancy of epoxy resins. They observed that the higher improvements—68% reduction in PHRR—did not correspond to the exfoliated structure (with just 38% of reduction). In addition, time-to-ignition remained unaltered with the presence of nanoclays. From the presented results, it is important to note that we are not able to conclude that the clay system tested by Camino is more effective than the LDH clay tested by Zammarano [46] since the cone heat fluxes used were different and, as it is known, they have a great influence on the flammability and on PHRR reduction, as stated in Section 4 of this work review.

A large number of works report the use of cationic nanoclays combined with phosphorous groups, either present in the epoxy prepolymer structure [49], in the hardener structure [51], or in the nanoclays structure [51–53; 55–56]. The combination with other conventional phosphorous flame retardants is also an approach

Table 1 Resume of Papers in Open Literature Reporting the Use of Clays for Flame Retardancy

Composite	Reinforcement Fibers Type and Amount	Polymer Matrix	Clay Type and Amount	Thermal Improvements	References
Nano	No	Epoxy	Bentonite (ammonium and phosphonium modified) 4.7 wt%	−35% PHRR (70 kW m^{-2}) −20% PHRR (30 kW m^{-2}) +TTI nanocomposites LOI ≈ 20 nanocomposites LOI ≈ 18 pure resin	[48]
		Epoxy (modified with phosphorous)	MMT 5–7.5 wt%	−40% PHRR (DGEBA, clay) −38% PHRR (modified DGEBA + clay) −17% PHRR (TGDDM, clay) −48% PHRR (modified TGDDM + clay) 50 kW m^{-2}	[49]
		Epoxy	LDH 5 wt%	Self-extinguishing in UL94 test −51% PHRR (ABS) −40% PHRR (TS) 35 kW m^{-2}	[46]
			MMT 5 wt%	−27% PHRR 35 kW m^{-2}	
		Epoxy	MMT (different commercial solutions) 10 wt%	−68% PHRR (Cloisite 30B; 50 kW m^{-2})	[50]
		Epoxy	Bentonite and Sepiolite 1 wt% (with/without organo-phosphorous amine)	LOI = 21 pure resin LOI = 25 (Bent) LOI = 36 (Bent + phosphorous amine) LOI = 34 (Sep + phosphorous amine) V-0 classification (clay + phosphorous amine)	[51]

Continued

Table 1 Resume of Papers in Open Literature Reporting the Use of Clays for Flame Retardancy—cont'd

Composite	Reinforcement Fibers Type and Amount	Polymer Matrix	Clay Type and Amount	Thermal Improvements	References
		Epoxy	Bentonite (phosphonium modified) 5 wt% (with/without ATH and organo-phosphorous FR)	+Decomposition temperature (clays + ATH) LOI = 21 pure resin LOI = 23 (clays) LOI = 31 (clays + phosphorous FR) LOI = 22 (clays + ATH) −38% PHRR (clays, 70 kW m^{-2}) −61% PHRR (clays + ATH, 70 kW m^{-2}) −73% PHRR (clays + phosphorous FR, 70 kW m^{-2}) −25% PHRR (clays, 35 kW m^{-2}) −50% PHRR (clays + ATH, 35 kW m^{-2}) −67% PHRR (clays + phosphorous FR, 35 kW m^{-2})	[52]
		Epoxy	Bentonite (phosphonium modified) 5 wt% (with/without APP and MB)	LOI = 20.5 pure resin LOI = 21.2 (clays) LOI = 21.9 (clays + APP) LOI = 23.5 (clays + MB) −54% PHRR (clays, 70 kW m^{-2}) −65% PHRR (clays + APP, 70 kW m^{-2}) −30% PHRR (clays, 35 kW m^{-2}) −45% PHRR (clays + APP, 35 kW m^{-2})	[53]
		Epoxy	MMT (different organic treatments) 0.5, 1, 2, 3 wt%	LOI = 20.9 pure resin LOI = 23.2 (2% nanocomposite)	[54]

		Epoxy	Bentonite (phosphonium modified) (with/without low-melting silicate glass—Cepree) 5 wt%	LOI = 21 pure resin LOI = 25 nanocomposite −42% PHRR (clay, 70 kW m^{-2}) −36% PHRR (clay, 50 kW m^{-2}) −34% PHRR (clay, 35 kW m^{-2})	[55]
		Epoxy	LDH (different organic solvents) 1, 3% and 5%	Lower burning rate Self-extinguishing behavior (V-1 or V-2 classification)	[47]
		Epoxy	Bentonite (phosphonium modified) 5 wt% (influence of drying method)	−55% PHRR (freeze-dried clays) −47% PHRR (spray dried clays) −26% PHRR (unmodified clays)	[56]
		UP	MMT (different commercial solutions—Cloisite) 5, 7.5% and 10% (with/without traditional FR)	−39% PHRR (10% Cloisite 25A) −27% PHRR (5% Cloisite 10A) −64% PHRR (5% Cloisite25A + 20% APP) −70% PHRR (5% Cloisite10A + 20% APP)	[23]
		UP	LDH 1% and 5%	−47% PHRR (1% A-LDHs) −33% PHRR (5% S-LDHs)	[57]
FRP	Glass	Epoxy	MMT (with/without organic treatment) 1, 3, 6 and 10 wt%	+7% Tg (laminates, 6% organic treated MMT) −85% Average time of burning decreased (10wt% organoclay) −75% Average extent of burning decreased (10 wt% organoclay) Determined by UL94	[58]
	Carbon	Epoxy (DGEBA and TGDDM)	MMT 5 wt%	−19% PHRR −14% AHRR −24% smoke obscuration (TGDDM, 50 kW m^{-2})	[25]

Continued

Table 1 Resume of Papers in Open Literature Reporting the Use of Clays for Flame Retardancy—cont'd

Composite	Reinforcement Fibers Type and Amount	Polymer Matrix	Clay Type and Amount	Thermal Improvements	References
	Glass	UP	MMT (commercial Cloisite) 0.05 and 0.20 wt% (in clay-CNF sheets)	−60% PHRR of composite (0.20 wt% clay in CNF sheets) −14% weight loss of clay-CNF sheet at 1000 °C (when adding clay)	[59]
	Glass	UP	MMT (commercial Cloisites) 5, 15 and 25 wt% (in clay/CNF sheets)	+17 °C decomposition temperature (25 wt% clay) +40% residue after TGA (25 wt% clay) −41% PHRR (5 wt% clay Na^+) −51% PHRR (10 wt% clay Na^+) −70% PHRR (25 wt% clay Na^+)	[60]
	Glass	UP	MMT 25 wt% (in clay/CNF paper) (comparison with POSS)	−67% PHRR (50 kW m^{-2}) −52% THR (50 kW m^{-2}) −35% ASEA (50 kW m^{-2}) −26% AMLR (50 kW m^{-2})	[61]

PHRR, peak of heat release rate; LOI, limiting oxygen index; MMT, montmorillonite; ATH, aluminum trihydroxide; APP, ammonium polyphosphate; THR, total heat released; TTI, time to ignition; DGEBA, Diglycidyl Ether of Bisphenol A; TGDDM, Tetraglycidyl Diamino Diphenyl Methane; FR, Flame Retardant; MB, Melamine Borate; CNF, Carbon Nanofiber; AHRR, Average Heat Release Rate; ASEA, Average Specific Extinction Area; AMLR, Average Mass Loss Rate; POSS, Polyhedral Oligomer Silsesquioxanes; ABS, 3-aminobenzenesulfonic acid LDH; TS, 4-toluenesulfonic acid mono-hydrate LDH.

presented by the team of Shartel [52,53]. Hussain et al. [49] synthesized organo-phosphorous modified epoxy resins (DGEBA and TGDDM) and studied the effect of the addition of a commercial montmorillonite in the fire reaction, degradation, and thermal properties. The modified DGEBA epoxy resin and the resin incorporating nanoclays showed significant reductions in the PHRR, 50% and 40%, respectively. No synergistic effects were observed with the incorporation of nanoclays in modified DGEBA, PHRR just decreases in 38%. Similar results were found for the TGDDM epoxy resin. The authors suggest that improvements are a result of the higher polymer charring in fire conditions due to the presence of phosphorous groups.

Toldy et al. [51] tested a phosphorous modified amine hardener and nanoclays in order to improve the flame retardancy of epoxy resin. A V-0 classification was observed when phosphorous-based hardener (TEDAP) and 1% nanoclays were used. Concerning LOI results the best value, LOI = 36, was obtained when Bentonite and TEDAP were used. The authors observed that Sepiolite is more effective alone (without phosphorous groups) when compared with the other clays. On the other hand, Bentonite is more effective in the presence of phosphorous groups (due to the presence of iron ions present on the surface of this clay).

An antagonism effect in flame retardancy was verified by Schartel et al. [52] when using organo-phosphorous with clays, where reductions in PHRR were of 67% (35 kW m^{-2}) and 73% (70 kW m^{-2}). Similar results were obtained when using the phosphorous flame retardant alone. Further studies from this team [53] confirm the antagonism effect of another phosphorous system, namely APP, with clays. The strong antagonism was shown in properties such as HRR, THR, and LOI, because of inhibiting intumescence caused by the presence of clays. The authors explained that the addition of clays increases the viscosity of the liquid products formed in the pyrolysis zone, hindering the formation of the intumescent layer.

In 2011, the same team [56] determined the influence of the phosphonium-based clay's drying process on epoxy flame-retardant nanocomposites. They have tested different drying processes and verified the formation of different surface areas that have promoted different flame-retardancy effects. The BET surface of clays increased with spray drying and freeze-drying processes resulting in their homogeneous distribution in the resin. These differences in the morphology resulted in different reductions in the PHRR, higher BET surfaces give higher reductions in PHRR—about 55% at 50 kW m^{-2}. They have observed and concluded that a nanocomposite morphology indicates a promising flame-retardant structure, however, this behavior is very dependent of ablation and migration mechanisms that will occur at higher temperatures and are the key for protective barrier formation.

Another recent work on phosphonium modified clays (TPPMMT) was presented by Wu et al. [55]. They have studied the combination of these modified clays with low-melting silicate glass, Ceepree (CP), for epoxy flame retardancy. CP and TPPMMT together do not change the pyrolysis behavior of the epoxy resin, but increase, as expected, the inorganic residue. The PHRR is reduced by around 32–42% when 5 wt% TPPMMT is added, and 51–63% when 10 wt% CP is added. CP is more effective than clays due to an important inorganic-carbonaceous surface protection layer formation. An antagonism effect was observed through the combination of these two additives.

The combination of ATH with phosphonium modified clays was also approached by Schartel [52] who has considered this combination as a promising flame-retardant solution. They observed reductions in PHRR of 50% (heat flux of 35 kW m^{-2}) and 61% (heat flux of 70 kW m^{-2}) for the referred clay + ATH system in epoxy resin.

6.2.2 Unsaturated polyester resin

Despite the interest on clay-based polymer for flame retardancy, as previously shown for epoxy resins, a limited number of works studied the use of clays on UP resin [23,57]. Nazaré et al. [23] studied the effect of different organic modified nanoclays and their combinations with ATH and APP in the fire reaction properties of an unsaturated polyester resin. They observed that the addition of 10% of organoclay to the polyester matrix decreased PHRR in 39% and THR in 15%. The total smoke produced, however, was increased by 11%. These cone calorimetric results together with XRD analysis suggested to the authors that flammability improvements are not always dependent on the formation of nanocomposite structures. The authors found a nearly linear relationship between the PHRR and the clay content, stating that higher clay loadings result in the formation of thicker insulating layers on the surface of the polymer, therefore providing a barrier to mass and energy transport. They have also observed a synergetic effect between nanoclays and traditional flame retardants. Anionic clays, namely modified adipate LDHs, have also provided significant decreases in the flammability of a polyester resin [57], with decreases in PHRR of 47% (1 wt% clay) and 22% (5 wt% clay). The HRR and smoke production (SEA) curves of UP/adipate LDH nanocomposites are shown in Figure 8.

These results suggest that LDH clays can be more effective than conventional clays in generating a barrier structure which prevents the volatiles to escape. According to the authors, the process of barrier formation is composed of two steps: (1) an intercalated char/metal oxide structure is formed, and (2) further oxidation of the

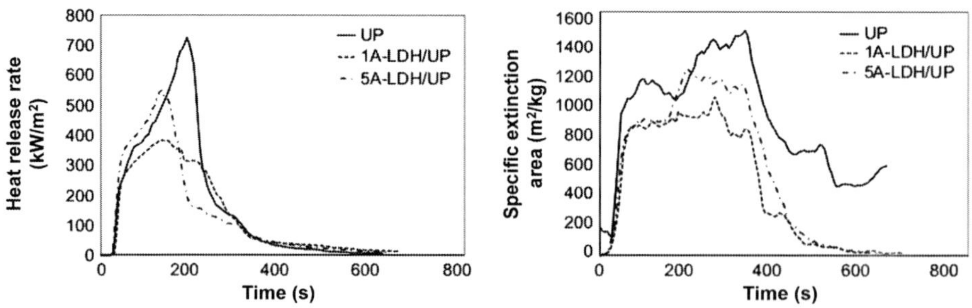

FIGURE 8

Heat release rate (left) and specific extinction area (right) curves for adipate LDH/UP nanocomposites [57].

residue is catalyzed by the metal oxides themselves. In addition, they decompose by an endothermic reaction, releasing water vapor, consequently promoting the dilution of the combustible gases and reducing the heat release of the nanocomposite. The authors concluded that the fire reaction behavior of the prepared nanocomposites is not always dependent on the formation of well dispersed and exfoliated structures. This enhancement of properties can be related to the thermal and physical linkage of the clay platelets and the polymer chains or to the physico-chemical adsorption of the degradation volatiles on the LDH, which depends on their organic modification.

6.2.3 Fiber-reinforced polymer

Several methods are currently used for providing flame-retardant properties to fiber-reinforced polymer composites, such as the use of mineral and ceramic wool, high-performance fibers, surface coatings, or modified polymer resins. In the aerospace industry, for instance, the use of surface coatings or the chemical or physical modification of the resin is seen as the most suitable solution, as the retention of mechanical properties after heat exposure is the fundamental requirement. On the other hand, the use of thermally insulative layers of mineral and ceramic wool is a considered an appropriate solution for naval and transportation industries. Surface coatings are passive fire protection systems that can be bonded onto the surface of fiber-reinforced composites in the form of paint, spray, or mat. They are considered as an efficient way to provide fire protection to composite materials without affecting the physical and mechanical properties. They also have the advantage of being applied after composite's manufacturing, therefore not requiring changes during the processing.

The use of intumescent surface coatings in GFRP composites has been reported by Lungille and Correia [62,63]. Kandola et al. reported the use of hybrid intumescent mats [64] and the simultaneous use of conventional fire-retardant additives added to the resin [65] as fire protection systems to UP/GFRP composites. However, in this work, focus will be given on fire protection systems based on modified polymeric resins, namely through the addition of nanoparticles.

6.2.4 FRP-epoxy resin

In a work from 2007, the fire performance of two different epoxy resins modified with nanoclay and phosphorous-based curing agent and their corresponding laminates produced with carbon fibers were reported by Liu et al. [25]. The use of nanoclays, phosphorous curing agent, or the combined use of both worsened the fire performance of the DGEBA resin composite. When using a TGGDM resin, there were slight improvements in the phosphorous containing sample, namely decreases of 19% in PHRR, 14% in the AHRR and 24% in the smoke obscuration. This work emphasized the close relationship between char yields obtained in TGA and cone calorimetry with the PHRR, highlighting the role of the condensed phase in minimizing combustion.

In the same year, Bozkurt et al. [58] published a work on the thermal and mechanical properties of glass fiber-reinforced clay/epoxy nanocomposites by adding clays to the epoxy resin. Addition of clay significantly reduced the flammability of polymer composites. The average extent of burning and average time of burning were reduced by 75% and 85%, respectively, due to the addition of 10 wt% organically modified MMT into epoxy. The improvements were higher with organo-modified clays due to enhanced dispersion in the polymeric matrix.

6.2.5 FRP-unsaturated polyester resin

Gou et al. have been one of the most active working in the flame retardancy of unsaturated polyester/glass fiber-reinforced composites [59–61]. In their first published work in this area, they have studied the production and incorporation of clay/carbon nanofiber hybrid sheets onto the surface of GFRP composites [59]. Significant improvements in the thermal stability of the carbon nanosheet with the addition of nanoclays were obtained. 0.20 wt% of nanoclay content in carbon nanosheet provided reductions in PHRR of 60%. Authors concluded that a combined effect of barrier/insulation from the nanoclays and reemitting effect of carbon nanofibers acted on the combustion of the composites. A continuous barrier to oxygen supply to the composite was effectively produced by combining these two nanomaterials.

The same approach was used by Zhuge et al. [60] but comparing two different commercial clays (Cloisite Na^+ and organically modified Cloisite 20A) in higher amounts. When unmodified MMT clay was incorporated into the CNF nanopaper, the PHRR decreased dramatically with the weight fraction of MMT and reductions in the PHRR of 70% were achieved with 25 wt% of clay content in the nanopaper.

When this team compared clays with POSS [61], they observed that clays were more effective in reducing not only the PHRR, but also the average mass loss, average smoke obscuration and total heat released. The PHHR of composite materials increased significantly with the addition of POSS-CNF nanopaper in the surface. On the contrary, for the CNF-clay nanopaper, significant decreases in PHRR (67%) were obtained. POSS particles were observed to migrate rapidly to the surface of polymer nanocomposites, leading the authors to believe that this could increase the HRR or even quickly ignite the polymer resin. Furthermore, clays have a layered structure, serving as a barrier to mass and heat transfer, while POSS has a cage-like structure with organic groups attaching to each corner of the cage, where no barrier effect can be attained.

6.3 Flame-retardancy mechanisms of nanoclays

Just as reference, the flame-retardant effect of clay mineral particles and its mechanism in thermoplastic matrices is well described in the literature and they can be summarized as follows:

1. Migration of the clays to the surface of the polymer during combustion, leading to the formation of a rich silicate layer which acts as a physical barrier that slows mass and heat transfer during combustion [66–68].

2. The silicate layer protects the polymer from heat and also limits the oxygen flow and the escape of volatiles; these actions interfere with the combustion process reducing the amount of available combustible [69].

In the case of a thermosetting matrix, the mechanism is essentially different from that described in thermoplastic matrices, since the cross-linked structure hinders the migration of clay particles to the polymer surface and the formation of a protective layer there. However, the barrier effect provided by the clays during the combustion of thermoset polymers has been pointed out as the main responsible flame-retardant mechanism [23,25,48,70].

The work done by Schartel in 2011 [56], though, has stated that there is a migration of rich silicate bubbles to the surface of the polymer during combustion. Bubbling is therefore reported as a mechanism responsible for the improved flame retardancy of epoxy/layered silicate nanocomposite. The formation of the bubbles, which are covered by silicate agglomerates, is important both in the creation of solid agglomerates and on its transportation process to the surface, thus contributing to the creation of a protective barrier. The formation of bubbles, their migration, reorientation, agglomeration, and ablation mechanisms have shown an effect on the development of a residue layer during the combustion of epoxy nanocomposites. Clay minerals are also claimed to act as flame retardants due to their catalytic effect on charring and/or reinforcement of the char structure. The formation of a carbonaceous layer changes the combustion products and promotes a reduction on the permeability of oxygen and mass transport [50,51,56].

In the case of anionic nanoclays, their main flame-retardant mechanisms are: LDHs release aqueous vapor and carbon dioxide during decomposition, promoting dilution of combustible gases of pyrolysis, increasing the ignition time and reducing the heat release during the combustion; porous thermal-decomposed products of LDHs with large specific surface area, absorbing the smoke and gases produced in the course of combustion [57]. The thermal decomposition of LDHs greatly depends on the nature of the layer cations and interlayer anions, and, as expected, on the experimental conditions during heating (oxidant, inert, or reducing conditions) [71,72]. Usually, decomposition takes place in four steps:

1. Loss of physically adsorbed water (50−200 °C).
2. Loss of interlayer water.
3. Removal of hydroxyl groups from the layers as water vapor (250−500 °C).
4. Loss of the interlayer anion, with the formation of metal oxides (e.g.: Al_2O_3, MgO) and carbon dioxide (500−800 °C).

From this behavior, it is expected that the metal oxides formed during the different decomposition stages increase the thermal stability of the yield char.

7. Polymer/CNT nano- and microcomposites

7.1 Carbon nanotubes

Carbon nanotubes (CNTs) are one dimensional carbon materials consisting of graphitic sheets rolled into seamless tubes. The properties of nanotubes depend on atomic arrangement (how the sheets of graphite are "rolled"), the diameter and length of the tubes, and the morphology, or nano structure. They have diameters ranging from about a nanometer to tens of nanometers with lengths up to centimeters, therefore aspect ratios greater than 1000. Due to their symmetric structure, carbon nanotubes usually have excellent properties [73].

For the synthesis of CNTs, three main synthetic methods are used: arc-discharge, laser ablation, and catalytic methods such as chemical vapor deposition. Depending on the process of fabrication, it is possible to obtain three types of CNTs: single-walled CNTs (SWCNTs), double-walled (DWCNTs), and multi-walled CNTs (MWCNTs). As the name implies, SWCNTs consist of a single graphene layer rolled up into a seamless cylinder, whereas DWCNTs and MWCNTs consist of two or more concentric cylindrical shells of graphene sheets. MWCNTs show a lower tensile strength and modulus, when compared to SWCNTs. SWCNTs have diameters in the range of 0.4 up to 5.6 nm and MWCNTs have much higher diameters, which may vary from 20 to 100 nm. MWNTs are also more rigid because their section is much larger compared to that of SWCNTs [74,75].

In 2007, the potential of CNTs has been investigated for many highly demanding applications such as aerospace applications [76,77]. The combination of high mechanical performances with extraordinary high electrical and thermal conductivities of CNT, allied with their high aspect ratio and low density, makes them a very interesting solution for multifunctional polymeric composites reinforcement. The increasing utilization of multifunctional composites in a wide range of technical applications will eventually lead them to encounter events where extreme thermal resistance is required, thus motivating the need of understanding the thermal and fire reaction behavior of these new composite systems. Furthermore, CNTs are interesting flame-retardant candidates due to their elongated shape and thermal stability.

7.2 Flame-retardant polymer/CNT nano- and microcomposites

Despite the reports of the exceptional physical and mechanical properties of epoxy nanocomposites with CNTs [78–81], there are only a very limited number of studies on the epoxy/CNT composites flammability and the reported results are dependent of raw materials (nanotubes and epoxy resin types) and sample preparation processes.

To our knowledge, one of the first works reporting the flammability of epoxy resins filled with CNTs was made by our team [82]. The addition of 1% CNT to a low viscous epoxy resin promoted a small increase in time-to-ignition and a decrease (17%) in PHRR, as it is possible to see in Figure 9.

Rahatekar et al. [83] reported reductions of almost 50% in peak of mass loss with a small addition (0.0025 wt%) of highly aligned MWCNTs to an epoxy resin. Tests

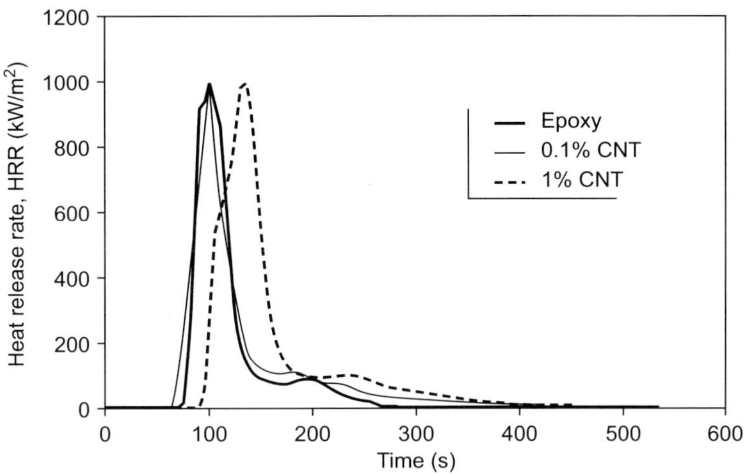

FIGURE 9

Heat release rate curves for epoxy/CNT nanocomposites [82].

were performed in a gasification apparatus, which measures the mass loss rate (MLR) of a sample exposed to a constant heat flux under an inert atmosphere. Comparison with an MMT/epoxy nanocomposite allowed the conclusion that it is necessary to use higher clay loading to obtain similar reductions. Authors claimed that these aligned nanotubes were easily exfoliated in the resin using high shear mixing, therefore contributing to enhanced properties. The combination of MMT and MWCNTs for improved flammability of an epoxy resin was approached by Lee et al. [84]. A significant increased LOI (33) was obtained with 10% MMT and 1% MWCNTs; the pure epoxy resin showed an LOI of 21. The same team [85] published a similar study, however using fluorinated MMT and MWCNTs in order to create hydrophobic functional groups for improved dispersion into the epoxy resin, once again resulting in increased LOI values and char yield when using nanotubes. However, LOI value obtained (31) was not as high as in the previous study, when using both MMT and MWCNTs. The use of 0.1 wt% of CNTs alone increased LOI from 21 (pure resin) to 23. In both the cases, authors concluded that the increased char yield formed in the presence of MWCNTs acts as both heat barrier and thermal insulation, in addition to decreasing the degradation rate of epoxy resin. Increased LOI values and a flame-retardant classification (V-0) were also obtained by Kuan et al. [86], although considerable higher amounts (up to 9%) of functionalized CNTs were added to a modified epoxy resin containing silicon to induce a sol–gel reaction. LOI increased up to 29 with increasing CNT content and V-0 classification was obtained for nanotube contents higher than 3%. Considerable increases in glass transition (42 °C) and char yields (47%) were also reported.

Hapuarachchi et al. also studied the combined effect of nanoclays (Sepiolite) and MWCNTs, but using unsaturated polyester resin as matrix [87]. CNTs alone hindered the thermo-oxidation of the resin in the early stages of degradation, by increasing the onset degradation temperature to 20 °C. An even higher increase in the onset degradation temperature (36 °C) was obtained when adding Sepiolite clays to the UP/MWCNT nanocomposites. A synergetic behavior was also found in terms of PHRR, with a decrease of 50% for Sepiolie/MWCNTs nanocomposites. The THR, however, did not change substantially, suggesting to authors that the physical barrier formed during combustion prolongs the burning times without decreasing the total amount of combustible material.

Two other works in 2010 and 2011, reported by Pereira [88] and Gérard teams [89], respectively, for imparting flame-retardant properties to an epoxy resin, was the combination of traditional flame-retardant APP with multi-walled CNTs. Pereira et al. report synergy in reducing PHRR and THR of an epoxy resin by using MWCNTs and APP. These authors refer that physical and chemical actions between CNT, APP, and the polymer produce a "fire blocking" barrier that prevents the release of degradation volatiles. This barrier was perceptible during cone calorimeter tests where some smoke escaped by the sample holder bottom face. On the other hand, antagonism effects were found on properties such as ignition times and PHRR by Gérard et al. [89]. Authors stated that the presence of nanotubes increases the material thermal conductivity, therefore resulting in lower ignition times. Furthermore, another suggested phenomenon is that the sample containing CNT can absorb more quickly the radiant heat, leading to the accumulation of heat at the surface of the material, resulting in a faster ignition.

7.2.1 Fiber-reinforced polymer

The use of CNT bucky papers has since 2010 emerged as a solution for fire protection of fiber-reinforced composites. Bucky papers are 3D structures consisting of sheets of entangled CNTs that can be produced by filtrating CNT suspensions. These sheets usually have thicknesses ranging between 30 and 500 μm. Bucky paper structures represent a path toward the production of polymer composites with high and well dispersed nanotube content [90].

Recently, Zhang's team published two works [91,92] regarding the use of CNT bucky papers (3D organized CNT structures)—as fire retardants in epoxy/carbon fiber composites. In their first work [91], they produced CNT bucky papers and incorporated them on the surface of carbon fiber epoxy composites for fire shielding. They have prepared carbon nanofiber, MWCNT and SWCNT papers, however, only MWCNT bucky paper acted as an effective flame-retardant. Important improvements on the ignition time, PHRR, THR, and the smoke emission were reported, as resumed in Table 2. The authors discussed and demonstrated that bucky papers act as barriers, reducing the release of polymer fragments from matrix into the air, therefore delaying time to the critical point of the concentration of combustible gases, despite their high thermal conductivity. Due to their dense structures and small pore size, MWCNT bucky papers were able to provide low gas and mass permeabilities, acting therefore as a fire shield. In their second work [92], they compared the effectiveness of SWCNT and MWCNT bucky papers as fire shield

Table 2 Resume of Papers in Open Literature Reporting the Use of Carbon Nanotubes for Flame Retardancy

Composite	Reinforcement Fibers Type and Amount	Polymer Matrix	CNT Type and Amount	Thermal Improvements	References
Nano	No	Epoxy	CNT 0.1 and 1 wt%	−17% PHRR Increase in TTI Decrease in smoke production (1 wt% CNT)	[82]
		Epoxy	MWCNT 1 wt% (combined effect with MMT)	LOI = 21 pure resin LOI = 33 (CNT + 10% MMT) +10.5% Char yield (CNT + 10% MMT)	[84]
		Epoxy	MWCNT 0.03, 0.25 and 0.5 wt% (comparison with MMT)	−50% PMLR (0.25 and 0.5 wt% CNT) −40% PMLR (0.03 wt% CNT) (determined in a gasification apparatus)	[83]
		Epoxy	Fluorinated MWCNT 0.1 wt% (with/without MMT)	LOI = 21 pure resin LOI = 23 (CNT) LOI = 31 (CNT + MMT) +2.5% Char yield (CNT) +6% Char yield (CNT + MMT)	[85]
		Epoxy	MWCNT 1, 3, 5, 7 and 9 wt%	+42 °C in Tg (7% CNT) +47% Char yield at 750 °C LOI = 22 pure resin LOI = 29 (9 wt% CNT) V-0 classification (3, 5, 7, 9 wt% CNT)	[86]
		Epoxy	MWCNT 0.5 wt% (combined effect with APP)	−28% PHRR (APP + CNT) −4% THR (APP + CNT) +8% Char yield (APP + CNT) (determined in mass loss calorimeter)	[89]

Continued

The use of nanoclays and carbon nanotubes may also be a strategic approach for producing new fiber-reinforced polymeric composites with enhanced thermal and mechanical properties.

In terms of flame retardancy, nanoclays and CNTs have shown to be efficient in reducing the mass loss rate, therefore promoting significant reductions in the peak of heat release rate. However, time-to-ignition usually decreases and smoke production increases. This effect is far below that achieved when using traditional flame retardants such as ammonium polyphosphate or aluminum trihydrate. Therefore, the possible synergistic effect of combining nanoclays with traditional flame retardants has been of increasing attention.

The variety of available nanomaterials and the possibility to change the composition during their synthesis continues to make these very interesting candidates for producing materials with new features. One of the main drawbacks, however, is that most work done in this area shows that it becomes crucial to produce well-dispersed polymer nanocomposites in order to achieve the best results. It becomes necessary to establish a dispersion procedures for each system used. An alternative solution to overcome the dispersion difficulties in FRP composites could be the use of preassembled stable nanomaterial structures such as bucky papers. This new approach must be investigated, as well as other several critical issues when using nanomaterials for flame retardancy as listed below:

- *Health safety concerns*. Either during nanoparticles's handling and during fire scenario (effect on smoke toxicity or asbestos).
- *Durability and aging*. Nanocomposites could suffer from hydrolytic, oxidative, and photo-oxidative instabilities. A deep understanding about these effects on the composite durability is needed.
- *Life cycle assessment (LCA)*. For example, the impact of introducing nanoparticles on the recyclability of composites must be evaluated.
- *Processing parameters*. When flame retardants and nanomaterials are incorporated into resins, knowledge about the rheology and cure kinetic is fundamental for process control and optimization.

Acknowledgments

Authors are gratefully acknowledged for financial support from the Portuguese Science and Technology Foundation (FCT) under the Program Science 2007 and FIRE-FRP Project (PTDC/ECM/100779/2008).

References

[1] Advani SG, Sozer EM. Process modelling in composites manufacturing. Marcel Dekker Inc; 2003.
[2] Horrocks AR, Price D, editors. Fire retardant materials. CRC Press; 2001.

[3] Mallick PK. Fibre-reinforced composites: materials, manufacturing and design. 3rd ed. CRC Press; 2008.
[4] Morgan P. Carbon fibers and their composites. Taylor & Francis; 2005.
[5] Mouritz AP, Gibson AG. Fire properties of polymer composite materials. Solid mechanics and its applications, vol. 143. Springer; 2006.
[6] McConnell VP. Decoding the 'stealth factor' in global carbon fibre supply and demand. Reinf Plast 2008;52:18–23.
[7] European Flame Retardants Association (EFRA).
[8] Einde LVD, Zhao L, Seible F. Use of FRP composites in civil structural applications. Constr Build Mater 2003;17:389–403.
[9] Hollaway LC. A review of the present and future utilisation of FRP composites in the civil infrastructure with reference to their important in-service properties. Constr Build Mater 2010;24:2419–45.
[10] Dodds N, Gibson AG, Dewhurst D, Davies JM. Fire behaviour of composite laminates. Composites Part A 2000;31:689–702.
[11] Correia JR, Branco FA, Ferreira JG, Bai Y, Keller T. Fire protection systems for building floors made of pultruded GFRP profiles: part 1: experimental investigations. Composites Part B 2010;41:617–29.
[12] Shartel B, Bartholmai M, Knoll U. Some comments on the main fire retardancy mechanisms in polymer nanocomposites. Polym Adv Technol 2006;17:772–7.
[13] Schartel B, Bartholmai M, Knoll U. Some comments on the use of cone calorimeter data. Polym Degrad Stab 2005;88:540–7.
[14] Schartel B, Hull TR. Development of fire-retarded materials—interpretation of cone calorimeter data. Fire Mater 2007;31:327–54.
[15] Kiliaris P, Papaspyrides CD. Polymer/layered silicate (clay) nanocomposites: an overview of flame retardancy. Prog Polym Sci 2010;35:902–5.
[16] Mouritz AP, Feih S, Kandare E, Mathys Z, Gibson AG, Des Jardin PE, et al. Review of fire structural modelling of polymer composites. Composites Part A 2009;40:1800–14.
[17] Laoutid F, Bonnaud L, Alexandre M, Lopez-Cuesta J-M, Dubois Ph. New prospects in flame retardant polymer materials: from fundamentals to nanocomposites. Mater Sci Eng R 2009;63:100–25.
[18] Mouritz AP, Mathys Z, Gibson AG. Heat release of polymer composites in fire. Composites Part A 2006;37:1040–54.
[19] Mouritz AP. Review of smoke toxicity of aerospace composites. In: Proceedings of the 5th international conference on composites in fire. United Kingdom; July 10–11, 2008.
[20] Brown JR, Mathys Z. Reinforcement and matrix effects on the combustion properties of glass reinforced polymer composites. Composites Part A 1997;28A:675–81.
[21] Avila MB, Dembsey NA, Dore C. Effect of resin type and glass content on the reaction to fire characteristics of typical FRP composites. Composites Part A 2008;39:1503–11.
[22] Kandola BK, Akonda MH, Horrocks AR. Use of high-performance fibres and intumescents as char promoters in glass-reinforced polyester composites. Polym Degrad Stab 2005;88:123–9.
[23] Nazaré S, Kandola BK, Horrocks AR. Flame-retardant unsaturated polyester resin incorporating nanoclays. Polym Adv Technol 2006;17:294–303.
[24] Kandare E, Kandola BK, Price D, Nazaré S, Horrocks AR. Study of the thermal decomposition of flame-retarded unsaturated polyester resins by thermogravimetric analysis and Py-GC/MS. Polym Degrad Stab 2008;93:1996–2006.

[25] Liu W, Varley RJ, Simon GP. Understanding the decomposition and fire performance processes in phosphorus and nanomodified high performance epoxy resins and composites. Polymer 2007;48:2345–54.
[26] Levchik SV, Weil ED. Thermal decomposition, combustion and flame-retardancy of epoxy resins—a review of the recent literature. Polymer International 2004;53: 1901–29.
[27] Eibl S. Influence of carbon fibre orientation on reaction-to-fire properties of polymer matrix composites. Fire Mater 2012;36:309–24.
[28] Bras ML, Bourbigot S, Mortaigne B, Cordellier G. Comparative study of the fire behavior of glass-fibre reinforced unsaturated polyesters using a cone calorimeter. Polym Polym Compos 1998;6:535–9.
[29] Morgan AB, Wilkie CA. Flame retardant polymer nanocomposites. Wiley Interscience; 2007.
[30] Kandola BK. Flammability issues with composites used in different sectors. Northwest Composites Centre.
[31] Horöld S. Phosphorous flame retardants in thermoset resins. Polym Degrad Stab 1999; 64:427–31.
[32] Braun U, Balabanovich AI, Schartel B, Knoll U, Artner J, Ciesielski M, et al. Influence of the oxidation state of phosphorus on the decomposition and fire behaviour of flame-retarded epoxy resin composites. Polymer 2006;47:8495–508.
[33] Alexandre M, Dubois P. Polymer-layered silicate nanocomposites: preparation, properties and uses of a new class of materials. Mater Sci Eng R 2000;28:1–63.
[34] Paul DR, Robeson LM. Polymer nanotechnology: nanocomposites. Polymer 2008;49: 3187–204.
[35] Pavlidou S, Papaspyrides CD. A review on polymer–layered silicate nanocomposites. Prog Polym Sci 2008;33:1119–98.
[36] Rives V, editor. Layered double hydroxides: present and future. New York: Nova Science Publishers, Inc; 2001.
[37] Azeez AA, Ree KY, Park SJ, Hui D. Epoxy clay nanocomposites – processing, properties and applications: a review. Composites Part B 2013;45:308–20.
[38] Kameda T, Saito S, Umetsu Y. Mg-Al layered double hydroxide intercalated with ethylene-diaminetetraacetate anion: synthesis and application to the uptake of heavy metal ions from an aqueous solution. Sep Purif Technol 2005;47:20–6.
[39] Evans DG, Duan X. Preparation of layered double hydroxides and their applications as additives in polymers, as precursors to magnetic materials and in biology and medicine. Chem Commun 2006;5:485–96.
[40] Utracki A, Sepehr M, Boccaleri E. Review: synthetic, layered nanoparticles for polymeric nanocomposites (PNCs). Polym Adv Technol 2007;18:1–37.
[41] Niambo C, Chen D, Su S, Wilkie CA. Variation of benzyl anions in MgAl-layered double hydroxides: fire and thermal properties in PMMA. Polym Degrad Stab 2009;94: 496–505.
[42] Duquesne S, Jama C, Bras ML, Delobel R, Recourt P, Gloaguen JM. Elaboration of EVA–nanoclay systems—characterization, thermal behaviour and fire performance. Compos Sci Technol 2003;63:1141–8.
[43] Wagenknecht U, Kretzschmar B, Reinhardt G. Investigations of fire retardant properties of polypropylene-clay-nanocomposites. Macromol Symp 2003;194: 207–12.

[44] Zhang J, Delichatsios MA, Bourbigot S. Experimental and numerical study of the effects of nanoparticles on pyrolysis of a polyamide 6 (PA6) nanocomposite in the cone calorimeter. Combust Flame 2009;156:2056–62.

[45] Zhao C, Quin H, Gong F, Feng M, Zhang S, Yang M. Mechanical, thermal and flammability properties of polyethylene/clay nanocomposites. Polym Degrad Stab 2005;87: 183–9.

[46] Zammarano M, Franceschi M, Bellayer S, Gilman JW, Meriani S. Preparation and flame resistance properties of revolutionary self-extinguishing epoxy nanocomposites based on layered double hydroxides. Polymer 2005;46:9314–28.

[47] Becker C, Gabbardo AD, Wypych F, Amico SC. Mechanical and flame-retardant properties of epoxy/Mg–Al LDH composites. Composites Part A 2011;42:196–202.

[48] Hartwig A, Putz D, Schartel B, Bartholmai M, Wendschuh-Josties M. Combustion behaviour of epoxide based nanocomposites with ammonium and phosphonium Bentonites. Macromol Chem Phys 2003;204:2247–57.

[49] Hussain M, Varley RJ, Mathys Z, Cheng YB, Simon GP. Effect of organo-phosphorus and nano-clay materials on the thermal and fire performance of epoxy resins. J Appl Polym Sci 2004;91:1233–53.

[50] Camino G, Tartaglione G, Frache A, Manferti C, Costa G. Thermal and combustion behaviour of layered silicate-epoxy nanocomposites. Polym Degrad Stab 2005;90: 354–62.

[51] Toldy A, Tóth N, Anna P, Keglevich Gy, Kiss K, Marosi Gy. Flame retardancy of epoxy resin with phosphorus containing reactive amine and clay minerals. Polym Adv Technol 2006;17:778–81.

[52] Schartel B, Knoll U, Hartwig A, Putz D. Phosphonium-modified layered silicate epoxy resins nanocomposites and their combinations with ATH and organo-phosphorus fire retardants. Polym Adv Technol 2006;17:281–93.

[53] Shartel B, Weiß A, Mohr F, Kleemeier M, Hartwig A, Braun U. Flame retarded epoxy resins by adding layered silicate in combination with the conventional protection-layer-building flame retardants melamine borate and ammonium polyphosphate. J Appl Polym Sci 2010;118:1134–43.

[54] Kaynak C, Nakas GI, Isitman NA. Mechanical properties, flammability and char morphology of epoxy resin/montmorillonite nanocomposites. Appl Clay Sci 2009;46: 319–24.

[55] Wu G, Schartel B, Kleemeier M, Hartwig A. Flammability of layered silicate epoxy nanocomposites combined with low-melting inorganic ceepree glass. Polym Eng Sci 2012;52:507–17.

[56] Schartel B, Weiß A, Sturm H, Kleemeier M, Hartwig A, Vogt C, et al. Layered silicate epoxy nanocomposites: formation of the inorganic-carbonaceous fire protection layer. Polym Adv Technol 2011;22:1581–92.

[57] Pereira CMC, Herrero M, Labajos FM, Marques AT, Rives V. Preparation and properties of new flame retardant unsaturated polyester nanocomposites based on layered double hydroxides. Polym Degrad Stab 2009;94:939–46.

[58] Bozkurt E, Kaya E, Tanoglu M. Mechanical and thermal behavior of non-crimp glass fibre reinforced layered clay/epoxy nanocomposites. Compos Sci Technol 2007;67: 3394–403.

[59] Zhao Z, Gou J, Bietto S, Ibeh C, Hui D. Fire retardancy of clay/carbon nanofibre hybrid sheet in fibre reinforced polymer composites. Compos Sci Technol 2009;69:2081–7.

[60] Zhuge J, Tang Y, Gou J, Chen R-H, Ibeh C, Hu Y. Flammability of carbon nanofibre–clay nanopaper based polymer composites. Polym Adv Technol 2011;22:2250–6.
[61] Tang Y, Zhuge J, Gou J, Chen R-H, Ibeh C, Hu Y. Morphology, thermal stability, and flammability of polymer matrix composites coated with hybrid nanopapers. Polym Adv Technol 2011;22:1403–13.
[62] Lungille K, Nguyen D, Veinot DE. Inorganic intumescent coatings for improved fire protection of GRP. Fire Technol 1999;35:99–110.
[63] Correia JR, Branco FA, Ferreira JG. The effect of different passive fire protection systems on the fire reaction properties of GFRP pultruded profiles for civil construction. Composites Part A 2010;41:441–52.
[64] Kandare E, Chukwudolue C, Kandola BK. The use of fire-retardant intumescent mats for fire and heat protection of glass fibre-reinforced polyester composites: thermal barrier properties. Fire Mater 2010;34:21–38.
[65] Kandare E, Chukwudolue C, Kandola BK. The effect of fire-retardant additives and a surface insulative fabric on fire performance and mechanical property retention of polyester composites. Fire Mater 2011;35:143–55.
[66] Ogata N, Jimenez G, Kawai H, Ogihara T. Structure and Thermal/Mechanical properties of poly(-lactide)-clay blend. J Polym Sci Polym Phys 1997;35:389–96.
[67] Beyer G. Nanocomposites: a new class of flame retardants for polymers. Plast Addit 2002;4:22–8.
[68] Zanetti M, Bracco P, Costa L. Thermal degradation behaviour of PE/clay nanocomposites. Polym Degrad Stab 2004;85:657–65.
[69] Porter D, Meycalfe E, Thomas MJK. Nanocomposite fire retardants = a review. Fire Mater 2000;24:45–52.
[70] Wang L, Su S, Chen D, Wilkie CA. Variation of anions in layered double hydroxides: effects on dispersion and fire properties. Polym Degrad Stab 2009;94:770–81.
[71] Rives V. Characterisation of layered double hydroxides and their decomposition products. Mater Chem Phys 2002;75:19–25.
[72] Zubitur M, Gómez MA, Cortázar M. Structural characterization and thermal decomposition of layered double hydroxide/poly(p-dioxanone) nanocomposites. Polym Degrad Stab 2009;94:804–9.
[73] Thosthenson ET, Ren Z, Chou T-W. Advances in the science and technology of carbon nanotubes and their composites: a review. Compos Sci Technol 2001;61:1899–912.
[74] Coleman JN, Khan U, Blau WJ, Gun YK. Small but strong: a review of the mechanical properties of carbon nanotube–polymer composites. Carbon 2006;44:1624–52.
[75] Paradise M, Goswami T. Carbon nanotubes – production and industrial applications. Mater Des 2007;28:1477–89.
[76] Pambaguian L, Edtmaier C, Janhsen T, Ferrato M, Chereau P, Forero S. Non- organic matrix materials reinforced with carbon nanotubes for space applications. In: Proceedings of the 2nd international conference on micro and nanotechnology – Viennano '07. March 14–16. Vienna, Austria; 2007.
[77] Fidelus JD, Lojkowski W, Lewandowska M, Bielinski D, Malsh I, Hoffschulz H, et al. 9th Nanoforum report nanotechnology in aerospace, European nanotechnology gateway; February 2007.
[78] Ci L, Bai J. The reinforcement role of carbon nanotubes in epoxy composites with different matrix stiffness. Compos Sci Technol 2006;66:599–603.

[79] Abdalla M, Dean D, Theodore M, Fielding J, Nyairo E, Price G. Magnetically processed carbon nanotube/epoxy nanocomposites: morphology, thermal, and mechanical properties. Polymer 2010;51:1614–20.

[80] Hernández-Pérez A, Avilés F, May-Pat A, Valadez-González A, Herrera-Franco PJ, Bartolo-Pérez P. Effective properties of multiwalled carbon nanotube/epoxy composites using two different tube. Compos Sci Technol 2008;68:1422–31.

[81] Moisala A, Li Q, Kinloch IA, Windle AH. Thermal and electrical conductivity of single- and multi-walled carbon nanotube-epoxy composites. Compos Sci Technol 2006;66:1285–8.

[82] Pereira CMC. Flame retardancy of thermoset polymers based on nanoparticles and carbon nanotubes. Solid State Phenom 2009;151:79–87.

[83] Rahatekar SS, Zammarano M, Matko S, Koziol KK, Windle AH, Nyden M, et al. Effect of carbon nanotubes and montmorillonite on the flammability of epoxy nanocomposites. Polym Degrad Stab 2010;95:870–9.

[84] Lee SK, Bai BC, Im JS, In SJ, Lee Y-S. Flame retardant epoxy complex produced by addition of montmorillonite and carbon nanotube. J Ind Eng Chem 2010;16:891–5.

[85] Im JS, Lee SK, In SJ, Lee Y-S. Improved flame retardant properties of epoxy resin by fluorinated MMT/MWCNT additives. J Anal Appl Pyrol 2010;89:225–32.

[86] Kuan C-F, Chen W-J, Li Y-L, Chen C-H, Kuan H-C, Chiang C-L. Flame retardance and thermal stability of carbon nanotube epoxy composite prepared from sol–gel method. J Phys Chem Solids 2012;71:539–43.

[87] Hapuarachchi TD, Bilotti E, Reynolds CT, Peijs T. The synergistic performance of multiwalled carbon nanotubes and sepiolite nanoclays as flame retardants for unsaturated polyester. Fire Mater 2011;35:157–69.

[88] Pereira CMC, Dias D, Costa I. Mechanical, thermal and fire behavior of high performance epoxy/CNT composites. In: 6th Conference on modification, degradation and stabilization of polymers; September 5–9, 2010. Athens, Greece.

[89] Gérard C, Fontaine G, Bourbigot S. Synergistic and antagonistic effects in flame retardancy of an intumescent epoxy resin. Polym Adv Technol 2011;22:1085–90.

[90] Pereira CMC, Nóvoa P, Martins M, Forero S, Hepp F, Pambaguian L. Characterization of carbon nanotube 3D-structures infused with low viscosity epoxy resin system. Compos Struct 2010;92:2252–7.

[91] Wu Q, Zhu W, Zhang C, Liang Z, Wang B. Study of fire retardant behavior of carbon nanotube membranes and carbon nanofibre paper in carbon fibre reinforced epoxy composites. Carbon 2010;48:1799–806.

[92] Wu Q, Bao J, Zhang C, Liang R, Wang B. The effect of thermal stability of carbon nanotubes on the flame retardancy of epoxy and bismaleimide/carbon fibre/buckypaper composites. J Therm Anal Calorim 2011;103:237–42.

[93] Kashiwagi T, Grulke E, Hilding J, Harris R, Awad W, Douglas J. Thermal degradation and flammability properties of poly(propylene)/carbon nanotube composite. Macromol Rapid Commun 2002;23:761–5.

[94] Kashiwagi T, Grulke E, Hilding J, Groth K, Harris R, Butler K, et al. Thermal and flammability properties of polypropylene/carbon nanotube nanocomposites. Polymer 2004; 45:227–39.

[95] Beyer G. Short communication: carbon nanotubes as flame retardants for polymers. Fire Mater 2002;26:291–3.

CHAPTER

Fire Retardancy of Elastomers and Elastomer Nanocomposites

18

S.K. Srivastava

Department of Chemistry, Indian Institute of Technology Kharagpur, Kharagpur, India

T. Kuila

Surface Engineering and Tribology Division, CSIR-Central Mechanical Engineering Research Institute, Durgapur, India

CHAPTER OUTLINE

1. Introduction

Back in 1990, the Toyota research group [1] showed that the use of montmorillonite (MMT) can improve the mechanical, thermal, gas barrier, and flame retardant properties of polymeric materials without hampering the optical translucency behavior of the matrix. Since then, the majority of the research has been focused in improving the fire retardance properties of polymer nanocomposites using cost-effective and environmental friendly nanofillers with the aim of extending the applications of these materials in automotive, aerospace, construction, electronic industries, etc.

Polymer Green Flame Retardants. http://dx.doi.org/10.1016/B978-0-444-53808-6.00018-4

as well as their use in day-to-day life [2–34]. The most investigated nanofillers are natural layered silicate (e.g. MMT), synthetic clay (e.g. layered double hydroxides, LDHs), expanded graphite (EG), graphite nanoplatelet (GNP), carbon nanotubes (CNTs), graphene, magnesium hydroxide (MH), alumina trihydrate (ATH), and hydrotalcite [2–34]. The thermal degradation of these polymer nanocomposites results in the formation of a protective oxide layer on the burning surfaces, which in turn reduces the burning temperature and prevents the oxygen supply from the atmosphere to the bulk of the material [16,34]. The classic route to enhance the fire retardance characteristics of polymers involves the incorporation of flame retardants (FR) either during the polymerization process or more frequently after polymerization (during processing) [7,13,22–24]. These compounds should be resistant toward aging, hydrolysis, and corrosion and should not generate any toxic gases (other than that produced by the polymer itself) or increase the smoke density of the burning polymer undergoing degradation. Flame retardants are inorganic, organophosphorus, or nitrogen-based halogenated compounds; intumescent comprises also a widely used class of flame retardants.

In Table 1, the estimated worldwide consumption of flame retardants as reported in the year 2006 is presented [7]. It is apparent that aluminum trihydrate and MH are the most used ones among all kinds of inorganic flame retardants. Although, these materials are essentially nontoxic, easy to handle, and relatively inexpensive, the high incorporation levels required for effective flame retardancy results in processing difficulties and in the deterioration of the mechanical, physical, and electrical properties of the host polymer. These properties can often be dramatically improved by the application of surface treatments that improve the compatibility of the inorganic filler with the hydrophobic polymer matrix, thus resulting in improved dispersion and enhanced properties. Since high loadings of inorganic hydroxides are required for flame retardancy, studies have been focused on developing synergistic systems, in which aluminum trihydrate, MH, or other inorganic fillers are used in conjunction with other flame retardants, in order to reduce overall addition levels. A wide range

Table 1 Estimated Worldwide Flame Retardant Consumption (Year 2006) [7]

Flame Retardant Type	Estimated Consumption (tonnes)	Percentage
ATH	600,000	38
MH	34,000	2
Brominated compounds	290,000	18
Chlorinated compounds	170,000	11
Phosphorus compounds	260,000	16
Nitrogen compounds	53,000	3
Antimony trichloride	90,000	6
Others	93,000	6
Total	1,590,000	100

ATH, Alumina trihydrate; MH, Magnesium hydroxide.

of conventional flame retardant additives have been reported to show synergistic effects such as antimony trioxide, zinc borate (ZB), melamine and its derivatives, and a number of phosphorus-based flame retardants, including red phosphorus (RP) itself. The halogenated (chlorine and bromine) flame retardants comprise another class of FRs that reacts with flammable gases to slow down or prevent the burning process. Some of the representatives are polybrominated diphenylethers and hexabromocyclododecane. The brominated and chlorinated compounds cover 18% and 11% by volume of the total global production, respectively. Organophosphorus compounds, on the other hand, cover about 20% (by volume) of the total global production. The most important nitrogen-based flame retardants are melamine cyanurate, other melamine salts, and guanidine. The presence of these nitrogen-based flame retardants inhibits the formation of flammable gases and are primarily used in polymers containing nitrogen such as polyurethane (PU) and polyamide. Antimony compounds, zinc hydroxystannate (ZHS) and zinc stannate, and other materials have also been used as flame retardant fillers in polymers. Zinc stannates and nanoclay flame retardants exhibit a pronounced synergistic behavior when used in conjunction with fillers. However, in recent years, the development of halogen-free, low smoke, and fume flame retardant polymers remains one of the ideal choices, which could otherwise cause problems such as toxicity, corrosion, and smoke [32].

Elastomers find extensive applications in a variety of commercial as well as domestic products on account of their unique combination of mechanical strength and flexibility, which no other material can match. While the technical specifications of only a few of these products such as cable sheathing, conveyor belt, and fire-fighting hoses require fire resistance, the rest including commonly used tires, power transmission belts, etc. do not demand flame retardancy. In reality, all of them are susceptible to flame, and therefore, from the fire safety point of view, it is good to include one or more chemicals (mostly nanofillers) in their formulation in order to improve flame retardance. This would make these products more resistant to any sort of unwanted flame or fire exposure during production, transportation, and usage. Therefore, investigation of flame retardant properties of these elastomers and their blends and composites remains an interesting area of research toward the development of such elastomeric products in automotive, aerospace, construction, and electronic applications.

The present chapter reviews the FR behavior of industrially important elastomers and elastomer nanocomposites, presenting the reported finding on cone calorimetric studies, limiting oxygen index (LOI), and UL-94 tests. The elastomers addressed are ethylene vinyl acetate copolymer (EVA), ethylene–propylene diene rubber (EPDM), polyurethane (PU), silicone rubber (SR), styrene–butadiene rubber (SBR), acrylonitrile butadiene rubber (NBR), natural rubber (NR), and polychloroprene rubber (CR).

2. Preparative methods of elastomer nanocomposites

Polymer–inorganic filler nanocomposites can be prepared by in situ polymerization and by blending processes. The blending processes are of basically two types: solution blending and melt blending [4,8–10,34–36]. Depending on the dispersion of the inorganic filler in a polymer matrix, intercalated nanocomposites, exfoliated nanocomposites, and immiscible nanocomposites can be prepared. Nanocomposites can be synthesized by three techniques: in situ polymerization, ex situ solution mixing, and melt mixing [1–17]. In the case of in situ polymerization, polymeric chains are grown staring from the corresponding monomer, in the presence of the fillers. Generally, the fillers are dispersed in the monomer (bulk polymerization), in an organic solvent (solution polymerization), or in water (suspension or emulsion polymerization). The polymerization can be initiated by either heat or photoirradiation. The unreacted monomers and initiators are separated from the nanocomposites to obtain the purified product. Waterborne PU-based nanocomposites can be synthesized by this technique. In the case of ex situ solution mixing, nanofillers are first swollen in the solvent, and the polymer matrix is dissolved in the same solvent. The swollen dispersion of the filer is then added into the jelly-like solution of the polymer followed by stirring at a certain temperature. The excess solvent is removed under reduced pressure, and the nanocomposites are dried under vacuum to obtain the dry product [37–40]. Different kinds of thermoplastic polymers such as EVA, EPDM, low-density polyethylene (LDPE), Linear low-density polyethylene (LLDPE), SR, PU, and NBR, can be used to obtain nanocomposites. However, this method is not ecofriendly and cost effective due to the use of an excess amount of organic solvents. Moreover, thermosetting-based nanocomposites cannot be prepared by this method due to the insolubility of these polymers in the organic solvent. On the contrary, melt mixing of a polymer with a nanofiller is the most effective way to obtain nanocomposites for commercial applications. Nanofillers are added to the molten polymer under a shearing force to obtain homogeneously dispersed inorganic filler–polymer nanocomposites. The method is totally environmentally friendly and does not require solvents. Any kind of polymers such as thermoplastic or thermosetting can be used to prepare nanocomposites by this techniques.

3. Inorganic nanofillers used in the formation of polymer nanocomposites

Different types of inorganic fillers, e.g. clay minerals, layered double hydroxides (LDH), carbon nanotubes (CNTs), and expanded graphite (EG) have very often been used for the preparation of elastomeric nanocomposites. Clay and LDH are hydrophilic, while CNT and EG are purely carbonaceous materials [8,9,14,19,34–36]. On the contrary, polymeric materials are hydrophobic. Therefore, all the above-mentioned nanofillers, if incorporated in their natural form, will lead to phase-separated composites. In order to enhance the compatibility of a polymer with these fillers, surface modification is very essential. This is most often accomplished by ion exchanging the typical sodium or calcium cations (natural clay minerals) with organophilic ammonium or other "onium" ions [37–42]. In the case of anionic clays, the interlayer anions are replaced by surfactant molecules [43–59]. On the contrary, surface modification of CNT can be done by the oxidation of pure CNT in the presence of an acid mixture followed by covalent or noncovalent attachment of long chain organic molecules [60–63]. The most common surface-modifying agents contain at least one long alkyl chain. The identity of this organic modification is dependent on both the polymer to be used and the mode of preparation.

4. Characterization of elastomer nanocomposites

Characterization of polymer nanocomposites is the prerequisite before its use for specific applications. Due to the nanolevel dispersion of the inorganic filler in the polymer matrix, nanocomposite materials not only show improved flame retardant properties but also enhanced physicochemical properties. This is a key advantage, because many FRs are used at relatively high loadings, leading to the deterioration of the material's mechanical properties. The characterization of composites provides an idea about its nanostructure. This usually involves a combination of X-ray diffraction (XRD) and transmission electron microscopy (TEM).

Figure 1 shows the XRD patterns of MMT, EVA + MMTa, EVA + MMTa (5%) + C16 and EVA + MMTb (5%) + hexadecyl trimethyl ammonium bromide (C16) [64]. The d_{001} peak of pristine clay at $2\theta = 5.8°$ corresponds to a 1.4-nm

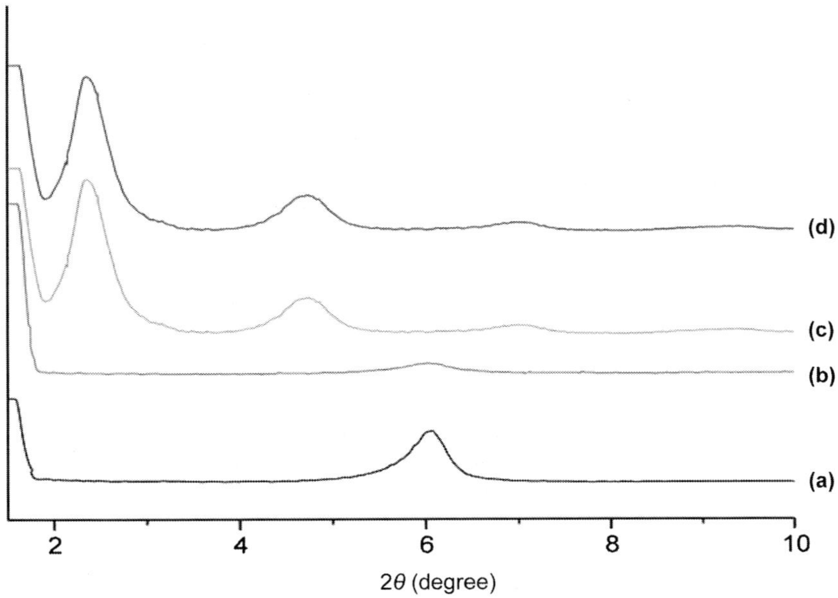

FIGURE 1

XRD patterns of MMT and EVA–MMT hybrids: (a) MMT; (b) EVA + MMTa; (c) EVA + MMTa (5%) + C16; (d) EVA + MMTb (5%) + C16 (C16: hexadecyl trimethyl ammonium bromide (C16) hexadecyl trimethyl ammonium bromide) [64].

(Figure 1(a)) interlayer spacing. The d_{001} peak of EVA–pristine clay composite without C16 (Figure 1(b)) is almost the same as that of pristine clay (Figure 1(a)) suggesting no intercalation of EVA into the silicate layers. However, the d_{001} peaks of the mixture (Figure 1(c,d)) are observed at a lower angle ($2\theta = 2.3°$) in the presence of C16 in hybrids than that of pristine clay, and corresponds to an increase in the basal spacing from 1.4 to 3.78 nm. These results indicate that the EVA with C16 could intercalate into the silicate layers and expand the basal spacing.

Acharya et al. [65] prepared ethylene–propylene diene monomer (EPDM) nanocomposites filled with dodecyl sulfate (DS) intercalated LDH (DS-LDH) by the solution method. Figure 2 displays the XRD patterns of DS-LDH and EPDM/DS-LDH in the angle range of 2–15°. It is shown that the basal spacing of DS-LDH has increased to 2.56 nm due to the intercalation of monolayer DS molecules between the hydrotalcite sheets compared to the basal spacing of 0.77 nm in pristine LDH. However, the XRD patterns of EPDM/LDH nanocomposites are characterized by the absence of the 0 0 *l* diffraction peaks corresponding to the DS-LDH, suggesting the formation of a partially exfoliated structure.

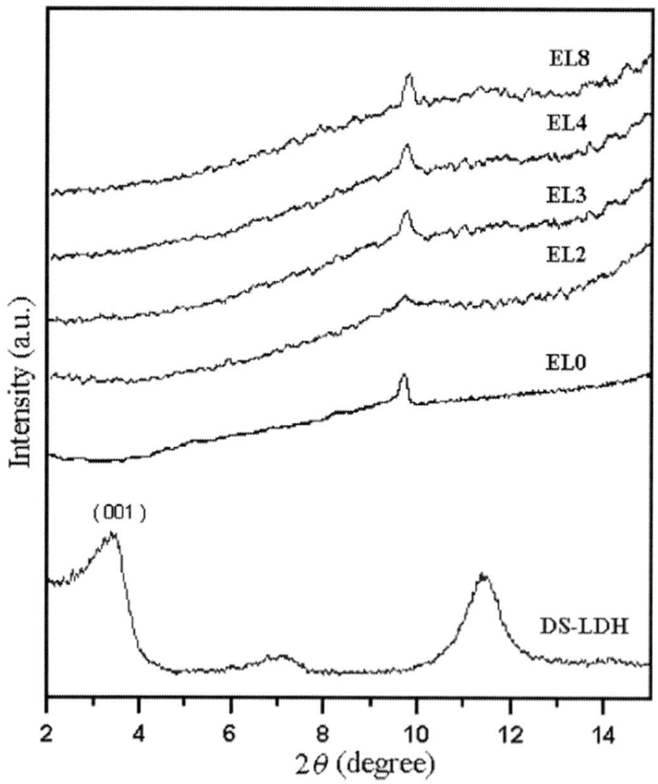

FIGURE 2

XRD spectra of EPDM/LDH composites with varying LDH contents [65].

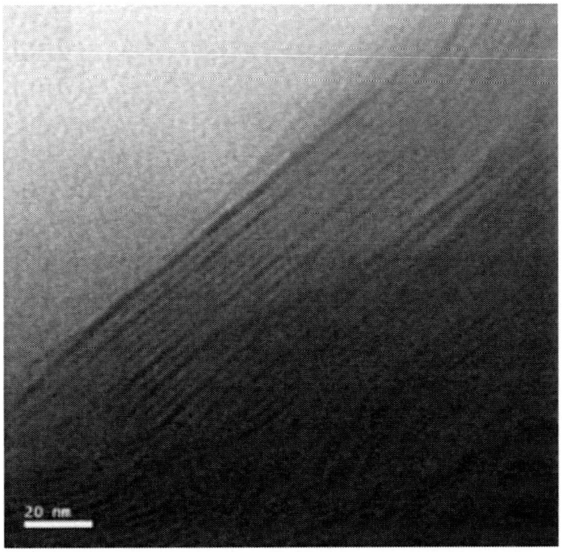

FIGURE 3

TEM image of EVA/LDH [40].

Reproduced with permission from Wiley.

Although XRD provides a partial picture about the distribution of a nanofiller and disappearance of the peak corresponding to *d* spacings, it does not always confirm the exfoliated nanocomposites; a complete characterization of nanocomposite morphology requires microscopic investigation. In this regard, TEM is the perfect tool to determine the actual dispersion of clay in the polymer. However, the analysis of TEM images also presents problems because the actual area that is imaged is very small compared to that of the whole material. Therefore, it is very essential to perform an analysis of the entire material so that enough images are taken in order to estimate the exact morphology of the system. The intercalated and exfoliated morphologies of EVA/LDH [40], EPDM/LDH [19], and SR/LDH [63] nanocomposites are shown in Figures 3–5 , respectively.

Fourier transform infrared (FTIR) spectra are also very informative when attempting to investigate the presence of functional groups in nanocomposites. Pramanik et al. [43] showed that dodecyl amine modified clay (12Me-MMT) have been successfully doped on to the EVA matrix. FTIR spectra of Na^+-MMT, 12Me-MMT, and EVA-28 and its hybrids with 4 wt% 12Me-MMT are shown in Figure 6, and the important bands are shown in Table 2. Appearance of new bands in the nanocomposites confirms the presence of 12Me-MMT in the EVA-28 nanocomposites.

Scanning electron microscopy (SEM) is also an important tool that gives important information about the surface morphology and dispersion of fillers in

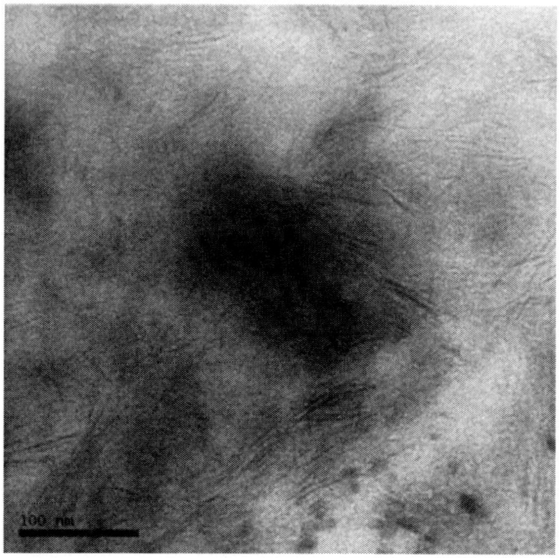

FIGURE 4

TEM image of EPDM/LDH [19].

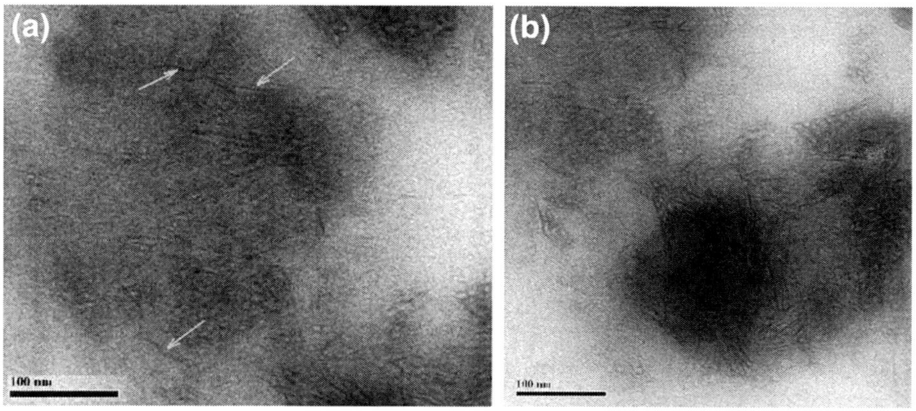

FIGURE 5

TEM images of (a) SR/DS-LDH (5 wt%) and (b) SR/DS-LDH (8 wt%) nanocomposites [63].

Reproduced with permission from Wiley.

nanocomposites. Figure 7(a,b) shows the SEM images of the EVA/clay immiscible composites at lower and higher magnifications [65]. It shows that a significant portion of the nanoclay had formed clusters of a 0.01- to 0.05-mm diameter and that the clusters are composed of sheets of material, in which the layers are no longer parallel to one another, but crumpled into small lumps. However, Pramanik et al. [9]

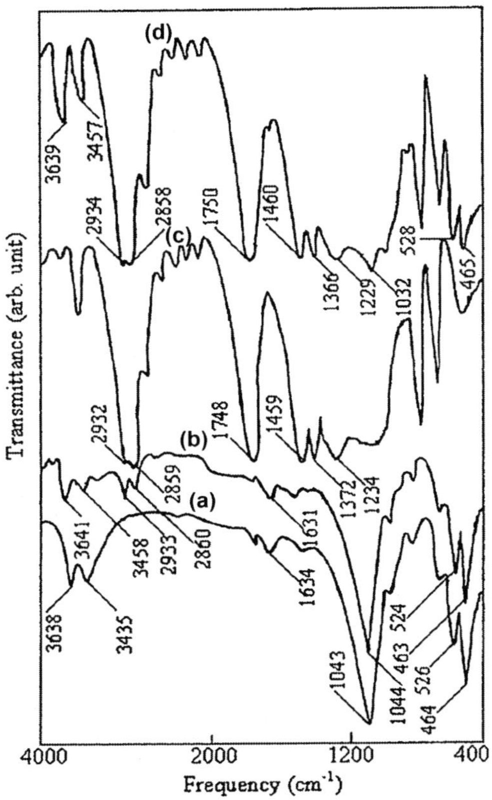

FIGURE 6

FTIR spectra of (a) Na-MMT, (b) 12Me-MMT, (c) EVA-28, and (d) EVA-28 and 4-wt % 12Me-MMT [37].

Reproduced with permission.

showed that clay minerals can be distributed throughout the matrix after appropriate surface modification.

5. Evaluation of flame retardancy of polymer nanocomposites

The assessment of the LOI is a preliminary tool and has been used extensively for evaluating the relative flammability of rubbers, textiles, paper, coatings, and other materials [66,67]. Furthermore, there are two types of preselection tests conducted on plastic materials to measure flammability characteristics. The first determines the material's tendency either to extinguish or to spread the flame during ignition. The second test measures the ignition resistance of plastics against electrical ignition

Table 2 Assignment of important bands in the IR spectra of Na^{+}-MMT, 12Me-MMT, pure EVA-28, and EVA-28/12Me-MMT hybrids. [37]

Observed Band (cm^{-1})	Band Assignment
3641/3639/3638	O—H stretching of structural hydroxyl groups
3435	O—H stretching of water
3458/3457	N—H stretching of NH_3^+
2934/2933/2932	C—H asymmetric stretching of CH_2 or CH_3
2860/2859/2858	C—H symmetric stretching of CH_2 or CH_3
1750/1748	C=O stretching of ester
1634/1631	O—H deformation of water
1460/1459	CH_2 scissoring
1372/1366–1234/1229	C—O stretching of ester
1044/1043/1032	Si—O stretching
528/526/524	Al—O—Si deformation
465/464/463	Si—O—Si deformation

Reproduced with permission.

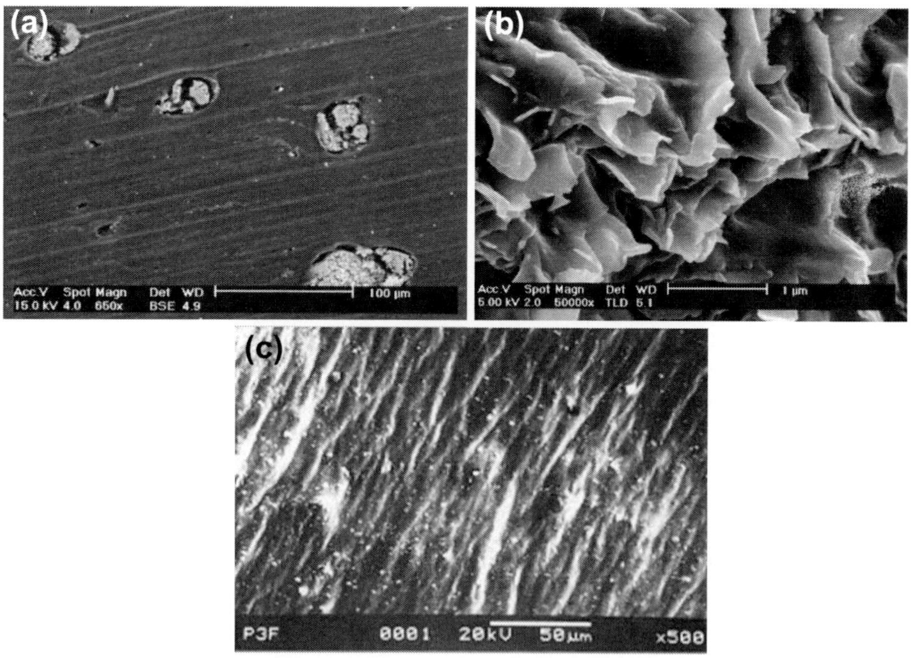

FIGURE 7

SEM of (a) EVA/pure clay at a low magnification, (b) at a higher magnification, and (c) EVA/organoclay nanocomposites [65].

sources. The material's resistance to ignition and surface tracking characteristics is described in UL 746A, which is similar to the test procedures described in IEC 60112, 60695, and 60950 [66,67].

6. Fire retardancy of elastomers and elastomer nanocomposites

Elastomer and elastomer nanocomposites are some of the most widely used commercial polymers due to the excellent combination of chemical and physical properties along with low cost, superior processability, and recyclability. Quite recently, investigations on the fire retardance characteristics of elastomers became intense due to the great potential of these materials in applications when flame retardancy is required. As a result, significant efforts have been made toward the development of the FR properties of the following elastomers and elastomer nanocomposites [68–84].

6.1 EVA copolymer

EVA copolymers are available as rubbers, thermoplastic elastomers, and plastics depending on the vinyl acetate content. This indeed, offers a wide spectrum of applications in different fields: electrical insulation, cable jacketing and repair, component encapsulation and water proofing, corrosion protection, packaging of components and shoes; these applications altogether dictate the extent of the industrial importance of these polymers [37–40,43,44]. However, bulk EVA does not often fulfill the requirements in terms of its thermal stability behavior and mechanical properties in some specific areas. To improve these properties, nanomaterials can be added as fillers. Zanetti et al. [4] dispersed octadecylammonium or aminododecanoic acid-exchanged fluorohectorite in EVA (19 wt% VA content) in an internal mixer. They studied the combustion behavior of the nanocomposites by using a mass loss calorimeter. Silicate clay minerals slow down the degradation of nanocomposites as compared to pure EVA. Beyer [69] also synthesized flame retardant nanocomposites by the melt blending of EVA with modified layered silicates; in this case, compared to neat EVA, an about 47% decrease in peak of heat release as well as a shift toward longer times was recorded for the burning time of a nanocomposite containing 5 wt% of nanofiller. Further addition of the nanofiller up to 10 wt% loadings could not lead to any significant decrease in the peak heat release rate (PHRR). It was suggested that the reduced flammability of the nanocomposites is due to the formation of a protective charred layer on nanocomposite surfaces during combustion. They also prepared two sets of composites containing 65 wt% ATH + 35 wt% of EVA-28 and 60 wt% ATH + 5 wt% MMT + 35 wt% of EVA-28 and investigated the effect of ATH on the flame retardancy of EVA by a cone calorimeter at 50 kW m^{-2}. It was noticed that the PHRR in the case of EVA/ATH/nanofiller nanocomposites shifted to a value (100 kW m^{-2}) lower than that of the EVA/ATH (200 kW m^{-2}) sample. It was suggested that the content of the flame retardant filler, ATH, had to be increased to 78 wt% within the EVA/ATH to obtain the same decrease in the PHRR values.

Zhang and coworkers studied the effect of nanoclays on the FR properties of the EVA/LDPE blend [71]. They observed that the nanoclay alone has little effect on the degradation of the polymer blend, whereas the use of inorganic fillers such as aluminum trihydroxide and MH generally decreased the onset degradation temperature and mass loss rate (MLR). It was also found (from cone calorimeter tests) that the addition of nanoclay reduced the heat release rate (HRR), while it increased smoke and CO yields. The addition of ATH improved the flame retardancy of the nanocomposites.

Shi et al. [72] prepared EVA (39 wt% VA content)/layered silicate nanocomposites by the melt method and observed that the heat release capacity (HRC) and total heat release (THR) were reduced by 21—24% and 16%, respectively. Figure 8 shows that the HRR of the exfoliated EVA nanocomposite containing 5 wt% of clay was reduced by 80% and that the MLR plots spread over a much longer period of time. The MLRs of pure EVA and the nanocomposites are shown in Figure 9. It shows that the MLR of pure EVA was spread over 482 s. On the contrary, the MLRs of the nanocomposites were only 335 s. Tang et al. [64] also prepared EVA/MMT nanocomposites by melt compounding and studied the flame retardant

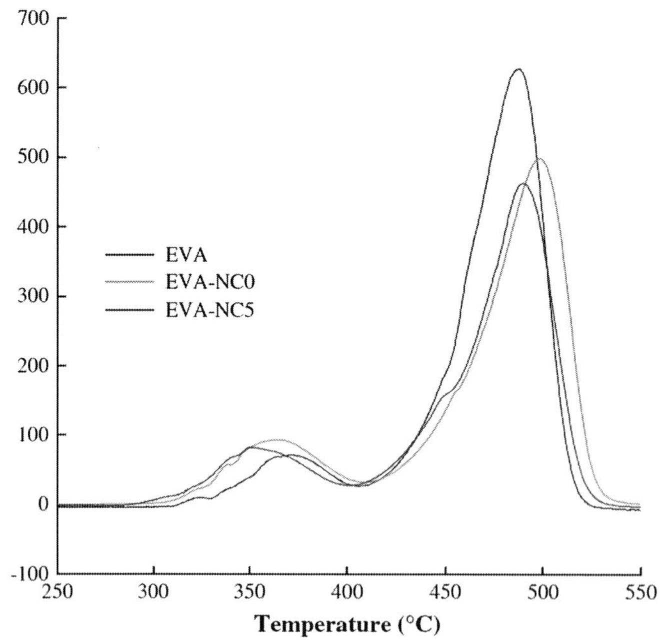

FIGURE 8

HRR versus temperature (degrees centigrade) for pure EVA, EVA-NC5 (PVAc-1 20 wt% + 5.6 wt% MMT), and EVA-NC0 (PVAc-1 20 wt%) [72].

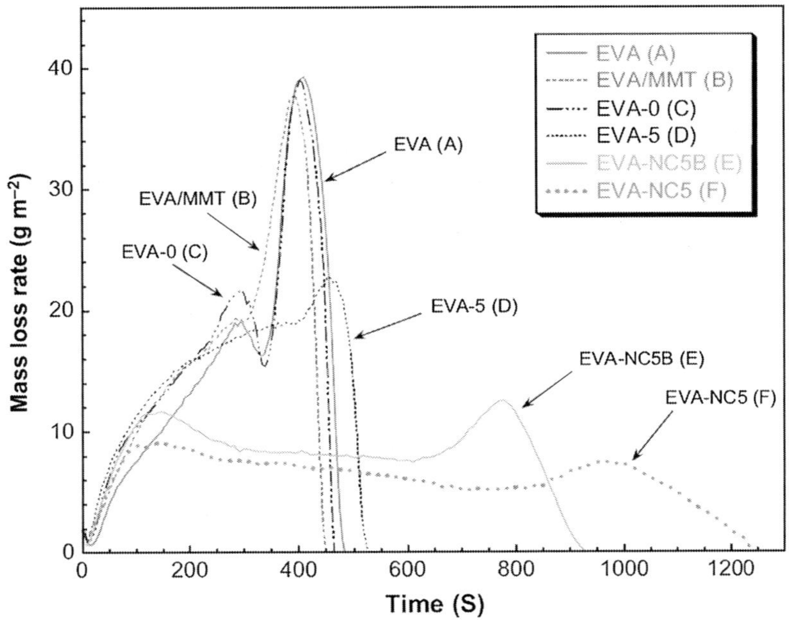

FIGURE 9

Mass loss rate plots recorded during gasification at a heat flux of 50 kW m^{-2} [72].

characteristics through the cone calorimeter test. They noticed that the HRR of the nanocomposite was 40% lower than that of pure EVA.

Duquesne et al. [73] prepared EVA-19/cloisite-Na^+ and EVA-19/cloisite-30B nanocomposites by the melt-mixing process. A significant decrease in the PHRR is observed as well as a widening of the peak when the loading of cloisite 30B increases. Cai and coworkers [74] prepared high-density polyethylene (HDPE)/EVA/organophilic MMT nanocomposites and observed that the HRR value decreased remarkably with the loading of organophilic MMT (OMMT). The PHRR of the HDPE-EVA/OMT/paraffin composites with 5 and 10 wt% of OMT loadings decreased by about 34.4% and 41.5% compared to that of HDPE–EVA/paraffin composites. These observations indicate that the prepared composites are stable, contributing to improved flammability properties.

DS-intercalated LDH (DS-LDH) are one of the important inorganic nanofillers whose effects on polymers especially polyethylene and EVA copolymer have been studied. Some findings regarding the LOI of neat EVA (EVA-18, EVA-28, EVA-45, and EVA-60) and its nanocomposites with DS-LDH content are presented in Table 3 [75]. The basic difference between neat EVA and EVA/DS-LDH nanocomposites is the duration of burning. The pure EVA burned more quickly compared to nanocomposites containing DS-LDH. Table 3 shows that the LOI values of all the neat EVA were in the range of 18–19. On the contrary, the nanocomposites showed

Table 3 LOI Data EVA/DS-LDH Nanocomposites [75]

DS-LDH (wt%)	EVA-18	EVA-28	EVA-45	EVA-60
0	18.6	19.3	18.0	19.3
1	21.3	21.3	19.3	22.2
3	23.4	23.0	21.3	23.4
5	24.2	24.2	22.2	24.2
8	24.4	25.5	25.5	24.8

EVA, ethylene vinyl acetate copolymer; DS-LDH, dodecyl sulfate intercalated layered double hydroxide.

improved LOI values; they increased by increasing the concentration of DS-LDH in EVA. This is mainly attributed to the presence of a charred layer in the nanocomposites, which impedes burning, acting as a barrier between the burning surface and supplied oxygen [28,30,76]. For the nanocomposites with a DS-LDH content of $\leq$3 wt%, the thickness of the charred layer appeared to be very low to prevent the burning of the composites. However, beyond 3 wt% of the DS-LDH loading, the formation of a thick charred layer suppresses the propagating downward flame by disrupting the oxygen supply to the burning specimen. In addition, the endothermic decomposition of LDH produced sufficient smoke and water vapor, which also accounted for the reduction of flammable characteristics in the case of nanocomposites [34,58,60,77,78].

Figure 10 shows the effects of different amounts of DS-LDH on the LOI values of the EVA/LDPE/DS-LDH nanocomposites with varying EVA/LDPE ratios of 30:70, 50:50, and 70:30 [75]. It is evident that the LOI values of the nanocomposites were relatively much higher with respect to those of neat EVA/LDPE blends and that the LOI values were increased with increasing DS-LDH contents [75]. The maximum LOI value was recorded for the nanocomposites with 8 wt% of DS-LDH content. During the LOI test, the flame propagated vertically downward along the burning sample. The propagation first took place through the surface layer and then finally reached the core of the sample. The basic difference between the burning of pure blend and nanocomposites was that in the case of the former no char residue was formed. The pure EVA/LDPE blend burned like a candle till the whole sample was consumed. On the contrary, the EVA/LDPE/DS-LDH nanocomposites burned quite slowly and formed three distinct regions during burning. These were the top or the skin layers consisting of char residue, the melt region, and the unburned solid region. Both the melt and the charred layers were together supported on the solid region. The formation of a charred layer on the burning surface acted as a physical barrier against the (downward) propagation of the flame along the LOI sample. However, on increasing the DS-LDH content in the EVA/LDPE blend, the thickness of the charred layer was sufficient to prevent burning. In addition, the formation of water vapors due to the endothermic decomposition of LDH was also likely to produce sufficient smoke thereby impeding the burning of

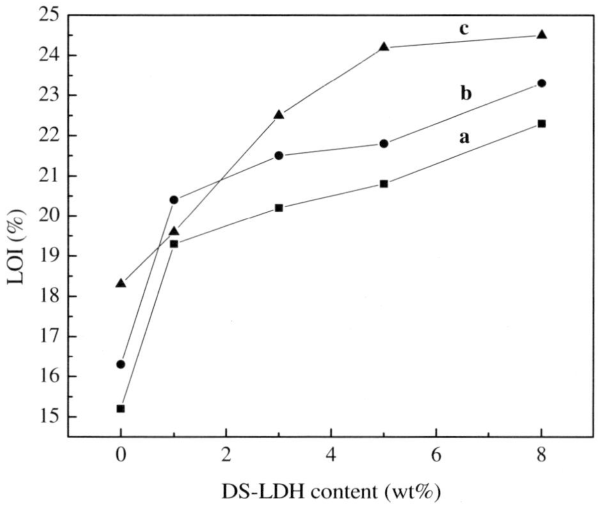

FIGURE 10

Effect of DS-LDH content on the LOI of EVA/LDPE/DS-LDH nanocomposites with EVA/LDPE ratio of (a) 30:70, (b) 50:50, and (c) 70:30 [75].

nanocomposites. This makes the self-sustained burning of the sample more and more difficult at lower oxygen concentrations, thus increasing the LOI value with LDH concentration.

Wilkie et al. [68] melt blended EVA (18 wt% VA content) with oleate intercalated MgAl LDH and ZnAl LDH. In both cases, the dispersion at the micrometer level was very good, yet relatively poor at the nanometer level. However, a considerable reduction in the PHRR was observed at 10 wt% of LDH loading. EVA/hydrotalcite nanocomposites have also been prepared by melt blending to investigate their flame retardant characteristics [28–30,79,80]. It was observed that the LOI values of the composites were gradually increased with increasing hydrotalcite content in EVA. The cone calorimeter characterization showed that the control sample of EVA showed a very sharp peak with a PHRR value of 1813 kW m^{-2}, and its ignition time was 92 s. On the contrary, the peaks of all nanocomposites were relatively smooth; the PHRR values below were 597 kW m^{-2}, and their ignition times became longer. Moreover, the PHRR values decreased on increasing the hydrotalcite content [28]. In another study, it was shown that, while pure EVA is flammable presenting an LOI value of 17.0, composites with 60 wt% of untreated ATH loading exhibited an LOI of 30.6 [29]. Very recently, Zhang et al. [79] showed that the addition of LDH could raise the thermal degradation temperatures of EVA (28 wt% VA content)/hyperfine magnesium hydroxide (HFMH)/LDH nanocomposite samples with 5–15 phr organomodified-LDH by 5–18 °C compared to control EVA/HFMH sample, when a 50% weight loss is selected as the point of comparison. The LOI tests showed that the LDH could act as a flame retardant synergist and compatibilizer to apparently increase the LOI value of EVA/HFMH/LDH nanocomposites. Du et al. [30] investigated the flame retardancy of an EVA/hydrotalcite composite and EVA/ hydrotalcite and concluded that the EVA blends with 60 wt% ultrasound-synthesized hydrotalcites have better flame retardant properties. The flammability characteristics of hydrotalcite with microencapsulated RP in halogen-free flame retardant EVA composites have also been studied [76]. Accordingly, the HRR and MLR of EVA composites blended with hydrotalcite greatly decreased compared with those blended with MH and aluminum hydroxide. The LOI values of EVA/hydrotalcite composites were 3–4% higher than those of the corresponding MH composites at 40–60 wt% loading levels, and 6% higher than that of the corresponding aluminum hydroxide composite at a 40 wt% loading level [76]. Lv et al. [80] noted that microcapsulated red phosphorus (MRP) has a good FR synergistic effect with MH nanoparticles in the EVA (28 wt% VA content)/MH/MRP blends [80].

CNT has also been used as a nanofiller for enhancing the fire retardancy of EVA nanocomposites [81–84]. Gao et al. [81] reported that CNTs play an important role in the reduction of PHRR by forming low permeability char containing graphitic carbon. Peeterbroeck et al. [22] used HDPE-coated MWNT as a filler in EVA and studied the effect of the filler on fire behavior in comparison with virgin EVA and EVA/MWNT nanocomposites. Their findings (based on cone calorimetry measurements) showed that the time to ignition (TTI) of EVA/MWNT nanocomposites was delayed in comparison with the unfilled EVA. This was probably due to the presence of acidic functions formed on the surface of MWNTs during the purification method. The cone calorimeter experimental results for EVA and the two corresponding nanocomposites filled with either 3 wt% MWNT or HDPE-coated MWNT showed that their HRR values were 707, 315, and 305 kW m^{-2}, respectively. Beyer and coworkers [85] studied the influence of CNT, organically modified MMTs, and LDH on the thermal degradation and fire retardancy of EVA (19 wt% vinyl acetate content); their findings are presented in Table 4. These investigations showed that the PHRR values of the nanocomposites were decreased significantly as compared to that of pure EVA. It was suggested that a barrier was formed, which inhibited mass transfer and provided thermal insulation to shield the underlying polymer from the fire source.

Ye et al. [86] studied the effect of MH in halogen-free flame retardant EVA (28 wt% VA content)/MH/MWNT nanocomposites. Table 5 provides the change in LOI values with the MWNT content of the EVA/MH/MWNT samples. It is noted that the LOI value increases on increasing the contents of MWNTs and reaches a maximum of 39% in the case of EVA/MH/MWNT (50:48:2 wt ratio) nanocomposite due to the synergistic effect of MH. Table 6 lists out the time to ignition (TTI), PHRR, and their ratio Fire performance index (FPI) values and the combustion time of pure EVA, EVA/MH, and EVA/MH/MWNT samples. It shows that the TTI values of the EVA/MH/MWNT samples are lower than those of the EVA/MH sample without MWNTs. But their combustion times increase on increasing the

Table 4 Cone Calorimetric Results for EVA and Its Nanocomposites (Heat Flux of 35 kW m^{-2}) [85]

Composition	PHRR (kW m^{-2})	Reduction (%)	THR (MJ m^{-2})	ASEA (m^2 kg^{-1})	AMLR (gs^{-1} m^2)	t_{ign} (s)
EVA	1772 ± 170		112 ± 4	399 ± 7	23.5 ± 0.3	75 ± 3.6
EVA/CNT (3 wt%)	597 ± 30	66	101 ± 1	553 ± 37	13.9 ± 0.6	63 ± 1.1
EVA/LDH (3 wt%)	1090 ± 58	39	106 ± 3	468 ± 23	18.0 ± 0.3	57 ± 1.2
EVA/MMT (3 wt%)	903 ± 24	49	99 ± 4	515 ± 43	14.6 ± 0.7	60 ± 2.1

PHRR, peak heat release rate; THR, total heat release; ASEA, average specific extinction area, a measure of smoke; AMLR, average mass loss rate; EVA, ethylene vinyl acetate copolymer; CNT, carbon nanotubes; LDH, layered double hydroxide; MMT, montmorillonite; t_{ig}*, time to ignition.*

Table 5 Formulations and LOI Values of EVA/MH and EVA/MH/MWNT Samples [86]

Samples	EVA (wt%)	MH (wt%)	MWNTs (wt%)	LOI (%)
EM-0	50	50	0	34 ± 0.5
EM-1	50	49	1	37 ± 0.5
EM-2	50	48	2	39 ± 0.5
EM-3	50	47	3	38 ± 0.5
EM-4	50	46	4	37 ± 0.5

EVA, ethylene vinyl acetate copolymer; MH, magnesium hydroxide; LOI, limiting oxygen index.

Table 6 Flammability Data of Pure EVA, EVA/MH, and EVA/MH/MWNT Samples with Different Amounts of MWNTs Determined from Cone Calorimeter Tests [86]

Sample	TTI (s)	pk-HRR (kW m^{-2})	FPI (m^{2}s kW^{-1})	pk-MLR (gs^{-1} m^{-2})	Combustion Time (s)
Pure EVA	43	791.7 ± 35	0.054	0.197 ± 0.009	285
EM-0	55	548.2 ± 27	0.100	0.165 ± 0.008	407
EM-1	54	392.7 ± 19	0.138	0.127 ± 0.006	496
EM-2	53	252.6 ± 12	0.210	0.085 ± 0.004	735
EM-3	52	278.5 ± 14	0.190	0.094 ± 0.004	768

EVA, ethylene vinyl acetate copolymer; TTI, time to ignition.

amounts of MWNTs and have prolonged to about 496–768 s from 258 s of pure EVA and 407 s of EVA/MH sample without MWNTs. Table 6 also shows that the EVA/MH/MWNT (50:48:2 wt ratio) has the greatest FPI (FPI is defined as the ratio of TTI to pk-HRR) value of 0.210 among these samples, which means it has the best fire resistance.

6.2 Ethylene–propylene diene monomer

EPDM rubber (ethylene–propylene diene Monomer (M-class) rubber) [87,88] is a type of synthetic rubber, with a wide range of applications. The E refers to Ethylene, P to Propylene, D to diene, and M refers to its classification in American Society for Testing and Materials (ASTM) standard D-1418. The "M" class includes rubbers having a saturated chain of the polymethylene type. The diene(s) currently used in the manufacture of EPDM rubbers are dicyclopentadiene, ethylidene norbornene, and vinyl norbornene. It is widely used in cold room doors for sealing purpose, industrial respirators, automotive paint spray environments, cable insulation, solar pool panels, and as a covering for water proof roofs [87,88].

A novel flame retardant system composed of nanokaolin and nano-HAO (nanosized hydroxyl aluminum oxalate) have been used for LDPE/EPDM blends; the findings are displayed in Figures 11 and 12 [89]. It was accordingly noted that

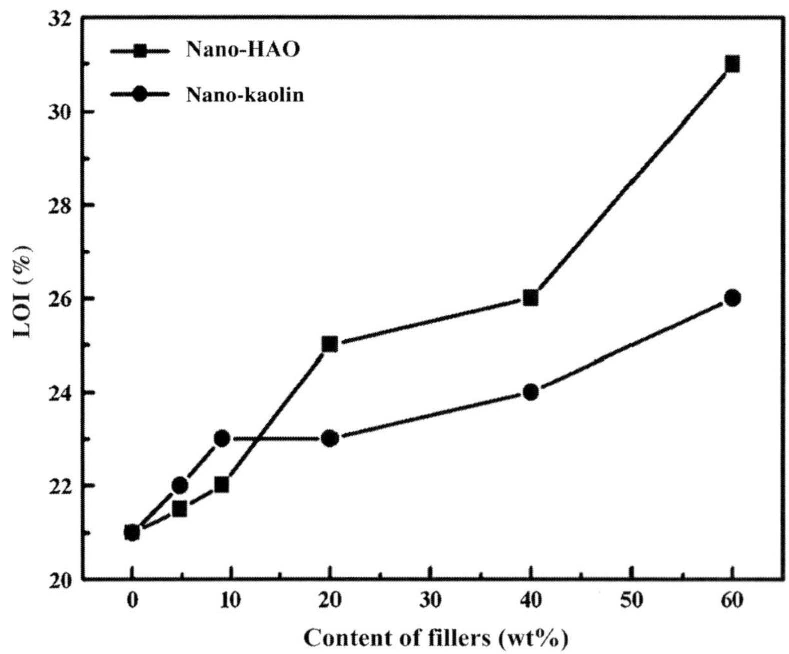

FIGURE 11

Effects of nano-HAO or nanokaolin on the LOI of LDPE/EPDM composites [89].

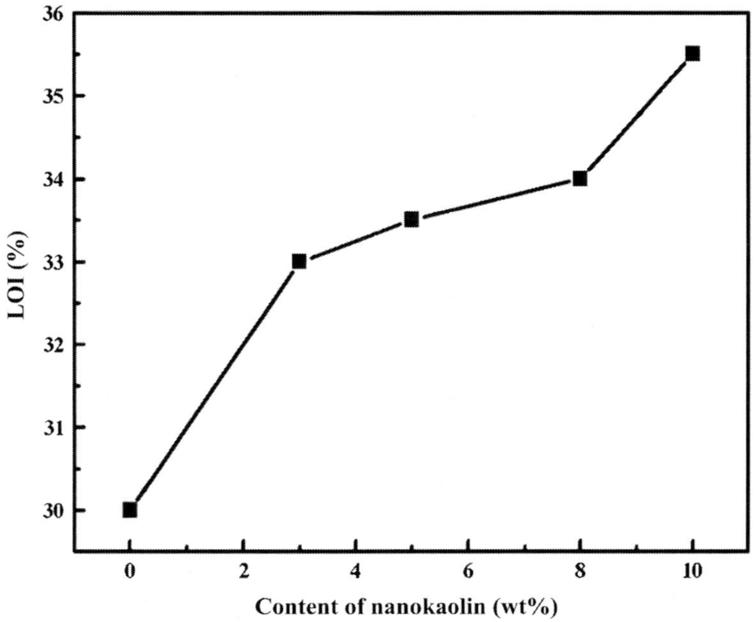

FIGURE 12

Effects of the addition of nanokaolin on the LOI of nano-HAO/LDPE/EPDM composites [89].

the addition of 12 wt% nanokaolin could improve the LOI values by 35.5%, whereas, in the presence of 12 wt% nano-HAO, an improvement of only 31.0% was achieved. UL-94 test revealed that the nano-HAO exhibited a better performance on stopping the combustion than that exhibited by nanokaolin. The flame on the horizontal specimen could cease when the content of nano-HAO increased to 60 wt%. These observations revealed the low efficiency of the nano-HAO on flame retarding the LDPE/EPDM system. It is observed that when (12 wt% of) nanokaolin took the place of (12 wt%) nano-HAO in the composites with 60 wt% flame retardants, the materials passed the UL94 V-0 standard. Cone calorimetic analysis showed that the addition of nanokaolin resulted in the decrease of HRR. In other words, the addition of nanokaolin into the nano-HAO/LDPE/EPDM composites resulted in the enhancement of the barrier properties of the charred layer, and thus, the transfer rate was retarded, resulting in the improvement of the flame retardancy of the composites.

Ismail et al. [27] prepared EPDM/halloysite nanotube (HNT) nanocomposites by mixing 0—100 parts per hundred rubber (phr) of HNTs with EPDM in a two-roll mill, and the findings of UL-94 tests are presented in Table 7 [27]. Accordingly, the unfilled EPDM is unclassified because the total ignition of five specimens had passed 250 s, as specified by the standard. Although, adding 5—15 phr of HNT strongly decreased the dripping and total ignition time, the EPDM/HNT remains still

Table 7 Classification of EPDM/HNT Nanocomposites According to the UL-94 Test [27]

Composites	Classified	Dripping Observed	Specimen Burns up to Holding Clamp	Total Flaming Combustion for All 5 Specimen (s)
C.1	Unclassified	Yes	Yes	396
C.2	Unclassified	Yes	Yes	342
C.3	Unclassified	No	Yes	288
C.4	Unclassified	No	Yes	231
C.5	V-2	No	No	191
C.6	V-1	No	No	154
C.7	V-1	No	No	99
C.8	V-0	No	No	46

unclassified. Table 7 also suggests that the addition of HNTs from 0 to 100 phr reduced the flammability of EPDM/HNT nanocomposites. It was found that the total ignition time of five specimens of unfilled EPDM had passed 250 s. On the contrary, addition of 50–70 phr of HNT strongly decreased the dripping and TTI to 100 s.

Tang and coworkers [90] studied the effect of paraffin, nano-MH, and red phosphorous on the flame retardant behavior of EPDM nanocomposites. The flammability measurements revealed that the LOI of the form-stable composite blend without any nano-MH flame retardant was only 17%. However, the addition of 38.46 wt% of nano-MH could increase the LOI value up to 28%. The enhanced fire resistance of the composite blends could be explained by the fire suppression mechanism of nano-MH. The thermal decomposition of MH is an endothermic process, which can release water to decrease temperature and dilute oxygen and flammable gases concentrated near the flame. Kang et al. [91] prepared EPDM/organoclay nanocomposites via melt mixing; cone calorimeter tests showed that the addition of organoclays helps to reduce the flammability of nanocomposites.

6.3 Polyurethane

PU constitutes an important class of functional polymers, whose properties can be tailormade by simply adjusting their compositions, i.e. by changing their molecular chain structure of soft and hard segments [59,92–98]. The hard segment corresponds to isocyanate and triol contributing stiffness and strength while the soft segment refers to the polyester or polyether unit contributing elasticity, flexibility, and damping ability. In order to meet the diverse demands of modern technologies, PU exists as coating, adhesive, foam, and thermoplastic elastomers. They find numerous commercial applications like adhesives, sealants, coatings, textiles, medical devices, automobiles, electronics, impact modifiers for other plastics, rollers, wheels, drive belts, and in constructions. The repeating units of PU consist of urethane linkage which can be produced from the reaction of an isocyanate (−N=C=O) with an alcohol (−OH). Although, it has an excellent resistance to oxygen, ozone, sunlight, oil, solvent, and fat, its combustibility and lack of thermal stability increase fire risks [99]. Therefore, the studies of PU related to the enhancement of its high resistance to burning in the presence of suitable environment-friendly halogen-free high performance flame retardant additives have become very important [100,101].

The flame retardancy of PU can be improved through the synthesis of phosphorus-containing PU [102,103]. This is due to the fact that phosphorus-containing flame retardants act through a condensed phase mechanism. Accordingly, the combustion of the outer layer of the polymeric material containing flame retardants leads to intumescent carbonaceous char, which acts as a barrier by impeding heat transfer and thereby decreasing the release of flammable volatile products.

Hosgor et al. [104] prepared carbonate-modified bis(4-glycidyloxy phenyl) phenyl phosphine oxide (CBGPPO)-based PU/silica nanocomposites (via the nonisocyanate route). They reported that CBGPPO used in PU synthesis facilitates the dispersion of silica particles, thus enhancing the flame retardancy of PU nanocomposites. Zinc borate is another example of an environment friendly flame retardant, smoke-suppressant, and antibacterial material used by Yildiz et al. [105] in the preparation of the PU nanocomposites by mechanical blending. They observed that the burning time of PU/0.5 wt% ZB nanocomposites increased dramatically by 160% compared to that of neat PU, implying that a significant improvement is brought by ZB to the flame retardancy of PU. Gao et al. [106] prepared PU/ZB nanocomposites by in situ polymerization; ZB nanoparticles are modified with poly(propylene glycol) phosphate ester and oleic acid in order to improve the dispersion of ZB in PU. The barrier effect of the dispersed (ZB) nanoparticles hindered the release of volatile degradation products from the polymer as well as restricted the diffusion of oxygen leading to the reduction of the degradation of PU nanocomposites. Berta et al. [107] investigated the effect of the molecular weight and functionality of polyol on the flame retardancy of PU/clay nanocomposites. The compositions of the PU nanocomposites produced are given in Table 8. Table 9 clearly shows that the PU nanocomposites exhibited a much better performance in the cone calorimeter than did their PU reference analogs. The best results were obtained with a polyol of 4000 MW; as shown in Figure 13, both the PHRR and average heat release rate (AHRR) were reduced to 80% and 43%, respectively, due to the barrier effect of the char-forming condensed phase.

Song et al. [108] synthesized PU/OMT nanocomposites by the in situ polymerization in the presence of the melamine polyphosphate (MPP), as a FR, and measured the flame retardant characteristics using cone calorimetry. Figure 14 shows that the PHRR values for PU/MPP (6 wt%), PU/OMT (5 wt%) and PU/OMT (5 wt%)/MPP (6 wt%) are higher (563, 472 and 243 kW m^{-2}) compared to that of neat PU (923 kW m^{-2}). It is apparent that the neat PU burns very fast after ignition and reaches a sharp peak on the HRR curve, whereas the incorporation of organophilic MMT (OMT) or MPP into PU results in a great decline of the HRR. Figure 15 shows that the MLR (gs^{-1} m^{-2}) of U/OMT, PU/MMP, and PU/OMT/MMP were

Table 8 Polyurethane (PU) Characteristics [107]

PU Reference	PU Nanocomposites	Polyol			
		Name	Calculated Mn	Functionality[a]	Wt% EO[b] End
I	NC-I	Acclaim 2220	2000	2	15
II	NC-II	Acclaim 4220	4000	2	15
III	NC-III	Daltocel F435	4000	2.5	17
IV	NC-IV	Arcol 1374	6000	2.3	15

[a] *Approximated.*
[b] *Ethylene Oxide.*

Table 9 Cone Calorimetry Results for PU Nanocomposites and Reference Materials [107]

	t_{ign} (s)		HRR_{peak} ($kW\ m^{-2}$)		PI (HRR_{peak}/t_{ign})		HRR_{ave} ($kW\ m^{-2}$)		SEA_{ave} ($m^2\ kg^{-1}$)	
PU	PU	NC-PU	PU	NC-PU	PU	NC-PU	PU	NC-PU	PU	NC-PU
I	29	29	2561	918	88	32	741	344	176	305
II	35	33	2254	641	64	19	637	363	235	412
III	26	17	2647	848	102	50	768	444	165	172
IV	22	25	2664	797	102	50	775	435	235	412

also found to be lower (0.24, 0.27, and 0.14, respectively) than that of pure PU (0.4), implying that the flame retardant mechanism of MPP-OMT is dominant in the condensed phase. These data measured by cone calorimetry are listed in Table 10. All these findings clearly suggest that the flame retardant nanocomposite exhibited enhanced flame retardant properties compared to those of neat PU due to the synergistic effects between the organoclay, PU and flame retardant.

Very recently, a significant improvement in the flame retardancy of PU was observed when MMT organomodified by 2-(2-(5,5-dimethyl-1,3,2-dioxaphosphinyl-2-ylamino)ethy-amino)-*N*,*N*,*N*-triethyl-2-oxoethanaminium chloride was used as a filler for the preparation PU nanocomposites by in situ polymerization [109]. It was found that the PHRRs of 1% 2-(2-(5,5-dimethyl-1,3,2-dioxaphosphinyl-2-alamino) ethy-amino)-*N*,*N*,*N*-triethyl-2-oxoethanaminium chloride/PU and 4% Na-MMT/PU were reduced by 16–17%. However, the AHRR and THR in both cases remained more or less the same; time of ignition (t_{ign}), on the other hand, changed significantly.

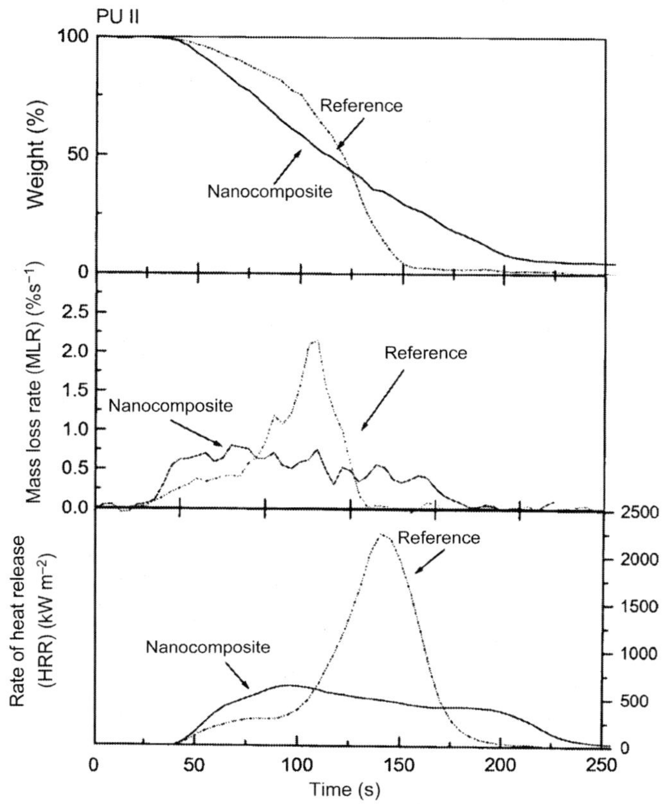

FIGURE 13

Cone calorimeter profiles of PU-II reference and NC-II nanocomposites [107].

It was also reported that 20% modified MMT/PU showed a 25% reduction of PHRR, a notably lower AHRR, and a longer t_{ign} (8 s) compared to that of the neat PU. It was, thus, proposed that the synergistic effect of 2-(2-(5,5-dimethyl-1,3,2-dioxaphosphinyl-2-ylamino)ethy-amino)-*N*,*N*,*N*-triethyl-2-oxoethanaminium chloride and MMT is responsible for the improvement of the flammability properties of PU. The modified polyhedral oligomeric silsesquioxanes (POSS) and coloisite 30B have also been used to enhance the flame retardancy of PU [110]. Accordingly, it has been found that octamethyl POSS-coated polyester or cotton fabrics do not contribute at all to the FR properties when used as an additive in the PU matrix. In contrast, a 55% decrease in the PHRR and a 50% increase in the time of ignition (t_{ign}) has been recorded in the case of the poly(vinylsilsesquioxane)/PU nanocomposites. Moreover, when cloisite 30B was used as a filler in PU, the ignition time (t_{ign}) increased, while the PHRR was reduced by 18%.

The fire retardancy of thermoplastic PU (TPU) has also received attention. TPU/10 wt% POSS nanocomposites prepared by melt mixing showed an 80%

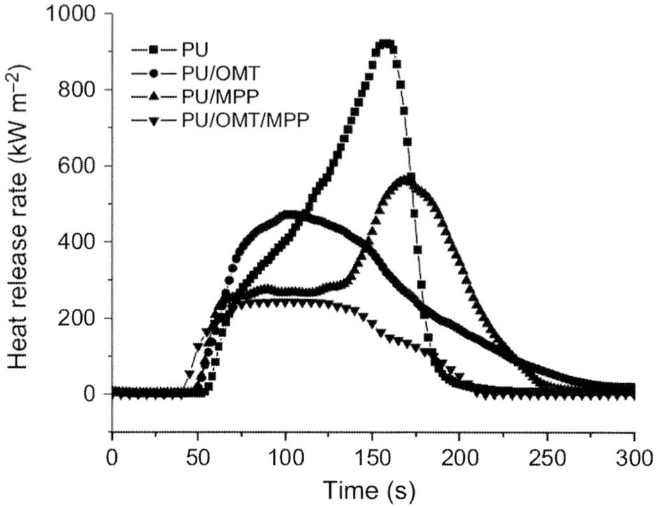

FIGURE 14

The patterns of HRR of PU, PU/OMT, PU/MPP, and PU/OMT/MPP nanocomposites [108].

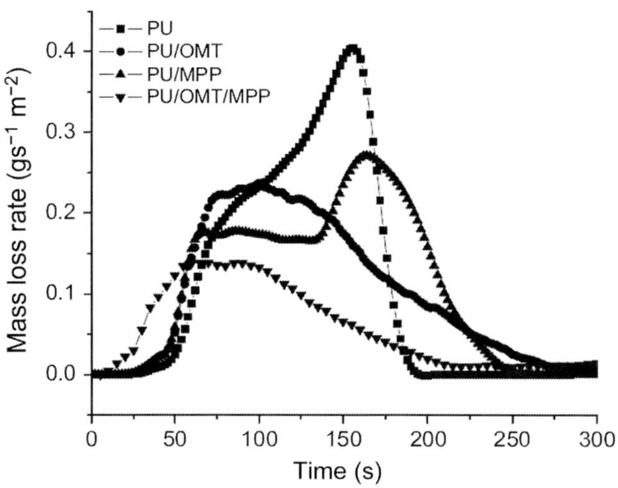

FIGURE 15

The patterns of MLR of PU, PU/OMT, PU/MPP, and PU/OMT/MPP nanocomposites [108].

decrease of PHRR compared to that of neat TPU, whereas the ignition time (t_{ign}) was found to be two times shorter than that of neat TPU [111]. The formation of a large intumescent char made of a silicon network in a polyaromatic structure acting as an efficient insulating material seemed to be responsible for the hindrance of the heat transfer and thus for the improved flame retardancy of these nanocomposites.

Table 10 The Data for Flammability of PU, PU/OMT, PU/MMP, and PU/OMT/MMP Nanocomposites [108]

Sample	PU	PU/OMT	PU/MPP	PU/OMT/MPP
Heat release rate (kW m^{-2})	923	472	563	243
Mass loss rate (gs^{-1} m^{-2})	0.4	0.24	0.27	0.14
Specific extinction area (m^2 kg^{-1})	1399	473	488	415
CO release amount (kg kg^{-1})	2.33	0.37	2.33	0.33
CO_2 release amount (kg kg^{-1})	3.67	1.91	3.67	1.71

PU, polyurethane; MPP, melamine polyphosphate.

Recently, (2 wt%) multiwalled carbon nanotube (MWNT) and (30 wt%) ammonium polyphosphate (APP) were used for the preparation of TPU nanocomposites by melt mixing, and the flame retardancy of the resulting materials were evaluated by cone calorimetry [112]. Compared to neat TPU, the nanocomposites of TPU/MWNT presented a 50% lower PHRR due to the formation of an efficient char layer. This char layer acted as an insulating barrier and reduced the release of volatile degradation products. They also reported that the development of an intumescent coating of TPU/APP and TPU/APP-MWNT permitted the decrease in the PHRR by 75% in either case compared to that for TPU (Figure 16), suggesting that the substitution of APP by MWNT is not beneficial in terms of PHRR.

GNPs also attracted attention for the preparation of PU nanocomposites. GNPs when present as a filler in PU/GNP nanocomposites act as intumescent flame retardants and significantly reduce the HRR of the nanocomposites [113] as shown in

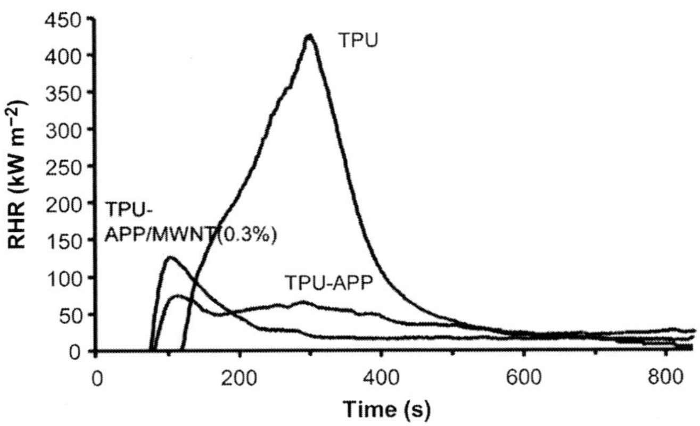

FIGURE 16

HRR as a function of time of pure TPU and TPU/APP-MWNT (external heat flux = 35 kW m^{-2}) [112].

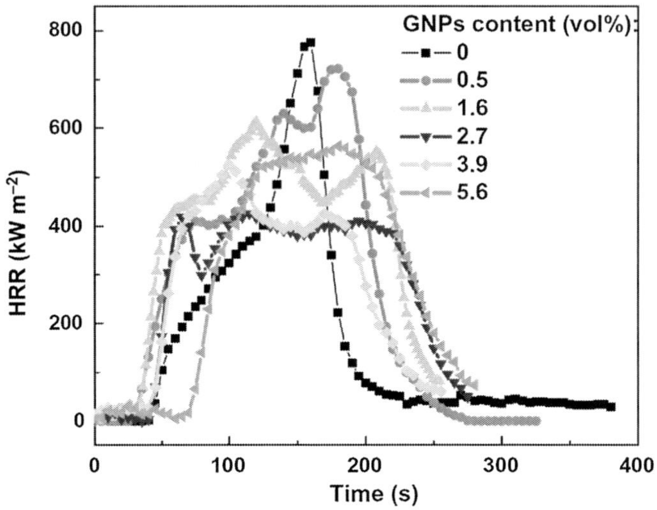

FIGURE 17

HRR curves for pure TPU and GNPs/TPU nanocomposites with different compositions [113].

Figure 17. In Table 11, the GNP content (vol%), TTI (s), flame out (FO) (s), PHRR (kW m^{-2}), THR (MJ m^{-2}) and residues remained (%) are tabulated. This table shows the maximum increase in time of ignition (63 s) and decrease in the PHRR (425.2 kW m^{-2}) at 5.6 and 2.7 vol % of GNP loadings in the TPU, respectively. The data in Table 12 refer to the maximum improvements in ignition time (t_{ign}), PHRR (kW m^{-2}), AHRR (kW m^{-2}), and LOI values of PU and its nanocomposites.

In the study of Song et al. [108], LOI measurements were performed to evaluate the flame retardant properties of PU/OMT nanocomposites in the presence of MPP; the LOI values were found to be 19, 20.5, 24.0, and 27.5 for neat PU, PU/OMT

Table 11 Combustion Testing Data of GNPs/TPU Nanocomposites [113]

GNP content (vol%)	0	0.5	1.6	2.7	3.9	5.6
TTI (t_{ign}) (s)	32	22	25	32	36	63
FO (s)	225	275	259	279	254	284
PHRR (kW m^{-2})	774.9	722.0	613.1	425.2	525.2	563.3
THR (MJ m^{-2})	62.2	89.2	95.6	77.2	70.9	81.7
Residues remaining (%)	24.6	6.1	6.8	9.4	10.8	13.7

GNP, graphite nanoplatelet; TTI, time to ignition; FO, flame out; PHHR, peak heat release rate; THR, total heat release.

Table 12 Cone Calorimetric and LOI Data of Various PU Nanocomposites (in Parentheses, the Percentage Increase (+) or Decrease (−) of the Value Compared with that of the Neat PU is Reported)

Samples	t_{ign} (s)	PHRR (kW m^{-2})	AHRR (kW m^{-2})	LOI	References
PU from polyol acclaim 4220 of molecular weight 4000 and 2.5 wt% cloisite 30B	33 (−5.7%)	641 (−71.5%)	363 (−43%)	–	107
PU/5 wt% OMT/6 wt% MPP	–	243 (−73.6%)	–	27.5 (44.7%)	108
PU/1% 2-(2-(5,5-dimethyl-1,3,2-dioxaphosphinyl-2-ylamino)ethy-amino)-*N,N,N*-triethyl-2-oxoethanaminium chloride	78 ± 1 (16.4%)	435 ± 9 (−16.8%)	252 ± 4 (−0.8%)	–	109
PU/20%2-(2-(5,5-dimethyl-1,3,2-dioxaphosphinyl-2-ylamino)ethy-amino)-*N,N,N*-triethyl-2-oxoethanaminium chloride-MMT	75 ± 2 (47%)	391 ± 5 (−10.3%)	242 ± 3 (−3%)	–	109
PU/8 wt% DS-LDH	–	–	–	23 (21%)	97
PU/NBR/DS-LDH	–	–	–	24 (23%)	99
MHBPU/2.5 wt% OMMT	–	–	–	28 (3.7%)	114
MHBPU/2.5 wt% modified OMMT by exchange process	–	–	–	32 (18.5%)	114
MHBPU/2.5 wt% modified OMMT by sonication process	–	–	–	33 (22.2%)	114
MHBPU/5 wt% modified OMMT by sonication process	–	–	–	36 (33.3)	114

PHRR, peak heat release rate; AHRR, average heat release rate.

(5 wt%), PU/MPP (6 wt%), and PU/OMT (5 wt%)/MPP (6 wt%), respectively. Thus, it was inferred that OMT exerts a synergistic effect with MPP reducing the flammability of PU.

LDHs are considered as a new emerging class of the most favorable layered crystals having efficient flame retardant properties. Kotal et al. [96] prepared PU/DS-intercalated LDH (DS-LDH) nanocomposites by the solution intercalation process. The LOI values were gradually improved with DS-LDH loading, and the maximum improvement (21%) was recorded for 8 wt% DS-LDH loading. It was therefore suggested that the LDH during combustion forms water vapor leaving a metal oxide residue, which obstructs the burning process by reducing oxygen supply toward the bulk specimen and thereby forms a char layer. The thickness of the char layer gradually increases with DS-LDH loading and acts as a barrier against the propagation of the flame. Kotal et al. [99] also prepared PU/NBR/DS-LDH nanocomposites (by applying a solution intercalation process). Similarly, the LOI values were found to gradually increase with DS-LDH loading; comparatively a higher improvement (23%) occurred at an 8 wt% DS-LDH loading in the PU/NBR blend rather than in PU. At lower DS-LDH loadings, the thickness of the charred layer was less, thus providing a milder barrier effect. The flame retardant behavior of PU/NBR/DS-LDH blend nanocomposites was attributed to be better by nanodispersion of DS-LDH in PU/NBR as shown in Figure 18. At 5 and 8 wt% of the DS-LDH loading in the PU/NBR blend, a partially exfoliated and intercalated morphology is developed. Some of the tactoids with a thickness of 30–50 nm and a lateral dimension of a few hundred nanometers are dispersed in the PU/NBR blend, which provides a sufficient barrier effect to give an improved LOI value.

Deka et al. [114] recently prepared epoxy resin-modified (*Mesua ferrea* L. seed oil based) hyperbranched PU (MHBPU) nanocomposites via an in situ technique using s-triazine-based highly branched poly(amido amine) (HBPAA) modified organonanoclay. According to their findings, MHBPU and MHBPU/OMMT (2.5 wt%) nanocomposites acquired a V-2 rating, whereas the MHBPU/2.5 wt% HBPAA–OMMT nanocomposites gained a V-1 rating. It was also observed that the LOI values increased from 28 to 33 for MHBPU/OMMT (2.5 wt%) and MHBPU/2.5 wt%HBPAA–OMMT nanocomposites, respectively, due to the homogeneous dispersion of HBPAA-modified OMMT in MHBPU. On further increasing the content of the modified OMMT in MHBPU, the formation of a continuous OMMT-rich carbonaceous surface was further facilitated, and as a result the LOI was further enhanced. It has been also reported that the LOIs of the PU/Epoxy resin/clay nanocomposites are improved with the addition of organophilic palygorskite or epoxy resin in PU; the maximum value (24.47%) was recorded at 3 wt% of organophilic palygorskite and 20 wt% epoxy resin content compared to that of neat PU [115].

UL94 testing of microcomposite and nanocomposite PU elastomers has shown that the former dripped heavily upon ignition with the rapid extinction of the flame and ignition of the underlying cotton wool by the burning fragments, while PU nanocomposites initially exhibit no dripping, but the flame eventually propagated up to the top of the specimen leading to an unclassified ranking [116]. It has been also reported that the nanocomposites of TPU show no significant improvement of LOI and UL-94 during the addition of 10 wt% POSS in TPU [111].

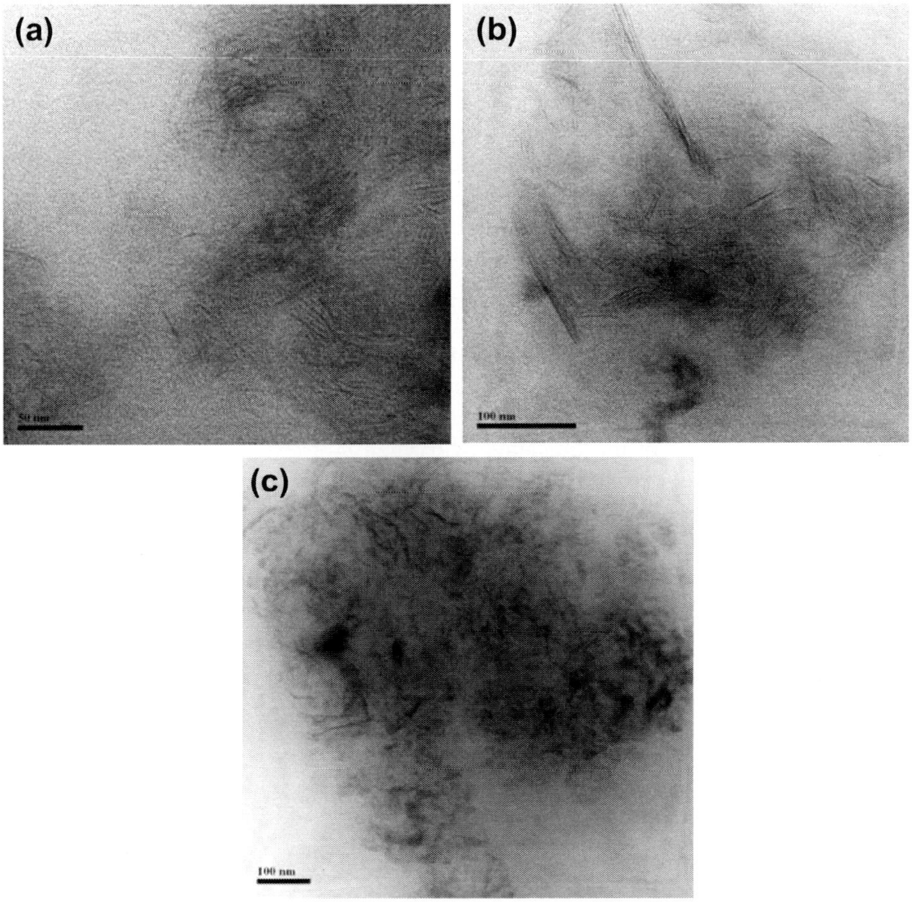

FIGURE 18

TEM images of (a) PU/NBR/DS-LDH (1 wt%) nanocomposite, (b) PU/NBR/DS-LDH (3 wt%) nanocomposite [99].

Reproduced with permission from Wiley.

6.4 Silicone rubber

Silicones are inert, synthetic compounds found in a wide variety of forms including silicone oil, silicone grease, SR, and silicone resin. Silicones are one of the most useful synthetic polymers widely used in many applications in civil engineering, construction building, electrical, transportation, aerospace, textiles, defence, and cosmetics industries. SR belongs to a family of thermoset elastomers that have a backbone of alternating silicone and oxygen atoms and methyl or vinyl side groups. They can be classified according to the polymer employed and the vulcanization process adopted for their production, as low temperature vulcanizable rubbers (low temperature vulcanizate rubber, LTV and room temperature vulcanizate rubber, RTV) and high temperature vulcanizable rubbers (HTV). SRs maintain their mechanical and electrical properties over a wide range of temperatures, and therefore comprise a choice for every application, from aerospace applications to medical devices [117,118]. They are used for the production of seals in the automotive industry, aerospace industry, connectors, cables for appliances and telecommunications, implants and devices for medical purpose, and packaging and baking pans for the food industry [119,120]. Silicones are greatly acknowledged for their better fire resistance properties compared to most carbon-based polymers. The use of silicones as flame retardant agents in other polymers offers an environmentally safer alternative to the halogenated FRs. The significant advantages of these silicones are attributed to the low HRR, minimal sensitivity to external heat flux, low yields of carbon monoxide release, and no emission of toxic smokes. The fire retardancy of silicones when exposed to elevated temperatures in the presence of oxygen is ascribed to the formation of a silica residue acting as "insulating blanket" that delays the volatilization of decomposed products. The flame retardancy of SR can be improved further by the addition of different fillers such as silica, calcium carbonate, wollastonite, mica, kaolin, MMT, carbon black, alumina trihydrate, magnesium dihydrate, and platinum compounds [121−127]. In silicone, calcium carbonate, wollastonite, and mica, the improvement in flame retardancy is governed by ceramization phenomena, whereas silica layer formation takes place in the case of MMT. Ohtani et al. [128] were the first to explain the flame retardant mechanism of SR in the presence of Pt compounds. According to the proposed mechanism, the thermal decomposition of the polydimethylsiloxane (PDMS-Pt) compound is strongly suppressed by the formation of the crosslinking structure induced by the Pt compound during the combustion, which is responsible for flame retardancy.

In several studies, cone calorimetric experiments have been performed to evaluate the flammability of silicone and its composites filled with inorganic fillers [121−124]. Mansouriet et al. [129] reported that the addition of 20 wt% of mica (muscovite)-reinforcing filler in silicone gum increases the TTI from 75 to 92 s, whereas the peak rate of release (kW m^{-2}) from 144 to 98, respectively. Upon firing, the cable (Silicone + 20% mica) (SNG20) formed a strong ceramic with no visible cracks. The residue was a coherent and strong ceramic that was able to withstand small mechanical shocks from the water spray applied at the end of firing. Ceramic microstructure analysis by field emission scanning electron microscopy (FESEM) revealed mica plates; the matrix in the cable insulation formed a multilayer structure (Figure 19). The continuous phase was mainly silica derived from SR pyrolysis and was dense, with pores as small as 30 nm, as shown in Figure 19. These findings clearly demonstrated the improved fire performance in the case of silicone/mica composites.

The formation of silicone polymer composites have also been reported using high-temperature vulcanizable (HTV) silicone gum with two chemically different micas (muscovite and phlogopite) in addition with glass frit and/or ferric oxide as

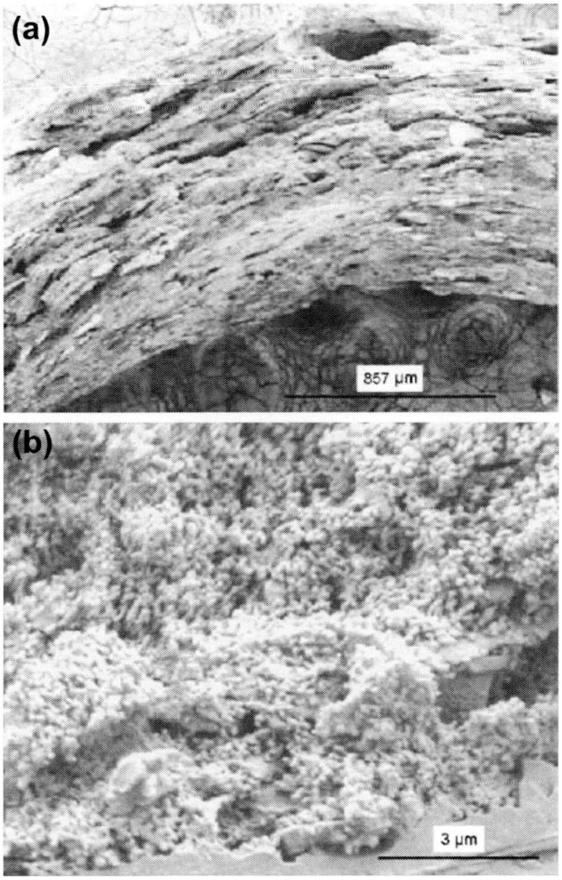

FIGURE 19

Scanning electron micrograph of a cross-section of SNG20 insulation layer after firing at 1050 °C at (a) low and (b) high magnifications [129].

With permission from Springer Science.

flame retardants by compounding on a two-roll mill [130]. Subsequently, the cone calorimetric method has been applied to calculate transthyretin (TTR), AHRR, and PHRR of the neat HTV and its composites. The findings showed that muscovite mica is the most successful filler to inhibit combustion delaying the TTI from 67 to 87 s and reduced the PHRR from 144 to 98 kW m^{-2}. It was also seen that both ferric oxide and glass frit effectively reduce the TTI to 55 and 62 s, respectively. From Table 13, it can be seen that both ferric oxide and glass frit in fact reduced the TTI (an undesirable effect in the field of passive fire protection); however, ferric oxide operated more effectively than glass frit did because, unlike glass frit, it actually lowered the HRR of silicone polymer.

Table 13 Flammability of Silicone Polymer Composites [130]

Sample	Filler	TTI (s)	AHRR (kW m^{-2})	PHRR (kW m^{-2})
HTVSi	None	67	117	144
SiEAP1	Glass frit	62	117	130
SiMO	Ferric oxide	55	102	117
SiGA1	Muscovite	87	93	98
SiGAEAP1	Muscovite + glass frit	66	121	134
SiGAMO	Muscovite + ferric oxide	69	87	95

TTI, time to ignition; AHRR, average heat release rate; PHRR, peak heat release rate.

Cone calorimetry tests using a flux of 30 kW m^{-2} on silicate/siloxane composites derived from vermiculite by reaction with hydroxy-terminated poly(dimethylsiloxane) have shown that the HRR of polyorganosiloxanes/vermiculite (20 wt%) is 70 kW m^{-2} over a 3-min burning period and the TTI period is 33 s [131]. Genevose and Shanks prepared PDMS foam composites with kaolin, MH, ZB, and firebrake 500 and investigated their fire performance [132]. The cone calorimetric results are shown in Table 14, and the HRR curves obtained with an incident heat flux of 35 kW m^{-2} are shown in Figure 20. From Table 14, it can be seen that the TTI of PDMS was found to be 123 s; the maximum value (155 s) was recorded in the case of the PDMS/Kaolin/MH/ZB; 73/16/6/5 composite due to the retarding effect of the mineral to the degradation of the polymer. The formation of the fine ash layer

Table 14 PDMS Composite Formulation and Fire Performance of PDMS Composites [132]

Sample	Composition (wt%)	TTI (s)	PHRR (kW m^{-2})	Time at PHRR (s)	Mean HRR (kW m^{-2})
1	PDMS	123	74/72	180/435	66.7
2	PDMS:Kaolin (72:28)	148	80/71	200/555	67.6
3	PDMS:Kaolin:MH (73:16:11)	144	78/118	200/605	75.1
4	PDMS:Kaolin:MH:ZB (73:16:10:1)	155	82/116	210/675	75.0
5	PDMS:Kaolin:MH:ZB (73:16:6:5)	109	84/108	170/690	74.2
6	PDMS:Kaolin:MH: anhZB (73:16:6:5)	121	76/96	200/610	68.3
7	PDMS:Kaolin:MH: anhZB (73:16:10:1)	120	76/114	170/710	69.7

PHRR, peak heat release rate; HRR, heat release rate.

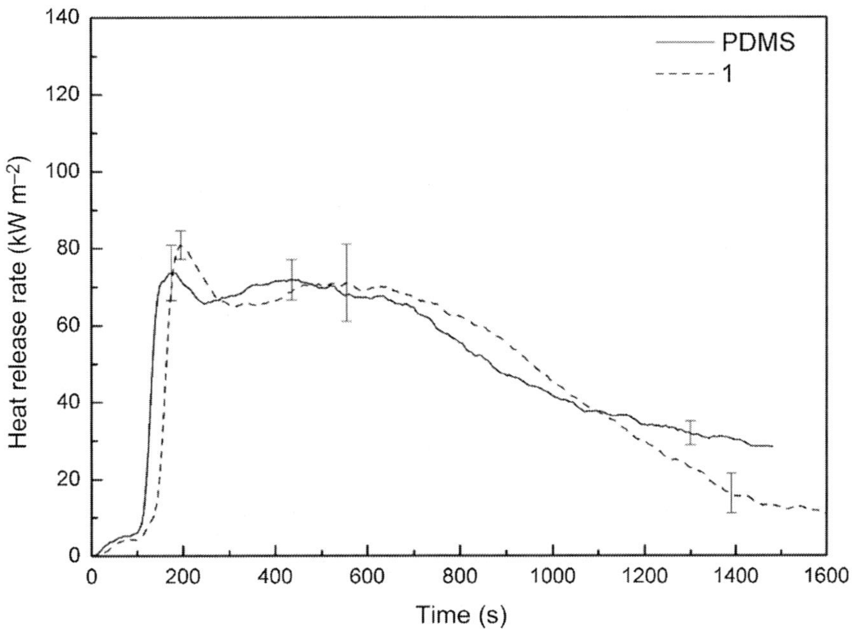

FIGURE 20

Heat release rates curves for PDMS and composite 1 at 35 kW m^{-2} heat flux [132].

on the surface exposed to the radiant heat resulted in the decrease of HRR to a minimum of about 66 kW m^{-2}; this layer reduced the oxygen permeation to the substrate and slowed down the diffusion of the pyrolysis gases emanating from the composite. The HRR curves of the composites in Figure 20 show that the first peak is similar to that of the PDMS; the second peak is associated with higher HRRs. Both peaks are indicative of the char formation and breakdown, attributed to fragmentation/cracks exposing an increased surface area to the incident heat flux. After the test, the residual char surface exhibited a thicker silica ash deposit, shown in Figure 21, suggesting that kaolin encouraged the decomposition of the PDMS polymer. However, composites containing MH and ZB provided a residual char surface with thicker silica ash and fewer cracks. These findings indicated that the presence of MH and ZB as FRs is not sufficient to improve the fire performance in the foam composites. The cone calorimetry test proved that silicone elastomers containing chalk also improved the flame retardancy of LDPE [133].

Yang et al. [134] prepared methyl vinyl silicone rubber (MVMQ)/MMT modified by hexadecyl trimethyl ammonium bromide (OMMT) nanocomposite by solution intercalation and flame retardant MVMQ/OMMT materials were prepared by mixing with flame retardant additives (MH and RP) at 25 °C, using a twin-roll mill. Their flammability performances were evaluated through UL-94 vertical burning tests. It was observed that MVMQ/OMMT nanocomposites containing various amounts of OMMT with the same content of SiO_2 as reinforcement filler

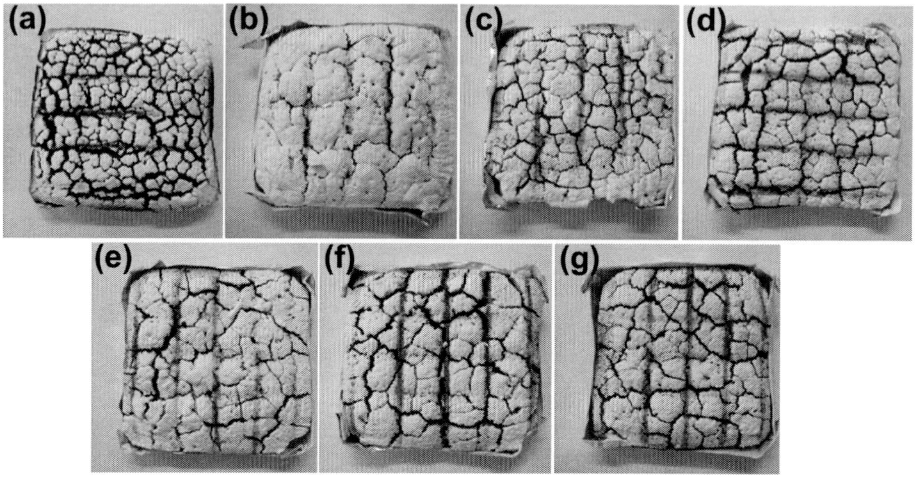

FIGURE 21

Photographs of the PDMS composite residue after cone calorimetry experiments (incident heat flux 35 kW m^{-2}). The composites: (a) PDMS foam, (b) composite 1, (c) composite 2, (d) composite 3, (e) composite 4, (f) composite 5, and (g) composite 6 [132].

and flame retardant additives have superior flammability properties. At a low filler loading of OMMT (1, 3, 5, and 7 phr), the flame retardant properties were considerably improved. The nanocomposite of MVMQ-OMMT-SiO_2 with different flame retardant additives (MH, RP) with a ratio of 100:1:20:20:5, respectively, showed the highest LOI than that of MVMQ. This is due to the synergistic flame retardant effect of MH, RP, and OMMT.

Chalk (30 wt% of stearate-coated calcium carbonate having an average particle size of 1.5 microns) and silicone (5 wt% of trimethylsilyl chain-ended PDMS gum containing nominally 0.2 mol% vinyl groups) have been proved to impart an improved flame retardancy to acrylate-based copolymers (65 wt%) in wire and cable applications [135]. This formulation successfully increased the LOI value of EVA copolymers from 18 to 34. They also reported that the addition of 5 wt% trimethylsilyl chain-ended PDMS gum and a stearate-coated calcium carbonate (average particle size of 1.5 mm) leads to the improvement of the flame retardancy of EVA copolymer as well. Andreasson et al. [136] reported that the composition of silicone (6.25 wt%) together with calcium carbonate (29 wt%) and magnesium hydroxide (29 wt%) improved the flame retardancy of EVA in terms of the LOI. Hermansson et al. [130] reported that a silicone elastomer containing chalk improves the flame retardancy of LDPE. In particular, the LOI value increased from 18 to 24.5 by adding 12.5 wt% of silicone elastomer and 30 wt% of chalk. The enhancement of the LDPE flame was suggested to be due to the synergetic formation from silicone and chalk of a stable intumescent structure covered by chars that acted as a heat barrier, thus preventing combustible gases from maintaining the flame.

6.5 Styrene–butadiene rubber

SBR, a random copolymer of styrene and butadiene, comprises one of the important classes of elastomers [84]. It is also commonly used in the preparation of blend composites as it constitutes one of the important polymers used in tire and rubber industries. Hence, there has been a constant effort to improve the performance characteristics with regard to the flame retardant properties of SBR to reduce the potential hazards due to the combustion of this material [84,137–139]. The fire hazards and smoke generation of SBR can also be reduced by plasticization with chlorinated paraffin in the presence of $Al(OH)_3$ [140]. Patents have also been filed on the utility of SBR in FR applications such as in pressure-sensitive adhesive tapes for television parts [141] and in impact resistant materials in combination with bisphenol A polycarbonate, fibrous polytetrafluoro ethylene, and other suitable ingredients [142]. Grazyna and others [143] reported that the ignition temperature of SBR corresponds to 348 °C.

Khattab [144] observed that the LOI value is 20.6 in the case of SBR, whereas in the presence of 10 phr of decabromobiphenyl oxide and $Al(OH_3)$, it increased to 37.7 and 21.8, respectively, suggesting the efficient flame retardant action of decabromobiphenyl oxide (DBBO) compared to that of $Al(OH)_3$ in SBR. Zhang et al. [145] reported that the LOI of the pure SBR (styrene unit to butadiene unit: 23.5%, trans 1,4 structure content:55% and cis 1,4 structure content:9.5%) corresponds to 19 and is not significantly altered even in the presence of MMT loadings of 10, 20, 30, and 40 phr. However, the combinations of $Mg(OH)_2$ and clay are effective for the improvement of oxygen index as compared to either only MMT or MH. Mishra et al. [139,146] studied the effect of nanodimensional calcium carbonate of 9, 15, and 21 nm on the flame retardancy properties of SBR. From Figure 22, it can be observed that by decreasing the size of nano-$CaCO_3$, the improvement in flame retardancy becomes stronger and better in all cases compared to that observed in the case of composite with commercial calcium carbonate [146]. At 2 wt% loading, the flammability rates were 1.44, 1.54, and 1.75 s mm^{-1}, respectively, with 21-, 15-, and 9-nm sizes of $CaCO_3$, while at 10 wt% loading, they were 1.63, 1.76, and 2.31 s mm^{-1}. However, the corresponding values for 2 and 10 wt% loadings of commercial $CaCO_3$ are 1.38 and 1.57 s mm^{-1}. The observed improvements in the flammability of rubber nanocomposites as compared to commercial $CaCO_3$-filled SBR and pristine SBR is attributed to the formation of char layers acting as an excellent insulator and mass transfer barrier.

Efforts have been made to improve the fire retardancy of SBR by blending them with the polymers of lower flammability. Studies on NBR/SBR elastomer blends have shown that the critical oxygen index (at above room temperature) is a function of the overall acrylonitrile content, and does not necessarily vary linearly with temperature [147]. The effect of low flammability polymer, chlorosulfonated polyethylene (CSM), on thermal properties and combustibility of SBR containing ZnO or nano-ZnO has also been investigated [148]. It has been observed that the oxygen index of the neat SBR (with 23.5% styrene content) is 0.293, while in the

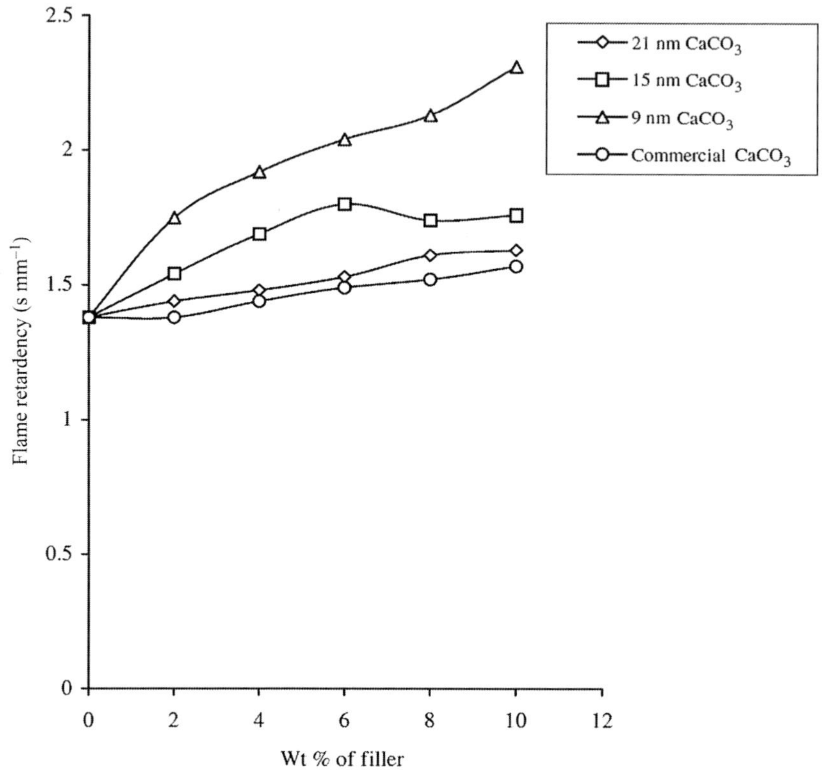

FIGURE 22

Flame retardancy of SBR filled with various sizes of $CaCO_3$ [146].

Reprinted with permission from Springer.

case of SBR/CSM (24% combined chlorine)-ZnO and SBR/CSM (24% combined chlorine)-nano-ZnO blends to >0.375 reaches higher values due to the better flame retardant properties of the CSM. In another study, the flammability properties of dichlorocarbene-modified styrene–butadiene rubber (DCSBR) with NR blends have been explored [149]. It can be observed from Figure 23 that the flame resistance of the blend increased with increasing DCSBR contents and decreased with increasing NR concentrations in the blend. It was suggested that an intumescent effect is a consequence of the presence of halogen-containing elastomers that restrict the flow of heat; the time to sustained ignition, the AHRR, the peak heat release, the total HRR, and the average MLR were 70.96 s, 975.65 kW m^{-2}, 2248.28 kW m^{-2}, 87.77 MJ m^{-2}, and 50.40 gs m^{-2}, respectively [140].

Zhang et al. [145] investigated the flammability behavior of MMT/SBR nanocomposites at 50-kW m^{-2} heat flux. They observed that the HRR by adding 20 phr MMT decreased from 1987 to 1442 kw m^{-2} in the case of SBR nanocomposites, whereas it was reduced only to 1693 kW m^{-2} when SBR microcomposites

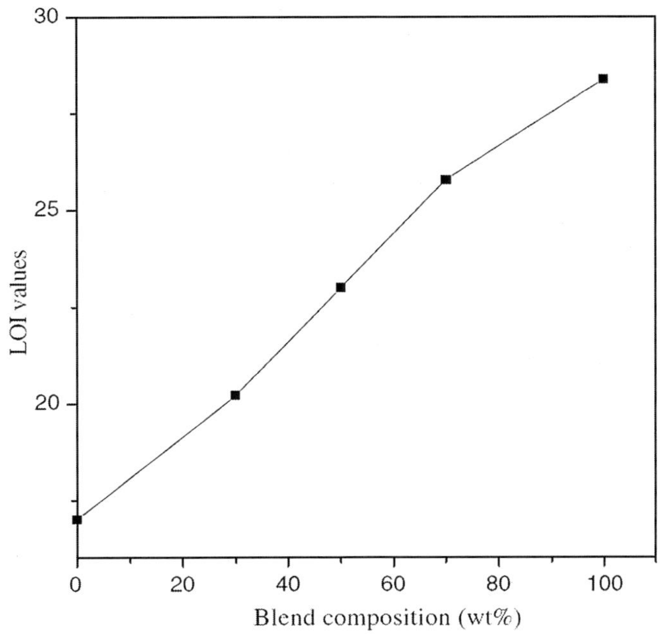

FIGURE 23

Variation of the limiting oxygen index value (LOI) of NR, DCSBR, and NR/DCSBR blends with different compositions [149].

were formed. The ignition time of the filled SBR/20 phr MMT nanocomposite was found to be 44 s, which is higher compared to that (18 s) recorded for pure SBR. Such a reduction in the HRR or MLR is the consequence of char formation during the combustion of composites, which acts as an excellent barrier toward heat and mass transfer.

6.6 Acrylonitrile–butadiene rubber

Nitrile Rubber is a synthetic rubber produced by the polymerization of acrylonitrile with butadiene. This rubber is also known as NBR, acrylonitrile–butadiene rubber, acrylonitrile rubber and nitrile-butadiene rubber. The properties of acrylonitrile–butadiene rubber (NBR), e.g. high tensile strength, high compression set, and poorer cold flexibility depends on the acrylonitrile content. It has an excellent resistance toward oils and fuels, and it is widely used in oil seals, packaging, diaphragms, blankets, etc. [84]. NBR has a wide variety of applications; therefore, an improvement of its flame resistance has become of major importance.

A great deal of research has been carried out on the improvement of the flame retardant characteristics of NBR by the addition of flame retardant materials preferably by using environmental friendly flame retardant/fillers or both [150–154].

Silica constitutes important fillers used in industrial applications. Hence, the flame retardant aspects of NBR with silica as the filler have also been investigated. Rybinski et al. [151] used two different grades of silica (Zeosil 175C and Ultrasil VN-3) to decrease the flammability of NBR (18% AN content). The fillers were loaded at 20, 30, 40, and 50 phr levels in unmodified and cryogenic modified states. It was observed that neat NBR exhibits an OI of 0.270. With increasing amounts of silica (grade Ultrasil VN-3) in an unmodified state, the value of OI increased as well; at 50 phr loading, the OI value was found to be 0.298. A further increment in the OI value was obtained with the cryogenic modification of the filler, OI value 0.305. In the case of a filler of a different grade, Zeosil 175C silica, at 50 phr loading, the OI was 0.285 and on modification of this filler, the OI raised to 0.310. The time of burning in air also increased for the NBR vulcanizates containing silica (ultraseal VN-3) at 20 and 30 phr, whereas the vulcanizates with 40 and 50 phr of silica showed a self-extinguishing behavior in air. An improvement in the OI values of sulfur and peroxide NBR vulcanizates modified with tetraethoxysilane compared to the corresponding NBR vulcanizates has been also reported. The presence of silica was proposed to restrict the passage of flammable decomposition products to inner layers and thereby reduce further combustion. Also, the char formation reduces the probability of further burning. In another study, NBR (18% AN content) was modified with hybrid poly(methyl siloxane) [152]. OI measurements showed that the OI value of NBR was 0.285, and it varied between 0.310 and 0.365, when NBR was modified by poly(siloxanes). This improvement in flame retardancy is due to the formation of silica on combustion, which improves the thermal stability and restricts the flow of energy and volatile products to inner layers thereby protecting the polymer.

Organomodified silicates have been also found to be highly effective in retarding the combustion of polymeric materials due to their layered structures. Very recently, the effect of modified MMT on the flame retardant characteristics of NBR with a 22% AN content was investigated [153]. The vulcanazites were prepared in the presence of sulfur and peroxide. Fillers like clay improve the flame retardant properties of polymers by the restriction of chain movements, thus increasing the degradation temperatures. The modified MMT used were nanobentonites, commercially known as Nanobent and Nanofil. The OI values for NBR were shown to be 0.205 and 0.215 when produced through peroxide and sulfur vulcanization, respectively. In the case of the peroxide vulcanazite, OI values with various grades of nanobents were observed to be >0.34; however, with nanofill grades, the values of OI were lower (in the range of 0.269–0.303). The OI values with sulfur vulcanazites have also been reported. These results indicated that both the grades of fillers are successful in improving the OI values. They reduce the flammability through the labyrinth effect: the volatile products are trapped in the layers of MMT, and thus, further combustion of inner layers of polymer is restricted.

NBR–clay composites, prepared by the cocoagulation process, exhibit also improved FR properties; with respect to unfilled NBR (24–26% AN content), which shows an OI value of 18.4 [154], the OI values of the composites with 10, 20, and 30 phr of silica loadings have been found to be 19.2, 19.5, and 20.7, respectively, indicating that the OI is improved with silica loading. Also, with clay as the flame retardant filler at 10, 20 and 30 phr loading, the OI values obtained are 10, 20.2, and 21.2, respectively. Therefore, it can be concluded that at similar phr loadings of silica and clay, the OI values were of the same range. In both cases, the char formation during burning restricted the flow of mass and heat to inner layers.

Moon and coworkers [150] prepared NBR foams compounded with different amounts of halogen-free flame retardants, namely RP, APP, and EG (10, 20, and 30 parts added too 100 parts of NBR) and a constant amount (195 parts) of other flame retardants, primarily alumina trihydrate (ATH) and studied their flame resistance properties. It was noted that the HRC, AHRR, and the THR ranged from 10 to 74 $J\,g^{-1}\,K^{-1}$, 8–60 $kW\,m^{-2}$, and 2.6–7.3 $MJ\,m^{-2}$ for the nonhalogenated NBR foams with a closed-cell structure; these values were significantly decreased upon increasing the amounts of the flame retardants. This reduction is attributed to the hard char formation and production of water arising from the interaction with aluminium trihydrate. It was also reported that the LOI values of NBR foams compounded with various halogen-free flame retardants were always higher compared to that of NBR. Cone calorimetric experiments have also been preformed in order to investigate the flame retardancy of modified MMT-filled NBR (22% AN content) composites [151]. The findings show that the TTI, PHRR (HRR_{max}), THR, the average effective heat of combustion, and the average MLR are improved compared to that for NBR vulcanazites.

6.7 Natural rubber

Natural rubber (NR), which is nothing but *Cis*-1,4-polyisoprene, has proved its utility as a "commodity polymer" as well as an "engineering elastomer" by virtue of its unique combination of physicomechanical properties i.e. excellent mechanical strength, very good abrasion resistance, and high resilience. The fact that its inherent high flammability restricts its usage in many possible critical applications necessitates the development of means for reducing its flammability. In this context, the introduction of flame retardant additives into the NR matrix remains a high priority option, and quite a good number of research papers have been published dealing with the development of flame retardant microcomposites and nanocomposites of NR [155–157]. To improve the flame retardancy of NR, flame retardant additives such as bromo derivatives of phosphorylated cashew nut shell liquid (PCNSL), antimony trioxide, and alumina trihydrate (ATH) have been used [156]. In spite of the concerns on the corrosiveness and toxicity of smoke and other emission products of rubber compounds based on decabromodiphenyl oxide and antimony trioxide, these two FR additives continue to be widely used. However, halogen-free flame retardant fillers, such as aluminum hydroxide, organoclay, and ZHS are also popular. Nonetheless, the partial replacement of flame retardant additives in NR compounds by ZHS produces no synergistic improvement in FR behavior [158]. Layered silicates or clays are suitable alternatives to (the usually expensive) halogen based, nitrogen-based, phosphorous-based, and intumescent FR (IFR) additives. Among the inexpensive inorganic fillers, talc and mica powder retard the flammability of NR satisfactorily without adversely affecting other vulcanizate properties [159]. A study on the flammability of poly [ethylene-co-(vinyl acetate)] (EVA)/(Std. Malaysian NR) (SMR L)/organoclay nanocomposites showed the positive effect of organoclay, as well [160].

Sodium MMT (NaMMT) at a dosage of 5% by weight has been reported to decrease the PHHR of neat NR by 38% [161]. What is more, in another study, organoclay (TMT)–NR nanocomposites were found to exhibit lower HRRs than those of MMT-NR composites [162]. Compared to neat NR (as well as filler type kaolin/NR composite), an intercalated kaolin/NR nanocomposite prepared by a two-step intercalation method has been reported to have a higher flame retardancy [163]. The flammability properties of NR nanocomposites filled with CNTs and organoclay have also been investigated and also compared with carbon black and pure NR composites [164]. Cone calorimetric studies have shown that the nanocomposites filled with organoclay and CNTs show a lower PHRR. A slight synergistic effect was found for NR nanocomposite having equal amounts of organoclay and CNTs.

Apart from sodium MMT, the effect of organically modified MMT on the flame retardant properties of the NR nanocomposites has been also studied [165]. The comparison of the HRR of pristine rubber (NR0) and a typical nanocomposite (3 wt% of org-MMT filled NR) at 50 kW m^{-2} heat flux showed that the NR3 nanocomposite had a 54.28% lower HRR peak value than the NR0, but NR5 (5 wt% of org-MMT filled NR) and NR7 (7 wt% of org-MMT filled NR) exhibited a >55% and 58% decrease in the HRR peak value, respectively. The MLR value of the nanocomposite was found to be significantly lower than that of the pristine rubber. Moreover, It is also noted that the ignition time is delayed by about 150% when 3 wt% of org-MMT is added to NR. In general, its flame retardant character is traced to the response accredited to a char layer, which developed on the outer surface of the sample during combustion [16]. Impressive flammability properties have been also reported in the case of natural NR containing organoclay modified by tributyl phosphate (TMMT), and show in particular the significant decrease in the HRR, MLR, and smoke produce rate [166].

The flammability of NR (expressed in the LOI) also improved when IFR additives are incorporated in NR [167]. Figure 24 shows that by adding 80 phr IFR, the LOI of NR is increased by about 70%, reaching an acceptable value of 27.0; the LOI increases further (by about 110%) when 0.5 phr of 4A zeolite is additionally included in the formulation. These observations give a clear indication of the synergistic effect of IFR and 4A zeolite. Natural rubber, when filled with 40 phr of carbon black, china clay, or aluminum powder is observed to show the same LOI value [168].

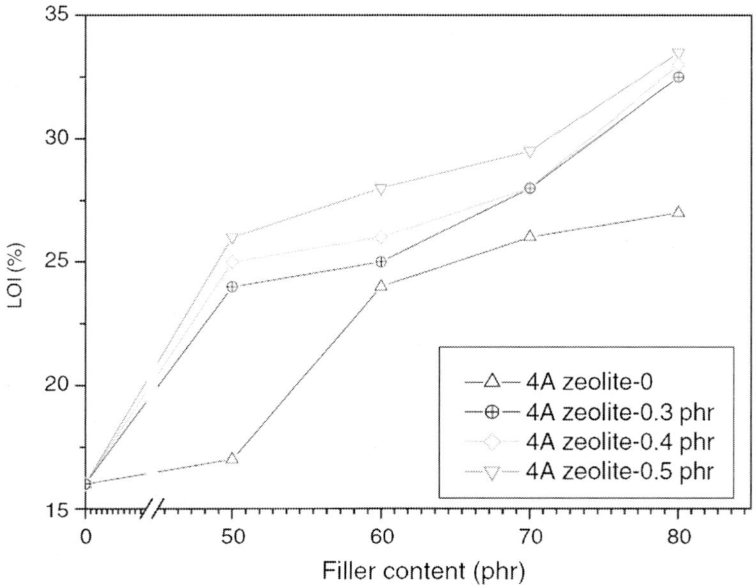

FIGURE 24

LOI value of different IFR and IFR-4a zeolite filled NR systems [167].

Reproduced with permission from Springer.

Nanocomposites of NR and polystyrene (PS)-encapsulated nanosilica prepared by the latex compounding method show higher LOI values as compared to those of neat NR [169]. On the other hand, a study on the effect of PCNSL on the LOI of NR indicated that this additive does not affect flammability to a great extent, too [170]. In Table 15, several data are tabulated implying routes for referring to the maximum improvements in the LOI and PHRR of NR composites.

Table 15 Data Referring to the Maximum Improvements in the LOI and PHRR of NR Composites (in Parentheses, the Percentage Increase (+) or Decrease (−) of the Value Compared with that of Neat NR)

FR Additive	LOI	PHRR (kW m^{-2})	References
NaMMT (5%)	–	1068 (−38%)	[161]
Org-MMT (3%)	–	16 (−53%)	[165]
TMMT (20%)	–	577 (−53%)	[166]
PS-silica (9%)	17.6 (+16%)	–	[169]
IFR (80%)	27 (+69%)	–	[167]
IFR (80%) + 4A zeolite (0.5%)	33.5 (+109%)	–	

6.8 Polychloroprene rubber

Polychloroprene Rubber (CR) possesses a combination of properties such as good heat resistance, very good resistance to oxygen, ozone and sunlight, self-extinguishing nature, good resistance to hydrocarbon oils, high tensile strength, and abrasion resistance making this polymer a special purpose rubber. It finds many applications e.g. joint seals, bridge bearings, all-purpose hoses, conveyer belting, cable sheathing, automotive molded parts, and electrical connectors, etc. [157,171] and demanded increased flammability resistance [172]. CR presents self-extinguishing behavior, attributed to the presence of the chlorine atom in the polychloroprene chain that slows down its burning process. In this, the released hydrogen chloride lowers the concentration of flammable gases produced by the pyrolysis of polychloroprene thereby hindering the burning processes [173]. However, in several cases, the addition of a flame retardant is needed so that the polymer can fulfill the application requirements. According to Barruel [174], efforts to improve flame resistance and to limit flame spreading rate can increase vulcanizate smoke production during pyrolysis or combustion of chloroprene rubber. In this regard, alumina trihydrate has beneficial effect on oxygen index and its capacity to limit smoke emission is very useful.

A detailed investigation by Paul Cusack [7] of ITRI Limited showed that there is a synergistic effect between ZHS and ATH and high effectiveness of ZHS-coated ATH in lowering HRR of CR. As shown in Table 16, that PHRR of CR vulcanizate with 50 parts ATH is 128 kW m^{-2}, almost 60% less than that of the gum vulcanizate (314 kW m^{-2}). When five parts of ATH is replaced by five parts of ZHS, the PHRR value further reduces to 95 kW m^{-2}, indicating a synergistic effect between the ATH and ZHS for flame resistance. But the most promising interesting result is obtained when ZHS-coated ATH is used instead of ATH. At 50 parts, ZHS-coated ATH decreases the PHRR to an extremely low value of (26 kW m^{-2}) confirming the higher effectiveness of this filler compared to ZHS-coated ATH.

Hornsby et al. [175] have shown that tin (IV) oxide and ZHS (even at low levels) increase the flame retardancy of CR, whereas chlorinated paraffin is effective only at a higher dosage. The results shown in Table 17 indicate that it is difficult to have a

Table 16 Cone Calorimeter Data for Polychloroprene Formulations [7]

Additive	Peak Rate of Heat Release (kW m^{-2})	Smoke Parameter (MW kg^{-1})
None	314	312
50 phr ATH	128	115
5 phr ZHS + 45 phr ATH	95	89
50 phr ZHS-coated ATH	26	15

ATH, Alumina trihydrate.

Table 17 Summary of Modified Polychloroprene Compositions and Their Limited Oxygen Indices [175]

Specimen Code	Polychloroprene Compound (phr)	Zinc Hydroxystannate (phr)	Tin (IV) Oxide (phr)	Chlorinated Paraffin (phr)	Total Chlorine Content (%)	LOI
A	100	–	–	–	40	39.7
B	100	–	1	–	40	43.1
C	100	–	3	–	40	44.8
D	100	–	5	–	40	42.4
E	100	1	–	–	40	41.9
F	100	3	–	–	40	46.9
G	100	5	–	–	40	42.7
H	100	–	–	20	45	40.0
J	100	–	5	20	45	43.4
K	100	–	–	50	50	45.6
L	100	–	5	50	50	54.7

LOI, limiting oxygen index.

direct correlation between the filler content and the observed LOI. But for both types of tin compounds, 3 phr was the optimum level (the increase in the LOI) with regard to gum vulcanizate was 4.9 and 7.2 for tin (IV) oxide and ZHS, respectively. This also confirmed the superiority of ZHS over tin (IV) oxide. The effect of CR on the flame retardancy properties of poly(ethylene(vinyl acetate)) has shown that the LOI increases linearly with increasing proportions of CR in the blend [176]. CR has also been blended with acrylonitrile–butadiene–styrene copolymer (ABS) and acrylonitrile–ethylene–propylene–diene terpolymer–styrene copolymer (AES) and their flammability characteristics have been studied. The LOI results indicated that CR marginally improves flame resistance of both the copolymers. In both the cases (of ABS and AES), the maximum improvement in the LOI was achieved at 40 parts of CR, and it has been found to be 3.5 units only [177]. NR/CR blend shows that the LOI increases with an increase in the CR content due to the formation of a thermally insulating char that inhibits its combustion [178]. The LOI is also found to increase from 36 to 41 on blending CR with eight parts of a polyphosphonate [179]. Flammability study on the flame retardancy of CR blended with a plant polymer indicated that 10 parts of the plant polymer is more effective than five parts of a commonly used FR additive "antimony trioxide" [180].

The effect of antimony trioxide and carbon black on the burning process of polychloroprene vulcanizates has been studied in detail by Nasybullin et al. [173]. This study showed that OI decreases from 0.324 to 0.302 with the incorporation of metal oxides. Such a decrease in the OI and hence flame resistance has been attributed to the capture of hydrogen chloride by the metal oxides. Vulcanization of the same compound is observed to have a minimal effect on the OI. On the other hand, antimony trioxide, antimony (III) oxide chloride and antimony trichloride have been observed to increase the OI of vulcanizate to 0.434, 0.425, and 0.473, respectively (at a constant level of 21 phr). This positive effect of antimony trioxide on the fire resistance of CR has its root in its reaction with the released HCl forming volatile chlorinated antimony compounds.

7. CONCLUSIONS

Recently, much more interest has been focused worldwide in developing flame retardant polymeric materials. This is because thousands of people are injured or killed in fires, and fire-related property loss reached annually amounts to billions of dollars invariably due to the fire in electrical distribution systems, which includes electrical wiring. The present article deals with the developments related to the fire retardancy of EVAC, ethylene–propylene diene rubber, PU, SR, SBR, NBR, NR, and CR. It can be concluded that LDH and ATH are the most effective fillers among all types of mineral clays, or CNTs and many other available nanofillers. The cost of LDH is also very less compared to that of CNT or clay. It can easily be synthesized in the laboratory at various compositions. The endothermic decomposition of LDH produces water vapor and smoke, which are nontoxic in nature. However, there is a challenge to meet flame retardancy requirements without compromising its mechanical performance and at an acceptable cost. Therefore, it is believed that the flame retardancy of elastomers is a field that will comprise the focal point of research in the near future.

References

[1] Okada A, Kawasumi M, Usuki A, Kojima Y, Kurauchi T, Kamigaito O. Synthesis and properties of nylon-6/clay hybrids. In: Schaefer DW, Mark JE, editors. Polymer based molecular composites. MRS symposium proceedings, vol. 171. Pittsburgh; 1990. pp. 45–50.

[2] Giannelis EP. Polymer layered silicate nanocomposites. Adv Mater 1996;8:29–35.

[3] Giannelis EP, Krishnamoorti R, Manias E. Polymer-silicate nanocomposites: model systems for confined polymers and polymer brushes. Adv Polym Sci 1999;138:107–47.

[4] Zanetti M, Camino G, Mulhaupt R. Combustion behaviour of EVA/fluorohectorite nanocomposites. Polym Degrad Stab 2001;74:413–7.

[5] Alexandre M, Dubois P. Polymer-layered silicate nanocomposites: preparation, properties and uses of a new class of materials. Mater Sci Eng:R 2000;28:1–63.

[6] Godovsky DY. Device applications of polymer-nanocomposites. Adv Polym Sci 2000;153:163–205.

[7] Cusak PA. Novel coated fillers enhance flame-retardant properties. Plast Addit Compd 2007;9:26–9.

[8] Ray SS, Okamoto M. Polymer/layered silicate nanocomposites: a review from preparation to processing. Prog Polym Sci 2003;28:1539–641.

[9] Pramanik M, Srivastava SK, Samantaray BK, Bhowmick AK. Rubber–clay nanocomposites by solution blending. J Appl Polym Sci 2003;87:221620.

[10] Srivastava SK, Pramanik M, Acharya H. Ethylene vinyl acetate copolymer/clay nanocomposites. J Polym Sci Part B: Polym Phys 2006;44:471–80.

[11] Barbosa R, Araújo EM, Melo TJA, Ito EN. Comparison of flammability behavior of polyethylene/Brazilian clay nanocomposites and polyethylene/flame retardants. Mater Lett 2007;61:2575–8.

[12] Maiti M, Bhowmick AK. New fluoroelastomer nanocomposites from synthetic montmorillonite. Compos Sci Technol 2008;68:1–9.

[13] Yi D, Yang R. Ammonium polyphosphate/montmorillonite nanocompounds in polypropylene. J Appl Polym Sci 2010;118:834–40.

[14] Leroux F, Besse JP. Polymer intercalated layered double hydroxide: a new emerging class of nanocomposites. Chem Mater 2001;13:3507–15.

[15] Costa FR, Saphiannikova M, Wagenknecht U, Heinrich G. Layered double hydroxide based polymer nanocomposites. Adv Polym Sci 2008;210:101–68.

[16] Srivastava SK. Barrier properties of ethylene vinyl acetate nanocomposites. In: Mittal V, editor. Barrier properties of polymer clay nanocomposites. USA: NOVA Pub; 2010. pp. 189–212.

[17] Srivastava SK, Pramanik M. Nanocomposites of ethylene vinyl acetate copolymers. In: Nalva HS, editor. Encyclopaedia of nanoscience and nanotechnology, vol. 17. USA: American Scientific publishers; 2011. pp. 1–25.

[18] Srivastava SK, Acharya H. Ageing and degradation behaviour of rubber nanocomposites. In: Thomas S, Stephen R, editors. Rubber nanocomposites: preparation, properties and applications. Singapore: John Wiley and Sons (Asia) Pvt. Ltd.; 2010. pp. 551–93.
[19] Acharya H, Srivastava SK, Bhowmick AK. Synthesis of partially exfoliated EPDM/LDH nanocomposites by solution intercalation: structural characterization and properties. Compos Sci Technol 2007;67:2807–16.
[20] Chen G. An exfoliated polyamide-6/layered double hydroxide nanocomposite via in situ intercalative polymerization between positively charged inorganic platelets. J Nanosci Nanotechnol 2006;6:1155–8.
[21] Kiliaris P, Papaspyrides CD. Polymer/layered silicate (clay) nanocomposites: an overview of flame retardancy. Prog Polym Sci 2010;35:902–58.
[22] Peeterbroeck S, Laoutid F, Taulemesse JN, Monteverde F, Lopez-Cuesta JM, Nagy JB, et al. Mechanical properties and flame-retardant behavior of ethylene vinyl acetate/high-density polyethylene coated carbon nanotube nanocomposites. Adv Funct Mater 2007;17:2787–91.
[23] Yu H, Zhang Z, Wang Z, Jiang Z, Liu J, Wang L, et al. Double functions of chlorinated carbon nanotubes in its combination with Ni_2O_3 for reducing flammability of polypropylene. J Phys Chem C 2010;114:13226–33.
[24] Im JS, Lee SK, In SJ, Lee YS. Improved flame retardant properties of epoxy resin by fluorinated MMT/MWCNT additives. J Ann Appl Pyrol 2010;89:225–32.
[25] Kim H, Miuram Y, Macosko CW. Graphene/polyurethane nanocomposites for improved gas barrier and electrical conductivity. Chem Mater 2010;22:3441–50.
[26] Ye L, Meng X, Ji X Y, ZM Li, Tang JH. Synthesis and characterization of expandable graphite–poly(methyl methacrylate) composite particles and their application to flame retardation of rigid polyurethane foams. Polym Degrad Stab 2009;94:971–9.
[27] Ismail H, Pasbakhsh P, Fauzi MNA, Bakar AA. Morphological, thermal and tensile properties of halloysite nanotubes filled ethylene propylene diene monomer (EPDM) nanocomposites. Polym Test 2008;27:841–50.
[28] Jiao CM, Wang ZZ, Ye Z, Hu Y, Fan WC. Flame retardation of ethylene–vinyl acetate copolymer using nano magnesium hydroxide and nano hydrotalcite. J Fire Sci 2006; 24:47–63.
[29] Zhang X, Guo F, Chen J, Wang G, Liu H. Investigation of interfacial modification for flame retardant ethylene vinyl acetate copolymer/alumina trihydrate nanocomposites. Polym Degrad Stab 2005;87:411–8.
[30] Du L, Qu B. Effects of synthesis condition on crystal morphological structure and thermal degradation behavior of hydrotalcite and flame retardant and mechanical properties of EVA/hydrotalcite blends. Polym Compos 2007;28:131–8.
[31] Prestona CML, Amarasinghea G, Hopewell JL, Shanksa RA, Mathys Z. Evaluation of polar ethylene copolymers as fire retardant nanocomposite matrices. Polym Degrad Stab 2004;84:533–44.
[32] Lu SY, Hamerton I. Recent developments in the chemistry of halogen-free flame retardant polymers. Prog Polym Sci 2002;27:1661–712.
[33] Zhang S, Horrocks AR. A review of flame retardant polypropylene fibres. Prog Polym Sci 2003;28:1517–38.
[34] Costa FR, Wagenknecht U, Heinrich G. LDPE/MgAl layered double hydroxide nanocomposite: thermal and flammability properties. Polym Degrad Stab 2007;92: 1813–23.

[35] Isitman NA, Kaynak A. Nanoclay and carbon nanotubes as potential synergists of an organophosphorus flame-retardant in poly(methyl methacrylate). Polym Degrad Stabil 2010;95:1523–32.
[36] Jeevananda T, Jang YK, Lee JH, Deepa S, Urs MV, Ranganathaiah C. Investigation of multi-walled carbon nanotube reinforced high-density polyethylene/carbon black nanocomposites using electrical, DSC and positron lifetime spectroscopy techniques. Polym Int 2009;58:755–80.
[37] Pramanik M, Srivastava SK, Samantaray BK, Bhowmick AK. Synthesis and characterization of organosoluble, thermoplastic elastomer/clay nanocomposites. J Polym Sci Part B: Polym Phys 2002;40:2065–72.
[38] Pramanik M, Srivastava SK, Samantaray BK, Bhowmick AK. EVA/Clay nanocomposites by solution blending: effect of aluminosilicate layers on mechanical and thermal properties. Macromol Res 2003;11:260–6.
[39] Pramanik M, Acharya H, Srivastava SK. Exertion of inhibiting effect by aluminosilicate layers on swelling of solution blended EVA/clay nanocomposites. Macromol Mater Eng 2004;289:562–7.
[40] Kuila T, Acharya H, Srivastava SK, Bhowmick AK. Synthesis and characterization of ethylene vinyl acetate/Mg-Al layered double hydroxide nanocomposites. J Appl Polym Sci 2007;104:1845–51.
[41] Huang JC, Zhu ZK, Ma XD, Qian XF, Yin J. Preparation and properties of montmorillonite/organo-soluble polyimide hybrid materials prepared by a one-step approach. J Mater Sci 2001;36:871–7.
[42] Bala P, Samantaray BK, Srivastava SK. Synthesis and characterization of Na-montmorillonite-alkylammonium intercalation compounds. Mater Res Bull 2000;35:1717–24.
[43] Pramanik M. *Studies on layered materials and polymer nanocomposites* [Ph.D. thesis]. India: I.I.T. Kharagpur; 2004.
[44] Kuila T, Acharya H, Srivastava SK, Bhowmick AK. Effect of vinyl acetate content on the mechanical and thermal properties of ethylene vinyl acetate/MgAl layered double hydroxide nanocomposites. J Appl Polym Sci 2008;108:1329–35.
[45] Newman SP, Jones W. Synthesis, characterization and applications of layered double hydroxides containing organic guests. New J Chem; 1998:105–15.
[46] Vucelic M, Moggridge GD, Jones W. Thermal properties of terephthalate- and benzoate-intercalated LDH. J Phys Chem 1995;99:8328–37.
[47] Hussain F, Hojjati M, Okamoto M, Gorga RE. Review article: polymer–matrix nanocomposites, processing, manufacturing, and application: an overview. J Compos Mater 2006;40:1511–75.
[48] Costa FR. *Mg-Al layered double hydroxide: a potential nanofiller and flame-retardant for polyethylene* [Ph.D. thesis]. Germany: Technischen Universit ät Dresden; 2007.
[49] Chen G. An exfoliated polyamide-6/layered double hydroxide nanocomposite via in situ intercalative polymerization between positively charged inorganic platelets. J Nanosci Nanotechnol 2006;6:1155–8.
[50] Wang GA, Wang CC, Chen CY. The disorderly exfoliated LDHs/PMMA nanocomposite synthesized by in situ bulk polymerization. Polymer 2005;46:5065–74.
[51] Moujahid EM, Besse JP, Leroux F. Poly(styrene sulfonate) layered double hydroxide nanocomposites. Stability and subsequent structural transformation with changes in temperature. J Mater Chem 2003;13:258–64.

[52] Chen W, Qu B. Structural characteristics and thermal properties of PE-g-A/MgAl-LDH exfoliation nanocomposites synthesized by solution intercalation. Chem Mater 2003;15:3208–13.
[53] Hsueh HB, Chen CY. Preparation and properties of LDHs/epoxy nanocomposites. Polymer 2003;44:5275–83.
[54] Zammarano M, Franceschi M, Bellayer S, Gilman GW, Meriani S. Preparation and flame resistance properties of revolutionary self-extinguishing epoxy nanocomposites based on layered double hydroxides. Polymer 2005;46:9314–28.
[55] Du L, Qu B, Meng Y, Zhu Q. Structural characterization and thermal and mechanical properties of poly(propylene carbonate)/MgAl-LDH exfoliation nanocomposite via solution intercalation. Compos Sci Technol 2006;66:913–8.
[56] Du L, Qu B. Structural characterization and thermal oxidation properties of LLDPE/MgAl-LDH nanocomposites. J Mater Chem 2006;16:1549–54.
[57] Lee WD, Im SS, Lim HM, Kim K. Preparation and properties of layered double hydroxide/poly(ethylene terephthalate) nanocomposites by direct melt compounding. Polymer 2006;47:1364–71.
[58] Zammarano M, Bellayer S, Gilman JW, Franceschi M, Beyer FL, Harris RH, et al. Delamination of organo-modified layered double hydroxides in polyamide 6 by melt processing. Polymer 2006;47:652–62.
[59] Kotal M, Srivastava SK, Manu SK, Saxena AK, Pandey KN. Preparation and properties of in-situ polymerized polyurethane/stearate intercalated layer double hydroxide nanocomposites. Polym Int 2013;62:728–35.
[60] Banerjee S, Benny TH, Wong SS. Covalent surface chemistry of single walled carbon nanotubes. Adv Mater 2005;17:17–29.
[61] Chen J, Hamon MA, Hu H, Chen YS, Rao AM, Eklund PC, et al. Solution properties of single-walled carbon nanotubes. Science 1998;282:95–8.
[62] Dyke CA, Tour JM. Solvent-free functionalization of carbon nanotubes. J Am Chem Soc 2003;125:1156–7.
[63] Pradhan B, Srivastava SK, Ananthakrishnan R, Saxena AK. Preparation and characterization of exfoliated layered double hydroxide/silicone rubber nanocomposites. J Appl Polym Sci 2011;119:343–51.
[64] Tang Y, Hu Y, Wang S, Gui Z, Chen Z, Fan W. Preparation and flammability of ethylene-vinyl acetate copolymer/montmorillonite nanocomposites. Polym Degrad Stab 2002;78:555–9.
[65] Hull TR, Price D, Liu Y, Wills CL, Brady J. An investigation into the decomposition and burning behaviour of ethylene-vinyl acetate copolymer nanocomposite materials. Polym Degrad Stab 2003;82:365–71.
[66] Morgan AB, Wilkie CA. Flame retardant polymer nanocomposites. Wiley Interscience; 2007, ISBN: 978-0-471-73426-0.
[67] Wilkie CA, Morgan AB. Fire retardancy of polymeric materials. CRC Press; 2009, ISBN: 978-1-4200-8400-9.
[68] Wanga X, Rathore R, Songtipya P, Jimenez-Gascod MM, Maniasc E, Wilkie CA. EVA-layered double hydroxide (nano) composites: mechanism of fire retardancy. Polym Degrad Stab 2011;96:301–13.
[69] Beyer G. Flame retardant properties of EVA-nanocomposites and improvements by combination of nanofillers with aluminium trihydrate. Fire Mater 2001;25:193–7.

[70] Jiao C, Wang Z, Chen X, Yu B, Hu Y. Irradiation crosslinking and halogen-free flame retardation of EVA using hydrotalcite and red phosphorus. Radiat Chem Phys 2006;75: 557–63.

[71] Zhang J, Hereid J, Hagen M, Bakirtzis D, Delichatsios MA, Fina A, et al. Effects of nano clay and fire retardants on fire retardancy of a polymer blend of EVA and LDPE. Fire Saf J 2009;44:504–13.

[72] Shi Y, Kashiwagi Y, Walters RN, Gilman JW, Lyon RE, Sogah DY. Ethylene vinyl acetate/layered silicate nanocomposites prepared by a surfactant free method: enhanced flame retardant and mechanical properties. Polymer 2009;50: 3478–87.

[73] Duquesne S, Jama C, Bras ML, Delobel R, Recourt P, Gloaguen JM. Elaboration of EVA-nanoclay systems—characterization, thermal behaviour and fire performance. Compos Sci Technol 2003;63:1141–8.

[74] Cai Y, Song L, He Q, Yang D, Hua Y. Preparation, thermal and flammability properties of a novel form-stable phase change materials based on high density polyethylene/poly(ethylene-co-vinyl acetate)/organophilic montmorillonite nanocomposites/paraffin compounds. Energy Convers Manage 2008;49:2055–62.

[75] Kuila T. *Preparation, characterization and properties of ethylene vinyl acetate copolymer/Mg-Al layered double hydroxide nanocomposites* [Ph.D. thesis]. India: IIT Kharagpur; 2009.

[76] Du L, Qu B, Xu Z. Flammability characteristics and synergistic effect of hydrotalcite with microencapsulated red phosphorus in halogen-free flame retardant EVA composite. Polym Degrad Stab 2006;91:995–1001.

[77] Costa FR, Leuteritz A, Wagenknecht U, Jehnichen D, Häußler L, Heinrich G. Intercalation of Mg–Al layered double hydroxide by anionic surfactants: preparation and characterization. Appl Clay Sci 2008;38:153–64.

[78] Costa FR, Saphiannikova M, Wagenknecht U, Heinrich G. Layered double hydroxide nanocomposites. Adv Polym Sci 2008;210:101–68.

[79] Zhang G, Ding P, Zhang M, Qu B. Synergistic effect of layered double hydroxide with hyperfine magnesium hydroxide in halogen-free-flame retardant EVA/HFMH/LDH nanocomposites. Polym Degrad Stab 2007;92:1715–20.

[80] Lv JP, Liu WH. Flame retardancy and mechanical properties of EVA nanocomposites based on magnesium hydroxide nanoparticles/microencapsulated red phosphorus. J Appl Polym Sci 2007;105:333–40.

[81] Gao F, Beyer B, Yuana O. A mechanistic study of fire retardancy of carbon nanotube/ethylene vinyl acetate copolymers and their clay composites. Polym Degrad Stab 2005;89:559–64.

[82] Peeterbroeck S, Alexandre M, Nagy JB, Pirlot C, Fonseca A, Moreau N, et al. Polymer-layered silicate–carbon nanotube nanocomposites: unique nanofiller synergistic effect. Compos Sci Technol 2004;64:2317–23.

[83] Peeterbroeck S, Laoutid F, Swoboda B, Lopez-Cuesta JM, Moreau N, Nagy JB, et al. How carbon nanotube crushing can improve flame retardant behaviour in polymer nanocomposites. Macromol Rapid Commun 2007;28:260–4.

[84] Peeterbroeck S, Breugelmans L, Alexandre M, Nagy JB, Viville P, Lazzaroni R, et al. The influence of the matrix polarity on the morphology and properties of ethylene vinyl acetate copolymers-carbon nanotube nanocomposites. Compos Sci Technol 2007;67:1659–65.

[85] Costache MC, Heidecker MJ, Manias E, Camino G, Frache A, Beyer G, et al. The influence of carbon nanotubes, organically modified montmorillonites and layered double hydroxides on the thermal degradation and fire retardancy of polyethylene, ethylene vinyl acetate copolymer and polystyrene. Polymer 2007;48:6532–45.
[86] Ye L, Wu Q, Qu B. Synergistic effects and mechanism of multiwalled carbon nanotubes with magnesium hydroxide in halogen-free flame retardant EVA/MH/MWNT nanocomposites. Polym Degrad Stab 2009;94:751–6.
[87] Bhowmick AK, Stephens HL. Handbook of elastomer. 2nd ed. New York: Marcel Dekker; 2001.
[88] Acharya H. *Synthesis, characterization and properties of polyolefinic elastomer nanocomposites* [Ph.D. thesis]. India: I.I.T. Kharagpur; 2007.
[89] Chang ZH, Guo F, Chen JF, Yu JH, Wang GQ. Synergistic flame retardant effects of nano-kaolin and nano-HAO on LDPE/EPDM composites. Polym Degrad Stab 2007;92:1204–12.
[90] Song G, Ma S, Tang G, Yin Z, Wang X. Preparation and characterization of flame retardant form-stable phase change materials composed by EPDM, paraffin and nano magnesium hydroxide. Energy 2010;35:2179–83.
[91] Kang D, Kim D, Yoon SH, Kim D, Barry C, Mead J. Properties and dispersion of EPDM/modified-organoclay nanocomposites. Macromol Mater Eng 2007;292:329–38.
[92] Hepburn C. Polyurethane elastomers. England: Applied Science Publishers Ltd.; 1982.
[93] Madbouly Samy A, Otaigbe Joshua U. Recent advances in synthesis, characterization and rheological properties of polyurethanes and POSS/polyurethane nanocomposites dispersions and films. Prog Polym Sci 2009;34:1283–332.
[94] Kotal M, Srivastava SK, Paramanik B. Enhancements in conductivity and thermal stabilities of polypyrrole/polyurethane nanoblends. J Phys Chem C 2011;115:1496–505.
[95] Kotal M, Srivastava SK, Bhowmick AK, Chakraborty SK. Morphology and properties of stearate-intercalated layered double hydroxide nanoplatelet-reinforced thermoplastic polyurethane. Polym Int 2011;60:772–80.
[96] Kotal M, Kuila T, Srivastava SK, Bhowmick AK. Synthesis and characterization of polyurethane/Mg-Al layered double hydroxide nanocomposites. J Appl Polym Sci 2009;114:2691–9.
[97] Kotal M, Srivastava SK, Manu SK. Layered double hydroxide as nanofiller in the development of polyurethane nanocomposites. J Nanosci Nanotechnol 2010;10:5730–40.
[98] Chattopadhyay DK, Raju KVSN. Structural engineering of polyurethane coatings for high performance application. Prog Polym Sci 2007;32:352–418.
[99] Kotal M, Srivastava SK, Bhowmick AK. Thermoplastic polyurethane and nitrile butadiene rubber blends with layered double hydroxide nanocomposites by solution blending. Polym Int 2010;59:2–10.
[100] Nyambo C, Kandare E, Wilkie CA. Thermal stability and flammability characteristics of ethylene vinyl acetate (EVA) composites blended with a phenyl phosphonate-intercalated layered double hydroxide (LDH), melamine polyphosphate and/or boric acid. Polym Degrad Stab 2009;94:513–20.
[101] Zhao CX, Liu Y, Wang DY, Wang DL, Wang YZ. Synergistic effect of ammonium polyphosphate and layered double hydroxide on flame retardant properties of poly(vinyl alcohol). Polym Degrad Stab 2008;93:1323–31.

[102] Alcon MJ, Ribera G, Galia M, Cadiz V. Synthesis, characterization and polymerization of isobutylbis(glycidylpropylether) phosphine oxide. Polymer 2003;44:7291–8.

[103] Liaw D-J, Lin S-P. Phosphorus-containing polyurethanes based on bisphenol-AF, prepared by N-akylation. J Appl Polym Sci 1996;61:315–9.

[104] Hosgor Z, Kayaman-Apohan N, Karatas S, Menceloglu Y, Gungor A. Preparation and characterization of phosphine oxide based polyurethane/silica nanocomposite via non-isocyanate route. Prog Org Coat 2010;69:366–75.

[105] Yildiz B, Seydibeyoglu MO, Guner FS. Polyurethane-zinc borate composites with high oxidative stability and flame retardancy. Polym Degrad Stab 2009;94:1072–5.

[106] Gao X, Guo Y, Tian Y, Li S, Zhou S, Wang Z. Synthesis and characterization of polyurethane/zinc borate nanocomposites. Colloids Surf A 2011;384:2–8.

[107] Berta M, Lindsay C, Pans G, Camino G. Effect of chemical structure on combustion and thermal behaviour of polyurethane elastomer layered silicate nanocomposites. Polym Degrad Stab 2006;91:1179–91.

[108] Song L, Hu Y, Tang Y, Zhang R, Chen Z, Fan W. Study on the properties of flame retardant polyurethane/organoclay nanocomposite. Polym Degrad Stab 2005;87:111–6.

[109] Huang G, Gao J, Li Y, Han L, Wang X. Functionalizing nano-montmorillonites by modified with intumescent flame retardant: preparation and application in polyurethane. Polym Degrad Stab 2010;95:245–53.

[110] Devaux E, Rochery M, Bourbigot S. Polyurethane/Clay and polyurethane/POSS nanocomposites as fame retarded coating for polyester and cotton fabrics. Fire Mater 2002;26:149–54.

[111] Bourbigot S, Turf T, Bellayer S, Duquesne S. Polyhedral oligomeric silsesquioxane as flame retardant for thermoplastic polyurethane. Polym Degrad Stab 2009;94:1230–7.

[112] Bourbigot S, Samyn F, Turf T, Duquesne S. Nanomorphology and reaction to fire of polyurethane and polyamide nanocomposites containing flame retardants. Polym Degrad Stab 2010;95:320–6.

[113] Quan H, Zhang B, Zhao Q, Yuen RKK, Li RKY. Facile preparation and thermal degradation studies of graphite nanoplatelets (GNPs) filled thermoplastic polyurethane (TPU) nanocomposites. Compos Part A 2009;40:1506–13.

[114] Deka H, Karak N. Influence of highly branched poly(amido amine) on the properties of hyperbranched polyurethane/clay nanocomposites. Mater Chem Phys 2010;124:120–8.

[115] Lei Z, Yang Q, Wu S, Song X. Reinforcement of polyurethane/epoxy interpenetrating network nanocomposites with an organically modified palygorskite. J Appl Polym Sci 2009;111:3150–62.

[116] Berta M, Saiani A, Lindsay C, Gunaratne R. Effect of cay dispersion on the rheological properties and flammability of polyurethane-clay nanocomposite elastomers. J Appl Polym Sci 2009;112:2847–53.

[117] Koerner G, Schulze M, Weis J. Silicones, chemistry and technology. Boca Raton: CRC Press; 1991.

[118] Noll W. Chemistry and technology of silicones. New York: Academic Press; 1968.

[119] Tomanek A. Silicones and Industry: a compendium for practical use, instruction and reference. Munich: Wacker-Chemie; 1991.

[120] Lopez LM, Cosgrove AB, Hernandez-Ortiz JP, Osswald TA. Modeling the vulcanization reaction of silicone rubber. Polym Eng Sci 2007;47:675–83.

[121] Changrok Y, Jonghyeok L, Daesuk B, Jongpil W, Ilyoung J, Wooyoung P. Preparation and characterization of fire-resistant silicone polymer composites containing inorganic flame retardants. Elastomers Compos 2010;45:87–93.
[122] Hamdani S, Longuet C, Lopez-Cuesta JM, Ganachaud F. Calcium and aluminium-based fillers as flame-retardant additives in silicone matrices. I. Blend preparation and thermal properties. Polym Degrad Stab 2010;95:1911–9.
[123] Dong-jing L, Jun-ming Q, Zu-min Q. Preparation and properties of flame-retardant silicone rubber materials. Gongkeban 2009;31:317–20.
[124] Timofeyeva SV, Osipov A Ye, Helevina OG. Materials with law fire hazard parameters based on hardened liquid siloxane rubbers. Pozharovzryvobezopasnost 2009;18: 25–30.
[125] Pradhan B, Srivastava SK, Bhowmick AK, Saxena A. Effect of bilayered stearate ion-modified Mg–Al layered double hydroxide on the thermal and mechanical properties of silicone rubber nanocomposites. Polym Int 2012;61:458–65.
[126] Pradhan B, Srivastava SK, Saxena A. Mechanical and thermal properties of silane grafted organomodified montmorillonite reinforced silicone rubber nanocomposites. J Nanosci and Nanotechnol 2012;12:8975–84.
[127] Pradhan B, Srivastava SK. Novel silicone rubber/layered double hydroxide nanocomposite: preparation and characterisation. In: Giri PK, et al., editors. Advanced nanomaterials and nanotechnology, Springer proceedings in physics 143. Berlin Heidelberg: Springer-Verlag; 2013. http://dx.doi.org/10.1007/978-3-642-34216-5_36.
[128] Hayashida K, Tsuge S, Ohtani H. Flame retardant mechanism of polydimethylsiloxane material containing platinum compound studied by analytical pyrolysis techniques and alkaline hydrolysis gas chromatography. Polymer 2003;44:5611–6.
[129] Mansouri J, Burfordy RP, Cheng YB, Hanu L. Formation of strong ceramified ash from silicone-based compositions. J Mater Sci 2005;40:5741–9.
[130] Hanu LG, Simon GP, Cheng YB. Thermal stability and flammability of silicone polymer composites. Polym Degrad Stab 2006;91:1373–9.
[131] Connell JE, Edwin M, Thomas MJK. Silicate-siloxane fire retardant composites. Polym Int 2000;49:1092–4.
[132] Genovese A, Shanks RA. Fire performance of poly(dimethyl siloxane) composites evaluated by cone calorimetry. Compos Part A 2008;39:398–405.
[133] Hermansson A, Hjertberg T, Sultan BA. The flame retardant mechanism of polyolefins modified with chalk and silicone elastomer. Fire Mater 2003;27:51–70.
[134] Yang L, Hu Y, Lu H, Song L. Morphology, thermal, and mechanical properties of flame-retardant silicone rubber/montmorillonite nanocomposites. J Appl Polym Sci 2006;99:3275–80.
[135] Davidson NS, Wilkinson K. *Flame retardant polymer composition*, European Patent 393,959; 1990.
[136] Andreasson U, Fagrell O, Sultan BA. *Flame retardant polymer composition*, European Patent 1,396,865; 2004.
[137] Hecker KC, Fruzzetti RE, Sinclair EA. Flammability and smoke properties. Rubber Age (N.Y.) 1973;105:25–32.
[138] Yang CP, Chen WT. Effects of brominated flame retardants and cross-linking agents on the flame retardancy of rubbers. J Appl Polym Sci 1988;36:963–78.
[139] Mishra S, Shimpi NG. Mechanical and flame retarding properties of styrene-butadiene rubber filled with nano-$CaCO_3$ as a filler and linseed oil as an extender. J Appl Polym Sci 2005;98:2563–71.

[140] Ushkov VA, Kulev D Kh, Tsigel'naya TI, Korotkevich S Kh. Ignitability and smoke generation capacity of butadiene-styrene rubber vulcanizates. Kauchuk i Rezina 1988;11:40–2.
[141] Kawaguchi T, Sato S, Fukushima Y. Flame retardant pressure-sensitive adhesive tapes for television parts and deflection yoke coil. U.S. Pat Appl Publ; 2004. US 20040126574 A1 20040701.
[142] Price RL, Lundy CE, Krishnan S. Flame-retardant impact resistant polycarbonate compositions. U.S; 1992. US 5109045 A 19920428.
[143] Grazyna J, Agnieszka K-J, Przemysław R, Dorota W, Iwona W. Flammability of diene rubbers. J Therm Anal Calorim 2010;102:1043–9.
[144] Khattab MA. Decabromobiphenyl oxide-aluminum hydroxide system as a flame retardant for styrene-butadiene rubber. J Appl Polym Sci 2000;78:2134–9.
[145] Zhang H, Wang Y, Wu Y, Zhang L, Yang J. Study on flammabililty of montmorillonite/styrene-butadiene rubber (SBR) nanocomposites. J Appl Polym Sci 2005;97:844–9.
[146] Mishra S, Shimpi NG, Patil UD. Effect of nano $CaCO_3$ on thermal properties of styrene butadiene rubber (SBR). J Polym Res 2007;14:449–59.
[147] Manley TR, Hampson FW. Flammability of polymeric products at various temperatures. Plast Rub Proc Appl 1983;3:13–6.
[148] Janowska G, Kucharska-Jastrzabek A. The effect of chlorosulphonated polyethylene on thermal properties and combustibility of butadiene-styrene rubber. J Therm Anal Calorim 2010;101:1093–9.
[149] Ramesan MT. Thermogravimetric analysis, flammability and oil resistance properties in natural rubber and dichlorocarbene modified styrene butadiene rubber blends. React Funct Polym 2004;59:267–74.
[150] Moon SC, Jo BW, Farris RJ. Flame resistance and foaming properties of NBR compounds with halogen-free flame retardants. Polym Compos 2009;30:1732–42.
[151] Rybinski P, Janowska G, Slusarski L. Influence of cryogenic modification of silica on thermal properties and flammability of cross-linked nitrile rubber. J Therm Anal Calorim 2010;101:665–70.
[152] Chrusciel J, Janowska G, Rybinski P, Slusarski L. Effect of the top layer modification of polymers on their thermostability and flammability. J Therm Anal Calorim 2006;84:339–44.
[153] Janowska G, Jastrzabek AK, Rybinski P. Thermal stability, flammability and fire hazard of butadiene-acrylonitrile rubber nanocomposites. J Therm Anal Calorim 2011;103:1039–46.
[154] Wu YP, Jia QX, Yu DS, Zhang LQ. Structure and properties of nitrile rubber (NBR) – clay nanocomposites by co-coagulating NBR latex and clay aqueous suspension. J Appl Polym Sci 2003;89:3855–8.
[155] Derouet D, Radhakrishnan N, Brosse JC, Boccaccio G. Phosphorus modification of epoxidized liquid natural rubber to improve flame resistance of vulcanized rubbers. J Appl Polym Sci 1994;52:1309–16.
[156] Menon ARR. Flame-retardant characteristics of natural rubber modified with a bromo derivative of phosphorylated cashew nut shell liquid. J Fire Sci 1997;15:3–13.
[157] Morton M. Rubber technology. New York: Van Nostrand Reinhold; 1987.
[158] Ismawi DHA, Harper JF, Ansarifar A. Influence of flame retardant additives on the flammability behaviour of natural rubber (NR). J Rubber Res (Kuala Lumpur, Malays) 2008;11:223–36.

CHAPTER

19 Self-extinguishing Polymer Blends Containing Organoclays

S. Pack, M.H. Rafailovich

Department of Materials Science and Engineering, Stony Brook University, Stony Brook, New York, USA

Marcia Simon

Professor and Director for Graduate Studies, Department of Oral Biology and Pathology School of Dental Medicine, Stony Brook University, Stony Brook, NY

CHAPTER OUTLINE

1. Introduction

For the past two decades, since it was first reported that the inorganic fillers could be incorporated into a nylon-6 matrix at the molecular level, polymer–clay nanocomposites have been extensively studied in both academic and industrial laboratories. Surprisingly, the nanocomposites exhibited superior material properties compared to the polymer itself, such as mechanical strength and modulus, as well as heat distortion temperature [1,2] and gas impermeability [3,4]. It was also established that the degree of exfoliation in the polymer matrices, which in turn depended on the interfacial interactions between the silicates and the polymer matrix, determined the extent to which the properties were optimized. A high degree of exfoliation was easily achieved when polymerization occurred from the surface of the silicates [1,2], but exfoliation

Polymer Green Flame Retardants. http://dx.doi.org/10.1016/B978-0-444-53808-6.00019-6

when silicates were extruded with polymers proved more challenging. One approach that has become very popular in recent years has been to modify the surface energy of clays via organic groups attached to alkylammonium ions which, cation-exchange with sodium ions present in the galleries of the clays [5,6]. In particular, it was shown that quaternary alkylammonium ions with ditallow hydrogenated molecules could be used to produce layered silicates, where the surface interactions could be tuned from hydrophobic through hydrophilic, to match those of the polymer matrix and achieve exfoliation in a broader array of polymers [7−10]. Later it was also demonstrated that when these functionalized silicates were added to immiscible polymer blends, chains of all the constituent polymers were able to absorb to the clay surfaces [11−14]. Owing to the high aspect ratio of clays platelet, polymer chains have potential to obtain physical contacts onto clay surfaces thereby reducing their surface energy. Recently, Si et al. [14] demonstrated that Cloisite 20A clays were segregated into interfaces between the immiscible polymer blends, forming in situ grafts with the polymers. Thus, the reduced interfacial energy at the interfaces could result in an improvement of mechanical and thermal ability properties.

Despite the advantages of using organophilic silicates, the manufacturing process has raised potential concerns since several toxic solvents are involved in mass production of the functionalized clays, which cause problems both in handling and disposal of the materials [15−17]. Moreover, the organophilic silicates tend to decompose at relatively low temperatures, which adversely impact the thermal stability of the nanocomposites and requires strict temperature control in the manufacturing process. Recently, Pack et al. [18] proposed a new technique for achieving exfoliation, which involves the adsorption of resorcinol bis(diphenyl phosphate) (RDP) into the clay galleries. This small molecule liquid was shown to intercalate effectively in the galleries of the clay, allowing them to be easily exfoliated under shear when mixed with polymers. RDP is nontoxic, thermally stable to 350 °C, and is easily degraded [19,20]. RDP-coated clays were shown to effectively rival the ditallow functionalized clays in effectiveness in nearly all applications when extruded together with homopolymers or polymer blends. The RDP clays were exfoliated in most polymers, including styrene and ethylene polymers, where the no ditallow coating formulation was effective [18]. In this chapter, we review possibilities of the phosphorus-based flame retardant (PFR) coated silicates to improve compatibilization, mechanical properties, and flame retardancy of polymer blends, which are not only immiscible but also biodegradable. We also demonstrate that these clays are not cytotoxic and in have been found effective in several biomedical applications.

2. PFR-coated aluminumsilicates

The addition of organophilic silicates is a good strategy to increase compatibilization of polymer blends and improve flame retardant (FR) properties. Si et al. [14] showed that the Cloisite clays treated with the ditallow surfactants could prevent further phase segregation because the clays favorably segregated to the interfaces between the immiscible polymer blends. They also reported that the addition of halogenated flame retardants (HFR) could render the homopolymer clay nanocomposites self-extinguishing, improving the degree of dispersion of the HFR particles [21]. Later, Pack et al. [22,23] developed the strategy in a complex system, where either the clays alone or a combination of the clays with carbon nanotubes (CNTs) was mixed with the HFR particles in the immiscible polymer blends. They found that the clays could be entangled with the CNTs as well as HFR particles at the interfaces in the immiscible polymer blends. Hence the addition of organoclays in the blends could enhance compatibility and thereby improve FR properties.

However, the HFR particles were recently phased out as new studies indicated that halogenated compounds were accumulating in living organisms, soil, ground water, and streams [24,25]. This environmental concern has led to seeking out alternatives. Nonhalogenated FR have been exclusively developed either as an additive or through in situ polymerization with polymers in order to satisfy the demand of environmental issues [26,27]. Pack et al. [18] recently demonstrated that resorcinol bis(diphenyl phosphate) (RDP) could be a good candidate as a surfactant to clays' surface as well as FR itself. The chemical structure of RDP is shown in Figure 1.

As we can see, the RDP has nonpolar and polar properties; the phenol groups are considered as nonpolar moieties, while the phosphoric acid groups can act as polar moieties. On the other hand, Koo et al. [28] reported that the ditallow molecules were not distributed uniformly on the clay's surface. Rather, the surfactants were formed as a patchy area on the clay surfaces. They proposed that the kind of patchy surfactants could render the organoclays as nonspecific fillers, which compatibilize immiscible polymer blends since the clays' surface had polar and nonpolar properties.

Considering the above, in order to study the properties of the clays, dispersions were formed in volatile solvents, which were then spread at the air—water interface of a Langmuir—Blodgett (LB) trough. A monolayer of the clays was then lifted onto silicon substrates and analyzed by various complementary techniques. In Figure 2, we show scanning force microscope topography and lateral force scans of the clay layer surfaces, from which one can see, the clay platelets are completely covered by the RDP, which is about 2.2 nm in thickness. The absorption of RDP molecules was confirmed in the X-ray scattering spectra of the clays, where diffraction peaks appear at 0.27 and 0.55 $\mathrm{\AA}^{-1}$, from which we obtain in the direction of [0,0,1] a value

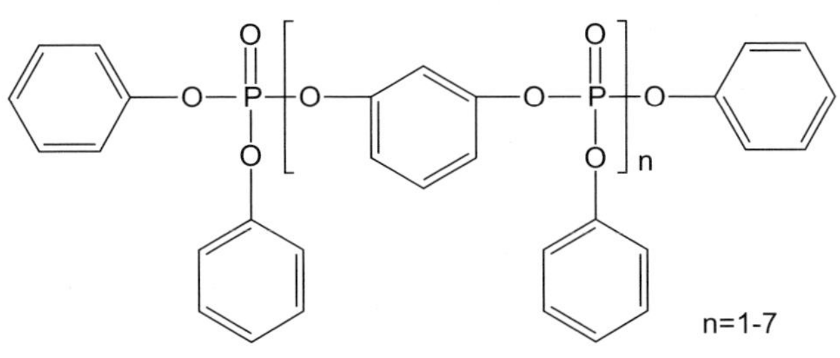

FIGURE 1

Chemical structure of resorcinol bis(diphenyl phosphate) (RDP) [18].

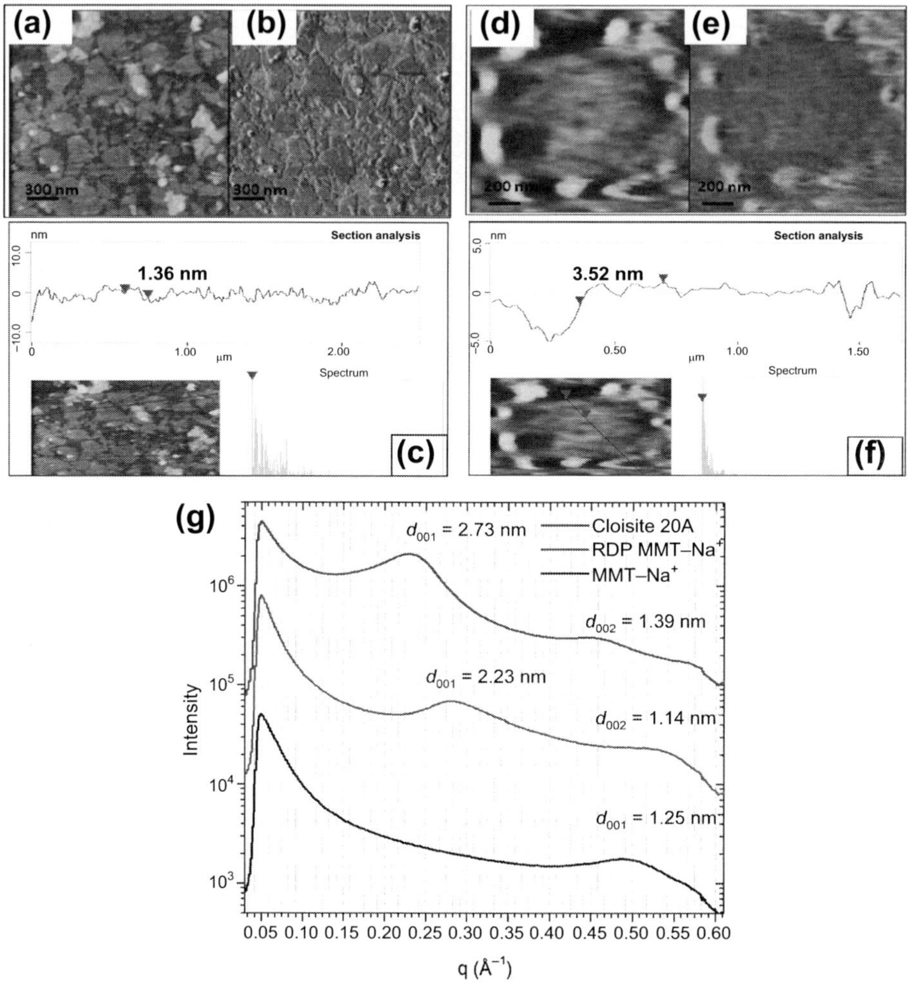

FIGURE 2

AFM images of uncoated MMT—Na^+ (a) Height and (b) friction, and (c) cross section of uncoated MMT—Na^+, AFM images of RDP-coated MMT—Na^+ (d) Height and (e) friction and (f) cross section of RDP-coated MMT—Na^+, (g) the SAXS of nanoclays [18].

for interlayer d-spacing of 2.23 nm which is consistent with the AFM results. Thus, the RDP is well intercalated in the clay galleries and is uniformly distributed on the clay platelets.

2.1 Homopolymer RDP-coated MMT—Na^+ nanocomposites

Since the RDP were well intercalated into the clay galleries, it could be exfoliated in Poly(methyl methacrylate) (PMMA) matrix. A cross section of PMMA/RDP-coated Montmorillonite (MMT)—Na^+ nanocomposite is shown in Figure 3. The TEM

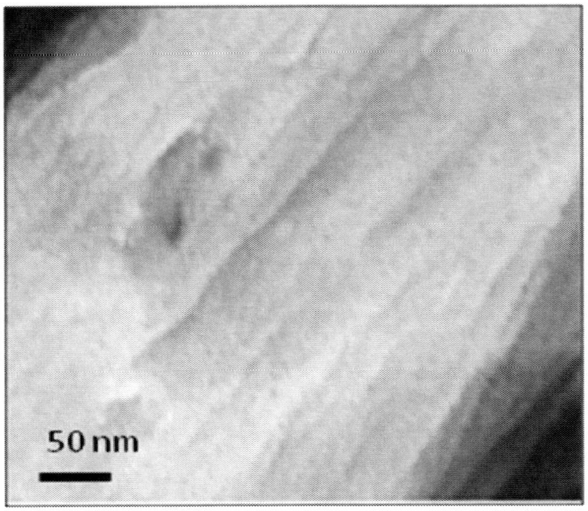

FIGURE 3

A cross section of TEM image of a PMMA/RDP-coated MMT—Na^+ (95/5 wt%) nanocomposite [18].

image shows that individual clay platelets are aligned in the normal direction of melt shearing. However, the Cloisite 20A is generally not compatible to styrenic polymers, such as polystyrene, unless in situ polymerization occurs with the monomers [7,29]. Pack et al. showed that the RDP-coated MMT—Na^+ could be fully exfoliated a polystyrene matrix in melt-blending under shear.

In Figure 4, as we can see that the exfoliated RDP-coated clays are well dispersed in a high impact polystyrene (HIPS) matrix, where no peak appears in the Small angle X-ray scattering (SAXS) spectrum, while the identified two peaks occur in the X-ray spectrum of HIPS/Closite 20A nanocomposite. This difference between the RDP-coated clays and Cloisite 20A can lead to a different morphology in TEM. As we can see in the figure, Cloisite 20A is highly intercalated in the polystyrene matrix. Compared to the Cloisite 20A, the RDP-coated clays are highly exfoliated in the same matrix. Subsequently, the different degree of compatibilization between the two organoclays in the PS matrix could lead to different storage modulus with temperature. In Figure 4(b), the storage modulus of HIPS/RDP-coated clays nanocomposite is higher than that of neat HIPS. Additionally, the modulus of the RDP-coated clay is almost two times of that of Cloisite 20A in the polymer matrix. Both the enhanced modulus and improved compatibility could be contributed to discrepancy of surface energy between the clays and polymers.

Contact angle measurements were conducted to explain the relationship between the polymer and RDP-coated clays in terms of the interfacial energy. In order to compare affinity of polymers to either the RDP-coated clays or the Cloisite 20A clays, first Polystyrene (PS) or PMMA monolayers were spun cast on glass cover slips and then the films were floated on top of the LB. The samples were annealed in oil free vacuum at $T = 180\ ^{\circ}C$, above Tg of the polymers, in order to allow the

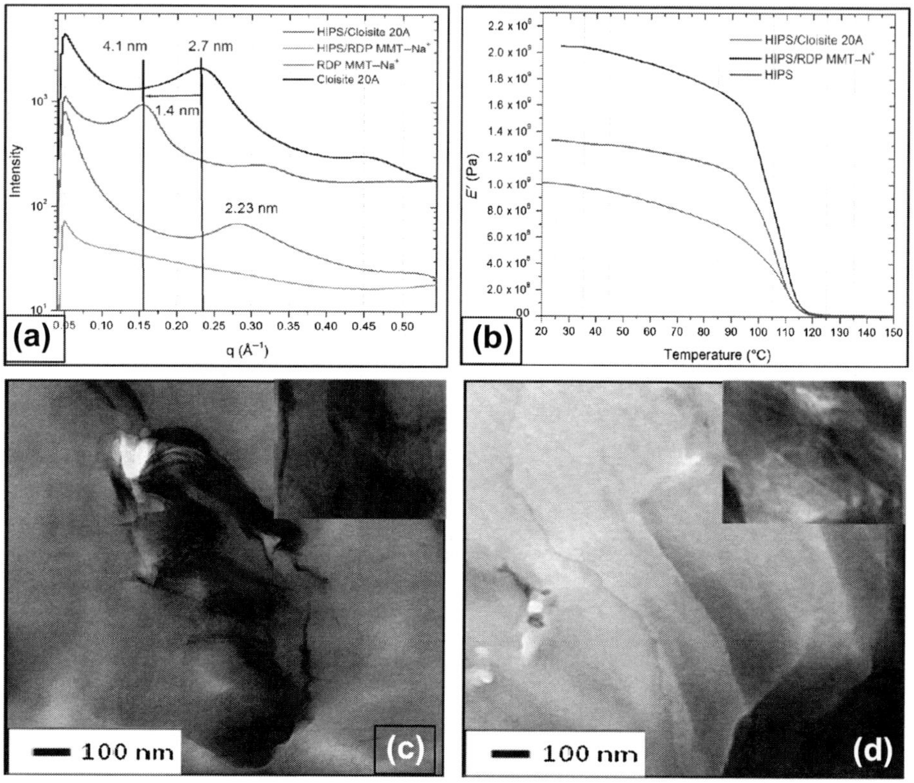

FIGURE 4

(a) SAXS spectra of HIPS/nanoclays nanocomposites; (b) storage modulus and tan δ of HIPS/RDP clays nanocomposite (the inserted image). TEM images of HIPS/nanoclays composites: (c) HIPS/Cloisite 20A, highly intercalated; and (d) HIPS/RDP MMT–Na^+, exfoliated [18].

chains to migrate and the system to reach equilibrium. Since the contact angle of the RDP remained relatively constant after an additional 24 hours of annealing, the system was assumed to be at equilibrium when the AFM measurements were performed. In Figure 5, we can see the contact angle of PS on RDP-coated clays is 7.7°, while the angle of the polymer on Cloisite 20A is 32. Furthermore, it can be seen that the contact angle of PMMA films on Cloisite 20A is about 2.5°, and on the RDP-coated clays it is 0.32°, or wetting occurs by the PMMA on the RDP-coated clays.

From the value of the contact angle, we can estimate the interfacial energy between the polymers and coated clays via Young's contact angle relation. In this case, surfactants are strongly absorbed onto the clay surfaces and are lifted onto the clay surfaces where they form very thin layers [18,30,31]. The Young's relationship with contact angle is,

$$\gamma_A = \gamma_B \cos\theta + \gamma_{A/B} \tag{1}$$

where γ_A and γ_B are the surface tension of the A and B phases, and $\gamma_{A/B}$ is the interfacial tension between the A and B phases.

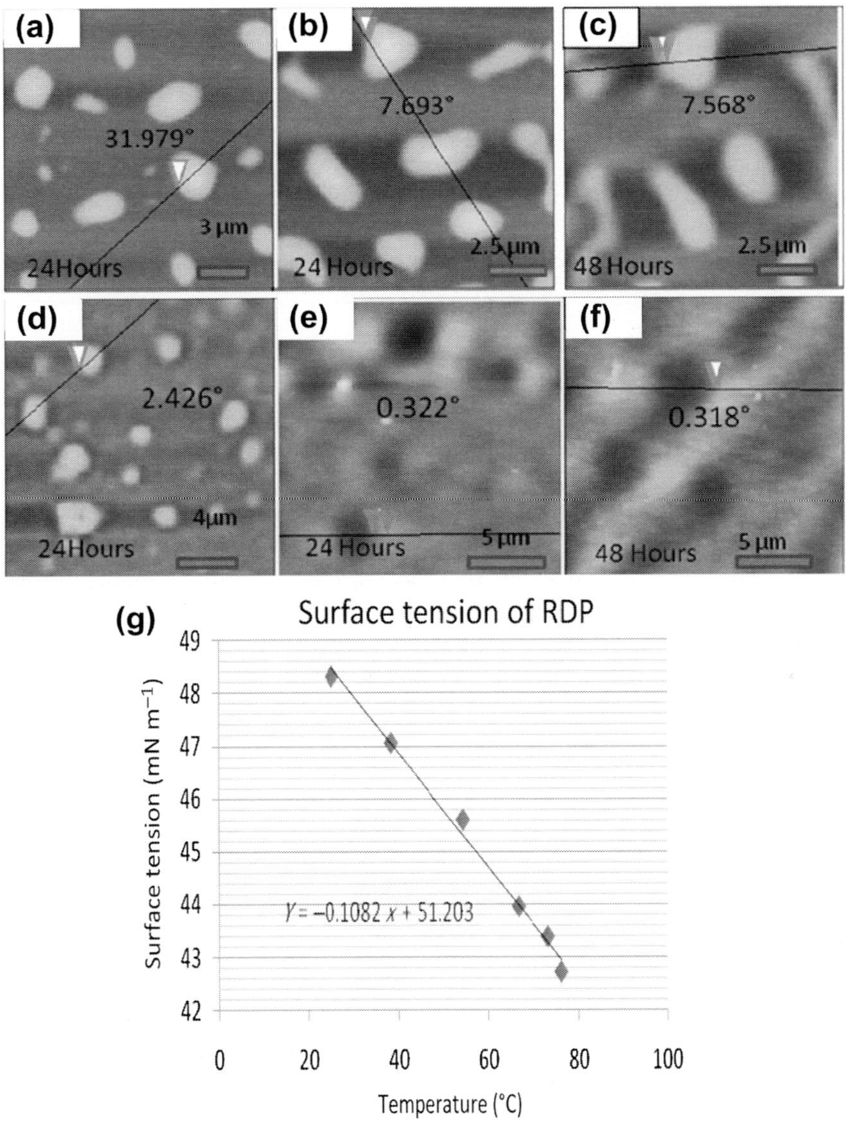

FIGURE 5

The contact angles of polymers (thin films) clays. (a) PS/Cloisite 20A and (b) PS/RDP MMT—Na^+ after 24 h annealing, and (c) PS/RDP MMT—Na^+ after 48 h annealing. (d) PMMA/Cloisite 20A and (e) PMMA/RDP MMT—Na^+ after 24 h annealing, and (f) PMMA/RDP MMT—Na^+ after 48 h annealing; and surface tension of RDP with temperature. (g): the surface tension of RDP as a function of temperature [18].

In order to determine the interfacial energy between polymers and RDP, the surface tension of RDP could be first measured by the Wilhelmy plate method. In Figure 5, the values of γ_{rdp} are shown as a function of temperature. Since the melt-blending process mainly operates at about 180 °C, we can estimate the surface tension of RDP at the temperature from the graph, which is 31.7 mN/m. On the other hand, the surface tensions of either PS or PMMA are nearly the same ($\gamma \sim 29$ mN/m) [30]. Therefore, from the Young's relationship, $\gamma_{\text{rdp/pmma}}$ is 2.75, while $\gamma_{\text{rdp/ps}}$ is 2.8 mN/m. Since we showed that PMMA did wet on RDP-coated clays, we could determine whether the polymers wet the RDP-coated clays through calculating the spreading coefficient *S*:

$$S = \gamma_{\text{rdp}} - \gamma_{\text{polymer}} - \gamma_{\text{rdp/polymer}} \tag{2}$$

where wetting occurs when $S > 0$. From the equation, $S = -0.25 < 0$ for PS and $S = 0-0$ for PMMA.

Therefore, in the case of PMMA, the RDP-coated clays are totally wetted by the polymers, which could lead to fully exfoliation. In contrast to the PMMA case, the spreading coefficient for PS is slightly negative thereby a high degree of exfoliation could occur, which is consistent with TEM images of HIPS/RDP-coated clays in Figure 4. However, the contact angle of PS on Cloisite 20A is relatively higher, which may indicate that the interaction between PS and Cloisite 20A is interrupted by unfavorable interfacial energy. This could lead to a large degree of intercalation, rather than exfoliation.

2.2 Immiscible polymer blends RDP-coated MMT$-$Na$^+$ nanocomposites

The addition of FR is often considered as a phase in a polymer matrix. This additive FR generally undergoes phase separation. Recently, Si et al. [21] utilization of organoclays with the halogen FR particles in PMMA matrix was achieved increasing the degree of dispersion of the FR particles. They found that the FR particles were segregated onto the clay surfaces thereby increasing the degree of dispersion of the FR particles. Later, Pack et al. [22] demonstrated that the organoclays segregated into the interfaces of polymer blends with the FR particles, without decreasing the degree of compatibilization of either PS/PMMA or Polycarbonate (PC)/Poly(styrene-co-acrylonitrile) (SAN)24 blends. In addition to the localization of the clays at interfaces, the intercalated clay structures with polymers could be stabilized until the polymers begun to degrade at about 250 °C. In spite of the improved stabilization of internal microstructures against heat treatment, there still remained the third phase in the blends. Therefore, FR had to be incorporated with either polymers or organoclays at the molecular level.

Pack et al. [18] found that the RDP-coated clays could be applied to the immiscible polymer blends. For instance, TEM images of a blend of 50% polycarbonate and 50% poly(styrene-co-acrylonitrile) (SAN24) with 5 wt% RDP-coated clays are shown in Figure 6. In the TEM images, apart from the Cloisite 20A, we can

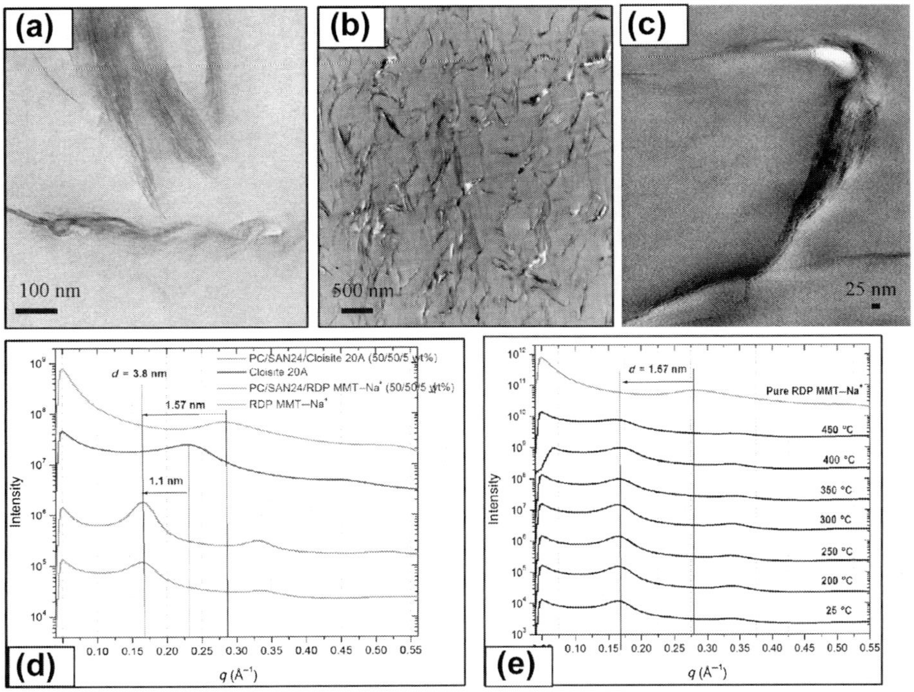

FIGURE 6

TEM images: (a) PC/SAN24/Cloisite 20A (50/50/5 wt%), (b) PC/SAN24/RDP MMT—Na^+ (50/50/5 wt%)—low magnification, and (c) PC/SAN24/RDP MMT—Na^+ (50/50/5 wt%)—high magnification. SAXS spectra: (d) PC/SAN24/RDP MMT-Na^+ and PC/SAN24/Cloisite 20A, and (e) PC/SAN24/RDP MMT—Na^+ as a function of temperature [18].

see that the RDP-coated clays are also well intercalated with polymer chains. Furthermore, the SAXS spectrum of RDP-coated clays has two identified peaks, which are relatively broad compared to the Cloisite 20A ones. This could be indicative of higher degree of dispersion of RDP-coated clays in the blend system. In particular, the intercalated microstructures of the clays with polymers were remained until 450 °C at which temperature is close to the decomposition temperature of polycarbonate. Moreover, in previous section, we show that the interfacial energy for either PMMA/RDP or PS/RDP clays is ~2.8 mN/m at 180 °C. This value is relatively high compared to that for PS/PMMA interfaces (~1.2 mN/m at 180 °C) [30]. Therefore, the addition of RDP-coated clays may not reduce the interfacial energy of the blend. Contrary, since interfacial tension for PC/RDP-coated clays is about 0.5 mN/m if $S = 0$, which is much lower than that for PC/SAN24 blend (~2.8 mN/m) [32,33], the segregation of RDP clays into the interfaces may reduce the interfacial energy of the blend significantly.

The increased compatibilization of PC/SAN24 blends when the RDP-coated clays are introduced can also result in a single peak and the improved storage

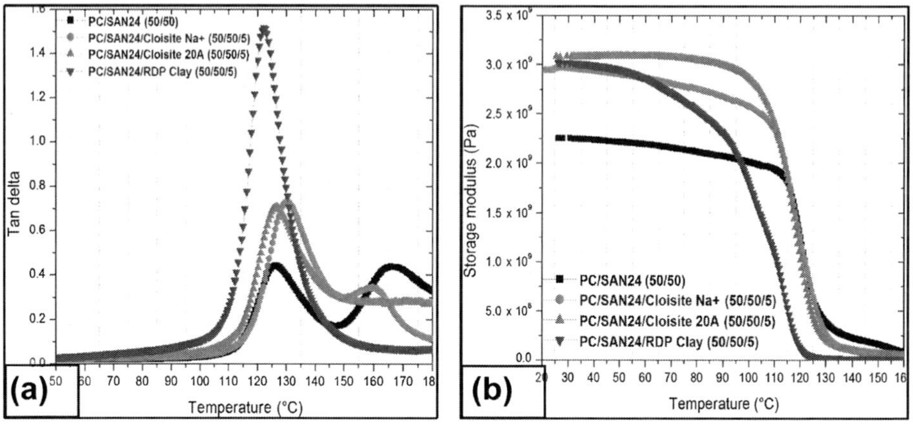

FIGURE 7

Tan delta and storage modulus of PC/SAN24 with different clays [18].

module in the dynamic mechanical properties, which is shown in Figure 7. Therefore, the addition of RDP-coated clays in polymer blends could lead to an enhancement of the material properties.

Since RDP has unique polarity to clays' surfaces, the strong absorption of RDP molecules to biodegradable polymers, such as starch, could be expected. Starch is not only inexpensive but can be completely biodegradable as well as instinctive FR. Despite its benefits, starch is brittle, and therefore difficult for the starch-based nanocomposites to be melt-blending without plasticizers [34,35]. Furthermore, very few polymers are compatible with starch. Recently, Pack et al. [36] investigated a series of blends of

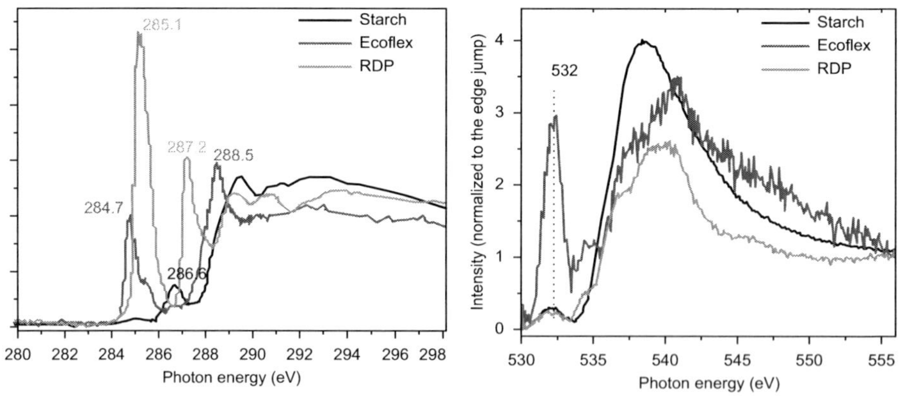

FIGURE 8

The spectra of Starch, Ecoflex, and RDP in a near-edge region: 280–310 eV is representative to carbon 1s transition, 530–570 eV is representative to oxygen 1s transition [36].

starch/poly(butyrate-adipate-terephthalate, or Ecoflex) with the phosphate oligomers as compatibilizers using a technique of near-edge X-ray absorption fine structures (NEXAFS), where starch does not have a very good electron contrast with Poly(butylenes adipate-co-teraphtalate) (PBAT). Therefore, the NEXAFS spectra of the polymer blends were used to analyze its chemical composition. In Figure 8, we can see that each spectrum of starch, Ecoflex, and RDP is very distinct to identify each other. Since Ecoflex has aromatic rings, a strong absorption peak appears at about 285 eV, which is unique compared to the peaks of starch. Moreover, these peaks can be utilized for assessing surface morphologies; dark phases have high absorption energy, while light phases have low absorption energy in the images of scanning transmission X-ray microscopy (STXM).

Accordingly, Pack et al. [36] demonstrated that the addition of RDP-coated starch could improve the degree of compatibilization of the immiscible biodegradable polymer blend. In Figure 9, we show the STXM images of a poly(lactic acid)/poly(butyrate-adipate-terephthalate) blend obtained by placing windows on the transition shown above. From the figure, we can see that the Ecoflex domains (the dark

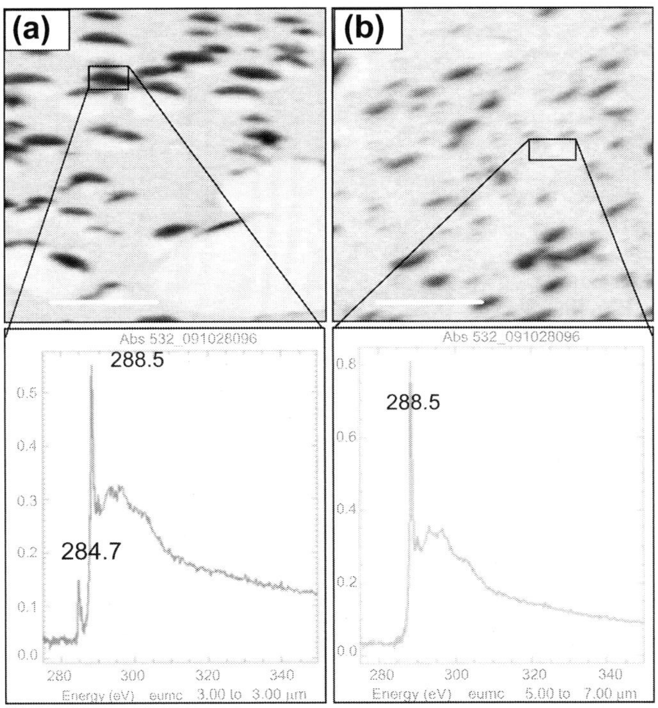

FIGURE 9

STXM images and NEXAF spectra: (a) PLA/Ecoflex/Starch (76/19/5 wt%), and (b) PLA/Ecoflex/RDP Starch (76/19/5 wt%). (The STXM images were taken at 284.7 eV and the scale bar is 5 mm [36].)

region) are aggregated in the Polylactic acid (PLA) matrix. However, the domains become smaller when the RDP-coated starch is added, while the interfaces get blunt thereby increasing the degree of dispersion of Ecoflex domains. Therefore, the improved compatibilization of the blend when the RDP-coated starch added could increase tensile properties and impact toughness, as reported in Ref. [36].

3. FR biodegradable polymers with nanoclays

Biodegradable polymers have been used in building construction, automobile, packing industries; it is increasing that more stringent flammable testing, such as UL94-V0, is required. PLA is the most biodegradable polymer being used in those industries since it has good mechanical properties and excellent transparency [37,38]. However, PLA is flammable and needs to be rendered FR before it can be used. Reti et al. [39] demonstrated that addition of both ammonium polyphosphate and either starch or lignin to PLA could achieve the vertical flammable designation. Zhan et al. [40] also showed that the addition of bisphosphorate diphosphoryl melamine had an impact on reducing flammability of PLA. Moreover, Wang et al. [37,41] developed a synthesis process, which involved attaching phosphorus oligomers to the backbone of PLA. In so doing, improved FR properties were obtained. However, it is still producing chemical wastes and involved with many synthesis steps to the polymerization of PLA. Another way to improve the flame retardancy of PLA is to blend it with starch. Owing to the fact that PLA is derived from starch, starch can be melt-blended with PLA, which is FR, where the carbonization process of the starch occurs at exposed surface from heat fronts [36,42]. However, crystallization of starch is a major drawback for melt-blending extrusion process with other polymers [43].

Interesting findings on the improvement of PLA flammability have been disclosed also by Pack et al. [18–36]. In fact Pack et al. [22,23] has published several research studies on FR polymers in the last few years. They showed that increased compatibilization of either PC/SAN or PS/PMMA blends could be achieved when small amounts of Cloisite 20A were added to the blends. Moreover, these blends with FR particles could be self-extinguishing since the clays helped increasing the degree of dispersion of the FR particles. However, aryl phosphate FR produce some problems in melt-blending extrusion processes, such as migration to the interior parts of the twin screw extruder; therefore they are difficult to be fed into the processing zones at low temperatures [44]. Pack et al. [18,36] showed a novel approach to overcome those problems: the direct absorption of resorcinol di(phenyl phosphate) (RDP) to clays' surfaces not only resulted in the reduction of the amount of the FR but also in the exfoliation of the clays in the polymer matrix. Therefore, they predicted that the absorption of RDP to either nanoparticles or starch might not only improve compatibilization for biodegradable polymer blends, such as PLA/Ecoflex, but also make the blends self-extinguishing. In Figure 10, we can see that the RDP-coated clays are well distributed and exfoliated in the Ecoflex domains. Furthermore, some of the RDP clays segregated at the interfaces between Ecoflex

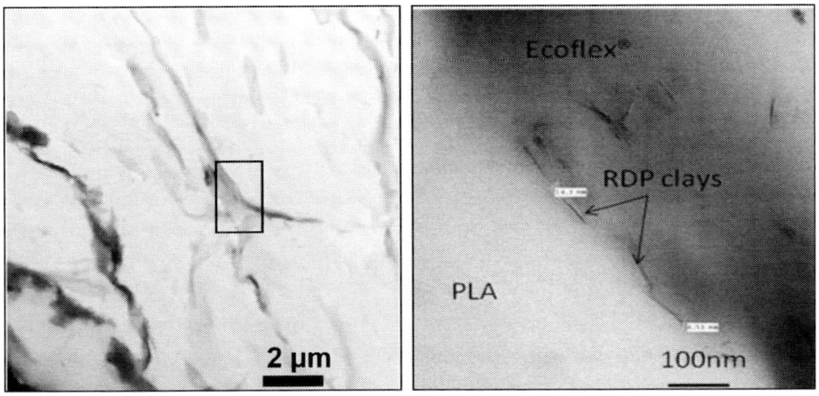

FIGURE 10

A TEM image of PLA/Ecoflex/RDP-coated clays (80/20/5 wt%): Dark phases are Ecoflex domains, and the RDP-coated clays segregated at interfaces (the scale bar is 2 μm) [36].

and PLA phases. Since Ecoflex is well known as a good ductile polymer, impact toughness of the blend may increase when 5% the RDP-coated clays are added.

However, they reported that the impact strength of the blend with RDP clays was comparable to that of the PLA/Ecoflex blend. It could be elucidated by the fact that the ductile domains became a brittle phase (which may be good for flammability of the blend). Since Ecoflex is relatively flammable compared to PLA [45], the flammability of Ecoflex may be retarded when the RDP-coated clays are added.

4. Relationship of surface energy and FR performance in immiscible polymer blends containing nanoclays

Relatively, it is easy to predict thermal responses of homopolymer clay nanocomposites since the inorganic fillers can be a major factor to produce FR performance in a fine geometry combustion process, or UL94-V0 vertical flame testing [46–53]. However, it is hard for immiscible polymer blends to achieve enhanced FR properties since microstructures of the polymer blends continuously change against heat fronts. Although intercalated and/or exfoliated clays exist in a polymer matrix, interfacial properties of polymer blends with the organoclays have to be considered in order to predict FR behaviors of the organoclays at the interfaces between the blends. Recently, Pack et al. [21] showed that when Cloisite 20A is added to PS/PMMA blends containing FR particles, the clays not only segregated into the interfaces of the blends but also facilitated the degree of distribution of the FR particles. Furthermore, the organoclays could have strong influence on the thermal response of the blends against heat fronts, where a formation of rigid layers containing clay platelets and carbonaceous chars was obtained at the exposed surface of melt polymers. This rigid char formation could act as a gas barrier to protect melt polymers from heat sources. Therefore, it is well established that the char formation occurs at a nearly exposed surface from heat fronts. The rate of mass loss from cone calorimetry has been typically used to interpret thermal behavior of polymers. In particular, charring polymers are well investigated since the mass loss rate dramatically reduces when chars are formed at an exposed surface against heat fronts. However, few research papers show that time to ignition can be correlated to the surface energy of polymer/fillers. Pack et al. [22] reported that the combination of CNTs with clays delayed the time to ignition since CNTs were entangled with clay platelets, thus increasing the heat capacity of the system. Therefore, the ignition time was enhanced.

The behavior of the time to ignition of biodegradable polymers can be comparable to that of petroleum-based polymers since most of biodegradable polymers have higher surface energy, altering thermal responses at heat-exposed surfaces. In Figure 11, the mass loss rate of two systems of biodegradable and immiscible polymer blends with different nanoparticles are presented. From the figure we can see that the addition of Closite 20A (C20A) leads to a delayed time to ignition of the starch/Ecoflex blend. The thermal response of that blend is different of that of the blend of PS/PMMA, where time to ignition of the blend was relatively increased to an earlier time [21,22]. Moreover, time to ignition of the PLA/Ecoflex blend with the addition of Cloisite Na^+ clays is almost same as that of the blend with no additives, which is shown in Figure 11(a). Since interfacial energy of each particle is different at interfaces, time to ignition may be controlled by interfacial penalty between polymers and nanoparticles.

The surface tension of C20A clays is relatively lower than that of starch [54], which may lead to segregation to either Ecoflex domains or at interfaces to reduce interfacial tension. However, the RDP-coated clays can cause an earlier ignition time because of its higher surface energy. Additionally, since PLA is derived from starch,

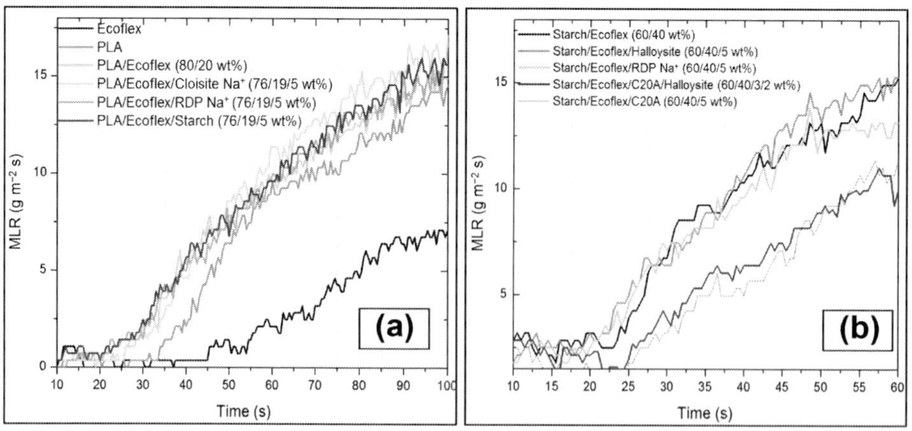

FIGURE 11

Time of ignition in the blends of (a) PLA/Ecoflex different nanoparticles and (b) Starch/Ecoflex different nanoparticles (b) within graphs of mass loss rate (MLR).

the behavior of time to ignition is similar when RDP clays are added. From Figure 10(a) we can see that when 5% RDP clays is added in PLA/Ecoflex blend, the time to ignition is earlier (about 25 s) than that of the neat polymer blend is ignited (around 30 s). Thus, the surface energy can be a crucial for the time to ignition in biodegradable polymers.

From the figure we can see that the time to ignition of the PLA is much shorter than that of Ecoflex. It is interesting to note that, in the case of the PLA/Ecoflex blend, the time to ignition is similar to that of pure PLA rather than intermediate to that of the two polymers. Since the surface energy of PLA is much lower than that of Ecoflex [45], it is reasonable to assume that the surface of the blend is enriched in the PLA component. Hence the time to ignition of all compounds is determined by the component at the surface, or the one having the lower surface energy. This can also be seen when starch is blended with Ecoflex. In Figure 10(b) we find that the time to ignition is determined by the starch content, even when RDP clay is present, Some RDP has higher surface energies than starch. On the other hand, the surface energy of neat Halloysite nanotube (HNT) is as yet unknown. On the other hand, when Cloisite 20A is added, the Cloisite becomes the component with the lowest surface energy and determines the time to ignition, even when HNT is present.

5. Mechanical properties of char formation

It is noteworthy to determine the quality of char formation since the char formation significantly plays an important role on time to ignition, mass loss rate, and heat release rate. In previous section, we explained that the exposed surface properties of biodegradable polymer blends can be governed by interfacial energy of the nanoparticles and polymers. This different surface energy can also lead to different char formation in microstructure. Pack et al. [18,22,23] reported that the flexible char layers could keep the decomposed gases from polymers before they are released throughout the porous chars surface, where surface energy of the charring polymer was lower than that of RDP-coated clays. Hence the chars formation had intermediated elastic modulus and hardness. On the other hand, since surface energy of starch (35−42 mN/m) is relatively lower than that of Ecoflex (53 mN/m) carbonized residues, formed as shell-like layers [36] starch may not be flexible to hold the volatile gases against heat fronts. Hence, the kind of chars layers could deteriorate FR properties, where the chars formation might be collapsed and thereby releasing heat and mass at a fast rate in cone calorimeter. Therefore, it is good to quantify elastic modulus and hardness of chars formation in order to understand a mechanism of flame retardancy, where nanoparticles can alter surface energy at the exposed surface. Chars formation from a series of starch/Ecoflex (60/40 wt %) blends with different nanoparticles was indented in a nanoindentation. The detail experimental method was described in ref.36. Elastic modulus and hardness are shown in Figure 12.

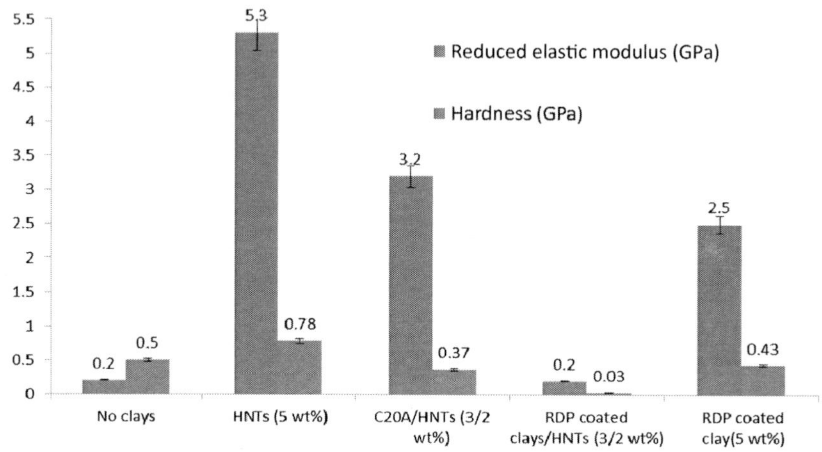

FIGURE 12

Elastic modulus and hardness of Starch/Ecoflex (60/40 wt%) blends with different nanoparticles [36].

From the figure we can see that when HNTs are added, both elastic modulus and hardness increase to 5.3 GPa and 0.78 GPa, respectively. The increased mechanical properties could be credited to good dispersion of HNTs in the polymer matrix. However, when Cloisite 20A are mixed with HNTs in the blends, both properties decrease to 3.2 GPa and 0.37 GPa, respectively. This reduction may be attributed to different surface energy of each nanoparticle. Since the surface energy of Cloisite 20A is lower than either the biodegradable polymers or HNTs, the exposed surface may mainly consist of the Cloisite 20A nanoparticles. However, when the RDP-coated clays were mixed with the HNTs, instead of the Cloisite 20A, the surface was not coated with nanoparticles since the surface energy of the nanoparticles might be higher than that of the polymers. Hence, the mechanical properties of the nanocomposites containing RDP clays/HNTs were similar to those of the clay-free matrix. Interestingly, when 5 wt% RDP-coated clays were solely added in the blend, the mechanical properties were comparable to the C20A/HNTs case. Therefore, a balance of surface energy of between polymers/particles can be crucial to determine mechanical properties of chars formation and time to ignition.

6. Biocompatibility of silicate–polymer nanocomposites

The status of polymer–silicate nanocomposites has recently been reviewed by Wu et al. [55]; in particular, they summarized the myriad of applications in the field of bioengineering where polymer–silicate nanocomposites could be used. Among these the most common appear to be cell growth scaffolds and drug delivery systems. In all cases, the most essential consideration is the biocompatibility of the

Table 1 Summary of Polymer–Silicate Nanocomposites Used for Biomedical Devices

Nanoparticles	Polymer	Key Experiment Results	References
Cloisite 20A	Poly(ethylene-co-vinyl acetate)	The addition of 10% clays improved cell proliferation	[58]
Silica	Collagen	Improved biocompatibility for osteogenic differentiation of human mesenchymal stem cells	[59]
Silica	Chitin	Silica-chitin composites provided biocompatible platforms for tissue engineering of bone implants	[60]
Somasif MTE	Fibrinogen	Enhanced cell adhesions on fibrinogen coated clay platelets with Trioctyl-ended trioctylmethylamine (TOMA)	[61]

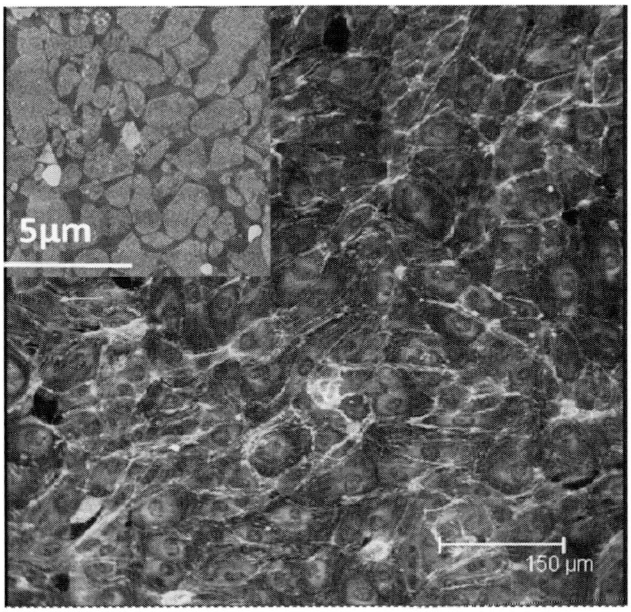

FIGURE 13

Endothelial tissue produced from cells plated on the TOMA substrate for 72 h. Inset: TOMA (Somasif) clay film lifted onto a Si wafer using the LB method and incubated with 0.5 mg/ml of fibrinogen.

clays. This question is raised since there has also been a great deal of research indicating that other nanoparticles can be detrimental to cells and tissue simply due to particle size [56,57]. Of most importance for these categories are titanium dioxide and gold, which are both approved for surgical implants in the bulk but appear to be toxic when produced in nanoparticle form. In contrast to these reports, the data summarized in Ref. [55] and the additional reports listed in Table 1 on silica nanocomposites does not list any incidence of toxicity to cells directly related to clay particles. In a few cases they even reported enhancement of cell adhesion and proliferation when clay platelets were added to the polymer matrix [58,61]. For example, recently Koo et al. [59] have demonstrated that clays can be beneficial in enhancing wound healing, facilitating blood clotting, and endothelialization, since as shown in Figure 13 clays can initiate fibrin fiber formation, cell adhesions, and migration of endothelial cells. Therefore, currently, with the present state of knowledge, there does not appear to be toxicological concerns regarding the addition of clays to nanocomposites.

7. CONCLUSIONS

In this chapter, we showed that ditallow functionalized clays can be replaced almost completely by RDP-coated clays. The RDP-coated clays can serve the same function in enhancing the mechanical, thermal, and FR properties of polymers and polymer blends. In fact we show that these clays may even perform better in styrenic polymers that the ditallow clays. A major advantage of the RDP-coated clays is the ease with which they could be produced simply via adsorption and mechanical agitation, without the need of toxic chemicals, solvents, or other reactions. The surface properties of the RDP-coated clays could be studied by making LB films and performing X-ray reflectivity, scanning force microscopy, and contact angle goniometry. The results showed that RDP coated the clay surfaces and intercalated into the galleries forming a layer of molecules approximately 2.2 nm thick within the clay galleries. Using TEM cross-sectional analysis, we demonstrated that the RDP coated achieved a high degree of exfoliation in either PS or PMMA, as predicted by contact angle goniometry. This technique was developed as a rapid screen for the interfacial energy between functionalized clays and polymers, which could be used to predict exfoliation. In this case we showed that the contact angle between PS/RDP clays was much lower than that of PS/C20A, indicating that in contrast to C20A, the RDP-coated clays had a lower energy penalty and therefore could be exfoliated in HIPS, SAN, and other styrenic polymers. We also showed that the addition of RDP-coated clays could compatibilize immiscible polymer blends. In the blend of PC/SAN24, the interfacial energy of the blend was relatively higher than that of the polymers with RDP clays, which resulted in the segregation of the RDP clays at the blend interfaces and a high degree of compatiblization.

The absorption of RDP molecules to biodegradable polymers was also examined aiming at the increase the degree of compatibilization of biodegradable polymer blends as well as FR properties. The RDP-coated starch could reduce the interfacial tension of starch/Ecoflex, since surface tension for RDP was relatively higher than that for starch. Hence the Ecoflex domains became smaller and were dispersed in the matrix. In the case of PLA/Ecoflex blend, the segregation of RDP-coated clays occurred at the interfaces or Ecoflex domains. Furthermore, a balance of interfacial energy of polymers/nanoparticles could play an important role on FR properties. We found that the surface energy of each comment determined time to ignition and mechanical properties of chars formation. Therefore, self-extinguishing formulations could be controlled by a precise energy balance on the exposed surfaces against heat fronts, where both chemical and physical interactions may occur between the rigid microstructures and decomposed molecules. Apart from the above mentioned merits, the RDP-coated clays are also not cytotoxic and hence they should not raise concerns when used as components of biodegradable polymers.

Acknowledgments

Funding from the NSF-DMR - 1344267 INSPIRE program, the New York State Center for Advance Technology (Sensor-CAT), and Israel Chemical Ltd is gratefully acknowledged. The authors thank Prof Chad Korach from Department of Mechanical Engineering at Stony Brook University for conducting the nanoindention on the char residues, Prof Harald Ade and Dr Sufal Swaraj from Department of Physics at North Carolina State University for conducting the NEXFS experiments with the biodegradable polymers. The contribution of numerous NSF-MRSEC Garcia High School and Undergraduate Research Scholars, (especially, Neil Muir, Ezra Bobo, and Rachel Davis) are also gratefully acknowledged.

References

[1] Kojima Y, Usuki A, Kawasumi M, Okada A, Fukushima Y, Kurauchi T, et al. Mechanical properties of nylon 6-clay hybrid. J Mater Res 1993;8:1185–9.

[2] Usuki A, Kojima Y, Kawasumi M, Okada A, Fukushima Y, Kurauchi T, et al. Synthesis of nylon 6-clay hybrid. J Mater Res 1993;8:1179–84.

[3] Messersmith PB, Giannelis EP. Synthesis and barrier properties of poly(ε-Caprolactone)-layered silicate nanocomposites. J Polym Sci A: Polym Chem 1995;33:1047–57.

[4] Giannelis EP. Polymer layered silicate nanocomposites. Adv Mater 1996;8:29–35.

[5] Vaia RA, Giannelis EP. Lattice model of polymer melt intercalation in organically-modified layered silicates. Macromolecules 1997;30:7990–9.

[6] Vaia RA, Giannelis EP. Polymer melt intercalation in organically-modified layered, silicates: model predictions and experiment. Macromolecules 1997;30:8000–9.

[7] Zeng C, Lee LJ. Poly(methyl methacrylate) and polystyrene/clay nanocomposites prepared by in-situ polymerization. Macromolecules 2001;34:4098–103.

[8] Park JH, Jana JC. Mechanism of exfoliation of nanoclay particles in epoxy-clay nanocomposites. Macromolecules 2003;36:2758–68.

[9] Barber GD, Calhoun BH, Moore RB. Poly(ethylene terephthalate) ionomer based clay nanocomposites produced via melt extrusion. Polymer 2005;46:6706–14.
[10] Si M, Goldman M, Rudomen G, Gelfer MY, Sokolov JC, Rafailovich MH. Effect of clay type on structure and properties of poly(methyl methacrylate)/clay nanocomposites. Macromol Mater Eng 2006;291:602–11.
[11] Wang Y, Zhang Q, Fu Q. Compatibilization of immiscible poly(propylene)/polystyrene blends using clay. Macromol Rapid Commun 2003;24:231–5.
[12] Ray SS, Bousmina M. Effect of organic modification on the compatibilization efficiency of clay in an immiscible polymer blend. Macromol Rapid Commun 2005;26:1639–46.
[13] Lee KY, Goettler LA. Structure-property relationships in polymer blend nanocomposites. Polym Eng Sci 2004;44:1103–11.
[14] Si M, Araki T, Ade H, Kilcoyne ALD, Fisher R, Sokolov JC, et al. Compatibilizing bulk polymer blends by using organoclays. Macromolecules 2006;39:4793–801.
[15] Vaia RA, Teukolsky RK, Giannelis EP. Interlayer structure and molecular environment of alkylammonium layered silicates. Chem Mater 1994;6:1017–22.
[16] Xie W, Gao Z, Pan W, Hunter D, Singh A, Vaia R. Thermal degradation chemistry of alkyl quaternary ammonium montmorillonite. Chem Mater 2001;13:2979–90.
[17] Jamshidian M, Tehrany EA, Imran M, Jacquot M, Desobry S. Poly-lactic acid: production, applications, nanocomposites, and release studies. Compr Rev Food Sci Food Saf 2010;9:552–71.
[18] Pack S, Kashiwagi T, Cao C, Korach C, Lewin M, Rafailovich M. Role of surface interactions in the synergizing polymer/clay flame retardant properties. Macromolecules 2010;43:5338–51.
[19] Levchik S, Bright DA, Alessio GR, Dashevsky S. New halogen-free fire retardant for engineering plastic applications. J Vinyl Addit Techn 2001;7:98–103.
[20] Kose T, Takigami H. International Symposium on flame retardants. www.dtsc.ca.gov; 2010.
[21] Si M, Zaitsev V, Goldman M, Frenkel A, Peiffer DG, Weil E, et al. Self-extinguishing polymer/organoclay nanocomposites. Polym Degrad Stabil 2007;92:86–93.
[22] Pack S, Si M, Koo J, Sokolov JC, Koga T, Kashiwagi T, et al. Mode-of-action of self-extinguishing polymer blends containing organoclays. Polym Degrad Stabil 2009;93: 306–26.
[23] Pack S, Kashiwagi T, Stemp D, Koo J, Si M, Sokolov JC, et al. Segregation of carbon nanotubes/organoclays rendering polymer blends self-extinguishing. Macromolecules 2009;42:6698–707.
[24] United States Environmental Protection Agency-Office of Solid Waste (5306P). EPA530-R-06–011. www.epa.gov; 2006.
[25] Soderstrom G, Sellstrom U, De Witt CA, Tyskind M. Photolytic debromination of decabromodiphenyl ether (BDE 209). Environ Sci Technol 2004;38:127–32.
[26] Lu S, Hamerton I. Recent developments in the chemistry of halogen-free flame retardant polymers. Prog Polym Sci 2002;27:1661–712.
[27] Levchik S, Weil E. Overview of recent developments in the flame retardancy of polycarbonates. Polym Int 2005;54:981–98.
[28] Koo J, Park S, Satija S, Tikhonov A, Sokolov JC, Rafailovich MH, et al. Characterization of Langmuir–Blodgett organoclay films using X-ray reflectivity and atomic force microscopy. J Colloid Interface Sci 2008;318:103–9.
[29] Fu X, Qutubuddin S. Synthesis of polystyrene-clay nanocomposites. Mater Lett 2000; 42:12–6.

[30] Wu S. Surface and interfacial tension of polymer melts. II. Poly(methyl methacrylate), poly(n-butyl methacrylate), and polystyrene. J Phys Chem 1970;74:632–8.
[31] Wyart FB, Martin P, Redon C. Liquid/liquid dewetting. Langmuir 1993;9:3682–90.
[32] Faldi A, Composto RJ, Winey KI. Unstable polymer bilayers. 1. Morphology of dewetting. Langmuir 1995;11:4855–61.
[33] Watkins VH, Hobbs SY. Determination of interfacial tensions between BPA polycarbonate and styrene-acrylonitrile copolymers from capillary thread instability measurements. Polymer 1993;34:3955–9.
[34] Walenta E, Fink H, Weigel P, Ganster J, Schaaf E. Structure-property relationships of extruded starch, 2a extrusion products from native starch. Macromol Mater Eng 2001;286:462–71.
[35] Ma XF, Yu JG, Wan JJ. Urea and ethanolamine as a mixed plasticizer for thermoplastic starch. Carbohydr Polym 2006;64:267–73.
[36] Pack S, Bobo E, Muir N, Yang K, Swaraj S, Ade H, et al. Engineering biodegradable polymer blends containing flame retardant-coated starch/nanoparticles. Polymer 2012;53:4787–99.
[37] Wei L, Wang D, Chen H, Chen L, Wang X, Wang Y. Effect of a phosphorus-containing flame retardant on the thermal properties and ease of ignition of poly(lactic acid). Polym Degrad Stabil 2011;96:1557–61.
[38] Wang D, Leuteritz A, Wang Y, Wagenknecht U, Heinrich G. Preparation and burning behaviors of flame retarding biodegradable poly(lactic acid) nanocomposite based on zinc aluminum layered double hydroxide. Polym Degrad Stabil 2010;95:2474–80.
[39] Reti C, Casetta M, Duquesne S, Bourbigot S, Delobel R. Flammability properties of intumescent PLA including starch and lignin. Polym Adv Technol 2008;19:628–35.
[40] Zhan J, Song L, Nie S, Hu Y. Combustion properties and thermal degradation behavior of polylactide with an effective intumescent flame retardant. Polym Degrad Stabil 2009; 94:291–6.
[41] Wang D, Song Y, Ling L, Wang X, Wang Y. A novel phosphorus-containing poly(lactic acid) toward its flame retardation. Polymer 2011;52:233–8.
[42] Wu K, Hu Y, Song L, Lu H, Wang Z. Flame retardancy and thermal degradation of intumescent flame retardant starch-based biodegradable composites. Ind Eng Chem Res 2009;48:3150–7.
[43] Liu H, Xie F, Yu L, Chen L, Lin L. Thermal processing of starch-based polymers. Prog Polym Sci 2009;34:1348–68.
[44] Levchik SV, Weill ED. Flame retardancy of thermoplastic polyesters—a review of the recent literature. Polym Int 2005;54:11–35.
[45] Pack S, Lewin M, Rafailovich M. A review of engineering biodegradable polymer blends: morphology, mechanical property, and flame retardancy. In: Morgan AB, Wilkie C, Nelson GL, editors. Fire and polymers VI: new advances in flame retardant chemistry and science. Washington, DC: ACS Symposium Series; American Chemical Society; 2012. pp. 427–43.
[46] Gilman JW. Flammability and thermal stability studies of polymer layered-silicate(clay) nanocomposites. Appl Clay Sci 1999;15:31–49.
[47] Gilman JW, Jackson CL, Morgan A, Harris R. Flammability properties of polymer-layered-silicate nanocomposites. Polypropylene and polystyrene nanocomposites. Chem Mater 2000;12:1866–73.
[48] Zhu J, Morgan AB, Lamelas FJ, Wilkie CA. Fire properties of polystyrene-clay nanocomposites. Chem Mater 2001;13:3774–80.

[49] Kashiwagi T, Harris R, Zhang X, Briber RM, Cipriano BH, Raghavan SR, et al. Flame retardant mechanism of polyamide 6_clay nanocomposites. Polymer 2004;45:881–91.
[50] Kashiwagi T, Mu M, Winey KI, Cipriano B, Raghavan S, Pack S, et al. Relation between the viscoelastic and flammability properties of polymer nanocomposites. Polymer 2008;49:4358–68.
[51] Tang Y, Lewin M, Pearce EM. Effects of annealing on the migration behavior of PA6/Clay nanocomposites. Macromol Rapid Commun 2006;27:1545–9.
[52] Lewin M, Pearce EM, Levon K, May-Marom A, Zammarano M, Willkie CA, et al. Nanocomposites at elevated temperatures: migration and structural changes. Polym Adv Technol 2006;17:226–34.
[53] Zammarano M, Gilman JW, Nyden M, Pearce EM, Lewin M. The role of oxidation in the migration mechanism of layered silicate in poly(propylene) nanocomposites. Macromol Rapid Commun 2006;27:693–6.
[54] Lewin M, Mey-Marom A, Frank R. Surface free energies of polymeric materials, additives and minerals. Polym Adv Technol 2005;16:429–41.
[55] Wu C, Gaharwar AK, Schexnailder PJ, Schmidt G. Development of biomedical polymer-silicate nanocomposites: a materials science perspective. Materials 2010;3:2986–3005.
[56] Vishwakarma V, Samal SS, Manoharan NJ. Safety and risk associated with nanoparticles—a review. J Mineral Mat Char Eng 2010;9:455–9.
[57] Barnard AS. One-to-one comparison of sunscreen efficacy, aesthetics and potential nanotoxicity. Nat Nanotechnol 2010;5:271–4.
[58] Lewkowitz-Shpuntoff HM, Wen NC, Singh A, Brenner N, Gambino R, Pernodet N, et al. The effect of organo clay and adsorbed FeO_3 nanoparticles on cells cultured on ethylene vinyl acetate substrates and fibers. Biomaterials 2009;30:8–18.
[59] Koo J, Rafailovich MH, Medved L, Tsurupa G, Kudryk BJ, Liu Y, et al. Evaluation of fibrinogen self-assembly: role of its αC region. J Thromb Haemost 2010;8:2727–35.
[60] Heinemann S, Heinemann C, Ehrlich H, Meyer M, Baltzer H, Worch H, et al. A novel biomimetic hybrid material made of silicified collagen: perspectives for bone replacement. Adv Eng Mater 2007;9:1061–8.
[61] Ehrlich H, Janussen D, Simon P, Bazhenov VV, Shapkin NP, Erler C, et al. Nanostructural organization of naturally occurring composites—part II: silica-chitin-based biocomposites. J Nanomater; 2008:1–8.

CHAPTER

Layered Double Hydroxides: An Emerging Class of Flame Retardants

20

K. Shanmuganathan, C.J. Ellison

McKetta Department of Chemical Engineering, The University of Texas at Austin, Austin, Texas, USA

CHAPTER OUTLINE

1. Introduction

Layered double hydroxides (LDHs) constitute a class of stacked inorganic sheets containing brucite-like $[Mg(OH)_2]$ layers. The structure of LDHs can be described by the general formula $[M^{II}_{1-x} \cdot M^{III}_x(OH)_2]_{intra}[(A^{n-})_{x/n} \cdot yH_2O]_{inter}$ where M^{II} and M^{III} represent divalent and trivalent cations, respectively, within the layer, while A represents intercalated anions in the hydrated interlayer space [1]. Substitution of divalent

Polymer Green Flame Retardants. http://dx.doi.org/10.1016/B978-0-444-53808-6.00020-2

cations with trivalent cations results in a net positive charge on the sheets. This charge is balanced by intercalation of anions in the hydrated interlayer regions. Owing to their layered structure, wide chemical compositions, variable layer charge density, ion-exchange properties, water swelling, and rheological properties, LDHs are often referred to as "anionic clays" [1]. They have been under intense research for use in diverse applications including catalysts and catalyst supports [2–4], chemical adsorbents [5–7], anion exchangers [8], bioactive nanocomposites [9–11], electroactive and photoactive materials [12–14], and flame retardant nanocomposites [15–17].

2. Structure and properties

2.1 Structural aspects of LDHs

Figure 1 shows the general structure of LDHs. The sheets can contain a range of metal cations including most of the first row transition metals as divalent cations and Fe, Al, and Ga as trivalent cations [11,19]. Recently, it has been found that tetravalent cations such as Zn^{4+} and Sn^{4+} can also be incorporated into the brucite-like layer [20,21]. Stacking of the layers in LDHs results in two different polymorphic forms: rhombohedral and hexagonal. The lattice parameter, c, in the rhombohedral

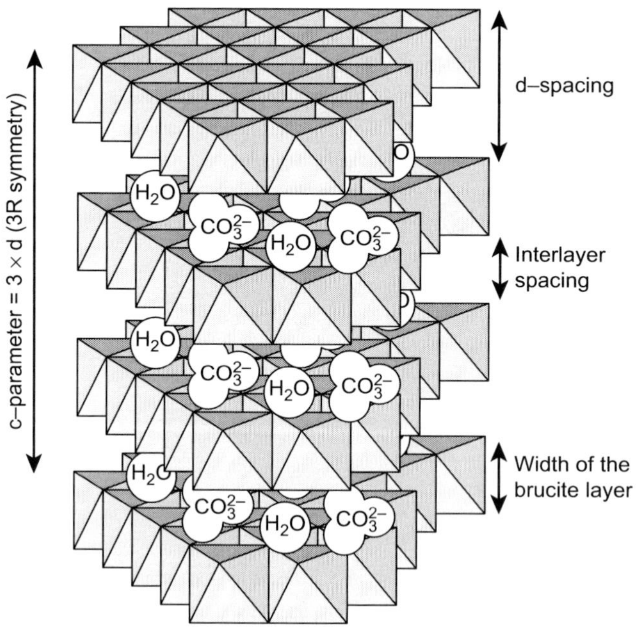

FIGURE 1

Ideal structure of an LDH, with interlayer carbonate anions [18].

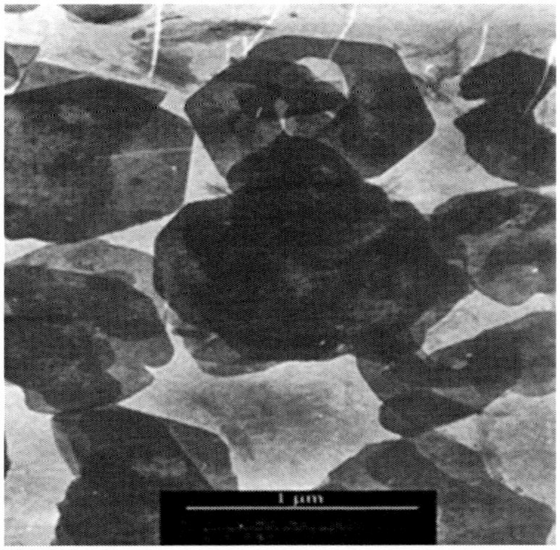

FIGURE 2

Transmission electron microscopy (TEM) image of an Mg/Al LDH [23].

form is three times the interlayer separation, while in the hexagonal form "c" is equal to twice the interlayer separation [22]. The particles are usually hexagonal shaped (Figure 2) and are found as aggregates ranging 1−10 μm in size [24]. When LDH particles are dispersed by sonication or delamination, the particle size reduces, and the hexagonal platelet width varies between 100 nm and a few microns [24,25]. The interlayer region is hydrated, and the amount of water in the interlayer region depends on the nature of the interlayer anions, the water vapor pressure, and temperature [22]. The water molecules are hydrogen bonded to both the metal hydroxide layers and interlayer anions. The interlayer region is very complex with continuous breaking and reforming of these hydrogen bonds [26,27], and the water molecules are in a continuous state of flux. The anions in the interlayer can be exchanged with other anions to facilitate processing and tailor the structure for particular applications. Apart from the chemical composition and degree of crystallinity, the crystallite size and its distribution are also important for certain applications.

2.2 Anion-exchange capacity

The anion-exchange capacity (AEC) is dependent on the proportion of divalent and trivalent cations and varies from 200 to 400 meq/100 g [28−30]. For comparison, sodium montmorillonite, a cationic clay, has an exchange capacity of 108 meq/100 g [31]. A higher AEC is consistent with tightly stacked layers due to the attractive forces between charged layers and intercalated anions in the gallery. It should be noted that the measured AEC values for LDHs are typically less than the theoretical values calculated from the structural formula, due to contamination by carbonate anions, which have a strong affinity for LDHs.

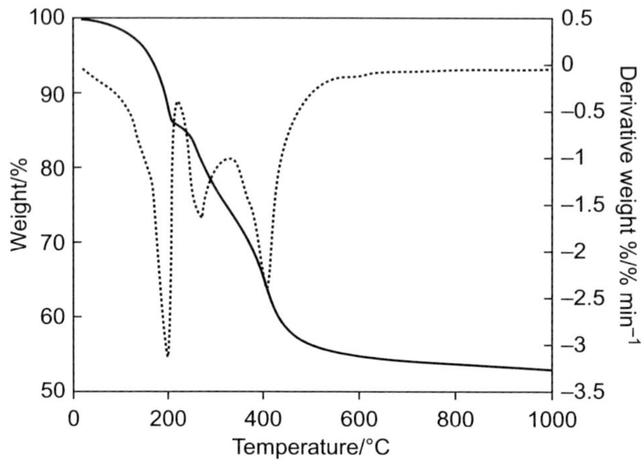

FIGURE 3

Thermogravimetric (solid line) and differential thermogravimetric (dotted line) analysis curves for a commercial Mg/Al LDH with intercalated carbonate [32].

2.3 Thermal behavior of LDHs

Figure 3 shows the multistep thermal degradation behavior of LDHs. The steps involve (1) removal of water physically adsorbed on the external surface of the crystallites; (2) removal of interlayer water; (3) removal of hydroxyl groups from the layers as water vapor; and (4) removal of the interlayer anion in some cases. The first step occurs below 100 °C and step (2) occurs between 100 and 200 °C. Steps (3) and (4) usually occur between 200 and 400 °C [32]. LDHs also undergo chemical and structural changes upon calcination at different temperatures. Calcination of LDHs at moderate temperatures yields a mixture of oxides, while at higher temperatures, corresponding spinels are formed [33]. Aramendia et al. [33] have shown that calcining Mg/Al LDH at 473 K produces no significant structural changes, while calcining them at 773 K leads to a periclase MgO phase and amorphous aluminum oxides. Calcination at 1073 K causes no additional phase changes, but the crystallinity is increased; when the temperature is further increased to 1273 K, spinels of $MgAl_2O_4$ are formed.

3. Synthesis of LDHs

LDHs can be synthesized on laboratory and industrial scales by simple and inexpensive methods with tailored physical and chemical properties suitable for many applications. Different synthesis routes result in significant changes in the structural features and physical properties of LDHs. A brief overview of typical methods of synthesis is provided below, and a list of references on synthesis of LDHs of varying chemistry is provided in Table 1. Additional details regarding the preparation, properties, and applications of LDHs can be found elsewhere [1,18,62].

Table 1 List of Reference Literature for Preparation of LDHs with Varying Chemical Composition

M^{2+}	M^{3+}	Interlayer Anion	Synthesis Method	References
Ba	Fe	OH	Coprecipitation	[34]
Ca	Al	CO_3	Coprecipitation	[11]
Ca	Al	Cl	Separate nucleation and aging	[35,36]
Ca	Al	NO_3	Coprecipitation	[37]
Ca	Fe	NO_3	Coprecipitation	[38]
Ca	Ga, Sc	Cl	Coprecipitation	[39]
Cd	Al	CO_3	Coprecipitation	[40]
Cd	Al	NO_3	Coprecipitation	[40]
Co	Al	Cl	Coprecipitation	[41]
Co	Al	CO_3	Urea hydrolysis	[42]
Cu	Al	CO_3, NO_3	Coprecipitation	[43]
Cu	Al	$[Fe(CN)_6]$	Coprecipitation	[44]
Cu	Cr	Cl	Coprecipitation	[45]
Li	Al	Br	Intercalation	[46]
Li	Al	Cl	Intercalation	[47]
Li	Al	CO_3	Coprecipitation	[48]
Li	Al	NO_3	Intercalation	[49]
Mg	Al	Cl	Coprecipitation	[29]
Mg	Al	CO_3	Coprecipitation	[50]
Mg	Al	CO_3	Sol–gel	[51]
Mg	Fe	CO_3	Coprecipitation	[52]
Mg	Al	Cl/NO_3	Sol–gel	[53]
Mg	Ca, Fe	Cl	Coprecipitation	[54]
Mg	Fe	Cl	Coprecipitation	[52]
Mn	Al	Cl	Coprecipitation	[55]
Ni	Al	Cl/NO_3	Sol–gel	[53]
Ni	Fe	CO_3	Urea hydrolysis	[56]
Ni	V	CO_3	Coprecipitation	[19]
Ni	Al	Cl	Coprecipitation	[57]
Zn	Al	NO_3	Coprecipitation	[58]
Zn	Al	NO_3	Urea hydrolysis	[59]
Zn	Al	CO_3	Sol–gel	[60]
Zn	Ti	CO_3	Coprecipitation	[61]

3.1 Coprecipitation

Coprecipitation is one of the most commonly used methods for the preparation of LDHs and can be performed at constant or variable pH. In general, the pH of the coprecipitation mixture has a crucial effect on the chemical, structural, and textural properties of the final material. In the variable pH method, solutions of metal salts are added to a solution of Na_2CO_3 until the pH reaches a specified value (typically around 10), and then, a solution of NaOH is added to maintain the pH value until the precipitation is complete. Thus, the metal hydroxides are formed first, and further addition of base results in coprecipitation [35]. The variable pH method has been shown to produce fine-grained crystals with rough surfaces and a relatively large surface area [63]. In the constant pH method, the mixed salt solutions (divalent and trivalent metal salts) and the base solution are simultaneously added dropwise to a reaction vessel at a rate such that the pH remains constant. The constant pH method involves low supersaturation and hence a relatively small number of nuclei lead to crystallites with a larger size and higher chemical homogeneity [63]. Subsequent aging processes at elevated temperatures or a hydrothermal treatment is often required to improve the crystallinity of LDHs prepared by the coprecipitation method [64,65]. It has been demonstrated that materials prepared by coprecipitation have many attractive properties for high technology applications, including high crystallinity, small particle size, high specific surface area, and high average pore diameter [66].

Reliably controlling the particle size distribution of LDHs is challenging with both coprecipitation methods. Due to long mixing times, nuclei formed at the beginning of the process have a much longer time to undergo crystal growth than those formed toward the end of the mixing process, and this leads to a wide distribution of crystallite sizes. However, Zhao et al. developed a process that involves a very rapid mixing and nucleation process in a colloid mill followed by a separate aging process. This has been found to yield well-formed crystallites with a narrow size distribution [35].

3.2 Urea hydrolysis

Highly crystalline LDHs can be obtained by homogeneous precipitation methods using the hydrolysis of urea [67–70]. In this method, mixed metal salt solutions and urea are combined with deionized water at 25 °C and then heated to 90–100 °C to start the hydrolysis reaction. The reaction is stopped by quenching it in a cold-water bath after reaction times varying from 18 to 48 h. The precipitates are extracted by centrifugation, washed with ethanol and deionized water, and dried by evaporation at room temperature or by freeze drying. Urea is a very weak Brønsted base ($pK_b \sim 13.8$) and highly soluble in water. Controlled hydrolysis in aqueous conditions yields ammonium cyanate, or its ionic form, while prolonged hydrolysis results in carbon dioxide or carbonate depending on the pH of the medium (Scheme 1) [23].

The nucleation rate can be controlled by changing parameters such as reaction temperature, urea amount, metal salt concentrations, and divalent to trivalent metal molar ratio. Temperature also has a very strong effect. The rate constant is about 200 times higher when the temperature is increased from 60 to 100 °C [70]. The reaction kinetics can be followed by monitoring the pH increase versus time. Higher temperature (90–100 °C) has been found to favor a high germination rate and using water/polar organic solvent mixtures helps to control the particle growth. High reaction

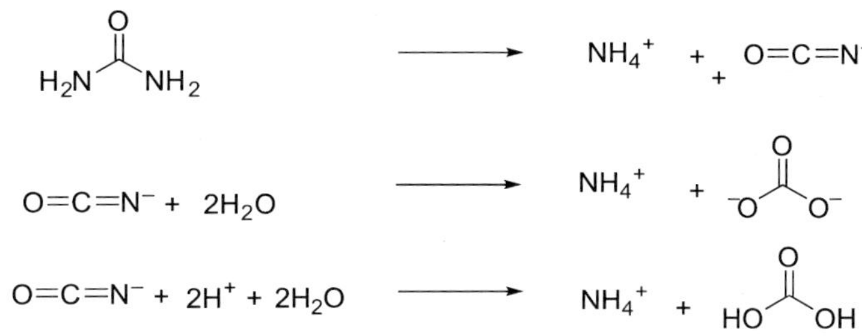

SCHEME 1

Reaction mechanism in urea hydrolysis [23].

temperatures, a low divalent to trivalent metal molar ratio, and high urea/metals ratio have been reported to yield a narrow particle size distribution [70].

3.3 Sol–gel approach

LDHs can also be prepared by a sol–gel route quite similar to the procedure used in the preparation of silica and other inorganic particles. Typical procedures involve dissolving the metal alkoxide in ethanol with the addition of HCl to adjust the pH to about 3. This solution is added dropwise into another metal alkoxide dissolved in ethanol, and the mixture is refluxed for 3 h under continuous stirring. Then, the pH of the solution is increased to approximately 10 by the addition of NaOH, and the solution is refluxed or aged until a gel is formed. The gel is then isolated by centrifugation or filtration, washed multiple times with ethanol and deionized water, and dried [53,60,71]. LDHs prepared by the sol–gel approach were found to exhibit a higher specific surface area than those prepared by coprecipitation procedures [60].

4. Preparation of polymer/LDH nanocomposites

In principle, LDHs can be blended into organic polymers through three different pathways, (1) intercalating the monomer between layers of LDHs followed by in situ polymerization, (2) direct polymer exchange, and (3) restacking of the exfoliated layers over the polymer [11]. Anionic monomers, such as acrylic acid, aniline-2-sulfonate, and vinyl benzene sulfonate, have been intercalated into LDHs and in situ polymerized to obtain polymer/LDH nanocomposites [72,73]. Although not very common, the in situ polymerization approach still appears to be an ideal route to blend inorganic layered structures into organic polymers. Melt compounding LDHs with polymers has been the most adopted approach to prepare polymer/LDH composites. Although exfoliation and homogeneous dispersion of LDHs in polymer matrices via this method pose a great challenge, using high shear mixing equipment and compatibilizing agents can facilitate the exfoliation of LDHs in polymers. In the restacking method, the exfoliated LDH sheets are brought into contact with a solution of a polymer and subsequently precipitated [74]. However, in many instances, this approach has yielded poorly defined materials. Messersmith and Stupp [75] have developed a novel approach to prepare well-ordered layered nanocomposites, where instead of using preformed LDHs, they rely on spontaneous self-assembly of the inorganic and organic phases from homogeneous aqueous solutions, yielding layered nanocomposites. This method could be generally applicable toward the introduction of preformed anionic polymers within LDHs.

5. Flame retardant behavior of LDHs

LDHs have been of significant interest in the last decade as potential flame retardant additives. In recent years, environmental concerns over the use of conventional halogenated flame retardants has accelerated the search for greener additives and a variety of inorganic additives such as silica, talc, clay, and various phosphorous and zinc based compounds are being investigated as flame retardants in polymeric materials [76–82]. In this regard, there has been tremendous interest in using LDHs as an environmentally benign flame retardant additive in many commercial polymers. Interestingly, in some cases, LDHs have exhibited superior flame retardancy compared to montomorillonite, a layered silicate well known for inhibiting the rapid burning of many polymeric materials [83]. LDHs display multiple modes of activity in impeding flame spread in polymers. Desorption and vaporization of interlayer and structural water in LDHs enable significant heat absorption and also dilute other combustible gases. Further formation of a metal oxide residue hinders oxygen and mass (fuel) transport to and from the bulk polymer phase beneath the burning surface. Thus, LDHs present a combined gas-phase and condensed-phase action on the burning material. In this section, we review some recent developments on the use of LDHs as flame retardants in various polymer matrices followed by a brief discussion on the synergistic role of LDHs with other flame retardants in various polymers.

5.1 Flame retardancy of LDHs in ethylene–vinyl acetate copolymers

Ethylene–vinyl acetate copolymers (EVA) are generally used in a broad range of applications, including wire and cable insulation, hot melt adhesives, packaging, and carpeting. Flame retardancy is an important requisite for EVA in most of these applications, particularly in cable insulation, and is typically achieved through the addition of conventional flame retardants. Thermal stability and flame retardancy of EVA in the presence of LDHs of varying chemistry have been studied extensively. In one of the earliest reports on the flame retardant effects of LDHs on EVA, Camino et al. [84] showed that hydrotalcites, which are naturally occurring Mg/Al LDHs, have a significant influence on the ignition and combustion behavior of EVA. Conducting a comparative study between hydrotalcites and other conventional hydrated fillers such as magnesium hydroxide (MH) and aluminum trihydrate (ATH), they found that hydrotalcites had the lowest mass loss rate in mass combustion calorimetry experiments conducted at an incident heat flux of 30.3 kW m^{-2} (Figure 4). Further, the ignition time (t_{ig}) of EVA/hydrotalcite composites was 150 s, which

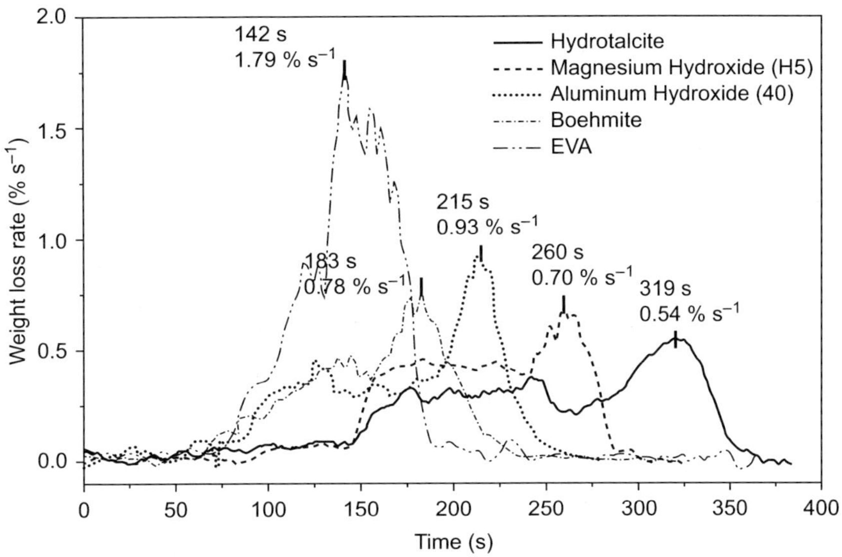

FIGURE 4

Mass loss calorimeter curves of EVA and EVA composites [84] obtained with an incident heat flux of 30.3 kW m^{-2}.

is almost twice that of neat EVA (78 s). Regardless of the type of filler, at least 50 wt % filler was required to have significant flame retardant effects on EVA.

Conventional fillers such as MH and ATH are known to have a significant water content that is endothermally released during decomposition. Apart from significant heat absorption, the released water vapor also dilutes the flammable gases during pyrolysis, thereby inhibiting flame spread in burning materials. LDHs have been reported to have significant quantities of water in the interlayer and can act in a similar way to retard flame spread. In addition, they can also form a mixed oxide residue upon burning leading to the formation of an insulative char layer that serves as further protection for the underlying polymer. The enhanced flame retardant effect of LDHs can be attributed to this combined physical and chemical action.

Ye and Qu [85] investigated the flame retardant effects of Mg/Al-PO_4 LDH in EVA and compared it with Mg/Al−CO_3. Phosphate intercalated Mg/Al LDH were prepared by anion exchange of Mg/Al-NO_3 LDH and incorporated into EVA by melt compounding. Mg/Al-PO_4 significantly influenced the thermal degradation and residual char formation in EVA. The thermal degradation of EVA is a two-step process. The first weight loss step occurs at 250−420 °C and is due to the evolution of acetic acid while the second weight loss step occurs between 420 and 600 °C and corresponds to the degradation of ethylene chains in EVA. Incorporation of 60 wt% Mg/Al-PO_4 did not enhance the thermal stability of EVA but resulted in 40−50% residual char. Phosphorous-containing compounds are known to catalyze the conversion of organic matter to charred layers by the formation of P-O-P and P-O-C complexes during combustion. This enhanced char formation of Mg/Al-PO_4 LDH can be expected to confer better flame retardance to EVA when compared to Mg/AlCO_3 LDH. As expected, phosphate intercalated LDH resulted in a more compact char layer compared to Mg/AlCO_3 thereby efficiently protecting the underlying polymer material from burning rapidly. The peak heat release rate (PHRR) of EVA during combustion in a cone calorimeter was reduced by 76% by incorporating 60 wt% of Mg/Al-PO_4, while incorporating a similar amount of Mg/Al–CO_3 LDH caused a reduction in the PHRR of only approximately 66%. The heat released during burning of a material is a driving force for further burning and can make the combustion self-sustaining. Hence, PHRR has been found to be the most important parameter to evaluate fire safety [77] and represents the point in a fire where the heat generation is sufficient to propagate the flame or ignite adjacent material. Therefore, a significant reduction in the PHRR by the addition of Mg/Al-PO_4 LDH indicates inhibition of the combustion process and factors associated with rapid flame spread. Phosphate intercalated LDH also revealed a better flame retardant performance than Mg/Al–CO_3 LDH in terms of reduction in mass loss rate and delayed ignition.

LDHs with intercalated borate anions were also found to be better flame retardants in EVA when compared to conventional flame retardant fillers such as ATH, MH, or zinc hydroxide (ZH). Nyambo and Wilkie [15] found that at 40 wt% loading, borate-intercalated Mg/Al and Zn/Al LDHs showed a significantly higher reduction of the PHRR (74% and 77%) in EVA/LDH composites than those consisting of ZH (47%), MH (65%), and their combinations. However, at similar loadings, ATH and zinc borate resulted in a higher reduction in the PHRR (89%) when compared to that of LDHs (Table 2). They also observed a significant variation in the char structure and form depending on the additive used. Zn/Al–borate, zinc borate, and ZH resulted in a hard compact char while Mg/Al–borate, MH, ATH, and combinations of MH and ATH yielded soft and powdery chars. The mixed oxide chars obtained after cone calorimetry were also determined to be amorphous.

In another study, Jiao et al. [86] compared the flame retardant effects of nano-MH and nanohydrotalcite (both treated with stearic acid with a median particle size of 50 nm) in EVA. While EVA composites with 150 phr of MH nanoparticles had a limiting oxygen index (LOI) of 40, hydrotalcite nanoparticles at a similar loading had an LOI of about 43. LOI is the minimum concentration of oxygen (in volume percent) that will just support combustion in a flowing mixture of oxygen and nitrogen and lead to downward burning of a vertically mounted test specimen (ASTM D2863). Since air comprises about 20.95% oxygen by volume, materials with LOI values below 20.95 are considered "flammable". It is suggested that materials with LOI values between 20.95 and 28 be classified as "slow burning" and those that have values >28 as "self-extinguishing" [87].

5.2 Flame retardancy of LDHs in epoxy

In a systematic investigation comparing montmorillonite, ATH, and Mg/Al-LDH modified with organic anions such as aminobenzene sulfonic acid (ABS),

Table 2 Cone Calorimeter Data [15] for Ethylene Vinyl Acetate Composites Obtained at an Incident Heat Flux of 35 kW m^{-2}

Material	PHRR (kW m^{-22})	% Reduction in PHRR	Avg. Mass Loss Rate (g s^{-1} m^{-2})	Time to Ignite, t_{ig} (s)	Char (%)
EVA	2027 ± 137	NA	48 ± 7	58 ± 4	0
EVA + 40% Mg/Al-borate	530 ± 51	74	15 ± 1	43 ± 1	26 ± 0
EVA + 40% Zn/Al-borate	460 ± 23	77	16 ± 4	51 ± 4	30 ± 1
EVA + 40% MH	703 ± 95	65	22 ± 3	63 ± 3	29 ± 1
EVA + 40% ATH	222 ± 29	89	9 ± 1	54 ± 5	31 ± 1
EVA + 40% ZH	1079 ± 106	47	27 ± 3	36 ± 4	34 ± 1
EVA + 40% zinc borate	231 ± 9	89	5 ± 1	50 ± 3	36 ± 1
EVA + 40% (2:1 MH:ATH)	577 ± 23	72	22 ± 1	63 ± 2	32 ± 1
EVA + 40% (2:1 ZH:ATH)	1052 ± 146	48	33 ± 4	44 ± 5	30 ± 1

4-toluenesulfonic acid monohydrate (TS), or 4-hydroxybenzenesulfonic acid, Zammarano et al. [88] showed for the first time that LDHs at 5 wt% loading can impart a self-extinguishing behavior to epoxy in a UL 94 horizontal flame test, while other flame retardants such as montmorillonite or ATH can only decrease the flame spread rate in epoxy. The fillers were stirred in epoxy under different conditions and subsequently cured with the aid of curing agents. Among the various fillers studied, only organically modified montmorillonite (Cloisite 30B) and LDHs modified with organic anions gave an intercalated nanocomposite structure (Figure 5) while unmodified LDHs, ATH and ammonium polyphosphate (APP) formed microcomposites in epoxy. The flame retardancy of these materials appeared to be a complex interplay of the level of dispersion of fillers, intrinsic properties of the fillers, their mode of action, and other factors. The superior flame resistance of organo-LDHs in polymers is a consequence of mixed metal oxide residue (confirmed by X-ray diffraction [XRD] experiments) and release of water. The higher flame retardant performance of organo-LDH/epoxy was also reflected in cone calorimeter tests where a 40–51% reduction in the PHRR of epoxy was observed with 5 wt% organo-LDH, while at a similar loading, organically modified montmorillonites yielded only a 27% reduction in the PHRR of epoxy. The difference in performance has been justified with a more compact and intumescent char obtained with LDH/epoxy as compared with fragmented residues in the case of the montmorillonite/epoxy system.

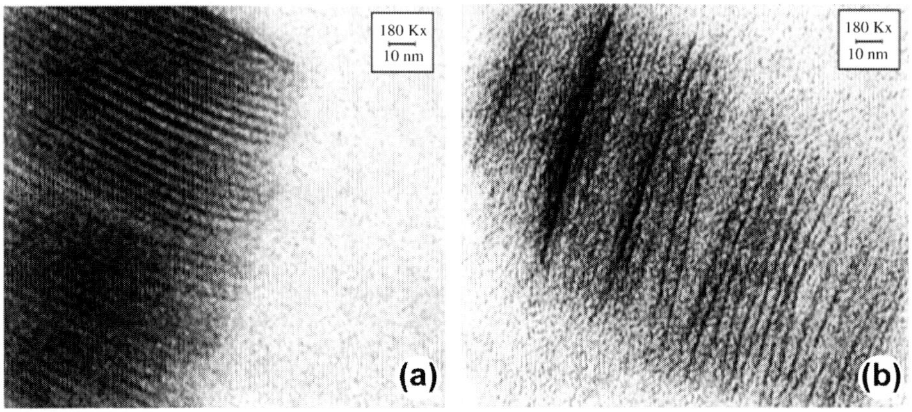

FIGURE 5

TEM micrographs of (a) Liquid epoxy monomer intercalated in Mg/Al-LDHs modified with ABS (6.4 wt%) after swelling; (b) same sample after curing with poly(propylene glycol) bis(2-aminopropyl) ether (Jeffamine D230) [88].

Thus, epoxy and LDHs modified with organic sulfonates (TS and ABS) evolve into an intumescent system upon burning (Figure 6), where the epoxy resin acts as a source of the char, sulfonate becomes the charring agent, water and CO_2 (evolved during the thermal decomposition of hydroxyl layers) serve to blow the char. The organo-LDH used in this study [88] also had synergistic effects with APP. For example, for 3-mm-thick samples, a 30 wt% loading of APP was necessary in D230 cured epoxy to attain the UL94-V0 rating (which indicates a

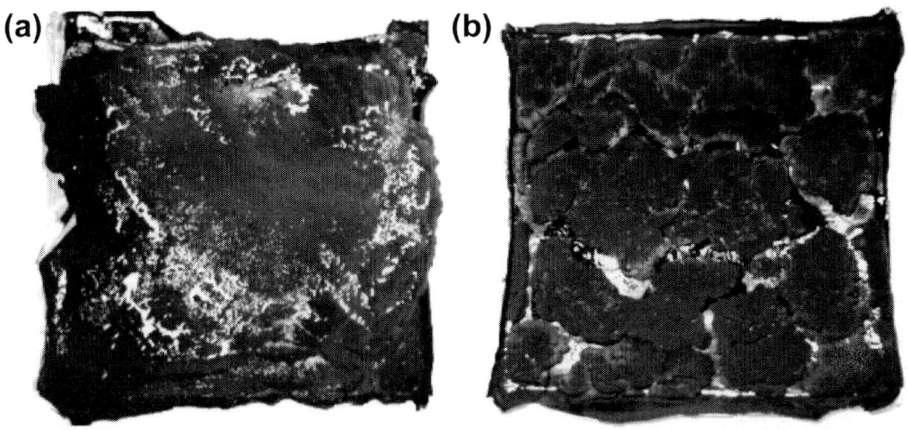

FIGURE 6

Images of carbonaceous residue obtained after cone calorimeter tests: (a) Compact and intumescent char of epoxy–LDH nanocomposite; (b) fragmented char of epoxy–montmorillonite nanocomposite at 5 wt% loading [88].

self-extinguishing behavior). However, when a 4 wt% Mg/Al-ABS was used, a similar rating was attained with an APP content of only 16–20 wt%.

5.3 Flame retardancy of LDHs in polyethylene

Polyethylenes (PEs) are one of the most widely used commodity polymers, primarily in packaging applications. Several investigations have been carried out on the flame retardant effects of LDHs in PEs. It is a challenging task to achieve a good dispersion of polar fillers such as layered silicates or silica in PE, which comprises a nonpolar hydrocarbon backbone. Hence, one of the foremost challenges in investigating the flammability of PE–LDH composites is to obtain a good dispersion of LDHs in the polymer matrix. Costantino et al. [89] prepared Zn/Al–CO_3 LDH by urea hydrolysis, exchanged the carbonate anions with nitrate to increase the interlayer distance, and then intercalated the LDHs with sodium stearate to favor mixing with PE. The modified LDHs had an interlayer spacing of 30.9 Å (XRD) due to stearate intercalation. Nanocomposites of PE with these modified LDHs, prepared by melt mixing in a Brabender mixer, revealed no diffraction peaks at lower angles indicating a partial exfoliation of LDHs in the PE matrix. In an early communication [89], they reported a 55% reduction in the PHRR of PE with 5 wt% (calculated from inorganic content) of modified LDHs.

In a recent report, Costa et al. modified Mg/Al–CO_3 LDH with sodium dodecylbenzene sulfonate (DBS) by solution intercalation [90]. They then prepared a master batch of organically modified LDHs (LDH-DBS) and maleic anhydride grafted PE (PE-g-MAH) by melt compounding in a 1:1 ratio. Subsequently, this master batch was diluted with appropriate amounts of low-density PE (LDPE) by melt mixing in a tightly intermeshing corotating twin-screw extruder. The use of such intensive mixing equipment has been reported to improve particle dispersion [91]. Figure 7(a)

FIGURE 7

(a) SEM image of organically modified Mg/Al–CO_3 (LDH–DBS), (b) TEM image of LDPE/LDH–DBS nanocomposite [91].

shows the scanning electron microscopy (SEM) of LDH–DBS particles before melt compounding where they appear as platelet-like structures stacked into particles ranging a few microns in width and a few hundred nanometers in thickness. Figure 7b is the TEM image of the LDPE/LDH–DBS nanocomposite, where the particles appear to be smaller after compounding into the polymer. It is hypothesized that the intensive shear force during mixing facilitates intercalation of polymer chains into the interlayers, and further delamination and rupture of layers results in particle exfoliation and size reduction as shown by other TEM images (Figure 8).

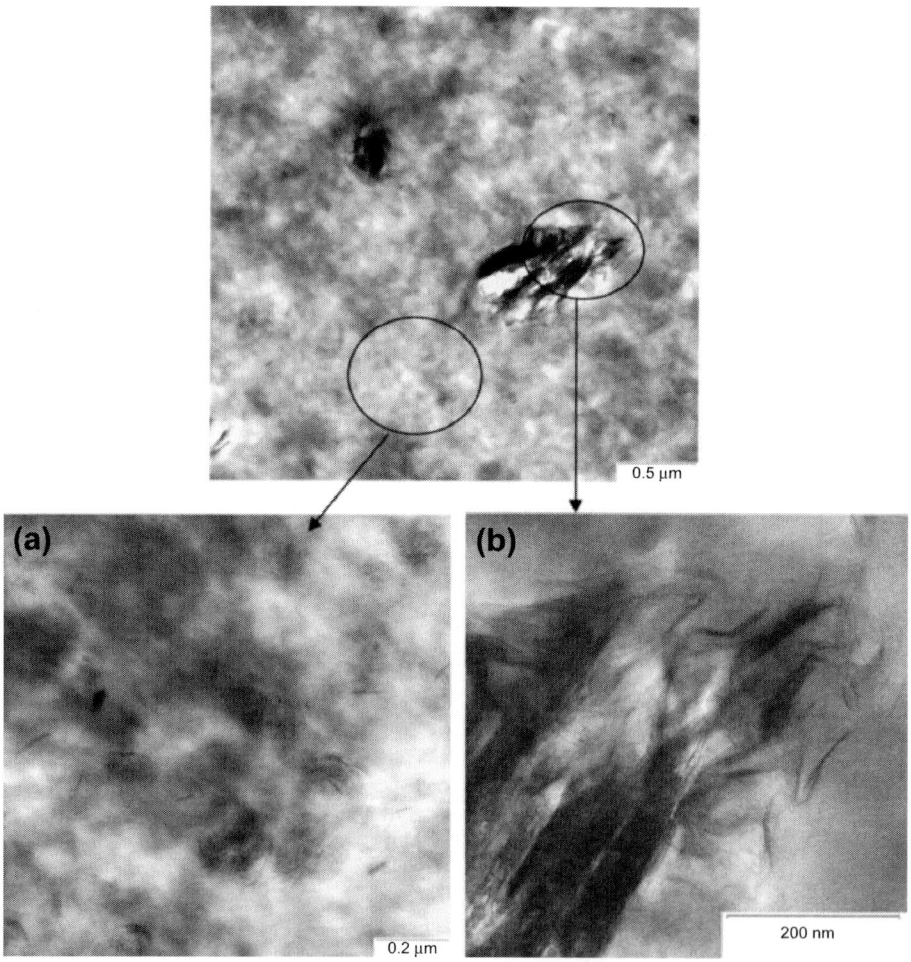

FIGURE 8

TEM micrographs showing that exfoliated LDH particles exist both (a) in the bulk matrix of LDPE and (b) also in the vicinity of the originating bigger particles [91].

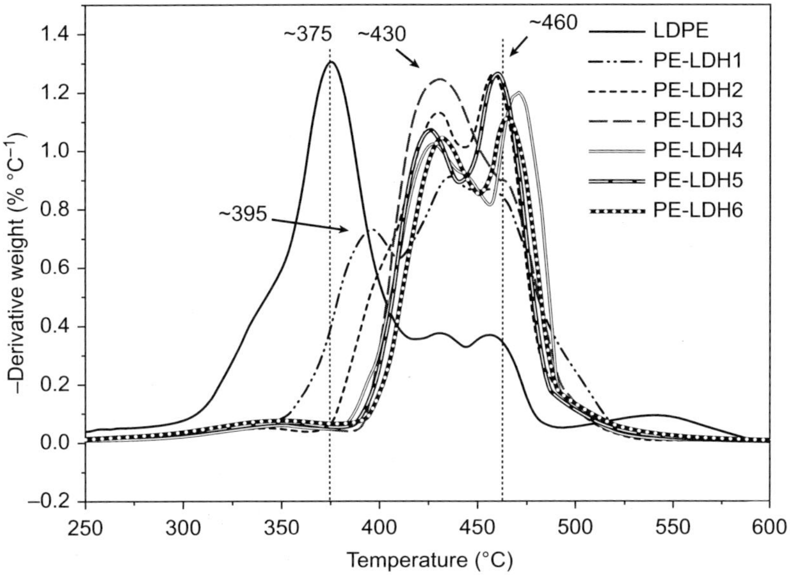

FIGURE 9

Derivative thermogravimetric analysis (DTGA) plots of LDPE/LDH–DBS nanocomposites [91] (LDH composition: PE–LDH1—2.43 wt%, PE–LDH2—4.72 wt%, PE–LDH3—6.89 wt%, PE–LDH4—8.95 wt%, PE–LDH5—12.75 wt%, PE–LDH6—16.2 wt%).

Costa et al. [91] determined that LDH–polymer composites prepared in this manner have a significant influence on the thermal stability of LDPE. As shown in Figure 9, the presence of LDHs in amounts as low as 2.43 wt% increased the onset decomposition temperature of LDPE by 20 °C. At higher concentrations of LDHs (6.89–16.20 wt%), low-temperature decomposition in LDPE is completely suppressed, and significant weight loss occurs only above 400 °C. The presence of LDHs also resulted in significant residual char at 600 °C for the LDPE composites. Under a radiant heat flux of 30 kW m^{-2} in a cone calorimeter, LDPE/LDH–DBS composites demonstrated significant flame retardance. While unfilled LDPE had a PHRR of about 800 kW m^{-2}, the composites had a PHRR as low as 300 kW m^{-2} with about 16.2 wt% filler concentration. The presence of LDHs also led to a delay in ignition time (t_{ig}) (time taken by the material to ignite at an incident heat flux of 30 kW m^{-2}) and a lower total heat release (THR). The THR is the amount of heat released during the entire combustion process. The fact that the THR is reduced by about 44% with 16.2 wt% LDHs is very significant.

The authors [91] highlight this as an advantage of LDHs over layered silicates, where no significant change in the THR is generally observed. The difference is likely due to the chemical action (endothermal release of water) of LDHs, which could function as a heat sink and reduce the total heat generated due to combustion. Layered silicates on the other hand function predominantly in a physical mode (protective char formation), unless chemically modified with other flame retardants.

Flammability of a material encompasses several aspects such as propensity to ignite, tendency for sustained combustion or rapid fire growth, and dripping of burning material. Hence, converting different results obtained in cone calorimeter tests into meaningful parameters or indices that define the propensity to cause fast fire growth has always been found to be more useful. The ratio of PHRR to ignition time (t_{ig}), PHRR/t_{ig} is often reported in the literature as an index for comparing the flammability of materials. Based on the cone calorimetry results, the authors found that LDPE/LDH–DBS nanocomposites have a decreasing PHRR/t_{ig} ratio with an increase in the concentration of LDH–DBS. When PHRR/t_{ig} ratio was plotted against THR, the position of LDPE/LDH–DBS nanocomposite shifted toward the origin indicating lower flammability potential of the composites (Figure 10). It is important to note here that the flammability of a material depends on the conditions of testing such as heat flux, source of heat flux and ignition, sample dimensions, its orientation and surface topography, and ambient conditions. Hence, the number and complexity of these variables can make comparisons across different investigations very difficult.

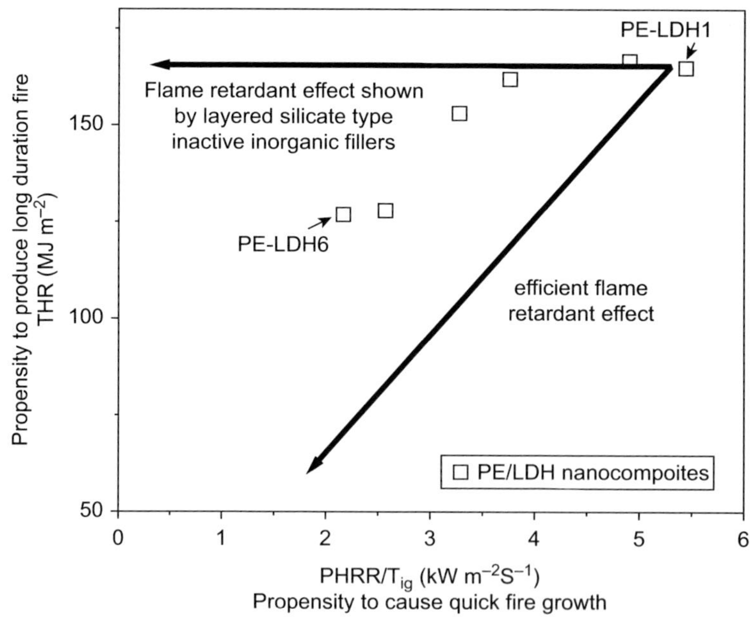

FIGURE 10

A plot of THR against PHRR/t_{ig} for different LDH concentrations indicating fire risk associated with LDPE/LDH nanocomposites (the LDH concentration increases from the right to the left as shown, PE–LDH1—2.43 wt%, PE–LDH2—4.72 wt%, PE–LDH3—6.89 wt%, PE–LDH4—8.95 wt%, PE–LDH5—12.75 wt%, PE–LDH6—16.2 wt%) [91].

5.4 Flame retardancy of LDHs in polypropylene

Polypropylene (PP) is a commercially important polymer used in a variety of applications owing to its low cost. However, this polymer is known for its thermooxidative degradation and high flammability. Hence, it has been of significant interest to find nonhalogenated flame retardants for PP. To this end, inorganic particulates including montmorillonite [92], carbon nanotubes (CNTs) [93], silica [76], MH [94], and other compounds such as APP [95] and melamine phosphate [96] have been investigated as flame retardant additives. In recent years, LDHs have also been explored as flame retardants for PP due to the chemical versatility of these "anionic clays" and their hydrated structure. Ding and Qu [97] compared the effects of montmorillonite and LDHs on the thermal degradation behavior properties of PP. In a TGA study, PP/LDH composites showed better thermal stability than did PP/montmorillonite composites at later stages of decomposition, which was ascribed to better char formation with LDHs. For example, the 80% weight loss temperature for PP/LDH nanocomposites was 416 °C and for PP/montmorillonite composites it was 399 °C. Significant enhancement in the thermal stability (higher decomposition temperature) of PP in the presence of LDHs has also been reported by other researchers [98,99].

Wang et al. [100] synthesized Co/Al-NO_3 LDHs modified with dodecylbenzene sulfonate. A master batch of this modified LDH and maleic anhydride grafted PP (PP-g-MA) was first melt compounded and later diluted with nonmaleated PP. These PP/LDH composites exhibited an enhanced thermal stability in TGA analysis. Activation energies at different stages of decomposition of PP determined by the Freidman method [101] or Flynn–Wall–Ozawa method [101] increased by 10–30% with the addition of 1.5–6 wt% LDHs.

While it has become customary to use organically modified LDHs and/or a compatibilizer such as PP-g-MA to overcome the difficulties in dispersing LDHs into hydrophobic PP, it is important to consider how these chemical entities affect the flame retardance of PP. In this regard, the influence of the LDH composition and the presence of PP-g-MA on the dispersion and flammability properties of the PP/LDH system have been systematically investigated by Manzi-Nshuti et al. [102] using a series of oleate-modified Zn/Mg/Al LDHs with different Zn/Mg contents. In this study, the reduction in the PHRR of nonmaleated PP was marginal (0–38%) with 1–4 wt% (inorganic mass fraction) of LDH. This was attributed to the relatively poor dispersion of LDHs in nonmaleated PP. To facilitate better dispersion of these oleate-modified LDHs in PP, different amounts of PP-g-MA were added as a compatibilizer. Cone calorimeter tests revealed a clear correlation between the amount of PP-g-MA and the flame retardance behavior of these PP/LDH systems containing maleated PP. For 4 wt% (inorganic mass fraction) LDH blends with a PP/PP-g-MA ratio of 1:1, a 68% reduction in the PHRR was observed compared to neat PP. Reduction in the amount of PP-g-MA from a PP/PP-g-MA ratio of 1:1 to 4:1 and 8:1 led to a 57% and 51% reduction in the PHRR, respectively, with 4 wt% oleate-modified LDH. Based on TEM and XRD analysis, the authors concluded that PP-g-MA facilitates better dispersion of

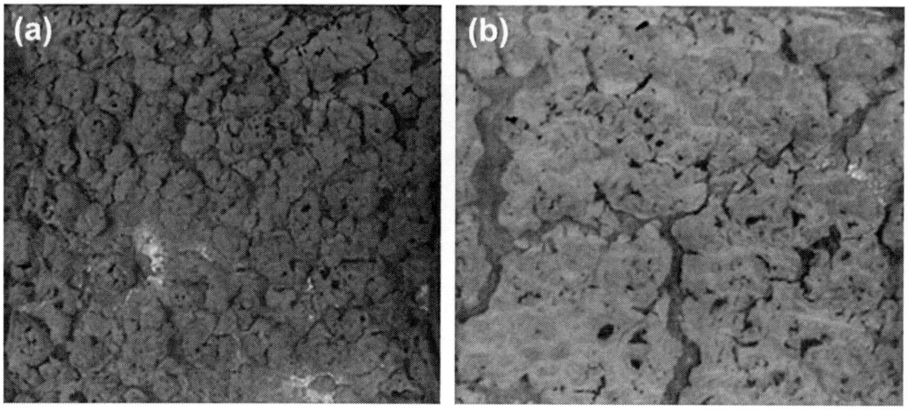

FIGURE 11

Char morphology of PP LDH composites combusted in a cone calorimeter (a) PP/PP-g-MA (1:1) blend with 4 wt% Mg/Al LDH, (b) PP/PP-g-MA (1:1) blend with 4 wt% Zn/Al LDH [102].

LDHs in PP yielding a better flame retardant behavior. Chemical composition of LDHs can also influence flame retardance behavior significantly. The PHRR reduction of PP/PP-g-MA (1:1) LDH nanocomposite was about 68% for Zn/Al LDH at 4 wt% loading while it was only 52% for Mg/Al LDH at similar conditions. The difference is likely due to differing char morphologies that are formed upon burning. Zn/Al LDH forms a more compact char upon burning as compared to that of Mg/Al LDH (Figure 11).

5.5 Flame retardancy of LDHs in poly(methyl methacrylate)

Nyambo et al. [103] reported on the flame retardant effects of alkyl carboxylate-modified LDHs in poly(methylmethacrylate) (PMMA). Mg/Al$-NO_3$ LDHs were synthesized by the coprecipitation method and intercalated with alkyl carboxylates of various chain lengths (decanoate ($C_{10}H_{19}O_2$ or C10), laurate (C12), myristate (C14), palmitate (C16), and stearate (C18)) through an anion-exchange process. The completion of the anion-exchange process was confirmed by the lack of an IR absorbance peak at 1384 cm^{-1} for NO_3 groups. XRD studies on the PMMA/modified LDH composites made by melt compounding revealed an increased *d* spacing with increasing alkyl chain lengths of the carboxylate anion leading to the conclusion that more organophilic LDHs expand the galleries to a greater extent by enhancing polymer intercalation. Thermal stability and char formation of PMMA were enhanced with these modified LDHs at all filler loadings. However, residual char formation was more than would be expected based on the amount of LDH added. This suggests that residual char is not just a mixed oxide residue and that LDHs may have additional mechanisms of inducing carbonaceous char formation in PMMA. The reduction in the PHRR of PMMA was between 49 and 58% for all the alkyl carboxylate-modified LDHs at 10 wt% loading. While studies in other polymers show a higher PHRR reduction with LDHs than with mineral fillers such as MH and ATH, in this PMMA study, the presence of the nanodispersed LDH at 10 wt% loading gives essentially the same reduction in the PHRR as the 10 wt% loading of MH or ATH. This suggests that further mechanistic investigation is needed to explain these similarities.

In a parallel study on polystyrene with the same alkyl carboxylate modified fillers, the authors found a higher PHRR reduction for Mg/Al-LDH modified with C10 carboxylate anions than with C22 carboxylate anions. This is unobvious considering the fact that organic modifiers with longer organophilic chains often yield better dispersion of LDHs in polymers. The authors propose that LDHs perform as flame retardants in a more complex way than montmorillonite fillers do and further investigation is necessary for a more complete understanding. The tensile strength of PMMA and polystyrene was found to be significantly reduced with the addition of 10 wt% of these modified LDHs, while elongation at break remained unchanged.

The role of divalent metal cations in LDHs on the flame retardant behavior of PMMA was studied by Manzi-Nshuti et al. [16]. Various LDHs with aluminum, as the trivalent cation, and zinc, cobalt, or nickel, as divalent cations, were prepared by coprecipitation and melt blended with PMMA. Among the different LDHs, cobalt-containing LDH displayed better flame retardance in PMMA. The reduction in the PHRR was 41%, 26%, and 16% for Co/Al-LDH, Zn/Al-LDH, and Ni/Al-LDH, respectively. Although PMMA nanocomposites with cobalt-containing LDHs were reported to have a better dispersion than other nanocomposites, a clear explanation has not been provided for the superior flame retardant performance of Co/Al-LDH over other LDHs. In another investigation that compared the role of trivalent metals in LDHs, Ca/Al-LDH gave a higher reduction in the PHRR of PMMA (54%) compared to that of Ca/Fe-LDH (34%) at a 10 wt% loading [104], which is likely due to differences in dispersion of these LDHs. Mg/Al-LDH intercalated with various benzyl anions of differing functionalities (i.e. carboxylate, sulfonate, and phosphonate) were prepared by coprecipitation and their flame retardant effects in PMMA were compared by Nyambo et al. [105]. The highest reduction in the PHRR of PMMA (46%) was achieved with Mg/Al-LDH modified with benzoic acid, while Mg/Al-LDH modified with benzene sulfonic acid produced the least effect (26% reduction in the PHRR) at 10 wt% loading. Different LDHs used in the same study yield different dispersion states in the polymer. However, a comprehensive analysis of all the available data does not facilitate clear correlations between various factors such as LDH chemistry, its dispersion state in the polymer, char structure, and PHRR.

5.6 Flame retardancy of LDHs in other polymers

LDHs have been examined as flame retardants in only a few other polymers such as nylon 6 [106], polylactic acid [107], and unsaturated polyesters [108]. Modified Mg/Al-LDH prepared by the coprecipitation of the metal salts and sodium dodecyl sulfate in aqueous solution [106] were incorporated into nylon 6, and their thermal

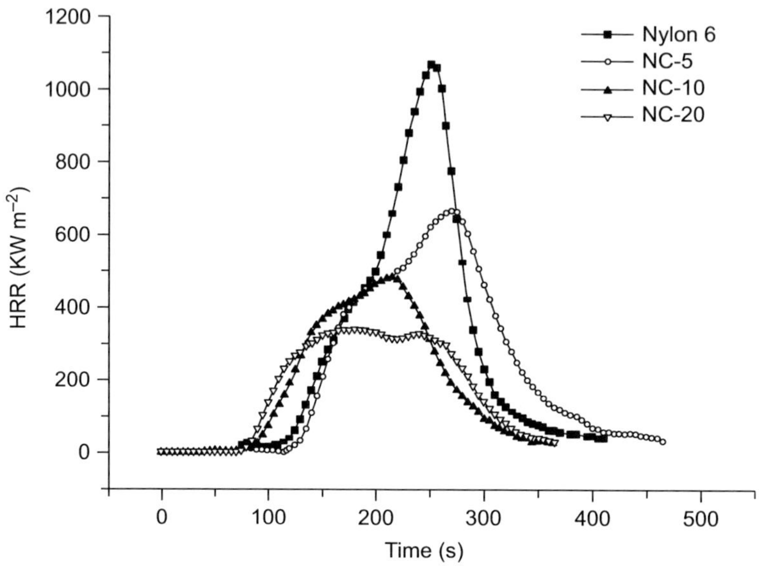

FIGURE 12

Heat release rate plots obtained by cone calorimeter investigations on nylon 6 and its nanocomposites with modified LDHs [106] (NC-5: Nylon 6/5 wt% LDH, NC-10: Nylon 6/10 wt% LDH, and NC-20: Nylon 6/20 wt% LDH).

and flammability properties were investigated. The thermal stability of nylon 6, as determined by TGA analysis, was reduced slightly in the presence of LDHs, which is likely due to catalytic degradation under alkaline conditions. In cone calorimeter combustion tests, the PHRR of nylon 6 was reduced by 37–68% with the addition of 5–20 wt% LDH (Figure 12). This is comparable to what has been achieved in nylon 6 with montmorillonite [77,80]. Pereira et al. [108] investigated the flammability of unsaturated polyester filled with LDH modified with either adipic acid or 2-methyl-2-propene-1-sulfonic acid. Combustion studies revealed a reduction in the PHRR of 32–46% with 1–5 wt% addition of LDH.

5.7 Flame retardancy of LDHs in combination with other flame retardants

The combustion of any material is a complex process with multiple physical and chemical events occurring simultaneously. In most instances, disrupting the combustion loop and realizing significant inhibition of flame spread demands multiple modes of action from additives of choice. Hence, it has been a rational approach to blend two or more flame retardant additives into a polymer. These additives generally differ in their mechanisms of action and are carefully chosen to complement one another and augment the overall flame retardance of polymers. Although LDHs themselves provide significant enhancement in the combustion behavior of polymers, they have also been used with other flame retardants with a view of establishing synergistic flame retardant effects. Several conventional flame retardants including MH, phosphorous, nitrogen, or boron-based compounds or inorganic fillers such as talc and fumed silica. have been used in combination with LDHs to enhance flame retardance. This section provides a brief overview on some of the recent developments in this direction.

Ye et al. [109] reported on the synergistic effects of exfoliated Mg/Al-LDH with other flame retardants such as hyperfine MH (HFMH), expanded graphite (EG), and microencapsulated red phosphorous (MRP) in LDPE/EVA blends. Since it has been a challenging task to disperse and exfoliate LDHs in nonpolar matrices such as LDPE, in addition to using sodium dodecyl sulfate-modified LDHs (OM-LDH), EVA was added as a compatibilizer. It was confirmed through XRD and TEM analysis of the dispersion state, that EVA is beneficial to the exfoliation of OM-LDH layers in the LDPE matrix. In this study, the exfoliated LDH layers acted as a synergistic compatibilizer for dispersing HFMH in LDPE/EVA blends. Flammability of these materials as evaluated by UL-94 and LOI tests is described in Table 3.

LDPE with HFMH (sample A) or LDPE/LDHs/HFMH (sample B) dripped during combustion and failed in the UL-94 vertical burning tests. Addition of EVA, which helped to exfoliate LDHs and/or disperse HFMH in LDPE, increased the LOI values slightly, but UL-94 ratings were not changed (samples C and F). Materials with HFMH and MRP or EG (samples G and I) had an LOI value of 34 and passed the UL-94 but still dripped during combustion. However, when LDHs were added to the above materials, the LOI values increased to 36 and 38, and

Table 3 List of LDPE/EVA Blends With Flame Retardant Additives and Their Flammability Behavior [109].

Sample	LDPE (phr)	EVA (phr)	LDH (phr)	HFMH (phr)	EG (phr)	MRP (phr)	LOI (%)	UL-94 Rating[a]	Dripping Behavior
A	100	–	–	100	–	–	29	Fail	Dripping
B	100	–	5	95	–	–	29	Fail	Dripping
C	80	20	–	100	–	–	31	Fail	Dripping
F	80	20	5	95	–	–	33	Fail	No dripping
G	80	20	–	95	–	5	34	V-O	Dripping
H	80	20	5	90	–	5	36	V-O	No dripping
I	80	20	–	95	5	–	34	V-O	Dripping
J	80	20	5	90	5	–	38	V-O	No dripping

[a] *V-O rating indicates self-extinguishing behavior, while 'Fail' indicates dripping and complete burn out of material.*

the materials burned without dripping in UL-94 with a V-0 rating, thus demonstrating the synergistic effects of LDHs with MRP and EG in LDPE/EVA/HFMH blends. Although MRP and EG are good char-forming agents, LDHs help to make the char more compact and homogeneous and also augment the gas-phase mechanism of HFMH. In another study by Du et al., the addition of 2–15 wt% EG to EVA/LDH blends helped to reduce the PHRR further while the addition of fumed silica caused no significant change in the combustion behavior of EVA/LDH composites [110].

LDHs were also found to augment the flame retardancy of APP in poly(vinyl alcohol) (PVA). APP decomposes into polyphosphoric acid, ammonia, and water upon heating. While polyphosphoric acid promotes char formation in polymers along with other char-forming agents, ammonia and water vapor tend to blow the char resulting in intumescence. Typically, higher loadings of APP (>30 wt%) are required to get significant flame retardance. In a study by Zhao et al. [111] addition of 15 wt% APP to PVA increased the LOI from 19.7 to 27.5 but the materials failed the UL-94 tests. However, replacing 0.1–0.3 wt% of APP with Zn/Al-LDH helped the materials to obtain a V-0 rating in the UL-94 tests in addition to increasing the LOI beyond 30. Adding LDHs >1.0 wt% to the PVA/APP system tends to decrease the flame retardance and the reasons for this are not clear. Jiao and Chen [112] found that ZnO can enhance the flame retardant effects of LDHs in EVA. In EVA/LDH composites, when LDH is partially replaced with ZnO, the UL-94 rating increases from V-1 to V-0 (i.e. self-extinguishing time drops from within 30 s (V-1) to within10 s (V-0)). ZnO is generally considered an inert additive, and it is unknown why this synergistic effect should be present when combined with LDHs. It has been suggested that ZnO enhances the viscosity of the EVA/LDH system and acts in a physical manner to yield better flame retardancy.

In another study, LDHs were found to enhance the intumescent flame retardance of APP/pentaerythritol(PER) in PP [99]. While a 100:20:10 ratio of PP:APP:PER revealed a PHRR of 506 kW m^{-2} as compared to 1275 kW m^{-2} of neat PP, a composite of PP:APP:PER:LDHs in the ratio 100:20:10:1 had a PHRR of 318 kW m^{-2}. Composites of PP/APP/PER (100:20:10) and PP/Zn/Al-LDHs (100:1) fail in the UL-94 test, but PP:APP:PER:LDH (100:20:10:1) passed the UL-94 test with a V-0 rating. Du and Fang [113] synthesized Ni/Al-LDH wrapped CNT hybrids by introducing CNTs into the hydrothermal synthesis of LDHs. Nucleation and in situ growth of LDH lamella on CNTs were allowed to occur during the coprecipitation of Ni^{2+} and Al^{3+} ions. Morphological analysis confirmed the wrapping of LDH lamellae on CNTs. They prepared PP composites with LDH, CNTs, and LDH wrapped CNTs by melt blending and investigated combustion behavior in a cone calorimeter. While the PHRR of PP composites with LDH (5 wt%) and CNTs (0.5 wt%) were 538 and 549 kW m^{-2}, respectively, incorporation of 5 wt % LDH-wrapped CNTs in PP led to a PHRR of 490 kW m^{-2}. It was proposed that the enhancement in flame retardancy could be due to a synergistic combination of barrier and free radical trapping effects of the CNTs along with the char forming behavior of LDHs.

6. CONCLUSIONS

In recent years, blending inorganic nanomaterials into organic polymers has become an attractive design approach, often yielding hybrid materials with significantly better mechanical, chemical, thermal, optical and other physical properties. Inorganic LDHs, also termed as "anionic clays", have gained increasing attention as multifunctional inorganic fillers. They have platelet morphology similar to that of cationic layered silicates but possess a much higher charge density and greater chemical tunability. New synthetic protocols including coprecipitation, urea hydrolysis, and sol–gel synthesis have been developed to synthesize LDHs from a variety of divalent and trivalent cations with narrow crystal size distributions and higher specific surface areas. The anions in the hydrated interlayer can be exchanged with long chain aliphatic anions either during synthesis or in a subsequent step to facilitate exfoliation of layers and homogeneous mixing in polymers.

High thermal stability, layered structure, and significant water content continue to motivate the use of LDHs as flame retardant additives in many different polymers (a summary is given in Table 4). Extensive research over the last decade has facilitated a better understanding on the flame retardation mechanisms of polymer/LDH composites. LDHs appear to have a dual mode of action to inhibit flame spread in polymers; (1) endothermic release of interlayer and structural water, which enables significant heat absorption and dilution of combustible gases, (2) formation of metal oxide residues that hinder oxygen transport to the bulk polymer phase under the burning surface providing a barrier effect. In many instances, LDHs have imparted better flame retardance to polymers than layered silicates such as montmorillonites and this is probably due to the combined gas phase and condensed-phase action of these anionic clays. It is also interesting to find that polymer/LDH nanocomposites have a lower PHRR and better LOI values when compared to polymers filled with conventional fillers such as ATH or MH, where the predominant mode of action is in the gas phase. The combined physical and chemical action of LDHs results in good performance both under radiant heat conditions and in standard flammability tests.

Although LDHs have been shown to significantly enhance the combustion behavior of polymers, in many instances, their addition alone cannot ensure that the materials will pass standard flammability tests such as UL-94, unless incorporated in significantly higher quantities (>50 wt%). However, this can compromise mechanical properties of the composite. A more useful approach has been to augment the flame retardant effects of LDHs with other fillers or additives. In this regard, the synergistic action of LDHs with other flame retardants such as APP, MRP, EG, and zinc oxide has been demonstrated.

The most interesting aspect of LDHs is their tunable chemistry, which has led many researchers to create LDHs with different metal compositions or to intercalate them with surfactant-like molecules. However, different LDHs systems in the same study almost assuredly yield different dispersion states, even in cases where the same polymer matrix is used. In most investigations, LDH chemistry and LDH dispersion in the polymer are confounded in flame retardance testing results.

Table 4 Summary of Significant Flame Retardant Effects of LDHs in Various Polymers

Polymer	Type of Filler	Concentration of Filler	Significant Observations in Flame-Retardancy	References
Poly(ethylene–co-vinyl acetate)	Mg/Al–CO_3	50 wt%	UL 94 rating: HB LOI: 29.5 ~100% increase in time to ignition	[84]
Poly(ethylene–co-vinyl acetate)	Mg/Al-PO_4	40–60 wt%	UL 94 rating: V-O LOI: 40 t_{ig} increased by 35% PHRR reduced by 76%	[85]
	Mg/Al–CO_3	40–60 wt%	UL 94 rating: V-O LOI: 38 t_{ig} increased by 14% PHRR reduced by 66%	
Poly(ethylene–co-vinyl acetate)	Mg/Al-borate Zn/Al-borate	40 wt% 40 wt%	PHRR reduced by 74% PHRR reduced by 77%	[15]
Poly(ethylene–co-vinyl acetate)	Hydrotalcite (Mg/Al–CO_3) treated with stearic acid	150 phr	UL 94 rating: V-O No dripping LOI: 43	[86]
Epoxy	Mg/Al–CO_3 (organically modified)	5 wt%	UL 94 horizontal flame test rating: self-extinguishing PHRR reduced by 40–51%	[88]
	Mg/Al–CO_3 (organically modified) Ammonium polyphosphate (APP)	4 wt% 16–20 wt%	UL 94 rating: V-O	
Poly(ethylene)	Zn/Al–CO_3 (organically modified)	5 wt%[a]	PHRR reduced by 55%	[89]

Poly(ethylene)	Mg/Al–CO_3 (organically modified)	16.2 wt%[a]	PHRR reduced by 63% THR reduced by 44% t_{ig} increased by 20%	[91]
Poly(propylene)	Zn/Mg/Al-nitrate (oleate modified)	4 wt%[a]	PHRR reduced by 38%	[102]
Poly(propylene): poly(propylene)-graft-maleic anhydride (1:1)	Zn/Mg/Al-nitrate (oleate modified)	4 wt%[a]	PHRR reduced by 68%	[102]
Poly(propylene): poly(propylene)-graft-maleic anhydride (1:1)	Mg/Al-nitrate (oleate modified)	4 wt%[a]	PHRR reduced by 52%	[102]
	Zn/Al-nitrate (oleate modified)	4 wt%[a]	PHRR reduced by 68%	
Poly(methyl methacrylate) (PMMA)	Mg/Al-nitrate (organically modified)	10 wt%	PHRR reduced by 49–58%	[103]
Poly(methyl methacrylate) (PMMA)	Ca/Al-nitrate (organically modified)	10 wt%	PHRR reduced 54%	[104]
Nylon 6	Mg/Al-nitrate (organically modified)	5–20 wt%	PHRR reduced 37–68%	[106]
LDHs with Other Flame Retardants				
LDPE: poly(ethylene-co-vinyl acetate) (80:20)	Mg/Al-NO_3 (organically modified) HFMH Microencapsulated red phosphorus	5 phr 90 phr 5 phr	UL 94 rating: V-O No dripping LOI: 36	[109]
	Mg/Al-NO_3 (organically modified) HFMH Expandable graphite	5 phr 90 phr 5 phr	UL 94 rating: V-O No dripping LOI: 38	[109]

Continued

Table 4 Summary of Significant Flame Retardant Effects of LDHs in Various Polymers—cont'd

Polymer	Type of Filler	Concentration of Filler	Significant Observations in Flame-Retardancy	References
PVA	APP Zn/Al nitrate + APP	15 wt% 0.1–0.5 wt% 14.5–14.9 wt%	UL 94 rating: fail LOI: 27.5 UL 94 rating: V-O LOI: 30.5	[111]
Poly(ethylene-co-vinyl acetate)	Mg/Al–CO_3 (organically modified)	120 phr	UL 94 rating: V-1	[112]
	Mg/Al–CO_3 (organically modified) Zinc oxide	115–118 phr 2–5 phr	UL 94 rating: V-O	
Poly(propylene)	APP PER	PP:APP:PER = 100:20:10	UL 94 rating: fail LOI: 30 PHRR reduced by 60%	[99]
	APP PER Zn/Al-NO_3 (organically modified)	PP:APP:PER:LDH = 100:20:10:1	UL 94 rating: V-O LOI: 33 PHRR reduced by 75%	

PHRR, peak heat release rate obtained from cone calorimeter tests; THR, Total heat released obtained from cone calorimeter tests.
[a] Calculated based on inorganic content in organically modified LDHs.

From a fundamental perspective, well-controlled investigations aimed toward understanding the effects of LDH dispersion on flame retardancy are necessary to delineate the effects of LDH chemistry. With this knowledge, more efforts to tune LDH chemistry, either by new synthetic protocols or intercalation processes, in conjunction with flame retardant formulation expertise should help to better establish the potential of LDHs as a promising green flame retardant.

References

[1] Forano C, Hibino T, Leroux F, Taviot-Gueho C. Layered double hydroxides. In: Bergaya F, Theng BKG, Lagaly G, editors. Handbook of clay science. Oxford: Elsevier Ltd,; 2006. pp. 1021–95.

[2] Li F, Tan Q, Evans D, Duan X. Synthesis of carbon nanotubes using a novel catalyst derived from hydrotalcite-like Co/Al layered double hydroxide precursor. Catal Lett 2005;99:151–6.

[3] Alejandre A, Medina F, Rodriguez X, Salagre P, Cesteros Y, Sueiras JE. Cu/Ni/Al layered double hydroxides as precursors of catalysts for the wet air oxidation of phenol aqueous solutions. Appl Catal B 2001;30:195–207.

[4] McKenzie AL, Fishel CT, Davis RJ. Investigation of the surface structure and basic properties of calcined hydrotalcites. J Catal 1992;138:547–61.

[5] Pavan PC, Gomes GA, Valim JB. Adsorption of sodium dodecyl sulfate on layered double hydroxides. Micropor Mesopor Mater 1998;21:659–65.

[6] Chitrakar R, Tezuka S, Sonoda A, Sakane K, Ooi K, Hirotsu T. Adsorption of phosphate from seawater on calcined Mg/Mn-layered double hydroxides. J Colloid Interface Sci 2005;290:45–51.

[7] Das J, Patra BS, Baliarsingh N, Parida KM. Adsorption of phosphate by layered double hydroxides in aqueous solutions. Appl Clay Sci 2006;32:252–60.

[8] Tagaya H, Ogata A, Kuwahara T, Ogata S, Karasu M, Kadokawa J-i, et al. Intercalation of colored organic anions into insulator host lattices of layered double hydroxides. Micropor Mater 1996;7:151–8.

[9] Choy J-H, Choi S-J, Oh J-M, Park T. Clay minerals and layered double hydroxides for novel biological applications. Appl Clay Sci 2007;36:122–32.

[10] Choy J-H, Kwak S-Y, Park J-S, Jeong Y-J, Portier J. Intercalative nanohybrids of nucleoside monophosphates and DNA in layered metal hydroxide. J Am Chem Soc 1999;121:1399–400.

[11] Leroux F, Besse J-P. Polymer interleaved layered double hydroxide: A new emerging class of nanocomposites. Chem Mat 2001;13:3507–15.

[12] Jayashree RS, Vishnu Kamath P. Layered double hydroxides of Ni with Cr and Mn as candidate electrode materials for alkaline secondary cells. J Power Sources 2002;107:120–4.

[13] Liu XM, Zhang YH, Zhang XG, Fu SY. Studies on Me/Al-layered double hydroxides (Me, Ni and Co) as electrode materials for electrochemical capacitors. Electrochim Acta 2004;49:3137–41.

[14] Giannelis EP, Nocera DG, Pinnavaia TJ. Anionic photocatalysts supported in layered double hydroxides: intercalation and photophysical properties of a ruthenium complex anion in synthetic hydrotalcite. Inorg Chem 1987;26:203–5.

[15] Nyambo C, Wilkie CA. Layered double hydroxides intercalated with borate anions: fire and thermal properties in ethylene vinyl acetate copolymer. Polym Degrad Stabil 2009;94:506–12.

[16] Manzi-Nshuti C, Wang D, Hossenlopp JM, Wilkie CA. Aluminum-containing layered double hydroxides: the thermal, mechanical, and fire properties of (nano)composites of poly(methyl methacrylate). J Mater Chem 2008;18:3091–102.

[17] Wang Z, Han E, Ke W. Influence of nano-LDHs on char formation and fire-resistant properties of flame-retardant coating. Prog Org Coat 2005;53:29–37.

[18] Rives V, Ulibarri MA. Layered double hydroxides (LDH) intercalated with metal coordination compounds and oxometalates. Coord Chem Rev 1999;181:61–120.

[19] Labajos MF, Sastre MD, Trujillano R, Rives V. New layered double hydroxides with the hydrotalcite structure containing Ni(II) and V(III). J Mater Chem 1999;9: 1033–9.

[20] Velu S, Satoh N, Gopinath CS, Suzuki K. Oxidative reforming of bio-ethanol over CuNiZnAl mixed oxide catalysts for hydrogen production. Catal Lett 2002;82: 145–52.

[21] Velu S, Suzuki K, Okazaki M, Osaki T, Tomura S, Ohashi F. Synthesis of new Sn-incorporated layered double hydroxides and their thermal evolution to mixed oxides. Chem Mat 1999;11:2163–72.

[22] Khan AI, O'Hare D. Intercalation chemistry of layered double hydroxides: recent developments and applications. J Mater Chem 2002;12:3191–8.

[23] Rao MM, Reddy BR, Jayalakshmi M, Jaya VS, Sridhar B. Hydrothermal synthesis of Mg–Al hydrotalcites by urea hydrolysis. Mater Res Bull 2005;40:347–59.

[24] Xu ZP, Stevenson GS, Lu CQ, Lu GQ, Bartlett PF, Gray PP. Stable suspension of layered double hydroxide nanoparticles in aqueous solution. J Am Chem Soc 2005; 128:36–7.

[25] Liu Z, Ma R, Ebina Y, Iyi N, Takada K, Sasaki T. General synthesis and delamination of highly crystalline transition-metal-bearing layered double hydroxides. Langmuir 2006;23:861–7.

[26] Marcelin G, Stockhausen NJ, Post JFM, Schutz A. Dynamics and ordering of intercalated water in layered metal-hydroxides. J Phys Chem 1989;93:4646–50.

[27] Pesic L, Salipurovic S, Markovic V, Vucelic D, Kagunya W, Jones W. Thermal-characteristics of a synthetic hydrotalcite-like material. J Mater Chem 1992;2: 1069–73.

[28] Parker LM, Milestone NB, Newman RH. The use of hydrotalcite as an anion absorbent. Ind Eng Chem Res 1995;34:1196–202.

[29] Inacio J, Taviot-Gueho C, Forano C, Besse JP. Adsorption of MCPA pesticide by MgAl-layered double hydroxides. Appl Clay Sci 2001;18:255–64.

[30] Ulibarri MA, Pavlovic I, Barriga C, Hermosín MC, Cornejo J. Adsorption of anionic species on hydrotalcite-like compounds: effect of interlayer anion and crystallinity. Appl Clay Sci 2001;18:17–27.

[31] Aranda P, Ruizhitzky E. Poly(ethylene oxide)-silicate intercalation materials. Chem Mat 1992;4:1395–403.

[32] Rives V. Characterisation of layered double hydroxides and their decomposition products. Mater Chem Phy 2002;75:19–25.

[33] Aramendia MA, Aviles Y, Borau V, Luque JM, Marinas JM, Ruiz JR, et al. Thermal decomposition of Mg/Al and Mg/Ga layered-double hydroxides: a spectroscopic study. J Mater Chem 1999;9:1603–7.

[34] Sranko D, Pallagi A, Kuzmann E, Canton SE, Walczak M, Sapi A, et al. Synthesis and characteristic properties of novel Ba(II)Fe(III) layered double hydroxides. Appl Clay Sci 2010;48:214–7.

[35] Zhao Y, Li F, Zhang R, Evans DG, Duan X. Preparation of layered double-hydroxide nanomaterials with a uniform crystallite size using a new method involving separate nucleation and aging steps. Chem Mat 2002;14:4286–91.

[36] Xu S, Chen Z, Zhang B, Yu J, Zhang F, Evans DG. Facile preparation of pure CaAl-layered double hydroxides and their application as a hardening accelerator in concrete. Chem Eng J 2009;155:881–5.

[37] Meyn M, Beneke K, Lagaly G. Anion-exchange reactions of layered double hydroxides. Inorg Chem 1990;29:5201–7.

[38] Woo MA, Woo Kim T, Paek MJ, Ha HW, Choy JH, Hwang SJ. Phosphate-intercalated Ca–Fe-layered double hydroxides: crystal structure, bonding character, and release kinetics of phosphate. J Solid State Chem 2011;184:171–6.

[39] Rousselot I, Taviot-Guého C, Leroux F, Léone P, Palvadeau P, Besse JP. Insights on the structural chemistry of hydrocalumite and hydrotalcite-like materials. Investigation of the series $Ca_2M^{3+}(OH)6Cl \cdot 2H_2O$ (M^{3+}: Al^{3+}, Ga^{3+}, Fe^{3+}, and Sc^{3+}) by x-ray powder diffraction. J Solid State Chem 2002;167:137–44.

[40] Vichi FM, Alves OL. Preparation of Cd/Al layered double hydroxides and their intercalation reactions with phosphonic acids. J Mater Chem 1997;7:1631–4.

[41] Prevot V, Casal B, Ruiz-Hitzky E. Intracrystalline alkylation of benzoate ions into layered double hydroxides. J Mater Chem 2001;11:554–60.

[42] Arai Y, Ogawa M. Preparation of Co–Al layered double hydroxides by the hydrothermal urea method for controlled particle size. Appl Clay Sci 2009;42:601–4.

[43] Lwin Y, Yarmo MA, Yaakob Z, Mohamad AB, Ramli Wan Daud W. Synthesis and characterization of Cu–Al layered double hydroxides. Mater Res Bull 2001;36: 193–8.

[44] Challier T, Slade RCT. Nanocomposite materials: polyaniline-intercalated layered double hydroxides. J Mater Chem 1994;4:367–71.

[45] Prévot V, Forano C, Besse JP. Hybrid derivatives of layered double hydroxides. Appl Clay Sci 2001;18:3–15.

[46] Besserguenev AV, Fogg AM, Francis RJ, Price SJ, O'Hare D, Isupov VP, et al. Synthesis and structure of the gibbsite intercalation compounds [LiAl2(OH)6]x {x = Cl, Br, NO_3} and $[LiAl2(OH)6]Cl \cdot H_2O$ using synchrotron X-ray and neutron powder diffraction. Chem Mat 1997;9:241–7.

[47] Fogg AM, Dunn JS, Shyu SG, Cary DR, O'Hare D. Selective ion-exchange intercalation of isomeric dicarboxylate anions into the layered double hydroxide $[LiAl_2(OH)_6]Cl \cdot H_2O$. Chem Mat 1998;10:351–5.

[48] Sissoko I, Iyagba ET, Sahai R, Biloen P. Anion intercalation and exchange in $Al(OH)_3$-derived compounds. J Solid State Chem 1985;60:283–8.

[49] Hou X, Kirkpatrick RJ. Thermal evolution of the Cl^-–$LiAl_2$ layered double hydroxide: a multinuclear MAS NMR and XRD perspective. Inorg Chem 2001;40:6397–404.

[50] Millange F, Walton RI, O'Hare D. Time-resolved X-ray diffraction study of the liquid-phase reconstruction of Mg–Al-carbonate hydrotalcite-like compounds. J Mater Chem 2000;10:1713–20.

[51] Yamaguchi N, Ando D, Tadanaga K, Tatsumisago M. Direct formation of Mg–Al-layered double-hydroxide films on glass substrate by the sol–gel method with hot water treatment. J Am Ceram Soc 2007;90:1940–2.

[52] Abdelkader NB, Bentouami A, Derriche Z, Bettahar N, de Ménorval LC. Synthesis and characterization of Mg–Fe layer double hydroxides and its application on adsorption of orange G from aqueous solution. Chem Eng J 2011;169:231–8.

[53] Prinetto F, Ghiotti G, Graffin P, Tichit D. Synthesis and characterization of sol–gel Mg/Al and Ni/Al layered double hydroxides and comparison with co-precipitated samples. Micropor Mesopor Mater 2000;39:229–47.

[54] Zhou J, Xu ZP, Qiao S, Liu Q, Xu Y, Qian G. Enhanced removal of triphosphate by MgCaFe-Cl-LDH: synergism of precipitation with intercalation and surface uptake. J Hazard Mater 2011;189:586–94.

[55] Malherbe F, Forano C, Besse JP. First coprecipitation of an LDH containing manganese as the divalent cation. J Mater Sci Lett 1999;18:1217–9.

[56] Li Y, Li HJ, Yang MY, He XX, Ni P, Kang LP, et al. Topochemical synthesis of Ni^{2+}/Fe^{3+} layered double hydroxides with large size. Appl Clay Sci 2011;52:51–5.

[57] Caravaggio GA, Detellier C, Wronski Z. Synthesis, stability and electrochemical properties of Ni/Al and Ni/V layered double hydroxides. J Mater Chem 2001;11:912–21.

[58] Lv L, Sun P, Gu Z, Du H, Pang X, Tao X, et al. Removal of chloride ion from aqueous solution by ZnAl-NO_3 layered double hydroxides as anion-exchanger. J Hazard Mater 2009;161:1444–9.

[59] Inayat A, Klumpp M, Schwieger W. The urea method for the direct synthesis of ZnAl layered double hydroxides with nitrate as the interlayer anion. Appl Clay Sci 2011;51:452–9.

[60] Tichit D, Lorret O, Coq B, Prinetto F, Ghiotti G. Synthesis and characterization of Zn/Al and Pt/Zn/Al layered double hydroxides obtained by the sol–gel method. Micropor Mesopor Mater 2005;80:213–20.

[61] Shao MF, Han JB, Wei M, Evans DG, Duan X. The synthesis of hierarchical Zn–Ti layered double hydroxide for efficient visible-light photocatalysis. Chem Eng J 2011;168:519–24.

[62] Mascolo G. Synthesis of anionic clays by hydrothermal crystallization of amorphous precursors. Appl Clay Sci 1995;10:21–30.

[63] Yun SK, Pinnavaia TJ. Water content and particle texture of synthetic hydrotalcite-like layered double hydroxides. Chem Mat 1995;7:348–54.

[64] Ulibarri MA, Fernandez JM, Labajos FM, Rives V. Anionic clays with variable valence cations: synthesis and characterization of cobalt aluminum hydroxide carbonate hydrate $[Co1\text{-}_xAl_x(OH)_2](CO_3)_{x/2}..nH_2O$. Chem Mat 1991;3:626–30.

[65] Ribet S, Tichit D, Coq B, Ducourant B, Morato F. Synthesis and activation of Co–Mg–Al layered double hydroxides. J Solid State Chem 1999;142:382–92.

[66] Crepaldi EL, Pavan PC, Valim JB. Anion exchange in layered double hydroxides by surfactant salt formation. J Mater Chem 2000;10:1337–43.

[67] Oh JM, Hwang SH, Choy JH. The effect of synthetic conditions on tailoring the size of hydrotalcite particles. Solid State Ionics 2002;151:285–91.

[68] Ogawa M, Kaiho H. Homogeneous precipitation of uniform hydrotalcite particles. Langmuir 2002;18:4240–2.

[69] Costantino U, Marmottini F, Nocchetti M, Vivani R. New synthetic routes to hydrotalcite-like compounds—characterisation and properties of the obtained materials. Eur J Inorg Chem 1998;1998:1439–46.

[70] Adachi-Pagano M, Forano C, Besse JP. Synthesis of Al-rich hydrotalcite-like compounds by using the urea hydrolysis reaction-control of size and morphology. J Mater Chem 2003;13:1988–93.

[71] Aramendía MA, Borau V, Jiménez C, Marinas JM, Ruiz JR, Urbano FJ. Comparative study of Mg/M(III) (M=Al, Ga, In) layered double hydroxides obtained by coprecipitation and the sol–gel method. J Solid State Chem 2002;168:156–61.

[72] Rey S, Mérida-Robles J, Han KS, Guerlou-Demourgues L, Delmas C, Duguet E. Acrylate intercalation and in situ polymerization in iron substituted nickel hydroxides. Polym Int 1999;48:277–82.

[73] Moujahid EM, Dubois M, Besse JP, Leroux F. Role of atmospheric oxygen for the polymerization of interleaved aniline sulfonic acid in LDH. Chem Mat 2002;14: 3799–807.

[74] Oriakhi CO, Farr IV, Lerner MM. Incorporation of poly(acrylic acid), poly(vinylsulfonate) and poly(styrenesulfonate) within layered double hydroxides. J Mater Chem 1996;6:103–7.

[75] Messersmith PB, Stupp SI. High-temperature chemical and microstructural transformations of a nanocomposite organoceramic. Chem Mat 1995;7:454–60.

[76] Kashiwagi T, Gilman JW, Butler KM, Harris RH, Shields JR, Asano A. Flame retardant mechanism of silica gel/silica. Fire Mater 2000;24:277–89.

[77] Gilman JW. Flammability and thermal stability studies of polymer layered-silicate (clay) nanocomposites. Appl Clay Sci 1999;15:31–49.

[78] Gilman JW, Jackson CL, Morgan AB, Harris R, Manias E, Giannelis EP, et al. Flammability properties of polymer/layered silicate nanocomposites. Polypropylene and polystyrene nanocomposites. Chem Mat 2000;12:1866–73.

[79] Shanmuganathan K, Deodhar S, Dembsey NA, Fan Q, Patra PK. Condensed-phase flame retardation in nylon 6-layered silicate nanocomposites: films, fibers, and fabrics. Poly Eng Sci 2008;48:662–75.

[80] Shanmuganathan K, Deodhar S, Dembsey N, Fan Q, Calvert PD, Warner SB, et al. Flame retardancy and char microstructure of nylon-6/layered silicate nanocomposites. J Appl Polym Sci 2007;104:1540–50.

[81] Deodhar S, Shanmuganathan K, Fan Q, Wilkie CA, Costache MC, Dembsey NA, et al. Calcium carbonate and ammonium polyphosphate-based flame retardant composition for polypropylene. J Appl Polym Sci 2011;120:1866–73.

[82] Fontaine G, Bourbigot S, Duquesne S. Neutralized flame retardant phosphorus agent: facile synthesis, reaction to fire in PP and synergy with zinc borate. Polym Degrad Stabil 2008;93:68–76.

[83] Qiu L, Chen W, Qu B. Morphology and thermal stabilization mechanism of LLDPE/MMT and LLDPE/LDH nanocomposites. Polymer 2006;47:922–30.

[84] Camino G, Maffezzoli A, Braglia M, De Lazzaro M, Zammarano M. Effect of hydroxides and hydroxycarbonate structure on fire retardant effectiveness and mechanical properties in ethylene-vinyl acetate copolymer. Polym Degrad Stabil 2001;74:457–64.

[85] Ye L, Qu BJ. Flammability characteristics and flame retardant mechanism of phosphate-intercalated hydrotalcite in halogen-free flame retardant EVA blends. Polym Degrad Stabil 2008;93:918–24.

[86] Jiao CM, Wang ZZ, Ye Z, Hu Y, Fan WC. Flame retardation of ethylene-vinyl acetate copolymer using nano magnesium hydroxide and nano hydrotalcite. J Fire Sci 2006; 24:47–64.

[87] Horrocks AR, Tunc M, Price D. The burning behavior of textiles and its assessment by oxygen-index methods. In: Harrison PW, editor. Textile progress. Manchester: Textile Institute; 1989. pp. 1–205.

[88] Zammarano M, Franceschi M, Bellayer S, Gilman JW, Meriani S. Preparation and flame resistance properties of revolutionary self-extinguishing epoxy nanocomposites based on layered double hydroxides. Polymer 2005;46:9314–28.

[89] Costantino U, Gallipoli A, Nocchetti M, Camino G, Bellucci F, Frache A. New nanocomposites constituted of polyethylene and organically modified Zn/Al-hydrotalcites. Polym Degrad Stabil 2005;90:586–90.

[90] Costa FR, Abdel-Goad M, Wagenknecht U, Heinrich G. Nanocomposites based on polyethylene and Mg–Al layered double hydroxide. I. Synthesis and characterization. Polymer 2005;46:4447–53.

[91] Costa FR, Wagenknecht U, Heinrich G. LDPE/Mg-Al layered double hydroxide nanocomposite: thermal and flammability properties. Polym Degrad Stabil 2007;92:1813–23.

[92] Qin H, Zhang S, Zhao C, Hu G, Yang M. Flame retardant mechanism of polymer/clay nanocomposites based on polypropylene. Polymer 2005;46:8386–95.

[93] Kashiwagi T, Grulke E, Hilding J, Groth K, Harris R, Butler K, et al. Thermal and flammability properties of polypropylene/carbon nanotube nanocomposites. Polymer 2004;45:4227–39.

[94] Mishra S, Sonawane SH, Singh RP, Bendale A, Patil K. Effect of nano-$Mg(OH)_2$ on the mechanical and flame-retarding properties of polypropylene composites. J Appl Polym Sci 2004;94:116–22.

[95] Chiu SH, Wang WK. Dynamic flame retardancy of polypropylene filled with ammonium polyphosphate, pentaerythritol and melamine additives. Polymer 1998;39:1951–5.

[96] Lv P, Wang Z, Hu K, Fan W. Flammability and thermal degradation of flame retarded polypropylene composites containing melamine phosphate and pentaerythritol derivatives. Polym Degrad Stabil 2005;90:523–34.

[97] Ding P, Qu B. Synthesis of exfoliated PP/LDH nanocomposites via melt-intercalation: structure, thermal properties, and photo-oxidative behavior in comparison with PP/MMT nanocomposites. Poly Eng Sci 2006;46:1153–9.

[98] Shi YQ, Chen F, Yang JT, Zhong MQ. Crystallinity and thermal stability of LDH/polypropylene nanocomposites. Appl Clay Sci 2010;50:87–91.

[99] Zhang M, Ding P, Qu B. Flammable, thermal, and mechanical properties of intumescent flame retardant PP/LDH nanocomposites with different divalent cations. Polym Compos 2009;30:1000–6.

[100] Wang DY, Das A, Leuteritz A, Boldt R, Häußler L, Wagenknecht U, et al. Thermal degradation behaviors of a novel nanocomposite based on polypropylene and Co–Al layered double hydroxide. Polym Degrad Stabil 2011;96:285–90.

[101] Starink MJ. The determination of activation energy from linear heating rate experiments: a comparison of the accuracy of isoconversion methods. Thermochim Acta 2003;404:163–76.

[102] Manzi-Nshuti C, Songtipya P, Manias E, Jimenez-Gasco MM, Hossenlopp JM, Wilkie CA. Polymer nanocomposites using zinc aluminum and magnesium aluminum oleate layered double hydroxides: effects of the polymeric compatibilizer and of composition on the thermal and fire properties of PP/LDH nanocomposites. Polym Degrad Stabil 2009;94:2042–54.

[103] Nyambo C, Songtipya P, Manias E, Jimenez-Gasco MM, Wilkie CA. Effect of MgAl-layered double hydroxide exchanged with linear alkyl carboxylates on fire-retardancy of PMMA and PS. J Mater Chem 2008;18:4827–38.

[104] Manzi-Nshuti C, Wang D, Hossenlopp JM, Wilkie CA. The role of the trivalent metal in an LDH: synthesis, characterization and fire properties of thermally stable PMMA/LDH systems. Polym Degrad Stabil 2009;94:705−11.

[105] Nyambo C, Chen D, Su S, Wilkie CA. Variation of benzyl anions in Mg/Al-layered double hydroxides: fire and thermal properties in PMMA. Polym Degrad Stabil 2009;94:496−505.

[106] Du L, Qu B, Zhang M. Thermal properties and combustion characterization of nylon 6/MgAl-LDH nanocomposites via organic modification and melt intercalation. Polym Degrad Stabil 2007;92:497−502.

[107] Wang DY, Leuteritz A, Wang YZ, Wagenknecht U, Heinrich G. Preparation and burning behaviors of flame retarding biodegradable poly(lactic acid) nanocomposite based on zinc aluminum layered double hydroxide. Polym Degrad Stabil 2010;95:2474−80.

[108] Pereira CMC, Herrero M, Labajos FM, Marques AT, Rives V. Preparation and properties of new flame retardant unsaturated polyester nanocomposites based on layered double hydroxides. Polym Degrad Stabil 2009;94:939−46.

[109] Ye L, Ding P, Zhang M, Qu BJ. Synergistic effects of exfoliated LDH with some halogen-free flame retardants in LDPE/EVA/HFMH/LDH nanocomposites. J Appl Polym Sci 2008;107:3694−701.

[110] Du LC, Zhang YC, Yuan XY, Chen JY. Combustion characteristics and synergistic effect of halogen-free flame-retarded EVA/hydrotalcite blends with expandable graphite and fumed silica. Polym-Plast Technol Eng 2009;48:1002−7.

[111] Zhao CX, Liu Y, Wang DY, Wang DL, Wang YZ. Synergistic effect of ammonium polyphosphate and layered double hydroxide on flame retardant properties of poly(vinyl alcohol). Polym Degrad Stabil 2008;93:1323−31.

[112] Jiao CM, Chen XL. Synergistic effects of zinc oxide with layered double hydroxides in EVA/LDH composites. J Therm Anal Calorim 2009;98:813−8.

[113] Du BX, Fang ZP. The preparation of layered double hydroxide wrapped carbon nanotubes and their application as a flame retardant for polypropylene. Nanotechnology 2010;21:1−6.

CHAPTER

21

Pathways to Biodegradable Flame Retardant Polymer (Nano)Composites

M. Murariu, F. Laoutid, Ph. Dubois
University of Mons & Materia Nova Research Center, Mons, Belgium

G. Fontaine, S. Bourbigot
Ecole Nationale Supérieure de Chimie de Lille, Villeneuve d'Ascq, France

E. Devaux, C. Campagne, M. Ferreira, S. Solarski
Université Lille Nord de France, Ecole Nationale Supérieure des Arts et Industries Textiles, Roubaix, France

CHAPTER OUTLINE

Polymer Green Flame Retardants. http://dx.doi.org/10.1016/B978-0-444-53808-6.00021-4

1. Introduction

(*Written by M. Murariu, F. Laoutid, and Ph. Dubois*)

The market for biodegradable polymers is growing every year, and important demands can be expected for those applications where biodegradability offers a clear advantage for the customers and the environment [1–3]. The extraordinary growth and interest for biopolymers is connected to a large number of factors including consumer demand for more environmentally sustainable products, development of biobased feedstocks, and the cumulative restrictions for the use of polymers with high "carbon footprint" of petrochemical origin, particularly in applications such as packaging, automotive, electrical, and electronics industry.

Moreover, as an illustration of the increased interest for utilization of biopolymers, a report from a US-based BCC Research [4] estimates that the total worldwide use of bioplastics, being approximately 570 ktons in 2010, will increase with a 41.4% compound annual growth rate from 2010 to 2015, to reach >3200 ktons in 2015 (Figure 1 [4,5]). It is anticipated that novel technical advances and innovations will lead to a biopolymer market boom, not only in the traditional sector as

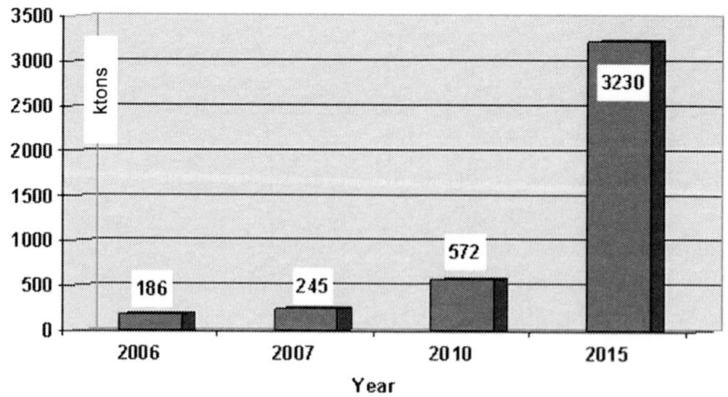

FIGURE 1

Total global biodegradable polymer market [4,5].

packaging but also in transportation, in the automotive and electronics industries, or in other applications with a higher profit potential.

In order to maximize the impact of biopolymers, it is noteworthy also mentioning a growing trend to compound the biobased plastics with petrochemical-based plastics to extend their market as durable products used in cars, cell phones, and elsewhere [4,6,7]. The focus has shifted to the total carbon footprint, and biobased polymers can not only replace existing polymers in various sectors but can also provide combinations of properties for new applications. Unfortunately, for certain applications such as in engineering sectors, biopolymers cannot be fully competitive with conventional thermoplastics since some of their properties are too weak. However, the development of special formulations of biopolymers (e.g. characterized by flame retardancy for application in electronics and in the electrical sector) is clearly in its early stages compared to conventional polymers. Generally speaking, many R&D works currently have as main objective the production using different techniques of new grades of biopolymers characterized by improved characteristics (processability, higher thermal resistance, flame retardancy, tailored electrical properties, impact strength, long durability, and stability at hydrolysis) to allow their utilization in engineering applications.

Selected aspects with focus on bionanocomposites (BNCs) characterized by FR properties will be discussed in this chapter. The design and synthesis of novel polymer BNCs using various nanofillers that are associated or not associated with traditional FR additives has gained an increased interest in polymer and materials science. BNCs obtained by adding nanofillers to biodegradable polymer matrices are very promising materials since they show not only improved performances but they also maintain the intrinsic proprieties of biodegradability.

Following the state of the art, it is also noteworthy mentioning that by considering the published results we can observe only few papers devoted to the flame retardancy of biopolymers (we will not include patents). We assumed that traditional applications of biopolymers such as disposable packaging or semidurable materials (e.g. textile materials) typically do not need FR characteristics. Nowadays, the market turns to more durable materials, and it is expected that some biopolymers, such as polylactic acid (PLA), will substitute thermoplastics such as polypropylene (PP), polyamide (PA), and polyethylene terephthalate (PET) even in those sectors, where fire hazard is an issue and flame retardancy is required [8]. Fire hazards result from the combination of different factors (ignition, amount of heat released during burning, rate of heat release (RHR), flame spread, etc.), and different testing procedures can be used at the laboratory scale for the evaluation of FR properties of the new formulations based on biopolymers. In this regard, the aim of this chapter is to present a screening with different approaches directed to the synthesis and characterization of BNC, "green" products characterized by specific end-use properties such as flame retardancy. The chapter also highlights the main developments in FR systems during the last decade using different nanofillers (clays, carbon nanotubes (CNTs), graphite derivatives or their combinations, etc.) or microfillers, associated or not associated with specific FR additives, and to present selected results in relation to the preparation and characterization of these novel BNCs, with a special focus on PLA-based nanocomposites. Finally, the chapter deals with the current researches and main results obtained by authors in the field of biopolymers characterized by flame retardancy, with special mention for the realization of biodegradable fibers especially designed with FR properties, and finally, the perspectives and future trends in the field of BNCs are discussed.

2. The state of the art: general considerations

(*Written by M. Murariu, F. Laoutid, and Ph. Dubois*)

2.1 Biodegradable polymer materials and BNCs

The terms "bioplastics" and "biodegradable plastics" have been used interchangeably, but there are important differences between the two categories of polymers. Bioplastics such as PLA, thermoplastic starch (TPS), polyhydroxyalkanoates (PHAs), polybutylene succinate (PBS), biopolyamides, and biopolyethylene are polymers that are made from renewable resources, such as food crops or biomass. Some of these bioplastics, but not all, are also biodegradable. Even though a generally recognized definition of the concept of "biopolymers" does not exist, we will accept that they are mainly as follows: (a) based on renewable resources and biodegradable; (b) based on renewable resources but not biodegradable; and (c) based on fossil resources and biodegradable.

Following some legislative restrictions, the "biodegradability" at the end-of-life cycle is a desired condition required by end users to increase the attractiveness and environmental impact of their products. As illustrated in Figure 2 [9,10], we would like to remind that even a number of petrochemical-based polymers are certified as biodegradable and compostable (e.g. polycaprolactone (PCL), polyvinylalcohol (PVOH), polyesteramides, and various copolyesters) [9].

Between the different biopolymers that can be developed as polymer matrices for the fabrication of BNCs, we will mainly focus on PLA, one of the most promising candidates for future developments as a potential alternative to petroleum-based polymers. PLA is not only biocompatible and biodegradable but it is also a thermoplastic aliphatic polyester produced from nonfossil renewable natural resources by the fermentation of polysaccharides or sugar, e.g. extracted from corn, potato, cane molasses, and sugar-beet, and therefore it allows the biological cycle to come to a full circle with PLA biodegradation, as well as the photosynthesis process [11−15]. Due to its properties (high tensile strength and rigidity, thermal stability, processing with traditional equipment, etc.) and biodegradability, PLA has a key position in the very promising bioplastic market. Moreover, the last tendencies evidence that improved PLA-based products are potentially interesting for engineering

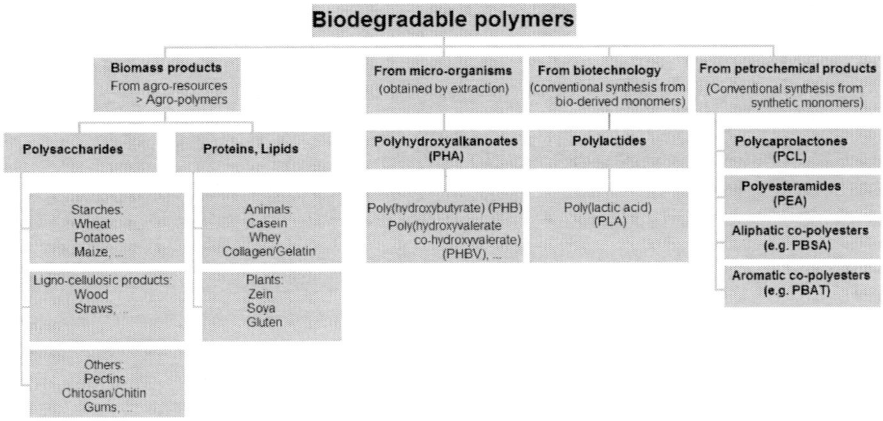

FIGURE 2

Classification of biodegradable polymers depending on their synthesis [9,10].

applications (transportation, electronic and electrical devices, mechanical and automotive parts, etc.).

In recent years, biobased polymers have triggered great interest since new sustainable development strategies are focused on their rapid development [10]. Consequently, to reach the end-user demands for further applications, the profile of biopolymers, such as PLA, can be tuned by combining the polymer matrix with different dispersed phases: microfillers and nanofillers, impact modifiers, plasticizers, polymers, other additives, etc. [16–20].

Polymer-based nanocomposites (PNC) are known to be radical alternatives to conventionally filled polymers due to their significantly improved characteristics (high strength and rigidity, thermal stability, barrier properties, flammability resistance, electrical and optical, or other specific properties) obtained at low nanofiller loadings [21–25]. Moreover, the transition from microparticles to nanoparticles yields dramatic changes in the physical properties of polymers since many important chemical and physical interactions are governed by the properties of high surface nanoscale-dispersed fillers [26].The real interest in nanotechnology is to create revolutionary properties and new functions by tailoring materials at the nanometer scale [27]. Therefore, to extend their applications, biopolymers have been formulated and associated with nanosized fillers, which could bring a large range of improved properties (increased strength, rigidity and heat resistance, decreased gas permeability and flammability, increased biodegradability, etc.). BNCs have been obtained using PLA or other biopolymers in combination with traditional nanofillers such as organomodified layered silicates (OMLS), CNTs, polyhedral oligomeric silsesquioxanes, graphite derivatives, ZnO, and other nanoparticles [24,28–32]. It is also noticeable that the utilization of new (bio)nanoparticles such as starch nanocrystals, or cellulose microfibrils and whiskers has recently received tremendous interest [33–35].

2.2 Short overview on polymer flame retardancy

Most polymers that are composed mainly of carbon and hydrogen are excellent fuels, and their combustion can be initiated by a small flame. In the presence of a combustive agent (oxygen or air) and a heat source, the combustion of these polymeric materials releases enough heat to maintain the flame. However, it is believed that the testing procedures and general concepts of polymer flame retardancy (briefly summarized below) are also applicable to achieve FR biopolymers. For more information on this subject, we recommend also one of our reviews [36].

When exposed to a heat source, the temperature of the substrate increases leading to polymer melting, followed by its thermal decomposition. At this early stage of combustion, the evolution of the process depends on the intensity of the heating source and polymer thermal properties (e.g. thermal conductivity, thermal diffusivity, heat capacity, enthalpy of fusion, and degree of crystallinity).

At higher temperatures, the weakest chemical bonds begin to break first, and polymer degradation starts, generating combustible and/or not combustible volatile products that ignite when the concentration of combustible gases reaches a critical value. The inflammation occurs under the action of an external source (flame or spark) or when the temperature is adequate, by autoignition. At this stage, a very exothermic radical mechanism takes place, generating products of combustion (CO, CO_2, soot, toxic gases or nontoxic gases, etc.) and a heat production, part of which is fed back to the substrate causing further degradation. If the evolved heat is enough to maintain the polymer degradation and the production of combustible volatile products, the combustion continues even in the absence of an external heating source: the phenomenon is therefore self-sustaining. This is one of the main inconveniences concerning the use of polymeric materials in some applications requiring flame retardancy, as they can easily contribute to the outbreak and the propagation of fire.

To improve the flame retardancy of polymeric materials i.e. limiting the risk of ignition in the presence of heat and to decrease the flame propagation during combustion, it is necessary to break the fire triangle (heat, fuel, oxidizing agent) using selected FRs that can be either physically mixed with the polymer during the transformation process (additive FRs), introduced before (monomers), or after the polymerization process by chemical grafting (reactive FRs). The selection of a suitable FR depends on the chemical structure of the biopolymer and requirements in terms of performance related to the application. There are different classes of FRs that can be classified according to their mode of action. We distinguish FRs acting in the gas phase from those acting in the condensed phase. FRs, such as mineral hydroxides and phosphorus compounds, may act in both the gas and condensed phases [36].

In the gas phase, the FR compound generates upon decomposition, in the flame zone, several active compounds that allow the reduction of the fire intensity. Halogenated compounds reduce the rate of the combustion reaction (flame poisoning) by producing active radicals that inhibit the action of the hot radicals, HO• and H• generated during the thermal decomposition of the polymer and that are responsible for the propagation of the flame.

Nonhalogen FRs, such as metal hydroxides, magnesium dihydroxide (MDH), and aluminum trihydroxide (ATH) or melamine derivatives, act either in the gas and condensed phase. They release inert gases upon endothermic decomposition, respectively, H_2O, CO_2, and NH_3, able to dilute the ignitable gas mixture. The endothermic decomposition of these compounds cools down the polymer and limits the heat rate released during the combustion. The residue formed during the thermal decomposition of metallic hydroxides forms a mineral protective barrier material providing a protection against the diffusion of heat and oxygen through the material and limits the volatilization of combustible products.

The use of phosphorus-based FR compounds allows the formation of a charred residue on the surface of the material through crosslinking reactions involving a part of the polymer and leading to the formation of a heat-resistant char structure that acts as a protective barrier. The expansion of this char could be enhanced by the incorporation of a suitable blowing agent leading to the formation of intumescent FR systems (details in Section 4).

Nanoparticles, due to their large aspect ratio, are capable, at a low incorporation level (typically 3–5 wt%) and when properly dispersed in the polymer matrix, of achieving very interesting FR properties. Their incorporation generally decreases the peak of heat release rate (PHRR), but does not reduce the total heat evolved (THE). The main fire retardant mechanisms in PNC is the formation of a protective layer during combustion that limits the heat transfer into the material, the volatilization of combustible degradation products, and the diffusion of oxygen into the material. This protective layer in the case of OMLS as a nanofiller is developed due to the accumulation of the clay that migrates to the surface upon heating and due to the thermally stable char formed thanks to the catalytic action of the Brønsted acid sites, generated during the thermal decomposition of the clay organomodifier [37,38]. These mechanisms modify the fire properties of PNC, sometimes improving them and in some other cases worsening them, depending on the type of fire test applied and the nature of the measurements recorded. However, to achieve good levels in terms of performances required by fire regulations, it is often useful to use nanoparticles in combination with conventional FRs. The polymer's flammability can be characterized by several tests [36,39,40] allowing the measurement of the appropriate flammability criteria, i.e. ignitability, time to ignition (TTI), flame-spread rate, heat release, etc.

3. Fire performances of biodegradable polymer nanocomposites

(*Written by M. Murariu, F. Laoutid, and Ph. Dubois*)

3.1 Clay nanoparticles

Clay nanocomposites have the largest market, with the main demand in packaging, FR applications, aerospace, and aviation (including military uses). According to different estimations, by 2016, clay nanocomposites are projected to increase their market share to approximately 58% [41]. In this context, it is not surprising that among the nanofillers considered for addition into biopolymers, clays (layered silicates) have been studied in the greatest detail. The great interest for the addition of clays into biopolymers can be attributed to their low costs and ability to dramatically increase polymer performances as compared to the neat matrix, including improved mechanical and flexural properties, elevated heat distortion temperature, enhanced barrier protection, accelerated biodegradation, and flame retardancy, as it comes out from an impressive list of reviews and publications, mainly concerning PLA as the polymer matrix; here, for additional information, only the studies performed by Okamoto, Sinha Ray, and their collaborators are cited [42–51].

The main mechanisms that govern the FR action of nanoclays in nonbiodegradable polymers (PE, PP, PA, PS, etc.) take place also in biodegradable thermoplastic polymers (e.g. protective mineral layer and char formation). As an illustration, in our group, we studied the effect of the incorporation of 3 wt% of two OMLS (Bentone 104 and Cloisite 30B, named below B104 and C30B, respectively) on the flame retardant (FR) properties of PLA. As expected, the PHRR of PLA during the cone calorimeter test was significantly reduced after the incorporation of the nanoclays (Figure 3). Slight differences between TTI and PHRR values were obtained depending on the nature of the nanoclay used. These differences could be attributed to the dispersion state of the nanofillers in the polymer matrix since they are treated with different organomodifiers: benzyl dimethyl hydrogenated tallow ammonium in the case of B104 and bis-(2-hydroxyethyl) methyl hydrogenated tallow ammonium for C30B. However, generally, PLA nanocomposites do not pass some traditional

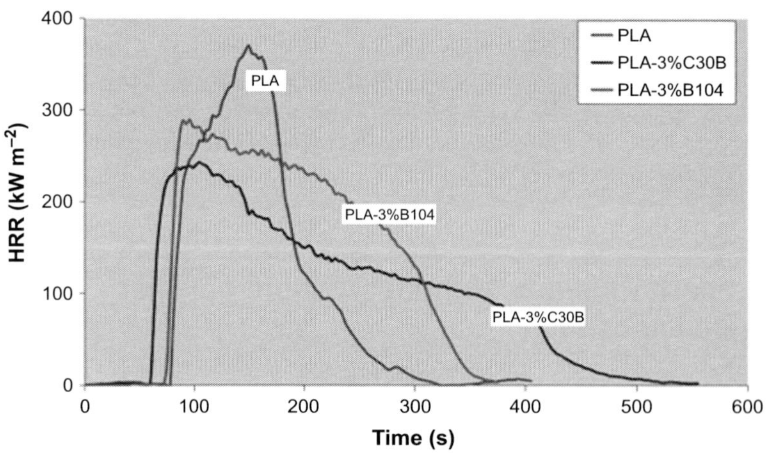

FIGURE 3

HRR versus time of pristine PLA and PLA–OMLS nanocomposites (heat flux: 35 kW m^{-2}).

Table 1 Results of the Horizontal Burning UL 94HB Test for Pristine PLA and Nanocomposites

Composition of Sample (% by Weight)	BR (mm min^{-1})	Category[a]	Remarks
PLA	25 (±1)	HB	Burning with drips
PLA—3% C30B	42 (±5)	Limit	No drips, charring
PLA—3% B104	47 (±13)	Limit	No drips, charring

[a] *Thickness of specimens of about 3.1 mm; a BR ≤ 40 mm min^{-1} allows the HB classification.*

burning tests such as the UL 94V, or even the UL 94HB, which usually give a lower UL-94 rating.

Table 1 gathers the results of UL94 HB characterization, including the information concerning the burning rate (BR), and also qualitative indications observed during testing, e.g. burning with or without drips or formation of char (Figure 4). It turns out that two different behaviors are evident for PLA and the corresponding nanocomposites. Due to combustion with drips, a low BR is recorded for pristine PLA (BR of 25 mm min^{-1}). In fact, this is generally explained by the removal of polymeric material from the pyrolysis zone, thus decreasing the fuel support rate. PLA nanocomposites (C30B/B104) burn without drips, but they show very high BR values (42—47 mm min^{-1}) attributed to the formation of a barrier layer related to the presence of nanoclay, which limits the propagation of fire inside the material but increases the flame-spread rate over the surface of the specimen.

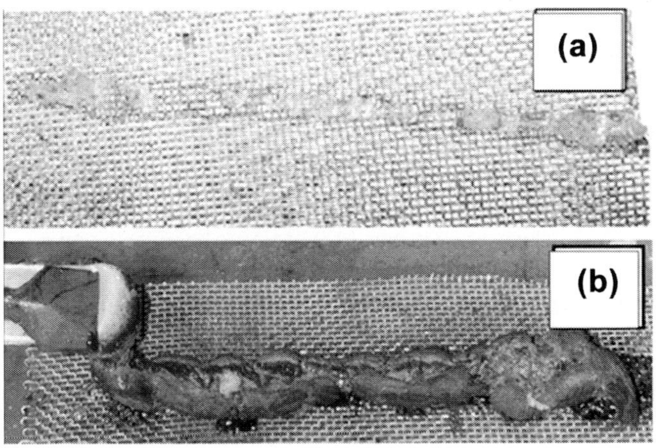

FIGURE 4

Illustration of the residual samples after UL 94HB testing: specimens burned with drips and without char—neat PLA (a) and without dripping and charring—PLA—3% C30B (b).

Finally, it is believed that the FR properties of BNC-containing OMLS are clearly influenced by the fire scenario. Moreover, it is generally accepted that the comparative evaluation of BNCs with respect to the neat polymer matrix by cone calorimetry provides positive and more beneficial information. From our experiments, it has been reconfirmed that the formation of a superficial carbonaceous-silicate charred layer can be considered as the principal mode of FR action leading to the reduction of PHRR, but usually TTI does not increase or the total heat release (THR) is not reduced (additional information in Sections 4 and 5). Because PNCs fail when evaluated according to other important flammability tests [52] such as the UL 94V test, it is believed that the association with traditional FR additives is remaining for further consideration, as one of the principal pathways leading to improved FR performances.

3.2 Carbonaceous nanofillers

3.2.1 Carbon nanotubes

CNTs have received much attention and are considered as key additives due to their positive impact, not only on the polymer mechanical and electrical properties (as reinforcing and as conductive nanofillers, respectively) but also in terms of thermal stability, flame retardancy, or other specific properties [26,53–55]. Furthermore, following the progress in their technology and as the price of CNTs decreases, they become more attractive for utilization in the production of PNC. However, the effective utilization of CNTs in PNCs strongly depends on the ability to homogeneously disperse them throughout the polymer matrix using different techniques: in situ polymerization, melt blending, solvent casting, etc. Thus, the improvements of polymer properties are strongly influenced by the good dispersion of CNTs and the existence of strong interactions between the polymer matrix and nanofiller [56,57].

In this context, several approaches for functionalization of CNTs including covalent sidewall coupling reactions and noncovalent interactions have been developed to overcome the solubility limitations. Rana et al. [58] reported the functionalization of multiwall CNTs (MWCNTs) with PCL using click chemistry, step followed by the investigation of MWCNT dispersion. However, according to the results recently published by our group, the dispersion of CNTs in different solvents can be improved in the presence of poly(2-(dimethylamino) ethyl methacrylate) (PDMAEMA) [59]. It was shown that PDMAEMA behaves as an anchoring agent toward CNTs. The same binding capability was ascribed to PLA-b-PDMAEMA block copolymers, where PDMAEMA acts as a linker toward CNTs. In this context, to obtain BNCs with improved properties, utilization of PLA-b-PDMAEMA block copolymers for the predispersion and noncovalent treatment of carbonaceous nanofillers (CNTs, graphite derivatives, etc.) appears as a reliable approach for the association of biosourced/biodegradable polyesters, such as PLA, to CNTs used as a reinforcement or conductive nanofiller.

CNTs were already used to reduce the flammability properties of numerous polymer materials, showing a general reduction of the combustion rate. The fire behavior of PP-MWCNT nanocomposites was studied in a typical cone calorimeter experiment by Kashiwagi et al. [60]. It was revealed that a network layer consisting of CNTs formed on the entire sample and acting as a physical shield is responsible for the decrease of the external heat flux. Schartel et al. [61] reported similar results for PA 6-MWCNT nanocomposites about the formation of a rigid MWCNT interconnected network structure. CNTs in the presence of oxygen are able to favor oxidative dehydrogenation of polymers, with the production of a stable surface layer of carbon char that provides an effective insulation to oxygen. This mechanism seems to be strictly related to the nature of the polymer, and it has been proved to be very effective for PE. It has been shown that the partial surface carbonization of the polymer matrix occurs subtracting part of the polymer from combustion and creating a char-supported protective structure [62].

Similarly, CNTs have been associated with biopolymers, and especially with PLA, due to the interest for its utilization in engineering applications. A large number of papers in the literature deal with different techniques of CNT dispersion, the improvements of mechanical, thermal, and electrical properties, the effect of CNTs on PLA crystallization, etc. [63–69].

Unfortunately, the number of studies regarding the FR properties of BNC-containing CNTs is relatively reduced; some explanations were presented before or in some reviews realized by some of us [8]. In a recent study conducted by the Bourbigot group, PLA–MWCNT nanocomposites were obtained by reactive extrusion via the ring opening polymerization of L,L-lactide using a continuous single-stage process [70]. Using mass loss calorimetry, it was experimentally established that the PLA–MWCNT sample exhibits lower HRR values than does the neat PLA (reduction by 20% of PHRR compared to virgin PLA) (Figure 5 [70]). It is also notable that the TTI of PLA–MWCNTs was not reduced compared to that of the virgin PLA (60 s versus 105 s, respectively), a behavior ascribed to the increase in the radiation in-depth absorption coefficient by the addition of CNTs. Moreover, due to radiation absorption and lower heat conductivity, accumulation of heat at the surface of the nanocomposite upon heating occurred, and the nanocomposite reached its ignition temperature more rapidly. It was assumed that the mechanism of protection involves the formation of a char layer covering the sample surface and acting as an insulated barrier, reducing volatiles escaping to the flame for a certain time.

As melamine-based compounds are known to be effective FRs, in the next experimental step, the melamine was grafted onto the surface of CNTs [8,71]. Then, MWCNTs and functionalized MWCNTs (f-MWCNTs) (functionalized multiwall carbon nanotubes) were incorporated into PLA via melt blending. Transmission electronic microscopy (TEM) analyses revealed that a high level of nanodispersion is achieved in PLA with f-MWCNTs, while the dispersion was poor using as-received MWCNTs. Hereafter, the PLA nanocomposite containing f-MWCNTs exhibited a PHRR reduction of 28% (Figure 6 [8]), ascribed to the formation of a char layer limiting mass/heat transfer.

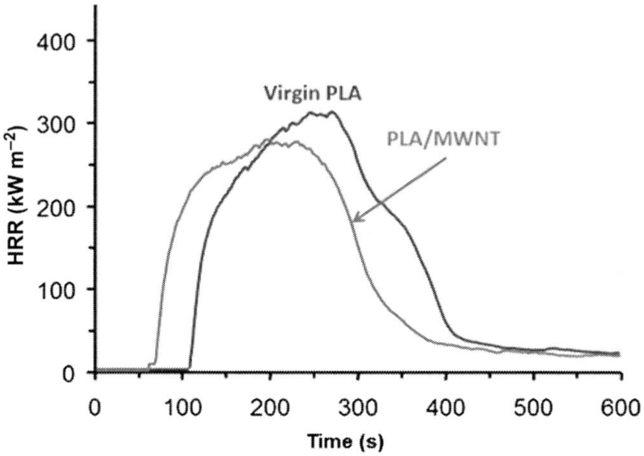

FIGURE 5

HRR curves as a function of time of PLA–1% MWCNTs compared to virgin PLA (heat flux: 35 kW m^{-2}) [70].

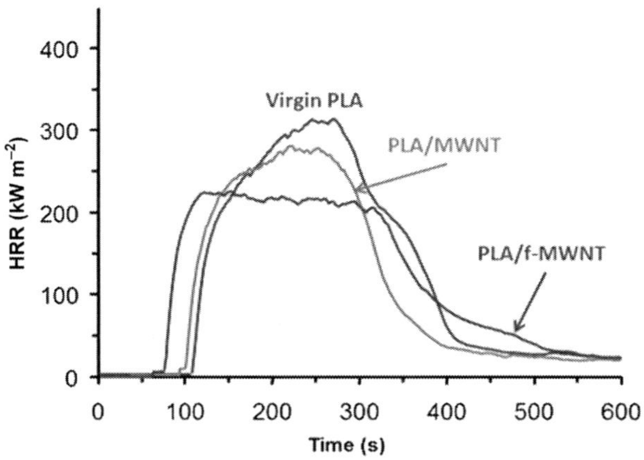

FIGURE 6

HRR curves as a function of time of PLA–MWCNTs and PLA–(f-MWCNTs) compared to virgin PLA (external heat flux: 35 kW m^{-2}) [8].

Furthermore, PLA–MWCNT nanocomposites were prepared by our group (laboratory of Ph. Dubois) using different approaches (e.g. the dry-mixing of PLA with MWCNTs followed by melt compounding, realization of masterbatch highly filled with CNTs and mixing it with the neat PLA, the predispersion of CNTs in solutions of (co)polymers, etc.). The production of PLA nanocomposites was followed by the evaluation of mechanical, thermal, electrical, and fire properties.

Regarding the FR properties, as was already mentioned, the PNCs that do not contain additional additives do not pass some traditional tests such as UL 94V test, required by the industry or by some applications, and in this context, special formulations are looked for. Furthermore, the testing of PLA–MWCNT nanocomposites according to the UL 94V test leads to a similar conclusion. Accordingly, as is shown in Figure 7, in this typical test (UL 94V), the specimens of PLA–MWCNT nanocomposites burn with flaming combustion up to the holding clamp, whereas interestingly enough, the residual specimens keep a rigid and cohesive consistence (Figure 7(b)). However, to have information about the antidripping properties and the BR, the UL 94HB horizontal test was used as the testing procedure. As illustrated in Figure 8, compared to the neat PLA that burns with drips (a behavior that could limit its utilization in electrical applications), the nanocomposites, i.e. PLA–(1–3)% MWCNTs, burn without dripping and form an important and cohesive rigid carbonaceous char, owing to the presence of CNTs.

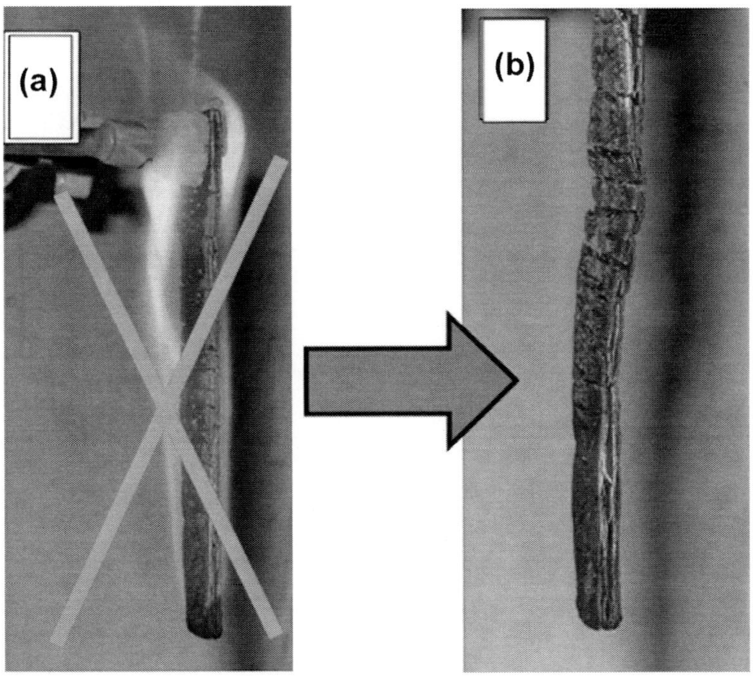

FIGURE 7

PLA–CNT nanocomposites cannot be classified according to the UL 94V test: illustration of the behavior during the testing of PLA (NW 6201D)–2% MWCNT nanocomposites (a) and the image of residual specimens (b).

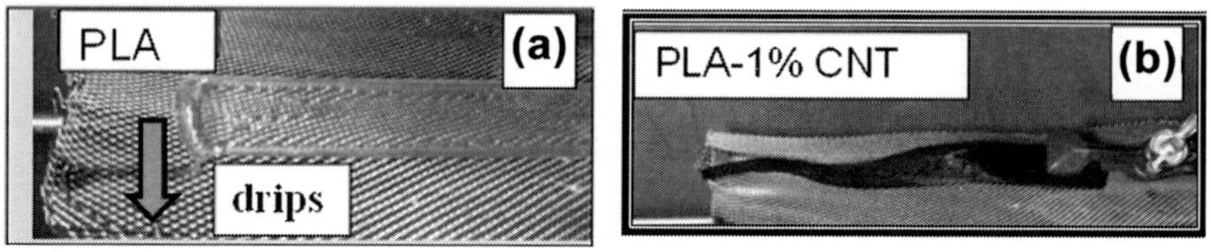

FIGURE 8

Illustration of the burning of PLA (a) and PLA–MWCNT nanocomposites (b) during the UL 94HB testing.

Table 2 gathers the results of the UL 94HB characterization of PLA–MWCNT nanocomposites compared to neat PLA. It turns out that two different behaviors are observed depending on sample formulation. Due to combustion with drips, on the one hand, a low BR is recorded for pristine PLA (BR about 17 mm min^{-1}), ascribed to the removal of polymeric material from the pyrolysis zone with an effect in decreasing the fuel support rate [72]. On the other hand, PLA–MWCNT specimens burn without dripping, showing low BR values (19–21 mm min^{-1}) that allow the HB classification, while maintaining charring and antidripping properties. It is supposed that the rigid network structure obtained during burning of PLA–MWCNTs is responsible for the increase in the melt viscosity, which eventually prevents dripping and flowing. The residual specimens keep more or less their original geometry (Figure 9), which is a main difference by comparing to other BNCs based on PLA, such as containing expanded graphite (EG) as a nanofiller, detailed in Section 3.2.2.

Interestingly, from the results presented in Table 2, most noticeable are the lower BR values obtained in the case of PLA–MWCNT, as compared to those of PLA–clay nanocomposites (Table 1, Section 3.1); this is ascribed to some differences between the FR mechanisms that can be strongly influenced by the nature

Table 2 Fire Properties (UL 94HB Test) of PLA and PLA–MWCNT Nanocomposites

Sample[a] (% by Weight)	BR (mm min^{-1})	Classification[b]	Remarks
PLA	17 (±2)	HB	Drips
PLA–1% MWCNT	21 (±1)	HB	No drips, charring
PLA–2% MWCNT	19 (±1)	HB	No drips, charring
PLA–3% MWCNT	21 (±1)	HB	No drips, charring

[a] *PLA 6201D, supplier NatureWorks: MWCNTs: Nanocyl NC7000, supplier Nanocyl SA Belgium.*
[b] *Thickness of specimens of approximately 3.1 mm; a BR <40 mm min^{-1} allows the HB classification.*

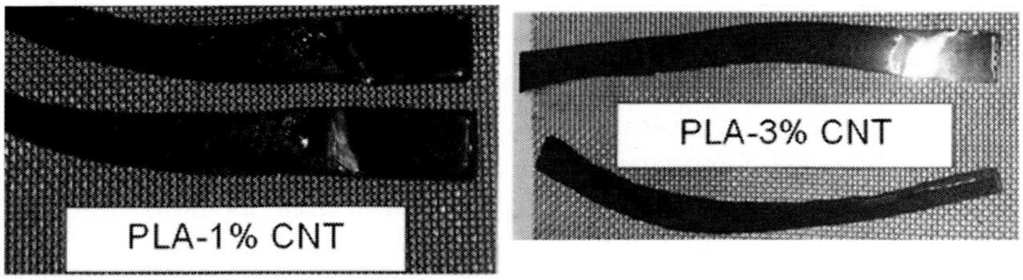

FIGURE 9

Residual specimens of PLA–MWCNT nanocomposites after the UL 94HB test.

of the nanofiller, without excluding the influence of some additional factors, such as those connected to the molecular characteristics of PLA.

As a preliminary conclusion, it can be stated that, during a fire, CNTs provide a protective char on the polymer surface that decreases the transfer of the combustible products to fire and simultaneously the diffusion of oxygen from the external environment toward the surface of the polymer. The char made of CNTs provides a thermal barrier to heat and decreases the rate of the heat transport, whereas the mass transfer is also decreased. Furthermore, it is believed that the combinations of CNTs with specific FRs can further pave the way to BNCs characterized by advanced flame retardancy.

3.2.2 Graphite derivative nanofillers

The high cost of CNTs has limited their extensive use in industrial sectors, and other options are to be considered. As an alternative to CNTs, nanofillers based on graphite derivatives (expandable graphite, EG, exfoliated graphite nanoplatelets (EGNPs), going to the extremely thin sheets of graphene of one atom thickness, etc.) are recently considered among the nanofillers that prove a reach potential due to their multifunctional properties (rigidity and high tensile strength, electrical conductivity, flame retardancy, and barrier properties) [26,73]. In this context, it is noteworthy mentioning that an important number of studies have been carried out on the dispersion of EG and EGNP in engineering or commodity polymers such as polyphenylene sulfide (PPS), polyamide 6 (PA 6), epoxy resins, poly(methyl methacrylate) (PMMA), or polypropylene (PP) [74–77].

Graphite derivatives as nanofillers combine the lower price and the layered structure of clays with the superior thermal and electrical properties of CNTs. In all cases, natural graphite must undergo chemical treatments to allow the polymer chains entrance between the nanofiller layers. Therefore, on the one hand, the dispersion of such graphite nanosheets in a matrix can play a key role in the improvement of both physical and mechanical properties of the resultant PNC. On the other hand, graphite is recognized for its self-lubricating and dry lubricating behavior, and it is expected that for engineering utilization, its addition into a biopolymer of interest, such as PLA, can simultaneously enhance the wear resistance (lower abrasion and friction coefficients) and electrical properties.

a. EG as a nanofiller

PNC-containing EG and expandable graphite as nanofillers have been prepared using various techniques: in situ polymerization, mixing in solution, melt compounding, etc. [74–78]. Simultaneously, many works have been devoted to the exfoliation of previously intercalated graphite using different methods: microwaves, radiofrequency wave treatment, sonication, milling, etc. [76,79,80].

Furthermore, to obtain "green" BNC with specific end-use properties and in response to the demand for extending PLA applications range, it has been recently demonstrated by some of us [72] that commercially available EG (i.e. TIMREX BNB90-TIMCAL Belgium SA) can be melt blended with PLA using different addition techniques. The BNCs were characterized by improved properties with respect to the neat PLA matrix: rigidity; improved crystallization kinetics, moderate increase of thermal stability, antistatic to conductive electrical properties, etc. Additionally, as demonstrated by Dynamic Mechanical Thermal Analysis (DMTA), melt blending PLA with EG leads to the enhancement of the storage modulus and offers the possibility to use these BNCs in applications requiring higher temperatures of utilization. Furthermore, using different techniques of investigation, it was revealed that the presence of EG significantly accelerates the crystallization process of PLA matrix. Currently, the use of PLA in technical applications (e.g. injection molded articles) is limited because PLA has a low crystallization rate when it is compared with many other thermoplastics.

In the context "fire properties", it is important to remind that some nanofillers such as CNTs, OMLS, and more recently, graphite derivatives have received much attention and are considered as potential additives due to their positive impact on principal thermomechanical properties and also in terms of flame retardancy. Generally, as already mentioned before, these nanofillers lead to the formation of a protective thermally stable surface layer during combustion, which limits the heat transfer from the flame to the substrate and mass transfer from the substrate to the flame. Thus, the overall rate of flame feeding by combustible products from polymer pyrolysis and thermo-oxidation is decreased [81,82]. According to some authors, it is likely that the FR mechanism of nanocomposites containing graphite derivatives can show some differences compared to those with OMLS [82]. In particular, even though both categories of nanofillers are characterized by a layered structure and they can lead to equivalent FR properties (e.g. at least as implied by the reduction of PHRR), it is noteworthy mentioning that the chemical composition and physico-mechanical properties of graphite are quite different from those of clays. The FR performances of OMLS are mainly ascribed to the structural collapse of the clays in nanocomposites during combustion, leading finally to the formation of a multilayered carbonaceous silicate that acts as a "physical" barrier [83]. In the case of graphite derivatives, during burning, the material swells and evolves gaseous products with consummation of heat to form a thicker porous foamed carbonaceous mass, which acts as an insulative barrier to interrupt or refrain the combustion. The larger amount of char

Table 3 Results of the Horizontal Burning Test (UL 94HB) for Different PLA Compositions [72]

Composition (wt%)	BR (mm min^{-1})	Classification[a]	Remarks
PLA	20 (±1)	HB	Burning with drips
PLA—3% EG	21 (±2)	HB	Burning with drips, low charring
PLA—6% EG	25 (±1)	HB	Burning without drips, charring
PLA—12% EG	21 (±3)	HB	Burning without drips, charring

[a] *Thickness of specimens is approximately 3.1 mm; a BR <40 mm min^{-1} allows the HB classification.*

formed during burning is ascribed to the additional expansion of graphite nanolayers [72]. Thus, the FR protection of graphite nanofillers exploits the benefits of both the layered structure ("physical" barrier mechanism) and intumescent/blowing effect ("chemical" mechanism), acting as a carbonization compound and as blowing agent [84].

Table 3 [72] gathers the results of UL 94HB characterization, while Figure 10(a)–(d) [72], illustrates different samples (without or with EG) during burning (Figure 10(a) and (b), respectively) or as recovered after the burning test (Figure 10(c) and (d)). PLA burns with intensive dripping and without any charring (Figure 10(a)). Due to its melting and dripping behavior, a low BR was recorded for pristine PLA (BR ≈ 20 mm min^{-1}). According to the results shown in Table 3, the investigated PLA–EG nanocomposites pass the UL 94HB test; the BR is

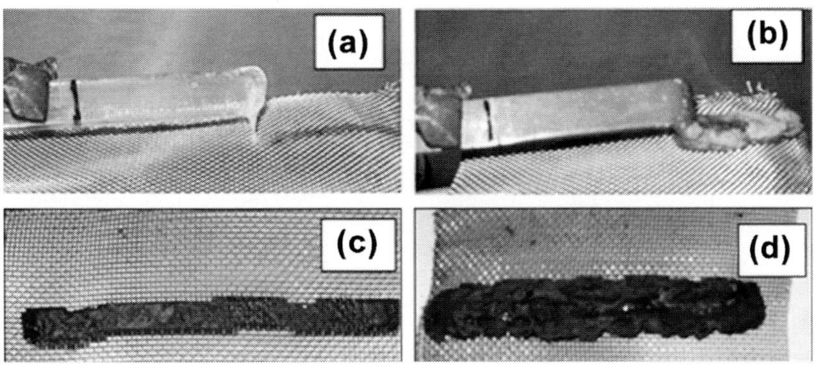

FIGURE 10

Illustration of the behavior during UL 94HB testing: specimens burned with dripping and without char formation: neat PLA (a) and without dripping: PLA–6% EG (b); Residual specimens of PLA–6% EG (c) and PLA–12% EG (d), the burning of nanocomposite samples lead to different amounts of char [72].

significantly <40 mm min^{-1}. The addition of EG into PLA significantly modifies the process of burning (Figure 10(b)). The PLA–EG nanocomposites burn with a decreased number of drips (i.e. PLA–3% EG) or even without any drips (PLA with 6% and 12% EG) and an intumescent char with an initial viscous consistency due to melting of PLA is formed. The formation of the (important) carbonaceous char is in direct correlation with the percentage of the EG nanofiller (consistent with cone calorimeter testing).

It is worth mentioning that the fire measurements by cone calorimetry on PLA–EG samples also prove the beneficial role of graphite in the decrease in the PHRR and similarly the direct dependence between the amount of charring and percentage of the EG nanofiller. For example, Figure 11 [72], shows comparative HRR traces recorded on unfilled PLA (which shows the highest PHRR) and selected PLA–EG nanocomposites. It can be seen that the addition of 6% EG leads to a reduction in the PHRR of about 30% with respect to neat PLA, whereas no significant difference between TTI was observed.

Interestingly enough, for all formulations containing EG, compared to neat PLA, an significant amount of char is formed (Figure 12(a) [72]), the thickness of which is directly correlated to the EG percentage (consistent with the observations previously achieved by the UL 94HB testing). During burning, the PLA–EG nanocomposite swells and evolves gaseous products with consummation of heat to form a porous foamed carbonaceous mass that acts as a barrier to interrupt/refrain the combustion. It is confirmed by these experiments that graphite is an efficient intumescent agent as it acts as a carbonization compound. The important amount of char formed during burning (UL 94HB and cone calorimeter testing)

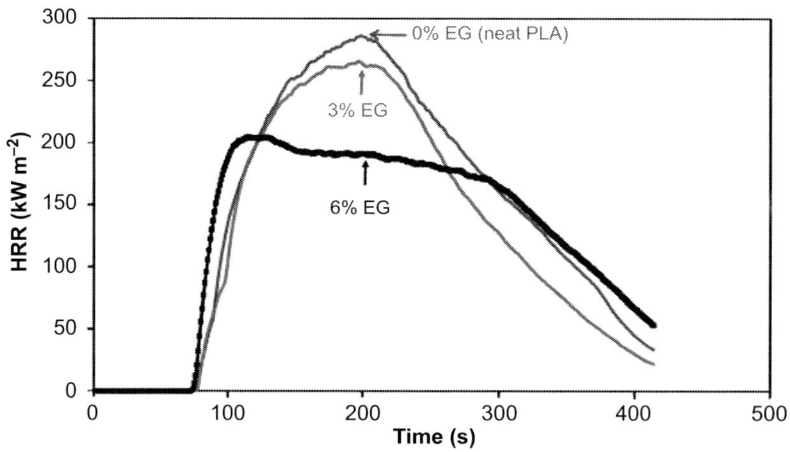

FIGURE 11

HRR plotted against time at a defined heat flux of 35 kW m^{-2}: neat PLA compared to PLA–EG nanocomposites with different loadings of the EG nanofiller [72].

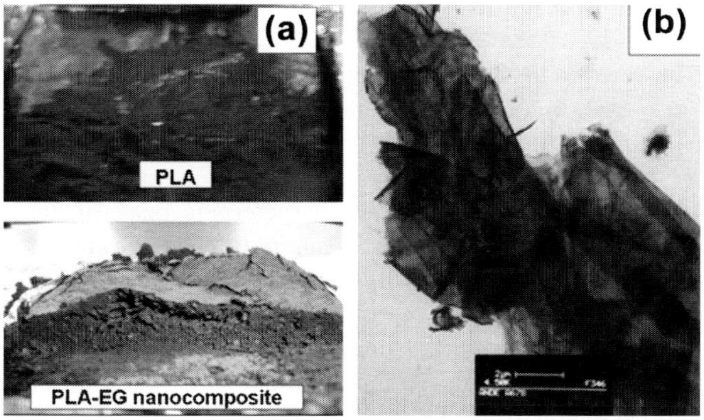

FIGURE 12

Illustration of the char formed in unfilled PLA and PLA–EG nanocomposites during cone calorimeter testing (a) and TEM image realized on the char of PLA–EG nanocomposite (b) [72].

was ascribed by us to the additional expansion of EG nanolayers provoked by both the heat and gases/products of combustion. In fact, this assumption is supported by TEM observations on the residue after burning (Figure 12(b)), obtaining images that revealed the presence of exfoliated graphene nanolayers of different dimensions. Furthermore, the formation of the char layer is believed to be principally responsible for the reduction of HRR values during cone calorimeter experiments.

b. PLA— EGNP

EGNPs due to their unique size and shape can lead to PNCs that show an improvement in desirable properties (electrical and thermal conductivity, stiffness, etc.). They were shown to nucleate different semicrystalline polymer matrices such as PP, polyhydroxybutyrate (PHB), PLA (e.g. investigations realized by the Drzal group [85]), causing dramatic changes in crystallization rates, polymer morphology, and properties of the resultant (nano)composites.

Following the interest for the utilization of graphite derivatives, we have evaluated different grades of graphite nanoplatelets such as xGnP(5) and xGnP(15) (supplier XG Sciences, Inc. USA). According to the supplier, these nanoparticles are characterized by about 10-nm thickness and mean diameters of 5 μm and 15 μm, respectively. For more information concerning these nanofillers, i.e. graphite nanoplatelets, we recommend as informative support additional references [79] [85], and [86]. Some comments concerning the principal properties of BNCs obtained using these nanofillers are summarized below.

By comparing both nanofillers, the addition of xGnP(5) to PLA led to nanocomposites characterized by a slightly improved thermal stability with respect to XGnP(15), ascribed to a somewhat better dispersion, and also to greater tensile strength values, differences that were more marked especially at higher amounts of the nanofiller (e.g. 12 wt%).

Moreover, interestingly enough in relation to the morphology of PLA−xGnP nanocomposites, even that each nanoplatelet has only consisting of about 10 graphene sheets, X-ray diffraction (XRD) analyses (results not shown here) have evidenced for nanocomposites a sharp peak at 2θ of approximately 26.5° that corresponds to the same d-spacing, the characteristic peak of natural graphite. Furthermore, additional information was obtained through TEM images (Figure 13(a) and (b)), which are fully consistent with the XRD analyses. On the one hand, clusters of xGnP exist throughout the polymer matrix, and their presence allows us to conclude that melt compounding cannot totally separate the multiple graphene layers, even when using nanofillers with an initially low dimension of the particles. On the other hand, from the TEM images at a higher magnification (Figure 13(b)), the individual graphene nanolayers with a significant length (even of hundreds of nanometers) can be observed in intercalated assemblies, as well as in dispersed single layers.

Concerning the FR properties, on the one hand, these nanocomposites cannot be classified according to UL 94V (vertical testing) due to the dripping and the flaming combustion of specimens to the holding clamp (as is illustrated in Figure 14). Interestingly as a remark, the formation of some cohesive and rigid char like in the case of CNTs has not been observed. On the other hand, these BNCs containing 3−12% xGnP were classified as HB materials according to UL 94HB horizontal test. The antidripping properties during burning were evidenced at about 6% nanofiller, whereas the important decrease in BR values were typically evidenced at higher

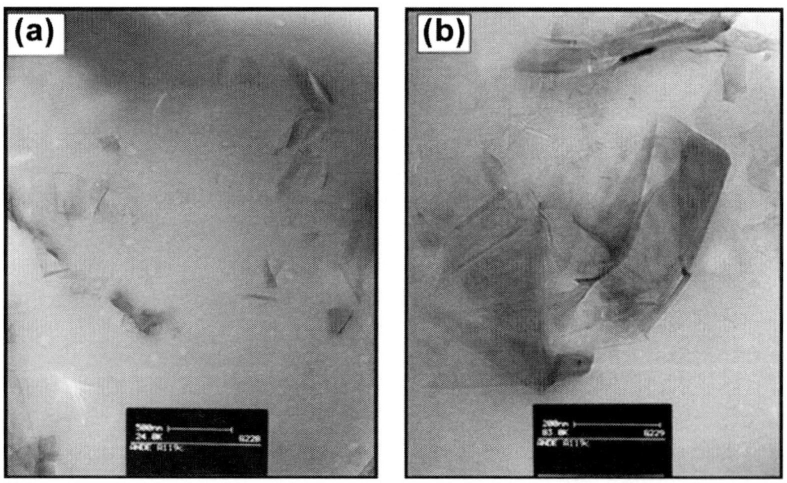

FIGURE 13

Typical TEM images of PLA-3% xGnP(15) nanocomposites showing the good dispersion of the nanofiller (a) and presence of EGNP (b).

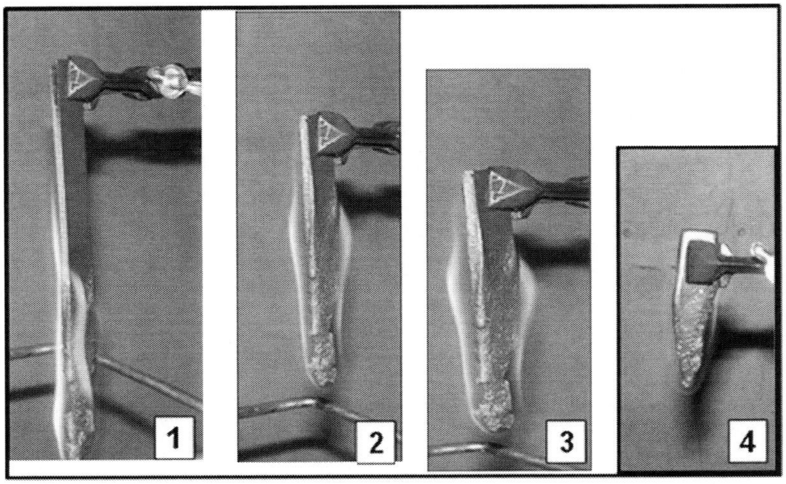

FIGURE 14

UL 94V testing on PLA-xGnP(5) nanocomposites.

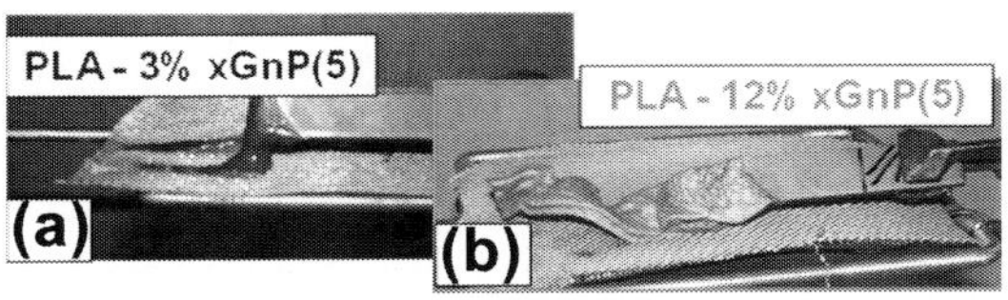

FIGURE 15

UL 94HB testing on PLA-xGnP(5) nanocomposites illustrating the formation of an important carbonaceous char in correlation with xGnP loading (3% (a) versus 12% (b)).

loadings (e.g. a nanocomposite containing 12% xGnP exhibited a BR of about 20 s). Noticeably, as is illustrated in Figure 15(a) and (b), the residual specimens obtained after the burning of PLA−xGnP nanocomposites were relatively more compact and did not show an expanded and porous structure as in the case of nanocomposites containing EG.

Finally, other confirmations of the improved FR properties have been provided by the additional investigations by cone calorimetry. As illustrated in Figure 16, compared to the neat PLA, an impressive reduction of PHRR of about 37% and 52% was obtained by the addition of 6% and 9% EGNP (xGnP(15)), respectively. Corroborated with other specific performances, these results again attest the high potential of graphite nanofillers in the realization of promising BNCs as interesting new materials for engineering applications.

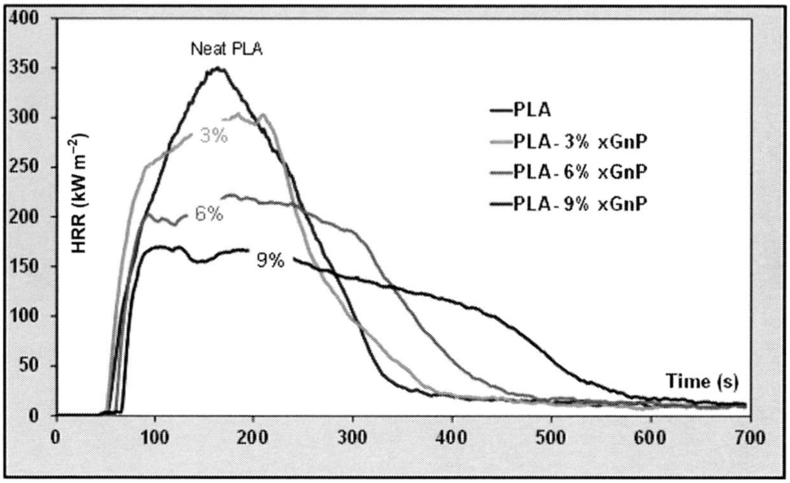

FIGURE 16

HRR plotted against time at a heating flux of 35 kW m^{-2}: neat PLA compared to PLA–xGnP(15) nanocomposites with different loadings of graphite nanoplatelets.

3.3 Combination of different nanoparticles

3.3.1 PLA nanocomposites containing EG and CNTs

The excellent mechanical properties and electrical conductivity of CNTs make them ideally suited as reinforcing and conductive nanofillers, capable of dissipating electrostatic charges and electromagnetic radiation. EG and EGNP produced by using natural graphite flakes [87] have been successfully used as a cheaper alternative to CNTs. In general, the reinforcing effectiveness and electrical properties depend on the aspect ratio of conducting fillers, uniformity of dispersion, the interfacial adhesion between nanofiller and matrix, etc. [88,89]. In addition to the important role of the nanofiller–polymer interface, the mechanical properties of nanocomposites can be influenced by other factors, such as the shape and aspect ratio of the nanofiller, as well as its orientation, distribution, and dispersion within the polymer matrix, which strongly affect the reinforcing efficiency [90].

Electrical characterization of PLA–EG nanocomposites revealed that an important decrease in the electrical resistivity is obtained at relatively high amounts of EG (i.e. 8–12 wt%), whereas some significant decrease in tensile strength was evidenced at this percentage of nanofiller. In this context, knowing the effectiveness of CNTs as conductive and as reinforcing nanofillers, it was considered of real interest to study the effect of the coaddition of CNTs into PLA–EG blends, aiming at the improvement of the electrical conductivity, thermomechanical and FR properties of these nanocomposites. In these experiments, the total loading of the nanofiller was maintained at a constant percentage of 6 wt%, whereas a part of EG was replaced by 1–3 wt% CNTs. Some short comments concerning the characterization of these hybrid nanocomposites (i.e. PLA–EG–CNT) are presented below.

First, in relation to the mechanical properties, it is noteworthy to mention that as has been reported elsewhere, in nanocomposites where the nanofiller is oriented, the mechanical reinforcing efficiency of CNTs is higher than those of nanoplatelets of the same aspect ratio [90]. However, at the same volume fraction, assuming an identical degree of dispersion and extent of influence on the surrounding polymer macromolecules, nanotubes can generate a significantly larger amount of interphase than the nanoplatelets. Moreover, by comparing to the PLA–6% EG nanocomposite (tensile strength >40 MPa), the hybrid nanocomposites were characterized by evident improvements of tensile strength (with up to 10 MPa) and a slight increase in the nominal strain at break, in direct correlation with the percentage of CNTs. The evident improvements of mechanical properties due to the coaddition of CNTs into PLA–EG blend were associated with better electrical properties obtained even where only 1% EG was replaced by a similar loading of CNTs. In all cases, as was confirmed by TEM images (Figure 17(a) and (b)) the nanofillers exhibited a rather good dispersion into the PLA matrix, which is fully consistent with the improvements of mechanical and electrical properties.

As was expected, the PLA–EG–CNT hybrid nanocomposites also showed improvements in FR properties with respect to neat PLA or to PLA–6% EG nanocomposites. By the addition of 2–3% CNTs into PLA–EG formulations, the cone calorimeter evaluations showed a reduction of the maximum peak of the RHR of >20–25% with respect to neat PLA (Figure 18), whereas UL 94HB tests allowed one to conclude that a low BR is evident in the case of the hybrid nanocomposites,

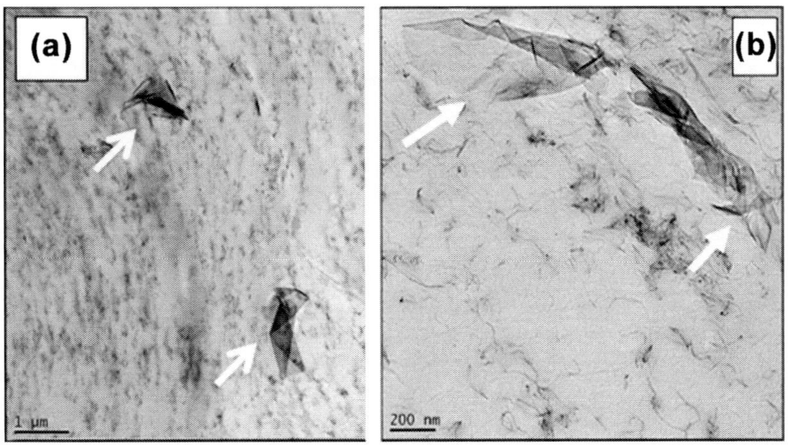

FIGURE 17

TEM images of PLA (NW 6201D)–4% EG–2% CNT hybrid nanocomposites at low (a) and high magnifications (b).

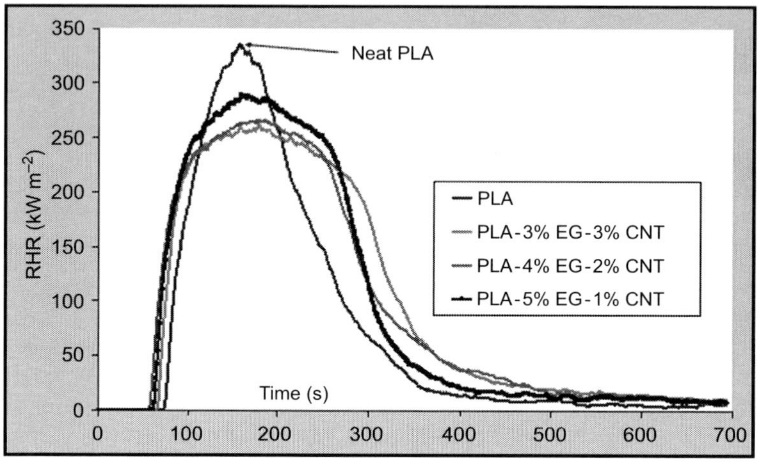

FIGURE 18

The RHR plotted against time: neat PLA (NW 6201D) compared to PLA–EG–CNT hybrid nanocomposites (heat flux: 35 kW m^{-2}).

i.e. values of 20–22 mm min^{-1}. Moreover, from Figure 19, it turns out that the coaddition of CNTs changed the behavior of the nanocomposites during burning (Figure 19(a) compared to Figure19(b), i.e. PLA–EG compared to PLA–EG–CNT, respectively), whereas the aspect/morphology of residual specimens is clearly changed (Figure 19(c) compared to Figure 19(d)). In fact, the presence of CNTs, even at a low loading (e.g. 1 wt%), led to a more compact, rigid, and cohesive char (Figure 19(d)), comportment that can be of real interest in applications

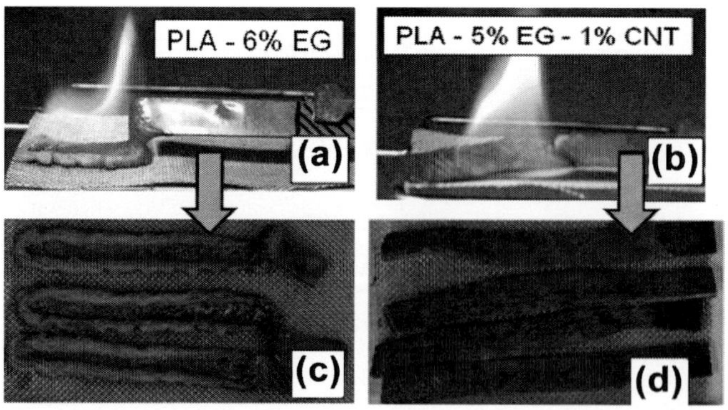

FIGURE 19

Illustration of burning during the UL 94HB testing of PLA–6% EG (a) vs. PLA–5% EG–1% CNT nanocomposites (b), and their residual specimens (c) and (d), respectively.

requiring flame retardancy and associated properties such as tensile strength and rigidity, thermal stability, and conductive electrical characteristics.

3.3.2 PLA nanocomposites containing EG and clays

The association of OMLS, which are among the cheapest and most effective nanoparticles, with various other nanofillers of different morphologies and specific properties, is not a new idea [91,92]. Following a collaboration with Camino's group [84], we have studied the effect of the coaddition of EG and a selected OMLS (C30B) on the thermal, mechanical, and FR properties of PLA, in order to develop new BNCs interesting for different applications related to electric and electronic engineering sectors, as well as in automotive parts. PLA nanocomposites containing OMLS (C30B was maintained at the same percentage, i.e. 3 wt%; its effectiveness at this loading was confirmed by us in previous experiments) and EG in the range 3–9% were obtained by melt compounding using internal kneaders.

Summarizing the principal effects ascribed to the coaddition of C30B into PLA–EG blends, it is first noteworthy mentioning that the ternary formulations (PLA–EG-C30B) exhibited better thermal stability that was mainly attributed to the effect of EG addition. In fact, as is seen from Table 4 [84], the onset of thermal degradation (considered as the temperature for 5% weight loss ($T_{5\%}$)) was as follows: PLA < PLA–EG < PLA–C30B ≈ PLA-(C30B-EG).

Moreover, the level of tensile strength was increased by the coaddition of clay, reaching a similar value to that of the neat PLA (about 63 MPa). It is noteworthy to mention that the rigidity, i.e. Young's modulus, has shown the highest increase for ternary systems (PLA–EG–clay), from 50% to 80% with respect to neat PLA, values in direct correlation to EG loadings.

Table 4 TGA Data of Neat PLA Compared to that of PLA Nanocomposites (Under Air Flow, 20 °C min^{-1}) [84]

Sample (% by Weight)	$T_{5\%}$ (°C)	T_{max} (°C)
PLA	340	377
PLA—3% C30B	352	391
PLA—3% EG	345	383
PLA—6% EG	344	383
PLA—9% EG	349	385
PLA—3% C30B - 3% EG	351	393
PLA—3% C30B - 6% EG	350	391
PLA—3% C30B - 9% EG	352	393

TGA, Thermogravimetric analysis.

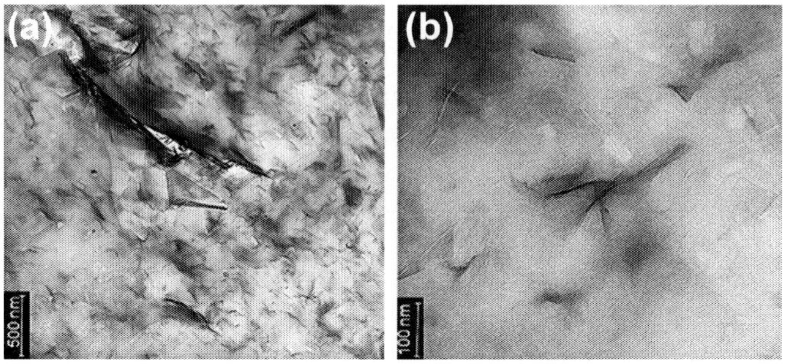

FIGURE 20

TEM patterns of PLA—3% C30B—6% EG at (a) 24 kX and (b) 100 kX of magnification [84].

Regarding the morphology of ternary compositions used for fire testing, typical TEM micrographs have proved that good dispersion is obtained by the coaddition of nanofillers (Figure 20(a) [84]). Single clay layers, bundles of graphite platelets, and well-dispersed graphite layers into the PLA matrix can be observed in high-magnification TEM images (Figure 20(b)). Following these observations, it was assumed that the presence of graphite further improves dispersion of the clay (consistent with the information supplied by XRD patterns) and also that the clay plays a similar role toward graphite.

Table 5 gathers the results of the UL 94HB characterization of different BNCs and of the neat PLA. In the case of PLA—3% C30B, as was expected and reported elsewhere [93], the burning occurs with charring, thus preventing the dripping, whereas the mean BR value is higher than that of the neat PLA, resulting in a "limit" flammability classification. The most interesting FR properties were attributed to the ternary compositions obtained by the coaddition of nanofillers, which, interestingly enough for potential utilization, present both antidripping properties, due to the

Table 5 Results of Horizontal Burning Test (UL94 HB) for PLA and nanocomposites [84]

Sample (% by Weight)	BR ($mm\ min^{-1}$)	Classification[a]	Remarks
PLA	31 (±2)	HB	Burning with drips
PLA–3%–C30B	39 (±3)	Limit	Burning without drips, charring
PLA–3% C30B–3% EG	30 (±3)	HB	Burning without drips, charring
PLA–3% C30B–6% EG	26 (±1)	HB	
PLA–3% C30B–9% EG	18 (±1)	HB	

[a] *For specimens with a thickness of approximately 3.2 mm; a BR <40 mm min*$^{-1}$ *allows the HB classification.*

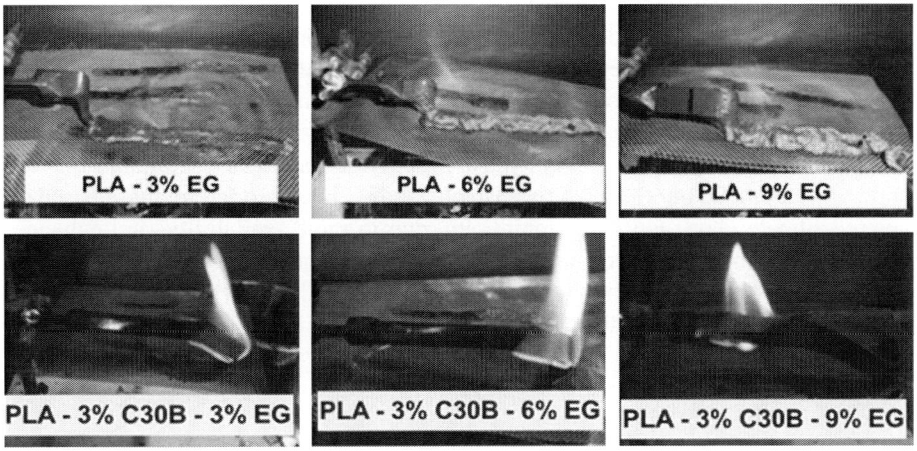

FIGURE 21

Illustration of the behavior during the UL 94HB testing: PLA—EG nanocomposites (burning with charring, carbonaceous char in correlation with EG amount) and PLA—C30B—EG nanocomposites (burning without dripping, formation of a rigid char) [84].

formation of a compact char (Figure 21 [84]), and an even lower BR as compared to the neat PLA and to the binary compositions allowing an HB classification. Thus, the ternary compositions owing to a favorable interaction between C30B and EG present the most advanced FR properties and can be considered for further technical applications.

3.4 Concluding remarks and further trends related to the utilization of nanofillers

As a preliminary conclusion, it is assumed that the FR properties of BNC-containing nanofillers of different morphologies are clearly influenced by the fire scenario and testing procedure. Furthermore, the comparative evaluation of BNCs and of the neat polymer matrix by cone calorimetry or using the UL 94HB horizontal test provided positive and more beneficial information. In the absence of additional FR additives, BNCs based on PLA as a polymer matrix and containing only nanofillers failed in other important flammability tests, such as the UL 94V (vertical flame test).

To further decrease the flammability of these BNCs, there is a need to find additional FR formulations of high effectiveness (see also Section 4). Research efforts have been and will be devoted to improve the FR properties of biopolymers by designing and synthesizing novel intumescent agents [94] and new FR additives. In other words, to obtain FR nanocomposites that are attractive for technical applications, products with low flammability are needed, characterized also by a high level of thermomechanical properties. Sometimes, the combination of biopolymers with microfillers or nanofillers, and/or traditional FR additives, is followed by the degradation of the polymer matrix. However, nanofillers like $Mg(OH)_2$, LDHs, and ZnO, combined with or without FR additives, can provoke more or less the degradation of a polymer matrix such as PLA [24,95,96].

Finally, various nanoreinforcements in biopolymer matrices (including PLA) are under development, but the most extensively studied BNCs so far are those based on layered-silicate clays, CNTs, graphite, etc. due to their availability, and enhancement that they induce in mechanical, thermal, barrier, and FR properties of the polymer matrix. We believe that some other nanofillers [97–99] such as sepiolite, boehmite, and halloysite, did not yet demonstrate their real potential. Associated with or without synergistic FR additives, it is assumed that their utilization can be regarded as possible new pathways in rendering BNCs attractive for a large field of applications.

4. Multicomponent FR systems

(*Written by G. Fontaine and S. Bourbigot*)

Depending on their nature, FRs can act chemically and/or physically in the solid, liquid, or gas phase. Some of the mechanisms have been already briefly described in Section 2.2. FRs interfere during a particular stage of the combustion process, e.g. during heating, decomposition, ignition, or flame spread. The burning of FR polymers should be considered as a complex process in which many individual stages occur simultaneously. As an example, hydroxides exhibit an endothermic decomposition, which cools down the substrate and dilutes the ignitable gas mixture due to the formation of inert gases associated with the formation of an oxide protective barrier. The FR action can be reinforced using nanoparticles. It is then one of the purposes of this section to describe the action of FRs with and without nanofillers in biodegradable polymers. In this section, the biodegradation of the polymers with FRs is also noted, a topic not always considered in the literature. We can assume as an explanation that on the one hand, these formulations need further optimization or, on the other hand, the interest for these products is relatively new and consequently it is needed some time to perform these specific studies.

As shown in Section 3, PNCs exhibit low flammability in terms of cone calorimetry, but they fail in other tests, in particular those with samples in the vertical position (e.g. limiting oxygen index (LOI), UL-94). The mechanism of protection involves the formation of a char layer covering the entire sample surface acting as an insulative barrier and reducing volatiles escaping to the flame. The formation of such a layer not forming cracks when burning is critical to obtain a low heat release rate (HRR) [100]. Nanoparticles have then been combined with conventional FRs in order to create synergistic effects in numerous polymers. The same approach can be applied to biodegradable polymers. It is the goal of this section to fully review the approaches appearing in the literature to flame retard those polymers. In order to show the benefit of using nanoparticles in biodegradable flame retarded polymers, the first two subsections will review their flame retardancy by combination of microfillers and by the concept of intumescence. Based on this, the combination of nanoparticles in microfiller-based formulations will be examined in the third subsection. The last subsection will address the different approaches and solutions suggested. Since the developments mainly concern PLA, its flame retardancy will be first considered and this will be extended to other biodegradable polymers.

4.1 Combination of microfillers

The market of FRs is mainly constituted of ATH, MDH, phosphorus-containing compounds, halogenated compounds, antimony trioxide, and nitrogenated compounds that act in the gas and/or condensed phase [100]. These conventional FRs were evaluated in PLA. Nishida et al. [101] used ATH to make PLA flame resistant and also to permit its depolymerization at high temperatures (300 °C). In this particular case, they exploited the depolymerization for recycling purposes. In this work, they considered the application of PLA for both durable use and also its recycling. No comment was given about the mechanism of action and the performance of the flame retarded PLA, but we can reasonably assume that the fire behavior of the material is similar to that of a typical thermoplastic flame retarded with ATH. Flame retardancy is due to physical effects requiring relatively large amounts of ATH (50−65 wt%). The action of ATH is a result of different effects [102]: (a) dilution of the polymer in the condensed phase; (b) decrease in the amount of available fuel; (c) increase of the amount of thermal energy needed to raise the temperature of the composition to the pyrolysis level, due to the high heat capacity of the fillers; (d) emission of water vapor (enthalpy of decomposition); (e) dilution of the gaseous phase by water vapor (decrease of the amount of fuel and oxygen in the flame); (f) possible endothermic interactions between the water and decomposition products in the flame; (g) decrease of feedback energy to the pyrolyzing polymer; (h) insulative effect of the oxides (Al_2O_3) remaining in the char; (i) charring of the materials.

More recently, Yanagisawa et al. [103] investigated the flame retardancy of PLA using ATH at a relatively high loading (50 wt%) combined with phenolic resins. The application was housings of electronic products where the required level of flame retardancy is high. It was reported that the mechanism of action was mainly due to the formation of homogeneous char layers produced by the phenol resins on the surface of the composite upon heating. The mechanism was not completely described by the authors, but we suspect that the phenolic resin should provide a high char yield that could be reinforced by the formation of alumina from ATH. This ceramified char then becomes protective enough to achieve a good FR performance.

Textiles are a major application area for PLA, which also requires to achieve flame retardancy. The flame retardancy of PLA fabrics was investigated by

Kubokawa et al. [104–106] using conventional FR treatments and measuring the LOI according to the ASTM Standard Method D2863. The LOI of untreated PLA fabric was 24 vol.%. When PLA was treated with several bromine-containing FRs and a phosphorus-containing FR, triphenylphosphate (TPhP), the LOI increases slightly to 26 vol.% with the bromine-containing FR (tetrabromobisphenol A, which was revealed to be the most efficient) and respectively to 28 vol.% with TPhP. It was found that the thermal stability of the FR fabrics was not improved by the FR treatment at lower temperatures, but it became much higher above 375 °C. They also noticed that the exothermal reactions were retarded upon heating in air, associated with the formation of a higher char yield. According to the explanations of the authors, we suspect that FR treatment promotes charring, suggesting a mechanism of action in the condensed phase. Note that in the case of the brominated treatment, action in the gaseous phase should also occur.

Our groups have investigated phosphinate-based FRs designed for PAs and polyesters (Exolit OP1230, an aluminum phosphinate and OP1311 of Clariant, Germany, which is the combination of OP1230 and MP200, a melamine polyphosphate (MP), in the weight ratio of 2:1) and compared their effect to that of "classical" FRs such as ammonium polyphosphate (APP) and melamine polyphosphate (MP200) [107]. The LOI of the flame retarded PLA increased as a function of the FR loading and jumped from 21 vol.% to 28 vol.% (PLA/OP1230) until 34 vol.% (PLA/APP) (Figure 22(a) [107]). In addition to this, all formulations exhibited a V-2 rating at the UL-94 test (3.2 mm), at a 10 wt% loading; a V-0 rating (3.2 mm) was achieved for 20 wt% of MP200 and 25 wt% of OP1311. The PHRR determined by cone calorimetry was decreased by 33% for MP200 while for the other FRs the reduction was less (Figure 22(b)). However, the ignition time is improved (longer) using MP200 and APP (100 s versus 50 s). Those results showed that the FR performance remains relatively poor in PLA. It was also observed that FR PLA compounds permitting to reach a high LOI can drip easily as the neat PLA, thus allowing only UL-94 V-2 rating. The poor reduction of PHRR was assigned to the low viscosity of the FR PLA upon heating. The charred material formed by the reaction between the additives and the polymer is not viscous enough to be expanded by the fuel consisting of the degradation gases and then to make a protective coating.

As for PLA, APP was used for flame retarding PBS [108]. A water-crosslinking technique was applied in order to limit the dripping of the material, thus keeping the structure of the composite during a vertical test such as the UL-94. The so-called water-crosslinking technique involves the combination of APP with tetraethoxysilane mixed together with PBS followed by crosslinking in water for an appropriate time. It results in the formation of a silsesquioxane network containing PBS chains in the material. Composites with APP alone (15 wt%) achieved a V-2 rating in the UL-94 test while those prepared with the water-crosslinking reaction were classified as UL-94 V-0. On the contrary, the LOI was significantly decreased from 37 to 27 vol.%. No comment and no description of the burning behavior were given, but it is suggested based on kinetic analysis, formation of phosphorus-rich residue

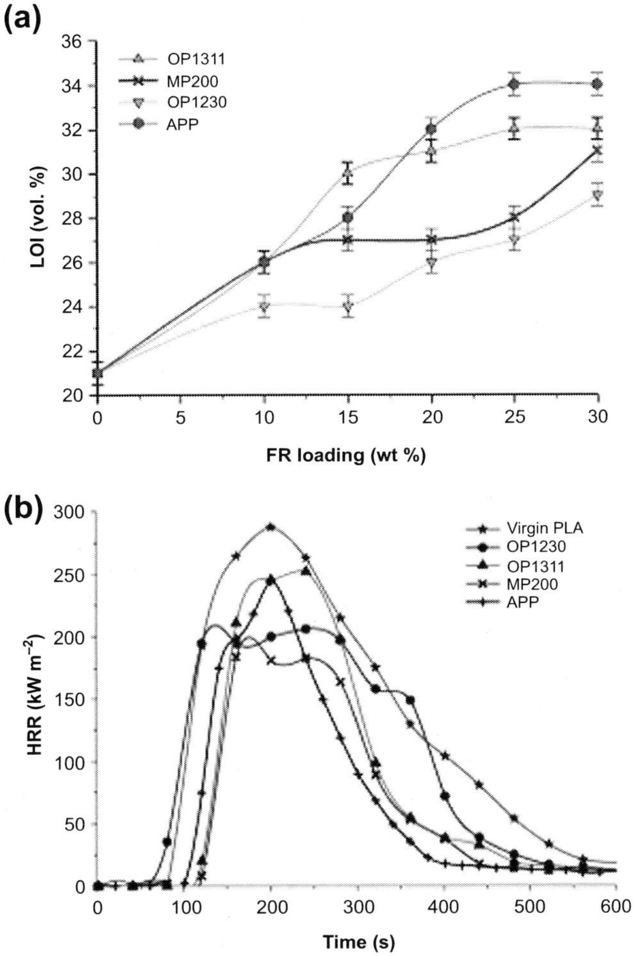

FIGURE 22

LOI as a function of the FR amount in PLA (a) and HRR as a function of time of flame retarded PLA at 30 wt% FR loading (external heat flux: 35 kW m^{-2}) (b) [107].

protecting PBS from further decomposition. From this, we suspect a mechanism first acting in the gaseous phase from the decomposition of APP diluting the flame, and in the second step, a mechanism in the condensed phase because of the formation of a protective residue. With the silsesquioxane network, the mechanism in the condensed phase should be favored.

PE-containing starch is another biodegradable polymer. Films of this polymer have been flame retarded using boron compounds (mixture of borax and boric acid) [109]. The LOI has only been used for estimating the flammability of the material. It is increased by 50% with borax or boric acid mixture (boron compounds in methanol) impregnated in the film. No comment is made about the mechanism of action. It has also been mentioned that this performance can be facilitated using chlorine/antimony oxide mixture. Nevertheless, this latter solution should be avoided because of the toxicity of antimony.

Natural fiber composites have been prepared by grafting hydrophobic monomer methyl methacrylate (MMA) onto chemically modified rice straw using a complex initiating system ($CuSO_4$/glycine/ammonium persulfate) in an aqueous medium with and without the additive, sodium silicate (SS) [110]. The LOI jumped from 20 to 24 vol.% with 10 wt% SS in the formulation. This result was confirmed by cone calorimetry; the PHRR reduced by 38%. The enhancement of the flame retardancy was assigned to the formation of charred residue with a silicate layer present in it. Because of the formation of a char residue, silicate acts as a protective barrier by reducing the heat and mass transfer between the flame and the samples as described in the case of nanocomposites [111].

Most of the mechanisms mentioned above concern the formation of protective barrier (sometimes also combined with a mechanism in the gas phase). Other mechanisms involving also the formation of a protective layer (intumescence) will be considered in the next subsection.

4.2 Intumescent formulations

The word "intumescence" comes from Latin "intumescere", which means "to swell up", that describes well the behavior of an intumescent material. When heating beyond a critical temperature, it begins to swell and then to expand. The result of this process is a foamed cellular charred layer on the surface, which protects the underlying material from the action of the heat flux or the flame [112,113]. This concept was applied for flame retarding polyolefins, PAs, and polyesters and was proven to be very efficient for those polymers. It should also be efficient in biodegradable polymers such as PLA, PA, PBS, and TPS.

Typically, the ingredients of intumescence are mainly composed of an inorganic acid or a material yielding acidic species upon heating, of a char former, and of a component that decomposes at the right temperature and at the right time to enable the blowing of the system. The compounds used as an acid source are acids, ammonium salts, and phosphates. Char formers are mainly hydroxyl-containing compounds, such as polyols and starch, and the blowing agents are compounds that can evolve gases upon heating, in the temperature range corresponding to the development of the intumescence process. Melamine, urea, and urea–formaldehyde resins are typical examples of blowing agents. A typical example of an intumescent system is the combination of APP with pentaerythritol (PER), where APP plays the role of both an acid source and a blowing agent (evolution of ammonia during the degradation of APP) and PER is the char former [114].

As PER comes from petrochemistry, Reti et al. [115] investigated the substitution of PER by lignin (LIG) or starch (ST), both of which are derived from renewable resources in the APP/char former system. The basic idea was to make an intumescent PLA with the highest percentage of biobased ingredients. At 40 wt% loading of APP/char former (PER, LIG, or ST), very high LOI values were observed (up to 58 vol.% with PER), and in all cases, an intumescent protective coating was formed at the surface of the material. The interesting feature of the "biobased systems" was the V-0 classification in the UL-94 test (3.2 mm) while PLA−APP/PER is only V-2 rated. By cone calorimetry (external heat flux of 35 kW m^{-2}), those systems exhibited a reduction of PHRR by 65% for APP/PER and by 40% for APP/LIG and APP/ST. The mode of action is the formation of an intumescent layer protecting the underlying material, limiting heat and mass transfer. The formulations were then optimized by an experimental design in order to have the highest amount of renewable resources (LIG and ST) while keeping acceptable fire performance (LOI >30 vol.% and V-0 rating in the UL-94 test). The authors showed that the amount of renewable resources can be increased to 12 wt% in APP/LIG and to 28 wt% in APP/ST (total loading was kept constant at 40 wt%).

Fibers can be incorporated into polymers in order to reinforce them, in particular, in terms of mechanical properties. Shumao et al. [116] used natural fibers (ramie fibers) combined with APP, and they made an intumescent system for PLA. At 40 wt % loading, the LOI reached 35 vol.%, and the formulation exhibited a V-0 rating. Thermal analysis and visual observation showed that the char yield of the intumescent PLA is greatly enhanced compared to that of the virgin polymer. The authors also reported an intumescent behavior of the material. They suggested that APP releases phosphoric acid, polyphosphoric acid, and nonflammable gases when exposed to a flame; the resulting acid contributes to intramolecular or intermolecular ramie dehydration, then dehydrogenizing, charring, and rupture of chemical bonds. Although it was not formally evidenced, this statement is probably true because it is a typical intumescent mechanism.

An intumescent FR (hereafter called SPDPM, Figure 23 displays its chemical structure) was synthesized from spirocyclic pentaerythritol bisphosphorate disphosphoryl chloride (SPDPC) and melamine by Zhan et al. [117]. SPDPM (spirocyclic pentaerythritol bisphosphorate disphosphoryl melamine) was evaluated at different loadings in PLA. The LOI reached a 38 vol.% at 25 wt% loading, and the corresponding material was V-0 rated. At a lower loading, the intumescent PLA was no longer V-0 rated (V-2 at 5 and 15 wt%) but LOI values were still quite high (28 and

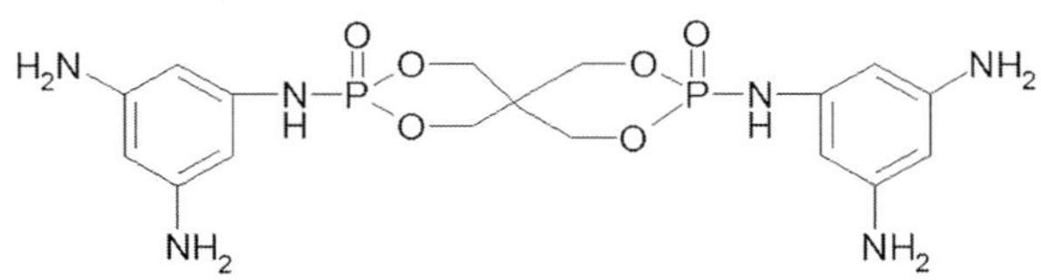

FIGURE 23

Intumescent FR (SPDPM) synthesized from SPDPC and melamine.

33 vol.% at 5 and 15 wt%, respectively). Visual observation showed an intumescent behavior with a high amount of char on the top of the material. The fire performance was confirmed by microscale combustion calorimetry (MCC); PHRR decreased by 33% at a 25 wt% loading. It was reported that the FR mechanism of SPDPM in PLA was not only due to the intumescent protective charred layer but also due to the change of the degradation process of PLA. This makes sense because the PHRR in the MCC experiment is significantly decreased indicating that the degradation pathway of the intumescent PLA was modified compared to that of the virgin PLA (in a typical MCC experiment only few milligrams of sample are used; heat and mass transfer in an intumescent system is not predominant and evolving gases entering into the combustor and burning with lower energy evidence different degradation pathways).

Intumescence has been also applied to PBS. Chen et al. [118] prepared intumescent PBS combining APP and melamine as was done in PLA [119]. The LOI jumped from 24 to 39 vol.% at a 25 wt% loading with the appropriate ratio of APP/Mel (5/1 wt/wt). The V-0 rating (3.2 mm) was also achieved with this formulation. The protection is provided by the formation of an expanded intumescent char at the surface of the sample. The mechanisms are similar to those described above in the case of PLA.

Starch-based materials are among the most popular biodegradable plastics, and are widely used due to the low cost and potential thermoplastic properties of starch [120]. It is reported that starch can be blended with thermoplastic polymers (e.g. PS and PE) to provide biodegradation properties to the composites [121,122]. Starch is not truly thermoplastic, but under certain shearing and temperature conditions, water or glycerol can act as a good plasticizer agent to disintegrate granules and overcome the strong interaction of starch molecules [123]. The efficiency of APP in those starch-based materials was evaluated by Marosi's group [124]. It was shown that plasticized TPS can be flame retarded efficiently by as low as 10 wt% APP since the LOI jumped from 23 to 33 vol.% and V-0 rating is achieved at UL-94 (3.2 mm). It is noteworthy that the LOI reached 60 vol.% at a 30 wt% APP loading. Intumescence is not strictly mentioned in the paper, but we can suspect such a mechanism because all ingredients are incorporated in the formulation (APP as acid source and blowing and starch as char former). Finally, it is suggested that this biocomposite could be utilized as a chemical fertilizer because of the presence of APP.

In a recent work, APP microencapsulated with melamine−formaldehyde (MCAPP) resin was used to improve the reaction of TPS to fire [125]. Here MCAPP was utilized as an intrinsic intumescent FR (APP is the acid source and melamine−formaldehyde resin is the char former) as suggested by Giraud [126−128] in order to restrain the reaction between APP and starch during processing. TPS/MCAPP composites with only 2 wt% MCAPP achieved a V-0 rating at UL 94 (3.2 mm) and the LOI was significantly increased (from 23 to 31 vol.%, Figure 24 [125]). Surprisingly, the incorporation of MCAPP could reduce the THR of TPS in the microscale combustion calorimetry (MCC) testing. This last experiment was done with low

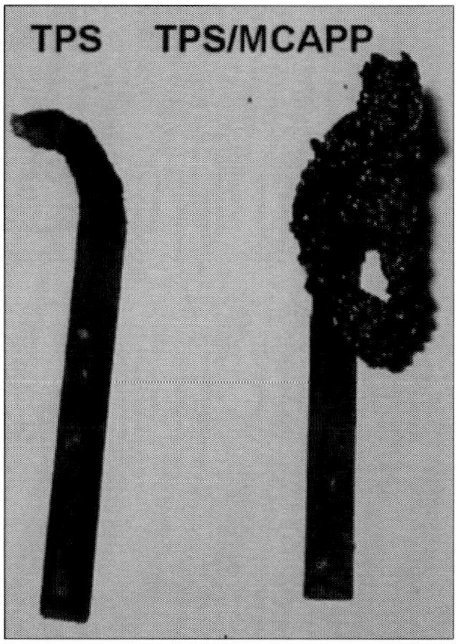

FIGURE 24

Intumescent TPS/MCAPP after the LOI test compared to virgin TPS [125].

amounts of sample (typically 5 mg) and in an MCC the sample underwent thermal degradation at a high heating rate as in TGA. The evolving gases were then burnt in a combustor, measuring HRR. In such a design, we do not believe that an efficient physical barrier (intumescent barrier) could be formed and the modification of HRR can only be due to the chemical changes of the evolving gases. If this is so, we should admit that the degradation pathway of the material is modified. It is not commented by the authors, but it should be investigated further.

4.3 Nanoparticles combined with conventional FRs

In the above section, we have shown that intumescence is a method to limit the flammability of biodegradable polymers (PLA, PBS, and TPS). Fontaine et al. [119] reported the combination of APP with melamine at an appropriate ratio (5−1, respectively) as efficient intumescent mixture for PLA at a loading as low as 10 wt%. The incorporation of an additional nanofiller such as organomodified montmorillonite (MMT; C30B) or MWCNTs showed large effects on the reaction of the materials to fire. A large synergistic effect was observed when using C30B (Figure 25(a)) while it was antagonistic when MWCNTs were used (Figure 25(b)). It can be observed that the intumescent PLA combined with C30B did not burn and the small PHRR was only due to a very short ignition of the material. At 30 wt%, extremely high LOI values were measured (up to 52 vol.%) and the V-0 rating (3.2 mm) was achieved. This jump in performance was partially

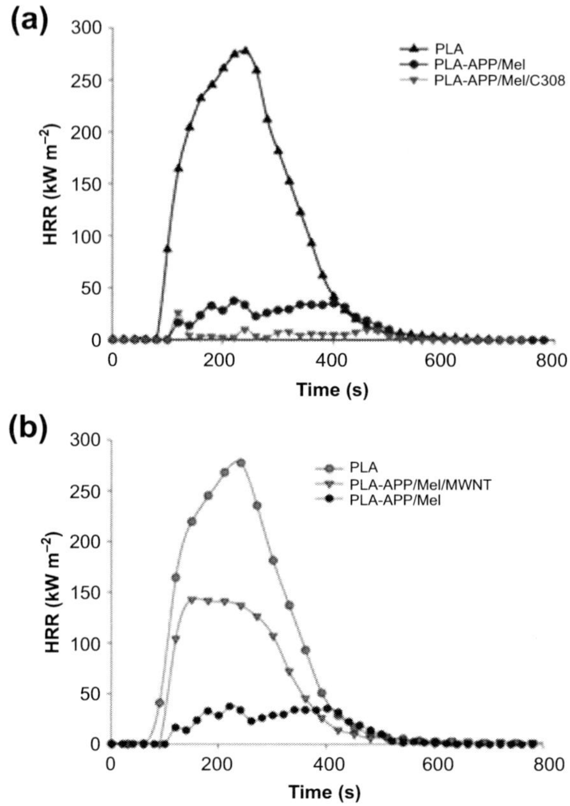

FIGURE 25

HRR curves as a function of time of PLA/APP/Mel/Nanoparticles (C30B (a) or MWCNT (b)) loaded at 30 wt%, compared to virgin PLA.

explained by an increase in the thermal stability of the formulation with C30B compared to that without. It might be justified according to previous works of our group [129,130] that C30B could react with APP to form aluminophosphate and silicophosphate stabilizing the intumescent structure at high temperatures. On the contrary, the incorporation of MWCNTs led to an antagonistic effect, which was not explained in the paper. According to the expertise of our group, this might be assigned to the accumulation of MWCNTs in the intumescent char. If this is the case, the apparent heat conductivity increases a lot and does not permit the intumescent coating to play its role as a heat barrier.

Intumescent PLA was also evaluated with a total loading of 10 wt%. This loading was still sufficient to get a relatively high LOI (33 vol.%) and a synergistic effect was observed with C30B (35 vol.%) while the LOI fell down when MWCNT is incorporated in the formulation (24 vol.%). It is noteworthy that in the UL-94 test, V-0 was achieved for formulations with and without C30B. As a conclusion, this kind of intumescent formulation offers an interesting way for designing flame retarded PLAs.

A similar approach has also been adopted by Wang et al. [131] in an intumescent PLA. The intumescent system is a very conventional one since it is the combination of APP, PER, and melamine cyanurate. The synergist was a zinc aluminum LDH (ZnAl-LDH) incorporated at 2 wt% in the intumescent PLA having 23 wt% intumescent ingredients. The authors claimed the formation of a nanocomposite structure, but according to their TEM pictures, only agglomerates of ZnAl-LDH are observed in PLA. No comment was made about the morphology of microfillers and nanofillers even though they can play a significant role in the flame retardancy of thermoplastics as shown by Samyn et al. [132]. Compared to the system without the synergist, the PHRR (cone calorimetry at 35 kW m^{-2}) was not reduced by incorporating it, but the two intumescent formulations exhibited a reduction by 65% against virgin PLA. The main benefit of ZnAl-LDH was to greatly decrease the THR (from 87 to 35 MJ m^{-2}). The role of ZnAl-LDH is not commented in the paper, but we suspect a reinforcement of the intumescent structure limiting the combustion time of the sample.

Starting from gypsum, a by-product of the lactic acid fabrication process, PLA composites were prepared by melt blending. The novelty of this approach is based on the selection of gypsum. It was made after a previous specific dehydration at 500 °C to obtain anhydrite II (AII) because it was shown that it permitted the preservation of PLA molecular weights, and it led to favorable key properties for the final composite products (other anhydrite forms, such as anhydrite III, obtained by drying gypsum hemihydrate at about 200 °C causes the contrary result) [133]. In order to obtain PLA—AII composites characterized by specific end-use FR properties, the addition of selected OMLS was considered [134]. The coaddition of AII and OMLS was found to lead to PLA nanocomposites characterized by good (nano) filler dispersion, thermal stability, and adequate mechanical resistance. The FR properties as evidenced by cone calorimetry showed a significant increase in the ignition time compared to neat PLA and a substantial decrease, i.e. approximately 40%, of PHRR; however, the UL94 HB test was successfully passed revealing a nondripping effect and extensive char formation.

Using the same approach as above for PLA, fumed silica was incorporated as a synergist in intumescent PBS. The performance of PBS containing the intumescent formulation APP/Mel has been already commented in the previous subsection. The dispersion of the fumed silica was acceptable, and the resulting material can be considered as a nanocomposite. The additional incorporation of fumed silica permitted the reduction of the total loading from 25 wt% to 17 wt% keeping the V-0 rating in the UL-94 test (3.2 mm) and an acceptable LOI (decrease from 39 to 33 vol.%). The suggested mechanism is the reinforcement of the intumescent char by silica migration. According to the work of our laboratories [135], silica can react with the degradation products of APP to yield silicophosphates. The formation of those silicophosphates permits the stabilization of phosphorus at a high temperature, and it reinforces the efficiency of the protective char. This is not considered in the paper, but we suspect that this is the mechanism involved in this case.

FR poly(butyl methacrylate)/SS/$Mg(OH)_2$ (MDH) nanocomposite has been prepared via in situ emulsion polymerization of butyl methacrylate, benzoyl peroxide, layered silicate, and conventional fire retardant additive, MDH [136]. It was shown that silicate is evenly dispersed in the polymeric matrix, so a polymer nanocomposite (PNC) can be synthesized. The LOI is the highest when combining MDH and layered silicate (30 vol.%) while it was not improved when incorporating the silicate alone (20 vol.% with and without silicate). In terms of cone calorimetry, the PHRR was decreased by 25% with the silicate and by 40% with the combination of MDH/silicate. Note that the TTI was decreased with the silicate alone while it was slightly increased with the combination of MDH/silicate because of the endothermic effect of MDH [102]. As a bonus, it was shown that the biodegradation rate of the nanocomposite is the fastest when containing MDH/layered silicate.

4.4 Concluding remarks

In this part, we have shown that the methods for flame retarding biodegradable polymers remain similar as those observed in other polymers. Nevertheless, no comprehensive study has been published to investigate the specific behavior of the thermal and fire degradation of FR biodegradable polymers. Some attempts have been made in the case of PBS [118] and PLA [119], but it is still limited and further work is needed.

The concept of intumescence has been widely used probably because it offers a compromise between the performance and the "green" aspects of the formulation. In this case, the basic mechanism seems to be the same as in the case of nonbiodegradable polymers (PP, PA etc.), but this approach permits the development of biodegradable polymers for durable applications. Hydrated minerals (ATH and MDH) have been also used and have been found to provide the same performance as in polyolefins but keeping their main disadvantage namely the high loading required to achieve acceptable flame retardancy. The good news of these two approaches is that they should help the biodegradation of the polymer compared to the unfilled one. Nanocomposite development is a relatively new technology in the field of flame retardancy, and it has been applied with relative success in the case of biodegradable polymers. PNCs are FR polymers in specific scenarios (high heat flux) but fail the flammability test (LOI, UL-94). The mechanism of protection is the formation of a mineral layer associated with char promotion, but it is not efficient enough to provide the highest standard of protection. This technology gives the best results when conventional FRs are also used, leading to synergistic effects. It was clearly shown with several examples in this section. Very promising developments in the synergy aspects are then expected for biodegradable polymers, and efforts should be made in this away.

5. Design of biodegradable fibers with FR Properties

(*Written by E. Devaux in collaboration with C. Campagne, M. Ferreira and S. Solarski*)

5.1 Spinning of multifilament yarns

For several years, PLA has been used in the surgical and biomedical domains in particular as stitches [137]. PLA was reserved mainly for these fields of application because of its high cost. But for some years, numerous studies were performed in order to improve the process of PLA synthesis, and this in the perspective to find new applications, particularly in the field of packaging and textiles where PLA would be a viable alternative to the commodity polymers obtained from petrochemistry. The production of PLA textile yarns for more general applications has been the object of numerous studies since the beginning of the 1980s [138]. Two main ways are considered in order to spin PLA: the solvent spinning on the one hand and melt spinning on the other hand. PLA yarns are henceforth available on the market, such as under the trademark of Ingeo [139].

Most of the synthetic or artificial fibers are produced by extrusion processes. The polymer is brought either to a molten state or to a viscous one by dissolving it using a solvent, or to the gel state. The polymer then passes through a die in order to obtain a continuous multifilament yarn. These dies are constituted of several holes of identical sizes to ensure the satisfactory regularity of the yarn. The shape of the holes can be modulated to vary the type of surface of the final product.

The oldest process is the so-called wet spinning. The polymer is put into solution in an appropriate solvent; this solution then passes through dies that are immersed in a chemical coagulating bath. This technique allows spinning multifilament yarns of acrylic, rayon, or aramide. The second process, called dry spinning, also involves dissolving the polymer in a solvent, but the yarns are solidified by evaporating the solvent by means of flowing air. It is typically used for the production of acetate, triacetate, or acrylic fibers. In both cases, a stage of drawing can be required in order to improve the mechanical properties of the yarns.

There are mainly two types of melt spinning processes. The first is high speed spinning where the polymer is brought to the molten state by means of an extruder, then it passes through dies to form continuous multifilament yarns that are directly collected with a very high speed (several thousand meters per minute). The second melt spinning process can be called spinning drawing (Figure 26). The molten polymer also passes through a specific die, but contrary to the first process, multifilament yarns are not directly collected. They undergo a more or less important drawing during the second stage. The yarns pass around two heated rolls having different rotation speeds, promoting the drawing of the fibers and the orientation of the constitutive macromolecules, thus improving the mechanical characteristics of the material. The spinning speeds can be beyond 1000 m min^{-1}. Numerous studies have been already realized on the spinning of

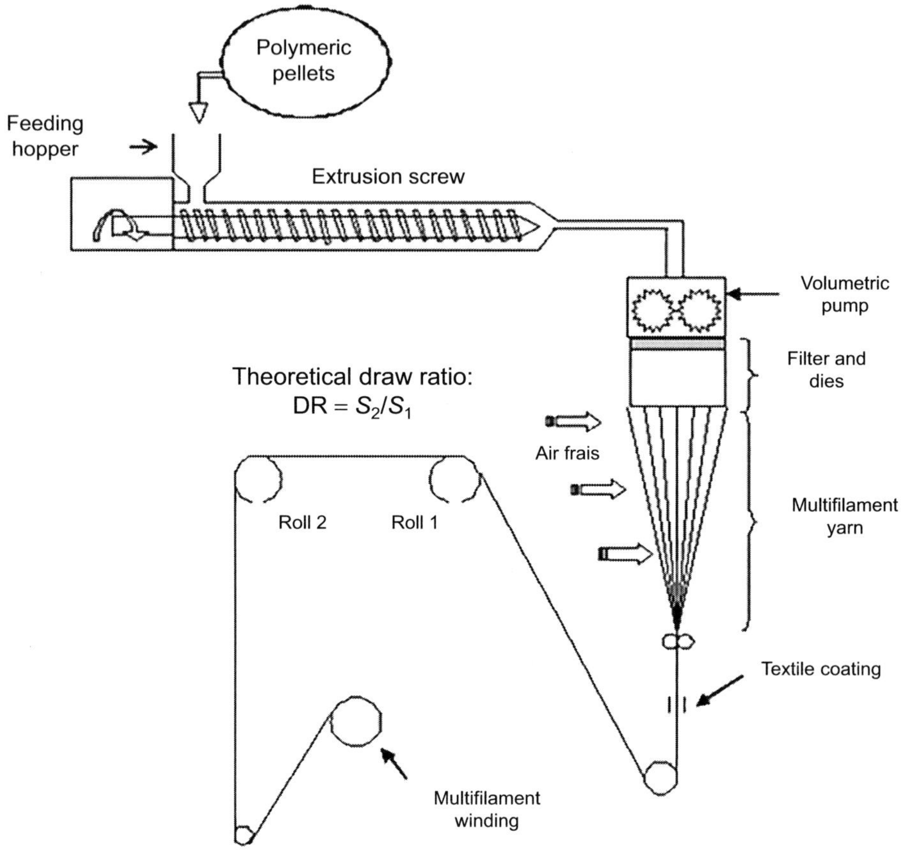

FIGURE 26

Typical melt spinning-drawing process.

PLA according to various processes [138]. Various types of PLA, presenting variable proportions of isomers L or D, have been transformed by dry and wet spinning in solution, as well as by both melt spinning processes (spinning with high speed or with a drawing).

5.1.1 The spinning of PLA in solution

PLA can be spun in solution by the dry or the wet way. As regards wet spinning, PLA can be solubilized in chloroform solution. Some quantity of toluene is used in the coagulation bath to precipitate the solvent and form the PLA filaments [140]. According to the molecular weight of the PLA, the concentration of polymer in the solvent must be between 6% and 12%. The lower the molecular weight of PLA is, the higher the concentration of polymer to reach a minimal viscosity must be. For the PLA multifilaments processed by dry spinning, the polymer is also solubilized in solvent, either in chloroform or in a chloroform/toluene mixture. The solvent is then eliminated by warm air imposed just after the extrusion step.

Different parameters influence the final properties of the PLA multifilaments [141]. First of all, the composition of the polymer solution has to be optimized. The PLA has to be dissolved first in chloroform. Toluene is then added to the solution in order to obtain the best dissolution of PLA. The maximal drawing ratio applicable for filaments increases with the density of entanglement of the macromolecular chains. This density of entanglement is optimal for a mixture of chloroform/toluene in the proportion of 40/60. In these conditions, the drawing ratio that can be applied to multifilaments is maximal, and the orientation of macromolecular chains of PLA polymer is sufficient to obtain optimized mechanical properties. The proportion of the solvent mixture has also an effect on the multifilament morphology [142]. For a solution of chloroform/toluene in the proportion of 50/50 containing PLA, multifilaments present an irregular structure, while multifilaments obtained from a 60/40 mixture present a regular fibrillar structure. This particular structure is obtained after drawing during the spinning process; the structure of undrawn multifilaments is macrofibrillar [142]. The concentration of PLA in solution has to be also optimized to obtain the best mechanical properties. The optimal concentration depends on the molecular weight of the PLA polymer and ensures the production of multifilaments with the best mechanical properties (Figure 27). To ensure good evaporation of the solvent during the dry spinning of PLA in solution, the temperature of the evaporation zone has to be optimized. The temperature allowing the complete evaporation and the achievement of multifilaments presenting a smooth structure is 60 °C [138].

5.1.2 Melt spinning of PLA

Different studies have been already carried out on the melt spinning of PLA with a high speed [143–146] or with a postdrawing [143,144,147–153]. These two techniques have the advantage that in process the use of solvents is avoided, contrary to

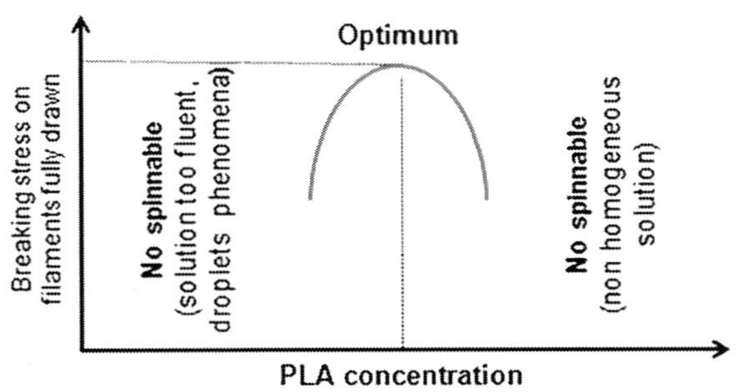

FIGURE 27

Influence of PLA concentration in the spinning solution on the breaking stress of the resulting filaments.

the spinning of the PLA in solution, and the production speeds are higher. For the spinning-drawing, it is possible to produce multifilaments to a speed close to 1000 m min^{-1}. As for the spinning with a high speed, the production speed can reach 5000 m min^{-1}.

As in the case of PET pellets, the melt spinning of PLA requires a preliminary step of drying [143,144], since according to principal suppliers, the recommended moisture content to prevent viscosity decrease and potential loss of properties must be <0.005% (50 ppm). Without the preliminary step of drying, the PLA would be subjected to a strong hydrolysis during the extrusion, which would decrease the macromolecular weight of the polymer. However, this drying step does not prevent the partial degradation of the PLA. So, a decrease in the molecular weight is generally observed, in particular for the PLA with a high macromolecular weight. For example, a PLA with a macromolecular weight equal to 330,000 g mol^{-1} presents a decrease of up to 105,000 g mol^{-1} after the spinning step [147]. Pellets of PLA are first molten in a single screw extruder, in which the temperature is adapted to obtain a polymer that is sufficiently fluid to pass through the dies, without causing an significant degradation.

5.2 Physico-chemical properties of PLA fibers

The spinning processes lead to modifications of PLA characteristics, and more particularly of mechanical, thermal properties, and topology of the fiber surface. Different parameters have detrimental effects on these modifications: the extrusion temperatures, the spinning speed of multifilaments (particularly in the case of melt spinning at high speed), the drawing rate, and the temperatures of the various rollers used during the drawing step of the multifilaments (for the process of spinning drawing).

The spinning speed for the melt spinning process at high speeds, and the drawing ratio applied to the multifilaments for the other processes are the main parameters, whose modification is associated with the variation of the thermophysical properties of PLA. The molten polymer passes through dies consisting of channels in order to produce a continuous multifilament yarn. The multifilament is covered with a spin finish that ensures monofilament cohesion during the process, then rolled up on two heated rolls with varying speeds (S_1 and S_2) to promote the drawing. The theoretical drawing of a multifilament is given by the draw ratio $DR = S_2/S_1$. Depending on the value of this parameter, different glass transition temperatures (T_g), and crystallization and melting temperatures (T_c and T_m, respectively), degree of crystallinity (χ_c) will be recorded. During the melt spinning process, the polymer is stretched to favor the orientation of the constitutive macromolecules in the direction of the drawing. This orientation of macromolecules increases the degree of crystallinity (cf. Figure 28). For the spinning with high speed, multifilaments produced at <1000 m min^{-1} are amorphous [143–145], because of the low kinetics of PLA crystallization. A speed superior to 1000 m min^{-1} is necessary to observe an increase in the crystallinity degree, which will lead to a variation of the glass transition. The increase in the crystallinity degree restricts the molecular mobility of

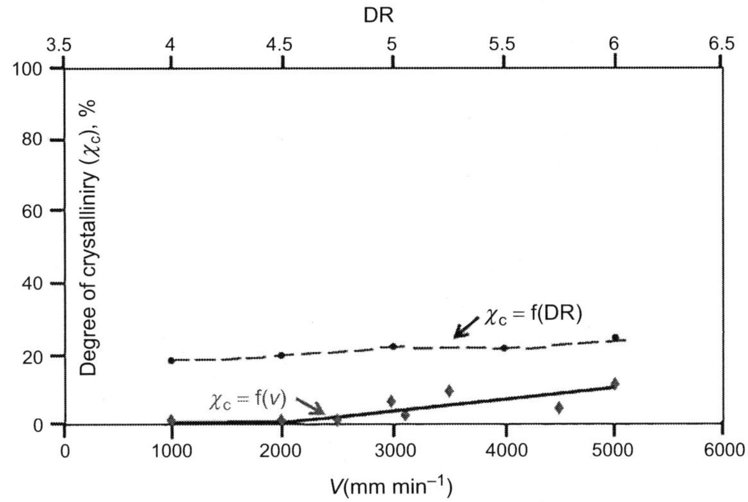

FIGURE 28

Influence of the spinning speed (high speed process) and of the draw ratio (spinning-drawing process) on the degree of crystallinity (χ_C) of the PLA filaments.

the PLA. The required energy to pass from the glassy to the rubbery state is thus superior. Cicero et al. [149] have shown an increase in T_g with the drawing ratio. Yuan et al. [150] observed a total disappearance of the glass transition after spinning. The increase in the molecular orientation during the spinning, and consequently of the crystallinity degree, will influence the cold crystallization of PLA. The higher the degree of crystallinity of the multifilament is, the lower is the temperature of cold crystallization [145].

The mechanical properties of PLA multifilaments are mainly dependent on the spinning conditions. So, for the spinning with high speed, the mechanical properties are improved up to a limit speed that depends on the type of PLA. For a PLA containing a copolymer of L-lactide (92%) and mesolactide (8%), the mechanical properties of filaments increase up to a spinning speed close to 4000 m min^{-1} [143,144]. On the other hand, for a PLA containing 99.3% of L-lactide and only 0.7% of D-lactide, the mechanical properties increase only for a speed equal to 2500 m min^{-1}. This phenomenon can be explained by the differences in crystallization kinetics. In the second case, the PLA will crystallize easily. A speed of 2500 m min^{-1} is then sufficient to obtain filaments with optimal mechanical properties. For the spinning drawing process, all the studies show that the mechanical properties increase strongly during drawing. Yuan et al. [150] demonstrated that for three different PLA (different molecular weights and contents in L-lactide), Young's modulus of unstretched filaments were between 1.2 and 2.4 GPa. After drawing, these values were between 3.6 and 5.4 GPa. Concerning the stress at break, it is supposed to vary between 42 and 103 MPa for filaments not stretched and between 300 and 600 MPa for the stretched ones. The optimal properties are thus obtained for the maximal drawing ratio. The mechanical properties are also influenced by the molecular weight of PLA. A high value allows one to obtain yarns having the best mechanical properties [145,150].

The PLA multifilaments obtained by spinning in solution present better mechanical properties than those produced using the molten method. As an example, yarns of PLA having a stress at break superior to 2 GPa were obtained by solution spinning [141]. As a comparison, the values of stress at break of PLA obtained by melt spinning vary from 0.3 to 0.9 GPa (with drawing) and from 0.15 to 0.4 GPa with high speeds. On the other hand, the speeds of fiber production in solution are lower (typically 10 m min^{-1}), whereas they reach several hundred m min^{-1} for the spinning drawing and several thousand m min^{-1} for the spinning at high speeds.

5.3 Nanocomposite PLA fibers

Different studies have already been carried out on the synthesis of PLA nanocomposites, including dispersion methods, morphology of the nanocomposite, or influence on the polymer properties. In order to disperse clays in a PLA matrix, various techniques are used. Firstly, intercalation by solvent [154] or in solution [155] allowed one to obtain various morphologies. Different commercial MMTs (Cloisite 15A, 25A, and 30B from Southern Clay Products, noted C15A, C25A, and C30B, respectively) were used to evaluate the influence of the organomodified agent on the dispersion of the silicate layers, as well as Bentone (noted B104, supplier Elementis Specialties, see hereinafter). For the nanocomposite containing C15A, the interlayer distance *(d)* increased very slightly. C25A possessed a mixture of intercalated and exfoliated structure in the PLA matrix. Exfoliation and a stable morphology were obtained just with C30B: the interactions of the carbonyl groups of PLA and hydroxyl of the organomodified agent in C30B seem to have an important role in the exfoliation of clay layers. The second method used to disperse clay in a PLA matrix was melt blending by extrusion [156–162]. On the one hand, this method allowed one to obtain generally an intercalated structure by using organomodified MMT. The organomodified agent contained hydroxyl functions (C30B), or long alkyl chains (C25A). On the other hand, the improvement of the properties, in particular in the thermal stability, was more significant for the PLA-C30B nanocomposite.

In situ polymerization is an interesting route for the realization of an appropriate dispersion of clay in a PLA matrix [163]. The synthesis of PLA nanocomposites has been carried out by in situ polymerization of L,L-lactide directly in the interlayer space of MMT. It has been found that if the organomodified agent possesses hydroxyl groups (C30B), it is possible to graft polymeric chains on the surface of the clay. Growing chains repel the layers and stabilize them; hence, an exfoliated structure is then obtained. In the absence of hydroxyl groups on the organomodified agent (C25A), an intercalated structure is obtained. The TGA has confirmed the improvement in the thermal stability, especially for nanocomposite containing C30B, thanks to the excellent dispersion of the layers (exfoliated structure). PLA nanocomposites are generally characterized by an increase in the crystallization rate with regard to pure PLA. The presence of clay does not modify the T_g and T_m [155,157,164]. Once more, the thermal stability is particularly improved when OMLS possesses hydroxyl groups (C30B) [156,164].

PLA nanocomposites have been also characterized by DMTA (Dynamic mechanical thermal analysis) [158]. The storage modulus is increased with the incorporation of nanofillers, and particularly above T_g. The motions of polymer chains are then restricted in the presence of clays. The loss modulus in the intercalated nanocomposites is significantly increased above T_g than below T_g, indicating that the plastic response during the deformation of PLA is more important in the presence of silicate, when the materials become soft.

The biodegradation of PLA is a multistep process. During the initial phase of degradation, the macromolecular chains with a high molecular weight are subjected to hydrolysis, creating oligomers with low molecular weights. This reaction can be accelerated by the presence of an acid or a base, and is affected by temperature and humidity. The deterioration of the material is then promoted, resulting in the decrease of the molecular weight. Microorganisms continue the process of degradation by converting the compounds with a low molecular weight in CO_2 and H_2O. The biodegradation of PLA is also affected by the presence of clay [48,158]. One of the major problems of the PLA matrix is the low kinetics of degradation compared to the dynamics of accumulation of wastes. The incorporation of nanofillers accelerates the degradation of PLA. Figure 29 [48], shows the biodegradation of a PLA sample and a PLA–OMLS nanocomposite [48]. However, the degradation kinetics of PLA is dependent on the nature of introduced nanofillers.

The effects of the addition of plasticizers on the properties of PLA and PLA nanocomposites have been widely studied. Different plasticizers can be used to

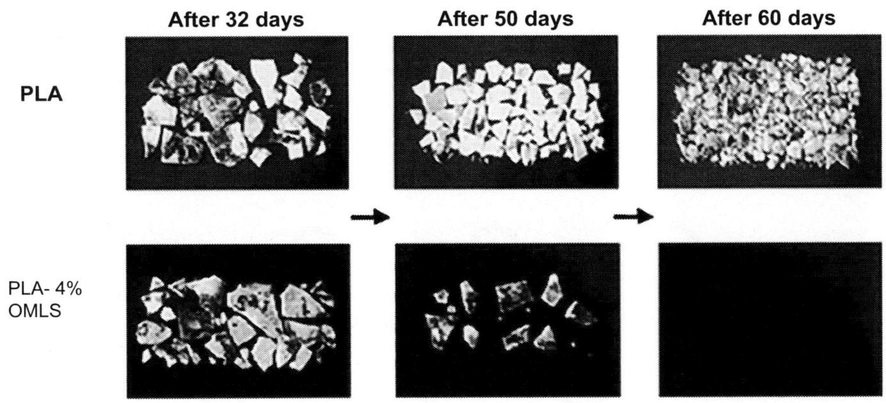

FIGURE 29

Pictures showing the biodegradability of neat PLA and PLA-4% OMLS nanocomposite recovered from compost (MMT modified with trimethyl octadecylammonium) [48].

modify the properties of PLA: polyethylene glycol with different molecular weights, some materials derived of citrate, glycerines, and oligomers of lactic acid [157,159,165–171]. The effects of the plasticizer on the properties of PLA are multiple. Generally, a decrease in T_g, in cold crystallization and melting temperature is observed. This decrease is linked to the improvement of the molecular mobility of PLA with the introduction of a plasticizer. Concerning the mechanical properties, plasticizers tend to decrease Young's modulus and the stress at break. On the other hand, the elongation at break of PLA increases in the presence of a plasticizer.

A melt spinning machine Spinboy I manufactured by Busschaert Engineering (Belgium) has been used to spin draw PLA nanocomposite multifilament yarns. They are rolled up on two heated rolls with varying speeds (S_1 and S_2) in order to ensure drawing. The theoretical drawing of a multifilament is given by the draw ratio $DR = S_2/S_1$. The speed of the feeding roll is fixed at 200 m min^{-1} and the speed of the drawing roll varies from 400 to 800 m min^{-1}. The temperature of the alimentation and drawing rolls are, respectively, 70 °C and 110 °C [172].

Multifilament yarns made with PLA and PLA nanocomposites can be then knitted manually on a rectilinear machine gauge 8. The yarns are processed with the maximal drawing ratio (3.5 for PLA multifilaments, PLA–1% C30B, PLA–4% C30B-10% DOA (Dioctyl adipate), PLA–4% B104, and 3 for PLA–2% C30B). The multifilaments with a yarn count of 1600 dtex are doubled before the knitting, in order to obtain a relatively thick knitting with a density of approximately 1000 g m^{-2}. A weft-knit Milano rib fabric has been used for the knitting fabric (Figure 30). This structure was used to obtain thick and tight knitting with a sufficient density to be tested with a cone calorimeter.

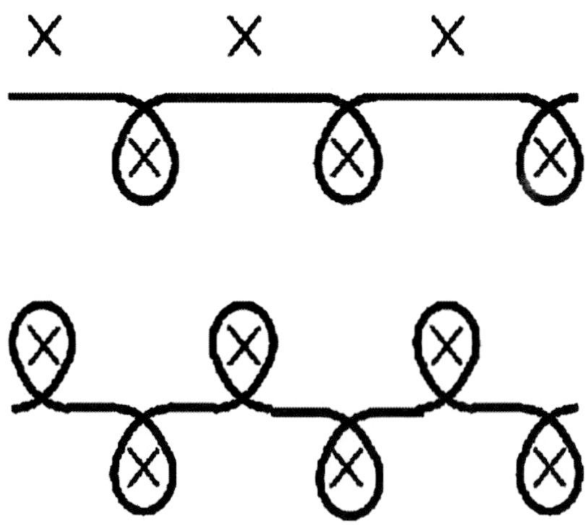

FIGURE 30

Knitted Milan point structure.

5.4 Nanocomposite PLA fiber properties

The properties of the fire resistance of PLA nanocomposite filaments have been estimated by oxygen consumption calorimetry. The sample is submitted to a given heat flow, and the device measures several data: the HRR, which represents the calorific flow of the material, the nature of emitted gases and their concentration, the time of ignition, etc. The heat flow is adjusted in order to simulate a weak, average, or important fire. The conception of the cone calorimeter by Vytenis Babraukas (National Institute of Standards and Technology) goes back to the beginning of the 1980s. Its functioning applies the principle of oxygen consumption [173], which leans on the study of Thornton. He showed in 1917 that for a large number of liquids and gases the ratio of heat evolved to the mass of oxygen consumed during combustion is equal to a constant [174]. Huggett extended this relation to most of the combustible, natural, or synthetic materials used in construction. Experimentally, he attributed a mean value of $13.1\ 10^3$ kJ kg^{-1} to this constant [175], which recently was confirmed for numerous polymers [176]. The knowledge of the mass depletion in oxygen promoted by the combustion of a material allows the direct calculation of the calorific flow.

The sample is submitted to a heat flow emitted by a truncated cone, in order to not perturb the flame. A paramagnetic analyzer allows the determination of the consummated oxygen quantity and thus, via the Huggett relation, gives access to the quantity of heat released by surface and time unit [177]: HRR (expressed in kilowatts per square meter). This parameter, in particular the value of its maximum, is usually taken into account for the evaluation of fire properties. It is indeed an indicator of the capacity of the material to propagate the flame to its environment. The integration of the HRR as a function of time gives the THE, expressed in kilojoules. The simultaneous dynamical acquisitions of additional combustion data are also realized:

- The Volume of Smoke Production in cubic meters per second, measured by laser detection based on the Beer–Lambert principle,
- The contents of carbon monoxide (CO, expressed in parts per million) and in carbon dioxide (CO_2, expressed as volume percentage) in the combustion gases thanks to specific analyzers. We deduct after calculation the total quantity of both gases produced (TCO and TCO_2 expressed in cubic meters).

This device simultaneously allows the qualification and the quantification of the phenomena connected to the fire (ignition, combustion, smokes, etc.). Trials have been carried out with the FTT cone calorimeter in the following conditions: horizontal orientation, interval of measurement = 3 s, forced ignition, and irradiance = 35 kW m^{-2}. These conditions allow the simulation of a beginning fire declared in ventilated premises.

Figure 31 presents the HRR curves of textile structures based on PLA and PLA nanocomposites. The ignition time of the PLA knitted fabric was 76 s. When the PLA is filled with C30B or B104, a slight decrease in TTI is observed. For knitted

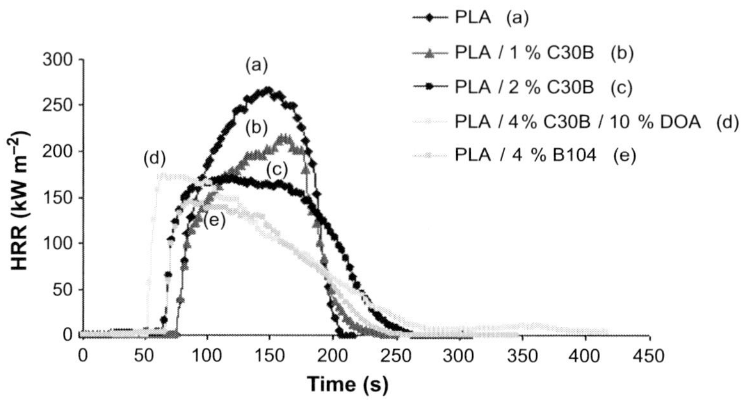

FIGURE 31

HRR curves of textile fabrics based on PLA and PLA nanocomposites (at 35 kW m^{-2}).

materials PLA−1% C30B, PLA−4% B104, PLA−2% C30B, PLA−4% C30B−10% DOA, the values of TTI are, respectively, 75, 66, 64, and 50 s. The decrease in TTI for a polymeric matrix filled with clay [178,179], but also for a polymer containing CNTs [180], has already been reported. This decrease can have several origins:

- The organomodifier present within the nanofiller degrades easily. Its thermal degradation could begin the degradation of the polymeric matrix.
- The presence of nanofillers promotes the modification of infrared radiation absorption. A more important part of the heat flow emitted by the cone calorimeter is then absorbed by the material, provoking a faster increase of the temperature on the surface of the sample. Thermal degradation can then be induced more quickly, and can generate flammable volatile degradation products that can promote the material ignition.
- The addition of clays entails a decrease in the mechanical properties of yarns. Thus, they become more fragile, and can break more easily during the knitting stage. Fibrils are indeed observed on the surface of knitted materials. These fibrils can then favor ignition.
- In the case of plasticized samples (PLA-4% C30B-10% DOA), the plasticizer can also play a role in the decrease of TTI. Thermogravimetry has shown that plasticizer volatilization starting toward 180 °C. This thus degrades more easily than the PLA whose degradation occurs at about 310 °C. Volatilization of the DOA can then produce gases, which can accelerate the ignition of the material.

The maximum value of HRR for textile structures with PLA is 265 kW m^{-2}. For samples containing 1% and 2% C30B, 4% C30B−10% DOA, the maximal values have been found to be, respectively, 213, 163, and 173 kW m^{-2}. It represents a respective decrease of 19, 38% and 34.7%. The most remarkable effect has been obtained for a nanofiller content of 2%. The HRR reach a maximum value (163 kW m^{-2}) between 80 and 180 s. This plateau is not observed for the textile structure with virgin PLA. A similar decrease of the HRR is obtained for the textile structure with PLA-4% C30B−10% DOA. Nevertheless, in this case, the plateau observed is less important. Moreover, the HRR tends to decrease gradually. Incandescence phenomena have been also observed at the end of the test, between 270 and 420 s. The strongest decrease of the HRR can be obtained when PLA−4% B104 yarns are used to produce knitted materials. The maximum value of HRR is 143 kW m^{-2}, which represents a 46% decrease with respect to the virgin PLA. B104 thus allows the decrease in a more significant way in the PHRR value compared with C30B.

Figure 32 presents the evolution of THE of the various knitted structures. Time of ignition for materials containing nanofillers is lower than that for virgin PLA, while the heat release arises more quickly. But according to the filler content in the fabric, the THE of the nanocomposite samples becomes lower than that of the PLA sample after a certain time (Table 6). When the sample combustion is over, we can observe that the THE of nanocomposite materials is lower than that of PLA. Nanocomposite samples thus possess a better reaction to fire.

The opacity of smoke is an important parameter to take into account because it hinders the evacuation of the persons present during a fire. If the visibility is indeed bad, a person trapped by fire cannot run away without panic; consequently, the risk

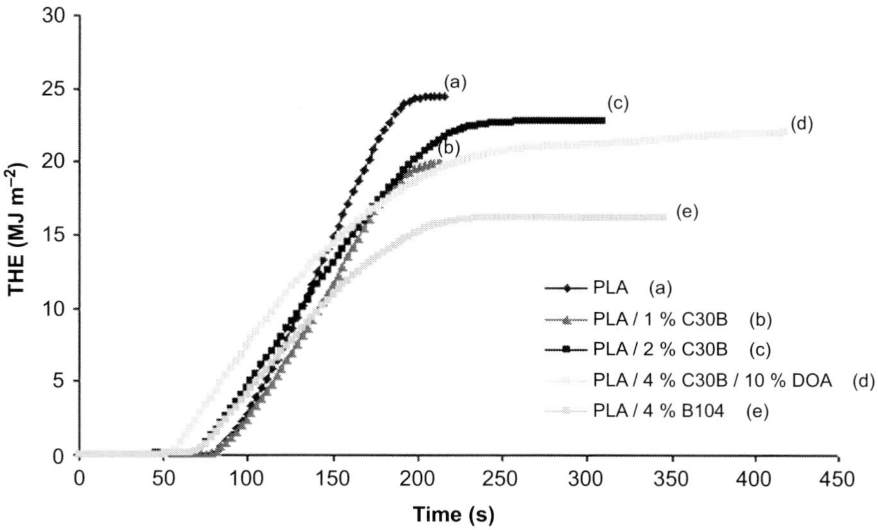

FIGURE 32

Evolution of the THE during burning of PLA and PLA nanocomposite-based knitted fabrics (at 35 kW m^{-2}).

Table 6 Time from Which the Total Heat Evolved (THE) becomes Lower Than That of Virgin PLA Fabric

Knitted Fabric (% by Weight)	Time from Which THE becomes Lower than That of the Virgin PLA Fabric (s)
PLA–1% C30B	101
PLA–2% C30B	138
PLA–4% C30B - 10% DOA	153
PLA–4% B104	123

of death, by being burned or by inhalation, increases. The analysis of smoke toxicity is of major importance because the statistics concerning fires show that the majority of the deaths are the consequence of inhalation of toxic compounds present in the smoke. In a fire, two kinds of exposure are to be considered regarding their toxic impact. The first one relates to a confined room and essentially concerns the intervention teams and the persons trapped by fire. The second concerns more particularly the surrounding population distant from 50 m to a few kilometers away from the disaster. The incurred risks are less important than in the previous case. However, it is important to note that these risks exist.

On the one hand, carbon monoxide (CO) is the major toxic element contained in smoke. It results from the incomplete combustion of materials. It is a colorless, odorless, and nonirritating gas. After inhalation, CO settles by reversible connection in the body and promotes cellular asphyxiation. A CO concentration of 100 ppm in the air provokes headaches, and 1500 ppm is the lethal threshold at more or less long term. So, the consideration of this parameter remains very important in the choice of a fireproof system. On the other hand, the carbon dioxide (CO_2) is the product of the complete combustion of the material. It is generally emitted in great quantities during a fire. The inhalation of CO_2 stimulates breath; this action in turn increases the oxygen inhalation of some gases or toxic vapors that can result from the fire. At a 5% concentration (50,000 ppm), stimulation is very important. An exposure for duration of 30 minutes produces signs of poisoning. Over 70,000 ppm, unconsciousness occurs in a few minutes. The CO/CO_2 ratio supplies information on the combustion mode of a material. When the combustion is complete, it produces only some water and CO_2. On the other hand, if it is incomplete, then CO and smoke are formed.

Figure 33 and Figure 34, respectively, show the emission of CO_2 and the total quantity of CO_2 evolved as a function of time, for several PLA nanocomposites. Samples possessing the lowest time of ignition (PLA-2% C30B, PLA-4% C30B-10% DOA, and PLA-4% B104) logically produce CO_2 faster. On the other hand, the peaks of CO_2 production of these samples are much lower than those of pure PLA (Figure 33). However, the total quantity of CO_2 evolved in the end does not

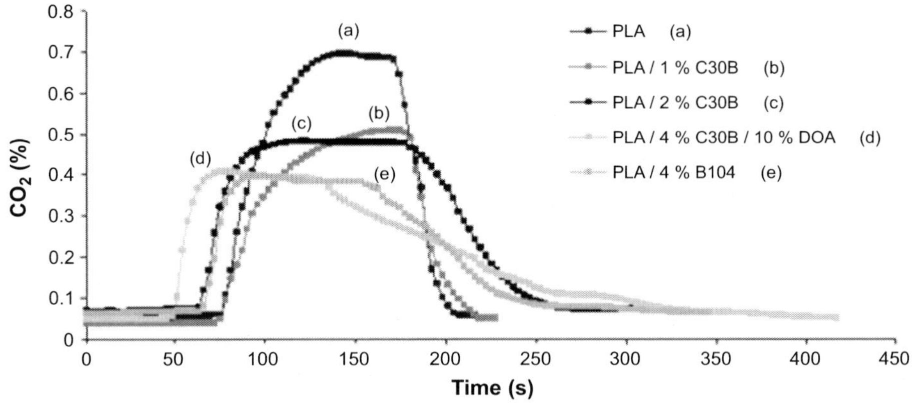

FIGURE 33

Production of CO_2 during burning of PLA and PLA nanocomposite-based knitted fabrics at 35 kW m^{-2}.

almost vary from one sample to another, except for the PLA-1% C30B sample where a slight decrease of CO_2 can be observed.

Figure 35 and Figure 36 respectively present CO emission and CO total quantity evolved as a function of time. A reduction of the peak of CO emission is observed for all nanocomposite samples (Figure 35). Only the PLA-4% B104 sample has its peak of CO emission appearing earlier; this then decreases rather quickly. Concerning the total quantity of CO evolved, we can observe that samples containing 4% of nanofiller (PLA−4% C30B−10% DOA and PLA−4% B104) finally emit more CO than

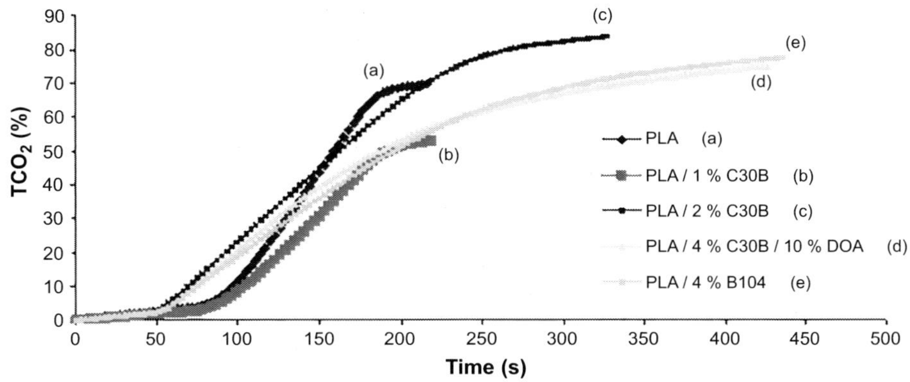

FIGURE 34

Total emission of CO_2 as a function of time during burning of PLA and PLA nanocomposite-based knitted fabrics at 35 kW m^{-2}.

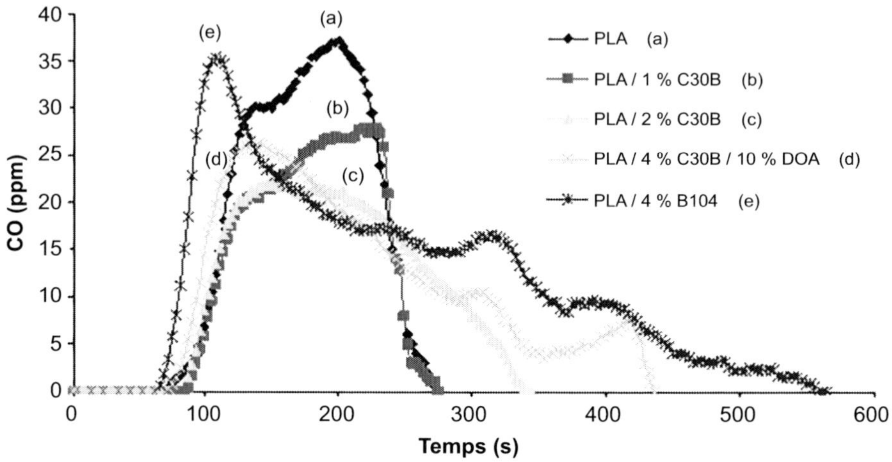

FIGURE 35

CO production during the burning of PLA and PLA nanocomposite-based knitted fabrics at 35 kW m^{-2}.

the weakly filled PLA or PLA samples (1 or 2% of C30B). This phenomenon can be explained by the incomplete combustion of the samples filled at 4%. Contrary to the samples of PLA, PLA-1% C30B and PLA-2% C30B where residues at the end of combustion are very low, those observed for the PLA-4% C30B-10% DOA, and PLA-4% B104 are much more. We can thus suppose that the protective coating formed by the fillers on the surface of the sample is more important in the case of nanocomposite samples filled at 4%. Combustion cannot then be complete, thus promoting the increase in the quantity of CO evolved.

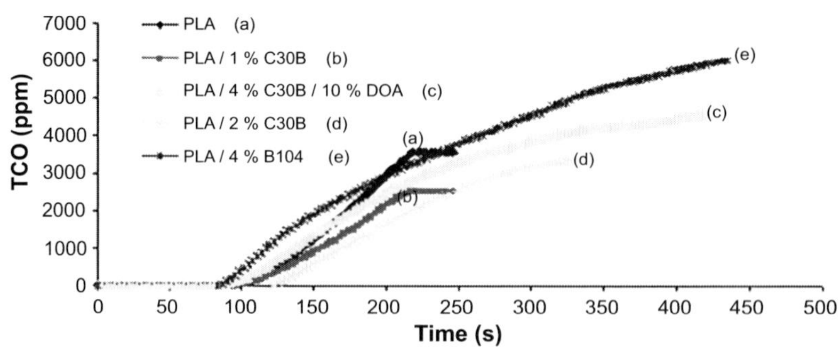

FIGURE 36

Total emission of CO as a function of time during burning of PLA and PLA nanocomposite-based knitted fabrics at 35 kW m^{-2}.

6. CONCLUSIONS AND FUTURE PERSPECTIVES

(*Joint contribution*)

The production of BNC using nanofillers of different morphologies, combined with or without traditional FRs, is certainly one of the most promising ways to improve the flame retardancy of biopolymers intended for engineering applications. Moreover, in competition with the petrochemical polymers, BNCs are very promising materials since they show improved performances (high strength and rigidity, thermal stability, low flammability, antistatic to conductive electrical characteristics, etc.), while maintaining the specific properties of the biodegradability of the polymer matrix.

Regarding the various nanoparticles addressed in this chapter (OMLS, CNTs, a special focus on graphite derivatives), it is generally assumed that the level of performance is typically associated with their good dispersion in the polymer matrix. Thus, it is believed that further work needs to be done for their surface treatment and adequate (pre)mixing. Following the state of the art and by considering the published results, it was recognized that only few papers are devoted to the flame retardancy of biopolymers. Moreover, a large number of studies are mainly found regarding FR compositions based on PLA, polyesters obtained from renewable resources and that are biodegradable, recently considered as environmental alternatives for certain technical applications. It is also noteworthy mentioning that the development of special FR formulations of biopolymers is clearly in its early stages compared to conventional nonbiodegradable petrochemical polymers. Various nanoreinforcements for biopolymer matrices are under development, but the most extensively studied BNCs, so far, are those based on OMLS, CNTs, graphite, etc. due to their availability, low cost, and enhancement induced in mechanical, thermal, barrier, and FR properties of the polymer matrix.

Generally speaking, methods to produce FR biodegradable BNCs remain similar to those used for traditional polymers. Moreover, the principal testing procedures for the determination of flame retardancy do not show (up to now) any differences. Based on our results and those reported in the literature, it is reconfirmed that the FR properties of BNCs containing different nanofillers or their synergistic combinations are clearly influenced by the fire scenario. On the one hand, in the absence of additional FR additives, the BNC e.g. based on PLA have failed when evaluated according to an important flammability test required by industry, i.e. UL 94V test. On the other hand, the evaluation of flame retardancy by cone calorimetry provides positive results along with more valuable information. The results show in almost every system a reduction of PHRR with respect to the neat polymer, whereas with a few exceptions, the TTI is almost always shorter. The main mechanism that governs the FR action in BNCs is similar to those reported for nonbiodegradable polymers: the formation of a protective thermally stable surface layer during combustion, which limits the heat transfer from the flame to the substrate and mass transfer from the substrate to the flame (supplemented by the formation of char). Additionally, useful information is obtained from the horizontal burning test (UL 94HB), i.e.

quantification of BR and qualitative indications, e.g. burning with or without drips or formation of char. Generally, it is quite possible, and this assumption is based on experimental results that the FR mechanism of the nanocomposites can show some differences related to the nature of the nanofiller (e.g. CNTs or graphite derivatives vs. OMLS). Coaddition of nanofillers, their combination with conventional FRs, utilization of a combination of microfillers, etc. represent possible approaches leading to "green" (nano)composites characterized by FR properties. As an example of the multiple applications, PLA-OMLS nanocomposites have been used for spinning and realization of textile fabrics, due to their improved FR properties.

Finally, it is assumed that utilization of nanofillers provides an opportunity for the enhancement of fire retardancy of biopolymers, but this protection is not enough to provide the highest standard of performance. Because BNCs failed when evaluated according to some tests (UL 94V, LOI); it is believed that the combination with traditional FR additives still remains for future consideration, as principal pathways leading to improved synergistic FR performances. To further decrease the flammability of biopolymers, there is a need to find additional FR formulations of high effectiveness by designing and synthesizing novel intumescent agents and new FR additives. We consider that the association of biopolymers with improved intumescent formulations can be a good compromise between the performances and the "green" aspects of the product. Very promising developments in synergy aspects are furthermore expected for biodegradable polymers and efforts should be made in this direction. It is also considered that other nanofillers did not yet show their real potential in polymer flame retardancy. Hence, nanoparticle dispersion still remains a challenge for advanced performance, whereas new pathways for different synergies and techniques of production that are easily applicable can undoubtedly increase the potential of BNCs as "green" products of high interest for engineering applications.

Abbreviations

AII	Anhydrite II
APP	Ammonium polyphosphate
APS	Ammonium persulfate
ATH	Aluminum trihydroxide
BMA	Butyl methacrylate
BNC	Bionanocomposites
BR	Burning rate
C30B	Cloisite 30B (organically-modified montmorillonite, from Southern Clay Products, USA)
CNTs	Carbon nanotubes
DMTA	Dynamic mechanical thermal analysis
DOA	Dioctyl adipate
DR	Draw ratio
EG	Expanded graphite
EGNP	Exfoliated graphite nanoplatelets

f-MWCNTs	Functionalized multiwall carbon nanotubes
FR	Flame retardant(s)
HRR (RHR)	Heat release rate (rate of heat release)
LDHs	Layered double hydroxides
LIG	Lignin
LOI	Limiting oxygen index
MB	Masterbatch
MC	Melamine cyanurate
MCAPP	Microencapsulated ammonium polyphosphate
MCC	Microscale combustion calorimetry
MDH	Magnesium dihydroxide
MMA	Methyl methacrylate
MMT	Montmorillonite
MP	Melamine polyphosphate
MWCNTs	Multiwall carbon nanotubes
OMLS	Organomodified layered silicates
PA 6	Polyamide 6
PBS	Polybutylene succinate
PCL	Polycaprolactone
PDMAEMA	Poly(2-(dimethylamino) ethyl methacrylate)
PE	Polyethylene
PEA	Polyesteramide
PER	Pentaerythritol
PET	Polyethylene terephthalate
PHA	Polyhydroxyalkanoate
PHB	Polyhydroxybutyrate
PHRR (PRHR)	Peak of heat release rate
PLA	Polylactic acid or polylactide
PMMA	Poly(methyl methacrylate)
PNC	Polymer nanocomposites
POSS	Polyhedral oligomeric silsesquioxanes
PP	Polypropylene
PPS	Poly(phenylene sulfide)
PVOH	Polyvinylalcohol
ROP	Ring opening polymerization
SPDPC	Spirocyclic pentaerythritol bisphosphorate disphosphoryl chloride
SPDPM	Spirocyclic pentaerythritol bisphosphorate disphosphoryl melamine
ST	Starch
TBP-A	Tetrabromobisphenol A
TEM	Transmission electronic microscopy
TEOS	Tetraethoxysilane
TGA	Thermogravimetric analysis
THE	Total heat evolved
THR	Total heat release
TPhP	Triphenylphosphate
TPS	Thermoplastic starch
TTI	Time to ignition
VSP	Volume of smoke production

XRD	X-ray diffraction
ZnAl-LDH	Zinc aluminum layered double hydroxide

Acknowledgments

The authors thank the Wallonia Region, Nord-Pas de Calais Region, and European Community for the financial support in the frame of the INTERREG IV—NANOLAC project. This work has been also supported by the European Commission and Région Wallonne FEDER program (Materia Nova) and OPTI^{2}MAT program of excellence, by the Interuniversity Attraction Pole program of the Belgian Federal Science Policy Office (PAI 6/27) and by FNRS-FRFC.

References

[1] Platt D. Biodegradable polymers—market report. Shawbury, Shrewsbury, Shropshire, UK: Smithers Rapra Limited; 2006.
[2] Madhavan Nampoothiri K, Nair NR, John RP. An overview of the recent developments in polylactide (PLA) research. Biores Technol 2010;101:8493–501.
[3] Dubois Ph, Murariu M. The "green" challenge: high performance PLA (nano) composites. JEC Compos Mag 2008;45:66–9.
[4] Smock D. Bioplastics: Technologies and global markets. BCC Research, report code PLS050A, http://www.bccresearch.com/report/bioplastics-technologies-markets-pls050a.html; September 2010.
[5] Rapid growth forecast for biodegradable polymers. Plast Addit Compd 2008;10:4.
[6] Shen L, Haufe J, Patel MK. Product overview and market projection of emerging bio-based plastics. PRO-BIP 2009. Utrecht: Copernicus Institute for Sustainable Development and Innovation, Utrecht University; 2009.
[7] Ravenstijn J. The state-of-the-art on bioplastics Products: markets, trends, and technologies. Polymedia Publ GmbH; 2010.
[8] Bourbigot S, Fontaine G. Flame retardancy of polylactide: an overview. Polym Chem 2010;1:1413–22.
[9] Averous L. Biodegradable multiphase systems based on plasticized starch: a review. J Macromol Sci Polym Rev 2004;C44:231–74.
[10] Bordes P, Pollet E, Avérous L. Nano-biocomposites: biodegradable polyester/nanoclay systems. Prog Polym Sci 2009;34:125–55.
[11] Auras R, Harte B, Selke S. An overview of polylactides as packaging materials. Macromol Biosci 2004;4:835–64.
[12] Garlotta D. A literature review of poly(lactic acid). J Polym Environ 2001;9:63–84.
[13] Sodergard A, Stolt M. Properties of lactic acid based polymers and their correlation with composition. Prog Polym Sci 2002;27:1123–63.
[14] Drumright RE, Gruber PR, Henton DE. Polylactic acid technology. Adv Mater 2000; 12:1841–6.
[15] Jamshidian M, Tehrany EA, Imran M, Jacquot M, Desobry S. Poly-lactic acid: production, applications, nanocomposites, and release studies. Compr Rev Food Sci F 2010;9: 552–71.

[16] Lim LT, Auras R, Rubino M. Processing technologies for poly (lactic acid). Prog Polym Sci 2008;33:820–52.

[17] Paul MA, Delcourt C, Alexandre M, Degée Ph, Monteverde F, Dubois Ph. Polylactide/montmorillonite nanocomposites. Study of the hydrolytic degradation. Polym Degrad Stabil 2005;87:535–42.

[18] Murariu M, Da Silva Ferreira A, Degée Ph, Alexandre M, Dubois Ph. Polylactide compositions. Part 1: effect of filler content and size on mechanical properties of PLA/calcium sulfate composites. Polymer 2007;48:2613–8.

[19] Yu L, Dean K, Li L. Polymer blends and composites from renewable resource. Prog Polym Sci 2006;31:576–602.

[20] Anderson KS, Schreck KM, Hillmyer MA. Toughening polylactide. Polym Rev 2008;48:85–108.

[21] Alexandre M, Dubois Ph. Polymer-layered silicate nanocomposites: preparation, properties and used of a new class of materials. Mater Sci Eng 2000;28:1–63.

[22] Wang L, He X, Wilkie C. The utility of nanocomposites in fire retardancy. Materials 2010;3:4580–606.

[23] Paul DR, Robeson LM. Polymer nanotechnology: nanocomposites. Polymer 2008;49:3187–204.

[24] Murariu M, Doumbia A, Bonnaud L, Dechief AL, Paint Y, Ferreira M, et al. High-performance polylactide/ZnO nanocomposites designed for films and fibers with special end-use properties. Biomacromolecules 2011;12:1762–71.

[25] Pavlidou S, Papaspyrides CD. A review on polymer–layered silicate nanocomposites. Prog Polym Sci 2008;33:1119–98.

[26] Hussain F, Hojjati M, Okamoto M, Gorga RE. Polymer–matrix nanocomposites, processing, manufacturing, and application: an overview. J Compos Mater 2006;40:1511–75.

[27] Gao F. Clay/Polymer composites: the Story. Mater Today 2004;7:50–5.

[28] Sinda Ray S, Bousmina M. Biodegradable polymers and their layered silicate nanocomposites: in greening the 21st century materials world. Prog Mater Sci 2005;50:962–1079.

[29] Sorrentino A, Gorrasi G, Vittoria V. Potential perspectives of bio-nanocomposites for food packaging applications. Trends Food Sci Tech 2007;18:84–95.

[30] Goffin AL, Duquesne E, Moins S, Alexandre M, Dubois Ph. New organic–inorganic nanohybrids via ring opening polymerization of (di)lactones initiated by functionalized polyhedral oligomeric silsesquioxane. Eur Polym J 2007;43:4103–13.

[31] Fortunati E, Armentano I, Iannoni A, Kenny JM. Development and thermal behaviour of ternary PLA matrix composites. Polym Degrad Stabil 2010;95:2200–6.

[32] Okamoto M. Biodegradable polymer/layered silicate nanocomposites: a review. In: Mallapragada S, Narasimhan B, editors. Handbook of biodegradable polymeric materials and their applications, vol. 1. American Scientific Publishers; 2005. pp. 1–45.

[33] Le Corre D, Bras J, Dufresne A. Starch nanoparticles: a review. Biomacromolecules 2010;11:1139–53.

[34] Oksman K, Mathew AP, Bondeson D, Kvien I. Manufacturing process of polylactic acid (PLA) cellulose whiskers nanocomposites. Compos Sci Technol 2006;66:2776–84.

[35] Petersson L, Oksman K. Biopolymer based nanocomposites: comparing layered silicates and microcrystalline cellulose as nanoreinforcement. Compos Sci Technol 2006;66:2187–96.

[36] Laoutid F, Bonnaud L, Alexandre M, Lopez-Cuesta JM, Dubois Ph. New prospects in FR polymer materials: from fundamentals to nanocomposites. Mater Sci Eng R 2009;63:100–25.

[37] Song R, Wang Z, Meng X, Zhang B, Tang T. Influences of catalysis and dispersion of organically modified montmorillonite on flame retardancy of polypropylene nanocomposites. J Appl Polym Sci 2007;106:3488–94.

[38] Jang BN, Costache M, Wilkie CA. The relationship between thermal degradation behavior of polymer and the fire retardancy of polymer/clay nanocomposite. Polymer 2005;46:10678–87.

[39] Troitzsch J. International plastics flammability handbook. 2nd ed. Munich, Vienna, New York: Hanser Publishers; 1990.

[40] Morgan AB, Wilkie CA. Flame retardant polymer nanocomposites. 1st ed. Hoboken, NJ, USA: John Wiley & Sons, Inc.; 2007.

[41] McWilliams A. Global markets for nanocomposites, nanoparticles, nanoclays, and nanotubes. Report NAN021E, BCC Research; January 2012. source: http://www.bccresearch.com/report/nanocomposites-global-markets-nan021e.html.

[42] Sinda Ray S, Yamada K, Okamoto M, Ogami A, Ueda K. New polylactide/layered silicate nanocomposites. 3. High-performance biodegradable materials. Chem Mater 2003;15:1456–65.

[43] Singh S, Sinda Ray S. Polylactide based nanostructured biomaterials and their applications. J Nanosci Nanotechnol 2007;7:2596–615.

[44] Maiti P, Yamada K, Okamoto M, Ueda K, Okamoto K. New polylactide/layered silicate nanocomposites: role of organoclays. Chem Mater 2002;14:4654–61.

[45] Sinha Ray S, Yamada K, Ogami A, Okamoto M, Ueda K. New polylactide/layered silicate nanocomposite: nanoscale control over multiple properties. Macromol Rapid Comm 2002;23:943–7.

[46] Sinha Ray S, Maiti P, Okamoto M, Yamada K, Ueda K. New polylactide/layered silicate nanocomposites. 1. Preparation, characterization and properties. Macromolecules 2002;35:3104–10.

[47] Sinha Ray S, Yamada K, Okamoto M, Ogami A, Ueda K. New polylactide/layered silicate nanocomposites. 4. Structure, properties and biodegradability. Compos Interfaces 2003;10:435–50.

[48] Sinha Ray S, Yamada K, Okamoto M, Ueda K. New polylactide-layered silicate nanocomposites. 2. Concurrent improvements of material properties, biodegradability and melt rheology. Polymer 2003;44:857–66.

[49] Okamoto M. Recent advances in polymer/layered silicate nanocomposites. Mater Sci Tech Ser 2006;22:756–79.

[50] Sinha Ray S, Yamada K, Okamoto M, Ueda K. Biodegradable polylactide/montmorillonite nanocomposites. J Nanosci Nanotechnol 2003;3:503–10.

[51] Yoshida O, Okamoto M. Direct melt intercalation of polylactide chains into nano-galleries: interlayer expansion and nano-composite structure. Macromol Rapid Comm 2006;27:751–7.

[52] Kiliaris P, Papaspyrides CD. Polymer/layered silicate (clay) nanocomposites: an overview of flame retardancy. Prog Polym Sci 2010;35:902–58.

[53] Szleifer I, Yerushalmi-Rozen R. Polymers and carbon nanotubes-dimensionality, interactions and nanotechnology. Polymer 2005;46:7803–18.

[54] Andrews R, Weisenberger MC. Carbon nanotube polymer composites. Curr Opin Solid State Mater Sci 2004;8:31–7.

[55] Stefanescu EA, Daranga C, Stefanescu C. Insight into the broad field of polymer nanocomposites: from carbon nanotubes to clay nanoplatelets, via metal nanoparticles. Materials 2009;2:2095−153.

[56] Bal S, Samal SS. Carbon nanotube reinforced polymer composites—a state of the art. Bull Mater Sci 2007;30:379−86.

[57] Bose S, Khare RA, Moldenaers P. Assessing the strengths and weaknesses of various types of pre-treatments of carbon nanotubes on the properties of polymer/carbon nanotubes composites: a critical review. Polymer 2010;51:975−93.

[58] Rana S, Yoo HJ, Cho JW, Chun BC, Park JS. Functionalization of multi-walled carbon nanotubes with poly(ε-caprolactone) using click chemistry. J Appl Polym Sci 2011; 119:31−7.

[59] Meyer F, Minoia A, Raquez JM, Spasova M, Lazzaroni R, Dubois Ph. Poly(aminomethacrylate) as versatile agent for carbon nanotube dispersion: an experimental, theoretical and application study. J Mater Chem 2010;20:6873−80.

[60] Kashiwagi T, Grulke E, Hilding J, Groth K, Harris R, Butler K, et al. Thermal and flammability properties of polypropylene/carbon nanotube nanocomposites. Polymer 2004;45:4227−39.

[61] Schartel B, Pötschke P, Knoll U, Abdel-Goad M. Fire behaviour of polyamide 6/multiwall carbon nanotube nanocomposites. Eur Polym J 2005;41:1061−70.

[62] Fina A, Bocchini S, Camino G. Catalytic fire retardant nanocomposites. Polym Degrad Stabil 2008;93:1647−55.

[63] Villmow T, Potschke P, Pegel S, Haussler L, Kretzschmar B. Influence of twin-screw extrusion conditions on the dispersion of multi-walled carbon nanotubes in a poly(lactic acid) matrix. Polymer 2008;49:3500−9.

[64] Song W, Zheng Z, Lu H, Wang X. Incorporation of multi-walled carbon nanotubes into biodegradable telechelic prepolymers. Macromol Chem Phys 2008;209:315−21.

[65] Song W, Zheng Z, Tang W, Wang X. A facile approach to covalently functionalized carbon nanotubes with biocompatible polymer. Polymer 2007;48:3658−63.

[66] Wu CS, Liao HT. Study on the preparation and characterization of biodegradable polylactide/multi-walled carbon nanotubes nanocomposites. Polymer 2007;48: 4449−58.

[67] Wu D, Wu L, Zhang M, Zhao Y. Viscoelasticity and thermal stability of polylactide composites with various functionalized carbon nanotubes. Polym Degrad Stabil 2008;93:1577−84.

[68] Tsuji H, Kawashima Y, Takikawa H, Tanaka S. Poly(L-lactide)/nano-structured carbon composites: conductivity, thermal properties, crystallization, and biodegradation. Polymer 2007;48:4213−25.

[69] Shieh YT, Twu YK, Su CC, Lin RH, Liu GL. Crystallization kinetics study of poly(L-lactic acid)/carbon nanotubes nanocomposites. J Polym Sci Pol Phys 2010; 48:983−9.

[70] Bourbigot S, Fontaine G, Gallos A, Bellayer S. Reactive extrusion of PLA and of PLA/carbon nanotubes nanocomposite: processing, characterization and flame retardancy. Polym Adv Technol 2011;22:30−7.

[71] Bourbigot S, Fontaine G, Gallos A, Gérard C, Bellayer S. Functionalized-carbon multiwall nanotube as flame retardant for polylactic acid. In: Wilkie CA, Morgan AB, Nelson GL, editors. Fire and polymers V—materials and concepts for fire retardancy, vol. 1013. Washington DC: American Chemical Society; 2009. pp. 25−34.

[72] Murariu M, Dechief AL, Bonnaud L, Paint Y, Gallos A, Fontaine G, et al. The production and properties of polylactide composites filled with expanded graphite. Polym Degrad Stabil 2010;95:889–900.
[73] Chung DDL. Review—graphite. J Mater Sci 2002;37:1475–89.
[74] Zhao YF, Xiao M, Wang SJ, Ge XC, Meng YZ. Preparation and properties of electrically conductive PPS/expanded graphite nanocomposites. Compos Sci Technol 2007;67:2528–34.
[75] Uhl FM, Yao Q, Nakajima H, Manias E, Wilkie CA. Expandable graphite/polyamide 6 nanocomposites. Polym Degrad Stabil 2005;89:70–84.
[76] Yasmin A, Luo JJ, Daniel IM. Processing of expanded graphite reinforced polymer nanocomposites. Compos Sci Technol 2006;66:1182–9.
[77] Kalaitzidou K, Fukushima H, Drzal LT. A new compounding method for exfoliated graphite–polypropylene nanocomposites with enhanced flexural properties and lower percolation threshold. Compos Sci Technol 2007;67:2045–51.
[78] Chen G, Chen X, Wang H, Wu D. Dispersion of graphite nanosheets in polymer resins via masterbatch technique. J Appl Polym Sci 2007;103:3470–5.
[79] Drzal LT, Fukushima H. Expanded graphite and products produced therefrom. US Patent 7,550,529 Board Trust Mich. State Univ. USA. 2009.
[80] Chen G, Wu D, Weng W, Wu C. Exfoliation of graphite flake and its nanocomposites. Carbon 2003;41:619–21.
[81] Castrovinci A, Camino G. Fire-retardant mechanisms in polymer nano-composite materials. In: Duquesne S, Magniez C, Camino G, editors. Multifunctional barriers for flexible structure-textile, series: Leather and paper, vol. 97. Springer Berlin Heidelberg; 2007. pp. 87–108.
[82] Wang D, Wilkie CA. Fire properties of polymer nanocomposites (Chapter 9). In: Mouritz AP, Gibson AG, editors. Fire properties of polymer composite materials, series: Solid mechanics and its applications, vol. 143. Springer Netherlands; 2006. pp. 287–312.
[83] Dasari A, Yu ZZ, Mai YW, Cai G, Song H. Roles of graphite oxide, clay and POSS during the combustion of polyamide 6. Polymer 2009;50:1577–87.
[84] Fukushima K, Murariu M, Camino G, Dubois Ph. Effect of expanded graphite/layered-silicate clay on thermal, mechanical and fire retardant properties of poly(lactic acid). Polym Degrad Stabil 2010;95:1063–76.
[85] Miloaga DG, Hosein H-AA, Rich JM, Kjoller K, Drzal LT. Scanning probe thermal analysis of polylactic acid/exfoliated graphite nanoplatelet (xGnP™) nanocomposites. J Biobased Mater Bioenergy 2008;2:78–84.
[86] Kalaitzidou K, Fukushima H, Drzal LT. Mechanical properties and morphological characterization of exfoliated graphite–polypropylene nanocomposites. Compos Part A-Appl S 2007;38:1675–82.
[87] Potts JR, Dreyer DR, Bielawski CW, Ruoff RS. Graphene-based polymer nanocomposites. Polymer 2011;52:5–25.
[88] Kim S, Drzal LT. Comparison of Exfoliated Graphite Nanoplatelets (xGnP) and CNTs for Reinforcement of EVA Nanocomposites Fabricated by Solution Compounding Method and Three Screw Rotating Systems. J Adhes Sci Technol 2009; 23:1623–38.
[89] Jing L, Pui-Shan W, Jang-Kyo K. Hybrid nanocomposites containing carbon nanotubes and graphite nanoplatelets. Mater Sci Eng A 2008;483–484:660–3.
[90] Liu H, Brinson LC. Reinforcing efficiency of nanoparticles: a simple comparison for polymer nanocomposites. Compos Sci Technol 2008;68:1502–12.

[91] Santangelo S, Gorrasi G, Di Lieto R, De Pasquale S, Patimo G, Piperopoulos E, et al. Polylactide and carbon nanotubes/smectite–clay nanocomposites: preparation, characterization, sorptive and electrical properties. Appl Clay Sci 2011;53:188–94.
[92] Zhang C, Tjiu WW, Liu TX, Lui WY, Phang IY, Zhang WD. Dramatically enhanced mechanical performance of Nylon-6 magnetic composites with nanostructured hybrid one-dimensional carbon nanotubes—two-dimensional clay nanoplatelet heterostructures. J Phys Chem B 2011;115:3392–9.
[93] Dubois Ph, Murariu M, Alexandre M, Degée Ph, Bourbigot S, Delobel R, et al. Polylactyde-based compositions. US Patent 8,119,713 (Materia Nova et al.). 2012.
[94] Zhu H, Zhu Q, Li J, Tao K, Xue L, Yan Q. Synergistic effect between expandable graphite and ammonium polyphosphate on flame retarded polylactide. Polym Degrad Stabil 2011;96:183–9.
[95] Katiyar V, Gerds N, Koch CB, Risbo J, Hansen HCB, Plackett D. Poly(L,L-lactide)-layered double hydroxide nanocomposites via *in situ* polymerization of L,L-lactide. Polym Degrad Stabil 2010;95:2563–73.
[96] Fan Y, Nishida H, Mori T, Shirai Y, Endo T. Thermal degradation of poly(L,L-lactide): effect of alkali earth metal oxides for selective L,L-lactide formation. Polymer 2004; 45:1197–205.
[97] Fukushima K, Tabuani D, Camino G. Nanocomposites of PLA and PCL based on montmorillonite and sepiolite. Mater Sci Eng C 2009;29:1433–41.
[98] Marney DCO, Russell LJ, Wu DY, Nguyen T, Cram DG, Rigopoulos N, et al. The suitability of halloysite nanotubes as a fire retardant for nylon 6. Polym Degrad Stabil 2008;93:1971–8.
[99] Özdilek C, Kazimierczak K, Van der Beek D, Picken SJ. Preparation and properties of polyamide-6-boehmite nanocomposites. Polymer 2004;45:5207–14.
[100] Bourbigot S, Duquesne S. Fire retardant polymers: recent developments and opportunities. J Mater Chem 2007;17:2283–300.
[101] Nishida H, Fan Y, Mori T, Oyagi N, Shirai Y, Endo T. Feedstock recycling of flame-resisting poly(lactic acid)/aluminum hydroxide composite to L,L-lactide. Ind Eng Chem Res 2005;44:1433–7.
[102] Bourbigot S, Le Bras M, Leeuwendal R, Shen KK, Schubert D. Recent advances in the use of zinc borates in flame retardancy of EVA. Polym Degrad Stabil 1999;64: 419–25.
[103] Yanagisawa T, Kiuchi Y, Iji M. Enhanced flame retardancy of polylactic acid with aluminum tri-hydroxide and phenolic resins. Kobunshi Ronbunshu 2009;66:49–54.
[104] Kubokawa H, Hatakeyama T. Thermal decomposition behavior of polylactide fabrics treated with flame retardants. Sen'i Gakkaishi 1999;55:349–55.
[105] Kubokawa H, Ohta M, Hatakeyama T. Flame-retarding of polylactide fabrics. Sen'i Gakkaishi 1999;55:290–7.
[106] Kubokawa H, Komatsu H. Flammability and thermal decomposition behavior of polylactide–cellulose union cloths. Sen'i Gakkaishi 2000;56:497–500.
[107] Bourbigot S, Duquesne S, Fontaine G, Bellayer S, Turf T, Samyn F. Characterization and reaction to fire of polymer nanocomposites with and without conventional flame retardants. Mol Cryst Liq Crys 2008;486:325/[1367]–339/[1381].
[108] Kuan CF, Kuan HC, Ma CCM, Chen CH. Flame retardancy and nondripping properties of ammonium polyphosphate/poly(butylene succinate) composites enhanced by water crosslinking. J Appl Polym Sci 2006;102:2935–45.
[109] Jana T, Roy BC, Maiti S. Biodegradable film: 7. Modification of the biodegradable film for fire retardancy. Polym Degrad Stabil 2000;69:79–82.

[110] Samal R, Sahoo PK. Development of a biodegradable rice straw-g-poly(methylmethacrylate)/sodium silicate composite flame retardant. J Appl Polym Sci 2009;113:3710–5.

[111] Bourbigot S, Duquesne S, Jama C. Polymer nanocomposites: how to reach low flammability? Macromol Symp 2006;233:180–90.

[112] Vandersall HJ. Intumescent coating systems, their development and chemistry. J Fire Flammabl 1971;2:97–140.

[113] Bourbigot S, Le Bras M, Duquesne S, Rochery M. Recent advances for intumescent polymers. Macromol Mater Eng 2004;289:499–511.

[114] Bourbigot S, Le Bras M, Delobel R. Carbonization mechanisms resulting from intumescence association with the ammonium polyphosphate-pentaerythritol fire retardant system. Carbon 1993;31:1219–30.

[115] Réti C, Casetta M, Duquesne S, Bourbigot S, Delobel R. Flammability properties of intumescent PLA starch and lignin. Polym Adv Technol 2008;19:628–35.

[116] Shumao L, Jie R, Hua Y, Tao Y, Weizhong Y. Influence of ammonium polyphosphate on the flame retardancy and mechanical properties of ramie fiber-reinforced poly(lactic acid) biocomposites. Polym Int 2010;59:242–8.

[117] Zhan J, Song L, Nie S, Hu Y. Combustion properties and thermal degradation behavior of polylactide with an effective intumescent flame retardant. Polym Degrad Stabil 2009;94:291–6.

[118] Chen Y, Zhan J, Zhang P, Nie S, Lu H, Song L, et al. Preparation of intumescent flame retardant poly(butylene succinate) using fumed silica as synergistic agent. Ind Eng Chem Res 2010;49:8200–8.

[119] Fontaine G, Bourbigot S. Intumescent polylactide: a nonflammable material. J Appl Polym Sci 2009;113:3860–5.

[120] Stewart R. Biopolymers: greater consumer awareness drives growth of biobased-resins market. Plast Eng 2007;63:24–30.

[121] Yu L, Petinakis S, Dean K, Bilyk A, Wu D. Green polymeric blends and composites from renewable resources. Macromol Symp 2007;249-250:535–9.

[122] Cao X, Chang PR, Huneault MA. Preparation and properties of plasticized starch modified with poly(ε-caprolactone) based waterborne polyurethane. Carbohyd Polym 2008;71:119–25.

[123] Avérous L, Halley PJ. Biocomposites based on plasticized starch. Biofuel Bioprod Bioref 2009;3:329–43.

[124] Matkó S, Toldy A, Keszei S, Anna P, Bertalan G, Marosi G. Flame retardancy of biodegradable polymers and biocomposites. Polym Degrad Stabil 2005;88:138–45.

[125] Wu K, Hu Y, Song HLL, Wang Z. Flame retardancy and thermal degradation of intumescent flame retardant starch-based biodegradable composites. Ind Eng Chem Res 2009;48:3150–7.

[126] Giraud S, Bourbigot S, Rochery M, Vroman I, Tighzert L, Delobel R. Flame behavior of cotton coated with polyurethane containing microencapsulated flame retardant agent. J Ind Text 2001;31:11–26.

[127] Giraud S, Bourbigot S, Rochery M, Vroman I, Tighzert L, Delobel R, et al. Flame retarded polyurea with microencapsulated ammonium phosphate for textile coating. Polym Degrad Stabil 2005;88:106–13.

[128] Giraud S, Salaün F, Bedek G, Vroman I, Bourbigot S. Influence of chemical shell structure on the thermal properties of microcapsules containing a flame retardant agent. Polym Degrad Stabil 2010;95:315–9.

[129] Bourbigot S, Le Bras M, Delobel R, Decressain R, Amoureux JP. Synergistic effect of zeolite in an intumescence process: study of the carbonaceous structures using solid-state NMR. J Chem Soc Faraday Trans 1996;92:149–58.
[130] Bourbigot S, Le Bras M, Dabrowski F, Gilman JW, Kashiwagi T. PA-6 clay nanocomposite hybrid as char forming agent in intumescent formulations. Fire Mater 2000;24:201–8.
[131] Wang DY, Leuteritz A, Wang YZ, Wagenknecht U, Heinrich G. Preparation and burning behaviors of flame retarding biodegradable poly(lactic acid) nanocomposite based on zinc aluminum layered double hydroxide. Polym Degrad Stabil 2010;95:2474–80.
[132] Samyn F, Bourbigot S, Jama C, Bellayer S, Nazare S, Hull R, et al. Characterisation of the dispersion in polymer flame retarded nanocomposites. Eur Polym J 2008;44:1631–41.
[133] Murariu M, Da Silva Ferreira A, Bonnaud L, Dubois Ph. Calcium sulfate as high-performance filler for polylactide (PLA) or how to recycle gypsum as by-product of lactic acid fermentation process. Compos Interfaces 2009;16:65–84.
[134] Murariu M, Bonnaud L, Yoann P, Fontaine G, Bourbigot S, Dubois Ph. New trends in polylactide (PLA)-based materials: "Green" PLA–calcium sulfate (nano)composites tailored with flame retardant properties. Polym Degrad Stabil 2010;95:374–81.
[135] Bourbigot S, Le Bras M, Bréant P, Trémillon JM, Delobel R. Zeolites: new synergistic agents for intumescent fire retardant thermoplastic formulations—criteria for the choice of the zeolite. Fire Mater 1996;20:145–54.
[136] Samal R, Rana PK, Mishra GP, Sahoo PK. Novel biodegradable flame retardant, poly(butyl methacrylate)/sodium silicate/$Mg(OH)_2$ nanocomposite. Polym Compos 2008;29:173–8.
[137] Schneider AK. Polylactide sutures. US Patent 3,636,956 (assigned Ethicon, Inc.). 1972.
[138] Agrawal AK, Bhalla R. Advances in the production of poly(lactic acid) fibers. A review. J Macromol Sci C 2003;C43:479–503.
[139] Vink ETH, Rábago KR, Glassner DA, Springs B, O'Connor RP, Kolstad J, et al. The sustainability of NatureWorks™ polylactide polymers and Ingeo™ polylactide fibers: an update of the future. Macromol Biosci 2004;4:551–64.
[140] Eling B, Gogolewski S, Pennings AJ. Biodegradable materials of poly(L-lactic acid): 1. Melt-spun and solution spun fibers. Polymer 1982;23:1587–93.
[141] Leenslag JW, Pennings AJ. High-strength poly(L-Lactide) fibers by a dry-spinning hot-drawing process. Polymer 1987;28:1695–702.
[142] Postema AR, Luiten AH, Oostra H, Pennings AJ. High-strength poly(L-lactide) fibers by a dry-spinning/hot-drawing process. I. Influence of the ambient temperature on the dry-spinning process. J Appl Polym Sci 1990;39:1265–74.
[143] Schmack G, Tändler B, Vogel R, Beyreuther R, Jacobsen S, Fritz HG. Biodegradable fibers of poly(L-lactide) produced by high-speed melt spinning and spin drawing. J Appl Polym Sci 1999;73:2785–97.
[144] Schmack G, Jehnichen D, Vogel R, Tändler B, Beyreuther R, Jacobsen S, et al. Biodegradable fibres spun from poly(lactide) generated by reactive extrusion. J Biotechnol 2001;86:151–60.
[145] Schmack G, Tändler B, Optiz G, Vogel R, Komber H, Häuβler L, et al. High-speed melt spinning of various grades of polylactides. J Appl Polym Sci 2004;91:800–6.
[146] Mezghani K, Spruiell JE. High speed melt spinning of poly(L-lactic acid) filaments. J Polym Sci Pol Phys 1998;36:1005–12.

[147] Fambri L, Pegoretti A, Fenner R, Incardona SD, Migliaseri C. Biodegradable fibres of poly(L-lactic acid) produced by melt spinning. Polymer 1997;38:79–85.
[148] Cicero JA, Dorgan JR, Dec SF, Knauss DM. Two-step melt-spun fibers of polylactide. Polym Degrad Stabil 2002;78:95–105.
[149] Cicero JA, Dorgan JR, Garrett J, Runt J, Lin JS. Effects of molecular architecture on two-step, melt spun poly(lactic acid) fibers. J Appl Polym Sci 2002;86:2839–46.
[150] Yuan X, Mak AFT, Kwok KW, Yung BKO, Yao K. Characterization of poly(L-lactic acid) fibers produced by melt spinning. J Appl Polym Sci 2001;81:251–60.
[151] Cicero JA, Dorgan JR, Janzen J, Garrett J, Runt J, Lin JS. Supramolecular morphology of two-step, melt-spun poly(lactic acid) fibers. J Appl Polym Sci 2002;86:2828–38.
[152] Okuzaki H, Kubota I, Kunugi T. Mechanical properties and structure of the zone-drawn poly(L-lactic acid) fibers. J Polym Sci Polym Phys 1999;37:991–6.
[153] Nishimura Y, Takasu A, Inai Y, Hirabayashi T. Melt spinning of poly(L-lactic acid) and its biodegradability. J Appl Polym Sci 2005;97:2118–24.
[154] Krikorian V, Pochan DJ. Poly (L-lactic acid)/layered silicate nanocomposite: fabrication, characterization, and properties. Chem Mater 2003;15:4317–24.
[155] Wu T, Chiang MF. Fabrication and characterization of biodegradable poly(lactic acid)/layered silicate nanocomposites. Polym Eng Sci 2005;45:1615–21.
[156] Pluta M, Galeski A, Alexandre M, Paul MA, Dubois Ph. Polylactide/montmorillonite nanocomposites and microcomposites prepared by melt blending: structure and some physical properties. J Appl Polym Sci 2002;86:1497–506.
[157] Paul MA, Alexandre M, Degée Ph, Henrist C, Rulmont A, Dubois Ph. New nanocomposite materials based on plasticized poly (L-lactide) and organo-modified montmorillonites: thermal and morphological study. Polymer 2003;44:443–50.
[158] Sinha Ray S, Okamoto M. Biodegradable polylactide and its nanocomposites: opening a new dimension for plastics and composites. Macromol Rapid Comm 2003;24: 815–40.
[159] Pluta M. Morphology and properties of polylactide modified by thermal treatment, filling with layered silicates and plasticization. Polymer 2004;45:8239–59.
[160] Nam PH, Fujimori A, Masuko T. The dispersion behaviour of clay particles in poly(L-lactide)/organo-modified montmorillonite hybrid systems. J Appl Polym Sci 2004;93:2711–20.
[161] Nam PH, Kaneko M, Ninomiya N, Fujimori A, Masuko T. Melt intercalation of poly(L-lactide) chains into clay galleries. Polymer 2005;46:7403–9.
[162] Di Y, Iannace S, Di Maio E, Nicolais L. Poly(lactic acid)/organoclay nanocomposites: thermal, rheological properties and foam processing. J Polym Sci Pol Phys 2005;43: 689–98.
[163] Paul MA, Alexandre M, Degée Ph, Calberg C, Jérôme R, Dubois Ph. Exfoliated polylactide/clay nanocomposites by in-situ coordination-insertion polymerization. Macromol Rapid Comm 2003;24:561–6.
[164] Paul MA, Delcourt C, Alexandre M, Degée Ph, Monteverde F, Rulmont A, et al. (Plasticized) polylactide/(organo-) clay nanocomposites by in situ intercalative polymerization. Macromol Chem Phys 2005;206:484–98.
[165] Martin O, Avérous L. Poly(lactic acid): plasticization and properties of biodegradable multiphase systems. Polymer 2001;42:6209–19.
[166] Baiardo M, Frisoni G, Scandola M, Rimelen M, Lips D, Ruffieux K, et al. Thermal and mechanical properties of plasticized poly(L-lactic acid). J Appl Polym Sci 2003;90: 1731–8.

[167] Labrecque LV, Kumar RA, Dave V, Gross RA, Mc-Carthy SP. Citrate esters as plasticizers for poly(lactic acid). J Appl Polym Sci 1997;66:1507−13.
[168] Jacobsen S, Fritz HG. Plasticizing polylactide—the effect of different plasticizers on the mechanical properties. Polym Eng Sci 1999;39:1303−13.
[169] Pillin I, Montrelay N, Grohens Y. Thermo-mechanical characterization of plasticized PLA: is the miscibility the only significant factor? Polymer 2006;47:4676−82.
[170] Younes H, Cohn D. Phase separation in poly(ethylene glycol)/poly(lactic acid) blends. Eur Polym J 1988;24:765−73.
[171] Ljungberg N, Wesslen B. The effects of plasticizers on the dynamic mechanical and thermal properties of poly(lactic acid). J Appl Polym Sci 2002;86:1227−34.
[172] Solarski S, Ferreira M, Devaux E. Influence of draw roll temperature on thermal and tensile properties of PLA filaments produced by melt spinning. Chem Fibers Int 2005;55:180−2.
[173] Babraukas V. Tests for flammability of plastics materials for part devices and appliances. Fire Mater 1984;8:81−95.
[174] Thornton W. The role of oxygen to the heat of combustion of organic compounds. Philos Mag 1917;33:196−205.
[175] Huggett C. Estimation of rate of heat release by means of oxygen consumption measurements. Fire Mater 1980;4:61−5.
[176] Walters RN, Hackett SM, Lyon RE. Heats of combustion of high temperature polymers. Fire Mater 2000;24:245−52.
[177] Grayson SJ. Heat release in fire. In: Babraukas V, Grayson SJ, editors. London: Elsevier Applied Science Pub.; 1992.
[178] Bourbigot S, Devaux E, Flambard X. Flammability of polyamide-6/clay hybrid nanocomposite textiles. Polym Degrad Stabil 2002;75:397−402.
[179] Gilman JW. Flammability and thermal stability studies of polymer layered-silicate (clay) nanocomposites. Appl Clay Sci 1999;15:31−49.
[180] Bellayer S. Développement de nouvelles techniques de caractérisation qualitative et quantitative de nanocomposites à matrice polymère: application au textile [PhD thesis]. Lille University, 2005. (http://ori.univ-lille1.fr/thematic-search.html?menuKey=these&submenuKey=authors&id=bellayer_severine)

CHAPTER

Improving the Flame Retardancy of Plant Oil Based Polymers

22

G. Lligadas, J.C. Ronda, M. Galià, V. Cádiz
Universitat Rovira i Virgili, Tarragona, Spain

CHAPTER OUTLINE

1. Introduction

One very desirable property for polymers is flame retardancy. The use of flame retardants to reduce combustibility of the polymers, and smoke or toxic fume production, therefore becomes a pivotal part of the development and applications of new materials. A flame retardant is a substance or treatment incorporated in or applied to a material that suppresses or delays its combustion under specific conditions. There exist two different approaches to achieve flame retardancy in polymers generally known as the additive type and the reactive type. Additive

Polymer Green Flame Retardants. http://dx.doi.org/10.1016/B978-0-444-53808-6.00022-6

type flame retardants, which are the mostly used, are generally incorporated into polymer by physical means whereas reactive ones are chemically linked into the polymer chain. The application of reactive flame retardants involves either the design of new, intrinsically flame retardant polymers or the modification of existing ones. Many kinds of both additive and reactive flame retardants have been tested to improve the flame resistance of polymers and have been applied to commercial products. Also, many efforts have been made to identify and collate information on the type of flame retardants used in consumer products, to review and assess the toxicity and environmental impact of the currently used principal flame retardants. Halogenated flame retardants have been the most common systems used up to date. However many of them, specially brominated flame retardants, are nowadays restricted in many countries due to the formation of hydrogen halides and dioxins, (which causes corrosion and toxicity), taking place during the combustion of such flame retardants containing polymers [1]. These drawbacks have increased the amount of research into the development of innovative and environmentally friendly halogen-free flame retardant polymers [2]. Phosphorus, silicon and boron-containing polymers are well recognized for their flame retardant properties, and they are increasingly becoming more popular than their halogen counterparts, as they generally give off non toxic combustion products.

Our approach to more environmentally friendly flame retardants not only involves the design of halogen-free systems but also the use of renewable resources. The growing demand for petroleum-based products and the resulting negative impact on the environment, plus the scarcity of nonrenewable resources, are some of the many factors that have recently raised the interest on the field of polymers derived from nonpetrochemical feedstock. In this context, increasing efforts have been focused on the use of plant oils and derivatives as monomers for polymer synthesis. Plant oils, made up of triglycerides molecules, consisting of glycerol and fatty acids, (Scheme 1), are considered to be one of the cheapest and most abundant biological sources available; furthermore, their use as annually renewable platform chemicals has numerous advantages, including low toxicity and inherent biodegradability.

Our research, thus, deals with the incorporation of reactive phosphorus, silicon or boron into different plant oil-derived polymeric structures as a more environmentally friendly technology.

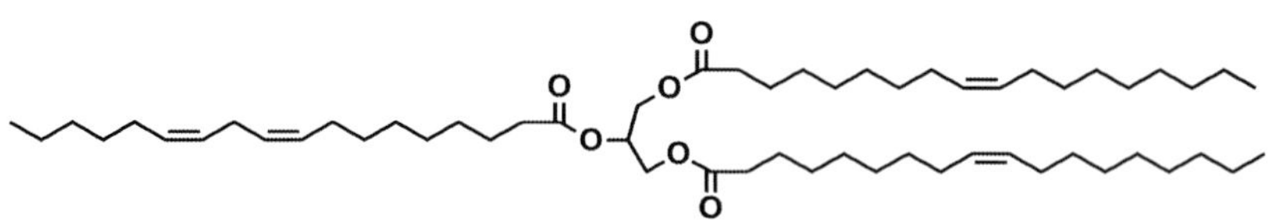

SCHEME 1

Representative chemical structure of a triglyceride molecule.

2. Phosphorus-containing plant oil based polymers

2.1 Phosphorus-containing epoxy resins from fatty acids

A new phosphorous-containing fatty acid diepoxide (DOPO III) (see Scheme 2) was obtained from 10-undecenoyl chloride and 10-(2′,5′-dihydroxyphenyl)-9,10-dihydro-9-oxa-10-phosphaphenanthrene-10-oxide (DOPO I) via esterification (DOPO II) and further epoxidation [3]. 10-undecenoyl chloride is a 10-undecenoic acid (UD) derivative. UD is a C11 fatty acid-derivative with a terminal carbon—carbon double bond. It can be easily obtained from castor oil. Up to 90% of the fatty acid fraction in castor oil is ricinoleic acid. UD can be obtained by heating ricinoleic acid under vacuum pyrolysis [4,5]. DOPO I is obtained from *p*-benzoquinone and 9,10-dihydro-9-oxa-10-phosphaphenanthrene-10-oxide (DOPO). Since 1972, this cyclic organic phosphorus compound, and its derivatives have been used to make many synthetic resins flame retardant [6]. This rigid and bulky group contain an unusually high thermally stable P—O—C bond, which can be attributed to the O=P—O group being protected by phenylene groups.

DOPO III was first crosslinked with 4,4′-diaminodiphenylmethane (DDM) and a phosphorylated hardener, bis(*m*-aminophenyl)methylphosphine oxide (BAMPO) [7]. DOPO III was also copolymerized with epoxidized 10-undecenoyl triglyceride

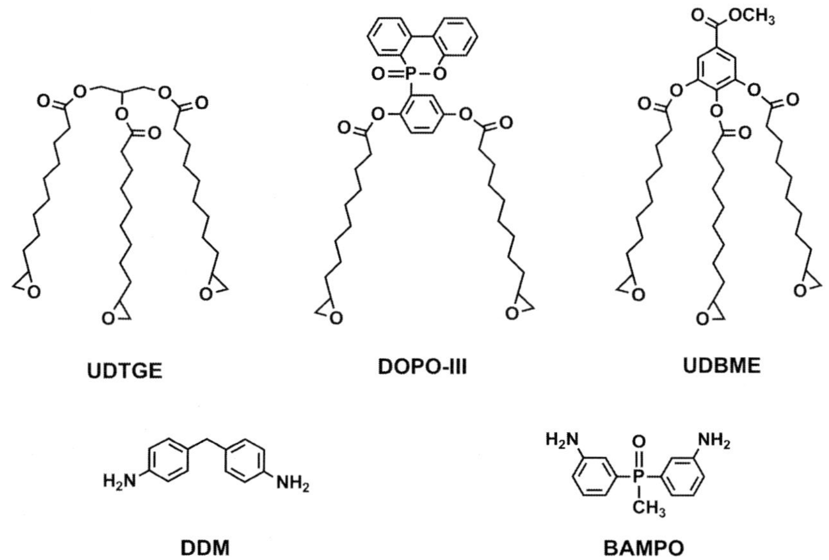

SCHEME 2

Chemical structures of epoxy monomers and hardeners. UDTGE: epoxidized 10-undecenoyl triglyceride; UDBME: epoxidized methyl 3,4,5-tris(10-undecenoyloxy)benzoate; DOPO-III: 10-(2′,5′-dihydroxyphenyl)-9,10-dihydro-9-oxa-10-phosphaphenanthrene-10-oxide; DDM: 4,4′-diaminodiphenylmethane; BAMPO: bis(*m*-aminophenyl)methylphosphine oxide.

(UDTGE) and epoxidized methyl 3,4,5-tris(10-undecenoyloxy)benzoate (UDBME) using the same hardeners to prepare epoxy resins with different phosphorus content [8]. The chemical structures are shown in Scheme 2.

Seven cured epoxy resins were prepared, with phosphorus content ranging from 0 to 5.7 wt% (entries 1–5 in Summary Table). The thermal properties of these thermosetting materials were evaluated by differential scanning calorimetry (DSC) and thermogravimetric analysis (TGA) (Table 1). Differences on the T_g values of the cured materials were observed, however it seemed that the presence of the bulky aromatic DOPO moiety, with a strong polar group (P=O), slightly increased the T_g values. To examine the effect of phosphorus on the thermal stability of the cured epoxy resins, TGA was carried out under nitrogen and air. The TGA data show that all the materials have good thermal stability both under nitrogen and air atmospheres, with initial temperatures of weight loss above 300 °C, regardless of the starting monomer or curing agent. Char yields at 800 °C under nitrogen were not influenced by phosphorus content, however under air the char yields for the phosphorus-containing resins increased with the phosphorus content. The phosphorus, therefore, plays an effective role in char formation, and because the char yield has been correlated to the flame retardancy [9], these resins were expected to have enhanced flame-retardant properties. The high char yield limits the production of combustion gases, decreases the exothermicity of the pyrolysis reactions, and inhibits the thermal conductivity of the burning material.

The flame retardancy of the cured polymers was evaluated by measuring the limiting oxygen index (LOI) values. It was found that phosphorus-containing resins show higher LOI values than phosphorus-free resins. The relationships between the phosphorus content and the LOI or the residue at 800 °C under air are plotted in Figure 1. The presence of phosphorus linearly increases the char yield at 800 °C under an air atmosphere. On the other hand, the LOI values increase with the phosphorus content increase, reaching a plateau value around 30.

To better understand the role of phosphorus in the flame-retardant properties of the polymer, element mapping was performed with energy dispersive X-ray spectroscopy (EDX) on the surfaces of the samples after the LOI test. Figure 2(a) shows the SEM micrograph of the top surface view of the DOPO-III/BAMPO specimen. The burned zone has the appearance of a black charred material with small cavities. The P distribution (Figure 2(b)) shows that the phosphorus density increased toward the top burned surface and that a phosphorus-rich layer formed. The white points in the figure denote P atoms. When they are heated, phosphorus compounds can form glass-like polyphosphoric acid, which protects the burning surface, or they can form inflammable phosphorus–carbon char by reacting with organic components. This protective layer is resistant to even higher temperatures and shields the underlying polymer from attack by oxygen and radiant heat, moreover, it prevents the combustible gases from transferring to the surface of the materials and feed the flame, thus improving the fire resistance.

Table 1 Thermal Data for the Cured Epoxy Resins

					TGA (N_2)		TGA (Air)		
Epoxy Monomer	**Molar Ratio**	**Curing Agent**	***P* (%)**	T_g[a]	$T_{5\%\ loss}$ (°C)[b]	$R_{800\ °C}$ (%)[c]	$T_{5\%\ loss}$ (°C)[b]	$R_{800\ °C}$ (%)[c]	**LOI**
UDTGE		DDM	0	56	363	11	321	0	21.9
UDTGE/DOPO-III	1/1	DDM	1.8	75	350	17	302	2	27.7
UDBME		DDM	0	85	299	20	283	0	23.5
UDBME/DOPO-III	1/1	DDM	1.9	80	313	18	294	4	28.5
UDBME		BAMPO	2.5	88	313	20	306	5	30.2
DOPO-III		DDM	3.9	98	350	18	333	10	31
DOPO-III		BAMPO	5.7	91	354	16	346	18	32

[a] *From the second heating scan of DSC measurements.*
[b] *Temperature of 5% weight loss.*
[c] *Residue at 800 °C.*

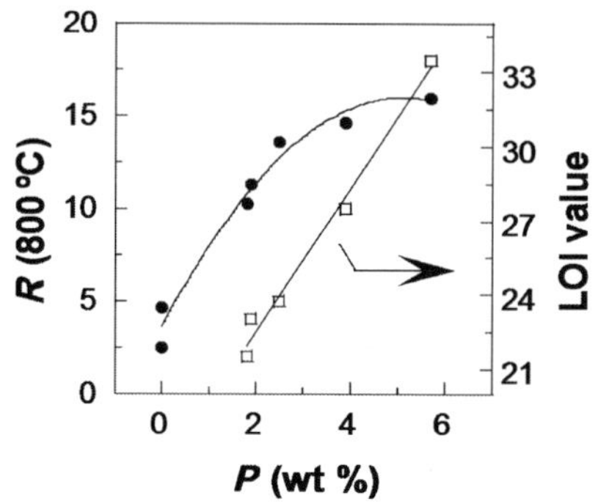

FIGURE 1

(□) Residue at 800 °C under an air atmosphere and (●) limiting oxygen index (LOI) values versus the phosphorus weight contents of the cured epoxy resins.

Reproduced from [8], with permission from Wiley.

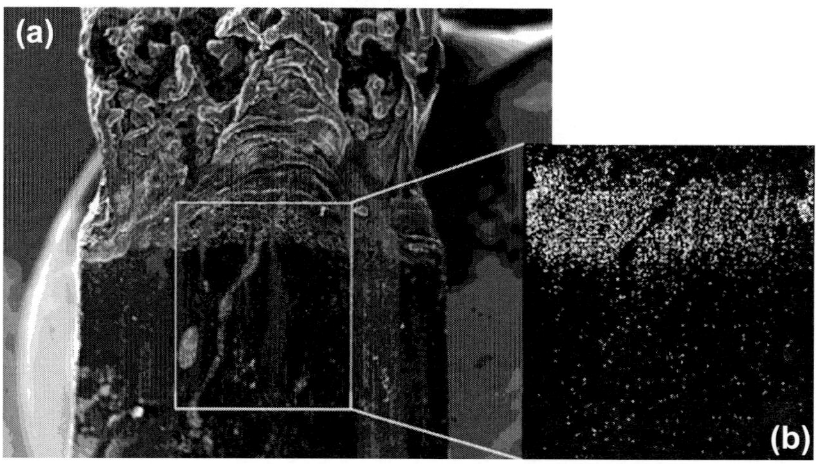

FIGURE 2

SEM (a) and SEM-EDX P mapping (b) micrographs of DOPO-III/BAMPO specimen after limiting oxygen index (LOI) test.

Reproduced from [3] , with permission from Wiley.

2.2 Acrylate and phosphine oxide-containing vegetable oils and their crosslinked materials

Acrylate triglycerides are usually low viscosity monomeric liquids, which can be free radically polymerized and can be easily copolymerized with other commercial comonomers, because of the high reactivity of the acrylate group. The special properties of these chemically modified triglycerides offer a widespread field of applications; so, they are excellent candidates as thermosetting liquid molding resins in techniques such as vacuum assisted resin transfer molding, composite fabrication processes or pressure sensitive adhesives manufacture. Among all the desirable properties thermosets and elastomers should have, flame retardancy is one of the most important concerning the security of the final users of these materials.

As can be seen in Scheme 3, phosphorus-containing acrylate sunflower oil was prepared from a new route that involves the singlet oxygen photooxygenation of the triglyceride double bonds, and further reduction of the resulting hydroperoxide derivatives to a mixture of secondary allylic alcohols (HSO). These allylic alcohols in the presence of chlorodiphenylphosphine give allylic phosphinites capable of undergoing a [2,3]-sigmatropic rearrangement [10] leading to tertiary phosphine oxides directly linked to the fatty acid chains (P–SO) in a one pot two step reaction. The obtained phosphorus-containing triglycerides with different hydroxyl content were activated for polymerization by acrylation [11].

These acrylate triglycerides (P-ASO) were radically crosslinked in the presence of different amounts of pentaerythritol tetraacrylate. In this way, four crosslinked materials with 1.4, 2.0, and 2.8% w/w of phosphorus content were synthesized (Entries 6–8 in Summary Table). The thermal properties of the final materials were evaluated. T_g values ranged from 14 °C to 25 °C. The temperatures of 5% weight loss of all samples were in the 305–330 °C range under nitrogen and between 270 and 300 °C under air. The char yields at 800 °C were below 5% both

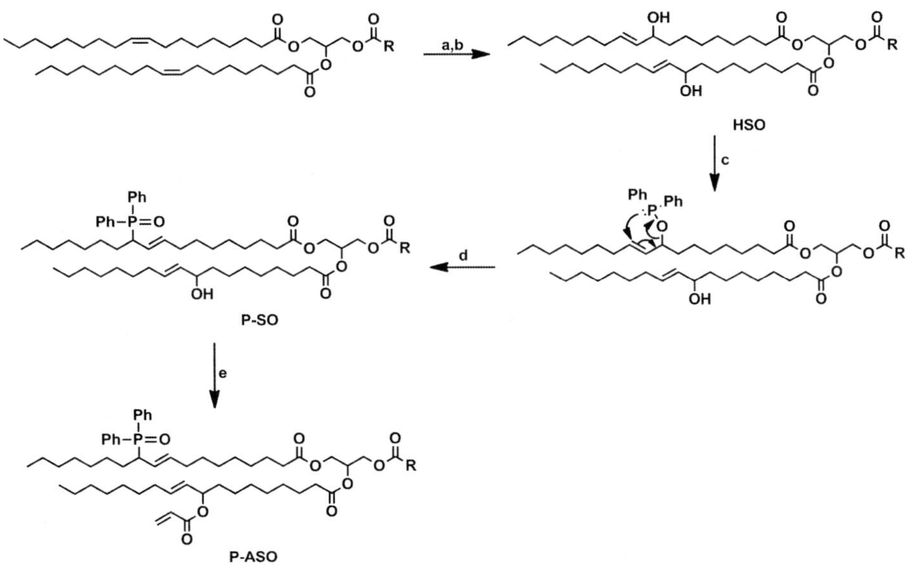

SCHEME 3

Synthesis of P-ASO derivatives from sunflower oils. Reagent and reaction conditions. (a) O_2, TPP, hν, CH_2Cl_2; (b) $NaBH_4$, CH_3OH, 0 °C; (c) $ClPPh_2$, DMAP, toluene, rt, 30 min; (d) reflux, 12 h; (e) Et_3N, acryloyl chloride, CH_2Cl_2, 0 °C to rt.

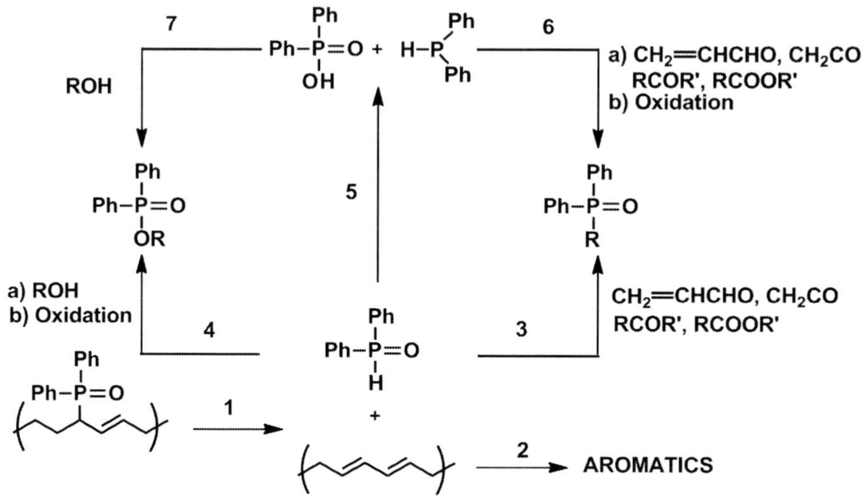

SCHEME 4

Proposed thermal degradation pathways of P-ASO thermosets. (1) P—C bond cleavage; (2) Diels-Alder aromatization; (3) P—H addition to electrophiles; (4) esterification and oxidation to alkyl diphenyl phosphinites; (5) thermal disproporcionation; (6) P—H addition to electrophiles; (7) esterification of diphenyl phosphinic acid.

in nitrogen and air. These low values could indicate that volatile phosphorus species are released upon heating. To prove this assumption, samples containing the highest phosphorus content were heated at high temperatures and volatiles were trapped and subsequently analyzed by GC/MS and ^{31}P NMR spectroscopy.

The analysis of the evolved gases, in all fractions, revealed the presence of several phosphorus aromatic compounds, such as diphenylphosphine oxide, methyldiphenyl phosphine oxide and methyl diphenylphosphinite. The formation of these compounds is probably a consequence of different chemical processes, which some of the compounds evolved during the decomposition of triglyceride moieties, undergo. These reactions are summarized in Scheme 4.

The flame retardancy of the resins was evaluated by applying the LOI test. Phosphorus-free resin gave a LOI of 19.6 which indicates that this material is not flame retardant. The addition of 1.4% of phosphorus gave a LOI of 21.2 and the maximum content of phosphorus (2.8%) gave a LOI of 22.4. This slight improvement on the flame retardant properties is related to the increase of the phosphorus content.

2.3 Fatty acid derived phosphorus-containing polyesters via acyclic diene metathesis polymerization

To obtain thermoplastic and thermosetting polyesters with improved flame retardancy, we used 10-undecenoic acid to design a set of α,ω-dienic monomers containing ester groups [12,13] (Scheme 5). So, 10-undecenoyl ester of the 10-undecenol,

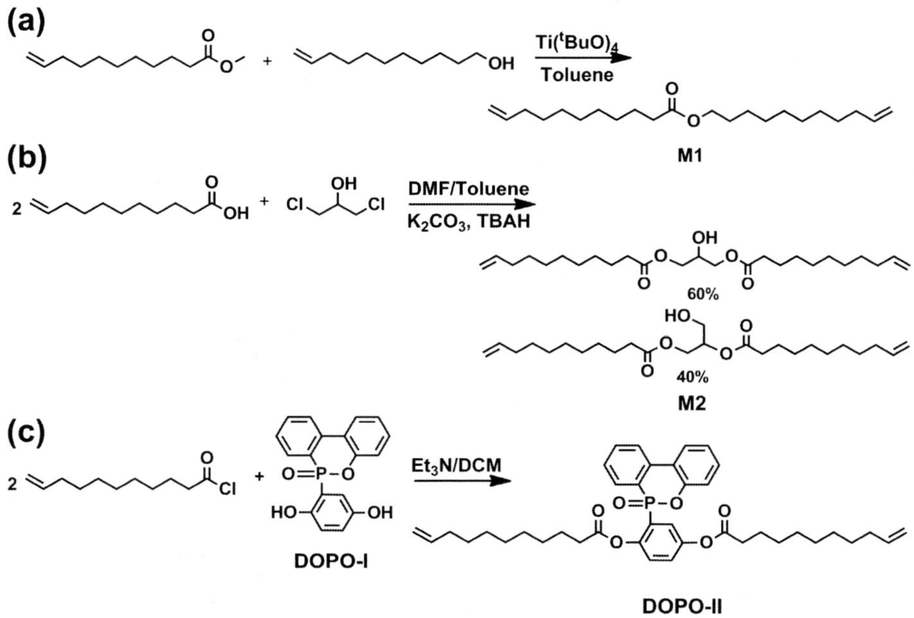

SCHEME 5

Routes (a-c) to prepare α,ω-dienes M1, M2 and DOPO-II from 10-undecenoic acid and derivatives.

M1, was synthesized by transesterification of the methyl 10-undecenoate with 10-undecenol (Scheme 5(a)). Bis-10-undecenoyl ester of the glycerol, M2, which is a mixture of two compounds with secondary and primary hydroxyl groups, was also prepared (Scheme 5(b)). Finally, to infer flame retardant properties to the system, the previously described α,ω-dienic phosphorus-containing monomer DOPO-II [3], was used (Scheme 5(c)). As it can be seen, all these monomers are difunctional, thus fulfilling the requirements to obtain lineal polymers via acyclic diene metathesis (ADMET) polymerization.

Metathesis of α,ω-dienes has been shown to be an efficient and powerful tool for the synthesis of a wide variety of linear polymers and polymer architectures that are not available using other polymerization methods [14]. The development of new, highly active and robust metathesis catalysts allows polymerizing dienic monomers in the presence of several functional groups. It has been demonstrated that ADMET polymerization can proceed in the presence of heteroatoms, as long as the terminal olefins are far enough from them [15–18].

Dienes M1 and DOPO II were copolymerized in the absence and in the presence of different amounts of methyl undecenoate (employed as chain stopper) using Grubbs 1st generation catalyst, Grubbs 2nd generation catalyst and Hoveyda-Grubbs 2nd generation catalyst working in bulk at 80 and 100 °C (Scheme 6).

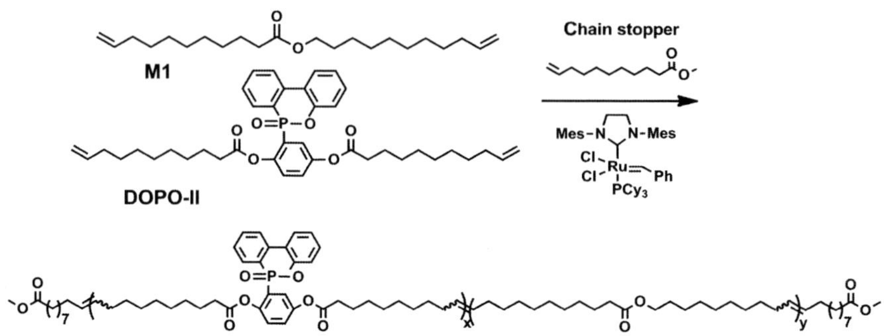

SCHEME 6

Acyclic diene metathesis copolymerization of M1 and DOPO-II.

The best results were obtained at 80 °C using the Grubbs second generation catalyst. In this way, by varying the comonomer molar ratio, a set of lineal polyesters with molecular weights ranging from 9×10^3 to 40×10^3 Da and with phosphorus contents ranging from 0% to 4.7% were obtained. These polyesters were semicrystalline or amorphous depending on the percentage of the bulky P-comonomer, (it impedes the crystallization of the polyester chains). So, the copolymers containing more than 50% of DOPO-II were amorphous whereas the rest were semicrystalline. The flame retardant behavior of these lineal polyesters was evaluated using the LOI test; improvement of the flame retardant properties was shown. LOI values were found to increase from 19 for the phosphorus-free polymers to 23.5 for polymers having 3% of P (entry 9 in Summary Table); at higher P content, the LOI value remained almost unchanged, which is a quite common phenomenon for P-based systems.

To prepare flame retardant polyester thermosets, the monomer M2 was also used which is a mixture of glycerol esters containing primary and secondary hydroxyl groups in a four to six ratio. This monomer was homopolymerized and copolymerized with the same phosphorus-containing monomer DOPO-II and the same chain stopper used before. In the case of the homopolymerization, the best results were achieved with the Hoveyda−Grubbs second generation catalyst; on the contrary, for the copolymerization the best results were obtained with the Grubbs second generation catalyst. Following this procedure and varying the molar ratios between monomers and the chain stopper, we were able to prepare a set of polyester-polyols with molecular weights ranging from 3.7×10^3 to 7.3×10^3 Da and containing different amounts of P (0, 1.6, 2.9, and 3.9%) (entries 10−12 in Summary Table). In this case, with the exception of the M2 homopolymer, these materials were amorphous and, due to their low molecular weight, presented a good solubility in solvents like dichloromethane (DCM), facilitating the introduction of radicallary crosslinkable functional groups in their backbone (Scheme 7).

Then, the hydroxyl groups of these polymers were esterified using acryloyl chloride and triethylamine in DCM to produce acrylate phosphorus-containing

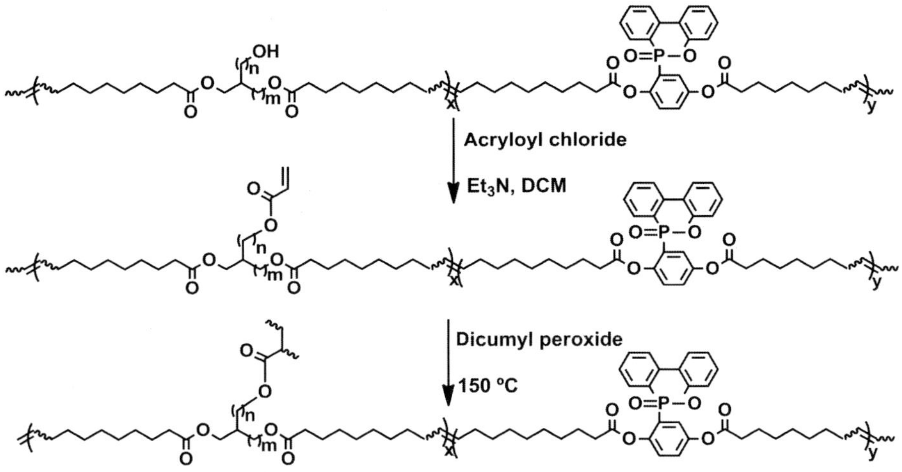

SCHEME 7

Chemical modification of hydroxyl-containing polyesters (n = 1, m = 0 and n = 0, m = 1) with acrylate moieties and further crosslinking.

polyesters. The acrylation step was monitored by NMR. Figure 3(b) shows a representative ^{1}H NMR spectrum of polyesters with the typical signals at 6.3, 6.1 and 5.7 ppm, characteristics of acrylate groups. Moreover, the enlargement of the 75–60 ppm region of ^{13}C NMR spectrum (Figure 3(a)) confirms full modification in terms of disappearance of methylene C5 and methine C2.

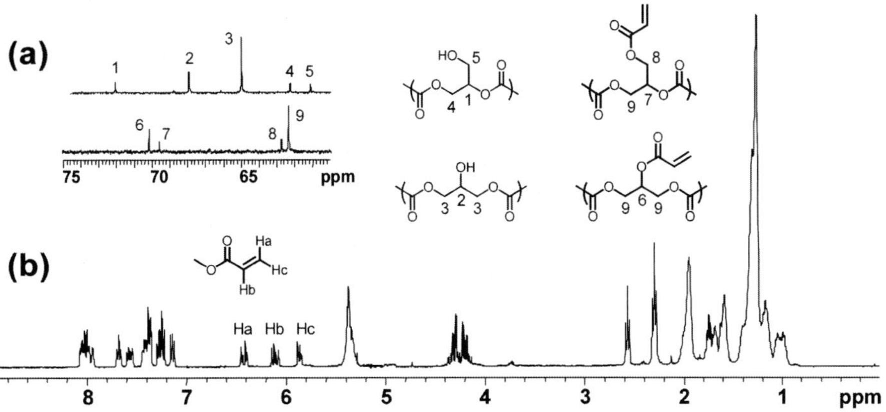

FIGURE 3

a) 75-60 ppm region of 13C NMR spectra of the hydroxyl polyester mixture (upper) and the resulting acrylated derivative (bottom) b) 1H NMR spectrum of acrylated polyester.

Reproduced from [13], with permission from Wiley.

The curing of these reactive materials with dicumyl peroxide at 150 °C led to a set of colorless materials. The tensile properties of the crosslinked materials showed to be dependent on the aromatic content, crosslinking density as well as the molecular weight of the parent polyesters. These plant oil based thermosets show T_g's ranging from 35 to 52 °C and good thermal stability. As expected, the amount of char yield at 800 °C during the TGA analysis under air atmosphere, increased proportionally to the phosphorus content, indicating how efficiently the burning surface would protect the rest of the polymer under real fire conditions. Moreover, the flame retardancy of these thermosets was evaluated using the LOI test which showed that in spite of its high aliphatic content, the presence of about 3% of phosphorus infers a notable fire resistance; as a result, LOI values of up to 25.6 were reached.

2.4 Cross-metathesis of fatty acid derivatives as an alternative way to phosphorus-containing reactive flame retardants

As previously mentioned, vegetable oils are triglycerides of different fatty acids with varying degrees of unsaturation; the C=C double bonds are capable of undergoing cationic polymerization. Among the Lewis acids capable of initiating the polymerization, $BF_3 \cdot OEt_2$ has been proven to be the most efficient initiator [19]. However, cationic homopolymerization of vegetable oils yields only low molecular weight viscous oils or soft rubbery materials of limited utility. Thus, olefinic monomers as styrene (ST), divinylbenzene (DVB), norbornadiene or dicyclopentadiene are usually copolymerized with vegetable oils to overcome these drawbacks [20].

Inspired by Larock's work [19], two different approaches to phosphorus-containing soybean oil copolymers were investigated [21]. First, two phosphorus-containing styrene derivatives, diphenyl styryl phosphine oxide and dimethyl-*p*-vinylbenzylphosphonate (M3), were tested as comonomers for the cationic copolymerization of soybean oil, ST and DVB, obtaining heterogeneous systems in all cases. There are different possible reasons for the heterogeneous nature of these copolymerizations. The first is the difference in the reactivity of soybean oil and styrenic monomers, which seems to be increased in the presence of M3 (homogeneous copolymers are obtained in absence of this monomer). Another disadvantage to be considered is that the cationic initiator, $BF_3 \cdot OEt_2$ is a strong Lewis acid that can interact with the basic P=O moieties. The strong coordination of the catalyst to the phosphoryl groups impedes the polymerization of the internal double bonds of the oil, favoring the homopolymerization of the more reactive styrenic monomers.

Aiming at the overcoming of this drawback, methyl 10-undecenoate was subjected to reaction with M3 using C1 and C2 to link the phosphorus moiety to the vegetable oil derivative (Scheme 8). As can be seen in Table 2, lower conversions in a clean reaction without large amounts of self-metathesis were obtained, when C1 was used. When C2 was employed, significant higher conversion was observed; the yield of self-metathesis also increased significantly. The highest cross-

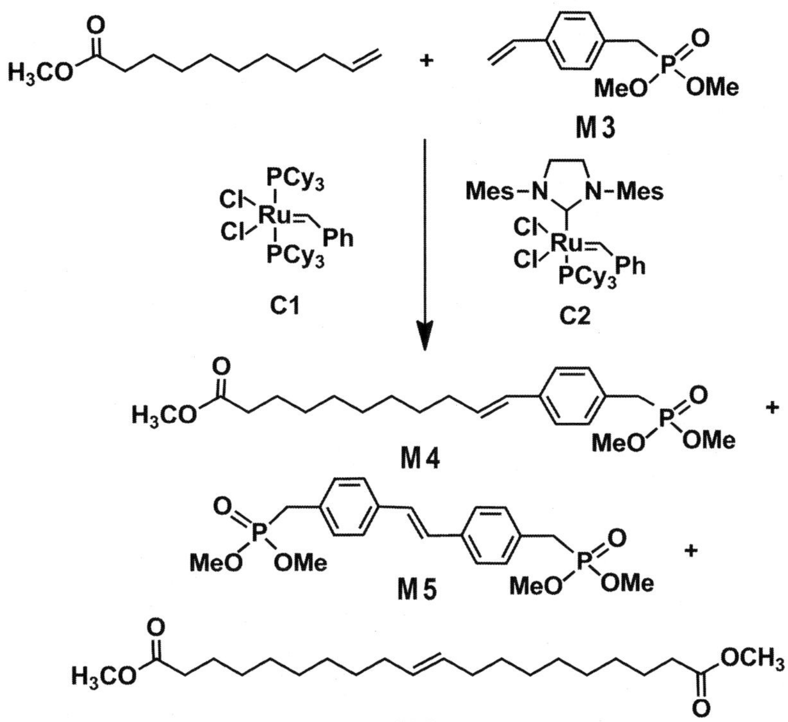

SCHEME 8

Metathesis reaction of methyl 10-undecenoate with M3 using first (C1) and second (C2) generation Grubbs catalysts.

metathesis yield was obtained by using stoichiometric amounts of methyl 10-undecenoate and M3, and 2% of C2. The cross-metathesis product (M4) was isolated and characterized by ^{1}H, ^{13}C and ^{31}P NMR spectroscopy. The ^{1}H and ^{13}C NMR spectra showed characteristic signals for the C=C double bond at 6.10–6.19 and 6.32 ppm, and at 129.42 and 131.50 ppm respectively.

M4 was mixed with soybean oil, ST and DVB, maintaining the amount of DVB constant and varying the amounts of M4, ST and oil, to obtain thermosets with a phosphorus content of 1 wt% (entry 13 in Summary Table). After cationic copolymerization, this second approach permitted the synthesis of homogeneous thermosets with T_g at around 60 °C. TGA analysis under air atmosphere showed a thermooxidative degradation process at temperatures higher than 500 °C for the phosphorus-containing resins. This behavior is in accordance with the mechanism of improved fire performance via phosphorus modification. The resulting thermosets had LOI values about 24.0, indicating an improvement in the fire retardant properties of the soybean oil-based copolymers.

Table 2 Self Metathesis (SM) and Cross Metathesis (CM) Results in the Reaction of Methyl 10-Undecenoate with M3 in Presence of C1 and C2 Catalysts. Reaction Conditions: [Methyl 10-Undecenoate + M3] = 1 M in CH_2Cl_2, [C] = 2 mol%, T = 40 °C, Time = 12 h

Catalyst	Molar Ratio (M3/Methyl 10-Undecenoate)	Conversion of Methyl 10-Undecenoate (%)[d]	CM[e] (%)	SM[f] (%)
C1	2.0	36.8	28.5	8.3
C1[a]	2.0	46.4	38.0	8.4
C1[b]	1.4	55.4	43.9	11.5
C1	1.0	53.7	38.0	15.7
C2[a,c]	1.0	70.9	42.6	28.3
C2[a]	1.0	90.4	60.4	30.0
C2	2.0	98.6	40.6	58.0

[a] *Without solvent.*
[b] *Reaction performed for 24 h. 0.4 mol of MU was added at 12 h of reaction.*
[c] *Catalyst concentration 1 mol%.*
[d] *Estimated by ^1H NMR.*
[e] *Yield of cross-metathesis products estimated by ^1H NMR.*
[f] *Yield of self-metathesis products estimated by ^1H NMR.*

3. Silicon-containing plant oil based polymers

3.1 Silicon-containing polyurethanes from vegetable oils

Polyurethanes are a versatile class of polymers which are used in a wide range of applications, and their structure can be tailored to specific requirements [22]. Depending on the hydroxyl value and other characteristics of the polyol, polyurethanes can be used for the development of adhesives, coatings, and flexible or rigid foams. In combination with isocyanates, vegetable oil-based polyols produce polyurethanes that can compete in many ways with polyurethanes derived from petrochemical polyols; thus, their preparation for general polyurethane use has been the subject of many studies [23].

Hydrosilylation of methyl 10-undecenoate with a trifunctional silane followed by a reduction of carboxylate groups was performed to prepare a novel silicon-containing fatty acid OH-terminated derivative with primary hydroxyl groups (M7) [24] (see Scheme 9). This polyol, together with epoxidized methyl oleate-based polyether polyol (M8) and 4,4′-methylenebis(phenyl isocyanate) (MDI), was used to prepare hybrid biobased-polyurethanes with a silicon content between 1.7% and 9.0% (entries 14–17 in Summary Table). The chemical structure of both polyols is shown in Scheme 9.

The most notable result of the progressive incorporation of M7 into M8 polyurethane matrix is the appearance of melting endotherms of variable enthalpy and position, and the downward shift of T_g. The prepared polyurethanes presented T_g's ranging from 9 to 39 °C, with lower T_g values found at higher M7 content.

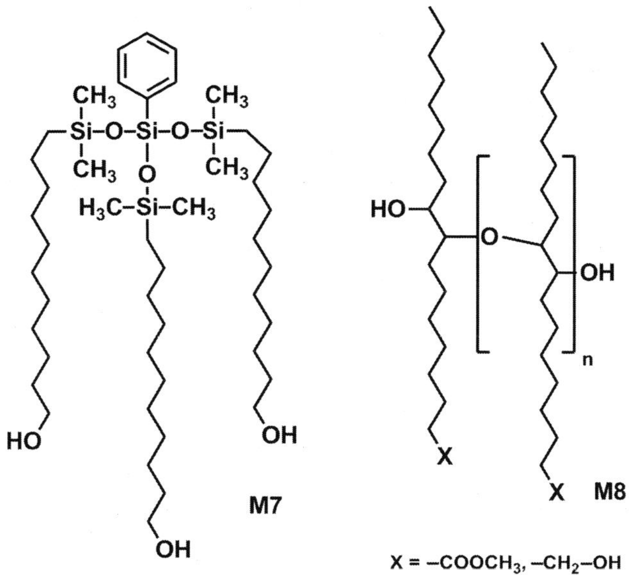

SCHEME 9

Chemical structure of polyols M7 and M8.

The samples with no crystallinity are transparent, while those semicrystalline are white. The lower T_g values of the silicon-containing polyurethanes can be related by the presence of Si—O and Si—C units in the backbone.

Thermal decomposition of the obtained polyurethanes was investigated by TGA under a nitrogen and air atmosphere. All polyurethanes were thermally stable over 250 °C. The char yields under nitrogen did not show significant differences among the polymers, but in air, the char yield of silicon-containing polymers was found to be slightly higher. Silicon has been proven to be one of the first choices as a flame retardant element. As soon as a silicon-containing polyol (M7) was introduced into M8 matrix (from 1.7% wt Si), the LOI was slightly improved, reaching a value of 19. LOI increased with the M7 ratio, achieving the maximum value in the case of neat M7 polyurethane. The significance of this result is that polyurethanes with 6.2% and 9.0% wt Si (whose LOI values are higher than 21) no longer burn in ambient air if there is no supplementary oxygen and are therefore very interesting materials for applications that require fire resistance. To figure out the role of silicon in the flammability enhancement of these polyurethanes, scanning electronic microscopy (SEM) and energy-dispersive X-ray spectroscopy (EDX) images of samples before and after the burning test (Figure 4) were taken.

The Si-mapping profile before the combustion test (Figure 4(b)) shows a homogeneous distribution of silicon, while the mapping after combustion (Figure 4(d)) shows that the burned surface was covered with char layers containing silicon atoms in very high densities. The relatively high oxygen density and low carbon density in

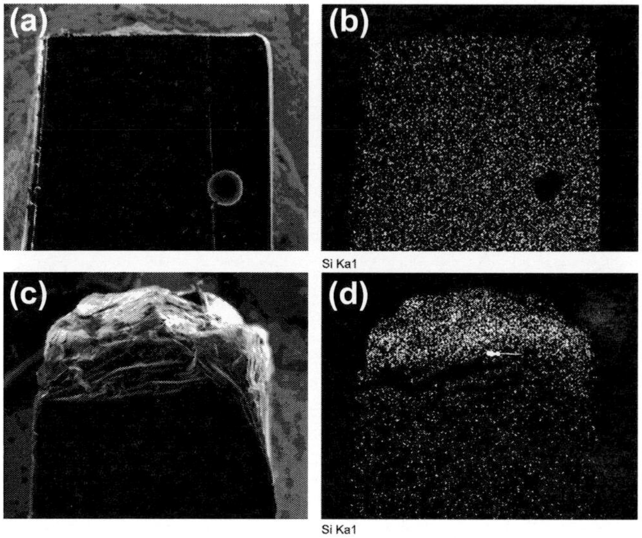

FIGURE 4

SEM and SEM-EDX Si mapping micrographs of polyurethane containing 9% wt Si before (a,b) and after (c,d) burning limiting oxygen index test: (left) SEM, and (right) SEM-EDX micrographs.

Reproduced from [24], with permission from ACS.

the top burned surface (O and C mapping profiles are not shown) demonstrated that the residue contained a certain amount of silica (SiO_2). These observations are in accordance with the mechanism of improved fire performance via silicon modification. Incorporating silicon into the polymers enhanced flame retardancy because it produced continuous layers of silica that retarded the oxidation of the char [25].

3.2 Silicon-containing soybean-oil based copolymers

With the objective of exploiting the possibilities to improve the flame retardancy of soybean oil/ST/DVB (SSD) thermosets and to preserve the environmentally friendly character of these materials, the influence of adding a silicon-containing comonomer into the system was studied [26]. Thus, soybean-oil-based copolymers with different silicon content, from 1.0 to 7.0 wt% (SSDSi), were prepared from soybean oil, ST, DVB, and *p*-trimethylsilylstyrene (SiST) by cationic polymerization using boron trifluoride etherate as initiator (entries 18–21 in Summary Table). The curing reaction was studied by dynamic DSC showing a broad exotherm that shifted at higher temperatures as the SiST increased (Table 3). The presence of this comonomer seemed to decrease the reactivity of the system.

Soxhlet extraction and NMR spectroscopy indicated that the copolymers consisted of a cross-linked network plasticized with varying amounts of oligomers and unreacted oil. This soluble fraction increased when the SiST content in the

Table 3 Compositions and Differential Scanning Calorimetric Data of the Silicon-Containing Soybean Oil Copolymers

Sample	Soluble Fraction (wt%)	Insoluble Fraction (wt%)	Aromatic Material (w%)			Si Soluble Fraction (wt%)
			Feed	Soluble Fraction	Insoluble Fraction	
SSD	8	92	47	13	87	–
SSDSi-1	17	83	48	14	86	–
SSDSi-3	23	77	51	16	84	0.7
SSDSi-5	25	75	54	23	77	3.0
SSDSi-7	32	68	56	27	73	3.5

feed increased, due to the lower reactivity of this monomer. Thermosets with glass transition temperatures ranging from 50 to 62 °C, which are thermally stable below 350 °C were obtained. Moreover DMTA analysis of the polymers indicated that the crosslinking density decreased as the SiST increased. The crosslinking density of the silicon-free sample, 6.6×10^2 mol m^{-3}, was found to be in the range of the previously described copolymers [27] and it was much higher than those of the silicon-containing samples. For example copolymer with 7 wt% Si had a crosslinking density of 1.5×10^2 mol m^{-3}.

To test the flame retardancy of the copolymers, LOI measurements were carried out. LOI values increased from 19.2 for the silicon-free system to 29.7 as the silicon content of the samples increased; so, the presence of silicon was shown to improve the flame-retardant properties of the copolymers. These results together with the fact that no significant differences were observed among char yield for the different copolymers, indicate that the condensed flame-retardant mechanism seems not to be responsible for the flame-retardant improvement which may suggest that a gas phase mechanism occurs.

4. Boron-containing plant oil based polymers

The use of several boron compounds for enhancing the flame retardant polymeric materials is well known since the twentieth century. Aromatic boronic acid based monomers are the most suitable because they are easily synthesized and they are known to release water during thermolysis, thus improving the flame retardant behavior of polymers (by diluting the flammable volatile substances and by providing a gas barrier to oxygen) [28]. Looking for new possibilities to improve the flame retardance of soybean oil/ST/DVD thermosets, a new heteroatom-containing comonomer was studied. In particular, boron-containing soybean oil based copolymers were prepared from soybean oil, ST, DVB and tris (4-vinylphenyl)boroxine (BST) (entries 22–24 in Summary Table) (Scheme 10) by cationic polymerization, using boron trifluoride etherate as initiator [29].

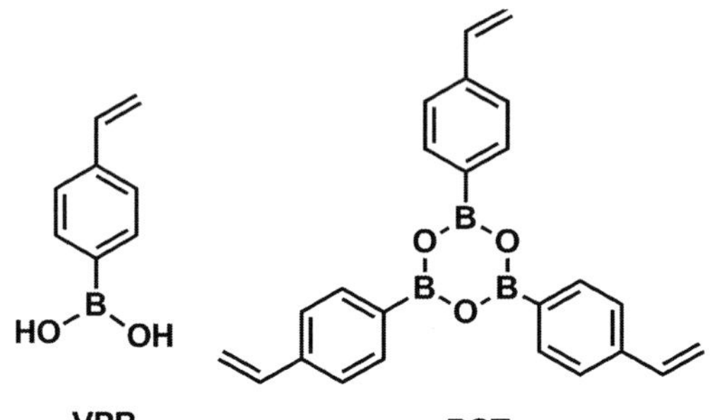

SCHEME 10

Structures of boron-containing styrenic monomers.

All the copolymers were obtained as brown, tough and ductile rubbers at room temperature. FTIR analyses confirmed the presence of the triglyceride carbonyl groups and aromatic moieties. However, the most significant observation was the presence of the B—O stretching band at 1350 cm^{-1} and of a broad band at c. 3300 cm^{-1} corresponding to boronic hydroxyl groups. This band indicates that in spite of using the boroxine as starting monomer, eventually the boronic acid was obtained in the cured material. Attempts to use 4-vinylphenyl boronic (VPB) (Scheme 10) acid as boron-containing comonomer were unsuccessful due to the poor solubility of the VPB in the polymerization mixture; in all cases heterogeneous materials were obtained.

To determine the thermal stability and decomposition behavior of the copolymers, thermogravimetric analysis was carried out under nitrogen and air atmospheres. Figure 5 shows the TGA curves and the first derivatives.

Under both nitrogen and air atmosphere, samples were thermally stable below 300 °C but a slight drifting starting around 100 °C was observed. This weight loss became more important as the temperature and the BST content increased; this can be attributed to the dehydration of boroxine rings occurring at higher temperatures [30]. The onset temperatures in air atmosphere were lower than those in nitrogen, mainly due to the mechanism of peroxidation that takes place when oxygen is present. In air, three weight loss stages could be seen. The first stage around 360 °C can be due to the decomposition of soluble components, while the second stage around 400—450 °C corresponds to the crosslinked polymer degradation that takes place at a faster rate. The third stage was observed at temperatures around 600 °C and corresponds to the thermooxidative degradation. The char yields increased with the boron content. The boron, therefore, played a positive role in char formation and because char yield has been correlated to flame retardancy [9], these polymers were expected to have good flame retardant properties.

For the purpose of further investigating the thermal degradation mechanism, samples were heated in an oven at 280 °C for 2 h with air as purge gas. For comparison, a soybean oil, ST and DVB copolymer containing BST as a fire retardant

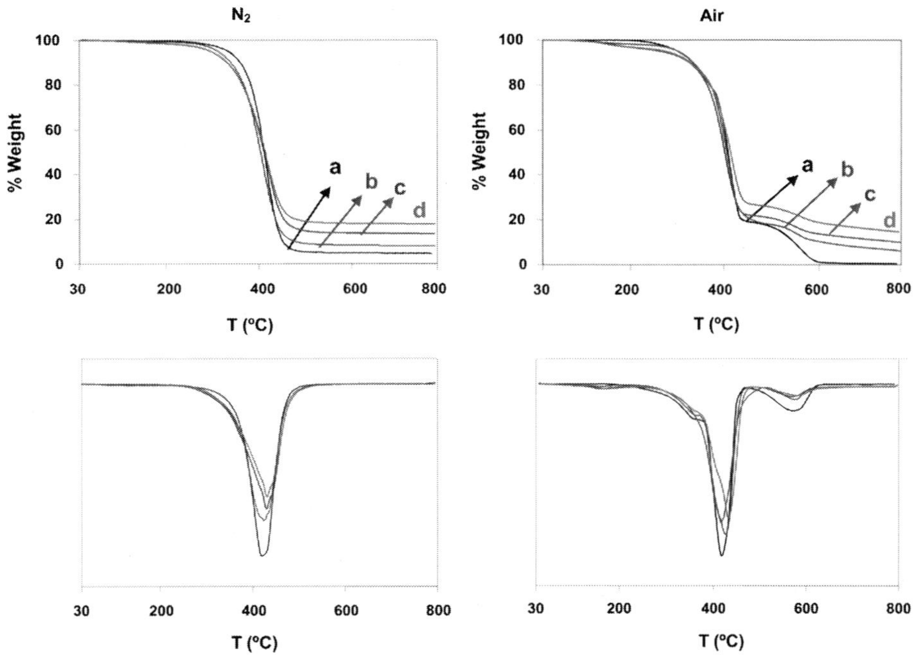

FIGURE 5

Thermogravimetric analysis plots and first derivative curves of the copolymers with 0 (a), 1.0 (b), 2.0 (c), and 3.0 (d) wt-% boron content.

Reproduced from reference [29], with permission from Elsevier.

additive (3% boron content) was also kept under investigation. The degradation temperature was selected at 280 °C. The analysis of the viscous residues by GC/MS showed the presence of various aliphatic and aromatic species (Scheme 11).

The mixture contained the expected degradation products for a soybean oil based polymer: fatty acids, fatty esters, alkanes and alkenes [31]. Moreover, the presence of aromatic products as ST, benzaldehyde, ethylbenzaldehyde, ethyl styrene, diethyl styrene and diphenyl butane and its isomers was also expected from an ST/DVD containing polymers [32,33]. The most noticeable feature of these analyses is that no boron-containing species were detected, thus implying that at low temperatures it remains in the solid residue. This result seems to be in accordance with a phase condensed mechanism. On the other hand, LOI values increased from 19.2 for the boron free system to 23.7 for the resin with the lowest boron content (1% B). As the boron content increased, the LOI value also increased up to 25.6. It has been described [34,35] for the case of boron-containing linear styrenic polymers that, to get small increases in the LOI values, boron contents higher that 3% are needed. These results show that when boron is chemically attached to the network, significant LOI increases can be obtained even when 50% of renewable materials of aliphatic structure are used.

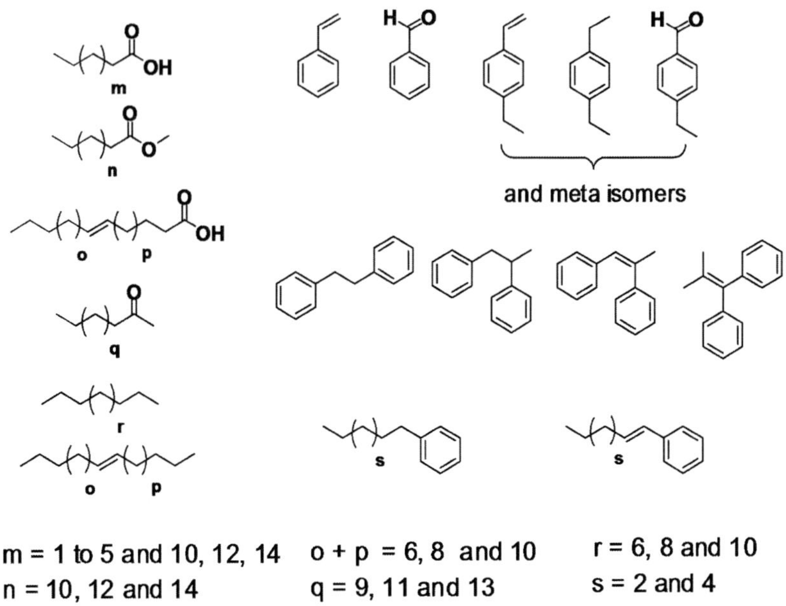

SCHEME 11

Products identified by GC–MS in the volatile pyrolysis fraction of boron-containing polymer.

5. Cone calorimerty studies of fire retardant soybean-oil-based copolymers containing silicon or boron: comparison of additive and reactive approaches

Because most fire retardant formulations are polymer specific, it is necessary to investigate both the burning and decomposition behavior of such new systems, to develop materials meeting current requirements. The incorporation of reactive fire retardant monomers containing boron and silicon has shown potential in these novel polymer systems. The fire retardancy and thermal stability of soybean-based copolymers reactively modified by copolymerization with SiST and BST have been compared with those of the samples prepared with equivalent amounts of the additive 1,3-diphenyl-1,1,3,3-tetramethyldisiloxane (SiAD) and tris-(phenylboroxine) (BAD) and with the heteroatom free soybean based copolymers [36]. Limiting oxygen index tests gave useful information about the ignitability of the materials, while cone calorimetry gave information on both the ignitability, and the burning behavior. The LOI values are summarized in Table 4.

Ignition occurs when the mass loss rate enables the production of sufficient volatiles while the effective heat of combustion is capable of sustaining ignition piloted by a spark. The heteroatom-containing samples had improved times to ignition (TTI), except for the CSiA system. While for the boron-containing copolymers, the presence of the heteroelement seemed to be a significant factor for the ignition time, for the

Table 4 Limiting Oxygen Index and Cone Calorimetric Data Measured with an Irradiance of 35 kW m^{-2}

Sample	LOI	TTI (s)	PHRR (kW m^{-2})	THR (mJ m^{-2})	FIGRA (kW $m^{-2}s^{-1}$)	MARHE (kW m^{-2})	Total Smoke Release (kW m^{-2})
C	19.2	25	509	36.7	7.3	235	742
CSiR	28.6	37	616	43.8	8.8	244	1189
CSiA	26.0	18	928	33.7	18.5	412	1403
CBR	25.6	44	409	32.5	3.7	178	371
CBA	25.2	52	527	32.8	6.6	278	986

C, heteroatom-free copolymer; CSiR, Si-containing reactive copolymer; CSiA, Si-containing additive copolymer; CBR, B-containing reactive copolymer; CBA, B-containing additive copolymer; LOI, limiting oxygen index; TTI, times to ignition; PHRR, peak heat release rate; THR, total heat release; FIGRA, fire growth rate index; MARHE, maximum average rate of heat emission.

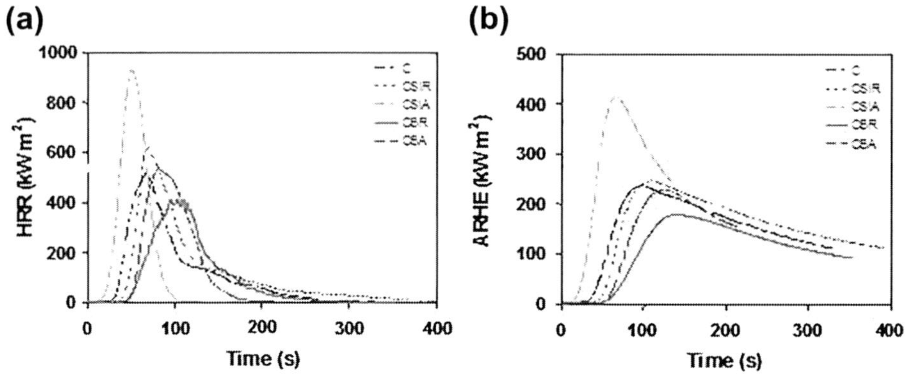

FIGURE 6

Heat release rates (HRR) versus time (a) and average rate of heat emission (ARHE) versus time (b) for the different systems.

Reproduced from reference [35] , with permission from Elsevier.

silicon-containing copolymers not only the presence of silicon but also the reactive nature of the fire retardant appeared to be important. This could indicate an early decomposition of SiAD, before it could play a role as a fire retardant. The heat release rates (HRR) versus time curves for the copolymers are presented in Figure 6(a).

The HRR of C increased rapidly after the initial 25 s and the peak heat release rate (PHRR) occurred at 70 s. Subsequently, the HRR decreased rapidly until it became negligible after 300 s. The shape of HRR curves was similar for all the copolymers. The results showed a significant decrease of 20% in the value of PHRR for the CBR system; the system with a similar boron content as a flame retardant additive showed a value similar to the heteroatom free system. For the silicon-containing copolymers, increase of the PHRR can be observed, especially in the case of the CSiA system. The values of the total heat evolved were very similar for all the samples. Unlike the time to ignition, the heat release rate and the total heat evolved did not correspond to LOI results, illustrating the difference of these parameters in fire retardancy studies. The average rate of heat emission (ARHE) curve is reported in Figure 6(b). This parameter, defined as the cumulative heat emission divided by time and its peak value (maximum average rate of heat emission, MARHE) is considered to be a good measure of the propensity for fire development under real scale conditions. MARHE for CBR system showed a notable reduction (24%) with respect to the C sample, while the incorporation of boron in additive form or silicon in reactive form led to a slight increase. The CSiA system, however, showed an increase of about 60%.

The cone calorimetry experiments gave much clearer evidence than the LOI measurements that incorporation of Si or B into the soybean oil copolymer resulted in fire retardation. This indicates the formation of a protective layer following ignition, reducing the rate of gas-phase fuel formation. The boron-containing copolymers were found to be more efficient flame retardants for this system than the silicon-containing copolymers. Moreover, the reactive or additive approach is a significant factor in terms of the level of fire retardancy achieved; the most significant improvements were obtained when reactive flame retardants were used.

6. CONCLUSIONS

Plant oils as well as fatty acids have been used in polymer technology for a long period of time. To extend the application possibilities of vegetable oil-based thermosets by improving their flame retardancy, the influence of introducing different heteroelements such as phosphorus, silicon or boron, into the polymeric structure was studied (Summary Table).

Summary Table Thermal and flame retardant characteristics of the described P (entries 1-13), Si (entries 14-21) and B (entries 22-24) containing polymers

Entry	Heteroatom (wt %)	AC[1] (wt %)	$T_{5\%}$[2] (°C)	Char[3] (%)	LOI	Reference
1	1.8	68	302	17	27.7	8
2	1.9	60	313	18	28.5	8
3	2.5	66	313	20	30.2	8
4	3.9	50	350	18	31.0	8
5	5.7	49	354	16	32.0	8
6	1.4	82	308	5	21.2	11
7	2.0	80	316	5	21.9	11
8	2.8	77	333	5	22.4	11
9	3.0	71	357	6	23.5	12
10	1.6	81	328	10	21.8	13
11	2.9	70	346	11	24.0	13
12	3.9	61	358	11	25.6	13
13	1.0	45	245	12	24.0	21
14	1.7	59	320	7	19.0	24
15	3.8	61	325	6	20.4	24
16	6.2	63	325	7	22.1	24
17	9.0	64	326	5	23.6	24
18	1.0	48	372	4	22.5	26
19	3.0	45	366	4	25.6	26
20	5.0	42	362	4	28.6	26
21	7.0	39	366	3	29.7	26
22	1.0	48	352	8	23.7	29
23	2.0	46	350	13	24.6	29
24	3.0	45	354	18	25.6	29

[1] *Aliphatic content from fatty acid or triglyceride moieties.*
[2] *Temperature of 5 % weight loss in nitrogen atmosphere*
[3] *Char yield at 800 °C in nitrogen atmosphere*

As a general trend the presence of the heteroelement increased the LOI values, leading to polymers with enhanced flame retardancy. In particular, the incorporation of phosphorus is the most fruitful approach to confer flame retardancy, as 2–3% P contents give high LOI values. Higher phosphorus amounts do not show a significant influence. On the contrary, higher amounts of silicon are necessary to increase the LOI. Another parameter that seems to have influence on the flammability is the aliphatic content (AC) from fatty acid and triglyceride moieties, as for similar heteroelement content, the higher the aliphatic content the lower the LOI values.

These results demonstrate the efficiency of heteroelement-containing plant oil-based polymers as flame retardants and show that it is possible to exploit renewable resources to manufacture original and useful materials.

References

[1] Stevens GC, Mann AH. Risks and benefits in the use of flame retardants in consumer products. London: DTI Report; 1999.
[2] Lu SY, Hamerton I. Recent developments in the chemistry of halogen-free flame retardant polymers. Prog Polym Sci 2002;27:1661–712.
[3] Lligadas G, Ronda JC, Galià M, Cádiz V. Synthesis and properties of thermosetting polymers from a phosphorous-containing fatty acid derivatives. J Polym Sci Part a: Polym Chem 2006;44:5630–44.
[4] Das G, Trivedi RK, Vasishtha AK. Heptaldehyde and undecylenic acid from castor oil. J Am Oil Chem Soc 1989;66:938–41.
[5] Van der Steen M, Stevens C. Undecylenic acid: a valuable and physiologically active renewable building block from castor oil. ChemSusChem 2009;2:692–713.
[6] Saito T. Cyclic organophosphorus compounds and process for making same. U.S. Patent 1972,3,702,878.
[7] Varma IK, Gupta U. Curing of epoxy resin with phosphorylated Diamine. J Macromol Sci Chem 1986;23:19–36.
[8] Lligadas G, Ronda JC, Galià M, Cádiz V. Development of novel phosphorus-containing epoxy resins from renewable resources. J Polym Sci Part A Polym Chem 2006;44: 6717–27.
[9] Van Krevelen DW. Some basic aspects of flame resistance of polymeric materials. Polymer 1975;16:615–20.
[10] Liron F, Knochel P. Stero Stereoselective [2,3] sigmatropic rearrangement of acyclic allylic phosphinites. Chem Commm; 2004:304–5.
[11] Montero de Espinosa L, Ronda JC, Galià M, Cádiz V. A straightforward strategies for the efficient synthesis of acrylate and phosphine oxide- containing vegetable oils and their crosslinked materials. J Polym Sci Part A Polym Chem 2009;47:4051–63.
[12] Montero de Espinosa L, Ronda JC, Galià M, Cádiz V, Meier MAR. Fatty acid derived phosphorus-containing polyesters via acyclic diene metathesis polymerization. J Polym Sci Part A Polym Chem 2009;47:5760–71.
[13] Montero de Espinosa L, Meier MAR, Ronda JC, Galià M, Cádiz V. Phosphorus-containing renewable polyester-polyols via ADMET polymerization: synthesis, functionalization, and radical crosslinking. J Polym Sci Part A Polym Chem 2010;48: 1649–60.

[14] Schwendeman JE, Church AC, Wagener KB. Synthesis and catalyst issues associated with ADMET polymerization. Adv Synth Catal 2002;344:597–613.
[15] Wagener KB, Brzezinska K, Anderson JD, Younkin TR, Steppe K, DeBoer W. Kinetics of acyclic diene metathesis (ADMET) polymerization. Influence of the negative neighboring group effect. Macromolecules 1997;30:7363–9.
[16] Baughman TW, Wagener KB. Recent advances in ADMET polymerization. Adv Polym Sci 2005;176:1–42.
[17] Shultz GV, Zakharov LN, Tyler DR. Transition-metal-containing polymers by ADMET: polymerization of cis-$Mo(CO)_4(Ph_2P(CH_2)_3CH{=}CH_2)_2$. Macromolecules 2008;4:5555–8.
[18] Terada K, Berda EB, Wagener KB, Sanda F, Masuda T. ADMET polycondensation of diketopiperazine-based dienes. Polymerization behavior and effect of diketopiperazine on the properties of the formed polymers. Macromolecules 2008;41:6041–6.
[19] Lu Y, Larock RC. Novel polymeric materials from vegetable oils and vinyl monomers: preparation, properties, and applications. ChemSusChem 2009;2:136–47.
[20] Li F, Larock RC. In: Mohanty AK, Misra M, Drzal LT, editors. Natural fibers, biopolymers and biocomposites. Boca Raton: CRC Press; 2005.
[21] Sacristán M, Ronda JC, Galià M, Cádiz V. Phosphorus-containing soybean oil copolymers: cross-metathesis of fatty acid derivatives as an alternative to phosphorus-containing reactive flame retardants. J Appl Polym Sci 2011;122:1649–58.
[22] Oertel G. Polyurethane handbook. 2nd ed. Munich: Hanser Publishers; 1993.
[23] Lligadas G, Ronda JC, Galià M, Cádiz V. Plant oils as platform chemicals for polyurethane synthesis: current state-of-the-art. Biomacromolecules 2010;11:2825–35.
[24] Lligadas G, Ronda JC, Galià M, Cádiz V. Novel silicon-containing polyurethanes from vegetable oils as renewable resources. Synthesis and properties. Biomacromolecules 2006;7:2420–6.
[25] Kambour RP, Lignon WV, Russell RR. Enhancement of the limiting oxygen index of an aromatic polycarbonate by the incorporation of silicone blocks. J Polym Sci Polym Lett 1978;16:327–33.
[26] Sacristán M, Ronda JC, Galià M, Cádiz V. Silicon-containing soybean-oil-based copolymers. Synthesis and properties. Biomacromolecules 2009;10:2678–85.
[27] Andjelkovic DD, Valverde M, Henna P, Lee F, Larock RC. Novel thermosets prepared by cationic copolymerization of various vegetable oils. Synthesis and their structure-property relationships. Polymer 2005;46:9674–85.
[28] Morgan AB, Jurs JL, Tour JM. Synthesis, flame-retardancy testing, and preliminary mechanism studies of nonhalogenated aromatic boronic acids: a new class of condensed-phase polymer flame-retardant additives for acrylonitrile–butadiene–styrene and polycarbonate. J Appl Polym Sci 2000;76:1257–68.
[29] Sacristán M, Ronda JC, Galià M, Cádiz V. Synthesis and properties of boron-containing soybean oil based thermosetting copolymers. Polymer 2010;51:6099–106.
[30] Kahraman G, Berskardesc O, Rzav ZMO, Piskin E. Bioengineering polyfunctional copolymers. VII. Synthesis and characterization of copolymers of *p*-vinylphenyl boronic acid with maleic and citraconic anhydrides and their self-assembled macrobranched supramolecular architectures. Polymer 2004;45:5813–28.
[31] Maher KD, Bressler DC. Pyrolysis of triglyceride materials for the production of renewable fuels and chemicals. Bioresour Technol 2007;98:2351–68.
[32] Risby TH, Yergey JA, Scocca JJ. Linear programmed thermal degradation mass spectrometry of polystyrene and poly(vinyl chloride). Anal Chem 1982;54:2228–33.

[33] McNeill IC, Zulfiqar M, Kousar T. A detailed investigation of the products of the thermal degradation of polystyrene. Polym Degrad Stab 1990;28:131–51.
[34] Armitage P, Ebdon JR, Hunt BJ, Jones MS, Thorpe FG. Chemical modification of polymers to improve flame retardance-I. The influence of boron-containing groups. Polym Degrad Stab 1996;54:387–93.
[35] Martín C, Hunt BJ, Ebdon JR, Ronda JC, Cádiz V. Synthesis, polymerization, and effects on the flame retardancy of boron-containing styrenic monomers. J Polym Sci Part A Polym Chem 2005;43:6419–30.
[36] Sacristán M, Ronda JC, Galià M, Cádiz V. Cone calorimetry studies of fire retardant soybean oil-based copolymers containing silicon or boron: comparison of additive and reactive approaches. Polym Degrad Stab 2010;95:1269–74.

CHAPTER

23

Flame Retardancy and Protection against Biodeterioration of Natural Fibers: State-of-Art and Future Prospects

R.M. Kozlowski

Institute for Engineering of Polymers Materials and Dyes, Piastow and ESCORENA Focal Point at the Institute of Natural Fibres and Medicinal Plants, Poznan, Poland

M. Muzyczek, J. Walentowska

The Institute of Natural Fibres and Medicinal Plants, Poznan, Poland

CHAPTER OUTLINE

Polymer Green Flame Retardants. http://dx.doi.org/10.1016/B978-0-444-53808-6.00023-8

1. Flammability

1.1 Introduction

Lignocellulosic and protein fibers comprise the two main groups of flammable natural fibers; these have been known and explored by humankind since 5000–7000 BC. The third group of natural fibers includes mineral and carbon fibers, which are generally fire resistant. The classification of all natural fibers is presented in Figure 1.

Lignocellulosic natural fibers, such as cotton, flax, hemp, jute, sisal, ramie, abaca, cabuya, henequen, viscose, acetate, triacetate, and some others, being cellulosic fibers present low flame resistance. In high humidity conditions and at elevated temperatures, they also undergo degradation i.e. biodeterioration by bacteria and fungi. Lignocellulosic fibers exhibit several virtues, the main of which are high hygroscopicity and ability to absorb moisture, they are nice and cool to handle, ability to remove electrostatic charges (prevents collecting electrostatic charges), nonallergenic properties, and soothing properties for skin, due to its natural pH balance. The smoothness of these cellulosic fibers, especially of linen, together with small hollows in the structure of the fiber ensures that the products made of these fibers will be highly resistant to soiling and are partially bacteriostatic.

Protein fibers, on the other hand (such as wool, silk, alpaca, and hair), contain nitrogen and sulfur, and thus, they behave differently under flame conditions; nitrogen is nonflammable and does not support combustion. Hence, these fibers ignite and burn with a much higher difficulty. Also, the development of a carbonaceous structure tends to inhibit flame. Furthermore, protein fibers are, in general, more resistant to biodeterioration. Bacteria and fungi do not directly attack these fibers in humid conditions. This is why (as everybody can see in the National Museum at Cairo in Egypt) mummies still have hair after >2000 years. However, there are some insects e.g. moths, which can destroy protein fibers.

Protein fibers are characterized by elasticity, excellent moisture sorption, inability to collect electrostatic charges, good thermal insulation, and satisfactory resistance to microorganism activity. The highest thermal insulation is observed in the case of wool; the stability and the structure of crimps are responsible for the high thermal insulation of wool. Among protein fibers, alpaca and vicuna wool present the highest thermal insulation properties, due to the inner structure of this wool (see the cross-section in Figure 2 and the longitudinal view in Figure 3) [1].

Wool is also characterized by high elasticity; crimps and keratin are responsible for specific "shock absorption" properties. The keratin structure under mechanical

FIGURE 1
Classification of natural fibers.

Source: Institute of Natural Fibres and Textile Institute. Textile terms and definitions.

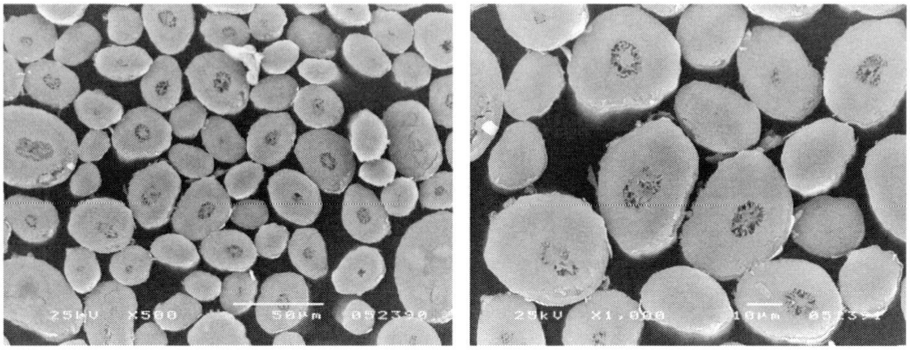

FIGURE 2

Scanning electron micrograph of the cross-sections of fibers of alpaca wool (Pinta S.A-beige/white diameter 29.1 μm).

pressure changes (i.e. stretching, bending) does not alter radically; it is only a reversible transformation of alpha-keratin into beta-keratin. Wool is also water repellent, which might appear to be somewhat of a contradiction. This feature results from the fact that a liquid water particle is larger than the molecular structure of the epicuticle, and thus, it tends to run off, while water vapor molecules are able to diffuse into the fiber through the microscopic pores in the surface. The above-mentioned phenomena explain the bacteriostatic properties of wool.

Natural fibers belong to the group of comfortable and healthy fibers with a positive influence on human physiology [2]. However, under extreme fire conditions and very high humidity, these fibers need flame retardant treatment and prevention against fungi mildew.

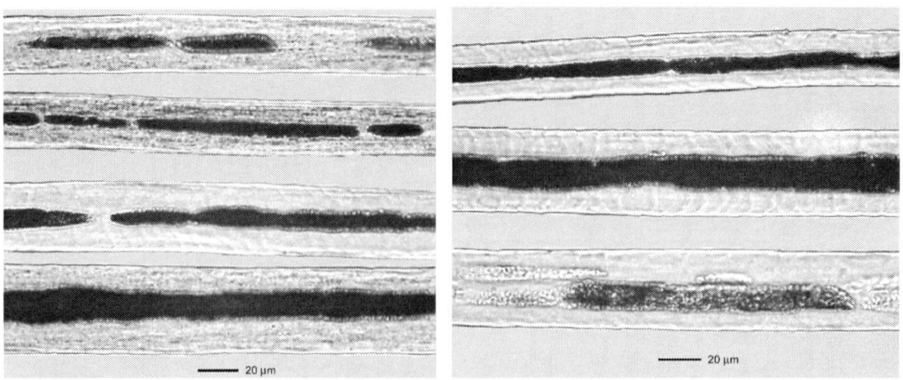

Pinta S.A. beige/white, $d = 29.1$ μm   606 S.A. white. $d = 35.4$ μm

FIGURE 3

Photographs (longitudinal view) of alpaca fibers.

1.2 Behavior in flame conditions of natural fibers versus other fibers

Natural lignocellulosic fibers (cotton, flax, hemp, ramie, etc.) are composed of carbon, hydrogen (fuels) and oxygen (supporter of combustion); thus, they are highly flammable and burn very easily.

Protein fibers (e.g. wool and silk), on the other hand, present a superior flame retardancy. The inherent flame retardancy of wool fibers, for example, originates from the keratin component and the presence of a large quantity of nitrogen (15–17% nitrogen) in wool along with its ability to absorb large amounts of moisture. This enables wool to pass some flammability tests without treatment e.g. high fabric surface mass—mainly in the horizontal test position. Investigations on combustibility have confirmed the higher fire resistance of wool compared to that of other fibers. In many cases, at respective large surface mass of the fabric (>800 g m^{-2}), wool is considered as a noncombustible material. Wool articles exposed to a sufficiently intense source of heat can ignite, but in most cases, they only smolder or carbonize and do not burn with a flame [3]. During combustion, wool burns to ash, which immediately gets cold and can be easily removed from the surface, for example from skin. In comparison with other fibers, wool releases smoke with less density. An interesting phenomenon is observed in some blends of wool with cellulosic fibers (linen, hemp); under flame conditions, cellulose changes from crystalline to amorphous, reducing the flammability of such blends (synergistic effect) [4,5]. This effect is sufficient in special areas of application where flame retardants cannot be used, but the requirement for flame retardancy is still important.

The flammability properties of selected natural fibers versus man-made fibers are presented in Table 1.

From the table, it becomes obvious that generally the behavior of natural fibers in flame conditions does not pose as much danger as some man-made fibers do. Under fire conditions, the combustion of natural lignocellulosic fibers leads to carbon dioxide and water. In some cases, only if oxygen delivery is not sufficient, the evolution of carbon monoxide is observed. In comparison, man-made fibers under flame conditions release more heat than do natural fibers. They also melt, and the dripping phenomenon is very often observed; dripping can propagate fire, and in the case of apparels, a melting polymer can seriously wound the skin. Depending on the chemical structure, man-made fibers also emit several toxic gases for example HCl, HCN, and NO_x. However, natural and man-made fibers differ in terms of glowing. The natural ones, especially wool, present a stronger tendency to glow in comparison to some man-made fibers.

The cone calorimetry results (ISO 5660) of natural fiber blends of wool and lignocellulosic fibers, flax and hemp, are presented in Table 2.

As arising from Table 2, blending wool with other raw natural materials (linen, hemp) in the ratio of 50:50, instead of using 100% wool, results in a lowering of the heat release rate (HRR) and the specific extinction area (SEA) at a heat flux of 35 kW m^{-2} and 50 kW m^{-2}.

Table 1 Flammability Properties of Various Fibers [3,6–8]

Fiber Type	LOI (ASTM/D-2863-70)	Heat of Com-bustion (kcal kg^{-1})	Self-Ignition Temp-erature (°C)	Flame Temp-erature (°C)	Flammability	Smoke Emission: Decrease in Visibility	Smoke Emission: Optical Density (%)	Gases Produced in Flaming Conditions
Cotton	18.4	3.9	400	860	Supports combustion; burns readily to char and afterglow	4	0.02	CO_2, CO, H_2O
Flax (linen)	17.4	3.8	390	–	–	4	0.02	CO_2, CO, H_2O
Wool	25.2	4.9	570–600	940	Supports combustion with difficulty; burns very slowly	18	0.09	CO_2, CO, HCN, NO_x, H_2O
Viscose	18.9	–	420	850	Supports combustion; burns readily to char and afterglow	–	–	–
Rayon	19.7	3.9	500	–	–	4	0.02	CO_2, CO, HCN, NO_x, H_2O
Polyamide	21.5	7.9	530	875	Supports combustion; melts	6	0.03	CO_2, CO, HCN, NO_x, H_2O
Polyester	20.6	5.7	450	695	Supports combustion; burns readily; melts and forms soot	28	0.14	CO_2, CO, H_2O
Polyvinyl chloride	37.1	5.1	–	–	Noncombustible and hence there is no combustion or flame	34	0.18	CO_2, CO, HCl, NO_x, H_2O
Polypropylene	18.6	6	570	–	–	66	0.39	CO_2, CO, H_2O
Polylactic acid	22.0	–	–	–	–	–	–	–
Acrylic	18.2	–	560	855	Supports combustion; burns readily; melts with sputtering	–	–	–
Modacrylic (Kanekaron)	30–35	–	–	–	–	–	–	–

Table 2 Cone Calorimetric Data of Natural Fluffy Nonwoven [9]

Code	Specimen Mass (g)	TTI (s)	HRR Max ($kW\ m^{-2}$)	SEA av ($m^2\ kg^{-1}$)
Wool fluffy nonwoven				
Heat flux of 35 $kW\ m^{-2}$	6.25	6.63	195.05	103.65
Heat flux of 50 $kW\ m^{-2}$	5.98	4.08	285.35	128.65
Wool/flax fluffy nonwoven				
Heat flux of 35 $kW\ m^{-2}$	3.96	8.78	116.27	34.65
Heat flux of 50 $kW\ m^{-2}$	3.49	4.51	175.13	96.96
Wool/hemp fluffy nonwoven				
Heat flux of 35 $kW\ m^{-2}$	7.25	7.87	155.91	36.58
Heat flux of 50 $kW\ m^{-2}$	6.35	3.83	186.31	94.28

HRR, heat release rate; SEA, specific extinction area; TTI: Time to ignition.

1.3 Systems of preventing lignocellulosic and protein fibers against fire

The main asset of applying of flame retardants to natural fibers are

- Delay of ignition time
- Reduction of heat released
- Reduction of surface spread of the flame
- Reduction of mass loss rate
- Reduction of fire propagation
- Protection of people's lives

All the above result in a higher possibility of extinguishing fire, longer time for evacuation, etc.

1.3.1 Cellulosic fibers

Nondurable finishes for lignocellulosic fibers (cotton, linen, hemp, ramie, jute, etc.): Nondurable finishes include a number of chemicals first described in the eighteenth century for this use, i.e. borates and ammonium phosphates. There are a number of proprietary formulations of ammonium phosphates or water—soluble ammonium polyphosphates combined with handling improving agents, or with agents such as ammonium bromide, which can give vapor phase action, especially useful for synthetic fibers. The water-insoluble version of ammonium polyphosphates has been used in conjunction with binding agents to make semidurable back coatings [7]. A fundamental study of this approach, which is particularly effective on cellulosics, was published by Horrocs et al. [10]. Other commercial nondurable treatments are phosphate salts of guanidine, dicyandiamide, or guanyl urea. Through cotton bearing, these finishes become phosphorylated, and the finish can then survive some washes. Some nondurable finishes contain ammonium bromide (especially

Table 3 (Nondurable) Flame Retardants for Fabric Protection [6]

No	Name	Producer	Composition
1	Aerotex NDC, NDS	American Cyjanamid Co	Composition of inorganic and organic compounds
2	Aflaman LS, GK	Thor chemic	Phosphorus–nitrogen reactive compound
3	Aflaman PS	Thor chemic	Ammonium salt
4	Akaustan A	BASF	Inorganic salts containing nitrogen
5	Amgard TR	Albright	Ammonium polyphosphate
6	Banfire	Lutex Chemical	Inorganic salts
7	Celluset 3, 7, 40	Apex Chemical	Composition of urea bromide
8	Consos flame retarder 270	Consos Inc	Combined resin
9	F/R P-26 solution	White Chemical Corp.	Organic bromic compound
10	Fire guard no 60	Polymer Research	Chlorinated nitrogen–phosphate compound
11	Fi-retard 5M	Arkansas Chemical Co.	Nitrogen compound
12	Firex 4100	Dr Th Bohme	Chlorinated phosphate salt
13	Firex 4160	Dr Th Bohme	Composition of inorganic and organic salts
14	Flacavon F 71, R, RL, PS	Schill & Seilacher	–
15	Flameproof 260		Phosphorus compound
16	Flame retardant 12, CM	DuPont Industrial Chem Dep	Composition of ammonium salts
17	Flameproof 2084	Apex Chemical Co.	Chlorinated phosphorus compound
18	Flameproof 311, 4227	Apex Chemical Co.	Inorganic salts
19	Flammentin AS, CN	Thor Chemie	–
20	Flammex CE	Zachimer und Schwarz	Diammonium phosphate and urea
21	Flamobin S	Henkel	Composition of inorganic and organic compounds and solvent
22	Flovan AS	Chemische Fabrik Pfersee	Inorganic ammonium compound
23	Flovan BU	Chemische Fabrik Pfersee	Composition of inorganic halogen compound and organic fixing agent
24	Flovan CG, Flovan CGN	Ciba	Composition of inorganic phosphorus compound and organic fixing agent
25	Flovan FD	Chemische Fabrik Pfersee	Composition of phosphorus inorganic compound and organic fixing agent

Table 3 (Nondurable) Flame Retardants for Fabric Protection [6]—cont'd

No	Name	Producer	Composition
26	Fobos M-2T	Instytut Włókien Naturalnych	Urea polyborates and polyphosphates
27	Fyran B	Crown-Metro Inc.	Ammonia compound
28	Fyran JA	Crown-Metro Inc.	Modified sulfonate
29	Fyran P	Crown-Metro Inc.	Inorganic compound without ammonium slats
30	Fyrol 99	Chemical Co.	Chlorinated phosphorus compound
31	Fyrol HNP	Chemical Co.	Phosphorus ester
32	Myflam	Mudrin	Composition of decabromodiphenyloxide and acrylic resin
33	Protenyl NE	Protex	Composition of metallic salts
34	Pyrex AM	Zachimer und Schwarz	Inorganic salt
35	Pyrosan LP	Laurel Products Co.	Inorganic salt
36	Pyroset BC	American Cyanamid Co.	Composition of dicyandiamide and phosphoric acid
37	Rucon Flammfest	Rudolf Chemische Fabrik	Organic phosphorus compound
38	Warconyl 175, 712	Sun Chemical	Composition of amine salt and nitrogen compounds

if intended for synthetics or blends) and several use sulfate and sulfamate salts, which can also intensify the action of phosphates. Some commercial (nondurable) flame retardants used for fabric protection are presented in Table 3 [6].

Semidurable finishes: Inexpensive phosphorus acids such as methylphosphonic can also provide the PO_4^{3-} anion in these salts such as cyanoguanidine (or possibly guanyl urea) and are believed to be a component of improved semidurable finishes such as Ciba's Flovan CGN. A few new semidurable systems (allowing limited washing or water soaking) using polyamine phosphonate salts have been patented by Ciba (now Huntsman), for example newly introduced Pyrovatim ®PBS.

Durable finishes: In the durable cellulosic finish area, the tetrakis(hydroxymethyl)phosphonium salt–urea–precondensate–ammonia cure system (known for about 50 years) is the leader in laundering durability. New developments in recent years are available on the market including some clever fine-tuning of the finish by small formulation changes, by mode of application, or by drying changes to obtain improved hand feel. The largest market for such treated fabrics is in work clothing. With improved esthetics, development of more customer applications will probably follow. Careful temperature control allows application to nonwovens, which cannot be protected from shrinking by a tender frame.

Blends of cotton with synthetics, such as those with high percentages of polyester and nylon continue to represent a challenge. It is well known that the finishes that can be effective on the meltable synthetics are defeated by the wick effect of cotton, and that the char-introducing finishes on the cotton do not do much to retard poorly charrable synthetics. It has long been possible to flame retard blends, in which cotton predominates, and in fact, some high-comfort blends of this sort are marketed such as 88% cotton with 12% high tenacity nylon in work clothing. Some progress has been made in flame-retarding blends with comparable or greater amounts of synthetic than cotton, but generally two-stage treatments and the application of a higher % P (such as by the use of methylophosphonate discussed below) were found to be necessary. Some commercially available fire retardants employed for durable (wet cleaning) fabric protection are presented in Table 4 [6].

Among durable flame retardant treatments, three phosphorous organic systems are the most important ones, worth mentioning:

- dialkylphosphonocarboxylic acid amide—Pyrowatex P, Ciba
- tetrakis(hydroxymethyl)phosphonium chloride—urea (THCP) commercial name Proban CC, Albright & Wilson and more recently,
- backcoating that often comprises a resin-bonded antimony—bromine system.

Table 4 Flame Retardants for Durable (Wet Cleaning) Fabric Protection [6]

No	Producer	Name
1	Thor Chemicals Ltd	Aflamman IST, PCS; Aflammit KWB; Alfamman AI
2	BASF	Akaustan Pc
3	Albright and Wilson	Antiblaze NW; Proban 210, 420 A, NX, Cl; Retardal AC, S
4	Yohimura	Apiros 307
5	White Chemical	Caliban F/R P 44; Caliban F/R P 53
6	Ugine Kuhlmann	Carbinul PA
7	Schill und Seilacher	Flacavon AR
8	Arkansas Chemical Co.	Flame Snub WND, WD + WDC
9	Dainippon Ink and Chemicals Inc.	Flameguard 5518, 5518conc, 5527 Flameguard 6188NS; 8115S
10	Stauffer Chem	Fyrol 76
11	Akzo Chemical	Fyrol CEF
12	Permagard products Ltd	Permaproof
13	Daj-Ichi Kogyo Seiyaku Co. Ltd	Pyroguard F-120
14	American Cyanamide Co.	Pyroset TKO, TKP, CP
15	Ciba	Pyrowatex CP, Pyrowatex CP-LP
16	Hooker Chemical Co.	Roxel
17	Rudolf Chemische	Rucon Flameproof S
38	Daikyo Chem Co. Ltd	Supervigol DS

The first system, being reactive, is bonded to the cellulose fiber during curing via a formaldehyde crosslinking reaction. The durability of the second flame retardant is not connected with the chemical reaction between cellulose and a phosphorus derivative, but it is a consequence of a curing process in ammonia atmosphere. However, both compounds decrease the mechanical strength of fibers and result in the creation of an additional amount of CO emission during combustion.

Recent progress in the flame retardancy of lignocellulosic fibers involves their chemical modification through the chemical reaction between hydroxyl groups of cellulose with a reactive fire retardant [8,11,12]. Cellulose, as a polyhydroxy compound, may be modified by the formation of chemical derivatives through the typical reactions of alcohols (substitution of hydroxyl groups). However, the physical structure of natural cellulosic fibers, especially crystalline regions, results in a limitation of chemical reactivity of cellulose. This crystalline structure renders the larger part of the molecule impenetrable by chemical reagents, so that the most important consideration in any reaction system is how to disrupt the crystalline regions so that they become as accessible to reagents as amorphous regions are. Therefore, pretreatment processes are often applied resulting in cellulose activation.

This is the reason for investigating safer chemicals that are able to bond to cellulose. Promising compounds that can be employed for this purpose are alkyl borates. An even more promising formulation has been developed by Foster−Miller Inc. They developed an intumescent system based on alumina trihydrate (ATH), and incorporated it into a cellulosic fabric using two different methods: microencapsulation and incorporation into a protective vinyl coating. The sources of alumina are aluminum alkoxides, which hydrolyze by water contained in fiber, creating ATH during in situ reaction. This method was tested for special fabrics in military tents.

Furthermore, the latest progress in genetic engineering gives another chance to obtain transgenic cellulose (cotton, flax, sisal, and hemp) with flame retardant groups built into the structure of cellulose or/and lignin: their hydroxyl groups can be substituted by nitrogen, which will then react with phosphorus groups, making cellulose inherently flame resistant.

Using polyperoxomolybdenate as a catalyst is a very common route for the functionalization and the flame retardancy enhancement of natural lignocellulosic fibers. This catalyst can promote durable flame retardancy in various natural cellulosic fibers by cationization and prevent further joining of the groups, a phenomenon responsible for the decrease of flammability.

1.3.2 Protein fibers

Wool fiber is a keratin protein fiber composed of amine acids linked by peptide bonds. The peptide units are crosslinked by disulfide bonds and have reactive groups such as amino hydroxyl and imidazol groups attached to them [13]. The presence of potentially reactive groups in wool allows the modification of some of its properties. However, it must be emphasized that this task is not easy due to the crosslinked structure of keratin. The inherent properties of wool allows its use for the production of clothes, furniture coverings, carpets, and blankets. Lately, wool received greater attention as a raw material able to fulfill requirements of antifire protection, both for buildings of public use and for the production of protective wear clothes. This is due to the inherent resistance of wool to fire as well as due to the fact that its flame retardancy can be improved through the use of fire retardants. Wool treated with flame retardants may be utilized for the production of protective clothes useful in conditions of maximum emergency (e.g. wool protective dresses used by explorers, fire brigades, military forces, and petrol industry). Also, flame retardant wool is used in the production of tapestry, floor coverings, carpets, and other goods for hotel equipment, restaurants, high buildings, airplanes, and ships [14].

In an International Wool Secretariat [15], notable scientific breakthrough was achieved with the discovery and development of a new mordanting (dye-assisting) process that markedly improves the natural flame resistance quality of wool. The process is known as Zirpro. Their output of treated wool textiles (carpets, aircraft furnishings, curtain fabrics, railway upholstery, and blankets) now exceeds 70 million lbs (35 million kilograms).

The Zirpro flame resistant treatment is based on the exhaustion of negatively charged complexes of titanium or zirconium with fluorides, citrates, or other carboxylic or hydroxyl—carboxylic acids on the positively charged wool fibers in acid conditions (pH <3) at a certain temperature. The complexing agent has to be present not only to avoid the precipitation of the extremely hydrolysis-prone titanium and zirconium compounds but also to ensure the negative charge of the metal complex by excess of the complexing agent [15]. The advantages of this method are

- Low add-on of chemicals keeps the cost down and ensures no increase in the weight of the treated material.
- The effective chemicals are exhausted into the fiber; as a result the handle and the natural soil resistance of wool fibers are in no way impaired.
- The physical properties of wool are not adversely affected.
- The treatment is simple; in the commonest method, the processor simply mixes chemicals together and adds them to the dye bath.
- The treatment has been successfully tested against 50 washings at 40 °C, and against dry cleaning.
- Color fastness of treated wool to water and perspiration is not affected.
- The use of organic flame retardants commonly increases smoke emission, but wool treated with the new mordanting process does not show a significant increase.
- Treatment can often be carried out concurrently with any one of the number of routine process applied to wool.
- Flexibility: various application methods have been devised to cover all possible requirements and products.
- The new treatment is compatible with easy-care treatments for wool including machine-wash ability.

High-performance fibers such as aramides (known as heat and flame resistant)—unfortunately with roughness touch—are blended with wool for achieving better handling and higher synergistic flame retardancy effects [16]. Blends of wool with high-performance fibers e.g. Kevlar (poly-*p*-phenylenediamine–terephthalamide) can lead to a synergistic effect (in terms of HRR), and it is much more cost effective [17]. Kevlar/Wool blends were studied in different structures (blends yarn by yarn and fiber by fiber, double woven rib structure and double faced structure). Also, Horrocks [18] expects desirable effects of manufacturing hybrid core yarns Kevlar/Wool blends, because they exhibit natural yarn aspects, good handling, dye ability, and high-performance in terms of flammability and mechanical properties.

Silk fiber is another natural protein fiber, and it is viewed as an environmentally favorable fiber since it contains only natural amino acid units. Silk fibers generally present a higher flame retardancy than the cellulose ones, as a result of the high nitrogen content of the former (~15–18%). Its LOI value is about 23%; however, further improvement of flame retardancy is needed to fulfill some commercial requirements. Silk fabrics treated with urea phosphoric acid salt by a pad dry process has flame retardancy with an LOI value >28%, and the flame retardancy finish is fast to dry cleaning while not fast to washing [19]. Recently, new reports were published on using *N*-hydroxymethyl (3-dimethyl phosphono)propionamide (Pyrovatex CP, Ciba–Geigy) to treat silk fabrics. In this case, silk fabrics have received excellent flame retardancy.

Leather is another excellent natural product widely used not only in the shoe industry but especially in more luxurious products e.g. in upholstery furniture and car and aircraft industries. For these applications, its fire behavior is very important. In comparison with man-made products and textiles used in car and upholstery applications, leather is a safer and also healthier choice [20].

Tanned leather is resistant to short-time exposure to temperatures up to 200 °C (higher temperatures result in pyrolysis). The exposure to temperatures in the range of 130 –170 °C for several minutes does not cause structural changes in leather. Resistance to higher temperatures requires appropriate finishing. In order to determine the flammability of natural and artificial leather, measurements have been carried out using ISO 3795:1989(E) and EN ISO 6940:2004 methods. The results of these measurements are presented in Tables 5 and 6.

Table 5 Leather Flammability in Compliance with ISO 3795:1989(E) [20]

Result	Natural Leather (For Furniture Applications)	Artificial PU Leather
Length of sample burnt (mm)	0	254
Combustion time (s)	0	230
Flammability degree (mm min^{-1})	0	66
Other observations	Ignition, glowing, flame going out before the first point of the measurement	Glowing: 1850 s causing a slow ashing of the sample

Table 6 Leather Flammability in Compliance with EN ISO 6940:2004 [20]

Material	Place of the Contact with Flame	Time of Contact with Flame (s)	Time of Sample Combustion(s)	Observations: Time of Glowing (s)	Observations: Length of Burnt (mm)
Natural leather	Surface (Grain)	18	0	0	Local
		19	0	593	28
	Edge	5	98	>600	200 (Whole)
		4	0	0	Local
Artificial PU leather	Surface	4	135	403	Ashing
		3	0	0	10
	Edge	1	71	313	Ashing

The results obtained by using the ISO 3795:1989 (E) method permit the classification of the tested leather as nonignitable materials (combustion time: 0 mm min^{-1}). On the other hand, artificial PU leather ignites and the flame spreads at a rate of 66 mm min^{-1}. The time until the ignition of the sample and until the beginning of glowing on the surface and on the edges exposed to the flame have been determined by the EN ISO 6940:2004 method. In the case of natural leather, the time of surface exposure to the flame was 18 s, whereas for artificial leather, it was 3 s (Table 6). When the flame was edge situated, natural leather ignited after 5 s and artificial leather ignited immediately after coming into contact with the flame, burned slowly, and then glowed until total ashing. Natural leather, when ignited, smoldered slowly, glowed for a long time, and shrinked. The results of the flammability measurements performed on a cone calorimeter are given in Table 7 [21].

Table 7 Leather Flammability Determined in a Cone Calorimeter (ISO 5660-1: 2002 (E)) [21]

Measurement Parameters (Heat Flux of 35 kW m^{-2})	Dimensional Unit	Natural Leather	Artificial PU Leather
Peak heat release rate (HRR)	kW m^{-2}	190.60	278.82
Time to sustained ignition (TI)	s	40.11	19.28
Total heat released (THR)	mJ m^{-2}	14.95	10.09
Average heat of combustion (HOC_{av})	mJ kg^{-1}	19.87	16.72
Average mass loss rate (MLR_{av})	g/s m^2	2.09	4.44
Average specific extinction area (SEA_{av})	m^2 kg^{-1}	137.66	238.32
Average CO emission (CO_{av})	kg kg^{-1}	0.028	0.043
Average CO_2 emission (CO_{2av})	kg kg^{-1}	2.110	2.059

In the case of natural leather, the HRR is 190 kW m^{-2} (on average) and is clearly lower than that observed for artificial leather (Table 7). Time to sustained ignition in the case of natural leather is 2.1 times longer than that of artificial leather and the mass loss rate is 2.1 times lower for natural leather. Smoke emission, expressed by SEA, in the case of natural leather is almost twice as low as that for artificial leather. Natural leather is also safer because of 1.5 times lower emission of carbon monoxide. A drawback of natural leather is its tendency to glowing, which can cause several-hour long smoldering of an upholstery system. Hence, there is a necessity of applying flame retardants or flame-retarding barriers for the upholstery system [21].

Recently, the most important role in fire retardancy is played by intumescent coatings. When exposed to high temperatures, they swell and increase their volume (50 times, and sometimes even 200 times), thus forming a protective barrier preventing heat transfer to the deeper layers of the material protected. Among the advantages of intumescent coatings are their low weight and the possibility of applying a relatively thin coat to obtain a heat insulating effect for a specified period of time. When swelling under the influence of heat, they form a carbonaceous or silicate layer of cellular structure with the simultaneous release of nonflammable gases, and thus effectively protect materials against oxygen access and flame spread.

The compounds most frequently used in the formulations of intumescent coatings are

- *A carbonizing substance* with a considerable number of carbon atoms, the thermal decomposition of which results in the formation of carbonaceous material having a large number hydroxyl groups, subject to esterification with acids. The carbonizing agents commonly used are polyhydric alcohols (erythritol, pentaerythritol, pentaerythritol dimer and trimer, arabitol, sorbitol, and inositol), polyhydric phenols (resorcinol), and sugars (glucose, maltose, arabinose, dextrin, starch, and cellulose).
- *A foam-producing substance*, which evolves copious quantities of nonflammable gases during its thermal decomposition, thus forming a foamed structure of carbonaceous layers. Dicyandiamide, melamine, polyphosphate, guanidine and derivates, glycine, urea, and chlorinated paraffins (70%) are some commonly employed foam-producing substances.
- *A dehydrating agent*, which during its thermal decomposition releases acid that esterifies the hydroxyl groups of the carbonizing substance. Some known dehydrating agents are phosphoric acid, monoammonium phosphate, diammonium phosphate, ammonium polyphosphate, melamine phosphate, guanyl urea phosphate, urea phosphate, boric acid, and ammonium tetraborate.
- *An esterification catalyst*, which accelerates the formation of esters with maximum utilization of a carbonizing substance. Catalysts used for this purpose are phosphoric acid, diammonium phosphate, monoammonium phosphate, melamine phosphate, polyphosphorylamide, tributyl phosphate, halogen-containing phosphoric ester, ammonium magnesium phosphate, boric acid, and borax.

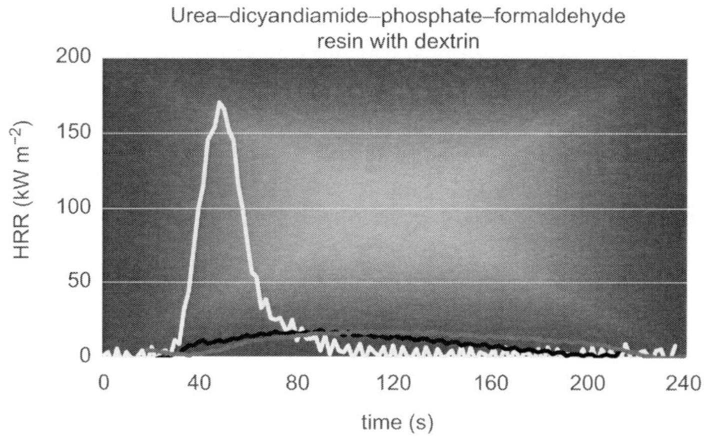

FIGURE 4

Cone calorimetry results of urea—dicyandiamide—phosphate—formaldehyde resin with dextrin-treated fabrics (…… Untreated fabric, —— FR-treated fabric, ------ FR-treated fabric after soaking) [6].

- *A bonding agent* such as phenol, urea, dicyandiamide—formaldehyde resins, polyurethane, chlorinated rubber resins, latexes of vinyl chloride—vinylidene chloride copolymers, polyvinyl acetate latex, polyacrylates, and silicates.

In one of our studies [6], the possibility of the utilization of intumescent fire retardants based on amino resins for fire retardant modification of upholstery fabrics through treating of the backside of the upholstery fabric was verified. The decorative fabric made of blended fibers was FR treated from the backside according to the following recipes: UDPD, urea—dicyandiamide—phosphate—formaldehyde resin with dextrin. The results of tests conducted in a cone calorimeter (ISO 5660, heat flux 35 kW m^{-2}) are presented in Figure 4.

The results of tests conducted in the cone calorimeter confirm the reduction of flammability intensity of decorative fabrics even after multiple leaching. The results of horizontal flammability tests (ISO 3795) are presented in Figure 5. These results show that the treatment has considerably limited the intensity of burning, even after several soaking actions. However, the flame combustion of the fabric from the unprotected side (fleece) was observed.

1.4 Demands that efficient flame retardants should meet

Flame retardants are used in many areas of applications from apparels, upholstery furniture, floor and wall coverings, modern transportation, composites, to insulation and building materials. Therefore, flame retardants must be carefully selected depending on the area of their application.

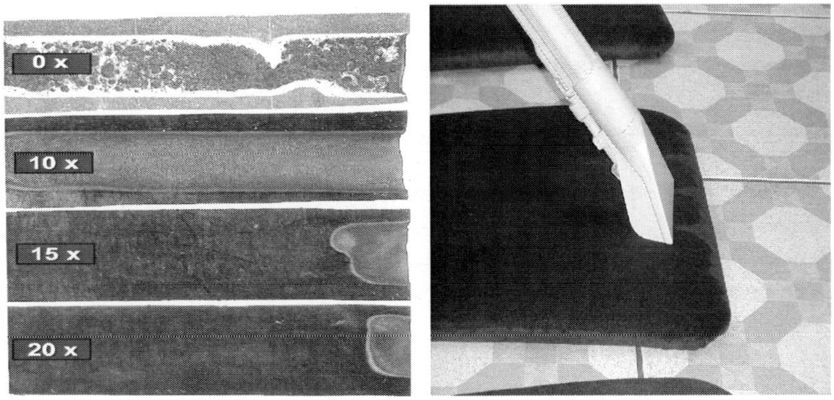

FIGURE 5

Photographs of urea–dicyandiamide–phosphate–formaldehyde resin with dextrin-treated fabrics subjected to the ISO 3795 horizontal test [6].

Flame retardants generally have to be characterized by

- high fire protection effectiveness
- no effect on mechanical strength and esthetics of protected fibers and textiles
- no emission of toxic and corrosive substances during their use
- no effect on the emission of toxic and corrosive gaseous products of the thermal decomposition and combustion of the matrix
- low bioavailability (no escaping and penetration of flame retardant through the skin from the FR apparel)
- ease of application
- high chemical stability
- insecticidal and fungicidal effectiveness (desirable)
- resistance to water (washing) and high humidity
- relatively low price

The above-listed requirements for features of flame retardants are very difficult to be obtained, and as a result, some of them have to be compromised.

1.5 Significant flame retardancy testing methods of fibers and textiles

The tests range in size from small scale and bench scale, utilizing equipment such as Bunsen burners or heat radiators of different types and shapes, up to full size tests on three-dimensional building components and structure [22].

Nowadays, fire tests can be divided into the following main groups [23]:

- Fire resistance tests concerning the ability of structures to withstand the effects of fire and prevent its spreading;

- Noncombustibility tests allowing one to the point to materials that burn and those that do not;
- Ignitability tests enabling the determination of the ease with which materials ignite;
- Reaction-to-fire tests determining important parameters of materials such as HRR and flame spread;
- Smoke production tests determining the ability of materials to produce smoke when on fire;
- Tests assessing the toxicity of gaseous products released during thermal decomposition and combustion. They provide qualitative and quantitative analytical data on gases emitted by materials involved in the fire.

Flammability tests, even when they measure the same parameters, differ to a greater or smaller extent, particularly as concerns

- details of test apparatus;
- specimen
  - size,
  - conditioning,
  - orientation during test;
- source of heat radiation;
- heat flux used, etc.

Flame retardant efficiency or the influence of diverse modifiers on fire performance of a material (e.g. wood, wood derivatives, textiles, plastics, rubbers, and composites) is usually determined by comparison of relevant flammability parameters obtained for the unprotected or the unmodified material and for protected or modified one.

Flammability or "reaction to fire" is a very imprecise term; however, it normally includes

- ignitability,
- flame spread,
- HRR.

However, it should be emphasized that two other important parameters, which occur during a real fire, namely

- smoke production
- toxicity of gaseous products of thermal degradation and combustion have to be taken into consideration when evaluating FR efficiency.

The method usually adopted to evaluate the flame retardancy of materials are

- Thermal analysis (thermogravimetric analysis (TGA), differential thermal analysis (DTA), and differential scanning calorimetry (DSC)) gives basic information on the mechanism of pyrolysis and combustion as well as data on the effect of fire retardants and different materials. However, the correlation between results of thermal analyses and real fires has not been established yet.

- The oxygen index test involves the measurement of the minimum concentration of oxygen in oxygen–nitrogen mixtures, which is necessary to sustain flame burning of a tested material.
- Reaction-to-fire tests for building products.
  Radiant panel test for floor coverings
  Standard: EN ISO 9239-1
  Measured and determined parameters: critical heat flux, smoke production
- Textiles: Determination of flammability and flame spread properties of vertically oriented specimens
  Tested materials: textile fabrics intended for apparel, curtains, and draperies in the form of single or multicomponent (coated, quilted, multilayered, sandwich construction, and similar combinations)
  Standards: EN ISO 6940:1998, EN ISO 6941:1998
  Measured and determined parameters: average time to ignition, minimal time to ignition, surface spread of flame
- Furniture. Assessment of the ignitability of upholstered furniture.
  Part 1: Ignition source: smoldering cigarette
  Part 2: Ignition source: match-flame equivalent
  Standards: EN 1021-1:1999, ISO 8191/1:1987, EN 1021-2:1999, and ISO 8191/2:1988
  Measured and determined parameters: appearance of ignition or smoldering, time to ignition, diameters of the destroyed area

Apart from flammability test methods applied for European classification, the most useful method, in our opinion, for determining the influence of different preservatives and modifiers on the flammability seems to be cone calorimetry (ISO 5660) (Figure 6).

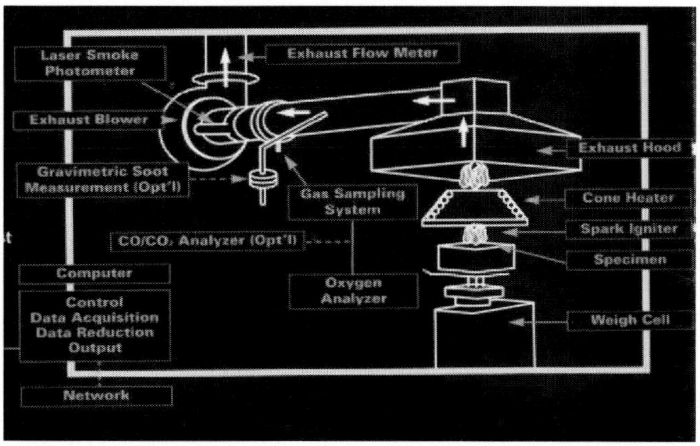

FIGURE 6

Cone calorimeter (ISO 5660).

Through cone calorimetric tests, the evaluation of most parameters is based on the oxygen consumption theory. The basis of oxygen consumption calorimetry is that for every kilogram of oxygen consumed by a burning object, 13.1 mJ of energy are released. At present, the O_2 consumption technique is used extensively in many laboratories all over the world, in both bench-scale and full-scale fire test applications (e.g. furniture calorimeter, room/corner test, and SBI (Single Burning Item test EN 13823) test).

1.5.1 Cone Calorimeter Testing Analyser

The connection between a cone calorimeter and a gas analyzer provides an advanced technique for measuring flammability, smoke generation, and toxicity of materials. The range of detection and quantification of different gases is presented in Table 8.

1.6 Recent progress in flame retardant fibers and other natural textile materials

Unfortunately, no revolutionary flame retardant system has been developed over the last decade, especially in the area of environmentally friendly durable finishes for all types of fibers and fabrics. On the other hand, the development has been responding to legislative procedures, specific test methods, cost, and environmental influences.

Under these circumstances, M. Levin [24] concluded the following: "Notwithstanding the considerable effort spent in the 50s, 60s and 70s of the 20th century on the flame retarding of cotton, no successful treatment has been evolved. There is still no treatment which would render flame-retardant cotton fabrics durable to multiple hot-alkaline, soft and hard—water launderings, without changing markedly the mechanical, physical and aesthetic properties with materials containing

Table 8 CTA System: Range of Detection and Quantification of Different Gases

Gas	Range of Detection
CO	1 ppm to 2%
CO_2	300 ppm to 10%
HCl	5–5000 ppm
HBr	5–1000 ppm
HF	5–5000 ppm
SO_2	15 ppm to 1%
HCN	5–500 ppm
NO	2 ppm to 2%
NO_2	15–500 ppm
NO_x	5 ppm to 2%
H_2O	5000 ppm to 10%
HC	25–5000 ppm

environmentally-friendly flame retardants, at a reasonable cost. Whereas for cotton fabrics, partial solutions exist, there are virtually no solutions for blends of cotton/polyesters, in spite of the enormous economic importance for such products. Similarly, no durable flame-retardant treatment that would preserve the mechanical and physical properties was developed for textiles. Existing treatments are non-durable to atmospheric conditions and suffer sizable losses in the mechanical properties. Developing new flame retardant systems for the automotive and aircraft industries is another important challenge. Two basic challenges for the future can be discerned: one challenge is the development of several new kinds of catalysts for FR systems, based on organometalics and macromolecular metal complexes on one hand and catalytic particles or catalysts precipitated on nanoparticles on the other hand. Another challenge is the application of nanocomposites."

The most important latest requirement for flame retardancy of different materials is the compliance with Registration, Evaluation and Authorisation of Chemicals (REACH) regulations and growing environmental awareness. Many flame retardants especially halogenated ones (Cl, Br) are not recommended [25,26]. Moreover, chemicals such as H_3BO_3, which was widely used as a flame retardant and fungicide can be harmful. The observed tendency is to opt for such combinations: fibrous materials that do not permit the use of flame retardants, or use safer substances improving flame retardancy.

Decomposition products are of high importance in the case of fire. If combustion products are only of low flammability (e.g. carbon and SiO_2), this mechanism is acceptable. Today, we demand from textile flame retardants low leachability, low emission of toxic gaseous substances, and low bioavailability. Some chemical and biochemical modification of natural fibrous polymers, for example the functionalization of cellulosic groups in lignocellulosic fibers and the exploration of enzymes, are promising routes in flame retardancy. Furthermore, nanomaterials are used as flame retardants; hence, there is a chance for significantly decreasing their amount needed for achieving satisfactory effectiveness [27].

A new trend in the flame retardancy of fibrous textiles involves using plasma and corona treatment along with very efficient intumescent systems and fire blockers with good elasticity and handleability of fabrics. The microcapsulation and introduction of the microcapsules into fibers (e.g. to hollow lignocellulosic and protein fibers) is also very promising [28]. The application of genetic engineering to obtain transgenic lignocellulosic and protein fibers with incorporated flame retardants groups "in statu nascendi" is currently being attempted, but this is still at the stage of fundamental research.

2. Biodeterioration

2.1 Introduction

At high humidity and elevated temperatures, natural fibers and textiles (nonwovens and fabrics) are susceptible to biodeterioration. The biodeterioration caused by microorganisms affects materials containing natural fibers, which are used in transport and building industries. Making these materials resistant to biodeterioration is very important not only in the case of textiles but also in the paper industry and in libraries, archives, and museums, where microbiological purity of air is essential. Hygienic finishing textiles used by humans, especially in medical applications, is also very important.

In unprotected cellulosic materials of natural fibers, the microorganisms cause the degradation of cellulose (which is the main component of natural fibers), decreasing the mechanical properties of these materials, also causing emission of odor, which leads to the worsening of microbiological purity of air. Lignin, the second main component of lignocellulosic fibrous materials, is more stable, due to the chemical structure of these phenolic compounds. On the other hand, the pectins, accompanying usually fibrous lignocellulosic materials, are easily decomposed by aerobic and anaerobic bacteria.

Fungi, especially mildew, play the biggest role in the biodeterioration of textile materials made from natural fibers, prompting the aerobic degradation of cellulose. The optimal conditions for their growth are a relative humidity of 70–90%, temperature at 24–30 °C, and a pH of about 6.0. Fungi, especially active in this process are *Aspergillus, Chaetomium, Alternaria, Trichoderma, Penicillium, Stachybotrys, Vertcilliu, and others.* The bacteria that prefer a very high relative humidity (96–99%) and a pH of 6.8–8.0 are of less importance. Bacterial strains decomposing cellulose mainly are *Clostridium*, *Bacillus*, and *Sporocytophaga*, *Cytophaga* [29].

The saprophytic mildew group of fungi plays both a positive and a negative role, in many branches of economy. Mildew is characterized by a wide spectrum of biochemical properties. Hence, they are utilized in the following areas:

- food industry (meat processing, cheese production)
- biotechnological production of antibiotics, enzymes, organic acids, lipids, and hormones
- proecological activity (biological processing of wastewater, composting of organic waste).

However, as already mentioned, there is also a negative aspect of mildew activity. They decompose food, paper, fibers (cellulosic, and protein), leather, and wood. The materials undergoing most considerable biodeterioration are textiles: fibers, fabrics, nonwovens, floor coverings, and carpets of both chemical and natural origin.

In studies on the textile resistance to mildew action, the following methods are used predominantly:

- Agar method (Standard EN 14,119:2003)
- Soil burial test (Standard EN ISO 11,721-1:2001).

Referring to bacteria, there are two basic methods used:

- Quality screening method (Standard AATCC 147-1998)
- Quantity method (Standard AATCC 100-1998).

Green natural fibers are biodegradable and recyclable. Nevertheless, the growth of fungi (especially mildew) is a considerable problem for natural fibers. The materials produced from natural fibers are frequently found in direct contact with human organisms (underwear, bed sheets, interior furnishing, hygienic products, and medical applications). The main role in the decomposition of natural textiles is played by mildew (e.g. *Chaetomium*, *Stachybotrys*, *Penicilium*, *Aspergillus*, and *Paecilomyces*). Among bacteria, the main strains responsible for deterioration of cellulose fibers are, e.g. *Clostridium*, *Cytophaga*, and also present on human skin *Corynebacterium*, *Bacillus*, *Micrococcus*, *Staphylococcus*, *Escherichia*, *Klebsiella*, and *Pseudomonas*. Their presence on natural fibers leads to the decrease in their mechanical performance.

A factor causing mildew growth is high air humidity, >70%. The main negative effect of microbiological decay of textiles is strength reduction. For example, the lignocellulosic bast fibers characterized by low resistance; after 16 days of the test, the tensile strength reduced by 83–96%. In the case of cotton fabric with a unit weight of 250 g m^{-2}, after 14 days of the test, the tensile strength also reduced by 82–98%, while in the case of cotton fabric with a unit weight of 910 g m^{-2}, after 6 days of the soil burial test, the tensile strength reduced by 92%. Viscose fibers completely lost their strength. At the same time, cotton fibers retained 50% of its initial strength. Cotton and viscose fabrics with a unit weight of 900 and 950 g m^{-2}, as a result of the soil test, showed a slower decomposition compared to that found in the case of raw cotton fibers. Higher resistance to mildew is shown by man-made fibers and fabrics. Polyacrylonitrile fibers after 240 days of testing presented a strength reduction of 5–13%; the elongation ability was also reduced (by 15–32%).

2.2 Systems and methods of textiles protection against biodeterioration

An ideal antimicrobial treatment of textiles should satisfy various requirements [30–32]: selective activity to undesirable microorganisms, low hazard (manufacturing, application, environment), durability to various cleaning and ironing processes, no deterioration of fabric quality, compatibility with textile chemical processes, and cost effectiveness. There are two basic methods to gain antimicrobial textile products: spinning antimicrobial (especially by electrospinning) fibers and treatment of fibers or fabrics in the finishing process; the latter comprises the most important one.

Antimicrobial agents can be applied to textile substrates, mainly fabrics, by exhaust, pad-dry cure, coating, spray, and foam techniques [33–36]. There are various methods to improve durability:

- Adsorption of the active agent on the fiber (e.g. ions, small molecules, colloids, nanoparticles, development of nanoparticles on the fiber by the sol–gel method); pretreatments of the fabric for the enhancement of adsorption

- Treating the fabric with carrier materials entrapping the antimicrobial agent (e.g. resin, zeolite, nanoparticles, microcapsules, cyclodextrin)
- Anchoring the antimicrobial agent or its carrier to the fiber by crosslinking (e.g. chlorotriazine or silane groups containing a crosslinker, chemicals of easy-care finishing)
- Application of polymeric antimicrobial material as a thin surface coating
- Development of antimicrobial groups directly on the fiber (e.g. quaternary ammonium groups by alkylation or high energy irradiation grafting).

Various chemicals can be used for the preparation of antimicrobial textiles. Nowadays, silver nanoparticles, quaternary ammonium compounds, chitosan, and active antimicrobial peptides (as a group of natural biocides) seem to be the most studied species. Silver is a representative of heavy metals, known for its toxicity for microorganisms. In ancient times, it was used to make water potable, while for centuries, it has been used for the treatment of wounds, etc.

Silver and other metals (and their ions) can be grafted to cellulosic fibers by various methods. For example, silver salt was incorporated in gel fiber used for wound healing [37]. In a study of Lee et al. [38], cotton fabric was padded with nanosized silver colloidal solution. In another study, socks simply impregnated with copper oxide were effectively used for wound healing on the feet of diabetic patients; copper fights against microorganism; moreover, copper is absorbed through the skin and thus improves the well-being of the skin [39]. Copper-based biocidal compounds have also been used as laundry additives: it has an antibacterial effect even at low temperature washing [40]. Cotton, pretreated with succinic acid anhydride, has been found to enhance the subsequent adsorption of silver and copper cations [41]. Metal ions embedded in various polymers have also been used on cellulosic fiber surfaces: alginate chains were crosslinked with copper ions on cotton fibers for dressing materials [42]; nanosilver was applied with polyvinyl alcohol [43]; poly(amidoamine) dendrimer was used with nanosilver on cotton/nylon blend fabric [44].

In protein fibers, the carboxyl groups are the active sites for metal cation adsorption. The adsorption capacity can be increased by various methods. Wool and silk, respectively, can be treated with a complexing agent such as tannic acid or ethylene diamine tetraacetic dianhydride. Modified fibers reacted easily with silver and copper [45—47]. Wool fabric treated with sulfur nanosilver colloid has been found to have various functionalities, such as mothproofing, antibiotic, and antistatic property [48]. In a study of Dubas et al. [49], silver nanoparticles were immobilized on silk fiber by following the layer-by-layer deposition method, exploiting the electrostatic interactions between oppositely charged polyelectrolytes; the following step involved dipping the fibers in diluted solutions of poly(diallyldimethylammonium chloride). Silver nanoparticles were capped with poly(methacrylic acid) (PMA), using an ultraviolet (UV) lamp; the reduction of silver ion by PMA led to the appearance of a pink color [49]. Silver-containing products are commercially available for wool [50].

Quaternary ammonium compounds (QACs) carry a positive charge at the N atom in solution. These chemicals, especially long alkyl group-containing species, such as cetylpyridinium chloride (bromide), have been used as disinfectants for a long time. QAC has been introduced into wool and silk fibers after the tosylation of protein fibers [51]. Thiol group-containing QAC can be adsorbed on wool by means of ionic and hydrophobic interactions, and it can also be linked covalently due its thiol functionality [52].

Apart from cellulose and protein fibers, there are some natural fibers believed inherently to be antimicrobial, such as kapok [53], hemp, flax, and bamboo. Essential oils of industrial hemp can significantly inhibit microbial growth, and its degree depends on variety and sowing time. [54].

However, as already mentioned, most natural fibers are sensitive to microbial assault. Hence, their protection is imperative. For this purpose, several methods have been developed; they are analyzed below.

2.2.1 Use of chemical biocides

Cooper fungicides, e.g. copper naphthenate, copper hydroxynaphthenate, copper 8-quinolinolate, in the 1940 and 1950s were widely used. In the 1960s, copper borate solubilized for aqueous application in a complex with zirconium was developed. Also, the tests on flax fibers and nonwovens protected by Cupramina B (a market product, on base of boron and copper compounds) have been conducted. In the case of these materials, a high efficiency was obtained—no mildew growth, no loss of breaking force, and no change in odor. Cupramina B is characterized by low toxicity for humans and warm-blooded animals and can be used in interiors where people stay permanently; thus, it finds application in protection of natural nonwovens and wood for automotive and building industries [55,56].

Studies have been carried out on fungi strains; the most frequently found ones as contaminants in these species are, i.e. *Aspergillus niger*, *Aspergillus ustus*, *Aspergillus flavus*, *Paecilomyces variotii*, *Trichoderma viride*, *Chaetomium globosum*, *Myrothecium verrucaria*, *Stachybotrys atra*, *Cladosporium herbarum*, *Penicillium chrysogenum*, and also their mixture. The following biocides have been also investigated: Econazole, Orthophenylphenol, Imazalil, and Thiabendazole. Orthophenylphenol and Imazalil at concentrations of 3% and 5%, respectively, have been found to be the most efficient with both individual fungi strains and the mixture of 10 tested strains. In addition, the effect of thiabendazole on paper properties has been positively evaluated, and no significant influence on the strength and level of oxidation has been observed [57].

2.2.2 Quaternary ammonium salts

Quaternary ammonium compounds (QACs) are low toxic effective in protection against biodeterioration. They are widely known to be bioactive and have a high antimicrobial activity. Anionic groups introduced into the cellulose backbone have been found to increase the QAC bonding ability of cellulose [58]. Antimicrobial polymer coating can be obtained on cotton surface using QAC-loaded alginate gel [59]. Polymers such as quaternary polyethyleneimines have also been used as antimicrobial agents on cotton and wool. Quaternary ammonium groups can be covalently bonded to cellulose by radiation-induced grafting as well [60,61]. An effective method of incorporating QACs in cellulose involves the use of ionic liquids. Ionic liquids usually consist of an organic cation and an inorganic or organic anion, which are liquids at a temperature below 100 °C and are nonvolatile even at 300 °C. An ionic liquid with a high thermal stability, high ionic conductivity, antielectrostatic and antimicrobial activity is 3-trimethoxy-silylpropyl-dimethyl-octadecyl-ammonium chloride. Ionic liquid application is also growing as a green solvent and biocide. The antimicrobial activity of ionic liquids strongly depends on their structure, especially of the anion type. $(NO_3)^-$ and $(NO_2)^-$ salts are very effective agents against bacteria and fungi [62]. Among quaternary ammonium compounds, quaternary silicones, such as 3-trimethoxy-silylpropyl-dimethyloctadecyl ammonium chloride, have been used as durable odor preventive agents for socks. However, it is <90% effective on bacteria and has limited activity against fungi [63].

The enhancement of cellulose antimicrobial stability can also be achieved through imidazolium salts with an array of N-substituted fatty acid side chains including versions with linear saturated chains, double bonds in different locations, and double bonds with different cis-trans geometries [61]. Imidazolium salts with C18 (stearic acid side chain has a melting point of 53.5° C and linoleic analog with a significant bend has a melting point of 46.8° C).

The research on ecofriendly antimicrobial finishing of natural textiles also includes silver nanoparticles. Silver, zinc, and also copper were effective against *Staphylococcus aureus* and *Klebsiella pneumoniae*—bacteria present on human skin.

2.2.3 2-Hydroxypropyl 3-piperasinyl quinoline methacrylate

2-Hydroxypropyl 3-piperasinyl Quinoline methacrylate (HPQM) is a biocide with a wide spectrum of activity against Gram-positive and Gram-negative bacteria, widely used for the protection of most thermoplastic polymers (including fibrous ones); its efficiency for the protection of natural fibers has been examined and compared to that of zinc dimethyl dithiocarbamate (TROYSAN-S82) and silver-substituted zeolite (ZEOMIC). The halo and plate-count-agar test showed that in the case of HPQM on a medium density polyethylene matrix, the % reduction of *Escherichia coli* and *S. aureus* bacteria reached a value of 99.9% [64].

2.2.4 Polyhexamethylenoguanidine copolymer

Modified cellulose fibers containing covalent bonded biocidal formulations based on polyhexamethylenoguanidine copolymer (PHMG) and its reactive derivatives are protected against *S. aureus*. This implies the possibility of using them in protective wear in health care. PHMG is a PHMG and guanidine copolymer of a relatively wide spectrum of biocidal activity. It belongs to the so-called AMMS Group (Antimicrobial Macromolecules), which is similar in structure and mode of action to natural proteins that fight infections i.e. host defence peptides (HDPS). Studies on the spectrum of activity of PHMG derivatives and their possible applications are also being conducted in Poland by the Institute for Engineering of Polymer Materials and Dyes.

2.2.5 Poly(hexamethylene biguanidine)

The commercial product on the base of cationic agents is a mixture of molecules of different sizes; it is used as a disinfectant. Poly(hexamethylene biguanidine) (PHMB) binds to cotton via weak ion—ion linkages with the carboxylate groups present on cotton; these bonds dissociate readily allowing the release of free PHMB. The attachment of PHMB on cotton can be increased by the introduction of anionic sites, e.g. by dyeing; however, due to strong ionic bonding, the release of PHMB from cotton becomes less facile; hence, antimicrobial efficiency decreases, a very large amount is needed [32,65] and potential bacterial resistance occurs.

2.2.6 Dyes as antimicrobial agents

The dye—bacteria relation has special importance. The two major classes of bacteria (Gram-positive and Gram-negative) are distinguished by their dye ability discovered by the Danish bacteriologist Gram at the end of the nineteenth century. Acridine dyes were used for battle wounds to prevent sepsis in both World Wars. After the discovery of penicillin, very effective antibiotics were used against bacteria, but nowadays, there is a global problem of microbial drug resistance; thus, antimicrobial dyes are becoming important again [66, 67]. Antimicrobial dyes belong to the following chemical groups: acridine, triphenylmethane, phenothiazine, azo, and fluorescence. They are used in dermatology; however, besides their strong antimicrobial activity, they have some other side effects as well [68]. In a study of Saji et al. [69], gentian violet (Crystal violet), the original Gram stain triphenylmethane type cationic dye was successfully used against Methicillin-resistant *S. aureus* after failing with some other agents [69]. However, recently synthesized antimicrobial anthraquinone dyes mostly contain a quaternary ammonium group [70—72].

2.2.7 TiO_2 nanoparticles

TiO_2 nanoparticles (mostly Degussa P-25, anatase—rutile powder) can be used for textile coating by applying dip-coat technology [32,73—77]. The antibacterial activity of TiO_2-coated fabric has been investigated under various conditions. UV radiation has been found to be more effective than white light. Although it is generally accepted that TiO_2 is bactericidally inactive in the absence of light, it has been found that the cellulose—TiO_2 composite possessed a bactericidal property in the presence and absence of light with different rates of cell destruction.

2.2.8 Triclosan (2,4,4′-trichloro-2′-hydroxydiphenyl ether)

Triclosan, used in many commercial products, breaks down to toxic dioxin; hence, it is banned in some European countries [78] and strongly criticized in the US as well. This criticism originated from experiments conducted on nursing mothers using Triclosan containing hygienic products. Triclosan and/or its metabolites were found in plasma and in breast milk.

2.2.9 Use of natural biocides

Among natural biocides, which have become the center of interest in recent years, the most significant ones are plant extract and oils, chitosan, and also antimicrobial peptides. Bacteriocidal and fungicidal properties of essentials oils have been known for ages. They find application especially in sanitary, cosmetic, pharmaceutical, food industries, and in plant protection. Active substances (alkaloids, flavonoids, terpenes, and tannins) present in essential oils such as oils of thyme, oregano, clove, sage, camomile, citrus (grapefruit), and mint are natural sources of antimicrobial properties [79]. The use of essentials oils from clove, cinnamon, oregano, and hemp as antimicrobial solutions in paper packaging have also been studied.

Chitosan is a deacetylated derivative of chitin, which comes from marine shells and molluscs. The antimicrobial activity of chitosan can be exploited along with its other functions such as dyeing improvement and antistatic and deodorant activity. Natural biocides include antimicrobial peptides (the so-called peptide antibiotics, AMPs). They are characterized by strong antimicrobial activity against a wide spectrum of microflora including bacteria, viruses, fungi, and some protozoan species. They are prevalent in nature as they can be found in most living organisms both proto- and eukaryotic. In human organisms, these peptides are an important element of the immune system. They develop on all body surfaces exposed to the environment, where they protect the organism against microorganisms that may penetrate. Their mechanism of activity relies on their ability to destroy cell membranes of microorganisms. Antibacterial peptides also have the ability to stimulate the immune system and the healing processes, suggesting their potential use for wound dressing in wounds prone to infections. Moreover, their natural origin implies that they would possibly exhibit low toxicity to people and environment.

Antimicrobial peptides as biocidal coating components for natural fibers can be applied in producing filtration, design and sterile materials, clothes and underwear for protection against contamination, in medicine, and defense. The peptides known for their antibacterial and antifungal activity are for example

- Cytropine 1 1—Gly-Len-Phe-Asp-Val-Ile-Lys-Lys-Val-Ala-Ser-Val-IIe-Gly-Gly-Len.NHz,
- Protegrine 1—Arg-Gly-Gly-Arg-Leu-Cys-Tyr-Cys-Arg-Arg-Arg-Phe-Cys-Val-Cys-Val-Gly-Gly-Arg-NHz,
- Temporine A—Phe-Len-Pro-Len-Ile-Gly-Arg-Val-Len-Ser-Gly-Ile-Len-NHz,
- Omiganane MBI-226—Ile-Len-Arg-Trp-Pro-Trp-Trp-Pro-Trp-Arg-Arg-Lys-NHz,
- Periganane MSI-78—Gly-Ile-Gly-Lys-Phe-Leu-Lys-Lys-Ala-Lys-Lys-Phe-Gly-Lys-AlaPhe-Val-Lys-Ile-Leu-Lys-Lys-NHz.

2.2.10 Chemical and genetic modification of natural fibers

Cellulose, the main component of natural fibers lignocellulosic, can be modified by chemical reaction, making it no longer susceptible to microbiological attack.

The chemical modification of natural fibers included for example the following methods:

- Acetylation: this method is based on the reaction of lignocellulosic material with acetic anhydride at elevated temperatures, usually without a catalyst. The acetylation strongly reduces hydrophilicity, change to hydrophobicity. In addition, such modification causes the bulking of the cell wall and renders the material less susceptible to biological decay. Biological protection is afforded by a lower equilibrium moisture content. Moreover, modified cell wall polymer chemistry so that the fibers are not easy recognized by enzymes [11,80,81].
- Mercerization: leads to an increase in the amount of amorphous cellulose at the expense of crystalline cellulose.
- Furfurylation: In a study of Kebony et al. [81], this process based on the treatment with furfuryl alcohol, enhanced the strength and durability of lignocellulosic materials. After treatment, the lignocellulosic fibers or wood is heated at a higher pressure for curing. The resultant lignocellulosic modified polymers became very stable and resistant to mildew and fungi.

The genetic modification of natural fibers is a new direction in the research area (e.g. production "in statu nascendi") of fibrous plants—alkanate, e.g. polyhydroxy butyric acid might allow for achieving more genetic resistance of cellulose fibers to the destructive action of microorganisms. At the moment, the results concerning the resistance of polyhydroxyalkanoates to mildew and bacteria are not available.

2.2.11 Bioavailability of compounds used for the protection of textile material from natural fibers

Biocides (bacteria, fungi including mildew, insects) applied in fibers and textiles may eventually migrate from textiles to the human body, posing their harmful activity on human physiology. An example of easy bioavailability materials are ZnO and Ag nanoparticles. The barrier effect of clothing is a key factor in protection against infection in many working environments, especially in hospitals. Bacteria are transported by carriers (vehicles) such as fluids or dust (air); hence, limitation or even prevention of vehicle transfer is an indirect protection against microbes.

2.3 Future trends

New regulations definitely limit the application of many commonly used biocides. Therefore, there is a growing demand for replacing old and not always environmentally safe compounds with new more ecological ones. Further to the implementation of EU Directive 98/8/EC on the introduction of biocidal products into the market and withdrawal of products based on substances exceptionally toxic to humans and the environment. The importance of ecological biocides is especially high because of the REACH system introduced in EU (Regulation (EC) No 1907/2006 of the European Parliament and of the Council of December 18, 2006, concerning the Registration, Evaluation, and Authorization of Chemicals).

A new class of antibacterial drugs is emerging from reviving an old antibiotic target [82]. An example is reexamining an established one-bacterial DNA gyrase, also known as type IIA topisomerase. This enzyme is critical in bacteria, where it cuts and reseals DNA to regulate a variety of important tasks. Quinolone antibiotics that target gyrase have been available since 1962. An antibiotic called GSK299423 that works against DNA gyrase in an entirely new way and could therefore exemplify a new class of antibiotics.

Silver, quaternary ammonium compounds, triclosan (in spite of its possible toxicity), and poly(hexamethylene biguanide) (PHMB) can be found as commercial products on the market. Valuable results on chitosan that might lead to the development of commercial products can already be found in the open literature. Dyes are common chemicals of the textile industry; hence, their possible application for antimicrobial finishing shows several advantages. Photomicrobiocidal agents are multifaceted materials, their properties and possible applications, like antimicrobial finishing, are the focus of recent research. Bandages for fighting pathogenic bacteria created phospholipid vesicles filled with microbiocide sodium azide (NaN_3). When exposed to toxins released by pathogenic bacteria, vesicles leak or burst [83].

Bursting vesicles easily knock *S. aureus* and *Pseudomonas aeruginosa*, and *E. coli*. The authors attached the vesicles to nonwoven polypropylene fabric. The aim of the work was to attempt to engineer a smart wound-dressing system that only releases an encapsulated antimicrobial agent in the presence of pathogenic bacteria without responding to harmless bacteria. The concept was based on the fact that the majority of pathogenic bacteria secrete virulence factors such as toxins and lipases that actively damage cell membranes, typically observed as tissue damage around infected wounds, while nonpathogenic bacteria do not have this effect [84].

3. CONCLUSIONS

Natural fibers such as lignocellulosics and protein are fully sustainable, renewable, and biodegradable. They have several advantages in comparison with man-made fibers and are widely used in the apparel sector and interior design. Unfortunately, they are flammable and not resistant to biodegradation. For this reason, their use is very often limited. For improving flame retardancy, delaying the time to ignition, decreasing HRR and surface spread of the flame, as well as significantly reducing fire propagation and release of smoke and gases are of major importance. There are three main ways to protect natural fibers against flammability. The methods described in this chapter include nondurable finishes that are very often based on simple compounds like ammonium phosphate, ammonium polyphosphate, melamine phosphate, and melamine borate. Durable flame retardants, which make natural textiles flame resistant up to 50 washes, are mainly Pyrowatex CP and THPC Proban. The mentioned flame retardants are called reactive systems of protection. A new development in the flame retardancy of lignocellulosic fibers is based on the chemical reaction between hydroxyl groups of cellulose and reactive polyphosphate, polyborate, or polyperoxymolybdate. Another promising method is the application of intumescent coatings that were mainly developed for backcoating of upholstery furniture, curtains, and draperies. In the case of protein fibers and leather, which ignite and burn with more difficulty than do cellulosic fibers, the most popular durable system is Zirpro that is based on the exhaustion of negatively charged complexes of titanium or zirconium onto positively charged protein fibers under acid conditions.

A very important matter is also the protection of natural cellulosic and protein fibers against biodeterioration, because they are susceptible to it. The decomposition of natural fibers can occur mainly in high humidity conditions at a relevant temperature in the presence of bacteria and fungi. The protection of these fibrous materials can be obtained by using nanocopper, nanosilver, quaternary ammonium salts (QACs), 2-hydroxypropyl-3-piperazinyl quinoline carboxylic acid methacrylate (HPQM), PHMG, some dyes, widely used compounds such as triclosan, chitosan, and lately developed antimicrobial peptides. Among the future trends are chemical functionalization of cellulosic fibers and genetic modification.

Unfortunately, enhancing the flame retardancy of natural fibers and making them resistant to biodeterioration create up to now significant problems related to toxicity and increased bioavailability of chemicals, which can be hazardous to human health and the environment. The awareness of these problems resulted in the creation of a controversial opinion that flame retardants and biocides used in consumer goods are toxic and ineffective, and some voices even demanded a ban on certain flame retardants and some biocides. These voices bring signals to researchers that they should pay more attention to the search for new safer flame retardants and biocides.

References

[1] Czaplicki Z, Kozlowski R. Properties and structure of Polish Alpaca wool. 7th International Conference of Textile Research Division. Cairo, Egypt: National Research Centre; 2010.

[2] Zimniewska M, Kozłowski R. Natural and man-made fibres and their role in creation of physiological parameters of human body. In: Molecular crystals and liquid crystals, vol. 418. Taylor & Francis; 2004. pp. 113–30.

[3] Kozlowski R, Muzyczek M. Improving the flame retardancy of natural fibres. In: Kozlowski R, editor. Handbook of natural fibres: processing and applications, vol. 2. Cambridge: Woodhead Publishing Limited; 2011. pp. 30–62.

[4] Kozlowski R, Mieleniak B, Muzyczek M, Kubacki A. Flexible fire barriers based on natural nonwoven textiles. Fire Mater 2002;(26):243–6.

[5] Kozlowski R, Muzyczek M, Mieleniak B. Upholstery fire barriers based on natural fibres. In: Goloshchapov A, Zaikov G, Ivanov V, editors. Essential results in chemical physics and physical chemistry. N.Y.: Nova Science Publishing; 2005. pp. 87–97.

[6] Kozlowski R, Muzyczek M, Mieleniak B, Wesolek D, Przybylak M. Available flame retardant treatment for natural fibres. General Assembly of 6 EU FP Multifunctional

barrier for flexible structure (textile, leather and paper) Flexifunbar, March 10–11, Poznan; 2005.
[7] Weil E, Yang Ch. Textile flame retardancy—state of the art and recent progress. Stamford: BCC Conference on Flame Retardancy; 2008.
[8] Kozłowski R, Helwig M. Progress in fire retardants for lignocellulosic materials. Proceedings of the 6th Arab International Conference on Materials Science: Plastics Materials For Agriculture Applications. Alexandria; 2000.
[9] Kozłowski R, Muzyczek M, Mieleniak B. Flexible non-woven upholstery composite based on wool and FR bast fibers. Conference on Natural Fibres: Vision 2020, December, 8–9. New Delhi; 2006.
[10] Horrocks AR. Safer textiles through flame retardant treatment: application of intumescents. 5th Arab International Conference of Materials Science: Materials and Fire. Alexandria; 1998.
[11] Rowell RM. In: Hon DNS, editor. Chemical modification of nonwood lignocellulosics. New York: Chemical Modification of Lignocellulosic Materials; 1996. pp. 229–46 [Chapter 9].
[12] Kozlowski R, Wladyka-Przybylak M. Natural polymers, wood and lignocellulosic materials. In: Horrocks AR, Price D, editors. Fire retardant materials. Cambridge: Woodhead Publishing Ltd; 2001. pp. 83–93 [Chapter 9].
[13] Kozlowski R, Muzyczek M. Flammability of wool and its fireproofing problems. International Conference of Wool. Sliven; 1977.
[14] Kozłowski R, Kicińska-Jakubowska A, Muzyczek M. Types of natural textiles used in interiors. In: Rowe T, editor. Interior textiles: design and developments. Cambridge: Woodhead Publishing Ltd; 2009. pp. 3–38.
[15] Wool blend mark pure new wool. IWS flame resist treatments. Leaflet of Inrenational Wool Secretary; 2000.
[16] Bourbigot S, Duquesne S, Bellayer S, Flambard X, Rochery M, Devaux E. Advances in the flame retardancy of the polymeric materials. Current perspectives presented at FRPM'05. In: Schartel B, editor. Novel developments in flame retardancy of textiles. Norderstedt: Broschiert – Books on Demand GmbH (BoD); 2007. pp. 159–80.
[17] Bourbigot S, Duquesne S, Flambard X, Bellayer S, Rochery M, Giraud S, et al. Flame retardancy of textiles: recent advances. FRPM'05, BAM, Berlin; 2005.
[18] Horrocks AR, Kandola BK. Flame retardant cellulose textiles. 6th European meeting on fire retardancy of polymeric materials. Lille; 1997.
[19] Guan JP, Chen GQ. Flame retardancy finish with an organophosphorus retardant on silk fabrics. Fire Mater 2006;30(6):415–24.
[20] Kozlowski R, Mieleniak B, Muzyczek M, Fiedorow R. Flammability and flame retardancy of leather. Leather Int; November 2006:24–6.
[21] Hjohlman M, Anderson P. Cone calorimeter testing of INCA leather and flax products. Test SP Raport within Flexifunbar Project. Boras: SP Technical Research Institute of Sweden; 2008.
[22] Babrauskas V. From Bunsen burner to heat release rate calorimeter [Chapter 2]. In: Babrauskas V, Grayson SJ, editors. Heat release in fires. London: E & FN Spon; 1992. pp. 7–29.
[23] Hall GS, Saunders RC, Allcorn RT. Fire tests methods and standards. In: The fire performance of timber—a literature survey. Timber Research and Development Association; 1980.

[24] Lewin M. Fire retardancy: challenges of the past, present and future. FRPM '05 BAM Conference. Berlin; September 7–9, 2005.
[25] Commission directive 2001/58/EC. Off. J. Eur. Communities.
[26] DiGangi J, Blum A, Bergman A, de Wit C. The San Antonio statement on brominated and chlorinated flame retardants. Environ Health Perspect 2010;118/12:516.
[27] Rochery M, Talk M, Giraud S, Lewandowski M, Devaux E, Bourbigot S, et al. Optimization of the structure of a polyurethane clay nanocomposite used as flame-retardant textile coating. FRPM' 05 BAM Conference. Berlin; September 7–9, 2005.
[28] Vromaz I, Dumont V, Giraud S, Saihi D, Bourbigot S. Nanoencapsulation of ammoniun phosphate as fire retardant for textile: effect of surfactants on the properties of nanocapsules. FRPM' 05 BAM Conference. Berlin; September 7–9, 2005.
[29] Szostak-Kotowa J. Biodeterioration of textiles. Int Biodeterior Biodegrad 2004;53: 165–70.
[30] Purwar R, Joshi M. Recent developments in antimicrobial finishing of textiles—a review. AATCC Rev 2004;4:22–6.
[31] Ramachandran T, Rajendrakumar K, Rajendran R. Antimicrobial textiles—an overview. Inst Eng (I) J-TX. Available from: http://www.ieindia.org/publish/tx/0204/feb04tx4.pdf; 2004 [accessed 09.03.10].
[32] Gao Y, Cranston R. Recent advances in antimicrobial treatments of textiles. Text Res J 2008;78:60–71.
[33] Anon. New multifunctional textiles: antimicrobial treatments. Available from: http://ebook-free-downloads.com/ebook-ppt-free-ppt-download-centrum-page-2.htm; 2005.
[34] Kozlowski R, Walentowska J. Antifungal agent on linen-cotton blended fabric and leather. Flexifunbar Project, 24.09.08 Lille, Final Meeting.
[35] Abdel-Halim ES, Fouda MMG, Hamdy I, Abdel-Mohdy FA, El-Sawy SM. Incorporation of chlorohexidin diacetate into cotton fabrics grafted with glycidyl methacrylate and cyclodextrin. Carbohyd Polym 2010;79:47–53.
[36] Newman GR, Walker M, Hobot JA, Bowler PG. Visualisaion of bacterial sequestration and bactericidal activity within Hydrofiber® wound dressings. Biomaterials 2006;27: 1129–39.
[37] Lee HJ, Yeo YS, Jeong SH. Antibacterial effect of nanosized silver colloidal solution on textile fabrics. J Mat Sci 2003;38:2199–204.
[38] Borkow G, Zatcoff RC, Gabbay J. Reducing the risk of skin pathologies in diabetics by using copper impregnated socks. Med Hypothesis 2009;73:883–6.
[39] Hall TJ, Wren MWD, Jeanes A, Gant VA. Decontamination of laundry at low temperature wit $CuWB_{50}$, a novel copper-based biocidal compound. Am J Infect Control 2009; 37:478–83.
[40] Nakashima T, Sakagami Y, Ito H, Matsuo M. Antibacterial activity of cellulose fabrics modified with metallic salts. Text Res J 2001;71:688–94.
[41] Grace M, Chand N, Bajpai SK. Copper alginate–cotton cellulose (CACC) fibers with excellent antibacterial properties. J Eng Fibers Fabr 2009;4:24–34. Available from: http://www.jeffjournal.org [accessed 18.01.10].
[42] Wasif AI, Laga SK. Use of nano silver as antimicrobial agent for cotton. AUTEX Res J 2009;9:5–13. Available from: http://www.autexrj.org [accessed 18.01.10].
[43] Ghosh S, Yadav S, Vasanthan N, Sekosan G. A study of antimicrobial property of textile fabric treated with modified dendrimers. J Appl Polym Sci 2010;115:716–22.

[44] Arai T, Freddi G, Colonna GM, Scotti E, Boschi A, Murakami R, et al. Absorption of metal cations by modified *B. mori* silk and preparation of fabrics with antimicrobial activity. J Appl Polym Sci 2001;80:297–303.

[45] Freddi G, Arai T, Colonna GM, Boschi A, Tsukada M. Binding of metal cations to chemically modified wool and antimicrobial properties of the wool–metal complexes. J Appl Polym Sci 2001;82:3513–9.

[46] Tsukada M, Arai T, Colonna GM, Boschi A, Freddi G. Preparation of metal-containing protein fibers and their antimicrobial properties. J Appl Polym Sci 2003;89:638–44.

[47] Ki HY, Kim JH, Kwon SC, Jeong SH. Study of multifunctional wool textile treated with nano-sized silver. J Mat Sci 2007;42:8020–4.

[48] Dubas ST, Kumlangdudsana P, Potiyaraj P. Layer-by-layer deposition of antimicrobial silver nanoparticles on textile fibers. Colloid Surf A Physicochem Eng Asp 2006;289: 105–9.

[49] NanoHorizons' Smart Silver Guards Wool Fabrics Against Microbes that Cause Odor. Available from http://www.azonano.com/news.asp?NewsID=7080 [Accessed 13.03.10].

[50] Abel T, Cohen JLI, Escalera J, Engel R, Filshtinskaya M, Fincher R, et al. Preparation and investigation of antibacterial protein-based surfaces. JTATM 2003;(2):3. Available from: http://www.tx.ncsu.edu/jtatm/volume3issue2/Articles/Engel/engel_full_55_03.pdf; 2003 [accessed 18.01.10].

[51] Diz M, Infante MR, Erra P. Antimicrobial activity of wool treated with a new thiol cationic surfactant. Text Res J 2001;71:695–700.

[52] Wollina U, Heide M, Müller-Litz W, Obenauf D, Ash J. Functional textiles in prevention of chronic wounds, wound healing and tissue engineering. In: Elsner P, Hatch K, Wigger-Alberti W, editors. Textiles and the skin. Curr Probl Dermatol, vol. 31. Basel: Karger; 2003. pp. 82–97. Available from: http://www.karger.com/gazette/67/elsner/wollina.pdf [accessed 23.01.10].

[53] Nissen L, Zatta A, Stefanini I, Grandi S, Sgorbati B, Biavati B, et al. Characterization and antimicrobial activity of essential oils of industrial hemp varieties (*Cannabis sativa* L.). Fitoterapia 2010;81:413–9.

[54] Kozlowski R, Walentowska J. Efficiency of biocides in protection of lignocellulosic non-wovens against biodeterioration. Ann Warsaw Agric Univ, For Wood Technol 2006;59:6–11.

[55] Walentowska J, Kozlowski R. Protection of natural fibres against biodeterioration caused by mildew. Corrosion Protection 9s/A/2006. pp. 328–31.

[56] Rakotonirainy MS, Fohrer F, Flieder F. Research on fungicides for aerial disinfection by thermal fogging in libraries and archives. Int Biodeterior Biodegrad 1999;44: 133–9.

[57] Son YA, Kim BS, Ravikumar K, Lee SG. Imparting durable antimicrobial properties to cotton fabrics using quaternary ammonium salts through 4-aminobenzenesulfonic acid-chloro-triazine adduct. Eur Polym J 2006;42:3059–67.

[58] Kim HW, Kim BR, Rhee YH. Imparting durable antimicrobial properties to cotton fabrics using alginate-quaternary ammonium complex nanoparticles. Carbohyd Polym 2010;79:1057–62.

[59] Kumar V, Bhardwaj YK, Rawat KP, Sabharwal S. Radiation induced grafting of vinylbenzyltrimethylammonium chloride (VBT) onto cotton fabric and study of its antimicrobial activities. Rad Phys Chem 2005;73:175–82.

[60] Ritter SK. New Chains For Ionic Liquids. Chem Eng News March 15 2010;88:11.

[61] Pernak J, Śmiglak M, Griffin ST, Hough WL, Wilson TB, Pernak A, et al. Long alkyl chain quaternary ammonium-based ionic liquids and potential applications. Green Chem 2006;8:798–806.
[62] Sivaramakrishnan CN. Antimicrobial finishes. Colourage; September 2007:50–4.
[63] Chammanee P, Sombatsompop K, Kositchaiyong A, Sombatsompop N. Effects of Anti-Bacterial Agents, Sample Preparation and Contact Time on AntiBacterial Efficacy in MDPE Film. J Macromol Sci Part B July 2009;48(4):755–65.
[64] Kawabata A, Taylor JA. The effect of reactive dyes upon the uptake and antibacterial action of poly(hexamethylene biguanide) on cotton. Part 2: uptake of poly(hexamethylene biguanide) on cotton dyed with β-sulphatoethylsulphonyl reactive dyes. Dyes Pigm 2006;68:197–204.
[65] Kawabata A, Taylor JA. The effect of reactive dyes upon the uptake and antibacterial action of poly(hexamethylene biguanide) on cotton. Part 3: reduction in the bactericidal efficacy of poly(hexamethylene biguanide) on cotton dyed with bis(monochlorotriazinyl) reactive dyes. Carbohyd Polym 2007;67:375–89.
[66] Wainwright M. Dyes in the development of drugs and pharmaceuticals. Dyes Pigm 2008;76:582–9.
[67] Balabanova M, Popova L, Tchipeva R. Dyes in dermatology. Clin Dermatol 2003;21: 2–6.
[68] Saji M, Taguchi S, Uchiyama K, Osono E, Hayama N, Ohkuni H. Efficacy of gentian-violet in the eradication of methicillin-resistant *Staphylococcus aureus* from skin lesions. J Hosp Infect 1995;31:225–8.
[69] Ma M, Sun Y, Sun G. Antimicrobial cationic dyes: part 1: synthesis and characterization. Dyes Pigm 2003;58:27–35.
[70] Liu S, Ma J, Zhao D. Synthesis and characterization of cationic monoazo dyes incorporating quaternary ammonium salts. Dyes Pigm 2007;75:255–62.
[71] Liu J, Sun G. The synthesis of novel cationic anthraquinone dyes with high potent antimicrobial activity. Dyes Pigm 2008a;77:380–6.
[72] Liu J, Sun G. The biocidal properties of anthraquininoid dyes. Dyes Pigm 2009;81: 231–4.
[73] Bozzi A, Yuranova T, Guasaquillo I, Laub D, Kiwi J. Self-cleaning of modified cotton textiles by TiO_2 at low temperatures under daylight irradiation. J Photochem Photobiol A 2005;174:156–64.
[74] Bozzi A, Yuranova T, Kiwi J. Self-cleaning of wool–polyamide and polyester textiles by TiO_2-rutile modification under daylight irradiation at ambient temperature. J Photochem Photobiol A 2005;172:27–34.
[75] Meilert KT, Laub D, Kiwi J. Photocatalytic self-cleaning of modified cotton textiles by TiO_2 clusters attached by chemical spacers. J Mol Catal A Chem 2005;237:101–8.
[76] Yuranova T, Rincon AG, Pulgarin C, Laub D, Xantopoulos N, Mathieu HJ, et al. Performance and characterization of Ag-cotton and Ag/TiO_2 loaded textiles during the abatement of *E. coli*. J Photochem Photobiol A Chem 2006;181:363–9.
[77] Kangwansupamonkon W, Lauruengtana V, Surassmo S, Ruktanonchai U. Antibacterial effect of apatite-coated titanium dioxide for textile application. Nanomedicine 2009;5:240–9.
[78] Environmental Working Group. Pesticide in soap, toothpaste and breast Milk—is it kid-safe? Available from: http://www.ewg.org [accessed 19.01.10].
[79] Rios JL, Recio MC. Medicinal plants and antimicrobial activity. J Ethnopharmacol 2005;100:80–4.

[80] Pott GT. Natural fibres with low moisture sensitivity. In: Wallenberger FT, Weston NE, editors. Natural fibres, plastics and composites. Kluwer Academic Publishers; 2004. pp. 106–10 [chapter 8].
[81] What is Kebony? Available from: http://www.kebony.com.
[82] Drahl C. Reviving an old antibiotic target. Chem Eng News 2010;88(33):11.
[83] Halford B. Bandages for fighting bad bacteria. Sci Technol Concentrates 2010;88(10):37.
[84] Zhou J, Loftus AL, Mulley G, Jenkins, Toby A. Thin film detection/response system for pathogenic bacteria. J Am Chem Soc 2010;132:6566–70.

CHAPTER

Influence of Fire Retardants and Nanofillers on Fire Toxicity

24

A.A. Stec, T.R. Hull

Centre for Fire and Hazard Science, University of Central Lancashire, Preston, UK

CHAPTER OUTLINE

1. Introduction

Most fire deaths and most fire injuries result from the inhalation of toxic smoke [1], both of which have increased with the widespread use of plastic materials. Fire toxicity is most important in areas where escape is restricted. Thus, most mass transport applications, such as airlines, railways, and passenger ships include requirements

Polymer Green Flame Retardants. http://dx.doi.org/10.1016/B978-0-444-53808-6.00024-X

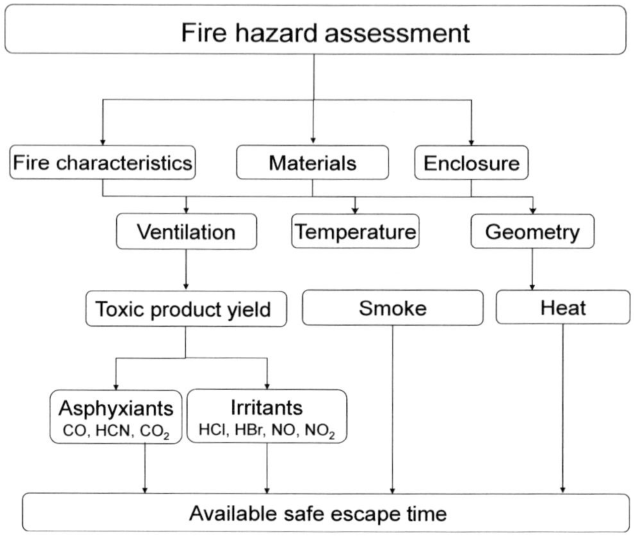

FIGURE 1

Schematic of factors required for fire hazard assessment relating to fire toxicity.

to quantify the fire toxicity of internal components. Fire hazard assessment requires consideration of the most probable fire scenarios, prediction of the rate of fire growth, the amount of fuel present, and its impact on the occupants and their ability to escape safely [2]. Figure 1 shows a schematic relationship between the factors required to assess the fire hazard, in order to ensure that the time required for escape is less than the time before escape becomes impossible (available safe escape time) [3].

The burning of an organic material produces a cocktail of products that vary with fire conditions. In addition to carbon dioxide (CO_2), water (H_2O), and oxygen (O_2) depletion these include asphyxiant gases, carbon monoxide (CO), and hydrogen cyanide (HCN); irritant gases, hydrogen chloride (HCl), hydrogen bromide (HBr), nitrogen oxides (NO, NO_2), and organoirritants, such as acrolein and formaldehyde, and particulates. They can cause death directly by asphyxiation or indirectly by inhibiting breathing (gaseous irritants and particulate irritants) or by visual obscuration (smoke), in each case preventing escape.

Prediction of toxic fire hazard depends on two parameters:

1. Time–concentration profiles for major products. These depend on the fire growth curve and the yields of toxic products.
2. Toxic potency of the products, based on estimates of doses likely to impair escape efficiency, cause incapacitation, or death.

While some real-life fires may be represented by a single fire stage, most fires progress through several different stages [4]. Burning behavior and particularly toxic product yields depend most strongly on a few factors, especially material composition, temperature, and oxygen concentration [5]. The generalized development of a fire has been recognized, and used to classify fire growth into a number of stages, from smoldering combustion and early well-ventilated flaming, through to fully developed underventilated flaming [6].

1.1 Fire statistics

Over the last 50 years, there has been a continuous change in the materials used for construction and content of buildings and transport with an increasing dependence on synthetic polymers and their composites with different reaction-to-fire properties. Compared with natural materials (wood, wool, cotton, leather, etc.), widely used synthetic polymers such as polyethylene (PE), polypropylene (PP), polyamide (PA), polystyrene (PS), and polyurethane (PU) burn more quickly, and generate more smoke and toxic effluents particularly in low-density, open structures such as fabrics and foams. Although the overall number of deaths has decreased [1] as smoke alarms and fire detection systems became available, UK Fire Statistics, with the best time series data in the world, show a progressive shift in the cause of death from "burns" to "overcome by toxic gas or smoke" from 1955 to 2008. More remarkable is the very large rise in fire toxicity injuries [1]. These are shown in Chapter 4 in Figures 1 and 2.

In Europe, most victims are found in the room of fire origin, often close to an exit, where they collapsed trying to escape. Usually, these are living rooms or bedrooms, with upholstery or bedding being the first thing to catch fire. EU directives for zero energy homes (e.g. the Passive House level (2012—2015) and the Zero Carbon level (2016—2020)) may result in millions of homes being lined in flammable insulation materials such as PS and PU foam (PUR) or equally flammable "environmental" materials such as flax or shredded paper, and this may have a dramatic impact on domestic fire deaths.

1.2 Fire scenarios and combustion conditions

Toxic product yields depend on the interactions between the material composition and the fire conditions; particularly temperature and ventilation and whether decomposition is nonflaming or flaming [7]. As the fire develops, the conditions change: the temperature increases and oxygen concentration decreases. The temperature and oxygen concentration vary significantly during a fire, and between different fires. Different fire scenarios are shown in Figure 2.

The most important fire stages identified by the ISO, from nonflaming to well-ventilated flaming to underventilated flaming, have been classified in terms of heat flux, temperature, oxygen availability, and CO_2 to CO ratio, equivalence ratio φ and combustion efficiency (the % conversion of fuel to fully oxygenated products, such as CO_2 and water).

The most significant differences in terms of toxic product yields with fire conditions arise between flaming and nonflaming combustion. Typically, nonflaming combustion produces small volumes of highly toxic, partially burnt effluent, well-ventilated fires produce larger volumes of less toxic effluent, while developed underventilated fires

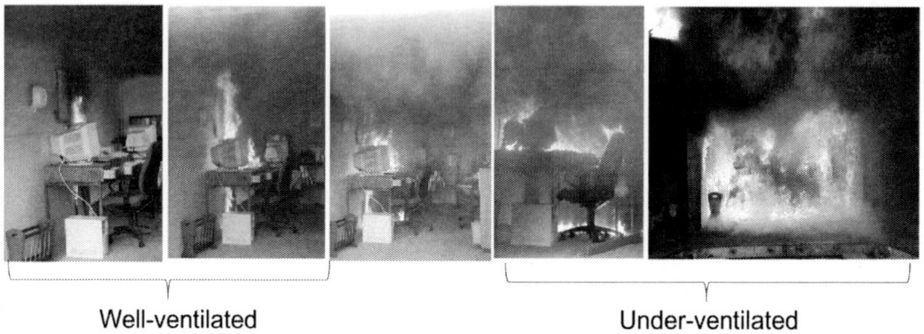

FIGURE 2

Stages of fire growth.

produce large volumes of toxic effluent, and therefore represent the greatest hazard. For flaming combustion, the fuel/air ratio has the greatest effect on the yields.

Research predicting the carbon monoxide evolution from flames of simple hydrocarbons, reviewed by Pitts [8], has shown the importance of the equivalence ratio, ϕ, for predicting the CO yield from the oxygen depletion in flaming conditions, presented in Table 1.

Stoichiometric combustion describes burning a material in its nominal chemical oxygen requirement ($\phi = 1$) providing just enough oxygen for full oxidation to CO_2 and H_2O. In practice, for condensed phase fuels, very little air usually results in incomplete combustion because mixing is never perfect. In well-ventilated fires, combustion is fuel lean with excess air (ϕ between 0.5 and 0.75), while in the later stages of a fire, when there is not enough air available and the conditions are underventilated, the equivalence ratio will be typically between 1.5 and 2.5. Each fire stage has a characteristic temperature and equivalence ratio, as summarized in Table 2.

Table 1 Equivalence Ratio

$$\phi = \frac{\text{actual fuel to air ratio}}{\text{stoichiometric fuel to air ratio}}$$

$\phi = 1$	"Stoichiometric" combustion
$\phi \ll 1$	Well-ventilated fires (fuel lean flames)
$\phi > 1$	Underventilated fires (fuel-rich flames)

Table 2 Classification of Fire Stages

ISO Fire Stage	Temperature/°C	Equivalence Ratio φ	CO_2/CO Ratio
1a. Oxidative pyrolysis	350	Not applicable	1–5
2. Well-ventilated flaming	650	$\phi < 0.75$	2–20
3. Under ventilated flaming:			
3a. Small fires	650	$\phi > 1.5$	2–20
3b. Post-flashover fires	825	$\phi > 1.5$	2–20

Fire can also be characterized by the CO_2/CO ratio as an indicator of combustion efficiency. However, this only correlates with the equivalence ratio (and degree of ventilation) when there is no gas phase quenching e.g. by hydrogen halides. Therefore, this parameter also corresponds to combustion efficiency.

1.3 Experimental methodology

Ideally, all bench-scale fire effluent toxicity test methods should be capable of reproducing the conditions in each of the stages of actual fires, including incipient, growing, and fully developed fires. It is therefore essential for the assessment of toxic hazard from fire that each fire stage can be adequately replicated, and preferably the individual fire stages treated separately [2,9].

Several test methods have been used to generate products for the purpose of evaluating the toxic product yields from burning polymers. However, many of them fail to relate this to a particular fire scenario [2,10,11]. In addition, although room- and larger-scale fire tests have been also conducted and the results published [4,12–14], only a few of these have attempted to segregate the fire stages, allowing the complexities of full-scale burning behavior to be addressed using a bench-scale model.

The steady-state tube furnace, ISO 19700 [15,16], has been developed specifically to replicate individual fire stages by decomposing materials under the full range of fire conditions from oxidative pyrolysis to fully developed, underventilated flaming [17,18]. By controlling the fuel feed rate and air flow into a tubular furnace, steady burning can be achieved for different fuel/air ratios even forcing combustion in oxygen-depleted atmospheres [19]. The apparatus is shown in Figure 3.

The apparatus typically consists of a tube furnace (600- to 800-mm heating zone) and a quartz tube (1600–1700 mm) that passes through the furnace and into a mixing and measurement chamber (27–30-l capacity). The standard procedure uses around 20 g of material uniformly placed in a silica boat (800 mm) to give a

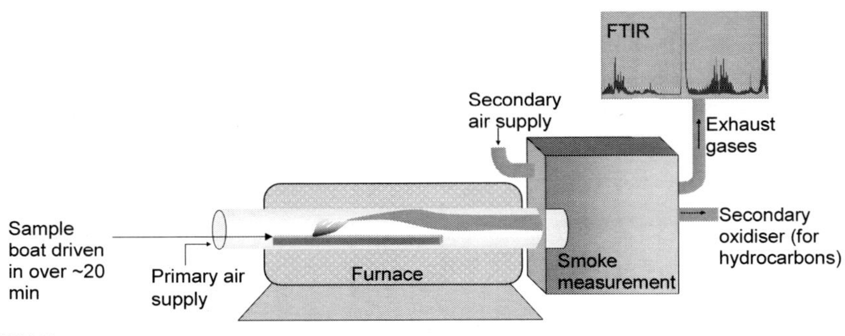

FIGURE 3

The steady-state tube furnace, ISO 19700.

linear density of 25 mg mm^{-1}. A drive mechanism pushes the specimen boat into the furnace tube at a rate of 40 mm min^{-1} to give a fuel introduction rate of 1.00 g min^{-1}. A constant stream of primary air is provided at the furnace tube entry, and secondary air is supplied into the mixing chamber to give a total air flow through the apparatus of 50 l min^{-1}. This provides a steady-state mass charge concentration of 20 mg l^{-1} (or 20 g m^{-3}) in the chamber. After dilution with secondary air, this effluent is equivalent to a fuel mass of 1 kg in 50 m^3 in a room. For materials that leave a residue or form a char, the residual mass is measured, and product yields based on mass loss concentrations may be also calculated. The requirement in each test run is to obtain a steady state of at least 5 min during which the concentrations of effluent gases and particulates can be measured. A light/photo cell system is used to determine smoke density across the mixing and measurement chamber. Organics (unburnt and partially burnt hydrocarbons) are determined as products of incomplete combustion using a secondary oxidizer for further oxidation at 900 °C in excess air, over silica wool, as the difference between secondary CO_2 and primary CO and CO_2 measured using a nondispersive infrared analyzer. The toxicity of organic species in the fire effluent may be quantified as a ratio of the actual organic yield to the organic yield of 10 mg l resulting in incapacitation, as described by Purser [4].

All scale-up of fire is difficult, but particularly in combustion toxicity where product yields may differ by two orders of magnitude, depending on fire scenarios [14]. It was shown that for many common materials, the yield of toxic products such as carbon monoxide, hydrogen cyanide, organoirritants, and smoke increases by a factor between 10 and 50 as the fire changes from well ventilated ($0.5 < \phi < 0.7$) to underventilated ($1 < \phi < 5$). The steady-state tube furnace has been shown to replicate a range of large-scale fire stages or conditions, characterizing the fire behavior of materials under controlled and well-defined laboratory conditions, in terms of the equivalence ratio (ϕ) or the CO_2/CO ratio [20]. Further, since each test run represents the burning behavior for a particular fire stage, the results are better defined than those of a single large-scale test, where individual fire stages may coexist so the transition is indistinguishable [21–23]. An example for CO from PP is presented in Figure 4.

The large-scale test data (from the ISO room corner test) show a good agreement with the steady-state tube furnace data; a significant improvement on any other toxicity test. In well-ventilated conditions, both the tube furnace and the large-scale fire give CO yields of around 0.02–0.03 g g^{-1}, then following the same rising trend to 0.1 g g^{-1} at $\phi = 1.3$ in the large-scale test and 0.17 g g^{-1} at $\phi = 1.5$ in the tube furnace and for underventilated flaming.

1.4 Toxic potency of fire effluent

Death or incapacitation may be predicted by quantifying the fire effluents in different fire conditions in small-scale tests, using chemical analysis. Lethality may be predicted using equations, based on rat lethality data, presented in ISO

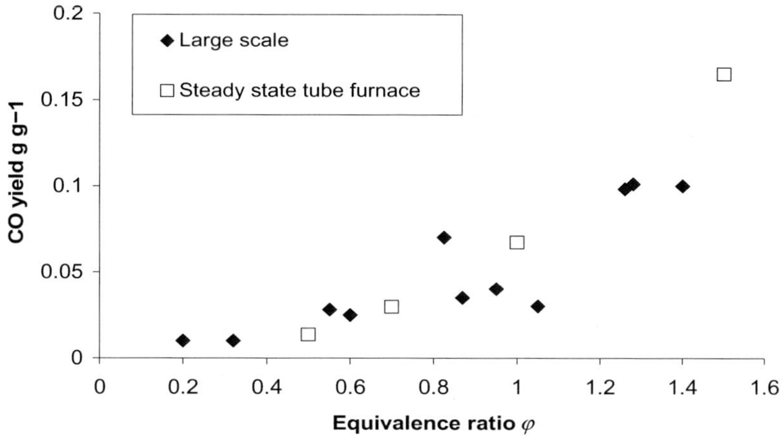

FIGURE 4

Comparison of tube furnace CO yields with large scale data for polypropylene [2].

13344 [24]. Incapacitation (the inability to effect one's own escape) may be predicted using methodology and consensus estimate data in ISO 13571 [25].

Toxic product data from chemical analysis may be expressed in various ways, including effluent gas concentrations, effluent gas yields, toxicity indices, Fractional Effective Dose (FED), Fractional Effective Concentration (FEC), and LC_{50} (lethal concentration to 50% of the population). The general approach in generating toxic potency data from chemical analysis is to assume additive behavior of individual toxicants, and to express the concentration of each as its fraction of the lethal concentration for 50% of the population for a 30-min exposure (LC_{50}) [4,24]. Thus, an FED = 1 indicates that the sum of the actual/lethality concentration ratios of individual species will be lethal to 50% of the population over a 30-in exposure. Since CO_2 increases the respiration rate, the Purser model, presented in Eqn (1), uses a multiplication factor for CO_2 driven by hyperventilation, V_{CO_2}, to increase the FED contribution from all the toxic species, and incorporates an acidosis factor A to account for toxicity of CO_2 in its own right [24].

$$\mathrm{FED} = \left\{\frac{[\mathrm{CO}]}{\mathrm{LC}_{50,\mathrm{CO}}} + \frac{[\mathrm{HCN}]}{\mathrm{LC}_{50,\mathrm{HCN}}} + \frac{[\mathrm{HCl}]}{\mathrm{LC}_{50,\mathrm{HCl}}} + \frac{[\mathrm{NO_2}]}{\mathrm{LC}_{50,\mathrm{NO_2}}} + \ldots + \mathrm{organics}\right\} \times V_{\mathrm{CO_2}} + A + \frac{21-[\mathrm{O_2}]}{21-5.4}, \qquad (1)$$

$$V_{\mathrm{CO_2}} = 1 + \frac{\exp(0.14[\mathrm{CO_2}]) - 1}{2},$$

where A is an acidosis factor equal to $[\mathrm{CO_2}] \times 0.05$.

Equation (1) is the Purser model for estimation of toxicity of fire effluents.

This equation is related only to lethality, or "cause of death". However, many people fail to escape from fires because of the incapacitating effect of smoke (obscuring visibility) and its irritant components that cause pain, inhibiting breathing, actually being the "reason for death".

ISO 13571 considers the four major hazards from fire that may prevent escape (toxic gases, irritant gases, heat, and smoke obscuration). It includes a separate calculation for prediction of incapacitation by each of the four hazards for humans exposed to fire effluents (indicating, in a nonnormative appendix that the effects of heat, smoke, and toxicants may be estimated independently).

Equations (2) and (3) have been taken from ISO 13571. They calculate the FED of asphyxiants, CO and HCN, and the FEC of sensory irritants in the fire effluent that limit escape.

$$\mathrm{FED} = \sum_{t_1}^{t_2} \frac{[\mathrm{CO}]}{35,000} \Delta t + \sum_{t_1}^{t_2} \frac{\exp([\mathrm{HCN}]/43)}{220} \Delta t. \tag{2}$$

Equation (2) is the FED model for ISO 13571.

Equation (2) considers the two significant asphyxiant fire gases CO and HCN. The FED value is calculated using the exposed dose relationship (concentration—time product, C•t) for CO. The lethal C•t product corresponds to the incapacitating dose (C•t) for CO of 35 000 μl l^{-1} min, equal to around 1170 ppm for a 30-min exposure) and an exponential relationship for HCN.

$$\mathrm{FEC} = \frac{[\mathrm{HCl}]}{\mathrm{IC}_{50,\mathrm{HCl}}} + \frac{[\mathrm{HBr}]}{\mathrm{IC}_{50,\mathrm{HBr}}} + \frac{[\mathrm{HF}]}{\mathrm{IC}_{50,\mathrm{HF}}} + \frac{[\mathrm{SO_2}]}{\mathrm{IC}_{50,\mathrm{SO_2}}} + \frac{[\mathrm{NO_2}]}{\mathrm{IC}_{50,\mathrm{NO_2}}} + \frac{[\mathrm{acrolein}]}{\mathrm{IC}_{50,\mathrm{acrolein}}} + \frac{[\mathrm{fomaldehyde}]}{\mathrm{IC}_{50,\mathrm{fomaldehyde}}} + \sum \frac{[\mathrm{irritant}]}{\mathrm{IC}_{50,\mathrm{irritant}}}. \tag{3}$$

Equation (3) is the FEC model from ISO 13571.

Equation (3) uses the same additive of Purser's principle in ISO 13344 to estimate the combined effect of all irritant gases (IC is the concentration resulting in incapacitation of 50% of the population).

ISO 13571 is a more robust methodology for ensuring the safety of potential fire victims. It is included here for completeness, to show the steps needed to avoid incapacitation, but subsequent results are presented using the simpler FED calculation of ISO 13344 (rat lethality) in order to illustrate the relative and absolute contributions of individual toxicants.

In order to relate the fire effluent toxicity to a maximum permissible loading, the FED can be related to the mass of material in a unit volume, which would cause 50% lethality or incapacitation for a given fire condition; as an LC_{50}, a specimen mass M of a burning polymeric material that would yield an FED equal to one within a volume of 1 m^3 according to Eqn (4) [24].

$$\mathrm{LC}_{50} = \frac{M}{\mathrm{FED} \times V}. \tag{4}$$

Equation (4) is the relation of LC_{50} to FED.Here V is the total volume of diluted fire effluent in cubic meters at standard temperature and pressure (STP). Comparing the toxic potencies of different materials, the lower the LC_{50} (the smaller the amount of materials necessary to reach the toxic potency) the greater its fire toxicity. LC_{50} values should be referenced to the fire condition under which they were measured.

1.5 Quantification of environmental pollution from fire effluents, and hazards from particulates and their impact on human health

Unwanted fires are characterized by incomplete combustion, often occurring at lower temperatures, and producing a rich cocktail of toxic and pollutant compounds compared to the small well-ventilated flaming and complete combustion of many laboratory test scenarios. In particular, real fires will produce more persistent bioaccumulative and toxic (PBT) products, such as polycyclic aromatic hydrocarbons (PAHs), and polychloro- and polybromo-dibenzo-dioxins and furans (PCDD/F and PBDD/F) from fuels containing halogens. In addition, research concerning exposure to respirable particles (from PM_{10}s that are <10 μm diameter down to nanoparticles, ~100 nm) from man-made sources indicates that exposure to background levels of particulates can cause irritation or damage to the respiratory system [26]. The World Health Organization estimates a 0.5% increase in daily mortality per 10 μg m^{-3} of PM_{10} (particulate matter smaller than about 10 μm) and smaller particles [27]. The general effect of particulates is to cause fluid release and inflammation in the lungs, preventing gas exchange. The toxicity and particulate formation mechanisms present during accidental fires are not well understood, but the toxic gaseous species present in the smoke are known to attach themselves to nanoparticles, which act as vehicles taking toxicants deep into the lung.

The particle size distribution is dependent on the material, temperature, and fire conditions. The particle size of the spherical droplets from smoldering combustion is generally of the order of 1 μm, while the size of the irregular soot particulates from flaming combustion is often larger, but much harder to quantify and dependent on the measuring technique and sampling position [20]. The deposition areas for humans as a function of particle size are shown in Figure 5.

The PAHs produced from fires agglomerate together into small spherical particles, which then adhere to one another, like a tangled string of beads. Both the volatile PAHs and the spherical particles will remain airborne almost indefinitely, but have the potential to cause significant damage to the lungs. The low-molecular-weight PAHs show a moderate potential to be adsorbed onto particulates whereas high molecular weight PAHs have stronger tendencies to adsorb to organic carbon, especially small particles (<2.5 μm) [28]. Of particular concern are the decomposition products from halogenated materials, including PVC and brominated flame retardants, which have been shown to produce large

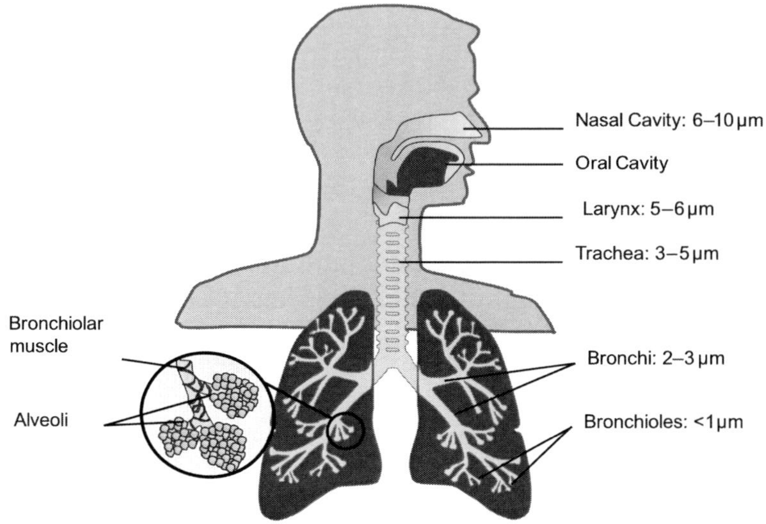

FIGURE 5

Particulate distribution in the respiratory tract.

quantities of PCDD/F or PBDD/F in unwanted fires. PCDDs and PBDDs are extremely toxic, chemically and thermally stable, and have a tendency to be strongly adsorbed on the surface of particulate matter. The rate of formation of PCDD/Fs is a function of temperature, and the quantity of unburnt carbon, which is dictated by the oxygen level [29]. Numerous studies on animals have also confirmed that some dioxin congeners are carcinogenic and produce mutagenic effects in certain species. However, it is difficult to extrapolate data from different fire test methods as there are significant differences in both the combustion conditions (such as large variations in temperature and oxygen environments), and also varying concentrations of precursors and chlorine in the product gases [30]. Therefore, at present, our knowledge of the distribution and quantities of these species in fire gases is fairly limited.

General awareness of the fact that fires may present dramatic and persistent adverse effects on the environment has been accentuated by a number of high impact incidents over the past half century [31]. Since quantitative data on environmentally hazardous components of fire effluent cannot be obtained from accidental fires, appropriate data must come from real-scale fire tests and simulations involving small-scale physical fire models. Such data are almost entirely absent from literature reports or enquiries relating to these incidents. Similar to PAHs, depending on the temperatures and oxygen regimes, the presence of unburned carbon particles in the postcombustion zone significantly increases the yield of PCDD/Fs.

2. Analysis of fire toxicity for different polymers and products

Estimation of fire toxicity is generally limited to those products considered most significant in causing incapacitation and death in fire victims. These consist predominantly of asphyxiant gases (CO, HCN, CO_2, and lowered oxygen) and irritants, which include acid gases (HBr, HCl, NO_2). In addition, examples of yield of carbon as organic carbon and carbonaceous particulates and other specific classes of compounds are also reported.

2.1 Toxic product yields and estimated fire toxicity of common polymers

The yields of toxicants from polymethylmethacrylate (PMMA), low-density polyethylene (LDPE), PVC, polyamide 6.6 (PA 6.6), and PS are presented in Figure 6 as a function of equivalence ratio, ϕ [2,32,33].

For PMMA, very efficient combustion at equivalence ratios below 1 is observed, with high CO_2 yields and very low yields of other carbon-containing products. In underventilated combustion conditions, there is a large increase in yields of organic gases and CO with increasing ϕ. For underventilated combustion, the yields of organics and CO are both high compared to well-ventilated flaming.

A similar pattern of product yields for LDPE is observed. The theoretical yield (100% conversion of fuel carbon to CO_2) for LDPE is 3.14 g/g under well-ventilated conditions. It can be seen that almost all the fuel carbon is converted to CO_2. The CO_2 yield follows an almost linear decrease as the ventilation is decreased, while the CO yield increases with decrease in ventilation from $\phi = 0.5$ to $\phi = 1.5$ then starts to decrease slowly, presumably due to limited oxygen availability. At an equivalence ratio of 1.6, the results show some increases in yields of hydrocarbons.

PS has a similar carbon content to that of LDPE, but a higher carbon/hydrogen ratio. The combustion product yield pattern is similar to that of LDPE, but with less sensitivity to the ventilation conditions. It shows a high CO_2 yield at low equivalence ratios, decreasing as $\phi > 1$, but there is a greater propensity to form carbon-rich soot throughout the ϕ range and especially at high equivalence ratios [34]. The relatively high CO yields in well-ventilated combustion conditions and low CO yields in underventilated conditions suggest the presence of stable aromatic molecules, and low hydrogen ratio, resulting in inefficient oxidation. The higher CO yield and correspondingly higher soot yield for aromatics and unsaturated fuels burning in well-ventilated conditions is well known; the lower CO yield under fuel-rich conditions is more interesting; the CO yield of 0.11 g g^{-1} for toluene under fuel-rich conditions, and attributed it to the thermal stability of the molecule [35]. This results in a further reduction in combustion efficiency as the aromatic hydrocarbons are not converted to CO. Since the main product of decomposition of PS is the monomer, with smaller quantities of dimer, trimer, and tetramer, these are also likely to

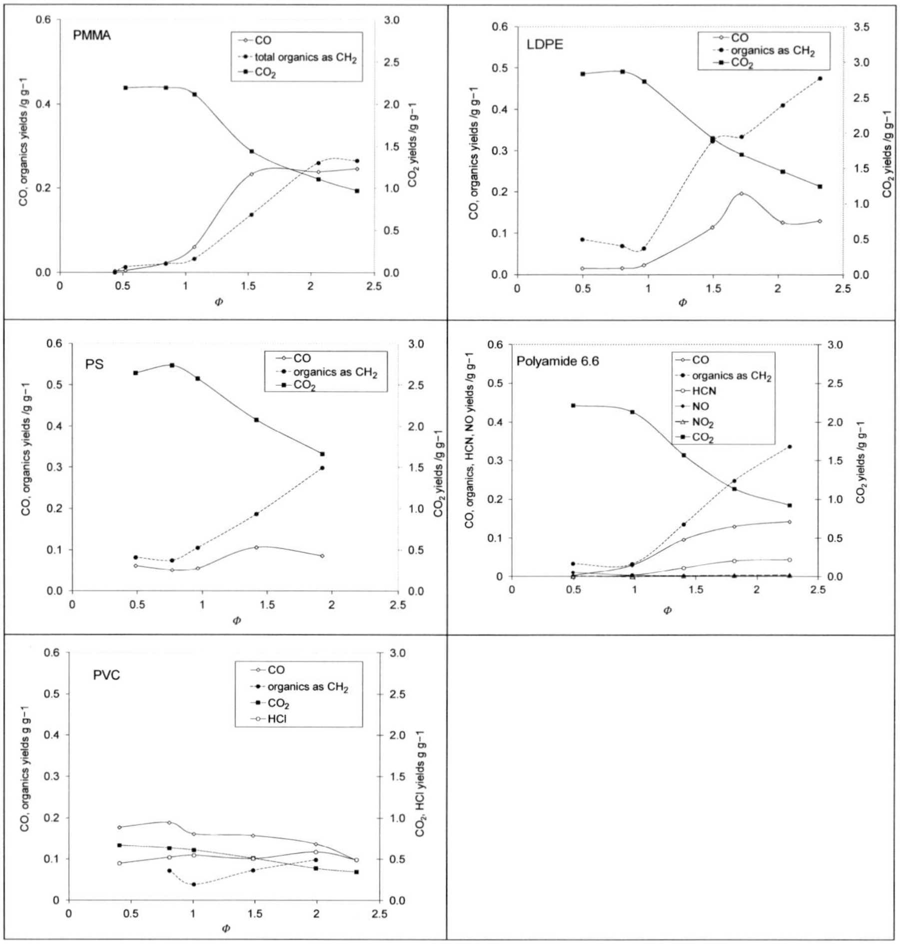

FIGURE 6

Fire effluent yields from polymethylmethacrylate (PMMA), low density polyethylene (LDPE), polystyrene (PS), polyamide (PA) 6.6, and polyvinyl chloride (PVC).

show similar enhancements in thermal stability, limiting the availability of OH· radicals, responsible for converting CO to CO_2.

PA 6.6 shows a similar trend to PP and PMMA. Combustion is efficient at low equivalence ratios with low yields of CO and organics, all of which increase at equivalence ratios >1. The efficiency of conversion of fuel carbon to CO is lower than that for PMMA, which may be related to the lower oxygen content of the polymer. The yields for CO and HCN show an increase with increase in ϕ. In well-ventilated combustion conditions, the yield was low, but increased steeply to a maximum of 0.44 g g^{-1} at $\phi = 2.5$. Under well-ventilated combustion conditions, the main toxic species produced was NO at a maximum yield of 0.012 g g^{-1} at

$\phi = 0.5$, decreasing as ϕ increased to a yield of $0.0026\ g\ g^{-1}$ in underventilated combustion conditions. The yield of NO_2 was approximately a factor of 10 lower than that of NO [10,17].

PVC burns with a low heat release rate, because the halogen atoms in the structure are released as HCl, accounting for almost 60% of its mass, which then inhibits the conversion of CO to CO_2 and combustion is inefficient, even under well-ventilated conditions [36]. The yield pattern is very different from that of all the other polymers described, in that the yields of all products are relatively similar across the whole ϕ range from 0.5 to 2.5. Combustion is very inefficient across the range, with relatively low CO_2 yields and high yields of CO, and organics. The CO yields in the steady-state tube furnace, are almost constant with increase in ϕ. The theoretical yield of HCl is $0.585\ g\ g^{-1}$, and it can be seen that most of the chlorine is released as HCl.

The major LDPE, PMMA, PS, PA 6.6, and PVC product yields, obtained from the steady-state tube furnace, have been translated into an overall estimation of the fire effluent toxicity, using the methods described in ISO 13344 indicating the contribution of each toxicant towards the overall fire hazard. Hypoxia is presented here as a decrease in oxygen supplied to, or utilized by, body tissue. The toxicity is expressed as FED for a fuel mass charge concentration of $20\ g\ m^{-3}$.

Most polymers without heteroatoms follow the trend shown by LDPE and PMMA of fire toxicity increasing from a very low value in well-ventilated conditions, to a much higher value in underventilated flaming. As can be seen from Figure 7, there is a large variation in FED values for materials containing chlorine or nitrogen. For well-ventilated tests, the largest FED value is for PVC. The FED

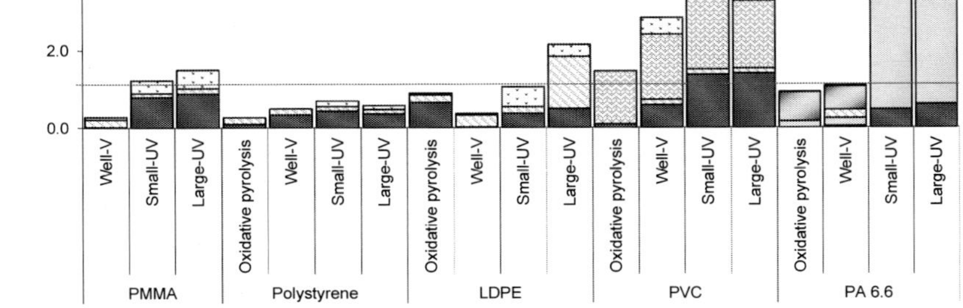

FIGURE 7

Fractional effective dose (FED) for common polymers [2].

is >1, indicating the lethality of the diluted fire atmosphere over 30 min. PVC is one of the few materials to show a predicted combustion toxicity that is almost independent of the equivalence ratio, showing unusually high toxicity under well-ventilated conditions. As an irritant gas, HCl will have the greatest effect of any toxic species. For underventilated fires, HCN from PA 6.6 makes the most significant contribution to the toxicity, and a high dependency on fire conditions is observed. HCN generated during small and large underventilated flaming tests makes the most significant contribution to the toxicity. For PA 6.6 in underventilated conditions, and for PVC, HCN, and HCl make the greatest contribution to the final FED, showing the importance of two toxicants other than CO. PMMA, LDPE, and PS are hydrocarbons without any halogens or nitrogen, and this is reflected in FED values well below 1 for the well-ventilated fire scenario. The fire effluent from PS has a relatively low toxicity, and shows a characteristically low dependence on fire conditions. The toxic contribution of CO under fuel-rich conditions is remarkably similar to those generated under fuel lean conditions, presented in section 2.2.2.

2.2 Effect of fire retardants and nanofillers on fire toxicity

2.2.1 Effect of fire retardants and nanoclays on fire toxicity of polypropylene

Samples of PP containing 5% PP grafted with maleic anhydride, to allow dispersion of nanoclay (NC), were compounded with 30% ammonium polyphosphate fire retardant and/or 5% Cloisite 20A NC. The toxicity of the PP materials is much lower than that for the PA samples in underventilated conditions (Figure 8). Again there is a progressive increase in the toxicity from small underventilated to large underventilated, which is shown consistently across the samples. Although the addition of either fire retardant (FR) or NC has no significant effect on the CO yield, it appears that the combination of both NC and FR results in higher CO yields for underventilated flaming [37,38].

2.2.2 Effect of fire retardants on ethylene-vinyl acetate copolymer

Ethylene-vinyl acetate copolymer (EVA) is widely used with mineral fillers in the cable industry, as a material for avoiding the toxic, smoky, corrosive effluent of PVC-sheathed cables. EVA containing 27% vinyl acetate was compounded into low smoke and fume formulation. Fire retarded composites were formulated with 30% by weight of the EVA and 70% aluminum hydroxide (ATH) or 65% ATH and either 5% zinc hydroxystannate (ZS) ($ZnSn(OH)_6$), 5% magnesium borate (MgB) ($MgO\ (B_2O_3)_2\ H_2O$), or 5% zinc borate (ZB) ($2ZnO.3B_2O_3.3.5H_2O$) [34]. In order to compare the effect of fillers on the combustion chemistry of composite EVA materials, the same mass of EVA was used for each experiment; thus, the filled composites containing 30% polymer and 70% fire retardant filler had a sample mass 3.33 times greater than those of pure EVA [35].

The contributions to FED, shown in Figure 9, indicate that except as a diluent filler, ATH and ATH with MgB or ZS have only a modest effect on fire toxicity,

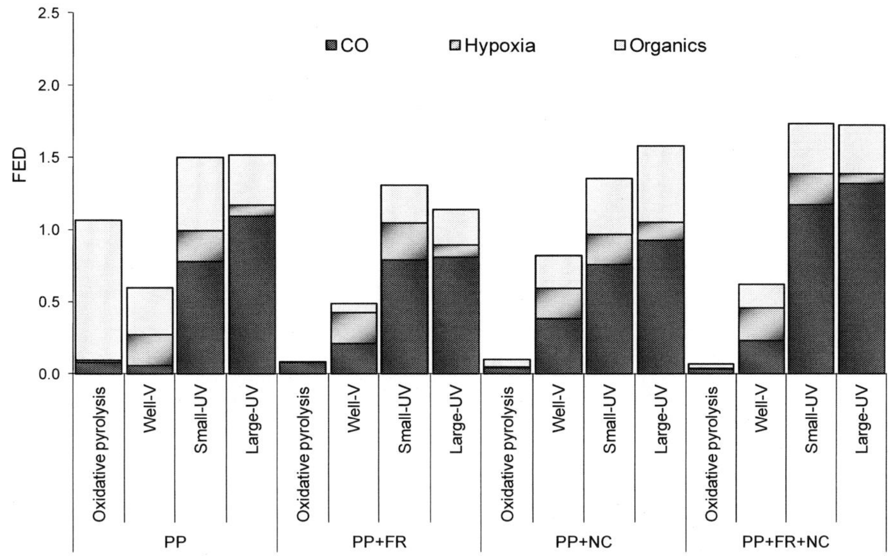

FIGURE 8

Fractional effective dose (FED) for polypropylene (PP) with fire retardant and nanoclays (NCs) [37,38].

increasing the CO contribution during underventilated burning compared with EVA alone. This is believed [34] to result from catalytic oxidation of char (after-glow) by the freshly formed alumina surface. In contrast, ATH with ZB has a dramatic effect on reducing the carbon monoxide contribution to fire toxicity especially in underventilated conditions. It is thought that ZB forms a glass, which destroys the char-oxidizing catalytic properties of the freshly formed alumina, leaving more of the carbonaceous residue in the condensed phase [35].

A separate study, using different EVA, with 68% ATH or magnesium hydroxide, or 63% when used with 5% NC (Figure 10) shows that ATH and magnesium hydroxide enhance the CO yield compared to the pure polymer, or polymer plus NC, which had little effect [37].

2.2.3 Effect of fire retardants on polyester materials

Polybutylene terephthalate was fire retarded with 18% aluminum phosphinate Exolit OP1240 and/or 5% of Cloisite 30B or sepiolite NC. Figure 11 shows the consistently higher toxicity from burning PBT under different fire conditions and the additional increase in toxicity resulting from the use of fire retardant presumably through gas phase inhibition. It is interesting to see this effect disappear on incorporation of either sepiolite or cloisite. A slight increase in CO yield from well-ventilated to underventilated conditions for PBT and PBT with sepiolite, and PBT + FR combinations is observed. As aluminum phosphinate is a gas phase flame retardant, the

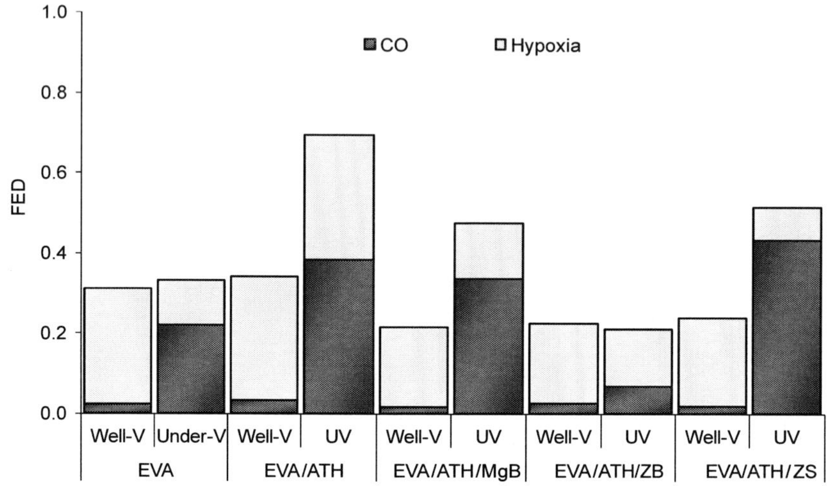

FIGURE 9

Contribution of each component to toxicity at different ventilation conditions for (ethylene-vinyl acetate copolymer) EVA fire retarded composites [2,35].

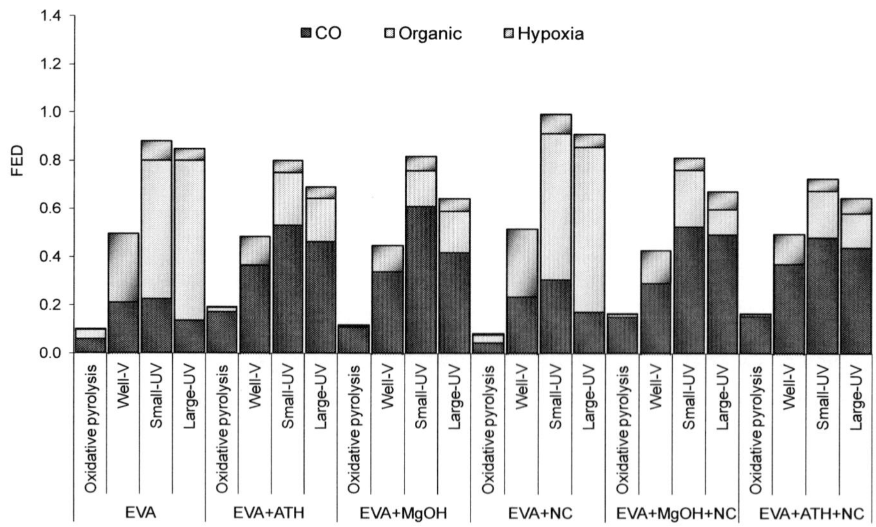

FIGURE 10

Contributions to (fractional effective dose) FED for ethylene-vinyl acetate copolymer (EVA) with fire retardant and nanoclay (NC) [37].

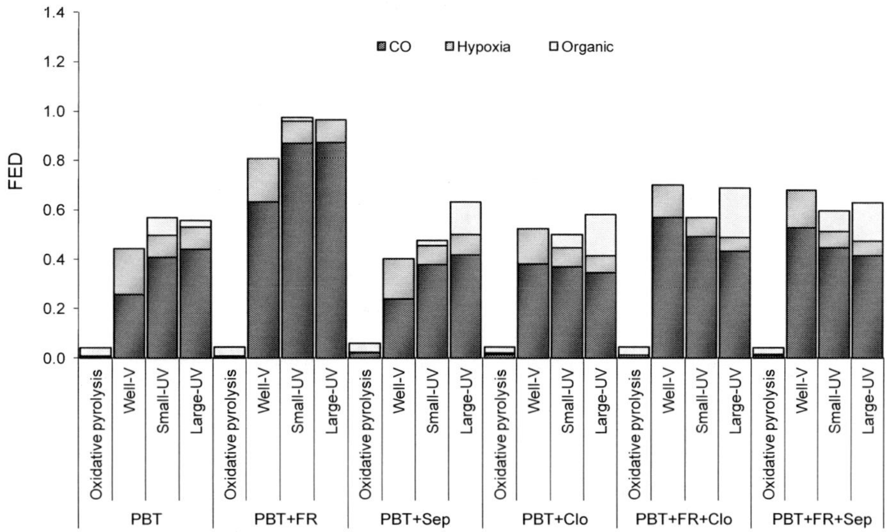

FIGURE 11

Contributions to (fractional effective dose) FED for (persistent bioaccumulative and toxic) PBT with fire retardant and nanoclays (NCs) [2,37].

increase with FR would be expected, but this effect seems to diminish in the presence of the NCs. The NC alone has little effect on the CO yield [37].

In a separate study (Figure 12), fire retarded glass-reinforced polyester (GRP) roof light material, supplied as a translucent flat sheet formulated to achieve a "class 1" rating to BS476 Part 7 was investigated [39]. This material contained both chlorine and bromine. The major effect on the fire toxicity of the presence of HCl and HBr was to increase the yield of carbon monoxide increasing its contribution to the toxicity, especially in well-ventilated combustion. CO makes the most significant contribution to the toxicity under all conditions, contributing around 80% of the FED. The CO contribution to the FED is hardly changed in different flaming scenarios, resulting in FEDs with a remarkably low sensitivity to fire conditions [2].

2.2.4 Effect of fire retardants and NCs on PAs

Samples of PA 6 also containing 18% aluminum phosphinate melamine polyphosphate mixture (Clariant's OP 1311) and/or 5% Cloisite 30B NC shared modest reduction in fire toxicity when compared to the virgin polymer. For oxidative pyrolysis, the only significant contribution to the toxicity comes from NO_2. The total value of the FED is higher for well-ventilated, small and large underventilated flaming. Similar to PA 6.6, the most toxicologically significant product from PA 6 is hydrogen cyanide (Figure 13). As the ventilation becomes limited, both the CO and HCN contributions increase while the oxygen depletion, by the nature of the experiment, decreases. In addition, a relationship between hypoxia and

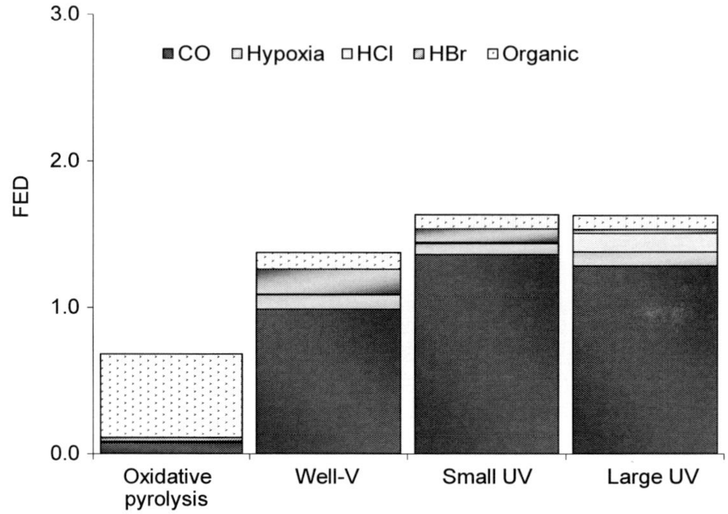

FIGURE 12

Contributions to (fractional effective dose) FED for (glass-reinforced polyester) GRP [2].

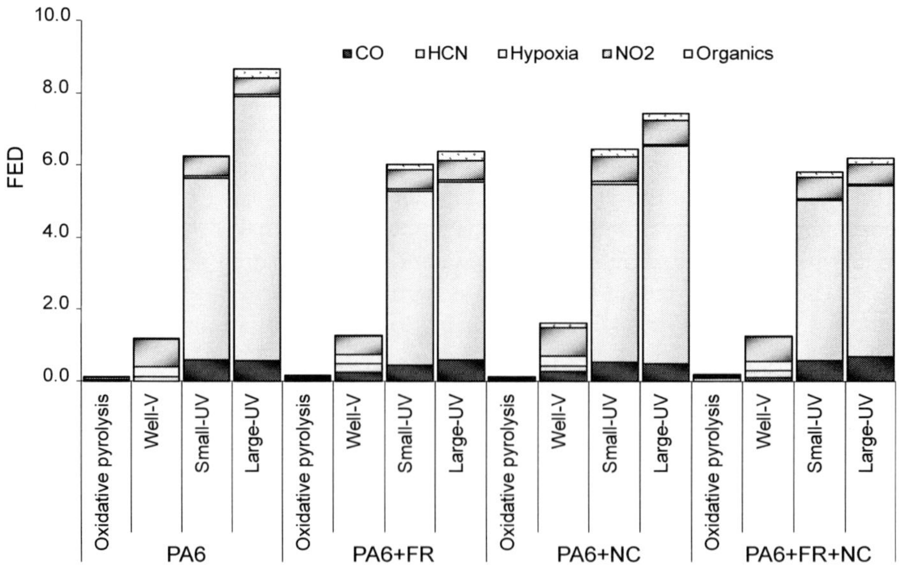

FIGURE 13

Contributions to fractional effective dose (FED) for polyamide (PA) 6 with fire retardant and nanoclay (NC) [2,37].

unburned hydrocarbons has been observed, as the production of both species is favored by underventilated conditions. The increase of HCN concentration and organic species, and CO and oxygen depletion are independent of the temperature of the furnace and give higher yields with reduced ventilation [38,40]. For well-ventilated burning, where the major toxicant is NO_2, there are only minor differences between the FR and NC materials compared to that for the virgin polymer. There is a decrease in CO for all the materials in the most dangerous, underventilated conditions on incorporation of either NC or FR, and only a slight increase when both are present [2,37].

Samples of PA 6, formulated to industry standards to meet UL94 V-0 at 0.8 mm, containing 30% glass fiber (abbreviated to PA6 + GF), for electrical applications, and fire retarded with either brominated PS and antimony trioxide (20%) (abbreviated to BrSb), or a blend of aluminum phosphinate and melamine polyphosphate (AlPiM), were burned under well-ventilated flaming conditions [41]. The relative contribution to the toxicity of individual components using FED, based on the Purser model is presented in Figure 14.

Figure 14 shows estimates of toxicity expressed as FED (assuming a loading of 1 kg of material in 50 m^3). They show a slight change in toxicity resulting from the presence of AlPiM (more HCN, but less NO_x), but a dramatic increase in both CO and HCN yields. It is notable that the fire toxicity of the PA 6 + GF + BrSb materials is a factor of 10 larger than that of PA 6 + GF and a factor of 5 larger than the

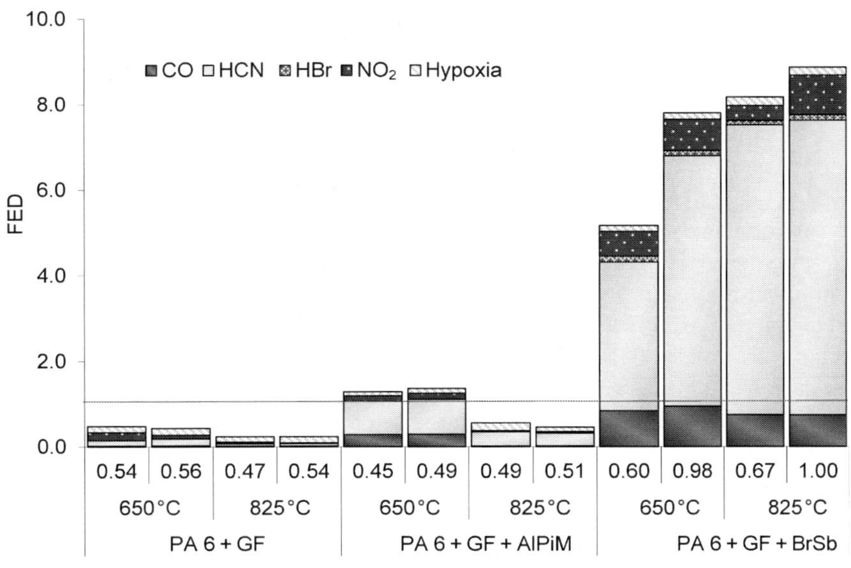

FIGURE 14

Fractional effective dose (FED) comparison for two different flame retardant systems in PA6 containing 30% glass fiber [42].

PA6 + GF + AlPiM material at 650 °C; at 825 °C it is a factor of 30 larger than the PA 6 + GF, and 17 larger than the AlPiM material. The BrSb materials all show high fire toxicity, even under the normally least toxic, well-ventilated fire conditions. The lower fire toxicity of the AlPiM materials at higher temperatures suggests a reduction in fire hazard. It appears that the AlPiM system "switches off" this flame quenching process as the fire becomes more severe, reducing the toxicity of the effluent.

2.3 Quantification of effluent toxicity during burning of fire retarded commercial products (precise formulation unknown)

2.3.1 Medium density fiber board

Medium density fiber board (MDF) is a wood based composite, composed of around 50% cellulose, 25% hemicellulose, and 25% lignin. MDF is a generic name used to define sheets composed mainly of lignin-cellulosic fibers glued with a synthetic resin or some other suitable adhesive system, combined together by means of pressure and heat. Samples of commercial MDF without and with flame retardant (MDF-FR) were investigated for fire toxicity. MDF was a BS476 Part 7 "class 2" rated non-flame-retarded construction material. Fire-resistant MDF (MDF-FR) was an MDF construction material treated with a fire retardant, which contained nitrogen, bromine, and phosphorus. The product met BS476 Part 7 "class 1" rated MDF. Both materials had a nominal density of 740 kg m^{-3}.

Although the relative yields are similar to LDPE and PS, the CO_2 yield from MDF under well-ventilated combustion conditions is considerably lower, corresponding to the lower carbon and higher oxygen content. The combustion is also more efficient than for the CH polymers, perhaps due to the additional oxygen in the molecular structure. Another important feature of cellulosic materials is that they are char formers, so that, especially in underventilated conditions, a significant residue remains, most of which is carbon. The yield of organic carbon is less than for the CH polymers, but the conversion of fuel carbon to CO is greater. MDF also contains approximately 3% nitrogen presumably from the glue or fire retardant, so a small amount of HCN is produced, especially in underventilated conditions. Small quantities of hydrogen chloride and hydrogen bromide were generated from MDF combustion and MDF-FR.

The toxic product yields obtained from the MDF and MDF-FR combustion under different fire conditions are presented in Figure 15. There are significant differences between MDF and MDF-FR. For both furnace temperatures for MDF, CO, and HCN yields increase with the increase of equivalence ratio, while NO_x yields decrease. However, organics contribute more to the FED. A different pattern is observed in the case of MDF-FR, which could be explained by the presence of the brominated flame retardant, which increases the CO, HCN, and NO_x yields. For MDF-FR, the most significant contributions to the toxicity are NO_x and HCN. These differences are the greatest at the higher furnace temperature. However, they seem to be

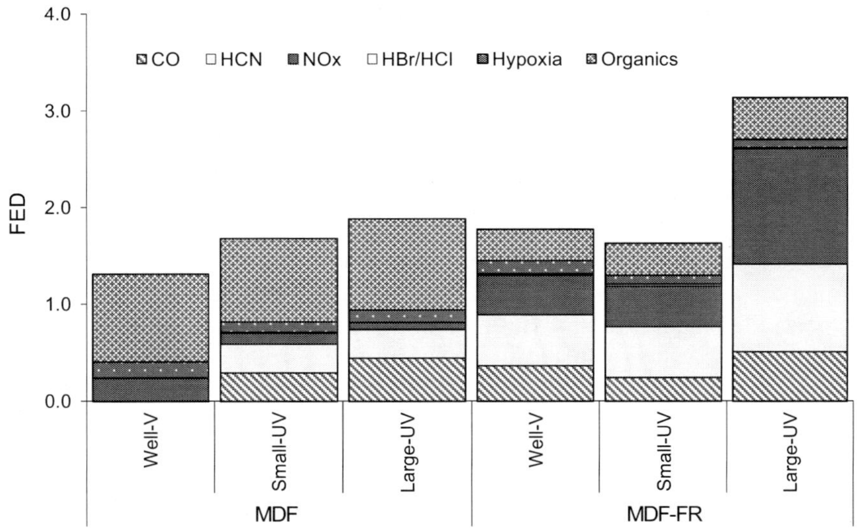

FIGURE 15

Contributions to fractional effective dose (FED) comparison for medium density fiber (MDF) and medium density fiber-flame retardant (MDF-FR).

independent of the temperature and ventilation conditions. Overall, the predicted toxicity is higher for MDF-FR than for MDF.

2.3.2 Insulation products

Modern, lightweight building materials are cheaper to produce, transport, and erect, and offer improved thermal insulation, allowing more efficient temperature control. In comparison to traditional materials, many insulation materials present a greater fire hazard, being less effective fire barriers, more combustible and having higher fire toxicity. Most common insulation materials comprise gases trapped in a matrix to inhibit convective heat transfer. The fire toxicity of six insulation materials: glass wool (GW), stone wool (SW), expanded PS foam (EPS), phenolic foam (PhF), PUR, and polyisocyanurate foam (PIR) were investigated under a range of fire conditions, oxidative pyrolysis, and well-ventilated flaming to underventilated flaming [42]. Both GW and SW are classified as noncombustible or limited combustibility depending on the binder content. While both lose small ($\sim$5%) quantities of pyrolyzable binders, most of the mass will not burn, and there is insufficient fuel for a flame to propagate through the bulk of the material, so their contribution to the fuel load is negligible. The foamed materials are organic polymer based, and depending on the material and fire conditions, a significant part of their mass is lost as fuel, and may contribute to the overall size of the fire. The thermal insulation materials were selected to have comparable densities [43].

For the two fibrous materials in the flaming condition, SW and GW, ignition was not observed even above 800 °C (nf). For the four foams, PUR, PIR, PhF, and EPS, ignition and steady flaming was achieved for the two flaming fire conditions and for intermediate ventilation conditions [42].

Figure 16 shows the increase in the carbon monoxide yield as the ventilation changes from well-ventilated to underventilated. In comparison to polymers without flame retardants, the CO yields for PIR and PUR in well-ventilated conditions are high. This suggests the presence of gas phase free radical quenchers, such as halogens or volatile phosphorus compounds, preventing the conversion of CO to CO_2 by reducing the availability of the OH· radical [36]. For the two nitrogen-containing polymers, PUR and PIR, the HCN yield is significant, both in well-ventilated flaming, and for underventilated flaming, and those materials have much greater fire toxicity than EPS or PhF. The GW and SW products show very low fire toxicity. The variation of NO_2 with fire conditions for the three materials shows less consistent trends although the PhF seems to show a progressive increase as the fire condition becomes underventilated. For PIR and PUR, there is a slight decreasing trend with underventilation, which corresponds to the increased yields of HCN, and reduced availability of oxygen. HCl and HBr were close to limits of detection.

2.3.2.1 Isocyanates from insulation materials

Although not specifically included in the normal lists of fire effluents for quantification of fire toxicity, it is been suggested that isocyanates (molecules with functional group —NCO, used in PUs and some binders) may pose a hitherto unquantified

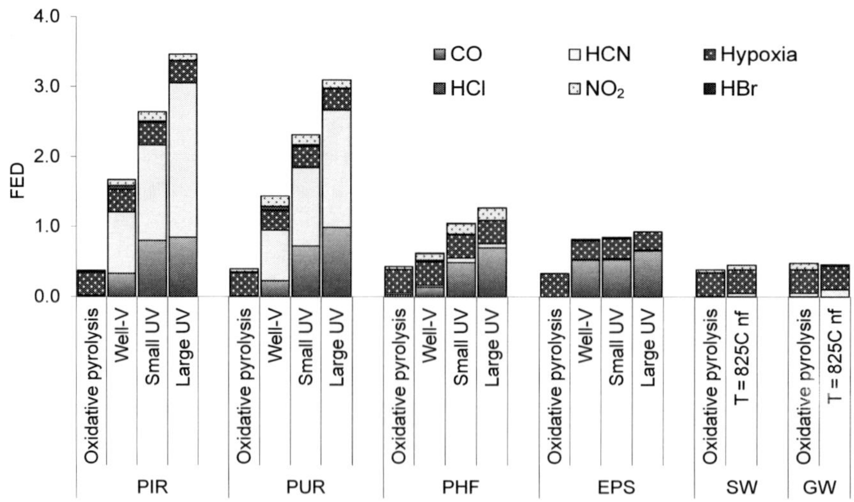

FIGURE 16

Fractional effective dose (FED) for insulation materials (for oxidative pyrolysis and flaming conditions, except stone wool (SW) and glass wool (GW) nonflaming (nf)) [42].

hazard in fire effluents. In ISO TC92 SC3, the main fire toxicants identified do not include isocyanates, although there is some evidence that they are present in fire effluents in toxicologicially significant concentrations. A cone calorimeter study [44] included five insulation materials, GW, SW, EPS foam, PUR foam, and PIR foam, as part of a larger project to investigate the presence of isocyanates in fire effluents. Each sample was exposed to an intermediate heat flux of 35 kW m^{-2}. It was found that isocyanate production was favored in the early well-ventilated stages of flaming, while hydrogen cyanide was favored in the more toxicologically significant under-ventilated stages [42].

For noncombustible samples, GW and SW, the materials were subject to a full 15-min pyrolysis, during which isocyanate collection continued. PS foam was only subjected to 10-min pyrolysis during which it did not ignite, while the PUR and PIR foams burnt for 5.5 and 9 min, respectively [45]. The isocyanates were only collected during these times. Thus, the sample collection time was greatest for the least flammable samples.

From the reported data, isocyanate yields have been calculated as shown in Table 3. In the original paper, the mass loss data for PUR or PIR is not given—for the calculation, data from experiments reported elsewhere [42] have been used. Despite the 37.5% mass loss from PS, no data are presented on the composition of the volatile products, and no explanation is provided, but it may be that a large amount of soot blocked the sampling lines. The yields are calculated on a mass charge basis (favored by engineers as it indicates the total amount of product that may be formed per unit mass present in a building) and as originally reported, on a mass loss basis (materials such as GW which were found to be 88.5% nonvolatile (glass fiber) and 11.5% organic binder), which is the yield from the organic binder alone, as though there was no glass fiber present [42].

The use of the mass loss yield results in a misleading impression of large isocyanate yields from the inorganic fiber insulation materials. In effect, it compares the yield from 1 kg of PUR foam with that from 9 kg of GW insulation, despite the fact

Table 3 Isocyanate Yield and Calculation Data [42]

	Concentration μg m^{-3}	Mass of Sample/g	Mass Loss/g	Mass of Isocyanate mg^{-1}	Mass Charge Yield mg g^{-1}	Mass Loss Yield mg g^{-1}
GW	8100	20	2.3	175	8.7	82.1
SW	990	14	0.5	21	1.5	68.9
PS	0	8	3.0	0	0	0
PUR	4500	17	13.0	36	2.1	2.7
PIR	3350	14	11.5	43	3.1	7

GW, glass wool; PS, polystyrene; PUR, polyurethane foam; PIR, polyisocyanurate foam; SW, stone wool.

that the latter is much less flammable, and would therefore only be volatilized in the presence of a much stronger external heat source [42].

2.3.3 PVC carpet

Another commercial material that has been investigated is PVC carpet, commonly found in domestic buildings [28,46,47]. The PVC carpet had a thickness of 2.0 mm, and contained almost as much plasticizer and mineral fillers as PVC polymer. The contribution of carbon monoxide and hydrogen chloride to FED is shown in Figure 17. For all fire conditions, HCl makes the greatest contribution to the toxicity. In common with the other halogenated materials investigated, increased yields for CO are observed. The HCl yields found from well-ventilated and underventilated combustion were close to the theoretical maximum. This supports an earlier study where the HCl yields are independent of fire stages and temperature effects [17,36].

2.3.3.1 Particulates, PAHs, PCDD/Fs, isocyanates from PVC carpet

The PVC carpet was subjected to an in-depth study to look for other noxious substances in addition to the eight standard ISO fire gases. It was found that in well-ventilated and underventilated tests on PVC carpet both large and small particles are generated (of the order of 1−4 μm), whereas in oxidative pyrolysis, only a narrow distribution of large particle sizes was found. Compared to well-ventilated tests, greater quantities of smaller, more dangerous particle sizes were found for the underventilated fire scenario. As small particles have a higher proportion of surface area

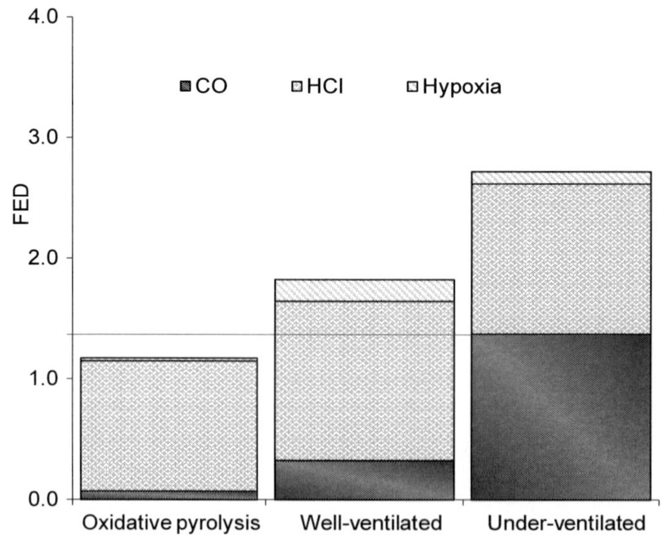

FIGURE 17

Contributions to Fractional effective dose (FED) for polyvinyl chloride (PVC) carpet [47].

per mass than do larger particles, it is possible that more toxic substances may be adsorbed onto smaller particles [28,46,47].

Although some chlorine will always appear as HCl, other chlorine containing gas or vapor species will be produced while, in some formulations, some chlorine may remain in the residue. A number of chlorine-containing species have been identified from large-scale fires burning a high proportion of PVC, including monochlorobenzene and dichlorobenzene and other chloroaromatic and chloroaliphatic hydrocarbons. Evidence exists to show that, depending on the fire situation, as much as 20% of the chlorine may exist in an organic form [48]. It was also found that the amount of chlorine from HCl could be very high in soot, even though the major part of the HCl produced was shown to be present in the gas phase [46,47]. In this case, for the PVC carpet, it is reported that the chlorine content adsorbed on the soot was in the range between 0.5 and 2.5 wt%, while in nonflaming fire conditions around approximately 18–20 wt% of chlorine is present in residue, compared to 10 wt% for flaming conditions which favor polychlorodibenzo furan and dioxin formation. These highly toxic materials are usually qualified by their toxic equivalence to 2,3,7,8-tetrachlorodibenzodioxin as toxic equivalency (TEQ). Total amounts of PCDD/F in the residue from PVC carpet shown in Figure 18 varies for different fire scenarios between 11 and 394 pg International toxic equivalents per gram (ITEQ/g) of burned material, corresponding to 41–1170 pg ITEQ/g residue [28].

From the studies, it is interesting to note that both well-ventilated and underventilated combustion of the PVC carpet produced the complete range of PAHs. In the

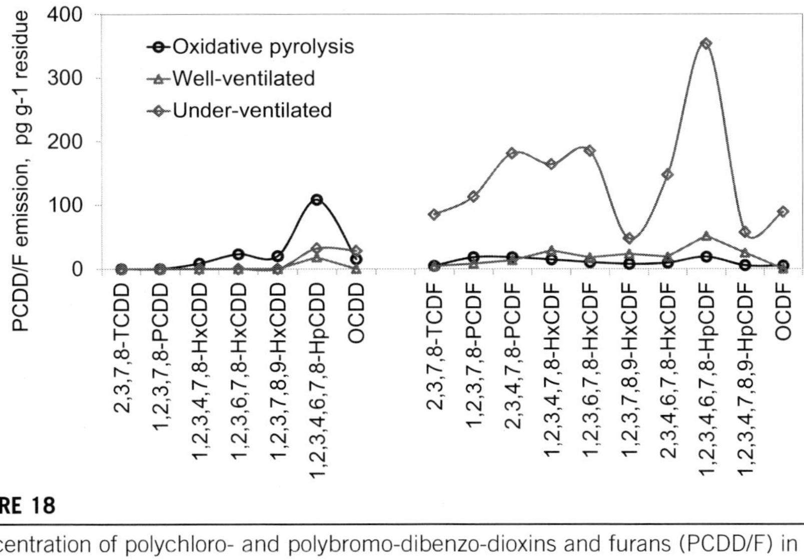

FIGURE 18

Concentration of polychloro- and polybromo-dibenzo-dioxins and furans (PCDD/F) in polyvinyl chloride (PVC) carpet residues under different fire conditions.

pyrolysis tests, however, only volatile and semivolatile PAHs were produced. For both well-ventilated and underventilated conditions, the toxicity weighted yields for the associated particle PAHs increased and generally dominated over the volatile species. The toxicity weighted yields for the underventilated tests were orders of magnitude higher than the yields from the well-ventilated tests, whereas for oxidative pyrolysis conditions, it is observed that the volatile part dominated the toxicity [46,47].

2.4 Cables and cable materials

Ten commercial cables (Figure 19) were investigated for fire toxicity using the conditions specified in the precursor International Electrotechnical Commission (IEC) standard, IEC 60695-7-50 [49], which uses the steady-state tube furnace, but a simplistic methodology for ensuring the setting up of the tube furnace ventilation condition to obtain a particular fire stage.

For each cable, the FED is shown for each of three fire stages. In most cases (8/10), oxidative pyrolysis is the least toxic, presumably due to the low mass loss of the polyolefin polymers at 350 °C [50]. However, this temperature is high enough to release HCl from PVC in two cases. In general, the FED, based on the limited range of gases analyzed, increases from oxidative pyrolysis, to well-ventilated, to developed flaming. In common with most materials, the most hazardous fire condition is confirmed to be the underventilated fully developed fire [2].

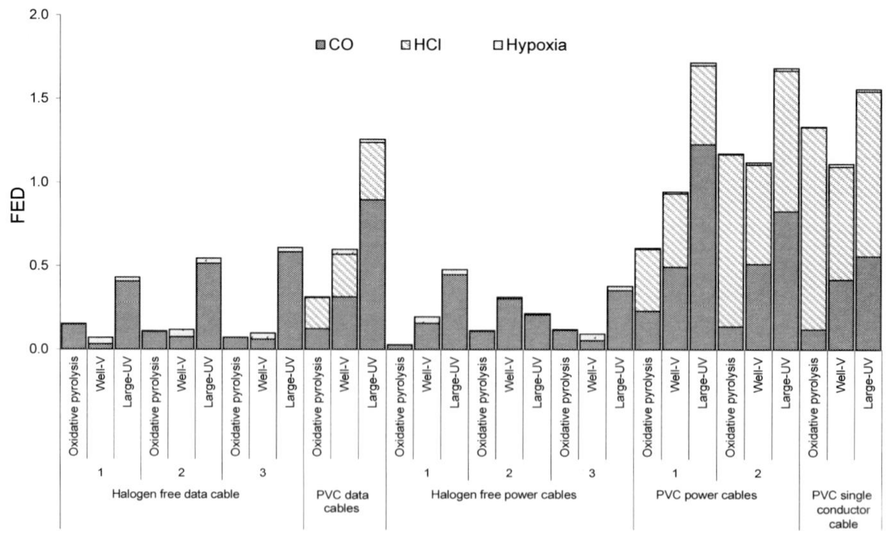

FIGURE 19

Contribution of individual gases to toxicity from burning cables.

The two major toxicants are seen to be CO and HCl. However, it is notable that the CO yield the PVC fires is greater than that in halogen-free fires, and it is also noted that the CO yield in PVC fires increases with underventilation, whereas for pure PVC, the yield is almost independent of ventilation [2].

3. CONCLUSIONS

Fire gases contain a mixture of fully oxidized products, such as CO_2, partially oxidized products, such as CO and aldehydes, fuel, or fuel degradation products, including aliphatic or aromatic hydrocarbons, and other stable gas molecules, such as nitrogen and hydrogen halides.

Most fire deaths arise from the inhalation of toxic gases, though the incapacitating effects of smoke and irritant gases also play an important role. Fire growth has been classified into a number of stages, from oxidative pyrolysis and early well-ventilated flaming, through to fully developed underventilated flaming. For many materials (such as CH- and CHO-containing polymers), the yield of the main toxicant, CO, has been shown to depend only on fire conditions, not on the nature of the polymer. It is therefore essential to the assessment of fire gas toxicity that these different fire stages can be adequately replicated.

The steady-state tube furnace (ISO 19700) provides an excellent method for exploring the relationship between combustion conditions and product yields. The yields of most of these species will depend on the material, the decomposition conditions (nonflaming or flaming), and for flaming, the ventilation conditions. For nonflaming decomposition conditions, some examples are presented of the effects of temperature and air flow on product yields, while flaming combustion conditions are expressed in terms of the fuel/air equivalence ratio, f.

For flaming conditions, examples are presented for five different natural and synthetic polymer classes demonstrating the relationship between the equivalence ratio and the product yields. The main carbon-containing products CO_2, CO, soot particulates (as total organic carbon in the gas phase and is expressed as the yield of CH_2) are reported. In addition, the yields of nitrogen-containing products (HCN, NO, and NO_2), and the HCl yield from a material with a high chlorine content (PVC), are presented. Yields for all gas phase products and residues are expressed as grams per gram (i.e. grams of product per gram of material on a mass charge basis). Equivalence ratios are varied from 0.5 to 2.5 in separate experiments representing well-ventilated to underventilated flaming.

For most natural and synthetic polymers, the yields of toxic products such as carbon monoxide are very low under well-ventilated combustion conditions, but increase steeply in underventilated combustion conditions as ϕ increases above 1. This results in a sigmoid relationship between equivalence ratio and yield, so that the yields of products of incomplete combustion can increase by factors of approximately 50 between well-ventilated flaming conditions ($\phi \sim 0.7$), and fuel-rich conditions ($\phi \sim 1.5$). The increase in products such as CO, HCN, organic irritants, and smoke particulates with φ is matched by a corresponding decrease in the products of more complete combustion—principally CO_2, H_2O, N_2, and NO_x. Superimposed on this basic pattern of decomposition and yields are lesser effects of decomposition temperature and oxygen concentration and specific effects related to the polymer composition. One important composition-related effect is residue formation in some natural and synthetic polymers, and in products containing fire retardants acting in the solid phase (such as phosphates), which leads to reduced yields of airborne products and particular trapping of fuel carbon. Other important yield modifiers are halogenated fire retardants, which act in the gas phase, resulting in increased yields of products of incomplete combustion particularly carbon monoxide and hydrogen cyanide throughout the range of equivalence ratios.

Assessment of fire gas toxicity is an essential component of fire hazard analysis. The toxicities of the effluents, showing the contribution of individual toxic components, are compared using the FED model. The predicted toxicities show variations of up to two orders of magnitude with change in fire scenario. They also show changes of at least one order of magnitude for different materials in the same fire scenario. Finally, they show that in many cases CO, which is often assumed to be the most, or even the only toxicologically significant fire gas, is of less importance than HCl or HCN when chlorine or nitrogen is present in the compound. For polyisocyanurate and PUR, there is a significant contribution from hydrogen cyanide resulting in a doubling of the overall toxicity, as the fire condition changes from well-ventilated to underventilated.

The presence of fire retardants and the incorporation of NC reduce the flammability, but for many fire retardants, there is no significant adverse effect of these additives on the toxicity of the material studies under most fire conditions. However, higher yields of CO under well-ventilated fire scenarios are often observed. The insulation materials showed an order of increasing fire toxicity, from SW (least toxic), GW, PS, phenolic, PU to polyisocyanurate foam (most toxic).

The current work shows lower carbon monoxide yields for all materials under well-ventilated conditions, compared to underventilated conditions, although the presence of halogens (presumably present as flame retardants) increases the CO and HCN yields in well-ventilated conditions.

References

[1] Fire Statistics United Kingdom 2008. Communities and local government; 2012.
[2] Stec AA, Hull TR, editors. Fire toxicity. Cambridge: Woodhead Publishing; 2010.
[3] ISO/TR 16738. Fire-safety engineering—technical information on methods for evaluating behaviour and movement of people; 2009.
[4] Purser DA. Assessment of hazards to occupants from smoke, toxic gases and heat. In: DiNenno PJ, editor. Chapter 2–6, SFPE handbook of fire protection engineering. 4th ed. Quincy (MA, USA): National Fire Protection Association; 2008. pp. 2-96–2-193.

[5] Stec AA, Hull TR, Lebek K, Purser JA, Purser DA. The effect of temperature and ventilation condition on the toxic product yields from burning polymers. Fire Mater 2008;32: 49–60.
[6] ISO TS 19706. Guidelines for assessing the fire threat to people; 2011.
[7] Purser DA. Toxic product yields and hazard assessment for fully enclosed design fires. Polym Int 2000;49:1232–55.
[8] Pitts WM. The global equivalence ratio concept and the formation mechanisms of carbon monoxide in enclosure fires. Prog Energy Combust Sci 1995;21:197–237.
[9] Hull TR, Paul KT. Bench-scale assessment of combustion toxicity—a critical analysis of current protocols. Fire Saf J 2007;42:340–65.
[10] Hull TR, Stec AA, Lebek K, Price D. Factors affecting the combustion toxicity of polymeric materials. Polym Degrad Stab 2007;92:2239–46.
[11] Stec AA, Hull TR. Fire Toxicity Assessment: Comparison of Asphyxiant Yields from Laboratory and Large Scale Flaming Fires Fire Safety Science, (in press 2014).
[12] ISO 9705. Fire tests—full scale room test for surface products; 1993.
[13] Blomqvist P, Lonnermark A. Characterization of the combustion products in large-scale fire tests: comparison of three experimental configurations. Fire Mater 2001; 25:71–81.
[14] Stec AA, Hull TR, Purser JA, Purser DA. Comparison of toxic product yields from bench-scale to ISO room. Fire Saf J 2009;44:62–70.
[15] ISO TS 19700. Controlled equivalence ratio method for the determination of hazardous components of fire effluents; 2007.
[16] BS 7990. Tube furnace method for the determination of toxic product yields in fire effluents; 2003.
[17] Stec AA, Hull TR, Purser JA, Blomqvist P, Lebek K. A comparison of toxic product yields obtained from five laboratories using the steady state tube furnace (ISO TS 19700). Fire Saf Sci 2008:653–64.
[18] Purser DA. The evolution of toxic effluents in fires and the assessment of toxic hazard. Toxicol Lett 1992;64-65:247–55.
[19] Stec AA, Hull TR, Lebek K. Characterisation of the steady state tube furnace (ISO TS 19700) for fire toxicity assessment. Polym Degrad Stab 2008;93:2058–65.
[20] Blomqvist P, Hertzberg T, Tuovinen H, Arrhenius K, Rosell L. Detailed determination of smoke gas contents using a small-scale controlled equivalence ratio tube furnace method. Fire Mater 2007;31:495–521.
[21] Blomqvist P, Hertzberg T, Tuovinen H. A small-scale controlled equivalence ratio tube furnace method—experience of the method and the link to large scale fires. In: Proceedings of the 10th international fire science and engineering conference (Interflam), Edinburgh; 2007.
[22] Hull TR, Lebek K, Stec AA, Paul KT, Price D, Schartel B. In: Bench-scale assessment of fire toxicity, advances in the flame retardancy of polymeric materials: current perspectives presented at FRPM'05. Norderstedt: Herstellung und Verlag; 2007. pp. 235–48.
[23] Hull TR, Lebek K, Pezzani M, Messa S. Comparison of toxic product yields of burning cables in bench and large-scale experiments. Fire Saf J 2008;43:140–50.
[24] ISO 13344. Estimation of the lethal toxic potency of fire effluents; 2004.
[25] ISO 13571. Life threat from fires – guidance on the estimation of time available for escape using fire data; 2012.

[26] Rhodes J, Smith C, Stec AA. Characterisation of soot particulates from fire retarded and nanocomposite materials, and their toxicological impact. Polym Degrad Stab 2011;96: 277–84.
[27] WHO, World Health Organisation. Protection of the human environment, outdoor air pollution. Geneva; 2004.
[28] Stec AA, Readman J, Blomqvist P, Gylestam D, Karlsson D, Wojtalewicz D, et al. Analysis of toxic effluents released from PVC carpet under different fire conditions. Chemosphere 2013;90:65–71.
[29] Huang H, Buekens A. On the mechanisms of dioxin formation in combustion processes. Chemosphere 1995;31:4099–117.
[30] Blomqvist P, Rosell L, Simonson M. Emissions from fires part II: simulated room fires. Fire Technol 2004;40:59–73.
[31] ISO 26367-1. Guidelines for assessing the adverse environmental impact of fire effluents—part I: General, 2011; 2011.
[32] Stec AA, Hull TR, Yields. Effects from common materials under different fire conditions. In: Bradley D, editor. Proceedings of the 6th international seminar on fire and explosion hazards, Leeds, UK; 2010. pp. 782–91.
[33] Purser DA. Influence of fire retardants on toxic and environmental hazards from fires [Chapter 24]. In: Hull TR, Kandola BK, editors. Fire retardancy of polymers—new strategies and mechanisms. RSC Publishing; 2009. pp. 381–404.
[34] Hull TR, Quinn RE, Areri IG, Purser DA. Combustion toxicity of fire retarded EVA. Polym Degrad Stab 2002;77:235–42.
[35] Hull TR, Carman JM, Purser DA. Prediction of CO evolution from small-scale polymer fires. Polym Int 2000;49:1259–65.
[36] Hull TR, Stec AA, Paul KT. Hydrogen chloride in fires. Fire Saf Sci 2008:665–76.
[37] Stec AA, Hull TR, Torero JL, Carvel R, Rein G, Bourbigot S, et al. Effects of fire retardants and nanofillers on the fire toxicity, fire and polymers V—materials and concepts for fire retardancy, Ch 21. In: ACS symposium series 1013. Oxford University Press; 2009. pp. 342–66.
[38] Stec AA, Hull TR. Assessment of fire toxicity from polymer nanocomposites, fire retardancy of polymers: new strategies and mechanisms [Chapter 25]. In: Hull TR, Kandola BK, editors. Cambridge (UK): Royal Society of Chemistry; 2009. pp. 405–18.
[39] BS 476-7: Fire tests on building materials and structures - Method of test to determine the classification of the surface spread of flame of products"; 1997.
[40] Stec AA, Hull TR. Yields of toxic combustion products from fire retarded nanocomposite polymers. In: Drysdale D, Bradley D, Molkov V, Carvel R, editors. Proceedings of the 5th international seminar on fire and explosion hazards, Edinburgh, UK, April 2007. University of Edinburgh; 2008. Prize winning papers, Section 2, pp. 125–35.
[41] Molyneux S, Stec AA, Hull TR. The effect of gas phase flame retardants on fire effluent toxicity. Polymer Degradation and Stability (in Press) 2013. http://dx.doi.org/10.1016/j.polymdegradstab.2013.09.013.
[42] Stec AA, Hull TR. Assessment of the fire toxicity of building insulation materials. Energy Build 2011;43:498–506.
[43] Papadopoulos AM. State of the art in thermal insulation materials and aims for future developments. Energy Build 2005;37:77–86.
[44] Blomqvist P, Hertzberg T, Dalene M, Skarping G. Isocyanates, aminoisocyanates and amines from fires—a screening of common materials found in buildings. Fire Mater 2003;27:275–94.

[45] Hertzberg T, Blomqvist P, Dalene M, Skarping G. Particles and isocyanates from fires, SP report 2003:05. Boras: SP Swedish National Testing and Research Institute; 2003.
[46] Blomqvist P, Simonson McNamee M, Stec A, Gylenstam D, Karlsson D. Characterisation of fire generated particles. Sp Rep 2010;01:2010.
[47] Blomqvist P, Simonson McNamee M, Stec AA, Gylestam D, Karlsson D. Detailed study of distribution patterns of polycyclic aromatic hydrocarbons and isocyanates under different fire conditions. Fire and Materials 2014;38:125–44.
[48] Lebek K, Hull TR, Price D. Products of rigid PVC burning under various fire conditions, fire and polymers: materials and concepts for hazard prevention. ACS Symposium Series No.922. Oxford University Press; 2005. pp. 334–47.
[49] IEC TS 60695-7-50 Fire hazard testing Part 7-50: toxicity of fire effluent estimation of toxic potency apparatus and test method.
[50] Hull TR, Lebek K, Robinson JE. Acidity, toxicity and European cable regulation, Transactions of the international wire and cable symposium (Trans IWCS); 2006.

CHAPTER

Recycling of Postindustrial and Postconsumer Plastics Containing Flame Retardants

25

M. Schlummer

Fraunhofer Institute for Process Engineering and Packaging, Freising, Germany

CHAPTER OUTLINE

1. Introduction

Unequivocally, the use of flame retarded polymers is a successful means to reduce the number of fire cases, and therefore, they have been used in various applications [1]. At the end of their use phase, flame retarded polymers become part of waste streams and may be subjected to recycling activities. If this is probably not the major pathway these days, its importance will increase in the future, since in the long run, production

Polymer Green Flame Retardants. http://dx.doi.org/10.1016/B978-0-444-53808-6.00025-1

of virgin polymers will no longer be based on crude oil only and will apply secondary resources and renewable raw materials fo biobased polymers with increasing shares.

Technically, most flame retarded polymers can be recycled without substantial loss of quality, and from this point of view, there are references that prove recyclability [2]. However, recyclability of a single polymer/FR system does not prove that recycling of FR polymers from real waste streams can be realized in industrial practice. In reality, FR polymers are mostly mixed with other waste polymers and foreign materials (such as dirt, dust, and foreign materials from production and waste reclamation activities) [3]. They require a comprehensive set of techniques in order to be reclaimed into grades of sufficient pureness, since only such grades will pass the industrial requirements, which are set for a secondary use in polymeric products [4,5].

As some older systems of brominated flame retardants (BFR) turned out to be hazardous after some decades of use, as polybrominated biphenyls (PBB) and diphenyl ethers (PBDE) may be degraded to highly toxic brominated dibenzo-*p*-dioxins and dibenzofurans (PBDD/F), there have been controversial discussions on the treatment of flame retarded plastics at the end of their use phase [2,6–11]. Legal activities have put pressure on old BFR systems in the past and hinder the use of polymers containing substantial amounts of PBB, PBDE, and PBDD/F in the production of recycled polymers [12]. But even if the discussions had started in 1993, there are still significant amounts of PBDE and PBDD/F in current waste streams [10,13,14].

Since the beginning of these discussions, there has been much development in the field of flame retarded polymers resulting in a comprehensive picture of old BFR, new BFR with a much lower PBDD/F building potential, and mainly phosphorous-based flame retardants [1,9,15–20].

Despite these obstacles, there have been successful approaches to realize the recycling of FR polymers. This has been demonstrated for pure polymer/FR systems, however, a number of studies could prove the recyclability of flame retarded polymers from real waste plastic mixtures, mainly obtained from waste electric and electronic equipment [5,21–25].

This paper will focus on the recycling of real waste streams containing BFR polymers and will only partly treat studies on the recyclability of virgin polymer/FR systems. However, a discussion of polymers with intrinsic flame retardancy (as polyvinyl chloride, PVC) is out of scope of this work.

2. FR polymers in waste streams

Mainly two types of waste streams have to be discussed with respect to the recycling of FR polymers, postindustrial waste, characterized by well-defined and mostly clean BFR–polymer systems, opposed to waste plastics, recovered from postconsumer waste streams, which contain nonpolymeric constituents, mixtures of several polymer types and mixtures of flame retardants.

2.1 Postindustrial waste

Postindustrial waste comprises plastic waste produced by plastic-processing companies. This waste is for example generated when extrusion or injection molding lines are started or stopped, or if the quality of the plastic parts does not comply with internal standards. Postindustrial waste can be collected separately for each line and therefore mostly provides a well-defined composition of a polymer type and a defined additive composition, e.g. a flame retardant. So, there is no issue of further sorting or separating of such waste fractions.

Significant amounts of waste FR polymers may be produced in companies producing

1. casings for electric and electronic equipment (printers, TV sets, notebooks, etc.),
2. flame retarded textiles (automotive, furniture) [26], or
3. construction materials such as flame retarded expanded polystyrene (FR-EPS), or flame retarded roofing membranes.

Technically, recycling of these fractions require a grinding step and a recompounding. In case that disturbing particles have entered the waste stream (e.g. dust, foreign products materials such as glass or metals), a melt filtration may be included in the reclamation process. Due to this comparably easy design of such recycling aggregates, postindustrial plastics are often recycled in plastic-processing companies by themselves. However, smaller companies use the service of specialized external recycling companies. This, however, may lead to mixtures of well-defined postindustrial plastics, if waste batch volumes are too small.

2.2 Postconsumer waste

Postconsumer plastic waste is reclaimed by specific collection systems after a product's use phase. Reclamation systems differ from country to country and may even be different regionally. The presence of FR plastics has been reported for construction waste [27], end-of-life vehicles (ELV) [21,28], Waste Electrical and Electronic Equipment (WEEE) [14,29], and upholstery textiles [26]. However, only a few collection systems have established a waste treatment that separates plastic fractions for recycling. This is the case for WEEE and ELV or EPS from packaging waste.

FR plastics in ELV are mainly due to additives in upholstery textiles, which are not separated for material recycling. Similarly, FR plastics in construction and demolition waste like flame retarded roof membranes or FR-EPS in insulation panels are normally not separated and applied to plastic recycling. Thus, WEEE is the main postconsumer plastic waste stream, which contains FR on the one hand and which is subjected to material recycling on the other hand [4,10,13,14,23,29].

Flame retardants in WEEE have been studied by a variety of authors [10,13,14,29], who revealed the presence of additive and reactively bound FR. The latter are mainly used in resins of printed circuit boards, whereas additive FRs are mainly found in thermoplastics. Among others, polybrominated diphenyl

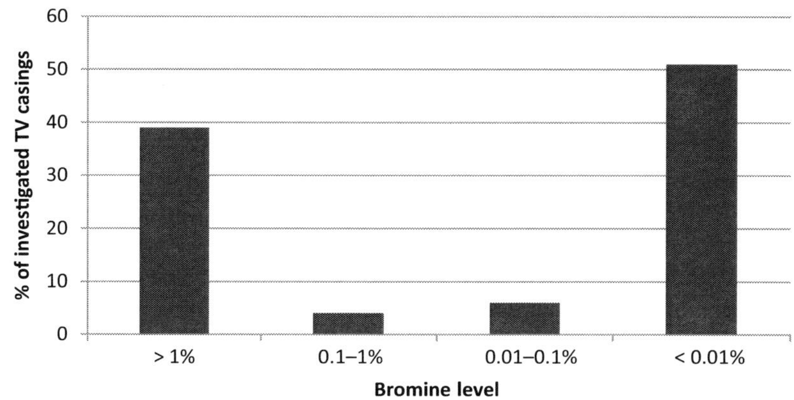

FIGURE 3

In 10 out of 100 TV-set casings, reduced bromine levels were identified, which would not be suitable as intentional FR equipment. This indicates that BFR plastics have entered the production of electrical and electronic equipment (EEE) in the past.

The main disadvantage of such processes is that hazardous contaminants such as PBDE may be diluted into a wide range of plastics. If bromine levels drop below an industrial control level, regranulated plastics might enter more sensitive product chains (toys, food contact materials). The sustainability of that approach remains questionable.

5.3 Separation techniques

In a German research project (SpectroDense, 2008–2010), a centrifugal force-assisted density separation (Sorticanter®) was applied to 28 different fractions of ground TV set and monitor casings derived from German dismantling plants. Bromine levels in these input fractions ranged from 30.000 to 40.000 mg kg^{-1}. In ABS and PS density fractions of the Sorticanter®, much lower bromine levels were analyzed. As depicted in Figure 5, the cumulative percentages of target fractions with bromine levels below 100, 500 and 1000 mg kg^{-1} accounted for 36%, 68%, and 93%. Furthermore, none of these fractions failed the RoHS compliance test, since bromine levels >1000 mg kg^{-1} could not be attributed to PBDE only.

In the framework of the same project, a manual dismantling campaign was supported by hand scanners (SSS and NIR). As displayed in Figure 6, out of 133 plastic parts from monitor casings, 79 weight% was made of BFR-free ABS, PS, or PC/ABS. These may be subjected to material recycling, considering the readdition of additives. However, a share of 25% of FR polymers is not reclaimed and has to be treated thermally.

Another German project investigated the combination of an automated NIR sorting device run at a WEEE recycling plant with a subsequent separation of BFR-equipped polymers by means of XRT in a university technical area [36]. NIR was used to separate gray ABS and PS fractions, which covered just a minority

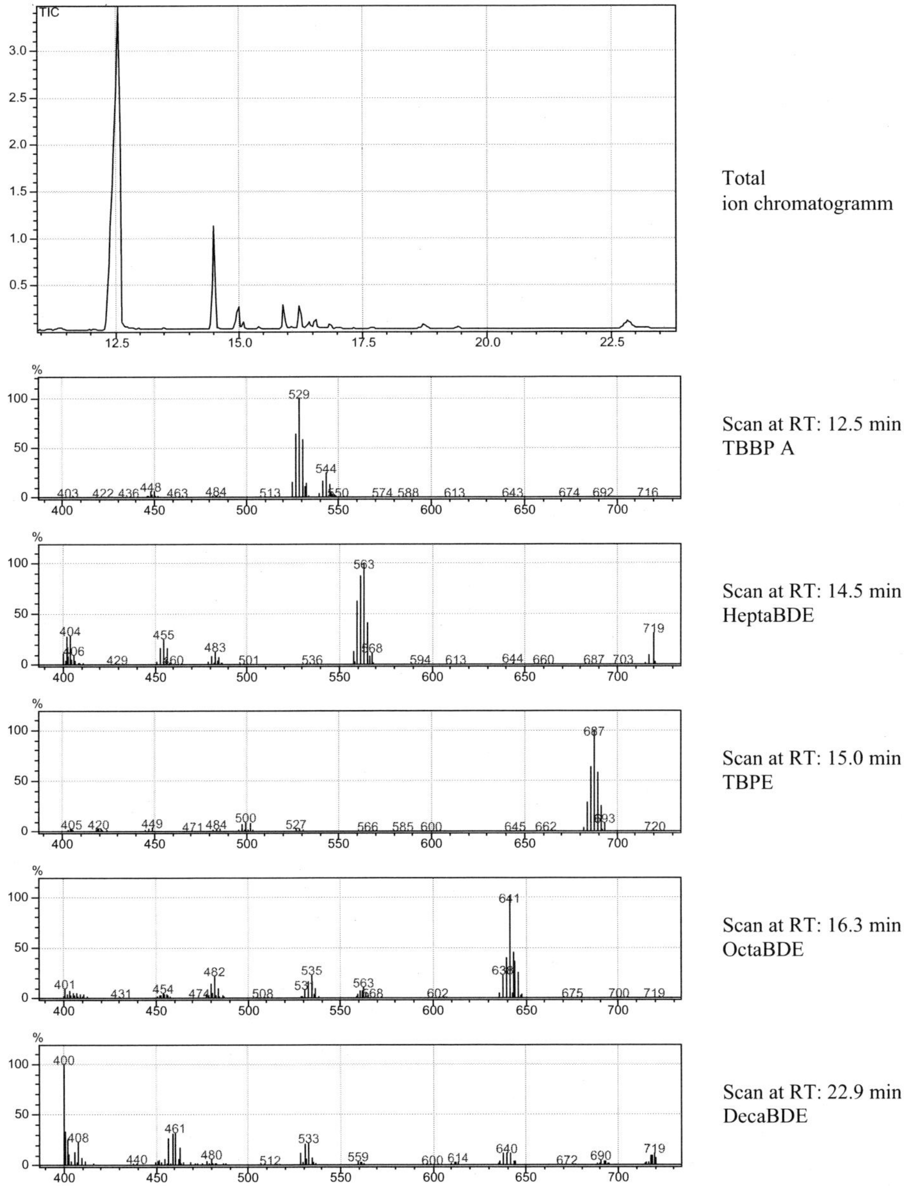

FIGURE 4

Mixture of BFRs identified in an AC adapter [35].

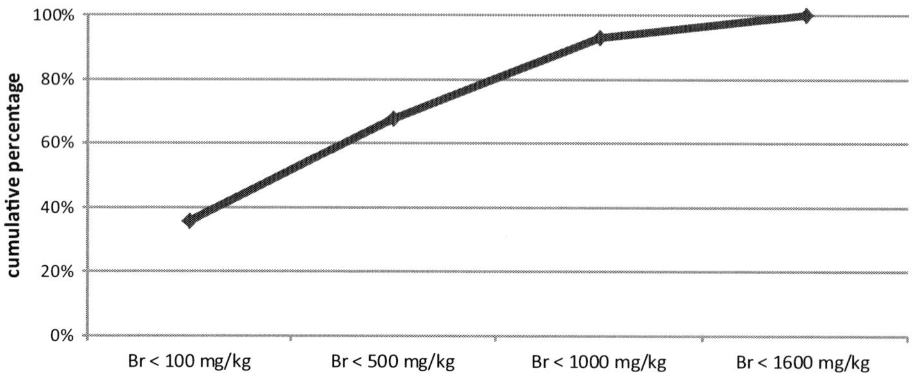

FIGURE 5

Cumulative percentage of bromine levels in 28 PS and ABS fractions separated from shreds of TV casing by sink and float. The input fraction of the technology contained about 40% of BFR-equipped casings, i.e. an average bromine level of 30.000–40.000 mg kg^{-1}.

of the plastics present in the treated WEEE. Subsequent XRT sorting, however, was rather effective, since none of the investigated samples of BFR-free ABS and PS exceeded a level of 1000 mg kg^{-1}.

5.4 Extractive approaches

The above-described extractive CreaSolv® process has been applied to WEEE plastics in several studies. Initially, Schlummer et al. [37] subjected density presorted BFR rich plastic shred from small WEEE to the solvent-based technique in a small technical batch trial. An extractive cleaning of such polymer solutions reduced amounts of initially present BFR by 98%.

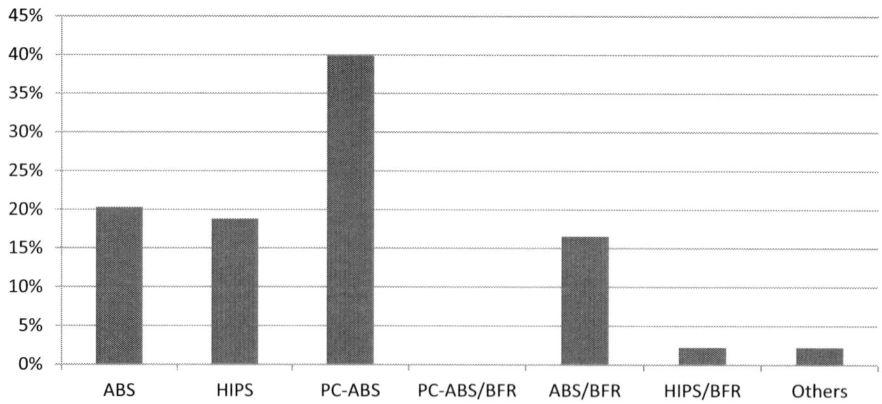

FIGURE 6

Manual sorting of 133 monitor casings from a German dismantling plant applying NIR for polymer identification and SSS for the detection of BFR.

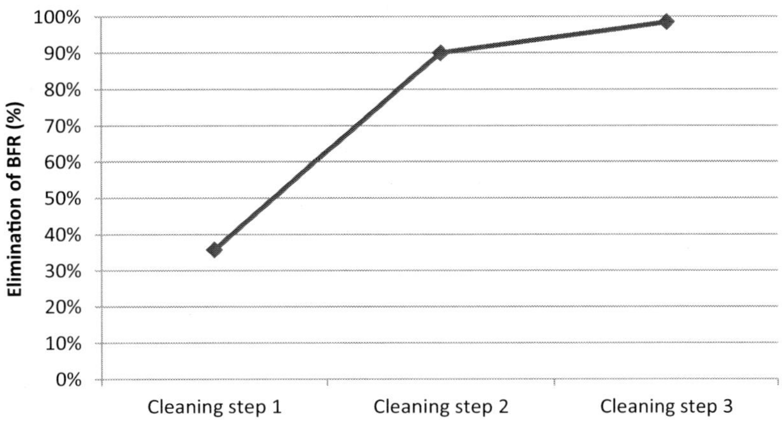

FIGURE 7

Within three cleaning steps, the CreaSolv® reaches an elimination of BFR >98%.

As the results indicated good mechanical properties of products and compatibility with RoHS and ChemVV, there were further trials in laboratory and in small technical scale within the German PolyResource Project [25,38,39]. These trials optimized single process steps in order to improve the process economy. As a main approach, batch process steps were reviewed to be replaced by continuous ones and were implemented for the extractive cleaning of dissolved polymers from contaminants as well as for the precipitation step. In addition, a stepwise dissolution of more than one target polymer was tested successfully.

The economically driven optimization of the process, however, did not reduce the cleaning efficiency. Figure 7 presents the most recent data on the cleaning potential of the CreaSolv® process. With only three consecutive cleaning steps, BFRs were eliminated by 98.4%. However, the CreaSolv® process is not yet available on a technical scale, despite the positive evaluation of its technical, economic, and environmental performance [23,25].

6. CONCLUSIONS

In industrial practice, polymer recycling is driven by monetary values of waste materials. Costs of input waste fractions and of conversion into recycled polymers have to be covered by the market price of the recycled polymer. In the case of commodity polymers such as PS and polypropylene, these prices are rather low (<1000 €/t) and do not allow many conversion processes.

In consequence, recycling activities focus on postindustrial plastics as these do not require much processing (grinding and remelting) and the applied additive polymer systems are normally not mixed with other waste plastics and they comply with legal limits as the related virgin polymers do. Postindustrial FR plastics can therefore most easily be recycled.

Recycling of flame retarded polymers from postconsumer waste, however, requires much more technical efforts and results in higher conversion cost. Since the quality of recycled products is in most cases not higher than that of postindustrial input fractions, market prices of recycled products are mainly governed by postindustrial recyclates. The probability of the presence of legally regulated FR causes further material testing costs. Therefore, at least plastics with a significant BFR share are normally withdrawn from plastic recycling.

Results of case studies presented in this paper, however, indicate that it is possible to recycle FR and even BFR-equipped plastic waste. Approaches include the sorting of FR-free fractions from mixed inputs and extractive techniques (CreaSolv®). Recyclates fulfill technical and legal requirements and may be applied in the production of new plastics [25].

Economic calculations revealed that related conversion cost can be covered by market prices of recycled polymers, but they indicated the importance of choosing input fractions. Sorting approaches should handle fractions with a high share of BFR-free plastics (>60%), whereas extractive approaches may focus on high shares of BFR. Therefore, these techniques do not compete on the same waste fractions and should be applied consecutively, i.e. sorting rejects are subjected to extraction technologies.

In consequence, it can be expressed that suitable technical approaches to recycle FR and BFR plastics are developed and may be put into large-scale industrial practice. Today, only sorting approaches have been realized. BFR-rich sorting residues, however, are thermally treated or exported to developing countries, where their fate is unclear. As discussed above, dilution of BFR plastics into recycled plastics may take place and these recycled polymers may return to developed countries. Therefore, stricter rules on *trans*-boundary transport of BFR-containing waste streams are required in order to prevent that practice.

FR and BFR polymers are evaluated as important materials in electric and electrical equipment [29,40], and therefore, they will stay in WEEE for the next decades. The increasing cost of crude oil will further raise the importance of recovering secondary polymeric raw materials from waste streams. Integrated approaches as the above-mentioned combination of sorting and extractive techniques would reclaim the maximum amount of polymers. For this reason, the approach has been included in a recent guide on guidelines on the best available techniques and best environmental practice for the recycling and disposal of articles containing PBDEs under the Stockholm Convention on persistent organic pollutants [41,42].

References

[1] Manor-Orit GP. Flame retardants and the environment. Spec Chem 2005;25(7):36.

[2] Dawson RB, Landry SD. Recyclability of flame retardant HIPS, PC/ABS, and PPO/HIPS used in electrical and electronic equipment. In: Proceedings of the international symposium on electronics and the environment. New Orleans; 2005.

[3] Delgado C, Barruetabena L, Salas O, Wolf O, editors. Assessment of the environmental advantages and drawbacks of existing and emerging polymers recovery processes. Luxembourg: Office for Official Publications of European Communities; 2007.

[4] Nnorom IC, Osibanjo O. Sound management of brominated flame retarded (BFR) plastics from electronic wastes: state-of-the art and options in Nigeria. Resour Conserv Recycl 2008;52(12):1362−72.

[5] Schlummer M, Mäurer A. Recycling of styrene polymers from shredded screen housings containing brominated flame retardants. J Appl Polym Sci 2006;102(2): 1262−73.

[6] Ebert J, Bahadir M. Formation of PBDD/F from flame-retarded plastic materials under thermal stress. Environ Int 2003;29(6):711−6.

[7] Hamm S, Strikkeling M, Ranken PF, Rothenbacher KP. Determination of polybrominated diphenyl ethers and PBDD/Fs during the recycling of high impact polystyrene containing decabromodiphenyl ether and antimony oxide. Chemosphere 2001;44(6): 1353−60.

[8] Imai T, Hamm S, Rothenbacher KP. Comparison of the recyclability of flame-retarded plastics. Environ Sci Technol 2003;37(3):652−6.

[9] Meyer H, Neupert M, Pump W, Willenberg W. Flammschutzmittel entscheiden über Wiederverwertbarkeit. (Flame retardants are decisive for material recycling). Kunststoffe 1993;83:253−7.

[10] Schlummer M, Gruber L, Mäurer A, Wolz G, van Eldik R. Characterisation of polymer fractions from waste electrical and electronic equipment (WEEE) and implications for waste management. Chemosphere 2007;67(9):1866−76.

[11] Zelinski V, Lorenz W, Bahadir M. Brominated flame retardants and resulting PBDD/F in accidental fire residues from private residences. Chemosphere 1993;27(8): 1519−28.

[12] EU. Directive 2002/95/EC on the restriction of the use of certain hazardous substances in electrical and electronic equipment (RoHS). OJ, vol. L37; 2006. pp. 19−23.

[13] Morf LS, Tremp J, Gloor R, Huber Y, Stengele M, Zennegg M. Brominated flame retardants in waste electrical and electronic equipment: substance flows in a recycling plant. Environ Sci Technol 2005;39(22):8691−9.

[14] Wager PA, Schluep M, Muller E, Gloor R. RoHS regulated substances in mixed plastics from waste electrical and electronic equipment. Environ Sci Technol 2012;46(2): 628−35.

[15] Levchik S, Weil ED. Flame retardants in commercial use or in advanced development in polycarbonates and polycarbonate blends. J Fire Sci 2006;24:137−51.

[16] Lu SY, Hamerton I. Recent developments in the chemistry of halogen-free flame retardant polymers. Prog Polym Sci 2002;27(8):1661−712.

[17] Morgan AB, Gilman JW. An overview of flame retardancy of polymeric materials: application, technology, and future directions. Fire Mater 2013;37:259−79.

[18] Nishihara H, Suda Y, Sakuma T. Halogen- and phosphorus-free flame retardant PC plastic with excellent moldability and recyclability. J Fire Sci 2003;21(6):451−64.

[19] Roth T, Urpi Bertran R, Pöhlein M, Wolf M, van Eldik R. Gas chromatographic determination of phosphate-based flame retardants in styrene-based polymers from waste electrical and electronic equipment. J Chromatogr A 2012;1262:188−95.

[20] Weil ED, Levchik SV. Flame retardants in commercial use or development for polyolefins. J Fire Sci 2008;26(1):5−43.

[21] Bocz K, Toldy A, Kmetty Á, Bárány T, Igricz T, Marosi G. Development of flame retarded self-reinforced composites from automotive shredder plastic waste. Polym Degrad Stab 2012;97(3):221–7.
[22] Brennan LB, Isaac DH, Arnold JC. Recycling of acrylonitrile–butadiene–styrene and high-impact polystyrene from waste computer equipment. J Appl Polym Sci 2002; 86(3):572–8.
[23] Freegard K, Tan G, Morton M. Final report on WRAP project PLA-037. Develop a process to separate brominated flame retardants from WEEE polymers. Banbury: Wrap; 2006.
[24] Park CH, Jeon HS, Yu HS, Han OH, Park JK. Application of electrostatic separation to the recycling of plastic wastes: separation of PVC, PET, and ABS. Environ Sci Technol 2008;42(1):249–55.
[25] Schlummer M, Mäurer A, Altnau G. Recycling of high performance polymers from electro(nic) scrap. In: Proceedings electronics goes green 2012+. Berlin: Fraunhofer Verlag; 2012.
[26] Kajiwara N, Sueoka M, Ohiwa T, Takigami H. Determination of flame-retardant hexabromocyclododecane diastereomers in textiles. Chemosphere 2009;74(11):1485–9.
[27] Mäurer A, Knauf U. EPS-Loop. Recycling of EPS-waste to expandable polystyrene. In: FAKUMA forum 2004. Germany: Friedrichshafen; 2004.
[28] Osada M, Tanigaki N, Takahashi S, Sakai S. Brominated flame retardants and heavy metals in automobile shredder residue (ASR) and their behavior in the melting process. J Mater Cycles Waste Manage 2008;10(2):93–101.
[29] Herat S. Environmental impacts and use of brominated flame retardants in electrical and electronic equipment. Environmentalist 2008;28(4):348–57.
[30] Riess M, Ernst T, Popp R, Muller B, Thoma H, Vierle O, et al. Analysis of flame retarded polymers and recycling materials. Chemosphere 2000;40:937–41.
[31] Dimitrakakis E, Janz A, Bilitewski B, Gidarakos E. Small WEEE: determining recyclables and hazardous substances in plastics. J Hazard Mater 2009;161:913–9.
[32] Weber R, Kuch B. Relevance of BFRs and thermal conditions on the formation pathways of brominated and brominated-chlorinated dibenzodioxins and dibenzofurans. Environ Int 2003;29:699–710.
[33] Gent MR, Menendez M, Toraño J, Torno S. DMS cyclone separation processes for optimization of plastic wastes recycling and their implications. Waste Manage Res 2011;29: 644–55.
[34] Sindiku O, Babayemi OJ, Osibanjo O, Schlummer M, Schluep M, Weber R. Screening e-waste plastics in Nigeria for brominated flame retardants using XRF—towards a methodology for assessing POPs BDEs in e-waste exports. Organohalogen Compd 2011;73:785–8.
[35] Schlummer M, Mäurer A, Gruber L, Wolz G. Coincidence of different brominated flame retardants (BFR) in electronic equipment—indication for inadequate material recycling of WEEE polymers. Organohalogen Compd 2006;68:2011–4.
[36] Krämer P, Walther G, Flamme S. Waste electrical and electronic equipment. Are potentials for increasing the creation of value sufficiently used? Müll Abfall 2009;12(09): 623–7.
[37] Schlummer M, Maurer A, Leitner T, Spruzina W. Report: recycling of flame-retarded plastics from waste electric and electronic equipment (WEEE). Waste Manage Res 2006;24:573–83.

[38] Dürr B. Demonstration plant for investigating the precipitation of polymer solutions in a continuous process. Diploma thesis. Munich: University of Applied Science; 2009.
[39] Stürzl S. Construction, measurement instrumentation and test operation of a small-scale continuous construction for precipitation. Diploma thesis. Weihenstephan-Triesdorf: University of Applied Science; 2012.
[40] Shaw SD, Blum A, Weber R, Kannan K, Rich D, Lucas D, et al. Halogenated flame retardants: do the fire safety benefits justify the risks? Rev Environ Health 2010;25(4): 261–305.
[41] UNEP. Technical review of the implications of recycling commercial penta and octabromodiphenyl ethers. Geneva. In: Stockholm convention document for 6th POP reviewing committee meeting (UNEP/POPS/POPRC.6/2); October 11–15, 2010.
[42] UNEP. Supporting document for technical review of the implications of recycling commercial penta and octabromodiphenyl ethers. Geneva. In: Stockholm convention document for 6th POP reviewing committee meeting (UNEP/POPS/POPRC.6/INF/6); October 11–15, 2010.
[43] Arends D. Laser Separation for the Sortation of Plastics from WEEE Shredder Light Fractions. Schriftenreihe der GDMB 2014;135:203–11.

CHAPTER

Methodology for Testing the Life Cycle Sustainability of Flame Retardant Chemicals and Nanomaterials

26

Marta Escamilla, Roshan Paul

LEITAT Technological Center, Terrassa (Barcelona), Spain

CHAPTER OUTLINE

1. Introduction

Fires are a common cause of harm to people and property and also generate a host of acute and chronic pollutants, including acid gases and persistent organic pollutants. Flame retardant (FR) chemicals are in great demand due to their extensive applications in various industrial sectors, including transportation and construction. There are six main families of FR chemicals that can be added to various kinds of polymers or materials, to enhance their FR properties. They include phosphorous (phosphonates,

Polymer Green Flame Retardants. http://dx.doi.org/10.1016/B978-0-444-53808-6.00026-3

phosphine oxides, phosphates, etc.), silicone (linear silanes, silsesquioxanes, etc.), boron (boric acids, borates, carboranes, etc.), nitrogen (melamine and derivatives), metal hydroxides (aluminum and magnesium hydroxides), and nanomaterials.

Most of these FR chemicals and also the halogenated ones have undesirable health and environmental effects. Various studies have shown that the concentrations of halogenated FRs in the environment and in human tissues are rising because of their increased presence in the environment. Their presence keeps on increasing in the manufacture, use, and after they have been discarded as waste. In this context, there is great interest in developing alternatives of halogenated FRs, and many new FR products containing nanoparticles are being developed. The life cycle assessment (LCA) of conventional FR chemicals is well discussed in the literature, but little information is available on nanomaterials, so this chapter will focus more on FR nanomaterials.

Carrying out an LCA of FR chemicals allows one to have a global knowledge of its environmental performance in all life cycle stages. LCA is a tool for the systematic evaluation of the environmental aspects of a product, and its main objective is to identify which materials pose greater impacts, where these impacts occur in the chemical life cycle, and their impact on environment and human health (HH).

The LCA is an environmental management tool used to assess the environmental impacts of a product over the entire life cycle, "from the cradle to the grave". According to ISO 14040 [1–3], LCA is defined as

> *"A technique for assessing the environmental aspects and potential impacts associated with a product by: compiling an inventory of relevant inputs and outputs of a product system, evaluating the potential environmental impacts associated with those inputs and outputs, interpreting the results of the inventory analysis and impact assessment phases in relation to the objectives of the study".*

The LCA exhaustively examines every stage of the whole life cycle, from the obtainment of raw materials, through its manufacture, distribution, use, and final disposal. In the case of FR chemicals and nanomaterials, the following stages should be taken into account:

- Synthesis/Production: Considering the wide variety of FR chemicals existing today, there cannot be any generalization. Depending on the nature of the FR chemical to be analyzed, the LCA strategy should be adopted starting from its raw materials, synthesis, functionalization, and final industrial production.
- Transport: The second important parameter to be considered is the packaging, storage, and transportation to end users. The stability and shelf life of the packed final product is also an important consideration.
- Use/Fire: It is quite possible that during the combustion of a material, it disintegrates due to the generated heat and emits gases. These gases are flammable and react with the oxygen present in the air and generate an increase in heat and additional combustion wastes. The increase in heat leads to the combustion of the remaining material, and a combustion cycle is produced. This chain reaction is repeated until the material totally disappears.

The operation method of FR chemicals consists of eliminating some products of the combustion cycle. FRs can take part in different phases of the combustion cycle, such as heating, decomposition, and smoke process. FRs can act physically as a protective layer or chemically in two different phases: the condensed phase and the vapor (or gas) phase. It is important to know how the FR chemical acts, its mode of action during a fire, how it protects the material, and to what extent. Also, it is needed to know whether the material or the FR chemical itself emits toxic gases.

- End of life: After a fire, the materials containing FR chemicals and the FR chemical itself should be recycled or disposed safely.

In the LCA analysis, inputs (resources) and outputs (emissions) are collected at each stage, in order to aggregate them over the life cycle, and then convert into the corresponding potential impacts on the environment. The sum of such environmental impacts represents the overall environmental effect of the life cycle of the product. In general terms, the LCA detects the hot spots in the whole life cycle and carries out environmental comparison of different actions that can improve the environmentally friendly nature of the production and use of chemicals. Furthermore, LCA can help to implement manufacturer strategies, identify opportunities, and to improve the environmental aspects of FRs along its life cycle.

A life cycle approach avoids shifting environmental burdens between life cycle phases. The essential characteristic and major strength of the LCA is its holistic nature, bringing the environmental impacts into one consistent framework. All necessary inputs and outputs in many stages and operations of the life cycle are considered to be within the system boundaries. This includes not only direct inputs and outputs for production, distribution, use, and final disposal, but also indirect inputs and outputs (e.g. from the initial production of the energy used) regardless of when or where they occur. Unfortunately, this holistic nature of LCA itself is its main limitation. In fact, a cradle-to-grave analysis of a product can be only achieved at the expense of simplifying other aspects [4—6].

Nowadays, nanomaterials are increasingly used in many production processes, and FR nanomaterials are now incorporated in a rising number of products. But there is a lack of knowledge regarding the environmental and HH impacts of these FR nanomaterials. Reference data are needed on the impact of nanomaterials in HH and environment at the different stages of their entire life cycle (production process, use phase, or end of life). Although LCA methodologies are being successfully used for many years in different fields, there is still a lack of LCA models for complex systems such as nanomaterials possessing novel properties, which are different from those of conventional chemicals.

Therefore, the existing impact assessment methods and databases of LCA methodologies should be adapted in order to include specific and new data of the potential impacts of FR nanomaterials on the environment and HH.

2. LCA approach

The International Organization for Standardization (ISO), an international federation of national standards bodies, has standardized the LCA framework within the series ISO 14040, establishing four interrelated basic phases for LCA studies, as shown in Figure 1.

The first phase of the LCA is the goal and scope definition, which defines the general context of the study. In the goal definition, parameters such as the functional unit, the intended application, the motivations for carrying out the study, the target audience or the limitations and assumptions are described.

The next step is the inventory analysis, where each life cycle stage is analyzed to determine the relevant inputs and outputs of the system. Extraction and consumption of resources (water, energy, raw materials), as well as releases to air, water, and soil are quantified throughout the whole life cycle phases.

Special efforts should be made to determine the potential release of nanomaterials and FR chemicals into the environment. The third phase is the impact evaluation, which also has to be fully adapted to nanomaterials.

The final step is the interpretation of results, where a critical evaluation of the results is done in order to verify its reliability. At this stage, the completeness, sensitivity, and consistency of data gathered and results obtained are essential.

The execution of iterative analysis is foreseen during the course of the LCA process to internally control the quality of the data. So whenever needed, the proposed methodology should be redefined and adapted during the LCA study development.

When carrying out a LCA study, an integrated approach is needed, taking into consideration, for instance:

- which impacts have to be covered
- how the impacts will be described
- how the data will be collected
- how the demand of resources and emissions will be allocated to the products

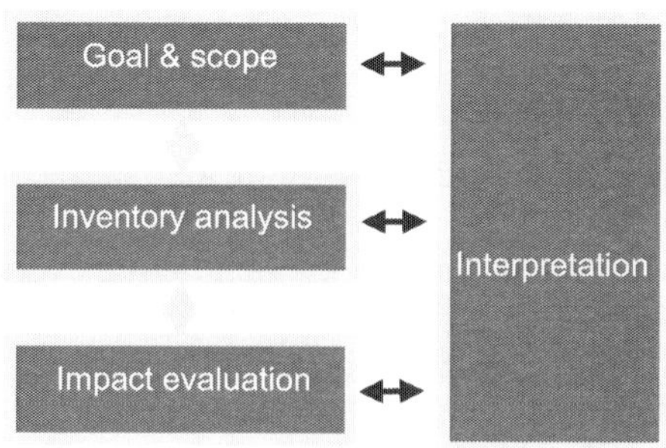

FIGURE 1

LCA diagram.

- how to deal with insecurities
- how to aggregate (if necessary) and to visualize the results
- which reference will be used for comparison

A methodological approach to achieve the entire comprehension of the environmental impact of FR nanomaterials, based on the LCA and risk assessment, is further explained in the following section of this chapter.

3. Methodology

3.1 Life cycle assessment

3.1.1 Goal and scope

As mentioned before, the goal and scope is the first stage of LCA and probably the most important one, as the scope is defined here and the manner in which the results have to be used is also stated. The goal and scope will determine the working plan of the entire study.

This first step also comprises the limitation of the system where the system boundaries are set, and a suitable functional unit is established. The functional unit will serve as a reference unit to be considered along the life cycle. FR nanomaterials should be studied using an approach that will model the different scenarios representative of their entire life cycle, so that different parameters that may affect its behavior such as the type of production process or materials used will be assessed.

Some basic assumptions should be made, for example:

- Will the study focuses only on the environmental aspects? Or it will also take into account their economic and social characteristics?
- Will the LCA be based on linear modeling?
- Will it address localized impacts? Will it be typically a steady state rather than a dynamic approach?
- The environmental impacts are described as "potential impacts" and not as real impacts because they are not actually specified in time and space.
- To avoid arbitrariness, an important aim will be to make transparent decisions and give clear and detailed information.

Finally, uncertainties and limitations need to be defined regarding data availability and quality (existing studies, databases, in vitro and in vivo tests) and those inherent to the FR nanomaterial models to assess toxicity, environmental fate, and ecotoxicity.

3.1.2 Inventory analysis

The inventory analysis of a LCA study comprises the data collection and the calculation procedures to quantify the inputs and outputs through the system boundaries. To make the analysis easier, the system can be divided into several interconnected subsystems.

In order to obtain the required information, different kinds of data can be used: (i) Primary data provided by manufacturers, suppliers, or the LCA analyst team;

(ii) Other information sources, such as literature and existing databases; or (iii) Data derived from estimations and educated guesses.

It is important to know the geographical site, time period, and type of technology to which the data refer and its representativeness.

3.1.3 Impact assessment

The impact assessment is the stage where the results derived from the inventory analysis are further processed in terms of potential environmental impacts. A base methodology that can be adapted and transformed for impact assessment of FR nanomaterials is the ReCiPe Mid/Endpoint method [7,8]. This methodology combines both midpoints and endpoints in a consistent framework (cause—effect chain), which allows a step-by-step interpretation and revision of the results. In order to calculate characterization factors (CFs) for FR nanomaterials, toxicity, fate, and exposure models can be adapted to the specific needs in this field [9].

The impact assessment consists of four distinct steps:

1. Selection and classification of impact categories

The environmental impact categories, which are of relevance to the study, are defined by their impact pathway and impact indicator, and the elementary flows from the inventory are assigned to the impact categories according to the ability of the substance to contribute to environmental problems.

The categories gather the different impact information collected by the indicators. The ReCiPe methodology has eighteen midpoint indicators, which are robust and with a low level of uncertainty, and three endpoint indicators, aimed to make the result interpretation easier.

The midpoint categories are (i) climate change; (ii) ozone depletion (OD); (iii) terrestrial acidification; (iv) freshwater eutrophication; (v) marine eutrophication; (vi) human toxicity (HT); (vii) photochemical oxidant formation; (viii) particulate matter formation; (ix) terrestrial ecotoxicity; (x) freshwater ecotoxicity; (xi) marine ecotoxicity; (xii) ionizing radiation; (xiii) agricultural land occupation (ALO); (xiv) urban land occupation (ULO); (xv) natural land transformation (NLT); (xvi) water depletion (WD); (xvii) mineral resource depletion (MRD); and (xviii) fossil resource depletion (FD).

At the endpoint level, the three categories to be considered are (i) damage to HH; (ii) damage to ecosystem diversity (ED); and (iii) damage to resource availability (RA). The first category comprises all the toxic effects of nanoparticles on the human body, both acute and chronic, and can be assessed by means of the in vitro studies performed in human cell lines and in vivo biodistribution and accumulation studies. The second category gathers the effects that nanoparticles have on nature and the biota of a given ecosystem (ecotoxicity as well as nontoxic interactions). The third category will take into account the emission of pollutants and the contribution to natural resource depletion during the manufacture process and the remaining life cycle of the particles.

To achieve the entire comprehension of the impact of FR nanomaterials in the environment, the full set of impact categories are to be considered, as specified in Table 1.

Table 1 Overview of the Midpoint Categories and the Connection between Midpoint and Endpoint Categories

	Indicator Name	Unit	Midpoint Impact Category	Endpoint Impact Category		
				HH	ED	RA
1. Climate change	Infrared radiative forcing	W × year m^{-2}	kg (CO_2 to air)	+	+	
2. OD	Stratospheric ozone concentration	ppt per year	kg (CFC-11 to air)	+	–	
3. Terrestrial acidification	Base saturation	year × m^2	kg (SO_2 to air)		+	
4. Freshwater eutrophication	Phosphorus concentration	year × kg m^{-3}	kg (P to freshwater)		+	
5. Marine eutrophication	Nitrogen concentration	year × kg m^{-3}	kg (N to freshwater)		–	
6. HT	HT hazard-weighted dose	–	kg (14DCB to urban air)	+		
7. Photochemical oxidant formation	Photochemical ozone concentration	kg	kg (NMVOC to air)	+	–	
8. Particulate matter formation	PM10 intake	kg	kg (PM10 to air)	+		
9. Terrestrial ecotoxicity	Hazard-weighted concentration	m^2 × year	kg (14DCB to industrial soil)		+	
10. Freshwater ecotoxicity	Hazard-weighted concentration	m^2 × year	kg (14DCB to freshwater)		+	
11. Marine ecotoxicity	Hazard-weighted concentration	m^2 × year	kg (14DCB to marine water)		+	
12. Ionizing radiation	Absorbed dose	man × Sv	kg (U^{235} to air)	+		
13. ALO	Occupation	m^2 × year	m^2 × year (agricultural land)		+	–
14. ULO	Occupation	m^2 × year	m^2 × year (urban land)		+	–
15. NLT	Transformation	m^2	m^2 (natural land)		+	–
16. WD	Amount of water	m^3	m^3 (water)			–
17. MRD	Grade decrease	kg^{-1}	kg (Fe)			+
18. FD	Upper heating value	MJ	kg (oil)			+

"+": A quantitative connection has been established in ReCiPe."–": Although it is an important link, no quantitative connection could be established.

2. Characterization

The goal of characterization in Life Cycle Impact Assessment (LCIA) is to estimate the potential contribution of different interventions (emissions, resource extractions, and land use) to different impact categories and to sum up the amounts of interventions into a single number within each impact category. The impact is expressed as an impact score in a common unit to all contributions within the impact category by applying characterization factors.

Characterization factors allow the conversion of results from the Life Cycle Inventory study (LCI) to impact indicators belonging to the previously described impact categories and fit well into a predefined mathematical equation (simplifying, Impact Indicator = Inventory data × Characterization factor). For the elements (inputs and outputs) of the inventory that have CFs associated for the different impact categories of the ReCiPe method, the impact assessment can be carried out and the impact loads can be calculated.

A suitable characterization factor adapted to FR nanomaterials has to be developed in the beginning of the LCIA studies. The ReCiPe model does not contain CFs specific for all FR chemicals or nanoparticles. In order to incorporate their impacts released into the environment, CFs will need to be generated, taking into account the most relevant categories in terms of the possible impact of FR nanomaterials, for instance HT and Freshwater Ecotoxicity.

3. Normalization

The characterized impact scores are related to a common reference in order to facilitate comparisons across impact categories. A normalizing parameter can be optionally defined attending to the properties of each type of FR chemicals and nanomaterials in order to carry out a comparison between the relative impacts of the different chemicals. The selection of an easily mutable reference value for normalization will be sought so as to make the Impact Analysis more adaptable to FR nanoparticles of different nature.

4. Weighting

If necessary, a ranking and/or weighting is performed of the environmental impact categories reflecting the relative importance they have in the study. Since this is a significant step in the LCIA models, mainly due to its subjectivity, the weighting of the different impact categories needs to be iteratively evaluated considering the relative importance for each of the three main sets of indicators. This aspect is also very important for the adaptation of the LCIA model to different scenarios in which FRs might already be present in variable amounts. The different weighting calculations performed have to be fully documented in the LCIA study, clearly defining the validity conditions for the performed assessment.

3.1.4 Interpretation

It is a systematic technique to identify, check, and to evaluate the information from the results of the LCI and LCIA. The task of interpretation aims at identifying significant environmental issues and evaluating the completeness, sensitivity, and consistency of both the data gathered to perform the life cycle study and the data generated by the study itself. The output of a decision model depends critically on the data input, and these data should therefore be examined critically.

The final results are analyzed in the interpretation phase according to the goals and the scope of the study, involving a review of all the LCA stages to check the consistency of the assumptions and the quality of the data. They are discussed in terms of the critical sources of impacts, providing transparent conclusions and recommendations to reduce them.

An uncertainty analysis could be done following Monte Carlo simulations in order to determine the uncertain aspects of the models and also the results of these models. A sensitive analysis can also be carried out, by varying key parameters in order to analyze the variation these parameters could cause in the final results and the relative influence they can have over the results.

In order to guarantee that the study has been performed according to the established standards and rules and that the conclusions are in accordance with the preset goal and scope, the method used, and the data quality, a peer review can be done. Specifications for a suitable peer review and the reviewer skills required for the verification can be found in the International Reference Life Cycle Data System (ILCD) Handbook: Review schemes for LCA. The reviewers perform a review and supervise the scheme application comprising of the tasks such as checking the completeness, internal consistency, transparency of the reports, data quality, and system boundaries.

Once the evaluation and interpretation of the data are finished, new processes can be created so as to include them in existing LCA databases following the existing protocols and LCIA methods. It is important to guarantee the representativeness and suitability of the results to be incorporated in the created processes for data sets and impact assessment methods.

3.2 Risk assessment

Although the LCA framework is suitable for the conventional FR chemicals and also for the FR nanomaterials, there are certain limits to its application and use, in particular regarding the assessment of toxicity impacts of nanomaterials. In order to fully assess potential risks and environmental impacts of these FRs, a combination of different methods needs to be applied, joining LCA with risk assessment approaches [10,11].

The risk assessment focuses on the toxicity impact and takes into account the existing data and bibliography, read across, and testing [12]. It includes the steps detailed in Figure 2.

3.2.1 Hazard identification

It is the qualitative determination of whether or not a particular hazardous agent is associated with environment or health effects of sufficient importance. Different tools, such as Quantitative Structure—Activity Relationship or read across, can also be used to estimate the damage, when no data are available.

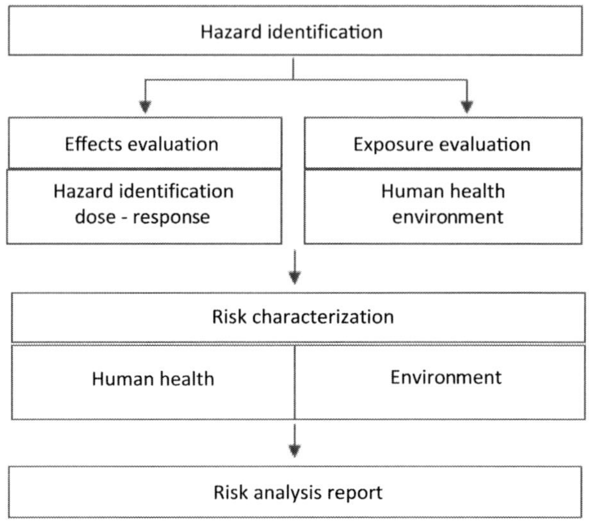

FIGURE 2

Risk assessment diagram.

3.2.2 Dose–response to level of exposure

The exposure scenarios in the different stages of the FR life cycle have to be defined. An exposure assessment is performed by undertaking a quantitative determination of the duration, the amount, and pattern of exposure. This stage determines the estimation of the concentrations/doses to which the human population and/or the environment are or may be exposed.

3.2.3 Exposure assessment in relevant exposure scenarios

A detailed evaluation of the effects, including risk identification (adverse effects that a substance with an inherent capability can produce), and dose–response evaluation (estimation of the relationship between dose or level of exposure to a substance, and incidence and severity effect) has to be developed. The quantitative relationship between the magnitude of the exposure and the probability of occurrence of adverse effects in the human population needs to be examined. The dose–response assessment is based on current available data and also extrapolations from the literature and bibliography.

3.2.4 Characterization of human health risk

The risk characterization estimates the incidence and the severity of adverse effects that may occur in the human population due to actual or predicted exposure. The exposure levels are compared with no effect levels for calculating risk characterization ratios. The analysis of HH risk has to be based on standardized methodology. Models and factors for toxicity effects (HT and ecotoxicity) in the LCA are based on the relative risk and associated consequences of FR nanomaterials that are released into the environment, accounting to environmental and biological side effects, human exposure, and the toxicological responses.

Characterization factors (CF) can be estimated, for instance, following the principles of the USEtox model [13]. This model can be analyzed in order to check the suitability of its use for FR nanomaterial CF. USEtox is the United Nations Environment Program (UNEP)-Society of Environmental Toxicology and Chemistry (SETAC) approved toxicity model, as the preferred model for characterization modeling of human and ecotoxic impacts in the LCIA. USEtox provides a parsimonious and transparent tool for HH and ecosystem CF estimates. Based on a referenced database, it has been used to calculate CFs for several thousand substances, and forms the basis of the recommendations from UNEP−SETAC's Life Cycle Initiative regarding the characterization of toxic impacts in the LCA [14]. This method allows accounting for fate, exposure, and effects of FR nanomaterials in order to provide impact measures.

The HT characterization factor in USEtox focuses on risks to the general population derived from environmental exposures. In addition to such analyses, a separate evaluation for a worker's toxicity can be performed and finally incorporated within the HT characterization factor. In order to estimate the environmental and HT characterization factors of chemicals, the following factors are needed: fate factors (FF), exposure factors (XF), and effect factors (EFs).

The FF and XF of a chemical in a certain compartment can be calculated by solving a set of mass balance equations that describe processes such as degradation and intercompartment transfer. For soluble compounds, the environmental XF is the fraction of the chemical dissolved in freshwater. In the case of nanomaterials, the environmental XF is the fraction of nanomaterials that are stable in suspension in the water column.

Ecotoxicological EF is defined in USEtox as 0.5/HC50, where HC50 (hazardous concentration) is the concentration at which 50% of the species are exposed above their EC50 (effective concentration). Human EF is derived for carcinogenic and noncarcinogenic effects, and for inhalation or oral exposure. The EFs reflect the change in lifetime disease probability due to the change in lifetime intake of a pollutant. These EFs are calculated as 0.5/ED50, where ED50 (effective dose) is calculated from animal toxicity data and extrapolation factors to account for exposure duration and interspecies differences.

In order to assess the effect of exposure to nanomaterials on workers, there are some specific tools that could be adapted to FR nanomaterials. These tools could be used for estimating some important parameters of the working environment.

- Stoffenmanager: Tool to assess the risk of exposure via inhalation (quantitative) and via the skin (qualitative). The exposure to powders via inhalation can be adapted to nanomaterials. The tool already has a nanomaterial module (Stoffenmanager Nano), although it does not provide a quantitative assessment but only a simplified qualitative solution based on control banding [15].

- Advanced Reach Tool: Tool to assess the exposure on workers. It is applicable for exposure to dust and powder. This model determines the exposure through parameters such as substance emission potential based on dustiness and activities [16].
- ECETOC: Tool for the risk assessment of the European Center for Ecotoxicology and Toxicity of Chemicals. The ECETOC Targeted Risk Assessment (TRA) Worker is a tool that allows the calculation of predictions for inhalation and dermal exposure on workers [17].
- EUSES: The European Union System for the Evaluation of Substances [18] is a decision-support instrument that enables one to carry out rapid and efficient assessments of the general risks posed by chemical substances.

4. European research projects

Some of the European research projects where LCA and RA are being carried out on different nanomaterials are listed below. It is expected that the LCA models of nanomaterials, in general and those used for FR purposes, will be fully established as an outcome of these projects.

- NANOPOLYTOX (www.nanopolytox.eu): The main goal of this project is to improve the understanding of the potential environmental/health impact of nanotechnology based products over their life cycle: use, reuse, recycling, and/or final treatment and disposal of products containing engineered nanomaterials. The LCA studies carried out within the project are aimed at providing a new framework capable of providing data on nanomaterial LCIA and allowing the comparison of nanomaterials with different properties.
- ENFIRO (www.enfiro.eu): This project offers a prototypical case study on substitution options for specific brominated FRs. The project delivers a comprehensive data set on the viability of production and application, environmental safety, and the LCA of the alternative FRs.
- NEPHH (www.nephh-fp7.eu): The main objective of the project is to identify and rate the important forms of nanotechnology related environmental pollution and health hazards that could result from activities involved in the life cycle of silicon based polymer nanocomposites currently used in a variety of industrial sectors and also to suggest means that might reduce or eliminate these impacts.
- NANOFATE (www.nanofate.eu): The objective of this project is to develop the knowledge and tools to enable a better assessment of the environmental fate and risk from high-volume nanoenabled products for which recycling is not an option. Particles from fuel additives, personal care, and antibacterial products will be studied through their postproduction life cycles i.e. from environmental entry as "spent product", through waste treatment to their final fates and potential toxic effects. This will test the applicability of the current fate and risk assessment methods and identify improvements required for better early stage assessment.

- NANOSUSTAIN (www.nanosustain.eu): The main objective is to explore, examine, and develop new solutions for the sustainable design, use, reuse, recycling, and final treatment and/or disposal of specific nanomaterials and associated products. This will be based on a comprehensive hazard characterization and impact assessment of selected environmentally and economically relevant products.
- ENPRA (www.enpra.eu): The aim of ENPRA is to implement a impact assessment approach for engineered nanoparticles used in vitro/in vivo toxicology tests, exposure assessment, and quantitative risk assessment.

5. CONCLUSIONS

Even though the LCA analysis of conventional FR chemicals is well studied, LCA methods of nanomaterials are not fully established. In this chapter, the possible LCA methodologies for FR nanomaterials are discussed, which may help in implementing manufacturer strategies and improving the environmental aspects along their life cycle. Thus, it is proposed that the LCA can be used as an effective tool to detect the hot spots in the whole life cycle of FR nanomaterials, just as in the case of conventional FR chemicals.

The combination of the LCA and risk assessment results will enable a quantitative and qualitative assessment of the overall impacts and trade-offs. This pragmatic approach of combining LCA with risk assessment will allow one to achieve good results, taking into account the specific characteristics of FR nanomaterials, such as their surface properties and chemical nature.

Abbreviations

LCA	Life cycle assessment
LCIA	Life cycle impact assessment
LCI	Life cycle inventory
SETAC	Society of environmental toxicology and chemistry
QSAR	Quantitative structure–activity relationship
CC	Climate change
OD	Ozone depletion
TA	Terrestrial acidification
FE	Freshwater eutrophication
ME	Marine eutrophication
HT	Human toxicity
POF	Photochemical oxidant formation
PMF	Particulate matter formation
TET	Terrestrial ecotoxicity
FET	Freshwater ecotoxicity
MET	Marine ecotoxicity
IR	Ionizing radiation

ALO	Agricultural land occupation
ULO	Urban land occupation
NLT	Natural land transformation
WD	Water depletion
MRD	Mineral resource depletion
FD	Fossil fuel depletion
FF	Fate factors
XF	Exposure factors
EF	Effect factors

References

[1] ISO 14040. Environmental management—life cycle assessment—principles and framework. International organization for standardization; 2006.
[2] ISO 14044. Environmental management—life cycle assessment—requirements and guidelines. International organization for standardization; 2006.
[3] European Commission—Joint Research Centre—Institute for Environment and Sustainability. International reference life cycle data system (ILCD) handbook of the European platform on life cycle assessment—general guide for life cycle assessment—detailed guidance (EUR 24708 EN). Luxembourg: Publications Office of the European Union; 2010.
[4] Guinée JB. Handbook on life cycle assessment: operational guide to the ISO standards. London (UK): Kluwer Academic Publishers; 2002.
[5] Klöpffer W. Life cycle sustainability assessment of products. Int J Life Cycle Assess 2008;13:89–95.
[6] Fullana P, Betz M, Hischier R, Puig R. Life cycle assessment applications: results from COST action 530. Madrid (Spain): AENOR/Emerald Group Publishing; 2008.
[7] ReCiPe Mid/Endpoint method, version 1.05. www.lcia-recipe.net; July 2010.
[8] Goedkoop MJ, Heijungs R, Huijbregts M, De Schryver A, Struijs J, Van Zelm R. ReCiPe 2008: a life cycle impact assessment method which comprises harmonised category indicators at the midpoint and the endpoint level. 1st ed. Report I: characterisation. Den Haag (The Netherlands): VROM; 2009.
[9] Hidalgo C, Janer G. NANOPOLYTOX (FP7-NMP-ENV-2009 project 247899): reports on the LCI data and LCIA analysis describing the new methodology used for nanomaterials. Barcelona (Spain): LEITAT Technological Center; 2011.
[10] Klöpffer W, Curran MA, Frankl P, Heijungs R, Köhler A, Olsen SI. Nanotechnology and life cycle assessment. A systems approach to nanotechnology and the environment: synthesis of results obtained at a workshop Washington, DC October 2–3, 2006. In: European Commission, DG research and project on Emerging Nanotechnologies, of the Woodrow Wilson International Center for Scholars; 2007.
[11] International stakeholder engagement meeting "towards global guidance for LCA databases". Boston (USA); September 30, 2009.
[12] European commission, European chemicals bureau, Joint Research Center. Technical guidance document on risk assessment. Ispra (Italy): European communities; 2003.
[13] www.USEtox.org.
[14] www.unep.fr/scp/lcinitiative.

[15] www.stoffenmanager.nl.
[16] www.advancedreachtool.com.
[17] www.ecetoc.org/tra.
[18] http://ihcp.jrc.ec.europa.eu/our_activities/public-health/risk_assessment_of_Biocides/euses.

Index

Note: Page numbers followed by f indicate figures; t, tables; b, boxes.

B

C

D

F

I

K

L

M

Q

R

U

Edwards Brothers, Inc.
Thorofare, NJ USA
August 27, 2014